NUMERICAL MATHEMATICS AND SCIENTIFIC COMPUTATION

General Editors

G.H. GOLUB R. JELTSCH
W.A. LIGHT E. SÜLI

NUMERICAL MATHEMATICS AND SCIENTIFIC COMPUTATION

---

*P. Dierckx: *Curve and surface fittings with splines*
*H. Wilkinson: *The algebraic eigenvalue problem*
*I. Duff, A. Erisman, and J. Reid: *Direct methods for sparse matrices*
*M. J. Baines: *Moving finite elements*
*J. D. Pryce: *Numerical solution of Sturm-Liouville problems*
K. Burrage: *Parallel and sequential methods for ordinary differential equations*
Y. Censor and S. A. Zenios: *Parallel optimization: theory, algorithms and applications*
M. Ainsworth, J. Levesley, M. Marletta and W. Light: *Wavelets, multilevel methods and elliptic PDEs*
W. Freeden, T. Gervens, and M. Schreiner: *Constructive approximation on the sphere: theory and applications to geomathematics*
Ch. Schwab: *p- and hp-finite element methods: theory and applications to solid and fluid mechanics*
J. W. Jerome: *Modelling and computation for applications in mathematics, science, and engineering*
Alfio Quarteroni and Alberto Valli: *Domain decomposition methods for partial differential equations*
G. E. Karniadakis and S. J. Sherwin: *Spectral/hp element methods for CFD*
I. Babuška and T. Strouboulis: *The finite element method and its reliability*
B. Mohammadi and O. Pironneau: *Applied shape optimization for fluids*
S. Succi: *The Lattice Boltzmann equation for fluid dynamics and beyond*

Monographs marked with an asterix (*) appeared in the series 'Monographs in Numerical Analysis' which has been folded into, and is continued by, the current series.

The Finite Element Method and its Reliability

IVO BABUŠKA
University of Texas at Austin

and

THEOFANIS STROUBOULIS
Texas A&M University

CLARENDON PRESS • OXFORD

2001

OXFORD
UNIVERSITY PRESS

Great Clarendon Street, Oxford OX2 6DP

Oxford University Press is a department of the University of Oxford
It furthers the University's objective of excellence in research, scholarship,
and education by publishing worldwide in

Oxford New York

Athens Auckland Bangkok Bogotá Buenos Aires Cape Town
Chennai Dar es Salaam Delhi Florence Hong Kong Istanbul Karachi
Kolkata Kuala Lumpur Madrid Melbourne Mexico City Mumbai Nairobi
Paris São Paulo Shanghai Singapore Taipei Tokyo Toronto Warsaw

with associated companies in Berlin Ibadan

Oxford is a registered trade mark of Oxford University Press
in the UK and in certain other countries

Published in the United States
by Oxford University Press Inc., New York

First published 2001

A catalogue record for this book is available from the British Library

Library of Congress Cataloging in Publication Data

Babuska, Ivo.
The finite element method and its reliability / Ivo Babuska and Theofanis Strouboulis.
(Numerical mathematics and scientific computation)
1. Finite element method 2. Reliability (Engineering) I. Strouboulis, Theofanis
II. Title III. Series.
TA347.F5 B33 2001 620′.001′51535–dc21 2001021269
ISBN 0 19 850276 1

Typeset by Newgen Imaging Systems (P) Ltd, Chennai, India
Printed in Great Britain by
T.J. International Ltd, Padstow, Cornwall

Ivo Babuška dedicates this book to his parents
Theofanis Strouboulis dedicates this book with love to Laurel Claire Easton

PREFACE

During the last 30 years a multitude of books on the Finite Element Method (FEM) have appeared. Many of these books address only the engineering or the mathematical aspects of the method, at an elementary or advanced level. The present book differs from all previous books especially in the following aspects.

1. It emphasizes the question of the reliability of FEM. By reliability we mean the analysis of the difference between the FE solution and the exact solution of the problem, i.e., the analysis of the error in the computed FE solution. The error is addressed first by employing a-priori estimates, then in terms of computable a-posteriori estimates including guaranteed upper and lower estimates. A main feature is the analysis of the reliability of the computed engineering quantities of interest which are the goal of the analysis. Further, the problem of the pollution in the FEM, and the improvement of the FE solution by post-processing are discussed.

2. The objective of the book is to give the reader a solid basis for understanding and the right feeling for the question of the reliability of FEM. The FEM is taken as a tool for practical computations but with emphasis on the understanding of the basic ideas and results which can be expected. Our approach to learning is based on a rigorous understanding of the *one-dimensional* FEM based on arguments which *are not* one-dimensional, and the understanding of the strong similarities and differences of the one-dimensional case with the two- and three-dimensional cases. In the one-dimensional setting, the ideas are most visible and the theoretical aspects are still accessible for the reader without prerequisites in advanced mathematics (functional analysis, etc). Nevertheless, there are also sometimes essential differences between one- and two-dimensional problems. To achieve our goals, every section has two parts, the one-dimensional (e.g. Section 4.2), and the two-dimensional (e.g. Section 4.2*). The two-dimensional problem is of course more important but also technically more difficult. For this reason, in contrast to the one-dimensional section where rigorous proofs are given, in the two-dimensional section we mainly use analogies and precise statements. We include many numerical results directly related to the presented theory. Also, we include many remarks elaborating on the presented results which illuminate still more the presented material.

3. The book is written for readers who are interested in understanding the FEM and questions about its reliability, errors, etc., especially in relation to practical computations. The book should be useful to engineers as well as applied mathematicians. The book addresses only the FEM for linear stationary strongly elliptic problems like, for example, the problems of linear stationary heat conduction, and elasticity, for which the theory and practice of the FEM is most advanced. There are no prerequisites for reading the book. Nevertheless, some very *elementary* knowledge of FEM on an engineering or

mathematical level could be useful. This book does not address various computational aspects because they are covered in the standard FE books.

4. The book is not a textbook because of its size; nevertheless, it could be used in graduate courses in engineering and applied mathematics programs. A textbook for a one-semester FEM course based on this book will be available next year from the same publisher, namely: I. Babuška, J. Whiteman, and T. Strouboulis (2002), Finite elements: an introduction to the method, error estimation, and adaptivity.

5. The book is mostly based on the results obtained by the authors and their students (especially Dibyendu K. Datta, Lin Zhang, Chandra S. Upadhyay, and Kevin Copps), which appeared in various journals and in the doctoral dissertations and the master's theses of D.K. Datta (1997, 2001), C.S. Upadhyay (1993, 1997), and the master's thesis of S.K. Gangaraj (1994), all at Texas A&M University, College Station, Texas.

The book has six chapters. The first one is an introduction with a brief historical background of the FEM and a brief description of the role of the FEM in engineering. The second chapter addresses the mathematical formulation of the model problems. Chapter 3 presents the basic theory of the FEM. Chapter 4 addresses questions related to the local properties of the FEM, namely pollution, and superconvergence. Chapter 5 is devoted to the a-posteriori estimates for the energy norm of the FEM error. Chapter 6 discusses the problem of guaranteed a-posteriori estimation of the error in the computed engineering outputs, and the a-posteriori estimation of the pollution.

The book was influenced by the discussions we had with many colleagues working in Finite Elements, in theory and practice. Among others we would like to thank Professors Tinsley Oden, Pierre Ladeveze, Olek Zienkiewicz, Leszek Demkowicz, Lars Wahlbin, Al Schatz, Bruce Kellog, Manil Suri, Benqi Guo, Barna Szabo, Carsten Carstensen, and John Osborn. Many thanks are also due to Dibyendu Datta, Lin Zhang, Chandra Upadhyay, and Kevin Copps for all the help and contributions to this book. We would like also to thank Frank Samstag for the excellent technical typing and manuscript construction using LaTeX typesetting software. Finally, we would like to express our appreciation to Sönke Adlung and the technical staff of Oxford University Press for their excellent effort which made this book a reality.

Austin, Texas I. B.
College Station, Texas T. S.
December, 2000

CONTENTS

1

INTRODUCTION

1.0 Introduction

The finite element method is today one of the major tools of CAE (Computer Aided Engineering). The growth of computer power and the availability of software allows engineers to solve more and more complex linear and nonlinear problems rapidly and with diminishing computational and human cost. More precisely the availability of solid based CAD geometry, automatic meshing, associativity between geometry and finite elements, the development of efficient solvers for system of linear equations, etc. has drastically reduced the time needed to solve complex problems. For a survey of various aspects of the use of the FEM in industry, refer to [Brown, 1997]. Many commercial and research codes are available and widely used, for example, MSC Nastran, ANSYS, ABAQUS, STRESS CHECK, MECCANICA, etc. These codes are used more and more as black boxes. Many times the detailed descriptions of the numerical algorithms employed in commercial FEM codes are not available to the user because they are proprietary. The graphical display of very high quality and sophistication often obscures the quality and reliability of the computed data and gives the user unsubstantiated confidence in the results. In addition, there is the trend of using the various commercial FEM codes as 'black' boxes by inexperienced analysts who may be unable to detect gross errors in the computed results. One has to ask the question: What is the purpose of computation? The answer is not simple; nevertheless, stated broadly the purpose is *to obtain reliable data leading to the proper engineering decisions.* Hence the emphasis on the problem of reliability, or a-posteriori error estimation, etc., is clearly coming to the forefront.

Today various versions of the FEM are available. The classical h version approximates the solution by piecewise polynomials of low order p, typically $p = 1, 2$. The p version utilizes high degrees of p, while the hp version combines the h and p versions; see, for example [Szabo and Babuška, 1991], and [Schwab, 1998]. Also, very recently, so called mesh free methods and the generalized FEM, have appeared; see, for example, [Babuška and Melenk, 1996], [Duarte and Oden, 1996], and [Strouboulis, Copps, and Babuška, 2001]. In this book only the h version will be addressed. There are dozens of books on finite elements focusing on different aspects, elementary or advanced, with engineering or mathematical character. Let us mention a few: [Babuška and Aziz, 1972], [Ciarlet, 1978], [Zienkiewicz and Taylor, 1989, 1991, 2000], [Brenner and Scott, 1994], [Eriksson *et al.*, 1996], [Bathe, 1996], [Braess, 1997], and [Ainsworth and Oden, 2000] for the h version, and [Szabo and Babuška, 1991] and [Schwab, 1998], which emphasize the p version. This present book has a different character than the aforementioned ones; it will focus on the problem of the reliability of the finite element method.

1.1 A brief history of the finite element method

To understand the progress in Finite Elements it is worthwhile to briefly review the history of the method, which was developed more or less in *parallel* in engineering and mathematics.

Finite Elements in engineering

The major development in engineering FEM was related to computers. In 1952 a great effort was made at Boeing to analyze aircraft structures. A procedure was developed, and appeared in the literature only by 1956 in [Turner *et al.*, 1956]. The method was essentially based on classical ideas of matrix structural mechanics cast in the framework of digital computers. In 1961, it was recognized that the FEM is a special version of the variational approach of Ritz [Ritz, 1909] and Galerkin [Galerkin, 1915]. By 1964 several aircraft companies had developed their own finite element programs and in 1965 the ASKA program, written by the group under J.H. Argyris, appeared on the market. The book by Zienkiewicz and Cheung [Zienkiewicz and Cheung, 1967] which appeared in 1967 stimulated further development and played an important role. In the late 1960s the program NASTRAN written by MacNeal–Schwendler Corporation appeared on the market. Since then, dozens of various commercial and research codes were written and are available.

Finite Elements in mathematics

If the FEM is understood as an approximate method for solving differential equations utilizing a variational principle and piecewise polynomial approximation (see Chapter 3), then likely G.W. Leibnitz (1646–1716) in 1696 was a user of FEM. At the same time L. Euler (1707–1783) introduced the variational methods with the approximation approach being essentially the main tool employed for the derivation of the Euler equation. Leibnitz, in a letter to Johann Bernoulli (1667–1748) [Leibnitz, 1962], addressed the problem of the brachistochrone, replacing the general descent curve with a piecewise linear curve with one and two internal nodes, he determined the coordinates of nodal points to achieve the brachistochrone.

In the nineteenth century the variational approaches were tools for proving the existence of a solution of differential equations and various discretizations were also used as mathematical tools. As an example let us mention the history of the variational method for solving the Poisson problem. G.F.B. Riemann (1826–1866) in his 1851 thesis used the variational principle of energy minimization, which he called the Dirichlet Principle, to prove the existence of the solution to this problem, essentially assuming that the *minimizer exists*. L.T.W. Weierstrass (1815–1897) found an essential flaw in Riemann's argument about the existence of the minimizer. This error was corrected much later by Hilbert [Hilbert, 1901] using essentially the principles of functional analysis. The use of the variational method as a tool for obtaining approximate solutions was first proposed by Ritz [Ritz, 1909], and Galerkin [Galerkin, 1915]. Courant touched on the subject of the variational method in six papers with the last one in 1943 [Courant, 1943] where he explicitly proposed a piecewise linear function for approximate solution of the Poisson problem via minimization of the energy. For this reason, the linear element is sometimes called the Courant element. Courant proposed the method in 1922 [Hurwitz and Courant, 1922], where he gave the proof of existence of the conformal mapping via minimization of the Dirichlet integral, by employing a sequence of functions, which he named the minimal sequence, which has the property that the potential energy decreases and converges to the infimum. A long footnote is given on the construction of the minimal sequence, where subdivision of the domain into triangles and the use of piecewise linear functions is proposed. It is interesting that this footnote appeared only in the first edition of the book, and was omitted in the next one, seemingly because it was an unimportant comment.

There are many other instances of discovery of the FEM. For example, K.O. Friedrichs in [Friedrichs, 1962] and [Friedrichs and Keller, 1966] uses piecewise linear functions on triangles and proves the convergence in the spaces $H^1(\Omega)$ and $L^2(\Omega)$ in today's standard way. He does not cite [Courant, 1943] in [Friedrichs, 1962], but does so in [Friedrichs and Keller, 1966]. Oganesjan [Oganesjan, 1963] uses piecewise linear functions on a triangular mesh and proves the classical estimates in the H^1 norm. He does not cite [Courant, 1943]. Feng Kang [Kang, 1965] proposed the same method for scalar second-order problems and elasticity, proving convergence but not giving a-priori error estimates, citing neither [Courant, 1943], nor [Friedrichs, 1962], nor [Oganesjan, 1963].

From the end of the 1960s to the beginning of the 1970s the basis of the rigorous mathematical theory was laid down [Babuška and Aziz, 1972], [Ciarlet, 1978], [Aziz, 1972]. For more about the history of FEM and additional references, see [Babuška, 1994], [Clough, 1990], [Oden, 1990], [Williamson, 1976]. After the end of the 1970s, much progress was made in all aspects of FEM. Many papers and books on FEM, engineering and mathematical, have been published. In 1978 two papers [Babuška and Rheinboldt, 1978a,b] showed that it is possible to estimate quantitatively the errors in the finite element solution a-posteriori, i.e. using the computed finite element solution and the data. Since then, various forms of a-posteriori error estimation have appeared in the engineering and mathematical literature, and various aspects of the reliability of FEM are coming to the forefront of research and computational practice; many references citing contributions to this latest development will be given in Chapter 5.

The history of FEM shows that it originated from different sources and today it is a very mature method where engineering, mathematics, and computer science are interwoven. On the other hand, the number and richness of the results are so large that it is hard for a single person to comprehend it. This has led to the development of various specialties concentrating only on a few particular aspects. For example, focusing on the computational aspect leads one to emphasize only 'recipes' whilst neglecting ideas behind them, while concentrating on mathematical aspects leads sometimes one to lose sight of the purpose of FEM, which is the computational analysis of engineering problems. Focusing on the computer-science aspects, such as the parallelization of FEM computations, also leads to an one sided view. The philosophy of this book relates to this situation and its various aspects.

1.2 On computational engineering

As we have already mentioned, the purpose of FEM is to provide tools for engineering decisions, which are related to safety and durability, such as, for example, questions concerning safety of nuclear powerplants, airplanes, dams, offshore drilling platforms, aircraft, automobiles, etc. Questions of this type have not only technical but also ethical, social, political, and economic aspects (see e.g. [Schwing and Alberto, 1980]). As a simple illustration, consider the safety of a bridge, which has to resist loads caused by traffic, high wind, and earthquakes. The following questions immediately arise: What is the worst combination of loads, the traffic load, wind, and earthquakes that the bridge should be designed to resist simultaneously? What kind of damages are acceptable under the various loading conditions? The formulation of the problems of safety and durability has evolved, depending on various circumstances, experience, progress in engineering (e.g. material sciences), etc. For example, in the 1970s the US Air Force changed its design requirements to employ criteria based on fracture mechanics [USAF, 1974], which were accepted in other environments much later or not at all.

It is necessary to realize that the FEM is a numerical method for constructing *approximate* solutions of a well defined mathematical problem. A necessary prerequisite is that the mathematical problem has 'good' properties. These are: existence and possibly uniqueness of the solution, and the continuous dependence of the solution on the input data, the coefficients of the equation, boundary conditions, and the domain. A mathematical problem with all these properties is then called 'well posed'. It is also important to keep in mind that there is uncertainty in all the employed input data and hence it may be meaningless to attempt to compute outputs which are very sensitive to these uncertainties. The need for well posedness was underlined by Hadamard [Hadamard, 1952]. Nevertheless Hadamard did not consider the need of continuous dependence on changes in the domain, which is very important in today's problems because of the approximate description of the domain, used, for example, in digital images, etc.

It is necessary to realize that the mathematical formulation is a simplification of reality, and the existence of a physical solution which can be observed in an experiment, does not guarantee that the solution of the mathematical problem exists and has the expected properties following from physical intuition. Let us give an example. Consider a simply supported circular uniformly loaded plate formulated by the Kirchoff model; this problem is addressed in classical books like, for example, in [Timoshenko and Woinowsky-Krieger, 1959]. Let us now consider a regular polygonal plate with n vertices, inscribed in the circular plate. One would expect that the solutions of the polygonal plates would converge to the solution of the circular plate for sufficiently large n. However, it can be shown that this is not the case (see [Babuška, 1962], [Babuška and Pitkaranta, 1990]). This paradox is the effect of the mathematical formulation of the boundary condition of simple support which does not capture sufficiently well the engineering problem. The need for a proper formulation of the boundary conditions is addressed also in [Babuška and Chleboun, 2001], where it is shown that when the traction boundary condition is used in the classical sense, the solution of Laplace's equation on the pixel approximation of a circular disk converges to a solution different than the solution of Laplace's equation for the exact domain. This difficulty may be avoided by reformulating the boundary condition.

The finite element method leads to approximate solutions of a *mathematical problem*, and their quality has to be judged by their difference from the exact solution of the mathematical problem and not an experiment. Whether the finite element solution, which we assume to be sufficiently accurate, is relevant to the engineering problem is a completely different question. Thanks to the developments of FEM, a *proper* use of it with a-posteriori error estimation could guarantee high quality of the solution and any discrepancy to the physical reality is due to the formulation of the problem and the data used. This also brings the problem of the validation of the modeling to the forefront; see [Roache, 1998]. In Europe, research into the methodology of the model validation was recently undertaken, see e.g. [Marczyk, 1997]. Nevertheless, any validation has to assume that the finite element approximation is sufficiently accurate such that there is no interference of the errors in the FE solution, and that the agreement or disagreement with the observed data is due only to the mathematical model. Hence the guaranteed quality of the FEM is of utmost importance for validation of the model. Let us also note that sometimes the meaning of the words validation and verification may be confused. Validation is related to the reliability of the mathematical model, while verification addresses the reliability of the approximate solution in comparison with the exact solution of the mathematical problem. In this book we will address exclusively the problem of verification.

The goal of computation has to be well specified so that the high quality of the finite element solution can be achieved with respect to this goal. The accuracy of the finite element solution has to be sufficient with respect to the goal of the analysis. Let us note that this is not always the case in engineering practice. We mention, as an example, the Sleipner Field development project in the North Sea which was scheduled to meet about 10% of the western Europe gas demand for a period of about 30 years. The cornerstone of the Sleipner development was the gravity base structure (GBS), a platform which was going to operate in 82 m of water and had 24 caisson cells over a base area of about 16 000 m^2, with a total height of 110 m and a concrete volume of about 75 000 m^3. During a test in August 1991, one of the Sleipner GBS sank in Gardsfjorden outside Stavanger. In the investigation of the accident it was found that the computed stresses which were critical to the design of the thickness of the wall of the caissons had a 47% error and this led to an inaccurate design and shear failure of the walls when the design loads were realized. Although there were also other deficiencies, the major reason for the accident was insufficient accuracy of the employed FEM solution for the purposes of the design. Proper use of quality control of the finite element solution could have avoided this accident. For more about the Sleipner accident, see [Jakobsen and Rasendahl, 1994] and [Rettedal, 1993].

References

M. Ainsworth and J.T. Oden (2000) *A-posteriori error estimation in Finite Element Analysis.* Wiley, New York.

A.K. Aziz (ed.) (1972) *The mathematical foundations of the Finite Element Method with applications to Partial Differential Equations, Proceedings of a Symposium held at the University of Maryland, Baltimore campus, June 26–30, 1972*, Academic Press, New York.

I. Babuška (1962) The theory of small changes in domain of definition in the theory of partial differential equations and its applications. In: *Proceedings of the Conference EQUADIFF 1962, Prague*, Academia, pp. 13–26.

I. Babuška and A.K. Aziz (1972) Survey lectures on the mathematical foundations of the Finite Element Method. In: A.K. Aziz (ed.) *The mathematical foundations of the Finite Element Method with applications to Partial Differential Equations, Proceedings of a Symposium held at the University of Maryland, Baltimore campus, June 26–30, 1972*, Academic Press, New York, pp. 3–363.

I. Babuška and W.C. Rheinboldt (1978a) Error estimates for adaptive finite element computations. *SIAM Journal of Numerical Analysis*, **15**, 736–754.

I. Babuška and W.C. Rheinboldt (1978b) A-posteriori error estimates for the finite element method. *International Journal for Numerical Methods in Engineering*, **12**, 1597–1615.

I. Babuška and J. Pitkaranta (1990) The plate paradox for hard and soft simple support. *SIAM Journal of Mathematical Analysis*, **21**, 551–576.

I. Babuška (1994) Courant element: before and after. In: M. Krizek, P. Neittaanmäki, and R. Sternberg (eds) *Finite Element Methods: fifty years of the Courant element*, Marcel Dekker, Inc.

I. Babuška and J.M. Melenk (1996) The partition of unity finite element method: basic theory and applications. *Computer Methods in Applied Mechanics and Engineering*, **139**, 289–314.

I. Babuška and J. Chleboun (2000) Solutions of Neumann boundary value problems defined on a sequence of 2D domains convergence, stability estimates (to appear in Mathematics of Computation).

K.J. Bathe (1996) *Finite element procedures.* Prentice Hall, Englewood Cliffs.

D. Braess (1997) *Finite Elements, theory, fast solvers and applications in solid mechanics.* Cambridge University Press, Cambridge.

S.C. Brenner and R.L. Scott (1994) *The mathematical theory of Finite Element Methods.* Springer, Berlin, New York.

D.H. Brown (1997) *CAE: ready for the next leap forward; a survey of trends.* D.H. Brown Assoc., Inc., 222 Grace Church St., Port Chester, NY 10573.

P.G. Ciarlet (1978) *The Finite Element Method for elliptic problems.* North-Holland, Amsterdam.

R.W. Clough (1990) Original formulation of the Finite Element Method. *Finite Elements in Analysis and Design*, **7**, 89–101.

R. Courant (1943) Variational methods for the solutions of equilibrium and vibrations. *Bulletin of the American Mathematical Society*, **49**, 1–23.

C.A.M. Duarte and J.T. Oden (1996) An *hp* adaptive method using clouds. *Computer Methods in Applied Mechanics and Engineering*, **139**, 237–262.

K. Eriksson, D. Estep, P. Hansbo, and C. Johnson (1996) *Computational differential equations*, Cambridge University Press, Cambridge.

K.O. Friedrichs (1962) A finite difference scheme for Neumann and Dirichlet problems, AEC Computing and Appl. Math. Center, Courant Institute of Math. Sci., REP TID 4500 NYU 9760.

K.O. Friedrichs and H.B. Keller (1966) A finite difference scheme for generalized Neumann problems. In: *Numerical solutions of Partial Differential Equations, Proceedings of the Symposium at University of Maryland, May 3–8, 1965*, Academic Press, New York, pp. 1–21.

B.G. Galerkin (1915) On some problems of elastic equilibration of rods and plates. *Vestnik Inzenerov Petrograd*, **19**, 879–908 (in Russian).

J. Hadamard (1952) *Lectures on Cauchy's problem in linear Partial Differential Equations.* Yale University Press, New Haven, CT (1923). Reprinted Dover.

D. Hilbert (1901) Über das Dirichletsche Princip in *Festschrift der Königlichen Gesselshaft der Wissenschaften zur Gottingen, Mathematische Physik Klasse*, Berlin, **17**.

A. Hurwitz and R. Courant (1922) *Vorlesungen über allgemeine Functionen Theorie und elliptische Functionen.* Springer, Berlin.

B. Jakobsen and F. Rasendahl (1994) The Sleipner platform accident. *Structural Engineering International*, **3**, 190–194 .

Kang Feng (1965) A difference formulation based on variational principle. *Applied Mathematics and Mathematical Computation*, **2**, 138–162 (in Chinese).

G. Leibnitz (1962) The letter to Johann Bernoulli. In: G. Hildeshein and G. Olms (eds) *G.W. Leibnitz, Mathematische Schriften Band 111/2*, Verlagsbuchhandlung.

J. Marczyk (1997) Meta-computing and computational stochastic mechanics. In: *Computational stochastic mechanics in a meta-computing perspective*, CIMNE 1–19, International Center for Numerical Methods in Engineering, Barcelona Spain.

J.T. Oden (1990) Historical comments on finite elements. In: S.G. Nash (ed.) *A history of scientific computing*, Adison-Wesley Publishing Company, pp. 152–167.

L.A. Oganesjan (1963) Numerical solutions of plates in solution of engineering problems on computer, *Lensovnarchoz CBTI, Sbornik*, 1–21 (in Russian).

W.K. Rettedal, O.T. Gudmestad, and T. Aarum (1993) Design of concrete platforms after Sleipner A-1 sinking. In: *Proceedings of the 12th international conference on offshore mechanics and arctic engineering, OMAE*, Vol. I, The American Society of Mechanical Engineering, pp. 309–319.

W. Ritz (1909) Uber eine neue Methode zur Lösung gewisser Variations Probleme der mathematischen Physik. *Journal für die Reine und Angewandte Mathematik de Gruyter, Berlin*, **35**, 1–61.

P.J. Roache (1998) *Verification and validation in Computational Science and Engineering*. Hermosa Publishers, Albuquerque.

C. Schwab (1998) *p- and hp-Finite Element Methods*, Oxford Science Publications.

R.C. Schwing and W.A. Alberto (eds) (1980) *Societal risk assessment: how safe is safe enough?* Plenum Press.

T. Strouboulis, K. Copps, and I. Babuška (2001) The generalized finite element method. In: State of the art surveys in computational mechanics. *Computer Methods in Applied Mechanics and Engineering*, **190**, 4081–4193.

B.A. Szabo and I. Babuška (1991) *Finite Element Analysis*, John Wiley & Sons, Inc., New York.

S.P. Timoshenko and S. Woinowsky-Krieger (1959) Theory of plates and shells, 2nd edition, McGraw-Hill Book Company, New York.

M. Turner, R.W. Clough, H.C. Martin, and L.J. Topp (1956) Stiffness and deflection analysis of complex structures. *Journal of Aeronautical Science*, **23**, 805–823.

US Air Force (1974) Airplane damage tolerance requirement MIL-A-83444 (USAF).

C.F. Williamson (1976) A history of Finite Element Method to the middle 1960s. Ph.D. Thesis, Boston University.

O.C. Zienkiewicz and Y.K. Cheung (1967) *The Finite Element Method in structural and continuum mechanics*. McGraw-Hill, New York.

O.C. Zienkiewicz and R.L. Taylor (1989, 1991) *The Finite Element Method, Fourth Edition*, Vol. 1 (1989), Vol. 2 (1991). McGraw-Hill, New York.

O.C. Zienkiewicz and R.L. Taylor (2000) *The Finite Element Method, Fifth Edition*, Vol. 1–3, Butterworth Heinemann, Oxford, Auckland, Boston.

2

MATHEMATICAL FORMULATION OF THE MODEL PROBLEM

2.0 Introduction

In this chapter, we introduce the model problems which will be utilized throughout the book to illustrate the main ideas. We give the classical and the variational formulation of the model problems, we establish conditions which guarantee the existence of a unique solution, and we explain the dependence of the smoothness of the solution on the smoothness of the input data. The same model problems will be utilized in the subsequent chapters to illustrate the Finite Element Method and the analysis of its properties.

2.1 The one-dimensional model problem

Let us consider an elastic bar of length l and cross-sectional area $A = A(x)$, as is, for example, shown in Fig. 2.1.1. The bar is loaded by a distributed axial load $f = f(x)$, and by concentrated axial loads a_1 and a_2 at the right and the left end points, respectively. Let $u = u(x)$, $x \in I \stackrel{\text{def}}{=} (0, l)$, denote the axial displacement of the bar, and assume that the bar is supported by a distributed axial elastic support, $c = c(x) \geq 0$, which exerts a distributed reaction $-cu$ and by elastic springs of stiffness $\alpha_1 \geq 0$, and $\alpha_2 \geq 0$, which exert axial reactions $-\alpha_1 u(0)$ and $-\alpha_2 u(l)$ at the left and the right end points, respectively. We will be interested in determining the axial displacement, u, the axial

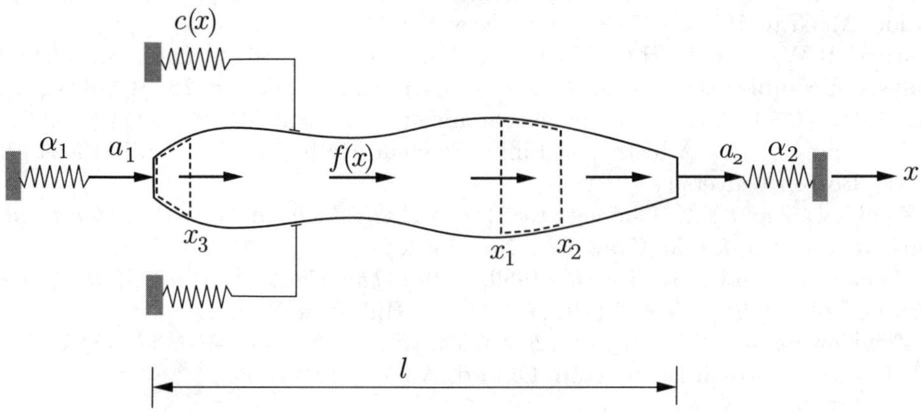

FIG. 2.1.1. The one-dimensional model problem: Deformation of an axially loaded elastic bar. A bar with variable cross section with applied distributed load f, concentrated end loads a_1 and a_2, distributed elastic support c, and elastic end supports α_1 and α_2

strain, du/dx, the axial stress, σ, and the axial force

$$P(x) = A(x)\sigma(x), \quad x \in I \tag{2.1.1a}$$

where we assumed that the stress σ is constant over the cross section. We will assume that the bar is linearly elastic, namely, that

$$\sigma(x) = E(x)\frac{du}{dx}(x) \tag{2.1.1b}$$

where $E = E(x) > 0$ is known as the elastic modulus, and (2.1.1b) is known as Hooke's law. Combining (2.1.1a) and (2.1.1b) we get

$$P(x) = A(x)E(x)\frac{du}{dx}(x) = AE(x)\frac{du}{dx}(x), \quad x \in I \tag{2.1.1c}$$

where $AE(x) = A(x)E(x)$ is known as the axial rigidity of the bar at x.

The mathematical formulation of the problem of the axially loaded bar will be based on the principle of axial equilibrium, which states that the sum of the forces on any segment (x_1, x_2) of the bar is zero, namely

$$P(x_2) - P(x_1) + \int_{x_1}^{x_2} (f - cu) = 0. \tag{2.1.2a}$$

Dividing by $x_2 - x_1$ we obtain

$$\frac{P(x_2) - P(x_1)}{x_2 - x_1} + \frac{1}{x_2 - x_1}\int_{x_1}^{x_2} (f - cu) = 0 \tag{2.1.2b}$$

and letting $x_1, x_2 \to x$, we get the *differential equation of equilibrium*

$$-\frac{dP}{dx}(x) + c(x)u(x) = f(x). \tag{2.1.3a}$$

This equation holds at any interior point x, $0 < x < l$. Further, under the assumption that $AE \in C^1(\bar{I})$, the axial force P can be eliminated by employing (2.1.1c), and we get

$$-\frac{d}{dx}\left(AE(x)\frac{du}{dx}(x)\right) + c(x)u(x) = f(x), \quad x \in I = (0, l). \tag{2.1.3b}$$

This is the *differential equation of equilibrium in terms of the axial displacement* u.

The equations (2.1.3a) and (2.1.3b) are local expressions of the equilibrium equation (2.1.2a) for interior points x, $0 < x < l$. Let us now derive the local expressions of equilibrium at the end points $x = 0$ and $x = l$. The equilibrium equation of a segment $(0, x_3)$, $x_3 > 0$, shown in Fig. 2.1.2b, is

$$P(x_3) - (\alpha_1 u(0) - a_1) + \int_0^{x_3} (f - cu) = 0 \tag{2.1.4a}$$

and noting that

$$\lim_{x_3 \to 0} \int_0^{x_3} (f - cu) = 0 \tag{2.1.4b}$$

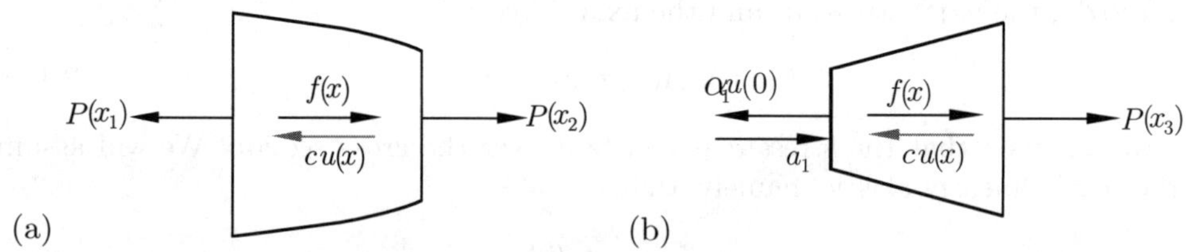

FIG. 2.1.2. Derivation of the differential equation of equilibrium and the natural boundary conditions for the one-dimensional model problem. (a) An interior segment (x_1, x_2) of the bar under the action of the applied load $\int_{x_1}^{x_2} f$, the elastic reaction $\int_{x_1}^{x_2} cu$, and the axial forces $P(x_1)$ and $P(x_2)$. (b) An end-segment $(0, x_3)$ at the left end of the bar under the action of the applied load $\int_0^{x_3} f$, the elastic reaction $\int_0^{x_3} cu$, the elastic reaction $\alpha_1 u(0)$, the applied force a_1, and the force $P(x_3)$

we get

$$P(0) = \alpha_1 u(0) - a_1. \tag{2.1.4c}$$

This is called the *natural boundary condition* at $x = 0$. Similarly, for $x_4 < l$ we have

$$a_2 - \alpha_2 u(l) - P(x_4) + \int_{x_4}^{l} (f - cu) = 0 \tag{2.1.5a}$$

and taking the limit as $x_4 \to l$, we get

$$-P(l) = \alpha_2 u(l) - a_2 \tag{2.1.5b}$$

where a_1, a_2, are the axial forces acting at $x = 0$ and $x = l$, respectively. Eliminating $P(0)$ and $P(l)$ using (2.1.1c) we get

$$AE(0)\frac{du}{dx}(0) = \alpha_1 u(0) - a_1 \tag{2.1.6a}$$

and

$$-AE(l)\frac{du}{dx}(l) = \alpha_2 u(l) - a_2. \tag{2.1.6b}$$

These are the *natural boundary conditions in terms of the displacement* u. The differential equation (2.1.3b) and the boundary conditions (2.1.6a), (2.1.6b) give the complete mathematical formulation of the model problem.

Remark 2.1.1 Note that (2.1.6a) and (2.1.6b) have the same form except for the difference in sign in the left-hand side. This difference can be eliminated by employing instead of du/dx, the derivative du/dn, where n is the outward unit normal with respect to $(0, l)$ at $x = 0$, and $x = l$. We then have

$$-AE(x)\frac{du}{dn}(x) = \alpha_{k(x)} u(x) - a_{k(x)}, \quad x \in I \tag{2.1.7}$$

where we let $k(0) = 1$, $k(l) = 2$. This is how the natural boundary condition is formulated in higher dimensions; see Remark 2.1*.1.

Let us now consider the case in which the stiffness of the spring support α_1 is much greater than the axial rigidity $AE(0)$ of the bar, namely, $AE(0)/\alpha_1 \approx 0$; we then have

$$u(0) = \frac{AE(0)}{\alpha_1}\frac{du}{dx}(0) + \frac{a_1}{\alpha_1} \approx \frac{a_1}{\alpha_1}. \tag{2.1.8a}$$

Hence, a special case of the natural boundary-condition at $x = 0$ is the boundary condition

$$u(0) = \frac{a_1}{\alpha_1}. \tag{2.1.8b}$$

This is the *essential*, or *kinematic*, or *Dirichlet* boundary condition at $x = 0$. Similarly, the Dirichlet boundary condition at $x = l$ is

$$u(l) = \frac{a_2}{\alpha_2} \tag{2.1.9}$$

and may be derived from the natural boundary condition at $x = l$.

Remark 2.1.2 Let us underline that the differential equation of equilibrium was derived under the assumption that f, c are continuous over $\bar{I} = [0, l]$, namely, $f, c \in C^0(\bar{I})$ (see A2.4.3). In the more general case where f, c are only piecewise continuous, the differential equation of equilibrium holds in the interior of the subintervals where f, c are continuous.

Remark 2.1.3 Let $\hat{x}$ be a point where f, c, are continuous, or suffer at most a jump discontinuity. Then, the local expression of equilibrium at $\hat{x}$ is

$$P(\hat{x}+0) - P(\hat{x}-0) = -\lim_{\varepsilon\to 0}\int_{\hat{x}-\varepsilon}^{\hat{x}+\varepsilon}(f - cu) = 0 \tag{2.1.10a}$$

and hence P is continuous. Equation (2.1.10a) can also be written as

$$AE(\hat{x}+0)\frac{du}{dx}(\hat{x}+0) = AE(\hat{x}-0)\frac{du}{dx}(\hat{x}-0) \tag{2.1.10b}$$

and hence du/dx is discontinuous at the points of discontinuity of AE. In the more general case in which f includes a concentrated force with intensity $\hat{f}$ at $x = \hat{x}$, we can write

$$f(x) = \hat{f}\delta(x;\hat{x}) + \tilde{f}(x), \quad x \in I \tag{2.1.11a}$$

where $\tilde{f}$ is piecewise continuous, and the symbol $\delta(\cdot;\hat{x})$ is Dirac's delta function at $\hat{x}$, characterized by the property

$$\int_0^l g\delta(\cdot;\hat{x}) = g(\hat{x}), \quad \forall g \in C^0(I) \tag{2.1.11b}$$

and the local expression of equilibrium at $\hat{x}$ is

$$P(\hat{x}+0) - P(\hat{x}-0) = -\lim_{\varepsilon \to 0} \left(\int_{\hat{x}-\varepsilon}^{\hat{x}+\varepsilon} (\hat{f}\delta(\cdot;\hat{x}) + \tilde{f} - cu) \right) = -\hat{f}. \tag{2.1.11c}$$

This means that P has a jump equal to $-\hat{f}$ at $\hat{x}$.

Remark 2.1.4 The differential equation of equilibrium in terms of the displacement (2.1.3b) holds in the subintervals of continuity of $(d/dx)(AE)$.

Remark 2.1.5 The mathematical formulation given above, can also be understood as the formulation of the problem of one-dimensional heat conduction in a bar, in which case u denotes the temperature, E is the coefficient of thermal conductivity, A is the area of the normal cross section of the bar, σ denotes the heat flux, the constitutive equation $\sigma(x) = -AE(x)(du/dx)(x)$ is Fourier's law, $c = c(x)$ is the coefficient of heat absorption, $f = f(x)$ is the distributed heat source, and α_i, $i = 1, 2$, are the heat-absorption coefficients at the ends of the bar. There are many other interpretations of the above model problem (e.g., one-dimensional flow through a porous medium, equilibrium of a laterally loaded stretched string, etc.).

Above, we have given the mathematical formulation of the one-dimensional model problem, which will be employed throughout this book, and we gave some physical interpretations. The main result is that the model problem consists of a second-order differential equation and boundary conditions which include derivatives up to first order, and are derived from a conservation principle.

Problems

P2.1.1. Let $AE = 1$, $l = 1$, $u(0) = 1$, $c = 0$, $\alpha_2 = 3$, $a_2 = 1$. Find the solution of (2.1.3b) in the case that $f = x$.

P2.1.2. Find the solution of P2.1.1 in the case $f(x) = 2\delta(x;\bar{x})$, $\bar{x} = \frac{1}{2}$.

P2.1.3. Find the solution of the model problem in the case that

$$AE = \begin{cases} 1 & \text{for } 0 < x < \frac{1}{2} \\ 2 & \text{for } \frac{1}{2} < x < 1 \end{cases} \tag{2.1.8}$$

$c \equiv 0$, $f \equiv 1$, and $u(0) = u(1) = 0$.

P2.1.4. Show that the solution of the model problem with $\alpha_1 = \alpha_2 = a_1 = a_2 = 0$, $f \equiv 1$, and $c \equiv 0$ does not exist.

P2.1.5. Determine $u = u(x;c)$, the solution of the model problem with $\alpha_1 = \alpha_2 = a_1 = a_2 = 0$, $AE \equiv 1$, $f \equiv 1$, $c(x) = c > 0$, as a function of c, and show that the limit $\lim_{c\to 0} u(\cdot;c)$ does not exist.

P2.1.6. Show that the solution of the model problem with $\alpha_1 = \alpha_2 = a_1 = a_2 = 0$, $c = 0$, and $f = 0$ is $u = \textit{constant}$.

P2.1.7. Determine the solution of the model problem with $\alpha_1 = \alpha_2 = a_1 = a_2 = 0$, $c = 0$, $AE = 1$, $l = 1$, and $f(x) = (x - \frac{1}{2})$.

P2.1.8. Consider the model problem with $\alpha_1 = \alpha_2 = a_1 = a_2 = 0$, $c = 0$, $AE = 1$, $l = 1$, and $f \in C(I)$. What is the condition that f must satisfy for the problem to have a solution?

2.1* The two-dimensional model problem

We will now introduce the two-dimensional model problems of heat conduction and elasticity, and discuss their similarities and differences with the one-dimensional case.

Let $\Omega \subseteq \mathbb{R}^2$ be a bounded two-dimensional *domain* with *boundary* $\Gamma = \partial\Omega$ consisting of m curvilinear polygonal boundaries $\Gamma^{(j)}$, $j = 1, \ldots, m$, with vertices $A_i^{(j)}, i = 1, \ldots, N(j)$, connected by smooth ($C^\infty$) curvilinear arcs $\Gamma_i^{(j)} = (A_i^{(j)}, A_{i+1}^{(j)})$, as is, for example, shown in Fig. 2.1*.1. We will denote by $\Gamma_i^{(j)} = (A_i^{(j)}, A_{i+1}^{(j)})$ the arc without its end points, and by $\bar{\Gamma}_i^{(j)} = \Gamma_i^{(j)} \cup \{A_i^{(j)}\} \cup \{A_{i+1}^{(j)}\}$ the arc together with its end points, and will let $N(j)$ denote the number of arcs in $\Gamma^{(j)}$ and $A_{N(j)+1}^{(j)} \equiv A_1^{(j)}$. We will denote by $\alpha_i^{(j)}$ the internal angle in $A_i^{(j)}$, and we will assume that $0 < \alpha_i^{(j)} \leq 2\pi$. We allow $\alpha_i^{(j)} = 2\pi$, and in this case the domain has a crack as shown in Fig. 2.1*.1. We will assume that

$$\partial\Omega = \Gamma = \bar{\Gamma}_{\mathrm{D}} \cup \Gamma_{\mathrm{N}}, \qquad \Gamma_{\mathrm{D}} \cap \Gamma_{\mathrm{N}} = \emptyset \tag{2.1*.1}$$

Here Γ_{D} (Γ_{N}) is the part of the boundary where Dirichlet (Neumann) boundary conditions are imposed and we will call Γ_{D} (Γ_{N}) the Dirichlet (Neumann) part of the domain boundary $\partial\Omega$.

First, we will decribe the model problem of steady-state heat conduction. Let u be the *temperature*, $\boldsymbol{\nabla} u \stackrel{\text{def}}{=} [\partial u/\partial x_1, \partial u/\partial x_2]^{\mathrm{T}}$, the *temperature gradient*, and $\mathbf{q}$ the *heat flux*.

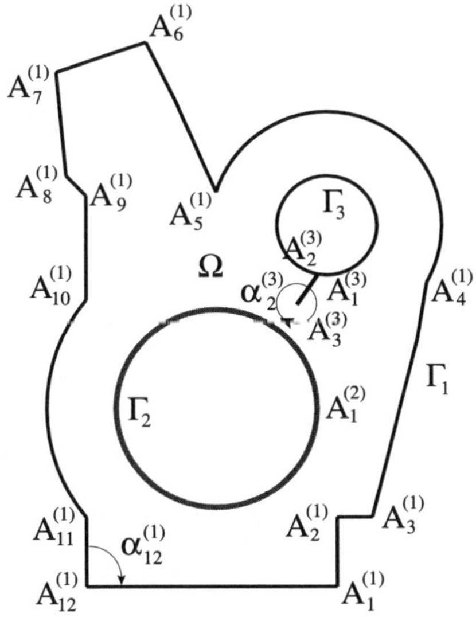

FIG. 2.1*.1. An example of a curvilinear polygonal domain with piecewise smooth boundary with $m = 3$, $N(1) = 12$, $N(2) = 1$, $N(3) = 2$, $\alpha_2^{(3)} = 2\pi$

We will assume a linear heat-conducting medium for which we have

$$\mathbf{q} = \mathbf{A}\boldsymbol{\nabla}u \tag{2.1*.2a}$$

or in component form

$$q_i = \sum_{j=1}^{2} a_{ij}\frac{\partial u}{\partial x_j}, \quad i = 1, 2 \tag{2.1*.2b}$$

where $\mathbf{A} = [a_{ij},\ i, j = 1, 2]$ is a symmetric positive definite matrix known as the *thermal conductivity matrix*. Equation (2.1*.2) is known as *Fourier's law.*

The mathematical formulation of the problem of heat conduction is based on the principle of conservation of heat flux, which for any arbitrary subdomain $\omega \subseteq \Omega$, states that

$$\int_{\partial\omega} \mathbf{q}\cdot\mathbf{n} + \int_{\omega}(f - cu) = \int_{\partial\omega}\sum_{i=1}^{2} q_i n_i + \int_{\omega}(f - cu) = 0 \tag{2.1*.3a}$$

where $\mathbf{n}$ is the outward unit normal on $\partial\omega$, f is the intensity of the heat source, and c is the heat absorption coefficient. Employing the Gauss theorem

$$\int_{\partial\omega}\sum_{i=1}^{2} q_i n_i = \int_{\omega}\sum_{i=1}^{2}\frac{\partial q_i}{\partial x_i}, \tag{2.1*.3b}$$

equation (2.1*.3a) can be written as

$$\int_{\omega}\left(\sum_{i=1}^{2}\frac{\partial q_i}{\partial x_i} - cu + f\right) = 0. \tag{2.1*.3c}$$

As in the one-dimensional case, assuming that f, c are continuous functions, $f, c \in C(\bar{\Omega})$, employing the mean value theorem and shrinking ω to a point $\mathbf{x}$, we get

$$-\boldsymbol{\nabla}\cdot\mathbf{q}(\mathbf{x}) + c(\mathbf{x})u(\mathbf{x}) = \sum_{i=1}^{2}\frac{\partial q_i}{\partial x_i}(\mathbf{x}) + c(\mathbf{x})u(\mathbf{x}) = f(\mathbf{x}), \quad \mathbf{x}\in\Omega. \tag{2.1*.3d}$$

This is the differential form of the principle of conservation of heat flux. Further, employing (2.1*.2) to eliminate $\mathbf{q}$, we get

$$-\sum_{i,j=1}^{2}\frac{\partial}{\partial x_i}\left(a_{ij}(\mathbf{x})\frac{\partial u}{\partial x_j}(\mathbf{x})\right) + c(\mathbf{x})u(\mathbf{x}) = f(\mathbf{x}), \quad \mathbf{x}\in\Omega \tag{2.1*.4a}$$

which can also be written as

$$-\boldsymbol{\nabla}\cdot(\mathbf{A}(\mathbf{x})\boldsymbol{\nabla}u(\mathbf{x})) + c(\mathbf{x})u(\mathbf{x}) = f(\mathbf{x}), \quad \mathbf{x}\in\Omega. \tag{2.1*.4b}$$

This is the partial differential equation satisfied by the temperature, which will be our two-dimensional model problem. For $\mathbf{x} \in \partial\Omega$, choosing ω such that $\mathbf{x} \in \partial\omega \cap \partial\Omega$, and shrinking ω to $\mathbf{x}$, we obtain

$$\mathbf{n}^{\mathrm{T}}(\mathbf{x})\mathbf{q}(\mathbf{x}) \stackrel{\text{def}}{=} \sum_{i=1}^{2} q_i(\mathbf{x})n_i(\mathbf{x}) = -\alpha(\mathbf{x})u(\mathbf{x}) + g(\mathbf{x}), \quad \mathbf{x} \in \partial\Omega \tag{2.1*.5a}$$

where $\mathbf{n} = (n_1, n_2)$ is the outward unit normal on $\partial\Omega$. This is the natural boundary condition at $\mathbf{x} \in \partial\Omega$, which can also be written as

$$-\mathbf{n}^{\mathrm{T}}(\mathbf{x})\mathbf{A}(\mathbf{x})\boldsymbol{\nabla}u(\mathbf{x}) \stackrel{\text{def}}{=} -\sum_{i,j=1}^{2} a_{ij}(\mathbf{x})\frac{\partial u}{\partial x_j}(\mathbf{x})n_i(\mathbf{x}) = -\alpha(\mathbf{x})u(\mathbf{x}) + g(\mathbf{x}). \tag{2.1*.5b}$$

In the special case that $\alpha \equiv 0$, (2.1*.5b) takes the form

$$\mathbf{n}^{\mathrm{T}}(\mathbf{x})\mathbf{A}(\mathbf{x})\boldsymbol{\nabla}u(\mathbf{x}) \stackrel{\text{def}}{=} \sum_{i,j=1}^{2} a_{ij}(\mathbf{x})\frac{\partial u}{\partial x_j}(\mathbf{x})n_i(\mathbf{x}) = g(\mathbf{x}). \tag{2.1*.6}$$

Equation (2.1*.6) (respectively (2.1*.5)) is known as the *Neumann* (respectively Newton) boundary condition at $\mathbf{x} \in \partial\Omega$. As in the one-dimensional case, the Dirichlet boundary condition can be obtained as a limiting case of the natural boundary condition. We have

$$u(\mathbf{x}) = \frac{g(\mathbf{x})}{\alpha(\mathbf{x})} - \sum_{i,j=1}^{2} \frac{a_{ij}(\mathbf{x})}{\alpha(\mathbf{x})}\frac{\partial u}{\partial x_j}(\mathbf{x})n_j(\mathbf{x}). \tag{2.1*.7a}$$

Letting $\mathbf{x} \in \Gamma_{\mathrm{D}}$, taking the limit as $a_{ij}(\mathbf{x})/\alpha(\mathbf{x}) \to 0$, and letting $g(\mathbf{x}) = \alpha(\mathbf{x})\bar{u}(\mathbf{x})$, we obtain

$$u(\mathbf{x}) = \bar{u}(\mathbf{x}), \quad \mathbf{x} \in \Gamma_{\mathrm{D}}. \tag{2.1*.7b}$$

This is the essential or *Dirichlet* boundary condition on Γ_{D}.

In the special case $\alpha_{ij} = \delta_{ij}$, where δ_{ij} is Kronecker's delta, the partial differential equation (2.1*.4) takes the form

$$-\Delta u(\mathbf{x}) + c(\mathbf{x})u(\mathbf{x}) = f(\mathbf{x}), \quad \mathbf{x} \in \Omega \tag{2.1*.8a}$$

and the Neumann boundary condition reads

$$\frac{\partial u}{\partial n}(\mathbf{x}) \stackrel{\text{def}}{=} \mathbf{n}^{\mathrm{T}}\boldsymbol{\nabla}u(\mathbf{x}) = g(\mathbf{x}), \quad \mathbf{x} \in \Gamma_{\mathrm{N}} \tag{2.1*.8b}$$

where

$$\Delta \stackrel{\text{def}}{=} \frac{\partial^2}{\partial x_1^2} + \frac{\partial^2}{\partial x_2^2} \tag{2.1*.8c}$$

is the *Laplacian operator*, and

$$\frac{\partial}{\partial n} \stackrel{\text{def}}{=} \mathbf{n}^{\mathrm{T}}\boldsymbol{\nabla} = \sum_{i=1}^{2} \frac{\partial}{\partial x_i}n_i \tag{2.1*.8d}$$

is the *outward normal derivative operator*.

Remark 2.1*.1 The Neumann or Newton boundary condition in the one-dimensional case may be formulated as a special case of the corresponding two-dimensional boundary condition.

Let us also formulate the model problem of plane elasticity. Let $\mathbf{u} = [u_1, u_2]^{\mathrm{T}}$ be the *displacement vector field*,

$$\boldsymbol{\epsilon} = \begin{Bmatrix} \epsilon_{11} \\ \epsilon_{22} \\ \gamma_{12} \end{Bmatrix} = \mathbf{D}\mathbf{u} = \begin{bmatrix} \dfrac{\partial}{\partial x_1} & 0 \\ 0 & \dfrac{\partial}{\partial x_2} \\ \dfrac{\partial}{\partial x_2} & \dfrac{\partial}{\partial x_1} \end{bmatrix} \begin{Bmatrix} u_1 \\ u_2 \end{Bmatrix} = \begin{Bmatrix} \dfrac{\partial u_1}{\partial x_1} \\ \dfrac{\partial u_2}{\partial x_2} \\ \dfrac{\partial u_1}{\partial x_2} + \dfrac{\partial u_2}{\partial x_1} \end{Bmatrix} \tag{2.1*.9a}$$

the *engineering strain vector*, where

$$\mathbf{D} = \begin{bmatrix} \dfrac{\partial}{\partial x_1} & 0 \\ 0 & \dfrac{\partial}{\partial x_2} \\ \dfrac{\partial}{\partial x_2} & \dfrac{\partial}{\partial x_1} \end{bmatrix} \tag{2.1*.9b}$$

and $\boldsymbol{\sigma} = [\sigma_{11}, \sigma_{22}, \sigma_{12}]^{\mathrm{T}}$ is the *engineering stress vector*, which is related to $\boldsymbol{\epsilon}$ by Hooke's law

$$\boldsymbol{\sigma} = \mathbf{E}\boldsymbol{\epsilon}. \tag{2.1*.9c}$$

Here $\mathbf{E} = [E_{ij},\ i,j = 1,2,3]$ is a symmetric positive definite elasticity matrix, which, in the case of isotropic material, is given by

$$\mathbf{E} = \begin{bmatrix} \lambda + 2\mu & \lambda & 0 \\ \lambda & \lambda + 2\mu & 0 \\ 0 & 0 & \mu \end{bmatrix} \tag{2.1*.9d}$$

where λ, μ are known as the *Lamé* constants. Note that

$$\begin{aligned} \boldsymbol{\sigma} \equiv \mathbf{0} \quad &\Longleftrightarrow \quad \boldsymbol{\varepsilon} = \mathbf{D}\mathbf{u} \equiv \mathbf{0} \\ &\Longleftrightarrow \quad \mathbf{u}(\mathbf{x}) = a_1 \begin{Bmatrix} 1 \\ 0 \end{Bmatrix} + a_2 \begin{Bmatrix} 0 \\ 1 \end{Bmatrix} + a_3 \begin{Bmatrix} -x_2 \\ x_1 \end{Bmatrix} \end{aligned} \tag{2.1*.9e}$$

where $\mathbf{u}_1 = [1, 0]^{\mathrm{T}}$, $\mathbf{u}_2 = [0, 1]^{\mathrm{T}}$, and $\mathbf{u}_3 = [-x_2, x_1]^{\mathrm{T}}$ are the *rigid-body modes.* The mathematical formulation is based on the principle of equilibrium of forces

$$\int_{\partial\omega} \mathbf{N}\boldsymbol{\sigma} + \int_{\omega} \mathbf{f} = \mathbf{0} \tag{2.1*.10a}$$

and on the principle of equilibrium of moments

$$\int_{\partial\omega} \mathbf{x} \times \mathbf{N}\boldsymbol{\sigma} + \int_{\omega} \mathbf{x} \times \mathbf{f} = 0 \tag{2.1*.10b}$$

where we used the notation

$$\mathbf{N} = \begin{bmatrix} n_1 & 0 & n_2 \\ 0 & n_2 & n_1 \end{bmatrix}. \tag{2.1*.10c}$$

Employing the Gauss theorem, and shrinking ω to a point $\mathbf{x}$ we get

$$\begin{bmatrix} \frac{\partial}{\partial x_1} & 0 & \frac{\partial}{\partial x_2} \\ 0 & \frac{\partial}{\partial x_2} & \frac{\partial}{\partial x_1} \end{bmatrix} \begin{Bmatrix} \sigma_{11}(\mathbf{x}) \\ \sigma_{22}(\mathbf{x}) \\ \sigma_{12}(\mathbf{x}) \end{Bmatrix} = \begin{Bmatrix} \frac{\partial \sigma_{11}}{\partial x_1}(\mathbf{x}) + \frac{\partial \sigma_{12}}{\partial x_2}(\mathbf{x}) \\ \frac{\partial \sigma_{12}}{\partial x_1}(\mathbf{x}) + \frac{\partial \sigma_{22}}{\partial x_2}(\mathbf{x}) \end{Bmatrix} = -\begin{Bmatrix} f_1(\mathbf{x}) \\ f_2(\mathbf{x}) \end{Bmatrix} \tag{2.1*.11a}$$

for any $\mathbf{x} \in \Omega$, which can also be written in the form

$$\mathbf{D}^{\mathrm{T}} \boldsymbol{\sigma}(\mathbf{x}) = \mathbf{f}(\mathbf{x}), \quad \mathbf{x} \in \Omega. \tag{2.1*.11b}$$

This is the differential equation of equilibrium. Furthermore, in the case of isotropic material, we have

$$\begin{bmatrix} (\lambda+\mu)\frac{\partial}{\partial x_1}\left(\frac{\partial}{\partial x_1} + \frac{\partial}{\partial x_2}\right) + \mu\Delta \\ (\lambda+\mu)\frac{\partial}{\partial x_2}\left(\frac{\partial}{\partial x_1} + \frac{\partial}{\partial x_2}\right) + \mu\Delta \end{bmatrix} \begin{Bmatrix} u_1 \\ u_2 \end{Bmatrix} = -\begin{Bmatrix} f_1 \\ f_2 \end{Bmatrix} \tag{2.1*.12a}$$

which can also be written as

$$(\lambda + \mu)\, \boldsymbol{\nabla}(\boldsymbol{\nabla} \cdot \mathbf{u}) + \mu \Delta \mathbf{u} = -\mathbf{f}. \tag{2.1*.12b}$$

This is the differential equation of equilibrium in the case of isotropic material.

Let us also derive the natural boundary conditions. Applying the equilibrium of forces at any $\mathbf{x} \in \partial\Omega$, we get

$$\begin{bmatrix} n_1 & 0 & n_2 \\ 0 & n_1 & n_2 \end{bmatrix} \begin{Bmatrix} \sigma_{11}(\mathbf{x}) \\ \sigma_{22}(\mathbf{x}) \\ \sigma_{12}(\mathbf{x}) \end{Bmatrix} = \begin{bmatrix} \alpha_{11} & \alpha_{12} \\ \alpha_{12} & \alpha_{22} \end{bmatrix} \begin{bmatrix} u_1(\mathbf{x}) \\ u_2(\mathbf{x}) \end{bmatrix} - \begin{bmatrix} g_1(\mathbf{x}) \\ g_2(\mathbf{x}) \end{bmatrix} \tag{2.1*.13a}$$

which can also be written in the form

$$\mathbf{N}\boldsymbol{\sigma}(\mathbf{x}) = \boldsymbol{\alpha}(\mathbf{x})\mathbf{u}(\mathbf{x}) - \mathbf{g}(\mathbf{x}). \tag{2.1*.13b}$$

This is the natural (or traction) boundary condition at $\mathbf{x} \in \partial\Omega$. Here $\boldsymbol{\alpha}$ is a symmetric positive semidefinite matrix which represents the action of an elastic support on $\partial\Omega$. Employing Hooke's law, $\boldsymbol{\sigma} = \mathbf{EDu}$, we obtain

$$-\mathbf{N}(\mathbf{x})\mathbf{E}(\mathbf{x})\mathbf{D}(\mathbf{x})\mathbf{u}(\mathbf{x}) = \boldsymbol{\alpha}(\mathbf{x})\mathbf{u}(\mathbf{x}) - \mathbf{g}. \tag{2.1*.13c}$$

This is the natural boundary condition in terms of the displacement. As before, the Dirichlet boundary condition at $\mathbf{x} \in \Gamma_{\mathrm{D}}$ can be derived from the natural boundary condition by noting that

$$\mathbf{u}(\mathbf{x}) = \boldsymbol{\alpha}^{-1}(\mathbf{x})\mathbf{g}(\mathbf{x}) + \boldsymbol{\alpha}^{-1}(\mathbf{x})\, \mathbf{N}(\mathbf{x})\mathbf{E}(\mathbf{x})\mathbf{D}(\mathbf{x})\mathbf{u}(\mathbf{x}) \tag{2.1*.14a}$$

letting $\bar{\mathbf{u}} = \boldsymbol{\alpha}^{-1}\mathbf{g}$, and taking the limit as $\boldsymbol{\alpha}^{-1}\mathbf{NEDu} \to 0$, to obtain

$$\mathbf{u}(\mathbf{x}) = \bar{\mathbf{u}}(\mathbf{x}), \quad \mathbf{x} \in \Gamma_{\mathrm{D}}. \tag{2.1*.14b}$$

Remark 2.1*.2 Several other types of boundary conditions may be encountered in practice. For example, the component of the displacement normal to the boundary and the tangential component of traction may be specified, or vice versa, etc.

Remark 2.1*.3 In the case of the scalar model problem, which in our case is the problem of heat conduction, the 'rigid-body mode' is the constant function. In the case of the vector-valued model problem, namely, the elasticity problem, there exist three 'rigid-body modes,' two constant vectors, and one which varies linearly in space and has the physical meaning of *rotation*. The rigid body mode of rotation sometimes poses difficulties in the FEM.

The model problems which we have addressed above belong to the category of elliptic boundary value problems. For an explanation of the basic ideas, see [Rektorys, 1998]. For the general theory, the reader may refer to [Courant and Hilbert, 1953], [Agmon, 1965], [Ladyzhenskaja and Uraltseva, 1968], [Lions and Magenes, 1972], [Egorov and Shubin, 1998], [Gilbarg and Trudinger, 1998], and [Taylor, 1999]. The theory of elasticity is, for example, addressed in [Malvern, 1969], [Gurtin, 1972], [de Veubeke, 1979], [Nečas and Hlavaček, 1981], [Marsden and Hughes, 1983], and [Kondratév and Oleijnik, 1990].

In this section we derived the mathematical formulation of the problems of heat conduction and elasticity in two dimensions.

Problems

P2.1*.1. Let r, θ be polar coordinates $x_1 = r\cos\theta$, $x_2 = r\sin\theta$. Show that the Laplacian operator Δ, can be written in the form

$$\Delta = \frac{\partial^2}{\partial x_1^2} + \frac{\partial^2}{\partial x_2^2} = \frac{\partial^2}{\partial r^2} + \frac{1}{r}\frac{\partial}{\partial r} + \frac{1}{r^2}\frac{\partial^2}{\partial \theta^2} = \frac{1}{r}\left(\frac{\partial}{\partial r}\left(r\frac{\partial}{\partial r}\right) + \frac{1}{r}\frac{\partial^2}{\partial \theta^2}\right). \tag{2.1*.15}$$

P2.1*.2. Consider the homogeneous Dirichlet problem for Poisson's equation

$$-\Delta u|_{\Omega_R} = 1, \qquad u|_{\partial\Omega_R} = 0 \tag{2.1*.16a}$$

where $\Omega_R \stackrel{\text{def}}{=} \{\mathbf{x} \mid x_1^2 + x_2^2 < R^2\}$. Show that $u(r,\theta) = u(r)$, can be obtained as the solution of the one-dimensional boundary value problem

$$-\frac{d}{dr}\left(r\frac{du}{dr}\right) = 1, \quad 0 < r < R, \qquad u(R) = 0, \qquad \frac{\partial u}{\partial r}(0) = 0. \tag{2.1*.16b}$$

Find an explicit form for u in terms of x_1, x_2.

P2.1*.3. Employ polar coordinates to obtain the explicit form of the solution of the nonhomogeneous Dirichlet problem

$$-\Delta u|_{\Omega_R} = 0, \qquad u|_{\partial\Omega_R} = \sin(k\theta). \tag{2.1*.17}$$

P2.1*.4. Consider the boundary value problem

$$-\left(\frac{\partial}{\partial x_1}\left(a_{11}\frac{\partial u}{\partial x_1}\right)+\frac{\partial}{\partial x_2}\left(a_{22}\frac{\partial u}{\partial x_2}\right)\right)\Big|_{\Omega}=1, \qquad u|_{\partial\Omega}=0 \tag{2.1*.18a}$$

where $\Omega = (-a,a)\times(-b,b)$, and a_{11}, a_{22} are constant functions. Employ the transformation

$$\xi_1 = c_1 x_1, \qquad \xi_2 = c_2 x_2 \tag{2.1}$$

to transform (2.1*.18a) into the standard Poisson equation

$$-\Delta\hat{u}|_{\hat{\Omega}} = \hat{C}, \qquad \hat{u}|_{\partial\hat{\Omega}} = 0 \tag{2.1*.18c}$$

where $\hat{u}(\xi_1,\xi_2) \stackrel{\text{def}}{=} u(x_1,x_2)$. Specify c_i, C, and $\hat{\Omega}$.

P2.1*.5. Let a_{ij} be the entries of a symmetric positive definite matrix. Construct a linear transformation $\boldsymbol{\xi} = \mathbf{C}\mathbf{x}$, which transforms the differential equation

$$\sum_{i,j=1}^{2}\frac{\partial}{\partial x_i}\left(a_{ij}\frac{\partial u}{\partial x_j}\right) \equiv 0 \tag{2.1*.19}$$

into the Laplace equation $\Delta\hat{u} \equiv 0$.

2.2 The variational formulation of the model problem

In the previous section, we gave the mathematical formulation of the model problem. Given AE, c, α_2, f, and a_2, such that

$$AE \in C^1(\bar{I}), \quad 0 < \beta_1 \le AE \le \beta_2 < \infty \tag{2.2.1a}$$

and

$$f, c \in C^0(\bar{I}), \quad c \ge 0, \ \alpha_2 \ge 0 \tag{2.2.1b}$$

we seek the solution of the boundary value problem:

Find $u \in C^2(0,l)$ such that

$$-\frac{d}{dx}\left(AE(x)\frac{du}{dx}(x)\right) + c(x)u(x) = f(x), \quad x \in I \tag{2.2.1c}$$

and

$$u(0) = 0, \quad -AE(l)\frac{du}{dx}(l) = \alpha_2 u(l) - a_2. \tag{2.2.1d}$$

Everywhere below, we will employ the homogeneous (zero) Dirichlet boundary condition at the left end point, $u(0) = 0$, except if it is stated otherwise. The boundary value problem (2.2.1) is often referred to as the *classical* or *strong* formulation of the boundary value problem for the model problem. It seeks a solution which satisfies the differential equation (2.2.1c) at every interior point, and the boundary conditions (2.2.1d) at the end points.

Remark 2.2.1 In the case that AE, f, and c are only piecewise smooth, u is only piecewise $C^2(\bar{I})$, and the jump conditions at the points of discontinuity of AE, f, and c must be added to (2.2.1c) and (2.2.1d).

We will now construct an alternative formulation. Let v be a *sufficiently smooth* function which is defined over $\bar{I} = [0, l]$, and satisfies the condition $v(0) = 0$. Multiplying (2.2.1c) by v, integrating over $(0, l)$, and then employing integration by parts

$$\int_0^l \frac{d}{dx}\left(AE\frac{du}{dx}\right)v = \left[AE\frac{du}{dx}v\right]_0^l - \int_0^l AE\frac{du}{dx}\frac{dv}{dx} \tag{2.2.2a}$$

we obtain

$$\int_0^l \left(AE\frac{du}{dx}\frac{dv}{dx} + cuv\right) - AE(l)\frac{du}{dx}(l)v(l) = \int_0^l fv.$$

Substituting $AE(l)(du/dx)(l)$ from (2.2.1d), the natural boundary condition at $x = l$, we get

$$\int_0^l \left(AE\frac{du}{dx}\frac{dv}{dx} + cuv\right) + \alpha_2 u(l)v(l) = \int_0^l fv + a_2 v(l). \tag{2.2.2b}$$

Equation (2.2.2b) will be used to construct a new formulation, which seeks a solution u which satisfies (2.2.2b) and the essential boundary condition $u(0) = 0$, for all sufficiently smooth functions v which satisfy $v(0) = 0$. This formulation is known as the *variational* or *weak* formulation of the model problem.

Before we can write the precise mathematical statement of the variational formulation, we have to define what we mean by the term 'sufficiently smooth.' Let us recall the assumptions

$$0 < \beta_1 \leq AE(x) \leq \beta_2 < \infty, \qquad c(x) \geq 0, \;\; x \in I, \qquad \alpha_2 \geq 0. \tag{2.2.3a}$$

We will say that a function v is sufficiently smooth if we have

$$\|v\|_{\mathcal{U}} \stackrel{\text{def}}{=} \sqrt{\int_0^l \left(AE\left(\frac{dv}{dx}\right)^2 + cv^2\right) + \alpha_2(v(l))^2} < \infty. \tag{2.2.3b}$$

The symbol $\|\cdot\|_{\mathcal{U}}$ is a mapping of functions into real numbers, and is known as an *energy norm.* In the cases that either $c \not\equiv 0$, or $\alpha_2 \neq 0$, or u is constrained to satisfy the boundary condition $u(0) = 0$, $\|\cdot\|_{\mathcal{U}}$ also has the property

$$\|u\|_{\mathcal{U}} = 0 \quad \Longleftrightarrow \quad u \equiv 0 \tag{2.2.3c}$$

and it is a *norm.* For the definitions of norm and seminorm, see Appendix A2.1.1. The set

$$\mathcal{U} = \mathcal{U}(I) \stackrel{\text{def}}{=} \{v = v(x), \; x \in I \mid \|v\|_{\mathcal{U}} < \infty, \; v(0) = 0\} \tag{2.2.4}$$

will be called the *energy space*. The energy space is the set of all functions v which have finite energy norm, $\|v\|_{\mathcal{U}}$, and satisfy the zero essential boundary conditions $v(0) = 0$. The set $\mathcal{U}$ is a linear space of functions (it satisfies the properties given in Appendix A2.1.1). Note that the functions $v \in \mathcal{U}$ are not required to satisfy the natural boundary conditions.

Remark 2.2.2 The terminology stems from structural mechanics where the quantity $\frac{1}{2}\|v\|_{\mathcal{U}}^2$ is known as the *strain energy* which corresponds to the axial displacement v.

Remark 2.2.3 The energy space $\mathcal{U}$ is the completion of the space of smooth functions with respect to the energy norm. More precisely, a function v belongs to $\mathcal{U}$ if there exists a sequence of functions $v_j \in C^1(\bar{I})$, $v(0) = 0$, which is a Cauchy sequence in $\mathcal{U}$ (see A2.2.2), and converges to v in $\mathcal{U}$, namely, $\lim_{j\to\infty} \|v - v_j\|_{\mathcal{U}} = 0$. This means that $\mathcal{U}$ is complete, which is an essential property of the energy-space.

Remark 2.2.4 Under the assumptions (2.2.3a), we have $\mathcal{U} \subseteq C^0(\bar{I})$, and the boundary condition $u(0) = 0$ has good meaning. The essential boundary conditions are preserved when limits of smooth functions are taken in the $\|\cdot\|_{\mathcal{U}}$ norm. For any $u \in C^1(I)$, with $u(0) \neq 0$, there exists no Cauchy sequence $u_n \in \mathcal{U}$, with $u_n(0) = 0$, such that $\|u - u_n\|_{\mathcal{U}(I)} \to 0$ as $n \to \infty$ (see P2.2.23).

Remark 2.2.5 Given any $u \in \mathcal{U}$ with $(du/dx)(l) \neq 0$, there exists a Cauchy sequence of functions $u_n \in C^1(I) \subseteq \mathcal{U}$, such that $(d/dx)(u_n)(l) = 0$, while $\|u - u_n\|_{\mathcal{U}} \to 0$ as $n \to \infty$ (see P2.2.21). Hence, the natural boundary conditions are not preserved when the limits are taken in the norm $\|\cdot\|_{\mathcal{U}}$. This is the reason why the natural boundary conditions are not included in the definition of $\mathcal{U}$.

Let $\mathcal{B} : \mathcal{U} \times \mathcal{U} \mapsto \mathbb{R}$ be given by

$$\mathcal{B}(u, v) \stackrel{\text{def}}{=} \int_0^l \left(AE\frac{du}{dx}\frac{dv}{dx} + cuv \right) + \alpha_2 u(l)v(l) \tag{2.2.5a}$$

and $\mathcal{L} : \mathcal{U} \mapsto \mathbb{R}$ be given by

$$\mathcal{L}(v) \stackrel{\text{def}}{=} \int_0^l fv + a_2 v(l). \tag{2.2.5b}$$

The mapping $\mathcal{B}$ is linear with respect to each of its arguments, namely

$$\mathcal{B}(\alpha_1 u_1 + \alpha_2 u_2, v) = \alpha_1 \mathcal{B}(u_1, v) + \alpha_2 \mathcal{B}(u_2, v) \tag{2.2.6a}$$

and

$$\mathcal{B}(u, \alpha_1 v_1 + \alpha_2 v_2) = \alpha_1 \mathcal{B}(u, v_1) + \alpha_2 \mathcal{B}(u, v_2). \tag{2.2.6b}$$

$\mathcal{B}(\cdot,\cdot)$ is known as a *bilinear* form. Further, under the assumptions (2.2.1), $\mathcal{B}(\cdot,\cdot)$ defines an *inner product* on $\mathcal{U}$, with the corresponding norm

$$\|v\|_{\mathcal{U}} = \sqrt{\mathcal{B}(v, v)}. \tag{2.2.6c}$$

An important property which we will use often below is the *Cauchy–Schwarz inequality*

$$|\mathcal{B}(u, v)| \leq \|u\|_{\mathcal{U}} \, \|v\|_{\mathcal{U}}. \tag{2.2.6d}$$

It follows that $|\mathcal{B}(u, v)| < \infty$, for any u, $v \in \mathcal{U}$ (see P2.2.2).

The mapping $\mathscr{L}$ is linear, namely

$$\mathscr{L}(\alpha_1 v_1 + \alpha_2 v_2) = \alpha_1 \mathscr{L}(v_1) + \alpha_2 \mathscr{L}(v_2) \tag{2.2.7a}$$

and it is often called a linear form. We will only be interested in the cases where

$$|\mathscr{L}(v)| \leq \mathscr{C}\, \|v\|_{\mathscr{U}} \quad \forall v \in \mathscr{U} \tag{2.2.7b}$$

with $\mathscr{C}$ independent of v. When (2.2.7b) holds, we will say that $\mathscr{L}$ is a *bounded* linear form, or a *bounded linear functional* on $\mathscr{U}$ (see Appendix A2.1.2). For example, assuming that

$$\|f\|_{L^2(I)} < \infty, \qquad |a_2| < \infty \tag{2.2.7c}$$

and employing the Cauchy–Schwarz inequality we get

$$|\mathscr{L}(v)| = \left| \int_0^l fv + a_2 v(l) \right| \leq \|f\|_{L^2(I)} \|v\|_{L^2(I)} + |a_2|\, |v(l)| \tag{2.2.7d}$$

where $\|v\|_{L^2(I)} \stackrel{\text{def}}{=} \sqrt{\int_I v^2}$, and employing the inequalities

$$\|v\|_{L^2(I)} \leq \mathscr{C} \|v\|_{\mathscr{U}} \tag{2.2.7e}$$

and

$$|v(l)| \leq \mathscr{C}\, \|v\|_{\mathscr{U}} \tag{2.2.7f}$$

which will be proven in Lemma 2.2.1 (see (2.2.18b,c)) below, we get

$$|\mathscr{L}(v)| \leq \mathscr{C}\big(\|f\|_{L^2(I)} + |a_2|\big) \|v\|_{\mathscr{U}} \quad \forall v \in \mathscr{U}. \tag{2.2.7g}$$

Using the above notations, the *generalized or variational formulation* of the mathematical problem can be written as follows:

Find $u_{EX} \in \mathscr{U}$ such that

$$\mathscr{B}(u_{EX}, v) = \mathscr{L}(v) \quad \forall v \in \mathscr{U}. \tag{2.2.8}$$

The solutions of (2.2.8) are called *generalized* or *weak* solutions because they need not satisfy the differential equation in the classical *pointwise* sense. The functions u are often called *trial solutions* or *trial functions* because they can be tried to see if they satisfy (2.2.8). The functions v are called *test functions* because they are used to test if a particular trial solution satisfies the variational formulation.

Note that if the classical solution exists, then it is also the weak solution, but the converse is not true, in general.

Remark 2.2.6 As we will see below, the boundedness of $\mathscr{L}$ is needed to guarantee the existence of $u_{EX} \in \mathscr{U}$.

Remark 2.2.7 The variational equation (2.2.8) may be replaced by

$$\mathscr{B}(u_{EX}, v) = \mathscr{L}(v) \quad \forall v \in C^\infty_{(0}(\bar{I}) \stackrel{\text{def}}{=} \{w \in C^\infty(\bar{I}) \mid w(0) = 0\}. \tag{2.2.9a}$$

This is because $C^\infty_{(0}(\bar{I})$ is *dense* in $\mathcal{U}$, i.e., any function $v \in \mathcal{U}$ may be approximated arbitrarily close by functions from $C^\infty_{(0}(\bar{I})$, namely

$$\forall v \in \mathcal{U}, \ \forall \epsilon > 0 \quad \exists w_\epsilon(v) \in C^\infty_{(0}(\bar{I}), \quad \|v - w_\epsilon(v)\|_{\mathcal{U}} < \epsilon. \tag{2.2.9b}$$

We will now show that the generalized formulation (2.2.8) has at most one solution. We will address the case with more general (weaker) assumptions imposed on AE than (2.2.1a) and (2.2.3a), employed above.

Theorem 2.2.1 *Under the assumption $c \geq 0$, $\alpha_2 \geq 0$, and*

$$\gamma x^\vartheta \leq AE, \quad 0 \leq \vartheta < 1, \ \gamma > 0 \tag{2.2.10}$$

the solution of (2.2.8), if it exists, is unique.

Proof Assume that there exist $u_1, u_2 \in \mathcal{U}$, $u_1 \not\equiv u_2$, such that

$$\mathcal{B}(u_1, v) = \mathcal{L}(v) \quad \forall v \in \mathcal{U} \tag{2.2.11a}$$

and

$$\mathcal{B}(u_2, v) = \mathcal{L}(v) \quad \forall v \in \mathcal{U}. \tag{2.2.11b}$$

Subtracting, we get $u_1 - u_2 \in \mathcal{U}$, and

$$\mathcal{B}(u_1 - u_2, v) = 0 \quad \forall v \in \mathcal{U} \tag{2.2.11c}$$

and letting $v = u_1 - u_2 \in \mathcal{U}$, we obtain

$$\|u_1 - u_2\|^2_{\mathcal{U}} = \mathcal{B}(u_1 - u_2, u_1 - u_2) = 0. \tag{2.2.11d}$$

Now, because of the assumptions, $\|\cdot\|_{\mathcal{U}}$ is a norm on $\mathcal{U}$ (see P2.2.4), and we have

$$\|u_1 - u_2\|_{\mathcal{U}} = 0 \quad \Longrightarrow \quad u_1 - u_2 \equiv 0. \tag{2.2.11e}$$

Note that in the case $\vartheta \geq 1$, assumption (2.2.10) is not sufficient to guarantee that $\|\cdot\|_{\mathcal{U}}$ is a norm (see P2.2.5). □

Hence, *if there exists a solution of (2.2.8), and if the classical solution belongs to $\mathcal{U}$, these two solutions are identical*, and (2.2.8) is a generalization of the classical formulation.

We will now address the existence of solutions of (2.2.8), under the assumption that $\mathcal{L}$ is bounded. We will say that:

A linear functional $\mathcal{M} : \mathcal{U} \mapsto \mathbb{R}$ is bounded if there exists a $\mathcal{C}$, $0 \leq \mathcal{C} < \infty$, such that

$$|\mathcal{M}(v)| \leq \mathcal{C}\, \|v\|_{\mathcal{U}} \quad \forall v \in \mathcal{U}. \tag{2.2.12a}$$

We will denote *the set of bounded linear functionals on* $\mathcal{U}$, by $L(\mathcal{U}, \mathbb{R})$. It can be proven that $L(\mathcal{U}, \mathbb{R})$ is a linear space, and it can be normed by the norm $\|\cdot\|_{L(\mathcal{U},\mathbb{R})}$, defined by

$$\|\mathcal{M}\|_{L(\mathcal{U},\mathbb{R})} \stackrel{\text{def}}{=} \sup_{v \in \mathcal{U}} \frac{|\mathcal{M}(v)|}{\|v\|_{\mathcal{U}}} = \sup_{v \in \mathcal{U}} \left|\mathcal{M}\left(\frac{v}{\|v\|_{\mathcal{U}}}\right)\right| = \sup_{\substack{v \in \mathcal{U} \\ \|v\|_{\mathcal{U}} = 1}} |\mathcal{M}(v)|. \tag{2.2.12b}$$

Note that every $\mathcal{M} \in L(\mathcal{U}, \mathbb{R})$ is continuous. This follows from the observation that for any $v_1, v_2 \in \mathcal{U}$, we have

$$|\mathcal{M}(v_1) - \mathcal{M}(v_2)| = |\mathcal{M}(v_1 - v_2)| \leq \|\mathcal{M}\|_{L(\mathcal{U},\mathbb{R})} \|v_1 - v_2\|_{\mathcal{U}}. \tag{2.2.12c}$$

Let us now state the main result about the existence of solutions of (2.2.8).

Theorem 2.2.2 *Under the assumption that*

$$0 < \beta_1 \leq AE(x) \leq \beta_2 < \infty, \qquad 0 \leq c(x) \leq \gamma < \infty, \qquad 0 \leq \alpha_2 \tag{2.2.13a}$$

and that f is such that

$$\|\mathcal{L}\|_{L(\mathcal{U},\mathbb{R})} = \sup_{v \in \mathcal{U}} \frac{|\mathcal{L}(v)|}{\|v\|_{\mathcal{U}}} = \sup_{v \in \mathcal{U}} \frac{\left|\int_0^l fv + a_2 v(l)\right|}{\|v\|_{\mathcal{U}}} < \infty \tag{2.2.13b}$$

there exists a $u_{EX} \in \mathcal{U}$ such that

$$\mathcal{B}(u_{EX}, v) = \mathcal{L}(v) \quad \forall v \in \mathcal{U}. \tag{2.2.13c}$$

The proof follows from the Lax–Milgram lemma (see Appendix A2.3.1).

Remark 2.2.8 Instead of $0 < \beta_1 \leq AE(x) \leq \beta_2 < \infty$ we can assume $\beta_1 x^\theta \leq AE(x) \leq \beta_2 < \infty$, $0 \leq \theta < 1$ (see P2.2.24).

We will now show that the solution to the variational problem (2.2.8) is also the minimizer of a quadratic functional.

Theorem 2.2.3 *Let*

$$\Pi(u) \overset{\text{def}}{=} \tfrac{1}{2}\mathcal{B}(u,u) - \mathcal{L}(u) = \tfrac{1}{2}\|u\|_{\mathcal{U}}^2 - \mathcal{L}(u). \tag{2.2.14a}$$

Then u_{EX}, the solution of (2.2.13c), also satisfies

$$\Pi(u_{EX}) = \min_{u \in \mathcal{U}} \Pi(u) = -\tfrac{1}{2}\|u_{EX}\|_{\mathcal{U}}^2. \tag{2.2.14b}$$

Proof From the assumption that $\mathcal{L}$ is bounded, we have

$$|\mathcal{L}(u)| \leq \|\mathcal{L}\|_{L(\mathcal{U},\mathbb{R})} \|u\|_{\mathcal{U}} \quad \forall u \in \mathcal{U} \tag{2.2.15a}$$

and hence

$$\tfrac{1}{2}\|u\|_{\mathcal{U}}^2 - \|\mathcal{L}\|_{L(\mathcal{U},\mathbb{R})}\|u\|_{\mathcal{U}} \leq \tfrac{1}{2}\|u\|_{\mathcal{U}}^2 - |\mathcal{L}(u)| \leq \Pi(u) \quad \forall u \in \mathcal{U} \tag{2.2.15b}$$

which proves that Π is bounded from below, and

$$-\infty < \inf_{u \in \mathcal{U}} \Pi(u). \tag{2.2.15c}$$

Letting

$$u = u_{EX} + \chi, \quad \chi \in \mathcal{U} \tag{2.2.16a}$$

we get

$$\Pi(u) = \tfrac{1}{2}\mathscr{B}(u_{EX} + \chi, u_{EX} + \chi) - \mathscr{L}(u_{EX} + \chi) \tag{2.2.16b}$$

and expanding we obtain

$$\Pi(u) = \Pi(u_{EX}) + \mathscr{B}(u_{EX}, \chi) - \mathscr{L}(\chi) + \tfrac{1}{2}\mathscr{B}(\chi, \chi) \tag{2.2.16c}$$

and taking into account that

$$\mathscr{B}(u_{EX}, \chi) = \mathscr{L}(\chi) \quad \forall \chi \in \mathscr{U} \tag{2.2.16d}$$

and that

$$\mathscr{B}(u_{EX}, u_{EX}) = \|u_{EX}\|^2_{\mathscr{U}} = \mathscr{L}(u_{EX}) \tag{2.2.16e}$$

we get $\Pi(u_{EX}) = -\frac{1}{2}\|(u_{EX}\}^2_u$, and

$$\Pi(u) = \Pi(u_{EX}) + \tfrac{1}{2}\|\chi\|^2_{\mathscr{U}}. \tag{2.2.16f}$$

This proves (2.2.14b). □

Remark 2.2.9 In 1851, Riemann proved the existence of a solution of (2.2.8) under the assumption that there exists a minimizer of Π. As it was pointed out by Weierstrass in 1870, however that assumption had yet to be proven. This was done 31 years later by employing the theory introduced by Hilbert in 1901.

Let us now show that the boundedness of $\mathscr{L}$ on $\mathscr{U}$ can be proven under relatively weak assumptions about the smoothness of AE, c, and f.

Lemma 2.2.1 *Under the assumptions that*

$$\begin{cases} 0 < \beta_1 \overset{\text{def}}{=} \inf\limits_{0 \le x \le l}(AE(x)), & \beta_2 \overset{\text{def}}{=} \sup\limits_{0 \le x \le l}(AE(x)) < \infty, \\ 0 \le \inf\limits_{0 \le x \le l}(c(x)), & \gamma \overset{\text{def}}{=} \sup\limits_{0 \le x \le l}(c(x)) < \infty \end{cases} \tag{2.2.17a}$$

and

$$\|f\|_{L^2} < \infty, \qquad 0 \le \alpha_2 < \infty, \qquad |a_2| < \infty \tag{2.2.17b}$$

we have

$$\|\mathscr{L}\|_{L(\mathscr{U},\mathbb{R})} \overset{\text{def}}{=} \sup_{v \in \mathscr{U}} \frac{|\int_0^l fv + a_2 v(l)|}{\|v\|_{\mathscr{U}}} < \infty. \tag{2.2.17c}$$

Proof Noting that any $v \in \mathscr{U}$ is constrained to satisfy $v(0) = 0$, we obtain

$$|v(x)| = \left|\int_0^x \frac{dv}{dx}\right| \le \sqrt{x}\sqrt{\int_0^x \left(\frac{dv}{dx}\right)^2} \quad \forall v \in \mathscr{U} \tag{2.2.18a}$$

and employing that $0 < \beta_1 \leq AE$, we obtain

$$|v(x)| \leq \sqrt{x}\frac{1}{\sqrt{\beta_1}}\sqrt{\int_0^x AE\Big(\frac{dv}{dx}\Big)^2} \leq \sqrt{x}\frac{1}{\sqrt{\beta_1}}\|v\|_{\mathcal{U}} \quad \forall v \in \mathcal{U} \tag{2.2.18b}$$

from which it follows that

$$\|v\|^2_{L^2(I)} = \int_0^l v^2 \leq \Big(\int_0^l x\Big)\frac{1}{\beta_1}\|v\|^2_{\mathcal{U}} = \frac{l^2}{2\beta_1}\|v\|^2_{\mathcal{U}} \quad \forall v \in \mathcal{U}. \tag{2.2.18c}$$

Using the Cauchy–Schwarz inequality and (2.2.18c), we get

$$\Big|\int_0^l fv\Big| \leq \|f\|_{L^2(I)}\|v\|_{L^2(I)} \leq \|f\|_{L^2(I)}\frac{l}{\sqrt{2}}\frac{1}{\sqrt{\beta_1}}\|v\|_{\mathcal{U}} \quad \forall v \in \mathcal{U} \tag{2.2.19a}$$

and employing that

$$|\mathcal{L}(v)| \leq \Big|\int_0^l fv\Big| + |a_2|\,|v(l)| \tag{2.2.19b}$$

we obtain

$$|\mathcal{L}(v)| \leq \frac{1}{\sqrt{\beta_1}}\Big(\frac{l}{\sqrt{2}}\|f\|_{L^2(I)} + \sqrt{l}|a_2|\Big)\|v\|_{\mathcal{U}} \quad \forall v \in \mathcal{U} \tag{2.2.19c}$$

from which it follows that

$$\|\mathcal{L}\|_{L(\mathcal{U},\mathbb{R})} \overset{\text{def}}{=} \sup_{v\in\mathcal{U}}\frac{|\mathcal{L}(v)|}{\|v\|_{\mathcal{U}}} \leq \frac{1}{\sqrt{\beta_1}}\Big(\frac{l}{\sqrt{2}}\|f\|_{L^2(I)} + \sqrt{l}\,|a_2|\Big) < \infty. \tag{2.2.19d}$$

□

Remark 2.2.10 Lemma 2.2.1 holds also under the weaker assumption that AE satisfies (2.2.10); see P2.2.4.

Remark 2.2.11 The above proof was based on the inequality

$$|v(x)| \leq \sqrt{x}\sqrt{\int_0^x \Big(\frac{dv}{dx}\Big)^2} \leq \sqrt{l}\sqrt{\int_0^l \Big(\frac{dv}{dx}\Big)^2} \quad \forall v \in H^1_{(0}(I) \tag{2.2.20a}$$

and

$$\|v\|_{L^2(I)} = \sqrt{\int_0^l v^2} \leq l\sqrt{\int_0^l \Big(\frac{dv}{dx}\Big)^2} = l\Big\|\frac{dv}{dx}\Big\|_{L^2(I)} \quad \forall v \in H^1_{(0}(I) \tag{2.2.20b}$$

where we let

$$v \in H^1_{(0}(I) \overset{\text{def}}{=} \Big\{v \Big| \int_0^l \Big(\frac{dv}{dx}\Big)^2 < \infty,\ v(0) = 0\Big\}. \tag{2.2.20c}$$

Inequalities of this type are known as *Poincaré* inequalities.

Remark 2.2.12 Note that we have $\mathcal{U} \subseteq C^0(I)$, i.e., the functions in $\mathcal{U}$ are continuous. This follows from the observation that

$$|v(x_2) - v(x_1)| = \left| \int_{x_1}^{x_2} \frac{dv}{dx} \right| \le \sqrt{|x_2 - x_1|} \frac{1}{\sqrt{\beta_1}} \|v\|_{\mathcal{U}} \quad \forall v \in \mathcal{U}. \tag{2.2.21}$$

This property of the energy space $\mathcal{U}$, does not hold in higher dimensions.

The condition $\|f\|_{L^2(I)} < \infty$ is sufficient *but not necessary* for the boundedness of $\mathcal{L}$. For example, we have

$$\left| \int_0^l x^{-5/4} v \right| \le \left(\int_0^l x^{-5/4} \sqrt{x} \frac{1}{\sqrt{\beta_1}} \|v\|_{\mathcal{U}} \right) = \left(\int_0^l x^{-3/4} \right) \frac{1}{\sqrt{\beta_1}} \|v\|_{\mathcal{U}} \tag{2.2.22}$$

and hence $\mathcal{L}$ is bounded for $f = x^{-5/4}$, even though $\|f\|_{L^2(I)} = \|x^{-5/4}\|_{L^2(I)}$ does not exist.

We will now specify more precisely a class of loads f for which $\mathcal{L}$ is bounded.

Lemma 2.2.2 *Let f be such that*

$$\left| \int_x^l f \right| \le \mathcal{C} \frac{1}{\sqrt{x}} x^\delta (l - x)^\delta \tag{2.2.23a}$$

for some $\delta > 0$. Then

$$\sup_{v \in \mathcal{U}} \frac{1}{\|v\|_{\mathcal{U}}} \left| \int_0^l fv \right| \le \frac{1}{\sqrt{\beta_1}} \left\| \int_x^l f \right\|_{L^2(I)}. \tag{2.2.23b}$$

Proof Let

$$F(x) \stackrel{\text{def}}{=} \int_x^l f. \tag{2.2.24a}$$

Then

$$\int_0^l v \frac{dF}{dx} = \lim_{\epsilon \to 0} \left(- \int_\epsilon^{l-\epsilon} F \frac{dv}{dx} + v(l-\epsilon)F(l-\epsilon) - v(\epsilon)F(\epsilon) \right) \quad \forall v \in \mathcal{U}. \tag{2.2.24b}$$

Noting that $v \in \mathcal{U}$ satisfies the condition $v(0) = 0$, and hence

$$|v(\epsilon)| \le \frac{\sqrt{\epsilon}}{\sqrt{\beta_1}} \|v\|_{\mathcal{U}} \tag{2.2.25a}$$

and we get

$$|v(\epsilon)F(\epsilon)| = |v(\epsilon)|\,|F(\epsilon)| \le \frac{\sqrt{\epsilon}}{\sqrt{\beta_1}} \|v\|_{\mathcal{U}}\, |F(\epsilon)| \quad \forall v \in \mathcal{U} \tag{2.2.25b}$$

and similarly

$$|v(l-\epsilon)F(l-\epsilon)| = |v(l-\epsilon)|\,|F(l-\epsilon)| \le \frac{\sqrt{l-\epsilon}}{\sqrt{\beta_1}} \|v\|_{\mathcal{U}}\, |F(l-\epsilon)| \quad \forall v \in \mathcal{U}. \tag{2.2.25c}$$

From assumption (2.2.23a), we have

$$|F(\epsilon)| \leq \mathscr{C}\frac{1}{\sqrt{\epsilon}}\epsilon^{\delta}(l-\epsilon)^{\delta} \tag{2.2.26a}$$

and hence

$$|v(\epsilon)F(\epsilon)| \leq \frac{\sqrt{\epsilon}}{\sqrt{\beta_1}}\,\|v\|_{\mathscr{U}}\mathscr{C}\epsilon^{-(1/2)+\delta}(l-\epsilon)^{\delta} = \frac{\mathscr{C}}{\sqrt{\beta_1}}\epsilon^{\delta}(l-\epsilon)^{\delta}\|v\|_{\mathscr{U}} \quad \forall v \in \mathscr{U} \tag{2.2.26b}$$

and similarly

$$|v(l-\epsilon)F(l-\epsilon)| \leq \frac{\sqrt{l-\epsilon}}{\sqrt{\beta_1}}\,\|v\|_{\mathscr{U}}\mathscr{C}(l-\epsilon)^{-(1/2)+\delta}\epsilon^{\delta} = \frac{\mathscr{C}}{\sqrt{\beta_1}}\epsilon^{\delta}(l-\epsilon)^{\delta}\,\|v\|_{\mathscr{U}} \tag{2.2.26c}$$

for every $v \in \mathscr{U}$. Hence we have proven that

$$\max(|v(\epsilon)F(\epsilon)|, |v(l-\epsilon)F(l-\epsilon)|) \leq \frac{\mathscr{C}}{\sqrt{\beta_1}}\epsilon^{\delta}(l-\epsilon)^{\delta}\,\|v\|_{\mathscr{U}} \quad \forall v \in \mathscr{U} \tag{2.2.27a}$$

from which it follows that

$$\int_0^l v\frac{dF}{dx} = \lim_{\epsilon\to 0}\left(-\int_{\epsilon}^{l-\epsilon} F\frac{dv}{dx} + v(l-\epsilon)F(l-\epsilon) - v(\epsilon)F(\epsilon)\right) = -\int_0^l F\frac{dv}{dx} \tag{2.2.27b}$$

and recalling that $f = -dF/dx$, we obtain

$$\begin{aligned}\left|\int_0^l fv\right| = \left|\int_0^l F\frac{dv}{dx}\right| &\leq \|F\|_{L^2(I)}\left\|\frac{dv}{dx}\right\|_{L^2(I)} \\ &\leq \frac{1}{\sqrt{\beta_1}}\,\|F\|_{L^2(I)}\,\|v\|_{\mathscr{U}} \quad \forall v \in \mathscr{U}.\end{aligned} \tag{2.2.27c}$$

□

Remark 2.2.13 The above proof reflects the fact that integral $\int_0^l fv$ is understood in the following sense: Given f such that

$$F \overset{\text{def}}{=} \int_x^l f \in L^2(I) \tag{2.2.28a}$$

let $F_n \in C^{\infty}(I)$, $n = 1, 2, \ldots$, be such that

$$|F_n(x)| \leq \mathscr{C}x^{(-1/2)+\delta}(l-x)^{\delta}, \qquad \delta > 0 \tag{2.2.28b}$$

and

$$\|F_n - F\|_{L^2(I)} \to 0 \quad \text{as } n \to \infty. \tag{2.2.28c}$$

Then, we have

$$\int_0^l fv \overset{\text{def}}{=} \lim_{n\to\infty}\int_0^l F_n\frac{dv}{dx} = \int_0^l F\frac{dv}{dx} \quad \forall v \in H^1_{(0}(I). \tag{2.2.28d}$$

Remark 2.2.14 Note that $F(x) = \int_x^l f$ is allowed to have jump discontinuities, and to diverge to infinity in the neighborhood of $x = 0$, but with growth slower than $1/\sqrt{x}$.

Remark 2.2.15 Given any $\psi \in \mathcal{U}$, we may define the linear functional

$$\mathcal{L}(v) = \mathcal{B}(\psi, v), \quad v \in \mathcal{U} \tag{2.2.29a}$$

and we have

$$\|\mathcal{L}\|_{L(\mathcal{U},\mathbb{R})} = \sup_{v\in\mathcal{U}} \frac{|\mathcal{L}(v)|}{\|v\|_{\mathcal{U}}} = \sup_{v\in\mathcal{U}} \frac{|\mathcal{B}(\psi, v)|}{\|v\|_{\mathcal{U}}} = \|\psi\|_{\mathcal{U}}. \tag{2.2.29b}$$

Let us now summarize the above results regarding the existence and uniqueness of solutions for the model problem.

Theorem 2.2.3 *Under the assumptions*

$$0 < \beta_1 \le AE(x) \le \beta_2 < \infty, \qquad 0 \le c(x) \le \gamma < \infty, x \in I,$$
$$0 \le \alpha_2 < \infty \tag{2.2.30a}$$

and

$$\left| \int_x^l f \right| \le \mathcal{C} x^{-1/2+\delta}(l - x)^{\delta}, \quad \delta > 0 \tag{2.2.30b}$$

and $|a_2| < \infty$, *there exists a unique* $u_{EX} \in \mathcal{U}$ *which satisfies (2.2.8), and we have*

$$\|u_{EX}\|_{\mathcal{U}} = \|\mathcal{L}\|_{L(\mathcal{U},\mathbb{R})}. \tag{2.2.30c}$$

Proof The uniqueness of the solution was proven in Theorem 2.2.1. The existence follows from the Lax–Milgram lemma (see Appendix A2.3.1) and Lemma 2.2.2. The equality (2.2.30c) follows directly from the definition of the energy norm using duality, namely

$$\|u_{EX}\|_{\mathcal{U}} = \sup_{v\in\mathcal{U}} \frac{\mathcal{B}(u_{EX}, v)}{\|v\|_{\mathcal{U}}} = \sup_{v\in\mathcal{U}} \frac{\mathcal{L}(v)}{\|v\|_{\mathcal{U}}} = \|\mathcal{L}\|_{L(\mathcal{U},\mathbb{R})}. \tag{2.2.31}$$

□

We will call the model problem with AE, c, and a_2, satisfying (2.2.30a), the *regular* model problem.

Note that above the assumption

$$0 < \beta_1 = \inf_{0\le x\le l} AE(x) \tag{2.2.32}$$

was critical for establishing the boundedness of $\mathcal{L}$. Let us now consider the case of the model problem with

$$AE \equiv x^{\vartheta}, \quad 0 \le \vartheta < 1, \qquad c \equiv 0, \qquad 0 \le \alpha_2 \tag{2.2.33}$$

for which assumption (2.2.32) is violated. This problem will be called the *singular* model problem, which is of interest because its solution has an algebraic singularity, $x^{1-\vartheta}$, at

$x = 0$, even in the case $f \equiv 0$, similarly as the solutions of boundary value problems in two dimensions (three dimensions) have algebraic singularities, r^α, at corner points (edges or vertices). The variational formulation of the singular model problem is given by (2.2.8) with bilinear form

$$\mathcal{B}(u, v) = \int_0^l x^\vartheta \frac{du}{dx}\frac{dv}{dx}\, dx + \alpha_2 u(l)v(l), \quad u, v \in \mathcal{U}^\vartheta \tag{2.2.34a}$$

where

$$\mathcal{U}^\vartheta \stackrel{\text{def}}{=} \{w = w(x),\ x \in I \mid \|w\|_{\mathcal{U}^\vartheta} < \infty,\ w(0) = 0\} \tag{2.2.34b}$$

is the energy space and

$$\|v\|_{\mathcal{U}^\vartheta} \stackrel{\text{def}}{=} \sqrt{\mathcal{B}(v,v)} = \sqrt{\int_0^l x^\vartheta \left(\frac{dv}{dx}\right)^2 + \alpha_2 (v(l))^2} \tag{2.2.34c}$$

is the energy norm for the singular model problem; see P2.2.4 and P2.2.5.

Let us analyze the assumptions on $\mathcal{L}$, under which we can prove the existence of solutions for the singular model problem. We will show that the assumption

$$\left|\int_x^l f\right| \le \mathcal{C}\frac{1}{\sqrt{x}} x^{\bar\delta}(l - x)^{\bar\delta} \tag{2.2.35}$$

for some $\bar\delta > 0$, is not sufficient for guaranteeing the boundedness of $\mathcal{L}$ on $\mathcal{U}^\vartheta$, and hence the existence of solutions for the singular model problem. Consider the case that $f \equiv x^{-1.31}$. Then

$$|F(x)| = \left|\int_x^l t^{-1.31}\right| = \frac{1}{0.31}|x^{-0.31} - l^{-0.31}| \le \mathcal{C}\frac{1}{\sqrt{x}} x^{\bar\delta}(l-x)^{\bar\delta} \tag{2.2.36a}$$

and assumption (2.2.35) is satisfied for $\bar\delta = 0.19$. Further, we have

$$\begin{aligned}\|x^{0.31}\|_{\mathcal{U}^{0.4}} &= \int_0^l x^{0.4}\left(\frac{d}{dx}(x^{0.31})\right)^2 + \alpha_2 l^{0.31} \\ &= \int_0^l (0.31)^2 x^{-0.98} + \alpha_2 l^{0.31} < \infty \end{aligned} \tag{2.2.36b}$$

and hence

$$x^{0.31} \in \mathcal{U}^{0.4}. \tag{2.2.36c}$$

Letting $v = x^{0.31}$, we get

$$\int_\epsilon^l fv = \int_\epsilon^l x^{-1.31}x^{0.31} = \int_\epsilon^l \frac{1}{x} \tag{2.2.36d}$$

and hence the functional $\int_0^l x^{-1.31} v$ is not bounded on $\mathcal{U}^{0.4}$.

From the above example it follows that the conditions which must be imposed on f (or $F = \int_x^l f$) in order to guarantee that $\|\mathscr{L}\|_{L(\mathscr{U}^\vartheta,\mathbb{R})}$ is finite, are more restrictive than (2.2.35) with arbitrary $\bar{\delta}$. Employing the Cauchy–Schwarz inequality, we get

$$\left|\int_0^l F\frac{dv}{dx}\right| = \left|\int_0^l x^{-\vartheta/2}Fx^{\vartheta/2}\frac{dv}{dx}\right| \leq \sqrt{\int_0^l x^{-\vartheta}F^2}\,\|v\|_{\mathscr{U}^\vartheta}. \tag{2.2.37a}$$

For the boundedness of $\mathscr{L}$, it is sufficient to have

$$\int_0^l x^{-\vartheta}F^2 < \infty. \tag{2.2.37b}$$

Working as in Lemma 2.2.2 we get that the condition

$$|F(x)| = \left|\int_x^l f\right| \leq \mathscr{C}\frac{1}{\sqrt{x}}x^{\vartheta/2}x^{\delta}(l-x)^{\delta} \tag{2.2.37c}$$

is sufficient for existence of solutions for the singular model problem.

Another sufficient condition may be obtained by noting that

$$|v(x)| = \left|\int_0^x \frac{dv}{dx}\right| \leq \sqrt{\int_0^x x^{-\vartheta}}\sqrt{\int_0^x x^{\vartheta}\left(\frac{dv}{dx}\right)^2} \leq \frac{x^{(1-\vartheta)/2}}{\sqrt{1-\vartheta}}\|v\|_{\mathscr{U}^\vartheta} \tag{2.2.38a}$$

and hence

$$\left|\int_0^l fv\right| \leq \left|\int_0^l \frac{x^{(1-\vartheta)/2}}{\sqrt{1-\vartheta}}\,|f|\right|\|v\|_{\mathscr{U}^\vartheta}. \tag{2.2.38b}$$

It follows that for the boundedness of $\mathscr{L}$, it is sufficient to have

$$\left|\int_0^l x^{(1-\vartheta)/2}\,|f|\right| < \infty \tag{2.2.38c}$$

where we recall that $0 < \vartheta < 1$.

Remark 2.2.16 In the special case $\vartheta = 1$, we have $AE(x) = x$. This choice of AE corresponds to the one-dimensional formulation of the two-dimensional axisymmetric model problem; see P2.1*.2. Note that in this case the value $u(0)$ cannot be imposed as an essential boundary condition because it is not preserved in the limit with respect to the energy norm; see P2.2.20.

Above, we always employed the homogeneous (zero) essential boundary condition $u(0) = 0$. The nonhomogeneous case

$$u(0) = a \neq 0 \tag{2.2.39a}$$

can be reduced to the homogeneous one by letting

$$u_{EX} = u_{Dir} + \tilde{u}_{EX} \tag{2.2.39b}$$

where u_{Dir} is such that

$$\int_0^l \left(\frac{d}{dx}(u_{Dir})\right)^2 < \infty, \quad u_{Dir}(0) = a \tag{2.2.39c}$$

and by seeking $\tilde{u}_{EX}$ as the solution of the problem:

Find $\tilde{u}_{EX} \in \mathcal{U}$ such that

$$\mathcal{B}(\tilde{u}_{EX}, v) = \tilde{\mathcal{L}}(v) \stackrel{\text{def}}{=} \mathcal{L}(v) - \mathcal{B}(u_{Dir}, v) \quad \forall v \in \mathcal{U}. \tag{2.2.39d}$$

Note that $u_{Dir} \notin \mathcal{U}$ because $u_{Dir}(0) = a \neq 0$; nevertheless, $\mathcal{B}(u_{Dir}, v)$ is a bounded linear functional on $\mathcal{U}$, and hence $\tilde{\mathcal{L}}$ is bounded. From Theorems 2.2.1 and 2.2.2 it follows that $\tilde{u}_{EX}$ exists and it is unique. Note also that $\tilde{u}_{EX}$ depends on u_{Dir}. Letting

$$\mathcal{U}_{u_{Dir}} = \{v \mid v = u_{Dir} + \tilde{v}, \ \tilde{v} \in \mathcal{U}\} \tag{2.2.39e}$$

and, working analogously as above, we can show that

$$\Pi(u_{Dir} + \tilde{u}_{EX}) = \min_{u \in \mathcal{U}_{u_{Dir}}} \Pi(u) \tag{2.2.39f}$$

where Π is the quadratic functional defined in (2.2.14a).

Let us also consider the case where natural boundary conditions are imposed at both ends, namely

$$AE(0)\frac{du}{dx}(0) = \alpha_1 u(0) - a_1 \tag{2.2.40a}$$

and

$$-AE(l)\frac{du}{dx}(l) = \alpha_2 u(l) - a_2. \tag{2.2.40b}$$

In this case, we let

$$\mathcal{B}(u, v) \stackrel{\text{def}}{=} \int_0^l \left(AE\frac{du}{dx}\frac{dv}{dx} + cuv\right) + \alpha_1 u(0)v(0) + \alpha_2 u(l)v(l) \tag{2.2.40c}$$

and

$$\mathcal{L}(v) \stackrel{\text{def}}{=} \int_0^l fv + a_2 v(l) - a_1 v(0) \tag{2.2.40d}$$

and

$$\mathcal{U} \stackrel{\text{def}}{=} \{v \mid \|v\|_{\mathcal{U}} \stackrel{\text{def}}{=} \sqrt{\mathcal{B}(v, v)} < \infty\}. \tag{2.2.40e}$$

In the cases that $\alpha_1 > 0$, or $\alpha_2 > 0$, or $\int_0^l c > 0$, $\|\cdot\|_{\mathcal{U}}$ is a norm, and the above theory holds with some small technical changes to account for the fact that we no longer have the condition $v(0) = 0$ (see P2.2.15 and P2.2.16). In the special case that $\alpha_1 = \alpha_2 = 0$, $c \equiv 0$, $\|\cdot\|_{\mathcal{U}}$ is only a seminorm (see A.2.4.3). To address this case, let

$$\mathcal{U}_Q \stackrel{\text{def}}{=} \{v \in \mathcal{U} \mid Q(v) = 0\} \tag{2.2.41a}$$

where Q is a linear functional such that $Q(1) \neq 0$, e.g., $Q(v) = v(0)$, or $Q(v) = \int_0^l v$, etc. Then, $\|\cdot\|_{\mathcal{U}}$ is a norm on $\mathcal{U}_Q$, and there exists a unique $u_{EX}^Q \in \mathcal{U}_Q$, such that

$$\mathcal{B}(u_{EX}^Q, v) = \mathcal{L}(v) \quad \forall v \in \mathcal{U}_Q. \tag{2.2.41b}$$

Note that there may exist $v \in \mathcal{U} - \mathcal{U}_Q$, for which

$$\mathcal{B}(u_{EX}^Q, v) \neq \mathcal{L}(v) \tag{2.2.41c}$$

and u_{EX}^Q may not satisfy (2.2.8). However, in the cases that

$$\mathcal{L}(1) = 0 \tag{2.2.41d}$$

u_{EX}^Q is also a solution of (2.2.8). This can be proven by noting that any $v \in \mathcal{U}$ can be uniquely decomposed into the form

$$v = v_1 + v_2 = (v - Q(v)) + Q(v) \tag{2.2.41e}$$

with

$$v_1 = v - Q(v) \in \mathcal{U}_Q, \qquad v_2 = Q(v) \equiv \mathit{constant}. \tag{2.2.41f}$$

Then, noting that

$$\mathcal{L}(v_2) = \mathcal{L}(Q(v)) = 0 \quad \forall v \in \mathcal{U}, \tag{2.2.41g}$$

we have

$$\mathcal{B}(u_{EX}^Q, v) - \mathcal{L}(v) = \mathcal{B}(u_{EX}^Q, v_1) - \mathcal{L}(v_1) = 0 \quad \forall v \in \mathcal{U} \tag{2.2.41h}$$

and hence, $u_{EX}^Q + \mathcal{C}$ satisfies (2.2.8). A unique solution can be selected by imposing the additional condition $Q(u_{EX}^Q) = 0$. See also [Rektorys, 1998].

Remark 2.2.17 Note that u_{EX}^Q is the solution of the variational problem:
Find $u_{EX}^Q \in \mathcal{U}_Q$ such that

$$\mathcal{B}(u_{EX}^Q, v) = \mathcal{L}(v) - \frac{\mathcal{L}(1)}{Q(1)} Q(v) \quad \forall v \in \mathcal{U}. \tag{2.2.42}$$

In the cases that $\mathcal{L}(1) \neq 0$, u_{EX} does not exist, while u_{EX}^Q does. See also P2.1.4, P2.1.5, and P2.1.6.

In this section, we presented the variational formulation of the mathematical problem together with conditions for the existence and uniqueness of generalized solutions for the regular and the singular model problems. The existence and uniqueness of the solution must be addressed before an attempt is made to construct numerical approximations.

Problems

P2.2.1. Consider the regular model problem and assume that $AE \in C^1(\bar{I})$, and $c, f \in C^0(\bar{I})$. Prove that u_{EX} exists, and $u_{EX} \in C^2(\bar{I})$.
Hint: Use the fact that the solution of any linear second-order differential equation can be written in the form $u_{EX} = \mathscr{C}_1\psi_1 + \mathscr{C}_2\psi_2 + \psi_p$, where ψ_1, ψ_2 are two linearly independent solutions of the homogeneous differential equation, and ψ_p is a particular solution of the nonhomogeneous differential equation.

P2.2.2. Show that the energy space $\mathscr{U}$, defined in (2.2.4), is a linear space and $\|\cdot\|_{\mathscr{U}}$, defined in (2.2.3b), has the properties of a norm. Show further that the bilinear form (2.2.5a) satisfies conditions given in A2.1.3 for $X = Y = \mathscr{U}$.

P2.2.3. Consider the regular model problem with

$$\mathscr{B}(u,v) = \int_0^l \left(AE\frac{du}{dx}\frac{dv}{dx} + cuv\right) + \alpha_2 u(l)v(l), \qquad \mathscr{L}(v) = \int_0^l fv \tag{2.2.43a}$$

and

$$\mathscr{U} \stackrel{\text{def}}{=} \{v \mid \mathscr{B}(v,v) < \infty,\ v(0) = 0\} \tag{2.2.43b}$$

and assume that $\inf_{x\in I} c(x) > 0$. What is the corresponding classical form of the boundary value problem? What is the classical form in the case that $\mathscr{U} \stackrel{\text{def}}{=} \{v \mid \mathscr{B}(v,v) < \infty\}$, instead of $\mathscr{U}$ defined in (2.2.4)?

P2.2.4. Show that $\mathscr{U}^\vartheta \subseteq C^0(\bar{I})$, for any $0 \le \vartheta < 1$.
Hint: For any $v \in \mathscr{U}$, we have

$$|v(\delta)| = \left|\int_0^\delta \frac{dv}{dx}\right| = \left|\int_0^\delta x^{-\vartheta/2}x^{\vartheta/2}\frac{dv}{dx}\right| \le \sqrt{\int_0^\delta x^{-\vartheta}}\ \|v\|_{\mathscr{U}(I)}$$
$$\le \frac{\sqrt{\delta^{1-\vartheta}}}{\sqrt{1-\vartheta}}\|v\|_{\mathscr{U}(I)}. \tag{2.2.44}$$

P2.2.5. Show that $\|\cdot\|_{\mathscr{U}^\vartheta}$, has all properties of a norm on $\mathscr{U}^\vartheta$, for $0 \le \vartheta < 1$, but not for $\vartheta = 1$.

P2.2.6. Find the exact solution $u_{EX} = u_{EX}(c_0)$ of the regular model problem with $AE \equiv 1$, $\alpha_2 = a_2 = 0$, $f \equiv 1$, $c \equiv c_0$, with c_0 not necessarily nonnegative. For what values of c_0 does u_{EX} exist?

P2.2.7. Find the exact solution $u_{EX} = u_{EX}(\alpha)$ of the regular model problem with $AE \equiv 1$, $c \equiv 1$, $f \equiv 1$, $\alpha_2 = \alpha$, with α not necessarily nonnegative, $a_2 = 0$. For what values of α does u_{EX} exist?

P2.2.8. Consider the regular model problem with $AE \equiv 1$, $c \equiv 0$, $\alpha_2 = a_2 = 0$, let $F(x) = \int_0^x f$, and assume that $F(x) = 0$, for $0 < x < \frac{1}{3}l$, and $\frac{2}{3}l < x < l$. Obtain the general form of the solution u_{EX} of this problem. Under what conditions on f (respectively F) does u_{EX} exist in $\mathscr{U}$?

P2.2.9. Let $u_{EX}^{(n)} \in \mathscr{U}$, $n = 1, 2, 3, \ldots$, be such that

$$\mathscr{B}(u_{EX}^{(n)}, v) = \mathscr{L}^{(n)}(v) = \int_0^l f^{(n)}v \quad \forall v \in \mathscr{U} \tag{2.2.45a}$$

where

$$f^{(n)}(x) = \begin{cases} 0, & 0 < x < \frac{1}{2}(l - n^{-1}),\ \frac{1}{2}(l + n^{-1}) < x < l \\ n, & \frac{1}{2}(l - n^{-1}) < x < \frac{1}{2}(l + n^{-1}) \end{cases} \tag{2.2.45b}$$

Determine $u_{EX}^{(n)}$, and let $u_{EX}^{(\infty)}$ be defined by

$$u_{EX}^{(\infty)}(x) \overset{\text{def}}{=} \lim_{n\to\infty} u_{EX}^{(n)}(x), \quad x \in I. \tag{2.2.45c}$$

Prove that

$$\lim_{n\to\infty} \|u_{EX}^{(n)} - u_{EX}^{(\infty)}\|_{\mathcal{U}} = 0. \tag{2.2.45d}$$

P2.2.10. Consider the same model problem as in P2.2.9, except with

$$f^{(n)}(x) = \begin{cases} 0, & 0 < x < \frac{1}{2}(l - n^{-1}),\ \frac{1}{2}(l + n^{-1}) < x < l \\ 4n^2, & \frac{1}{2}(l - n^{-1}) < x < \frac{1}{2}l \\ -4n^2, & \frac{1}{2}l < x < \frac{1}{2}(l + n^{-1}) \end{cases} \tag{2.2.46}$$

and determine $u_{EX}^{(n)}$ and $u_{EX}^{(\infty)}$. Show that although $u_{EX}^{(n)} \in \mathcal{U}$, we have $u_{EX}^{(\infty)} \notin \mathcal{U}$, and the sequence $\|u_{EX}^{(n)} - u_{EX}^{(\infty)}\|_{\mathcal{U}}$ diverges.

P2.2.11. Determine the exact solution $u_{EX} = u_{EX}(\gamma)$ of the regular model problem with $AE \equiv 1$, $c \equiv 0$, $f \equiv x^\gamma$, $\alpha_2 = 1$, $a_2 = 0$. For what values of γ is $u_{EX}(\gamma) \in \mathcal{U}$?

P2.2.12. Let $Q : H^1(I) \mapsto \mathbb{R}$, $Q(1) \neq 0$, be a bounded linear functional and let

$$|||v||| \overset{\text{def}}{=} \sqrt{\left\|\frac{dv}{dx}\right\|_{L^2(I)}^2 + \|Q(v)\|_{L^2(I)}^2}. \tag{2.2.47}$$

Prove that $|||\cdot|||$ is a norm on $H^1(I)$.

P2.2.13. Show that $|\cdot|_{H^1(I)} \overset{\text{def}}{=} \|d\cdot/dx\|_{L^2(I)}$ is a norm on

$$H^1_{\int_0^\ell}(I) \overset{\text{def}}{=} \{u \in H^1(I) \mid \textstyle\int_0^l u = 0\}. \tag{2.2.48}$$

P2.2.14. Prove the existence and uniqueness of the solution of the variational problem: Find $u_{EX}^{\int_0^l} \in H^1_{\int_0^\ell}(I)$ which satisfies

$$\int_0^l \frac{d}{dx}(u_{EX}^{\int_0^l})\frac{dv}{dx} = \mathcal{L}(v) \overset{\text{def}}{=} \int_0^l fv + a_1 v(0) + a_2 v(l) \quad \forall v \in H^1_{\int_0^\ell}(I). \tag{2.2.49}$$

P2.2.15. Consider the setting of the previous problem and assume that f, a_1, and a_2 are such that $\mathcal{L}(1) = 0$. Show that the solution $u_{EX}^{\int_0^l}$ of P2.2.14 also satisfies

$$\int_0^l \frac{d}{dx}(u_{EX}^{\int_0^l})\frac{dv}{dx} = \mathcal{L}(v) \quad \forall v \in H^1(I). \tag{2.2.50}$$

P2.2.16. Consider the setting of P2.2.14 and assume that $f \in L^2(I)$ and $\mathscr{L}(1) \neq 0$. Show that $u_{EX}^{\int_0^l}$ exists in the sense of P2.2.14, but not P2.2.15.

P2.2.17. Determine $u_{EX}^{\int_0^l}$ in the case that $f \equiv 1$, $a_1 = 0$, $a_2 = 2$.

P2.2.18. Prove

$$\Pi(\tilde{u}_{EX} + u_{Dir}) = \min_{u \in \mathfrak{U}_{u_{Dir}}} \Pi(u) \tag{2.2.51a}$$

and

$$\Pi(\tilde{u}_{EX} + u_{Dir} + \chi) = \Pi(\tilde{u}_{EX} + u_{Dir}) + \tfrac{1}{2} \|\chi\|_{\mathfrak{U}}^2 \quad \forall \chi \in \mathfrak{U}. \tag{2.2.51b}$$

P2.2.19. Construct examples of where $\Pi(\tilde{u}_{EX} + u_{Dir}) > 0$, and $\Pi(\tilde{u}_{EX} + u_{Dir}) < 0$.

P2.2.20. Let $u_{EX}^{(\infty)} = 1 - x$, and show that it is possible to construct a sequence of smooth functions $u_{EX}^{(j)}$ such that

$$u_{EX}^{(j)}(0) = u_{EX}^{(j)}(1) = 0 \tag{2.2.52a}$$

and

$$\lim_{j \to \infty} \int_0^1 x \Big(\frac{d}{dx} (u_{EX}^{(\infty)} - u_{EX}^{(j)}) \Big)^2 = 0. \tag{2.2.52b}$$

Hint: Use $\int_0^{1/e} x((d/dx)|\ln|\ln x||)^2 < \infty$.

P2.2.21. Let $u_{EX}^{(\infty)} = x$ and show that it is possible to construct a sequence of smooth functions $u_{EX}^{(j)} \in \mathfrak{U}$, such that

$$\frac{d}{dx}(u_{EX}^{(j)})(1) = 0 \tag{2.2.53a}$$

and

$$\lim_{j \to \infty} \|u_{EX}^{(\infty)} - u_{EX}^{(j)}\|_{\mathfrak{U}} = 0 \tag{2.2.53b}$$

where $\mathfrak{U}$ is the energy space and $\|\cdot\|_{\mathfrak{U}}$ the energy norm for the regular model problem.

P2.2.22. Prove Theorem 2.2.2 by using the Lax–Milgram lemma.

P2.2.23. Let $u_{EX}^{(\infty)} = 1 - x$, and prove that there is no sequence $u_{EX}^{(n)}$ which is Cauchy in $\mathfrak{U}$, such that

$$u_{EX}^{(n)}(0) = 0 \tag{2.2.54a}$$

and

$$\lim_{n \to \infty} \|u_{EX}^{(\infty)} - u_{EX}^{(n)}\|_{L^2(I)} = 0. \tag{2.2.54b}$$

Hint: Use the fact that for any $v \in \mathfrak{U}$, we have $|v(x)| \leq \sqrt{x}\,\|v\|_{\mathfrak{U}}$.

P2.2.24. Prove Theorem 2.2.2 under the assumption given in Remark 2.2.13.

2.2* The variational formulation of the model problem in two dimensions

We will now give the variational formulation of the model problems of heat conduction and plane elasticity introduced in Section 2.1*. Let us first consider the problem of heat conduction. The two-dimensional analogues of the assumptions (2.2.17a) are

$$0 < \beta_1(y_1^2 + y_2^2) \le \sum_{i,j=1}^{2} a_{ij}(\mathbf{x}) y_i y_j \le \beta_2(y_1^2 + y_2^2) \quad \forall \mathbf{y} \in \mathbb{R}^2, \;\; \mathbf{y} \ne \mathbf{0} \tag{2.2*.1a}$$

which can also be written in the form

$$0 < \beta_1 \mathbf{y}^{\mathrm{T}}\mathbf{y} \le \mathbf{y}^{\mathrm{T}}\mathbf{A}(\mathbf{x})\mathbf{y} \le \beta_2 \mathbf{y}^{\mathrm{T}}\mathbf{y} \quad \forall \mathbf{y} \in \mathbb{R}^2, \;\; \mathbf{y} \ne \mathbf{0} \tag{2.2*.1b}$$

for any $\mathbf{x} \in \Omega$, and

$$0 \le c(\mathbf{x}) \le \gamma, \quad \mathbf{x} \in \Omega, \qquad 0 \le \alpha(\mathbf{x}) < \infty, \quad \mathbf{x} \in \Gamma_{\mathrm{N}}. \tag{2.2*.1c}$$

Further, we will assume that Γ_{D} has positive length, on which homogeneous Dirichlet boundary conditions are imposed, namely

$$u|_{\Gamma_{\mathrm{D}}} \equiv 0. \tag{2.2*.2}$$

Analogously as in the one-dimensional case, we will define the energy norm by

$$\|u\|_{\mathcal{U}} \stackrel{\text{def}}{=} \sqrt{\int_{\Omega} \left(\sum_{i,j=1}^{2} a_{ij} \frac{\partial u}{\partial x_i} \frac{\partial u}{\partial x_j} + cu^2 \right) + \int_{\Gamma_{\mathrm{N}}} \alpha u^2} \tag{2.2*.3a}$$

which can also be written in the form

$$\|v\|_{\mathcal{U}} \stackrel{\text{def}}{=} \sqrt{\int_{\Omega} ((\boldsymbol{\nabla} u)^{\mathrm{T}} \mathbf{A} \boldsymbol{\nabla} u + cu^2) + \int_{\Gamma_{\mathrm{N}}} \alpha u^2} \tag{2.2*.3b}$$

and the energy space by

$$\mathcal{U} \stackrel{\text{def}}{=} \mathcal{U}(\Omega) \stackrel{\text{def}}{=} \{ v = v(x), \; x \in \Omega \,|\, \|v\|_{\mathcal{U}} < \infty, \; v|_{\Gamma_{\mathrm{D}}} \equiv 0 \}. \tag{2.2*.3c}$$

In contrast to the one-dimensional case, $\mathcal{U} \not\subseteq C^0(\Omega)$, which means that *the functions in* $\mathcal{U}$ *are not necessarily continuous and the meaning of the boundary value* $u|_{\Gamma_{\mathrm{D}}}$ *has to be properly defined*; see for example [Adams, 1975].

Analogously as in the one-dimensional case, we will introduce the bilinear form $\mathcal{B}$: $\mathcal{U}(\Omega) \times \mathcal{U}(\Omega) \mapsto \mathbb{R}$

$$\mathcal{B}(u, v) \stackrel{\text{def}}{=} \int_{\Omega} \left(\sum_{i,j=1}^{2} a_{ij} \frac{\partial u}{\partial x_i} \frac{\partial v}{\partial x_j} + cuv \right) + \int_{\Gamma_{\mathrm{N}}} \alpha uv \tag{2.2*.4a}$$

which can also be written in the form

$$\mathscr{B}(u,v) = \int_\Omega ((\boldsymbol{\nabla} u)^{\mathrm{T}} \mathbf{A} \boldsymbol{\nabla} v + cuv) + \int \alpha uv \tag{2.2*.4b}$$

and the linear form $\mathscr{L} : \mathscr{U}(\Omega) \mapsto \mathbb{R}$,

$$\mathscr{L}(v) \stackrel{\text{def}}{=} \int_\Omega fv + \int_{\Gamma_{\mathrm{N}}} gv. \tag{2.2*.4c}$$

In most instances below, we will omit the subscript Ω from $\mathscr{B}_\Omega$ and $\mathscr{L}_\Omega$, except where needed for clarity.

As in the one-dimensional case, (2.2*.1) and (2.2*.2) ensure that the bilinear form (2.2*.4a), the energy space (2.2*.3c), and the norm $\|\cdot\|_{\mathscr{U}}$ have all the properties needed for the existence and uniqueness of the solution.

As in the one-dimensional case, the case of non-zero essential boundary condition $u|_{\Gamma_{\mathrm{D}}} = \bar{u} \not\equiv 0$, can be reduced to the homogeneous case $u|_{\Gamma_{\mathrm{D}}} \equiv 0$, by letting $u = u_{Dir} + \tilde{u}$, where u_{Dir} is a sufficiently smooth function which satisfies the boundary condition $u_{Dir}|_{\Gamma_{\mathrm{D}}} = \bar{u}$.

As in the one-dimensional case, in the special case that $\Gamma_{\mathrm{D}} = \emptyset$, $\alpha \equiv 0$ and $c \equiv 0$, $\|\cdot\|_{\mathscr{U}} = \sqrt{\mathscr{B}(\cdot,\cdot)}$ is a norm modulo constant function (see A.2.4.3), and the solution u_{EX} exists and is unique up to an additive constant if

$$\mathscr{L}(1) = \int_\Omega f + \int_{\partial\Omega} g = 0. \tag{2.2*.5}$$

Let us now consider the case of the elasticity problem. The analogue of assumption $0 < \beta_1 \leq AE \leq \beta_2$, is

$$0 < \beta_1 \, |\boldsymbol{\epsilon}|^2 \leq \boldsymbol{\epsilon}^{\mathrm{T}} \mathbf{E} \boldsymbol{\epsilon} \leq \beta_2 \, |\boldsymbol{\epsilon}|^2 \tag{2.2*.6a}$$

where $|\boldsymbol{\epsilon}|^2 = \boldsymbol{\epsilon}^{\mathrm{T}} \boldsymbol{\epsilon}$, and the analogue of assumption $0 < \alpha_2$ is

$$0 \leq \mathbf{u}^{\mathrm{T}} \boldsymbol{\alpha} \mathbf{u}. \tag{2.2*.6b}$$

The bilinear form $\mathscr{B}$ is given by

$$\mathscr{B}(\mathbf{u},\mathbf{v}) = \int_\Omega ((\mathbf{D}\mathbf{u})^{\mathrm{T}} \mathbf{E} \mathbf{D} \mathbf{v} + \mathbf{u}^{\mathrm{T}} \mathbf{C} \mathbf{u}) + \int_{\Gamma_{\mathrm{N}}} \mathbf{u}^{\mathrm{T}} \boldsymbol{\alpha} \mathbf{v} \tag{2.2*.7a}$$

the linear form $\mathscr{L}$ by

$$\mathscr{L}(\mathbf{v}) = \int_\Omega \mathbf{v}^{\mathrm{T}} \mathbf{f} + \int_{\Gamma_{\mathrm{N}}} \mathbf{v}^{\mathrm{T}} \mathbf{g} \tag{2.2*.7b}$$

the energy norm by $\|\mathbf{u}\|_{\mathscr{U}} = \sqrt{\mathscr{B}(\mathbf{u},\mathbf{u})}$, where we assume that $\mathbf{u}|_{\Gamma_{\mathrm{D}}} \equiv \mathbf{0}$ and that Γ_{D} has positive length. In the special case that $\Gamma_{\mathrm{D}} = \emptyset$ and $\boldsymbol{\alpha} = \mathbf{0}$, the solution exists only if

$$\mathscr{L}(\mathbf{u}) = 0 \quad \forall \mathbf{u} \equiv \alpha_1 \begin{Bmatrix} 1 \\ 0 \end{Bmatrix} + \alpha_2 \begin{Bmatrix} 0 \\ 1 \end{Bmatrix} + \alpha_3 \begin{Bmatrix} -x_2 \\ x_1 \end{Bmatrix} \tag{2.2*.8}$$

and $\mathbf{u}_{EX}$ is determined up to constants $\alpha_1, \alpha_2, \alpha_3$, and $\|\cdot\|_{\mathcal{U}} = \sqrt{\mathcal{B}(\cdot,\cdot)}$ is only a seminorm modulo these constants, which may be selected by imposing the additional conditions

$$\int_\Omega u_1 = \int_\Omega u_2 = \int_\Omega (-x_2 u_1 + x_1 u_2) = 0. \tag{2.2*.9a}$$

These conditions, where $u = (u_1, u_2)$, always have good meaning. In contrast, the condition

$$u_1(\bar{\mathbf{x}}) = u_2(\bar{\mathbf{x}}) = u_2(\bar{\mathbf{y}}) = 0. \tag{2.2*.9b}$$

where $\mathbf{y} - \mathbf{x}$ is not parallel to the x_2 direction, which is often imposed in practice, may be meaningless in some cases. For example, it can happen that of $u_1(\bar{\mathbf{x}})$, $u_2(\bar{\mathbf{x}})$, and/or $u_2(\bar{\mathbf{y}})$ do not exist (they are infinite), although $\mathbf{u} \in \mathcal{U}$, and if this is the case, the corresponding condition cannot be imposed; see P2.2*.1.

Remark 2.2*.1 The condition (2.2*.9b) has good meaning in most cases of interest because in practice we know more about the smoothness of u_{EX} than the minimal condition $u_{EX} \in \mathcal{U}$.

In the case of heat conduction, under the assumption that either $\alpha > 0$ or $\Gamma_{\mathrm{D}} \neq 0$, the energy norm $\|\cdot\|_{\mathcal{U}}$ is equivalent to the $H^1(\Omega)$ norm, namely, there exist constants $\mathcal{C}_1$, $\mathcal{C}_2$, such that

$$\mathcal{C}_1 \|u\|_{H^1(\Omega)} \leq \|u\|_{\mathcal{U}} \leq \mathcal{C}_2 \|u\|_{H^1(\Omega)} \quad \forall v \in \mathcal{U}. \tag{2.2*.10}$$

In the cases that $\alpha = 0$ and $\Gamma_{\mathrm{D}} = \emptyset$, $\|\cdot\|_{\mathcal{U}}$ is only a seminorm on $\mathcal{U}$, and hence not equivalent to $\|\cdot\|_{H^1(\Omega)}$. However, $\|\cdot\|_{\mathcal{U}}$ is still a norm on

$$\mathcal{U}_{\int_\Omega} \stackrel{\text{def}}{=} \{v \in \mathcal{U} |\ \textstyle\int_\Omega v = 0\} \tag{2.2*.11}$$

as in the one-dimensional case.

In the case of elasticity, the situation is analogous, namely, if either $\boldsymbol{\alpha} \not\equiv \mathbf{0}$, or $\Gamma_{\mathrm{D}} \neq \emptyset$, then we have the equivalence

$$\mathcal{C}_1 \|\mathbf{v}\|_{H^1(\Omega)} \leq \|\mathbf{v}\|_{\mathcal{U}} \leq \mathcal{C}_2 \|\mathbf{v}\|_{H^1(\Omega)} \quad \mathbf{v} \in \boldsymbol{\mathcal{U}}. \tag{2.2*.12}$$

The inequality

$$\mathcal{C}_1 \|\mathbf{v}\|_{H^1(\Omega)} \leq \|\mathbf{v}\|_{\mathcal{U}} \tag{2.2*.13}$$

is known as Korn's inequality, and $\mathcal{C}_1$ as Korn's constant (see [Nečas, 1967], [Kondratév and Olejnik, 1990]). In the cases that $\Gamma_{\mathrm{D}} = \emptyset$ and $\boldsymbol{\alpha} \equiv \mathbf{0}$, then $\|\cdot\|_{\mathcal{U}}$ is only a seminorm on $\mathcal{U}$; nevertheless, it is still a norm on

$$\boldsymbol{\mathcal{U}}_{\int_\Omega} \stackrel{\text{def}}{=} \{\mathbf{v} \in \boldsymbol{\mathcal{U}} | \textstyle\int_\Omega \mathbf{v} = 0,\ \int_\Omega (-x_2 v_1 + x_1 v_2) = 0\} \tag{2.2*.14}$$

and the equivalence $\|\cdot\|_{\mathcal{U}} \approx \|\cdot\|_{H^1}$ still holds.

For further details on variational methods, see [Rektorys, 1977], [Rektorys, 1998], [Nečas, 1967], [Mikhlin, 1964], and [Mikhlin, 1971]. For the basics of functional analysis referred here the reader may consult [Adams, 1975], [Mason, 1985], [Oden, 1979], [Oden and Demkowicz, 1996].

Remark 2.2*.2 Above we were interested only in the case where the solution u_{EX} has finite energy and the variational form utilized the same spaces for the trial and test functions. The major tool for proving the existence was the Lax–Milgram lemma. It is possible to generalize it (see [Babuška and Aziz, 1972]), which would allow the use of different trial and test functions. This would allow solutions which have infinite energy to be tested.

Remark 2.2*.3 There are many generalizations to the notion of the solution, e.g. a very weak solution which allows us to deal with the solutions under very weak assumptions (see, e.g., [Nečas, 1967]).

In this section we have given the variational formulation of the steady heat conduction and the elasticity model problems in two dimensions, and we underlined the similarities and differences with the one-dimensional case.

Problems

P2.2*.1. Show that

$$u = \ln(x_1^2 + x_2^2) \notin \mathcal{U}(\Omega_R) = \{u \mid \int_{\Omega_R} |\boldsymbol{\nabla} u|^2 < \infty, u|_{\partial\Omega_R} \equiv 0\} \tag{2.2*.15a}$$

where $\Omega_R \overset{\text{def}}{=} \{x \mid x_1^2 + x_2^2 < R^2\}$. On the other hand, the function

$$w = \ln\left(\ln\left(e\frac{x_1^2 + x_2^2}{R^2}\right)\right) \notin C^0(\Omega_R) \tag{2.2*.15b}$$

while $w \in \mathcal{U}(\Omega_R)$.

P2.2*.2. Let $\Omega = \Omega_R$, and prove that $\mathcal{L}(v) = \int_\Omega fv$, where $f \in L^2(\Omega)$, is a bounded linear functional on $\mathcal{U} = \{u \mid \int_\Omega |\boldsymbol{\nabla} u|^2 < \infty, u|_{\partial\Omega} \equiv 0\}$.

P2.2*.3. Prove that the functional $\mathcal{L}(v) = v_1(\bar{\mathbf{x}})$, $\bar{\mathbf{x}} \in \Omega$, is not bounded on $\mathcal{U}$, the energy space for the elasticity problem. For simplicity, consider the case $\Gamma_{\mathrm{D}} \equiv \partial\Omega$.

P2.2*.4. Consider the elasticity problem on $\Omega = (0,1) \times (0,l)$ with prescribed traction g on $\partial\Omega$, $\alpha = 0$ and $f = 0$. Assume that λ and μ are constants such that

$$\lambda + 2\mu = \frac{E}{1-\nu^2}, \quad \lambda = \frac{E\nu}{1-\nu^2}, \qquad 0 \leq \nu < 0.5. \tag{2.2*.16a}$$

Show that (2.2*.7a) holds. Show that $\mathbf{u} = (u_1, u_2)$, with

$$u_1(x_1, x_2) = -\frac{x_1^2 x_2}{2} + \frac{\nu x_2^3}{6} + \frac{x_2^3}{3} + ay + b \tag{2.2*.16b}$$

and

$$u_2(x_1, x_2) = -\frac{\nu x_1 x_2^2}{2} + \frac{\nu x_2^3}{6} - ax + c \tag{2.2*.16c}$$

the solution of the elasticity problem with $f = 0$ and any a, b, c. Determine the traction $\mathbf{g}$ on $\partial\Omega$, and show that

$$\mathcal{L}(v) = \int_{\partial\Omega} \mathbf{v}^{\mathrm{T}}\mathbf{g} = 0, \quad \forall v = (d - rx_2, e + rx_1) \tag{2.2*.16d}$$

where d, e, r, are arbitrary constants.

P2.2*.5. Show that

$$\mathbf{u}_{EX} = \left(\ln\left(\ln\left(e\frac{x_1^2 + x_2^2}{R^2}\right)\right), 0\right) \in \mathcal{U}(\Omega_R). \tag{2.2*.17}$$

Determine $\mathbf{f}$, and $\mathbf{g}$ corresponding to $\mathbf{u}_{EX}$, assuming that $\Gamma_{\mathrm{N}} = \partial\Omega_R$.

2.3 Smoothness of the solution of the model problem

In the previous section, the existence of solutions for the regular problem was established under the assumptions

$$0 < \beta_1 \leq AE \leq \beta_2, \qquad 0 \leq c \leq \gamma, \qquad F \in L^2(I), \qquad 0 \leq \alpha_2 \tag{2.3.1a}$$

where F is the primitive of f from which it follows that

$$u_{EX} \in \mathcal{U} \subseteq C^0(\bar{I}). \tag{2.3.1b}$$

When we know additional information about the smoothness of AE, c, and f, we can say more about the smoothness of u_{EX}. For example, if $c, f \in C^0(\bar{I})$, $AE \in C^1(\bar{I})$, we have $u_{EX} \in C^2(\bar{I})$. In general, we can prove that u_{EX} has one or two more derivatives than AE, c, and f, namely:

Theorem 2.3.1 *Let AE, c, and f be such that*

$$\left\|\frac{d^{q'}}{dx^{q'}}(AE)\right\|_{L^\infty(I)} < A, \qquad \left\|\frac{d^{r'}c}{dx^{r'}}\right\|_{L^\infty(I)} < B, \qquad \left\|\frac{d^{t'}f}{dx^{t'}}\right\|_{L^\infty(I)} < C \tag{2.3.2a}$$

for all integers q', r', t', such that

$$0 \leq q' \leq q, \qquad 0 \leq r' \leq r, \qquad 0 \leq t' \leq t \tag{2.3.2b}$$

where $\|\cdot\|_{L^\infty(I)}$ is the L^∞ norm defined in A2.4.1. Then, for the regular model problem under the assumptions (2.2.17a,b) we have

$$\left\|\frac{d^j}{dx^j}(u_{EX})\right\|_{L^\infty(I)} \leq D \tag{2.3.2c}$$

for all integers j, such that $0 \leq j \leq s$, with D dependent upon the bounds A, B, C, the input-data parameters β_1, a_2, α_2, and

$$s = \min(r + 2, t + 2, q + 1) \tag{2.3.2d}$$

If the assumptions (2.3.2a,b) hold piecewise, then u_{EX} has s piecewise-bounded derivatives.

Proof Let us first consider the case where $c \equiv 0$. Then,

$$-\frac{d}{dx}\Big(AE\frac{d}{dx}(u_{EX})\Big) = f \tag{2.3.3a}$$

and integrating we get

$$-AE(x)\frac{d}{dx}(u_{EX})(x) = F(x) = \int_0^x f - \mathscr{C}$$

and hence

$$\frac{d}{dx}(u_{EX})(x) = \frac{1}{AE(x)}\Big(\mathscr{C} - \int_0^x f\Big) = \frac{1}{AE(x)}(\mathscr{C} - F(x)) \tag{2.3.3b}$$

where we let $F(x) = \int_0^x f$, and integrating again, we obtain

$$u_{EX}(x) = \mathscr{C}\int_0^x \frac{1}{AE(x)} - \int_0^x \frac{F(x)}{AE(x)}. \tag{2.3.3c}$$

Note that (2.3.3c) satisfies the essential boundary condition $u_{EX}(0) = 0$. The constant $\mathscr{C}$ is determined from the natural boundary condition at $x = l$, namely

$$\alpha_2 u_{EX}(l) + AE(l)\frac{d}{dx}(u_{EX})(l) = a_2 \tag{2.3.3d}$$

which gives

$$\alpha_2\Big(-\int_0^l \frac{F}{AE} + \mathscr{C}\int_0^l \frac{1}{AE}\Big) + (-F(l) + \mathscr{C}) = a_2. \tag{2.3.3e}$$

Noting that the conditions $\alpha_2 \geq 0$ and $\int_0^l (1/AE) > 0$ imply that $1 + \alpha_2\int_0^l (1/AE) > 0$, we have

$$\mathscr{C} = \frac{a_2 + F(l) + \alpha_2\int_0^l (1/AE)F}{1 + \alpha_2\int_0^l (1/AE)} \tag{2.3.3f}$$

with $\mathscr{C} \leq \mathscr{C}_2\,\|f\|_{L^2(I)}$, where $\mathscr{C}_2$ depends on β_1, a_2, α_2, l, and A.

The assumption $\|d^t f/dx^t\|_{L^\infty(I)} < C$ implies that

$$\left\|\frac{d^{t'}F}{dx^{t'}}\right\|_{L^\infty(I)} \leq \mathscr{C}_1, \quad 0 \leq t' \leq t+1 \tag{2.3.4a}$$

where $\mathscr{C}_1$ depends on C and l. Further, from (2.3.2a) and (2.2.17), we have

$$\left\|\frac{d^{q'}}{dx^{q'}}\Big(\frac{1}{AE}\Big)\right\|_{L^\infty(I)} \leq \mathscr{C}_3, \quad 0 \leq q' \leq q \tag{2.3.4b}$$

and

$$\left\| \frac{d^{q'}}{dx^{q'}} \left(\frac{1}{AE} F \right) \right\|_{L^\infty(I)} \leq \mathscr{C}_4, \quad 0 \leq q' \leq \min((t+1), q) \tag{2.3.4c}$$

where $\mathscr{C}_4$ depends on β_1, a_2, α_2, l, A, C, q, and t. Then, (2.3.2c,d) follow by employing the above estimates in (2.3.3b) and (2.3.3c).

Let us now consider the general case where $c \not\equiv 0$. From Theorem 2.2.3 we know that u_{EX} exists, and it is bounded, namely

$$\|u_{EX}\|_{\mathscr{U}} \leq \tilde{\mathscr{C}} < \infty. \tag{2.3.5a}$$

Using inequality (2.2.18b) we get

$$\|u_{EX}\|_{L^\infty(I)} \leq \mathscr{C}_5 < \infty. \tag{2.3.5b}$$

Let

$$g = f - c u_{EX} \tag{2.3.6a}$$

then from (2.3.5b) and (2.3.2a) we get

$$\|g\|_{L^\infty(I)} \leq \mathscr{C}_6 < \infty \tag{2.3.6b}$$

and we have

$$-\frac{d}{dx}\left(AE \frac{d}{dx}(u_{EX}) \right) = g \tag{2.3.6c}$$

which is equivalent to the problem with $c \equiv 0$. Hence

$$\left\| \frac{d^{s'}}{dx^{s'}}(u_{EX}) \right\|_{L^\infty(I)} \leq \mathscr{C}_7 < \infty, \quad 0 \leq s' \leq \min(q+1, t+2, 2) \tag{2.3.7a}$$

where $\mathscr{C}_7$ depends on β_1, a_2, α_2, l, A, B, C, q, and t. It follows that

$$\left\| \frac{d^k}{dx^k}(c u_{EX}) \right\|_{L^\infty(I)} < \mathscr{C}_8 < \infty, \quad k = \min(s', r). \tag{2.3.7b}$$

Using, once more, the above argument, we get

$$\left\| \frac{d^{s''}}{dx^{s''}}(u_{EX}) \right\|_{L^\infty(I)} < \mathscr{C}_9 < \infty, \quad 0 \leq s'' \leq \min(q+1, t+2, k+2) \tag{2.3.7c}$$

and hence

$$\left\| \frac{d^{s'}}{dx^{s'}}(c u_{EX}) \right\|_{L^\infty(I)} \leq \mathscr{C}_{10} < \infty, \quad 0 \leq s' \leq \min(r+2, t+2, q+1, s''+2) \tag{2.3.7d}$$

where the constants $\mathscr{C}_5, \ldots, \mathscr{C}_{10}$ depend on A, B, C, and β_1. The proof of (2.3.2c) can now be completed by *induction*. □

Theorem 2.3.1 states that u_{EX}, the exact solution of the model problem, is guaranteed to have bounded derivatives up to a maximal order s which depends on the smoothness of the coefficients AE, c, and the right-hand side f. The bounds depend on the parameters $\beta_1 = \min_{x \in I}(AE)$, a_2, α_2, the bounds for the derivatives of AE, c, and f and the maximal order of derivative of the exact solution; see P2.3.1–P2.3.3.

Remark 2.3.1 Theorem 2.3.1 is often called the *shift* theorem, and the technique employed in its proof is known as the *boot-strapping approach.*

Remark 2.3.2 In the proof of Theorem 2.3.1, we assumed that the solution satisfies the homogeneous Dirichlet condition $u(0) = 0$. The proof can be easily modified to apply to other types of boundary conditions.

Remark 2.3.3 In the case that AE and c are smooth, Theorem 2.3.1 states that the exact solution of the model problem has two more bounded derivatives than the right-hand side f.

Let us now give two applications of the above theorem.

Example 2.3.1 Let $G(\cdot;\bar{x}) \in \mathcal{U}$ be the exact solution of (2.2.8) for the choice $f \equiv \delta(\cdot;\bar{x})$, $a_2 = 0$, for which we have

$$\mathcal{L}(v) = \int_0^l \delta(\cdot;\bar{x})v = v(\bar{x}). \tag{2.3.8a}$$

Note that

$$|\mathcal{L}(v)| = |v(\bar{x})| \leq \frac{1}{\sqrt{\beta_1}}\sqrt{l}\,\|v\|_{\mathcal{U}} \tag{2.3.8b}$$

and hence $G(\cdot;\bar{x})$ exists. From Theorem 2.3.1, it follows that $G(\cdot;\bar{x})$ has s-bounded derivatives with respect to its first argument, with $s = \min(r+2, q+1)$ in the intervals $(0,\bar{x})$, $(\bar{x}, l)$.

Example 2.3.2 Let us consider the singular model problem for $0 \leq \vartheta < 1$ with $a_2 = 0$ and $\alpha_2 = 0$, namely

$$-\frac{d}{dx}\left(x^{\vartheta}\frac{du}{dx}(x)\right) = f(x), \qquad 0 \leq x \leq l \tag{2.3.9a}$$

$$u(0) = 0, \quad l^{\vartheta}\frac{du}{dx}(l) = 0. \tag{2.3.9b}$$

Integrating the differential equation and letting $F(x) = -\int_x^l f$, we get

$$-x^{\vartheta}\frac{d}{dx}(u_{EX})(x) = -\int_x^l f + \mathscr{C} = F(x) + \mathscr{C} \tag{2.3.9c}$$

and

$$-\frac{d}{dx}(u_{EX})(x) = x^{-\vartheta}F(x) + \mathscr{C}x^{-\vartheta} \tag{2.3.9d}$$

and applying the boundary condition at $x = l$ we obtain $\mathscr{C} = 0$. Thus, if $\vartheta > 0$, the first derivative of the solution is not bounded, in general, even for smooth f, e.g., $f \equiv 1$. For $\frac{1}{2} \le \vartheta$ the integral $\int_0^1((d/dx)(u_{EX}))^2$ does not exist and hence $u_{EX} \notin H^1_{(0}(I)$. Note that this does not contradict the statement $u_{EX} \in \mathscr{U}^\vartheta$ since

$$\|u_{EX}\|_{\mathscr{U}^\vartheta} = \sqrt{\int_0^l x^\vartheta \Big(\frac{d}{dx}(u_{EX})\Big)^2} = \sqrt{\int_0^l x^{-\vartheta} F^2} < \infty, \quad \text{for } \vartheta < 1. \tag{2.3.9e}$$

Similarly, the Green's function $G(\cdot;\bar{x})$, for the singular model problem has unbounded derivative at $x = 0$.

Remark 2.3.4 Let us underline that *it makes sense to attempt to compute a quantity of interest only if it has a finite value.* For example, it is meaningless to attempt to determine $(d/dx)(u_{EX})(0)$ or $\max_{x\in I}((d/dx)(u_{EX}))$ in the case of the singular model problem.

In this section we have determined relations between the smoothness of the input data and the smoothness of the exact solution for the model problem, where the smoothness of a function is measured by the maximum order of bounded derivatives of the function in the interval of interest. In Theorem 2.3.1, we have proven that the solution of the model problem has bounded derivatives of one or two orders higher than the input data. An analogous result, but with some important differences, holds in higher dimensions.

Problems

P2.3.1. Let f be such that

$$\|f\|_{L^2(I)} \le B, \qquad \Big\|\frac{d^t}{dx^t}(f)\Big\|_{L^\infty(I)} \le C \tag{2.3.10}$$

for some integers $t > 0$. Bound $\|d^j f/dx^j\|_{L^\infty(I)}$ for integer j, $0 \le j < t$, in terms of B, C. Let u_{EX} denote the solution of the regular model problem with $AE \equiv 1$, $c \equiv 0$, $\alpha_2 = 0$, $a_2 = 0$, and load f, where we have

$$\|f\|_{L^2(I)} \le B, \qquad \Big\|\frac{d^t}{dx^t}(f)\Big\|_{L^\infty(I)} \le C \tag{2.3.11}$$

for some integer $t \ge 0$. Estimate $\|(d^j/dx^j)(u_{EX})\|_{L^\infty(I)}$ for integer j, $0 \le j \le t+2$, in terms of B, C.

P2.3.2. Let u_{EX} be the solution of the regular model problem with the same input data as in P2.3.1 and assume that

$$\Big\|\frac{d^{t'} f}{dx^{t'}}\Big\|_{L^\infty(I)} \le B, \quad 0 \le t' \le t. \tag{2.3.12}$$

Estimate $\|(d^j/dx^j)(u_{EX})\|_{L^\infty(I)}$ for integer j, $0 \le j \le t+2$, in terms of B.

P2.3.3. Let u_{EX} be the solution of the regular model problem with $c \equiv 0$, $\alpha_2 = 0$, $a_2 = 0$, $f \equiv 1$, and assume that

$$\Big\|\frac{d^{q'}}{dx^{q'}}(AE)\Big\|_{L^\infty(I)} \le B, \quad 0 \le q' \le q. \tag{2.3.13}$$

Estimate $\|(d^j/dx^j)(u_{EX})\|_{L^\infty(I)}$ for integer j, $0 \le j \le q+1$ in terms of B.

P2.3.4. Let u_{EX} be the solution of the singular model problem with $c \equiv 0$, $\alpha_2 = 0$, $a_2 = 0$, $f \equiv 1$. Estimate $\|(d^j/dx^j)(u_{EX})\|_{L^\infty(I_\epsilon)}$ and $\|(d^j/dx^j)(x^\vartheta(d/dx)(u_{EX}))\|_{L^\infty(I_\epsilon)}$, where $I_\epsilon = (\epsilon, l)$, $\epsilon > 0$, for $j = 0, 1, 2$.

P2.3.5. Prove Theorem 2.3.1 for the case that the natural boundary conditions

$$AE(0)\frac{du}{dx}(0) = \alpha_1 u(0) - a_1, \qquad -AE(l)\frac{du}{dx}(l) = \alpha_2 u(l) - a_2 \tag{2.3.14}$$

are imposed, assuming $\alpha_1 > 0$.

2.3* Smoothness of the solution of the two-dimensional model problem

As in the one-dimensional case, the smoothness of the solution of the two-dimensional model problem depends on the smoothness of the coefficients, the loads, the boundary conditions (e.g., a_{ij}, c, α, f, $\bar{u}$, g), and, in addition, *on the smoothness of the boundary* $\partial\Omega$. In the very special case that:

(i) The boundary $\partial\Omega$ is a smooth function of the arc length,
(ii) Only homogeneous essential boundary conditions $u|_{\partial\Omega} \equiv 0$ are imposed

the smoothness of the solution depends only on the smoothness of the coefficients a_{ij}, c, and the right-hand side f of the differential equation and an analogue of Theorem 2.3.1 can be formulated, as follows:

Theorem 2.3*.1 *Under the assumptions*

$$\left\|\frac{\partial^{k_1+k_2} a_{ij}}{\partial x_1^{k_1}\partial x_2^{k_2}}\right\|_{L^\infty(\Omega)} \le A, \quad 0 \le k_1+k_2 \le q, \;\; k_1, k_2 \ge 0 \tag{2.3*.1a}$$

and

$$\left\|\frac{\partial^{k_1+k_2} c}{\partial x_1^{k_1}\partial x_2^{k_2}}\right\|_{L^\infty(\Omega)} \le B, \quad 0 \le k_1+k_2 \le t \tag{2.3*.1b}$$

and

$$\left\|\frac{\partial^{k_1+k_2} f}{\partial x_1^{k_1}\partial x_2^{k_2}}\right\|_{L^\infty(\Omega)} \le C, \quad 0 \le k_1+k_2 \le t, \;\; k_1, k_2 \ge 0 \tag{2.3*.1c}$$

we have

$$\left\|\frac{\partial^{k_1+k_2} u_{EX}}{\partial x_1^{k_1}\partial x_2^{k_2}}\right\|_{L^\infty(\Omega)} \le K, \quad 0 \le k_1+k_2 \le \min(q+1, t+2), \;\; k_1, k_2 \ge 0. \tag{2.3*.1d}$$

The above statement is 'almost' true in the sense that the smoothness of u_{EX} is only slightly worse than indicated above. Nevertheless the bound (2.3*.1d) holds for the L^2 norm, i.e. we have

$$\left\|\frac{\partial^{k_1+k_2} u_{EX}}{\partial x_1^{k_1}\partial x_2^{k_2}}\right\|_{L^2(\Omega)} \le K, \quad 0 \le k_1+k_2 \le \min(q+1, t+2), \;\; k_1, k_2 \ge 0. \tag{2.3*.1e}$$

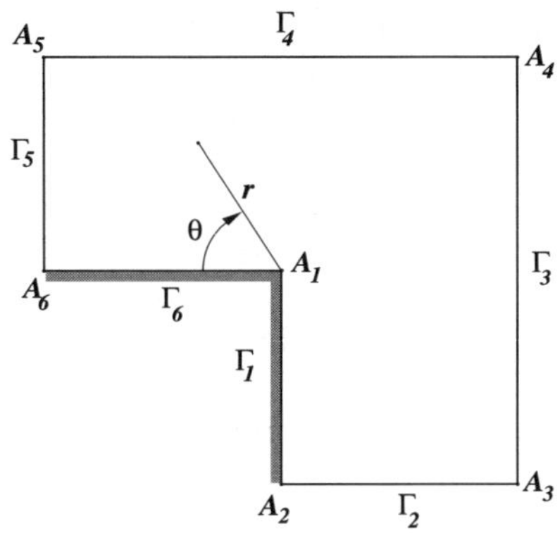

FIG. 2.3*.1. The L-shaped domain and the definition of the polar coordinates with respect to vertex A_1

More precisely, it can be shown that for sufficiently smooth a_{ij} and c, we have

$$\|u_{EX}\|_{H^{k+2}(\Omega)} \leq \mathscr{C} \|f\|_{H^k(\Omega)} \tag{2.3*.1f}$$

where $H^k(\Omega)$ is the Sobolev space defined in Appendix A2.4.2, and $\mathscr{C}$ depends on Ω and the coefficients a_{ij}, but is independent of f. Thus we see the typical 'shift' between the $H^k(\Omega)$ space, which characterizes the smoothness of f, and the $H^{k+2}(\Omega)$ space which characterizes the smoothness of u_{EX}. We will not elaborate any further. The corresponding result for the elasticity problem can be stated analogously. Further details can be found in [Lions and Magenes, 1972], [Ladyzhenskaja and Uraltseva, 1968] , [Nečas, 1967], and [Babuška and Aziz, 1972].

In practice, the boundary $\partial\Omega$ has one or several *corner points*, which are points of discontinuity of the slope of the boundary or points where the type of boundary condition changes (e.g. from Dirichlet to Neumann), and then *the smoothness of the solution is influenced by these points.* In the neighborhood of the corner points with internal angle greater than π, the derivatives of the solution are usually unbounded even if the coefficients and the right-hand side of the differential equation and the boundary conditions are very smooth. Let us for example consider the problem of Laplace's equation in the L-shaped domain Ω shown in Fig. 2.3*.1, namely

$$\Delta u_{EX} = 0 \quad \text{in } \Omega, \tag{2.3*.2a}$$

$$u_{EX} = 0 \quad \text{on } \Gamma_{\mathrm{D}} = A_1A_6 \cup A_1A_2, \tag{2.3*.2b}$$

$$\frac{\partial}{\partial n}(u_{EX}) = g \quad \text{on } \Gamma_{\mathrm{N}} = \Gamma - \Gamma_{\mathrm{D}}. \tag{2.3*.2c}$$

Then we can select g such that

$$u_{EX}(r,\theta) = Ar^{2/3}\sin\left(\tfrac{2}{3}\theta\right). \tag{2.3*.2d}$$

In the more general case, for any smooth g, the exact solution u_{EX} can be expressed in the form

$$u_{EX}(r,\theta) = Ar^{2/3}\sin\left(\tfrac{2}{3}\theta\right) + \tilde{u}_{EX}(r,\theta) \tag{2.3*.2e}$$

where $\tilde{u}_{EX}$ is smoother than the term $r^{2/3}\sin(\frac{2}{3}\theta)$. Note that although all the input data are smooth, the derivatives of u_{EX} are unbounded in the neighborhood of A_1. Analogous results also hold for the elasticity problem. For details, see [Grisvard, 1985, 1992], [Nazarov and Plamenevsky, 1992], [Dauge, 1988].

Hence, if the domain Ω has corners and/or points where the type of boundary conditions change, then the solution may be (and usually is) singular in the neighborhood of these points. Assume, for example, that Ω is a polygonal domain with sides Γ_i and internal angles α_i at the vertices A_i, that the type of the boundary conditions changes only at the corner points of the domain ($\alpha_i = 2\pi$ can be formally a corner), and let us be interested in the mixed boundary value problem for the Poisson equation

$$-\Delta u_{EX} = f \quad \text{in } \Omega, \tag{2.3*.3a}$$

$$u_{EX} = 0 \quad \text{on } \Gamma_{\mathrm{D}}, \tag{2.3*.3b}$$

$$\frac{\partial}{\partial n}(u_{EX}) = g \quad \text{on } \Gamma_{\mathrm{N}} = \Gamma - \Gamma_{\mathrm{D}} = \bigcup_{j=1}^{Q} \Gamma_j. \tag{2.3*.3c}$$

We then have

$$\left\| u_{EX} - u_{EX}^{sing} \right\|_{H^k} \leq \mathscr{C} \|f\|_{H^l(\Omega)} + \sum_{i=1}^{Q} \|g\|_{H^{s_i}(\Gamma_i)} \tag{2.3*.4a}$$

where $k = \min(l+2, s_i+1)$, and

$$u_{EX}^{sing} = \sum_{i=1}^{Q} \sum_{j=1}^{M_i} a_{i,j} r_i^{\alpha_{i,j}} \varphi_{i,j}(\theta_i) \, |\ln r_i|^{\beta_{i,j}} \chi_i(r_i) \tag{2.3*.5}$$

where (r_i, θ_i) are polar coordinates in the vertices A_i, $\varphi_{i,j}(\theta_i)$ are analytic functions in θ_i, $\varphi_{i,j}$ and $\alpha_{i,j}$ depend on α_i and the boundary condition, and $\beta_{i,j} = 0$ or 1; $\alpha_{i,j+1} \geq \alpha_{i,j}$, $\chi_i(r_i)$ is a C^∞ function such that $\chi_i = 1$ for $r_i < r_{i,0}$ and $\chi_i = 0$ for $r_i > 2r_{i,0}$ when $r_{i,0}$ is properly chosen. A similar, yet more complicated situation exists for the elasticity problem; see [Grisvard, 1992, 1985], [Nazarov and Plamenevsky, 1992], [Kozlov *et al.*, 1977], and [Dauge, 1988]. In the case of elasticity the coefficients $a_{i,j}$ are called the stress-intensity factors and are often of major engineering interest, especially for $j = 1$, because they relate to the fracture of material. For computation of the singular functions, we refer to [Papadakis and Babuška, 1995], [Leguillon and Sanchez-Palencia, 1987], and [Vasilopoulos, 1988].

Remark 2.3*.1 The singular model problem was introduced as a one-dimensional analogue of the two-dimensional problem where the boundary of the domain has corner points.

The purpose of this section was to address the similarities and differences on the dependence of the smoothness of the exact solution on the smoothness of the input data in the one and the two-dimensional cases. When the boundary is sufficiently smooth, the smoothness of the solution of the two-dimensional model problem depends on the smoothness of the coefficients and the right-hand side of the differential equation, in analogue to the one-dimensional case. Most often in practice, the boundary has corner points or points where the type of boundary condition changes, and the shift theorem does not hold, and there is no one-dimensional analogue. Nevertheless, there is some similarity with the singular model problem which was introduced exactly for this purpose.

Problems

P2.3*.1. Show that $u_{EX} \equiv r^{2/3} \sin(\frac{2}{3}\theta)$ has finite energy, i.e. $\int_\Omega |\boldsymbol{\nabla} u_{EX}|^2 < \infty$.

P2.3*.2. Show that the function $u_{EX} \equiv r^{-2/3} \sin(\frac{2}{3}\theta)$ satisfies (2.3*.2a,b) (except at the point A_1), but the integral $\int_\Omega |\boldsymbol{\nabla} u_{EX}|^2$ does not exist, and $u_{EX} \notin \mathcal{U}$.

P2.3*.3. Show that $u_{EX} = r^{2/3} \cos(\frac{2}{3}\theta)$ satisfies (2.3*.2a–c) with (2.3*.2b) replaced by $(\partial/\partial n)(u_{EX}) = 0$ on $A_1A_0 \cup A_1A_2$ ($(\partial/\partial n)(u_{EX})$ at A_1 does not have meaning). Show that $\int_\Omega |\boldsymbol{\nabla} u_{EX}|^2 < \infty$, and hence $u_{EX} \in \mathcal{U}$. Show that $u_{EX} = r^{-2/3} \cos(\frac{2}{3}\theta)$ also satisfies the boundary value problem, but $u_{EX} \notin \mathcal{U}$.

2.4 Historical and bibliographical remarks

Differential equations have been at the center of interest for a relatively long time. Today we often distinguish between the 'classical' and modern theories of partial differential equations. The classical approach concentrates on the strong formulation, classical definition of derivatives, etc. (for the history of calculus, we refer to [Boyer, 1959]). The modern approach is based on the use of functional analysis; it concentrates on the weak formulations, Sobolev spaces, etc. For the history of differential equations in the 20th Century, we refer to [Brezis and Browder, 1998].

There are today hundreds of books on differential equations, the classical and modern theories, as well as on functional analysis, elementary and sophisticated. Only relatively recently (1967) has the theory of the solutions on the domains with piecewise smooth boundaries, which is very important for practical applications, been initiated by V.A. Kondratév [Kondratév, 1967], and today the theory is well progressed, although far from complete.

This book is based on the 'modern' theory of PDEs, its weak solution, variational principles, etc. The relation between the minimization of the energy and the problem of partial differential equations was discovered in 1832 by G. Green (1793–1841). The rigorous relation was obtained only at the beginning of the twentieth century by D. Hilbert and further developed by B. Levi, G. Fubini, S. Zaremba, H. Lebesgue, and others. Hilbert (1862–1943) formulated the relation between the variational formulation and the solution as the most outstanding mathematical problem (Problem No. 20) in his celebrated address to the International Mathematical Congress in Paris 1900. In that address Hilbert formulated 23 problems which had a strong influence on mathematics in the first half of the twentieth century.

In Chapter 2 we have used certain names in connection with various mathematical notions. Let us mention these scientists and the time when they lived:

Dirichlet (Peter Gustav Lejeune) 1805–1859

Fourier (Jean Baptiste Joseph) 1768–1830

Gauss (Johann Carl Friedrich) 1777–1855

Laplace (Pierre Simon) 1749–1827

Neumann (Franz Ernst) 1798–1895.

From this list, some old roots of the theory of PDEs are clearly visible.

References

R.A. Adams (1975) *Sobolev spaces.* Academic Press, New York.

S. Agmon (1965) *Lectures on elliptic boundary value problems.* Van Nostrand, Princeton, NJ.

I. Babuška and A.K. Aziz (1972) Survey lectures on the mathematical foundations of the Finite Element Method. In: A.K. Aziz (ed.) *The mathematical foundations of the Finite Element Method with applications to Partial Differential Equations, Proceedings of a symposium held at the University of Maryland, Baltimore campus, June 26–30, 1972*, Academic Press, New York, pp. 47–82.

C.B. Boyer (1959) *The history of calculus and its conceptual development.* Dover, New York.

H. Brezis and F. Browder (1998) Partial Differential Equations in the twentieth century. *Advances in Mathematics*, **135**, 76–144.

R. Courant and D. Hilbert (1953) *Methods of mathematical physics I.* Wiley.

M. Dauge (1988) *Elliptic boundary value problems in corner domains: smoothness and asymptotics of solutions.* Lecture Notes in Mathematics, Vol. **1341**, Springer, Berlin.

Y.V. Egorov and M.A. Shubin (1998) *Foundation of the classical theory of Partial Differential Equations.* Second Edition. Springer, Berlin.

B.F. de Veubeke (1979) *A course in elasticity.* Springer, Berlin.

D. Gilbarg and N. Trudinger (1998) *Elliptic Partial Differential Equations of second order*, Third printing. Springer, Berlin.

P. Grisvard (1985) *Elliptic problems in nonsmooth domains*, Monographs and Studies in Mathematics, **21**, Pitman Boston.

P. Grisvard (1992) *Singularities in boundary value problems.* Masson Springer.

M.E. Gurtin (1972) The linear theory of elasticity. In: C. Truesdell (ed.) *Handbuch der physik VI a/2, mechanics of solids II*, Springer, Berlin.

V.A. Kondratév (1967) Boundary value problems for elliptic equations in domains with conical or angular points. *Trudy Moskovskogo Matematicheskogo Obschetsva*, **16**, 209–292.

V.A. Kondratév and O.A. Olejnik (1990) Hardy's and Korn's type inequalities and their applications. *Rendiconti di Matematica e della sue Applicazioni, Serie VII*, **7** (*10*), 641–666.

V.A. Kozlov, V.G. Mazja and J. Rossmann (1977) *Elliptic boundary value problems in domains with point singularities.* Mathematical Surveys and Monographs, **52**, American Mathematical Society.

O. Ladyzhenskaja and N. Uraltseva (1968) *Linear and quasilinear elliptic equations.* Academic Press, New York.

D. Leguillon and E. Sanchez-Palencia (1987) *Computation of singular solutions in elliptic problems and elasticity.* Masson.

J.L. Lions and E. Magenes (1972) *Nonhomogenous boundary value problems and applications*, Vol. 1, Springer, Berlin.

L.E. Malvern (1969) *Introduction to the mechanics of continuum.* Prentice Hall, Englewood Cliffs.

J.E. Marsden and T.J.R. Hughes (1983) *Mathematical foundation of elasticity.* Prentice Hall, Englewood Cliffs.

J. Mason (1985) *Methods of functional analysis for applications in solid mechanics*, Elsevier North Holland.

S.G. Mikhlin (1964) *Variational methods in mathematical physics.* Pergammon Press, Oxford.

S.G. Mikhlin (1971) *The numerical performance of variational methods.* Wolters-Noordhoff, Groningen.

S. Nazarov and B.A. Plamenevsky (1992) *Elliptic problems in domains with piecewise smooth boundaries.* Walter de Grupter.

J. Nečas (1967) *Les méthodes diréctes en théorie des équations élliptiques.* Academia, Prague.

J. Nečas and I. Hlavaček (1981) *Mathematical theory of elastic and elastoplastic bodies: an introduction.* Elsevier, Amsterdam.

J.T. Oden (1979) *Applied functional analysis: a first course for students of mechanics and engineering science.* Prentice Hall, Englewood Cliffs.

J.T. Oden and L.F. Demkowicz (1996) *Applied functional analysis.* CRC Press, New York.

P.J. Papadakis and I. Babuška (1995) A numerical procedure for the determination of certain quantities related to the stress intensity factors in two dimensional elasticity. *Computer Methods in Applied Mechanics and Engineering*, **122**, 69–92.

K. Rektorys (1977) *Variational methods in mathematics, science and engineering.* D. Reidel Publishing Co., Dordrecht.

K. Rektorys (1998) *Solving ordinary and boundary value problems in science and engineering.* CRC Press, Boca Raton.

M.E. Taylor (1999) *Partial differential equations, basic theory*, Second Edition. Springer, Berlin.

D. Vasilopoulos (1988) On the determination of higher order terms of singular elastic stress fields near corners. *Numerische Mathematik*, **52**, 51–97.

3

THE FINITE ELEMENT METHOD

3.0 Introduction

In this chapter, we present the finite element method for our model problem. We first describe the Galerkin method which constructs approximate solutions $u_{S^{(n)}}$, by satisfying the variational problem on finite-dimensional spaces $S^{(n)} \subseteq \mathcal{U}$, namely:

Find $u_{S^{(n)}} \in S^{(n)}$ such that

$$\mathcal{B}(u_{S^{(n)}}, v) = \mathcal{L}(v) \quad \forall v \in S^{(n)}. \tag{3.0.1}$$

We also prove the best-approximation property of the Galerkin method, namely, that

$$\|u_{EX} - u_{S^{(n)}}\|_{\mathcal{U}} = \min_{\chi \in S^{(n)}} \|u_{EX} - \chi\|_{\mathcal{U}}. \tag{3.0.2}$$

The finite element method is introduced as a special case of the Galerkin method in which the finite-dimensional spaces $S^{(n)}$, are constructed by employing a basis of continuous, piecewise polynomial functions defined over a *mesh* Δ, which is a partition of the domain I into non-overlapping subdomains I_j^{Δ}, $j = 1, \ldots, N(\Delta)$, called *elements*, namely

$$\bar{I} = \bigcup_{j=1}^{N(\Delta)} \overline{I_j^{\Delta}}, \qquad I_j^{\Delta} \cap I_k^{\Delta} = \emptyset \quad k \neq j. \tag{3.0.3}$$

We denote the finite-dimensional space $S^{(n)}$, by S_{Δ}^{p}, the space of continuous functions which are polynomials of degree p over the elements I_j^{Δ}, of the mesh Δ, and satisfy the Dirichlet boundary conditions. The finite element solution denoted by $u_{S_{\Delta}^{p}}$, is defined as the solution of the discrete problem:

Find $u_{S_{\Delta}^{p}} \in S_{\Delta}^{p}$ such that

$$\mathcal{B}(u_{S_{\Delta}^{p}}, v) = \mathcal{L}(v) \quad \forall v \in S_{\Delta}^{p}. \tag{3.0.4}$$

We give detailed constructions of the finite element approximations, and derive estimates for the error

$$e_{S_{\Delta}^{p}} \stackrel{\text{def}}{=} u_{EX} - u_{S_{\Delta}^{p}}. \tag{3.0.5}$$

The typical error estimate reads

$$\|e_{S_{\Delta}^{p}}\|_{\mathcal{U}} = \min_{\chi \in S_{\Delta}^{p}} \|u_{EX} - \chi\|_{\mathcal{U}} \leq \mathcal{C}(h^{\Delta})^{\mu}, \tag{3.0.6}$$

where h^{Δ} is the size of the maximal element in Δ, and the exponent μ depends on p and the smoothness of u_{EX}.

Further, we introduce the *Lagrangian* interpolant $\mathcal{I}^p_\Delta u \in S^p_\Delta$, which is well defined for any continuous function u, and derive estimates for various norms of the interpolation error, under various assumptions about the smoothness of u. For example, under the assumption that

$$\left\| \frac{d^r u}{dx^r} \right\|_{L^q(I)} < \infty \tag{3.0.7a}$$

we prove that

$$\|u - \mathcal{I}^p_\Delta u\|_{\mathcal{U}} \leq \mathscr{C}(h^\Delta)^\mu \left\| \frac{d^r u}{dx^r} \right\|_{L^q(I)} \tag{3.0.7b}$$

for $q = 2$ or ∞, where

$$\mu = \min(p, r-1), \quad 1 \leq r \leq p+1. \tag{3.0.7c}$$

We also introduce the spaces of functions with *fractional derivatives*, and construct more general versions of the interpolation estimates.

Further, we analyze the error in the finite element solution in the special case that

$$u_{EX}(x) = x^\alpha, \quad \alpha > \tfrac{1}{2}, \ \alpha \neq \text{ integer}. \tag{3.0.8}$$

This solution is of particular interest because it is a one-dimensional analogue of the solutions in polygonal domains in higher dimensions. We consider the class of *radical* meshes Δ^γ_n, with nodes at

$$x_j^{\Delta^\gamma_n} = \frac{j^\gamma}{N^\gamma}, \quad j = 0, 1, \ldots, N(\Delta^\gamma_n) = n, \tag{3.0.9a}$$

and prove the estimate

$$\left\| \frac{d}{dx}\left(x^\alpha - (x^\alpha)_{S^p_\Delta}\right) \right\|_{L^2(I)} \leq \begin{cases} \mathscr{C}N^{-\mu}, & p \neq \gamma(\alpha - \frac{1}{2}) \\ \mathscr{C}N^{-p}\sqrt{\ln N}, & p = \gamma(\alpha - \frac{1}{2}) \end{cases} \tag{3.0.9b}$$

where

$$\mu = \min\left(p, \gamma\left(\alpha - \tfrac{1}{2}\right)\right). \tag{3.0.9c}$$

It follows that by choosing

$$\gamma \geq \gamma_{cr} = \frac{p}{\alpha - \frac{1}{2}}, \quad \alpha > \frac{1}{2}, \tag{3.0.9d}$$

which results in a clustering of the meshes Δ^γ_n towards $x = 0$, we obtain the optimal rate of convergence, for any value of the exponent α. However, for α close to $\frac{1}{2}$, and high p, the meshes Δ^γ_n are so highly graded that problems with roundoff error may occur.

Finally, we address the question of optimal meshes in a general setting. The main idea of our approach is to employ an estimate of the interpolation error for an entire sequence of meshes represented by a *density* function, which describes the density of elements in all the meshes in the sequence. For example, we have

$$\left\| \frac{d}{dx}(u_{EX} - \mathcal{I}^p_{\Delta^\Phi_n} u_{EX}) \right\|_{L^2(I)} \approx \frac{\mathscr{C}(p)}{(N(\Delta^\Phi_n))^p} \left\| \frac{1}{\Phi^p} \frac{d^{p+1}}{dx^{p+1}}(u_{EX}) \right\|_{L^2(I)}, \tag{3.0.10a}$$

where Φ is the density function which is such that

$$\int_a^b \Phi > 0, \quad a < b, \qquad \int_0^1 \Phi = 1, \tag{3.0.10b}$$

and Δ^Φ_n is the corresponding density mesh of n elements which is characterized by the property

$$\int_{I_j^{\Delta^\Phi_n}} \Phi = \frac{1}{n}, \quad j = 1, 2, \ldots, N(\Delta^\Phi_n) = n. \tag{3.0.10c}$$

This leads us to the definition of the optimal density function Φ^{opt}, which is the one which minimizes the above interpolation-error estimate, and the principle of *equidistribution of error* which states that *a mesh is asymptotically optimal in a given norm if it equidistributes the error among the elements in that norm.* We also show that, unless some assumptions about the smoothness of the solution are imposed, this principle is false. However, the needed assumptions are rather mild and hold for all the problems encountered in engineering.

3.1 The Galerkin method and the properties of the approximate solution

We will now present the Galerkin method for constructing approximate solutions of the variational problem (2.2.8).

Let $\varphi_i \in \mathscr{U}$, $i = 1, \ldots, n$, be *linearly independent* functions, namely

$$\sum_{i=1}^{n} c_i \varphi_i \equiv 0 \quad \Longrightarrow \quad c_i = 0, \quad i = 1, \ldots, n, \tag{3.1.1a}$$

and denote by $S^{(n)}$ the space of functions which are linear combinations of the φ_i, $i = 1, \ldots, n$, namely

$$S^{(n)} \stackrel{\text{def}}{=} \left\{ u \in \mathscr{U} \,\middle|\, u = \sum_{i=1}^{n} c_i \varphi_i \right\}. \tag{3.1.1b}$$

Then $S^{(n)}$ is the n-dimensional subspace of $\mathscr{U}$, generated by the *basis* $\{\varphi_i\}_{i=1}^n$.

Given $S^{(n)} \subseteq \mathscr{U}$, we will construct $u_{S^{(n)}} \in S^{(n)}$ as the solution of the following *discrete* problem:

Find $u_{S^{(n)}} \in S^{(n)}$ such that

$$\mathscr{B}(u_{S^{(n)}}, v) = \mathscr{L}(v) \quad \forall v \in S^{(n)}. \tag{3.1.2a}$$

We will call $u_{S^{(n)}}$ the *Galerkin approximation of $u_{EX} \in \mathcal{U}$ from $S^{(n)}$*. Employing the expansion

$$u_{S^{(n)}} = \sum_{j=1}^{n} c_j \varphi_j \tag{3.1.2b}$$

and applying (3.1.2a) for $v = \varphi_i$, for $i = 1, \ldots, n$, we obtain the system of linear equations

$$\sum_{j=1}^{n} \mathcal{B}(\varphi_j, \varphi_i) c_j = \mathcal{L}(\varphi_i), \quad i = 1, \ldots, n. \tag{3.1.2c}$$

Denoting $\mathbf{K} = [K_{ij}, i, j = 1, \ldots, n]$, and $\mathbf{f} = [f_1, \ldots, f_n]^T$, where

$$K_{ij} \stackrel{\text{def}}{=} \mathcal{B}(\varphi_j, \varphi_i), \qquad f_i \stackrel{\text{def}}{=} \mathcal{L}(\varphi_i) \tag{3.1.2d}$$

and $\mathbf{c} = [c_1, \ldots, c_n]^T$, (3.1.2c) can be rewritten in the form

$$\mathbf{K}\mathbf{c} = \mathbf{f}. \tag{3.1.2e}$$

The matrix $\mathbf{K}$ (respectively the vector $\mathbf{f}$) is known as the *stiffness matrix* (respectively the *load vector*) and $\mathbf{c}$ is the *solution vector.*

Let us address the existence and uniqueness of $u_{S^{(n)}}$.

Theorem 3.1.1 *There exists a unique $u_{S^{(n)}}$ which satisfies (3.1.2a).*

Proof The existence and uniqueness of $u_{S^{(n)}}$ is equivalent to the existence and uniqueness of a $\mathbf{c} \in \mathbb{R}^n$, which satisfies $\mathbf{K}\mathbf{c} = \mathbf{f}$. To prove this, it is sufficient to prove that the homogeneous problem $\mathbf{K}\mathbf{c} = \mathbf{0}$ has only the trivial (zero) solution $\mathbf{c} = \mathbf{0}$, namely

$$\mathbf{K}\mathbf{c} = \mathbf{0} \quad \Longrightarrow \quad \mathbf{c} = \mathbf{0} \tag{3.1.3a}$$

or equivalently

$$\mathcal{B}(u_{S^{(n)}}, v) = 0 \quad \forall v \in S^{(n)} \quad \Longrightarrow \quad u_{S^{(n)}} \equiv 0. \tag{3.1.3b}$$

Noting that

$$\|u_{S^{(n)}}\|_{\mathcal{U}}^2 = \mathcal{B}(u_{S^{(n)}}, u_{S^{(n)}}) = \mathbf{c}^{\mathrm{T}}\mathbf{K}\mathbf{c} \tag{3.1.3c}$$

and because $\|\cdot\|_{\mathcal{U}}$ is a norm, we have

$$\begin{aligned} \|u_{S^{(n)}}\|_{\mathcal{U}}^2 = \mathbf{c}^{\mathrm{T}}\mathbf{K}\mathbf{c} = 0 \quad &\Longrightarrow \quad u_{S^{(n)}} = \sum_{i=1}^{n} c_i \varphi_i \equiv 0 \\ &\Longrightarrow \quad c_i = 0, \quad i = 1, \ldots, n \end{aligned} \tag{3.1.3d}$$

where the first implication follows because $\|\cdot\|_{\mathcal{U}}$ is a norm, and the second implication follows from the linear independence of the φ_i. □

The coefficients c_j are known as the *degrees of freedom* of the approximation.

Let us now address the properties of the matrix $\mathbf{K}$.

Corollary 3.1.1 *We have*

$$\mathbf{K} = \mathbf{K}^{\mathrm{T}} \tag{3.1.4}$$

$$\mathbf{d}^{\mathrm{T}}\mathbf{K}\mathbf{d} \geq 0 \quad \forall \mathbf{d} \in \mathbb{R}^n \tag{3.1.5a}$$

$$\mathbf{d}^{\mathrm{T}}\mathbf{K}\mathbf{d} = 0 \quad \Longleftrightarrow \quad \mathbf{d} = \mathbf{0}. \tag{3.1.5b}$$

Proof Recalling the symmetry of $\mathcal{B}(\cdot,\cdot)$, we obtain

$$K_{ij} = \mathcal{B}(\varphi_i, \varphi_j) = \mathcal{B}(\varphi_j, \varphi_i) = K_{ji}. \tag{3.1.6}$$

Letting

$$w = \sum_{i=1}^{n} d_i \varphi_i \tag{3.1.7a}$$

we get

$$\mathbf{d}^{\mathrm{T}}\mathbf{K}\mathbf{d} = \sum_{i=1}^{n} d_i \left(\sum_{j=1}^{n} d_j \mathcal{B}(\varphi_i, \varphi_j) \right) = \mathcal{B}\left(\underbrace{\sum_{i=1}^{n} d_i \varphi_i}_{w}, \underbrace{\sum_{j=1}^{n} d_j \varphi_j}_{w} \right) = \|w\|_{\mathcal{U}}^2 > 0 \tag{3.1.7b}$$

for $\mathbf{d} \neq \mathbf{0}$, because $\|\cdot\|_{\mathcal{U}}$ is a norm, and

$$\mathbf{d}^{\mathrm{T}}\mathbf{K}\mathbf{d} = \|w\|_{\mathcal{U}}^2 = 0 \quad \Longleftrightarrow \quad w = \sum_{i=1}^{n} d_i \varphi_i \equiv 0 \quad \Longleftrightarrow \quad \mathbf{d} = \mathbf{0}. \tag{3.1.7c}$$

□

Hence we have proven that $\mathbf{K}$ is a *symmetric positive definite* matrix.

Remark 3.1.1 The symmetry and positive-definiteness of $\mathbf{K}$ guarantees good performance of many algorithms for solving the linear system $\mathbf{Kc} = \mathbf{f}$ (e.g., the conjugate-gradient algorithm, Gauss elimination without pivoting, Cholesky factorization, etc.); see e.g. [Atkinson, 1989].

Let

$$e_{S^{(n)}} \stackrel{\text{def}}{=} u_{EX} - u_{S^{(n)}} \tag{3.1.8}$$

denote the *error* in $u_{S^{(n)}}$. We will now prove that $u_{S^{(n)}}$ is the best approximation of u_{EX} from $S^{(n)}$, when the error is measured in the energy norm, namely

$$\|e_{S^{(n)}}\|_{\mathcal{U}} = \min_{\chi \in S^{(n)}} \|u_{EX} - \chi\|_{\mathcal{U}}. \tag{3.1.9}$$

The proof is based on the following *orthogonality condition* for the error.

Lemma 3.1.1 *We have*

$$\mathcal{B}(e_{S^{(n)}}, v) = 0 \quad \forall v \in S^{(n)}. \tag{3.1.10}$$

Proof We have

$$\mathcal{B}(u_{EX}, v) = \mathcal{L}(v) \quad \forall v \in \mathcal{U} \tag{3.1.11a}$$

and

$$\mathcal{B}(u_{S^{(n)}}, v) = \mathcal{L}(v) \quad \forall v \in S^{(n)}. \tag{3.1.11b}$$

Now because

$$S^{(n)} \subseteq \mathcal{U} \tag{3.1.11c}$$

we can employ the same test function $v \in S^{(n)}$ into the first and the second equality, and subtracting we get (3.1.10). □

Remark 3.1.2 Equation (3.1.10) is known as *the orthogonality condition for the error*, and is a basic property of the Galerkin method.

Let us now prove that $u_{S^{(n)}}$ is the best approximation of u_{EX} from $S^{(n)}$ with respect to the energy norm.

Theorem 3.1.2 *Let $u_{S^{(n)}} \in S^{(n)}$ be the solution of (3.1.2a). Then*

$$\|e_{S^{(n)}}\|_{\mathcal{U}} = \min_{\chi \in S^{(n)}} \|u_{EX} - \chi\|_{\mathcal{U}} = \sqrt{\|u_{EX}\|_{\mathcal{U}}^2 - \|u_{S^{(n)}}\|_{\mathcal{U}}^2}. \tag{3.1.12}$$

Proof Let $\mathcal{B}u_{EX} \in S^{(n)}$ be *the best approximation of u_{EX} in the energy norm from $S^{(n)}$*, namely

$$\|u_{EX} - \mathcal{B}u_{EX}\|_{\mathcal{U}} = \min_{\chi \in S^{(n)}} \|u_{EX} - \chi\|_{\mathcal{U}}. \tag{3.1.13}$$

Note that $\mathcal{B}u_{EX}$ exists because $S^{(n)}$ is finite dimensional. For any $\chi \in S^{(n)}$, we have

$$u_{EX} - \chi = u_{EX} - u_{S^{(n)}} + u_{S^{(n)}} - \chi = e_{S^{(n)}} + v_{S^{(n)}} \tag{3.1.14a}$$

where we let

$$v_{S^{(n)}} \stackrel{\text{def}}{=} u_{S^{(n)}} - \chi \in S^{(n)} \tag{3.1.14b}$$

and hence

$$\|u_{EX} - \chi\|_{\mathcal{U}}^2 = \|e_{S^{(n)}} + v_{S^{(n)}}\|_{\mathcal{U}}^2 = \mathcal{B}(e_{S^{(n)}} + v_{S^{(n)}}, e_{S^{(n)}} + v_{S^{(n)}}) \tag{3.1.14c}$$

and expanding we get

$$\|u_{EX} - \chi\|_{\mathcal{U}}^2 = \|e_{S^{(n)}}\|_{\mathcal{U}}^2 + \|v_{S^{(n)}}\|_{\mathcal{U}}^2 + 2\mathcal{B}(e_{S^{(n)}}, v_{S^{(n)}}). \tag{3.1.14d}$$

Noting that $v_{S^{(n)}} \in S^{(n)}$, and recalling the orthogonality condition (3.1.10)

$$\mathcal{B}(e_{S^{(n)}}, v_{S^{(n)}}) = 0 \tag{3.1.14e}$$

we obtain

$$\|u_{EX} - \chi\|_{\mathcal{U}}^2 = \|e_{S^{(n)}}\|_{\mathcal{U}}^2 + \|v_{S^{(n)}}\|_{\mathcal{U}}^2. \tag{3.1.14f}$$

Taking the minimum of both sides over $S^{(n)}$, we get

$$\|u_{EX} - \mathcal{B}u_{EX}\|_{\mathcal{U}}^2 = \min_{\chi\in S^{(n)}} \|u_{EX} - \chi\|_{\mathcal{U}}^2 = \|e_{S^{(n)}}\|_{\mathcal{U}}^2 + \min_{v_{S^{(n)}}\in S^{(n)}} \|v_{S^{(n)}}\|_{\mathcal{U}}^2 \tag{3.1.14g}$$

and noting that

$$\min_{v_{S^{(n)}}\in S^{(n)}} \|v_{S^{(n)}}\|_{\mathcal{U}}^2 = 0 \tag{3.1.14h}$$

we get

$$\|e_{S^{(n)}}\|_{\mathcal{U}} = \min_{\chi\in S^{(n)}} \|u_{EX} - \chi\|_{\mathcal{U}} = \|u_{EX} - \mathcal{B}u_{EX}\|_{\mathcal{U}}. \tag{3.1.14i}$$

Letting $\chi \equiv 0$ into (3.1.14f) we obtain

$$\|u_{EX}\|_{\mathcal{U}}^2 = \|e_{S^{(n)}}\|_{\mathcal{U}}^2 + \|u_{S^{(n)}}\|_{\mathcal{U}}^2 \tag{3.1.14j}$$

which proves the second equality in (3.1.12). □

Remark 3.1.3 In the case that $u_{EX} \notin \mathcal{U}$, *the best approximation property (3.1.12) has no meaning,* although $u_{S^{(n)}}$ can be formally determined even in that case; see, for example, P3.1.5.

Let us state another important consequence of the orthogonality condition:

Theorem 3.1.3 *Let* $\Pi : \mathcal{U} \mapsto \mathbb{R}$ *be defined by*

$$\Pi(v) \stackrel{\text{def}}{=} \tfrac{1}{2}\mathcal{B}(v,v) - \mathcal{L}(v) = \tfrac{1}{2}\|v\|_{\mathcal{U}}^2 - \mathcal{L}(v), \quad v \in \mathcal{U}. \tag{3.1.15a}$$

Then we have

$$\Pi(u_{S^{(n)}}) = \min_{v\in S^{(n)}} \Pi(v) = -\tfrac{1}{2}\|u_{S^{(n)}}\|_{\mathcal{U}}^2 \tag{3.1.15b}$$

and

$$\|e_{S^{(n)}}\|_{\mathcal{U}}^2 = 2\big(\Pi(u_{S^{(n)}}) - \Pi(u_{EX})\big) = \|u_{EX}\|_{\mathcal{U}}^2 - \|u_{S^{(n)}}\|_{\mathcal{U}}^2. \tag{3.1.15c}$$

Proof For any $\chi \in S^{(n)}$, we have

$$\Pi(u_{S^{(n)}} + \chi) = \tfrac{1}{2}\mathcal{B}(u_{S^{(n)}} + \chi, u_{S^{(n)}} + \chi) - \mathcal{L}(u_{S^{(n)}} + \chi) \tag{3.1.16a}$$

and hence

$$\Pi(u_{S^{(n)}} + \chi) = \underbrace{\tfrac{1}{2}\|u_{S^{(n)}}\|_{\mathcal{U}}^2 - \mathcal{L}(u_{S^{(n)}})}_{\Pi(u_{S^{(n)}})} + \tfrac{1}{2}\|\chi\|_{\mathcal{U}}^2 + \underbrace{\mathcal{B}(u_{S^{(n)}}, \chi) - \mathcal{L}(\chi)}_{0}. \tag{3.1.16b}$$

where we have used that

$$\mathscr{B}(u_{S^{(n)}}, \chi) = \mathscr{L}(\chi) \quad \forall \chi \in S^{(n)}. \tag{3.1.16c}$$

We have thus proven that

$$\Pi(u_{S^{(n)}}) + \tfrac{1}{2}\|\chi\|_{\mathscr{U}}^2 = \Pi(u_{S^{(n)}} + \chi) \quad \forall \chi \in S^{(n)} \tag{3.1.16d}$$

from which it follows that

$$\Pi(u_{S^{(n)}}) \leq \Pi(u_{S^{(n)}} + \chi) \quad \forall \chi \in S^{(n)} \tag{3.1.16e}$$

which means that $u_{S^{(n)}}$ minimizes Π over $S^{(n)}$. Recalling now that

$$\|u_{S^{(n)}}\|_{\mathscr{U}}^2 = \mathscr{B}(u_{S^{(n)}}, u_{S^{(n)}}) = \mathscr{L}(u_{S^{(n)}}) \tag{3.1.16f}$$

we obtain

$$\Pi(u_{S^{(n)}}) = \tfrac{1}{2}\|u_{S^{(n)}}\|_{\mathscr{U}}^2 - \mathscr{L}(u_{S^{(n)}}) = \tfrac{1}{2}\|u_{S^{(n)}}\|_{\mathscr{U}}^2 - \|u_{S^{(n)}}\|_{\mathscr{U}}^2 = -\tfrac{1}{2}\|u_{S^{(n)}}\|_{\mathscr{U}}^2. \tag{3.1.16g}$$

□

Let us now give an important consequence of Theorem 3.1.3.

Corollary 3.1.2 *Let $\{S^{(n_i)}\}_{i=1}^{\infty}$ be a nested sequence of finite dimensional subspaces of $\mathscr{U}$, namely*

$$S^{(n_1)} \subset S^{(n_2)} \subset \cdots \subset S^{(n_i)} \subset S^{(n_{i+1})} \subset \cdots \subseteq \mathscr{U} \tag{3.1.17a}$$

and let $\{u_{S^{(n_i)}}\}_{i=1}^{\infty}$ be the corresponding sequence of Galerkin approximations. Then

$$\cdots \leq \|e_{S^{(n_{i+1})}}\|_{\mathscr{U}} \leq \|e_{S^{(n_i)}}\|_{\mathscr{U}} \leq \cdots \leq \|e_{S^{(n_2)}}\|_{\mathscr{U}} \leq \|e_{S^{(n_1)}}\|_{\mathscr{U}}. \tag{3.1.17b}$$

Proof Recalling that

$$\|u_{S^{(n)}}\|_{\mathscr{U}} = \sqrt{-2\Pi(u_{S^{(n)}})} \tag{3.1.18a}$$

and noting that

$$S^{(n_1)} \subseteq S^{(n_2)} \quad \Longrightarrow \quad \min_{v \in S^{(n_2)}} \Pi(v) \leq \min_{v \in S^{(n_1)}} \Pi(v) \tag{3.1.18b}$$

and hence

$$-\tfrac{1}{2}\|u_{S^{(n_2)}}\|_{\mathscr{U}}^2 = \min_{v \in S^{(n_2)}} \Pi(v) \leq \min_{v \in S^{(n_1)}} \Pi(v) = -\tfrac{1}{2}\|u_{S^{(n_1)}}\|_{\mathscr{U}}^2. \tag{3.1.18c}$$

It follows that

$$\|e_{S^{(n_2)}}\|_{\mathscr{U}}^2 = \|u_{EX}\|_{\mathscr{U}}^2 - \|u_{S^{(n_2)}}\|_{\mathscr{U}}^2 \leq \|u_{EX}\|_{\mathscr{U}}^2 - \|u_{S^{(n_1)}}\|_{\mathscr{U}}^2 = \|e_{S^{(n_1)}}\|_{\mathscr{U}}^2 \tag{3.1.18d}$$

which proves (3.1.17b). □

Remark 3.1.4 Corollary 3.1.2 states that, in the case that a sequence of finite-dimensional subspaces $S^{(n_i)} \subseteq \mathscr{U}$, has the *telescopic property* (3.1.17a), the sequence

$\|e_{S^{(n_i)}}\|_{\mathcal{U}}$ is not increasing. Note also that the inequalities in (3.1.17b) do not hold in the strict sense ($<$), in general.

Remark 3.1.5 If $S^{(n_i)} \not\subseteq S^{(n_{i+1})}$, we could have $\|e_{S^{(n_i)}}\|_{\mathcal{U}} < \|e_{S^{(n_{i+1})}}\|_{\mathcal{U}}$, even though $n_{i+1} > n_i$.

Remark 3.1.6 Let us underline that *Corollary 3.1.2 was proven for the global energy norm* $\|\cdot\|_{\mathcal{U}}$. The employed argument does not hold for any other norm, such as the L^∞ norm, the L^2 norm, or the L^∞ norm of the derivative, etc.

Above we addressed the case where u_{EX} satisfies the homogeneous Dirichlet boundary condition

$$u_{EX}(0) = 0. \tag{3.1.19a}$$

Let us now consider the case that u_{EX} satisfies the nonhomogeneous Dirichlet boundary condition

$$u_{EX}(0) = a \neq 0. \tag{3.1.19b}$$

As we have already seen in Section 2.2, this case can be reduced to the homogeneous one by letting

$$u_{EX} = \tilde{u}_{EX} + u_{Dir} \tag{3.1.19c}$$

where u_{Dir} is a function which satisfies

$$u_{Dir}(0) = a, \qquad \mathcal{B}(u_{Dir}, u_{Dir}) < \infty. \tag{3.1.19d}$$

It follows that $\tilde{u}_{EX}(0) = 0$, and $\tilde{u}_{EX}$ is the solution of the following variational problem:

Find $\tilde{u}_{EX} \in \mathcal{U}$ such that

$$\mathcal{B}(\tilde{u}_{EX}, v) = \mathcal{L}(v) - \mathcal{B}(u_{Dir}, v) \quad \forall v \in \mathcal{U}. \tag{3.1.19e}$$

Letting $\tilde{u}_{S^{(n)}} \in S^{(n)}$ denote the Galerkin approximation of $\tilde{u}_{EX}$, and applying Theorem 3.1.1, we obtain

$$\|\tilde{u}_{EX} - \tilde{u}_{S^{(n)}}\|_{\mathcal{U}} = \min_{\chi \in S^{(n)}} \|(u_{EX} - u_{Dir}) - \chi\|_{\mathcal{U}}. \tag{3.1.19f}$$

Note that $\tilde{u}_{EX}$, $\tilde{u}_{S^{(n)}}$, and $u_{S^{(n)}} = \tilde{u}_{S^{(n)}} + u_{Dir}$ depend on u_{Dir}, which should be selected such that it facilitates the numerical treatment.

Let us also prove that

$$\Pi(\tilde{u}_{S^{(n)}} + u_{Dir}) = \min_{v \in S^{(n)}} \Pi(v + u_{Dir}). \tag{3.1.20}$$

Expanding $\Pi(\tilde{u}_{S^{(n)}} + u_{Dir} + \chi)$ for $\chi \in S^{(n)}$, and taking into account that

$$\mathcal{B}(\tilde{u}_{S^{(n)}}, \chi) = \mathcal{L}(\chi) - \mathcal{B}(u_{Dir}, \chi) \quad \forall \chi \in S^{(n)} \tag{3.1.21a}$$

we get

$$\Pi(\tilde{u}_{S^{(n)}} + u_{Dir} + \chi) = \tfrac{1}{2}\|\tilde{u}_{S^{(n)}} + u_{Dir}\|^2_{\mathcal{U}} - \mathcal{L}(\tilde{u}_{S^{(n)}} + u_{Dir}) + \tfrac{1}{2}\|\chi\|^2_{\mathcal{U}} \tag{3.1.21b}$$

and hence

$$\Pi(\tilde{u}_{S^{(n)}} + u_{Dir} + \chi) = \Pi(\tilde{u}_{S^{(n)}} + u_{Dir}) + \tfrac{1}{2}\|\chi\|^2_{\mathcal{U}} \quad \forall \chi \in S^{(n)} \tag{3.1.21c}$$

which proves (3.1.20).

Remark 3.1.7 Hence, in both cases of homogeneous and nonhomogeneous Dirichlet boundary conditions, using that $\Pi(u) = \Pi(u_{EX}) + \chi$, and letting $\chi = u_{S^{(n)}} - u_{EX} \in \mathcal{U}$, we have

$$\|e_{S^{(n)}}\|^2_{\mathcal{U}} = 2(\Pi(u_{S^{(n)}}) - \Pi(u_{EX})) \tag{3.1.22a}$$

and

$$\Pi(u_{EX}) \le \Pi(u_{S^{(n)}}). \tag{3.1.22b}$$

Further, in the case of the homogeneous Dirichlet boundary condition, $u_{EX}(0) = 0$, we also have

$$\Pi(u_{EX}) = -\tfrac{1}{2}\|u_{EX}\|^2_{\mathcal{U}}, \qquad \Pi(u_{S^{(n)}}) = -\tfrac{1}{2}\|u_{S^{(n)}}\|^2_{\mathcal{U}} \tag{3.1.23a}$$

and (3.1.22a) can be also written in the form

$$\|e_{S^{(n)}}\|^2_{\mathcal{U}} = \|u_{EX}\|^2_{\mathcal{U}} - \|u_{S^{(n)}}\|^2_{\mathcal{U}} \tag{3.1.23b}$$

and hence

$$\|u_{S^{(n)}}\|^2_{\mathcal{U}} \le \|u_{EX}\|^2_{\mathcal{U}} \tag{3.1.23c}$$

and

$$\Pi(e_{S^{(n)}}) = -\tfrac{1}{2}\|e_{S^{(n)}}\|^2_{\mathcal{U}} = \Pi(u_{EX}) - \Pi(u_{S^{(n)}}). \tag{3.1.23d}$$

Let us now give an example which illustrates the convergence of the error $e_{S^{(n)}}$, in the simplest possible setting.

Example 3.1.1 Consider the regular model problem with $AE \equiv 1$, $c \equiv 0$, $f \equiv -2$, $\alpha_2 = 0$, $a_2 = 0$, and $l = 1$, namely the boundary value problem

$$-\frac{d^2u}{dx^2}(x) = -2 \quad 0 < x < 1, \qquad u(0) = 0, \qquad \frac{du}{dx}(1) = 0 \tag{3.1.24a}$$

with exact solution

$$u_{EX}(x) = x^2 - 2x, \qquad 0 \le x \le 1, \tag{3.1.24b}$$

and let

$$u_{S^{(n)}} = \sum_{j=1}^{n} c_j \varphi_j \tag{3.1.25a}$$

with $h = 1/n$, and

$$\varphi_j(x) = \begin{cases} \frac{1}{h}(x-(j-1)h) & (j-1)h \le x \le jh \\ \frac{1}{h}((j+1)h - x) & jh \le x \le j+1 \\ 0 & \text{otherwise.} \end{cases} \tag{3.1.25b}$$

for $j = 1, \ldots, n-1$, and $\varphi_n(x) = \frac{1}{h}(x-(n-1)h)$, $(n-1)h \le x \le nh$, and zero otherwise. As it will be proven in Corollary 4.1.1, we have

$$c_j = u_{EX}\Big(\frac{j}{n}\Big) = \Big(\frac{j}{n}\Big)^2 - 2\Big(\frac{j}{n}\Big), \quad j = 1, \ldots, n. \tag{3.1.25c}$$

It follows that

$$\Big\| \frac{d}{dx}(u_{EX} - u_{S(n)}) \Big\|_{L^\infty(I)} = \frac{1}{n} \tag{3.1.26a}$$

and

$$\|u_{EX} - u_{S(n)}\|_{\mathcal{U}} = \frac{1}{\sqrt{3}}\frac{1}{n} \tag{3.1.26b}$$

and hence

$$\Pi(u_{S(n)}) - \Pi(u_{EX}) = \frac{1}{2}\|u_{EX} - u_{S(n)}\|_{\mathcal{U}}^2 = \frac{1}{2}\big(\|u_{EX}\|_{\mathcal{U}}^2 - \|u_{S(n)}\|_{\mathcal{U}}^2\big) = \frac{1}{6}\frac{1}{n^2}. \tag{3.1.26c}$$

Remark 3.1.8 As we have seen in the preceding example, and as we will prove below in the general case, we have

$$\Big\| \frac{d}{dx}(e_{S(n)}) \Big\|_{L^\infty(I)} \approx \|e_{S(n)}\|_{\mathcal{U}} \approx \frac{1}{n} \tag{3.1.27a}$$

and hence a reduction of $\|e_{S(n)}\|_{\mathcal{U}}$ implies a proportional reduction of $\|(d/dx)(e_{S(n)})\|_{L^\infty(I)}$, the maximum error in the derivative. Further, in Chapter 5, we will see that, under certain assumptions, $\|e_{S(n)}\|_{\mathcal{U}}$ may be also estimated by a *computable estimator* $\mathcal{E}$, and we have

$$\Big\| \frac{d}{dx}(e_{S(n)}) \Big\|_{L^\infty(I)} \approx \|e_{S(n)}\|_{\mathcal{U}} \approx \mathcal{E}. \tag{3.1.27b}$$

Nevertheless, sometimes in the engineering codes $\mathcal{E}^2$ is reported instead of $\mathcal{E}$, the justification being that $\mathcal{E}^2$ estimates the energy of the error which has a physical meaning. This could however lead to misunderstandings because when $\|e_{S(n)}\|_{\mathcal{U}}^2/\|u_{EX}\|_{\mathcal{U}}^2 \approx 4\%$, we have $\|e_{S(n)}\|_{\mathcal{U}}/\|u_{EX}\|_{\mathcal{U}} \approx 20\%$, which is hardly negligible.

In this section, we presented the basic properties of the Galerkin method.

Problems

P3.1.1. Consider the regular model problem with $AE \equiv 1$, $c \equiv 0$, $\alpha_2 = 0$, $a_2 = 0$, and $l = \pi/2$. Using $\varphi_i(x) = \sin((2i-1)x)$, $i = 1, 2, \ldots$, as the basis functions compute the entries of the stiffness matrix K_{ij} for arbitrary n.

P3.1.2. Consider the model problem of P3.1.1 with

$$f(x) = \sum_{k=1}^{\infty} \frac{1}{k} \sin((2k-1)x). \tag{3.1.28a}$$

Show that

$$u_{EX}(x) = \sum_{k=1}^{\infty} c_k \sin((2k-1)x) \tag{3.1.28b}$$

and

$$u_{S(n)} = \sum_{k=1}^{n} c_k \sin((2k-1)x). \tag{3.1.28c}$$

Determine the general form of c_k, and prove that

$$\lim_{n\to\infty} \|u_{EX} - u_{S(n)}\|_{\mathcal{U}} = 0. \tag{3.1.28d}$$

This means that *the series converges in the energy norm.*

P3.1.3. For the model problem and the input data given in P3.1.1, determine the entries of the load vector, the approximate solution $u_{S(n)}$, and the energy norm of the error $\|u_{EX} - u_{S(n)}\|_{\mathcal{U}}$ as functions of n.

P3.1.4. Consider the model problem of P3.1.1 with

$$\mathcal{L}(v) = v\left(\frac{\pi}{4}\right) \quad \forall v \in \mathcal{U}. \tag{3.1.29}$$

Employ the series expansions (3.1.28b) and (3.1.28c), determine the general form of c_k, and estimate $\|u_{EX} - u_{S(n)}\|_{\mathcal{U}}$ as function of n.

P3.1.5. Consider the model problem of P3.1.1 with

$$\mathcal{L}(v) = \frac{dv}{dx}\left(\frac{\pi}{4}\right). \tag{3.1.30a}$$

Prove that

$$\lim_{n\to\infty} \|u_{EX} - u_{S(n)}\|_{L^2(I)} = 0 \tag{3.1.30b}$$

while $\|u_{EX} - u_{S(n)}\|_{\mathcal{U}}$ diverges with n. This means that $u_{S(n)}$ *converges to* u_{EX} *in the* L^2 *norm, but not in the energy norm.*

P3.1.6. Let $\Pi : \mathbb{R}^n \mapsto \mathbb{R}$ be defined by

$$\Pi(\mathbf{a}) \stackrel{\text{def}}{=} \tfrac{1}{2}\mathbf{a}^T\mathbf{K}\mathbf{a} - \mathbf{a}^T\mathbf{f}, \quad \mathbf{a} \in \mathbb{R}^n, \tag{3.1.31a}$$

and let $\mathbf{c} \in \mathbb{R}^n$ be the solution of $\mathbf{K}\mathbf{c} = \mathbf{f}$. Prove that

$$\Pi(\mathbf{c}) = -\min_{\mathbf{a}\in\mathbb{R}^n} \Pi(\mathbf{a}). \tag{3.1.31b}$$

P3.1.7. Consider the regular problem with $AE \equiv 1$, $c \equiv 0$, and f, corresponding to

$$u_{EX}(x) = x^\alpha - x \qquad 0 \le x \le \tfrac{\pi}{2}, \quad \frac{1}{2} < \alpha < 1 \tag{3.1.32a}$$

and let

$$u_{S^{(n)}} = \sum_{j=1}^{n} u_{EX}\Big(\frac{j}{n}\Big)\varphi_j \tag{3.1.32b}$$

where φ_j is given in (3.1.25b), and determine $\|e_{S^{(n)}}\|_{\mathcal{U}}$ by employing the identity

$$\|e_{S^{(n)}}\|_{\mathcal{U}} = \sqrt{2\big(\Pi(u_{S^{(n)}}) - \Pi(u_{EX})\big)} = \sqrt{\|u_{EX}\|^2_{\mathcal{U}} - \|u_{S^{(n)}}\|^2_{\mathcal{U}}}. \tag{3.1.32c}$$

What is the minimal number m for which

$$\|e_{S^{(n)}}\|_{\mathcal{U}} - I_m(\|e_{S^{(n)}}\|_{\mathcal{U}}) \le 1\%\|e_{S^{(n)}}\|_{\mathcal{U}}, \tag{3.1.32d}$$

where I_m denotes the Gauss–Legendre numerical integration using m integration points, in the cases that $\alpha = 0.75$, $\alpha = 0.65$, and $\alpha = 0.55$?

P3.1.8. Restate and prove P3.1.6 in the case of the nonhomogeneous Dirichlet boundary conditions, $u(0) = a \ne 0$.

3.1* The Galerkin method in two dimensions

The Galerkin method and its properties for the two-dimensional case can be stated analogously as in the one-dimensional case. Given a set of linearly independent functions $\varphi_i \in \mathcal{U}$, $i = 1, \dots, n$, then $u_{S^{(n)}}$ can be constructed as in (3.1.2). The best approximation property and the case of nonhomogeneous essential boundary conditions can then be addressed analogously as in the one-dimensional case (see e.g. [Kantorovich and Krylov, 1958], [Rektorys, 1977], [Rektorys, 1998], [Mikhlin, 1964], [Mikhlin, 1971], [Nečas, 1967]).

Problems

P3.1*.1. Let $\Omega = (-1,1) \times (-1,1)$, and consider the Dirichlet problem

$$-\Delta u|_\Omega = f, \qquad u|_{\partial\Omega} = 0 \tag{3.1*.1a}$$

with

$$u_{EX}(x_1, x_2) = (1 - x_1^2)(1 - x_2^2) \tag{3.1*.1b}$$

and hence

$$f(x_1, x_2) = 4 - 2(x_1^2 + x_2^2). \tag{3.1*.1c}$$

a) Derive the weak formulation of the problem, namely, the expressions for the bilinear form $\mathcal{B}(\cdot,\cdot)$, the linear form $\mathcal{L}(\cdot)$, and describe the energy space $\mathcal{U}$ for the problem.
b) Let

$$u_{S^{(1)}}(x_1, x_2) = c_1 \sin \pi x_1 \sin \pi x_2 \tag{3.1*.1d}$$

and determine c_1.

c) Determine the energy norm of the error, $\|u_{EX} - u_{S^{(1)}}\|_{\mathcal{U}}$, directly from the definition

$$\|u_{EX} - u_{S^{(1)}}\|_{\mathcal{U}} = \sqrt{\mathcal{B}(u_{EX} - u_{S^{(1)}}, u_{EX} - u_{S^{(1)}})} \tag{3.1*.2a}$$

and by employing the identities

$$\|e_{S^{(1)}}\|_{\mathcal{U}} = \sqrt{\|u_{EX}\|_{\mathcal{U}}^2 - \|u_{S^{(n)}}\|_{\mathcal{U}}^2} \tag{3.1*.2b}$$

and

$$\|e_{S^{(1)}}\|_{\mathcal{U}} = \sqrt{2}\sqrt{\Pi(u_{S^{(1)}}) - \Pi(u_{EX})}. \tag{3.1*.2c}$$

P3.1*.2. Let $\Omega = (-1,1) \times (-1,1)$, and consider the Dirichlet problem for the Poisson equation $-\Delta u|_{\Omega} = f$, with Dirichlet boundary conditions

$$u|_{\Gamma_1} = 4(1 - x_2^2), \qquad u|_{\Gamma - \Gamma_1} = 0 \tag{3.1*.3a}$$

where

$$\Gamma_1 = \{x_1, x_2 \mid x_1 = -1, \ |x_2| < 1\} \tag{3.1*.3b}$$

and let f be such that

$$u_{EX}(x_1, x_2) = (1 - x_2^2)(1 - x_1)^2. \tag{3.1*.3c}$$

a) Determine the Galerkin solution $u_{S^{(1)}} = c_1\varphi_1$, for $\varphi_1 = \sin\pi x_1 \sin\pi x_2$, and for $\varphi_1 = 4(1 - x_2^2)(1 - x_1)$, and $\|u_{S^{(1)}}\|_{\mathcal{U}}$ in each case.

b) Determine $\|e_{S^{(1)}}\|_{\mathcal{U}}$ directly from the definition of the norm, and by employing the identity

$$\|e_{S^{(n)}}\|_{\mathcal{U}} = \sqrt{2}\sqrt{\Pi(u_{S^{(1)}}) - \Pi(u_{EX})}. \tag{3.1*.4}$$

P3.1*.3. Let $\Omega = (-1,1) \times (-1,1)$ and consider the Neumann boundary value problem

$$(-\Delta u + cu)|_{\Omega} = f, \qquad \frac{\partial u}{\partial n}\bigg|_{\partial\Omega} = 0 \tag{3.1*.5}$$

with $c \equiv 1$ and $f \equiv 1$. Derive the variational formulation of the problem, and define $\mathcal{B}$, $\mathcal{L}$, and $\mathcal{U}$. Is the function

$$v = \ln(x_1^2 + x_2^2) \tag{3.1*.6}$$

admissible as a basis function?

In the special case $c \equiv 0$,

(a) Does the Neumann problem have a solution for $f \equiv 1$?

(b) Does the Neumann problem have a solution for $f = x_1$? If yes, find it.

(c) Write down the system of linear equations $\mathbf{Kc} = \mathbf{f}$, obtained by formally employing the Galerkin method using $n = 3$ and the basis functions $\varphi_1 = 1$, $\varphi_2 = x_1$, $\varphi_3 = x_2$

for the case $f \equiv 1$ and $f = x_1$. What is the main difference between these two systems of linear equations?

P3.1*.4. Consider the isotropic elasticity problem in $\Omega = (-1,1) \times (-1,1)$ with body force $\mathbf{f} = (1,0)$, $u|_{\partial\Omega} \equiv 0$, $\lambda = 1$, and $\mu = 2$. Derive the weak formulation and compute the Galerkin approximation $\mathbf{u}_{S^{(2)}}$ using

$$\varphi_1 = ((1-x_1^2)(1-x_2^2), 0), \qquad \varphi_2 = (0, (1-x_1^2)(1-x_2^2)). \tag{3.1*.7}$$

P3.1*.5. Consider the elasticity problem introduced in P3.1*.4 with

$$\mathscr{L}(\mathbf{v}) = v_1(0,0). \tag{3.1*.8}$$

Does there exist an exact solution $\mathbf{u}_{EX} \in \mathscr{U}$? Does the Galerkin approximation $\mathbf{u}_{S^{(2)}} \in S^{(2)}$ exist?

3.2 The finite element method

Let Δ be a *mesh*, a partition of the domain I into a set of $N(\Delta)$ non-overlapping subdomains, I_j^Δ, $j = 1, \ldots, N(\Delta)$, which are called *elements*, namely

$$\Delta = \{I_j^\Delta\}_{j=1}^{N(\Delta)}, \qquad \bar{I} = \bigcup_{j=1}^{N(\Delta)} \bar{I}_j^\Delta, \qquad I_j^\Delta \cap I_k^\Delta = \emptyset, \quad j \neq k. \tag{3.2.1}$$

In one dimension we will always understand Δ in terms of an increasing sequence of *nodes* x_j^Δ

$$0 = x_0^\Delta < x_1^\Delta < x_2^\Delta < \cdots < x_{N-1}^\Delta < x_N^\Delta = l \tag{3.2.2a}$$

by letting

$$I_j^\Delta = (x_{j-1}^\Delta, x_j^\Delta). \tag{3.2.2b}$$

We will denote the size of the element I_j^Δ by

$$h_{I_j^\Delta} \stackrel{\text{def}}{=} |I_j^\Delta| = x_j^\Delta - x_{j-1}^\Delta \tag{3.2.2c}$$

and the maximal element size for the mesh Δ by

$$h^\Delta \stackrel{\text{def}}{=} \max_{I_j^\Delta \in \Delta} h_{I_j^\Delta}. \tag{3.2.2d}$$

We will refer to h^Δ as the *mesh size*. In many instances we will write h instead of h^Δ, and I_i instead of I_i^Δ.

Given a mesh Δ and a positive integer p, we will define S_Δ^p as the space of continuous functions, which are polynomials of degree p in each element, and satisfy the Dirichlet boundary condition, namely

$$S_\Delta^p \stackrel{\text{def}}{=} \{v \in C^0(I) \,|\, v(0) = 0,\ v|_{I_i^\Delta} \in \mathscr{P}^p(I_i^\Delta)\} \tag{3.2.3}$$

where $C^0(I)$ is the space of continuous functions defined over $\bar{I}$, and $\mathcal{P}^p(I_i^\Delta)$ is the space of polynomials of degree p defined over I_i^Δ. The notation $v|_{I_i^\Delta}$ means the *restriction* of the function v in the element I_i^Δ, namely

$$v|_{I_i^\Delta}(x) \stackrel{\text{def}}{=} v(x), \quad x \in I_i^\Delta. \tag{3.2.4}$$

We will call S_Δ^p the finite element space of degree p on the mesh Δ, or simply the finite element space. The *dimension* of S_Δ^p, which is equal to $pN(\Delta)$, is known as *the number of degrees of freedom* of the finite element approximation.

The finite element space S_Δ^p is the finite-dimensional subspace of $\mathcal{U}$ which consists of all the continuous functions which vanish at the left end point of the domain at $x = 0$, and are polynomials of degree p in each element of Δ. For example, the space S_Δ^1 is the space of continuous functions which are linear over each element I_i^Δ, and vanish at $x = 0$.

Let us now construct various bases for S_Δ^p. The functions

$$\varphi_j^\Delta(x) \stackrel{\text{def}}{=} \begin{cases} \dfrac{x - x_{j-1}^\Delta}{h_{I_j^\Delta}}, & x \in I_j^\Delta \\ \dfrac{x_{j+1}^\Delta - x}{h_{I_{j+1}^\Delta}}, & x \in I_{j+1}^\Delta \\ 0, & x \notin I_j^\Delta \cup I_{j+1}^\Delta \end{cases} \tag{3.2.5a}$$

for $j = 1, \ldots, N(\Delta) - 1$, and

$$\varphi_N^\Delta(x) = \begin{cases} \dfrac{x - x_{N-1}^\Delta}{h_{I_N^\Delta}}, & x \in I_N^\Delta \\ 0, & x \notin I_N^\Delta \end{cases} \tag{3.2.5b}$$

are known as *hat* functions and form a basis for S_Δ^1. A basis for S_Δ^p, for $p > 1$, may be constructed by adding to the hat functions φ_j^Δ, $j = 1, \ldots, N(\Delta)$, given above, polynomial *bubble* functions φ_k^Δ, $k = N(\Delta) + 1, \ldots, pN(\Delta)$, which are such

$$\varphi_{N+qj}^\Delta|_{I_j^\Delta} \in \mathcal{P}_0^p(I_j^\Delta), \qquad \varphi_{N+qj}^\Delta|_{I-I_j^\Delta} \equiv 0, \qquad q = 1, \ldots, p-1 \tag{3.2.5c}$$

where $\mathcal{P}_0^p(I_j^\Delta)$ is the space of polynomials of degree p which vanish at the end points of I_j^Δ. One possible set of such functions are the *Peano* bubble functions

$$\tilde{\phi}_q(\tilde{x}) = \begin{cases} \dfrac{1}{q!}(\tilde{x}^q - 1) & \text{for } q \text{ even} \\ \dfrac{1}{q!}\tilde{x}(\tilde{x}^{q-1} - 1) & \text{for } q \text{ odd} \end{cases} \tag{3.2.5d}$$

where $\tilde{x} \in (-1, 1)$, $2 \le q \le p$, and the φ_j^Δ are obtained by letting

$$\varphi_{N+qj}^\Delta|_{I_j^\Delta}(x) = \tilde{\phi}_q(\tilde{x}_{I_j^\Delta}(x)), \qquad q = 1, \ldots, p-1, \quad j = 1, \ldots, N(\Delta), \tag{3.2.5e}$$

where we employed the transformation

$$\tilde{x}_{I_j^\Delta}(x) = \frac{2}{h_{I_j^\Delta}}\Big(x - \frac{x_{j-1}^\Delta + x_j^\Delta}{2}\Big). \tag{3.2.5f}$$

Another possible choice are the primitives of the Legendre polynomials

$$\tilde{\phi}_q(\hat{x}) = \int_{-1}^{\hat{x}} \tilde{P}_{q-1}, \quad q \geq 2 \tag{3.2.5g}$$

where $\hat{P}_l$ is the Legendre polynomial of degree l.

The *finite element solution* $u_{S_\Delta^p}$ is defined as the Galerkin approximation of u_{EX} from S_Δ^p, namely, the solution of the discrete variational problem:

Find $u_{S_\Delta^p} \in S_\Delta^p$ which satisfies

$$\mathcal{B}\big(u_{S_\Delta^p}, v\big) = \mathcal{L}(v) \quad \forall v \in S_\Delta^p. \tag{3.2.6}$$

The error in the finite element solution $u_{S_\Delta^p}$ will be denoted by

$$e_{S_\Delta^p} \stackrel{\text{def}}{=} u_{EX} - u_{S_\Delta^p}. \tag{3.2.7}$$

Remark 3.2.1 In this book we consider only *conforming* finite element methods which are characterized by the property

$$S_\Delta^p \subseteq \mathcal{U}. \tag{3.2.8}$$

Remark 3.2.2 The finite element basis functions φ_j^Δ are zero everywhere, except over the two elements connected to a node for the hat functions, and one element, for the bubble functions. A consequence of this is that the matrix $\mathbf{K}$ is sparse.

Remark 3.2.3 Using the best-approximation property of the Galerkin method

$$\|e_{S_\Delta^p}\|_{\mathcal{U}} = \|u_{EX} - u_{S_\Delta^p}\|_{\mathcal{U}} \leq \|u_{EX} - \chi\|_{\mathcal{U}} \quad \forall \chi \in S_\Delta^p \tag{3.2.9a}$$

and letting $\chi = \mathcal{I}_\Delta^p u_{EX}$, where $\mathcal{I}_\Delta^p u_{EX}$ is a finite element interpolant defined in Section 3.3 below, for smooth u_{EX}, we get

$$\|e_{S_\Delta^p}\|_{\mathcal{U}} \leq \|u_{EX} - \mathcal{I}_\Delta^p u_{EX}\|_{\mathcal{U}} \leq \mathcal{C}(h^\Delta)^p. \tag{3.2.9b}$$

This is a standard *a-priori* estimate for the finite element method. The above argument is based on the best-approximation property, and holds only for the global energy norm of the error $\|e_{S_\Delta^p}\|_{\mathcal{U}}$. Similar estimates can also be proven for other norms, e.g., $\|e_{S_\Delta^p}\|_{L^\infty(I)}$, $\|e_{S_\Delta^p}\|_{H^1(I)}$, etc., by employing different arguments.

Remark 3.2.4 In Section 4.1 we will prove that, in the special case that

$$\frac{1}{AE}\Big|_{I_j^\Delta} \in \mathcal{P}^{p-1}(I_j^\Delta), \quad j = 1, \ldots, N(\Delta), \qquad c \equiv 0 \tag{3.2.10a}$$

the finite element solution $u_{S^p_\Delta}$, is exact at the nodes, namely

$$u_{S^p_\Delta}(x_j^\Delta) = u_{EX}(x_j^\Delta), \quad j = 1, \dots, N(\Delta), \tag{3.2.10b}$$

and we also have the elementwise estimate

$$\|u_{EX} - u_{S^p_\Delta}\|_{\mathcal{U}(I_j^\Delta)} = \min_{\substack{v \in \mathcal{P}^p(I_j^\Delta) \\ v(x_k^\Delta) = u_{EX}(x_k^\Delta) \\ k = j-1, j}} \|u_{EX} - v\|_{\mathcal{U}(I_j^\Delta)} \le \mathcal{C}(h_{I_j^\Delta})^{p+1/2}. \tag{3.2.10c}$$

A more general definition of the finite element space is given by

$$S(\Delta; \mathbf{p}^\Delta; \mathcal{F}^\Delta) \stackrel{\text{def}}{=} \{w \in C^0(I) \mid w(0) = 0, \ w|_{I_j^\Delta} \in \mathcal{P}^{p_j}(I_j^\Delta; F_{I_j^\Delta})\} \tag{3.2.11a}$$

where $\mathbf{p}^\Delta = [p_1, \dots, p_{N(\Delta)}]^{\mathrm{T}}$ is the vector of element degrees, $\mathcal{F}^\Delta = \{F_{I_j^\Delta}\}^{N(\Delta)}$ is the set of smooth bijective element transformations, $F_{I_j^\Delta} : \tilde{I} \stackrel{\text{def}}{=} (-1, 1) \mapsto I_j^\Delta$, (see A3.2.1) which are such that $F_{I_j^\Delta}(-1) = x_{j-1}^\Delta$, $F_{I_{j-1}^\Delta}(1) = x_j^\Delta$, and

$$S^p(I_j^\Delta) = \mathcal{P}^{p_j}(I_j^\Delta; F_{I_j^\Delta}) \stackrel{\text{def}}{=} \{v = v(x), x \in I_j^\Delta \mid \exists \tilde{v} \in \mathcal{P}^{p_j}(\tilde{I}) : \ v = \tilde{v} \circ F_{I_j^\Delta}^{-1}\} \tag{3.2.11b}$$

is the *element space* for element I_j^Δ, which is also known as the space of *mapped (or pull-back) polynomials* of degree p_j over I_j^Δ. Hence, $S(\Delta; \mathbf{p}^\Delta; \mathcal{F}^\Delta)$ is *the linear space of continuous functions which satisfy the Dirichlet boundary conditions and are pull-back polynomials of degree p_j over each element I_j^Δ.*

Remark 3.2.5 In the special case that $F_{I_j^\Delta}$ is *affine*, namely

$$F_{I_j^\Delta}(\tilde{x}) = \frac{x_j^\Delta + x_{j-1}^\Delta}{2} + \frac{h_j^\Delta}{2}\tilde{x}, \quad \tilde{x} \in \tilde{I} = (-1, 1) \tag{3.2.12a}$$

we have

$$\mathcal{P}^{p_j}(I_j^\Delta; F_{I_j^\Delta}) \equiv \mathcal{P}^{p_j}(I_j^\Delta). \tag{3.2.12b}$$

Let $\tilde{\varphi}_m$, $m = 1, \dots, p_j + 1$ be a basis for $\mathcal{P}^{p_j}(\tilde{I})$. The functions $\tilde{\varphi}_m$ are known as the *element shape functions*. In practice the following choices are often employed.

1. *The Lagrangian shape functions*:

Given a set of points $\{\tilde{\xi}_j\}_{j=1}^{(p+1)}$, with $\tilde{\xi}_i \neq \tilde{\xi}_j$, $i \neq j$, and $\tilde{\xi}_1 = -1$, $\tilde{\xi}_{p+1} = 1$, we will let

$$\tilde{\mathcal{L}}_i(\tilde{x}; \{\tilde{\xi}_j\}_{j=1}^{(p+1)}) \stackrel{\text{def}}{=} \prod_{\substack{j=1 \\ j \neq i}}^{(p+1)} \frac{(\tilde{x} - \tilde{\xi}_j)}{(\tilde{\xi}_i - \tilde{\xi}_j)}, \quad \tilde{x} \in \tilde{I}, \ i = 1, \dots, (p+1) \tag{3.2.13a}$$

and

$$\tilde{\varphi}_1(\tilde{x}) = \tilde{\mathcal{L}}_1(\tilde{x}; \{\tilde{\xi}_j\}_{j=1}^{(p+1)}), \qquad \tilde{\varphi}_2(\tilde{x}) = \tilde{\mathcal{L}}_{p+1}(\tilde{x}; \{\tilde{\xi}_j\}_{j=1}^{(p+1)}) \tag{3.2.13b}$$

and

$$\tilde{\varphi}_s(\tilde{x}) = \tilde{\mathcal{L}}_{s-1}(\tilde{x}; \{\tilde{\xi}_j\}_{j=1}^{(p+1)}), \quad s = 3, \ldots, p+1. \tag{3.2.13c}$$

The polynomials $\tilde{\mathcal{L}}_i$, $i = 1, \ldots, (p+1)$, are known as the *Lagrangian* polynomials corresponding to the points $\{\tilde{\xi}_j\}_{j=1}^{p+1}$. Note that we have

$$\sum_{i=1}^{(p+1)} \tilde{\mathcal{L}}_i(\tilde{x}; \{\tilde{\xi}_j\}_{j=1}^{(p+1)}) = 1. \tag{3.2.13d}$$

In practice the set of equidistributed points

$$\tilde{\xi}_j = -1 + 2\frac{j-1}{p} \quad j = 1, \ldots, (p+1) \tag{3.2.13e}$$

is most often employed.

2. *Hierarchic shape functions*:
For an element of degree p, the hierarchic shape functions consist of the two affine functions

$$\tilde{\varphi}_1(\tilde{x}) = \frac{1-\tilde{x}}{2}, \qquad \tilde{\varphi}_2(\tilde{x}) = \frac{1+\tilde{x}}{2}, \tag{3.2.14a}$$

and $\tilde{\varphi}_i$, $i = 3, \ldots, p+1$, given either by the *Peano monomials*

$$\tilde{\varphi}_i(\tilde{x}) \stackrel{\text{def}}{=} \begin{cases} \dfrac{1}{(i-1)!}(\tilde{x}^{i-1} - 1), & \text{for } i = 3, 5, 7, \ldots \\ \dfrac{1}{(i-1)!}(\tilde{x}^{i-1} - \tilde{x}), & \text{for } i = 4, 6, 8, \ldots, \end{cases} \tag{3.2.14b}$$

or the *primitives of the Legendre polynomials*

$$\tilde{\varphi}_i(\tilde{x}) \stackrel{\text{def}}{=} \sqrt{\frac{2i-3}{2}} \int_{-1}^{\tilde{x}} \tilde{P}_{i-2}, \quad i = 3, 4, 5, \ldots \tag{3.2.14c}$$

where $\tilde{P}_i, i = 0, 1, 2, \ldots$ are the Legendre polynomials (see A3.1.1) of degree i which satisfy orthogonality condition

$$\int_{-1}^{1} \tilde{P}_i \tilde{P}_j = \begin{cases} \dfrac{2}{2i+1}, & \text{for } i = j \\ 0, & \text{for } i \neq j \end{cases} \tag{3.2.14d}$$

and $\tilde{P}_0 \equiv 1$. Note that the primitives of the Legendre polynomials satisfy the orthogonality conditions

$$\int_{-1}^{1} \frac{d}{d\tilde{x}}(\tilde{\varphi}_i) \frac{d}{d\tilde{x}}(\tilde{\varphi}_j) = \sqrt{\frac{2i-3}{2}} \sqrt{\frac{2j-3}{2}} \int_{-1}^{1} \tilde{P}_{i-2} \tilde{P}_{j-2} = \delta_{ij} \tag{3.2.15a}$$

for $i, j = 3, 4, 5, \ldots$, where δ_{ij} is Kronecker's delta (see A2.6.1), and

$$\int_{-1}^{1} \frac{d}{d\tilde{x}}(\tilde{\varphi}_i) = \sqrt{\frac{2i-3}{2}} \int_{-1}^{1} \tilde{P}_{i-2} = \sqrt{\frac{2i-3}{2}} \int_{-1}^{1} \tilde{P}_{i-2}\tilde{P}_0 = 0 \qquad (3.2.15b)$$

and hence

$$\tilde{\varphi}_i(-1) = \tilde{\varphi}_i(1) = 0, \quad i = 3, 4, 5, \ldots. \qquad (3.2.15c)$$

The Legendre polynomials satisfy the relation (see A3.1.3)

$$(2n+1)\tilde{P}_n = \frac{d}{d\tilde{x}}(\tilde{P}'_{n+1} - \tilde{P}'_{n-1}), \quad n = 1, 2, 3, \ldots \qquad (3.2.15d)$$

with $\tilde{P}_0 \equiv 1$, $\tilde{P}_1 \equiv \tilde{x}$, from which it follows that

$$\tilde{\varphi}_i(\tilde{x}) = \frac{1}{\sqrt{2(2i-3)}}(\tilde{P}_{i-1}(\tilde{x}) - \tilde{P}_{i-3}(\tilde{x})), \quad i = 3, 4, 5, \ldots \qquad (3.2.15e)$$

Remark 3.2.6 The choice of the $\tilde{\varphi}_i$, $i = 3, 4, \ldots$, as the primitives of the Legendre polynomials was motivated by the observation that, in the special case that the $F_{I_j^\Delta}$ are affine and

$$AE|_{I_j^\Delta} \equiv constant, \quad j = 1, \ldots, N(\Delta), \qquad c \equiv 0 \qquad (3.2.16a)$$

we have

$$K_{ij}^{(k)} = \widetilde{AE}|_{I_k^\Delta}\Big(\frac{d}{d\tilde{x}}(F_{I_j^\Delta})\Big)^{-1} \int_{\tilde{I}} \frac{d}{d\tilde{x}}(\tilde{\varphi}_i)\frac{d}{d\tilde{x}}(\tilde{\varphi}_j) = \widetilde{AE}|_{I_k^\Delta}\Big(\frac{d}{d\tilde{x}}(F_{I_j^\Delta})\Big)^{-1}\delta_{ij} \qquad (3.2.16b)$$

for $i, j = 3, \ldots, (p+1)$, where δ_{ij} is Kronecker's delta. Noting that $(d/d\tilde{x})(\tilde{\varphi}_1)$ and $(d/d\tilde{x})(\tilde{\varphi}_2)$ are constant, we also have

$$K_{ij}^{(k)} = \int_{\tilde{I}} AE|_{I_k^\Delta}\frac{d}{d\tilde{x}}(\tilde{\varphi}_i)\frac{d}{d\tilde{x}}(\tilde{\varphi}_j)\Big(\frac{d}{d\tilde{x}}(F_{I_j^\Delta})\Big)^{-1} = 0 \qquad (3.2.16c)$$

for $i = 1, 2$ and $j = 3, 4, \ldots$. It follows that the only non zero off-diagonal term in $\mathbf{K}^{(k)}$ is $K_{12}^{(k)}$. This choice also minimizes the effect of the roundoff error.

Given a basis $\tilde{\varphi}_j$, $j = 1, \ldots, p+1$, for $\mathcal{P}^p(\tilde{I})$ any $\tilde{u} \in \mathcal{P}^p(\tilde{I})$ can be written in the form

$$\tilde{u}(\tilde{x}) = \sum_{j=1}^{p+1} \tilde{Q}_j(\tilde{u})\tilde{\varphi}_j(\tilde{x}) \qquad (3.2.17a)$$

where $\tilde{Q}_j : \mathcal{P}^p(\tilde{I}) \mapsto \mathbb{R}$ is known as the jth *degree of freedom*. For example, in the case of the Lagrangian shape functions we have

$$\tilde{Q}_j(\tilde{u}) = \tilde{u}(\tilde{\xi}_j) \qquad (3.2.17b)$$

where $\tilde{\xi}_j$ is the jth Lagrangian node. In the case of the hierarchic basis which employs the primitives of the Legendre polynomials, we have

$$\tilde{Q}_1(\tilde{u}) = \tilde{u}(-1), \qquad \tilde{Q}_2(\tilde{u}) = \tilde{u}(1) \tag{3.2.18a}$$

and

$$\tilde{Q}_j(\tilde{u}) = \int_{-1}^{1} \frac{d\tilde{u}}{d\tilde{x}} \tilde{P}_{j-2}, \quad j \geq 3. \tag{3.2.18b}$$

The case of the nonhomogeneous Dirichlet boundary condition, $u_{EX}(0) = a \neq 0$, can be addressed by employing the function

$$u_{Dir}(x) = \begin{cases} a\dfrac{x_1^\Delta - x}{h_1^\Delta}, & 0 \leq x < x_1^\Delta \\ 0, & x \geq x_1^\Delta \end{cases} \tag{3.2.19a}$$

and by determining the finite element solution $\tilde{u}_{S^p_\Delta}$, which is the best approximation of $\tilde{u}_{EX} = u_{EX} - u_{Dir}$ from S^p_Δ, namely

$$\left\|\tilde{u}_{EX} - \tilde{u}_{S^p_\Delta}\right\|_{\mathcal{U}} = \left\|u_{EX} - \left(\tilde{u}_{S^p_\Delta} + u_{Dir}\right)\right\|_{\mathcal{U}} = \min_{\chi \in S^p_\Delta} \|u_{EX} - (\chi + u_{Dir})\|_{\mathcal{U}}. \tag{3.2.19b}$$

It follows that

$$\left\|u_{EX} - \left(\tilde{u}_{S^p_\Delta} + u_{Dir}\right)\right\|_{\mathcal{U}} \leq \|u_{EX} - (\mathcal{I}^p_\Delta \tilde{u}_{EX} + u_{Dir})\|_{\mathcal{U}} = \|u_{EX} - \mathcal{I}^p_\Delta u_{EX}\|_{\mathcal{U}} \tag{3.2.19c}$$

and hence the interpolation error estimates can be used also in this case.

As we have seen in Section 3.1, the convergence of the approximate solution $u_{S^{(n)}}$ to u_{EX} can be addressed by letting $n \to \infty$. For the Galerkin finite element method $n = pN(\Delta)$, and the convergence of $u_{S^p_\Delta}$ to u_{EX} can be achieved in one of the following three ways:

1. By refining the mesh such that the mesh size h^Δ tends to zero as the number of elements $N(\Delta)$ diverges to infinity, while keeping p fixed. This is the h *version* of the FEM.
2. By increasing p while keeping the mesh Δ fixed. This is the p *version* of the FEM (see e.g. [Szabo and Babuška, 1991], [Babuška and Suri, 1994], [Schwab, 1998]).
3. By letting h^Δ tend to zero, and increasing p simultaneously. This is the *hp version* of the FEM (see e.g. [Schwab, 1998] and [Babuška and Suri, 1994]).

Today all three versions are used in practice. Throughout this book, we mostly concentrate on the h version with $1 \leq p \leq 3$.

Let us now briefly elaborate on the construction of the entries of the stiffness matrix $\mathbf{K}$ and the load vector $\mathbf{f}$.

The finite element solution in the element I_j^Δ can be written in the form

$$u_{S^p_\Delta}(x) = \sum_{m=1}^{p+1} Q_m^{I_j^\Delta}\left(u_{S^p_\Delta}\right)\varphi_m^{I_j^\Delta}(x), \quad x \in I_j^\Delta \tag{3.2.20a}$$

where

$$\varphi_m^{I_j^\Delta}(x) \stackrel{\text{def}}{=} \tilde{\varphi}_m\big(F_{I_j^\Delta}^{-1}(x)\big) = \tilde{\varphi}_m \circ F_{I_j^\Delta}^{-1} \tag{3.2.20b}$$

are the mapped element shape functions, and $Q_m^{I_j^\Delta}$, $m = 1, \ldots, p{+}1$, are the corresponding degrees of freedom. Here we will employ the hierarchic basis with the primitives of the Legendre polynomials as bubble functions. In this case we have

$$Q_1^{I_j^\Delta}\big(u_{S_\Delta^p}\big) = u_{S_\Delta^p}(x_{j-1}^\Delta), \qquad Q_2^{I_j^\Delta}\big(u_{S_\Delta^p}\big) = u_{S_\Delta^p}(x_j^\Delta). \tag{3.2.20c}$$

The continuity of $u_{S_\Delta^p}$ at the interior nodes x_j^Δ, $j = 2, \ldots, N(\Delta) - 1$, is enforced by requiring that

$$Q_1^{I_j^\Delta}\big(u_{S_\Delta^p}\big) = Q_2^{I_{j-1}^\Delta}\big(u_{S_\Delta^p}\big), \quad j = 2, \ldots, N(\Delta) \tag{3.2.20d}$$

and the Dirichlet boundary condition $u_{S_\Delta^p}(0) = a$ by letting

$$Q_1^{I_1^\Delta}\big(u_{S_\Delta^p}\big) = a. \tag{3.2.20e}$$

Extending $\varphi_m^{I_j^\Delta}$ outside I_j^Δ by zero, we have

$$u_{S_\Delta^p}(x) = \sum_{j=1}^{N(\Delta)} \sum_{m=1}^{p_j+1} Q_m^{I_j^\Delta}\big(u_{S_\Delta^p}\big)\varphi_m^{I_j^\Delta}(x). \tag{3.2.20f}$$

Let us now elaborate on the construction of the linear system of finite element equations, $\mathbf{Kc} = \mathbf{f}$.

In the simplest case where we have *only one element* $N(\Delta) = 1$, we have

$$\begin{bmatrix} K_{11} & K_{12} & \cdots & K_{1,p+1} \\ K_{21} & K_{22} & \cdots & K_{2,p+1} \\ K_{31} & K_{32} & \cdots & K_{3,p+1} \\ \vdots & \vdots & \vdots & \vdots \\ K_{p+1,1} & K_{p+1,2} & \cdots & K_{p+1,p+1} \end{bmatrix} \begin{bmatrix} Q_1\big(u_{S_\Delta^p}\big) \\ Q_2\big(u_{S_\Delta^p}\big) \\ Q_3\big(u_{S_\Delta^p}\big) \\ \vdots \\ Q_{p+1}\big(u_{S_\Delta^p}\big) \end{bmatrix} = \begin{bmatrix} f_1 \\ f_2 \\ f_3 \\ \vdots \\ f_{p+1} \end{bmatrix} \tag{3.2.21a}$$

where

$$K_{ij} = \mathscr{B}\big(\varphi_i^{I_1^\Delta}, \varphi_j^{I_1^\Delta}\big), \qquad f_i = \mathscr{L}\big(\varphi_i^{I_1^\Delta}\big) \tag{3.2.21b}$$

Employing the Dirichlet condition

$$Q_1\big(u_{S_\Delta^p}\big) = u_{S_\Delta^p}(0) = u_{EX}(0) = a \tag{3.2.21c}$$

we get

$$\begin{bmatrix} K_{22} & \cdots & K_{2,p+1} \\ K_{32} & \cdots & K_{3,p+1} \\ \vdots & \vdots & \vdots \\ K_{p+1,2} & \cdots & K_{p+1,p+1} \end{bmatrix} \begin{bmatrix} Q_2(u_{S^p_\Delta}) \\ Q_3(u_{S^p_\Delta}) \\ \vdots \\ Q_{p+1}(u_{S^p_\Delta}) \end{bmatrix} = \begin{bmatrix} f_2 - K_{21}Q_1(u_{S^p_\Delta}) \\ f_3 - K_{31}Q_1(u_{S^p_\Delta}) \\ \vdots \\ f_{p+1} - K_{N1}Q_1(u_{S^p_\Delta}) \end{bmatrix}. \tag{3.2.21d}$$

The above elimination of $Q_1(u_{S^p_\Delta})$, is the simplest example of *condensation*, which is utilized, as above, for applying Dirichlet boundary conditions, or for eliminating the internal degrees of freedom corresponding to the element bubble functions.

To address the case $N(\Delta) > 1$, we will split $\mathscr{B}$ and $\mathscr{L}$ into element contributions, namely

$$\mathscr{B}(u,v) = \sum_{j=1}^{N(\Delta)} \mathscr{B}_{I_j^\Delta}(u|_{I_j^\Delta}, v|_{I_j^\Delta}), \qquad \mathscr{L}(v) = \sum_{j=1}^{N(\Delta)} \mathscr{L}_{I_j^\Delta}(v|_{I_j^\Delta}) \tag{3.2.22a}$$

where

$$\mathscr{B}_{I_j^\Delta}(u,v) \stackrel{\text{def}}{=} \int_{I_j^\Delta} \left(AE\frac{du}{dx}\frac{dv}{dx} + cuv \right) + \delta_{jN}\alpha_2 u(l)v(l) \tag{3.2.22b}$$

and

$$\mathscr{L}_{I_j^\Delta}(v) \stackrel{\text{def}}{=} \int_{I_j^\Delta} fv + \delta_{jN}\alpha_2 v(l) \tag{3.2.22c}$$

for $j = 1, \ldots, N(\Delta)$, where δ_{jN} is Kronecker's delta. We will then introduce the *element stiffness matrix* and *load vector*

$$\mathbf{K}^{I_j^\Delta} = \left[K_{mn}^{I_j^\Delta}, m, n = 1, \ldots, p_j + 1\right], \qquad \mathbf{f}^{I_j^\Delta} = \left[f_m^{I_j^\Delta}, m = 1, \ldots, p+1\right] \tag{3.2.23a}$$

with entries

$$K_{mn}^{I_j^\Delta} = \mathscr{B}_{I_j^\Delta}(\varphi_m^{I_j^\Delta}, \varphi_n^{I_j^\Delta}), \qquad f_m^{I_j^\Delta} = \mathscr{L}_{I_j^\Delta}(\varphi_m^{I_j^\Delta}). \tag{3.2.23b}$$

The *global stiffness matrix* $\mathbf{K}$ (respectively the global load vector $\mathbf{f}$) will be constructed by *assembling* the element stiffness matrices $\mathbf{K}^{I_j^\Delta}$ (respectively the element load vectors $\mathbf{f}^{I_j^\Delta}$). For this, we will employ the following ordering of the degrees of freedom

$$\mathbf{c} = \left[Q_2^{I_1^\Delta}, \ldots, Q_{p+1}^{I_1^\Delta}, Q_2^{I_2^\Delta}, \ldots, Q_{p+1}^{I_2^\Delta}, \ldots, Q_2^{I_{N(\Delta)}^\Delta}, \ldots, Q_{p+1}^{I_{N(\Delta)}^\Delta}\right]^{\mathrm{T}} \tag{3.2.24a}$$

where we omitted the argument $u_{S^p_\Delta}$ from $Q_k^{I_j^\Delta}(u_{S^p_\Delta})$ for simplicity. Note that in the above expression for $\mathbf{c}$ we have already employed the condition of continuity of $u_{S^p_\Delta}$ at the nodes, which reads

$$Q_2^{I_{j-1}^\Delta}(u_{S^p_\Delta}) = Q_1^{I_j^\Delta}(u_{S^p_\Delta}) \tag{3.2.24b}$$

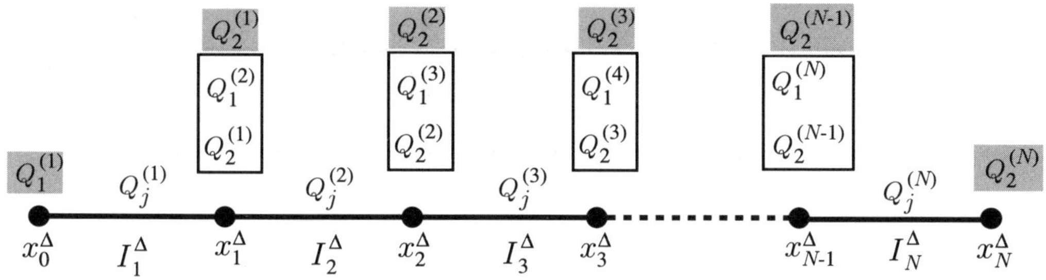

FIG. 3.2.1. The degrees of freedom over the finite element mesh. $Q_j^{(i)}$ are the bubble (internal) degrees of freedom in the element I_i^Δ. $\square$ indicates the constraint.

see Fig. 3.2.1, and the homogeneous Dirichlet boundary condition

$$Q_1^{I_1^\Delta}(u_{S_\Delta^p}) = 0. \tag{3.2.24c}$$

The correspondence between the global and the local degrees of freedom can be written in the compact form

$$c_{glob(k,j)} = c_{k+(j-1)(p+1)} = Q_{k+1}^{I_j^\Delta}, \quad k = 1, \ldots, p, \;\; j = 1, \ldots, N(\Delta) \tag{3.2.24d}$$

where

$$glob(k,j) = k + (j-1)(p+1) \tag{3.2.24e}$$

is the global number for the kth degree of freedom of the jth element. Using this enumeration, we have the following assembly equations

$$K_{glob(1,j),glob(1,j)} = K_{2,2}^{I_j^\Delta} + K_{1,1}^{I_{j+1}^\Delta}, \quad j = 1, \ldots, N-1 \tag{3.2.25a}$$

$$K_{glob(1,N),glob(1,N)} = K_{2,2}^{I_N^\Delta}, \tag{3.2.25b}$$

$$K_{glob(k,j),glob(l,j)} = K_{k+1,l+1}^{I_j^\Delta}, \quad 2 \le h \le l \le p, \;\; l,j = 1, \ldots, N \tag{3.2.25c}$$

$$K_{glob(k,j),glob(l,i)} = 0, \quad i,j = 1, \ldots, N, i \ne j, \;\; 2 \le k \le l \le p \tag{3.2.25d}$$

and

$$K_{glob(k,j),glob(l,i)} = K_{glob(l,i),glob(k,j)} \tag{3.2.25e}$$

$$f_{glob(1,j)} = f_2^{I_j^\Delta} + f_1^{I_{j+1}^\Delta}, \quad j = 1, \ldots, N \tag{3.2.25f}$$

$$f_{glob(1,N)} = f_2^{I_N^\Delta} \tag{3.2.25g}$$

$$f_{glob(k,j)} = f_{k+1}^{I_j^\Delta}, \quad k = 2, \ldots, p, \;\; j = 1, \ldots, N. \tag{3.2.25h}$$

Let us, for example, consider the mesh of $N(\Delta) = 4$, elements of degree $p = 2$. In this case, there is one internal degree of freedom in each element, $\mathbf{K}^{I_j^\Delta}$ is a 3×3 matrix, $\mathbf{f}^{I_j^\Delta}$ is a 3×1 vector. Employing the enumeration given in Fig. 3.2.2,

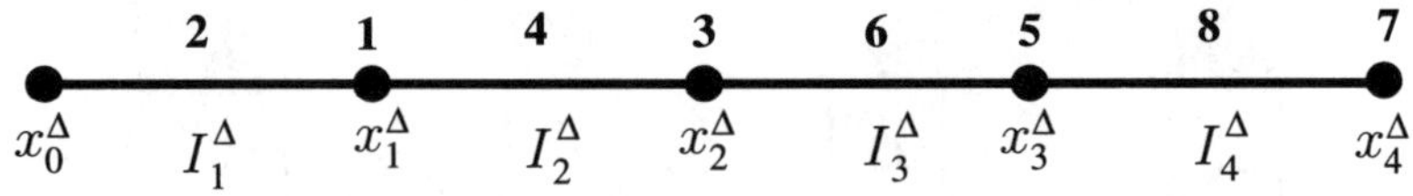

FIG. 3.2.2. Enumeration of the degrees of freedom for a four element mesh of quadratic elements.

we get

$$\mathbf{K} = \begin{bmatrix} K_{22}^{I_1^\Delta} + K_{11}^{I_2^\Delta} & & & & & & & \\ K_{32}^{I_1^\Delta} & K_{33}^{I_1^\Delta} & & & & & & \\ K_{21}^{I_2^\Delta} & 0 & K_{22}^{I_2^\Delta} + K_{11}^{I_3^\Delta} & & & & & \\ K_{31}^{I_2^\Delta} & 0 & K_{32}^{I_2^\Delta} & K_{33}^{I_2^\Delta} & & & & \\ 0 & 0 & K_{21}^{I_3^\Delta} & 0 & K_{22}^{I_3^\Delta} + K_{11}^{I_4^\Delta} & & & \\ & & K_{31}^{I_3^\Delta} & & K_{32}^{I_3^\Delta} & K_{33}^{I_3^\Delta} & & \\ 0 & 0 & 0 & 0 & K_{21}^{I_4^\Delta} & 0 & K_{22}^{I_4^\Delta} & \\ 0 & 0 & 0 & 0 & K_{31}^{I_4^\Delta} & & K_{32}^{4} & K_{33}^{I_4^\Delta} \end{bmatrix} \tag{3.2.26a}$$

and

$$\mathbf{f} = \left[f_2^{I_1^\Delta} + f_1^{I_2^\Delta}, f_3^{I_1^\Delta}, f_2^{I_2^\Delta} + f_1^{I_3^\Delta}, f_3^{I_2^\Delta}, f_2^{I_3^\Delta} + f_1^{I_4^\Delta}, f_3^{I_3^\Delta}, f_2^{I_4^\Delta}, f_3^{I_4^\Delta}\right]. \tag{3.2.26b}$$

Remark 3.2.7 Let us also give another way of arriving at the assembled equations. Recall that in each element I_j^Δ, u_{EX} satisfies the variational problem:

Find $u_{EX} \in \mathcal{U}(I_j^\Delta)$ such that

$$\mathcal{B}_{I_j^\Delta}(u_{EX}, v) = \mathcal{L}_{I_j^\Delta}^{EX}(v) \stackrel{\text{def}}{=} \int_{I_j^\Delta} fv + P_{EX}(x_j^{\Delta,-})v(x_j^\Delta) - P_{EX}(x_{j-1}^{\Delta,+})v(x_{j-1}^\Delta) \tag{3.2.27a}$$

for all $v \in \mathcal{U}(I_j^\Delta)$, where

$$P_{EX}(x_j^{\Delta,\pm}) \stackrel{\text{def}}{=} \lim_{\substack{\delta\to 0\\ \delta>0}} \left(AE(x_j^\Delta \pm \delta)\frac{d}{dx}(u_{EX})(x_j^\Delta \pm \delta)\right) \tag{3.2.27b}$$

is the exact axial force, and $\mathcal{U}(I_j^\Delta)$ is the energy space of I_j^Δ, where the functions v in $\mathcal{U}(I_1^\Delta)$ are constrained to satisfy $v(0) = 0$. In Chapter 5 we will denote $\mathcal{U}(I_1^\Delta)$ by $\mathcal{U}_{(0}(I_1^\Delta)$ to underline that the zero boundary condition is imposed at the left end. Let us for the moment assume that $P_{EX}(x_{j-1}^{\Delta,+})$ and $P_{EX}(x_j^{\Delta,-})$ are known, and let $u_{S^p(I_j^\Delta)}$ be the solution of the discrete variational problem:

Find $u_{S^p(I_j^\Delta)} = \sum_{k=1}^{p+1} Q_k^{I_j^\Delta} \varphi_k^\Delta \in S^p(I_j^\Delta)$ such that

$$\mathcal{B}_{I_j^\Delta}(u_{S^p(I_j^\Delta)}, v) = \mathcal{L}_{I_j^\Delta}^{EX}(v) \quad \forall v \in S^p(I_j^\Delta), \tag{3.2.28a}$$

where in the first element we utilize $S^p_{(0}(I_j^\Delta)$ instead of $S^p(I_j^\Delta)$. Letting $v = \varphi_l^\Delta$, $l = 1, \dots, p+1$, we obtain the linear system of equations

$$\begin{bmatrix} K_{11}^{I_j^\Delta} & K_{12}^{I_j^\Delta} & \cdots & K_{1(p+1)}^{I_j^\Delta} \\ K_{21}^{I_j^\Delta} & K_{22}^{I_j^\Delta} & \cdots & K_{2(p+1)}^{I_j^\Delta} \\ \vdots & \vdots & & \vdots \\ K_{(p+1)1}^{I_j^\Delta} & K_{(p+1)2}^{I_j^\Delta} & \cdots & K_{(p+1)(p+1)}^{I_j^\Delta} \end{bmatrix} \begin{Bmatrix} Q_1^{I_j^\Delta} \\ Q_2^{I_j^\Delta} \\ \vdots \\ Q_{p+1}^{I_j^\Delta} \end{Bmatrix} = \begin{Bmatrix} f_1^{I_j^\Delta} \\ f_2^{I_j^\Delta} \\ \vdots \\ f_{p+1}^{I_j^\Delta} \end{Bmatrix} + \begin{Bmatrix} -P_{EX}(x_{j-1}^{\Delta,+}) \\ P_{EX}(x_j^{\Delta,-}) \\ \vdots \\ 0 \end{Bmatrix}. \tag{3.2.28b}$$

Let us now eliminate the unknown end reactions $P_{EX}(x_{j-1}^{\Delta,+})$ and $P_{EX}(x_j^{\Delta,-})$ by employing the *nodal equilibrium* equations

$$P_{EX}(x_j^{\Delta,-}) = P_{EX}(x_j^{\Delta,+}) \tag{3.2.29a}$$

where we assumed that there is no concentrated load applied at node x_j^Δ. Adding the second equation from the element I_j^Δ

$$\sum_{k=1}^{p+1} K_{2k}^{I_j^\Delta} Q_k^{I_j^\Delta} = f_2^{I_j^\Delta} + P_{EX}(x_j^{\Delta,-}) \tag{3.2.29b}$$

to the first equation from the element I_{j+1}^Δ

$$\sum_{k=1}^{p+1} K_{1k}^{I_{j+1}^\Delta} Q_k^{I_{j+1}^\Delta} = f_1^{I_{j+1}^\Delta} - P_{EX}(x_j^{\Delta,+}) \tag{3.2.29c}$$

we get

$$(K_{22}^{I_j^\Delta} + K_{11}^{I_{j+1}^\Delta}) Q_2^{I_j^\Delta} + \sum_{\substack{k=1 \\ k\neq 2}}^{p+1} K_{2k}^{I_j^\Delta} Q_k^{I_j^\Delta} + \sum_{k=2}^{p+1} K_{1k}^{I_{j+1}^\Delta} Q_k^{I_{j+1}^\Delta} = f_2^{I_j^\Delta} + f_1^{I_{j+1}^\Delta} \tag{3.2.29d}$$

where we also took into account the continuity condition $Q_2^{I_j^\Delta} = Q_1^{I_{j+1}^\Delta}$. Equation (3.2.29d) is the assembled equation corresponding to the jth node.

Let us also note that the equations $k = 3, \dots, (p+1)$ are *internal* and can be eliminated at the element level. We have

$$\begin{bmatrix} K_{33}^{I_j^\Delta} & \cdots & K_{3(p+1)}^{I_j^\Delta} \\ \vdots & & \vdots \\ K_{(p+1)3}^{I_j^\Delta} & \cdots & K_{(p+1)(p+1)}^{I_j^\Delta} \end{bmatrix} \begin{Bmatrix} Q_3^{I_j^\Delta} \\ \vdots \\ Q_{p+1}^{I_j^\Delta} \end{Bmatrix}$$

$$= \begin{Bmatrix} f_3^{I_j^\Delta} \\ \vdots \\ f_{p+1}^{I_j^\Delta} \end{Bmatrix} - \begin{Bmatrix} -K_{31}^{I_j^\Delta} Q_1^{I_j^\Delta} - K_{32}^{I_j^\Delta} Q_2^{I_j^\Delta} \\ \vdots \\ -K_{(p+1)1}^{I_j^\Delta} Q_1^{I_j^\Delta} - K_{(p+1)2}^{I_j^\Delta} Q_2^{I_j^\Delta} \end{Bmatrix} \tag{3.2.30a}$$

and denoting the left-hand-side matrix by ${}^{\text{int}}\mathbf{K}^{I_j^\Delta}$, and its inverse by $({}^{\text{int}}\mathbf{K}^{I_j^\Delta})^{-1}$, we obtain

$$Q_{i+2}^{I_j^\Delta} = \sum_{l=1}^{p-1} \left({}^{\text{int}}K_{il}^{I_j^\Delta}\right)^{-1} f_{l+2}^{I_j^\Delta} - \sum_{l=1}^{p-1} \left({}^{\text{int}}K_{il}^{I_j^\Delta}\right)^{-1} \left(K_{(l+1)1}^{I_j^\Delta} Q_1^{I_j^\Delta} - K_{(l+1)2}^{I_j^\Delta} Q_2^{I_j^\Delta}\right) \quad (3.2.30b)$$

for $i = 1, \ldots, p-1$, and substituting into the original linear system of element equations, (3.2.27c), we get the *condensed* element equations

$$\begin{bmatrix} {}^{\text{cond}}K_{11}^{I_j^\Delta} & {}^{\text{cond}}K_{12}^{I_j^\Delta} \\ {}^{\text{cond}}K_{21}^{I_j^\Delta} & {}^{\text{cond}}K_{22}^{I_j^\Delta} \end{bmatrix} \begin{Bmatrix} Q_1^{I_j^\Delta} \\ Q_2^{I_j^\Delta} \end{Bmatrix} = \begin{Bmatrix} {}^{\text{cond}}f_1^{I_j^\Delta} \\ {}^{\text{cond}}f_2^{I_j^\Delta} \end{Bmatrix} + \begin{Bmatrix} -P_{EX}(x_{j-1}^{\Delta,+}) \\ P_{EX}(x_j^{\Delta,-}) \end{Bmatrix}. \quad (3.2.30c)$$

These equations can then be assembled into a global tridiagonal system by forming the nodal equilibrium equations. The same results could be obtained by rearranging the unknowns c_j so that the internal degree of freedom will be first, and by employing Gauss elimination without pivoting.

Remark 3.2.8 The determination of the entries of the element stiffness matrices is done by splitting the element bilinear form $\mathscr{B}_{I_j^\Delta}$ into three terms

$$\mathscr{B}_{I_j^\Delta}(u,v) = {}^1\mathscr{B}_{I_j^\Delta}(u,v) + {}^2\mathscr{B}_{I_j^\Delta}(u,v) + {}^3\mathscr{B}_{I_j^\Delta}(u,v), \quad j = 1, \ldots, N(\Delta), \quad (3.2.31a)$$

where

$${}^1\mathscr{B}_{I_j^\Delta}(u,v) = \int_{I_j^\Delta} AE \frac{du}{dx}\frac{dv}{dx}, \qquad {}^2\mathscr{B}_{I_j^\Delta}(u,v) = \int_{I_j^\Delta} cuv \quad (3.2.31b)$$

and

$${}^3\mathscr{B}_{I_j^\Delta}(u,v) = \alpha_2 u(l) v(l) \delta_{jN} \quad (3.2.31c)$$

and by summing the contributions

$$\mathbf{K}^{I_j^\Delta} = {}^1\mathbf{K}^{I_j^\Delta} + {}^2\mathbf{K}^{I_j^\Delta} + {}^3\mathbf{K}^{I_j^\Delta}. \quad (3.2.31d)$$

Similarly, the load vectors are determined by employing the splitting

$$\mathscr{L}_{I_j^\Delta}(v) = {}^1\mathscr{L}_{I_j^\Delta}(v) + {}^2\mathscr{L}_{I_j^\Delta}(v) \quad (3.2.32a)$$

where

$${}^1\mathscr{L}_{I_j^\Delta}(v) = \int_{I_j^\Delta} fv, \qquad {}^2\mathscr{L}_{I_j^\Delta}(v) = \alpha_2 v(l) \delta_{jN} \quad (3.2.32b)$$

and by summing the contributions

$$\mathbf{f}^{I_j^\Delta} = {}^1\mathbf{f}^{I_j^\Delta} + {}^2\mathbf{f}^{I_j^\Delta}. \quad (3.2.32c)$$

The contributions ${}^{i}K^{I_j^\Delta}_{m,n}$ and ${}^{i}f^{I_j^\Delta}_{m}$ are evaluated in the master element. For example, we have

$$ {}^{1}K^{I_j^\Delta}_{m,n} = \int_{\tilde{I}} AE(F_{I_j^\Delta}(\tilde{x})) \frac{d}{d\tilde{x}}(\tilde{\varphi}_m)(\tilde{x}) \frac{d}{d\tilde{x}}(\tilde{\varphi}_n)(\tilde{x}) \frac{1}{(F'_{I_j^\Delta}(\tilde{x}))^2} F'_{I_j^\Delta}(\tilde{x})\, d\tilde{x}. \tag{3.2.33} $$

Note that the integrand consists of the *polynomial* function $(d/d\tilde{x})(\tilde{\varphi}_n)(d/d\tilde{x})(\tilde{\varphi}_m)$, multiplied by the function $(AE \circ F_{I_j^\Delta})(1/F'_{I_j^\Delta})$, which, in general, is not a polynomial.

Remark 3.2.9 *Everywhere in the book we will assume that the integrals are determined with negligible error.* Note that in most cases of practical interest, especially in higher dimensions, the integrals must be computed numerically, and it is necessary to address the problem of the estimation and control of the error in the numerical quadrature. For the asymptotic results to hold, it is sufficient to employ Gauss–Legendre quadrature with p quadrature points (see [Ciarlet, 1991]), which integrates exactly polynomials of degree $2p-1$ (see [Atkinson, 1989]), and hence for $AE \equiv constant$, and for affine $F_{I_j^\Delta}$, the entries ${}^{1}K^{(j)}_{m,n}$ are computed exactly. In the cases where AE is unsmooth, e.g., the singular model problem, the Gauss–Legendre quadrature with p points is not sufficient, and special adaptive integration rules which account for the unsmooth character of the integrand are more appropriate (see [Espelid, 1994], [Pressens *et al.*, 1983], [Krommer and Ueberhuber, 1998], and [Schwab, 1994]). The tolerance in the adaptive integration depends on the desired accuracy of the finite element solution. Typically, if the desired accuracy in the energy norm is 1% and the stiffness matrix is scaled so that its diagonal terms are equal to one, then the admissible absolute tolerance for the numerical integration of all coefficients is about 10^{-4}.

Above we have seen that the construction of the finite element equations is done in four steps:

- mesh generation and identification of the correspondence between the element and the global degrees of freedom
- calculation of the element stiffness matrix and load vector
- assembly of the element into the global arrays
- constraining of the stiffness matrix and the load vector to enforce the essential boundary conditions.

Details about the implementation may be found, among others, in [Akin, 1995].

In this section we described the construction and the basic properties of the finite element approximation.

Problems

P3.2.1. Consider the regular model problem with input data

$$ AE \equiv 1, \qquad c \equiv 1, \qquad f \equiv 1, \qquad \alpha_2 = 1, \qquad a_2 = 1 \tag{3.2.34} $$

and let Δ be a uniform mesh of linear elements, namely, $h_{I_j^\Delta} = h^\Delta = N^{-1}$, $p = 1$. Compute the entries of the stiffness matrix and the load vector.

P3.2.2. Count the number of arithmetic operations required to solve a system of linear equations by the method of Gauss elimination without pivoting.

P3.2.3. Consider the regular model problem with $AE \equiv 1$, $c \equiv 0$, and a uniform mesh of quadratic elements ($h_{I_j^\Delta} = h^\Delta = N^{-1}$, $p = 2$). Show that the degrees of freedom corresponding to the bubble functions can be systematically eliminated from the element stiffness matrices $\mathbf{K}^{I_j^\Delta}$, $j = 1, \ldots, N(\Delta)$ before assembly (condensed out at the element level) and that the cost of solving the global system of linear equations is exactly the same as for the mesh of linear elements.

P3.2.4. Consider the regular model problem with

$$AE \equiv 1, \qquad c \equiv 0, \qquad f \equiv 1, \qquad \alpha_2 = 0, \qquad a_2 = 0 \tag{3.2.35a}$$

and a uniform mesh $\Delta = \{((j-1)h, jh)\}_{j=0}^{N(\Delta)}$, $h = N^{-1}$. For each node j define

$$\bar{\varphi}_j^\Delta(x) \stackrel{\text{def}}{=} \begin{cases} \dfrac{x - x_{j-1}^\Delta}{h} + \dfrac{1}{h^2}(x - x_j^\Delta)(x - x_{j-1}^\Delta), & x \in I_j^\Delta \\ \dfrac{x_{j+1}^\Delta - x}{h} + \dfrac{1}{h^2}(x - x_j^\Delta)(x - x_{j+1}^\Delta), & x \in I_{j+1}^\Delta \\ 0, & x \notin I_j^\Delta \cup I_{j+1}^\Delta \end{cases} \tag{3.2.35b}$$

and let $u_{S_\Delta} = \sum_{j=1}^{N(\Delta)} c_j \bar{\varphi}_j^\Delta$. Write down the stiffness matrix, the load vector, and determine the solution vector. Explain the difference between the solution vector c for the above choice of basis functions, and the solution vector $\mathbf{c}$ when $u_{S_\Delta^p}$ is constructed using the φ_j^Δ hat functions.

P3.2.5. Compute the analytic expressions for the solution $u_{S_\Delta^1}$, the norms $\|u_{EX}\|_{\mathcal{U}}$ and $\|u_{S_\Delta^1}\|_{\mathcal{U}}$ and the values of $\Pi(u_{EX})$ and $\Pi(u_{S_\Delta^1})$, for the problem and the mesh given in P3.2.1. Recall the identity

$$\|e_{S_\Delta^1}\|_{\mathcal{U}} = \sqrt{\sum_{j=1}^{N(\Delta)} \|e_{S_\Delta^1}\|^2_{\mathcal{U}(I_j^\Delta)}} = \sqrt{2}\sqrt{\Pi(u_{S_\Delta^1}) - \Pi(u_{EX})} \tag{3.2.36}$$

and determine $\|e_{S_\Delta^1}\|_{\mathcal{U}}$ by directly computing and summing the element contributions $\|e_{S_\Delta^1}\|^2_{\mathcal{U}(I_j^\Delta)}$, and also by evaluating $\Pi(u_{S_\Delta^1})$ and $\Pi(u_{EX})$.

P3.2.6. Repeat P3.2.5 in the case where a nonhomogeneous essential boundary condition $u(0) = a$ is employed, and address the case where $\|u_{EX}\|_{\mathcal{U}} > \|u_{S_\Delta^1}\|_{\mathcal{U}}$.

P3.2.7. Draw a figure of the finite element basis functions for $p = 1, 2, 3$, for a uniform mesh Δ of $N(\Delta) = 4$ elements in the case of the (a) Lagrangian shape functions, (b) Peano functions, and (c) primitives of Legendre polynomials.

3.2* The finite element method in two dimensions

Here we will describe the construction of the finite element approximation in two dimensions.

As in one dimension, a mesh $\Delta = \{\tau_j^\Delta\}_{j=1}^{N(\Delta)}$ is a partition of the domain Ω into non-overlapping subdomains τ_j^Δ, namely

$$\bar{\Omega} = \bigcup_{j=1}^{N(\Delta)} \overline{\tau_j^\Delta}, \qquad \tau_j^\Delta \cap \tau_i^\Delta = \emptyset, \quad j \neq k \tag{3.2*.1}$$

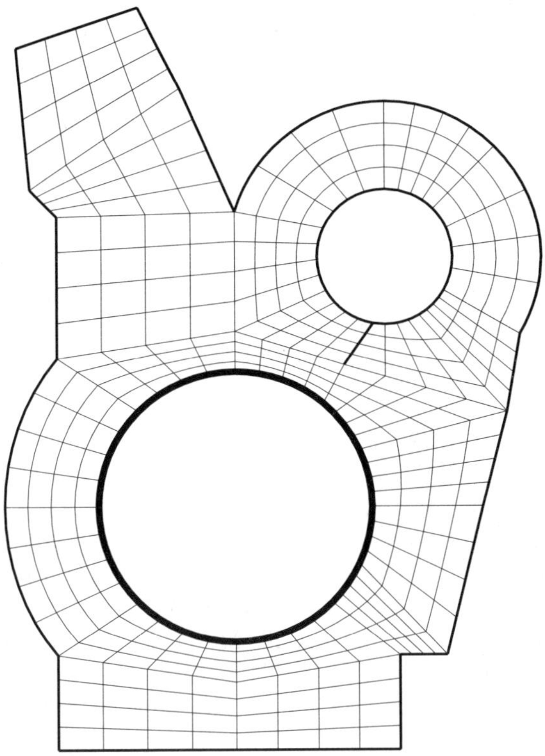

FIG. 3.2*.1. An example of a regular mesh of curvilinear quadrilaterals

where $\overline{\tau_j^\Delta} \stackrel{\text{def}}{=} \tau_j^\Delta \cup \partial\tau_j^\Delta$ is the element τ_j^Δ along with its boundary. For the meshes Δ employed in practical two-dimensional computations, τ_j^Δ is either a curvilinear triangle or quadrilateral.

A mesh is called *regular* if the intersection of the boundaries of any two elements $\partial\tau_i^\Delta \cap \partial\tau_j^\Delta$, $i \neq j$, is either empty, a common vertex, or entire edge for both elements. An example of a regular mesh of curvilinear quadrilaterals is shown in Fig. 3.2*.1.

As in the one-dimensional case we will denote by $\mathbf{F}_{\tau_j^\Delta}$, a smooth continuous bijective transformation which maps the *master* element $\tilde{\tau}$ onto τ_j^Δ, and is such that it takes the same values on an edge, or a vertex which is common between two elements, namely

$$\mathbf{F}_{\tau_j^\Delta}|_{\partial\tau_j^\Delta \cap \partial\tau_k^\Delta} = \mathbf{F}_{\tau_k^\Delta}|_{\partial\tau_j^\Delta \cap \partial\tau_k^\Delta} \quad \forall \tau_j^\Delta, \tau_k^\Delta \colon \partial\tau_j^\Delta \cap \partial\tau_k^\Delta \neq \emptyset. \tag{3.2*.2a}$$

As we will see below, such a transformation can be easily constructed under reasonable assumptions about the shape of the τ_j^Δ. In the case of triangular elements, the master-element $\tilde{\tau}$ will be taken to be the equilateral triangle $\tilde{A}_1\tilde{A}_2\tilde{A}_3$ with vertices $\tilde{A}_1 = (-1,0), \tilde{A}_2 = (1,0), \tilde{A}_3 = (0,\sqrt{3})$, while for quadrilateral elements $\tilde{\tau}$ will be taken to be the square $\tilde{A}_1\tilde{A}_2\tilde{A}_3\tilde{A}_4$ with vertices at $\tilde{A}_1 = (-1,-1), \tilde{A}_2 = (1,-1), \tilde{A}_3 = (1,1), \tilde{A}_4 = (-1,1)$.

The simplest possible mapping is the affine mapping,

$$\mathbf{F}(\tilde{\mathbf{x}}) = \mathbf{T}\tilde{\mathbf{x}} + \mathbf{b} \tag{3.2*.2b}$$

where $\mathbf{T} \in \mathbb{R}^2 \times \mathbb{R}^2$ is an *invertible* matrix, and $\mathbf{b} \in \mathbb{R}^2$, which transforms the master equilateral triangle $\tilde{\tau}$ into a skewed triangle with straight edges, and the master square $\tilde{\tau}$ into a parallelogram. In the general case, where τ_j^Δ is a skewed quadrilateral with straight edges (not a parallelogram), or a curvilinear triangle or quadrilateral $\mathbf{F}_{\tau_j^\Delta}$ is nonlinear. We will always assume (see A3.2.1–A3.2.3)

$$\min_{\tilde{\mathbf{x}} \in \tilde{\tau}} \det\left(\left(\nabla \mathbf{F}_{\tau_j^\Delta}\right)(\tilde{\mathbf{x}})\right) > 0. \tag{3.2*.2c}$$

Analogously as in the one-dimensional case, we will define the finite element space as the set of continuous functions which satisfy the Dirichlet boundary conditions, and are mapped polynomials in each element, namely

$$S(\Delta, \mathbf{p}^\Delta, \mathcal{F}^\Delta) \stackrel{\text{def}}{=} \left\{ w \in C^0(\Omega) \,\middle|\, w|_{\Gamma_D} \equiv 0,\ w|_{\tau_j^\Delta} \circ \mathbf{F}_{\tau_j^\Delta} \in \tilde{S}^{p_j}(\tilde{\tau}) \right\} \tag{3.2*.3}$$

where $\mathbf{p}^\Delta$ is the vector of element degrees, and $\mathcal{F}^\Delta = \{\mathbf{F}_{\tau_j^\Delta}\}_{j=1}^{N(\Delta)}$, the set of element mappings $\tilde{S}^{p_j} \equiv \tilde{S}^{p_j}(\tilde{\tau})$ is the master element space. Note that for triangular elements we have $\tilde{S}^{p_j} \equiv \tilde{\mathcal{P}}^{p_j}(\tilde{\tau})$, while for parallelogram elements $\tilde{\mathcal{P}}^{p_j}(\tilde{\tau}) \subset \tilde{S}^{p_j}$, where $\tilde{\mathcal{P}}^{p_j}(\tilde{\tau})$ is the space of polynomials of degree p_j over $\tilde{\tau}$.

Let us now elaborate on the span and the shape functions for the space $\tilde{S}^p$. The shape functions must be constructed in such a way that they can be patched together to form global basis functions which are continuous. Moreover, to ensure good approximation properties, we want to have

$$\tilde{\mathcal{P}}^p(\tilde{\tau}) \subseteq \tilde{S}^p. \tag{3.2*.4}$$

Let us now see how these requirements can be satisfied for triangles and quadrilaterals.

(a) *Triangular elements*:
A polynomial of degree p is specified in terms of $\frac{1}{2}(p+1)(p+2)$ degrees of freedom. In the case of triangular elements $3 + 3(p-1) = 3p$, of these degrees of freedom must be assigned to the boundary of the element (e.g., 3 vertex values, and $3(p-1)$ edge values), while the remaining degrees of freedom $\frac{1}{2}(p+1)(p+2) - 3p = \frac{1}{2}(p-1)(p-2)$, for $p \geq 3$, are assigned to the interior of the element.

The shape functions may be written in simple form in terms of the *barycentric* coordinates

$$L_1 = \frac{1}{2}\left(1 - \tilde{x}_1 - \frac{\tilde{x}_2}{\sqrt{3}}\right), \qquad L_2 = \frac{1}{2}\left(1 + \tilde{x}_1 - \frac{\tilde{x}_2}{\sqrt{3}}\right), \qquad L_3 = \frac{\tilde{x}_2}{\sqrt{3}}. \tag{3.2*.5a}$$

Note that L_i is the affine function which is equal to one at the vertex i and vanishes at the opposite to the vertex, and we have

$$L_1 + L_2 + L_3 = 1. \tag{3.2*.5b}$$

Figure 3.2*.2 shows the master element and its enumeration of vertices and edges.

(a1) *Lagrangian shape functions*:
Analogously as in the one-dimensional case, we will employ a set of interpolation points $\{\tilde{\boldsymbol{\xi}}_k\}_{k=1}^{(1/2)(p+1)(p+2)}$, and will construct the Lagrangian shape functions $\tilde{\mathcal{L}}_j = \tilde{\mathcal{L}}_j(\cdot\,; \{\tilde{\boldsymbol{\xi}}_k\}_{k=1}^{(1/2)(p+1)(p+2)}) \in \tilde{\mathcal{P}}^p(\tilde{\tau})$, as the polynomials of degree p, which are such that

$$\tilde{\mathcal{L}}_j\left(\tilde{\xi}_i\,; \{\tilde{\boldsymbol{\xi}}_k\}_{k=1}^{(1/2)(p+1)(p+2)}\right) = \delta_{ik}. \tag{3.2*.6}$$

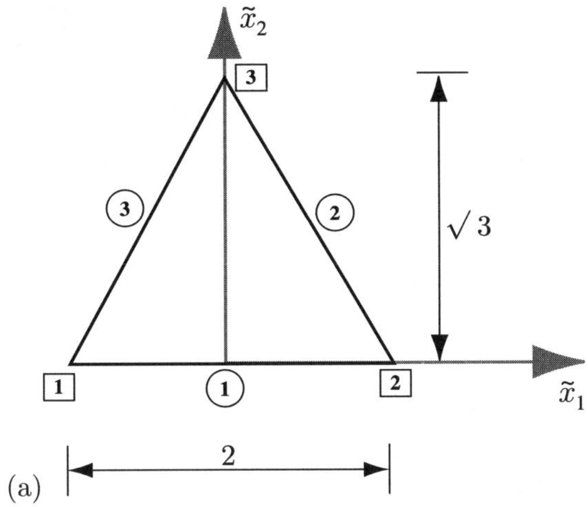

FIG. 3.2*.2. The triangular master element. The enumeration of the vertices [1], [2], [3], and the edges ①, ②, ③

Let us now give an example of the points $\tilde{\boldsymbol{\xi}}_k$, $k = 1, \ldots, \frac{1}{2}(p+1)(p+2)$, written in barycentric coordinates. The most commonly used sets of interpolation points $\tilde{\boldsymbol{\xi}}_k$, are the vertex points:

$$\tilde{\boldsymbol{\xi}}_1 = (1, 0, 0), \qquad \tilde{\boldsymbol{\xi}}_2 = (0, 1, 0), \qquad \tilde{\boldsymbol{\xi}}_3 = (0, 0, 1) \tag{3.2*.7}$$

the edge points

$$\begin{aligned} \tilde{\boldsymbol{\xi}}_m &= \left(\frac{i}{p}, \frac{j}{p}, 0\right), \quad 4 \le m \le p+2 \\ \tilde{\boldsymbol{\xi}}_m &= \left(0, \frac{i}{p}, \frac{j}{p}\right), \quad p+3 \le m \le 2p+1 \\ \tilde{\boldsymbol{\xi}}_m &= \left(\frac{i}{p}, 0, \frac{j}{p}\right), \quad 2p+2 \le m \le 3p \end{aligned} \tag{3.2*.8}$$

where i, j are integers which satisfy $i + j = p$ and $1 \le i, j \le p - 1$, and the interior points

$$\tilde{\boldsymbol{\xi}}_m = \left(\frac{i}{p}, \frac{j}{p}, \frac{k}{p}\right), \quad 3p+1 \le m < np \tag{3.2*.9}$$

where i, j, k are integers such that $i + j + k = p$, $1 \le i, j, k \le p - 2$. The edge points need to be employed for $p \ge 2$, and the interior points for $p \ge 3$. Figure 3.2*.3 shows the interpolation points for $p = 1, 2, 3, 4$. It can be shown that the Lagrangian shape functions $\mathcal{L}_j$ exist and are linearly independent for these sets of points. For $p = 1$, we have

$$\tilde{\mathcal{L}}_1 = L_1, \qquad \tilde{\mathcal{L}}_2 = L_2, \qquad \tilde{\mathcal{L}}_3 = L_3 \tag{3.2*.10}$$

and for $p = 2$

$$\tilde{\mathcal{L}}_1 = L_1(2L_1 - 1), \qquad \tilde{\mathcal{L}}_2 = L_2(2L_2 - 1), \qquad \tilde{\mathcal{L}}_3 = L_3(2L_3 - 1) \tag{3.2*.11a}$$

$$\tilde{\mathcal{L}}_4 = 4L_1L_2, \qquad \tilde{\mathcal{L}}_5 = 4L_2L_3, \qquad \tilde{\mathcal{L}}_6 = 4L_3L_1. \tag{3.2*.11b}$$

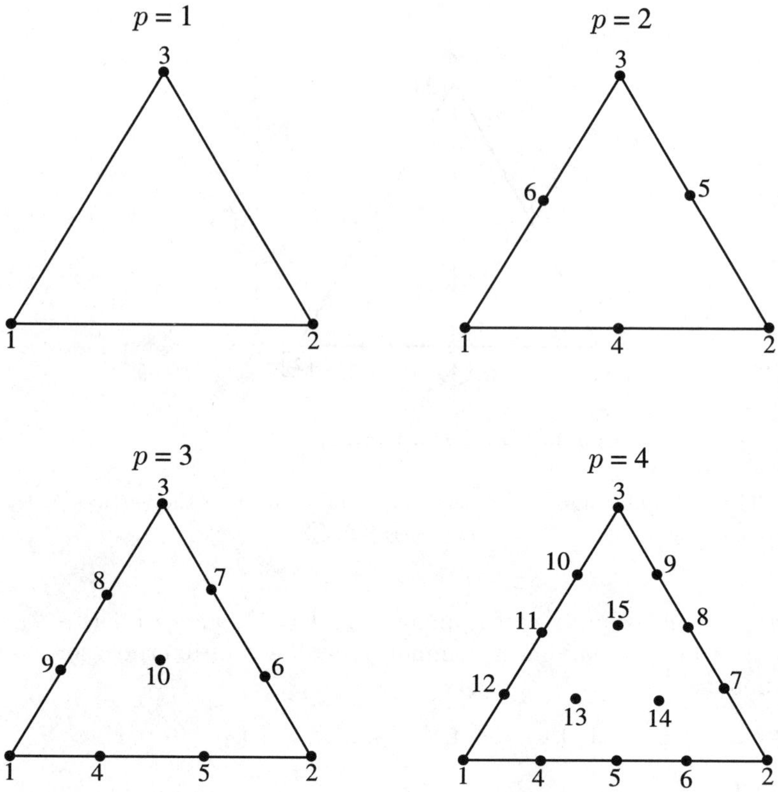

FIG. 3.2*.3. The interpolation points for the elements of degree (a) $p = 1$, (b) $p = 2$, (c) $p = 3$, (d) $p = 4$, and their enumeration.

Note that the restriction of the two-dimensional shape-functions $\tilde{\mathcal{L}}_j$, $j = 1, \ldots, \frac{1}{2}(p+1)(p+2)$, on each edge is exactly the set of one-dimensional Lagrangian shape functions corresponding to the interpolation points on the edge. This is important because it makes possible the patching of the element shape functions to form global basis functions which are continuous through the element boundaries.

(a2) *The hierarchic shape functions*:
The hierarchic shape functions also consist of vertex, edge, and interior functions. As in the one-dimensional case, the vertex functions are affine and are given by

$$\tilde{\varphi}_i(\tilde{x}) = L_i(\tilde{x}), \quad i = 1, 2, 3. \tag{3.2*.12a}$$

The edge shape functions are constructed by extending the one-dimensional bubble functions over each edge to the interior of the element. For example, for edge ①, of Fig. 3.2*.2, the one-dimensional bubble functions over the edge can be extended to the interior of the element by

$$\tilde{\varphi}_j(\tilde{x}) = \tilde{\varphi}_l^{(1-d)}(\tilde{x}_1)\frac{L_1(\tilde{x}_1, \tilde{x}_2)L_2(\tilde{x}_1, \tilde{x}_2)}{L_1(\tilde{x}_1, 0)L_2(\tilde{x}_1, 0)}, \quad j = 4, 5, \ldots, p+2 \tag{3.2*.12b}$$

where $l = 3, 4, \ldots, p+1$, $j = l+1$, and $\tilde{\varphi}_l^{(1-d)}$ denotes the one-dimensional edge bubble function. For elements of cubic degree or higher, $p \geq 3$, the element basis must also include internal shape functions. The cubic internal shape function is given by

$$\tilde{\varphi}_1^{int} = L_1 L_2 L_3 \tag{3.2*.13a}$$

and the higher-order internal shape functions can be generated by multiplying $\tilde{\varphi}_1^{int}$ by the Legendre polynomials P_i. For example, the quartic internal shape functions are given by

$$\tilde{\varphi}_2^{int} = L_1 L_2 L_3 P_1(L_2 - L_1), \qquad \tilde{\varphi}_3^{int} = L_1 L_2 L_3 P_1(2L_3 - 1) \tag{3.2*.13b}$$

and similarly for the internal shape functions of quintic order and higher.

Let us note that the sign of the one-dimensional bubble functions of even degree constructed by integration of the Legendre polynomials, depends on the employed orientation of the one-dimensional element. This has to be taken into account when assembling the entries corresponding to even degrees of freedom along the edges of the mesh.

Similarly, the Lagrangian shape-functions can be classified into vertex, edge, and internal shape-functions.

(b) *Quadrilateral elements*:
In this case the requirement of continuity of the global basis functions through the inter-element boundaries requires $4 + 4(p-1) = 4p$ degrees of freedom to be placed at the boundary of the element, namely, 4 vertex values and $4(p-1)$ edge values. There are various possibilities for constructing the basis of monomials in each element. The following families of shape functions are the ones most often used in practice.

(b1) *The tensor product shape functions*:
Also known as bi-p shape functions, these shape functions are constructed as polynomials of degree p in both variables $\tilde{x}_1$ and $\tilde{x}_2$ *separately*, by taking tensor products of the one-dimensional shape functions. For example

$$\tilde{\varphi}_{i,k}(\tilde{x}_1, \tilde{x}_2) = \tilde{\mathscr{L}}_i(\tilde{x}_1; \{\tilde{\boldsymbol{\xi}}_j\}_{j=1}^{p+1}) \tilde{\mathscr{L}}_k(\tilde{x}_2; \{\tilde{\boldsymbol{\xi}}_j\}_{j=1}^{p+1}) \tag{3.2*.14}$$

$i, k = 1, \ldots, p+1$, are the $(p+1)^2$ tensor product Lagrangian shape functions. For $p = 1$ four bilinear shape functions are obtained. The tensor product shape functions can also be based on the one-dimensional hierarchic shape functions, and in this case we get:

- Four bilinear vertex shape functions which are products of the one-dimensional nodal shape functions
- $4(p-1)$ edge shape functions which are products of the one-dimensional bubble functions in one variable, and the nodal shape functions in the other variable
- $(p-1)^2$ internal shape functions which are products of the one-dimensional bubble functions.

The span of the tensor product shape functions will be denoted by

$$\tilde{Q}^p \stackrel{\text{def}}{=} \left\{ \tilde{v} \,\middle|\, \tilde{v} = \sum_{i,j=0}^{p} a_{i,j} \tilde{x}_1^i \tilde{x}_2^j \right\}. \tag{3.2*.15a}$$

Note that

$$\tilde{\mathcal{P}}^p \subseteq \tilde{Q}^p \tag{3.2*.15b}$$

and $\tilde{Q}^p$ has the same asymptotic properties as $\tilde{\mathcal{P}}^p$.

(b2) *The serendipity shape functions*:
As we have seen above, $4p$ degrees of freedom must be assigned to the boundary of the element to achieve conformity. We have

$$\dim(\tilde{\mathcal{P}}^p) = \tfrac{1}{2}(p+1)(p+2) \leq \dim(\tilde{Q}^p) = (p+1)^2. \tag{3.2*.16}$$

For $p \geq 2$, several conforming element bases with dimension less than $(p+1)^2$ are possible. The most common example is the one corresponding to the *serendipity* or *trunc* space

$$\tilde{Q}'^p \stackrel{\text{def}}{=} \left\{ \tilde{v} \,\middle|\, \tilde{v} = a_{1,p+1}x_1x_2^{p+1} + a_{p+1,1}x_1^{p+1}x_2 + \sum_{\substack{i,j=0\\ i+j\leq p}}^{p} a_{i,j}x_1^ix_2^j \right\} \tag{3.2*.17a}$$

which has the minimal number of shape functions which are needed for conformity, and satisfies the requirement

$$\tilde{\mathcal{P}}^p \subseteq \tilde{Q}'^p. \tag{3.2*.17b}$$

For elements of quadratic and cubic degree ($p = 2$ and $p = 3$), we have $4p = \dim(\tilde{Q}'^p)$. For elements of quartic degree or higher ($p \geq 4$), interior shape functions must be also added in order to satisfy the requirement $\tilde{\mathcal{P}}^p \subseteq \tilde{Q}'^p$, and we have $4p < \dim(\tilde{Q}'^p)$. Further details on the character of the shape functions for straight and curvilinear elements and their graphical display may be found in [Szabo and Babuška, 1991].

Remark 3.2*.1 A vertex hat function in two dimensions is non-zero only on the patch of the elements connected to the vertex.

Remark 3.2*.2 The continuity of the basis functions at the vertices and across the inter-element edges is guaranteed by the fact that the element mappings can be patched together into a continuous global mapping, and from the construction of the degrees of freedom for the element.

Remark 3.2*.3 A mesh Δ may include both triangular and quadrilateral elements.

Thus far we have considered only regular meshes. In two and three dimensions, meshes with irregular connections are often used to facilitate the local refinement of the mesh, as is, for example, shown in Fig. 3.2*.4. The continuity of the global basis functions through the edges with irregular connections, may be ensured by imposing constraints. For example, in the case $p = 1$, the degree of freedom at node $\boxed{8}$ may be constrained to be the average of the degrees of freedom at $\boxed{2}$ and $\boxed{6}$, or a special transition shape function may be added to element τ_1. Another possibility is to employ special *transition* element shape functions.

As in the one-dimensional case, we will denote the element size by

$$h_\tau \stackrel{\text{def}}{=} \max_{\mathbf{x},\mathbf{y}\in\tau} |\mathbf{x}-\mathbf{y}|. \tag{3.2*.18}$$

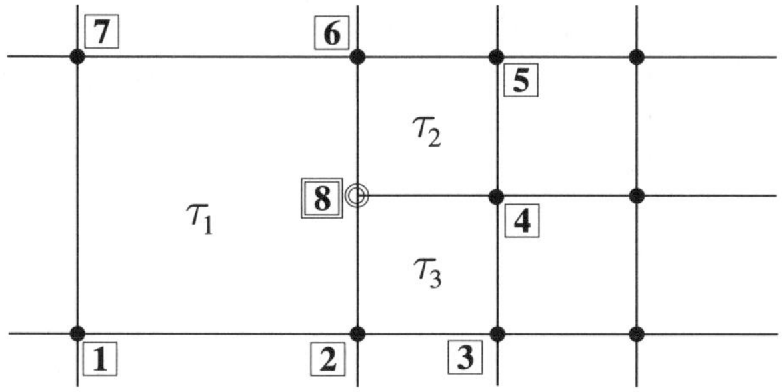

FIG. 3.2*.4. A detail from a 1-irregular mesh. The vertex at 8 is often referred to as a hanging node

Unlike the one-dimensional case where the mesh size $h_{I_j^\Delta}$ was sufficient to characterize the geometry of the elements, in two dimensions additional parameters are needed. For triangles with straight edges, in addition to h_τ, it is sufficient to employ the *aspect ratio*

$$\alpha_\tau \stackrel{\text{def}}{=} \frac{h_\tau}{\rho_\tau} \leq \alpha_{\max} < \infty \tag{3.2*.19}$$

where ρ_τ is the diameter of the largest circle which can be inscribed in τ. The parameters h_τ, ρ_τ are also sufficient to characterize the geometry of parallelograms. The triangular or rectangular elements are always required to satisfy the *maximal angle* condition

$$\max_{\tau\in\Delta}\left(\max_{i=1,\dots,n(\tau)} \beta_i^\tau\right) \leq \pi - \gamma, \quad \gamma > 0 \tag{3.2*.20a}$$

where β_i^τ is the ith interior angle of the element τ, and $n(\tau)$ is the number of vertices of τ. This condition is needed for ensuring that the approximation properties of the corresponding finite element spaces will not degenerate and the finite element method will converge; see [Babuška and Aziz, 1976] and [Apel, 1999]. Another condition which is often used is the *minimal angle* condition

$$\min_{\tau\in\Delta}\left(\min_{i=1,\dots,n(\tau)} \beta_i^\tau\right) \geq \gamma > 0 \tag{3.2*.20b}$$

which is equivalent to the aspect ratio condition and implies the maximal angle condition. Often, for the solution of problems with boundary-layers, meshes with needle elements which satisfy only the maximal but not the minimal angle condition are employed. In the case where the domain has curvilinear boundaries, like the domain shown in Fig. 3.2*.1, elements with curved edges must be used. Details on the properties of curvilinear elements and on sufficient conditions for their convergence can be found in [Ciarlet, 1991].

For various constructions of the mapping functions $\mathbf{F}_\tau$, refer to [Szabo and Babuška, 1991]. Two types of the mapping $\mathbf{F}_\tau$ are used in practice, the isoparametric and the blending-function mapping. In the case of isoparametric mapping, the components of $\mathbf{F}_\tau$ can be expressed as linear combinations of the element shape functions of degree p. More generally, the components of $\mathbf{F}_\tau$ can be expressed in terms of the element shape functions of degree q, and if $q < p$ (respectively $q > p$), we call the transformation subparametric (respectively superparametric). For example, the quadratic isoparametric transformation for triangular elements is given by

$$\mathbf{x} = \mathbf{F}(\tilde{\mathbf{x}}) = \sum_{j=1}^{6} \bar{\mathbf{x}}_j \tilde{\mathcal{L}}_j(\tilde{\mathbf{x}}) \tag{3.2*.21}$$

where $\bar{\mathbf{x}}_j$, $j = 1, 2, 3$, are the positions of the vertices, and $\bar{\mathbf{x}}_j$, $j = 4, 5, 6$ the positions of the mid-points, and $\tilde{\mathcal{L}}_j$ are the corresponding quadratic Lagrangian shape functions. Note that the above mapping is not guaranteed to be invertible. In the case that $p = 2$ and the midpoints of the curved edges $\mathbf{x}_k$, $k = 4, 5, 6$, are relatively close to the corresponding midpoints of the straight edges $\mathbf{x}_{k'}$, namely, if $(1/h_\tau^2)\,|\,\mathbf{x}_k - \mathbf{x}_{k'}|$ is sufficiently small and the element satisfies the minimal angle condition, then $\mathbf{F}_\tau$ is invertible; see [Ciarlet, 1991]. The isoparametric mapping for curvilinear quadrilaterals is constructed analogously. One important case is the general skewed quadrilateral in which the transformation is bilinear.

Remark 3.2*.4 Above, we have discussed conditions on the location of the mid-side nodes for isoparametric Lagrangian elements of quadratic degree ($p = 2$), which are needed for the invertibility of the isoparametric element transformation. Similar conditions are needed for the interior nodes which are present in Lagrangian elements of higher degree; see for example [Lenoir, 1986]. The problem of locating the interior nodes may be avoided by constructing the element transformation using the isoparametric map only on the edges of the element, and linear blending to extend the edge maps to the interior of the element.

Remark 3.2*.5 The span of the element space constructed using either subparametric or isoparametric transformation includes the linear polynomials, namely, $\mathcal{P}^1(\tau) \subseteq S^p(\tau)$. This is not the case, in general, for a superparametric or a blending-function transformation; see Remark 3.2*.8.

Remark 3.2*.6 The span of the tensor product (bi-p) element space of degree p, for quadrilateral elements with straight edges includes all polynomials of degree p, namely, $\mathcal{P}^p(\tau) \subseteq S^p(\tau)$. This is, however, not the case for the serendipity (trunc) element space of degree p, unless the elements are parallelograms. This is important for the convergence of the method, as it was recently shown by [Arnold *et al.*, 2001], where it was shown that the serendipity basis leads to the optimal rate of convergence for meshes of parallelograms, while it only leads to a sub-optimal rate of convergence for meshes of general quadrilaterals.

Remark 3.2*.7 The problem of checking the invertibility of the element transformations is not easy. In fact, it often happens that the 3D meshes constructed by commercial mesh generators include elements in which the element transformation is not invertible, although it is difficult to detect them.

Let us note that, unless the boundary of the domain is a piecewise polynomial curve, the use of the parametric mapping introduces errors in the representation of the boundary. Nevertheless an exact representation of the boundary can be achieved by employing a blending function mapping (see [Szabo and Babuška, 1991]). Let us now give an example of a blending-function mapping.

Example 3.2*.1 (*Blending-function mapping*) Consider the case of a quadrilateral element with one curved edge, as shown in Fig. 3.2*.5a, and let $\mathbf{x} = \varphi(\tilde{\mathbf{x}})$ be the one-dimensional transformation which defines the curved edge. The transformation

$$x_1 = \frac{1}{4}(1-\tilde{x}_1)(1-\tilde{x}_2)x_1^{(1)} + \frac{1}{4}(1-\tilde{x}_1)(1+\tilde{x}_2)x_1^{(4)} + \varphi_1(\tilde{x}_2)\frac{1+\tilde{x}_1}{2} \qquad (3.2^*.22a)$$

$$x_2 = \frac{1}{4}(1-\tilde{x}_1)(1-\tilde{x}_2)x_2^{(1)} + \frac{1}{4}(1-\tilde{x}_1)(1+\tilde{x}_2)x_2^{(4)} + \varphi_2(\tilde{x}_2)\frac{1+\tilde{x}_1}{2} \qquad (3.2^*.22b)$$

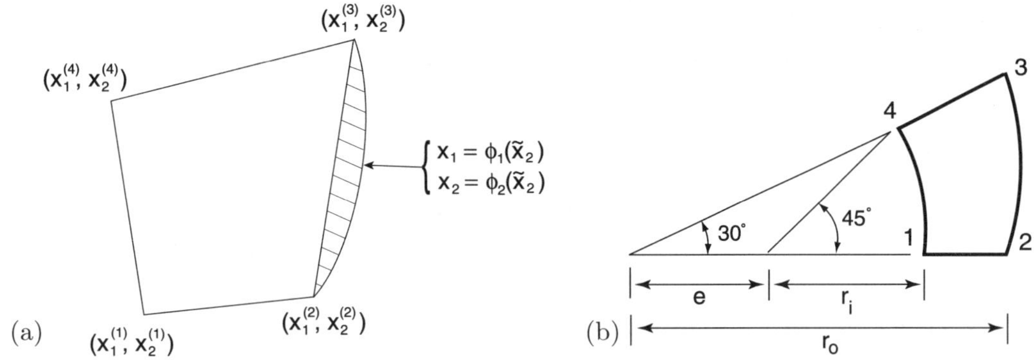

FIG. 3.2*.5. †Blending function transformation: (a) A quadrilateral with one curved edge. (b) A quadrilateral element bounded by two circular segments

is a linear blending-function map which represents the geometry exactly. The functions $\varphi_1(\tilde{x}_2)$, $\varphi_2(\tilde{x}_2)$ represent the difference between the curved side and the chord of the quadrilateral.

Let us now consider the case of the element shown in Fig. 3.2*.5b, which is bounded by two circular segments; see [Gordon, 1971] and [Szabo and Babuška, 1991, p. 109]. The center of the larger circle is located at the origin, while the center of the smaller circle is at the point $(l, 0)$. Letting†

$$\varphi^{(2)}(\tilde{x}_2) = r_0\Big(\cos\Big(\frac{\pi}{12}(1+\tilde{x}_2)\Big), \sin\Big(\frac{\pi}{12}(1+\tilde{x}_2)\Big)\Big) \tag{3.2*.23a}$$

$$\varphi^{(4)}(\tilde{x}_2) = \Big(l + r_1\cos\frac{\pi}{12}\Big(\frac{3}{2}+\tilde{x}_2\Big), r_1\sin\frac{\pi}{12}\Big(\frac{3}{2}+\tilde{x}_2\Big)\Big) \tag{3.2*.23b}$$

where $l = (\sqrt{2}/2)(\sqrt{3}-1)r_1$, we have

$$\begin{aligned}
x_1 &= \tfrac{1}{4}(1-\tilde{x}_1)(1-\tilde{x}_2)(l+r_1) + \tfrac{1}{4}(1+\tilde{x}_1)(1-\tilde{x}_2)r_0 \\
&\quad + \tfrac{1}{4}(1+\tilde{x}_1)(1+\tilde{x}_2)\tfrac{\sqrt{3}}{2}r_0 \\
&\quad + \tfrac{1}{4}(1-\tilde{x}_1)(1+\tilde{x}_2)\Big(l + \tfrac{\sqrt{2}}{2}r_1\Big) \\
&\quad + \frac{1+\tilde{x}_1}{2}\Big[\varphi_1^{(2)}(\tilde{x}_2) - \tfrac{1}{2}(1-\tilde{x}_2)r_0 - \tfrac{1}{2}(1+\tilde{x}_2)\tfrac{\sqrt{3}}{2}r_0\Big] \\
&\quad + \frac{1-\tilde{x}_1}{2}\Big[\varphi_1^{(4)}(\tilde{x}_2) - \tfrac{1}{2}(1-\tilde{x}_2)(l+r_1) - \tfrac{1}{2}(1+\tilde{x}_2)\Big(l + \tfrac{\sqrt{2}}{2}r_1\Big)\Big] \\
&= \tfrac{1}{2}(1+\tilde{x}_1)\varphi_1^{(2)}(\tilde{x}_2) + \tfrac{1}{2}(1-\tilde{x}_1)\varphi_1^{(4)}(\tilde{x}_2)
\end{aligned} \tag{3.2*.24a}$$

and similarly

$$x_2 = \tfrac{1}{2}(1+\tilde{x}_1)\varphi_2^{(2)}(\tilde{x}_2) + \tfrac{1}{2}(1-\tilde{x}_1)\varphi_2^{(4)}(\tilde{x}_2). \tag{3.2*.24b}$$

† Reprinted from [Szabo and Babuška, 1991], by permission of John Wiley & Sons, Inc.

Let us now summarize the properties of the element space $S^p(\tau)$. We have:

(a) When an iso, sub, super parametric mapping or a blending mapping $\mathbf{F}_\tau$ is used, the constant function belongs to $S^p(\tau)$, namely

$$1 \in S^p(\tau). \tag{3.2*.25a}$$

(b) For $\mathbf{F}_\tau$ affine, $S^p(\tau)$ includes all polynomials of degree p, namely

$$\mathcal{P}^p(\tau) \subseteq S^p(\tau). \tag{3.2*.25b}$$

(c) For $\mathbf{F}_\tau$ bilinear which is used when τ is a quadrilateral, we have

$$\mathcal{P}^p(\tau) \subset S^p_{\text{bi-}p}(\tau). \tag{3.2*.25c}$$

(d) For the sub or isoparametric mapping, we have

$$\mathcal{P}^1(\tau) \subseteq S^p(\tau) \tag{3.2*.25d}$$

while for superparametric or blending mapping,

$$\mathcal{P}^1(\tau) \not\subseteq S^p(\tau) \tag{3.2*.25e}$$

in general.

Remark 3.2*.8 The property $\mathcal{P}^1(\tau) \subseteq S^p(\tau)$ ensures that when solving problems of elasticity, the rigid-body modes, i.e. translation and rotation, have zero energy. This condition is often desirable, but is often satisfied at the expense of introducing errors in the approximation of the boundary. When the blending-function mapping is employed in finite element computations in elasticity the rigid rotation mode is not energy free, in general. Note however that with the exception of the special case of very thin domains, this does not affect the accuracy of the finite element method. On the other hand, the use of blending mappings eliminate the need for approximating the boundary.

Remark 3.2*.9 The construction of the finite element mesh is the most laborious part of finite element analysis. There has been considerable effort to automate mesh generation, and several commercial mesh generators have been developed; see, for example, [Thompson and Soni, 1999], [Carey, 1997], [George, 1996], [George and Borouhaki, 1998], [Liseikin, 1999], and the references in these works.

As in the one-dimensional case, the construction of the finite element equations is done in four steps:

- generation of the mesh and identification of the degrees of freedom
- calculation of the element arrays
- assembly of the element arrays into the global arrays
- application of the essential boundary conditions by constraining the global arrays.

Similarly as in one dimension we can speak about h, p, and hp versions of the finite element method; see e.g. [Szabo and Babuška, 1991].

Let us also give the two-dimensional analogue of Remark 3.2.7.

Remark 3.2*.10 Consider the case of the anisotropic Laplacian $(\partial/\partial x_j)(a_{ij}(\partial u/\partial x_j)) = f$. As in the one-dimensional case, the exact solution u_{EX} satisfies the following variational problem in each element τ:

Find $u_{EX}|_\tau \in \mathcal{U}_{\partial\tau_D}(\tau)$ such that

$$\mathcal{B}_\tau(u_{EX}|_\tau, v) = \mathcal{L}_\tau^{EX}(v) \stackrel{\text{def}}{=} \int_\tau fv + \oint_{\partial\tau} \sum_{j=1}^{2} a_{ij} \frac{\partial}{\partial x_i}(u_{EX}) n_i^{\partial\tau} v \quad \forall v \in \mathcal{U}_{\partial\tau_D}(\tau) \tag{3.2*.26a}$$

where $\partial\tau_D \stackrel{\text{def}}{=} \partial\tau \cap \Gamma_D$, and

$$\mathcal{U}_{\partial\tau_D}(\tau) \stackrel{\text{def}}{=} \{v \mid \| v\|_{\mathcal{U}(\tau)} < \infty,\ v|_{\partial\tau_D} \equiv 0\} \tag{3.2*.26b}$$

and $n_i^{\partial\tau}$ are the components of the exterior normal on $\partial\tau$. As in the one-dimensional case, let us, for the moment, assume that the boundary term $\sum_{j=1}^{2}(a_{ij}(\partial/\partial x_i)(u_{EX})n_i^{\partial\tau})|_{\partial\tau}$ exists and it is known, and let $u_{S^p(\tau)}$ be the solution of the discrete problem:

Find $u_{S^p(\tau)} = \sum_{k=1}^{ndf(\tau)} Q_k^\tau \varphi_k^\Delta \in S_{\partial\tau_D}^p(\tau)$ such that

$$\mathcal{B}_\tau(u_{S^p(\tau)}, v) = \mathcal{L}_\tau^{EX}(v) \quad \forall v \in S_{\partial\tau_D}^p(\tau) \tag{3.2*.27a}$$

where

$$S_{\partial\tau_D}^p(\tau) \stackrel{\text{def}}{=} \{v \in S^p(\tau) \mid v|_{\partial\tau} \equiv 0\} \tag{3.2*.27b}$$

and Q_k^τ denotes the kth degree of freedom of the element τ, φ_k^Δ the kth element shape function, $ndf(\tau)$, the number of degrees of freedom in the element τ. Letting $v = \varphi_l^\Delta$, $l = 1, \ldots, ndf(\tau)$, we obtain the linear system

$$\begin{bmatrix} K_{11}^\tau & K_{12}^\tau & \cdots & K_{1(ndf(\tau))}^\tau \\ K_{21}^\tau & K_{22}^\tau & \cdots & K_{2(ndf(\tau))}^\tau \\ K_{31}^\tau & K_{32}^\tau & \cdots & K_{3(ndf(\tau))}^\tau \\ \vdots & \vdots & & \vdots \\ K_{(ndf(\tau))1}^\tau & \cdots & & K_{(ndf(\tau))(ndf(\tau))}^\tau \end{bmatrix} \begin{Bmatrix} Q_1^\tau \\ Q_2^\tau \\ Q_3^\tau \\ \vdots \\ Q_{(ndf(\tau))}^\tau \end{Bmatrix}$$

$$= \begin{Bmatrix} F_1^\tau \\ \vdots \\ \vdots \\ \vdots \\ F_{(ndf(\tau))}^\tau \end{Bmatrix} + \begin{Bmatrix} s_1^{\partial\tau}(u_{EX}) \\ \vdots \\ \vdots \\ \vdots \\ s_{(ndf(\tau))}^{\partial\tau}(u_{EX}) \end{Bmatrix}. \tag{3.2*.27c}$$

Here K_{ij}^{τ} are the entries of the element stiffness matrix given by

$$K_{ij}^{\tau} = \int_{\tau} \sum_{k,l=1}^{n} a_{kl} \frac{\partial u_i}{\partial x_k} \frac{\partial u_j}{\partial x_l}. \tag{3.2*.28a}$$

The entries of the element load vector F_i^{τ} are

$$F_i^{\tau} = \int_{\tau} f \varphi_i^{\tau} \tag{3.2*.28b}$$

and $s_i^{\partial\tau}(u_{EX})$ are the entries of the boundary term

$$s_i^{\partial\tau}(u_{EX}) \stackrel{\text{def}}{=} \oint_{\partial\tau} \sum_{k,l=1}^{2} a_{kl} \frac{\partial}{\partial x_l}(u_{EX}) n_k^{\partial\tau} \varphi_i^{\Delta}. \tag{3.2*.28c}$$

Note that

$$s_i^{\partial\tau}(u_{EX}) = s_i^{\partial\tau_{\mathrm{N}}}(u_{EX}) + s_i^{\partial\tau - \partial\tau_{\mathrm{N}}}(u_{EX}) \tag{3.2*.29}$$

where $\partial\tau_{\mathrm{N}} \stackrel{\text{def}}{=} \partial\tau \cap \Gamma_{\mathrm{N}}$, and $s_i^{\partial\tau - \partial\tau_{\mathrm{N}}}(u_{EX})$ is unknown. Note that for φ_i^{τ} an interior shape function, we have $s_i^{\partial\tau} = 0$.

As in the one-dimensional case these unknowns can be eliminated by employing the nodal and edge equilibrium equations. Given a vertex X, let τ_X denote an element with a vertex at X. Let $i(\tau_X)$ denote the local number of the element vertex which is connected to X. We then have

$$\sum_{\tau_X} s_{i(\tau_X)}^{\partial\tau_X}(u_{EX}) = 0 \tag{3.2*.30a}$$

where the sum is taken over all τ_X connected to node X. In the cases that i is a degree of freedom on the edge ϵ, we have

$$s_{i(\tau_\epsilon)}^{\tau_\epsilon}(u_{EX}) + s_{i(\tau_\epsilon^*)}^{\tau_\epsilon^*}(u_{EX}) = 0 \tag{3.2*.30b}$$

where $\epsilon = \partial\tau_\epsilon \cap \partial\tau_\epsilon^*$, and $i(\tau_\epsilon)$, $i(\tau_\epsilon^*)$ is the local number for the degree of freedom in τ_ϵ, τ_ϵ^*, respectively. The global system of linear equations is obtained by employing (3.2*.30a), (3.2*.30b), and the element equations for the interior degrees of freedom.

Remark 3.2*.11 Let us also discuss the evaluation of the entries of the element stiffness matrices and load vectors. For the entries of the stiffness matrix we have

$$K_{mn}^{\tau} = \int_{\tau} \sum_{i,j=1}^{2} a_{ij} \frac{\partial}{\partial x_i}(\varphi_m^{\tau}) \frac{\partial}{\partial x_j}(\varphi_n^{\tau}) \tag{3.2*.31a}$$

and employing the element transformation $\mathbf{x} = \mathbf{F}_{\tau}(\tilde{\mathbf{x}})$, we have

$$\frac{\partial}{\partial x_i}(\varphi_m^{\tau}) = \frac{\partial}{\partial \tilde{x}_k}(\varphi_m^{\tau}) \frac{\partial \tilde{x}_k}{\partial x_i} = \frac{\partial}{\partial \tilde{x}_k}(\tilde{\varphi}_m^{\tau})(\boldsymbol{\nabla}\mathbf{F}_{\tau})_{ki}^{-1} \tag{3.2*.31b}$$

and we get

$$K_{mn}^{\tau} = \int_{\tau} \sum_{i,j=1}^{2} a_{ij} \frac{\partial \tilde{x}_k}{\partial x_i} \frac{\partial \tilde{x}_l}{\partial x_j} \frac{\partial}{\partial \tilde{x}_i}(\tilde{\varphi}_m^{\tau}) \frac{\partial}{\partial \tilde{x}_j}(\tilde{\varphi}_n^{\tau}) \det \left| \frac{\partial x_p}{\partial \tilde{x}_q} \right| \tag{3.2*.31c}$$

where $(\nabla \mathbf{F}_\tau)_{ki}^{-1}$ are the entries of the Jacobian of the inverse transformation $\mathbf{F}_\tau^{-1}$, and $\det | \partial x_p / \partial \tilde{x}_q | = \det | \nabla \mathbf{F}_\tau |$ is the determinant of $\nabla \mathbf{F}_\tau$. Similarly, for the entries of the load vector we have

$$F_m^{\tau} = \int_{\tau} f \varphi_m^{\tau} = \int_{\tilde{\tau}} f \circ \mathbf{F}_\tau \tilde{\varphi}_m^{\tau} \det | \nabla F_\tau |. \tag{3.2*.32a}$$

Let $\varepsilon = \partial \tau \cap \Gamma_{\mathrm{N}}$, and let us also be interested in the contribution $\int_\varepsilon g \varphi_m^{\Delta}$. Assuming that ε is the edge ① shown in Fig. 3.2*.2, we have

$$\int_{\varepsilon} g \varphi_m \, ds = \int_{-1}^{1} \tilde{g}(\tilde{x}) \tilde{\varphi}_m(\tilde{x}_1, 0) \sqrt{\left(\left(\frac{\partial x_1}{\partial \tilde{x}_1} \right)^2 + \left(\frac{\partial x_2}{\partial \tilde{x}_1} \right)^2 \right) (\tilde{x}_1, 0)} \, d\tilde{x}_1. \tag{3.2*.32b}$$

The assembly follows analogous steps as in one dimension. Further, details about the organization and computer implementation of finite element computations may be found, among others, in [Becker, Carey, and Oden, 1981], [Akin, 1995], and [Bathe, 1996].

Remark 3.2*.12 As in the one-dimensional case, the integrals over $\tilde{\tau}$ have to be computed numerically. In the case of quadrilateral elements, the product Gauss rule may be employed in the master square. In the case of triangular elements, Gaussian quadrature has only an indirect analog with the one-dimensional case; see also [Cowper, 1973], [Dunavant, 1985], [Hammer, Marlowe, and Stroud, 1956], [Espelid, 1994], [Lyness and Jespersen, 1975], [Schwab, 1994], and [Wandzura and Xiao, 1999].

In this section, we outlined the construction of the finite element approximation in two dimensions.

Problems

P3.2*.1. Derive the assembled global equation for the central node of each of one of the four meshes of triangles shown in Fig. 3.2*.6, for the Laplace equation, $\Delta u = 0$.

P3.2*.2. Let $\Omega = (-1, 1) \times (-1, 1)$ and consider the problem

$$(-\Delta u + u)|_{\Omega} = f, \qquad \frac{\partial u}{\partial n}\bigg|_{\Gamma} = g \tag{3.2*.33}$$

with f and g such that the exact solution is $u_{EX} = x_1^2 + x_2^2$. Determine the finite element solution for the meshes of triangles shown in Fig. 3.2*.6, and the mesh of the four bilinear $(p = 1)$ squares determined by the nine vertices.

3.3 The finite element interpolation and approximation error

In this section we will employ a *local approximation* of the exact solution, which is known as the *Lagrangian interpolant*, to study the asymptotic behavior of the error in the finite element solution. Let us first define the Lagrangian interpolant (see also [Atkinson, 1989]).

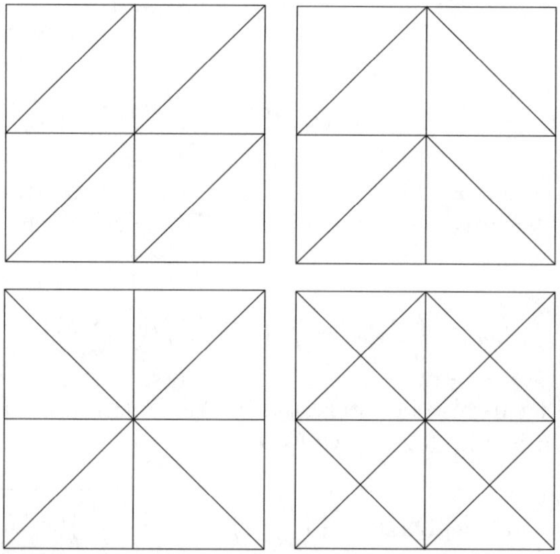

FIG. 3.2*.6. Patterns of meshes of triangles

Given any $u \in C^0(I_i^\Delta)$, we will define its *element Lagrangian interpolant* $\mathcal{I}^p_{I_i^\Delta} u \in \mathcal{P}^p(I_i^\Delta)$, by

$$\mathcal{I}^p_{I_i^\Delta} u \stackrel{\text{def}}{=} \sum_{j=0}^{p} \mathcal{L}_j(\cdot;\{\xi_l^{I_i^\Delta}\}_{l=0}^p) u(\xi_j^{I_i^\Delta}) \tag{3.3.1a}$$

where $\{\xi_l^{I_i^\Delta}\}_{l=0}^p$ is the set of the employed *interpolation points* which are such that

$$x_{i-1}^\Delta = \xi_0^{I_i^\Delta} < \xi_1^{I_i^\Delta} < \cdots < \xi_p^{I_i^\Delta} = x_i^\Delta \tag{3.3.1b}$$

and $\mathcal{L}_j(\cdot;\{\xi_l^{I_i^\Delta}\}_{l=0}^p)$ is the jth Lagrangian polynomial corresponding to these points.

It is customary to employ the equidistributed set of interpolation points, namely

$$\xi_l^{I_i^\Delta} = x_{i-1}^\Delta + \frac{l}{p} h_{I_i^\Delta}, \quad l = 0, \ldots, p. \tag{3.3.1c}$$

As we will see below, other sets of interpolation points are *preferable when* $p > 4$. We will call $\mathcal{I}^p_{I_i^\Delta} : C^0(I) \mapsto \mathcal{P}^p(I_i^\Delta)$ the Lagrangian interpolation operator of degree p over the element I_i^Δ.

Given any $u \in C^0(I)$, we define its *global Lagrangian interpolant* $\mathcal{I}^p_\Delta u \in C^0(I)$, by

$$\mathcal{I}^p_\Delta u|_{I_i^\Delta} \stackrel{\text{def}}{=} \mathcal{I}^p_{I_i^\Delta} u. \tag{3.3.1d}$$

Note that the continuity of $\mathcal{I}^p_\Delta u$ at the nodes is automatically guaranteed because we employed $\xi_0^{I_i^\Delta} = x_{i-1}^\Delta$ and $\xi_p^{I_i^\Delta} = x_i^\Delta$ in all the elements. We will call $\mathcal{I}^p_\Delta : C^0(I) \mapsto C^0(I)$ the global Lagrangian interpolation operator of degree p, over the mesh Δ.

We will now construct asymptotic estimates for the L^∞ norm of the *interpolation error* $u - \mathcal{I}^p_{I_i^\Delta} u$, and its derivative $(d/dx)(u - \mathcal{I}^p_{I_i^\Delta} u)$.

Theorem 3.3.1 *Let $u \in C^0(I_i^\Delta)$ be such that*

$$\left\| \frac{d^r u}{dx^r} \right\|_{L^\infty(I_i^\Delta)} < \infty. \tag{3.3.2a}$$

Then

$$\|u - \mathcal{I}^p_{I_i^\Delta} u\|_{L^\infty(I_i^\Delta)} \leq \mathscr{C}(h_{I_i^\Delta})^{\mu+1} \left\| \frac{d^r u}{dx^r} \right\|_{L^\infty(I_i^\Delta)} \tag{3.3.2b}$$

$$\left\| \frac{du}{dx} - \frac{d}{dx}(\mathcal{I}^p_{I_i^\Delta} u) \right\|_{L^\infty(I_i^\Delta)} \leq \mathscr{C}(h_{I_i^\Delta})^{\mu} \left\| \frac{d^r u}{dx^r} \right\|_{L^\infty(I_i^\Delta)} \tag{3.3.2c}$$

where

$$\mu = \min(p, r-1), \quad 1 \leq r \leq p+1 \tag{3.3.2d}$$

and $\mathscr{C}$ depends only on p and r.

Proof Let $\tilde{I} = (-1, 1)$, and $F_{I_i^\Delta} : \tilde{I} \mapsto I_i^\Delta$, be the affine transformation

$$F_{I_i^\Delta}(\tilde{x}) \stackrel{\text{def}}{=} \frac{1}{2}(x_{i-1}^\Delta + x_i^\Delta) + \frac{h_{I_i^\Delta}}{2}\tilde{x}, \quad \tilde{x} \in \tilde{I}. \tag{3.3.3}$$

Given a function $u : I_i^\Delta \mapsto \mathbb{R}$, we will denote by $\tilde{u}_{I_i^\Delta} : \tilde{I} \mapsto \mathbb{R}$ the corresponding function over $\tilde{I}$, defined by

$$\tilde{u}_{I_i^\Delta}(\tilde{x}) \stackrel{\text{def}}{=} u\big(F_{I_i^\Delta}(\tilde{x})\big) = u \circ F_{I_i^\Delta}(\tilde{x}), \quad \tilde{x} \in \tilde{I}. \tag{3.3.4a}$$

Below we will suppress the subscript I_i^Δ from $\tilde{u}_{I_i^\Delta}$. Expanding $\tilde{u}$ in Taylor series about $\tilde{x} = 0$, we get

$$\tilde{u}(\tilde{x}) = \sum_{j=0}^{r-1} \frac{1}{j!} \frac{d^j \tilde{u}}{d\tilde{x}^j}(0)\tilde{x}^j + \tilde{R}_r(\tilde{x}) = \tilde{U}^{0,r-1}(\tilde{x}) + \tilde{R}_r(\tilde{x}) \tag{3.3.4b}$$

where we let

$$\tilde{U}^{0,r-1}(\tilde{x}) \stackrel{\text{def}}{=} \sum_{j=0}^{r-1} \frac{1}{j!} \frac{d^j \tilde{u}}{d\tilde{x}^j}(0)\tilde{x}^j \tag{3.3.4c}$$

and

$$\tilde{R}_r \stackrel{\text{def}}{=} \tilde{u} - \tilde{U}^{0,r-1}. \tag{3.3.4d}$$

$\tilde{R}_r$ is the *remainder* of the Taylor series expansion. Noting that

$$\frac{d^r}{d\tilde{x}^r}(\tilde{R}_r) = \frac{d^r \tilde{u}}{d\tilde{x}^r}, \qquad \frac{d^j}{d\tilde{x}^j}(\tilde{R}_r)(0) = 0, \quad j = 0, \ldots, r-1 \tag{3.3.5a}$$

we get

$$\tilde{R}_r(\tilde{x}) = \underbrace{\int_0^{\tilde{x}} \cdots \int_0^{\tilde{t}}}_{r \text{ times}} \frac{d^r \tilde{u}}{d\tilde{x}^r} \tag{3.3.5b}$$

from which it follows that

$$|\tilde{R}_r(\tilde{x})| = \Bigg|\underbrace{\int_0^{\tilde{x}} \cdots \int_0^{\tilde{t}}}_{r \text{ times}} \frac{d^r \tilde{u}}{d\tilde{x}^r}\Bigg| \le \frac{|\tilde{x}|^r}{r!}\left\|\frac{d^r \tilde{u}}{d\tilde{x}^r}\right\|_{L^\infty(\tilde{I})} \le \frac{1}{r!}\left\|\frac{d^r \tilde{u}}{d\tilde{x}^r}\right\|_{L^\infty(\tilde{I})} \tag{3.3.6a}$$

and hence we have

$$\|\tilde{R}_r\|_{L^\infty(\tilde{I})} \le \frac{1}{r!}\left\|\frac{d^r \tilde{u}}{d\tilde{x}^r}\right\|_{L^\infty(\tilde{I})}. \tag{3.3.6b}$$

Similarly, we get

$$\left\|\frac{d}{d\tilde{x}}(\tilde{R}_r)\right\|_{L^\infty(\tilde{I})} \le \frac{1}{(r-1)!}\left\|\frac{d^r \tilde{u}}{d\tilde{x}^r}\right\|_{L^\infty(\tilde{I})}. \tag{3.3.6c}$$

Denoting by $\tilde{\mathcal{I}}^p : C^0(\tilde{I}) \mapsto \mathcal{P}^p(\tilde{I})$, the p-degree interpolation operator over the master element $\tilde{I}$, and interpolating both sides of (3.3.4b) we get

$$\tilde{\mathcal{I}}^p\tilde{u}(\tilde{x}) = \sum_{j=0}^{r-1} \frac{1}{j!}\frac{d^j \tilde{u}}{d\tilde{x}^j}(0)\tilde{\mathcal{I}}^p(\tilde{x}^j) + \tilde{\mathcal{I}}^p\tilde{R}_r(\tilde{x}) = \tilde{\mathcal{I}}^p U^{0,r-1}(\tilde{x}) + \tilde{\mathcal{I}}^p\tilde{R}_r(\tilde{x}). \tag{3.3.7}$$

Noting that

$$\tilde{\mathcal{I}}^p(\tilde{x}^j) \equiv \tilde{x}^j, \quad j = 0, 1, \ldots, r-1 \le p \tag{3.3.8a}$$

it follows that

$$\tilde{\mathcal{I}}^p U^{0,r-1} \equiv \sum_{j=0}^{r-1} \frac{1}{j!}\frac{d^j \tilde{u}}{d\tilde{x}^j}(0)\tilde{\mathcal{I}}^p(\tilde{x}^j) \equiv \tilde{U}^{0,r-1} \tag{3.3.8b}$$

and hence

$$\tilde{u} - \tilde{\mathcal{I}}^p\tilde{u} = \tilde{u} - \tilde{\mathcal{I}}^p\tilde{U}^{0,r-1} - \tilde{\mathcal{I}}^p\tilde{R}_r = \tilde{u} - \tilde{U}^{0,r-1} - \tilde{\mathcal{I}}^p\tilde{R}_r = \tilde{R}_r - \tilde{\mathcal{I}}^p\tilde{R}_r \tag{3.3.8c}$$

and using the triangle inequality we get

$$\|\tilde{u} - \tilde{\mathcal{I}}^p\tilde{u}\|_{L^\infty(\tilde{I})} = \|\tilde{R}_r - \tilde{\mathcal{I}}^p\tilde{R}_r\|_{L^\infty(\tilde{I})} \le \|\tilde{R}_r\|_{L^\infty(\tilde{I})} + \|\tilde{\mathcal{I}}^p\tilde{R}_r\|_{L^\infty(\tilde{I})}. \tag{3.3.8c}$$

Further, using that

$$\|\tilde{\mathcal{I}}^p\tilde{R}_r\|_{L^\infty(\tilde{I})} \le \mathscr{C}(p)\|\tilde{R}_r\|_{L^\infty(\tilde{I})} \tag{3.3.9}$$

we get

$$\|\tilde{u} - \tilde{\mathcal{I}}^p\tilde{u}\|_{L^\infty(\tilde{I})} \le (1 + \mathscr{C}(p))\|\tilde{R}_r\|_{L^\infty(\tilde{I})} \le (1 + \mathscr{C}(p))\,\frac{1}{r!}\left\|\frac{d^r \tilde{u}}{d\tilde{x}^r}\right\|_{L^\infty(\tilde{I})} \tag{3.3.10}$$

with $0 < r \le p + 1$. This is the estimate for the interpolation error over $\tilde{I}$.

Let us now rewrite (3.3.10) over I_i^Δ by employing the transformation

$$\frac{d^r\tilde{u}}{d\tilde{x}^r} = \left(\frac{dx}{d\tilde{x}}\right)^r \frac{d^r u}{dx^r} = \left(\frac{h_{I_i^\Delta}}{2}\right)^r \frac{d^r u}{dx^r}. \tag{3.3.11a}$$

which implies

$$\left\|\frac{d^r\tilde{u}}{d\tilde{x}^r}\right\|_{L^\infty(\tilde{I})} = \left(\frac{h_{I_i^\Delta}}{2}\right)^r \left\|\frac{d^r u}{dx^r}\right\|_{L^\infty(I_i^\Delta)} < \infty \tag{3.3.11b}$$

and

$$\|u - \mathcal{I}^p_{I_i^\Delta} u\|_{L^\infty(I_i^\Delta)} = \|\tilde{u} - \tilde{\mathcal{I}}^p\tilde{u}\|_{L^\infty(\tilde{I})} \tag{3.3.11c}$$

and substituting into (3.3.10) we obtain

$$\|u - \mathcal{I}^p_{I_i^\Delta} u\|_{L^\infty(I_i^\Delta)} \leq \frac{1}{r!2^r}(1 + \mathscr{C}(p))(h_{I_i^\Delta})^r \left\|\frac{d^r u}{dx^r}\right\|_{L^\infty(I_i^\Delta)} \tag{3.3.12}$$

for $1 \leq r \leq p+1$, and the interpolation error estimate (3.3.2b) has been proven.

Let us now prove (3.3.2c), the estimate for the derivative of the interpolation error. Recalling that

$$\tilde{u} - \tilde{\mathcal{I}}^p\tilde{u} = \tilde{R}_r - \tilde{\mathcal{I}}^p\tilde{R}_r \tag{3.3.13a}$$

differentiating with respect to $\tilde{x}$, and using the triangle inequality, we get

$$\left\|\frac{d\tilde{u}}{d\tilde{x}} - \frac{d}{d\tilde{x}}(\tilde{\mathcal{I}}^p\tilde{u})\right\|_{L^\infty(\tilde{I})} \leq \left\|\frac{d}{d\tilde{x}}(\tilde{R}_r)\right\|_{L^\infty(\tilde{I})} + \left\|\frac{d}{d\tilde{x}}(\tilde{\mathcal{I}}^p\tilde{R}_r)\right\|_{L^\infty(\tilde{I})}. \tag{3.3.13b}$$

The first term in the right-hand side has been estimated in (3.3.6c). For the second term, note that, because

$$\tilde{\mathcal{I}}^p\tilde{R}_r \in \mathcal{P}^p(\tilde{I}) \tag{3.3.14a}$$

it can be shown (see Remark 3.3.3 below) that

$$\left\|\frac{d}{d\tilde{x}}(\tilde{\mathcal{I}}^p\tilde{R}_r)\right\|_{L^\infty(\tilde{I})} \leq \bar{\mathscr{C}}(p)\|\tilde{\mathcal{I}}^p\tilde{R}_r\|_{L^\infty(\tilde{I})} \tag{3.3.14b}$$

and employing that

$$\|\tilde{\mathcal{I}}^p\tilde{R}_r\|_{L^\infty(\tilde{I})} \leq \mathscr{C}(p)\|\tilde{R}_r\|_{L^\infty(\tilde{I})} \tag{3.3.14c}$$

we get

$$\left\|\frac{d}{d\tilde{x}}(\tilde{\mathcal{I}}^p\tilde{R}_r)\right\|_{L^\infty(\tilde{I})} \leq \mathscr{C}'(p)\|\tilde{R}_r\|_{L^\infty(\tilde{I})} \leq \frac{1}{r!}\mathscr{C}'(p)\left\|\frac{d^r\tilde{u}}{d\tilde{x}^r}\right\|_{L^\infty(\tilde{I})} \tag{3.3.14d}$$

where we let $\mathscr{C}'(p) = \bar{\mathscr{C}}(p)\mathscr{C}(p)$. Employing (3.3.6c) and (3.3.14d) into (3.3.13b), we get

$$\left\|\frac{d}{d\tilde{x}}(\tilde{u} - \tilde{\mathcal{I}}^p\tilde{u})\right\|_{L^\infty(\tilde{I})} \leq \left(\frac{1}{(r-1)!} + \frac{1}{r!}\mathscr{C}'(p)\right)\left\|\frac{d^r\tilde{u}}{d\tilde{x}^r}\right\|_{L^\infty(\tilde{I})} \tag{3.3.15}$$

which is the estimate for the derivative of the interpolation error written over $\tilde{I}$. Employing now that

$$\left\|\frac{d}{dx}(u - \mathcal{I}^p_{I_i^\Delta} u)\right\|_{L^\infty(I_i^\Delta)} = \frac{2}{h_{I_i^\Delta}}\left\|\frac{d}{dx}(\tilde{u} - \tilde{\mathcal{I}}^p\tilde{u})\right\|_{L^\infty(\tilde{I})} \tag{3.3.16a}$$

and

$$\left\|\frac{d^r\tilde{u}}{d\tilde{x}^r}\right\|_{L^\infty(\tilde{I})} = \left(\frac{h_{I_i^\Delta}}{2}\right)^r \left\|\frac{d^r u}{dx^r}\right\|_{L^\infty(I_i^\Delta)} \tag{3.3.16b}$$

we get

$$\left\|\frac{d}{dx}(u - \mathcal{I}^p_{I_i^\Delta} u)\right\|_{L^\infty(I_i^\Delta)} \leq \frac{1}{2^{r-1}(r-1)!}\left(1 + \frac{1}{r}\mathscr{C}'(p)\right) \times (h_{I_i^\Delta})^{r-1}\left\|\frac{d^r u}{dx^r}\right\|_{L^\infty(I_i^\Delta)}. \tag{3.3.16c}$$

This proves (3.3.2c), the estimate for the derivative of the interpolation error. □

Remark 3.3.1 The above proof of the interpolation estimates is sufficiently detailed for the analysis of the h version of the finite element method. For the analysis of the p version, the optimal dependence of the constants $\mathscr{C}(p)$ and $\overline{\mathscr{C}}(p)$ on the degree p of the elements is also needed.

Remark 3.3.2 The main ingredients of the proof of the estimates of the interpolation error are:

(a) An affine mapping $F_{I_i^\Delta}$ is used to restate the estimates over the master element $\tilde{I}$.

(b) The interpolation operator reproduces all polynomials of degree p, and we have

$$\tilde{u} - \tilde{\mathcal{I}}^p\tilde{u} = \tilde{R}_r - \tilde{\mathcal{I}}^p\tilde{R}_r \tag{3.3.17a}$$

and hence

$$\|\tilde{u} - \tilde{\mathcal{I}}^p\tilde{u}\|_{L^\infty(\tilde{I})} \leq \|\tilde{R}_r\|_{L^\infty(\tilde{I})} + \|\tilde{\mathcal{I}}^p\tilde{R}_r\|_{L^\infty(\tilde{I})} \tag{3.3.17b}$$

and similarly

$$\left\|\frac{d}{d\tilde{x}}(\tilde{u} - \tilde{\mathcal{I}}^p\tilde{u})\right\|_{L^\infty(\tilde{I})} \leq \left\|\frac{d}{d\tilde{x}}(\tilde{R}_r)\right\|_{L^\infty(\tilde{I})} + \left\|\frac{d}{d\tilde{x}}(\tilde{\mathcal{I}}^p\tilde{R}_r)\right\|_{L^\infty(\tilde{I})}. \tag{3.3.17c}$$

(c) The remainder $\tilde{R}_r$ is estimated by

$$\|\tilde{R}_r\|_{L^\infty(\tilde{I})} \leq \frac{1}{r!}\left\|\frac{d^r\tilde{u}}{d\tilde{x}^r}\right\|_{L^\infty(\tilde{I})} \tag{3.3.18a*}$$

and its derivative $(d/d\tilde{x})(\tilde{R}_r)$ by

$$\left\|\frac{d}{d\tilde{x}}(\tilde{R}_r)\right\|_{L^\infty(\tilde{I})} \leq \frac{1}{(r-1)!}\left\|\frac{d^r\tilde{u}}{d\tilde{x}^r}\right\|_{L^\infty(\tilde{I})}. \tag{3.3.18b}$$

(d) The interpolant of the remainder $\tilde{\mathcal{I}}^p \tilde{R}_r$ is estimated by

$$\|\tilde{\mathcal{I}}^p \tilde{R}_r\|_{L^\infty(\tilde{I})} \le \mathscr{C}(p)\|\tilde{R}_r\|_{L^\infty(\tilde{I})} \tag{3.3.19a}$$

and its derivative $(d/d\tilde{x})(\tilde{\mathcal{I}}^p \tilde{R}_r)$, by

$$\left\|\frac{d}{d\tilde{x}}(\tilde{\mathcal{I}}^p \tilde{R}_r)\right\|_{L^\infty(\tilde{I})} \le \bar{\mathscr{C}}(p)\|\tilde{\mathcal{I}}^p \tilde{R}_r\|_{L^\infty(\tilde{I})} \le \bar{\mathscr{C}}'(p)\|\tilde{R}_r\|_{L^\infty(\tilde{I})}. \tag{3.3.19b}$$

(e) Combining the above, we get the interpolation estimates over $\tilde{I}$

$$\|\tilde{u} - \tilde{\mathcal{I}}^p_{I_i^\Delta}\tilde{u}\|_{L^\infty(I_i^\Delta)} \le \frac{1}{r!2^r}\,(1 + \mathscr{C}(p))(h_{I_i^\Delta})^r \left\|\frac{d^r u}{dx^r}\right\|_{L^\infty(I_i^\Delta)} \tag{3.3.20a}$$

and

$$\left\|\frac{d}{d\tilde{x}}(\tilde{u} - \tilde{\mathcal{I}}^p\tilde{u})\right\|_{L^\infty(\tilde{I})} \le \left(\frac{1}{(r-1)!} + \frac{1}{r!}\mathscr{C}'(p)\right)(h_{I_i^\Delta})^r \left\|\frac{d^r u}{dx^r}\right\|_{L^\infty(\tilde{I})} \tag{3.3.20b}$$

where $1 \le r \le p+1$.

(f) The estimates over I_i^Δ are obtained by transforming the norms from $\tilde{I}$ to I_i^Δ.

The derivation of the interpolation estimates in two dimensions is based on completely analogous arguments.

Remark 3.3.3 The inequality

$$\left\|\frac{d\tilde{v}}{d\tilde{x}}\right\|_{L^\infty(\tilde{I})} \le \bar{\mathscr{C}}(p)\|\tilde{v}\|_{L^\infty(\tilde{I})} \quad \forall \tilde{v} \in \mathscr{P}^p(\tilde{I}) \tag{3.3.21}$$

follows from the observations that: (a) the norms in both sides of the inequality can be rewritten as norms in a finite-dimensional space; and that (b) any two norms of a finite-dimensional space are equivalent.

We will now analyze the dependence of the constant $\mathscr{C}(p)$ in the estimate

$$\|\tilde{\mathcal{I}}^p \tilde{v}\|_{L^\infty(\tilde{I})} \le \mathscr{C}(p)\|\tilde{v}\|_{L^\infty(\tilde{I})} \tag{3.3.22}$$

on the selection of the interpolation points $\tilde{\xi}_k$, $k = 0, \ldots, p$. We will assume that the interpolation points $\tilde{\xi}_k$ are:

(i) distinct

$$\tilde{\xi}_j \ne \tilde{\xi}_k, \quad j \ne k, \ j,k = 0,\ldots,p \tag{3.3.23a}$$

and include the end points

$$\tilde{\xi}_0 = -1, \qquad \tilde{\xi}_p = 1 \tag{3.3.23b}$$

(ii) symmetric with respect to the mid-point of the master interval.

We will analyze $\mathscr{C}(p)$ for the following choices of the interpolation points ξ_k, $k = 0, \dots, p$.

(a) *Uniformly distributed points*:

$$\tilde{\xi}_k^{unif} = -1 + 2\frac{k}{p}, \quad k = 0, \dots, p. \tag{3.3.24a}$$

(b) *Canonical Chebyshev points*:

$$\tilde{\xi}_k^{Cheb} = -\frac{\cos(\pi(2k+1)/(-2p+2))}{\cos(\pi/(2p+2))}, \quad k = 0, \dots, p. \tag{3.3.24b}$$

These points are obtained by mapping the roots of the Chebyshev polynomial (see A3.1.4) of degree $(p+1)$, by a linear mapping which transforms the smallest and the biggest root to the left and the right end points of $\tilde{I}$, respectively.

(c) *Gauss–Lobatto points*:

$$\tilde{\xi}_0^{Lob} = -1, \qquad \tilde{\xi}_p^{Lob} = 1, \qquad \frac{d}{d\tilde{\xi}}(\tilde{P}_p)(\tilde{\xi}_k^{Lob}) = 0, \quad k = 1, \dots, p-1. \tag{3.3.24c}$$

These points are the roots of the derivative of the p-degree Legendre polynomial, together with the end points of the interval.

(d) *Gauss–Legendre points*:

$$\tilde{\xi}_0^{Leg} = -1, \qquad \tilde{\xi}_p^{Leg} = 1, \qquad \tilde{P}_{p-1}(\tilde{\xi}_k^{Leg}) = 0, \quad k = 1, \dots, p-1. \tag{3.3.24d}$$

These points are the zeros of the Legendre polynomial of degree $p-1$, together with the end points of the interval.

(e) *Fejer points*:

$$\tilde{\xi}_0^{Fej} = -1, \qquad \tilde{\xi}_p^{Fej} = 1, \qquad \int_{-1}^{\tilde{\xi}_k^{Fej}} \tilde{P}_p = 0, \quad k = 1, \dots, p-1. \tag{3.3.24e}$$

These points are the zeros of the primitive of the Legendre polynomial of degree p, together with the end points of the interval.

(f) *Optimal points*:

Let us first give some background needed for the definition of the optimal points. We will use the notation

$$\tilde{\mathcal{I}}^p(\{\tilde{\xi}_l\}_{l=0}^p) = \sum_{k=0}^{p} \tilde{v}(\tilde{\xi}_k)\tilde{\mathcal{L}}_k(\cdot; \{\tilde{\xi}_l\}_{l=0}^p) \tag{3.3.25a}$$

to underline the dependence of the interpolation operator $\tilde{\mathcal{I}}^p$ on $\{\tilde{\xi}_l\}_{l=0}^p$, the employed set of interpolation points. We will measure the effect of the choice of the set $\{\tilde{\xi}_l\}_{l=0}^p$, on the interpolation operator $\tilde{\mathcal{I}}^p(\{\tilde{\xi}_l\}_{l=0}^p)$, by employing the operator norm

$$|||\tilde{\mathcal{I}}^p(\{\tilde{\xi}_l\}_{l=0}^p)||| = \sup_{\substack{\tilde{v}\in C^0(\tilde{I})\\ \tilde{v}\neq 0}} \frac{|||\tilde{\mathcal{I}}^p v|||}{\|\tilde{v}\|_{L^\infty(\tilde{I})}} = \sup_{\substack{\tilde{v}\in C^0(\tilde{I})\\ \|\tilde{v}\|_{L^\infty(\tilde{I})}=1}} |||\tilde{\mathcal{I}}^p v|||. \tag{3.3.25b}$$

Note that the value of the norm of the interpolation operator, $|||\tilde{\mathcal{I}}^p(\{\tilde{\xi}_l\}_{l=0}^p)|||$, depends on the choice of the set of interpolation points $\{\tilde{\xi}_l\}_{l=0}^p$, and on the employed norm $|||\cdot|||$. Employing the operator norm $|||\cdot|||_\infty$ corresponding to the norm $|||\cdot||| = \|\cdot\|_{L^\infty(\tilde{I})}$ we get

$$|||\tilde{\mathcal{I}}^p(\{\tilde{\xi}_l\}_{l=0}^p)|||_\infty = \left\| \sum_{k=0}^p |\mathcal{L}_k(\cdot\,; \{\tilde{\xi}_l\}_{l=0}^p)| \right\|_{L^\infty(\tilde{I})} \tag{3.3.26a}$$

and hence

$$\mathscr{C}(p; \{\tilde{\xi}_l\}_{l=0}^p) \stackrel{\text{def}}{=} |||\tilde{\mathcal{I}}^p(\{\tilde{\xi}_l\}_{l=0}^p)|||_\infty = \left\| \sum_{k=0}^p |\mathcal{L}_k(\cdot; \{\tilde{\xi}_l\}_{l=0}^p)| \right\|_{L^\infty(\tilde{I})}. \tag{3.3.26b}$$

The term $|||\tilde{\mathcal{I}}^p(\{\tilde{\xi}_l\}_{l=0}^p)|||_\infty$ is known as the *Lebesgue constant.*

Let us give the definition of the optimal points:

A set of points $\{\tilde{\xi}_l^{\text{opt}}\}_{l=0}^p$ is optimal, with respect to the operator norm $|||\cdot|||$, if

$$|||\tilde{\mathcal{I}}^p(\{\tilde{\xi}_l^{\text{opt}}\}_{l=0}^p)||| \leq |||\tilde{\mathcal{I}}^p(\{\tilde{\xi}_l\}_{l=0}^p)||| \tag{3.3.27}$$

for any admissible set of interpolation points $\{\tilde{\xi}_l\}_{l=0}^p$.

Let us now compare the values of the Lebesgue constant, $|||\tilde{\mathcal{I}}^p(\{\tilde{\xi}_l\}_{l=0}^p)|||_\infty$, for the various choices of the interpolation points, namely, the set of equidistributed points, $\{\tilde{\xi}_l^{unif}\}_{l=0}^p$, the set of canonical Chebyshev points, $\{\tilde{\xi}_l^{Cheb}\}_{l=0}^p$, the set of Gauss–Lobatto points, $\{\tilde{\xi}_l^{Lob}\}_{l=0}^p$, and the set of Gauss–Legendre points, $\{\tilde{\xi}_l^{Leg}\}_{l=0}^p$. We will use the notation

$$\mathscr{C}^{unif}(p) \stackrel{\text{def}}{=} \mathscr{C}(p; \{\tilde{\xi}_l^{unif}\}_{l=0}^p) \tag{3.3.28a}$$

and similarly for the other sets of points. For the equidistributed set of points, it can be shown that

$$\mathscr{C}e^{p/2} \leq \mathscr{C}^{unif}(p) \tag{3.3.28b}$$

which means that $\mathscr{C}^{unif}(p)$ *diverges at least exponentially with* p. The exponential growth of $\mathscr{C}^{unif}(p)$ can be seen in Fig. 3.3.1. In contrast, $\mathscr{C}^{Cheb}(p)$, $\mathscr{C}^{Lob}(p)$, $\mathscr{C}^{Leg}(p)$, and $\mathscr{C}^{opt}(p)$ grow much slower with p, as it can be seen from Fig. 3.3.2.

The optimal set of points $\{\tilde{\xi}_l^{opt}\}_{l=0}^p$ is characterized by the property

$$\left\| \sum_{k=0}^p |\mathcal{L}_k(\cdot\,; \{\tilde{\xi}_l^{opt}\}_{l=0}^p)| \right\|_{L^\infty\left(\tilde{\xi}_{m-1}^{opt}, \tilde{\xi}_m^{opt}\right)} = \left\| \sum_{k=0}^p |\mathcal{L}_k(\cdot\,; \{\tilde{\xi}_l^{opt}\}_{l=0}^p)| \right\|_{L^\infty\left(\tilde{\xi}_m^{opt}, \tilde{\xi}_{m+1}^{opt}\right)} \tag{3.3.29}$$

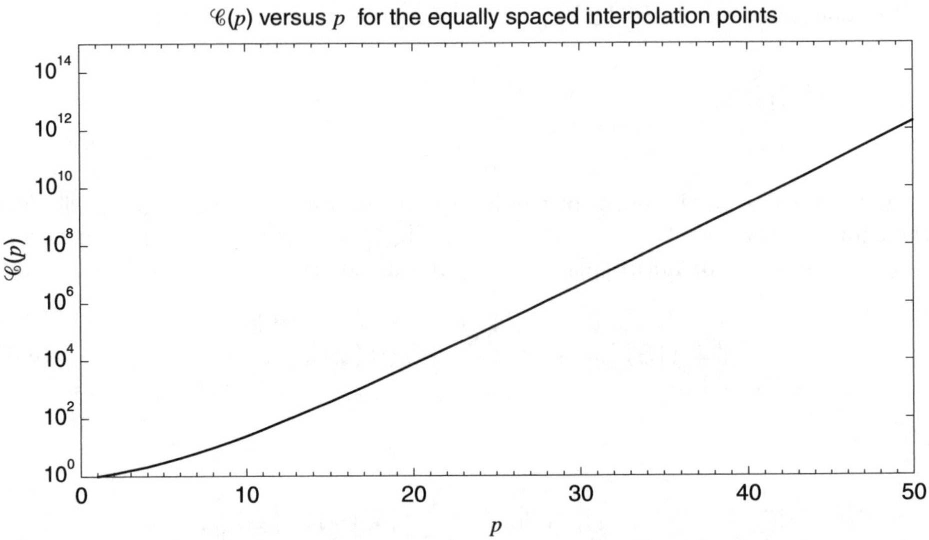

FIG. 3.3.1. The growth of the constant $\mathscr{C}^{unif}(p)$ with increasing p. Note that $\mathscr{C}^{unif}(p)$ grows at least exponentially with p namely $\mathscr{C}e^{p/2} \leq \mathscr{C}^{unif}(p)$

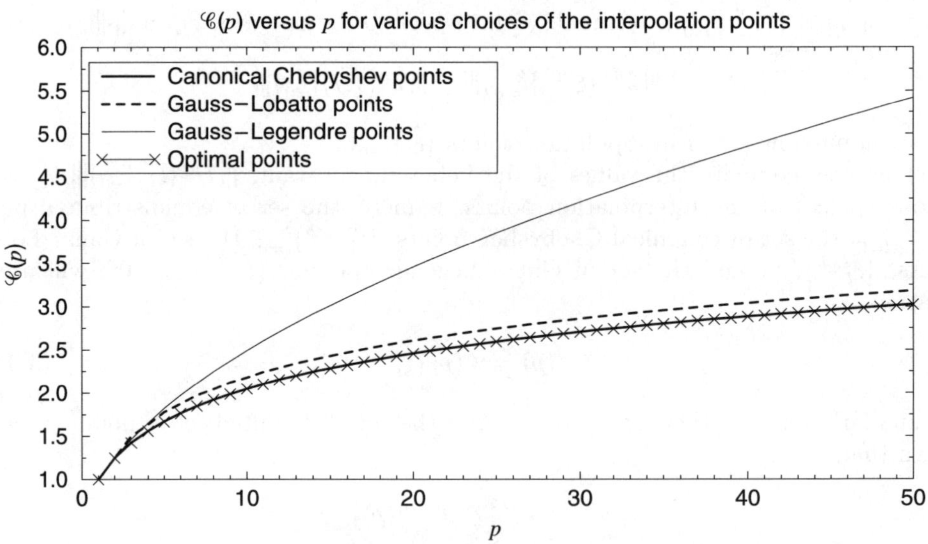

FIG. 3.3.2. The growth of $\mathscr{C}^{opt}(p)$ and the value of $\mathscr{C}(p)$ for various choices of the interpolation points with increasing p. Note that, for the considered choices of interpolation points, the values of $\mathscr{C}(p)$ are small for the entire practical range of p, and we have $\mathscr{C}^{opt}(p) \approx \mathscr{C}^{Cheb}(p)$, and $\mathscr{C}^{opt}(50) \approx \mathscr{C}^{Cheb}(50) \approx 3$

for $m = 1, \ldots, p-1$, namely, that *the maximum error is equidistributed in the subintervals when the end points are consecutive optimal interpolation points.* This property is known as *the Erdös conjecture* [Erdös, 1961], and it was proven in [De Boor and Pinkus, 1978].

TABLE 3.3.1. Coordinates of the optimal interpolation points $\tilde{\xi}_i^{opt}$ in $[0, 1]$, the right half of the master interval for $p = 3, 4, 5, 6, 7, 8, 9, 10$. The coordinates of the points in $[-1, 0]$ can be obtained by symmetry

$p = 3$	$p = 4$	$p = 5$	$p = 6$
0.4177913013559897	0	0.2689070447719729	0
1.0	0.620911304689210	0.7341266671891750	0.4461215299911069
	1.0	1.0	0.8034402382691069
			1.0
$p = 7$	$p = 8$	$p = 9$	$p = 10$
0.1992877299056662	0	0.1585652886576400	0
0.5674306027472531	0.3477879716116667	0.4601498259228992	0.2848880010669260
0.8488719610366556	0.6535334790799031	0.7166138606253076	0.5466676961746041
1.0	0.8802308527184542	0.9027709752917723	0.7640984545671448
	1.0	1.0	0.9195087517942989
			1.0

TABLE 3.3.2. The values $\mathscr{C}^{opt}(p)$ and $\mathscr{C}^{Cheb}(p)$, for $1 \leq p \leq 20$, which correspond to the optimal interpolation points and the canonical Chebyshev points. Note that $\mathscr{C}^{opt}(p) \leq \mathscr{C}^{Cheb}(p)$ for all p

p	$\mathscr{C}^{opt}(p)$	$\mathscr{C}^{Cheb}(p)$	p	$\mathscr{C}^{opt}(p)$	$\mathscr{C}^{Cheb}(p)$
1	1.000000000	1.000000000	11	2.106580259	2.123920344
2	1.250000000	1.250000000	12	2.157118966	2.174698976
3	1.422919573	1.429869531	13	2.203955206	2.222172555
4	1.559490210	1.570170269	14	2.247593208	2.259988185
5	1.672210365	1.685140331	15	2.288440923	2.302446068
6	1.768134581	1.782534592	16	2.326833039	2.344922856
7	1.851599391	1.866990314	17	2.363047517	2.381276788
8	1.925457621	1.941577624	18	2.397317710	2.410477129
9	1.991684987	2.008324917	19	2.429842418	2.448201482
10	2.051705759	2.068745729	20	2.460787749	2.479197705

Employing the Erdös property (3.3.29), we determined the optimal points $\{\tilde{\xi}_l^{opt}\}_{l=0}^p$ by using the Lobatto–Legendre points of degree p as initial guess, and adjusting their locations iteratively, one at a time, to satisfy (3.3.29). Table 3.3.1 lists the points for $p = 3, 4, 5, 6, 7, 8, 9, 10$ in the interval $[0, 1]$, the right half of the master element, and Table 3.3.2 compares the values of $\mathscr{C}^{opt}(p)$ and $\mathscr{C}^{Cheb}(p)$, for $1 \leq p \leq 20$. It can be seen that $\mathscr{C}^{opt}(p)$, $\mathscr{C}^{Cheb}(p)$ practically coincide. Note also that for $p = 50$, which is considerably higher than the interpolation order employed in any practical situation, we have $\mathscr{C}^{opt}(50) \approx \mathscr{C}^{Cheb}(50) \approx 3$, while $\mathscr{C}^{unif}(20) > 10^4$.

We will now illustrate the dependence of the magnitude of the interpolation error on the selection of the interpolation points by employing a classical example.

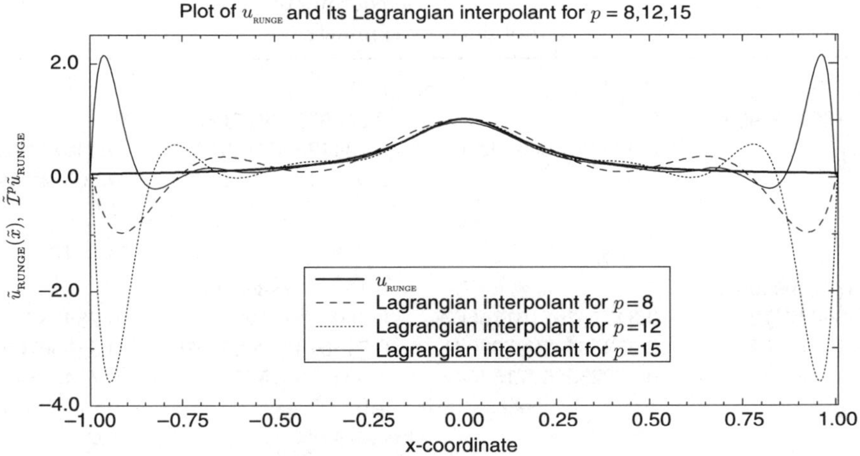

FIG. 3.3.3. Runge's example. The graph of the function $\tilde{u}_{\text{RUNGE}}(\tilde{x}) = 1/(1+25\tilde{x}^2)$, $-1 < \tilde{x} < 1$ and its Lagrangian interpolants $\tilde{\mathcal{I}}^p\tilde{u}_{\text{RUNGE}}$, for $p = 8$, 12, 15, for uniform distribution of the interpolation points. Runge proved that the maximum value of the interpolation error in intervals $0.728 < |\tilde{x}| < 1$ diverges with p. As it can be seen from the graph, although the interpolated function $\tilde{u}_{\text{RUNGE}}$ is very smooth, the interpolation error is very large near the end points and the absolute value of its maximum increases with the degree p of the interpolant, although not monotonically

Example 3.3.1 Consider the C^∞ function

$$\tilde{u}_{\text{RUNGE}}(\tilde{x}) = \frac{1}{1+25\tilde{x}^2}, \quad -1 \le \tilde{x} \le 1. \tag{3.3.30a}$$

Runge proved that *the maximum value of the interpolation error,* $(\tilde{u}_{\text{RUNGE}} - \tilde{\mathcal{I}}^p\tilde{u}_{\text{RUNGE}})(\tilde{x})$, *diverges with* p, *for* $0.728 < |\tilde{x}| < 1$. This can be seen from Fig. 3.3.3 which gives the graph of $\tilde{u}_{\text{RUNGE}}$ and its Lagrangian interpolant $\tilde{\mathcal{I}}^{p,unif}\tilde{u}_{\text{RUNGE}}$ for $p = 8$, 12, 15, for a uniform set of $p+1$ interpolation points, and from Fig. 3.3.4 which compares the graphs of the interpolation error $(\tilde{u}_{\text{RUNGE}} - \tilde{\mathcal{I}}^{12}\tilde{u}_{\text{RUNGE}})$, for the sets of uniformly spaced points and the canonical Chebyshev points. We have

$$\|\tilde{u}_{\text{RUNGE}} - \tilde{\mathcal{I}}^{12,Cheb}\tilde{u}_{\text{RUNGE}}\|_{L^\infty(\tilde{I})} = 0.0707 \tag{3.3.30b}$$

and

$$\|\tilde{u}_{\text{RUNGE}} - \tilde{\mathcal{I}}^{12,opt}\tilde{u}_{\text{RUNGE}}\|_{L^\infty(\tilde{I})} = 0.0705 \tag{3.3.30c}$$

which are much smaller than the error corresponding to the set of equidistributed points

$$\|\tilde{u}_{\text{RUNGE}} - \tilde{\mathcal{I}}^{12,unif}\tilde{u}_{\text{RUNGE}}\|_{L^\infty(\tilde{I})} = 3.6577. \tag{3.3.30d}$$

This is a clear illustration that *Lagrangian interpolation using a uniform distribution of the interpolation points in each element could lead to very large interpolation errors for*

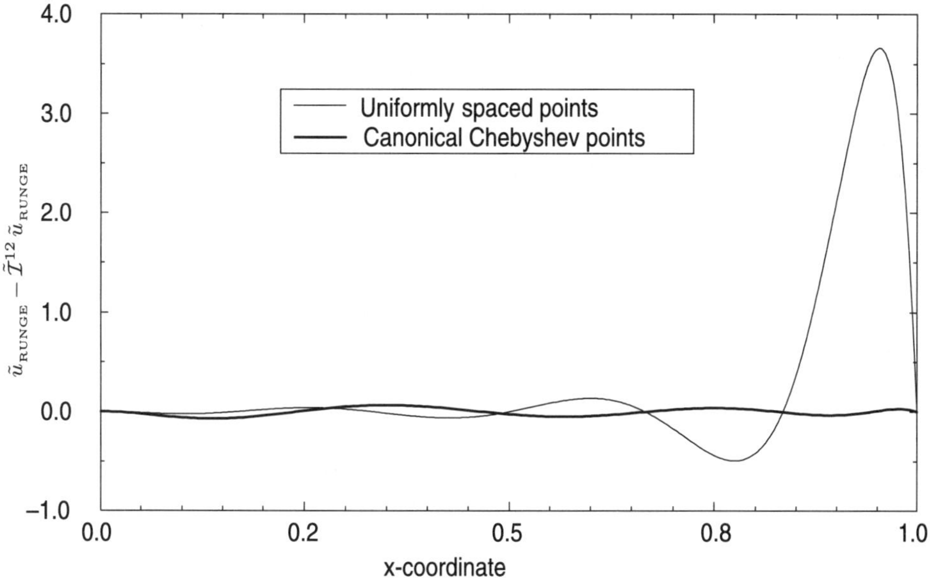

FIG. 3.3.4. Runge's example. The graph of the error $\tilde{u}_{\text{RUNGE}} - \tilde{\mathcal{I}}^{12}\tilde{u}_{\text{RUNGE}}$ for the sets of uniformly spaced points and the canonical Chebyshev points, over the half interval (0, 1). Note that in the case that the canonical Chebyshev points are employed, the maximum interpolation error is practically negligible in comparison with the maximum error for uniformly spaced interpolation points

high p, and that *the use of the canonical Chebyshev points or the optimal points is much preferred.* A complete discussion of the problem of the optimal interpolation points in one, two, and three dimensions is given in [Chen and Babuška, 1995, 1996]. For the properties of the various families of polynomials, see [Borwein and Erdelyi, 1995].

Remark 3.3.4 The constant $\bar{\mathscr{C}}(p)$, in the inequality

$$\left\| \frac{d\tilde{v}}{dx} \right\|_{L^\infty(I)} \leq \bar{\mathscr{C}}(p) \|\tilde{v}\|_{L^\infty(I)}, \quad \tilde{v} \in \mathscr{P}^p(\tilde{I}) \tag{3.3.31}$$

also depends on the choice of interpolation points $\{\tilde{\xi}_k\}_{k=0}^p$, and can be analyzed similarly.

Remark 3.3.5 For the h and p version, the choice of the interpolation points is not critical because the finite element method uses the projection and the Lagrangian interpolation is only a tool for constructing the basis of the finite element space.

Remark 3.3.6 Lagrangian interpolation is often used in practice to describe the boundary of curvilinear domains in two and three dimensions, and to construct the element mapping for elements adjacent to the boundary, and, in this case the use of the optimal points is essential because, the use of a uniformly distributed set of points could introduce excessive error, which could affect the overall performance of the finite element method, especially when the p version with high p (e.g. $p = 8$) is employed.

In the remainder of this chapter, we will suppress the dependence of the constants on the choice of the interpolation points.

Above we constructed estimates for the L^∞ norm of the interpolation error and its derivative. We will now construct the corresponding estimates in the L^2 norm.

Theorem 3.3.2 *Let $u \in C^0(I_i^\Delta)$ be such that*

$$\left\| \frac{d^r u}{dx^r} \right\|_{L^2(I_i^\Delta)} < \infty. \tag{3.3.32a}$$

for some integer r, such that

$$1 \leq r \leq p+1 \tag{3.3.32b}$$

Then

$$\|u - \mathcal{I}^p_{I_i^\Delta} u\|_{L^2(I_i^\Delta)} \leq \mathscr{C}_1 (h_{I_i^\Delta})^{\mu+1} \left\| \frac{d^r u}{dx^r} \right\|_{L^2(I_i^\Delta)} \tag{3.3.32c}$$

and

$$\left\| \frac{du}{dx} - \frac{d}{dx}(\mathcal{I}^p_{I_i^\Delta} u) \right\|_{L^2(I_i^\Delta)} \leq \mathscr{C}_2 (h_{I_i^\Delta})^{\mu} \left\| \frac{d^r u}{dx^r} \right\|_{L^2(I_i^\Delta)} \tag{3.3.32d}$$

with

$$\mu = \min(p, r-1) = r-1 \tag{3.3.32e}$$

where $\mathscr{C}_1$ and $\mathscr{C}_2$ depend only on r and p.

Proof As in the proof of Theorem 3.3.1, we will first prove the estimates over the master element. Assumption (3.3.32a) implies that

$$\frac{d^{r-1}\tilde{u}}{d\tilde{x}^{r-1}} \in C^0(\tilde{I}) \tag{3.3.33a}$$

and hence we can once more employ the Taylor series expansion

$$\tilde{u}(\tilde{x}) = \sum_{j=0}^{r-1} \frac{1}{j!} \frac{d^j \tilde{u}}{d\tilde{x}^j}(0)\tilde{x}^j + \tilde{R}_r(\tilde{x}) \tag{3.3.33b}$$

and we have

$$\|\tilde{R}_r\|_{L^2(\tilde{I})} \leq \mathscr{C}(r) \left\| \frac{d^r \tilde{u}}{d\tilde{x}^r} \right\|_{L^2(\tilde{I})}. \tag{3.3.33c}$$

Recalling that

$$\tilde{u} - \tilde{\mathcal{I}}^p \tilde{u} = \tilde{R}_r - \tilde{\mathcal{I}}^p \tilde{R}_r \tag{3.3.34a}$$

and using the triangle inequality we get

$$\|\tilde{u} - \tilde{\mathcal{I}}^p \tilde{u}\|_{L^2(\tilde{I})} \leq \|\tilde{R}_r\|_{L^2(\tilde{I})} + \|\tilde{\mathcal{I}}^p \tilde{R}_r\|_{L^2(\tilde{I})}. \tag{3.3.34b}$$

Further, because

$$\tilde{R}_r \in C^0(\tilde{I}) \tag{3.3.35a}$$

we have

$$\|\tilde{\mathcal{I}}^p \tilde{R}_r\|_{L^2(\tilde{I})} \leq \mathscr{C}(p)\|\tilde{R}_r\|_{L^2(\tilde{I})}. \tag{3.3.35b}$$

The rest of the proof, and the proof of (3.3.32c), follows exactly the same lines as the proof of Theorem 3.3.1. □

In all the above estimates $\mathscr{C}$ is a generic constant, which takes on different values for each estimate, and is independent of the parameters that it multiplies.

Remark 3.3.7 Replacing the semi-norm $\|d^r u/dx^r\|_{L^2(I_i^\Delta)}$ by the norm $\|u\|_{H^r(I_i^\Delta)} \overset{\text{def}}{=} \sqrt{\sum_{j=0}^{r} \|d^j u/dx^j\|^2_{L^2(I_i^\Delta)}}$ in the right-hand side of the estimates (3.3.32c) and (3.3.32d), we obtain the interpolation estimates

$$\|u - \mathcal{I}^p_{I_i^\Delta} u\|_{L^2(I_i^\Delta)} \leq \mathscr{C}(h_{I_i^\Delta})^{\mu+1}\|u\|_{H^r(I_i^\Delta)} \tag{3.3.36a}$$

and

$$\left\|\frac{du}{dx} - \frac{d}{dx}(\mathcal{I}^p_{I_i^\Delta} u)\right\|_{L^2(I_i^\Delta)} \leq \mathscr{C}(h_{I_i^\Delta})^{\mu}\|u\|_{H^r(I_i^\Delta)} \tag{3.3.36b}$$

with

$$\mu = \min(p, r-1). \tag{3.3.36c}$$

These estimates hold under the assumption that u satisfies (3.3.32a) for some integer r.

Using the above results, we will now construct estimates for the global energy norm of the interpolation error $\|u - \mathcal{I}^p_\Delta u\|_{\mathcal{U}}$.

Theorem 3.3.3 *Let $u \in C^0(I)$ be such that*

$$\left\|\frac{d^r u}{dx^r}\right\|_{L^\infty(I)} < \infty \tag{3.3.37a}$$

with

$$1 \leq r \leq p+1. \tag{3.3.37b}$$

Then

$$\|u - \mathcal{I}^p_\Delta u\|_{\mathcal{U}} \leq \mathscr{C}(h^\Delta)^{\mu}\left\|\frac{d^r u}{dx^r}\right\|_{L^\infty(I)} \tag{3.3.37c}$$

where

$$\mu = \min(p, r-1) = r-1 \tag{3.3.37d}$$

and $\mathscr{C}$ depends only on r, p, β_2, and γ.

Proof We have

$$\|u - \mathcal{I}^p_\Delta u\|^2_{\mathcal{U}} \leq \mathscr{C} \sum_{j=1}^{N(\Delta)} h_{I^\Delta_j} \left(\beta_2 \left\| \frac{d}{dx}(u - \mathcal{I}^p_{I^\Delta_j} u) \right\|^2_{L^\infty(I^\Delta_j)} + \gamma \|u - \mathcal{I}^p_{I^\Delta_j} u\|^2_{L^\infty(I^\Delta_j)} \right) \tag{3.3.38a}$$

where we used that $\beta_2 = \|AE\|_{L^\infty(I)}$ and $\gamma = \|c\|_{L^\infty(I)}$. Employing the estimates (3.3.2b) and (3.3.2c) we get

$$\|u - \mathcal{I}^p_\Delta u\|^2_{\mathcal{U}} \leq \mathscr{C} \sum_{j=1}^{N(\Delta)} h_{I^\Delta_j} (\beta_2 (h_{I^\Delta_j})^{2\mu} + \gamma (h_{I^\Delta_j})^{2\mu+2}) \left\| \frac{d^r u}{dx^r} \right\|^2_{L^\infty(I)} \tag{3.3.38b}$$

where $\mathscr{C}$ is the biggest of the constants in (3.3.2b) and (3.3.2c). Further, realizing that

$$\beta_2 (h_{I^\Delta_j})^{2\mu} + \gamma (h_{I^\Delta_j})^{2\mu+2} \leq \mathscr{C} (h^\Delta)^{2\mu} \tag{3.3.38c}$$

and that $\sum_{j=1}^{N(\Delta)} h_{I^\Delta_j} = l$ we obtain

$$\|u - \mathcal{I}^p_\Delta u\|_{\mathcal{U}} \leq \mathscr{C} (h^\Delta)^\mu \left\| \frac{d^r u}{dx^r} \right\|_{L^\infty(I)} \tag{3.3.38d}$$

with $\mu = \min(p, r-1)$, $1 \leq r \leq p+1$, and $\mathscr{C}$ depending only on r, p, β_2, and γ. □

Theorem 3.3.4 *Let $u \in C^0(I)$ be such that*

$$\left\| \frac{d^r u}{dx^r} \right\|_{L^2(I)} < \infty \tag{3.3.39a}$$

with

$$1 \leq r \leq p+1. \tag{3.3.39b}$$

Then

$$\|u - \mathcal{I}^p_\Delta u\|_{\mathcal{U}} \leq \mathscr{C} (h^\Delta)^\mu \left\| \frac{d^r u}{dx^r} \right\|_{L^2(I)} \tag{3.3.39c}$$

where

$$\mu = \min(p, r-1) = r - 1 \tag{3.3.39d}$$

and $\mathscr{C}$ depends only on β_2, γ, p, and r.

Let us now use the above results to derive an a-priori estimate for $\|e_{S^p_\Delta}\|_{\mathcal{U}}$, the energy norm of the error in the finite element solution.

Corollary 3.3.1 *Under the assumption that*

$$\left\| \frac{d^r}{dx^r}(u_{EX}) \right\|_{L^2(I)} < \infty, \tag{3.3.40a}$$

with $1 \le r \le p+1$, *we have*

$$\|e_{S^p_\Delta}\|_{\mathcal{U}} \le \mathscr{C}(h^\Delta)^\mu \left\| \frac{d^r}{dx^r}(u_{EX}) \right\|_{L^2(I)}, \tag{3.3.40b}$$

where $\mu = \min(p, r-1) = r-1$.

Proof Recalling the best-approximation property of the finite element method (Theorem 3.1.2), we have

$$\|e_{S^p_\Delta}\|_{\mathcal{U}} = \|u_{EX} - u_{S^p_\Delta}\|_{\mathcal{U}} = \min_{\chi \in S^p_\Delta} \|u_{EX} - \chi\|_{\mathcal{U}} \le \|u_{EX} - \mathcal{I}^p_\Delta u_{EX}\|_{\mathcal{U}}, \tag{3.3.41}$$

and (3.3.40b) follows from (3.3.39c). □

The estimate (3.3.40b) is called a-priori because it is known prior to the determination of the finite element solution for the particular case of interest. It proves that the finite element method converges as h^Δ tends to zero, with rate μ dependent on the degree p of the elements, and the smoothness of u_{EX}.

As in Remark 3.3.7, we can also obtain the estimate

$$\|e_{S^p_\Delta}\|_{\mathcal{U}} \le \mathscr{C}(h^\Delta)^\mu \|u_{EX}\|_{H^r(I)}, \tag{3.3.42}$$

which holds under the assumption that u_{EX} satisfies (3.3.40a) for any integer r.

Remark 3.3.8 Corollary 3.3.1 can also be understood as an approximation error estimate, namely:

Given a u_{EX} which satisfies (3.3.40a), for any sequence of meshes Δ, and any positive integer p, there exists a function $u_{S^p_\Delta} \in S^p_\Delta$, such that

$$\|u_{EX} - u_{S^p_\Delta}\|_{\mathcal{U}} \le \mathscr{C}(h^\Delta)^\mu \left\| \frac{d^r}{dx^r}(u_{EX}) \right\|_{L^2(I)}. \tag{3.3.43}$$

Note that, in general, $u_{S^p_\Delta}$ *is different to the Lagrangian interpolant.*

From Corollary 3.3.1, it follows that if u_{EX} is sufficiently smooth, namely, if

$$\left\| \frac{d^{p+1}}{dx^{p+1}}(u_{EX}) \right\|_{L^2(I)} < \infty \tag{3.3.44a}$$

we have

$$\|u_{EX} - u_{S^p_\Delta}\|_{\mathcal{U}} \le \mathscr{C}(h^\Delta)^p. \tag{3.3.44b}$$

This means that the rate of convergence of the energy norm of the error $\|e_{S^p_\Delta}\|_{\mathcal{U}}$, is at least p. Let us now prove that, with the exception of the case that $u_{EX} \in S^p_\Delta$, the rate of convergence of $\|e_{S^p_\Delta}\|_{\mathcal{U}}$ cannot exceed p.

Theorem 3.3.5 *Let* u_{EX} *be such that*

$$\left\| \frac{d^{p+2}}{dx^{p+2}}(u_{EX}) \right\|_{L^2(I)} < \infty \tag{3.3.45a}$$

and

$$u_{EX} \notin S^p_\Delta. \tag{3.3.45b}$$

Then for any sequence of uniform meshes Δ *we have*

$$\mathscr{C}(h^\Delta)^p \leq \|e_{S^p_\Delta}\|_{\mathscr{U}} \tag{3.3.45c}$$

where $\mathscr{C}$ *depends on* u_{EX}.

Proof We will prove (3.3.45c) in the case of linear elements ($p = 1$). From the assumption that

$$\left\| \frac{d^3}{dx^3}(u_{EX}) \right\|_{L^2(I)} < \infty \tag{3.3.46a}$$

it follows that

$$\frac{d^2}{dx^2}(u_{EX}) \in C^0(I). \tag{3.3.46b}$$

Further, because $u_{EX} \notin S^1_\Delta$, there exists a $J = [a, b] \subset I$ such that

$$0 < \alpha < \min_{x \in J} \left| \frac{d^2}{dx^2}(u_{EX}) \right|. \tag{3.3.46c}$$

Let us now take a mesh with mesh size $h^\Delta < \frac{1}{2}(b - a)$, such that there exists at least one element $I^\Delta_j \subset J$. Then we have

$$\mathscr{C}\left(\alpha(h_{I^\Delta_j})^{3/2}\right)^2 \leq \min_{v \in \mathscr{P}^0(I^\Delta_j)} \left\| \frac{d}{dx}(u_{EX}) - v \right\|^2_{L^2(I^\Delta_j)} \tag{3.3.46d}$$

and hence for any $\chi \in S^1_\Delta$, we have

$$\mathscr{C}\alpha h^\Delta \leq \sqrt{\sum_{I^\Delta_j \subset J} \left\| \frac{d}{dx}(u_{EX}) - \frac{d\chi}{dx} \right\|^2_{L^2(I^\Delta_j)}} \leq \left\| \frac{d}{dx}(u_{EX}) - \frac{d\chi}{dx} \right\|_{L^2(I)}. \tag{3.3.46e}$$

□

Remark 3.3.9 The estimate

$$\|e_{S^p_\Delta}\|_{\mathscr{U}} \leq \mathscr{C}(h^\Delta)^\mu \tag{3.3.47}$$

is a typical *a-priori upper bound* for the error in the finite element method. It is called a-priori because it utilizes information about the input data, the meshes, and the smoothness of the solution, which is known prior to the computation of the finite element solution. A-priori estimates are important because they indicate the factors which affect the accuracy of the finite element solution.

Remark 3.3.10 The lower a-priori estimate (3.3.45c) also holds for arbitrary mesh with $h^\Delta = \max_{I^\Delta_j \subseteq J} h_{I^\Delta_j}$.

Remark 3.3.11 A-priori estimates are sometimes used for the estimation of the error in a concrete case of interest by replacing the unknown function $(d^r/dx^r)(u_{EX})$ by an approximation $\mathcal{A}^*_{d^r u/dx^r}(u_{S^p_\Delta})$, which is determined by a post-processing of the finite element solution, namely

$$\|e_{S^p_\Delta}\|_{\mathcal{U}} \approx \mathcal{E} = \mathcal{C} h^\mu \big\|\mathcal{A}^*_{d^r u/dx^r}(u_{S^p_\Delta})\big\|_{L^2(I)}. \tag{3.3.48}$$

In general, estimates of this type are not accurate because:

- the constant $\mathcal{C}$ can only be specified a-priori for an entire class of solutions and may be pessimistic for the solution of interest
- the approximation of $d^r/dx^r(u_{EX})$ by $\mathcal{A}^*_{d^r u/dx^r}(u_{S^p_\Delta})$ may be very poor.

Above we derived estimates for the interpolation error for any function u, under the general assumption that it has bounded or square integrable derivatives of a certain order. We will now consider the special case of the function x^α, which has analogous smoothness as the solutions of the two-dimensional model problems in polygonal domains.

Lemma 3.3.1 *For any $\alpha > 0$, $\alpha \neq$ integer, we have*

$$\|x^\alpha - \mathcal{I}^p_{I_1^\Delta} x^\alpha\|_{L^\infty(I_1^\Delta)} \leq \mathcal{C}(h_{I_1^\Delta})^{\bar{\mu}+1} \tag{3.3.49a}$$

and

$$\|x^\alpha - \mathcal{I}^p_{I_1^\Delta} x^\alpha\|_{L^2(I_1^\Delta)} \leq \mathcal{C}\sqrt{h_{I_1^\Delta}}\,(h_{I_1^\Delta})^{\bar{\mu}+1} \tag{3.3.49b}$$

where

$$\bar{\mu} = \min(p, \alpha - 1). \tag{3.3.49c}$$

Further, for $\alpha > 1$ we have

$$\left\|\frac{d}{dx}(x^\alpha) - \frac{d}{dx}(\mathcal{I}^p_{I_1^\Delta} x^\alpha)\right\|_{L^\infty(I_1^\Delta)} \leq \mathcal{C}(h_{I_1^\Delta})^{\bar{\mu}} \tag{3.3.49d}$$

and for $\alpha > \frac{1}{2}$

$$\left\|\frac{d}{dx}(x^\alpha) - \frac{d}{dx}(\mathcal{I}^p_{I_1^\Delta} x^\alpha)\right\|_{L^2(I_1^\Delta)} \leq \mathcal{C}\sqrt{h_{I_1^\Delta}}\,(h_{I_1^\Delta})^{\bar{\mu}}. \tag{3.3.49e}$$

In all the above estimates $\mathcal{C}$ depends only on α and p.

Proof Let

$$x = \frac{h_{I_1^\Delta}}{2}(1 + \tilde{x}), \quad \tilde{x} \in \tilde{I} = (-1, 1) \tag{3.3.50a}$$

be the mapping which maps $\tilde{I}$ onto I_1^Δ. Then

$$x^\alpha = \left(\frac{h_{I_1^\Delta}}{2}\right)^\alpha (1 + \tilde{x})^\alpha \tag{3.3.50b}$$

and hence

$$\|x^\alpha - \mathcal{I}^p_{I_1^\Delta} x^\alpha\|_{L^\infty(I_1^\Delta)} = \frac{1}{2^\alpha}(h_{I_1^\Delta})^\alpha \|(1+\tilde{x})^\alpha - \tilde{\mathcal{I}}^p(1+\tilde{x})^\alpha\|_{L^\infty(\tilde{I})} \tag{3.3.50c}$$

where $\tilde{\mathcal{I}}^p$ is the interpolation operator over $\tilde{I}$. It follows that

$$\|x^\alpha - \mathcal{I}^p_{I_1^\Delta} x^\alpha\|_{L^\infty(I_1^\Delta)} = \mathscr{C}(\alpha,p)(h_{I_1^\Delta})^\alpha \tag{3.3.51a}$$

where

$$\mathscr{C}(\alpha,p) = \frac{1}{2^\alpha}\left\|(1+\tilde{x})^\alpha - \tilde{\mathcal{I}}^p\big((1+\tilde{x})^\alpha\big)\right\|_{L^\infty(\tilde{I})}. \tag{3.3.51b}$$

□

The proof of (3.3.49b), (3.3.49d), and (3.3.49e) follow similar steps.

Remark 3.3.12 Note that for $\frac{1}{2} < \alpha < 1$, $(d/dx)(x^\alpha)$ is unbounded at $x = 0$, and (3.3.49d) has no meaning, while $(d/dx)(x^\alpha) \in L^2(I)$ for $\frac{1}{2} < \alpha$, and hence (3.3.49e) holds for $\frac{1}{2} < \alpha$.

Let us now derive an estimate for the global energy norm of the interpolation error $\|x^\alpha - \mathcal{I}^p_\Delta x^\alpha\|_{\mathscr{U}}$.

Theorem 3.3.6 *Let Δ be a sequence of uniform meshes. For $\alpha > \frac{1}{2}$, $\alpha \not\equiv$ integer, we have*

$$\|x^\alpha - \mathcal{I}^p_\Delta x^\alpha\|_{\mathscr{U}} \le \begin{cases} \mathscr{C}h^\mu, & \textit{for } p \ne \alpha - \frac{1}{2} \\ \mathscr{C}h^\mu\sqrt{|\ln h|}, & \textit{for } p = \alpha - \frac{1}{2} \end{cases} \tag{3.3.52a}$$

where

$$\mu = \min\big(p, \alpha - \tfrac{1}{2}\big). \tag{3.3.52b}$$

Proof Employing Lemma 3.3.1, and Theorem 3.3.1, we have

$$\begin{aligned}\|x^\alpha - \mathcal{I}^p_\Delta x^\alpha\|^2_{\mathscr{U}} \le{}& \beta_{I_1^\Delta} \underbrace{\left\|\frac{d}{dx}(x^\alpha - \mathcal{I}^p_{I_1^\Delta} x^\alpha)\right\|^2_{L^2(I_1^\Delta)}}_{\le \mathscr{C}(h_{I_1^\Delta})^{2\bar{\mu}+1}} + \gamma_{I_1^\Delta} \underbrace{\|x^\alpha - \mathcal{I}^p_{I_1^\Delta} u\|^2_{L^2(I_1^\Delta)}}_{\le \mathscr{C}(h_{I_1^\Delta})^{2\bar{\mu}+3}} \\ &+ \sum_{j=2}^{N(\Delta)} \Bigg(\beta_{I_j^\Delta} \underbrace{\left\|\frac{d}{dx}(x^\alpha - \mathcal{I}^p_{I_j^\Delta} x^\alpha)\right\|^2_{L^2(I_j^\Delta)}}_{\le \mathscr{C}(h_{I_j^\Delta})^{2p+1}\left\|\frac{d^{p+1}}{dx^{p+1}}(x^\alpha)\right\|^2_{L^\infty(I_j^\Delta)}} \\ &+ \gamma_{I_j^\Delta} \underbrace{\|x^\alpha - \mathcal{I}^p_{I_j^\Delta} x^\alpha\|^2_{L^2(I_j^\Delta)}}_{\le \mathscr{C}(h_{I_j^\Delta})^{2p+3}\left\|\frac{d^{p+1}}{dx^{p+1}}(x^\alpha)\right\|^2_{L^\infty(I_j^\Delta)}} \Bigg)\end{aligned} \tag{3.3.53a}$$

where we used the notations $\beta_{I_j^\Delta} = \|AE\|_{L^\infty(I_j^\Delta)}$ and $\gamma_{I_j^\Delta} = \|c\|_{L^\infty(I_j^\Delta)}$. Assuming that

$$p < \alpha - \tfrac{1}{2} \tag{3.3.53b}$$

we obtain

$$\|x^\alpha - \mathcal{I}_\Delta^p x^\alpha\|_{\mathcal{U}}^2 \le \mathscr{C}\left(h^{2\min(p+1/2,\alpha-1/2)}(1+h^2) \right.$$
$$\left. + \sum_{j=2}^{N(\Delta)} \left(((j-1)h)^{\alpha-(p+1)}\right)^2 h(h^{2p}+h^{2p+2}) \right) \tag{3.3.53c}$$

and neglecting the higher-order terms we get

$$\|x^\alpha - \mathcal{I}_\Delta^p x^\alpha\|_{\mathcal{U}}^2 \le \mathscr{C}h^{2\mu} \tag{3.3.53d}$$

where $\mu = \min(p, \alpha - \frac{1}{2})$. The same result also follows for the case $p > \alpha - \frac{1}{2}$.

The estimate in the case $p = \alpha - \frac{1}{2}$ follows by employing that

$$\sum_{j=1}^{N} \frac{1}{j} \le \mathscr{C} \ln(N). \tag{3.3.54}$$

□

The estimate

$$\|x^\alpha - \mathcal{I}_\Delta^p x^\alpha\|_{\mathcal{U}} \le \mathscr{C}h^\mu \tag{3.3.55a}$$

can be derived for $p \ne \alpha - \frac{1}{2}$ by a formal application of the estimate

$$\|u_{EX} - \mathcal{I}_\Delta^p u_{EX}\|_{\mathcal{U}} \le \mathscr{C}h^\mu \left\| \frac{d^r}{dx^r}(u_{EX}) \right\|_{L^2(I)} \tag{3.3.55b}$$

as if the function $u_{EX} = x^\alpha$ *had* $\mu = \alpha + \frac{1}{2}$ *square-integrable derivatives.* This approach can be made rigorous by constructing a more general estimate which employs *non-integer* derivatives of u_{EX}. Let us explain how non-integer derivatives of a function can be defined.

We start by noting that any

$$u \in H_0^1(I) = \left\{ u \in H^1(I) \,\middle|\, u(0) = u(1) = 0 \right\} \tag{3.3.56a}$$

where $I \overset{\text{def}}{=} (0,1)$ can be written in the form of its sine series expansion, namely

$$u \equiv \sum_{k=1}^{\infty} a_k \sin(k\pi x) \tag{3.3.56b}$$

and we have

$$\|u\|_{H^0(I)} \overset{\text{def}}{=} \|u\|_{L^2(I)} \overset{\text{def}}{=} \sqrt{\frac{1}{2}\sum_{k=1}^{\infty} a_k^2} = \sqrt{\frac{1}{2}\sum_{k=1}^{\infty} a_k^2 (1+\pi^2 k^2)^0} < \infty \tag{3.3.56c}$$

and

$$\|u\|_{H^1(I)} \stackrel{\text{def}}{=} \sqrt{\|u\|^2_{L^2(I)} + \left\|\frac{du}{dx}\right\|^2_{L^2(I)}} \stackrel{\text{def}}{=} \sqrt{\frac{1}{2}\sum_{k=1}^{\infty} a_k^2(1+\pi^2k^2)^1} < \infty. \tag{3.3.56d}$$

The above could be understood, respectively, as the definition of the H^0-norm and the H^1-norm for functions in $H_0^1(I)$, namely

$$\|u\|_{H^l(I)} \stackrel{\text{def}}{=} \sqrt{\frac{1}{2}\sum_{k=1}^{\infty} a_k^2(1+\pi^2k^2)^l} \tag{3.3.56e}$$

for $l = 0$ and $l = 1$. We will generalize this definition to non-integer l, by letting

$$\|u\|_{H^\theta(I)} \stackrel{\text{def}}{=} \sqrt{\frac{1}{2}\sum_{k=1}^{\infty} a_k^2(1+\pi^2k^2)^\theta}, \quad 0 \le \theta \le 1. \tag{3.3.57}$$

It can be shown that $\|\cdot\|_{H^\theta(I)}$ is a norm on $H_0^1(I)$, and we will let $H_0^\theta(I)$ denote the completion of $H_0^1(I)$ with respect to this norm. We will call $H_0^\theta(I)$ *the space of functions with square-integrable fractional derivatives of order* θ.

Let us now gain some understanding of the character of functions which belong to $H_0^\theta(I)$. The following example shows that

$$H_0^\theta(I) \not\subseteq C^0(I), \quad \theta < \tfrac{1}{2}. \tag{3.3.58}$$

Example 3.3.2 We will let

$$\hbar(x) = \begin{cases} 1, & 0 \le x < \frac{1}{2} \\ -1, & \frac{1}{2} < x \le 1 \end{cases} \tag{3.3.59a}$$

and we will prove that

$$\hbar \in H_0^\theta(I), \quad \theta < \tfrac{1}{2}. \tag{3.3.59b}$$

We have

$$\hbar(x) = \sum_{\substack{k=1 \\ k \text{ odd}}}^{\infty} \frac{4}{\pi k}\sin(k\pi x) \tag{3.3.59c}$$

and noting that the series

$$\|\hbar\|^2_{H_0^\theta(I)} = \sum_{\substack{k=1 \\ k \text{ odd}}}^{\infty} \frac{8}{\pi^2k^2}(1+\pi^2k^2)^\theta \tag{3.3.59d}$$

converges for $\theta < \frac{1}{2}$, and hence (3.3.59b) holds.

Let us now prove that *for* $\frac{1}{2} < \theta < 1$, *the functions in* $H_0^\theta(I)$ *are continuous.*

Theorem 3.3.7 *Let $H_0^\theta(I)$ denote the completion of $H_0^1(I)$ in the $\|\cdot\|_{H^\theta(I)}$ norm. Then*

$$H^\theta(I) \subseteq C^0(I), \quad \tfrac{1}{2} < \theta < 1. \tag{3.3.60}$$

Proof To prove $H^\theta(I) \subseteq C^0(I)$, we must show that for any

$$u = \sum_{k=1}^{\infty} a_k \sin(kx) \tag{3.3.61a}$$

with

$$\|u\|^2_{H^\theta(I)} = \frac{1}{2}\sum_{k=1}^{\infty} a_k^2 (1+\pi^2 k^2)^\theta < \infty \tag{3.3.61b}$$

we have

$$\lim_{\delta\to 0} |u(x+\delta) - u(x)| = 0 \tag{3.3.61c}$$

at every point $x \in I$. We have

$$|u(x) - u(x+\delta)| = \left|\sum_{k=1}^{\infty} a_k(\sin(k\pi x)(1-\cos(k\pi\delta)) - \cos(k\pi x)\sin(k\pi\delta))\right|. \tag{3.3.62a}$$

Noting that for any ϵ, $0 < \epsilon \leq 1$ we have

$$|\sin(k\pi\delta)| \leq (k\pi\delta)^\epsilon, \qquad |1-\cos(k\pi\delta)| \leq 2(k\pi\delta)^\epsilon \tag{3.3.62b}$$

we obtain

$$\begin{aligned} |u(x) - u(x+\delta)| &\leq \left|\sum_{k=1}^{\infty} a_k 3(k\pi\delta)^\epsilon\right| \\ &= 3\delta^\epsilon \left|\sum_{k=1}^{\infty} a_k (1+\pi^2k^2)^{\theta/2} \frac{(\pi k)^\epsilon}{(1+\pi^2k^2)^{\theta/2}}\right| \end{aligned} \tag{3.3.62c}$$

and employing the Cauchy-Schwarz inequality, we get

$$|u(x) - u(x+\delta)| \leq 3\delta^\epsilon \underbrace{\sqrt{\sum_{k=1}^{\infty} a_k^2(1+\pi^2k^2)^\theta}}_{\|u\|_{H^\theta(I)}} \sqrt{\sum_{k=1}^{\infty} \underbrace{\frac{(\pi k)^{2\epsilon}}{(1+(\pi k)^2)^\theta}}_{\leq (\pi k)^{2(\epsilon-\theta)}}} \tag{3.3.62d}$$

and hence we have proven that

$$|u(x) - u(x+\delta)| \leq 3\delta^\epsilon \|u\|_{H^\theta(I)} \sqrt{\sum_{k=1}^{\infty} (\pi k)^{2(\epsilon-\theta)}}. \tag{3.3.62e}$$

The series $\sum_{k=1}^{\infty}(\pi k)^{2(\epsilon-\theta)}$ converges if

$$2(\epsilon-\theta) < -1 \quad\Longrightarrow\quad \epsilon < \theta - \tfrac{1}{2}. \tag{3.3.63a}$$

Because $\frac{1}{2} < \theta < 1$, we have $\theta - \frac{1}{2} > 0$, and hence it is possible to choose ϵ, such that

$$0 < \epsilon < \theta - \tfrac{1}{2} \tag{3.3.63b}$$

and letting

$$\mathscr{C}(\theta) \stackrel{\text{def}}{=} \sqrt{\sum_{k=1}^{\infty} \pi^2 k^{2(\epsilon-\theta)}} \tag{3.3.63c}$$

which exists because of (3.3.63b), we have

$$|u(x) - u(x+\delta)| \leq \sqrt{\sum_{k=1}^{\infty}(\pi k)^{2(\epsilon-\theta)}}\|u\|_{H^\theta(I)}\delta^\epsilon \leq \mathscr{C}(\theta)\|u\|_{H^\theta(I)}\delta^\epsilon \tag{3.3.63d}$$

and the continuity of u has been proven. □

In Theorem 3.3.2 we have proven that

$$\|u - \mathcal{I}_\Delta^1 u\|_{L^2(I)} \leq \mathscr{C}h\|u\|_{H^1(I)} \quad \forall u \in H^1(I). \tag{3.3.64a}$$

Note that

$$H_0^1(I) \subseteq C^0(I) \tag{3.3.64b}$$

and the interpolant $\mathcal{I}_\Delta^1 u$ is well defined for any $u \in H^1(I)$. We will now let $\mathcal{B}_\Delta^p u \in S_\Delta^p$ denote the best approximation of u from $H_0^\theta(I)$ in the L^2 norm, and we will extend the interpolation estimate (3.3.64a) to functions from $H^\theta(I)$, $0 \leq \theta \leq 1$, as an approximation estimate, namely:

For any $u \in H_0^\theta(I)$, we have

$$\|u - \mathcal{B}_\Delta^p u\|_{L^2(I)} \leq \mathscr{C}h^\theta\|u\|_{H^\theta(I)}. \tag{3.3.65}$$

We will employ the following lemma.

Lemma 3.3.2 *Given $u \in H_0^\theta(I)$, $0 < \theta < 1$, and $0 < t < 1$, there exists a splitting*

$$u = u_1^{(t)} + u_2^{(t)} \tag{3.3.66a}$$

with $u_1^{(t)}$, $u_2^{(t)}$, depending on t, and such that

$$u_1^{(t)} \in L^2(I) \equiv H^0(I): \quad \|u_1^{(t)}\|_{L^2(I)} \leq \mathscr{C}t^\theta\|u\|_{H^\theta(I)} \tag{3.3.66b}$$

and

$$u_2^{(t)} \in H^1(I): \quad \|u_2^{(t)}\|_{H^1(I)} \leq \mathscr{C}t^{\theta-1}\|u\|_{H^\theta(I)} \tag{3.3.66c}$$

with $\mathscr{C}$ independent of u, t, and θ.

Proof Letting

$$u \equiv \sum_{k=1}^{\infty} a_k \sin(k\pi x) \tag{3.3.67a}$$

and employing the splitting

$$u(x) = \underbrace{\sum_{k=1}^{[1/t]} a_k \sin(k\pi x)}_{u_2^{(t)}(x)} + \underbrace{\sum_{k=[1/t]+1}^{\infty} a_k \sin(k\pi x)}_{u_1^{(t)}(x)} \tag{3.3.67b}$$

where $[1/t]$ is the integer part of $1/t$, and $0 < t < 1$ is given, we have

$$\|u_1^{(t)}\|^2_{L^2(I)} = \frac{1}{2}\sum_{k=[1/t]+1}^{\infty} a_k^2 \leq \underbrace{\frac{1}{(1+\pi^2([1/t]+1)^2)^\theta}}_{\leq \mathscr{C}t^{2\theta}} \underbrace{\frac{1}{2}\sum_{k=[1/t]+1}^{\infty} a_k^2(1+\pi^2k^2)^\theta}_{\leq \|u\|^2_{H^\theta(I)}} \tag{3.3.67c}$$

and hence

$$\|u_1^{(t)}\|^2_{L^2(I)} \leq \mathscr{C}t^{2\theta}\|u\|^2_{H^\theta(I)} \tag{3.3.67d}$$

and

$$\|u_2^{(t)}\|^2_{H^1(I)} = \frac{1}{2}\sum_{k=1}^{[1/t]} a_k^2(1+\pi^2k^2) = \frac{1}{2}\sum_{k=1}^{[1/t]} a_k^2(1+\pi^2k^2)^\theta(1+\pi^2k^2)^{1-\theta} \tag{3.3.67e}$$

from which we get

$$\|u_2^{(t)}\|^2_{H^1(I)} \leq \underbrace{\left(1+\frac{\pi^2}{t^2}\right)^{1-\theta}}_{\leq \mathscr{C}t^{-2(1-\theta)}} \|u\|^2_{H^\theta(I)} \leq \mathscr{C}t^{-2(1-\theta)}\|u\|^2_{H^\theta(I)} \tag{3.3.67f}$$

and the lemma has been proven. □

We will now use the splitting $u = u_1^{(t)} + u_2^{(t)}$ to generalize the interpolation estimate to $H_0^\theta(I)$. Given any $u \in H_0^\theta(I)$, $0 < \theta < 1$, note that

$$u_2^{(t)} \in H^1(I) \subseteq C^0(I) \tag{3.3.68a}$$

and $\mathcal{I}_\Delta^1 u_2^{(t)}$ is well defined for any t, $0 < t < 1$, and we have

$$\|u_2^{(t)} - \mathcal{I}_\Delta^p u_2^{(t)}\|_{L^2(I)} \leq \mathscr{C}h^\Delta\|u_2^{(t)}\|_{H^1(I)}. \tag{3.3.68b}$$

Employing the triangle inequality

$$\|u - \mathcal{I}_\Delta^p u_2^{(t)}\|_{L^2(I)} \leq \|u_1^{(t)}\|_{L^2(I)} + \|u_2^{(t)} - \mathcal{I}_\Delta^p u_2^{(t)}\|_{L^2(I)} \tag{3.3.68c}$$

we get

$$\|u - \mathcal{I}^p_\Delta u_2^{(t)}\|_{L^2(I)} \le \|u_1^{(t)}\|_{L^2(I)} + \mathcal{C}h^\Delta \|u_2^{(t)}\|_{H^1(I)}. \tag{3.3.68d}$$

Using now that

$$\|u_1^{(t)}\|_{L^2(I)} \le \mathcal{C}t^\theta \|u\|_{H^\theta(I)} \qquad \|u_2^{(t)}\|_{H^1(I)} \le \mathcal{C}t^{\theta-1}\|u\|_{H^\theta(I)} \tag{3.3.68e}$$

we get

$$\|u - \mathcal{I}^p_\Delta u_2^{(t)}\|_{L^2(I)} \le \mathcal{C}\big(t^\theta + \mathcal{C}h^\Delta t^{\theta-1}\big)\|u\|_{H^\theta(I)} \tag{3.3.68f}$$

and choosing $t = h^\Delta$ we obtain

$$\|u - \mathcal{I}^p_\Delta u_2^{(t)}\|_{L^2(I)} \le \mathcal{C}(h^\Delta)^\theta \|u\|_{H^\theta(I)}. \tag{3.3.69a}$$

It follows that for any $u \in H^\theta(I)$, $0 < \theta < 1$, we have

$$\|u - \mathcal{B}^p_\Delta u\|_{L^2(I)} = \min_{\chi \in S^p_\Delta} \|u - \chi\|_{L^2(I)} \le \mathcal{C}(h^\Delta)^\theta \|u\|_{H^\theta(I)}. \tag{3.3.69b}$$

The main steps employed in the definition of $H_0^\theta(I)$, and the generalization of the interpolation estimate to $H^\theta(I)$, are:

(a) Any

$$u \in H_0^1(I) \subseteq H^0(I) \equiv L^2(I) \tag{3.3.70a}$$

can be uniquely represented by the sine series

$$u \equiv \sum_{k=1}^\infty a_k \sin(k\pi x) \tag{3.3.70b}$$

and we have

$$\|u\|_{H^l(I)} = \sqrt{\frac{1}{2}\sum_{k=1}^\infty a_k^2(1+\pi^2k^2)^l} \tag{3.3.70c}$$

for $l = 0$ and $l = 1$. Extending this expression to non-integer l, we define

$$\|u\|_{H^\theta(I)} \overset{\text{def}}{=} \sqrt{\frac{1}{2}\sum_{k=1}^\infty a_k^2(1+\pi^2k^2)^\theta}. \tag{3.3.70d}$$

$\|\cdot\|_{H^\theta(I)}$ is a norm in $H_0^1(I)$, and we let $H^\theta(I)$ denote the completion of $H_0^1(I)$ in $L^2(I)$ in this norm.

(b) We then have

$$H_0^1(I) \subseteq H^\theta(I) \subseteq H^0(I) \equiv L^2(I) \tag{3.3.71a}$$

and that

$$H^\theta(I) \not\subseteq H_0^1(I). \tag{3.3.71b}$$

Further, given any h, $0 < h < 1$, any $u \in H_0^\theta(I)$ can be split as follows:

$$u = u_1^{(h)} + u_2^{(h)} \tag{3.3.72a}$$

with

$$u_1^{(h)} \in H^0(I): \quad \|u_1^{(h)}\|_{H^0(I)} \leq \mathscr{C}h^\theta \|u\|_{H^\theta(I)} \tag{3.3.72b}$$

and

$$u_2^{(h)} \in H_0^1(I): \quad \|u_2^{(h)}\|_{H^1(I)} \leq \mathscr{C}h^{\theta-1} \|u\|_{H^\theta(I)}. \tag{3.3.72c}$$

(c) Employing this splitting we get the best-approximation estimate

$$\min_{\chi \in S_\Delta^p} \|u - \chi\|_{H^0(I)} \leq \|u_1^{(h)}\|_{H^0(I)} + \underbrace{\min_{\chi \in S_\Delta^p} \|u_2^{(h)} - \chi\|_{H^0(I)}}_{\mathscr{C}h^\Delta \|u_2^{(h)}\|_{H^1(I)}} \leq \mathscr{C}(h^\Delta)^\theta \|u\|_{H^\theta(I)}. \tag{3.3.73}$$

We will now give a more general approach for the construction of spaces of functions with fractional derivatives. This approach is known as the K method (see [Bergh and Löfstrom, 1976]).

Let H_1 and H_2, $H_2 \subset H_1$, be normed spaces with corresponding norms $\|\cdot\|_{H_1}$ and $\|\cdot\|_{H_2}$ which satisfy

$$\|u\|_{H_1} \leq \|u\|_{H_2} \quad \forall u \in H_2 \subset H_1, \tag{3.3.74a}$$

and let

$$K^2(u,t) \stackrel{\text{def}}{=} \inf_{\substack{u_i \in H_i, i=1,2 \\ u = u_1^{(t)} + u_2^{(t)}}} \left(\|u_1^{(t)}\|_{H_1}^2 + t^2 \|u_2^{(t)}\|_{H_2}^2 \right) \tag{3.3.74b}$$

where $0 < t < \infty$. Note that

$$K^2(u,t) \leq K^2(u,0) = \|u\|_{H_1}^2 \tag{3.3.74c}$$

and hence $K(u,t)$ exists for any $u \in H_1$. Next, for any θ, $0 < \theta < 1$, let

$$\|u\|_{H^\theta(H_1,H_2)} \stackrel{\text{def}}{=} \sqrt{\int_0^\infty (t^{-2\theta} K^2(u,t)) \frac{1}{t}\, dt}. \tag{3.3.74d}$$

$\|\cdot\|_{H^\theta(H_1,H_2)}$ is a norm on H_2, and we will denote by $H^\theta(H_1, H_2)$ the completion of H_2 in this norm. The space $H^\theta(H_1, H_2)$ is known as the *interpolated space between* H_1 *and* H_2.

As an example, we will consider

$$H_2 = H_0^1(I) \subset H_1 = H^0(I) \equiv L^2(I) \tag{3.3.75}$$

with $\|\cdot\|_{H_1} = \|\cdot\|_{H^0(I)}$ and $\|\cdot\|_{H_2} = \|\cdot\|_{H^1(I)}$, and we will prove that the norm $\|\cdot\|_{H^\theta(H^0(I),H_0^1(I))}$ is equivalent to the norm $\|\cdot\|_{H^\theta(I)}$.

Theorem 3.3.8 *For any $0 < \theta < 1$, there exist $\mathscr{C}_1(\theta)$, $\mathscr{C}_2(\theta)v$, such that*

$$\mathscr{C}_1(\theta)\|u\|_{H^\theta(I)} \leq \|u\|_{H^\theta(H_1,H_2)} \leq \mathscr{C}_2(\theta)\|u\|_{H^\theta(I)} \quad \forall u \in H_0^\theta(I). \tag{3.3.76}$$

Proof Recall that any $u \in H_0^\theta(I)$ has a unique representation in the form $u \equiv \sum_{k=1}^\infty a_k \sin(k\pi x)$, and let

$$b_k^{(1)} + b_k^{(2)} = 1, \quad k = 1, 2, \ldots. \tag{3.3.77a}$$

Employing the splitting

$$u = u_1 + u_2, \qquad u_i = \sum_{k=1}^\infty b_k^{(i)} a_k \sin(k\pi x), \quad i = 1, 2 \tag{3.3.77b}$$

we obtain

$$K^2(u,t) = \inf_{b_k^{(1)}+b_k^{(2)}=1} \frac{1}{2}\left(\sum_{k=1}^\infty a_k^2((b_k^{(1)})^2 + t^2(b_k^{(2)})^2(1+\pi^2k^2))\right). \tag{3.3.77c}$$

The above infimum is achieved for

$$b_k^{(2)} = \frac{1}{1+t^2(1+\pi^2k^2)}, \qquad b_k^{(1)} = 1 - b_k^{(2)} \tag{3.3.78a}$$

and we get

$$K^2(u,t) = \frac{1}{2}\sum_{k=1}^\infty a_k^2 \frac{t^2(1+\pi^2k^2)}{1+t^2(1+\pi^2k^2)}. \tag{3.3.78b}$$

Therefore we have

$$\begin{aligned}\|u\|^2_{H^\theta(L^2(I),H_0^1(I))} &= \int_0^\infty (t^{-2\theta}K^2(u,t))\frac{1}{t}\,dt \\ &= \frac{1}{2}\sum_{k=1}^\infty a_k^2 \int_0^\infty \frac{t^{1-2\theta}(1+\pi^2k^2)}{1+t^2(1+\pi^2k^2)}\,dt.\end{aligned} \tag{3.3.78c}$$

To prove the equivalence $\|u\|_{H^\theta(I)} \approx \|u\|_{H^\theta(L^2(I),H^1(I))}$, it is sufficient to show that there exist positive constants $\mathscr{C}_1$ and $\mathscr{C}_2$ such that

$$\begin{aligned}\mathscr{C}_1\sum_{k=1}^\infty a_k^2(1+\pi^2k^2)^\theta &\leq \sum_{k=1}^\infty a_k^2 \int_0^\infty \frac{t^{1-2\theta}(1+\pi^2k^2)}{1+t^2(1+\pi^2k^2)}\,dt \\ &\leq \mathscr{C}_2\sum_{k=1}^\infty a_k^2(1+\pi^2k^2)^\theta.\end{aligned} \tag{3.3.79}$$

This is proven in Lemma 3.3.3. □

Lemma 3.3.3 *For any θ, $0 < \theta < 1$, and integer k, $1 \le k < \infty$ we have*

$$\mathscr{C}_1(\theta)(1+\pi^2k^2)^\theta \le \int_0^\infty \frac{t^{1-2\theta}(1+\pi^2k^2)}{1+t^2(1+\pi^2k^2)}dt \le \mathscr{C}_2(\theta)(1+\pi^2k^2)^\theta \tag{3.3.80}$$

with $\mathscr{C}_1$ and $\mathscr{C}_2$ independent of k.

Proof Let

$$\varphi(t;\theta,k) = \frac{t^{1-2\theta}(1+\pi^2k^2)}{1+t^2(1+\pi^2k^2)} \tag{3.3.81a}$$

and let us consider the partition

$$\int_0^\infty \varphi(t;\theta,k)\,dt = I_1(\theta,k) + I_2(\theta,k) \tag{3.3.81b}$$

where

$$I_1(\theta,k) = \int_0^{1/\sqrt{1+\pi^2k^2}} \varphi(t;\theta,k)\,dt, \qquad I_2(\theta,k) = \int_{1/\sqrt{1+\pi^2k^2}}^\infty \varphi(t;\theta,k)\,dt. \tag{3.3.81c}$$

Then, for $0 \le t \le 1/\sqrt{1+\pi^2k^2}$, we have

$$\frac{1}{2}t^{1-2\theta}(1+\pi^2k^2) \le \frac{t^{1-2\theta}(1+\pi^2k^2)}{1+t^2(1+\pi^2k^2)} \le t^{1-2\theta}(1+\pi^2k^2) \tag{3.3.81d}$$

and, for $1/\sqrt{1+\pi^2k^2} \le t < \infty$,

$$\frac{1}{2}t^{-1-2\theta} \le \frac{t^{1-2\theta}(1+\pi^2k^2)}{1+t^2(1+\pi^2k^2)} \le t^{-1-2\theta}. \tag{3.3.81e}$$

Hence

$$\mathscr{C}_1(1+\pi^2k^2)^\theta \le I_1(\theta,k) \le \mathscr{C}_2(1+\pi^2k^2)^\theta \tag{3.3.81f}$$

and

$$\mathscr{C}_1(1+\pi^2k^2)^\theta \le I_2(\theta,k) \le \mathscr{C}_2(1+\pi^2k^2)^\theta \tag{3.3.81g}$$

and the lemma is proven by adding the last two inequalities. □

Remark 3.3.13 Various other properties of the norm $\|\cdot\|_{H^\theta(H_1,H_2)}$, and, equivalently, of the norm $\|\cdot\|_{H^\theta}$, may be found in [Bergh and Löfstrom, 1976]. For example, we have

$$\|u\|_{H^\theta(H_1,H_2)} \le \mathscr{C}(\theta)\|u\|_{H_1}^{1-\theta}\|u\|_{H_2}^\theta \quad \forall u \in H_2. \tag{3.3.82}$$

Remark 3.3.14 Thus far, we discussed the interpolation spaces between $H^0(I)$ and $H^1(I)$. Similarly, we can define interpolated spaces between $H^1(I)$ and $H^2(I)$, $H_0^1(I)$ and $H_0^1(I) \cap H^2(I)$, $H_0^1(I)$ and $H_0^2(I)$, etc. The case $\theta = \frac{1}{2}$ is exceptional, and it requires special attention. A complete discussion of the theory of interpolated spaces is given in [Bergh and Löfstrom, 1976].

The norm $\|\cdot\|_{H^\theta(H_1,H_2)}$ is known as the *fractional Sobolev norm.* We will now introduce another norm which is more suitable for analyzing finite element computations.

For any θ, $0 < \theta < 1$, we will let

$$\|u\|_{B^\theta(H_1,H_2)} \stackrel{\text{def}}{=} \sqrt{\sup_{0<t<\infty} \left(t^{-\theta}K(u,t)\right)^2}. \tag{3.3.83}$$

It can be proven that $\|\cdot\|_{B^\theta(H_1,H_2)}$ is a norm on H_2, and is known as the *Besov norm.* The completion of $H_2 = H^1(I)$ with respect to $\|\cdot\|_{B^\theta(H_1,H_2)}$ will be denoted by $B^\theta(H_1, H_2)$, and is known as the *Besov space.*

We will now address the relationship between the Besov norm $\|\cdot\|_{B^\theta(H_1,H_2)}$ and the fractional Sobolev norm $\|\cdot\|_{H^\theta(H_1,H_2)}$.

Theorem 3.3.9 *Let H_1, H_2 be normed spaces with $H_2 \subset H_1$, and let $u \in H^\theta(H_1, H_2)$, $0 < \theta < 1$. Then, we have*

$$\|u\|_{B^\theta(H_1,H_2)} \le \|u\|_{H^\theta(H_1,H_2)} \tag{3.3.84a}$$

and

$$\mathscr{C}(\theta,\theta')\|u\|_{H^\theta(H_1,H_2)} \le \|u\|_{B^{\theta'}(H_1,H_2)}, \quad \theta' > \theta \tag{3.3.84b}$$

and hence

$$H^\theta(H_1, H_2) \subset B^\theta(H_1, H_2) \subset H^{\theta-\epsilon}(H_1, H_2). \tag{3.3.84c}$$

The proof is given in [Bergh and Löfstrom, 1976].

Remark 3.3.15 The norm $\|\cdot\|_{H^\theta(I)}$ was introduced to motivate the discussion, and has meaning only for functions which can be represented in the form $u \equiv \sum_{k=1}^{\infty} a_k \sin(k\pi x)$. On the other hand, the norms $\|\cdot\|_{H^\theta(H_1,H_2)}$ and $\|\cdot\|_{B^\theta(H_1,H_2)}$ have more general meaning.

Remark 3.3.16 There exist various norms which can be used to define Sobolev spaces with fractional derivatives. One of the oldest norms is the Aronszajn-Slobodetskii norm, which is given by

$$\|v\|_{H^\theta_{\text{AS}}(H^0(I),H^1(I))} \stackrel{\text{def}}{=} \sqrt{\int_0^1\int_0^1 \frac{(v(x)-v(y))^2}{|x-y|^{2\theta+1}}\,dx\,dy} \tag{3.3.85}$$

and is equivalent to the fractional Sobolev norm $\|\cdot\|_{H^\theta(H_1,H_2)}$, given above. As we will see below, the evaluation of some of these norms for the functions of interest may be rather difficult.

We will now give some insight about the properties of the functions in $H^\theta(H^0(I), H^1(I))$ and $B^\theta(H^0(I), H^1(I))$.

Example 3.3.3 Let us consider the *Heaviside step function*

$$\mathscr{h}(x;x_0) = \begin{cases} 0, & 0 < x < x_0 \\ 1, & x_0 \le x < 1 \end{cases} \tag{3.3.86a}$$

where $0 < x_0 < 1$, and let us determine the value of θ for which we have

$$\hbar(\cdot\,; x_0) \in B^\theta(H^0(I), H^1(I)). \tag{3.3.86b}$$

Let us introduce the *regularized Heaviside step function*

$$\hbar_\delta(x; x_0) = \begin{cases} 0, & 0 < x \le x_0 - \frac{1}{2}\delta \\ 1, & x_0 + \frac{1}{2}\delta < x < 1 \\ \dfrac{1}{\delta}\Big(x - \Big(x_0 - \dfrac{1}{2}\delta\Big)\Big), & x_0 - \frac{1}{2}\delta < x < x_0 + \frac{1}{2}\delta \end{cases} \tag{3.3.87a}$$

and employ the splitting

$$\hbar(\cdot\,; x_0) = \underbrace{(\hbar(\cdot\,; x_0) - \hbar_\delta(\cdot\,; x_0))}_{u_1^{(\delta)} \in H^0(I)} + \underbrace{\hbar_\delta(\cdot\,; x_0)}_{u_2^{(\delta)} \in H^1(I)} = u_1^{(\delta)} + u_2^{(\delta)}. \tag{3.3.87b}$$

Figure 3.3.5 shows the graphs of $\hbar(\cdot\,; \frac{1}{2})$ and $\hbar_\delta(\cdot\,; \frac{1}{2})$.

Letting

$$u_1^{(\delta)} = \hbar(\cdot\,; x_0) - \hbar_\delta(\cdot\,; x_0) \in H^0(I) \tag{3.3.87c}$$

and

$$u_2^{(\delta)} = \hbar_\delta(\cdot\,; x_0) \in H^1(I) \tag{3.3.87d}$$

we have

$$\|u_1^{(\delta)}\|^2_{H^0(I)} = \frac{\delta}{12} \le \delta \tag{3.3.87e}$$

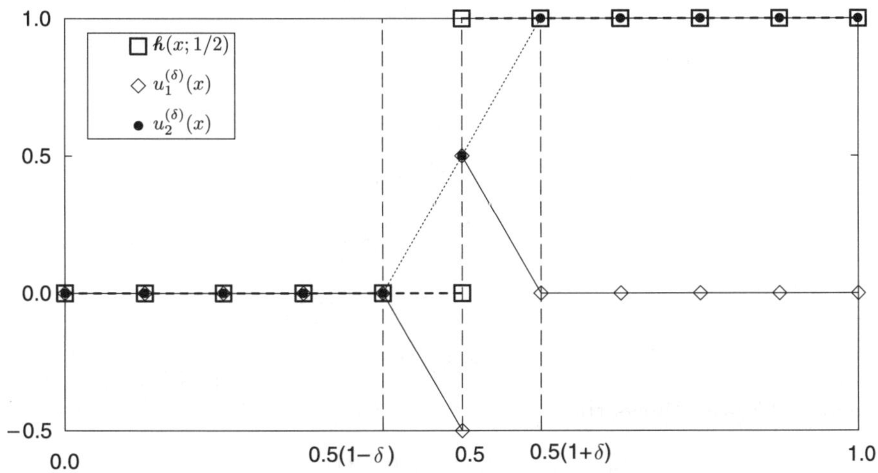

FIG. 3.3.5. The step function $\hbar(\cdot\,; \frac{1}{2})$ and its regularization $\hbar_\delta(\cdot\,; \frac{1}{2})$

and

$$\|u_2^{(\delta)}\|^2_{H^1(I)} = \frac{1}{\delta} + \frac{1}{2} - \frac{\delta}{6} \le \frac{2}{\delta} \tag{3.3.87f}$$

for $0 \le \delta \le 1$. For $t \le 1$, selecting δ such that

$$\delta = \frac{t^2}{\delta} \quad \Longrightarrow \quad \delta = t \tag{3.3.88a}$$

we get

$$K^2(u,t) \le 3t. \tag{3.3.88b}$$

For $t \ge 1$ we let

$$u_1 \equiv \hbar(\cdot\,; x_0), \qquad u_2 \equiv 0 \tag{3.3.89a}$$

and we get

$$\|u_1\|_{H^0(I)} \le 1 \tag{3.3.89b}$$

and hence

$$K^2(u,t) \le 1. \tag{3.3.89c}$$

Therefore we have

$$\|\hbar(\cdot\,; x_0)\|_{B^{1/2}(H^0(I),H^1(I))} = \sqrt{\sup_{0<t<\infty} \left(K(\hbar(\cdot\,; x_0), t)t^{-1/2}\right)^2} < \infty, \tag{3.3.90a}$$

and hence

$$\hbar(\cdot\,; x_0) \in B^{1/2}(H^0(I), H^1(I)). \tag{3.3.90b}$$

On the other hand, we have

$$\hbar(\cdot\,; x_0) \notin H^{1/2}(H^0(I), H^1(I)). \tag{3.3.91a}$$

This can be shown by the approach used in Example 3.3.2. Note that

$$\hbar(\cdot\,; x_0) \in H^{(1/2)-\epsilon}(H^0(I), H^1(I)) \tag{3.3.91b}$$

for any $\epsilon > 0$, but

$$\|\hbar(\cdot\,; x_0)\|_{H^{(1/2)-\epsilon}(H^0(I),H^1(I))} \to \infty \quad \text{as } \epsilon \to 0. \tag{3.3.91c}$$

From Theorem 3.3.6 it follows that

$$\hbar(\cdot\,; x_0) \notin B^{\alpha}(H^0(I), H^1(I)) \tag{3.3.91d}$$

for any $\alpha > \frac{1}{2}$.

Let us give an example of the evaluation of the Aronszajn-Slobodetskii norm.

Example 3.3.4 We will determine the value of $\|\hbar(\cdot;\tfrac{1}{2})\|_{H^{\alpha}_{\mathrm{AS}}(H^0(I),H^1(I))}$. For $x \le \frac{1}{2}$ we have

$$\left|\hbar(x;\tfrac{1}{2}) - \hbar(y;\tfrac{1}{2})\right| = \begin{cases} 0 & \text{for } y \le \frac{1}{2} \\ 1 & \text{for } y > \frac{1}{2} \end{cases} \tag{3.3.92a}$$

and for $x > \frac{1}{2}$

$$\left|\hbar(x;\tfrac{1}{2}) - \hbar(y;\tfrac{1}{2})\right| = \begin{cases} 1 & \text{for } y \le \frac{1}{2} \\ 0 & \text{for } y > \frac{1}{2}. \end{cases} \tag{3.3.92b}$$

Hence for $x \le \frac{1}{2}$ we have

$$\begin{aligned} \int_0^1 \frac{\left|\hbar(x;\frac{1}{2}) - \hbar(y;\frac{1}{2})\right|^2}{|x-y|^{2\theta+1}}\,dy &= \int_{1/2}^1 \frac{dy}{|x-y|^{2\theta+1}} \\ &= \frac{1}{2\theta}\left(-(1-x)^{-2\theta} + \left(\tfrac{1}{2}-x\right)^{-2\theta}\right) \end{aligned} \tag{3.3.92c}$$

and for $x > \frac{1}{2}$

$$\begin{aligned} \int_0^1 \frac{\left|\hbar(x;\frac{1}{2}) - \hbar(y;\frac{1}{2})\right|^2}{|x-y|^{2\theta+1}}\,dy &= \int_0^{1/2} \frac{dy}{|x-y|^{2\theta+1}} \\ &= \frac{1}{2\theta}\left(-x^{-2\theta} + \left(x-\tfrac{1}{2}\right)^{-2\theta}\right). \end{aligned} \tag{3.3.92d}$$

Therefore we have

$$\left\|\hbar(\cdot;\tfrac{1}{2})\right\|_{H^{\theta}_{\mathrm{AS}}(H^0(I),H^1(I))} = \frac{1}{\sqrt{\theta}}\frac{\sqrt{4(\frac{1}{2})^{1-2\theta} - 2}}{\sqrt{1-2\theta}}, \quad \theta < \frac{1}{2} \tag{3.3.93a}$$

and hence

$$\hbar(\cdot;\tfrac{1}{2}) \in H^{(1/2)-\epsilon}(H^0(I), H^1(I)), \quad 0 < \epsilon < \tfrac{1}{2}. \tag{3.3.93b}$$

Note that

$$\left\|\hbar(\cdot;\tfrac{1}{2})\right\|_{H^{\theta}_{\mathrm{AS}}(H^0(I),H^1(I))} \approx \frac{1}{\sqrt{1-2\theta}} \quad \to \infty \tag{3.3.94a}$$

as $\theta \to \frac{1}{2} - 0$, and hence

$$\hbar(\cdot;\tfrac{1}{2}) \notin H^{1/2}(H^0(I), H^1(I)). \tag{3.3.94b}$$

From the above example, it can be seen that the evaluation of the Aronszajn-Slobodetskii norm could be rather difficult, especially in higher dimensions.

For the fractional Sobolev space norm $\|\cdot\|_{H^\theta(H_1,H_2)}$, we have

$$\|\hbar(\cdot;\tfrac{1}{2})\|_{H^\theta(H^0(I),H^1(I))} = \int_0^\infty t^{-2\theta} K^2(\hbar(\cdot;x_0),t)\frac{1}{t}dt \tag{3.3.95a}$$

and hence

$$\|\hbar(\cdot;\tfrac{1}{2})\|_{H^\theta(H^0(I),H^1(I))} \le 3\int_0^1 t^{-2\theta}\,dt + \int_1^\infty t^{-2\theta}\frac{1}{t}\,dt \le 3\frac{1}{1-2\theta} + \frac{1}{2\theta} \tag{3.3.95b}$$

and hence the norm is finite for $\theta < \frac{1}{2}$. By constructing a lower bound, it can be shown that $\|\hbar(\cdot;\frac{1}{2})\|_{H^{1/2}(H^0(I),H^1(I))}$ does not exist.

Let us now consider the case of the special function x^α.

Example 3.3.5 Let us determine the values of θ for which we have

$$x^\alpha \in B^\theta(H^0(I), H^1_{(0}(I)), \quad -\tfrac{1}{2} < \alpha < \tfrac{1}{2} \tag{3.3.96}$$

where we recall that $H^1_{(0}(I) \stackrel{\text{def}}{=} \{u \in H^1(I) \mid u(0) = 0\}$.

Letting x^α_δ denote the regularization of x^α, given by

$$x^\alpha_\delta(x) = \begin{cases} \delta^{\alpha-1}x, & 0 < x < \delta \\ x^\alpha, & \delta \le x \le 1 \end{cases} \tag{3.3.97a}$$

and

$$u_1^{(\delta)} \equiv x^\alpha - x^\alpha_\delta, \qquad u_2^{(\delta)} \equiv x^\alpha \tag{3.3.97b}$$

we have

$$\|u_1^{(\delta)}\|^2_{H^0(I)} \le \mathscr{C}\delta^{2\alpha+1}, \qquad \|u_2^{(\delta)}\|^2_{H^1(I)} \le \mathscr{C}\delta^{2\alpha-1}. \tag{3.3.97c}$$

For $t < 1$, selecting $\delta = t$ we get

$$K^2(t,u) \le \mathscr{C}t^{2\alpha+1}. \tag{3.3.97d}$$

For $t \ge 1$, letting $u_1 \equiv x^\alpha$, $u_2 \equiv 0$, we get $\|u_1\|_{H^0(I)} < \infty$.

Hence we have proven that

$$x^\alpha \in B^{\alpha+1/2}(H^0(I), H^1_{(0}(I)), \quad -\tfrac{1}{2} < \alpha < \tfrac{1}{2}. \tag{3.3.98}$$

Following similar steps we get

$$x^\alpha \in B^{\alpha-1/2}(H^1_{(0}(I), H^1_{(0}(I)\cap H^2(I)), \quad \tfrac{1}{2} < \alpha < 1. \tag{3.3.99}$$

Further, a computation of the Aronszajn-Slobodetskii norm shows that

$$x^\alpha \in H^{\alpha+(1/2)-\epsilon}(H^0(I), H^1(I)), \quad \text{for } \epsilon > 0 \tag{3.3.100a}$$

while

$$x^\alpha \notin H^{\alpha+1/2}(H^0(I), H^1(I)). \tag{3.3.100b}$$

Remark 3.3.17 The meaning of

$$x^\alpha \in B^{\alpha+1/2}(H^0(I), H^1_{(0}(I)) \tag{3.3.101a}$$

is that *the function* x^α, $-\frac{1}{2} < \alpha < \frac{1}{2}$, *has* $\alpha + \frac{1}{2}$ *derivatives which are square integrable in the Besov sense.* Similarly,

$$x^\alpha \in B^{\alpha-1/2}(H^1_{(0}(I), H^1_{(0}(I) \cap H^2(I)) \tag{3.3.101b}$$

has the meaning that x^α, $\frac{1}{2} < \alpha < 1$ has $\alpha + \frac{1}{2}$ derivatives which are square integrable in the Besov space.

The Besov spaces are useful for characterizing the precise smoothness of the exact solution in a-priori estimates of the error in the finite element method. Let us state the main result:

Theorem 3.3.10 *Given a function*

$$u \in B^{1+\theta}_{(0}(I) \stackrel{\text{def}}{=} B^\theta(H^1_{(0}(I), H^1_{(0}(I) \cap H^2(I)), \quad 0 < \theta < 1 \tag{3.3.102a}$$

we have

$$\min_{\chi \in S^p_\Delta} \|u - \chi\|_{\mathcal{U}} \le \mathscr{C} h^\theta \|u\|_{B^{1+\theta}_{(0}(I)} \tag{3.3.102b}$$

where $\mathscr{C}$ *depends only on* θ *and* p.

Proof For any given t, $0 < t < \infty$, let $u = u_1^{(t)} + u_2^{(t)}$, with $u_1^{(t)} \in H^1_{(0}(I)$, such that

$$|u_1^{(t)}|_{H^1(I)} \le t^\theta \|u\|_{B^{1+\theta}_{(0}(I)} \tag{3.3.103a}$$

where $|\cdot|_{H^1(I)}$ is the H^1 seminorm, and $u_2^{(t)} \in H^1_{(0}(I) \cap H^2(I)$, such that

$$\|u_2^{(t)}\|_{H^2(I)} \le \frac{1}{t} t^\theta \|u\|_{B^{1+\theta}_{(0}(I)} = t^{\theta-1}\|u\|_{B^{1+\theta}_{(0}(I)}. \tag{3.3.103b}$$

Using the triangle inequality

$$\min_{\chi \in S^p_\Delta} \|u - \chi\|_{\mathcal{U}} \le \|u_1^{(t)}\|_{\mathcal{U}} + \min_{\chi \in S^p_\Delta} \|u_2^{(t)} - \chi\|_{\mathcal{U}} \tag{3.3.104a}$$

and the estimates

$$\|u_1^{(t)}\|_{\mathcal{U}} \le \mathscr{C} |u_1^{(t)}|_{H^1(I)} \le \mathscr{C} t^\theta \|u\|_{B^{1+\theta}_{(0}(I)} \tag{3.3.104b}$$

and

$$\min_{\chi \in S^p_\Delta} \|u_2^{(t)} - \chi\|_{\mathcal{U}} \le \mathscr{C} h^\Delta \|u_2^{(t)}\|_{H^2(I)} \le \mathscr{C} h^\Delta t^{\theta-1} \|u_2^{(t)}\|_{H^2(I)} \tag{3.3.104c}$$

and choosing $t = h$, (3.3.102b) follows.

□

Let us also state, without proof, a more general version of the above theorem.

Theorem 3.3.11 *Given a* $u \in B^{\theta}(H^{\bar{\alpha}}(I), H^{\bar{\alpha}+1}(I))$, *where* $\bar{\alpha} = 0, 1, 2, \ldots$, $0 < \theta < 1$, *we have*

$$\min_{\chi \in S_{\Delta}^{p}} \|u - \chi\|_{\mathcal{U}} \leq \mathcal{C} h^{\mu} \|u\|_{B^{\theta}(H^{\bar{\alpha}}(I), H^{\bar{\alpha}+1}(I))}, \tag{3.3.105a}$$

with

$$\mu = \min(p, \bar{\alpha} + \theta - 1). \tag{3.3.105b}$$

Remark 3.3.18 The exponent μ in (3.3.105) is optimal in the sense that it cannot be improved.

An application of (3.3.105) for the function

$$x^{\alpha} \in B^{\alpha - 1/2}(H^1_{(0}(I), H^1_{(0}(I) \cap H^2(I)), \quad \tfrac{1}{2} < \alpha < 1 \tag{3.3.106a}$$

gives

$$\min_{\chi \in S_{\Delta}^{p}} \left\| \frac{d}{dx}(x^{\alpha} - \chi) \right\|_{L^2(I)} \leq \mathcal{C} h^{\alpha - 1/2} \|x^{\alpha}\|_{B^{1+(\alpha-1/2)}(H^1_{(0}(I), H^1_{(0}(I) \cap H^2(I))} \tag{3.3.106b)}$$

where we note that $\|x^{\alpha}\|_{B^{1+(\alpha-1/2)}} < \infty$.

Remark 3.3.19 The fractional Sobolev spaces cannot be used to characterize precisely the smoothness of x^{α}. Note that while

$$x^{\alpha} \in H^{\alpha - 1/2 - \epsilon}(I) \quad \forall \epsilon > 0 \tag{3.3.107a}$$

we have

$$\|x^{\alpha}\|_{H^{\alpha-1/2-\epsilon}(I)} \approx \frac{1}{\sqrt{\epsilon}} \longrightarrow \infty \quad \text{as } \epsilon \longrightarrow 0 \tag{3.3.107b}$$

and hence

$$x^{\alpha} \notin H^{\alpha - 1/2}(I). \tag{3.3.107c}$$

The estimate

$$\min_{\chi \in S_{\Delta}^{p}} \|u - \chi\|_{\mathcal{U}} \leq \mathcal{C} h^{\mu - \epsilon} \|u\|_{H^{\theta}(H^{\bar{\alpha}}(I), H^{\bar{\alpha}+1}(I))} \approx \mathcal{C} \frac{h^{\mu - \epsilon}}{\sqrt{\epsilon}} \tag{3.3.108a}$$

where

$$\mu = \min(p, \alpha - 1), \quad \alpha = \bar{\alpha} + \theta \tag{3.3.108b}$$

is *misleading* in the sense that a better rate is observed in practice. Although it is possible to obtain a better estimate by selecting ϵ depending on h, the Besov space approach is preferable because it leads to a more accurate estimate.

Remark 3.3.20 In the above estimates, we have proven that

$$\min_{\chi \in S^p_\Delta} \|u - \chi\|_{\mathscr{U}} \leq \mathscr{C}(u,p)h^\mu, \quad \mu > 0. \tag{3.3.109}$$

In [Babuška and Osborn, 2000] it is shown that there exists a coefficient AE, which satisfies the requirement $0 < \beta_1 \leq AE \leq \beta_2$, such that the convergence is arbitrarily slow.

In this section, we derived estimates for the interpolation error. These estimates can be used to obtain estimates of the error in the finite element solution measured in the global energy norm, and in certain cases, where the nodal values of the finite element solution are exact, to get asymptotic estimates of the error over one element.

Problems

P3.3.1. Let u be a function with r bounded derivatives defined over $\bar{I}_i^\Delta$ and let $\mathcal{I}^3_{I_i^\Delta} u$ be its cubic Lagrangian interpolant which satisfies

$$\mathcal{I}^3_{I_i^\Delta} u(x_{i-1}^\Delta + \xi_j h_{I_i^\Delta}) = u(x_{i-1}^\Delta + \xi_j h_{I_i^\Delta}), \quad 0 \leq j \leq 3 \tag{3.3.110a}$$

with $\xi_0 = 0$, $\xi_1 = \beta$, $\xi_2 = 1 - \beta$, $\xi_3 = 1$, $0 < \beta < \frac{1}{2}$. Derive the asymptotic error estimate

$$\|u - \mathcal{I}^3_{I_i^\Delta} u\|_{L^\infty(I_i^\Delta)} \leq \mathscr{C}_1(\beta)(h_{I_i^\Delta})^{\mu+1} \left\| \frac{d^r u}{dx^r} \right\|_{L^\infty(I_i^\Delta)} \tag{3.3.110b}$$

and

$$\left\| \frac{du}{dx} - \frac{d}{dx}(\mathcal{I}^3_{I_i^\Delta} u) \right\|_{L^\infty(I_i^\Delta)} \leq \mathscr{C}_2(\beta)(h_{I_i^\Delta})^{\mu} \left\| \frac{d^r u}{dx^r} \right\|_{L^\infty(I_i^\Delta)} \tag{3.3.110c}$$

where

$$\mu = \min(3, r-1), \quad 1 \leq r \leq 4 \tag{3.3.110d}$$

and determine the explicit form of the constants $\mathscr{C}_1(\beta)$, $\mathscr{C}_2(\beta)$. Find the values of β which minimize $\mathscr{C}_1(\beta)$ and $\mathscr{C}_2(\beta)$.

P3.3.2. Let u be such that

$$\left\| \frac{d^r u}{dx^r} \right\|_{L^\infty(I_i^\Delta)} \leq C^r, \quad C^r \geq 1, \ 1 \leq r \leq 4. \tag{3.3.111}$$

Use the asymptotic estimates given in P3.3.1 with any r, $1 \leq r \leq 4$, and find the values of r which lead to the minimal values of the asymptotic error bounds as functions of C^r and $h_{I_i^\Delta}$.

P3.3.3. For the function $u \equiv e^{\gamma(x - x_i^\Delta)}$ compute the exact values of $\|u - \mathcal{I}^3_{I_i^\Delta} u\|_{L^\infty(I_i^\Delta)}$, and $\|(d/dx)(u - \mathcal{I}^3_{I_i^\Delta} u)\|_{L^\infty(I_i^\Delta)}$, as functions of $h_{I_i^\Delta}$, β, and γ, where $\mathcal{I}^3_{I_i^\Delta}$ is the interpolation operator given in P3.3.1. Compare these values with the corresponding asymptotic error estimates derived in P3.3.1.

P3.3.4. Let u be a function with bounded first derivative. Derive the asymptotic error estimate

$$\|u - \mathcal{I}^1_{I_i^\Delta} u\|_{L^\infty(I_i^\Delta)} \leq \mathscr{C}_0(h_{I_i^\Delta})^{\delta_0} \left\| \frac{du}{dx} \right\|_{L^\infty(I_i^\Delta)} \tag{3.3.112a}$$

and

$$\|u - \mathcal{I}^1_{I_i^\Delta} u\|_{L^2(I_i^\Delta)} \le \mathscr{C}_0 (h_{I_i^\Delta})^{\delta_1} \left\| \frac{du}{dx} \right\|_{L^2(I_i^\Delta)} \tag{3.3.112b}$$

and determine the values of $\mathscr{C}_0$, δ_0, $\mathscr{C}_1$, and δ_1.

P3.3.5. Consider the regular model problem with input data corresponding to the exact solution $u_{EX} = g(x; x_0)$, where

$$g(x; x_0) = \begin{cases} 0, & 0 \le x \le x_0 \\ x - x_0, & x_0 < x \le 1 \end{cases} \tag{3.3.113}$$

with $0 < x_0 < 1$. Employ Theorem 3.3.4 to construct an estimate for the energy norm or the error in the Lagrangian interpolant over a uniform mesh Δ, and use this result to bound the energy norm of the finite element solution. Utilize the construction of Besov space given in Example 3.3.1. Let $x_0 = \frac{1}{2}$ and consider the cases of uniform meshes with odd and even number of elements. What can you say about the error in each case?

P3.3.6. Find the value of θ for which we have

$$x^\alpha \in B^\theta(\mathscr{U}^\vartheta, \mathscr{U}^\vartheta \cap H^2(I)) \tag{3.3.114}$$

for $\frac{1}{2} < \alpha < 1$, where $\mathscr{U}^\vartheta$, $0 < \vartheta < 1$, is the energy space for the singular model problem.

P3.3.7. Utilize the result of P3.3.6 to construct an estimate for the energy norm of the error in the finite element solution of the singular model problem.

P3.3.8. Let u_{EX} be the solution of the variational problem:

Find $u_{EX} \in \mathscr{U}$ such that

$$\mathscr{B}(u_{EX}, v) = v(\tfrac{1}{2}), \quad \forall v \in \mathscr{U} \tag{3.3.115}$$

where $\mathscr{B}$ is the bilinear form for the regular model problem for $AE \equiv 1$, $c \equiv 1$, $\alpha_2 = a_2 = 0$, $l = 1$. Estimate the energy norm of the error $\|e_{S^p_\Delta}\|_{\mathscr{U}}$ from above and below, where Δ is a uniform mesh, for odd and even number of elements. For the upper estimate utilize the Besov norm.

P3.3.9. Prove Theorem 3.3.5 for all meshes Δ with the property

$$\alpha \max_{i=1,\dots,N(\Delta)} h_{I_i^\Delta} \le \min_{i=1,\dots,N(\Delta)} h_{I_i^\Delta}, \quad \alpha > 0. \tag{3.3.116}$$

P3.3.10. Show by a counterexample that Theorem 3.3.5 does not hold for arbitrary meshes and functions u_{EX}.

Hint: Select u_{EX} such that it is zero in some subinterval, and construct meshes with $h^\Delta \not\to 0$.

3.3* The finite element interpolation and approximation error in two dimensions

We will now construct the finite element interpolation error estimates in two dimensions. Let us first summarize the main ideas involved in the construction of the estimates in one dimension.

Given a function u with $d^r u/dx^r \in L^\infty(I_j^\Delta)$, we:

(a) Construct $\tilde{u}_{I_i^\Delta} : \tilde{I} = (-1,1) \mapsto \mathbb{R}$, the corresponding function over the master element $\tilde{I}$ by

$$\tilde{u}_{I_i^\Delta}(\tilde{x}) \stackrel{\text{def}}{=} u\big(F_{I_i^\Delta}(\tilde{x})\big) \tag{3.3*.1a}$$

using the affine mapping

$$F_{I_i^\Delta}(\tilde{x}) = \tfrac{1}{2}(x_i^\Delta + x_{i+1}^\Delta) + \tfrac{1}{2}h_{I_i^\Delta}\tilde{x}. \tag{3.3*.1b}$$

It follows that

$$\frac{d}{d\tilde{x}}(\tilde{u}_{I_i^\Delta})(\tilde{x}) = \frac{du}{dx}(x)\frac{d}{d\tilde{x}}(F_{I_i^\Delta})(\tilde{x}) = \frac{h_{I_i^\Delta}}{2}\frac{du}{dx}(x) \tag{3.3*.1c}$$

and

$$\left\|\frac{d^r}{d\tilde{x}^r}(\tilde{u}_{I_i^\Delta})\right\|_{L^\infty(\tilde{I})} = \left(\frac{h_{I_i^\Delta}}{2}\right)^r \left\|\frac{d^r u}{dx^r}\right\|_{L^\infty(I_i^\Delta)} < \infty. \tag{3.3*.1d}$$

(b) Construct the Lagrangian interpolant $\tilde{\mathcal{I}}^p \tilde{u}_{I_i^\Delta}$ over $\tilde{I}$

$$\tilde{\mathcal{I}}^p \tilde{u}_{I_i^\Delta} = \sum_{j=0}^{p} \mathcal{L}_j(\tilde{x}; \{\tilde{\xi}_l\}_{l=0}^p)\tilde{u}_{I_i^\Delta}(\tilde{\xi}_j) \tag{3.3*.2a}$$

where $\tilde{\xi}_l$, $l = 0, \ldots, p$, is a set of distinct points, with $\tilde{\xi}_0 = -1$, and $\tilde{\xi}_p = 1$, and prove the estimates

$$\|\tilde{u}_{I_i^\Delta} - \tilde{\mathcal{I}}^p \tilde{u}_{I_i^\Delta}\|_{L^\infty(\tilde{I})} \leq \mathscr{C}(p) \left\|\frac{d^r}{d\tilde{x}^r}(\tilde{u}_{I_i^\Delta})\right\|_{L^\infty(\tilde{I})} \tag{3.3*.2b}$$

$$\left\|\frac{d}{d\tilde{x}}(\tilde{u}_{I_i^\Delta} - \tilde{\mathcal{I}}^p \tilde{u}_{I_i^\Delta})\right\|_{L^\infty(\tilde{I})} \leq \mathscr{C}(p)\left\|\frac{d^r}{d\tilde{x}^r}(\tilde{u}_{I_i^\Delta})\right\|_{L^\infty(\tilde{I})} \tag{3.3*.2c}$$

where

$$1 \leq r \leq p+1 \tag{3.3*.2d}$$

and $\mathscr{C}(p)$ also depends on $\{\tilde{\xi}_l\}_{l=0}^p$, the employed set of points.

(c) Construct the Lagrangian interpolant $\mathcal{I}_{I_i^\Delta}^p u$ over I_i^Δ by

$$\mathcal{I}_{I_i^\Delta}^p u(x) \stackrel{\text{def}}{=} \tilde{\mathcal{I}}^p \tilde{u}_{I_i^\Delta}\big(F_{I_i^\Delta}^{-1}(x)\big) \tag{3.3*.3a}$$

and transform the norms by

$$\|u - \mathcal{I}_{I_i^\Delta}^p u\|_{L^\infty(I_j^\Delta)} = \|\tilde{u}_{I_i^\Delta} - \tilde{\mathcal{I}}^p \tilde{u}_{I_i^\Delta}\|_{L^\infty(\tilde{I})} \tag{3.3*.3b}$$

and

$$\left\|\frac{d}{dx}(u - \mathcal{I}_{I_i^\Delta}^p u)\right\|_{L^\infty(I_j^\Delta)} = \frac{2}{h_{I_i^\Delta}}\left\|\frac{d}{d\tilde{x}}(\tilde{u}_{I_i^\Delta} - \tilde{\mathcal{I}}^p \tilde{u}_{I_i^\Delta})\right\|_{L^\infty(\tilde{I})}. \tag{3.3*.3c}$$

(d) Combine (3.3*.1b), (3.3*.2b), and (3.3*.3b) to get the estimate.

$$\|u - \mathcal{I}^p_{I_i^\Delta} u\|_{L^\infty(I_i^\Delta)} \le \mathscr{C}(p,r)(h_{I_i^\Delta})^{\mu+1} \left\| \frac{d^r u}{dx^r} \right\|_{L^\infty(I_i^\Delta)} \tag{3.3*.4a}$$

and (3.3*.1b), (3.3*.2c), and (3.3*.3c) to obtain

$$\left\| \frac{du}{dx} - \frac{d}{dx}(\mathcal{I}^p_{I_i^\Delta} u) \right\|_{L^\infty(I_i^\Delta)} \le \mathscr{C}(p,r)(h_{I_i^\Delta})^{\mu} \left\| \frac{d^r u}{dx^r} \right\|_{L^\infty(I_i^\Delta)} \tag{3.3*.4b}$$

where

$$\mu = \min(p, r-1), \quad 0 \le r \le p+1. \tag{3.3*.4c}$$

(e) Follow similar steps, under the assumption that $d^r u/dx^r \in L^2(I)$, to obtain

$$\|u - \mathcal{I}^p_{I_i^\Delta} u\|_{L^2(I_i^\Delta)} \le \mathscr{C}(p,r)(h_{I_i^\Delta})^{\mu+1} \left\| \frac{d^r u}{dx^r} \right\|_{L^2(I_i^\Delta)} \tag{3.3*.5a}$$

and

$$\left\| \frac{du}{dx} - \frac{d}{dx}(\mathcal{I}^p_{I_i^\Delta} u) \right\|_{L^2(I_i^\Delta)} \le \mathscr{C}(p,r)(h_{I_i^\Delta})^{\mu} \left\| \frac{d^r u}{dx^r} \right\|_{L^2(I_i^\Delta)} \tag{3.3*.5b}$$

with

$$\mu = \min(p, r-1), \quad 0 \le r \le p+1. \tag{3.3*.5c}$$

Remark 3.3*.1 The interpolant $\mathcal{I}^p_{I_i^\Delta} u$ has meaning only in the cases that $u \in C^0(I_i^\Delta)$; this is the case when $r \ge 1$. The interpolation estimates may be extended to the case $0 \le r < 1$ by employing a *generalized interpolant*; see also Remark 3.3*.4 below.

Let us now give the construction of the interpolation error estimates in two dimensions. For concreteness, let us consider the case of triangular elements with straight edges. As in the one-dimensional case, given a u such that

$$\max_{\substack{m+n=r \\ m,n\ge 0}} \left\| \frac{\partial^{m+n} u}{\partial x_1^m \partial x_2^n} \right\|_{L^\infty(\tau_j^\Delta)} < \infty \tag{3.3*.6}$$

we:

(a) Construct $\tilde{u}_{\tau_j^\Delta} : \tilde{\tau} \mapsto \mathbb{R}$, the corresponding function over the master element τ, by

$$\tilde{u}_{\tau_j^\Delta}(\tilde{\mathbf{x}}) \stackrel{\text{def}}{=} u(\mathbf{F}_{\tau_j^\Delta}(\tilde{\mathbf{x}})) \tag{3.3*.7a}$$

where $\mathbf{F}_{\tau_j^\Delta} : \tilde{\tau} \mapsto \tau_j^\Delta$ is an affine mapping which transforms the master triangle $\tilde{\tau}$ into τ_j^Δ. It follows that

$$\left\| \frac{\partial^k \tilde{u}_{\tau_j^\Delta}}{\partial \tilde{x}_1^m \partial \tilde{x}_2^n} \right\|_{L^\infty(\tilde{\tau})} \le \mathscr{C}_1 (h_{\tau_j^\Delta})^k \left\| \frac{\partial^k u}{\partial x_1^m \partial x_2^n} \right\|_{L^\infty(\tau_j^\Delta)}, \qquad m,n \ge 0, \quad m+n=k \tag{3.3*.7b}$$

where $\mathscr{C}_1$ depends on the the interior angles of τ_j^Δ, but not on its size. The inequality (3.3*.7b) is the two-dimensional analogue of the equality (3.3*.1d).

(b) Construct the Lagrangian interpolant $\tilde{\mathcal{I}}^p\tilde{u}_{\tau_j^\Delta}$ over $\tilde{\tau}$ by

$$\tilde{\mathcal{I}}^p\tilde{u}_{\tau_j^\Delta}(\mathbf{x}) = \sum_{j=1}^{np} \tilde{\mathcal{L}}_j(\mathbf{x}; \{\tilde{\xi}_l\}_{l=1}^{np})\tilde{u}_{\tau_j^\Delta}(\tilde{\xi}_j) \tag{3.3*.8a}$$

where $\tilde{\xi}_l$, $l = 1, \ldots, np = \frac{1}{2}(p+1)(p+2)$ are the interpolation points given in Section 3.2*, and $\tilde{\mathcal{L}}_j(\cdot; \{\tilde{\xi}_l\}_{l=1}^{np}) \in \mathcal{P}^p(\tilde{\tau})$ is the jth Lagrangian shape function over $\tilde{\tau}$. We can now prove estimates which are analogous to (3.3*.2b) and (3.3*.2c), namely

$$\|\tilde{u}_{\tau_j^\Delta} - \tilde{\mathcal{I}}^p\tilde{u}_{\tau_j^\Delta}\|_{L^\infty(\tilde{\tau})} \leq \mathscr{C}(p) \max_{\substack{m+n=r\\ m,n\geq 0}} \left\|\frac{\partial^r \tilde{u}_{\tau_j^\Delta}}{\partial\tilde{x}_1^m \partial\tilde{x}_2^n}\right\|_{L^\infty(\tilde{\tau})} \tag{3.3*.8b}$$

and

$$\left\|\frac{\partial}{\partial\tilde{x}_i}(\tilde{u}_{\tau_j^\Delta} - \tilde{\mathcal{I}}^p\tilde{u}_{\tau_j^\Delta})\right\|_{L^\infty(\tilde{\tau})} \leq \mathscr{C}(p) \max_{\substack{m+n=r\\ m,n\geq 0}} \left\|\frac{\partial^r \tilde{u}_{\tau_j^\Delta}}{\partial\tilde{x}_1^m \partial\tilde{x}_2^n}\right\|_{L^\infty(\tilde{\tau})}. \tag{3.3*.8c}$$

(c) Construct the Lagrangian interpolant $\mathcal{I}^p_{\tau_j^\Delta}u$ over τ_j^Δ by

$$\mathcal{I}^p_{\tau_j^\Delta}u(\mathbf{x}) \stackrel{\text{def}}{=} \tilde{\mathcal{I}}^p\tilde{u}_{\tau_j^\Delta}(F^{-1}_{\tau_j^\Delta}(\mathbf{x})) \tag{3.3*.9a}$$

for which we have

$$\|u - \mathcal{I}^p_{\tau_j^\Delta}u\|_{L^\infty(\tau_j^\Delta)} \leq \mathscr{C}_2\|\tilde{u}_{\tau_j^\Delta} - \tilde{\mathcal{I}}^p\tilde{u}_{\tau_j^\Delta}\|_{L^\infty(\tilde{\tau})} \tag{3.3*.9b}$$

and

$$\left\|\frac{\partial}{\partial x_i}(u - \mathcal{I}^p_{\tau_j^\Delta}u)\right\|_{L^\infty(\tau_j^\Delta)} \leq \frac{\mathscr{C}_3}{h_{\tau_j^\Delta}} \max_{i=1,2} \left\|\frac{\partial}{\partial\tilde{x}_i}(\tilde{u}_{\tau_j^\Delta} - \tilde{\mathcal{I}}^p\tilde{u}_{\tau_j^\Delta})\right\|_{L^\infty(\tilde{\tau})} \tag{3.3*.9c}$$

where $\mathscr{C}_2$ and $\mathscr{C}_3$ depend on the interior angles of τ_j^Δ. The inequalities (3.3*.8b) and (3.3*.8c) are analogous to the equalities (3.3*.3b) and (3.3*.3c). The dependence of the constants $\mathscr{C}_i$, $i = 1, 2, 3$, on the interior angles of τ_j^Δ can be expressed in terms of the *minimal interior angle* of τ_j^Δ, or, equivalently, in terms of the aspect ratio $\alpha(\tau_j^\Delta)$.

(d) Combine (3.3*.7b), (3.3*.8b), (3.3*.9b), to get

$$\|u - \mathcal{I}^p_{\tau_j^\Delta}u\|_{L^\infty(\tau_j^\Delta)} \leq \mathscr{C}(h_{\tau_j^\Delta})^{\mu+1} \max_{\substack{m+n=r\\ m,n\geq 0}} \left\|\frac{\partial^{m+n} u}{\partial x_1^m \partial x_2^n}\right\|_{L^\infty(\tau_j^\Delta)} \tag{3.3*.10a}$$

and (3.3*.7b), (3.3*.8c), (3.3*.9c) to obtain

$$\left\|\frac{\partial}{\partial x_i}(u - \mathcal{I}^p_{\tau_j^\Delta}u)\right\|_{L^\infty(\tau_j^\Delta)} \leq \mathscr{C}(h_{\tau_j^\Delta})^{\mu} \max_{\substack{m+n=r\\ m,n\geq 0}} \left\|\frac{\partial^{m+n} u}{\partial x_1^m \partial x_2^n}\right\|_{L^\infty(\tau_j^\Delta)} \tag{3.3*.10b}$$

where

$$\mu = \min(p, r-1), \quad 0 < r \le p+1 \tag{3.3*.10c}$$

and the constants $\mathscr{C}$ depend on the *mesh aspect ratio* $\alpha_\Delta \stackrel{\text{def}}{=} \max_{j=1,\ldots,N(\Delta)} \alpha(\tau_j^\Delta)$, or, equivalently, the minimum of the minimal interior angle of the elements over the mesh.

(e) Following similar steps, under the assumption that

$$\max_{\substack{m+n=r\\ m,n\ge 0}} \left\| \frac{\partial^{m+n} u}{\partial x_1^m \partial x_2^n} \right\|_{L^2(\tau_j^\Delta)} < \infty \tag{3.3*.11}$$

we get

$$\|u - \mathcal{I}^p_{\tau_j^\Delta} u\|_{L^2(\tau_j^\Delta)} \le \mathscr{C}(h_{\tau_j^\Delta})^{\mu+1} \max_{\substack{m+n=r\\ m,n\ge 0}} \left\| \frac{\partial^{m+n} u}{\partial x_1^m \partial x_2^n} \right\|_{L^2(\tau_j^\Delta)} \tag{3.3*.12a}$$

$$\left\| \frac{\partial}{\partial x_i}(u - \mathcal{I}^p_{\tau_j^\Delta} u) \right\|_{L^2(\tau_j^\Delta)} \le \mathscr{C}(h_{\tau_j^\Delta})^{\mu} \max_{\substack{m+n=r\\ m,n\ge 0}} \left\| \frac{\partial^{m+n} u}{\partial x_1^m \partial x_2^n} \right\|_{L^2(\tau_j^\Delta)}. \tag{3.3*.12b}$$

For further details see [Ciarlet, 1991] and [Scott and Zhang, 1990].

Remark 3.3*.2 In order to keep the constants $\mathscr{C}$ bounded, uniformly, for all meshes in a mesh sequence, it is sufficient to keep the mesh aspect ratios bounded for all the meshes in the sequence or, equivalently, the minimal interior angles for all the meshes in the sequence bounded from below. This is the reason why it is important that the mesh generator avoids the construction of slender triangular elements.

Remark 3.3*.3 Analogous estimates hold for curvilinear triangles and for quadrilaterals with straight or curved edges, under additional assumptions about the geometry of the elements (see [Lenoir, 1986] and [Ciarlet, 1991]). These assumptions state that, as the size h_τ of the element converges to zero, the curvilinear triangles converge to triangles with straight edges, and the curvilinear quadrilaterals to parallelograms. These assumptions may not be satisfied in many practical cases, like, for example, the so-called *crown elements* which have two straight edges and one circular edge with radius proportional to h_τ. Also in the more general case in which a curved quadrilateral converges to a straight-edge quadrilateral which is not a parallelogram, the above estimates hold only when the tensor-product basis is used. A detailed analysis may be found in [Arnold *et al.*, 2000]. In the general case, in which the elements τ do not converge to a straight edge element, there is a loss in the order of convergence.

Let us also give the global estimates for the error in $\mathcal{I}^p_\Delta u$ and its derivatives. For the L^∞ norm we have

$$\|u - \mathcal{I}^p_\Delta u\|_{L^\infty(\Omega)} \le \mathscr{C} \max_{\tau\in\Delta} \left((h_\tau)^{\mu+1} \max_{m+n=r} \left\| \frac{\partial^{m+n} u}{\partial x_1^m \partial x_2^n} \right\|_{L^\infty(\tau)} \right) \tag{3.3*.13a}$$

and

$$\left\| \frac{\partial}{\partial x_i}(u - \mathcal{I}^p_\Delta u) \right\|_{L^\infty(\Omega)} \le \mathscr{C} \max_{\tau\in\Delta} \left((h_\tau)^{\mu} \max_{m+n=r} \left\| \frac{\partial^{m+n} u}{\partial x_1^m \partial x_2^n} \right\|_{L^\infty(\tau)} \right) \tag{3.3*.13b}$$

and for the L^2 norm

$$\|u - \mathcal{I}^p_\Delta u\|_{L^2(\Omega)} \le \mathscr{C} \sqrt{\sum_{\tau\in\Delta} (h_\tau)^{2\mu+2} \Big(\max_{m+n=r} \Big\| \frac{\partial^{m+n} u}{\partial x_1^m \partial x_2^n} \Big\|_{L^2(\tau)} \Big)^2} \tag{3.3*.14a}$$

and

$$\Big\| \frac{\partial}{\partial x_i}(u - \mathcal{I}^p_\Delta u) \Big\|_{L^2(\Omega)} \le \mathscr{C} \sqrt{\sum_{\tau\in\Delta} (h_\tau)^{2\mu} \Big(\max_{m+n=r} \Big\| \frac{\partial^{m+n} u}{\partial x_1^m \partial x_2^n} \Big\|_{L^2(\tau)} \Big)^2} \tag{3.3*.14b}$$

where, in both cases, we let

$$\mu = \min(p, r-1), \quad 1 < r \le p+1. \tag{3.3*.15}$$

For uniform meshes, the L^∞ estimates become

$$\|u - \mathcal{I}^p_\Delta u\|_{L^\infty(\Omega)} \le \mathscr{C} h^{\mu+1} \max_{m+n=r} \Big\| \frac{\partial^{m+n} u}{\partial x_1^m \partial x_2^n} \Big\|_{L^\infty(\Omega)} \tag{3.3*.16a}$$

$$\Big\| \frac{\partial}{\partial x_i}(u - \mathcal{I}^p_\Delta u) \Big\|_{L^\infty(\Omega)} \le \mathscr{C} h^{\mu} \max_{m+n=r} \Big\| \frac{\partial^{m+n} u}{\partial x_1^m \partial x_2^n} \Big\|_{L^\infty(\Omega)} \tag{3.3*.16b}$$

and similarly for the L^2 estimates.

Remark 3.3*.4 Note that the assumption

$$\max_{\substack{m+n=r\\ m,n\ge 0}} \Big\| \frac{\partial^{m+n} u}{\partial x_1^m \partial x_2^n} \Big\|_{L^\infty(\Omega)} < \infty \tag{3.3*.17}$$

guarantees the continuity of u only if $r \ge 2$. Nevertheless, the interpolation-error estimates can also be extended to the case $0 \le r < 2$, by employing a generalized interpolant $\hat{\mathcal{I}}^p_\Delta u$ (see [Clément, 1975], [Brenner and Scott, 1994], and [Scott and Zhang, 1990]), and we have

$$\|u - \hat{\mathcal{I}}^p_\Delta u\|_{L^2(\Omega)} \le \mathscr{C} \sqrt{\sum_{\tau\in\Delta} (h_\tau)^{2\mu+2} \Big(\max_{m+n=r} \Big\| \frac{\partial^{m+n} u}{\partial x_1^m \partial x_2^n} \Big\|_{L^2(\omega_\tau)} \Big)^2} \tag{3.3*.18a}$$

$$\Big\| \frac{\partial}{\partial x_i}(u - \hat{\mathcal{I}}^p_\Delta u) \Big\|_{L^2(\Omega)} \le \mathscr{C} \sqrt{\sum_{\tau\in\Delta} (h_\tau)^{2\mu} \Big(\max_{m+n=r} \Big\| \frac{\partial^{m+n} u}{\partial x_1^m \partial x_2^n} \Big\|_{L^2(\omega_\tau)} \Big)^2} \tag{3.3*.18b}$$

where

$$\mu = \min(p, r-1), \quad 0 < r \le p+1 \tag{3.3*.18c}$$

and ω_τ is the union of all elements which have a common edge or vertex with τ, including τ. The generalized interpolant $\hat{\mathcal{I}}^p_\Delta u$, may be understood as $\mathcal{I}^p_\Delta \hat{u}$, the standard interpolant of a function $\hat{u}$, obtained by locally averaging u.

Remark 3.3*.5 A further investigation of the angle condition (see [Babuška and Aziz, 1976]) reveals that the minimal-angle condition (or the mesh aspect-ratio condition) is a sufficient but not a necessary condition for the constant $\mathscr{C}_3$ to remain uniformly bounded. The necessary condition is that the maximal interior angle of all the elements in the mesh must be bounded away from the value of π radians. This is known as the *maximal-angle condition.* Hence very slender elements (*needle elements*) may also be employed, if they satisfy the maximal-angle condition, although they may violate the minimal angle condition. Such slender elements are often needed for the optimal approximation of solutions which exhibit strong one-dimensional features, e.g., boundary layers, edge singularities in three dimensions. For an analysis of anisotropic meshes, see [Apel, 1999].

Remark 3.3*.6 The construction of the Lagrangian interpolant using a set of uniformly spaced interpolation points in two or three dimensions can lead to large errors, as it was illustrated in the one-dimensional case. As we have already mentioned in Section 3.3, Lagrangian interpolation is often used to represent the geometry of the boundary of the domain and large interpolation errors could lead to the creation of spurious corners and edges. For certain problems (e.g. the Helmholtz equation) the solution of the mathematical problem can be very sensitive to small changes in the domain, and it is very important to represent the geometry with high accuracy. The optimal interpolation points which were introduced in the one-dimensional case can also be determined in the two-dimensional case. For a rectangular element or a right parallelepiped, the optimal points can be obtained as tensor products of the optimal points determined in one dimension. The optimal points for triangles and tetrahedra cannot be obtained from the one-dimensional case, and are given in [Chen and Babuška, 1996].

Remark 3.3*.7 The above interpolation results also hold for parallelogram elements, with the same maximal-angle condition as in the case of triangular elements. In the general case, where the elements are generally quadrilateral with straight or curved edges, additional conditions are needed to guarantee the non-degeneracy of the approximation.

The interpolated spaces can be constructed in two dimensions analogously as in the one-dimensional case. For example, we may employ the two-dimensional version of the Aronszajn-Slobodetskii norm

$$\|v\|_{H^\theta} = \sqrt{\int_\Omega \int_\Omega \frac{|v(\mathbf{x}) - v(\mathbf{y})|^2}{|\mathbf{x} - \mathbf{y}|_{\mathbb{R}^2}^{2\theta+2}} dx\, dy} \tag{3.3*.19}$$

where $|\mathbf{x} - \mathbf{y}|_{\mathbb{R}^2} = \sqrt{(x_1 - y_1)^2 + (x_2 - y_2)^2}$. Similarly, we can define the Besov norms, the corresponding Besov spaces, and we have

$$\min_{\chi \in S^p_\Delta} \|u - \chi\|_{\mathscr{U}} \leq \mathscr{C} h^\theta \|u\|_{B^{1+\theta}_{(I)}(\Omega)}. \tag{3.3*.20}$$

As we have already underlined in several instances, the great majority of the boundary value problems employed in engineering are set in polygonal domains with several corner points in the neighborhood of which the exact solution u_{EX} has algebraic singularities. For example, in the case of the Laplacian in a domain which has only one corner point with angle π/α with homogeneous Neumann boundary conditions in the neighborhood of the corner point, we have

$$u_{EX} = r^\alpha \sin(\alpha\theta) + \tilde{u}_{EX} \in B^\alpha(H^1(\Omega), H^2(\Omega)) \tag{3.3*.21a}$$

where we assumed that the boundary conditions are smooth and we have $\tilde{u}_{EX} \in H^2(\Omega)$, and for quasi-uniform meshes we get

$$\|r^\alpha \sin(\alpha\theta) - \mathcal{B}^p_\Delta(r^\alpha \sin(\alpha\theta))\|_{\mathcal{U}} \leq \mathcal{C} h^\alpha \tag{3.3*.21b}$$

and similarly for $\|e_{S^p_\Delta}\|_{\mathcal{U}}$. Note the difference with the one-dimensional analogue where

$$u_{EX} \equiv x^\alpha \in B^{\alpha-1/2}(H^1(\Omega), H^2(\Omega)). \tag{3.3*.21c}$$

It can be also shown that

$$u_{EX} = r^\alpha \sin(\alpha\theta) + \tilde{u}_{EX} \in H^{\alpha-\epsilon}(H^1(\Omega), H^2(\Omega)) \tag{3.3*.22a}$$

for $\epsilon > 0$, and that

$$u_{EX} = r^\alpha \sin(\alpha\theta) + \tilde{u}_{EX} \notin H^{\alpha}(H^1(\Omega), H^2(\Omega)). \tag{3.3*.22b}$$

This shows the advantage of using the Besov spaces to describe the smoothness of u_{EX} instead of the Sobolev spaces of fractional degree. For estimates in other norms, see [Ciarlet, 1991], [Brenner and Scott, 1994]. For a detailed analysis of the approximation for the p version, we refer to [Babuška and Guo, 2000a,b] and references therein.

Remark 3.3*.8 Let us give an example which underlines the need for accurate numerical integration when norms of the error are evaluated in benchmark computations. Consider the model problem of heat conduction in the heterogeneous square domain $\Omega = (-1, 1)^2$, namely, the problem

$$\boldsymbol{\nabla} \cdot (k(\mathbf{x})\boldsymbol{\nabla} u)(\mathbf{x}) = 0, \quad \mathbf{x} \in \Omega \tag{3.3*.23a}$$

with

$$k(\mathbf{x}) = \begin{cases} 100, & \mathbf{x} \in \Omega_1 \cup \Omega_3 \\ 1, & \mathbf{x} \in \Omega_2 \cup \Omega_4 \end{cases} \tag{3.3*.23b}$$

where we let $\Omega_1 = (0,1)^2$, $\Omega_3 = (-1,0)^2$, $\Omega_2 = (0,1) \times (-1,0)$, and $\Omega_4 = (-1,0) \times (0,1)$, with Neumann boundary condition

$$\left.\frac{\partial u}{\partial n}\right|_{\partial\Omega} = g \tag{3.3*.23c}$$

where g is determined to correspond to the exact solution

$$u_{EX}(\mathbf{x}) = r^\alpha (a_i \sin(\alpha\theta) + b_i \cos(\alpha\theta)), \quad x \in \Omega_i \tag{3.3*.24a}$$

where (r, θ) are the polar coordinates of the point $\mathbf{x}$, and the polar coordinate system is located at $(0, 0)$, with exponent

$$\alpha = 0.126902 \tag{3.3*.24b}$$

and coefficients

$$a_1 = 0.100000, \qquad a_2 = -9.60396, \qquad a_3 = -0.480355, \qquad a_4 = 7.701564 \tag{3.3*.24c}$$

$$b_1 = 1.0000, \qquad b_2 = 2.960396, \qquad b_3 = -0.882756, \qquad b_4 = -6.456461. \tag{3.3*.24d}$$

Let us now employ a subdivision of Ω into a uniform 6×6 mesh of squares, which we will denote by Δ, and attempt to evaluate the energy norm by employing a product Gauss rule in each square, namely

$$\|u_{EX}\|_{\mathcal{U}}^2 \cong \sum_{\tau \in \Delta} \left(\sum_{l=1}^{np(\tau)} k(\mathbf{x}_l^\tau) |\nabla u_{EX}|(\mathbf{x}_l^\tau) w_l^\tau |J_\tau| \right) \tag{3.3*.25}$$

where $\mathbf{x}_l^\tau$, w_l^τ are the lth integration point and weight in the element τ, and $np(\tau)$ is the number of integration points used in the element τ. We considered three cases:

(a) The 3×3 Gauss rule is used in each element.

(b) The 12×12 Gauss rule is used in each element.

(c) The 12×12 Gauss rule is used in the key elements, i.e. the four squares with a vertex at the origin $(0, 0)$, while the 3×3 Gauss rule was used in the rest of the elements.

The results are given in Table 3.3*.1.

From the computed results given in Table 3.3*.1, it can be seen that it is practically impossible to obtain an accurate value for the energy norm $\|u_{EX}\|_{\mathcal{U}}$ for this problem by employing the product Gauss rule on a uniform mesh of squares. Better results can be obtained by employing the Gauss rule on a graded mesh obtained by employing several nested refinements of the key elements. A better approach is to employ integration by parts to obtain

$$\|u_{EX}\|_{\mathcal{U}}^2 = \int_{\partial\Omega} u_{EX} \Big(k \frac{\partial}{\partial n}(u_{EX}) \Big). \tag{3.3*.26}$$

TABLE 3.3*.1. The values of the energy norm $\|u_{EX}\|_{\mathcal{U}}$, computed by numerical integration

Integration rule	Computed value
a: 3×3	27.53832
b: 12×12	34.22664
c: $12 \times \frac{12}{3} \times 3$	34.22621
Exact value	41.42021

In this integral, the integrand is smooth, and it can be evaluated with 8 correct significant digits by employing a one-dimensional Gauss rule on each element edge which lies on the boundary $\partial\Omega$. Having evaluated the energy norm of the exact solution, the energy norm of the error in a finite element approximation $u_{S^p_\Delta}$, may be evaluated accurately by employing the identity

$$\|e_{S^p_\Delta}\|^2_{\mathcal{U}} = 2(\Pi(u_{S^p_\Delta}) - \Pi(u_{EX})) = \|u_{EX}\|^2_{\mathcal{U}} - \|u_{S^p_\Delta}\|^2_{\mathcal{U}} \qquad (3.3^*.27)$$

where $\|u_{S^p_\Delta}\|_{\mathcal{U}}$ can be evaluated accurately by employing a low order (e.g. $(p+1)$ order) product Gauss rule in each element. Using inaccurate values for $\|u_{EX}\|^2_{\mathcal{U}}$, or $\|u_{EX}-u_{S^p_\Delta}\|^2_{\mathcal{U}}$, could lead to gross underestimation of the error in $u_{S^p_\Delta}$; see also P3.1.7.

In this section we have seen that the interpolation estimates in two dimensions, are analogous to the estimates given for one dimension, with some important differences stemming from the fact that in higher dimensions the functions in the energy space are not necessarily continuous.

Problems

P3.3*.1. Let Ω be the L-shaped domain of Fig. 2.3*.1, and

$$u_{EX} \equiv r^{2/3} \sin\left(\tfrac{2}{3}\theta\right). \qquad (3.3^*.28a)$$

Show that $\|u_{EX}\|_{B^{2/3}(\Omega)} < \infty$.

Hint: Let $\varphi_\delta(r) = \varphi(r/\delta)$, be a smooth function with $\varphi(r) = 0$ for $r \leq \frac{1}{2}$ and $\varphi(r) = 1$ for $r > 1$, and use the splitting

$$u_{EX} = \varphi_\delta(r)u_{EX} + (1 - \varphi_\delta(r))u_{EX}. \qquad (3.3^*.28b)$$

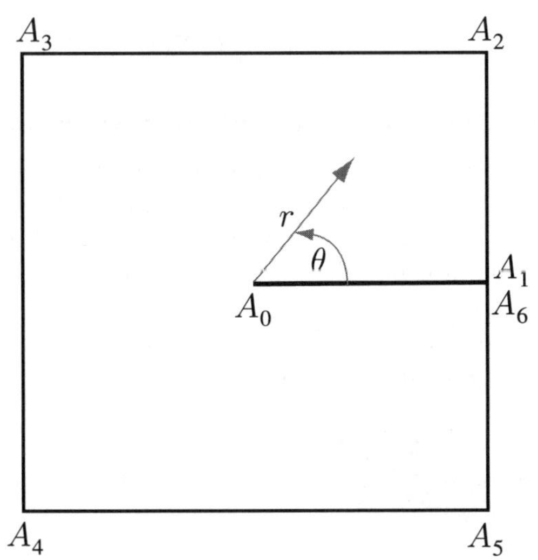

FIG. 3.3*.1. The domain for problem P3.3*.3

P3.3*.2. Let Ω be the L-shaped domain shown in Fig. 2.3*.1, and consider the problem

$$-\Delta u|_{\Omega} = 0 \qquad u|_{\Gamma_D} = 0, \quad \Gamma_D = \overline{A_1A_2} \cup \overline{A_6A_1} \tag{3.3*.29a}$$

$$\left.\frac{\partial u}{\partial n}\right|_{A_4A_3} = 1 \qquad \left.\frac{\partial u}{\partial n}\right|_{\Gamma-\Gamma_D-A_4A_3} = 0. \tag{3.3*.29b}$$

Then we have

$$u_{EX} = a_1 r^{2/3}\sin\left(\tfrac{2}{3}\theta\right) + a_2 r^{4/3}\sin\left(\tfrac{4}{3}\theta\right) + \hat{u} \tag{3.3*.29c}$$

where $\hat{u} \in H^3(\Omega)$. Let $\{\Delta_j\}_{j=1}^{\infty}$ be a sequence of quasi-uniform meshes, $j = 1, 2, \ldots$, i.e., meshes such that $\max_{\tau\in\Delta_j} h_\tau < \mathscr{C} \min_{\tau\in\Delta_j} h_\tau$, and prove that

$$\left\|u_{EX} - u_{S^p_{\Delta_j}}\right\| \le \mathscr{C}(N(\Delta_j))^{\mu} \tag{3.3*.29d}$$

where the rate of convergence μ is:

(1) for $a_1 \neq 0$, with $\mu = \frac{1}{3}$ for $p = 1$ and $p = 2$
(2) for $a_1 = 0$ and $a_2 \neq 0$, $\mu = \frac{1}{2}$, for $p = 1$, and $\mu = \frac{2}{3}$, for $p = 2$
(3) for $a_1 = a_2 = 0$, $\mu = \frac{1}{2}$, for $p = 1$, and $\mu = 1$, for $p = 2$.

P3.3*.3. Consider the elasticity problem in the cracked domain Ω shown in Fig. 3.3*.1, and assume that the edges $\overline{A_1A_0}$ and $\overline{A_0A_6}$ of the crack are traction free. Under the assumption that the boundary conditions on $\Gamma - \overline{A_1A_0} \cup \overline{A_0A_6}$ are sufficiently smooth, the solution can be written in the form

$$u_1^{EX} = ar^{1/2}\psi_1(\theta) + \tilde{u}_1^{EX}, \qquad u_2^{EX} = br^{1/2}\psi_2(\theta) + \tilde{u}_2^{EX} \tag{3.3*.30a}$$

where $\psi_1(\theta)$ and $\psi_2(\theta)$ are analytic functions in θ, and $\tilde{u}_1^{EX}, \tilde{u}_2^{EX} \in H^2(\Omega)$. Let $\{\Delta_j\}_{j=1}^{\infty}$ be a sequence of quasi-uniform meshes, and show that

$$\left\|u_{EX} - u_{S^p_{\Delta_j}}\right\|_{\mathscr{U}} \le \mathscr{C}(N(\Delta_j))^{1/4}. \tag{3.3*.30b}$$

3.4 The finite element interpolation and approximation: a special case

In Section 3.3 we derived asymptotic estimates for the intepolation error under rather general assumptions about the smoothness of the exact solution expressed in terms of various norms (the L^2 or the L^∞ norm of $(d^r/dx^r)(u_{EX})$, the Besov norm, etc.). In this section we analyze the interpolation-error estimates for the special solution x^α. We consider the special case $AE \equiv 1$, $c \equiv 0$, in which the nodal values are exact, and the finite element solution is very close to the interpolant.

Let us consider the regular model problem

$$-\frac{d^2u}{dx^2}(x) = f(x), \quad x \in (0,1) \tag{3.4.1a}$$

with boundary conditions

$$u(0) = 0, \qquad \frac{du}{dx}(1) = \alpha \tag{3.4.1b}$$

and right-hand side

$$f(x) = -\alpha(\alpha - 1)x^{\alpha-2}, \quad x \in (0, 1) \tag{3.4.1c}$$

chosen to correspond to the exact solution

$$u_{EX}(x) = x^{\alpha} \tag{3.4.1d}$$

where we will assume that

$$\alpha > \tfrac{1}{2}, \quad \alpha \neq \text{integer}. \tag{3.4.1e}$$

Note that for this model problem the energy norm coincides with the L^2 norm of the derivative, which is also known as the H^1 seminorm, namely

$$\|u\|_{\mathcal{U}} = |u|_{H^1(I)} = \left\|\frac{du}{dx}\right\|_{L^2(I)}. \tag{3.4.2a}$$

The condition $\alpha > \frac{1}{2}$ is needed to guarantee that $\|x^{\alpha}\|_{\mathcal{U}}$ is finite, namely

$$\left\|\frac{d}{dx}(x^{\alpha})\right\|_{L^2(I)} = \sqrt{\int_0^1 \left(\frac{d}{dx}(x^{\alpha})\right)^2} = \alpha\sqrt{\int_0^1 x^{2\alpha-2}} = \frac{\alpha}{\sqrt{2\alpha-1}} < \infty. \tag{3.4.2b}$$

The variational formulation of (3.4.1) is given by:

Find $u \in H^1_{(0}(I)$ such that

$$\mathcal{B}(u, v) = (u, v)_{H^1_0(I)} = \int_0^1 \frac{du}{dx}\frac{dv}{dx} = \mathcal{L}(v) = \mathcal{B}(x^{\alpha}, v) \quad \forall v \in H^1_{(0}(I). \tag{3.4.3a}$$

Note that

$$|\mathcal{L}(v)| = \left|\int_0^1 \frac{d}{dx}(x^{\alpha})\frac{dv}{dx}\right| \leq \frac{\alpha}{\sqrt{2\alpha-1}}\left\|\frac{dv}{dx}\right\|_{L^2(I)} \tag{3.4.3b}$$

and hence, for $\alpha > \frac{1}{2}$, $\mathcal{L}$ is bounded, and $u_{EX} \equiv x^{\alpha}$ is the unique solution of the above variational problem.

Below we will analyze the L^2 norm of the derivative of the error in the finite element approximation of the function x^{α} for the special class of *radical* (or γ-*radical*) meshes $\{\Delta_n^{\gamma}\}_{n=1}^{\infty}$, with nodes specified by

$$x_j^{\Delta_n^{\gamma}} = \left(\frac{j}{N}\right)^{\gamma}, \quad j = 0, 1, \ldots, N = N(\Delta_n^{\gamma}) = n \tag{3.4.4}$$

where $0 < \gamma < \infty$. For $\gamma = 1$, the sequence of uniform meshes $\{\Delta_n^1\}_{n=1}^{\infty}$ is obtained. For $\gamma < 1$ (respectively $\gamma > 1$), the sequence of meshes which are graded towards $x = 1$ (respectively $x = 0$) is obtained. For example, for $N = 10$, and $\gamma = \frac{1}{10}, \frac{1}{2}$, and 2, respectively, the second node of the mesh is at $x_1^{\Delta_{10}^{0.1}} = 0.79433$, $x_1^{\Delta_{10}^{0.5}} = 0.31623$, and $x_1^{\Delta_{10}^{2}} = 0.01$, respectively. Below, we will often use Δ instead of Δ_n^{γ}, unless we need to use the more precise notation for clarity.

Let us now state some useful estimates of the size of the elements in Δ_n^γ.

Lemma 3.4.1 *For* $j \geq 2$

$$\mathscr{C}_1 x_j^{1-1/\gamma} N^{-1} \leq h_{I_j^{\Delta_n^\gamma}} \leq \mathscr{C}_2 x_j^{1-1/\gamma} N^{-1}, \quad j = 1, 2, \ldots \tag{3.4.5a}$$

$$\mathscr{C}_3 x_{j-1}^{1-1/\gamma} N^{-1} \leq h_{I_j^{\Delta_n^\gamma}} \leq \mathscr{C}_4 x_{j-1}^{1-1/\gamma} N^{-1}, \quad j = 2, 3, \ldots \tag{3.4.5b}$$

where the constants $\mathscr{C}_i$, $i = 1, \ldots, 4$, *depend only on* γ.

Proof We have

$$\frac{h_{I_j^{\Delta_n^\gamma}}}{N^{-1} x_j^{1-1/\gamma}} = N x_j^{1/\gamma}\Big(1 - \frac{x_{j-1}}{x_j}\Big) = j\Big(1 - \Big(\frac{j-1}{j}\Big)^\gamma\Big) = \frac{1 - (1 - 1/j)^\gamma}{1/j}. \tag{3.4.6a}$$

Hence, in order to prove (3.4.5a), it is sufficient to show that there exist constants $\mathscr{C}_1$, $\mathscr{C}_2$ such that

$$\mathscr{C}_1 \leq \frac{1 - (1 - 1/j)^\gamma}{1/j} \leq \mathscr{C}_2, \quad j = 1, 2, 3, \ldots. \tag{3.4.6b}$$

Noting that $1/j \in (0, 1]$, for $j = 1, 2, 3, \ldots$, it follows that the constants $\mathscr{C}_1$, $\mathscr{C}_2$, can be determined from the condition

$$\mathscr{C}_1 \leq \frac{1 - (1 - x)^\gamma}{x} \leq \mathscr{C}_2, \quad x \in (0, 1] \tag{3.4.6c}$$

or

$$\mathscr{C}_1 x \leq 1 - (1 - x)^\gamma \leq \mathscr{C}_2 x, \quad x \in (0, 1]. \tag{3.4.6d}$$

Let

$$\varphi(x) = 1 - (1 - x)^\gamma, \quad x \in (0, 1] \tag{3.4.7a}$$

and note that

$$\frac{d\varphi}{dx}(x) = \gamma(1 - x)^{\gamma - 1}. \tag{3.4.7b}$$

Thus, for $\gamma < 1$, φ is a *monotonically increasing function* with *monotonically increasing derivative*, and it follows that

$$\varphi(0) + \frac{d\varphi}{dx}(0)x \leq \varphi(x) \leq \varphi(0) + (\varphi(1) - \varphi(0))x, \quad \gamma < 1. \tag{3.4.7c}$$

For $\gamma > 1$, φ is a *monotonically increasing function* with *monotonically decreasing derivative*, and we have

$$\varphi(0) + (\varphi(1) - \varphi(0))x \leq \varphi(x) \leq \varphi(0) + \frac{d\varphi}{dx}(0)x, \quad \gamma > 1. \tag{3.4.7d}$$

Taking into account that $\varphi(0) = 0$, $\varphi(1) = 1$, and $(d\varphi/dx)(0) = \gamma$, the above inequalities give

$$\gamma x \leq \varphi(x) \leq x, \quad \gamma < 1 \tag{3.4.8a}$$

and

$$x \leq \varphi(x) \leq \gamma x, \quad \gamma > 1 \tag{3.4.8b}$$

and hence

$$\mathscr{C}_1 = \begin{cases} \gamma, & \gamma < 1 \\ 1, & \gamma \geq 1 \end{cases} \qquad \mathscr{C}_2 = \begin{cases} 1, & \gamma < 1 \\ \gamma, & \gamma \geq 1. \end{cases} \tag{3.4.8c}$$

In order to prove (3.4.5b) we note that

$$\begin{aligned} \frac{h_{I_j^{\Delta_n^\gamma}}}{N^{-1} x_{j-1}^{1-1/\gamma}} &= N x_{j-1}^{-1/\gamma}\Big(\frac{x_j}{x_{j-1}} - 1\Big) = (j-1)\Big(\Big(\frac{j}{j-1}\Big)^\gamma - 1\Big) \\ &= \frac{(1 + 1/(j-1))^\gamma - 1}{1/(j-1)} \end{aligned} \tag{3.4.9a}$$

and the constants $\mathscr{C}_3$, $\mathscr{C}_4$, can be determined from the condition

$$\mathscr{C}_3 x \leq (1+x)^\gamma - 1 \leq \mathscr{C}_4 x, \quad x \in (0, 1]. \tag{3.4.9b}$$

Letting $\varphi(x) = (1+x)^\gamma - 1$, and following the same procedure as above, we get

$$\mathscr{C}_3 = \begin{cases} 2^\gamma - 1, & \gamma < 1 \\ \gamma, & \gamma > 1 \end{cases} \qquad \mathscr{C}_4 = \begin{cases} \gamma, & \gamma < 1 \\ 2^\gamma - 1, & \gamma > 1. \end{cases} \tag{3.4.9c}$$

□

Let us now state the asymptotic estimates for the L^2 norm of the derivative of the interpolation error.

Lemma 3.4.2 *Let $\{\Delta_n^\gamma\}_{n=1}^\infty$ be a sequence of γ-radical meshes, and let*

$$\alpha > \tfrac{1}{2}, \quad \alpha \neq \textit{integer} \tag{3.4.10a}$$

and

$$\delta \overset{\text{def}}{=} \alpha - \frac{p}{\gamma} - 1. \tag{3.4.10b}$$

Then, in the cases that $2\delta \neq -1$, we have

$$\sum_{j=2}^{n} \Big\| \frac{d}{dx}(x^\alpha - \mathcal{I}^p_{I_j^\Delta} x^\alpha) \Big\|^2_{L^2(I_j^{\Delta_n^\gamma})} \leq \mathscr{C}(\alpha, p, \gamma) \Big| \frac{1 - (x_1^{\Delta_n^\gamma})^{2\delta+1}}{2\delta + 1} \Big| N^{-2p} \tag{3.4.10c}$$

while in the case $2\delta = -1$, we have

$$\sum_{j=2}^{n} \left\| \frac{d}{dx}(x^\alpha - \mathcal{I}^p_{I_j^\Delta} x^\alpha) \right\|^2_{L^2(I_j^{\Delta^\gamma_n})} \leq \mathscr{C}(\alpha, p, \gamma) \ln\left(\frac{1}{x_1^{\Delta^\gamma_n}}\right) N^{-2p}. \tag{3.4.10d}$$

Moreover

$$\left\| \frac{d}{dx}(x^\alpha - \mathcal{I}^p_{I_1^\Delta} x^\alpha) \right\|^2_{L^2(I_1^{\Delta^\gamma_n})} \leq \mathscr{C}(\alpha, p) h_{I_1^{\Delta^\gamma_n}} (h_{I_1^{\Delta^\gamma_n}})^{2\mu} \leq \mathscr{C}(\alpha, p) N^{-2\gamma(\mu+1/2)} \tag{3.4.10e}$$

with

$$\mu = \min(p, \alpha - 1). \tag{3.4.10f}$$

Proof Note that the function

$$\left| \frac{d^{p+1} x^\alpha}{dx^{p+1}} \right| = |\alpha(\alpha-1)\ldots(\alpha-p)| x^{\alpha-p-1} \tag{3.4.11a}$$

is monotonically increasing if $\alpha > p+1$, monotonically decreasing if $\alpha < p+1$, and hence

$$\left\| \frac{d^{p+1} x^\alpha}{dx^{p+1}} \right\|_{L^\infty(I_j)} \leq |\alpha(\alpha-1)\cdots(\alpha-p)| x_{\tilde{j}}^{\alpha-p-1} \tag{3.4.11b}$$

where

$$\tilde{j} = \begin{cases} j, & \alpha - p - 1 > 0 \\ j-1, & \alpha - p - 1 < 0. \end{cases} \tag{3.4.11c}$$

Employing the interpolation-error estimate given in Theorem 3.3.2, and taking into account that

$$\left\| \frac{d^{p+1} x^\alpha}{dx^{p+1}} \right\|^2_{L^2(I_j)} \leq h_{I_j^\Delta} (\alpha(\alpha-1)\cdots(\alpha-p))^2 x_{\tilde{j}}^{2(\alpha-p-1)} \tag{3.4.11d}$$

we obtain for $j \geq 2$

$$\begin{aligned} \left\| \frac{d}{dx}(x^\alpha - \mathcal{I}^p_{I_j^\Delta} x^\alpha) \right\|^2_{L^2(I_j^{\Delta^\gamma_n})} &\leq \mathscr{C}(h_{I_j^\Delta})^{2p} \left\| \frac{d^{p+1} x^\alpha}{dx^{p+1}} \right\|^2_{L^2(I_j^{\Delta^\gamma_n})} \\ &\leq \mathscr{C}(h_{I_j^\Delta})^{2p+1} x_{\tilde{j}}^{2(\alpha-p-1)} \end{aligned} \tag{3.4.11e}$$

where $\alpha - p \neq 0$, and $\mathscr{C}$ is independent of j and $h_{I_j^{\Delta^\gamma_n}}$.

Employing Lemma 3.4.1 we get

$$(h_{I_{\tilde{j}}^\Delta})^{2p} x_{\tilde{j}}^{2(\alpha-p-1)} \leq \mathscr{C}^{2p} x_{\tilde{j}}^{2(\alpha-p-1)} x_{\tilde{j}}^{(1-1/\gamma)2p} N^{-2p} \leq \mathscr{C}^{2p} N^{-2p} x_{\tilde{j}}^{2\alpha-2(p/\gamma)-2} \tag{3.4.12a}$$

where $x_{\tilde{j}}$ is either $x_j^{\Delta^\gamma_n}$ or $x_{j-1}^{\Delta^\gamma_n}$, and $\mathscr{C}$ is either the constant $\mathscr{C}_2$ or $\mathscr{C}_4$, which is given in the proof of Lemma 3.4.1. Hence

$$\sum_{j=2}^{n} \left\| \frac{d}{dx}(x^\alpha - \mathcal{I}^p_{I_j^\Delta} x^\alpha) \right\|^2_{L^2(I_j^{\Delta^\gamma_n})} \leq \mathscr{C}^{2p} \sum_{j=2}^{n} h_{I_j^\Delta} x_{\tilde{j}}^{2\delta} N^{-2p} \leq \mathscr{C}^{2p} N^{-2p} \int_{x_1^{\Delta^\gamma_n}}^{1} x^{2\delta} \tag{3.4.12b}$$

and estimates (3.4.10c), (3.4.10d) follow. The estimate (3.4.10e) has already been proven in Lemma 3.3.1. □

We will now employ the above results to construct an asymptotic estimate for the global L^2 norm of the derivative of the interpolation error. We will first consider the case of uniform meshes, which has been already addressed in Theorem 3.3.6.

Theorem 3.4.1 *Let $\alpha > \frac{1}{2}$, $\alpha \neq$ integer. Then, for $\gamma = 1$, we have*

$$\left\|\frac{d}{dx}(x^\alpha - \mathcal{I}^p_{\Delta^1_n} x^\alpha)\right\|_{L^2(I)} \leq E_n(\alpha, p; 1) \stackrel{\text{def}}{=} \begin{cases} \mathcal{C}N^{-p}, & \alpha > p + \frac{1}{2} \\ \mathcal{C}N^{-p}\sqrt{\ln(N)}, & \alpha = p + \frac{1}{2} \\ \mathcal{C}N^{-\alpha+1/2}, & \alpha < p + \frac{1}{2} \end{cases} \tag{3.4.13}$$

where $N = N(\Delta^1_n) = n$.

Proof We will employ the estimates given in Lemma 3.4.2 above, for the special case where $\gamma = 1$ (uniform mesh). We have the following cases:

(1) $\alpha - 1 > p$: Then $\delta = (\alpha - 1) - p > 0$, and (3.4.10c) gives

$$\sum_{j=2}^{N} \left\|\frac{d}{dx}(x^\alpha - \mathcal{I}^p_{I^\Delta_j} x^\alpha)\right\|^2_{L^2(I^\Delta_j)} \leq \mathcal{C}N^{-2p} \tag{3.4.14a}$$

and (3.4.10e) holds with $\mu = p$, namely

$$\left\|\frac{d}{dx}(x^\alpha - \mathcal{I}^p_{I^\Delta_1} x^\alpha)\right\|^2_{L^2(I^\Delta_1)} \leq \mathcal{C}N^{-2(p+1/2)} \leq \mathcal{C}N^{-2p}. \tag{3.4.14b}$$

Hence

$$\left\|\frac{d}{dx}(x^\alpha - \mathcal{I}^p_{\Delta^1_n} x^\alpha)\right\|^2_{L^2(I)} \approx \left\|\frac{d}{dx}(x^\alpha - \mathcal{I}^p_{\Delta^1_n} x^\alpha)\right\|^2_{L^2(I - I_1^{\Delta^1_n})} \leq \mathcal{C}N^{-2p} \tag{3.4.14c}$$

and the energy norm of the error converges to zero as N^{-p}, as the number of elements N diverges to infinity.

(2) $\alpha - 1 < p$: Then we have to distinguish three cases:

(a) $2\delta + 1 > 0$, which means that

$$p + \tfrac{1}{2} < \alpha. \tag{3.4.15a}$$

Then

$$\left\|\frac{d}{dx}(x^\alpha - \mathcal{I}^p_{I^\Delta_1} x^\alpha)\right\|^2_{L^2(I^\Delta_1)} \leq \mathcal{C}N^{-2(\alpha-1/2)} \leq \mathcal{C}N^{-2p} \tag{3.4.15b}$$

and

$$\sum_{j=2}^{N} \left\|\frac{d}{dx}(x^\alpha - \mathcal{I}^p_{I^\Delta_j} x^\alpha)\right\|^2_{L^2(I^\Delta_j)} \leq \mathcal{C}N^{-2p} \tag{3.4.15c}$$

and the energy norm of the error converges to zero as N^{-p}.

(b) $2\delta + 1 = 0$ which means that

$$\alpha = p + \tfrac{1}{2}. \tag{3.4.16a}$$

Then

$$\left\| \frac{d}{dx}(x^\alpha - \mathcal{I}^p_{I_1^\Delta} x^\alpha) \right\|^2_{L^2(I_1^\Delta)} \leq \mathscr{C} N^{-2(\alpha-1/2)} = \mathscr{C} N^{-2p} \tag{3.4.16b}$$

and

$$\sum_{j=2}^{N} \left\| \frac{d}{dx}(x^\alpha - \mathcal{I}^p_{I_j^\Delta} x^\alpha) \right\|^2_{L^2(I_j^\Delta)} \leq \mathscr{C} N^{-2p} \ln(N) \tag{3.4.16c}$$

and we get

$$\begin{aligned} \left\| \frac{d}{dx}(x^\alpha - \mathcal{I}^p_{\Delta} x^\alpha) \right\|_{L^2(I)} &\approx \left\| \frac{d}{dx}(x^\alpha - \mathcal{I}^p_{\Delta} x^\alpha) \right\|_{L^2(I - I_1^\Delta)} \\ &\leq \mathscr{C} N^{-p} \sqrt{\ln(N)} \end{aligned} \tag{3.4.16d}$$

and the energy norm of the error converges to zero as $N^{-p}\sqrt{\ln(N)}$.

(c) $2\delta + 1 < 0$ which implies

$$\alpha < p + \tfrac{1}{2}. \tag{3.4.17a}$$

Then

$$\left\| \frac{d}{dx}(x^\alpha - \mathcal{I}^p_{I_1^\Delta} x^\alpha) \right\|^2_{L^2(I_1^\Delta)} \leq \mathscr{C} N^{-2(\alpha-1/2)} \tag{3.4.17b}$$

and

$$\sum_{j=2}^{N} \left\| \frac{d}{dx}(x^\alpha - \mathcal{I}^p_{I_j^\Delta} x^\alpha) \right\|^2_{L^2(I_j^\Delta)} \leq \mathscr{C} N^{-2(2p-(\alpha-1/2))} \tag{3.4.17c}$$

and because, in this case, $4p - 2\alpha + 1 > 2\alpha - 1$, we get

$$\begin{aligned} \left\| \frac{d}{dx}(x^\alpha - \mathcal{I}^p_{\Delta} x^\alpha) \right\|_{L^2(I)} &\approx \left\| \frac{d}{dx}(x^\alpha - \mathcal{I}^p_{\Delta} x^\alpha) \right\|_{L^2(I_1^\Delta)} \\ &\leq \mathscr{C} N^{-(\alpha-1/2)} \end{aligned} \tag{3.4.17d}$$

and the energy norm of the error converges to zero as $N^{-(\alpha-1/2)}$. □

Let us now consider the case of the radical meshes with $\gamma > 1$, which are clustered toward the singular point at $x = 0$.

Theorem 3.4.2 *For $\alpha > \frac{1}{2}$, and $\alpha \not\equiv$ integer, we have*

$$\left\| \frac{d}{dx}(x^\alpha - \mathcal{I}^p_{\Delta_n^\gamma} x^\alpha) \right\|_{L^2(I)} \leq E_n(\alpha, p; \gamma) = \begin{cases} \mathscr{C} N^{-p}, & \gamma > \gamma_{cr} \\ \mathscr{C} N^{-p}\sqrt{\ln(N)}, & \gamma = \gamma_{cr} \\ \mathscr{C} N^{\gamma(-\alpha+1/2)}, & \gamma < \gamma_{cr} \end{cases} \tag{3.4.18a}$$

where

$$\gamma_{cr} \stackrel{\text{def}}{=} \frac{p}{\alpha - \frac{1}{2}}. \tag{3.4.18b}$$

Proof We have

$$\left\| \frac{d}{dx}(x^\alpha - \mathcal{I}^p_{I_1^\Delta} x^\alpha) \right\|^2_{L^2(I_1^{\Delta_n^\gamma})} \le \mathscr{C}(\alpha, p) N^{-\gamma(2\alpha-1)} \tag{3.4.19a}$$

and in order to have

$$N^{-\gamma(2\alpha-1)} \le N^{-2p} \tag{3.4.19b}$$

we must choose γ such that $\gamma(2\alpha - 1) \ge 2p$, which gives

$$\gamma = \frac{p\epsilon}{\alpha - \frac{1}{2}} \ge \gamma_{cr} \stackrel{\text{def}}{=} \frac{p}{\alpha - \frac{1}{2}}, \quad \epsilon \ge 1. \tag{3.4.20a}$$

For this choice of γ we have

$$2\delta + 1 = 2\alpha - \frac{2}{\epsilon}\left(\alpha - \frac{1}{2}\right) - 1 = 2\alpha\left(1 - \frac{1}{\epsilon}\right) - \left(1 - \frac{1}{\epsilon}\right) = (2\alpha - 1)\left(1 - \frac{1}{\epsilon}\right). \tag{3.4.20b}$$

Noting that $2\alpha - 1 > 0$, for $\epsilon > 1$ we get $2\delta + 1 > 0$, and hence

$$\sum_{j=2}^{N} \left\| \frac{d}{dx}(x^\alpha - \mathcal{I}^p_{I_j^\Delta} x^\alpha) \right\|^2_{L^2(I_j^{\Delta_n^\gamma})} \le \mathscr{C} \left| \frac{1 - x_1^{2\delta+1}}{2\delta + 1} \right| N^{-2p} \le \mathscr{C} N^{-2p}. \tag{3.4.21a}$$

In the case $\epsilon = 1$, we have $\gamma = \gamma_{cr}$ and $2\delta + 1 = 0$, and we get

$$\sum_{j=2}^{N} \left\| \frac{d}{dx}(x^\alpha - \mathcal{I}^p_{I_j^\Delta} x^\alpha) \right\|^2_{L^2(I_j^{\Delta_n^\gamma})} \le \mathscr{C} \ln\left(\frac{1}{x_1^{\Delta_n^\gamma}}\right) N^{-2p} = \mathscr{C} \ln(N^\gamma) N^{-2p} \tag{3.4.21b}$$

and hence we have

$$\left\| \frac{d}{dx}(x^\alpha - \mathcal{I}^p_{\Delta_n^\gamma} x^\alpha) \right\|_{L^2(I)} \le \mathscr{C}(\sqrt{\ln(N)} N^{-p} + N^{-p}) \le \mathscr{C} N^{-p} \sqrt{\ln(N)}. \tag{3.4.22}$$

On the other hand, for $\gamma < \gamma_{cr}$ we have

$$\left\| \frac{d}{dx}(x^\alpha - \mathcal{I}^p_{\Delta_n^\gamma} x^\alpha) \right\|_{L^2(I)} \le \mathscr{C} N^{-\gamma(\alpha - 1/2)}. \tag{3.4.23}$$

□

Recall that from the best-approximation property of the finite element solution, it follows that

$$\|e_{S^p_\Delta}\|_{\mathscr{U}} = \min_{\chi \in S^p_\Delta} \left\| \frac{d}{dx}(x^\alpha - \chi) \right\|_{L^2(I)} \le \left\| \frac{d}{dx}(x^\alpha - \mathcal{I}^p_{\Delta_n^\gamma} x^\alpha) \right\|_{L^2(I)}. \tag{3.4.24}$$

Note that for $p = 1$, we have $u_{S^1_\Delta} \equiv \mathcal{I}^1_\Delta u_{EX}$; however, for $p > 1$, $u_{S^p_\Delta} \not\equiv \mathcal{I}^p_\Delta u_{EX}$, and the inequality in (3.4.24) is strict. In Section 4.3 we will prove that these estimates are optimal in the sense that they also give the asymptotic behavior of the finite element solution, possibly with different constants.

The asymptotic estimates for the energy norm of the error in the finite element solution are summarized by the following Theorem [Gui and Babuška, 1986]:

Theorem 3.4.3 *Let* $u_{EX} \equiv x^\alpha$, *with* $\alpha > \frac{1}{2}$, $\alpha \neq$ *integer, and let* Δ *denote a mesh from a sequence of* γ*-radical meshes. Then,*

$$\lim_{N\to\infty} N^p \|e_{S^p_\Delta}\|_{\mathscr{U}} = \mathscr{C}(\alpha,p)\frac{\gamma^{p+1/2}}{\sqrt{(2\alpha-1)\gamma-2p}}, \quad \gamma > \gamma_{cr} \tag{3.4.25a}$$

and

$$\lim_{N\to\infty} \frac{N^p}{\sqrt{\ln(N)}}\|e_{S^p_\Delta}\|_{\mathscr{U}} = \mathscr{C}(\alpha,p)\gamma^{p+1/2}, \quad \gamma = \gamma_{cr} \tag{3.4.25b}$$

where

$$\mathscr{C}(\alpha,p) = \frac{\alpha\Gamma(\alpha)|\sin(\pi\alpha)|}{\sqrt{\pi}} \frac{\Gamma(p-\alpha+1)}{4^p\sqrt{2p+1}\,\Gamma(p+\frac{1}{2})} \tag{3.4.25c}$$

where Γ *denotes the Gamma function, and*

$$\lim_{N\to\infty} N^{\gamma(\alpha-1/2)}\|e_{S^p_\Delta}\|_{\mathscr{U}} = \mathscr{C}_1(\alpha,\gamma,p), \quad \gamma < \gamma_{cr} \tag{3.4.25d}$$

where $0 < \mathscr{C}_1(\alpha,\gamma,p) < \infty$, *cannot be written in terms of elementary functions.*

We will denote by γ_{opt} the value of γ which minimizes the right-hand side of (3.4.25a). We have:

$$\gamma_{opt} = \frac{p+\frac{1}{2}}{\alpha-\frac{1}{2}}. \tag{3.4.25e}$$

It follows that:

Theorem 3.4.4 *The sequence* $\{\Delta_n^{\gamma_{opt}}\}_{n=1}^\infty$ *leads to the minimal asymptotic value of the energy norm of the error, namely*

$$\lim_{N\to\infty} \frac{\|e_{S_{\Delta_N^{\gamma_{opt}}}}\|_{\mathscr{U}}}{\|e_{S_{\Delta_N^\gamma}}\|_{\mathscr{U}}} \leq 1, \quad \gamma > \gamma_{cr}. \tag{3.4.26}$$

Proof The minimum of the right-hand side of

$$\lim_{N\to\infty} N^p \|e_{S^p_\Delta}\|_{\mathscr{U}} = \mathscr{C}(\alpha,p)\frac{\gamma^{p+1/2}}{\sqrt{(2\alpha-1)\gamma-2p}} \tag{3.4.27a}$$

occurs at the root of

$$\frac{d}{d\gamma}\Big(\frac{\mathscr{C}(\alpha,p)\gamma^{p+1/2}}{\sqrt{(2\alpha-1)\gamma-2p}}\Big) = 0 \tag{3.4.27b}$$

which is given by (3.4.25e), and (3.4.26) follows. □

Above we have proven that for any sequence of *radical* meshes the energy norm of the error converges to zero at best as N^{-p}. Hence, the question arises whether there exist other sequences of meshes for which the energy norm of the error converges to zero faster than N^{-p}. We will now prove that this is impossible.

Theorem 3.4.5 *Let $\alpha > \frac{1}{2}$, $\alpha \neq$ integer. Given any sequence of meshes $\{\Delta_n\}_{n=1}^{\infty}$ there exists a constant $\mathscr{C} > 0$, independent of Δ_n, such that*

$$\mathscr{C} N^{-p} \leq \|e_{S^p_\Delta}\|_{\mathscr{U}} = \min_{\chi \in S^p_{\Delta_n}} \left\| \frac{d}{dx}(x^\alpha - \chi) \right\|_{L^2(I)}. \tag{3.4.28}$$

Proof For simplicity we will discuss only the case of linear elements ($p = 1$), and $\alpha < 2$. Noting that $(d^2/dx^2)(x^\alpha)$ is monotone, and

$$0 < \mathscr{C}_0 = \alpha|\alpha - 1| \leq \min_{x \in I} \left| \frac{d^2}{dx^2}(x^\alpha) \right| \tag{3.4.29a}$$

and that the derivative of the finite element solution is constant in each element, namely

$$\frac{d}{dx}\left((x^\alpha)_{S^1_{\Delta_n}}\right)\Big|_{I^\Delta_j} \equiv \textit{constant} \tag{3.4.29b}$$

it follows that

$$\mathscr{C}_0^2 (h_{I_j^{\Delta_n}})^3 \frac{1}{12} \leq \min_{\chi \in S^1(I_j^{\Delta_n})} \left\| \frac{d}{dx}(x^\alpha - \chi) \right\|^2_{L^2(I_j^{\Delta_n})}. \tag{3.4.29c}$$

Employing $\chi \in S^p_\Delta$, and summing over all the elements, we get

$$\mathscr{C} \sum_{j=1}^{N(\Delta_n)} (h_{I_j^{\Delta_n}})^3 \leq \min_{\chi \in S^p_\Delta} \left\| \frac{d}{dx}(x^\alpha - \chi) \right\|^2_{L^2(I)} \tag{3.4.29d}$$

where the element sizes $h_{I_j^{\Delta_n}}$, $j = 1, \ldots, N(\Delta_n) = n$ are arbitrary.

To estimate the sum $\sum_{j=1}^{N(\Delta_n)} (h_{I_j^{\Delta_n}})^3$, we will partition the elements in two sets

$$J(\Delta_n) \stackrel{\text{def}}{=} \left\{ j \,\middle|\, h_{I_j^{\Delta_n}} \geq \frac{1}{2N(\Delta_n)} \right\} \tag{3.4.30a}$$

and

$$J^*(\Delta_n) \stackrel{\text{def}}{=} \left\{ j \,\middle|\, h_{I_j^{\Delta_n}} < \frac{1}{2N(\Delta_n)} \right\}. \tag{3.4.30b}$$

Then we have

$$\sum_{j \in J^*(\Delta_n)} h_{I_j^{\Delta_n}} \leq N \frac{1}{2N} = \frac{1}{2} \tag{3.4.30c}$$

and since $|I| = 1$, we have

$$\frac{1}{2} \leq \sum_{j \in J(\Delta_n)} h_{I_j^{\Delta_n}} = 1 - \sum_{j \in J^*(\Delta_n)} h_{I_j^{\Delta_n}}. \tag{3.4.30d}$$

Therefore

$$\frac{1}{8} N^{-2} \leq \left(\frac{1}{2N}\right)^2 \sum_{j \in J(\Delta_n)} h_{I_j^{\Delta_n}} \leq \sum_{j \in J(\Delta_n)} (h_{I_j^{\Delta_n}})^3 \leq \sum_{j=1}^{N(\Delta_n)} (h_{I_j^{\Delta_n}})^3 \tag{3.4.30e}$$

and therefore we have

$$\mathscr{C} N^{-2} \leq \mathscr{C} \sum_{j=1}^{N(\Delta_n)} (h_{I_j^{\Delta_n}})^3 \leq \min_{\chi \in S_\Delta^p} \left\| \frac{d}{dx}(x^\alpha - \chi) \right\|^2_{L^2(I)} \tag{3.4.31}$$

which proves (3.4.28) for $p = 1$.

The proof for $p > 1$ follows exactly the same steps, where $(d^{p+1}/dx^{p+1})(x^\alpha)$ must be employed instead of $(d^2/dx^2)(x^\alpha)$. □

Let us now discuss the meaning of Theorems 3.4.1–3.4.4.

1. We have seen that for sequences of radical meshes $\{\Delta_n^\gamma\}_{n=1}^\infty$, with $\gamma > \gamma_{cr}$, the energy norm of the error in the finite element solution $\|e_{S_\Delta^p}\|_{\mathscr{U}}$ converges to zero as N^{-p}, which is the maximal possible rate, independently of the value of the exponent α, i.e., the optimal rate can be achieved for any α, $\alpha = \frac{1}{2} + \epsilon$, with $\epsilon > 0$ arbitrarily small.
2. The value of γ_{cr}, which is the minimal value of γ required to achieve the optimal rate of convergence, increases as the degree p of the elements is increased, and as the singularity becomes stronger as α tends to the value $\frac{1}{2}$ from above.
3. Over-refinement, i.e., the use of sequences of radical meshes with $\gamma > \gamma_{cr}$ is better than under-refinement, namely, the use of sequences of radical meshes with $\gamma < \gamma_{cr}$.
4. The level of over-refinement, i.e., the value of $\gamma > \gamma_{cr}$, does not influence the rate of convergence, but it does influence the value of the constant in the estimate.

Let us now show that the results of the above asymptotic analysis are also useful for rather coarse meshes.

Example 3.4.1 Let us employ the radical meshes $\Delta = \Delta_n^\gamma(x_j^{\Delta_n^\gamma} = N^{-\gamma} j^\gamma, j = 0, 1, \ldots, N(\Delta_n^\gamma) = n)$ and elements of degree p, and compute the relative value of the L^2 norm of the derivative of the error in the finite element solution

$$e_{\text{REL}} = e_{\text{REL}}(u; \gamma, N, p) \stackrel{\text{def}}{=} \left\| \frac{d}{dx}(x^\alpha - (x^\alpha)_{S_\Delta^p}) \right\|_{L^2(I)} \Big/ \left\| \frac{d}{dx}(x^\alpha) \right\|_{L^2(I)} \tag{3.4.32a}$$

and the values of $\mathscr{C}_N = e_{\text{REL}}/E_N$, where

$$E_N = E_N(\alpha, p; \gamma) \stackrel{\text{def}}{=} \begin{cases} N^{-\mu}, & p \neq \gamma(\alpha - \frac{1}{2}) \\ N^{-p}\sqrt{\ln N}, & p = \gamma(\alpha - \frac{1}{2}) \end{cases} \tag{3.4.32b}$$

TABLE 3.4.1a. Values of the relative energy norm of the error in the finite element solution, e_{REL}, and of $\mathscr{C}_N = e_{\mathrm{REL}} N^{\alpha-1/2}$ for the function x^α ($\alpha = 0.7$) obtained using uniform meshes of N elements of degree p ($p = 1$, 2, 3, 4). It can be seen that $\mathscr{C}_N$ practically achieves its asymptotic value for $N \geq 2$. Note that the accuracy achieved by uniform meshes is rather low, e.g., even for $p = 4$ and $N = 512$ only 8% accuracy has been achieved ($\alpha = 0.7$, $\gamma = 1$)

$e_{\mathrm{REL}}(N(\Delta), p)$		$p = 1$	$p = 2$	$p = 3$	$p = 4$
$N = 2$	e_{REL}	0.3742	0.2853	0.2430	0.2168
	$\mathscr{C}_2$	0.42982	0.32776	0.27918	0.24906
$N = 16$	e_{REL}	0.2473	0.1882	0.1603	0.1430
	$\mathscr{C}_{16}$	0.43064	0.32766	0.27908	0.24903
$N = 512$	e_{REL}	0.1237	0.09412	0.08017	0.07151
	$\mathscr{C}_{512}$	0.43078	0.32775	0.27917	0.24901

and

$$\mu = \min\left(p, \gamma\left(\alpha - \tfrac{1}{2}\right)\right). \tag{3.4.32c}$$

We are interested to see *how big N should be so that the approximation is practically in the asymptotic range*, or, in other words, how quickly $\mathscr{C}_N$ approaches its asymptotic value $\mathscr{C}_\infty = \mathscr{C}(\alpha, \gamma, p)$.

Table 3.4.1a gives the values of the relative error in the finite element solution for $\alpha = 0.7$, and $\gamma = 1$ (uniform meshes), for $N = 2, 16, 512$, and the corresponding values of $\mathscr{C}_N$. It can be seen that the asymptotic behavior is practically achieved for $N \geq 2$. From the magnitude of the values of the relative error we conclude that, in general, it is not practical to employ uniform meshes in order to obtain high accuracy (i.e., $e_{\mathrm{REL}} < 1\%$). For example, only 8% accuracy was obtained for a uniform mesh of $N = 512$ quartic ($p = 4$) elements or $M = 2025$ degrees of freedom. Table 3.4.1b gives the values of the relative error for the Lagrangian interpolant. Comparing Tables 3.4.1a and 3.4.1b, it can be seen that, as expected, for $p > 1$ the energy norm of the interpolation error is bigger than the energy norm of the error in the finite element solution. This difference can be minimized by employing the optimal interpolation points to construct the Lagrangian interpolant; in the reported results a uniform set of points was used. Figure 3.4.1 shows the graph of the relative error, e_{REL}, versus the number of degrees of freedom, $M = Np$, in log-log scale for $\alpha = 2.9$. We have

$$e_{\mathrm{REL}}(N, p) \simeq \mathscr{C} N^{-\mu} \tag{3.4.33}$$

where $\mu = \min(p, \alpha - \frac{1}{2})$. For $p \leq 2$ the rate of convergence μ, is equal to p, while for $p \geq 3$ the rate of convergence is equal to 2.4. Table 3.4.2 gives the values of e_{REL} and $\mathscr{C}_N$ for $\alpha = 1.1$, $p = 2$, and various radical meshes ($N = 2^k$, $k = 1, 2, \ldots, 9$, $\gamma = 2, \frac{10}{3}, \frac{25}{6}, 5$). Note that the asymptotic behavior of the error is achieved for rather coarse meshes

TABLE 3.4.1b. Values of the relative error in the Lagrangian interpolant constructed using a uniformly distributed set of interpolation points in each element. The corresponding values of the relative error in the finite element solution are given in Table 3.4.1a. It can be seen that the interpolation error is different than the error in the finite element solution and this difference increases with p ($\alpha = 0.7$, $\gamma = 1$)

$N(\Delta)$	e_{REL}			
	$p = 1$	$p = 2$	$p = 3$	$p = 4$
2	0.3742	0.2870	0.2480	0.2255
16	0.2473	0.1893	0.1636	0.1488
512	0.1237	0.09466	0.08180	0.07439

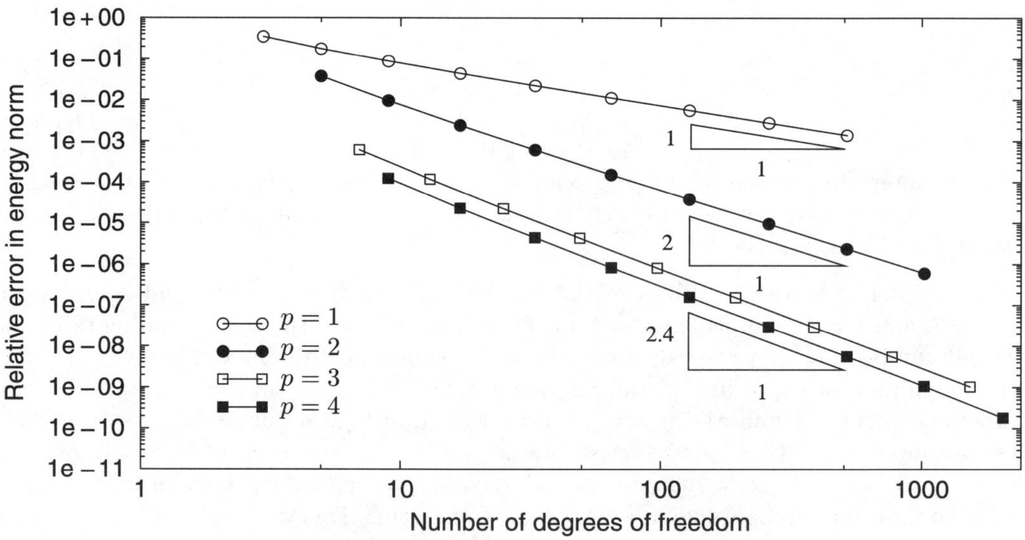

FIG. 3.4.1. Graph of the relative energy norm of the error in the finite element solution, e_{REL}, versus the number of degrees of freedom, $M = Np$, for the exact solution x^α ($\alpha = 2.9$), and uniform meshes of elements of degree p ($p = 1, 2, 3, 4$). Note that for $p = 1$ or 2 we have $p < \alpha - \frac{1}{2} = 2.4$, and the rate of convergence is equal to p, while for $p \geq 3$, the rate of convergence is equal to $\alpha - \frac{1}{2} = 2.4$

($N \geq 4$) for all the γs employed. Figure 3.4.2 shows the graph of the relative error, versus the number of degrees of freedom for $\alpha = 1.1$, and for various radical meshes. We note that the convergence graphs are rather insensitive to the choice of $\gamma \geq \gamma_{cr} = 3.33$, in the neighborhood of the optimal radical mesh ($\gamma_{opt} = 4.167$). Figure 3.4.3 gives the graphs of the relative error versus the number of degrees of freedom in the mesh (e_{REL} versus $M = Np$) for $\alpha = 1.1$ and the optimal radical meshes with elements of degree

TABLE 3.4.2. Values of the relative energy norm of the error in the finite element solution, e_{REL}, and the values of the constant $\mathscr{C}_N = e_{\mathrm{REL}}/E_N$ for the function x^α ($\alpha = 1.1$) obtained using various radical meshes ($x_j^\Delta = N^{-\gamma}j^\gamma$ with $\gamma = 2, \frac{10}{3}, \frac{25}{6}, 5$ and $N = 2^k$, $k = 1, 2, \ldots, 9$) of quadratic elements ($p = 2$). Note that $\mathscr{C}_N$ practically achieves its asymptotic value $\mathscr{C}_\infty$, and the asymptotic rate of convergence (equal to $\gamma(\alpha - \frac{1}{2})$ for $\gamma < \gamma_{cr}$, or equal to p for $\gamma > \gamma_{cr}$) is achieved for $N \geq 16$

$N(\Delta)$	e_{REL}	$\mathscr{C}_N$	e_{REL}	$\mathscr{C}_N$
	$\gamma = 2.0$		$\gamma_{cr} = 3.3333$	
2	0.1776×10^{-1}	0.4080×10^{-1}	0.1658×10^{-1}	0.7966×10^{-1}
4	0.7859×10^{-2}	0.4148×10^{-1}	0.5570×10^{-2}	0.7569×10^{-1}
16	0.1500×10^{-2}	0.4179×10^{-1}	0.4869×10^{-3}	0.7486×10^{-1}
64	0.2844×10^{-3}	0.4182×10^{-1}	0.3721×10^{-4}	0.7474×10^{-1}
	$\gamma_{opt} = 4.167$		$\gamma = 5.0$	
2	0.1998×10^{-1}	0.799×10^{-1}	0.2410×10^{-1}	0.964×10^{-1}
4	0.6480×10^{-2}	1.037×10^{-1}	0.7865×10^{-2}	1.258×10^{-1}
16	0.4818×10^{-3}	1.233×10^{-1}	0.5585×10^{-3}	1.430×10^{-1}
64	0.3137×10^{-4}	1.285×10^{-1}	0.3541×10^{-4}	1.450×10^{-1}

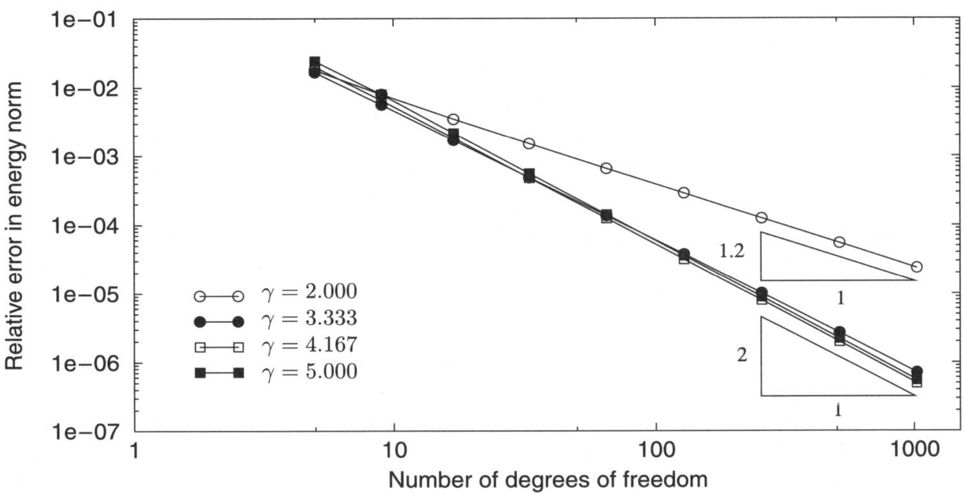

FIG. 3.4.2. Graph of the relative energy norm of the error in the finite element solution, e_{REL}, versus the number of degrees of freedom, $M = Np$, for the exact solution x^α ($\alpha = 1.1$) and various radical meshes of quadratic elements ($p = 2$). The optimal rate of convergence, which is equal to two, and is obtained for $\gamma_{opt} = 4.167$. It can be seen that the error curves are rather insensitive to the selection of γ in the neighborhood of γ_{opt}

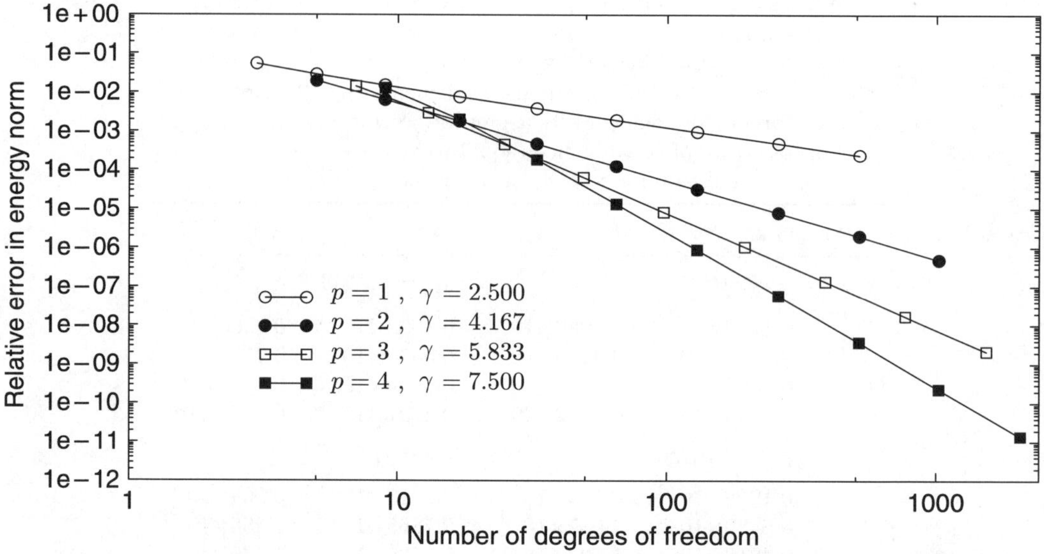

FIG. 3.4.3. Graph of the relative energy norm of the error in the finite element solution, e_{REL}, versus the number of degrees of freedom, $M = Np$, for the exact solution x^α ($\alpha = 1.1$) and the optimal radical meshes for elements of degree p ($p = 1, 2, 3, 4$). Note that each graph attains its corresponding optimal asymptotic rate equal to p

p ($p = 1, 2, 3, 4$). Note that every graph attains rate of convergence equal to p, as the number of degrees of freedom diverges to infinity.

We also examined the question of *optimality of the radical meshes with* γ_{opt} *among all radical meshes with the same number of degrees of freedom, when the meshes are rather coarse.* Table 3.4.3 gives the values of the ratio of the error for various radical meshes to the error for the asymptotically optimal radical mesh. Note that the optimal radical mesh gives the smallest values of the relative error (among all other radical meshes) for sufficiently large N. For coarse meshes, the smallest error may be achieved for some other radical mesh. Nevertheless the error for the mesh which corresponds to γ_{opt} is close to the minimum value of the error, even for very coarse meshes.

Another important question is, *can the optimal radical meshes be used in practical computations?* Table 3.4.4 gives the values of the minimum and the maximum mesh size for the critical and optimal radical meshes corresponding to $\alpha = 0.7$, $N = 256$, for $1 \leq p \leq 5$. From these results it can be seen that, in the case of strong singularities and high p, the optimal radical meshes are highly graded and, unless higher precision is employed, some problems with round-off could occur; see also [Babuška, Strouboulis, and Copps, 1997], [Babuška and Rheinboldt, 1979].

Remark 3.4.1 The objective of the analysis given above is to show the existence of properly refined meshes for which the optimal rate of convergence can be achieved. In practical computations we are often interested in achieving engineering accuracy, e.g., 5%, which could be achieved before the approximation enters its asymptotic range, and reducing the value of the constant $\mathscr{C}$ may be more important than achieving the optimal rate.

TABLE 3.4.3. Values of the ratio of the relative energy norm of the error in the finite element solution for various radical meshes with respect to the corresponding error in the finite element solution for the optimal radical mesh, $e_{\text{REL}}/e_{\text{REL}}^{opt}$. Note that asymptotically the error is minimum for the optimal radical mesh. For coarse meshes the minimum error over the entire class of radical meshes (which may be achieved for a radical mesh different than the asymptotically optimal one) is rather close to the error for the asymptotically optimal radical mesh ($\alpha = 0.7$, $p = 2$)

$N(\Delta)$	$e_{\text{REL}}/e_{\text{REL}}^{opt}$						e_{REL}^{opt}
	$\gamma = 1$	$\gamma = 4$	$\gamma = 7$	$\gamma_{cr} = 10$	$\gamma_{opt} = 12.5$	$\gamma = 15$	
2	0.98	0.69	0.74	0.90	1.0	1.06	2.915×10^{-1}
4	1.60	0.76	0.67	0.83	1.0	1.17	1.548×10^{-1}
8	3.84	1.21	0.77	0.83	1.0	1.22	0.5630×10^{-1}
16	11.19	2.33	1.02	0.89	1.0	1.20	0.1681×10^{-1}
32	35.85	4.92	1.44	0.96	1.0	1.17	0.4572×10^{-2}
64	119.81	10.84	2.12	1.05	1.0	1.15	0.1191×10^{-2}
128	408.82	24.41	3.15	1.14	1.0	1.13	0.3038×10^{-3}
256	1409.01	55.51	4.73	1.22	1.0	1.12	0.7672×10^{-4}
512	4881.74	126.87	7.14	1.31	1.0	1.12	0.1928×10^{-4}

TABLE 3.4.4. The minimum and maximum size of the elements for the critical and optimal radical meshes of $n = 256$ elements for element degree p ($p = 1, 2, 3, 4, 5$). Note that for $p > 1$ the ratio of the maximum to the minimum element size in the optimal mesh is more than 10^{28} (more than 10^{52} for $p = 4$) for the optimal mesh and close to that for the critical mesh. It follows that for high p (e.g., $p \geq 3$) it could be difficult to employ the optimal or the critical mesh (or any other mesh for which the optimal rate of convergence is achieved) because of the round-off error ($\alpha = 0.7$, $n = 256$)

p	γ_{cr}	γ_{opt}	γ_{cr}		γ_{opt}	
			$\min_{j=1,\ldots,N(\Delta)} \lvert I_j^\Delta \rvert$	$\max_{j=1,\ldots,N(\Delta)} \lvert I_j^\Delta \rvert$	$\min_{j=1,\ldots,N(\Delta)} \lvert I_j^\Delta \rvert$	$\max_{j=1,\ldots,N(\Delta)} \lvert I_j^\Delta \rvert$
1	5	7.5	9.09×10^{-13}	1.94×10^{-2}	8.67×10^{-19}	2.89×10^{-2}
2	10	12.5	8.27×10^{-25}	3.84×10^{-2}	7.89×10^{-31}	4.77×10^{-2}
3	15	17.5	7.52×10^{-37}	5.70×10^{-2}	7.17×10^{-43}	6.62×10^{-2}
4	20	22.5	6.84×10^{-49}	7.53×10^{-2}	6.53×10^{-55}	8.43×10^{-2}
5	25	27.5	6.22×10^{-61}	9.32×10^{-2}	5.93×10^{-67}	1.02×10^{-1}

In this section, we addressed the interpolation error for the special function x^α which is an one-dimensional analogue of the singularities of solutions of elliptic partial differential equations in two-dimensional domains with corners. We have seen that the rate of convergence depends on the refinement of the mesh in the neighborhood of the singular point. For elements of cubic degree or higher ($p \geq 3$) and strong singularities (e.g. $0.5 < \alpha \leq 0.6$), it may be practically impossible to use the meshes for which the optimal rate of convergence is achieved, because the relative size of the maximum to the minimum element for the properly refined meshes is very large and difficulties due to the round-off error could occur.

Problems

P3.4.1. Consider the regular model problem given in (3.4.1) with exact solution

$$u_{\alpha,\delta}(x) = (x+\delta)^{\alpha} - \alpha(1+\delta)^{\alpha-1}x \tag{3.4.34}$$

and let Δ denote a uniform mesh. Derive an asymptotic estimate for the energy norm of the error, $\|e_{S^p_\Delta}\|_{\mathcal{U}}$, as function of α, δ, and p.

P3.4.2. Consider P3.4.1 for the exact solution $u_{EX}(x) = x\ln x - x$, and elements of degree $p = 1, 2, 3$.

P3.4.3. Consider the singular model problem

$$\frac{d}{dx}\left(x^{\vartheta}\frac{du}{dx}(x)\right) = f(x), \quad x \in (0,1), \qquad u(0) = 0, \qquad \frac{du}{dx}(1) = 0 \tag{3.4.35}$$

with $0 \le \vartheta < 1$, and let f be such that the exact solution $u_{\alpha,\delta}$ is given in P3.4.1, and let $\{\Delta_n\}_{n=1}^{\infty}$ be the sequence of uniform meshes. Derive an asymptotic estimate for the energy norm, $\|e_{S^p_{\Delta_n}}\|_{\mathcal{U}^{\vartheta}}$, as a function of $h, \vartheta, \alpha, \delta$, and p.

P3.4.4. Let $\{\Delta_n\}$ be a sequence of radical meshes in (0, 1). Show that there exist constants $\mathscr{C}_1 \le \mathscr{C}_2$ such that

$$\mathscr{C}_1 \le N(\Delta_n)\int_{I_j^{\Delta}} x^{1/\gamma-1} \le \mathscr{C}_2. \tag{3.4.36}$$

3.4* The interpolation error for the function $r^{\alpha}\varphi(\theta)$ in two dimensions

In Section 2.3* we have seen that the exact solution in the neighborhood of each corner point, where there is a discontinuity in the tangent and/or the type of boundary condition along the boundary, has the same behavior as the singular solution

$$u_{EX}(r,\theta) = r^{\alpha}|\ln r|^{\beta}\chi(\theta) \tag{3.4*.1}$$

where r, θ are polar coordinates centered at the corner point, the exponents α, β depend on the magnitude of the interior angle, and the type of boundary conditions applied on the edges emanating from the corner point, and χ is a smooth function. For example, in the case of the Laplacian in an L-shaped domain with homogeneous essential boundary conditions applied on the edges emanating from the re-entrant corner, we have

$$u_{EX}(r,\theta) = Ar^{2/3}\sin\left(\frac{2\theta}{3}\right) + \tilde{u}_{EX} \tag{3.4*.2}$$

where $\tilde{u}_{EX}$ is a function which is smoother than $r^{2/3}\sin(2\theta/3)$. Therefore, the analysis of the convergence of the finite element method for solutions of the type (3.4*.1) and the corresponding domains is very relevant to the understanding of the behavior of the finite element method in practical problems.

In two dimensions, it is necessary to weaken the notion of radical mesh so that it applies to the meshes that are constructed by employing practical mesh generators.

A sequence of meshes $\{\Delta_n^{\gamma}\}_{n=1}^{\infty}$ is *almost γ-radical* or *$(\mathscr{C}_1, \mathscr{C}_2; \gamma)$-radical* if there exist constants $0 < \mathscr{C}_1 \le \mathscr{C}_2 < \infty$, independent of n, such that

$$\mathscr{C}_1 \le N(\Delta_n^{\gamma})\int_{\tau^{\Delta_n^{\gamma}}} r^{2(1/\gamma-1)} \le \mathscr{C}_2 \tag{3.4*.3}$$

for all elements $\tau^{\Delta_n^{\gamma}}$.

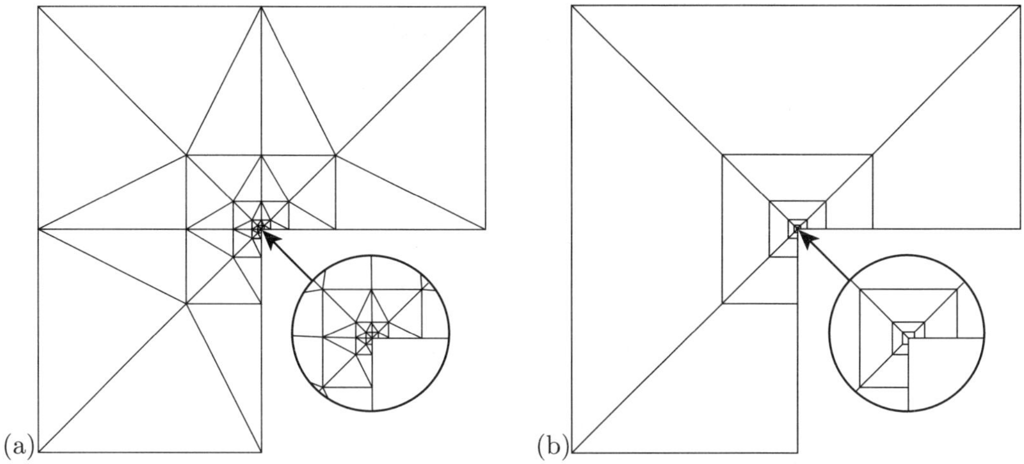

FIG. 3.4*.1. Examples of γ-radical meshes in two dimensions

Analogously as in the one-dimensional case, by employing a proper selection of γ, depending on α in (3.4*.1) we can obtain a rate of convergence $O(N^{-p/2})$, i.e., the same rate of convergence as in the case of the smooth solution. For $\gamma = 0$, the rate of convergence is $O(N^{-\mu/2}(\ln N)^{\beta})$, $\mu = \min(\alpha, p)$. Further details may be found in [Babuška, Kellogg, and Pitkaranta, 1979].

There are many ways to create the almost γ-radical meshes in two dimensions. Figure 3.4*.1a shows an example of a radical mesh of triangles and quadrilaterals for the L-shaped domain.

Here we have outlined the setting for the analysis of solutions with algebraic singularities at corner points in two dimensions. Although the analysis is somewhat technical and was not presented, the conclusions are analogous as in the one-dimensional case.

Problems

P3.4*.1. Does the mesh Figure 3.4*.1a satisfy the aspect-ratio condition?

P3.4*.2. Let Ω be the L-shaped domain with boundary conditions corresponding to $u_{EX} = r^{2/3}\sin(2\theta/3)$, where r, θ are polar coordinates centered at the re-entrant corner. Using the results of Section 3.3*, estimate $|u_{EX} - \mathcal{I}^p_{\Delta_n} u_{EX}|_{H^1(\Omega)}$ for elements of degree p, and the almost γ-radical meshes constructed in P3.4*.2. Show that there exists $\beta > 0$ such that

$$\|u_{EX} - \mathcal{I}^p_{\Delta_n} u_{EX}\|_{\mathcal{U}} \leq \mathcal{C}(N(\Delta_n))^{-\beta} \tag{3.4*.4}$$

where $\mathcal{C} = \mathcal{C}(\gamma)$, is independent of n. Show that the optimal rate $\beta = p/2$ can be achieved for $\gamma \geq \gamma_{cr}$. Determine γ_{cr} and compare it with the corresponding value in the one-dimensional case.

P3.4*.3. Using an analogous argument as in Theorem 3.4.5, prove that no mesh could give better asymptotic convergence than $O(N^{-p/2})$.

P3.4*.4. Formulate and prove analogs to Lemma 3.4.2 and Theorem 3.4.1 for the meshes shown in Fig. 3.4*.1.

3.5 The description of the mesh and analyses of its optimality

In the previous section, we analyzed the asymptotic behavior of the energy norm of the interpolation error for a special family of meshes $\{\Delta_n^\gamma\}_{n=1}^\infty$, the radical meshes with nodes given by

$$x_j^{\Delta_n^\gamma} = \left(\frac{j}{N}\right)^\gamma, \quad j = 0, 1, \ldots, N = N(\Delta_n^\gamma). \tag{3.5.1a}$$

A more general construction is given by

$$x_j^{\Delta_n} = g\left(\frac{j}{N}\right), \quad j = 0, 1, \ldots, N = N(\Delta_n) \tag{3.5.1b}$$

where $g : [0,1] \mapsto [0,l]$ is a *continuous, strictly increasing function*, which is known as the *mesh-grading function.* The radical meshes are generated by employing $g \equiv x^\gamma$ as the mesh-grading function. The idea of using a grading function to characterize the meshes *cannot be generalized to more than one dimension.*

Another way to describe sequences of meshes is in terms of a *mesh-density function.*

We will call any function $\Phi : I = (0,l) \mapsto \mathbb{R}_0^+$, which is such that for any $0 \leq a < b \leq l$, we have

$$\int_a^b \Phi > 0 \tag{3.5.2a}$$

and

$$\int_0^l \Phi = 1 \tag{3.5.2b}$$

a *mesh-density* or *density function.*

To any density function Φ, there is a unique corresponding sequence of *density meshes* $\{\Delta_n^\Phi\}_{n=1}^\infty$, which are specified by the condition

$$\int_{I_j^{\Delta_n^\Phi}} \Phi = \frac{1}{n}, \quad j = 1, 2, \ldots, N(\Delta_n^\Phi) = n. \tag{3.5.2c}$$

We will call Φ and Δ_n^Φ the corresponding *density mesh* of n elements.

Let us determine the relation between the grading and the density function. Given a density function Φ, let $\Gamma : [0,l] \to [0,1]$ be its primitive, namely

$$\Gamma(x) \stackrel{\text{def}}{=} \int_0^x \Phi. \tag{3.5.3a}$$

It follows that $\Gamma \in C^0([0,l])$, and that for any $0 \leq a < b \leq l$, we have

$$\Gamma(b) - \Gamma(a) = \int_a^b \Phi > 0. \tag{3.5.3b}$$

Thus Γ is a continuous and strictly increasing function, and hence it is invertible with Γ^{-1} continuous, strictly increasing, and we have

$$\Gamma\big(x_j^{\Delta_n^\Phi}\big) = \int_0^{x_j^{\Delta_n^\Phi}} \Phi = \sum_{k=1}^{j} \int_{I_k^{\Delta_n^\Phi}} \Phi = \frac{j}{N} \quad \Longrightarrow \quad x_j^{\Delta_n^\Phi} = \Gamma^{-1}\Big(\frac{j}{N}\Big). \tag{3.5.3c}$$

It follows that Γ^{-1} is the grading-function which generates the sequence of density meshes $\{\Delta_n^\Phi\}_{n=1}^\infty$.

Remark 3.5.1 Let us give the motivation leading to the idea of the density function. Given any sequence of meshes $\{\Delta_n\}_{n=1}^\infty$, $N(\Delta_n) = n$, and any interval $J = (a, b)$, $0 \le a < b \le l$, let $N(\Delta_n, J)$ be the number of elements of Δ_n which intersect J, and

$$\varphi_n(J) = \frac{N(\Delta_n, J)}{N(\Delta_n)}. \tag{3.5.4a}$$

Let Φ be the function which satisfies

$$\int_J \Phi = \varphi(J) = \lim_{n\to\infty} \varphi_n(J) \tag{3.5.4b}$$

provided that the limit φ exists. We then have

$$\Phi(x) = \lim_{|J_x|\to 0} \frac{\varphi(J_x)}{|J_x|} \tag{3.5.4c}$$

where J_x are intervals of length $|J_x|$ such that $x \in J_x$. The function Φ, if it exists, is the density function corresponding to the given sequence of meshes $\{\Delta_n\}_{n=1}^\infty$.

Remark 3.5.2 In the one-dimensional case, it is always possible to construct the meshes Δ_n which achieve the equidistribution

$$\int_{I_j^{\Delta_n}} \Phi = \frac{1}{n}, \quad j = 1, \ldots, N(\Delta_n) \tag{3.5.5a}$$

for any admissible Φ. This may not however be possible in more than one dimension, where there are restrictions imposed by the available mesh generator. For this reason, we will broaden the definition of the density meshes:

We will call a sequence $\{\Delta_n\}_{n=1}^\infty$, $(\mathscr{C}_1, \mathscr{C}_2)$-equilibrated with respect to the density Φ, if

$$0 < \mathscr{C}_1 \frac{1}{N(\Delta_n)} \le \int_{I_j^{\Delta_n}} \Phi \le \mathscr{C}_2 \frac{1}{N(\Delta_n)} < \infty, \quad j = 1, \ldots, N(\Delta_n). \tag{3.5.5b}$$

Note that the density meshes which satisfy (3.5.2c) are (1,1)-equilibrated.

Below we will let $l = 1$ and $I = (0, 1)$.

Let us now show that the estimates for the interpolation error can be expressed in terms of the density function.

Theorem 3.5.1 *Let $u_{EX} \in C^0(I)$ with*

$$\frac{d^{p+1}}{dx^{p+1}}(u_{EX}) \in L^2(I). \tag{3.5.6a}$$

Given a sequence of density meshes $\{\Delta_n^\Phi\}_{n=1}^\infty$, we have

$$\left\|\frac{d}{dx}(u_{EX} - \mathcal{I}^p_{\Delta_n^\Phi} u_{EX})\right\|_{L^2(I)} \approx \frac{\mathscr{C}(p)}{(N(\Delta_n^\Phi))^p}\left\|\frac{1}{\Phi^p}\frac{d^{p+1}}{dx^{p+1}}(u_{EX})\right\|_{L^2(I)} \tag{3.5.6b}$$

where the symbol $\approx$ denotes equality up to higher-order terms in $N(\Delta_n^\Phi)$.

Proof Employing the approximation

$$\int_{I_j^\Delta}\left(\frac{d^{p+1}}{dx^{p+1}}(u_{EX})\right)^2 \approx h_{I_j^\Delta}\left(\frac{d^{p+1}}{dx^{p+1}}(u_{EX})(x_j^\Delta)\right)^2 \tag{3.5.7a}$$

the interpolation estimate for the L^2 norm, gives

$$\left\|\frac{d}{dx}(u_{EX} - \mathcal{I}^p_{\Delta_n^\Phi} u_{EX})\right\|^2_{L^2(I_j^\Delta)} \approx \mathscr{C} h^{2p+1}_{I_j^{\Delta_n^\Phi}}\left(\frac{d^{p+1}}{dx^{p+1}}(u_{EX})\big(x_j^{\Delta_n^\Phi}\big)\right)^2 \tag{3.5.7b}$$

and summing over all the elements in the mesh, we get

$$\left\|\frac{d}{dx}(u_{EX} - \mathcal{I}^p_{\Delta_n^\Phi} u_{EX})\right\|^2_{L^2(I)} \approx \mathscr{C}\sum_{j=1}^{N(\Delta_n^\Phi)} h^{2p+1}_{I_j^{\Delta_n^\Phi}}\left(\frac{d^{p+1}}{dx^{p+1}}(u_{EX})\big(x_j^{\Delta_n^\Phi}\big)\right)^2. \tag{3.5.7c}$$

From the definition of the density meshes, (3.5.2b), we get

$$\frac{1}{N(\Delta_n^\Phi)} = \int_{I_j^{\Delta_n^\Phi}} \Phi \approx h_{I_j^{\Delta_n^\Phi}}\Phi\big(x_j^{\Delta_n^\Phi}\big) \tag{3.5.8a}$$

and hence

$$h_{I_j^{\Delta_n^\Phi}} \approx \frac{1}{N(\Delta_n^\Phi)\Phi\big(x_j^{\Delta_n^\Phi}\big)}. \tag{3.5.8b}$$

Substituting into (3.5.7c) we obtain

$$\left\|\frac{d}{dx}(u_{EX} - \mathcal{I}^p_{\Delta_n^\Phi} u_{EX})\right\|^2_{L^2(I)}$$

$$\approx \frac{\mathscr{C}}{(N(\Delta_n^\Phi))^{2p}}\sum_{j=1}^{N(\Delta_n^\Phi)} \frac{h_j^{\Delta_n^\Phi}}{(\Phi\big(x_j^{\Delta_n^\Phi}\big))^{2p}}\left(\frac{d^{p+1}}{dx^{p+1}}(u_{EX})\big(x_j^{\Delta_n^\Phi}\big)\right)^2 \tag{3.5.9a}$$

and noting that asymptotically

$$\sum_{j=1}^{N(\Delta_n^\Phi)} \frac{1}{(\Phi\big(x_j^{\Delta_n^\Phi}\big))^{2p}}\left(\frac{d^{p+1}}{dx^{p+1}}(u_{EX})\big(x_j^{\Delta_n^\Phi}\big)\right)^2 h_j^{\Delta_n^\Phi} \approx \int_I \frac{1}{\Phi^{2p}}\left(\frac{d^{p+1}}{dx^{p+1}}(u_{EX})\right)^2 \tag{3.5.9b}$$

(3.5.6b) follows. □

Remark 3.5.3 Equation (3.5.6b) is an *asymptotic estimate* in the sense that it holds as the number of elements $N(\Delta_n^\Phi)$ in the mesh Δ_n^Φ diverges to infinity. Note that *because the meshes Δ_n^Φ are density meshes, it is guaranteed that the mesh size $h^{\Delta_n^\Phi}$, tends to zero as the number of elements, $N(\Delta_n^\Phi)$, diverges to infinity.*

Based on the asymptotic estimate (3.5.6b), we will now define the optimal density function, $\Phi_{opt}^{H^1}$, as the solution of the minimization problem:

Find

$$\Phi_{opt}^{H^1} \in \mathcal{D} \overset{\text{def}}{=} \{\Phi \mid \textstyle\int_I \Phi = 1,\ \int_a^b \Phi > 0,\ \forall a, b : 0 \leq a < b < 1\} \tag{3.5.10a}$$

and

$$\left\| \left(\frac{1}{\Phi_{opt}^{H^1}}\right)^p \frac{d^{p+1}}{dx^{p+1}}(u_{EX}) \right\|_{L^2(I)} \leq \min_{\Phi \in \mathcal{D}} \left\| \left(\frac{1}{\Phi}\right)^p \frac{d^{p+1}}{dx^{p+1}}(u_{EX}) \right\|_{L^2(I)}. \tag{3.5.10b}$$

Using the method of Lagrange multipliers, we obtain

$$\Phi_{opt}^{H^1} = K \left(\frac{d^{p+1}}{dx^{p+1}}(u_{EX}) \right)^{2/(2p+1)} \tag{3.5.10c}$$

with

$$K = \left(\int_I \left(\frac{d^{p+1}}{dx^{p+1}}(u_{EX}) \right)^{2/(2p+1)} \right)^{-1}. \tag{3.5.10d}$$

Following similar steps we can construct the asymptotic estimates and define the optimal density function with respect to other norms. For example, the asymptotic estimate for the L^2 norm of the interpolation error is

$$\|u_{EX} - \mathcal{I}_\Delta^p u_{EX}\|_{L^2(I)} \approx \frac{\mathscr{C}}{N^{p+1}} \left\| \frac{1}{\Phi^{p+1}} \frac{d^{p+1}}{dx^{p+1}}(u_{EX}) \right\|_{L^2(I)} \tag{3.5.11a}$$

and the corresponding optimal density function is given by

$$\Phi_{opt}^{L^2} = \left(\int_I \left(\frac{d^{p+1}}{dx^{p+1}}(u_{EX}) \right)^{2/(2p+3)} \right)^{-1} \left(\frac{d^{p+1}}{dx^{p+1}}(u_{EX}) \right)^{2/(2p+3)}. \tag{3.5.11b}$$

Remark 3.5.4 The derivation of the optimal density function Φ_{opt}, holds under the assumption that $(d^{p+1}u/dx^{p+1})(x)$, is equal to zero only in a set of measure zero, which usually is the case in practice.

Let us give some examples of optimal density functions.

Example 3.5.1 Let $u_{EX}(x) = \sin((\pi/2)x)$, $l = 1$. Then

$$\frac{d^{p+1}}{dx^{p+1}}(u_{EX})(x) = \begin{cases} (-1)^{(p-1)/2} \left(\frac{\pi}{2}\right)^{p+1} \sin\left(\frac{\pi}{2}x\right), & \text{for } p \text{ odd} \\ (-1)^{p/2} \left(\frac{\pi}{2}\right)^{p+1} \cos\left(\frac{\pi}{2}x\right), & \text{for } p \text{ even} \end{cases} \tag{3.5.12a}$$

From

$$\Phi_{opt}^{H^1} = K\Big(\frac{d^{p+1}}{dx^{p+1}}(u_{EX})\Big)^{2/(2p+1)} \tag{3.5.12b}$$

it follows that $\Phi_{opt}^{H^1}$ is maximum at $x = 1$ for p odd (respectively at $x = 0$ for p even), and hence the local mesh size for the optimal mesh is minimal at $x = 1$ for p odd (respectively $x = 0$ for p even). Hence *the optimal mesh is completely different for elements of different degrees.*

For the L^2 norm we have

$$\Phi_{opt}^{L^2} = K\Big(\frac{d^{p+1}}{dx^{p+1}}(u_{EX})\Big)^{2/(2p+3)} \tag{3.5.12c}$$

and hence *the optimal meshes are different for different norms.*

Example 3.5.2 Let $u_{EX} \equiv x^\alpha$, $\alpha > \frac{1}{2}$. Note that, in the case $\alpha < p + 1$,

$$\frac{d^{p+1}}{dx^{p+1}}(u_{EX})(x) = \alpha(\alpha-1)\cdots(\alpha-p)x^{\alpha-p-1} \tag{3.5.13a}$$

is unbounded at $x = 0$. Nevertheless the asymptotic estimate used in the derivation of the optimal density is still valid, and we have

$$\Phi_{opt}(x) = \frac{x^{2(\alpha-p-1)/(2p+1)}}{\int_I x^{2(\alpha-p-1)/(2p+1)}} \tag{3.5.13b}$$

and the inverse of the corresponding optimal grading function g_{opt}, is

$$g_{opt}^{-1}(x) = \frac{\int_0^x \Phi_{opt}}{\int_I x^{2(\alpha-p-1)/(2p+1)}} = \frac{x^{(\alpha-1/2)/(p+1/2)}}{\int_I x^{2(\alpha-p-1)/(2p+1)}} \tag{3.5.13c}$$

and hence

$$g_{opt}(x) = x^{(p+1/2)/(\alpha-1/2)}. \tag{3.5.13d}$$

In Section 3.4 we analyzed the interpolation error for the mesh with grading function $g \equiv x^\gamma$, and in Theorem 3.4.3 we obtained the optimal value of the exponent $\gamma_{opt} = (p+\frac{1}{2})/(\alpha-\frac{1}{2})$. Note that here we obtained the same optimal mesh by employing the density function.

Let us underline that *the meshes which are constructed using an optimal density are optimal, with respect to the particular norm, only asymptotically as the number of elements diverges to infinity.* More precisely:

Let $\{\Delta_n\}_{n=1}^\infty$ be a sequence of meshes with the property that the number of elements $N(\Delta_n)$ diverges to infinity with n. We will say that $\{\Delta_n^{\text{as. opt.}}\}_{n=1}^\infty$ is *asymptotically optimal* with respect to the norm $|||\cdot|||$ if

$$\lim_{n\to\infty} \sup_{\Delta_n^*} e_{REL}(u_{EX}; |||\cdot|||; \Delta_n^{\text{as. opt.}}, \Delta_n^*) \stackrel{\text{def}}{=} \lim_{n\to\infty} \sup_{\Delta_n^*} \frac{|||u - \mathcal{I}^p_{\Delta_n^{\text{as. opt.}}} u|||_I}{|||u - \mathcal{I}^p_{\Delta_n^*} u|||_I} \le 1 \tag{3.5.14a}$$

for every sequence $\{\Delta_n^*\}_{n=1}^{\infty}$ such that

$$\lim_{n\to\infty} \frac{N(\Delta_n^{\text{as. opt.}})}{N(\Delta_n^*)} = 1. \tag{3.5.14b}$$

Let us also define what we mean by optimal mesh for a fixed number of elements.

A mesh $\Delta_{\bar{N}}^{opt} = \Delta_{\bar{N}}^{opt}(u_{EX}; |||\cdot|||; p; \mathcal{I}_{\Delta}^{p})$ is an optimal mesh of $\bar{N}$ elements for the solution u_{EX} and the norm $|||\cdot|||$, if

$$\left|\left|\left|u_{EX} - \mathcal{I}^p_{\Delta^{opt}_{\bar{N}}} u_{EX}\right|\right|\right| \le |||u_{EX} - \mathcal{I}^p_{\Delta} u_{EX}||| \quad \forall \Delta, N(\Delta) \le \bar{N}. \tag{3.5.15}$$

Note that the optimal mesh of $\bar{N}$ elements $\Delta_{\bar{N}}^{opt}$ may not be unique.

Another type of meshes which, are close to the optimal meshes, are the *asymptotically equilibrated meshes.*

A sequence of meshes $\{\Delta_n^{\text{as. eq.}}\}_{n=1}^{\infty}$ is *asymptotically equilibrated* with respect to the exact solution u_{EX} and the norm $|||\cdot|||$ if it satisfies

$$\lim_{n\to\infty} \frac{\left|\left|\left|u_{EX} - \mathcal{I}^p_{\Delta_n^{\text{as. eq.}}} u_{EX}\right|\right|\right|_{I_j^{\Delta_n}}}{\left|\left|\left|u_{EX} - \mathcal{I}^p_{\Delta_n^{\text{as. eq.}}} u_{EX}\right|\right|\right|_{I_k^{\Delta_n}}} = 1, \quad \forall j,k = 1,\ldots,N(\Delta_n). \tag{3.5.16}$$

Remark 3.5.5 As in the case of the density meshes, the definitions of the asymptotically optimal and equilibrated meshes may be weakened. For example:

A sequence of meshes $\{\Delta_n\}_{n=1}^{\infty}$ is $((\mathscr{C}_1, \mathscr{C}_2); (\mathscr{C}_3, \mathscr{C}_4))$-asymptotically optimal with respect to the exact solution u_{EX}, and the norm $|||\cdot|||$, if there exist constants $0 < \mathscr{C}_1 \le \mathscr{C}_2 < \infty$ and $0 < \mathscr{C}_3 \le \mathscr{C}_4 < \infty$, independent of n, such that

$$\mathscr{C}_1 \le e_{REL}(u_{EX}; |||\cdot|||; \Delta_n, \Delta_n^*) \le \mathscr{C}_2 \tag{3.5.17a}$$

for every sequence $\{\Delta_n^*\}_{n=1}^{\infty}$, such that

$$\mathscr{C}_3 \le \frac{N(\Delta_n)}{N(\Delta_n^*)} \le \mathscr{C}_4. \tag{3.5.17b}$$

A sequence of meshes $\{\Delta_n\}_{n=1}^{\infty}$ is $\mathscr{C}_5$-asymptotically equilibrated with respect to the exact solution u_{EX} and the norm $|||\cdot|||$, if there exists a constant $\mathscr{C}_5$, independent of n, such that for every mesh Δ_n in the sequence we have

$$\frac{\max_{j=1,\ldots,N(\Delta_n)} \left|\left|\left|u_{EX} - \mathcal{I}^p_{\Delta_n} u_{EX}\right|\right|\right|_{I_j^{\Delta_n}}}{\min_{k=1,\ldots,N(\Delta_n)} \left|\left|\left|u_{EX} - \mathcal{I}^p_{\Delta_n} u_{EX}\right|\right|\right|_{I_k^{\Delta_n}}} \le \mathscr{C}_5. \tag{3.5.18}$$

Next we will prove that the sequence of density meshes $\{\Delta_n^{\Phi_{opt}^{H^1}}\}_{n=1}^{\infty}$, which is asymptotically optimal with respect to the H^1 seminorm, is also asymptotically equilibrated.

Theorem 3.5.2 *Assume that u_{EX} is sufficiently smooth so that*

$$\Phi_{opt}^{H^1} \stackrel{\text{def}}{=} \left(\int_I \left(\frac{d^{p+1}}{dx^{p+1}}(u_{EX}) \right)^{2/(2p+1)} \right)^{-1} \left(\frac{d^{p+1}}{dx^{p+1}}(u_{EX}) \right)^{2/(2p+1)} \tag{3.5.19a}$$

is well defined. Then the sequence $\{\Delta_n^{\Phi_{opt}^{H^1}}\}_{n=1}^{\infty}$ *is asymptotically equilibrated with respect to the* H^1 *seminorm, namely*

$$\lim_{n\to\infty} \frac{\left\| \frac{d}{dx}\left(u_{EX} - \mathcal{I}^p_{\Delta_n^{\Phi_{opt}^{H^1}}} u_{EX}\right) \right\|_{L^2(I_j^{\Delta_n^{\Phi_{opt}^{H^1}}})}}{\left\| \frac{d}{dx}\left(u_{EX} - \mathcal{I}^p_{\Delta_n^{\Phi_{opt}^{H^1}}} u_{EX}\right) \right\|_{L^2(I_k^{\Delta_n^{\Phi_{opt}^{H^1}}})}} = 1. \tag{3.5.19b}$$

Proof Denoting $\Delta_n^{\Phi_{opt}^{H^1}}$ by Δ^{opt}, we have

$$\left\| \frac{d}{dx}(u_{EX} - \mathcal{I}^p_{I_j^{\Delta}} u_{EX}) \right\|^2_{L^2(I_j^{\Delta})}$$
$$\approx \left(\frac{\mathscr{C}}{N\Phi^{H^1}_{opt}(x_j^{\Delta^{opt}})} \right)^{2p+1} \left(\frac{d^{p+1}}{dx^{p+1}}(u_{EX})(x_j^{\Delta^{opt}}) \right)^2 \tag{3.5.20a}$$

and recalling that

$$\Phi^{H^1}_{opt} = K\left(\frac{d^{p+1}}{dx^{p+1}}(u_{EX}) \right)^{2/(2p+1)} \tag{3.5.20b}$$

we get

$$\left\| \frac{d}{dx}(u_{EX} - \mathcal{I}^p_{I_j^{\Delta}} u_{EX}) \right\|^2_{L^2(I_j^{\Delta})} \approx \frac{\mathscr{C}}{N^{2p+1}} = constant \tag{3.5.20c}$$

for $j = 1, 2, \ldots, N(\Delta^{opt})$, and hence the sequence of density meshes $\{\Delta_n^{\Phi_{opt}^{H^1}}\}_{n=1}^{\infty}$ is asymptotically equilibrated. □

Remark 3.5.6 Theorem 3.5.2 holds also for the optimal density meshes in other norms, e.g., the L^2 norm and the L^∞ norm of the derivative, etc., and its proof follows analogous steps.

Remark 3.5.7 The converse of Theorem 3.5.1 is known as the *principle of equidistribution of error*, and states: *The sequence of equilibrated meshes is asymptotically optimal with respect to the norm in which the mesh is equilibrated.*

This principle is the basis of any adaptive finite element method in one and many dimensions. As we will see below, in Counterexample 3.5.1, the minimal requirement $u_{EX} \in \mathcal{U}$ is not sufficient for the validity of the principle.

Next we show that the error is insensitive with respect to small perturbations of the optimal mesh, when it is measured in the norm in which the mesh is optimal.

Theorem 3.5.3 *Let* Φ_{opt} *denote the optimal density function with respect to the* $|||\cdot|||$ *norm, and let* $\{\Delta_n^{\Phi_{opt}}\}_{n=1}^{\infty}$ *be the corresponding sequence of density meshes. Then, for any* $\Psi : I \to \mathbb{R}$, *with*

$$\int_I \Psi = 0 \tag{3.5.21a}$$

and any ϵ which satisfies

$$\int_a^b \Phi_{opt} + \epsilon \int_a^b \Psi > 0 \quad \forall 0 \le a < b \le 1 \tag{3.5.21b}$$

we have

$$\left|\left|\left| u_{EX} - \mathcal{I}^p_{\Delta_n^{\Phi_{opt}+\epsilon\Psi}} u_{EX} \right|\right|\right|^2 \approx \left|\left|\left| u_{EX} - \mathcal{I}^p_{\Delta_n^{\Phi_{opt}}} u_{EX} \right|\right|\right|^2 + \mathscr{C}\epsilon^2. \tag{3.5.21c}$$

Proof We will prove (3.5.21c) for the L^2 norm of the derivative; the proof for other norms is completely analogous.

Employing the asymptotic estimate of the interpolation error in terms of the density function, we have

$$\left\| \frac{d}{dx}\left(u_{EX} - \mathcal{I}^p_{\Delta_n^{\Phi_{opt}+\epsilon\Psi}} u_{EX}\right) \right\|^2_{L^2(I)} \approx \frac{\mathscr{C}}{N^{2p}} \int_I \frac{1}{(\Phi_{opt}+\epsilon\Psi)^{2p}} \left(\frac{d^{p+1}u_{EX}}{dx^{p+1}}\right)^2 \tag{3.5.22a}$$

and employing the second order Taylor-series expansion

$$\frac{1}{(\Phi_{opt}+\epsilon\Psi)^{2p}} = \frac{1}{\Phi_{opt}^{2p}} - \frac{2p}{\Phi_{opt}^{2p+1}}\epsilon + \frac{(2p+1)p}{2}\frac{1}{\Phi_{opt}^{2p+2}}\epsilon^2 + O(\epsilon^3) \tag{3.5.22b}$$

we get

$$\begin{aligned} &\left\| \frac{d}{dx}\left(u_{EX} - \mathcal{I}^p_{\Delta_n^{\Phi_{opt}+\epsilon\Psi}} u_{EX}\right) \right\|^2_{L^2(I)} \\ &\approx \frac{\mathscr{C}}{N^{2p}}\Bigg(\int_I \frac{1}{\Phi_{opt}^{2p}}\left(\frac{d^{p+1}}{dx^{p+1}}(u_{EX})\right)^2 - 2\epsilon p \int_I \frac{\Psi}{\Phi_{opt}^{2p+1}}\left(\frac{d^{p+1}}{dx^{p+1}}(u_{EX})\right)^2 \\ &\quad + \epsilon^2 \frac{(2p+1)p}{2} \int_I \frac{\Psi^2}{\Phi_{opt}^{2p+2}}\left(\frac{d^{p+1}}{dx^{p+1}}(u_{EX})\right)^2 \Bigg). \end{aligned} \tag{3.5.22c}$$

Recalling that $\Phi_{opt}^{H^1} = K(d^{p+1}u_{EX}/dx^{p+1})^{2/(2p+1)}$, we get

$$\int_I \frac{\Psi}{\Phi_{opt}^{2p+1}}\left(\frac{d^{p+1}}{dx^{p+1}}(u_{EX})\right)^2 = K\int_I \Psi = 0 \tag{3.5.22d}$$

and hence

$$\begin{aligned} \left\| \frac{d}{dx}\left(u_{EX} - \mathcal{I}^p_{\Delta_n^{\Phi_{opt}+\epsilon\Psi}} u_{EX}\right) \right\|^2_{L^2(I)} &= \left\| \frac{d}{dx}\left(u_{EX} - \mathcal{I}^p_{\Delta_n^{\Phi_{opt}}} u_{EX}\right) \right\|^2_{L^2(I)} \\ &\quad + \mathscr{C}\frac{(2p+1)p}{2N^{2p}} \left\| \frac{\Psi}{\Phi_{opt}^{p+1}} \frac{d^{p+1}}{dx^{p+1}}(u_{EX}) \right\|^2_{L^2(I)} \epsilon^2 \end{aligned} \tag{3.5.22e}$$

which is (3.5.21c). The constant $\mathscr{C}$ in the estimate (3.5.21c) is given by

$$\mathscr{C} = \frac{(2p+1)p}{2N^{2p}} \int_I \frac{\Psi^2}{\Phi_{opt}^{2p+2}}\left(\frac{d^{p+1}}{dx^{p+1}}(u_{EX})\right)^2. \tag{3.5.22f}$$

□

The above result shows that *the error, measured in the norm in which the density is optimal, is not sensitive to small perturbations of the mesh around the optimal mesh.* As a consequence, the determination of the mesh which minimizes a given norm of the error is not easy. In fact it is an ill-posed problem. Nevertheless we are interested in the norm of the error and not the mesh, and the above result implies that the precise determination of the optimal mesh is not necessary.

Remark 3.5.8 In practice, the computation of the optimal density function involves the approximate evaluation of $((d^{p+1}/dx^{p+1})(u_{EX}))^{2/(2p+1)}$ from a finite element solution computed using a relatively coarse mesh. A consequence of Theorem 3.5.2 is that the error in the extraction of $(d^{p+1}/dx^{p+1})(u_{EX})$ will not cause large variation of the norm of the error from the value which corresponds to the optimal mesh.

Let us now compare the asymptotically optimal, and equilibrated, meshes with the meshes which are *optimal* or *equilibrated for a fixed number of elements.* Recall that

(a) A mesh Δ^{eq} is *equilibrated* with respect to the $|||\cdot|||$ norm if

$$|||u_{EX} - \mathcal{I}^p_{\Delta^{eq}} u_{EX}|||_{I^\Delta_j} = \frac{1}{\sqrt{N(\Delta^{eq})}} |||u_{EX} - \mathcal{I}^p_{\Delta^{eq}} u_{EX}|||_I \tag{3.5.23a}$$

for $j = 1, \ldots, N(\Delta^{eq})$.

(b) A mesh Δ^{opt} is *optimal* with respect to the $|||\cdot|||$ norm if

$$|||u_{EX} - \mathcal{I}^p_{\Delta^{opt}} u_{EX}|||_I \leq |||u_{EX} - \mathcal{I}^p_{\Delta^*} u_{EX}|||_I \quad \forall \Delta^* : \; N(\Delta^*) \geq N(\Delta^{opt}). \tag{3.5.23b}$$

Note that, in general, $\Delta^{eq} \neq \Delta^{opt}$, and Δ^{eq}, and Δ^{opt}, may not be unique.

We will now investigate the following questions:

1. For relatively coarse meshes (when $N(\Delta)$ is relatively small), what is the difference between the norm of the error for the optimal density mesh, the norm of the error for the equilibrated mesh, and the norm of the error for the optimal mesh with the same number of elements?
2. How large should be $N(\Delta)$ in order for the error for the optimal density mesh to be practically the same as the error for the optimal mesh, for elements of degree p and for the norm of interest?

We will address these questions by employing the following example.

Example 3.5.3 We considered the exact solution

$$u_{EX}(x) = (x + \delta)^\alpha, \quad x \in I = (0, 1) \tag{3.5.24a}$$

with $\delta \geq 0$, $\alpha > 0.5$, and meshes of elements of degree $p = 1$ and 2. In each case we constructed density, equilibrated and optimal meshes for the interpolation error measured in the L^2 norm of the derivative (the H^1 seminorm) denoted, respectively, by $\Delta_n^{\Phi^{H^1}_{opt}}$, Δ_n^{eq,H^1}, Δ_n^{opt,H^1}, $n = 1, 2, \ldots$, or the corresponding meshes for the L^2 norm of the value.

The density meshes were constructed by employing the condition

$$\int_{I_j^\Delta} \Phi_{opt} = \frac{1}{N(\Delta)} \tag{3.5.24b}$$

which gives

$$x_i^{\Delta_n^{\Phi_{opt}^{H^1}}} = \left(\frac{i}{N}\left((1+\delta)^{1/\gamma_{opt}} - \delta^{1/\gamma_{opt}}\right) + \delta^{1/\gamma_{opt}}\right)^{\gamma_{opt}} - \delta, \quad i = 0, \ldots, N \tag{3.5.24c}$$

where $\gamma_{opt} = (p + \frac{1}{2})/(\alpha - \frac{1}{2})$. Note that for $\delta = 0$ the optimal density mesh $\Delta_N^{\Phi_{opt}^{H^1}}$ coincides with the optimal radical mesh $\Delta_N^{\gamma_{opt}}$. In each case an equilibrated mesh Δ_N^{eq,H^1} was determined starting from the optimal density mesh $\Delta_N^{\Phi_{opt}^{H^1}}$ by adjusting the location of the interior nodes of the mesh, one node at a time, in order to equalize the error norms for the two elements attached to the node, repeating this procedure until convergence was achieved. The iterations were terminated when the difference between the maximum and minimum values of the element norms of the error normalized by the maximum element norm of the error was less than 0.1%. The optimal mesh Δ_N^{opt,H^1} was *determined* by employing a global optimization algorithm using the equilibrated mesh Δ_N^{eq,H^1} as an initial guess. We do not claim to have actually determined the optimal mesh. Nevertheless we expect the norm of the error for the mesh which resulted from the global optimization to be practically minimal for the number of elements employed.

In all the examples below, we employed $\mathcal{I}_\Delta^1 u_{EX}$, the standard piecewise linear interpolant of u_{EX}, and $\mathcal{I}_\Delta^2 u_{EX}$, the piecewise quadratic Lagrangian interpolant with an interpolation node at the midpoint of each element.

Table 3.5.1 (respectively Table 3.5.2) gives the values of the relative error in the H^1 seminorm for the uniform, optimal density, equilibrated and optimal meshes for the H^1 seminorm for $\alpha = 0.75$, meshes of linear elements ($p = 1$) with $2 \leq N(\Delta) \leq 25$ (respectively meshes of quadratic elements ($p = 2$) with $2 \leq N(\Delta) \leq 9$). Note that for $p = 1$ the values of the relative error for the optimal density, equilibrated and optimal meshes practically coincide for $N(\Delta) > 5$. In contrast, for $p = 2$, the values of the relative error for the optimal density mesh differ by more than 20% (but less than 100%) from the corresponding values of the relative error for the equilibrated and the optimal meshes for $N(\Delta) \leq 9$. On the other hand, the values of the relative error for the equilibrated and the optimal mesh are practically identical for $N(\Delta) \geq 2$. In order to determine a minimal number of elements N for which the error for the optimal density mesh is practically the same as the error for the equilibrated mesh (and hence the error for the optimal mesh) we extended the results of Table 3.5.2 to $10 \leq N(\Delta) \leq 40$. These results are given in Table 3.5.3, where the calculation of the norm of the error of the optimal mesh was omitted because it is practically equal with the error for the equilibrated mesh. From Table 3.5.3 it can be seen that the optimal density meshes are practically optimal for $N(\Delta) \geq 40$. Let us underline that these conclusions were obtained for the rather unsmooth solution $x^{3/4}$.

Fig. 3.5.1 (respectively Fig. 3.5.2) gives the graphs of convergence of the relative value of the H^1-seminorm of the interpolation error for the uniform, optimal-density, and equilibrated meshes, for $\alpha = 0.65$, and $\alpha = 0.75$, in the case of linear ($p = 1$) (respectively quadratic ($p = 2$)) elements.

TABLE 3.5.1. Values of the relative interpolation error in the H^1 seminorm for the function $x^{3/4}$ obtained for the uniform, optimal density, equilibrated, optimal meshes (for the H^1 seminorm) of N elements of degree $p = 1$. Note that even for small values of N $(N > 5)$ the norm of the errors for the optimal density, equilibrated, and optimal mesh are practically the same

$N(\Delta)$	$\|(d/dx)(x^{3/4}) - (d/dx)(\mathcal{I}^1_\Delta x^{3/4})\|_{L^2(I)}/\|(d/dx)(x^{3/4})\|_{L^2(I)}$			
	Uniform Mesh	Density Mesh	Equilibrated Mesh	Optimal Mesh
2	0.2816	0.2529	0.2289	0.2262
4	0.2372	0.1545	0.1416	0.1397
6	0.2144	0.1100	0.1027	0.1014
9	0.1938	0.7658×10^{-1}	0.7274×10^{-1}	0.7207×10^{-1}
12	0.1804	0.5867×10^{-1}	0.5633×10^{-1}	0.5590×10^{-1}
16	0.1678	0.4472×10^{-1}	0.4331×10^{-1}	0.4305×10^{-1}
20	0.1588	0.3611×10^{-1}	0.3518×10^{-1}	0.3500×10^{-1}
25	0.1502	0.2911×10^{-1}	0.2849×10^{-1}	0.2838×10^{-1}

TABLE 3.5.2. Values of the relative interpolation error in the H^1 seminorm for the function $x^{3/4}$ obtained for the uniform, optimal density, equilibrated, optimal meshes (for the H^1 seminorm) of $N(\Delta)$ elements of degree $p = 2$. Note that (in contrast with the case $p = 1$, given in Table 3.5.1) the norm of the error for the optimal density mesh is different (by more than 20% but less than 100%) than the norm of the error for the equilibrated and the optimal meshes for $N(\Delta) \leq 10$ while it is much smaller than the norm of the error for the uniform mesh

$N(\Delta)$	$\|(d/dx)(x^{3/4}) - (d/dx)(\mathcal{I}^2_\Delta x^{3/4})\|_{L^2(I)}/\|(d/dx)(x^{3/4})\|_{L^2(I)}$			
	Uniform Mesh	Density Mesh	Equilibrated Mesh	Optimal Mesh
2	0.2016	0.2002	0.1369	0.1336
3	0.1822	0.1350	0.8940×10^{-1}	0.8671×10^{-1}
4	0.1695	0.9318×10^{-1}	0.6319×10^{-1}	0.6124×10^{-1}
5	0.1603	0.6723×10^{-1}	0.4710×10^{-1}	0.4568×10^{-1}
6	0.1531	0.5048×10^{-1}	0.3650×10^{-1}	0.3544×10^{-1}
7	0.1474	0.3917×10^{-1}	0.2912×10^{-1}	0.2832×10^{-1}
8	0.1426	0.3124×10^{-1}	0.2379×10^{-1}	0.2316×10^{-1}
9	0.1384	0.2546×10^{-1}	0.1980×10^{-1}	0.1931×10^{-1}

We also studied the effect of moving the singular point outside the interval on the asymptotic behavior of the error for the density meshes. The exact solution $(x+\delta)^\alpha$, with $\delta > 0$, is a one-dimensional analog of solutions of structural or thermal problems in two-dimensional domains, in the neighborhood of a fillet at a re-entrant corner. Table 3.5.4 (respectively Table 3.5.5) gives the values of the relative interpolation error in the H^1 seminorm for the function $(x + \delta)^{0.65}$, for $\delta = 0$, 0.000001×2^k, (respectively $\delta = 0$, 0.000001×5^k) $k = 0, 1, \ldots, 4$ and $p = 1$ (respectively $p = 2$). We see that even for very

TABLE 3.5.3. Values of the relative interpolation error in the H^1 seminorm for the function $x^{3/4}$ obtained for the uniform, optimal density, equilibrated meshes (for the H^1 seminorm) of N elements of degree $p = 2$. Note, that for $N(\Delta) \geq 40$ the norm of the error for the optimal density mesh is practically identical with the norm of error for the equilibrated mesh

$N(\Delta)$	$\|(d/dx)(x^{3/4}) - (d/dx)(\mathcal{I}^2_\Delta x^{3/4})\|_{L^2(I)}/\|(d/dx)(x^{3/4})\|_{L^2(I)}$		
	Uniform Mesh	Density Mesh	Equilibrated Mesh
11	0.1316	0.1781×10^{-1}	0.1434×10^{-1}
13	0.1262	0.1314×10^{-1}	0.1087×10^{-1}
16	0.1198	0.8950×10^{-2}	0.7622×10^{-2}
20	0.1133	0.5881×10^{-2}	0.5151×10^{-2}
25	0.1072	0.3843×10^{-2}	0.3432×10^{-2}
32	0.1008	0.2388×10^{-2}	0.2175×10^{-2}
40	0.9532×10^{-1}	0.1548×10^{-2}	0.1506×10^{-2}

TABLE 3.5.4. Values of the relative interpolation error in the H^1 seminorm for the function $(x+\delta)^{0.65}$, $\delta = 0,\ 0.000001\times2^k$, $k = 0,\dots,4$, obtained for the uniform, optimal density, equilibrated, optimal meshes (for the H^1 seminorm) of $N(\Delta) = 5$ elements of degree $p = 1$. Note that for $\delta > 0.000016$ the norms of error for the optimal density, equilibrated and optimal meshes are practically identical

δ	$\|(d/dx)((x+\delta)^{0.65}) - (d/dx)(\mathcal{I}^1_\Delta(x+\delta)^{0.65})\|_{L^2(I)}/\|(d/dx)((x+\delta)^{0.65})\|_{L^2(I)}$			
	Uniform Mesh	Density Mesh	Equilibrated Mesh	Optimal Mesh
0.000000	0.4241	0.2711	0.2397	0.2352
0.000001	0.4086	0.2092	0.2075	0.2071
0.000002	0.4050	0.2036	0.2023	0.2020
0.000004	0.4006	0.1975	0.1966	0.1963
0.000008	0.3951	0.1909	0.1902	0.1901
0.000016	0.3883	0.1837	0.1832	0.1831

small positive values of δ, i.e., when the singular point is displaced even very slightly outside the interval I, the norm of the error for the density mesh practically coincides with the norm of the error for the equilibrated and the optimal mesh.

We also studied the effect of the exponent α of the singularity. Table 3.5.6 (respectively Table 3.5.7) gives the relative interpolation errors for the uniform, optimal density, equilibrated and optimal meshes for the H^1 seminorm for $p = 1$ and $\alpha = 0.61 + 0.10k$, $k = 0, 1, 9$ (respectively for $p = 2$ and $\alpha = 0.61 + 0.10k$, $k = 0, 1, \dots, 19$). Note that for $p = 1$ (respectively $p = 2$) the errors for the optimal density, equilibrated and optimal meshes are practically the same for $\alpha > 1$ (respectively $\alpha > 2$). This can also be seen from Figs. 3.5.3, and 3.5.4.

Finally, Table 3.5.8 gives the relative interpolation error in the L^2 norm for the uniform Δ_N^{unif}, optimal density $\Delta_N^{\Phi_{opt}^{L^2}}$, equilibrated Δ_N^{eq,L^2}, and optimal mesh Δ_N^{opt,L^2} for

TABLE 3.5.5. Values of the relative interpolation error in the H^1 seminorm for the function $(x+\delta)^{0.65}$, $\delta = 0, 0.000001 \times 5^k$, $k = 0, \ldots, 4$, obtained for the uniform, optimal density, equilibrated, optimal meshes (for the H^1 seminorm) of $N(\Delta) = 5$ elements of degree $p = 2$. Note that for $\delta > 0.000625$ the norms of the error for the optimal density, equilibrated and optimal meshes are practically identical

δ	$\Vert(d/dx)((x+\delta)^{0.65}) - (d/dx)(\mathcal{I}^2_\Delta(x+\delta)^{0.65})\Vert_{L^2(I)}/\Vert(d/dx)((x+\delta)^{0.65})\Vert_{L^2(I)}$			
	Uniform Mesh	Density Mesh	Equilibrated Mesh	Optimal Mesh
0.000000	0.3477	0.2051	0.1305	0.1250
0.000001	0.3275	0.8651×10^{-1}	0.8544×10^{-1}	0.8533×10^{-1}
0.000005	0.3149	0.7369×10^{-1}	0.7323×10^{-1}	0.7319×10^{-1}
0.000025	0.2946	0.6023×10^{-1}	0.6006×10^{-1}	0.6004×10^{-1}
0.000125	0.2626	0.4660×10^{-1}	0.4655×10^{-1}	0.4654×10^{-1}
0.000625	0.2137	0.3346×10^{-1}	0.3344×10^{-1}	0.3344×10^{-1}

TABLE 3.5.6. Values of the relative interpolation error in the H^1 seminorm for the function $(x+\delta)^{0.65}$, $\alpha = 0.61 + 0.10k$, $k = 0, 1, \ldots, 9$, obtained for the uniform, optimal density, equilibrated, optimal meshes (for the H^1 seminorm) of $N(\Delta) = 8$ elements of degree $p = 1$. Note that for $\alpha > 1$ the norms of the error for the optimal density, equilibrated and optimal meshes are practically identical

δ	$\Vert(d/dx)((x+\delta)^{0.65}) - (d/dx)(\mathcal{I}^1_\Delta(x+\delta)^{0.65})\Vert_{L^2(I)}/\Vert(d/dx)((x+\delta)^{0.65})\Vert_{L^2(I)}$			
	Uniform Mesh	Density Mesh	Equilibrated Mesh	Optimal Mesh
0.61	0.5095	0.2677	0.2372	0.2332
0.71	0.2652	0.1156	0.1079	0.1074
0.81	0.1244	0.5320×10^{-1}	0.5102×10^{-1}	0.5096×10^{-1}
0.91	0.4303×10^{-1}	0.1942×10^{-1}	0.1891×10^{-1}	0.1888×10^{-1}
1.01	0.3553×10^{-2}	0.1756×10^{-2}	0.1727×10^{-2}	0.1727×10^{-2}
1.11	0.2956×10^{-1}	0.1630×10^{-1}	0.1612×10^{-1}	0.1611×10^{-1}
1.31	0.5059×10^{-1}	0.3504×10^{-1}	0.3488×10^{-1}	0.3488×10^{-1}
1.51	0.5560×10^{-1}	0.4665×10^{-1}	0.4658×10^{-1}	0.4658×10^{-1}

the L^2 norm for $\alpha = 0.75$, $2 \le N(\Delta) \le 10$, and $p = 1$. Note that the errors for the optimal density, equilibrated and optimal meshes are practically the same for $N(\Delta) \ge 2$.

From the above example, we see that although the theory given above has asymptotic character, i.e., it is valid as the number of elements in the mesh diverges to infinity, we can also expect to obtain meshes with global error close to that of the optimal mesh even for relatively coarse meshes and unsmooth solutions with strong algebraic singularities like the ones encountered in practical applications.

Remark 3.5.9 Note that in practice:

1. We do not know the exact solution u_{EX} and hence the needed information (e.g., $(d^{p+1}/dx^{p+1})(u_{EX})$) has to be extracted from the computed solution. Since the norm of the error is rather insensitive to perturbations around the optimal mesh,

TABLE 3.5.7. Values of the relative interpolation error in the H^1 seminorm for the function x^α, $\alpha = 0.61 + 0.10k$, $k = 0, 1, \ldots, 19$, obtained for the uniform, optimal density, equilibrated, optimal meshes (for the H^1 seminorm) of $N(\Delta) = 4$ elements of degree $p = 2$. Note that for $\alpha > 2$ the norms of the error for the optimal density, equilibrated and optimal meshes are practically identical

δ	$\|(d/dx)(x^\alpha) - (d/dx)(\mathcal{I}^2_\Delta x^\alpha)\|_{L^2(I)}\|(d/dx)(x^\alpha)\|_{L^2(I)}$			
	Uniform Mesh	Density Mesh	Equilibrated Mesh	Optimal Mesh
0.61	0.4756	0.3947	0.2503	0.2364
0.81	0.1011	0.4920×10^{-1}	0.3502×10^{-1}	0.3428×10^{-1}
1.01	0.2427×10^{-2}	0.9527×10^{-3}	0.7693×10^{-3}	0.7640×10^{-3}
1.21	0.2343×10^{-1}	0.8865×10^{-2}	0.7751×10^{-2}	0.7714×10^{-2}
1.51	0.1650×10^{-1}	0.6994×10^{-2}	0.6544×10^{-2}	0.6535×10^{-2}
1.81	0.5073×10^{-2}	0.2659×10^{-2}	0.2576×10^{-2}	0.2574×10^{-2}
2.01	0.2174×10^{-3}	0.1340×10^{-3}	0.1316×10^{-3}	0.1316×10^{-3}
2.21	0.3701×10^{-2}	0.2677×10^{-2}	0.2650×10^{-2}	0.2650×10^{-2}
2.51	0.6818×10^{-2}	0.6011×10^{-2}	0.5992×10^{-2}	0.5992×10^{-2}

TABLE 3.5.8. Values of the relative interpolation error in the L^2 norm for the function $x^{3/4}$, obtained using various types of meshes (uniform, optimal density, equilibrated, optimal meshes for the L^2 norm) of $N(\Delta) = 2, \ldots, 10$ elements of degree $p = 1$. Note, that in contrast with the results given in Table 3.5.1 (for the optimal density, equilibrated and optimal meshes for the H^1 seminorm) the norms of the error for the optimal density, equilibrated and optimal meshes for the L^2 norm are practically identical even for the coarsest possible mesh ($N(\Delta) = 2$)

$N(\Delta)$	$\|x^{3/4} - \mathcal{I}^1_\Delta x^{3/4}\|_{L^2(I)}/\|x^{3/4}\|_{L^2(I)}$			
	Uniform Mesh	Density Mesh	Equilibrated Mesh	Optimal Mesh
2	0.3115×10^{-1}	0.2052×10^{-1}	0.2029×10^{-1}	0.2028×10^{-1}
3	0.1881×10^{-1}	0.9456×10^{-2}	0.9357×10^{-2}	0.9354×10^{-2}
4	0.1314×10^{-1}	0.5414×10^{-2}	0.5366×10^{-2}	0.5365×10^{-2}
5	0.9947×10^{-2}	0.3502×10^{-2}	0.3475×10^{-2}	0.3474×10^{-2}
6	0.7923×10^{-2}	0.2448×10^{-2}	0.2433×10^{-2}	0.2432×10^{-2}
7	0.6536×10^{-2}	0.1808×10^{-2}	0.1798×10^{-2}	0.1798×10^{-2}
8	0.5532×10^{-2}	0.1390×10^{-2}	0.1383×10^{-2}	0.1383×10^{-2}
9	0.4775×10^{-2}	0.1101×10^{-2}	0.1096×10^{-2}	0.1096×10^{-2}
10	0.4187×10^{-2}	0.8941×10^{-3}	0.8905×10^{-3}	0.8910×10^{-3}

we expect small sensitivity of the norm of the error to the quality of the extraction employed in the construction of the density.

2. We are not interested in the interpolation error but in the error of the finite element solution. Hence we have to relate the interpolation error and the error of the finite element solution. This will be done in the next chapter.

3. We are not interested in the asymptotic behavior of the norms of the error as the number of elements diverges to infinity, but in obtaining the desired accuracy in the

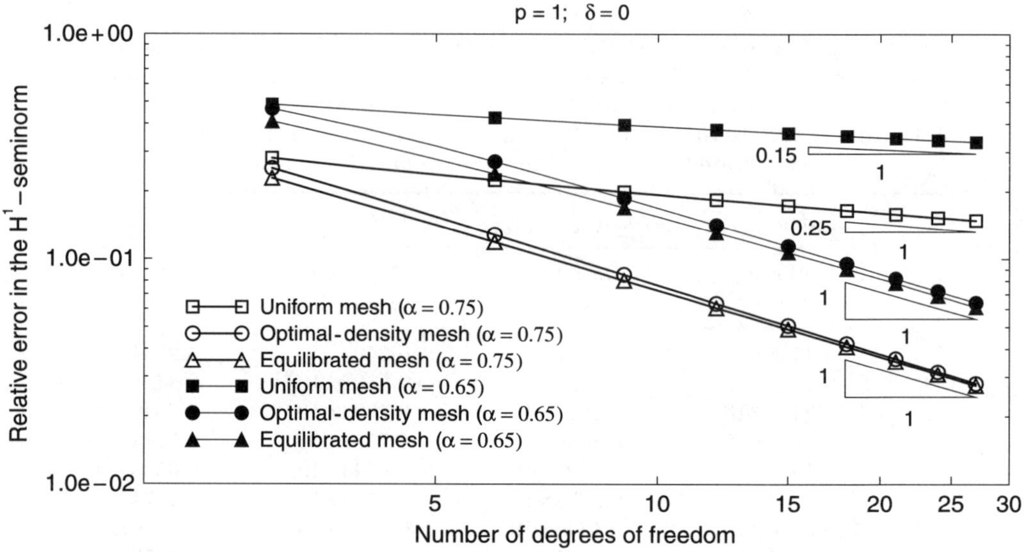

FIG. 3.5.1. Graphs of the relative interpolation error versus the number of degrees of freedom, $M = pN + 1$, for the exact solution x^α, $\alpha = 0.65, 0.75$, and several types of meshes (uniform, optimal density, equilibrated) of linear elements ($p = 1$). Note that the graphs corresponding to the optimal density and the equilibrated meshes practically coincide after a sufficiently large number of elements which depends on α and p

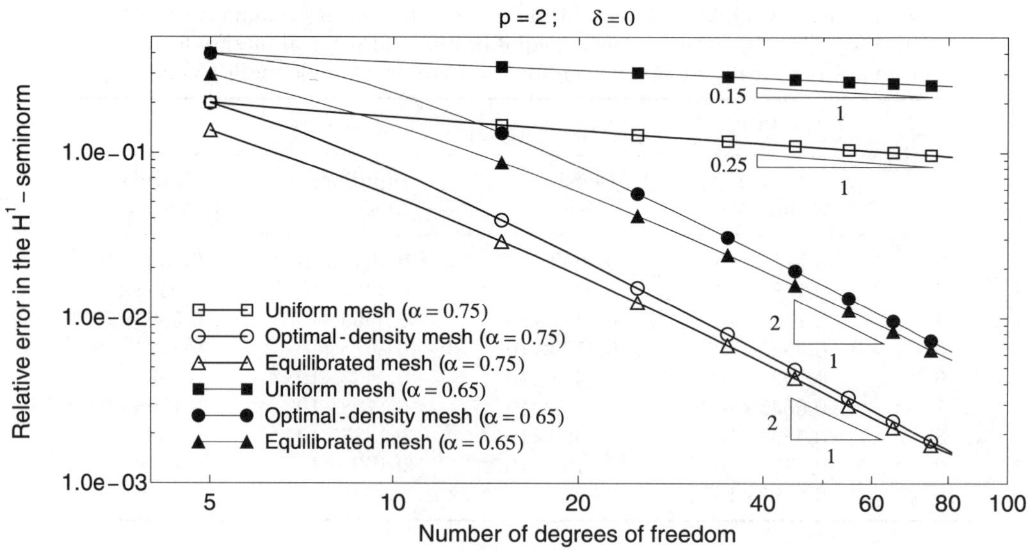

FIG. 3.5.2. Graphs of the relative interpolation error versus the number of degrees of freedom, $M = pN + 1$, for the exact solution x^α, $\alpha = 0.65, 0.75$, and several types of meshes (uniform, optimal density, equilibrated) of quadratic elements ($p = 2$). Comparing with the corresponding graphs for linear elements ($p = 1$) given in Fig 3.5.1, we see that in the case of quadratic ($p = 2$) elements the convergence of the H^1 seminorm of the error for the optimal density mesh to the H^1 seminorm of the error for the equilibrated mesh is slower than in the case of linear elements

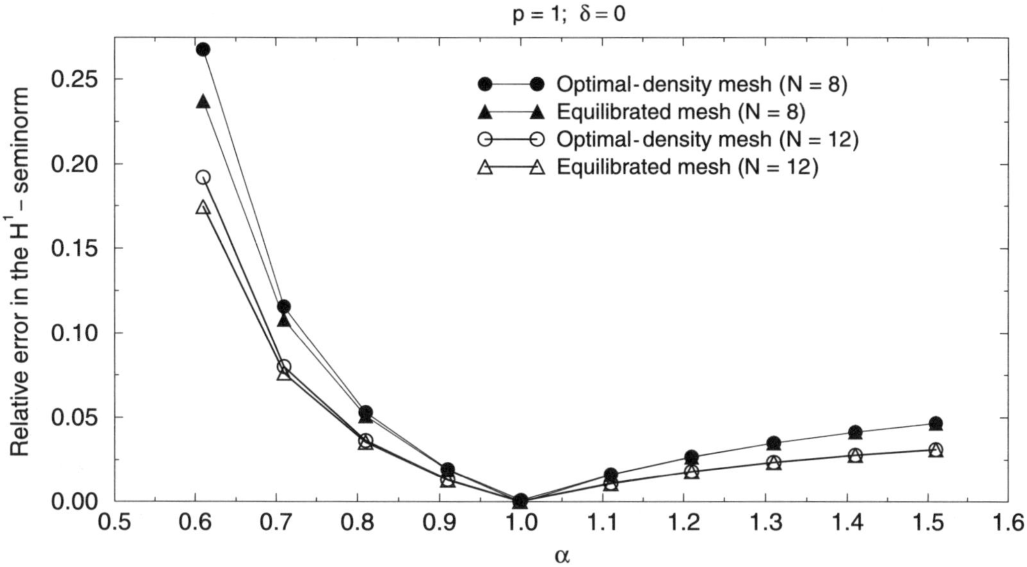

FIG. 3.5.3. Graphs of the relative interpolation error versus the strength of the singularity, α, for the exact solution x^α, $0.61 \leq \alpha \leq 1.51$, and several types of meshes (optimal density, equilibrated) of linear elements ($p = 1$). Note that for $\alpha > 1$ the values of the H^1 seminorm of the error for the optimal density mesh and the equilibrated mesh are practically the same for the range of α which is shown in the graph

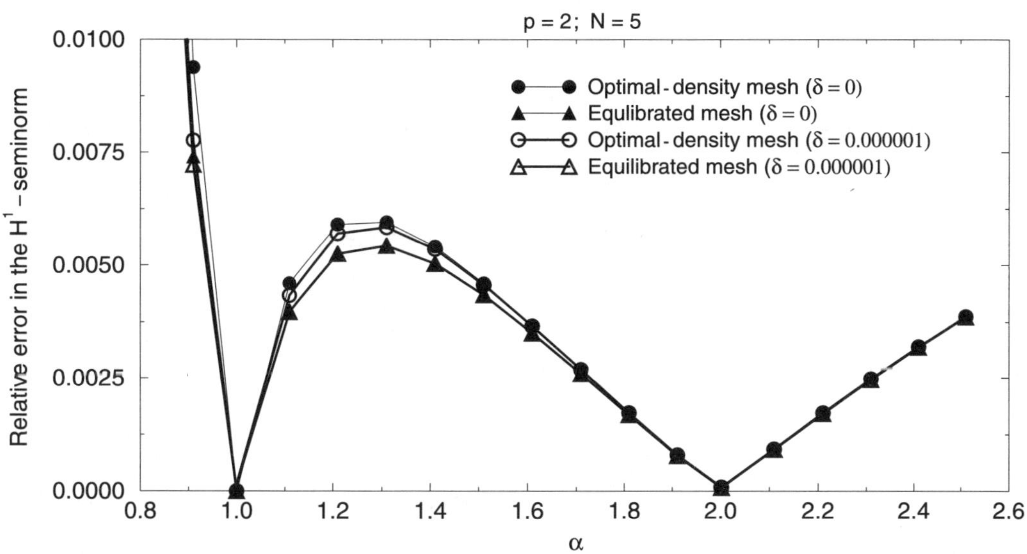

FIG. 3.5.4. Graphs of the relative interpolation error versus the strength of the singularity, α, for the exact solution $(x + 0.000001)^\alpha$, and several types of meshes (optimal density, equilibrated) of five quadratic elements ($N = 5, p = 2$) for $0.91 \leq \alpha \leq 2.51$. Note that the values of the H^1 seminorm of the error for both the optimal density mesh and the equilibrated mesh are practically identical for the considered range of α

most effective way. For this purpose we need to have an a-posteriori error estimator (see Chapter 5), and to have some information whether the asymptotic theory can be used for the class of problems of interest.

We will conclude this section by showing that the converse of Theorem 3.5.2, which we called the principle of equidistribution of error, does not hold for arbitrary $u_{EX} \in \mathcal{U}$.

Example 3.5.4 *There exist solutions u_{EX} for which the optimal meshes $\{\Delta_n^{opt}\}_{n=1}^{\infty}$ need not coincide, asymptotically, with the equilibrated meshes $\{\Delta_n^{EQ}\}_{n=1}^{\infty}$.*

Proof We will construct a $u_{EX} \in H^1(I)$, a corresponding sequence of meshes $\{\Delta_n^{EQ}\}_{n=1}^{\infty}$ which are equilibrated in the $|\cdot|_{H^1}$ norm, and a sequence of unequilibrated meshes $\{\Delta_n^*\}_{n=1}^{\infty}$ such that

$$\lim_{n\to\infty} \frac{N(\Delta_n^*)}{N(\Delta_n^{EQ})} = 1 \tag{3.5.25a}$$

and

$$\begin{aligned} \mathcal{C}n &\le e_{REL}\Big(u_{EX}; \Big\|\frac{d\,\cdot}{dx}\Big\|_{L^2(I)}; \Delta_n^{EQ}, \Delta_n^*; p\Big) \\ &\overset{\text{def}}{=} \frac{\big\|(d/dx)(u_{EX} - \mathcal{I}^1_{\Delta_n^{EQ}} u_{EX})\big\|_{L^2(I)}}{\big\|(d/dx)(u_{EX} - \mathcal{I}^1_{\Delta_n^*} u_{EX})\big\|_{L^2(I)}} \end{aligned} \tag{3.5.25b}$$

as $n \to \infty$, where $N(\Delta_n^{EQ})$ diverges to infinity with n.

Let $v_k \in H^1(I)$ be the function with $v_k(0) = 0$, and

$$\frac{d}{dx}(v_k)(x) = \begin{cases} 1, & x \in (0, 2^{-k-1}) \\ (-1)^j, & x \in ((j-\frac{1}{2})2^{-k}, (j+\frac{1}{2})2^{-k}), \quad 1 \le j \le 2^k - 1 \\ 1, & x \in (1 - 2^{-k-1}, 1) \end{cases} \tag{3.5.26}$$

The graph of v_k, for $k = 1, 2, 3$ is given in Fig. 3.5.5. Then let

$$u_{EX}(t) = \sum_{k=1}^{\infty} \alpha_k \int_0^t \frac{d}{dx}(v_k) = \sum_{k=1}^{\infty} \alpha_k v_k(t) \tag{3.5.27a}$$

and

$$\frac{d}{dx}(u_{EX})(x) = \sum_{k=1}^{\infty} \alpha_k \frac{d}{dx}(v_k)(x) \tag{3.5.27b}$$

where $\{\alpha_k\}_{k=1}^{\infty}$ is square summable ($\sum_{k=1}^{\infty} \alpha_k^2 < \infty$), and will be specified below. Let Δ_n^{EQ} be the mesh with nodes

$$x_j^{\Delta_n^{EQ}} = 2^{-n} j, \quad j = 0, 1, \ldots, N(\Delta_n^{EQ}) = 2^n. \tag{3.5.28}$$

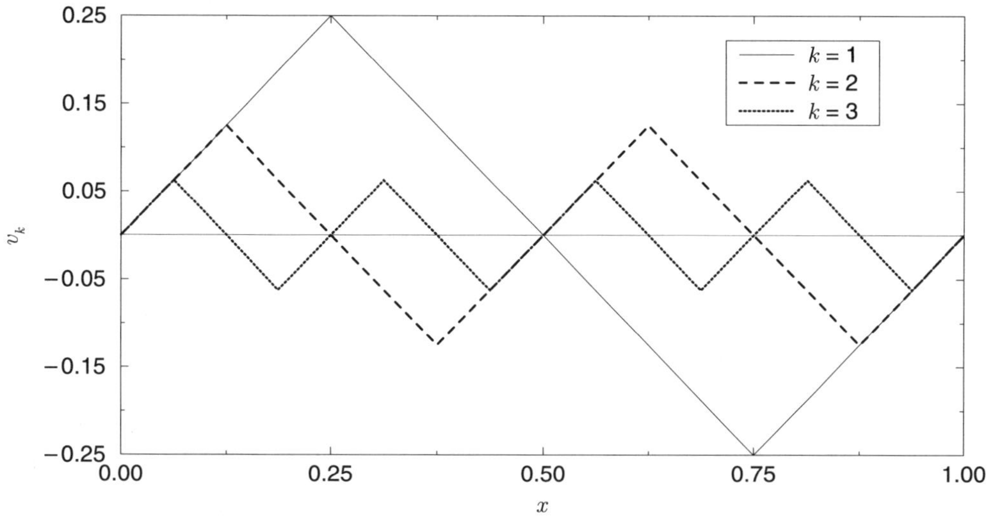

FIG. 3.5.5. Graphs of the functions v_k, for $k = 1, 2, 3$, employed in the construction of the counterexample for the principle of equidistribution of error

Noting that

$$\int_0^1 \frac{d}{dx}(v_k)\frac{d}{dx}(v_l) = 0, \quad l \neq k \tag{3.5.29a}$$

it follows that

$$\mathcal{I}^1_{\Delta_n^{EQ}} u_{EX} = \sum_{k=1}^{n-1} \alpha_k v_k \tag{3.5.29b}$$

is the piecewise-linear interpolant of u_{EX}, and hence

$$\frac{d}{dx}\left(u_{EX} - \mathcal{I}^1_{\Delta_l^{EQ}} u_{EX}\right) = \sum_{k=l}^{\infty} \alpha_k \frac{d}{dx}(v_k) \tag{3.5.30a}$$

and

$$E_{S^1_{\Delta_l}} = \left\| \frac{d}{dx}\left(u_{EX} - \mathcal{I}^1_{\Delta_l^{EQ}} u_{EX}\right) \right\|_{L^2(I)} = \sum_{k=l}^{\infty} \alpha_k^2. \tag{3.5.30b}$$

Let us now employ the coefficients

$$\alpha_k = \begin{cases} 2^{-k}, & \text{for } k = 2^m, \ m = 0, 1, 2, \ldots \\ 0, & \text{otherwise.} \end{cases} \tag{3.5.31a}$$

Figure 3.5.6a gives a graph of the partial sum function

$$\frac{d}{dx}(u_{EX}^{(2^m)})(x) = \sum_{k=0}^{2^m} \alpha_k \frac{d}{dx}(v_k)(x) \tag{3.5.31b}$$

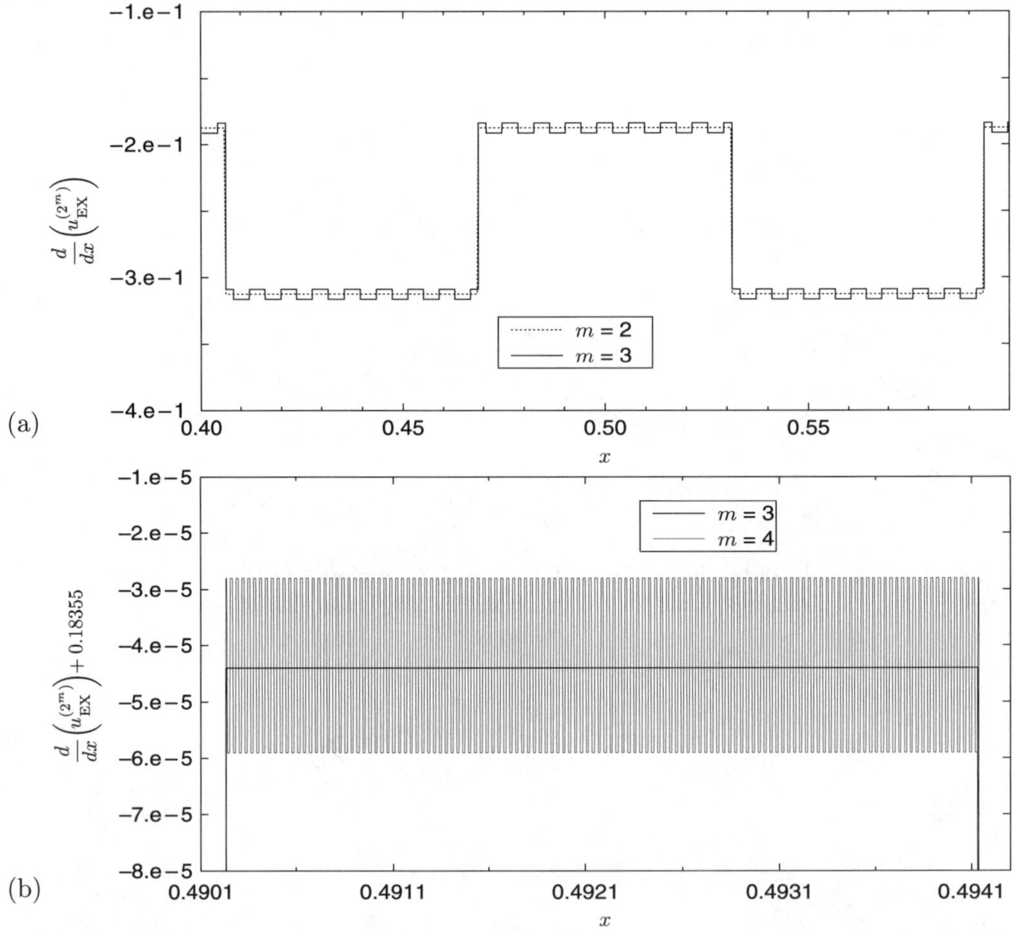

FIG. 3.5.6. The function which is employed in the construction of a counterexample for the principle of equidistribution of error. Graphs of $(d/dx)(u_{EX}^{(2^m)}) = \sum_{k=0}^{2^m} \alpha_k (d/dx)(v_k)$, where $u_{EX} = \sum_{k=0}^{\infty} \alpha_k v_k$ is the function employed in the construction of the counterexample. (a) $m = 2, 3$ and $0.40 \leq x \leq 0.60$; (b) $m = 3, 4$ and $0.4901 \leq x \leq 0.4941$. Note that $(d/dx)(u_{EX}^{(2^m)})$ is discontinuous at the points $(j - \frac{1}{2})2^{-k}$, $(j + \frac{1}{2})2^{-k}$ for $1 \leq j \leq 2^k - 1$

for $m = 2, 3$ and $0.40 \leq x \leq 0.60$, and Fig. 3.5.6b the graph of $(d/dx)(u_{EX}^{(2^m)})(x)$, for $m = 3, 4$ and $0.4901 \leq x \leq 0.4941$. Note that the oscillation of the graph of the function $(d/dx)(u_{EX}^{(2^m)})$ increases with m, and in the limit $(d/dx)(u_{EX})$ is discontinuous almost everywhere. Note also that *the error is symmetric with respect to the nodal points and hence it is equidistributed in any norm.* The value of the H^1 seminorm of the error in the elements of the mesh Δ_l^{EQ} is the same in every element and will be denoted by $\xi_{\Delta_l}^{EQ}$, namely

$$\xi_{\Delta_l^{EQ}} \overset{\text{def}}{=} \left\| \frac{d}{dx}\left(u_{EX} - \mathcal{I}^1_{\Delta_l^{EQ}} u_{EX}\right) \right\|_{L^2(I_j^{\Delta_l^{EQ}})} = \sqrt{2^{-l} \sum_{k=l}^{\infty} \alpha_k^2} \qquad (3.5.32)$$

for $j = 1, \ldots, N(\Delta_l^{EQ}) = 2^l$.

Let us now compare the error for Δ_l^{EQ} with the error for the mesh Δ_l^* with $N(\Delta_l^*) = 2^l - 1$ nodes given by

$$x_0^{\Delta_l^*} = 0, \qquad x_j^{\Delta_l^*} = \left(j + \tfrac{1}{2}\right)2^{-l}, \quad 1 \le j \le 2^l - 2, \ l > 1, \qquad x_{2^l-1}^{\Delta_l^*} = 1. \tag{3.5.33a}$$

Note that

$$N(\Delta_l^*) = N(\Delta_l^{EQ}) - 1. \tag{3.5.33b}$$

The element norms of the interpolation error for the interior elements of the mesh Δ_l^* are

$$\left\| \frac{d}{dx}(u_{EX} - \mathcal{I}^1_{\Delta_l^*} u_{EX}) \right\|^2_{L^2(I_j^{\Delta_l^*})} = 2^{-l} \sum_{k=l+1}^{\infty} \alpha_k^2, \quad j = 2, \ldots, 2^l - 2 \tag{3.5.34a}$$

while in the first and the last element we have

$$\left\| \frac{d}{dx}(u_{EX} - \mathcal{I}^1_{\Delta_l^*} u_{EX}) \right\|^2_{L^2(I_j^{\Delta_l^*})} \le \mathscr{C}2^{-l} \sum_{k=l-1}^{\infty} \alpha_k^2, \quad j = 1 \text{ or } j = N \tag{3.5.34b}$$

and hence

$$\begin{aligned} \left(E_{S^1_{\Delta_l^*}}\right)^2 &= \left\| \frac{d}{dx}(u_{EX} - \mathcal{I}^1_{\Delta_l^*} u_{EX}) \right\|^2_{L^2(I)} \\ &\le \left(\bar{E}_{S^1_{\Delta_l^*}}\right)^2 \stackrel{\text{def}}{=} \sum_{k=l+1}^{\infty} \alpha_k^2 + \mathscr{C}2^{-l} \sum_{k=l-1}^{\infty} \alpha_k^2. \end{aligned} \tag{3.5.35}$$

Let us now compare the values of $E_{S^1_{\Delta_l^{EQ}}}$ with the corresponding values of the upper bound $\bar{E}_{S^1_{\Delta_l^*}}$ for $E_{S^1_{\Delta_l^*}}$ given by the right-hand side of (3.5.35) for the choice of coefficients α_k given in (3.5.31a). Table 3.5.9 lists the values of $E_{S^1_{\Delta_l^{EQ}}}$ and $\bar{E}_{S^1_{\Delta_l^*}}$ for $1 \le l \le 16$. We see that for $l = 4, 8, 16, 32, 64, 128, 256, \ldots$, the H^1 seminorm of the interpolation error for the mesh Δ_l^* is smaller, by several orders of magnitude, than the H^1 seminorm of the interpolation error for the mesh Δ_l^{EQ}, which is equilibrated with respect to the H^1 seminorm. □

Remark 3.5.10 We based the above counterexample [Babuška and Vogelius, 1984] on the construction of a special function with discontinuous derivative almost everywhere. This function is very peculiar in the sense that there is total lack of correlation between the length of an element and the associated interpolation error. We can exclude such functions by assuming that the norm of the interpolation error in any element in the mesh, with possible exception of a finite number of elements, is bounded by the minimum of the norms of the interpolation error in the two subintervals which result from the bisection of the element multiplied by a constant which does not depend on the mesh. The purpose of the above counterexample is to demonstrate that the principle of equidistribution of the element norms of the error is correct only if some additional assumptions are imposed on the smoothness of the solution. The minimal assumption, namely that the exact solution belongs to the energy space, is not sufficient.

TABLE 3.5.9. A counterexample of the optimality of the equilibrated mesh. The values of the H^1 seminorm of the interpolation error $E_{S^1_{\Delta_l^{EQ}}}$ for the sequence of equilibrated meshes $\{\Delta_l^{EQ}\}_{l=1}^{\infty}$ for the specially constructed exact solution u_{EX}, and the upper bound $\bar{E}_{S^1_{\Delta_l^*}}$ of the H^1 seminorm of the interpolation error for a corresponding sequence of unequilibrated meshes $\{\Delta_l^*\}_{l=1}^{\infty}$. Here we use the symbol $**$ to indicate the values of l for which H^1 seminorm of the interpolation error for the equilibrated mesh Δ_l^{EQ} is much bigger than the H^1 seminorm of the interpolation error for the mesh Δ_l^*

l	$N(\Delta_l^{EQ})$	$N(\Delta_l^*)$	α_l	$\frac{3}{4}\left(E_{S^1_{\Delta_l^{EQ}}}\right)^2$	$\frac{3}{4}\left(\bar{E}_{S^1_{\Delta_l^*}}\right)^2$
1	2		2^{-1}	2^{-2}	$2^{-4}+\mathscr{C}2^{-1}$
2	4	3	2^{-2}	2^{-4}	$2^{-8}+\mathscr{C}2^{-4}$
3	8	7	0	2^{-8}	$2^{-8}+\mathscr{C}2^{-7}$
4**	16	15	2^{-4}	2^{-8}	$2^{-16}+\mathscr{C}2^{-12}$
5	32	31	0	2^{-16}	$2^{-16}+\mathscr{C}2^{-13}$
6	64	63	0	2^{-16}	$2^{-16}+\mathscr{C}2^{-22}$
7	128	127	0	2^{-16}	$2^{-16}+\mathscr{C}2^{-23}$
8**	256	255	2^{-8}	2^{-16}	$2^{-32}+\mathscr{C}2^{-24}$
9	512	511	0	2^{-32}	$2^{-32}+\mathscr{C}2^{-25}$
10	1024	1023	0	2^{-32}	$2^{-32}+\mathscr{C}2^{-42}$
11	2048	2047	0	2^{-32}	$2^{-32}+\mathscr{C}2^{-43}$
12	4096	4095	0	2^{-32}	$2^{-32}+\mathscr{C}2^{-44}$
13	8192	8191	0	2^{-32}	$2^{-32}+\mathscr{C}2^{-45}$
14	16384	16383	0	2^{-32}	$2^{-32}+\mathscr{C}2^{-46}$
15	32768	32767	0	2^{-32}	$2^{-32}+\mathscr{C}2^{-46}$
16**	65536	65535	2^{-16}	2^{-32}	$2^{-64}+\mathscr{C}2^{-47}$

In this section, we introduced the idea of the density function which is useful for characterizing sequences of meshes in one and higher dimensions, we obtained asymptotic expressions for various norms of the interpolation error in terms of the density function, and defined the asymptotically optimal density function in a norm as the one which minimizes the corresponding asymptotic estimate. The meshes obtained by employing the optimal density lead to errors, measured in the norm for which the density was constructed, which are very close to the errors for the optimal meshes. The equilibrated meshes perform even better than the density meshes and their errors are very close to the errors for the corresponding optimal meshes even for very coarse meshes. Thus the principle of equidistribution of error derived from asymptotic considerations is very robust and performs well in both the asymptotic and the pre-asymptotic range of practical computations. We have also seen that the error is rather insensitive to perturbations of the mesh around the optimal one, when the error is measured in the norm corresponding to the optimal mesh. This means that, in practical computations, there is no need to determine the optimal mesh very accurately. In practice, the problem of mesh optimization must be considered within a class of admissible meshes which are constructed by a mesh generator. The choice of the class of admissible meshes may influence the result of the mesh optimization substantially. The ideas of the density function and the equidistribution of error are general and can be employed in the problem of mesh optimization in many dimensions. We also gave a counterexample to show that the principle of equidistribution of error works only when the solution is 'reasonable' in some sense.

Problems

P3.5.1. In the general case, where $AE \not\equiv$ *constant* and $c \equiv 0$, we can write

$$\|u_{EX} - \mathcal{I}_\Delta^p u_{EX}\|_{\mathcal{U}}^2 \approx \mathcal{C} \sum_{j=1}^{N(\Delta)} AE(x_j^\Delta)(h_{I_j^\Delta})^{2p+1}\Big(\frac{d^{p+1}}{dx^{p+1}}(u_{EX})(x_j^\Delta)\Big)^2 \qquad (3.5.36a)$$

asymptotically as $h^\Delta \to 0$. Consider the case

$$AE \equiv x^\vartheta, \quad 0 \le \vartheta < 1, \qquad c \equiv 0, \qquad u_{EX} \equiv x^\alpha - \alpha x \qquad (3.5.36b)$$

and determine the optimal density for the energy norm. How does it compare with the corresponding optimal density for the case $AE \equiv 1$?

P3.5.2. Let $AE \equiv 1$, $c \equiv 0$, $u_{EX} \equiv x^\alpha - \alpha x$, and determine the optimal density for the L^2 norm. Compare the density of the asymptotically optimal meshes for the energy and the L^2 norm in the neighborhood of $x = 0$.

P3.5.3. Determine the value of $\mathcal{C}$ in the estimate

$$\|u_{EX} - \mathcal{I}_\Delta^p u_{EX}\|_{L^2(I)} \approx \frac{\mathcal{C}}{N^{p+1}}\Big\|\frac{1}{\Phi^{p+1}}\frac{d^{p+1}}{dx^{p+1}}(u_{EX})\Big\|_{L^2(I)} \qquad (3.5.37)$$

for the case $AE \equiv 1$, $c \equiv 0$, $p = 1$.

P3.5.4. Determine the density Φ_γ which corresponds to the radical mesh Δ with nodes $x_j^\Delta = (j/N)^\gamma$, $j = 0, \dots, N(\Delta)$. Assuming $u_{EX} \equiv x^\alpha - \alpha x$, determine the range of values of γ for which the integral

$$\mathcal{C}(\gamma, p) = \int_I \frac{1}{\Phi_\gamma^{2p}}\Big(\frac{d^{p+1}}{dx^{p+1}}(u_{EX})\Big)^2 \qquad (3.5.38a)$$

exists. For these values of γ, we have

$$\|u_{EX} - \mathcal{I}_\Delta^p u_{EX}\|_{\mathcal{U}}^2 \le \mathcal{C} N^{-2p}\mathcal{C}(\gamma, p). \qquad (3.5.38b)$$

This result implies that, *whenever $\mathcal{C}(\gamma, p)$ exists, the energy norm of the error converges with the optimal rate.* Compare these observations with the results stated in Theorem 3.4.2.

P3.5.5. Compare the values of the L^2 norm of the derivative of the error for the meshes corresponding to the optimal density, given in Tables 3.5.1–3.5.3 with the values computed from Lemma 3.4.2 for $\gamma = \gamma_{opt}$.

P3.5.6. Show that the condition

$$\int_{I_j^\Delta} \Phi = \frac{1}{N(\Delta)}, \quad j = 1, \dots, N(\Delta) \qquad (3.5.39)$$

determines uniquely the sequence of density meshes corresponding to a given density function.

3.5* The description of the mesh and analysis of its optimality in two dimensions

As we have already seen in Section 3.4*, the notion of the radical meshes can be generalized to two and three dimensions by employing a weakened definition. Then all the results presented in Section 3.5 also hold in the multi-dimensional case.

As we discussed earlier, we employed the solution x^α in one dimension as an analogue of the solution $r^\beta\varphi(\theta)$ which occurs in two-dimensional problems in the neighborhood of corner points or points where there is an abrupt change on the type of boundary condition. In the one-dimensional case, we proved that for $\frac{1}{2} < \alpha < 1$ we have $x^\alpha \in B^{\alpha-1/2}(I)$ and that the energy norm of the error in the finite element solution for uniform meshes converges to zero as $h^{\alpha-1/2}$ (or $N^{-(\alpha-1/2)}$). In the two-dimensional case, we can show that for $0 < \beta < 1$ we have $r^\beta \in B^\beta(\Omega)$ and that the energy norm of the error converges to zero as h^β (or $N^{-\beta/2}$). In the one-dimensional case, we proved that for a class of meshes which are sufficiently graded toward the singular point at $x = 0$, the energy norm of the error converges as N^{-p}. Analogously, in the two-dimensional case, we can construct meshes which are sufficiently graded in the radial direction toward the singular points, and we can obtain convergence for the energy norm of the error as $N^{-p/2}$. This is the best possible rate which can be achieved for any mesh; the proof of this result is analogous to the one-dimensional case. As in the one-dimensional case, under-refinement leads to a lower rate of convergence and hence it is always preferable to employ meshes which are slightly over-refined at the singularities. As in the one-dimensional case, as the degree of the elements is increased very strong grading of the mesh is required at the singular points. For more, see Chapter 6. See also [Hugger, 1993a], [Hugger, 1993b]. Also in two dimensions the equidistribution principle is the main tool to construct almost optimal meshes.

Problems

P3.5*.1. Consider the L-shaped domain shown in Fig. 2.3*.1 and let the density function be $\Phi(x_1, x_2) = r^\gamma$. Assume that $\Delta_n = \{\tau_j^{\Delta_n}\}_{j=1}^{N(\Delta_n)}$ which are such that

$$0 < \mathscr{C}_1 \le N(\Delta_n) \int_{\tau_j} \Phi \le \mathscr{C}_2 < \infty \tag{3.5*.1}$$

where $\mathscr{C}_1$ and $\mathscr{C}_2$ are constants independent of n. Assume that the elements are of degree p and satisfy the aspect ratio condition. Show that

$$\left\| r^\alpha \sin \alpha\theta - \mathcal{I}^p_{\Delta_n}(r^\alpha \sin \alpha\theta) \right\|_{\mathscr{U}} \le \mathscr{C}(N(\Delta_n))^{-\beta} \tag{3.5*.2}$$

where $\beta > 0$ depends on α and γ. Show that you can select γ so that $\beta = p/2$.

P3.5*.2. Consider the mesh constructed in Problem P3.4*.2. Does this mesh satisfy (3.5*.1) for some γ?

P3.5*.3. Discuss the optimal density function Φ in terms of $\sum_{i+j=p+1} |(\partial^{p+1}u/\partial x_1^i \partial x_2^j)|^2$. compare it with (3.5.8).

3.6 Historical and bibliographical comments

The convergence of the finite element method is based on the estimate

$$\|u_{EX} - u_{S^p_\Delta}\|_{\mathscr{U}} \le \mathscr{C} \inf_{\chi \in S^p_\Delta} \|u_{EX} - \chi\|_{\mathscr{U}} \tag{3.6.1}$$

where $\mathscr{C}$ is independent of p and Δ. Choosing χ the interpolant or approximant of u_{EX} we obtain an a-priori error estimate for the finite element solution.

One dimensional interpolation and approximation by polynomial functions has a long history. Interpolation was already used by H. Brigg (1561–1630) when constructing his tables of logarithms. The Newton interpolation formula was known to I. Newton (1642–1727) since 1676. W. Leibnitz (1646–1716) independently obtained similar results around 1672. Since then, many results on interpolation have been obtained. Approximation of functions by polynomials was studied first in a systematic way by P.L. Chebyshev (1821–1894). Results on the rate of convergence (with respect to degree of polynomials) was first analyzed by D. Jackson (1888–1946) in his 1911 dissertation. Since then, many sophisticated results have been obtained.

Much fewer results are available for higher dimensional (multivariate) interpolation. The first attempt to study multivariate interpolation was likely that of Narumi (1920). Most results today are related to the finite element method; see [Ciarlet and Raviart, 1972], [Babuška and Guo, 2000a,b], [Chen and Babuška, 1995, 1996].

The one dimensional approximation theorem proven by K.T.W. Weierstrass (1815–1897) in 1885, which showed that any continuous function can be approximated by polynomials, was extended in 1891 to several variables by C.E. Picard (1856–1941). Various theorems were proven recently, mostly in connection with finite elements.

So far, we mentioned only interpolation and approximation by polynomial functions. Of course other families of functions could be considered. For example, H. Pade in 1900 treated interpolation problems by using rational functions. Various other families of functions are of interest today, e.g., splines, wavelets, etc.

In higher dimensions, approximation by a family of functions satisfying a differential equation was also studied. For example, the problem of approximation satisfying the Laplace equation by harmonic functions is related to the approximating functions with complex variables; this is addressed in [Mergelyan, 1962]. An analogous problem is the approximation of functions satisfying elliptic partial differential equations with analytic coefficients, which was addressed in [Bergman, 1961] and [Vekua, 1967], and plays an important role in the generalized finite element method; see [Babuška, Caloz, and Osborn, 1994], [Melenk and Babuška, 1996], [Babuška and Melenk, 1997], [Strouboulis, Babuška, and Copps, 2000], [Strouboulis, Copps, and Babuška, 2000], and the references in these works. The approximation rate and smoothness of the approximated function are closely related. In fact, the Sobolev norm can be also defined by the approximation rate obtained using analytic functions. This is true also for the finite element approximation [Babuška, Kellogg, and Pitkaranta, 1979], [Babuška and Guo, 2000b].

The integrals appearing in the finite element method are usually computed numerically. A numerical integration for determining π was already used by Archimedes (287–212 BC). It seems that the first proposal of numerical quadrature was made by B. Cavalieri (1598–1647) in 1639. The well known Simpson formula was mentioned in Newton's book Principia Mathematika. The composite Simpson rule was proposed by T. Simpson (1710–1761) in 1743. The Gaussian quadrature was analyzed by C.F. Gauss (1777–1855) in 1814. Today many quadrature formulae are known and studied. In three dimensions the integration is called cubature. Likely the first cubature formula was proposed by J.C. Maxwell (1831–1879) in 1877. A connection between orthogonal polynomials and cubature, analogously as for Gaussian quadrature, was likely first studied by P. Appell (1855–1930) in 1890.

Today there are many books of various levels of sophistication on interpolation, approximation, and numerical integration. We mention, e.g., [Cheney, 1986], [Cheney, 1966], [Davis, 1963], [Davis and Rabinowitz, 1984], [Natanson, 1964] [Achieser, 1947], [Rivlin, 1969]. For a survey of the history of interpolation, approximation, and numerical integration, we refer to [Brezinski, 1994].

The best approximation property (3.6.1) is essential and also holds for various other differential equations and variational forms, such as mixed methods, etc, in which case (3.6.1) is equivalent to the inf sup condition, called today the BB condition. The inf sup condition was proven in a general form in [Babuška, 1972], [Babuška, 1973] (see also [Babuška and Aziz, 1972]), and, independently, in a form specially related to the mixed method, it was proven in [Brezzi, 1974]; see, e.g., [Brezzi and Fortin, 1991].

Today the finite element method is used in many different forms and for solving various partial differential equations, integral equations, etc.

References

N.I. Achieser (1947) *Theory of approximation.* Ungar, New York.

J.E. Akin (1995) *Finite elements for analysis and design.* Academic Press, New York.

T. Apel (1999) *Anisotropic finite elements, local estimates and applications.* Teubner.

D. Arnold, D. Boffi, and R. Falk (2000) *On serendipity elements.* Penn State University (preprint).

K.E. Atkinson (1989) *An introduction to numerical analysis.* John Wiley, New York

I. Babuška (1972) The rate of convergence for the finite element method. *SIAM Journal of Numerical Analysis*, **8**, 304–315.

I. Babuška and A.K. Aziz (1972) Survey lectures on the mathematical foundations of the finite element method. In: A.K. Aziz (ed.) *The mathematical foundations of the finite element method with applications to partial differential equations, Proceedings of a symposium held at the University of Maryland, Baltimore campus, June 26–30, 1972*, Academic Press, New York, pp. 3–359.

I. Babuška (1973) The finite element method with Lagrange multipliers. *Numerische Mathematik*, **20**, 179–192.

I. Babuška and A.K. Aziz (1976) On the angle condition in the finite element method. *SIAM Journal of Numerical Analysis*, **13**, 214–226.

I. Babuška, R.B. Kellogg, and J. Pitkaranta (1979) Direct and inverse error estimates for finite elements with mesh refinements. *Numerische Mathematik*, **33**, 447–471.

I. Babuška and W.C. Rheinboldt (1979) Analysis of optimal finite element meshes in R^1. *Mathematics of Computation*, **33**, 435–463.

I. Babuška and M. Vogelius (1984) Feedback and adaptive finite element solution for one-dimensional boundary value problems. *Numerische Mathematik*, **44**, 75–102.

I. Babuška, G. Caloz, and J. Osborn (1994) Special finite element methods in a class of second order elliptic problems with rough coefficients. *SIAM Journal of Numerical Analysis*, **31**, 945–981.

I. Babuška and M. Suri (1994) The p and h–p versions of the finite element method, basic principles and properties. *SIAM Review*, **36**, 578–632.

I. Babuška and J.M. Melenk (1997) The partition of unity method. *International Journal for Numerical Methods in Engineering*, **40**, 727–758.

I. Babuška, T. Strouboulis, and K. Copps (1997) h–p optimization of finite element approximation analysis of the optimal mesh sequence in one dimension. *Computer Methods in Applied Mechanics and Engineering*, **150**, 89–108.

I. Babuška and J. Osborn (1999) Can a finite element method perform arbitrarily badly? *Mathematics of Computation*, **230**, 443–462.

I. Babuška and B.Q. Guo (2000a) Optimal estimates for lower and upper bounds of approximation error in p version of the finite element method in two dimensions. *Numerische Mathematik*, **85**, 197–218.

I. Babuška and B.Q. Guo (2000b) Direct and inverse approximation theorems for the p version of the finite element method in the frame of weighted Besov spaces. Part I: Approximability of functions in the weighted Besov spaces. Part II: Optimal rate of convergence of the p version finite element solutions. Part III: Inverse approximation theorems (preprints).

K.J. Bathe (1996) *Finite element procedures.* Prentice Hall, Englewood Cliffs.

E. Becker, G.F. Carey, and J.T. Oden (1981) *Finite elements: an introduction.* Prentice Hall.

J. Bergh and J. Löfstrom (1976) *Interpolation spaces.* Springer, Berlin, New York.

S. Bergman (1961) *Integral operators in the theory of linear partial differential equations.* Springer, Berlin.

S.C. Brenner and R.L. Scott (1994) *The mathematical theory of finite element methods.* Springer, Berlin, New York.

C. Brezinski (1994) Historical perspective in interpolation, approximation, and quadrature. In: P.G. Ciarlet and J.L. Lions (eds) *Handbook of numerical analysis*, Vol. III, Elsevier, Amsterdam, pp. 3–46.

F. Brezzi (1974) On the existence, uniqueness, and approximation of saddle-point problems arising from Lagrange multipliers. *Revue francaise, d'automatique informatique, et reserche operationelle: Analyse numerique*, **8**, 129–151.

F. Brezzi, M. Fortin (1991) *Mixed and hybrid finite element methods.* Springer, Berlin.

G.F. Carey (1997) *Computational grids generation, adaptation and solution strategies.* Taylor and Francis, London.

Q. Chen and I. Babuška (1995) Approximate optimal points for polynomial interpolation of real functions in an interval and in a triangle. *Computer Methods in Applied Mechanics and Engineering*, **128**, 405–417.

Q. Chen and I. Babuška (1996) The optimal symmetrical points for polynomial interpolation of real functions in the tetrahedron. *Computer Methods in Applied Mechanics and Engineering*, **137**, 89–94.

E.W. Cheney (1966) *Introduction to approximation theory.* McGraw-Hill, New York.

E.W. Cheney (1986) *Multivariate approximation theory: selected topics.* SIAM.

P.G. Ciarlet and P.A. Raviart (1972) General Lagrange and Hermite interpolation in $\mathbb{R}^N$ with applications to Finite Element Methods. *Archives for Rational Mechanics and Analysis*, **46**, 177–199.

P.G. Ciarlet (1991) Basic error estimates for elliptic problems. In: P.G. Ciarlet and J.L. Lions (eds) *Handbook of numerical analysis, Vol. II; finite element methods* (Part 1). Elsevier Science Publishers B. V., North-Holland, Amsterdam, pp. 19–351.

P. Clément (1975) Approximation by finite element functions using local regularization. *RAIRO: Analyse numerique*, **R-2**, 77–84.

R.G. Cowper (1973) Gaussian quadrature formulas for triangles. *International Journal for Numerical Methods in Engineering*, **7**, 405–408.

J.P. Davis (1963) *Interpolation and approximation.* Blaisdell Waltham.

J.P. Davis and P. Rabinowitz (1984) *Methods of numerical integration.* Academic Press, New York.

C. De Boor and A. Pinkus (1978) Proof of Bernstein and Erdös concerning the optimal nodes for polynomial interpolation. *Journal of Approximation Theory*, **24**, 289–303.

D.A. Dunavant (1985) High-degree efficient symmetric Gaussian quadrature rules for the triangle. *International Journal for Numerical Methods in Engineering*, **21**, 1129–1148.

P. Erdös (1961) Problems and results on the theory of interpolation, II. *Acta Mathematica Scientia Hungary*, **12**, 235–241.

T.O. Espelid (1994) On integrating vertex singularities using extrapolation. *BIT Numerical Mathematics*, **34**, 62–79.

P.L. George (1996) Automatic mesh generation and finite element computations. In: P.G. Ciarlet and J.L. Lions (eds) *Handbook of numerical analysis*, North Holland, pp. 69–193.

P.L. George and H. Borouhaki (1998) *Delaunay triangulation and meshing.* Hermes, Paris.

W.J. Gordon (1971) Blending function method of bivariate and multivariate interpolation and approximation. *SIAM Journal of Numerical Analysis*, **8**, 158–177.

W. Gui and I. Babuška (1986) The *h*, *p* and *hp* versions of the Finite Element Method in one dimension. Part I: The error analysis of the *p* version. Part II: The error analysis of the *h* and *hp* versions. *Numerische Mathematik*, **49**, 577–621; 613–657.

P.C. Hammer, O.J. Marlowe, and A.H. Stroud (1956) Numerical integration over simplexes and cones. *Math. Tabl. Natn. Res. Connc. Washington*, **10**, 130–137.

J. Hugger (1993a) The theory of density representation of finite element meshes: examples of density operations with quadrilateral elements in the mapped domain. *Computer Methods in Applied Mechanics and Engineering*, **109**, 17–39.

J. Hugger (1993b) Recovery and few parameter representation of the optimal mesh density function for near optimal finite element meshes. *Computer Methods in Applied Mechanics and Engineering*, **109**, 41–71.

L.V. Kantorovich and V.I. Krylov (1958) *Approximate methods of higher analysis.* Interscience Publishers.

A.R. Krammer and C.W. Ueberhuber (1998) *Computational integration.* SIAM.

M. Lenoir (1986) Optimal isoparametric finite elements and error estimates for domains involving curved boundaries. *SIAM Journal of Numerical Analysis*, **23**, 562–580.

V.D. Liseikin (1999) *Grid generation methods.* Springer, Berlin.

J.N. Lyness and D. Jespersen (1975) Moderate Degree Symmetric Quadrature Rules for the Triangle. *Journal of the Institute of Mathematics and its Applications*, **15**, 19–32.

J.M. Melenk and I. Babuška (1996) The partition of unity Finite Element Method: basic theory and applications. *Computer Methods in Applied Mechanics and Engineering*, **139**, 289–314.

S.N. Mergelyan (1962) Uniform approximation to functions of a complex variable. Mathematical Surveys and Monographs, **3**, American Mathematical Society.

S.G. Mikhlin (1964) *Variational methods in mathematical physics.* Pergammon Press, Oxford, New York.

S.G. Mikhlin (1971) *The numerical performance of variational methods.* Wolters-Noordhoff, Groningen.

I.P. Natanson (1964) *Constructive function theory.* Ungar, New York.

J. Necas (1967) *Les méthodes diréctes en théorie des équations élliptiques.* Academia Prague.

R. Pressens, E. de Doncker-Kapenga, C. Uberhuber, and D. Kahaner (1983) *Quadpack: a subroutine package for automatic integration.* Springer, Berlin.

K. Rektorys (1977) *Variational methods in mathematics, science and engineering.* D. Reidel Publishing Co., Dordrecht.

K. Rektorys (1998) *Solving ordinary and boundary value problems in science and engineering.* CRC Press, Boca Raton.

T.J. Rivlin (1969) *An introduction to the approximation of functions.* Dover, New York.

C. Schwab (1994) Variable order composite quadrature of singular and nearly singular integrals. *Computing*, **53**, 173–194.

C. Schwab (1998) *p- and hp-Finite Element Methods.* Oxford University Press, Oxford.

L.R. Scott and S. Zhang (1990) Finite element interpolation of nonsmooth functions satisfying boundary conditions. *Mathematics of Computation*, **54**, 483–493.

T. Strouboulis, K. Copps, and I. Babuška (2000) The generalized finite element method: an example of its implementation and illustration of its performance. *International Journal for Numerical Methods in Engineering*, **47**, 1401–1419.

T. Strouboulis, I. Babuška, and K. Copps (2001) The design and analysis of the generalized finite element method. *Computer Methods in Applied Mechanics and Engineering* (to appear).

B.A. Szabo and I. Babuška (1991) *Finite element analysis.* John Wiley & Sons, Inc., New York.

J.F. Thompson and B.V. Soni (eds) (1999) *Handbook of grid generation.* CRC Press.

I.N. Vekua (1967) *New methods for solving elliptic equations.* North Holland.

S. Wandzura and H. Xiao (1999) Quadrature rules on triangles in $\mathbb{R}^2$. *Yale University, Department of Computer Science Report*, YALEU/DCS/RR-1168.

4

LOCAL BEHAVIOR IN THE FINITE ELEMENT METHOD

4.0 Introduction

In Chapter 3 we derived a-priori estimates $\|e_{S^p_\Delta}\|_{\mathcal{U}}$ for the global energy norm of the error. In this chapter, we will further analyze the properties of the error.

In Section 4.1, we construct estimates for the nodal errors of the finite element solution, which prove that, for sufficiently smooth input data and elements of quadratic degree or higher ($p \geq 2$), the nodal values of the finite element solution are *superconvergent*, namely, we have

$$|e_{S^p_\Delta}(x_j^\Delta)| \leq \|e_{S^p_\Delta}\|_{\mathcal{U}} \|G(\cdot; x_j^\Delta) - \mathcal{I}^p_\Delta G(\cdot; x_j^\Delta)\|_{\mathcal{U}} \leq \mathcal{C} h^{2p} \tag{4.0.1}$$

where $G(\cdot; x_j^\Delta)$ is the Green's function which corresponds to x_j^Δ, while we have only

$$\|e_{S^p_\Delta}\|_{L^\infty(I_j^\Delta)} \approx \mathcal{C} h^{p+1}. \tag{4.0.2}$$

This means that, as the mesh size h tends to zero, the nodal errors converge to zero with a rate higher than the rate of convergence of the maximum error.

In Section 4.2, we employ a generalization of the approach of Section 4.1, which allows us to construct estimates for the L^2 norm of the error, and other lower-order norms of the error which are known as *negative norms*. In particular, for sufficiently smooth input data, we have

$$\|e_{S^p_\Delta}\|_{L^2(I)} \leq \mathcal{C} h \|e_{S^p_\Delta}\|_{\mathcal{U}} \tag{4.0.3}$$

which means that the L^2 norm of the error converges to zero by a power of h faster than the energy norm. The corresponding result for the negative norms of the error reads

$$\|e_{S^p_\Delta}\|_{H^{-l}(I)} \stackrel{\text{def}}{=} \sup_{v \in H^l(I)} \frac{|\int_I e_{S^p_\Delta} v|}{\|v\|_{H^l(I)}} \leq \mathcal{C} h^{\min(p, l+1)} \|e_{S^p_\Delta}\|_{\mathcal{U}}. \tag{4.0.4}$$

Hence by increasing the negative order of the norm, the asymptotic rate of convergence increases up to a maximum threshold which is determined by the degree p of the elements, and the smoothness of the input data.

In Section 4.3, we introduce a splitting of the error $e_{S^p_\Delta}$ into the *local error* $e^{loc}_{S^p_\Delta}$, and the *global* or *pollution error* $e^{glob}_{S^p_\Delta}$, namely,

$$e_{S^p_\Delta} = \underbrace{u_{EX} - \mathcal{A}^p_\Delta u_{EX}}_{e^{loc}_{S^p_\Delta}} + \underbrace{\mathcal{A}^p_\Delta u_{EX} - u_{S^p_\Delta}}_{e^{glob}_{S^p_\Delta}} \tag{4.0.5}$$

where $\mathcal{A}^p_\Delta u_{EX}$ is a best approximation of the exact solution u_{EX}, which is characterized by the property that it practically coincides with the finite element solution $u_{S^p_\Delta}$ for sufficiently smooth input data. We prove that the local error has the same asymptotic behavior as the interpolation error, namely

$$\|e^{loc}_{S^p_\Delta}\|_{\mathcal{U}(I^\Delta_j)} \approx \|u_{EX} - \mathcal{I}^p_\Delta u_{EX}\|_{\mathcal{U}(I^\Delta_j)} \approx \mathcal{C}h^{\mu+1/2} \tag{4.0.6}$$

where μ depends on the smoothness of u_{EX} in the neighborhood of the element I^Δ_j, while for the pollution error, we have

$$\|e^{glob}_{S^p_\Delta}\|_{\mathcal{U}(I^\Delta_j)} \leq \mathcal{C}\sqrt{h}\sqrt{(e^\Delta_j + e^\Delta_{j-1})^2 + \left(\frac{e^\Delta_j - e^\Delta_{j-1}}{h}\right)^2} \tag{4.0.7}$$

where e^Δ_j denotes the error at the jth node. This estimate reflects the fact that the pollution error is practically affine in each element, and is due to the behavior of the approximation in the mesh outside the neighborhood of the element.

In Section 4.4 we analyze the pollution error and demonstrate that it is related to the non-uniformity of the mesh and the unsmoothness of the Green's function. We prove that when the Green's function is smooth and the meshes are quasi-uniform, the pollution error is negligible relative to the local error. On the other hand, if the Green's function is unsmooth, the pollution error may be the dominant component of the error in the majority of the elements, depending on the refinement of the mesh and the degree p of the elements.

The analysis of the pollution error given in Section 4.4 has essentially one-dimensional character and cannot be used in many dimensions. A more general approach which has a direct multi-dimensional analogue can be based on the interior estimates which are discussed in Section 4.5. The typical interior estimate reads

$$\begin{aligned}\|e_{S^p_\Delta}\|_{\mathcal{U}(I'')} \leq \mathcal{C} \inf_{\chi \in S^p_\Delta} \left(\|u_{EX} - \chi\|_{\mathcal{U}(I'_\Delta)} + \frac{1}{d}\|u_{EX} - \chi\|_{L^2(I'_\Delta)}\right) \\ + \frac{\mathcal{C}}{d}\|e_{S^p_\Delta}\|_{L^2(I'_\Delta)}\end{aligned} \tag{4.0.8a}$$

with

$$I'' = (a'', b'') \subset I'_\Delta = (x^\Delta_m, x^\Delta_n) \subseteq I' = (a', b') \tag{4.0.8b}$$

and

$$d \leq \min(a' - a'', b'' - b'). \tag{4.0.8c}$$

The first term in the right-hand side expresses the local error in I'', while the second term $(\mathcal{C}/d)\|e_{S^p_\Delta}\|_{L^2(I'_\Delta)}$ accounts for the pollution error in I'', and may also be replaced by $(\mathcal{C}/d^{l+1})\|e_{S^p_\Delta}\|_{H^{-l}(I')}$.

In Section 4.6, we analyze the properties of the local error. We show that the derivatives of the local error at the mapped Gauss–Legendre points of order p converge to zero as h^{p+1}, while the maximum absolute value of the derivative of the error in each element converges as h^p, namely

$$\left|\frac{d}{dx}(e^{loc}_{S^p_\Delta})(x_l^{G,I_j^\Delta})\right| \leq \mathscr{C}(h_{I_j^\Delta})^{p+1}\left\|\frac{d^{p+2}}{dx^{p+2}}(u_{EX})\right\|_{L^\infty(I_j^\Delta)} \tag{4.0.9a}$$

for $l = 1, \ldots, p$, where x_l^{G,I_j^Δ} are the mapped Gauss–Legendre points of order p in I_j^Δ, which are given by

$$x_l^{G,I_j^\Delta} \stackrel{\text{def}}{=} F_{I_j^\Delta}(\tilde{x}_l^G), \qquad \tilde{P}_p(\tilde{x}_l^G) = 0, \qquad l = 1, \ldots, p \tag{4.0.9b}$$

while we have

$$\mathscr{C}(h_{I_j^\Delta})^p\left\|\frac{d^{p+1}}{dx^{p+1}}(u_{EX})\right\|_{L^\infty(I_j^\Delta)} \leq \left\|\frac{d}{dx}(e^{loc}_{S^p_\Delta})\right\|_{L^\infty(I_j^\Delta)}. \tag{4.0.9c}$$

These results hold under the assumptions that:

(a) The $p+2$ derivative of the exact solution u_{EX} is bounded in the L^∞ norm.

(b) The absolute value of the $p+1$ derivative of u_{EX} is bounded from below away from zero.

If, in addition,

(c) The meshes are such that the pollution is negligible;

the above results also hold for the total error $e_{S^p_\Delta}$, and the Gauss–Legendre points of degree p are superconvergence points for the derivative of the finite element solution. In Section 4.6, we also show that, for elements of quadratic degree or higher ($p \geq 2$), the values of the local error at the mapped Lobatto–Legendre points of order p converge to zero as h^{p+2}, while the maximum absolute value of the local error in each element converges only as h^{p+1}. It follows that, if the meshes are such that the pollution in the value is negligible, for elements of quadratic degree or higher, the finite element solution is superconvergent at the Lobatto–Legendre points of order p, which include the end points of the element. We underline that the superconvergence points are useful for extracting higher-order accuracy from the finite element solution, or its derivative, only if the meshes are such that the pollution error is negligible relative to the local error for the extracted quantity.

The analysis of the local error given in Section 4.6 has one-dimensional character and can be used only in a few special cases in higher dimensions. In Section 4.7 we give a general approach for analyzing the local error, which is applicable to the most general cases in one and many dimensions. The main result is that, on locally uniform meshes, the error $e_{S^p_\Delta}$ coincides asymptotically with a function $\Psi^{\bar{x},H}_{S^p_\Delta}$, namely

$$\left\|\frac{d}{dx}(e_{S^p_\Delta} - \Psi^{\bar{x},H}_{S^p_\Delta})\right\|_{L^\infty(S(\bar{x},H/2))} \leq \mathscr{C}h^{\bar{\nu}}\left\|\frac{d}{dx}(e_{S^p_\Delta})\right\|_{L^\infty(S(\bar{x},H))} \tag{4.0.10}$$

where the function $\Psi^{\bar{x},H}_{S^p_\Delta}$ can be constructed by solving a periodic boundary value problem in a master mesh.

In Section 4.8 we introduce post-processing methods which employ local averaging of the finite element solution and/or its derivatives to construct superconvergent quantities (e.g., values or derivatives). We call these post-processing methods *recoveries*, or *superconvergent recoveries*, to underline that, in certain cases, the recovered quantities are indeed superconvergent. For example, for a recovery $\mathcal{A}^*_{du/dx}$, which samples the derivative of the finite element solution at the Gauss–Legendre points, in the cases that the meshes are such that the pollution is negligible, we have

$$\left\| \frac{d}{dx}(u_{EX}) - \mathcal{A}^*_{du/dx}(u_{S^p_\Delta}) \right\|_{L^\infty(I^\Delta_j)} \leq \mathscr{C}(h_{I^\Delta_j})^{p+1} \leq \mathscr{C} h_{I^\Delta_j} \left\| \frac{d}{dx}(e_{S^p_\Delta}) \right\|_{L^\infty(I^\Delta_j)}. \quad (4.0.11)$$

We show that a superconvergent recovery can be constructed in two ways:

- by averaging superconvergent values
- by averaging a properly chosen set of values, none of which may be superconvergent, for which *cancellation of error* occurs.

Let us note that in many dimensions we either have too few superconvergence points and/or these points may be cumbersome to use, and the construction of superconvergent recoveries is almost exclusively based on the principle of cancellation of error. Once more, we underline that superconvergent recovery is possible only if the pollution is negligible.

4.1 A-priori estimates for the nodal errors and the maximum error

In this section, we derive estimates of the convergence of the nodal errors of the finite element solution, as the mesh size h tends to zero, by employing the properties of the Green's function. Although this approach cannot be directly generalized to the multi-dimensional case, it is important because it provides us with insight.

Let us consider the model problem (2.2.8) with the exact solution u_{EX} and its finite element approximation $u_{S^p_\Delta} \in S^p_\Delta$. We will assume that the coefficients AE and c are sufficiently smooth, where the precise smoothness will be specified below at the proper places. The Green's function $G(\cdot;\bar{x})$ has been introduced in Section 2.2 and in Example 2.3.1 as the exact solution of the model problem with loads

$$f \equiv \delta(\cdot, \bar{x}), \qquad a_2 = 0. \quad (4.1.1a)$$

It follows that $G(\cdot;\bar{x})$ is the solution of the variational problem:

Find $G(\cdot;\bar{x}) \in \mathscr{U}$ such that

$$\mathscr{B}(G(\cdot;\bar{x}), v) = v(\bar{x}) \quad \forall v \in \mathscr{U}. \quad (4.1.1b)$$

We will now prove that there exists a unique G, and it is symmetric with respect to its arguments.

Lemma 4.1.1 *There exists a unique $G \in \mathscr{U}$ which satisfies (4.1.1b), and*

$$G(x;\bar{x}) = G(\bar{x};x), \quad \forall x, \bar{x} \in I. \quad (4.1.2)$$

Proof The existence and uniqueness of the Green's function G follows from the boundedness of the functional $\mathscr{L}(v) = v(\bar{x})$. In the case of the regular model problem

we have

$$\|v\|_{L^\infty(I)} \le \frac{\sqrt{l}}{\sqrt{\beta_1}}\|v\|_{\mathcal{U}} \quad \forall v \in \mathcal{U} \tag{4.1.3a}$$

and in the case of the singular model problem we have

$$\|v\|_{L^\infty(I)} \le \frac{l^{(1-\vartheta)/2}}{\sqrt{1-\vartheta}}\|v\|_{\mathcal{U}^\vartheta} \quad \forall v \in \mathcal{U}^\vartheta \tag{4.1.3b}$$

and hence in both cases we have

$$|v(\bar{x})| \le \|v\|_{L^\infty(I)} \le \mathcal{C}\|v\|_{\bar{\mathcal{U}}} \quad \forall v \in \bar{\mathcal{U}} \tag{4.1.3c}$$

with $\bar{\mathcal{U}} \equiv \mathcal{U}$, $\mathcal{C} = \sqrt{l}/\sqrt{\beta_1}$, in the case of the regular model problem, and $\bar{\mathcal{U}} \equiv \mathcal{U}^\vartheta$, $\mathcal{C} = l^{(1-\vartheta)/2}/\sqrt{1-\vartheta}$, in the case of the singular model problem.

The symmetry of the Green's function follows from the symmetry of $\mathcal{B}$. We have

$$G(\bar{x};x) = \mathcal{B}(G(\cdot;\bar{x}),\ G(\cdot;x)) = \mathcal{B}(G(\cdot;x),\ G(\cdot;\bar{x})) = G(x;\bar{x}). \tag{4.1.3d}$$

□

Remark 4.1.1 Note that *the symmetry of G does not hold in the cases that $\mathcal{B}$ is not symmetric.*

We will now employ the properties of the Green's function G in order to estimate the nodal errors of the finite element solution.

Theorem 4.1.1 *Consider the regular model problem, and let AE and c satisfy the additional assumptions*

$$\left\|\frac{d^{q'}}{dx^{q'}}(AE)\right\|_{L^\infty(I)} \le A, \qquad \left\|\frac{d^{r'}c}{dx^{r'}}\right\|_{L^\infty(I)} \le B \tag{4.1.4a}$$

for all integers q', r', such that $0 \le q' \le q$, $0 \le r' \le r$. Then, for any nodal point x_j^Δ, we have

$$|e_{S^p_\Delta}(x_j^\Delta)| \le \|e_{S^p_\Delta}\|_{\mathcal{U}}\|G(\cdot;x_j^\Delta) - \mathcal{I}^p_\Delta G(\cdot;x_j^\Delta)\|_{\mathcal{U}} \le \mathcal{C}(h^\Delta)^\mu\|e_{S^p_\Delta}\|_{\mathcal{U}} \tag{4.1.4b}$$

with

$$\mu = \min(q, r+1, p) \tag{4.1.4c}$$

and $\mathcal{C}$ depending only on q, r, and p.

Proof From the definition of the Green's function $G(\cdot;x_j^\Delta)$, and the symmetry of $\mathcal{B}$, we get

$$e_{S^p_\Delta}(x_j^\Delta) = \mathcal{B}(G(\cdot;x_j^\Delta), e_{S^p_\Delta}) = \mathcal{B}(e_{S^p_\Delta}, G(\cdot;x_j^\Delta)) \tag{4.1.5a}$$

and employing the orthogonality

$$\mathcal{B}(e_{S^p_\Delta}, \mathcal{I}^p_\Delta G(\cdot;x_j^\Delta)) = 0 \tag{4.1.5b}$$

we get

$$e_{S^p_\Delta}(x_j^\Delta) = \mathcal{B}(e_{S^p_\Delta}, G(\cdot; x_j^\Delta)) = \mathcal{B}(e_{S^p_\Delta}, G(\cdot; x_j^\Delta) - \mathcal{I}^p_\Delta G(\cdot; x_j^\Delta)) \tag{4.1.5c}$$

and employing the Cauchy–Schwarz inequality, we obtain

$$|e_{S^p_\Delta}(x_j^\Delta)| \leq \|e_{S^p_\Delta}\|_{\mathcal{U}} \|G(\cdot; x_j^\Delta) - \mathcal{I}^p_\Delta G(\cdot; x_j^\Delta)\|_{\mathcal{U}}. \tag{4.1.5d}$$

From the smoothness assumptions (4.1.4a), Theorem 2.3.1, and Example 2.3.1, it follows that

$$\max\left(\left\| \frac{d^s}{dx^s}(G(x; x_j^\Delta)) \right\|_{L^\infty((0, x_j^\Delta))}, \left\| \frac{d^s}{dx^s}(G(x; x_j^\Delta)) \right\|_{L^\infty((x_j^\Delta, l))} \right) \leq \mathcal{C}(A, B) \tag{4.1.6a}$$

with

$$s = \min(q + 1, r + 2). \tag{4.1.6b}$$

Recalling that x_j^Δ is a nodal point, we have

$$\|G(\cdot; x_j^\Delta) - \mathcal{I}^p_{I_k^\Delta} G(\cdot; x_j^\Delta)\|_{\mathcal{U}(I_k^\Delta)} \leq \mathcal{C}(h_{I_k^\Delta})^\mu \sqrt{h_{I_k^\Delta}} \tag{4.1.6c}$$

for all elements I_k^Δ with

$$\mu = \min(s - 1, p). \tag{4.1.6d}$$

Squaring (4.1.6c), and summing over the elements, we obtain

$$\underbrace{\sqrt{\sum_{k=1}^{N(\Delta)} \|G(\cdot; x_j^\Delta) - \mathcal{I}^p_{I_k^\Delta} G(\cdot; x_j^\Delta)\|^2_{\mathcal{U}(I_k^\Delta)}}}_{\|G(\cdot; x_j^\Delta) - \mathcal{I}^p_\Delta G(\cdot; x_j^\Delta)\|_{\mathcal{U}}} \leq \sqrt{(h^\Delta)^{2\mu} \sum_{k=1}^{N(\Delta)} h_{I_k^\Delta}} \leq \mathcal{C}(h^\Delta)^\mu \tag{4.1.6e}$$

and (4.1.4b) follows. □

Remark 4.1.2 Theorem 4.1.1 also holds under the more general assumption that AE and c satisfy (4.1.4a) piecewise in each element. Hence AE and c are allowed to have discontinuities at the nodal points.

We will now use the estimate (4.1.4b) to prove that, *in the special case that the inverse of AE is a polynomial of degree $(p-1)$, and $c \equiv 0$, the finite element solution $u_{S^p_\Delta}$ is exact at the nodes.*

Corollary 4.1.1 *Consider the regular model problem, and assume, in addition, that*

$$\frac{1}{AE} \in \mathcal{P}^{p-1}(I), \quad c \equiv 0. \tag{4.1.7a}$$

Then we have

$$u_{S^p_\Delta}(x_j^\Delta) = u_{EX}(x_j^\Delta), \quad j = 0, 1, \ldots, N(\Delta). \tag{4.1.7b}$$

Proof For any $\bar{x} \in I$, let $\bar{G}(\cdot;\bar{x}) \in C^0(I)$ be defined by

$$\bar{G}(x;\bar{x}) \stackrel{\text{def}}{=} \begin{cases} \mathcal{C}_1 \displaystyle\int_0^x \frac{1}{AE} + \mathcal{C}_2, & 0 \le x \le \bar{x} \\ \mathcal{C}_3 \displaystyle\int_x^l \frac{1}{AE} + \mathcal{C}_4, & \bar{x} \le x \le l \end{cases} \tag{4.1.8a}$$

where the constants $\mathcal{C}_i$, $i = 1, \dots, 4$, are determined from the conditions

$$\bar{G}(x;\bar{x}) = G(x,\bar{x}), \quad x = 0, \bar{x}, l. \tag{4.1.8b}$$

Note that $G(\cdot;\bar{x}) \in \mathcal{U} \subseteq C^0(I)$, and hence $G(x,\bar{x})$ are well defined at any point x. Further, note that because $1/AE \in \mathcal{P}^{p-1}(I)$, it follows that $\bar{G}(\cdot;\bar{x})$ is piecewise polynomial of degree p, namely

$$\bar{G}(\cdot;\bar{x})\big|_{(0,\bar{x})} \in \mathcal{P}^p((0,\bar{x})), \qquad \bar{G}(\cdot;\bar{x})\big|_{(\bar{x},l)} \in \mathcal{P}^p((\bar{x},l)). \tag{4.1.8c}$$

We will now show that G and $\bar{G}$ coincide. Let

$$\delta G = G(\cdot;\bar{x}) - \bar{G}(\cdot;\bar{x}). \tag{4.1.9a}$$

Then

$$\mathcal{B}(\delta G, \delta G) = \mathcal{B}(G(\cdot;\bar{x}), \delta G) - \mathcal{B}(\bar{G}(\cdot;\bar{x}), \delta G) = \delta G(\bar{x}) - \mathcal{B}(\bar{G}(\cdot;\bar{x}), \delta G) \tag{4.1.9b}$$

and employing that

$$\mathcal{B}(\bar{G}(\cdot;\bar{x}), v) = \int_0^l AE \frac{d}{dx}(\bar{G}(\cdot;\bar{x})) \frac{dv}{dx} = \mathcal{C}_1 \int_0^{\bar{x}} \frac{dv}{dx} + \mathcal{C}_3 \int_{\bar{x}}^l \frac{dv}{dx} \tag{4.1.9c}$$

we get

$$\|\delta G\|_{\mathcal{U}}^2 = \mathcal{B}(\delta G, \delta G) = \delta G(\bar{x}) - \mathcal{C}_1(\delta G(\bar{x}) - \delta G(0)) - \mathcal{C}_3(\delta G(l) - \delta G(\bar{x})). \tag{4.1.9d}$$

From the construction of $\bar{G}$ we have

$$\delta G(\bar{x}) = \delta G(0) = \delta G(l) = 0 \tag{4.1.9e}$$

and hence

$$\|\delta G\|_{\mathcal{U}}^2 = \mathcal{B}(\delta G, \delta G) = 0 \tag{4.1.9f}$$

which proves that $\delta G \equiv 0$. Hence we have $G(\cdot;\bar{x}) \equiv \bar{G}(\cdot;\bar{x})$, i.e., the Green's function $G(\cdot;\bar{x})$ is a piecewise polynomial of degree p, and hence

$$\mathcal{I}_\Delta^p G(\cdot; x_j^\Delta) \equiv G(\cdot; x_j^\Delta) \in S_\Delta^p \tag{4.1.10a}$$

and

$$|e_{S_\Delta^p}(x_j^\Delta)| \le \|e_{S_\Delta^p}\|_{\mathcal{U}} \|G(\cdot; x_j^\Delta) - \mathcal{I}_\Delta^p G(\cdot; x_j^\Delta)\|_{\mathcal{U}} = 0 \tag{4.1.10b}$$

which proves that $u_{S_\Delta^p}(x_j^\Delta) = u_{EX}(x_j^\Delta)$. □

Remark 4.1.3 Corollary 4.1.1 can also be proven under the more general assumption

$$\frac{1}{AE}\bigg|_{I_k^\Delta} \in \mathcal{P}^{p-1}(I_k^\Delta), \quad k = 1, 2, \ldots, N(\Delta). \tag{4.1.11}$$

Let us now prove a more general result about the convergence of the nodal errors $e_{S_\Delta^p}(x_j^\Delta)$.

Corollary 4.1.2 *Consider the regular model problem, and let AE, c, and f satisfy the additional assumptions*

$$\left\|\frac{d^{q'}}{dx^{q'}}(AE)\right\|_{L^\infty(I)} \leq A, \qquad \left\|\frac{d^{r'}c}{dx^{r'}}\right\|_{L^\infty(I)} \leq B, \qquad \left\|\frac{d^{t'}f}{dx^{t'}}\right\|_{L^\infty(I)} \leq C \tag{4.1.12a}$$

for any integers q', r', t', such that $0 \leq q' \leq p$, $0 \leq r', t' \leq p-1$. Then we have

$$|e_{S_\Delta^p}(x_j^\Delta)| \leq \mathcal{C}(h^\Delta)^{2p}, \quad j = 1, \ldots, N(\Delta). \tag{4.1.12b}$$

Proof From the smoothness assumptions (4.1.12a), it follows that

$$\|G(\cdot; x_j^\Delta) - \mathcal{I}_\Delta^p G(\cdot; x_j^\Delta)\|_{\mathcal{U}} \leq \mathcal{C}(K, B)(h^\Delta)^p \tag{4.1.13a}$$

and

$$\|e_{S_\Delta^p}\|_{\mathcal{U}} \leq \mathcal{C}(C)(h^\Delta)^p \tag{4.1.13b}$$

and substituting into (4.1.4b) we obtain

$$|e_{S_\Delta^p}(x_j^\Delta)| \leq \underbrace{\|e_{S_\Delta^p}\|_{\mathcal{U}}}_{\leq \mathcal{C}(C)(h^\Delta)^p} \underbrace{\|G(\cdot; x_j^\Delta) - \mathcal{I}_\Delta^p G(\cdot; x_j^\Delta)\|_{\mathcal{U}}}_{\leq \mathcal{C}(K,B)(h^\Delta)^p} \leq \mathcal{C}(h^\Delta)^{2p}. \tag{4.1.13c}$$

□

Remark 4.1.4 The above proof is based on the estimate

$$\min_{\chi \in S_\Delta^p} \|G(\cdot; x_j^\Delta) - \chi\|_{\mathcal{U}} \leq \mathcal{C}(h^\Delta)^p. \tag{4.1.14}$$

The assumptions (4.1.12a) which state that the input data AE, c, and f are sufficiently smooth, are essential for the validity of this estimate.

Remark 4.1.5 Note that if $\bar{x}$ is not a nodal point, then, we only have

$$\mathcal{C}_1\sqrt{h^\Delta} \leq \min_{\chi \in S_\Delta^p} \|G(\cdot; \bar{x}) - \chi\|_{\mathcal{U}} \leq \mathcal{C}_2\sqrt{h^\Delta} \tag{4.1.15}$$

and hence the argument in the proof of Corollary 4.1.2 cannot be used for points $\bar{x}$ which are not nodal points.

Remark 4.1.6 Corollaries 4.1.1 and 4.1.2 do not hold in the case that the finite element approximation is constructed to have continuous first or higher-order derivatives, namely, $S_\Delta^p \subseteq C^k(I)$, for $k \geq 1$. This is because

$$G(\cdot; \bar{x}) \notin C^1(I) \tag{4.1.16a}$$

and hence

$$\mathscr{C}h^p < \min_{\chi \in S_\Delta^p \subseteq C^1(I)} \|G(\cdot;\bar{x}) - \chi\|_{\mathscr{U}}. \tag{4.1.16b}$$

Remark 4.1.7 In the case of the regular model problem, for which AE and f satisfy the additional assumptions

$$\left\|\frac{d}{dx}(AE)\right\|_{L^\infty(\tilde{I})} \le K < \infty, \qquad \|f\|_{L^\infty(\tilde{I})} \le B < \infty \tag{4.1.17a}$$

and for meshes which are such that

$$\kappa h^\Delta \le h_{I_j^\Delta}^\Delta \le h^\Delta \tag{4.1.17b}$$

for $j = 1, \ldots, N(\Delta)$, with $\kappa > 0$ independent of Δ, it can be shown that

$$\left\|\frac{d}{dx}(e_{S_\Delta^p})\right\|_{L^\infty(I)} \le \mathscr{C}\|e_{S_\Delta^p}\|_{\mathscr{U}}. \tag{4.1.17c}$$

This relates to the observation that

$$|e_{S_\Delta^p}(x_j^\Delta)| \le \mathscr{C}h^\Delta \|e_{S_\Delta^p}\|_{\mathscr{U}}. \tag{4.1.17d}$$

We will now give an example which illustrates the above results.

Example 4.1.1 Consider the regular model problem with input data

$$AE \equiv 1, \qquad \alpha_2 = 10, \qquad c \equiv 0, 1, 5, 10, 50, \qquad f = \delta(\cdot; \tfrac{1}{2}), \qquad a_2 = 0. \tag{4.1.18a}$$

Figure 4.1.1 shows the graphs of $G(\cdot; \frac{1}{2})$ and $(d/dx)(G(x; \frac{1}{2}))$. In the case $c \equiv 0$, $G(\cdot; \bar{x})$ is piecewise linear, and, if the mesh Δ has a node at $\bar{x}$, we have $G(\cdot; \bar{x}) \in S_\Delta^1$. For $c \equiv$ *constant* $\neq 0$, $G(\cdot; \bar{x})$ is a piecewise exponential function, and, if the mesh has a node at $\bar{x}$, we have

$$\min_{\chi \in S_\Delta^p} \|G(\cdot; \bar{x}) - \chi\|_{\mathscr{U}} \le \mathscr{C}h^p \tag{4.1.18b}$$

and we get superconvergence at $\bar{x}$, namely,

$$|u_{EX}(\bar{x}) - u_{S_\Delta^p}(\bar{x})| \le (h^\Delta)^{2p}. \tag{4.1.18c}$$

To illustrate the superconvergence of the nodal values, $u_{S_\Delta^p}(x_j^\Delta)$, we considered the model problem with input data

$$AE \equiv 1, \qquad c \equiv 10, \qquad f \equiv 0, \qquad \alpha_2 = 10, \qquad a_2 = 1 \tag{4.1.19}$$

and computed the finite element solution $u_{S_\Delta^p}$, using uniform meshes with mesh size $h = \frac{1}{4}, \frac{1}{8}, \frac{1}{16}, \frac{1}{32}$, and elements of degree $p = 2, 3$, and 4. Table 4.1.1 gives the values of $h^{-2p} e_{S_\Delta^p}(\bar{x})$, for $\bar{x} = \frac{1}{4}$, and $\bar{x} = \frac{1}{2}$. We employed the uniform meshes Δ with $h = \frac{1}{4}$, $\frac{1}{2}$, $\frac{1}{16}$, and $\frac{1}{32}$, for which there is always a node at $\bar{x} = \frac{1}{4}$ and $\bar{x} = \frac{1}{2}$. It can be seen that

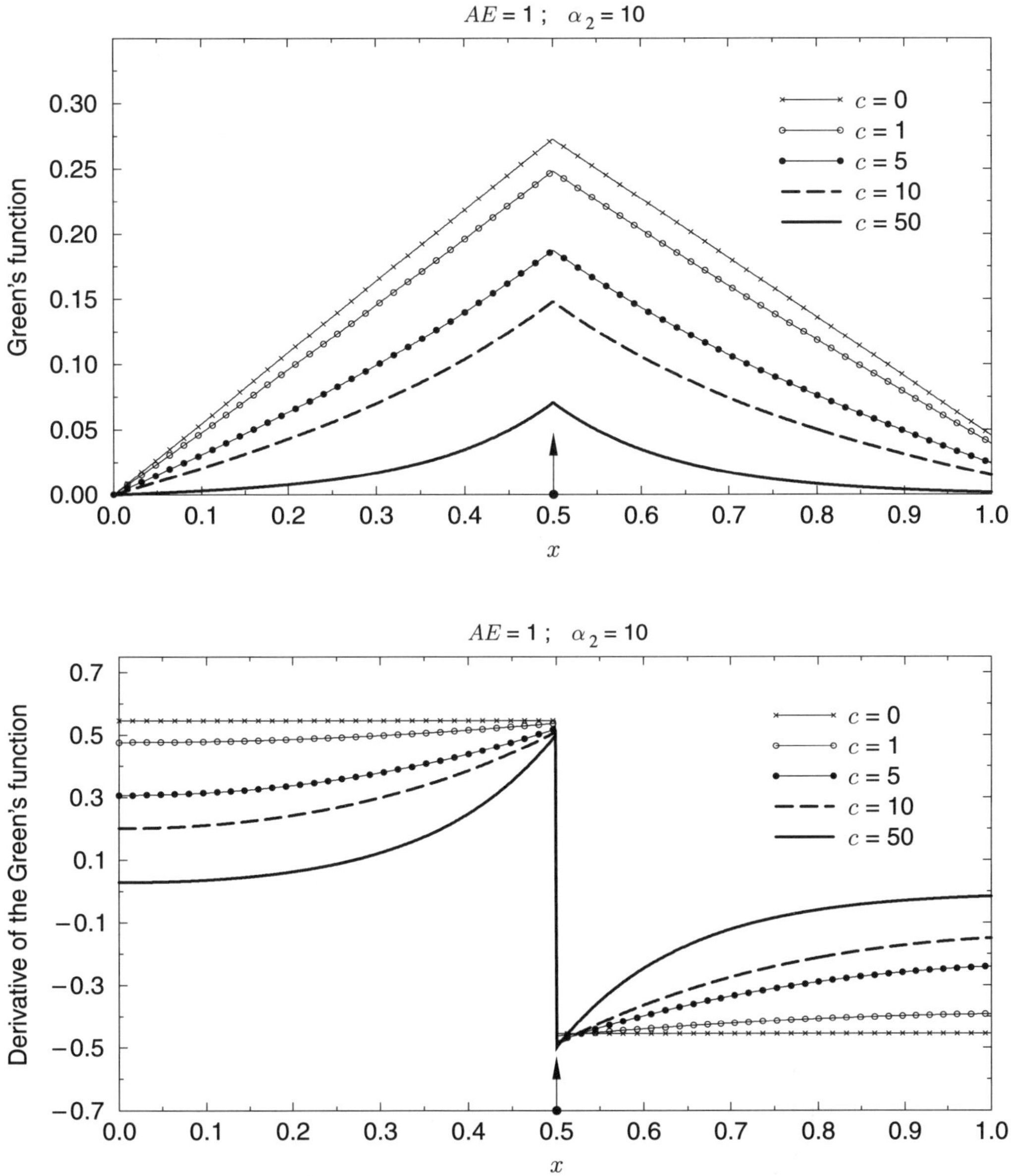

FIG. 4.1.1. An illustration of the Green's function for the regular model problem. Input data: $AE \equiv 1$, $c \equiv constant$, $f = \delta(\cdot; \frac{1}{2})$, $\alpha_2 = 10$, $a_2 = 0$. (a) Graph of the Green's function for $G(x; \frac{1}{2})$, $0 \le x \le 1$, $c \equiv 0, 1, 5, 10, 50$. (b) Graph of the derivative of the Green's function $(d/dx)(G(x; \frac{1}{2}))$, $0 \le x \le 1$, for $c \equiv 0, 1, 5, 10, 50$

the quantities $h^{-2p} e_{S^p_\Delta}(\frac{1}{4})$ and $h^{-2p} e_{S^p_\Delta}(\frac{1}{2})$ converge rather quickly to constant values, which means that the asymptotic rate of convergence $2p$ is achieved at the points $\bar{x} = \frac{1}{4}$ and $\bar{x} = \frac{1}{2}$, even for rather coarse meshes.

We will now address the convergence of $\|e_{S^p_\Delta}\|_{L^\infty(I^\Delta_j)}$, the maximum error in each element. We will need the following lemma.

TABLE 4.1.1. Values of $h^{-2p}\, e_{S^p_\Delta}(\bar{x})$ at $\bar{x} = \frac{1}{4}$ and $\bar{x} = \frac{1}{2}$ for uniform meshes of size h of elements of degree p. Input data: $AE \equiv 1$, $c \equiv 10$, $f \equiv 0$, $\alpha_2 = 10$, $a_2 = 1$. Mesh data: Uniform meshes with mesh size $h = \frac{1}{4}, \frac{1}{8}, \frac{1}{16}, \frac{1}{32}$, and elements of degree $p = 2, 3, 4$. The meshes have been chosen such that there is always a node at $\bar{x} = \frac{1}{4}$ and $\bar{x} = \frac{1}{2}$. From the above results, we can see that the asymptotic rate $2p$ is achieved at both points even for rather coarse meshes

h	$h^{-2p} e_{S^p_\Delta}(\frac{1}{4})$			$h^{-2p} e_{S^p_\Delta}(\frac{1}{2})$		
	$p = 2$	$p = 3$	$p = 4$	$p = 2$	$p = 3$	$p = 4$
1/4	0.7070×10^{-3}	0.6248×10^{-4}	0.1974×10^{-5}	0.1319×10^{-2}	0.1265×10^{-3}	0.3686×10^{-5}
1/8	0.6838×10^{-3}	0.6193×10^{-4}	0.1931×10^{-5}	0.1271×10^{-2}	0.1257×10^{-3}	0.3590×10^{-5}
1/16	0.6780×10^{-3}	0.6180×10^{-4}	0.1920×10^{-5}	0.1259×10^{-2}	0.1255×10^{-3}	0.3566×10^{-5}
1/32	0.6765×10^{-3}	0.6177×10^{-4}	0.1896×10^{-5}	0.1256×10^{-2}	0.1255×10^{-3}	0.3505×10^{-5}

Lemma 4.1.2 *Under the assumptions that*

$$\tilde{u}_{EX} \in C^{p+1}(\tilde{I}) \tag{4.1.20a}$$

and

$$\left\| \frac{d^{p+1}}{d\tilde{x}^{p+1}}(\tilde{u}_{EX}) \right\|_{L^\infty(\tilde{I})} \le A < \infty \tag{4.1.20b}$$

and

$$0 < \tilde{\mathscr{C}}_{EX} \le \min_{\tilde{x}\in\tilde{I}} \left| \frac{d^{p+1}}{d\tilde{x}^{p+1}}(\tilde{u}_{EX})(\tilde{x}) \right| \tag{4.1.20c}$$

we have

$$0 < \mathscr{C}(p)\tilde{\mathscr{C}}_{EX} \le \min_{\tilde{\chi}\in\mathscr{P}^p(\tilde{I})} \|\tilde{u}_{EX} - \tilde{\chi}\|_{L^\infty(\tilde{I})} \tag{4.1.21}$$

with $\mathscr{C}(p) > 0$ independent of $\tilde{u}_{EX}$.

Proof Let $\tilde{\mathscr{B}}^{p,\infty}\tilde{u}_{EX} \in \mathscr{P}^p(\tilde{I})$ be the best approximation of $\tilde{u}_{EX}$ in the L^∞ norm, namely

$$\|\tilde{u}_{EX} - \tilde{\mathscr{B}}^{p,\infty}\tilde{u}_{EX}\|_{L^\infty(\tilde{I})} = \min_{\tilde{\chi}\in\mathscr{P}^p(\tilde{I})} \|\tilde{u}_{EX} - \tilde{\chi}\|_{L^\infty(\tilde{I})}. \tag{4.1.22a}$$

From the Chebyshev *equioscillation theorem* (see [Atkinson, 1989], p. 224), it follows that there exist at least $(p+2)$ points

$$-1 \le \bar{x}_0 < \bar{x}_1 < \cdots < \bar{x}_{p+1} \le 1 \tag{4.1.22b}$$

such that

$$\tilde{u}_{EX}(\bar{x}_j) - \tilde{\mathscr{B}}^{p,\infty}\tilde{u}_{EX}(\bar{x}_j) = \sigma(-1)^j \|\tilde{u}_{EX} - \tilde{\mathscr{B}}^{p,\infty}\tilde{u}_{EX}\|_{L^\infty(I)} \tag{4.1.22c}$$

for $j = 0, 1, \ldots, (p + 1)$, with $\sigma = \pm 1$ depending only on $\tilde{u}_{EX}$ and p. Further, because $\tilde{u}_{EX}$, $\tilde{\mathcal{B}}^{p,\infty}\tilde{u}_{EX}$ are continuous, there exist $(p + 1)$ points $\tilde{x}_l \in \tilde{I} = [-1, 1]$, $l = 0, \ldots, p$, such that

$$\tilde{u}_{EX}(\tilde{x}_l) = \tilde{\mathcal{B}}^{p,\infty}\tilde{u}_{EX}(\tilde{x}_l), \quad l = 0, 1, \ldots, p. \tag{4.1.22d}$$

Hence $\tilde{\mathcal{B}}^{p,\infty}\tilde{u}_{EX} \in \mathcal{P}^p(\tilde{I})$ is the Lagrangian interpolant $\tilde{u}_{EX}$ with interpolation points $\tilde{x}_l$, $l = 0, 1, \ldots, p$, and for any $\tilde{x} \in \tilde{I}$, we have

$$(\tilde{u}_{EX} - \tilde{\mathcal{B}}^{p,\infty}\tilde{u}_{EX})(\tilde{x}) = \frac{1}{(p+1)!}\frac{d^{p+1}}{d\tilde{x}^{p+1}}(\tilde{u}_{EX})(\hat{x})\prod_{l=0}^{p}(\tilde{x} - \tilde{x}_l) \tag{4.1.22e}$$

with $\hat{x}$ depending on $\tilde{x}$. Further, there exists a constant $\tilde{\mathscr{C}}(p) > 0$ which is independent of the distribution of the $\tilde{x}_l$, such that

$$\tilde{\mathscr{C}}(p) \leq \max_{\tilde{x}\in\tilde{I}}\left|\prod_{l=0}^{p}(\tilde{x} - \tilde{x}_l)\right| \tag{4.1.22f}$$

and hence

$$\frac{1}{(p+1)!}\min_{\tilde{x}\in\tilde{I}}\left|\frac{d^{p+1}}{d\tilde{x}^{p+1}}(\tilde{u}_{EX})(\tilde{x})\right|\max_{\tilde{x}\in\tilde{I}}\left|\prod_{l=0}^{p}(\tilde{x} - \tilde{x}_l)\right| \leq \|\tilde{u}_{EX} - \tilde{\mathcal{B}}^{p,\infty}\tilde{u}_{EX}\|_{L^\infty(\tilde{I})} \tag{4.1.22g}$$

and we get

$$\frac{\tilde{\mathscr{C}}(p)}{(p+1)!}\tilde{\mathscr{C}}_{EX} \leq \|\tilde{u}_{EX} - \tilde{\mathcal{B}}^{p,\infty}\tilde{u}_{EX}\|_{L^\infty(\tilde{I})} \tag{4.1.22h}$$

which is (4.1.21). □

We are now ready to prove a lower estimate for $\|e_{S^p_\Delta}\|_{L^\infty(I^\Delta_j)}$:

Theorem 4.1.2 *Under the assumptions (4.1.20a)–(4.1.20c), we have*

$$\mathscr{C}(p)\,\mathscr{C}_{EX}(h_{I^\Delta_j})^{p+1} \leq \|e_{S^p_\Delta}\|_{L^\infty(I^\Delta_j)} \tag{4.1.23}$$

with $\mathscr{C}(p)$ dependent only on p.

Proof Let

$$F_{I^\Delta_j}(\tilde{x}) = x^\Delta_{j-1/2} + \frac{h_{I^\Delta_j}}{2}\tilde{x} \tag{4.1.24a}$$

be the affine mapping which maps $\tilde{I} = (-1, 1)$ into I^Δ_j, and let

$$\tilde{u}_{EX,I^\Delta_j} \stackrel{\text{def}}{=} u_{EX}|_{I^\Delta_j} \circ F_{I^\Delta_j} \tag{4.1.24b}$$

be the exact solution over $\tilde{I}$, and let $\tilde{\mathcal{B}}^{p,\infty}(\tilde{u}_{EX,I^\Delta_j})$ be the best L^∞ approximation of $\tilde{u}_{EX,I^\Delta_j}$ from $\mathcal{P}^p(\tilde{I})$. We then have

$$\mathcal{B}^{p,\infty}_{I^\Delta_j}(u_{EX}) \stackrel{\text{def}}{=} \tilde{\mathcal{B}}^{p,\infty}(u_{EX}|_{I^\Delta_j} \circ F_{I^\Delta_j}) \tag{4.1.24c}$$

and using Lemma 4.1.2., we obtain

$$\mathscr{C}(p)\tilde{\mathscr{C}}_{EX} \leq \|\tilde{u}_{EX,I_j^\Delta} - \tilde{\mathcal{B}}^{p,\infty}(\tilde{u}_{EX,I_j^\Delta})\|_{L^\infty(\tilde{I})} = \|u_{EX} - \mathcal{B}^{p,\infty}_{I_j^\Delta}(u_{EX})\|_{L^\infty(I_j^\Delta)}. \quad (4.1.25a)$$

Further, noting that

$$\tilde{\mathscr{C}}_{EX} = \min_{\tilde{x}\in\tilde{I}} \left|\frac{d^{p+1}}{d\tilde{x}^{p+1}}(\tilde{u}_{EX,I_j^\Delta})(\tilde{x})\right| = \left(\frac{dx}{d\tilde{x}}\right)^{p+1} \min_{x\in I_j^\Delta} \left|\frac{d^{p+1}}{dx^{p+1}}(u_{EX})(x)\right| \quad (4.1.25b)$$

using that $dx/d\tilde{x} = h_{I_j^\Delta}/2$, and employing (4.1.25b) into (4.1.25a), we get (4.1.23). □

We are now ready to prove that $\|e_{S^p_\Delta}\|_{L^\infty(I)}$ converges to zero as h^{p+1}.

Corollary 4.1.3 *Consider the regular model problem with the same assumptions as in Corollary 4.1.2, and assume, in addition, that there exists a subinterval $\hat{I} = (a, b) \subseteq I$, with $a < b$, such that*

$$u_{EX} \in C^{p+1}(\hat{I}) \quad (4.1.26a)$$

and

$$0 < \mathscr{C}_{EX} \leq \min_{x\in\hat{I}} \left|\frac{d^{p+1}}{dx^{p+1}}(u_{EX})\right|. \quad (4.1.26b)$$

Then, assuming that the meshes Δ are such that

$$\kappa h^\Delta \leq h_{I_j^\Delta} \leq h^\Delta \quad (4.1.26c)$$

for $j = 1, \ldots, N(\Delta)$, with $\kappa > 0$ independent of Δ, we have

$$\mathscr{C}_1 (h^\Delta)^{p+1} \leq \|e_{S^p_\Delta}\|_{L^\infty(I)} \leq \mathscr{C}_2\,(h^\Delta)^{p+1}. \quad (4.1.27)$$

Proof The left inequality follows from Theorem 4.1.2. Let us now prove the right inequality. In the special case that $AE \equiv 1$ and $c \equiv 0$, we have $\mathscr{B}_{I_j^\Delta}(u, v) = \int_{I_j^\Delta}(du/dx)(dv/dx)$, and for any $\bar{x} \in I_j^\Delta$, integrating by parts we get

$$|e_{S^p_\Delta}(\bar{x})| = \left|\int_{I_j^\Delta} \frac{d}{dx}(e_{S^p_\Delta})\frac{d}{dx}\left(G(\cdot;\bar{x}) - \mathcal{I}^1_\Delta G(\cdot;\bar{x})\right)\right| \quad (4.1.28a)$$

and employing the Cauchy–Schwarz inequality, we get

$$|e_{S^p_\Delta}(\bar{x})| \leq \underbrace{\left\|\frac{d}{dx}(e_{S^p_\Delta})\right\|_{L^2(I_j^\Delta)}}_{\leq \mathscr{C}(h^\Delta)^{p+\frac{1}{2}}} \underbrace{\left\|\frac{d}{dx}\left(G(\cdot;\bar{x}) - \mathcal{I}^1_\Delta G(\cdot;\bar{x})\right)\right\|_{L^2(I_j^\Delta)}}_{\leq \mathscr{C}(h^\Delta)^{\frac{1}{2}}} \leq \mathscr{C}(h^\Delta)^{p+1}. \quad (4.1.28b)$$

This proves the right inequality in the special case $AE \equiv 1$, $c \equiv 0$. Note that in (4.1.28a) we have employed that $e_{S^p_\Delta}(x^\Delta_{j-1}) = 0$, which follows from Corollary 4.1.1. The proof of right inequality in the general case $AE \not\equiv$ *constant*, and/or $c \not\equiv 0$, can be obtained by employing a similar approach as in the proof of Theorem 4.3.3 below. □

Remark 4.1.8 Under the assumptions that

(a) AE, c, and f satisfy the assumptions for the regular model problem and are sufficiently smooth

(b) $u_{EX} \in C^{p+1}(\bar{I})$, and $(d^{p+1}/dx^{p+1})(u_{EX})$ does not change sign in the element

(c) the meshes satisfy the quasi-uniformity condition (4.1.26c), we have

$$\max_{j=1,\dots,N(\Delta)} |e_{S^p_\Delta}(x_j^\Delta)| \leq \mathscr{C}(h^\Delta)^{2p} \tag{4.1.29a}$$

while

$$\|e_{S^p_\Delta}\|_{L^\infty(I)} \approx \mathscr{C}(h^\Delta)^{p+1} \tag{4.1.29b}$$

where h^Δ is the maximum size of the elements in the mesh. For elements of quadratic degree or higher, $p \geq 2$, we have $p+1 < 2p$, and hence

$$\max_{j=1,\dots,N(\Delta)} |e_{S^p_\Delta}(x_j^\Delta)| \leq \mathscr{C}(h^\Delta)^{2p} \ll \|e_{S^p_\Delta}\|_{L^\infty(I)} \approx \mathscr{C}(h^\Delta)^{p+1}. \tag{4.1.29c}$$

Hence, for $p \geq 2$, the nodal values of the finite element solution are *superconvergent* with rate $2p$.

In this section, we employed the Green's function $G(\cdot;\bar{x})$ to construct the following estimate for the pointwise error in the finite element solution

$$|e_{S^p_\Delta}(x_j^\Delta)| \leq \|e_{S^p_\Delta}\|_{\mathscr{U}} \|G(\cdot;x_j^\Delta) - \mathcal{I}^p_\Delta G(\cdot;x_j^\Delta)\|_{\mathscr{U}} \leq \mathscr{C}(h^\Delta)^\mu \|e_{S^p_\Delta}\|_{\mathscr{U}} \tag{4.1.30a}$$

with μ depending on p and the smoothness of AE and c. We used this estimate to prove that, for the regular model problem with sufficiently smooth input data, the error in the nodal values of the finite element solution converges to zero with the exceptional rate $2p$

$$|e_{S^p_\Delta}(x_j^\Delta)| \leq \mathscr{C}(h^\Delta)^{2p} \tag{4.1.30b}$$

at all the nodes x_j^Δ, while

$$\|e_{S^p_\Delta}\|_{L^\infty(I_j^\Delta)} \approx \mathscr{C}(h_{I_j^\Delta})^{p+1} \tag{4.1.30c}$$

in all the elements I_j^Δ. This means that for $p \geq 2$, the nodal values converge with higher rate than any other values. This is a first example of superconvergence.

Problems

P4.1.1. Construct the Green's function for the regular model problem in the case that $AE \equiv 1, c \equiv 1, \alpha_2 = 1$.

P4.1.2. Construct the Green's function $G_\vartheta(\cdot;\bar{x})$ for the singular model problem. Determine the rate of convergence μ for the energy norm of the interpolation error of $G_\vartheta(\cdot;\bar{x})$

$$\|G_\vartheta(\cdot;\bar{x}) - \mathcal{I}^p_\Delta G_\vartheta(\cdot;\bar{x})\|_{\mathcal{U}^\vartheta} \approx \mathcal{C}h^\mu \tag{4.1.31}$$

for a sequence of uniform meshes Δ.

P4.1.3. Consider the regular model problem and assume that AE and c are continuous with bounded derivatives up to order r and q, respectively, and let Δ be a uniform mesh with mesh size h. Construct an asymptotic estimate for $h^{-1}|e_{S^p_\Delta}(x^\Delta_j) - e_{S^p_\Delta}(x^\Delta_{j-1})|$.
Hint: Note that

$$\left|e_{S^p_\Delta}(x^\Delta_j) - e_{S^p_\Delta}(x^\Delta_{j-1})\right| = \left|\mathcal{B}(e_{S^p_\Delta}, G(\cdot;x^\Delta_j) - G(\cdot;x^\Delta_{j-1}))\right|. \tag{4.1.32}$$

4.1* A-priori estimates for the nodal errors and the maximum error in two dimensions

Here we will only underline the differences between the one- and the two-dimensional cases.

A major difference is that *the Green's function for the two-dimensional model problem does not belong to the energy space.* For example, the Green's function for the Laplace operator is the solution of the boundary value problem

$$\Delta G(\mathbf{x};\bar{\mathbf{x}}) = \delta(\mathbf{x}-\bar{\mathbf{x}}), \quad \mathbf{x} \in \Omega \tag{4.1*.1a}$$

$$G(\cdot;\bar{\mathbf{x}})|_{\Gamma_D} = 0, \qquad \frac{\partial}{\partial n}(G(\cdot;\bar{\mathbf{x}}))\Big|_{\Gamma_N} = 0 \tag{4.1*.1b}$$

and can be written in the form

$$G(\mathbf{x};\bar{\mathbf{x}}) = \frac{1}{2\pi}\ln|\mathbf{x}-\bar{\mathbf{x}}| + G_s(\mathbf{x};\bar{\mathbf{x}}), \quad \mathbf{x},\bar{\mathbf{x}} \in \Omega \tag{4.1*.1c}$$

where G_s is smooth in the neighborhood of $\bar{\mathbf{x}}$, and hence

$$G(\cdot;\bar{\mathbf{x}}) \notin \mathcal{U}. \tag{4.1*.1d}$$

It follows that the approach which was used for deriving estimates for the nodal errors in Section 4.1 cannot be used in higher dimensions. Nevertheless the analogue of the estimate (4.1.27) holds, namely

$$\mathcal{C}_1 h^{p+1} \le \|e_{S^p_\Delta}\|_{L^\infty(\Omega)} \le \mathcal{C}_2 h^{p+1}|\ln h|^r \tag{4.1*.2}$$

with $r = 1$ for $p = 1$, and $r = 0$ otherwise, provided that all of the input data, including the boundary $\partial\Omega$, are smooth. The left estimate can be proven analogously as in the one-dimensional case. The proof of the right estimate is rather technical and it is based on a special approximation of the Green's function. The logarithmic term, which appears in the right estimate in the case of linear elements, is not due to a technicality in the proof, and reflects the actual behavior of the error. For certain special meshes of elements of quadratic degree or higher, the nodal values are superconvergent. For more details we refer to Theorem 10.1 in [Wahlbin, 1991], and to [Ciarlet, 1991], [Schatz, 1998], and [Scott, 1976].

Problems

P4.1*.1. Let Ω be a domain with smooth boundary, let

$$\omega(\bar{x}) = B(\bar{\mathbf{x}}, R) \stackrel{\text{def}}{=} \Big\{\mathbf{x} \,\big|\, |\mathbf{x} - \bar{\mathbf{x}}| < R\Big\} \subseteq \Omega = B(\bar{\mathbf{x}}, 2R) \tag{4.1*.3}$$

and let $G_{\omega(\bar{x})}$ be the solution of the problem:
Find $G_{\omega(\bar{x})} \in \mathcal{U}$ such that

$$\int_\Omega (\boldsymbol{\nabla} G_{\omega(\bar{x})} \cdot \boldsymbol{\nabla} v + uv) = \frac{1}{|\omega(\bar{x})|} \int_{\omega(\bar{x})} v \quad \forall v \in \mathcal{U}. \tag{4.1*.4}$$

Estimate $\|G_{\omega(\bar{x})}\|_{H^2(\Omega)}$.

P4.1*.2. Let $\Omega = B(\mathbf{0}, 1)$, and $u(x) = \ln(x)$. Prove that $u \notin H^1(\Omega)$.

4.2 A-priori estimates for negative norms of the error

We will now construct estimates for the error in the finite element solution in the L^2 norm of the solution and its primitives, by extending the approach which was employed in Section 4.1.

In Section 4.1 we employed the Green's function $G(\cdot;\bar{x})$, which is the solution of the problem:

Find $G(\cdot;\bar{x}) \in \mathcal{U}$ such that

$$\mathcal{B}(G(\cdot;\bar{x}), v) = \int_0^l \delta(\cdot;\bar{x}) v = v(\bar{x}), \quad v \in \mathcal{U}. \tag{4.2.1}$$

In this section we will employ the function $H(\cdot;g)$ which is a generalization of the Green's function $G(\cdot;\bar{x})$, and is defined as the solution of the variational problem:

Find $H(\cdot;g) \in \mathcal{U}$ such that

$$\mathcal{B}(H(\cdot;g), v) = \int_0^l gv \quad \forall v \in \mathcal{U}. \tag{4.2.2}$$

Note that

$$G(\cdot;\bar{x}) \equiv H(\cdot;\delta(\cdot;\bar{x})). \tag{4.2.3}$$

Corollary 4.1.2 can now be generalized as follows:

Corollary 4.2.1 *Consider the regular model problem, and assume, in addition, that*

$$\left\|\frac{d^{q'}}{dx^{q'}}(AE)\right\|_{L^\infty(I)} \leq A, \qquad \left\|\frac{d^{r'}c}{dx^{r'}}\right\|_{L^\infty(I)} \leq B \tag{4.2.4a}$$

for all integers q', r', such that $0 \leq q' \leq q$, $0 \leq r' \leq r$. Then, for any $\mathscr{g} \in H^s(I)$, we have

$$\left|\int_I \mathscr{g} e_{S^p_\Delta}\right| \leq \mathscr{C}(h^\Delta)^\mu \|e_{S^p_\Delta}\|_{\mathscr{U}} \left\|\frac{d^s\mathscr{g}}{dx^s}\right\|_{L^2(I)} \tag{4.2.4b}$$

with

$$\mu = \min(s+1, q, r+1, p) \tag{4.2.4c}$$

with $\mathscr{C}$ depending only on s, r, and p.

Proof Employing Theorem 2.3.1, we get

$$\|H(\cdot;\mathscr{g})\|_{H^{\lambda+1}(I)} \leq \mathscr{C}\left\|\frac{d^s\mathscr{g}}{dx^s}\right\|_{L^2(I)} < \infty \tag{4.2.5a}$$

with

$$\lambda = \min(s+1, q, r+1). \tag{4.2.5b}$$

Letting $v = e_{S^p_\Delta}$ in (4.2.2), using the symmetry of the bilinear form, employing the orthogonality of the error, and the Cauchy–Schwarz inequality, we get

$$\left|\int_I \mathscr{g} e_{S^p_\Delta}\right| = |\mathscr{B}(e_{S^p_\Delta}, H(\cdot;\mathscr{g}) - \mathcal{I}^p_\Delta H(\cdot;\mathscr{g}))| \leq \|e_{S^p_\Delta}\|_{\mathscr{U}}\|H(\cdot;\mathscr{g}) - \mathcal{I}^p_\Delta H(\cdot;\mathscr{g})\|_{\mathscr{U}}. \tag{4.2.6}$$

Employing the interpolation-error estimate, and (4.2.5) we get

$$\|H(\cdot;\mathscr{g}) - \mathcal{I}^p_\Delta H(\cdot;\mathscr{g})\|_{\mathscr{U}} \leq \mathscr{C}(h^\Delta)^\mu\|H(\cdot;\mathscr{g})\|_{H^{\lambda+1}(I)} \leq \mathscr{C}(h^\Delta)^\mu\left\|\frac{d^s\mathscr{g}}{dx^s}\right\|_{L^2(I)} \tag{4.2.7a}$$

where

$$\mu = \min(\lambda, p). \tag{4.2.7b}$$

Using (4.2.7a) into (4.2.6), (4.2.4b) follows. □

We are now ready to construct an estimate for the L^2 norm of the error.

Corollary 4.2.2 *Consider the regular model problem, and assume, in addition, that*

$$\left\|\frac{d}{dx}(AE)\right\|_{L^\infty(I)} \leq K. \tag{4.2.8a}$$

Then we have

$$\|e_{S^p_\Delta}\|_{L^2(I)} \leq \mathscr{C}h\|e_{S^p_\Delta}\|_{\mathscr{U}}. \tag{4.2.8b}$$

Proof Employing $g = e_{S^p_\Delta}$, and $s = 0$, in (4.2.4b), we obtain

$$\|e_{S^p_\Delta}\|^2_{L^2(I)} \leq \mathscr{C}h\|e_{S^p_\Delta}\|_{\mathscr{U}}\|e_{S^p_\Delta}\|_{L^2(I)} \tag{4.2.9}$$

and (4.2.8b) follows. □

Remark 4.2.1 The argument employed in the proof of Corollary 4.2.2 was proposed by [Aubin, 1967], [Nitsche, 1968], and by [Oganesjan and Ruchovec, 1969].

Remark 4.2.2 In the special case that

$$e_{S^p_\Delta}(x^\Delta_{j-1}) = 0 \tag{4.2.10a}$$

we have

$$\|e_{S^p_\Delta}\|_{L^2(I^\Delta_j)} \leq \frac{1}{\sqrt{2\beta_1}} h_{I^\Delta_j}\|e_{S^p_\Delta}\|_{\mathscr{U}(I^\Delta_j)} \tag{4.2.10b}$$

and hence

$$\|e_{S^p_\Delta}\|_{L^2(I)} \leq \frac{1}{\sqrt{2\beta_1}}\sqrt{\sum_{j=1}^{N(\Delta)}(h_{I^\Delta_j})^2\|e_{S^p_\Delta}\|^2_{\mathscr{U}(I^\Delta_j)}} \leq \frac{1}{\kappa\sqrt{2\beta_1}}h^\Delta\|e_{S^p_\Delta}\|_{\mathscr{U}} \tag{4.2.10c}$$

where κ is the parameter which controls the quasi-uniformity of the mesh.

For sufficiently smooth AE, c, and f, we have

$$\|e_{S^p_\Delta}\|_{\mathscr{U}} \approx \mathscr{C}h^p \tag{4.2.11a}$$

and from (4.2.8b) we get

$$\|e_{S^p_\Delta}\|_{L^2(I)} \leq \mathscr{C}h\|e_{S^p_\Delta}\|_{\mathscr{U}} \leq \mathscr{C}h^{p+1}. \tag{4.2.11b}$$

Further, it can be proven that

$$\mathscr{C}h^{p+1+\varepsilon} \leq \|e_{S^p_\Delta}\|_{L^2(I)} \tag{4.2.11c}$$

for any $\varepsilon > 0$, and hence

$$\|e_{S^p_\Delta}\|_{L^2(I)} \approx \mathscr{C}h^{p+1}. \tag{4.2.11d}$$

Thus *the maximum possible rate of convergence in the L^2 norm is $(p+1)$.*

The assumption $\|(d/dx)(AE)\|_{L^\infty(I)} \leq K$ and hence Corollary 4.2.2 does not apply in the case of the singular model problem. Let us now derive an L^2 estimate for the singular model problem.

Corollary 4.2.3 *Consider the singular model problem with $0 < \vartheta < 1$, and assume, in addition, that*

$$f \in L^2(I). \tag{4.2.12a}$$

Then

$$\|e_{S^p_\Delta}\|_{L^2(I)} \leq \mathscr{C}(\vartheta) h^{(1-\vartheta)/2} \|e_{S^p_\Delta}\|_{\mathscr{U}^\vartheta}. \tag{4.2.12b}$$

Proof For simplicity we will prove (4.2.12b) only for uniform meshes, and $p = 1$. Let $\mathscr{g} \in L^2(I)$, and denote by $u^{\vartheta,\mathscr{g}}_{EX} \equiv H(\cdot;\mathscr{g})$ the solution of the singular problem

$$-\frac{d}{dx}\Big(x^\vartheta \frac{d}{dx}(u^{\vartheta,\mathscr{g}}_{EX})\Big)(x) = \mathscr{g}(x), \quad 0 < x < 1 \tag{4.2.13a}$$

$$u^{\vartheta,\mathscr{g}}_{EX}(0) = 0, \qquad \frac{d}{dx}(u^{\vartheta,\mathscr{g}}_{EX})(1) = 0. \tag{4.2.13b}$$

It follows that

$$u^{\vartheta,\mathscr{g}}_{EX}(x) = -\int_0^x \frac{1}{t^\vartheta} F(t) + \mathscr{C} \int_0^x \frac{1}{t^\vartheta} \tag{4.2.13c}$$

where

$$F(t) = \int_0^t \mathscr{g}(y)\, dy \tag{4.2.13d}$$

and

$$|\mathscr{C}| \leq A \overset{\text{def}}{=} \sqrt{\int_0^1 \mathscr{g}^2}. \tag{4.2.13e}$$

On each I^Δ_j, $j > 1$, we have

$$\|u^{\vartheta,\mathscr{g}}_{EX} - \mathcal{I}^1_\Delta u^{\vartheta,\mathscr{g}}_{EX}\|_{\mathscr{U}^\vartheta(I^\Delta_j)} \leq \mathscr{C} h^{1+\frac{\vartheta}{2}}_{I^\Delta_j} j^{\frac{\vartheta}{2}} \Big\| \frac{d^2}{dx^2}(u^{\vartheta,\mathscr{g}}_{EX}) \Big\|_{L^2(I^\Delta_j)}. \tag{4.2.14a}$$

Using that

$$\frac{d^2}{dx^2}(u^{\vartheta,\mathscr{g}}_{EX}) = -\frac{1}{x^\vartheta}\mathscr{g}(x) + \frac{\vartheta}{x^{\vartheta+1}} F(x) - \frac{\mathscr{C}\vartheta}{x^{\vartheta+1}} \tag{4.2.14b}$$

we get

$$\Big\| \frac{d^2}{dx^2}(u^{\vartheta,\mathscr{g}}_{EX}) \Big\|_{L^2(I^\Delta_j)} \leq \Big\| \frac{\mathscr{g}}{x^\vartheta} \Big\|_{L^2(I^\Delta_j)} + \vartheta \Big\| \frac{F}{x^{\vartheta+1}} \Big\|_{L^2(I^\Delta_j)} + \mathscr{C}\vartheta \Big\| \frac{1}{x^{\vartheta+1}} \Big\|_{L^2(I^\Delta_j)}. \tag{4.2.14c}$$

Realizing that for $j > 1$, we have

$$\Big\| \frac{1}{x^{\vartheta+1}} \Big\|_{L^2(I^\Delta_j)} \leq \mathscr{C}(j-1)^{-(\vartheta+1)} h^{-(\vartheta+1)} \sqrt{h} \tag{4.2.15a}$$

and

$$\Big\| \frac{\mathscr{g}}{x^\vartheta} \Big\|_{L^2(I^\Delta_j)} \leq \mathscr{C} \|\mathscr{g}\|_{L^2(I^\Delta_j)} (j-1)^{-\vartheta} h^{-\vartheta} \tag{4.2.15b}$$

and that

$$|F(x)| \le \sqrt{x}\sqrt{\int_0^1 \boldsymbol{g}^2} = \sqrt{x}A \tag{4.2.15c}$$

we get

$$\left\| \frac{d^2}{dx^2}(u_{EX}^{\vartheta,\boldsymbol{g}}) \right\|^2_{L^2(I_j^\Delta)} \le \mathscr{C}\big((j-1)^{-2(\vartheta+1)}\, h^{-2(\vartheta+1)}\, hA^2 + jh(j-1)^{-2(\vartheta+1)}\, h^{-2(\vartheta+1)} hA^2 + \|\boldsymbol{g}\|^2_{L^2(I_j^\Delta)} (j-1)^{-2\vartheta} h^{-2\vartheta}\big). \tag{4.2.15d}$$

Summing over all the elements except the first, we get

$$\sum_{j=2}^{N(\Delta)} \|u_{EX}^{\vartheta,\boldsymbol{g}} - \mathcal{I}_\Delta^1 u_{EX}^{\vartheta,\boldsymbol{g}}\|^2_{\mathcal{U}^\vartheta(I_j^\Delta)} \le \mathscr{C}\Bigg(A^2\, h^{(1-\vartheta)} \sum_{j=2}^{N(\Delta)} \frac{j^\vartheta}{(j-1)^{2(\vartheta+1)}} + A^2 h^{(2-\vartheta)} \sum_{j=2}^{N(\Delta)} \frac{j^{\vartheta+1}}{(j-1)^{2(\vartheta+1)}} + h^{-\vartheta+2}\, \|\boldsymbol{g}\|^2_{L^2(I)} \sum_{j=2}^{N(\Delta)} \frac{j^\vartheta}{(j-1)^{2\vartheta}} \Bigg) \le \mathscr{C} A^2 h^{1-\vartheta}. \tag{4.2.16}$$

Further, on I_1^Δ we have

$$\frac{d}{dx}(u_{EX}^{\vartheta,\boldsymbol{g}} - \mathcal{I}_\Delta^1 u_{EX}^{\vartheta,\boldsymbol{g}}) = \frac{d}{dx}(u_{EX}^{\vartheta,\boldsymbol{g}}) - \frac{1}{h}\, u_{EX}^{\vartheta,\boldsymbol{g}}(h) \tag{4.2.17a}$$

and hence

$$\|u_{EX}^{\vartheta,\boldsymbol{g}} - \mathcal{I}_\Delta^1 u_{EX}^{\vartheta,\boldsymbol{g}}\|^2_{\mathcal{U}^\vartheta(I_1^\Delta)} \le \mathscr{C}\Big(\int_{I_1^\Delta} x^\vartheta \Big(\frac{d}{dx}(u_{EX}^{\vartheta,\boldsymbol{g}})\Big)^2 + \frac{1}{h^2}\, h^{\vartheta+1}\, (u_{EX}^{\vartheta,\boldsymbol{g}}(h))^2 \Big) \le \mathscr{C} A^2\, (h^{2-\vartheta} + h^{1-\vartheta}). \tag{4.2.17b}$$

Combining (4.2.16) and (4.2.17b), we get

$$\|u_{EX}^{\vartheta,\boldsymbol{g}} - \mathcal{I}_\Delta^1 u_{EX}^{\vartheta,\boldsymbol{g}}\|_{\mathcal{U}^\vartheta} \le \mathscr{C} h^{(1-\vartheta)/2} \|\boldsymbol{g}\|_{L^2(I)}. \tag{4.2.18}$$

Using now the Cauchy–Schwarz inequality, and letting $\boldsymbol{g} \equiv e_{S_\Delta^p}$, (4.2.12b) follows. □

Let us also prove that the estimate $\|e_{S_\Delta^p}\|_{L^2(I)} \le \mathscr{C} h^{(1-\vartheta)/2} \|e_{S_\Delta^p}\|_{\mathcal{U}^\vartheta}$ is optimal.

Example 4.2.1 Consider the singular model problem with $f = 0$, $\alpha_2 = 0$ and $a_2 = 1$, namely,

$$-\frac{d}{dx}\Big(x^\vartheta \frac{du}{dx}\Big) \equiv 0, \qquad u(0) = 0, \qquad \frac{du}{dx}(1) = 1 \tag{4.2.19a}$$

for which

$$u_{EX} \equiv \frac{x^{1-\vartheta}}{1-\vartheta}. \tag{4.2.19b}$$

Let Δ be a uniform mesh, $p = 1$, and $u_i^\Delta = u_{S_\Delta^1}(x_i)$. Employing $v = v_{I_j^\Delta}$, where

$$v_{I_j^\Delta}(x) = \begin{cases} 0 & \text{for } 0 \le x \le x_{j-1}^\Delta \\ (x - x_{j-1}^\Delta) & \text{for } x_{j-1}^\Delta \le x \le x_j^\Delta \\ h_{I_j^\Delta} & \text{for } x_j^\Delta \le x \le 1 \end{cases} \tag{4.2.20a}$$

in the discrete variational statement

$$\int_I x^\vartheta \frac{d}{dx}(u_{S_\Delta^1}) \frac{dv}{dx} = v(1) \quad v \in S_\Delta^1 \tag{4.2.20b}$$

we obtain

$$\frac{u_j^\Delta - u_{j-1}^\Delta}{h} \int_{I_j^\Delta} x^\vartheta = h \tag{4.2.21a}$$

which can also be written as

$$u_j^\Delta - u_{j-1}^\Delta = h^2 \left(\int_{I_j^\Delta} x^\vartheta \right)^{-1}. \tag{4.2.21b}$$

On the other hand

$$u_{EX}(x_j) - u_{EX}(x_{j-1}) = \frac{1}{1-\vartheta}(x_j^{1-\vartheta} - x_{j-1}^{1-\vartheta}) = \int_{I_j^\Delta} x^{-\vartheta}. \tag{4.2.21c}$$

Employing now that

$$h^2 \left(\int_{I_j^\Delta} x^\vartheta \right)^{-1} \le \int_{I_j^\Delta} x^{-\vartheta} \tag{4.2.21d}$$

we get

$$u_j^\Delta - u_{j-1}^\Delta \le u_{EX}(x_j^\Delta) - u_{EX}(x_{j-1}^\Delta) \tag{4.2.21e}$$

and hence

$$u_{EX}(x_{j-1}^\Delta) - u_{j-1}^\Delta \le u_{EX}(x_j^\Delta) - u_j^\Delta \tag{4.2.21f}$$

from which it follows that

$$\frac{\vartheta^2}{1-\vartheta} h^{1-\vartheta} = u_{EX}(h) - u_1^\Delta \le u_{EX}(x_j^\Delta) - u_j^\Delta. \tag{4.2.21g}$$

From (4.2.21g) it follows that

$$\mathscr{C}h^{1-\vartheta} \leq \|e_{S^p_\Delta}\|_{L^2(I)}. \tag{4.2.22a}$$

Hence using Corollary 4.2.3 we have

$$\mathscr{C}h^{1-\vartheta} \leq \|e_{S^p_\Delta}\|_{L^2(I)} \leq \mathscr{C}h^{(1-\vartheta)/2}\|e_{S^p_\Delta}\|_{\mathscr{U}^\vartheta}. \tag{4.2.22b}$$

Analogously as in (4.2.18) we have

$$\|e_{S^p_\Delta}\|_{\mathscr{U}^\vartheta} \leq \mathscr{C}h^{(1-\vartheta)/2} \tag{4.2.22c}$$

It follows that the estimate (4.2.12b) cannot be improved.

Remark 4.2.3 Let us now compare the L^2 norm error estimates for the regular and the singular model problems. In the case of the regular model problem, with smooth AE, we have

$$\|e_{S^p_\Delta}\|_{L^2(I)} \leq \mathscr{C}h\|e_{S^p_\Delta}\|_{\mathscr{U}} \tag{4.2.23a}$$

which means that the L^2 norm of the error converges to zero faster than the energy norm by a full power of h. In the case of the singular model problem, with square integrable f, we have

$$\|e_{S^1_\Delta}\|_{L^2(I)} \leq \mathscr{C}h^{(1-\vartheta)/2}\|e_{S^p_\Delta}\|_{\mathscr{U}^\vartheta} \tag{4.2.23b}$$

and the L^2 norm of the error converges faster than the energy norm only by a factor of $h^{(1-\vartheta)/2}$.

Remark 4.2.4 Analogously, in the two-dimensional case for smooth input data, including the boundary $\partial\Omega$, we have

$$\|e_{S^p_\Delta}\|_{L^2(\Omega)} \leq \mathscr{C}h\|e_{S^p_\Delta}\|_{\mathscr{U}} \tag{4.2.24a}$$

while in the case that $\partial\Omega$ has corner points, we have

$$\|e_{S^p_\Delta}\|_{L^2(\Omega)} \leq \mathscr{C}h^\alpha\|e_{S^p_\Delta}\|_{\mathscr{U}} \tag{4.2.24b}$$

where the exponent α depends on the internal angles of the corner points and the boundary conditions on the edges emanating from the corner points.

Above we constructed a-priori estimates for the L^2 norm of the error in the finite element solution. Let us now recall that, from the *duality principle*, for any $w \in L^2(I)$, we have

$$\|w\|_{L^2(I)} = \sup_{v\in L^2(I)} \frac{|\int_I wv|}{\|v\|_{L^2(I)}} = \sup_{\substack{v\in L^2(I)\\ \|v\|_{L^2(I)}=1}} \left|\int_I wv\right|. \tag{4.2.25}$$

Taking now the supremum in (4.2.25) over the smaller space $H^l(I) \subset L^2(I)$, we obtain a functional $\|\cdot\|_{H^{-l}(I)} : L^2(I) \mapsto \mathbb{R}_0^+$, defined by

$$\|w\|_{H^{-l}(I)} \stackrel{\text{def}}{=} \sup_{v \in H^l(I)} \frac{|\int_I wv|}{\|v\|_{H^l(I)}} = \sup_{\substack{v \in H^l(I) \\ \|v\|_{H^l(I)}=1}} \left| \int_I wv \right| \tag{4.2.26a}$$

where we recall that

$$\|v\|_{H^l(I)} = \sqrt{\sum_{j=0}^{l} \left\| \frac{d^j v}{dx^j} \right\|_{L^2(I)}^2}. \tag{4.2.26b}$$

It can be proven that $\|\cdot\|_{H^{-l}(I)}$, $l > 0$, satisfies 12a–12d of A2.1.1, and hence it is a norm. We will denote by $H^{-l}(I)$ the Hilbert space obtained by completing $L^2(I)$ in this norm.

Similarly, taking the supremum in (4.2.25) over the space

$$H^l_{(0}(I) \stackrel{\text{def}}{=} \left\{ v \in H^l(I) \mid \frac{d^j v}{dx^j}(0) = 0,\ j = 0, \ldots, l-1 \right\} \tag{4.2.27a}$$

we obtain the norm

$$\|w\|_{H^{-l}_{(0}(I)} \stackrel{\text{def}}{=} \sup_{v \in H^l_{(0}(I)} \frac{|\int_I wv|}{\|d^l v/dx^l\|_{L^2(I)}} = \sup_{\substack{v \in H^l_{(0}(I) \\ \|d^l v/dx^l\|_{L^2(I)}=1}} \left| \int_I wv \right| \tag{4.2.27b}$$

and the space $H^{-l}_{(0}(I)$, as the completion of $L^2(I)$ in the norm $\|\cdot\|_{H^{-l}_{(0}(I)}$.

Let us now find the relationship between $H^{-l}_{(0}(I)$ and $H^{-l}(I)$. Noting that

$$\|v\|_{H^l(I)} \le \mathscr{C}(l) \left\| \frac{d^l v}{dx^l} \right\|_{L^2(I)} \quad \forall v \in H^l_{(0}(I) \tag{4.2.28a}$$

we have

$$\|w\|_{H^{-l}_{(0}(I)} = \sup_{v \in H^l_{(0}(I)} \left(\frac{\|v\|_{H^l(I)}}{\|d^l v/dx^l\|_{L^2(I)}} \frac{|\int_I wv|}{\|v\|_{H^l(I)}} \right) \le \mathscr{C}(l) \sup_{v \in H^l_{(0}(I)} \frac{|\int_I wv|}{\|v\|_{H^l(I)}} \tag{4.2.28b}$$

and noting that $H^l_{(0}(I) \subset H^l(I)$, we obtain

$$\sup_{v \in H^l_{(0}(I)} \frac{|\int_I wv|}{\|v\|_{H^l(I)}} \le \sup_{v \in H^l(I)} \frac{|\int_I wv|}{\|v\|_{H^l(I)}} = \|w\|_{H^{-l}(I)} \tag{4.2.28c}$$

and we get

$$\|w\|_{H^{-l}_{(0}(I)} < \mathscr{C}(l) \|w\|_{H^{-l}(I)} \quad \forall w \in H^{-l}(I) \tag{4.2.28d}$$

and hence

$$H^{-l}_{(0}(I) \supseteq H^{-l}(I). \tag{4.2.28e}$$

We will now construct an asymptotic estimate for $\|e_{S^p_\Delta}\|_{H^{-l}(I)}$.

Theorem 4.2.1 *Consider the regular model problem, and let AE, c satisfy the additional assumptions*

$$\left\|\frac{d^{q'}}{dx^{q'}}(AE)\right\|_{L^\infty(I)} \leq K, \qquad \left\|\frac{d^{r'}c}{dx^{r'}}\right\|_{L^\infty(I)} \leq B \tag{4.2.29a}$$

for q', r' integers such that $0 \leq q' \leq q$, $0 \leq r' < r$. Then we have

$$\|e_{S^p_\Delta}\|_{H^{-l}(I)} \leq \mathscr{C}(l)h^\mu \|e_{S^p_\Delta}\|_{\mathcal{U}} \tag{4.2.29b}$$

where

$$\mu = \min(p, l+1, q, r+1). \tag{4.2.29c}$$

Proof For any $g \in H^l(I)$, we have

$$\left|\int_I g e_{S^p_\Delta}\right| \leq \|e_{S^p_\Delta}\|_{\mathcal{U}} \underbrace{\|H(\cdot;g) - \mathcal{I}^p_\Delta H(\cdot;g)\|_{\mathcal{U}}}_{\leq \mathscr{C}h^\mu \|H(\cdot;g)\|_{H^{\lambda+1}(I)}} \leq \|e_{S^p_\Delta}\|_{\mathcal{U}} \mathscr{C}h^\mu \|g\|_{H^l(I)} \tag{4.2.30a}$$

where

$$\mu = \min(p, \lambda), \qquad \lambda = \min(l+1, q, r+1). \tag{4.2.30b}$$

Using this estimate in the definition of the $H^{-l}(I)$ norm, we get

$$\|e_{S^p_\Delta}\|_{H^{-l}(I)} = \sup_{\substack{g \in H^l(I) \\ \|g\|_{H^l(I)}=1}} \left|\int_I e_{S^p_\Delta} g\right| \leq \sup_{\substack{g \in H^l(I) \\ \|g\|_{H^l(I)}=1}} \left(\mathscr{C}h^\mu \|g\|_{H^l(I)} \|e_{S^p_\Delta}\|_{\mathcal{U}}\right) \tag{4.2.30c}$$

and (4.2.29b) has been proven. □

The same estimate also holds for $\|e_{S^p_\Delta}\|_{H^{-l}_{(0}(I)}$.

Remark 4.2.5 For sufficiently smooth input data, i.e., for $q \geq l+1 \geq 1$, and $r \geq l \geq 0$, we have $\mu = \min(p, l+1)$, and

$$\|e_{S^p_\Delta}\|_{H^{-l}(I)} \leq \mathscr{C}h^{\min(p,l+1)} \|e_{S^p_\Delta}\|_{\mathcal{U}}. \tag{4.2.31}$$

As we will see in Example 4.2.5 below, this estimate is optimal, and thus, *the rate of convergence of $\|e_{S^p_\Delta}\|_{H^{-l}(I)}$ increases with the order l of the negative norm up to the threshold* $l = p-1$. For example, for linear elements ($p = 1$), the threshold is $l = 0$.

Remark 4.2.6 The above approach for constructing estimates for negative norms of the error in the finite element solution is sometimes called the *Aubin–Nitsche method*, although it was also discovered independently by Oganesjan and Rukhovec; see [Aubin, 1967], [Nitsche, 1968], and [Oganesjan and Rukhovec, 1969].

We can also estimate $\|e_{S^p_\Delta}\|_{H^{-l}(I)}$ in terms of the norm of the load f.

Corollary 4.2.4 *Consider the regular model problem, and assume, in addition, that*

$$\left\|\frac{d^{q'}}{dx^{q'}}(AE)\right\|_{L^\infty(I)} \le K, \qquad \left\|\frac{d^{r'}c}{dx^{r'}}\right\|_{L^\infty(I)} \le B, \qquad \left\|\frac{d^{s'}f}{dx^{s'}}\right\|_{L^2(I)} \le C, \tag{4.2.32a}$$

for all integers q', r', s', such that $0 \le q' \le q$, $0 \le r' \le r$, $0 \le s' \le s$. Then we have

$$\|e_{S^p_\Delta}\|_{H^{-l}(I)} \le \mathscr{C} h^{\mu+\lambda}\|f\|_{H^s(I)} \tag{4.2.32b}$$

where

$$\mu = \min(p, l+1, q, r+1), \qquad \lambda = \min(p, q, r+1, s+1) \tag{4.2.32c}$$

and $\mathscr{C}$ depends only on q, r, s, p, and l.

Proof Employing the assumptions about the smoothness of AE, c, and f, and the best-approximation property of the finite element solution, we get

$$\|e_{S^p_\Delta}\|_{\mathscr{U}} \le \|u_{EX} - \mathcal{I}^p_\Delta u_{EX}\|_{\mathscr{U}} \le \mathscr{C} h^\lambda \left\|\frac{d^{\lambda+1}u_{EX}}{dx^{\lambda+1}}\right\|_{L^2(I)} \le \mathscr{C} h^\lambda \|f\|_{H^s(I)} \tag{4.2.33}$$

with $\lambda = \min(p, q, r+1, s+1)$, and using this result in $\|e_{S^p_\Delta}\|_{H^{-l}(I)} \le \mathscr{C} h^\mu \|e_{S^p_\Delta}\|_{\mathscr{U}}$, (4.2.30b) follows. □

Remark 4.2.7 For sufficiently smooth input data, we have $\lambda = p$, and we get

$$\|e_{S^p_\Delta}\|_{H^{-l}(I)} \le \mathscr{C} h^{2p}, \qquad l \ge p. \tag{4.2.34}$$

It can also be proven that this estimate is optimal, and hence *the maximum possible rate of convergence in the negative norm is equal to* $2p$.

We will now address the meaning of the negative norms. We will first address the $\|\cdot\|_{\mathscr{U}^{-1}}$ norm, which is defined by

$$\|w\|_{\mathscr{U}^{-1}} \stackrel{\text{def}}{=} \sup_{v\in\mathscr{U}} \left(\frac{1}{\|v\|_{\mathscr{U}}}\left|\int_I wv\right|\right) = \sup_{v\in\mathscr{U}}\left|\int_I w\frac{v}{\|v\|_{\mathscr{U}}}\right| = \sup_{\substack{v\in\mathscr{U}\\ \|v\|_{\mathscr{U}}=1}}\left|\int_I wv\right|. \tag{4.2.35a}$$

Integrating by parts, and employing that for any $v \in \mathscr{U}$, we have $v(0) = 0$, we get

$$\int_I wv = -\int_I \left(\int_x^l w\right)\frac{dv}{dx} = \int_I W\frac{dv}{dx}. \tag{4.2.35b}$$

Here we let

$$W(x) \stackrel{\text{def}}{=} -\int_x^l w \tag{4.2.35c}$$

which has meaning because $w \in L^2(I) \subset L^1(I)$, and hence it is integrable. Employing (4.2.35b) into, (4.2.35a), we get

$$\begin{aligned}\|w\|_{\mathcal{U}^{-1}} &= \sup_{\substack{v\in\mathcal{U}\\ \|v\|_{\mathcal{U}}=1}} \left|\int_0^l wv\right| = \sup_{\substack{v\in\mathcal{U}\\ \|v\|_{\mathcal{U}}=1}} \left|\int_I W\frac{dv}{dx}\right| \\ &\leq \sup_{\substack{v\in\mathcal{U}\\ \|v\|_{\mathcal{U}}=1}} \left\|\frac{dv}{dx}\right\|_{L^2(I)} \|W\|_{L^2(I)} \end{aligned} \tag{4.2.35d}$$

and hence we have proven

$$\|w\|_{\mathcal{U}^{-1}} \leq \sup_{\substack{v\in\mathcal{U}\\ \|v\|_{\mathcal{U}}=1}} \left\|\frac{dv}{dx}\right\|_{L^2(I)} \left\|\int_x^l w\right\|_{L^2(I)} \qquad \forall w \in \mathcal{U}^{-1}. \tag{4.2.35e}$$

Let us now show that *every function in* $\mathcal{U}^{-1}$ *can be interpreted as a generalized derivative of a function from* $L^2(I)$.

Theorem 4.2.2 *Given a* $w \in \mathcal{U}^{-1}$, *there exists a* $W \in L^2(I)$, *such that*

$$W = -\int_x^l w, \qquad \frac{d}{dx}(W) = w \tag{4.2.36}$$

where the derivative $(d/dx)(W)$ *is understood in a generalized sense.*

Proof Recall that $\mathcal{U}^{-1}$ is the completion of $L^2(I)$ with respect to the norm $\|\cdot\|_{\mathcal{U}^{-1}}$. Let $\{w_n^{(1)}\}_{n=1}^\infty$, $\{w_n^{(2)}\}_{n=1}^\infty \in L^2(I)$, be two Cauchy sequences which are equivalent in the sense that

$$\lim_{n\to\infty} \|w_n^{(1)} - w_n^{(2)}\|_{\mathcal{U}^{-1}} = 0 \tag{4.2.37a}$$

and let

$$W_n^{(k)} \equiv -\int_x^l w_n^{(k)} \tag{4.2.37b}$$

for $k = 1, 2$. We then have

$$\begin{aligned}\|W_n^{(k)} - W_m^{(k)}\|_{L^2(I)} &= \sup_{v\in\mathcal{U}} \frac{|\int_I (W_n^{(k)} - W_m^{(k)})(dv/dx)|}{\|dv/dx\|_{L^2(I)}} \\ &\leq \mathscr{C} \sup_{v\in\mathcal{U}} \frac{|\int_I (w_n^{(k)} - w_m^{(k)})v|}{\|v\|_{\mathcal{U}}}. \end{aligned} \tag{4.2.37c}$$

Thus $\{W_n^{(k)}\}_{n=1}^\infty$, for $k = 1, 2$, is a Cauchy sequence in $L^2(I)$, which is complete, and hence it converges. Denoting the limits by $W^{(k)} \in L^2(I)$ we have

$$\lim_{n\to\infty} \|W_n^{(k)} - W^{(k)}\|_{L^2(I)} = 0. \tag{4.2.37d}$$

Further, we have

$$\lim_{n\to\infty} \|W_n^{(1)} - W_n^{(2)}\|_{L^2(I)} = \lim_{n\to\infty} \|w_n^{(1)} - w_n^{(2)}\|_{\mathcal{U}^{-1}} = 0 \tag{4.2.37e}$$

and, the two limits are identical. It follows that $\mathcal{U}^{-1}$ can be identified with the space of equivalent Cauchy sequences in $L^2(I)$.

On the other hand, given any $W \in L^2(I)$, there exists a sequence $\{W_n\}_{n=1}^{\infty}$ of smooth functions which satisfy $W_n(l) = 0$, and converge to W in the L^2 sense. It follows that $w_n = (d/dx)(W_n)$ is a Cauchy sequence in $\mathcal{U}^{-1}$, and its limit can be identified with w. If W is differentiable, then $w = (d/dx)(W)$, where the derivative is understood in the classical sense. If W is not differentiable, then $(d/dx)(W)$ is understood in the sense of generalized functions (see [Gel'fand and Shilov, 1968]). □

Remark 4.2.8 We have

$$\|\mathcal{g}\|_{\mathcal{U}^{-1}} = \|H(\cdot;\mathcal{g})\|_{\mathcal{U}}. \tag{4.2.38}$$

This result can be used for computing $\|\mathcal{g}\|_{\mathcal{U}^{-1}}$.

Remark 4.2.9 Using the above notations we have

$$e_{S^p_\Delta} = H(\cdot;\mathfrak{R}) \tag{4.2.39a}$$

where

$$\mathfrak{R} \stackrel{\text{def}}{=} f - \mathfrak{D}\, u_{S^p_\Delta} \in \mathcal{U}^{-1} \tag{4.2.39b}$$

is the *residual load*, and we have

$$\|e_{S^p_\Delta}\|_{\mathcal{U}} = \|\mathfrak{R}\|_{\mathcal{U}^{-1}}. \tag{4.2.39c}$$

In Chapter 5, we will utilize (4.2.39c) to construct a-posteriori estimates for $\|e_{S^p_\Delta}\|_{\mathcal{U}}$.

Let us now give an example of how the above results can be utilized.

Example 4.2.2 Let $W_n \in H^1_{0)}(I)$ be given by

$$W_n(x) = \begin{cases} 1 & 0 < x < l\left(1 - \frac{1}{n}\right) \\ n(l-x), & l\left(1 - \frac{1}{n}\right) < x < l. \end{cases} \tag{4.2.40a}$$

Figure 4.2.1 shows the graphs of W_n for various values of n. We have

$$\lim_{n\to\infty} \|W_n - 1\|^2_{L^2(I)} = \lim_{n\to\infty} \int_{l(1-1/n)}^{l} (1 - n(l-x))^2\, dx = 0. \tag{4.2.40b}$$

Hence W_n converges to 1 in $L^2(I)$, and $\{W_n\}_{n=1}^{\infty}$ is a Cauchy sequence in $L^2(I)$. It follows that the sequence $\{w_n\}_{n=1}^{\infty}$, where

$$w_n = -\frac{d}{dx}(W_n) = \begin{cases} 0, & 0 < x < l\left(1 - \frac{1}{n}\right) \\ n, & l\left(1 - \frac{1}{n}\right) \le x < l \end{cases} \tag{4.2.40c}$$

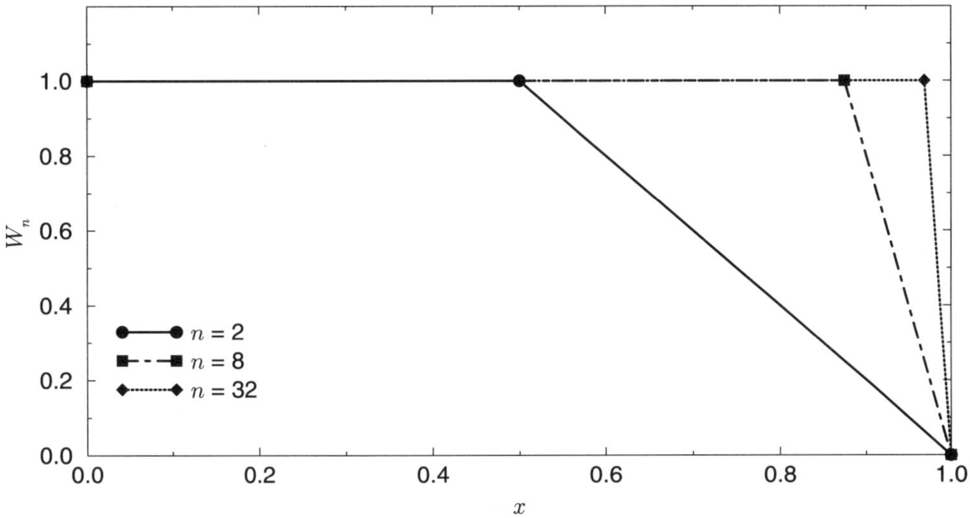

FIG. 4.2.1. Graphs of W_n for $n = 2, 8, 32$, for $l = 1$

is a Cauchy sequence in $\mathcal{U}^{-1}$, and

$$\lim_{n\to\infty} \|w_n - \delta(\cdot;l)\|_{\mathcal{U}^{-1}} = 0. \tag{4.2.40d}$$

Hence w_n converges to $(d/dx)(W) \equiv \delta(\cdot;l)$ in $\mathcal{U}^{-1}$.

Let us now consider the regular model problem with $AE \equiv 1$, $c \equiv 0$, $a_2 = 0$. We have

$$-H(\cdot;w_n)(x) = \begin{cases} x, & 0 < x < l\left(1 - \frac{1}{n}\right) \\ l\left(1 - \frac{1}{n}\right) - \frac{n}{2}\left((l-x)^2 - \frac{l^2}{n^2}\right), & l\left(1 - \frac{1}{n}\right) \le x < 1. \end{cases} \tag{4.2.41a}$$

It follows that

$$\lim_{n\to\infty} \left\| H\left(\cdot; \frac{d}{dx}(W_n)\right) - H(\cdot; -\delta(\cdot;l)) \right\|_{\mathcal{U}} = 0. \tag{4.2.41b}$$

This means that $H(\cdot;w_n)$ converges to $H(\cdot;\delta(\cdot;l)) \in \mathcal{U}$, the solution of the boundary value problem

$$\frac{d^2}{dx^2}(H(\cdot;\delta(\cdot;l)))(x) = \delta(x;l), \quad 0 < x < l \tag{4.2.41c}$$

$$H(\cdot;\delta(\cdot;l))(0) = 0, \quad \frac{d}{dx}(H(\cdot;\delta(\cdot;l)))(l) = 0. \tag{4.2.41d}$$

This means that a nonhomogeneous Neumann boundary condition $P(l) = a_2$ can be substituted by the homogeneous Neumann boundary condition, $P(l) = 0$, and a concentrated load $-a_2\delta(\cdot;l-0)$ which is added to the right-hand side f.

In the special case $AE \equiv 1$, $c \equiv 0$, $\alpha_2 = 0$, $a_2 = 0$, we have $\mathcal{U}^{-1} \equiv H_{(0}^{-1}(I)$. Let us now prove that this is also true in the general case where $AE \not\equiv$ *constant*, and/or $c \not\equiv 0$.

Theorem 4.2.3 *Let $\mathcal{U}$ be the energy space for the regular model problem. Then we have*

$$\mathcal{U}^{-1} \equiv H_{(0}^{-1}(I). \tag{4.2.42}$$

Proof From the definition of the regular model problem, we have

$$0 < \beta_1 \leq AE \leq \beta_2, \qquad 0 \leq c \leq \gamma. \tag{4.2.43a}$$

It follows that the norms $\|\cdot\|_{H_{(0}^1(I)}$ and $\|\cdot\|_{\mathcal{U}}$ are equivalent, namely, there exist positive constants $\mathcal{C}_3$ and $\mathcal{C}_4$, such that

$$\mathcal{C}_3\|v\|_{H_{(0}^1(I)} \leq \|v\|_{\mathcal{U}} \leq \mathcal{C}_4\|v\|_{H_{(0}^1(I)} \quad \forall v \in \mathcal{U}. \tag{4.2.43b}$$

From the definition of the norm $\|\cdot\|_{H_{(0}^{-1}(I)}$, we get

$$\frac{1}{\mathcal{C}_4}\|u\|_{H_{(0}^{-1}(I)} = \frac{1}{\mathcal{C}_4} \sup_{v \in H_{(0}^1(I)} \frac{|\int_I uv|}{\|v\|_{H_{(0}^1(I)}} \leq \sup_{v \in \mathcal{U}} \frac{|\int_I uv|}{\|v\|_{\mathcal{U}}} = \|u\|_{\mathcal{U}^{-1}} \tag{4.2.44a}$$

and hence

$$\mathcal{U}^{-1} \subseteq H_{(0}^{-1}(I). \tag{4.2.44b}$$

Further, from the definition of the norm $\|\cdot\|_{\mathcal{U}^{-1}}$, we get

$$\|u\|_{\mathcal{U}^{-1}} = \sup_{v \in \mathcal{U}} \frac{|\int_I uv|}{\|v\|_{\mathcal{U}}} \leq \frac{1}{\mathcal{C}_3} \sup_{v \in H_{(0}^1(I)} \frac{|\int_I uv|}{\|v\|_{H_{(0}^1(I)}} = \frac{1}{\mathcal{C}_3}\|u\|_{H_{(0}^{-1}(I)} \tag{4.2.45a}$$

and hence

$$H_{(0}^{-1}(I) \subseteq \mathcal{U}^{-1}. \tag{4.2.45b}$$

□

Remark 4.2.10 Analogous constructions can be given for $\|\cdot\|_{(\mathcal{U}^{\vartheta})^{-1}}$, and $(\mathcal{U}^{\vartheta})^{-1}$, where $\mathcal{U}^{\vartheta}$ is the energy space for the singular model problem.

Let us also consider the case that zero Dirichlet houndary conditions are imposed at both end points. In this case the energy space is

$$\mathcal{U}_0 \stackrel{\text{def}}{=} \{v \mid \|v\|_{\mathcal{U}} < \infty, \ v(0) = v(1) = 0\} \tag{4.2.46a}$$

and the elements of $\mathcal{U}_0^{-1}$ can be understood as generalized derivatives of functions from

$$L_0^2(I) \stackrel{\text{def}}{=} \{V | \|V\|_{L^2(I)} < \infty, \ \textstyle\int_I V = 0\}. \tag{4.2.46b}$$

The condition $\int_I V = 0$ follows from the observation that every $v \in \mathcal{U}_0$ satisfies

$$\int_I \frac{dv}{dx} = v(1) - v(0) = 0. \tag{4.2.46c}$$

Noting that $\mathcal{U}_0 \subset \mathcal{U}$, we have

$$\|w_n\|_{\mathcal{U}_0^{-1}} = \sup_{v \in \mathcal{U}_0} \frac{|\int_I w_n v|}{\|v\|_{\mathcal{U}}} \le \sup_{v \in \mathcal{U}} \frac{|\int_I w_n v|}{\|v\|_{\mathcal{U}}} = \|w_n\|_{\mathcal{U}^{-1}} \tag{4.2.46d}$$

which proves that

$$\mathcal{U}^{-1} \subset \mathcal{U}_0^{-1}. \tag{4.2.46e}$$

Hence the completions of $L^2(I)$ in the $\mathcal{U}^{-1}$ and the $\mathcal{U}_0^{-1}$ norm are different. For example, for the W_n defined in Example 4.2.2 we have

$$\lim_{n\to\infty} \left\|0 - \frac{d}{dx}(W_n)\right\|_{\mathcal{U}_0^{-1}} = \lim_{n\to\infty} \sup_{\substack{v \in \mathcal{U}_0 \\ \|v\|_{\mathcal{U}_0}=1}} \left| \int_{1-1/n}^{1} -nv \right| = 0 \tag{4.2.47}$$

where we have used that $|v(x)| \le \sqrt{1-x}\,\|v\|_{\mathcal{U}_0}$. Hence $w_n = (d/dx)(W_n)$ *converges to the zero function in* $\mathcal{U}_0^{-1}$*, while it converges to the Dirac function in* $\mathcal{U}^{-1}$.

Similarly $H^{-1}(I)$ and $H_0^{-1}(I)$, the completions of $L^2(I)$ in the $\|\cdot\|_{H^{-1}(I)}$ and $\|\cdot\|_{H_0^{-1}(I)}$ norms, are different.

The meaning of the negative norm $\|\cdot\|_{H^{-1}(I)}$ can be also understood analogously. Given any $g \in H^{-1}(I)$, we have

$$\|g\|_{H^{-1}(I)} = \|H(\cdot; g)\|_{H^1(I)} \tag{4.2.48a}$$

where $H(\cdot; g)$ is the solution of the problem:

Find $H(\cdot; g) \in H^1(I)$ such that

$$\int_0^l \left(\frac{d}{dx}(H(\cdot; g))\,\frac{dw}{dx} + H(\cdot; g)w\right) = \int_0^l gw \quad \forall w \in H^1(I). \tag{4.2.48b}$$

Let us also address the meaning of the $H_{(0}^{-k}(I)$ norms for $k > 1$, where

$$H_{(0}^k(I) \stackrel{\text{def}}{=} \left\{ v \,\middle|\, \left\|\frac{d^k v}{dx^k}\right\|_{L^2(I)} < \infty,\ \frac{d^s v}{dx^s}(0) = 0,\ s = 0, \ldots, k-1 \right\}. \tag{4.2.49a}$$

For any $w \in L^2(I)$ we will let

$$A_{(0}^{(k)}(w)(x) \stackrel{\text{def}}{=} (-1)^k \underbrace{\int_x^l \cdots \int_s^l \int_t^l}_{k \text{ times}} w. \tag{4.2.49b}$$

We will call $A_{(0}^{(k)}(w)$ the k-th *antiderivative* of w. Integrating by parts, we get

$$\int_I wv = \int_I A_{(0}^{(k)}(w)\,\frac{d^k v}{dx^k} \quad \forall v \in H_{(0}^k(I) \tag{4.2.49c}$$

and hence

$$\|w\|_{H_{(0}^{-k}(I)} = \sup_{v\in H_{(0}^{k}(I)} \frac{|\int_I wv|}{\|d^kv/dx^k\|_{L^2(I)}} = \sup_{v\in H_{(0}^{k}(I)} \frac{\left|\int_I A_{(0}^{(k)}(w)(d^kv/dx^k)\right|}{\|d^kv/dx^k\|_{L^2(I)}}. \quad (4.2.49d)$$

This proves that

$$\|w\|_{H_{(0}^{-k}(I)} = \|A_{(0}^{(k)}(w)\|_{L^2(I)}. \quad (4.2.49e)$$

Hence $H_{(0}^{-k}(I)$ is the space of generalized functions w which are such that $A_{(0}^{(k)}(w) \in L^2(I)$.

Given a $w \in H_{(0}^{-k}(I)$, the norm $\|w\|_{H_{(0}^{-k}(I)}$ can be evaluated by employing that

$$\begin{aligned}\|w\|_{H_{(0}^{-k}(I)} &= \sup_{v\in H_{(0}^{k}(I)} \frac{|\int_I wv|}{\|v\|_{H_{(0}^{k}(I)}} = \sup_{v\in H_{(0}^{k}(I)} \frac{|\int_I (d^k/dx^k)(u_w)(d^kv/dx^k)|}{\|d^kv/dx^k\|_{L^2(I)}} \\ &\stackrel{\text{def}}{=} \left\|\frac{d^k}{dx^k}(u_w)\right\|_{L^2(I)}\end{aligned} \quad (4.2.50a)$$

where u_w is the solution of the problem:

Find $u_w \in H_{(0}^{k}(I)$ such that

$$\int_I \frac{d^k}{dx^k}(u_w)\,\frac{d^kv}{dx^k} = \int_I wv \quad \forall v \in H_{(0}^{k}(I). \quad (4.2.50b)$$

Another argument shows that $A_{(0}^{(k)}(w)$ coincides with $(d^k/dx^k)(u_w)$.

Remark 4.2.11 From the above, it can be seen that *the negative norms and the corresponding spaces depend strongly on the norm and the space employed in the supremum.*

Let us now give an example which explains why the rate of convergence of the negative norms of the error cannot exceed $2p$.

Example 4.2.3 Let us consider the model problem with input data $AE \equiv 1$, $c \equiv 0$, $\alpha_2 = 0$, $l = 1$, and f, a_2, consistent with the exact solution $u_{EX}(x) = x^3$, and let $u_{S^2_\Delta}$ be the finite element solution computed using a uniform mesh Δ of quadratic elements ($p = 2$). For the case considered, *the error is periodic over the mesh*, and is given by

$$e_{S^2_\Delta}(x) = (x_j^\Delta - x)(x_{j+1/2}^\Delta - x)(x - x_{j-1}^\Delta), \quad x \in I_j^\Delta. \quad (4.2.51a)$$

Further, note that

$$\int_{I_j^\Delta} e_{S_\Delta^2} = 0 \tag{4.2.51b}$$

for each $j = 1, 2, \ldots, N(\Delta)$. It follows that for $x \in I_i^\Delta$ we have

$$A_{(0}^{(1)}(e_{S_\Delta^2})(x) = -\int_x^1 e_{S_\Delta^2} = -\int_x^{x_i^\Delta} e_{S_\Delta^2} - \sum_{j=i+1}^{N} \int_{I_j^\Delta} e_{S_\Delta^2} = -\int_x^{x_i^\Delta} e_{S_\Delta^2} \tag{4.2.51c}$$

and *hence* $A_{(0}^{(1)}(e_{S_\Delta^2})$ *is periodic over the mesh*, and

$$A_{(0}^{(1)}(e_{S_\Delta^2})(x) = -\frac{1}{2}h \int_{\tilde{x}(x)}^1 \tilde{e}_{S_\Delta^2} = -\int_{\tilde{x}(x)}^1 \frac{1}{8}h^3(1-s)^2 s \frac{1}{2}h\, ds = h^4 \tilde{g}(\tilde{x}(x)) \tag{4.2.51d}$$

where $\tilde{g}(\tilde{x}) = 12(-1 - 6\tilde{x}^2 + 8\tilde{x}^3 - 3\tilde{x}^4)$.

Let us now compute $\|e_{S_\Delta^2}\|_{H_{(0}^{-1}}$. We have

$$\|A_{(0}^{(1)}(e_{S_\Delta^2})\|_{L^2(I_i^\Delta)}^2 = \int_{-1}^1 \frac{h}{2}\left(\tilde{A}_{(0}^{(1)}(\tilde{e}_{S_\Delta^2})\right)^2 d\tilde{x} = \int_{-1}^1 \frac{h}{2} h^8 \tilde{g}^2(\tilde{x})\, d\tilde{x} = \frac{h^9}{2520} \tag{4.2.52a}$$

and summing over the elements we get

$$\|A_{(0}^{(1)}(e_{S_\Delta^2})\|_{L^2(I)}^2 = \sum_{i=1}^{N} \|A_{(0}^{(1)}(e_{S_\Delta^2})\|_{L^2(I_i^\Delta)}^2 = N\|A_{(0}^{(1)}(e_{S_\Delta^2})\|_{L^2(I_i^\Delta)}^2 = \frac{h^8}{2520} \tag{4.2.52b}$$

where we used that $N = 1/h$. Hence, we have

$$\|e_{S_\Delta^2}\|_{H_{(0}^{-1}(I)} = \|A_{(0}^{(1)}(e_{S_\Delta^2})\|_{L^2(I)} = \frac{1}{6\sqrt{70}}h^4. \tag{4.2.52c}$$

Thus $\int_{I_j^\Delta} e_{S_\Delta^2} = 0$ *implies that* $A_{(0}^{(1)}(e_{S_\Delta^2})$ *is periodic, and that* $\|e_{S_\Delta^2}\|_{H^{-1}(I)}$ *converges as* $h^{2p} = h^4$.

Let us now compute the second antiderivative $A_{(0}^{(2)}(e_{S_\Delta^2})(x)$, $x \in I_i^\Delta$. Note that

$$\int_{I_j^\Delta} A_{(0}^{(1)}(e_{S_\Delta^2}) > 0 \tag{4.2.53a}$$

for each $j = 1, \ldots, N(\Delta)$, and hence $A_{(0}^{(2)}$ *is not periodic*. For $x \in I_i^\Delta$, we have

$$\begin{aligned} A_{(0}^{(2)}(e_{S_\Delta^2})(x) &= -\int_x^1 A_{(0}^{(1)}(e_{S_\Delta^2}) = -\int_x^{x_i^\Delta} A_{(0}^{(1)}(e_{S_\Delta^2}) - \sum_{j=i+1}^{N} \int_{I_j^\Delta} A_{(0}^{(1)}(e_{S_\Delta^2}) \\ &= -\int_{\tilde{x}(x)}^1 h^4 \tilde{g}(s)\frac{h}{2}\, ds - \sum_{j=i+1}^{N} \int_{-1}^1 h^4 \tilde{g}(\tilde{x})\frac{h}{2}\, d\tilde{x} \end{aligned} \tag{4.2.53b}$$

and hence

$$A_{(0}^{(2)}(e_{S^2_\Delta})(x) = \frac{h^5}{2}\Big(\int_{\tilde{x}(x)}^{1} \tilde{g} + \frac{1}{30}(N-(i+1))\Big)$$
$$= \frac{h^5}{2}\Big(\tilde{G}(\tilde{x}) + \frac{1}{30}\Big(\frac{1}{h}-(i+1)\Big)\Big) \quad (4.2.53c)$$

where $\tilde{G}(\tilde{x}) = -\frac{12}{5}(-8-5\tilde{x}-10\tilde{x}^3+10\tilde{x}^4-3\tilde{x}^5)$, and we get

$$\|e_{S^2_\Delta}\|^2_{H^{-2}_{(0}(I)} = \sum_{i=1}^{N} \|A_{(0}^{(2)}(e_{S^2_\Delta})\|^2_{L^2(I_i^\Delta)} = 0.00025h^{10} + 0.00069h^9 + 0.00056h^8 \quad (4.2.53d)$$

which shows that

$$\|e_{S^2_\Delta}\|_{H^{-2}_{(0}(I)} \approx \mathscr{C}h^4 \approx \|e_{S^2_\Delta}\|_{H^{-1}_{(0}(I)}. \quad (4.2.53e)$$

Let us now consider the case of arbitrary p. Letting $u_{EX}(x) = x^{p+1}$, we get

$$\int_{I_j^\Delta} A_{(0}^{(l)}(e_{S^p_\Delta}) = 0, \quad l = 0, \ldots, (p-2) \quad (4.2.54a)$$

and, following similar steps as above, we obtain

$$\|A_{(0}^{(l)}(e_{S^p_\Delta})\|^2_{L^2(I_j^\Delta)} = \mathscr{C}(h_{I_j^\Delta})^{2p+3+2l} \quad (4.2.54b)$$

for $l = 0, \ldots, p-1$, and hence

$$\|A_{(0}^{(l)}(e_{S^p_\Delta})\|^2_{L^2(I)} = \sum_{j=1}^{N(\Delta)} \mathscr{C}(h_{I_j^\Delta})^{2p+3+2l} = \mathscr{C}h^{2(p+1+l)}. \quad (4.2.54c)$$

On the other hand, we have

$$A_{(0}^{(p)}(e_{S^p_\Delta})(x) = \sum_{j=1}^{i-1} \int_{I_j^\Delta} A_{(0}^{(p-1)}(e_{S^p_\Delta}) + \int_{x_{i-1}^\Delta}^{x} A_{(0}^{(p-1)}(e_{S^p_\Delta}) \quad (4.2.54d)$$

with $\sum_{j=1}^{i-1} \int_{I_j^\Delta} A_{(0}^{(p-1)}(e_{S^p_\Delta}) > 0$, and $x \in I_i^\Delta$. Hence

$$\|A_{(0}^{(2p)}(e_{S^p_\Delta})\|^2_{L^2(I_i^\Delta)} = \Big(\sum_{j=1}^{i-1} \int_{I_j^\Delta} A_{(0}^{(p-1)}(e_{S^p_\Delta})\Big)^2 + \Big(\int_{x_{i-1}^\Delta}^{x} A_{(0}^{(p-1)}(e_{S^p_\Delta})\Big)^2$$
$$+ 2\Big(\sum_{j=1}^{i-1} \int_{I_j^\Delta} A_{(0}^{(p-1)}(e_{S^p_\Delta})\Big) \int_{I_i^\Delta} A_{(0}^{(p-1)}(e_{S^p_\Delta}) \quad (4.2.54e)$$

and we have

$$\|A_{(0}^{(2p)}(e_{S^p_\Delta})\|^2_{L^2(I_i^\Delta)} \approx \mathscr{C}h^{4p} + \mathscr{C}h^{4p+1} + \mathscr{C}h^{4p+2} \quad (4.2.54f)$$

and neglecting the higher-order terms we obtain

$$\|A_{(0}^{(2p)}(e_{S_\Delta^p})\|_{L^2(I)}^2 \approx \left(\sum_{j=1}^{i-1} \int_{I_j^\Delta} A_{(0}^{(p-1)}(e_{S_\Delta^p})\right)^2 \approx \mathscr{C}h^{4p}. \tag{4.2.54g}$$

The above arguments can be generalized to quasi-uniform meshes.

Let us also give an example where the exact solution is not a polynomial.

Example 4.2.4 Consider the regular model problem with input data

$$AE \equiv 1, \qquad c \equiv 0, \qquad f \equiv \sin(\pi x), \qquad \alpha_2 = 0, \qquad a_2 = 0, \quad l = 1 \tag{4.2.55a}$$

and let $u_{S_\Delta^p}$ be the finite element solution which is obtained by employing a uniform mesh Δ of elements of degree $p = 1, 2, 3$. Figure 4.2.2 shows the graphs of $h^{-3}\, e_{S_\Delta^p}$, $h^{-4}\, A_{(0}^{(1)}(e_{S_\Delta^p})$, and $h^{-4}\, A_{(0}^{(2)}(e_{S_\Delta^p})$ for $p = 2$ and $h = \frac{1}{8}, \frac{1}{16}$. It can be seen that the local character of the error and its antiderivatives is exactly like the error and its antiderivatives corresponding to the local $(p+1)$ Taylor series expansion of the solution. Tables 4.2.1a–c report the values of $\|e_{S_\Delta^p}\|_{H_{(0}^{-l}(I)}$ and $h^{-\min(p+l+1,2p)}\|e_{S_\Delta^p}\|_{H^{-l}(I)}$ for $h = \frac{1}{2}, \frac{1}{4}, \frac{1}{8}, \frac{1}{16}, \frac{1}{32}$. We note that

$$\|e_{S_\Delta^p}\|_{H_{(0}^{-l}(I)} \approx \mathscr{C}(l)h^{2p}, \quad l \geq p \tag{4.2.55b}$$

where $\mathscr{C}(l)$ decreases for increasing values of l.

Remark 4.2.12 Above we addressed the model problem with the Dirichlet condition $u(0) = 0$, which ensures that the solution u_{EX} exists for every f. Let us now assume that

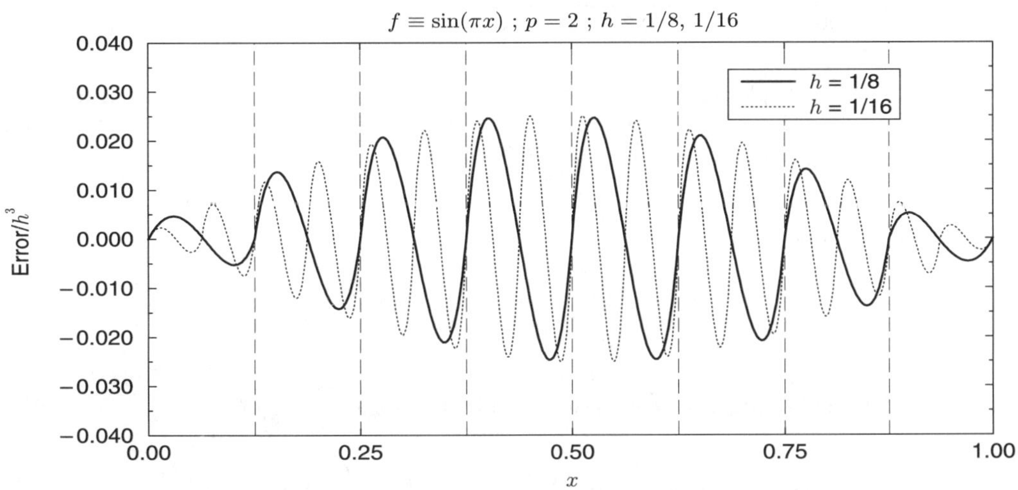

(a)

FIG. 4.2.2.

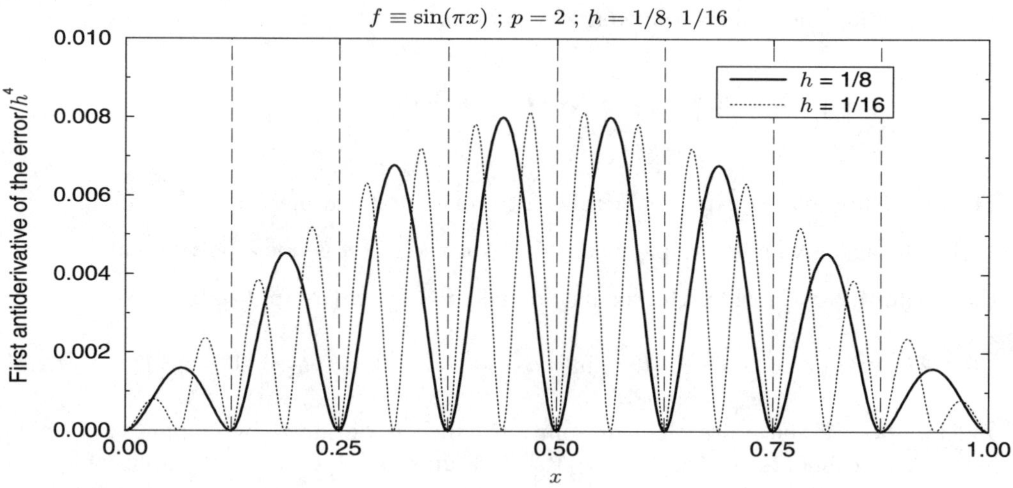

(b)

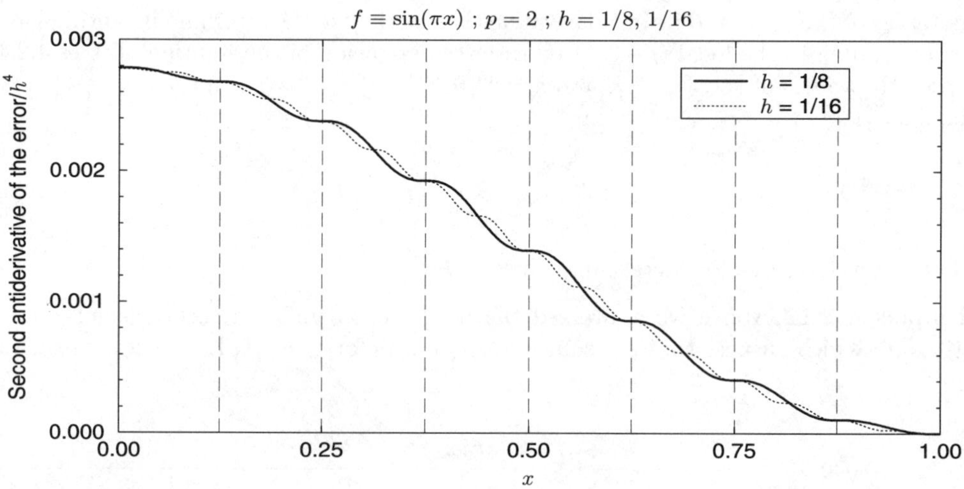

(c)

FIG. 4.2.2. An example which illustrates the convergence of the negative norms of the error. Input data: $AE \equiv 1$, $c \equiv 0$, $f \equiv \sin(\pi x)$, $\alpha_2 = 0$, $a_2 = 0$, $l = 1$. Mesh data: $h = \frac{1}{8}, \frac{1}{16}$, $p = 2$. Graphs of the functions (a) $e_{S^2_\Delta}/h^3$; (b) $A^{(1)}_{(0}(e_{S^2_\Delta})/h^4$; (c) $A^{(2)}_{(0}(e_{S^2_\Delta})/h^4$

f satisfies the condition

$$\int_0^1 f = 0 \tag{4.2.56a}$$

and consider the homogeneous Neumann problem

$$-\frac{d}{dx}\Big(AE\,\frac{du}{dx}\Big) = f, \qquad \Big(AE\,\frac{du}{dx}\Big)(0) = \Big(AE\,\frac{du}{dx}\Big)(1) = 0. \tag{4.2.56b}$$

Table 4.2.1. An example which illustrates the convergence of the negative norms of the error. Input data: $AE \equiv 1$, $c \equiv 0$, $f \equiv \sin(\pi x)$, $\alpha_2 = 0$, $a_2 = 0$, $l = 1$. Mesh data: Uniform mesh of elements of degree p. The values of $\|e_{S^p_\Delta}\|_{H^{-s}_{(0}(I)}$ for $s = 0, 1, 2, 3$. (a) Linear elements ($p = 1$). Note that $\|e_{S^1_\Delta}\|_{H^{-s}_{(0}(I)}$ converges as h^2 for all values of s. (b) Quadratic elements ($p = 2$). Note that $\|e_{S^2_\Delta}\|_{H^0(I)}$ converges as h^3, while $\|e_{S^2_\Delta}\|_{H^{-s}_{(0}(I)}$ converges as h^4 for $s \geq 1$. (c) Cubic elements ($p = 3$). Note that $\|e_{S^3_\Delta}\|_{H^0(I)}$ converges as h^4, while $\|e_{S^3_\Delta}\|_{H^{-s}_{(0}(I)}$ converges as h^5 for $s = 1$, and $\|e_{S^3_\Delta}\|_{H^{-s}_{(0}(I)}$ converges as h^6 for $s \geq 2$

h				
(a)	$\|e_{S^1_\Delta}\|_{H^0(I)}$	$h^{-2}\|e_{S^1_\Delta}\|_{H^0(I)}$	$\|e_{S^1_\Delta}\|_{H^{-1}_{(0}(I)}$	$h^{-2}\|e_{S^1_\Delta}\|_{H^{-1}_{(0}(I)}$
1/2	0.152870×10^{-1}	0.06115	0.437687×10^{-2}	0.01751
1/4	0.398034×10^{-2}	0.06369	0.115059×10^{-2}	0.01841
1/8	0.100520×10^{-2}	0.06433	0.291683×10^{-3}	0.01867
1/16	0.251935×10^{-3}	0.06450	0.731807×10^{-4}	0.01873
1/32	0.630236×10^{-3}	0.06454	0.183115×10^{-4}	0.01875
	$\|e_{S^1_\Delta}\|_{H^{-2}_{(0}(I)}$	$h^{-2}\|e_{S^1_\Delta}\|_{H^{-2}_{(0}(I)}$	$\|e_{S^1_\Delta}\|_{H^{-3}_{(0}(I)}$	$h^{-2}\|e_{S^1_\Delta}\|_{H^{-3}_{(0}(I)}$
1/2	0.228429×10^{-2}	0.00914	0.764919×10^{-3}	0.00306
1/4	0.628329×10^{-3}	0.01005	0.216944×10^{-3}	0.00347
1/8	0.160467×10^{-3}	0.01027	0.557055×10^{-4}	0.00357
1/16	0.403254×10^{-4}	0.01032	0.140159×10^{-4}	0.00359
1/32	0.100944×10^{-4}	0.01034	0.350954×10^{-5}	0.00359
(b)	$\|e_{S^2_\Delta}\|_{H^0(I)}$	$h^{-3}\|e_{S^2_\Delta}\|_{H^0(I)}$	$\|e_{S^2_\Delta}\|_{H^{-1}_{(0}(I)}$	$h^{-4}\|e_{S^2_\Delta}\|_{H^{-1}_{(0}(I)}$
1/2	0.153865×10^{-2}	0.01231	0.220990×10^{-3}	0.00354
1/4	0.197762×10^{-3}	0.01266	0.142548×10^{-4}	0.00365
1/8	0.248925×10^{-4}	0.01274	0.897959×10^{-6}	0.00368
1/16	0.311697×10^{-5}	0.01277	0.562327×10^{-7}	0.00369
1/32	0.389791×10^{-6}	0.01277	0.351627×10^{-8}	0.00369
	$\|e_{S^2_\Delta}\|_{H^{-2}_{(0}(I)}$	$h^{-4}\|e_{S^2_\Delta}\|_{H^{-2}_{(0}(I)}$	$\|e_{S^2_\Delta}\|_{H^{-3}_{(0}(I)}$	$h^{-4}\|e_{S^2_\Delta}\|_{H^{-3}_{(0}(I)}$
1/2	0.109279×10^{-3}	0.00175	0.403184×10^{-4}	0.0065
1/4	0.668125×10^{-5}	0.00171	0.236473×10^{-5}	0.0061
1/8	0.415826×10^{-6}	0.00170	0.145255×10^{-6}	0.0059
1/16	0.259639×10^{-7}	0.00170	0.903839×10^{-8}	0.0059
1/32	0.162236×10^{-8}	0.00170	0.564273×10^{-9}	0.0059
(c)	$\|e_{S^3_\Delta}\|_{H^0(I)}$	$h^{-4}\|e_{S^3_\Delta}\|_{H^0(I)}$	$\|e_{S^3_\Delta}\|_{H^{-1}_{(0}(I)}$	$h^{-5}\|e_{S^3_\Delta}\|_{H^{-1}_{(0}(I)}$
1/2	0.140642×10^{-3}	0.00225	0.105786×10^{-4}	0.00034
1/4	0.898511×10^{-5}	0.00230	0.338459×10^{-6}	0.00035
1/8	0.564652×10^{-6}	0.00231	0.106392×10^{-7}	0.00035
1/16	0.353391×10^{-7}	0.00232	0.332962×10^{-9}	0.00035
1/32	0.220945×10^{-8}	0.00232	0.104089×10^{-10}	0.00035
	$\|e_{S^3_\Delta}\|_{H^{-2}_{(0}(I)}$	$h^{-6}\|e_{S^3_\Delta}\|_{H^{-2}_{(0}(I)}$	$\|e_{S^3_\Delta}\|_{H^{-3}_{(0}(I)}$	$h^{-6}\|e_{S^3_\Delta}\|_{H^{-3}_{(0}(I)}$
1/2	0.133639×10^{-5}	0.00009	0.335149×10^{-6}	0.00002
1/4	0.214121×10^{-7}	0.00009	0.533177×10^{-8}	0.00002
1/8	0.336666×10^{-9}	0.00009	0.838671×10^{-10}	0.00002
1/16	0.526864×10^{-11}	0.00009	0.131278×10^{-11}	0.00002
1/32	0.823480×10^{-13}	0.00009	0.205256×10^{-13}	0.00002

In this case u_{EX} and $u_{S^p_\Delta}$ are unique up to additive constants, which can be fixed by imposing the conditions

$$\int_I u_{EX} = \int_I u_{S^p_\Delta} = 0. \tag{4.2.57a}$$

It follows that

$$\int_I e_{S^p_\Delta} = \int_I u_{EX} - \int_I u_{S^p_\Delta} = 0. \tag{4.2.57b}$$

This is the consistency condition which is necessary for the existence of $H(\cdot; e_{S^p_\Delta})$. It follows that the estimate

$$\|e_{S^p_\Delta}\|_{L^2(I)} \leq \mathscr{C} h \|e_{S^p_\Delta}\|_{\mathscr{U}} \tag{4.2.57c}$$

has meaning in this case as well. A similar interpretation can be employed for the estimates of the error measured in the negative norm.

In this section we introduced the spaces $H^l(I)$ for $l \leq 0$ and $\mathscr{U}^{-1}$. We have shown that for smooth input data and $l > 0$, we have

$$\|e_{S^p_\Delta}\|_{H^{-l}(I)} \leq \mathscr{C} h^{p+\min(p,l)} \tag{4.2.58}$$

i.e., the rate of convergence of the negative norm of the error in the finite element solution increases with the order l of the norm, up to a threshold of $2p$. We gave an interpretation of the 'negative' norm $\|\cdot\|_{H^{-l}(I)}$ as the L^2 norm of $(l-1)$-primitive function. These results will play an important role in the local analysis of the finite element error.

Problems

P4.2.1. Describe the functions in the space $(\mathscr{U}^\vartheta)^{-1}$ where $\mathscr{U}^\vartheta$ is the energy space for the singular model problem.

P4.2.2. Describe the functions in $\mathscr{U}^{-1}$ where $\mathscr{U}$ is the energy space for the regular model problem in the case that $\alpha_2 \neq 0$.

P4.2.3. In Example 4.2.1 compute $\|W_n\|_{H_0^{-k}(I)}$ and $\|W_n\|_{H_{0)}^{-k}(I)}$.

P4.2.4. Repeat the computations in Example 4.2.3 for linear elements ($p = 1$). Explain why for linear elements ($p = 1$),

$$\|e_{S^1_\Delta}\|_{H^{-l}(I)} \approx \mathscr{C} h^2 \tag{4.2.59}$$

for any $l \geq 0$.

P4.2.5. In Corollary 4.2.3 we have proven

$$\|e_{S^p_\Delta}\|_{L^2(I)} \leq \mathscr{C}(\vartheta) h^{(1-\vartheta)/2} \|e_{S^p_\Delta}\|_{\mathscr{U}^\vartheta} \tag{4.2.60}$$

for the singular model problem. Derive analogous estimates for $\|e_{S^p_\Delta}\|_{(\mathcal{U}^\vartheta)^{-1}}$ and $\|e_{S^p_\Delta}\|_{H^{-1}_{(0}(I)}$ in the case of quadratic elements ($p = 2$).

P4.2.6. Describe the functions in $H_0^{-k}(I)$.

P4.2.7. Consider the sequence of functions $w_n(x) = (1 - e^{-n(l-x)})^{k-1}$. Note that $w_n(x)$ converges to one for fixed x. Does w_n converge in $H^k_{(0}(I)$, and if yes, what is the limit?

P4.2.8. Consider the singular model problem with $f \equiv 0$, $\alpha_2 = 0$, $a_2 = 1$, and let Δ be quasi-uniform meshes. Show that

$$\mathcal{C}_1 h^{(1-\vartheta)/2} \le \|e_{S^1_\Delta}\|_{\mathcal{U}^\vartheta} \le \mathcal{C}_2 h^{(1-\vartheta)/2}. \tag{4.2.61}$$

Hint: The right inequality follows from the interpolation estimate. The proof of the left inequality employs the property $\int_{I^\Delta_j} x^\vartheta (d/dx)(e_{S^p_\Delta}) = 0$.

P4.2.9. Prove that $h^2(\int_{I^\Delta_j} x^\vartheta)^{-1} \le \int_{I^\Delta_j} x^{-\vartheta}$

P4.2.10. Consider the singular model problem with $f \equiv 0$, $\alpha_2 = 0$, $a_2 = 1$, and let Δ be quasi-uniform meshes. Prove that

$$\mathcal{C} h^{1-\vartheta} \le \|e_{S^1_\Delta}\|_{L^2(I)}. \tag{4.2.62}$$

P4.2.11. Let u_{EX} be the solution of the Problem P4.2.10. Estimate $\|u_{EX} - \mathcal{I}^1_\Delta u_{EX}\|_{L^2(I)}$ and compare this estimate with the result of P4.2.10. What can you conclude from this comparison?

4.2* A-priori estimates for negative norms of the error in two dimensions

The main ideas presented in Section 4.2 can be directly extended to the two-dimensional case. In particular:

1. Corollary 4.2.1 holds with $\mu = \min(s+1, p)$ under the assumption that the coefficients and the domain are smooth. The proof is based on the regularity of $H(\cdot; g)$ which depends on the regularity of g.
2. Corollary 4.2.2 also holds in the two-dimensional case under the assumption that the coefficients and the boundary $\partial\Omega$ are smooth. If this is not the case, e.g., *if the boundary $\partial\Omega$ has corners, then we cannot gain an entire power of h, analogously as in Corollary 4.2.3.*
3. Theorem 4.2.1 also holds in the two dimensional case, once more under the assumption of sufficient smoothness of the coefficients and the domain.
4. The error $e_{S^p_\Delta}$ has oscillatory character, similarly as in the one-dimensional case.

Let us address the $\mathcal{U}^{-1}$ norm for the model problem of the Poisson equation with homogeneous Dirichlet conditions:

$$-\Delta u|_\Omega = f \qquad u|_{\partial\Omega} = 0 \tag{4.2*.1a}$$

for which the energy space is

$$\mathcal{U} \equiv H^1_0(\Omega) = \left\{ v \mid \|v\|^2_{H^1_0(\Omega)} \stackrel{\text{def}}{=} \int_\Omega |\nabla v|^2 < \infty,\ v|_{\partial\Omega} = 0 \right\}. \tag{4.2*.1b}$$

As in the one-dimensional case, given a $w \in L^2(\Omega)$, we let

$$\|w\|_{H_0^{-1}(\Omega)} \stackrel{\text{def}}{=} \sup_{v \in H_0^1(\Omega)} \frac{|\int_\Omega wv|}{\|v\|_{H_0^1(\Omega)}} < \infty \tag{4.2*.2a}$$

which is a norm on $L^2(\Omega)$, and we denote by $H_0^{-1}(\Omega)$ the completion of $L^2(\Omega)$ in this norm.

The meaning of the norm $\|\cdot\|_{H_0^{-1}(\Omega)}$ can be understood by employing a similar construction as in the one-dimensional case. Given a $w \in H_0^{-1}(\Omega)$, we will let u_w be the solution of the variational problem:

Find $u_w \in H_0^1(\Omega)$ such that

$$\int_\Omega \left(\frac{d}{dx_1}(u_w) \frac{dv}{dx_1} + \frac{d}{dx_2}(u_w) \frac{dv}{dx_2} \right) = \int_\Omega wv \quad \forall v \in H_0^1(\Omega). \tag{4.2*.2b}$$

We then have

$$\|w\|_{H_0^{-1}(\Omega)} = \sup_{v \in H_0^1(\Omega)} \frac{|\int_\Omega wv|}{\|v\|_{H_0^1(\Omega)}} = \sup_{v \in H_0^1(\Omega)} \frac{|\int_\Omega \boldsymbol{\nabla} u_w \cdot \boldsymbol{\nabla} v|}{\|\boldsymbol{\nabla} v\|_{\mathbf{L}^2(I)}} \tag{4.2*.2c}$$

and letting $\mathbf{V} = \boldsymbol{\nabla} v$, we get

$$\|w\|_{H_0^{-1}(\Omega)} = \sup_{\mathbf{V} \in \mathbf{L}^2(\Omega)} \frac{|\int_\Omega \boldsymbol{\nabla} u_w \cdot \mathbf{V}|}{\|\mathbf{V}\|_{\mathbf{L}^2(I)}} = \|\boldsymbol{\nabla} u_w\|_{\mathbf{L}^2(I)}. \tag{4.2*.2d}$$

In the more general case where homogeneous Dirichlet boundary conditions are applied only on a part of the boundary, we have $\mathcal{U} \equiv H^1_{\Gamma_D}(\Omega)$, and the $\mathcal{U}^{-1}$ norm is determined by employing completely analogous steps with $H_0^1(\Omega)$ replaced by $H^1_{\Gamma_D}(\Omega)$.

For $k > 1$, analogously as in the one-dimensional case, given a $w \in L^2(\Omega)$, we let

$$\|w\|_{H_0^{-k}(\Omega)} \stackrel{\text{def}}{=} \sup_{v \in H_0^k(\Omega)} \frac{|\int_\Omega wv|}{\|v\|_{H_0^k(\Omega)}} < \infty \tag{4.2*.3a}$$

and we let $H_0^{-k}(\Omega)$ be the completion of $L^2(\Omega)$ in the $\|\cdot\|_{H_0^{-k}(\Omega)}$ norm.

Given a $w \in H_0^{-k}(\Omega)$, we will let u_w be the solution of the variational problem:
Find $u_w \in H_0^k(\Omega)$ such that

$$\int_\Omega \sum_{\substack{i,j=1 \\ i+j=k}}^{k} \frac{\partial^k u_w}{\partial x_1^i \partial x_2^j} \frac{\partial^k v}{\partial x_1^i \partial x_2^j} = \int_\Omega wv \quad \forall v \in H_0^k(\Omega) \tag{4.2*.3b}$$

and we have

$$\|w\|_{H_0^{-k}(\Omega)} = \|u_w\|_{H_0^k(\Omega)}. \tag{4.2*.3c}$$

Let us also address the case of the homogeneous Neumann problem

$$-\Delta u|_\Omega = f \qquad \left.\frac{\partial u}{\partial n}\right|_{\partial\Omega} = 0 \tag{4.2*.4a}$$

where f satisfies the consistency condition

$$\int_{\partial\Omega} f = 0. \qquad (4.2^*.4b)$$

It follows that u_{EX} and $u_{S^p_\Delta}$ are unique up to additive constants and, fixing these constants by imposing the conditions

$$\int_\Omega u_{EX} = \int_\Omega u_{S^p_\Delta} = 0 \qquad (4.2^*.4c)$$

we get

$$\int_\Omega e_{S^p_\Delta} = 0 \qquad (4.2^*.4d)$$

which is the consistency condition needed for the existence of $H(\cdot; e_{S^p_\Delta})$, employed in the proof of the estimate

$$\|e_{S^p_\Delta}\|_{L^2(\Omega)} \leq \mathscr{C} h^\beta \|e_{S^p_\Delta}\|_{\mathscr{U}}. \qquad (4.2^*.4e)$$

Hence, by imposing the conditions (4.2.4c), the negative-norm estimates have meaning also in the case of the pure Neumann problem.

In this section we outlined the similarities and the differences of the theory of the negative norms in one and two dimensions.

Problems

P4.2*.1. Consider the homogeneous Dirichlet problem

$$-\Delta u|_\Omega = f, \qquad u|_{\partial\Omega} = 0 \qquad (4.2^*.5a)$$

where Ω is a polygonal domain. Under the assumption that

$$\|u_{EX}\|_{B^{1+\alpha}(H^1_0(\Omega), H^2(\Omega)\cap H^1_0(\Omega))} \leq \mathscr{C}\|f\|_{L^2} \qquad (4.2^*.5b)$$

prove that, for quasi-uniform meshes, we have

$$\|e_{S^p_\Delta}\|_{\mathscr{U}} \leq \mathscr{C} h^\mu \|f\|_{L^2(\Omega)} \qquad (4.2^*.5c)$$

and

$$\|e_{S^p_\Delta}\|_{L^2(\Omega)} \leq \mathscr{C} h^{2\mu} \|f\|_{L^2(\Omega)} \qquad (4.2^*.5d)$$

and determine μ.

P4.2*.2. Let $\Omega = (0,\pi)\times(0,\pi)$, and consider the homogeneous Dirichlet problem

$$-\Delta u|_\Omega = f = \sum_{i,j=1}^{\infty} a_{ij}\sin(ix_1)\sin(jx_2), \quad u|_{\partial\Omega} = 0. \qquad (4.2^*.6a)$$

Show that

$$u_{EX} = \sum_{i,j=1}^{\infty} b_{ij} \sin(ix_1)\sin(jx_2) \tag{4.2*.6b}$$

and determine b_{ij}. Under the assumption that

$$\sum_{i,j=1}^{\infty} a_{ij}^2 (i^2 + j^2)^k < \infty \tag{4.2*.6c}$$

for which s does the norm $\|u_{EX}\|_{H^s(\Omega)}$ stay bounded? Discuss especially the case $k = 0$.

P4.2*.3. Let

$$u_{S^{(n)}} = \sum_{i,j=1}^{n} c_{ij} \sin(ix_1)\sin(jx_2) \tag{4.2*.7a}$$

be an approximate solution of the problem given in P4.2*.2, and determine the coefficients c_{ij}, $i, j \leq n$, by employing the Galerkin method. What can be said about $\|u_{EX} - u_{S^{(n)}}\|_{H^1(\Omega)}$ and $\|u_{EX} - u_{S^{(n)}}\|_{L^2(\Omega)}$? Show that there exists a $\beta > 0$ such that

$$\|u_{EX} - u_{S^{(n)}}\|_{L^2(\Omega)} \leq \mathscr{C} n^{-\beta} \|u_{EX} - u_{S^{(n)}}\|_{H^1(\Omega)}. \tag{4.2*.7b}$$

Determine the largest possible value of β.

P4.2*.4. Let Ω be an L-shaped domain, and consider the homogeneous Dirichlet problem (4.2*.5a). It is possible to show that

$$u_{EX} = \mathscr{C} r \sin\left(\tfrac{2}{3}\theta\right) + w \tag{4.2*.8a}$$

where

$$\max\left(|\mathscr{C}|, \|w\|_{H^2(\Omega)}\right) \leq \|f\|_{L^2(\Omega)}. \tag{4.2*.8b}$$

Prove that, for uniform meshes, we have

$$\|e_{S^p_\Delta}\|_{\mathscr{U}} \leq \mathscr{C} h^{2/3} \|f\|_{L^2(\Omega)} \tag{4.2*.8c}$$

and

$$\|e_{S^p_\Delta}\|_{L^2(\Omega)} \leq \mathscr{C} h^{4/3} \|f\|_{L^2(\Omega)}. \tag{4.2*.8d}$$

Hint: First show that

$$\|u_{EX}\|_{B^{2/3}(H_0^1(\Omega), H^2(\Omega)\cap H_0^1(\Omega))} \leq \mathscr{C}\|f\|_{L^2}. \tag{4.2*.8e}$$

P4.2*.5. Consider the same problem as in P4.2*.2 and P4.2*.3. Compute $\|e_{S^{(n)}}\|_{H_0^{-l}(\Omega)}$, and show that in contrast to the threshold we have seen for the finite element method which uses piecewise polynomial functions of degree p, a threshold is not present in this case.

P4.2*.6. For the problem given in P4.2*.4, show that, for a general $f \in H^k(\Omega)$, $k \geq 2$, and $p \geq 2$, we have

$$\mathscr{C} h^{4/3} \|f\|_{H^k(\Omega)} \leq \|e_{S^p_\Delta}\|_{L^2(\Omega)}. \tag{4.2*.9a}$$

Find the maximal β such that

$$\|e_{S^p_\Delta}\|_{\mathscr{U}^{-1}} \leq h^\beta \|f\|_{H^k(\Omega)} \tag{4.2*.9b}$$

for $k \geq 2$.

4.3 Local and pollution error in the finite element solution

In Section 3.2, we introduced the finite element space

$$S^p_\Delta = S^p_\Delta(I) = \{v \in C^0(I) \mid v|_{I^\Delta_j} \in S^p(I^\Delta_j),\ j = 1, \ldots, N(\Delta)\}. \tag{4.3.1a}$$

Here $S^p(I^\Delta_j)$ is the space of mapped polynomials over I^Δ_j. In the majority of the cases the element transformation $F_{I^\Delta_j}$ is affine and we have $S^p(I^\Delta_j) \equiv \mathscr{P}^p(I^\Delta_j)$. We will partition $S^p(I^\Delta_j) \equiv \mathscr{P}^p(I^\Delta_j)$ into the space of linear functions $\mathscr{P}^1(I^\Delta_j)$, and the space of *bubble* functions

$$S^p_0(I^\Delta_j) \equiv \mathscr{P}^p_0(I^\Delta_j) \stackrel{\text{def}}{=} \{v \in \mathscr{P}^p(I^\Delta_j) \mid v(x^\Delta_{j-1}) = v(x^\Delta_j) = 0\} \tag{4.3.1b}$$

and we will write

$$S^p(I^\Delta_j) \equiv \mathscr{P}^p(I^\Delta_j) = \mathscr{P}^1(I^\Delta_j) \oplus \mathscr{P}^p_0(I^\Delta_j). \tag{4.3.1c}$$

This means that every function $w \in \mathscr{P}^p(I^\Delta_j)$ can be written in the form

$$w = w_{\mathscr{P}^1(I^\Delta_j)} + w_{\mathscr{P}^p_0(I^\Delta_j)} \tag{4.3.1d}$$

with $w_{\mathscr{P}^1(I^\Delta_j)} \in \mathscr{P}^1(I^\Delta_j)$, and $w_{\mathscr{P}^p_0(I^\Delta_j)} \in \mathscr{P}^p_0(I^\Delta_j)$, uniquely defined. A similar partition,

$$S^p(I^\Delta_j) = S^1(I^\Delta_j) \oplus S^p_0(I^\Delta_j) \tag{4.3.1e}$$

can also be employed in the more general case that $S^p(I^\Delta_j)$ is the space of mapped polynomials in which $S^1(I^\Delta_j)$ is the space of mapped linear functions, and $S^p_0(I^\Delta_j)$ is the space of mapped polynomial bubbles.

We will write the bilinear form $\mathscr{B}(\cdot, \cdot)$ as a sum of element contributions $\mathscr{B}_{I^\Delta_j}(\cdot, \cdot)$, namely

$$\mathscr{B}(u, v) = \sum_{j=1}^{N(\Delta)} \mathscr{B}_{I^\Delta_j}(u|_{I^\Delta_j}, v|_{I^\Delta_j}) \tag{4.3.2a}$$

where $\mathscr{B}_{I^\Delta_j} : \mathscr{U}(I^\Delta_j) \times \mathscr{U}(I^\Delta_j) \mapsto \mathbb{R}$ is given by

$$\mathscr{B}_{I^\Delta_j}(v, w) \stackrel{\text{def}}{=} \int_{I^\Delta_j} \left(AE \frac{dv}{dx}\frac{dw}{dx} + cvw\right) + \delta_{jN}\alpha_2\, v(1)w(1), \quad v, w \in \mathscr{U}(I^\Delta_j) \tag{4.3.2b}$$

for $j = 1, 2, \ldots, N$, where $\delta_{jN} = 1$ for $j = N$, and zero otherwise.

We will now prove the element orthogonality condition

$$\mathcal{B}_{I_j^\Delta}(e_{S_\Delta^p}, v) = 0, \quad \forall v \in S_0^p(I_j^\Delta). \tag{4.3.3a}$$

For any $v \in S_0^p(I_j^\Delta)$, let $\tilde{v} \in S_\Delta^p$

$$\tilde{v}(x) = \begin{cases} v(x), & x \in I_j^\Delta \\ 0, & x \in I - I_j^\Delta \end{cases} \tag{4.3.3b}$$

be its *global extension* by zero outside I_j^Δ. We then have

$$\mathcal{B}_{I_j^\Delta}(e_{S_\Delta^p}, v) = \mathcal{B}(e_{S_\Delta^p}, \tilde{v}) = 0 \quad \forall v \in S_0^p(I_j^\Delta) \tag{4.3.3c}$$

where the second equality follows from the orthogonality of the error because $\tilde{v} \in S_\Delta^p$.

We will split the finite element solution $u_{S_\Delta^p}$ in the element I_j^Δ into a linear part $u_{S^1(I_j^\Delta)} \in S^1(I_j^\Delta)$, and a bubble part $u_{S_0^p(I_j^\Delta)} \in S_0^p(I_j^\Delta)$, namely,

$$u_{S_\Delta^p}|_{I_j^\Delta} = u_{S^1(I_j^\Delta)} + u_{S_0^p(I_j^\Delta)}. \tag{4.3.4a}$$

From the element orthogonality condition (4.3.3a), we get

$$0 = \mathcal{B}_{I_j^\Delta}(e_{S_\Delta^p}, v) = \mathcal{B}_{I_j^\Delta}\left(u_{EX} - u_{S^1(I_j^\Delta)}, v\right) - \mathcal{B}_{I_j^\Delta}\left(u_{S_0^p(I_j^\Delta)}, v\right) \quad \forall v \in S_0^p(I_j^\Delta) \tag{4.3.4b}$$

and hence

$$\mathcal{B}_{I_j^\Delta}\left(u_{S_0^p(I_j^\Delta)}, v\right) = \mathcal{B}_{I_j^\Delta}\left(u_{EX} - u_{S^1(I_j^\Delta)}, v\right) \quad \forall v \in S_0^p(I_j^\Delta). \tag{4.3.5}$$

It follows that $u_{S_0^p(I_j^\Delta)}$ is the best approximation of $u_{EX}|_{I_j^\Delta} - u_{S^1(I_j^\Delta)}$ from $S_0^p(I_j^\Delta)$, namely

$$\left\|\left(u_{EX} - u_{S^1(I_j^\Delta)}\right) - u_{S_0^p(I_j^\Delta)}\right\|_{\mathcal{U}(I_j^\Delta)} = \min_{\chi \in \mathcal{P}_0^p(I_j^\Delta)} \left\|\left(u_{EX} - u_{S^1(I_j^\Delta)}\right) - \chi\right\|_{\mathcal{U}(I_j^\Delta)}. \tag{4.3.6}$$

In Chapter 3 we introduced the Lagrangian interpolant, $\mathcal{I}_\Delta^p u_{EX}$, of the exact solution u_{EX}. We will now construct another interpolant $\mathcal{A}_\Delta^p u_{EX}$, which has the special property that it coincides with the finite element solution in the cases that the nodal values are exact.

Given $u_{EX} \in \mathcal{U}$, we will define $\mathcal{A}_{I_j^\Delta}^p u_{EX}$, as the solution of the minimization problem: Find $\mathcal{A}_{I_j^\Delta}^p u_{EX} \in S_{u_{EX}}^p(I_j^\Delta) \overset{\text{def}}{=} \{w \in S^p(I_j^\Delta) \mid w(x_k^\Delta) = u_{EX}(x_k^\Delta),\ k = j-1, j\}$ such that

$$\left\|\mathcal{A}_{I_j^\Delta}^p u_{EX} - u_{EX}\right\|_{\mathcal{U}(I_j^\Delta)} = \min_{v \in S_{u_{EX}}^p(I_j^\Delta)} \|v - u_{EX}\|_{\mathcal{U}(I_j^\Delta)}. \tag{4.3.7a}$$

It follows that $\mathcal{A}_{I_j^\Delta}^p u_{EX}$ is the solution of the following local Dirichlet problem:

Find $\mathcal{A}_{I_j^\Delta}^p u_{EX} \in S^p(I_j^\Delta)$ such that $\mathcal{A}_{I_j^\Delta}^p u_{EX}(x_k^\Delta) = u_{EX}(x_k^\Delta)$, $k = j-1, j$, and

$$\mathcal{B}_{I_j^\Delta}\left(\mathcal{A}_{I_j^\Delta}^p u_{EX}, v\right) = \mathcal{B}_{I_j^\Delta}(u_{EX}, v) \quad \forall v \in S_0^p(I_j^\Delta). \tag{4.3.7b}$$

The solution $\mathcal{A}^p_{I_j^\Delta} u_{EX}$ coincides with the finite element solution $u_{S^p_\Delta}|_{I_j^\Delta}$ in the special case that the nodal values are exact, and, for this reason, we will call it the *local finite element solution.*

We will denote by $\mathcal{A}^p_\Delta u_0 \in C^0(I)$ the continuous function which is obtained by patching together the local finite element solutions $\mathcal{A}^p_{I_j^\Delta} u_{EX}$, over the entire mesh, namely,

$$\mathcal{A}^p_\Delta u_{EX}|_{I_j^\Delta} = \mathcal{A}^p_{I_j^\Delta} u_{EX} \tag{4.3.8a}$$

for $j = 1, \ldots, N(\Delta)$. It follows that $\mathcal{A}^p_\Delta u_{EX}$ is the solution of the following problem:

Find $\mathcal{A}^p_\Delta u_{EX} \in S^p_{\Delta, u_{EX}} \stackrel{\text{def}}{=} \{w \in S^p_\Delta \mid w(x_j^\Delta) = u_{EX}(x_j^\Delta),\ j = 0, \ldots, N(\Delta)\}$, such that

$$\mathcal{B}(\mathcal{A}^p_\Delta u_{EX}, v) = \mathcal{B}(u_{EX}, v) \quad \forall v \in S^p_{\Delta,0} \tag{4.3.8b}$$

where

$$S^p_{\Delta,0} \stackrel{\text{def}}{=} \{v \in S^p_\Delta \mid v(x_j^\Delta) = 0,\ j = 0, \ldots, N(\Delta)\}. \tag{4.3.8c}$$

It follows that when the finite element solution $u_{S^p_\Delta}$ is exact at the nodes, we have $\mathcal{A}^p_\Delta u_{EX} \equiv u_{S^p_\Delta}$, while, in general, $u_{S^p_\Delta} \not\equiv \mathcal{I}^p_\Delta u_{EX}$. For example, in the case

$$\frac{1}{AE} \in \mathcal{P}^{p-1}(I), \qquad c \equiv 0, \qquad p > 1 \tag{4.3.8d}$$

in which we have

$$u_{S^p_\Delta} \equiv \mathcal{A}^p_\Delta u_{EX} \not\equiv \mathcal{I}^p_\Delta u_{EX}. \tag{4.3.8e}$$

Let us now give an example which illustrates the difference between $\mathcal{I}^p_\Delta u_{EX}$ and $\mathcal{A}^p_\Delta u_{EX}$.

Example 4.3.1 Consider the singular model problem with $AE \equiv x^\vartheta$, $c \equiv 0$, $a_2 = 0$, $u_{EX} \equiv \mathcal{C} x^{1-\vartheta}$, $\vartheta = \frac{3}{4}$. Figure 4.3.1 shows the graphs of the errors $u_{EX} - \mathcal{I}^2_{I_j^\Delta} u_{EX}$ and $u_{EX} - \mathcal{A}^2_{I_j^\Delta} u_{EX}$, for elements of quadratic degree ($p = 2$) over a uniform mesh of two elements: $I_1^\Delta = (0, \frac{1}{8})$ and $I_2^\Delta = (\frac{1}{8}, \frac{1}{4})$. The difference between the two errors $u_{EX} - \mathcal{A}^2_{I_1^\Delta} u_{EX}$ and $u_{EX} - \mathcal{I}^2_{I_1^\Delta} u_{EX}$ is clearly visible in the first element.

Although, in general, we have $\mathcal{A}^p_{I_j^\Delta} u_{EX} \not\equiv \mathcal{I}^p_{I_j^\Delta} u_{EX}$, it can be proven that $\mathcal{A}^p_{I_j^\Delta} u_{EX}$ is close to any Lagrangian interpolant $\mathcal{I}^p_{I_j^\Delta} u_{EX}$.

Theorem 4.3.1 *Given $u_{EX} \in \mathcal{U}$, there exists a $\mathcal{C}(p, K)$, where K is such that $|||\tilde{\mathcal{I}}^p(\{\tilde{\xi}_l\}_{l=0}^p)|||_\infty \le K$ (see (3.3.26b))*

$$\begin{aligned}\mathcal{C}(p,K)\|u_{EX} - \mathcal{I}^p_{I_j^\Delta} u_{EX}\|_{\mathcal{U}(I_j^\Delta)} &\le \|u_{EX} - \mathcal{A}^p_{I_j^\Delta} u_{EX}\|_{\mathcal{U}(I_j^\Delta)} \\ &\le \|u_{EX} - \mathcal{I}^p_{I_j^\Delta} u_{EX}\|_{\mathcal{U}(I_j^\Delta)}.\end{aligned} \tag{4.3.9}$$

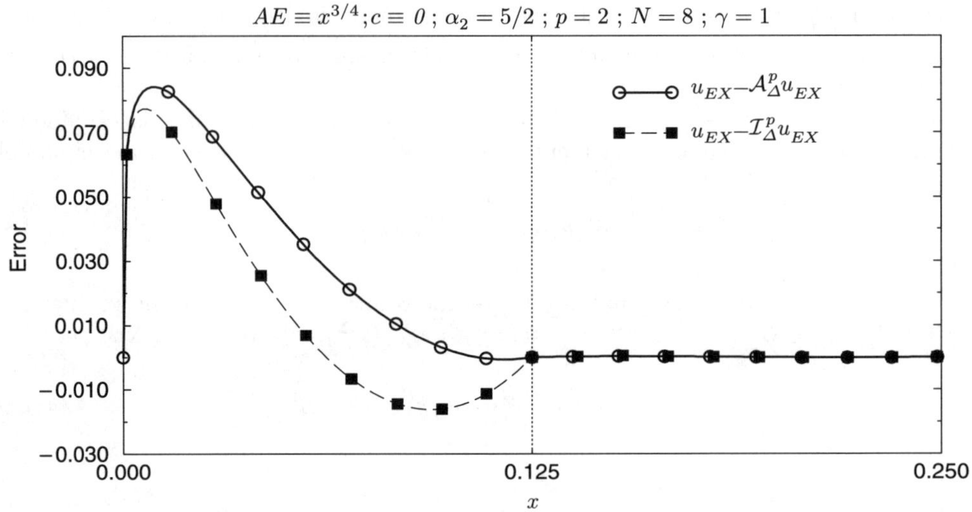

FIG. 4.3.1. An example which illustrates the difference between the Lagrangian interpolant $\mathcal{I}^2_{I_j^\Delta} u_{EX}$ and the local finite element solution $\mathcal{A}^2_{I_j^\Delta} u_{EX}$. Input data: $AE \equiv x^\vartheta$, $c \equiv 0$, $a_2 = 0$, $u_{EX} \equiv \mathscr{C} x^{1-\vartheta}$, $\vartheta = \frac{3}{4}$. Mesh data: Uniform mesh of eight quadratic elements ($N = 8$, $p = 2$). Graphs of the error in the local finite element solution $u_{EX} - \mathcal{A}^2_\Delta u_{EX}$, and the error in the Lagrangian interpolant $u_{EX} - \mathcal{I}^2_\Delta u_{EX}$, in the first and the second element. The difference between the two interpolants in the first element is clearly visible

Proof From the construction of $\mathcal{A}^p_{I_j^\Delta} u_{EX}$, it follows that

$$\begin{aligned} \left\| u_{EX} - \mathcal{A}^p_{I_j^\Delta} u_{EX} \right\|_{\mathfrak{U}(I_j^\Delta)} &= \min_{v \in S^p_{u_{EX}}(I_j^\Delta)} \| u_{EX} - v \|_{\mathfrak{U}(I_j^\Delta)} \\ &\leq \left\| u_{EX} - \mathcal{I}^p_{I_j^\Delta} u_{EX} \right\|_{\mathfrak{U}(I_j^\Delta)} \end{aligned} \tag{4.3.10a}$$

where we employed the notation

$$S^p_{u_{EX}}(I_j^\Delta) \stackrel{\text{def}}{=} \left\{ v \in S^p(I_j^\Delta) \mid v(x_l^\Delta) = u_{EX}(x_l^\Delta),\ l = j-1, j \right\} \tag{4.3.10b}$$

and the upper bound in (4.3.9) has been proven.

To prove the lower bound, we must show that

$$0 < \bar{\mathscr{C}}(p) = \inf_{w \in \mathfrak{U}(I_j^\Delta)} \frac{\left\| w - \mathcal{A}^p_{I_j^\Delta} w \right\|_{\mathfrak{U}(I_j^\Delta)}}{\left\| w - \mathcal{I}^p_{I_j^\Delta} w \right\|_{\mathfrak{U}(I_j^\Delta)}}. \tag{4.3.11a}$$

We will prove this by contradiction. Let us assume that $\bar{\mathscr{C}}(p) = 0$. Then, there exists a sequence $\{w_n\}_{n=1}^\infty$, $w_n \in \mathfrak{U}(I_j^\Delta)$, such that

$$\lim_{n\to\infty} \epsilon_n \stackrel{\text{def}}{=} \lim_{n\to\infty} \left\| w_n - \mathcal{A}^p_{I_j^\Delta} w_n \right\|_{\mathfrak{U}(I_j^\Delta)} = 0 \tag{4.3.11b}$$

where we let $\epsilon_n \stackrel{\text{def}}{=} \| w_n - \mathcal{A}^p_{I_j^\Delta} w_n \|_{\mathfrak{U}(I_j^\Delta)}$, while

$$\left\| w_n - \mathcal{I}^p_{I_j^\Delta} w_n \right\|_{\mathfrak{U}(I_j^\Delta)} = 1 \tag{4.3.11c}$$

for any n.

Let us now prove that such a sequence cannot exist. We have

$$(w_n - \mathcal{A}^p_{I^\Delta_j} w_n)(x^\Delta_{j-1}) = 0 \tag{4.3.12a}$$

and hence

$$\|w_n - \mathcal{A}^p_{I^\Delta_j} w_n\|_{L^\infty(I^\Delta_j)} \le \mathscr{C}\sqrt{h_{I^\Delta_j}}\,\|w_n - \mathcal{A}^p_{I^\Delta_j} w_n\|_{\mathscr{U}(I^\Delta_j)} = \mathscr{C}\sqrt{h_{I^\Delta_j}}\,\epsilon_n. \tag{4.3.12b}$$

This implies that

$$|w_n(x) - \mathcal{A}^p_{I^\Delta_j} w_n(x)| \le \mathscr{C}\sqrt{h_{I^\Delta_j}}\,\epsilon_n \quad \forall x \in I^\Delta_j. \tag{4.3.12c}$$

Applying (4.3.12c) at $\xi_l^{I^\Delta_j}$, $l = 0, \dots, p$, the interpolation points employed in the construction of $\mathcal{I}^p_{I^\Delta_j}$, and taking into account that

$$w_n(\xi_l^{I^\Delta_j}) = \mathcal{I}^p_{I^\Delta_j} w_n(\xi_l^{I^\Delta_j}), \quad l = 0, \dots, p \tag{4.3.12d}$$

we get

$$\left|\mathcal{I}^p_{I^\Delta_j} w_n(\xi_l^{I^\Delta_j}) - \mathcal{A}^p_{I^\Delta_j} w_n(\xi_l^{I^\Delta_j})\right| \le \mathscr{C}\sqrt{h_{I^\Delta_j}}\,\epsilon_n, \qquad l = 0, \dots, p. \tag{4.3.12e}$$

Further, because $(\mathcal{A}^p_{I^\Delta_j} w_n - \mathcal{I}^p_{I^\Delta_j} w_n) \in S^p_0(I^\Delta_j)$, we have

$$\begin{aligned} \|\mathcal{A}^p_{I^\Delta_j} w_n - \mathcal{I}^p_{I^\Delta_j} w_n\|_{L^\infty(I^\Delta_j)} &\le \mathscr{C} \max_{l=0,\dots,p} \left|\mathcal{A}^p_{I^\Delta_j} w_n(\xi_l^{I^\Delta_j}) - \mathcal{I}^p_{I^\Delta_j} w_n(\xi_l^{I^\Delta_j})\right| \\ &< \mathscr{C}\sqrt{h_{I^\Delta_j}}\,\epsilon_n \end{aligned} \tag{4.3.12f}$$

and

$$\|\mathcal{A}^p_{I^\Delta_j} w_n - \mathcal{I}^p_{I^\Delta_j} w_n\|_{\mathscr{U}(I^\Delta_j)} \le \frac{\mathscr{C}}{\sqrt{h_{I^\Delta_j}}} \|\mathcal{A}^p_{I^\Delta_j} w_n - \mathcal{I}^p_{I^\Delta_j} w_n\|_{L^\infty(I^\Delta_j)} \le \mathscr{C}\epsilon_n. \tag{4.3.12g}$$

Using now the triangle inequality

$$\underbrace{\|w_n - \mathcal{I}^p_{I^\Delta_j} w_n\|_{\mathscr{U}(I^\Delta_j)}}_{=1} \le \underbrace{\|w_n - \mathcal{A}^p_{I^\Delta_j} w_n\|_{\mathscr{U}(I^\Delta_j)}}_{\le \epsilon_n} + \underbrace{\|\mathcal{A}^p_{I^\Delta_j} w_n - \mathcal{I}^p_{I^\Delta_j} w_n\|_{\mathscr{U}(I^\Delta_j)}}_{\le \mathscr{C}\epsilon_n} \tag{4.3.12h}$$

we get

$$1 \le \epsilon_n + \mathscr{C}\,\epsilon_n \tag{4.3.12i}$$

which contradicts our assumption that $\lim_{n\to\infty} \epsilon_n = 0$. □

Using the above result, we can now prove an asymptotic estimate $\|u_{EX} - \mathcal{A}^p_{I_j^\Delta} u_{EX}\|_{\mathcal{U}(I_j^\Delta)}$.

Corollary 4.3.1 *Under the assumption*

$$\left\| \frac{d^q}{dx^q}(u_{EX}) \right\|_{L^\infty(I_j^\Delta)} \le K < \infty \tag{4.3.13a}$$

we have

$$\|u_{EX} - \mathcal{A}^p_{I_j^\Delta} u_{EX}\|_{\mathcal{U}(I_j^\Delta)} \le \mathscr{C} K h_{I_j^\Delta}^{\mu+1/2} \tag{4.3.13b}$$

where

$$\mu = \min(p, q-1), \quad 1 \le q \le p+1 \tag{4.3.13c}$$

and $\mathscr{C}$ depends only on p. Further, under the additional assumption that

$$\frac{d^{p+1}}{dx^{p+1}}(u_{EX}) \in C^0(I_j^\Delta) \tag{4.3.13d}$$

we have

$$\mathscr{C}' (h_{I_j^\Delta})^{p+1/2} \min_{x \in I_j^\Delta} \left| \frac{d^{p+1}}{dx^{p+1}}(u_{EX})(x) \right| \le \|u_{EX} - \mathcal{A}^p_{I_j^\Delta} u_{EX}\|_{\mathcal{U}(I_j^\Delta)} \tag{4.3.13e}$$

where $\mathscr{C}'$ depends only on p.

Proof The upper estimate (4.3.13b) follows from Theorem 3.3.1 and the best-approximation property

$$\begin{aligned} \|u_{EX} - \mathcal{A}^p_{I_j^\Delta} u_{EX}\|_{\mathcal{U}(I_j^\Delta)} &= \min_{\bar{\chi} \in S^p_{u_{EX}}(I_j^\Delta)} \|u_{EX} - \bar{\chi}\|_{\mathcal{U}(I_j^\Delta)} \\ &\le \|u_{EX} - \mathcal{I}^p_{I_j^\Delta} u_{EX}\|_{\mathcal{U}(I_j^\Delta)}. \end{aligned} \tag{4.3.14a}$$

The lower estimate follows from

$$\mathscr{C} (h_{I_j^\Delta})^{p+1/2} \min_{x \in I_j^\Delta} \left| \frac{d^{p+1}}{dx^{p+1}}(u_{EX})(x) \right| \le \min_{\chi \in S^{p-1}(I_j^\Delta)} \left\| \frac{d}{dx}(u_{EX}) - \chi \right\|_{L^2(I_j^\Delta)} \tag{4.3.14b}$$

and the observation that

$$\begin{aligned} \min_{\chi \in S^p(I_j^\Delta)} \|u_{EX} - \chi\|_{\mathcal{U}(I_j^\Delta)} &\le \min_{\chi \in S^p_{u_{EX}}(I_j^\Delta)} \|u_{EX} - \chi\|_{\mathcal{U}(I_j^\Delta)} \\ &= \|u_{EX} - \mathcal{A}^p_{I_j^\Delta} u_{EX}\|_{\mathcal{U}(I_j^\Delta)}. \end{aligned} \tag{4.3.14c}$$

To prove the inequality (4.3.14b) it is sufficient to show that

$$\min_{x \in \tilde{I}} \left| \frac{d^{p+1}}{d\tilde{x}^{p+1}}(\tilde{u}_{EX})(x) \right| \le \mathscr{C} \min_{\chi \in S^{p-1}(\tilde{I})} \left\| \frac{d}{d\tilde{x}}(\tilde{u}_{EX}) - \chi \right\|_{L^2(\tilde{I})}. \tag{4.3.15a}$$

To this end let us write

$$\frac{d}{d\tilde{x}}(\tilde{u}_{EX}) = \sum_{k=0}^{\infty} a_k\, P_k \tag{4.3.15b}$$

where P_k are the Legendre polynomials (see Appendix A3.1.1). Then

$$\min_{\chi \in S^{p-1}(\tilde{I})} \left\| \frac{d}{d\tilde{x}}(\tilde{u}_{EX}) - \chi \right\|_{L^2(\tilde{I})} = \sqrt{\sum_{k=p}^{\infty} a_k^2 \frac{2}{2k+1}}. \tag{4.3.15c}$$

Hence it is enough to prove that

$$a_p \geq \mathscr{C} \min_{x \in \tilde{I}} \left| \frac{d^{p+1}}{d\tilde{x}^{p+1}}(\tilde{u}_{EX})(x) \right|. \tag{4.3.15d}$$

This follows immediately from the observation that

$$\frac{2p+1}{2} a_p = \int_{\tilde{I}} \Big(\frac{d}{d\tilde{x}} u_{EX}(\tilde{x}) \Big) P_p(\tilde{x}) = \int_{\tilde{I}} \Big(\frac{d}{d\tilde{x}} u_{EX}(\tilde{x}) \Big) \frac{1}{2^p p!} \frac{d^p}{d\tilde{x}^p}(\tilde{x}^2 - 1)^p \tag{4.3.15e}$$

and p integrations by parts. □

Remark 4.3.1 It follows that for sufficiently smooth u_{EX}, we have

$$\|u_{EX} - \mathcal{I}^p_{I_j^\Delta} u_{EX}\|_{\mathcal{U}(I_j^\Delta)} \approx \|u_{EX} - \mathcal{A}^p_{I_j^\Delta} u_{EX}\|_{\mathcal{U}(I_j^\Delta)} \approx \mathscr{C}\, h_{I_j^\Delta}^{\mu+1/2} K \tag{4.3.16a}$$

for all elements I_j^Δ, for which we have

$$0 < K \leq \min_{x \in I_j^\Delta} \left| \frac{d^{p+1}}{dx^{p+1}}(u_{EX})(x) \right|. \tag{4.3.16b}$$

We will partition the error $e_{S^p_\Delta}$ into two components, namely,

$$e^{loc}_{S^p_\Delta} \stackrel{\text{def}}{=} u_{EX} - \mathcal{A}^p_\Delta u_{EX} \tag{4.3.17a}$$

the *local component of the error*, or the *local error*, and

$$e^{glob}_{S^p_\Delta} \stackrel{\text{def}}{=} \mathcal{A}^p_\Delta u_{EX} - u_{S^p_\Delta} \tag{4.3.17b}$$

the *global component of the error*, or the *pollution error*, and we have

$$e_{S^p_\Delta} = u_{EX} - u_{S^p_\Delta} = u_{EX} - \mathcal{A}^p_\Delta u_{EX} + \mathcal{A}^p_\Delta u_{EX} - u_{S^p_\Delta} = e^{loc}_{S^p_\Delta} + e^{glob}_{S^p_\Delta}. \tag{4.3.17c}$$

Remark 4.3.2 Corollary 4.3.1 can be stated as follows:

Under the assumption that $\|(d^r/dx^r)(u_{EX})\|_{L^\infty(I_j^\Delta)} < \infty$ we have

$$\|e^{loc}_{S^p(I_j^\Delta)}\|_{\mathcal{U}(I_j^\Delta)} \leq \mathscr{C} h_{I_j^\Delta}^{\mu+1/2} \left\| \frac{d^r}{dx^r}(u_{EX}) \right\|_{L^\infty(I_j^\Delta)} \tag{4.3.18a}$$

where $\mu = \min(p, r-1)$, $1 \le r \le p+1$. Further, assuming that $(d^{p+1}/dx^{p+1})(u_{EX}) \in C^0(I_j^\Delta)$, we have

$$\mathscr{C}(h_{I_j^\Delta})^{p+1/2} \min_{x \in I_j^\Delta} \left| \frac{d^{p+1}}{dx^{p+1}}(u_{EX}) \right| (x) \le \| e^{loc}_{S^p(I_j^\Delta)} \|_{\mathscr{U}(I_j^\Delta)}. \tag{4.3.18b}$$

Further, we will partition $e^{glob}_{S^p_\Delta}$ in each element into an affine and a bubble component, namely,

$$e^{glob}_{S^p_\Delta}|_{I_j^\Delta} = e^{glob}_{S^1(I_j^\Delta)} + e^{glob}_{S^p_0(I_j^\Delta)} \tag{4.3.19a}$$

and we have:

Theorem 4.3.2 *Under the minimal assumptions which guarantee the existence of solutions of the model problem, we have*

$$\mathscr{B}_{I_j^\Delta}(e^{glob}_{S^p_0(I_j^\Delta)} v) = -\mathscr{B}_{I_j^\Delta}(e^{glob}_{S^1(I_j^\Delta)}, v) \quad \forall v \in S^p_0(I_j^\Delta) \tag{4.3.19b}$$

and

$$\begin{aligned} \| e^{glob}_{S^p_0(I_j^\Delta)} \|_{\mathscr{U}(I_j^\Delta)} &\le \| e^{glob}_{S^1(I_j^\Delta)} \|_{\mathscr{U}(I_j^\Delta)} \\ &\le \mathscr{C}(1 + h_{I_j^\Delta}) \sqrt{h_{I_j^\Delta}} \sqrt{(e_j^\Delta + e_{j-1}^\Delta)^2 + \left(\frac{e_j^\Delta - e_{j-1}^\Delta}{h_{I_j^\Delta}} \right)^2} \end{aligned} \tag{4.3.19c}$$

where $\mathscr{C} = \max(\beta_2, \gamma/4)$, *and we employed the notation* $e_j^\Delta \stackrel{\text{def}}{=} e_{S^p_\Delta}(x_j^\Delta)$.

Proof For every bubble function $v \in S^p_0(I_j^\Delta)$, let $\tilde{v} \in \mathscr{U}$ be its global extension by zero outside I_j^Δ. Then, for any $v \in S^p_0(I_j^\Delta)$, we have

$$0 = \mathscr{B}(e_{S^p_\Delta}, \tilde{v}) = \mathscr{B}_{I_j^\Delta}(e_{S^p_\Delta}|_{I_j^\Delta}, v) = \mathscr{B}_{I_j^\Delta}(e^{loc}_{S^p(I_j^\Delta)}, v) + \mathscr{B}_{I_j^\Delta}(e^{glob}_{S^p(I_j^\Delta)}, v) \tag{4.3.20a}$$

and because

$$\mathscr{B}_{I_j^\Delta}(e^{loc}_{S^p(I_j^\Delta)}, v) = 0 \quad \forall v \in S^p_0(I_j^\Delta) \tag{4.3.20b}$$

we have

$$\mathscr{B}_{I_j^\Delta}(e^{glob}_{S^p(I_j^\Delta)}, v) = \mathscr{B}_{I_j^\Delta}(e^{glob}_{S^1(I_j^\Delta)} + e^{glob}_{S^p_0(I_j^\Delta)}, v) = 0 \quad \forall v \in S^p_0(I_j^\Delta) \tag{4.3.20c}$$

and we have proven that

$$\mathscr{B}_{I_j^\Delta}(e^{glob}_{S^p_0(I_j^\Delta)}, v) = -\mathscr{B}_{I_j^\Delta}(e^{glob}_{S^1(I_j^\Delta)}, v) \quad \forall v \in S^p_0(I_j^\Delta). \tag{4.3.20d}$$

Thus $e^{glob}_{S^p_0(I_j^\Delta)}$ is the energy projection of $e^{glob}_{S^1(I_j^\Delta)}$ into the bubble space $S^p_0(I_j^\Delta)$, and hence

$$\| e^{glob}_{S^p_0(I_j^\Delta)} \|_{\mathscr{U}(I_j^\Delta)} \le \| e^{glob}_{S^1(I_j^\Delta)} \|_{\mathscr{U}(I_j^\Delta)}. \tag{4.3.20e}$$

To estimate the energy norm of $e^{glob}_{S^1(I^\Delta_j)}$ in terms of the nodal errors, we note that

$$\|e^{glob}_{S^1(I^\Delta_j)}\|^2_{\mathcal{U}(I^\Delta_j)} \leq \beta_2 \Big\| \frac{d}{dx}(e^{glob}_{S^1(I^\Delta_j)}) \Big\|^2_{L^2(I^\Delta_j)} + \gamma \|e^{glob}_{S^1(I^\Delta_j)}\|^2_{L^2(I^\Delta_j)} \tag{4.3.21a}$$

and that

$$\Big\| \frac{d}{dx}(e^{glob}_{S^1(I^\Delta_j)}) \Big\|^2_{L^2(I^\Delta_j)} = \int_{I^\Delta_j} \left(\frac{e^\Delta_j - e^\Delta_{j-1}}{h_{I^\Delta_j}} \right)^2 = h_{I^\Delta_j} \left(\frac{e^\Delta_j - e^\Delta_{j-1}}{h_{I^\Delta_j}} \right)^2 \tag{4.3.21b}$$

and

$$\begin{aligned} \|e^{glob}_{S^1(I^\Delta_j)}\|^2_{L^2(I^\Delta_j)} &= \int_{I^\Delta_j} \left(\frac{e^\Delta_j + e^\Delta_{j-1}}{2} + \frac{e^\Delta_j - e^\Delta_{j-1}}{h_{I^\Delta_j}} (x - x^\Delta_{j-1/2}) \right)^2 \\ &= \tfrac{1}{4} h_{I^\Delta_j} (e^\Delta_j + e^\Delta_{j-1})^2 + \tfrac{1}{6} h_{I^\Delta_j} (e^\Delta_j - e^\Delta_{j-1})^2. \end{aligned} \tag{4.3.21c}$$

Employing (4.3.21b,c) into (4.3.21a), (4.3.19c) follows. □

Remark 4.3.3 We will utilize the partition $e_{S^p_\Delta} = e^{loc}_{S^p_\Delta} + e^{glob}_{S^p_\Delta}$ to construct a-priori estimates for the error of the finite element solution in an element, by constructing a-priori estimates for the local error $e^{loc}_{S^p_\Delta}$ and the pollution error $e^{glob}_{S^p_\Delta}$ separately. The partition will also have essential consequences in the study of the problem of a-posteriori estimation of the error.

In the cases that AE has bounded first derivative, the bubble part of the pollution error is negligible.

Theorem 4.3.3 *Consider the regular model problem, and let AE satisfy the additional assumption*

$$\Big\| \frac{d}{dx}(AE) \Big\|_{L^\infty(I^\Delta_j)} \leq K < \infty. \tag{4.3.22a}$$

Then

$$\|e^{glob}_{S^p_0(I^\Delta_j)}\|_{\mathcal{U}(I^\Delta_j)} \leq \mathcal{C}(K + \gamma) h_{I^\Delta_j} \|e^{glob}_{S^1(I^\Delta_j)}\|_{\mathcal{U}(I^\Delta_j)}. \tag{4.3.22b}$$

Proof Let $\bar{x} \in I^\Delta_j$ be a fixed point. Noting that

$$\int_{I^\Delta_j} \frac{d}{dx}(e^{glob}_{S^1(I^\Delta_j)}) \frac{dv}{dx} = \frac{d}{dx}(e^{glob}_{S^1(I^\Delta_j)}) \int_{I^\Delta_j} \frac{dv}{dx} = 0 \quad \forall v \in S^p_0(I^\Delta_j) \tag{4.3.23a}$$

we obtain

$$\mathcal{B}_{I^\Delta_j}(e^{glob}_{S^1(I^\Delta_j)}, v) = \int_{I^\Delta_j} (AE - AE(\bar{x})) \frac{d}{dx}(e^{glob}_{S^1(I^\Delta_j)}) \frac{dv}{dx} + \int_{I^\Delta_j} c e^{glob}_{S^1(I^\Delta_j)} v \tag{4.3.23b}$$

for any $v \in S_0^p(I_j^\Delta)$, and employing

$$|AE(x) - AE(\bar{x})| = \left| \int_{\bar{x}}^{x} \frac{d}{dx}(AE) \right| \leq \left\| \frac{d}{dx}(AE) \right\|_{L^\infty(I_j^\Delta)} |x - \bar{x}| \leq K\, h_{I_j^\Delta} \qquad (4.3.23c)$$

and the Cauchy–Schwarz inequality, we get

$$|\mathscr{B}_{I_j^\Delta}(e^{glob}_{S^1(I_j^\Delta)}, v)| \leq K h_{I_j^\Delta} \left\| \frac{d}{dx}(e^{glob}_{S^1(I_j^\Delta)}) \right\|_{L^2(I_j^\Delta)} \left\| \frac{dv}{dx} \right\|_{L^2(I_j^\Delta)} + \gamma \left\| e^{glob}_{S^1(I_j^\Delta)} \right\|_{L^2(I_j^\Delta)} \|v\|_{L^2(I_j^\Delta)} \qquad (4.3.23d)$$

for any $v \in S_0^p(I_j^\Delta)$. Further, every $v \in S_0^p(I_j^\Delta)$ satisfies $v(x_{j-1}^\Delta) = 0$, and hence

$$\|v\|_{L^2(I_j^\Delta)} \leq \mathscr{C} h_{I_j^\Delta} \|v\|_{\mathscr{U}(I_j^\Delta)} \qquad (4.3.23e)$$

and substituting in (4.3.23d) we get

$$\left|\mathscr{B}_{I_j^\Delta}(e^{glob}_{S^1(I_j^\Delta)}, v)\right| \leq \mathscr{C} h_{I_j^\Delta} (K + \gamma) \|e^{glob}_{S^1(I_j^\Delta)}\|_{\mathscr{U}(I_j^\Delta)} \|v\|_{\mathscr{U}(I_j^\Delta)} \qquad (4.3.23f)$$

for any $v \in S_0^p(I_j^\Delta)$, and letting $v = e^{glob}_{S_0^p(I_j^\Delta)}$, (4.3.22b) follows. □

As we will see in Section 4.3*, $\mathcal{A}_\Delta^p u_{EX}$ cannot be used in higher dimensions. We will now introduce another definition of the local finite element solution which can be used in the higher dimensional case.

Given $u_{EX} \in \mathscr{U}$, we will define $\mathcal{B}^p_{I_j^\Delta} u_{EX}$ as the solution of the minimization problem:

Find $\mathcal{B}^p_{I_j^\Delta} u_{EX} \in S^p(I_j^\Delta)$ such that

$$\left\|\mathcal{B}^p_{I_j^\Delta} u_{EX} - u_{EX}\right\|_{\mathscr{U}(I_j^\Delta)} = \min_{v \in S^p(I_j^\Delta)} \|v - u_{EX}\|_{\mathscr{U}(I_j^\Delta)}. \qquad (4.3.24a)$$

It follows that $\mathcal{B}^p_{I_j^\Delta} u_{EX}$ is the solution of the variational problem:

Find $\mathcal{B}^p_{I_j^\Delta} u_{EX} \in S^p(I_j^\Delta)$ such that

$$\mathscr{B}_{I_j^\Delta}\left(\mathcal{B}^p_{I_j^\Delta} u_{EX}, v\right) = \mathscr{B}_{I_j^\Delta}(u_{EX}, v) \quad \forall\, v \in S^p(I_j^\Delta). \qquad (4.3.24b)$$

When $c \equiv 0$, $\mathcal{B}^p_{I_j^\Delta} u_{EX}$ is defined only up to an arbitrary constant. This constant may be fixed by letting

$$\mathcal{B}^p_{I_j^\Delta} u_{EX}(x_j^\Delta) = u_{EX}(x_j^\Delta) \qquad (4.3.24c)$$

or

$$\int_{I_j^\Delta} \mathcal{B}^p_{I_j^\Delta} u_{EX} = \int_{I_j^\Delta} u_{EX} \qquad (4.3.24d)$$

etc. Note that, in general, we have

$$\mathcal{B}^p_{I^\Delta_j} u_{EX}(x^\Delta_{j-1}) \neq \mathcal{B}^p_{I^\Delta_{j-1}} u_{EX}(x^\Delta_{j-1}). \tag{4.3.24e}$$

Letting

$$\mathcal{B}^p_\Delta u_{EX}|_{I^\Delta_j} = \mathcal{B}^p_{I^\Delta_j} u_{EX} \tag{4.3.25a}$$

we will define the local component of the error by

$$\tilde{e}^{loc}_{S^p_\Delta} = u_{EX} - \mathcal{B}^p_\Delta u_{EX} \tag{4.3.25b}$$

and the global component by

$$\tilde{e}^{glob}_{S^p_\Delta} = \mathcal{B}^p_\Delta u_{EX} - u_{S^p_\Delta}. \tag{4.3.25c}$$

We will now prove that, when AE has bounded first derivative, the difference between $\mathcal{B}^p_{I^\Delta_j} u_{EX}$ and $\mathcal{A}^p_{I^\Delta_j} u_{EX}$ is negligible.

Theorem 4.3.4 *Consider the regular model problem, and assume, in addition, that*

$$\left\| \frac{d}{dx}(AE) \right\|_{L^\infty(I^\Delta_j)} \leq K < \infty. \tag{4.3.26a}$$

Then we have

$$\|\mathcal{B}^p_{I^\Delta_j} u_{EX} - \mathcal{A}^p_{I^\Delta_j} u_{EX}\|_{\mathfrak{U}(I^\Delta_j)} \leq \mathscr{C}(K,\gamma)\, h_{I^\Delta_j} \|\tilde{e}^{loc}_{S^p(I^\Delta_j)}\|_{\mathfrak{U}(I^\Delta_j)}. \tag{4.3.26b}$$

Proof Let $\delta\tilde{e}_{S^p(I^\Delta_j)}$ denote the difference between $\mathcal{B}^p_{I^\Delta_j} u_{EX}$ and $\mathcal{A}^p_{I^\Delta_j} u_{EX}$, namely,

$$\delta\tilde{e}_{S^p(I^\Delta_j)} \stackrel{\text{def}}{=} \mathcal{B}^p_{I^\Delta_j} u_{EX} - \mathcal{A}^p_{I^\Delta_j} u_{EX} \tag{4.3.27a}$$

with linear part

$$\delta\tilde{e}_{S^1(I^\Delta_j)}(x) \stackrel{\text{def}}{=} \frac{\tilde{e}^{\Delta,-}_j + \tilde{e}^{\Delta,+}_{j-1}}{2} + \frac{\tilde{e}^{\Delta,-}_j - \tilde{e}^{\Delta,+}_{j-1}}{h_{I^\Delta_j}}(x - x^\Delta_{j-1/2}) \tag{4.3.27b}$$

and bubble part

$$\delta\tilde{e}_{S^p_0(I^\Delta_j)} = \delta\tilde{e}_{S^p(I^\Delta_j)} - \delta\tilde{e}_{S^1(I^\Delta_j)} \tag{4.3.27c}$$

where we used the notations

$$\tilde{e}^{\Delta,+}_{j-1} \stackrel{\text{def}}{=} \mathcal{B}^p_{I^\Delta_j} u_{EX}(x^\Delta_{j-1}) - \mathcal{A}^p_{I^\Delta_j} u_{EX}(x^\Delta_{j-1}) \tag{4.3.27d}$$

and

$$\tilde{e}^{\Delta,-}_j \stackrel{\text{def}}{=} \mathcal{B}^p_{I^\Delta_j} u_{EX}(x^\Delta_j) - \mathcal{A}^p_{I^\Delta_j} u_{EX}(x^\Delta_j). \tag{4.3.27e}$$

Noting that

$$\mathcal{B}_{I_j^\Delta}\big(\delta\tilde{e}_{S^p(I_j^\Delta)}, v\big) = \mathcal{B}_{I_j^\Delta}\big(\mathcal{B}^p_{I_j^\Delta} u_{EX} - \mathcal{A}^p_{I_j^\Delta} u_{EX}, v\big) = 0 \quad \forall v \in S_0^p(I_j^\Delta) \tag{4.3.28a}$$

we get

$$\mathcal{B}_{I_j^\Delta}\big(\delta\tilde{e}_{S_0^p(I_j^\Delta)}, v\big) = -\mathcal{B}_{I_j^\Delta}\big(\delta\tilde{e}_{S^1(I_j^\Delta)}, v\big) \quad \forall v \in S_0^p(I_j^\Delta) \tag{4.3.28b}$$

and following similar steps as in the proof of Theorem 4.3.2, we get

$$\big\|\delta\tilde{e}_{S_0^p(I_j^\Delta)}\big\|_{\mathcal{U}(I_j^\Delta)} \le \big\|\delta\tilde{e}_{S^1(I_j^\Delta)}\big\|_{\mathcal{U}(I_j^\Delta)} \le \mathcal{C}\sqrt{h_{I_j^\Delta}}\max\big(|\tilde{\rho}_1^{I_j^\Delta}|, |\tilde{\rho}_2^{I_j^\Delta}|\big) \tag{4.3.29a}$$

where we let

$$\tilde{\rho}_1^{I_j^\Delta} \stackrel{\text{def}}{=} \frac{\tilde{e}_{j-1}^{\Delta,+} + \tilde{e}_j^{\Delta,-}}{2} \qquad \tilde{\rho}_2^{I_j^\Delta} \stackrel{\text{def}}{=} \frac{\tilde{e}_j^{\Delta,-} - \tilde{e}_{j-1}^{\Delta,+}}{h_{I_j^\Delta}}. \tag{4.3.29b}$$

Using the triangle inequality

$$\big\|\delta\tilde{e}_{S^p(I_j^\Delta)}\big\|_{\mathcal{U}(I_j^\Delta)} \le \big\|\delta\tilde{e}_{S^1(I_j^\Delta)}\big\|_{\mathcal{U}(I_j^\Delta)} + \big\|\delta\tilde{e}_{S_0^p(I_j^\Delta)}\big\|_{\mathcal{U}(I_j^\Delta)} \tag{4.3.29c}$$

and (4.3.29a), we obtain

$$\big\|\delta\tilde{e}_{S^p(I_j^\Delta)}\big\|_{\mathcal{U}(I_j^\Delta)} \le \mathcal{C}\sqrt{h_{I_j^\Delta}}\max\big(|\tilde{\rho}_1^{I_j^\Delta}|, |\tilde{\rho}_2^{I_j^\Delta}|\big). \tag{4.3.29d}$$

In the case that $c \equiv 0$, we have

$$\big\|\delta\tilde{e}_{S^p(I_j^\Delta)}\big\|_{\mathcal{U}(I_j^\Delta)} \le \mathcal{C}\sqrt{h_{I_j^\Delta}}\,|\tilde{\rho}_2^{I_j^\Delta}|. \tag{4.3.30}$$

Hence, in order to prove (4.3.26b), it is sufficient to show that

$$\max\big(|\tilde{\rho}_1^{I_j^\Delta}|, |\tilde{\rho}_2^{I_j^\Delta}|\big) \le \mathcal{C}\sqrt{h_{I_j^\Delta}}\,\big\|\tilde{e}^{loc}_{S^p(I_j^\Delta)}\big\|_{\mathcal{U}(I_j^\Delta)} \tag{4.3.31a}$$

and in the case that $c \equiv 0$, only that

$$|\tilde{\rho}_2^{I_j^\Delta}| \le \mathcal{C}\sqrt{h_{I_j^\Delta}}\,\|\tilde{e}^{loc}_{S^p(I_j^\Delta)}\|_{\mathcal{U}(I_j^\Delta)}. \tag{4.3.31b}$$

Let us first address the case $c|_{I_j^\Delta} \not\equiv 0$. Let $G_1^{I_j^\Delta}$ be the solution of the following problem:

Find $G_1^{I_j^\Delta} \in \mathcal{U}(I_j^\Delta)$ such that

$$\mathcal{B}_{I_j^\Delta}(G_1^{I_j^\Delta}, v) = \mathcal{L}_1^{I_j^\Delta}(v) \stackrel{\text{def}}{=} \tfrac{1}{2}\big(v(x_{j-1}^\Delta) + v(x_j^\Delta)\big) \quad \forall v \in \mathcal{U}(I_j^\Delta). \tag{4.3.32}$$

Now let $\bar{c} > 0$, and

$$\bar{G}_1^{I_j^\Delta} \stackrel{\text{def}}{=} G_1^{I_j^\Delta} - \frac{1}{\bar{c}\,h_{I_j^\Delta}}. \tag{4.3.33a}$$

From (4.3.32) it follows that

$$\mathcal{B}_{I_j^\Delta}(\bar{G}_1^{I_j^\Delta}, v) = \frac{1}{2}\big(v(x_{j-1}^\Delta) + v(x_j^\Delta)\big) - \frac{1}{\bar{c}\,h_{I_j^\Delta}} \int_{I_j^\Delta} cv \quad \forall v \in \mathcal{U}(I_j^\Delta). \tag{4.3.33b}$$

Employing $v \equiv \bar{G}_1^{I_j^\Delta}$ in (4.3.33b) we get

$$2\,\|\bar{G}_1^{I_j^\Delta}\|^2_{\mathcal{U}(I_j^\Delta)} = \bar{G}_1^{I_j^\Delta}(x_{j-1}^\Delta) + \bar{G}_1^{I_j^\Delta}(x_j^\Delta) - \frac{2}{\bar{c}\,h_{I_j^\Delta}} \int_{I_j^\Delta} c\,\bar{G}_1^{I_j^\Delta} \tag{4.3.33c}$$

which can also be written in the form

$$\|\bar{G}_1^{I_j^\Delta}\|^2_{\mathcal{U}(I_j^\Delta)} = \frac{1}{2\bar{c}\,h_{I_j^\Delta}} \sum_{k=0}^{1} \int_{I_j^\Delta} c \left(\frac{\bar{c}\,h_{I_j^\Delta}}{\int_{I_j^\Delta} c} \bar{G}_1^{I_j^\Delta}(x_{j-1+k}^\Delta) - \bar{G}_1^{I_j^\Delta} \right). \tag{4.3.33d}$$

Letting

$$\bar{c} = \frac{1}{h_{I_j^\Delta}} \int_{I_j^\Delta} c \tag{4.3.33e}$$

in (4.3.33d), we get

$$\|\bar{G}_1^{I_j^\Delta}\|^2_{\mathcal{U}(I_j^\Delta)} \leq \max_{k=j-1,j} \|\bar{G}_1^{I_j^\Delta} - \bar{G}_1^{I_j^\Delta}(x_k^\Delta)\|_{L^\infty(I_j^\Delta)} \tag{4.3.33f}$$

and employing that

$$\max_{k=j-1,j} \|\bar{G}_1^{I_j^\Delta} - \bar{G}_1^{I_j^\Delta}(x_k^\Delta)\|_{L^\infty(I_j^\Delta)} \leq \mathcal{C}\sqrt{h_{I_j^\Delta}}\,\|\bar{G}_1^{I_j^\Delta}\|_{\mathcal{U}(I_j^\Delta)} \tag{4.3.33g}$$

we obtain

$$\|\bar{G}_1^{I_j^\Delta}\|_{\mathcal{U}(I_j^\Delta)} \leq \mathcal{C}\sqrt{h_{I_j^\Delta}}. \tag{4.3.33h}$$

Noting that $\tilde{e}^{loc}_{S^p(I_j^\Delta)}$ satisfies the orthogonality condition

$$\mathcal{B}_{I_j^\Delta}\big(\tilde{e}^{loc}_{S^p(I_j^\Delta)}, 1\big) = 0 \tag{4.3.34a}$$

and using the symmetry of $\mathcal{B}$, we get

$$|\tilde{\rho}_1^{I_j^\Delta}| = \big|\mathcal{B}_{I_j^\Delta}(G_1^{I_j^\Delta}, \tilde{e}^{loc}_{S^p(I_j^\Delta)})\big| = \Big|\mathcal{B}_{I_j^\Delta}\big(\underbrace{G_1^{I_j^\Delta} - (\bar{c}h_{I_j^\Delta})^{-1}}_{\bar{G}_1^{I_j^\Delta}}, \tilde{e}^{loc}_{S^p(I_j^\Delta)}\big)\Big|. \tag{4.3.34b}$$

Employing the Cauchy–Schwarz inequality and the estimate (4.3.33h), we obtain

$$|\tilde{\rho}_1^{I_j^\Delta}| \leq \|\bar{G}_1^{I_j^\Delta}\|_{\mathcal{U}(I_j^\Delta)}\|\tilde{e}^{loc}_{S^p(I_j^\Delta)}\|_{\mathcal{U}(I_j^\Delta)} \leq \mathcal{C}\sqrt{h_{I_j^\Delta}}\|\tilde{e}^{loc}_{S^p(I_j^\Delta)}\|_{\mathcal{U}(I_j^\Delta)}. \tag{4.3.34c}$$

This proves the desired estimate for $|\tilde{\rho}_1^{I_j^\Delta}|$, with $\mathcal{C}$ independent of c. This completes the proof of the estimate (4.3.31a) for $|\tilde{\rho}_1^{I_j^\Delta}|$.

We will now estimate $|\tilde{\rho}_2^{I_j^\Delta}|$. Let $G_2^{I_j^\Delta}$ be the solution of the following problem: Find $G_2^{I_j^\Delta} \in \mathcal{U}(I_j^\Delta)$ such that

$$\mathcal{B}_{I_j^\Delta}(G_2^{I_j^\Delta}, v) = \mathcal{L}_2^{I_j^\Delta}(v) \stackrel{\text{def}}{=} \frac{v(x_j^\Delta) - v(x_{j-1}^\Delta)}{h_{I_j^\Delta}} \quad \forall v \in \mathcal{U}(I_j^\Delta). \tag{4.3.35}$$

Note that the consistency condition $\mathcal{L}_2^{I_j^\Delta}(1) = 0$, is satisfied and hence $G_2^{I_j^\Delta}$ exists also for $c \equiv 0$, in which case $G_2^{I_j^\Delta}$ may be fixed by employing the additional condition $G_2^{I_j^\Delta}(x_j^\Delta) = 0$. Employing the definition of $G_2^{I_j^\Delta}$, we get

$$\mathcal{B}_{I_j^\Delta}\left(G_2^{I_j^\Delta}, \tilde{e}^{loc}_{S^p(I_j^\Delta)}\right) = \frac{\tilde{e}^{loc}_{S^p(I_j^\Delta)}(x_j^\Delta) - \tilde{e}^{loc}_{S^p(I_j^\Delta)}(x_{j-1}^\Delta)}{h_{I_j^\Delta}} = \tilde{\rho}_2^{I_j^\Delta} \tag{4.3.36a}$$

and using the orthogonality

$$\mathcal{B}_{I_j^\Delta}\left(\tilde{e}^{loc}_{S^p(I_j^\Delta)}, v\right) = 0 \quad \forall v \in S^1(I_j^\Delta) \tag{4.3.36b}$$

with $v = \mathcal{I}^1_{I_j^\Delta} G_2^{I_j^\Delta}$, and the symmetry of $\mathcal{B}$, we get

$$\begin{aligned}|\tilde{\rho}_2^{I_j^\Delta}| &= \left|\mathcal{B}_{I_j^\Delta}\left(G_2^{I_j^\Delta} - \mathcal{I}^1_{I_j^\Delta} G_2^{I_j^\Delta}, \tilde{e}^{loc}_{S^p(I_j^\Delta)}\right)\right| \\ &\leq \|G_2^{I_j^\Delta} - \mathcal{I}^1_{I_j^\Delta} G_2^{I_j^\Delta}\|_{\mathcal{U}(I_j^\Delta)}\|\tilde{e}^{loc}_{S^p(I_j^\Delta)}\|_{\mathcal{U}(I_j^\Delta)}\end{aligned} \tag{4.3.36c}$$

where we used the Cauchy–Schwarz inequality. From the assumptions that AE has bounded derivative, and c is bounded, it follows that

$$\|G_2^{I_j^\Delta} - \mathcal{I}^1_{I_j^\Delta} G_2^{I_j^\Delta}\|_{\mathcal{U}(I_j^\Delta)} \leq \mathcal{C}(h_{I_j^\Delta})^{3/2}\left\|\frac{d^2}{dx^2}(G_2^{I_j^\Delta})\right\|_{L^\infty(I)} \leq \mathcal{C}\sqrt{h_{I_j^\Delta}} \tag{4.3.36d}$$

and hence

$$|\tilde{\rho}_2^{I_j^\Delta}| \leq \|G_2^{I_j^\Delta} - \mathcal{I}^1_{I_j^\Delta} G_2^{I_j^\Delta}\|_{\mathcal{U}(I_j^\Delta)}\|\tilde{e}^{loc}_{S^p(I_j^\Delta)}\|_{\mathcal{U}(I_j^\Delta)} \leq \mathcal{C}\sqrt{h_{I_j^\Delta}}\|\tilde{e}^{loc}_{S^p(I_j^\Delta)}\|_{\mathcal{U}(I_j^\Delta)}. \tag{4.3.36e}$$

This completes the proof of the estimates (4.3.31a) and (4.3.31b) for $|\tilde{\rho}_2^{I_j^\Delta}|$, and hence (4.3.26b) has been proven. □

The pollution error $e^{glob}_{S^p(I_j^\Delta)}$ expresses the global effect of the finite element method and will be analyzed in detail in the next section. A question that arises is how different are the two definitions of the local error, $e^{loc}_{S^p_\Delta}$ and $\tilde{e}^{loc}_{S^p_\Delta}$. As it will be seen in the following example, the difference between $e^{loc}_{S^p_\Delta}$ and $\tilde{e}^{loc}_{S^p_\Delta}$ is often practically negligible.

Example 4.3.2 Consider the regular model problem with input data

$$AE \equiv 1, \qquad c \equiv 50, \qquad f \equiv 0, \qquad \alpha_2 = 10, \qquad a_2 = 1. \tag{4.3.38a}$$

Then we have

$$u_{EX}(x) = \frac{e^{2\sqrt{c}}\,(e^{\sqrt{c}x} - e^{-\sqrt{c}x})}{\sqrt{c}\,e^{2\sqrt{c}} + \alpha_2\, e^{2\sqrt{c}} - \alpha_2 + \sqrt{c}}. \tag{4.3.38b}$$

Figure 4.3.2a (respectively Fig. 4.3.2b) gives the graphs of the errors $e^{loc}_{S^p_\Delta} = u_{EX} - \mathcal{A}^p_\Delta u_{EX}$, and $\tilde{e}^{loc}_{S^p_\Delta} = u_{EX} - \mathcal{B}^p_\Delta u_{EX}$ (respectively the derivatives of these errors) for a uniform grid of $N = 8$ linear elements ($p = 1$). It can be seen that the derivatives of the local finite element solutions $(d/dx)(\mathcal{A}^1_\Delta u_{EX})$ and $(d/dx)(\mathcal{B}^1_\Delta u_{EX})$ are practically identical even for the relatively coarse uniform mesh of eight elements.

Above, we analyzed the error in the energy norm. In practice, however, one is often interested in the pointwise error in the derivative (e.g., the heat flux, the stress, etc.). We will now show that the pointwise error in the derivative can be estimated by employing the energy norm estimates which were derived above.

Lemma 4.3.1 *Consider the setting of the regular model problem, and assume that there exists a* $\hat{u}_{S^p(I_j^\Delta)} \in S^p(I_j^\Delta)$ *such that*

$$\|u_{EX} - \hat{u}_{S^p(I_j^\Delta)}\|_{\mathcal{U}(I_j^\Delta)} \le \epsilon_1 \tag{4.3.39a}$$

and

$$\left\| \frac{d}{dx}(u_{EX} - \hat{u}_{S^p(I_j^\Delta)}) \right\|_{L^\infty(I_j^\Delta)} \le \frac{\epsilon_2}{\sqrt{h_{I_j^\Delta}}}. \tag{4.3.39b}$$

Then, for any $u_{S^p(I_j^\Delta)} \in S^p(I_j^\Delta)$, *we have*

$$\left\| \frac{d}{dx}(u_{EX} - u_{S^p(I_j^\Delta)}) \right\|_{L^\infty(I_j^\Delta)} \le \frac{\mathcal{C}}{\sqrt{h_{I_j^\Delta}}} \left(\|u_{EX} - u_{S^p(I_j^\Delta)}\|_{\mathcal{U}(I_j^\Delta)} + \epsilon_1 + \epsilon_2 \right) \tag{4.3.39c}$$

with $\mathcal{C}$ *depending only on* β_1 *and* p.

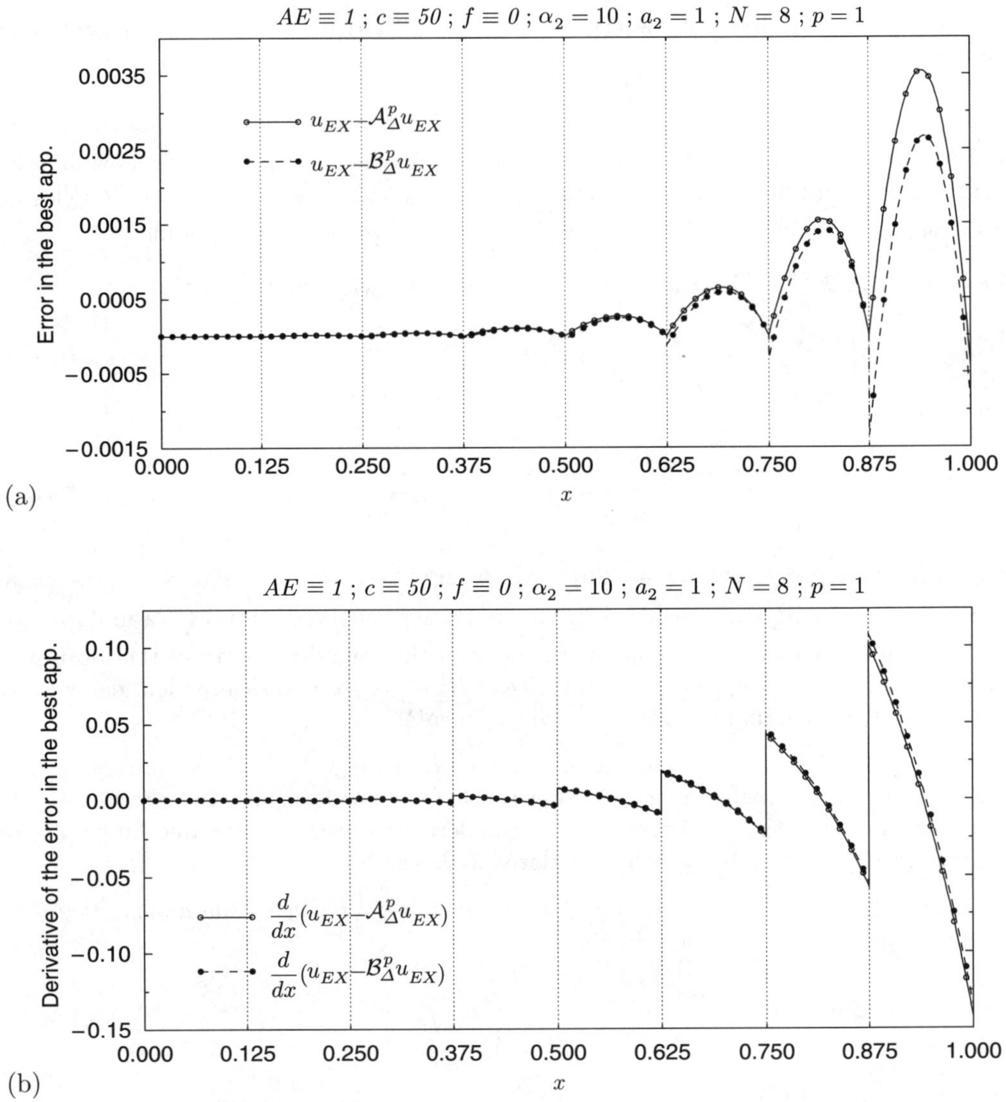

FIG. 4.3.2. An example which illustrates the difference between the two types of local finite element solution. Input data: $AE \equiv 1$, $c \equiv 50$, $f \equiv 0$, $\alpha_2 = 10$, $a_2 = 1$. Uniform mesh of $N = 8$ linear elements ($p = 1$). (a) Graphs of $(u_{EX} - \mathcal{A}^1_\Delta u_{EX})$, and $(u_{EX} - \mathcal{B}^1_\Delta u_{EX})$. (b) Graphs of $(d/dx)(u_{EX} - \mathcal{A}^1_\Delta u_{EX})$, and $(d/dx)(u_{EX} - \mathcal{B}^1_\Delta u_{EX})$. Note that $(d/dx)(u_{EX} - \mathcal{A}^1_\Delta u_{EX})$ and $(d/dx)(u_{EX} - \mathcal{B}^1_\Delta u_{EX})$ are practically identical

Proof Employing the triangle inequality we get

$$\left\| \frac{d}{dx}(u_{EX} - u_{S^p(I_j^\Delta)}) \right\|_{L^\infty(I_j^\Delta)} \leq \left\| \frac{d}{dx}(u_{EX} - \hat{u}_{S^p(I_j^\Delta)}) \right\|_{L^\infty(I_j^\Delta)} + \left\| \frac{d}{dx}(\delta u_{S^p(I_j^\Delta)}) \right\|_{L^\infty(I_j^\Delta)} \tag{4.3.40a}$$

where we let

$$\delta u_{S^p(I_j^\Delta)} \stackrel{\text{def}}{=} \hat{u}_{S^p(I_j^\Delta)} - u_{S^p(I_j^\Delta)}. \tag{4.3.40b}$$

Further, noting that $\delta u_{S^p(I_j^\Delta)} \in S^p(I_j^\Delta)$, we have

$$\left\| \frac{d}{dx}(\delta u_{S^p(I_j^\Delta)}) \right\|_{L^\infty(I_j^\Delta)} \leq \frac{\mathscr{C}(p)}{\sqrt{h_{I_j^\Delta}}} \|\delta u_{S^p(I_j^\Delta)}\|_{\mathscr{U}(I_j^\Delta)} \tag{4.3.40c}$$

and employing the triangle inequality we get

$$\left\| \frac{d}{dx}(\delta u_{S^p(I_j^\Delta)}) \right\|_{L^\infty(I_j^\Delta)} \leq \frac{\mathscr{C}(p)}{\sqrt{h_{I_j^\Delta}}} \Big(\|u_{EX} - u_{S^p(I_j^\Delta)}\|_{\mathscr{U}(I_j^\Delta)} + \|u_{EX} - \hat{u}_{S^p(I_j^\Delta)}\|_{\mathscr{U}(I_j^\Delta)} \Big) \tag{4.3.40d}$$

and (4.3.39c) follows. □

We will now use Lemma 4.3.1 with $u_{S^p(I_j^\Delta)} \equiv \mathcal{A}^p_{I_j^\Delta} u_{EX}$, and $\hat{u}_{S^p(I_j^\Delta)} \equiv \mathcal{I}^p_{I_j^\Delta} u_{EX}$, to prove an L^∞ norm estimate for the derivative of the local error.

Corollary 4.3.2 *Consider the setting of the regular model problem, and assume, in addition, that*

$$\left\| \frac{d^r}{dx^r}(u_{EX}) \right\|_{L^\infty(I_j^\Delta)} < \infty. \tag{4.3.41a}$$

Then we have

$$\left\| \frac{d}{dx}(u_{EX} - \mathcal{A}^p_{I_j^\Delta} u_{EX}) \right\|_{L^\infty(I_j^\Delta)} \leq \mathscr{C}\,(h_{I_j^\Delta})^\mu \left\| \frac{d^r}{dx^r}(u_{EX}) \right\|_{L^\infty(I_j^\Delta)} \tag{4.3.41b}$$

with $\mu = \min(p, r-1)$.

Proof In Corollary 4.3.1 we proved that

$$\|u_{EX} - \mathcal{A}^p_{I_j^\Delta} u_{EX}\|_{\mathscr{U}(I_j^\Delta)} \leq \mathscr{C} h_{I_j^\Delta}^{\mu+1/2} \left\| \frac{d^r}{dx^r}(u_{EX}) \right\|_{L^\infty(I_j^\Delta)}. \tag{4.3.42a}$$

From the standard estimates for the Lagrangian interpolant, we have

$$\|u_{EX} - \mathcal{I}^p_{I_j^\Delta} u_{EX}\|_{\mathscr{U}(I_j^\Delta)} \leq \mathscr{C} h_{I_j^\Delta}^{\mu+1/2} \left\| \frac{d^r}{dx^r}(u_{EX}) \right\|_{L^\infty(I_j^\Delta)} \approx \epsilon_1 \tag{4.3.42b}$$

and

$$\left\| \frac{d}{dx}(u_{EX} - \mathcal{I}^p_{I_j^\Delta} u_{EX}) \right\|_{L^\infty(I_j^\Delta)} \leq \mathscr{C}\,(h_{I_j^\Delta})^\mu \left\| \frac{d^r}{dx^r}(u_{EX}) \right\|_{L^\infty(I_j^\Delta)} \approx \frac{\epsilon_2}{\sqrt{h_{I_j^\Delta}}} \tag{4.3.42c}$$

where

$$\epsilon_1 \approx \epsilon_2 \approx \mathscr{C} h_{I_j^\Delta}^{\mu+1/2} \left\| \frac{d^r}{dx^r}(u_{EX}) \right\|_{L^\infty(I_j^\Delta)}. \tag{4.3.42d}$$

Applying Lemma 4.3.1 with $u_{S^p(I_j^\Delta)} \equiv \mathcal{A}^p_{I_j^\Delta} u_{EX}$, and $\hat{u}_{S^p(I_j^\Delta)} \equiv \mathcal{I}^p_{I_j^\Delta} u_{EX}$, (4.3.41b) follows. □

Let us also give an error estimate for the L^∞ norm of the derivative of $e^{glob}_{S^p(I_j^\Delta)}$, the global part of the error.

Corollary 4.3.3 *Consider the regular model problem with the additional assumption (4.3.41a). Then*

$$\left\| \frac{d}{dx}(e^{glob}_{S^p(I_j^\Delta)}) \right\|_{L^\infty(I_j^\Delta)} \leq \mathscr{C} \sqrt{(e_j^\Delta + e_{j-1}^\Delta)^2 + \left(\frac{e_j^\Delta - e_{j-1}^\Delta}{h_{I_j^\Delta}} \right)^2}. \tag{4.3.43}$$

Proof We have $e^{glob}_{S^p(I_j^\Delta)} = e^{glob}_{S^1(I_j^\Delta)} + e^{glob}_{S_0^p(I_j^\Delta)}$, and

$$\left\| \frac{d}{dx}(e^{glob}_{S^1(I_j^\Delta)}) \right\|_{L^\infty(I_j^\Delta)} = \frac{e_j^\Delta - e_{j-1}^\Delta}{h_{I_j^\Delta}} \tag{4.3.44a}$$

and because $e^{glob}_{S_0^p(I_j^\Delta)} \in S_0^p(I_j^\Delta)$, we have

$$\left\| \frac{d}{dx}(e^{glob}_{S_0^p(I_j^\Delta)}) \right\|_{L^\infty(I_j^\Delta)} \leq \frac{\mathscr{C}}{\sqrt{h_{I_j^\Delta}}} \| e^{glob}_{S_0^p(I_j^\Delta)} \|_{\mathfrak{U}(I_j^\Delta)} \tag{4.3.44b}$$

and employing that

$$\begin{aligned} \| e^{glob}_{S_0^p(I_j^\Delta)} \|_{\mathfrak{U}(I_j^\Delta)} &\leq \| e^{glob}_{S^1(I_j^\Delta)} \|_{\mathfrak{U}(I_j^\Delta)} \\ &\leq \mathscr{C} \sqrt{h_{I_j^\Delta}} \sqrt{(e_j^\Delta + e_{j-1}^\Delta)^2 + \left(\frac{e_j^\Delta - e_{j-1}^\Delta}{h_{I_j^\Delta}} \right)^2} \end{aligned} \tag{4.3.44c}$$

we get

$$\left\| \frac{d}{dx}(e^{glob}_{S_0^p(I_j^\Delta)}) \right\|_{L^\infty(I_j^\Delta)} \leq \mathscr{C} \sqrt{(e_j^\Delta + e_{j-1}^\Delta)^2 + \left(\frac{e_j^\Delta - e_{j-1}^\Delta}{h_{I_j^\Delta}} \right)^2}. \tag{4.3.44d}$$

Using now the triangle inequality

$$\left\| \frac{d}{dx}(e^{glob}_{S^p(I_j^\Delta)}) \right\|_{L^\infty(I_j^\Delta)} \leq \left\| \frac{d}{dx}(e^{glob}_{S^1(I_j^\Delta)}) \right\|_{L^\infty(I_j^\Delta)} + \left\| \frac{d}{dx}(e^{glob}_{S_0^p(I_j^\Delta)}) \right\|_{L^\infty(I_j^\Delta)} \tag{4.3.44e}$$

(4.3.43) follows. □

We will also show that the difference $(d/dx)(\mathcal{B}^p_{I_j^\Delta} u_{EX} - \mathcal{A}^p_{I_j^\Delta} u_{EX})$, measured in the L^∞ norm, is negligible.

Corollary 4.3.4 *Consider the setting of the regular model problem, and assume, in addition, that*

$$\left\| \frac{d}{dx}(AE) \right\|_{L^\infty(I_j^\Delta)} \le K < \infty \tag{4.3.45a}$$

we have

$$\left\| \frac{d}{dx}(\mathcal{B}^p_{I_j^\Delta} u_{EX} - \mathcal{A}^p_{I_j^\Delta} u_{EX}) \right\|_{L^\infty(I_j^\Delta)} \le \mathscr{C}\,(h_{I_j^\Delta})^{\mu+1} \left\| \frac{d^r}{dx^r}(u_{EX}) \right\|_{L^\infty(I_j^\Delta)}. \tag{4.3.45b}$$

Proof Noting that $(\mathcal{B}^p_{I_j^\Delta} u_{EX} - \mathcal{A}^p_{I_j^\Delta} u_{EX}) \in S^p(I_j^\Delta)$ we have

$$\begin{aligned} &\left\| \frac{d}{dx}(\mathcal{B}^p_{I_j^\Delta} u_{EX} - \mathcal{A}^p_{I_j^\Delta} u_{EX}) \right\|_{L^\infty(I_j^\Delta)} \\ &\quad \le \frac{\mathscr{C}}{\sqrt{h_{I_j^\Delta}}} \left\| \frac{d}{dx}(\mathcal{B}^p_{I_j^\Delta} u_{EX} - \mathcal{A}^p_{I_j^\Delta} u_{EX}) \right\|_{L^2(I_j^\Delta)} \end{aligned} \tag{4.3.46a}$$

and hence

$$\left\| \frac{d}{dx}(\mathcal{B}^p_{I_j^\Delta} u_{EX} - \mathcal{A}^p_{I_j^\Delta} u_{EX}) \right\|_{L^\infty(I_j^\Delta)} \le \frac{\mathscr{C}}{\sqrt{h_{I_j^\Delta}}} \|\mathcal{B}^p_{I_j^\Delta} u_{EX} - \mathcal{A}^p_{I_j^\Delta} u_{EX}\|_{\mathcal{U}(I_j^\Delta)} \tag{4.3.46b}$$

and recalling (4.3.26b), namely,

$$\begin{aligned} \|\mathcal{B}^p_{I_j^\Delta} u_{EX} - \mathcal{A}^p_{I_j^\Delta} u_{EX}\|_{\mathcal{U}(I_j^\Delta)} &\le \mathscr{C} h_{I_j^\Delta} \|\tilde{e}^{loc}_{S^p(I_j^\Delta)}\|_{\mathcal{U}(I_j^\Delta)} \\ &\le \mathscr{C}(h_{I_j^\Delta})^{\mu+3/2} \left\| \frac{d^r}{dx^r}(u_{EX}) \right\|_{L^\infty(I_j^\Delta)} \end{aligned} \tag{4.3.47}$$

(4.3.45b) follows. □

Remark 4.3.4 Corollaries 4.3.2–4.3.4 show that, if the pollution error is negligible, the error in the finite element solution is essentially the same as the error in the best approximation of the exact solution from the finite element space.

Remark 4.3.5 If the pollution is not negligible, the finite element solution is rather different than the local best approximation, and the accuracy of the finite element solution at a point can be established only if we have separate estimates of the local error and the pollution error.

Let us now give an example where the pollution error is not negligible.

Example 4.3.3 Consider the singular model problem

$$-\frac{d}{dx}\left(x^\vartheta \frac{du}{dx}\right) = 0, \qquad u(0) = 0, \qquad \left(x^\vartheta \frac{du}{dx}\right)(1) = -\alpha u(1) + 1 \tag{4.3.48}$$

with $0 < \vartheta < 1$. Let u^0_{EX} be the exact solution, and $u^0_{S^1_\Delta}$ the finite element solution, in the case $\alpha = 0$. Then we have

$$u_{EX} = Q_{EX}(\alpha) u^0_{EX}, \qquad Q_{EX}(\alpha) = \frac{1}{1 + \alpha u^0_{EX}(1)} \tag{4.3.49a}$$

and

$$u_{S^1_\Delta} = Q_h(\alpha) u^0_{S^1_\Delta}, \qquad Q_h(\alpha) = \frac{1}{1 + \alpha u^0_{S^1_\Delta}(1)}. \tag{4.3.49b}$$

Further

$$\mathscr{C}(\vartheta)\, h^{1-\vartheta} \leq \min_{\chi \in \mathscr{P}^p(I^\Delta_1)} \|u^0_{EX} - \chi\|^2_{\mathscr{U}^\vartheta(I^\Delta_1)} \leq \|e^0_{S^1_\Delta}\|^2_{\mathscr{U}^\vartheta(I^\Delta_1)} \leq \|e^0_{S^1_\Delta}\|^2_{\mathscr{U}^\vartheta} \tag{4.3.50a}$$

and recalling that $\mathscr{B}(u, v) = \int_0^1 x^\vartheta (du/dx)\,(dv/dx)$, we have

$$\|e^0_{S^1_\Delta}\|^2_{\mathscr{U}} = \mathscr{B}(u^0_{EX} - u^0_{S^1_\Delta}, e^0_{S^1_\Delta}) = \mathscr{B}(u^0_{EX}, e^0_{S^1_\Delta}) = e^0_{S^1_\Delta}(1). \tag{4.3.50b}$$

Hence

$$\mathscr{C}(\vartheta) h^{1-\vartheta} \leq |e^0_{S^1_\Delta}(1)|. \tag{4.3.50c}$$

We then have

$$\mathscr{C}(\vartheta)\, h^{1-\vartheta}\, \alpha \leq |Q_{EX}(\alpha) - Q_h(\alpha)| = \frac{\alpha |u^0_{S^1_\Delta}(1) - u^0_{EX}(1)|}{|1 + \alpha u^0_{EX}(1)|\,|1 + \alpha\, u^0_{S^1_\Delta}(1)|} \tag{4.3.51}$$

from which we get

$$\mathscr{C}(\vartheta)\, h^{1-\vartheta}\, \alpha \leq |e^\Delta_n + e^\Delta_{n-1}| \tag{4.3.52a}$$

and

$$\mathscr{C}(\vartheta)\, h^{1-\vartheta}\, \alpha \leq \left| \frac{e^\Delta_n - e^\Delta_{n-1}}{h} \right|. \tag{4.3.52b}$$

Hence

$$\mathscr{C}(\vartheta)\, \sqrt{h}\, h^{1-\vartheta}\, \alpha \leq \|e^{glob}_{S^1_{I^\Delta_n}}\|_{\mathscr{U}^\vartheta(I^\Delta_n)}. \tag{4.3.52c}$$

On the other hand we have

$$\|e^{loc}_{S^p_{I^\Delta_n}}\|_{\mathscr{U}^\vartheta(I^\Delta_n)} \leq \mathscr{C}(\vartheta)\, \sqrt{h}\, h. \tag{4.3.53}$$

From (4.3.52c) and (4.3.53) we see that the pollution cannot be neglected for any $\vartheta > 0$, and increases with ϑ. The above arguments hold for $\alpha \neq 0$; in the case that $\alpha = 0$ there is no pollution. Although we addressed only the pollution in I^Δ_n, from (4.3.51) it follows that the pollution is present in any I_j^Δ, $j = 2, \ldots, n$. A more detailed construction of the above is given in Example 4.4.3.

In this section, we introduced the local finite element solutions $\mathcal{A}_\Delta^p u_{EX}$ and $\mathcal{B}_\Delta^p u_{EX}$, as best approximations of the exact solution with respect to the energy norm in each element I_j^Δ. We used these local solutions to split the error $e_{S_\Delta^p}$, into a local and a global part

$$e_{S_\Delta^p}^{loc} = u_{EX} - \mathcal{A}_\Delta^p u_{EX} \qquad e_{S_\Delta^p}^{glob} = \mathcal{A}_\Delta^p u_{EX} - u_{S_\Delta^p} \tag{4.3.54a}$$

or

$$\tilde{e}_{S_\Delta^p}^{loc} = u_{EX} - \mathcal{B}_\Delta^p u_{EX} \qquad \tilde{e}_{S_\Delta^p}^{glob} = \mathcal{B}_\Delta^p u_{EX} - u_{S_\Delta^p}. \tag{4.3.54b}$$

Further, we employed the splitting of $e_{S_\Delta^p}^{glob}$

$$e_{S_\Delta^p}^{glob} = e_{S_\Delta^1}^{glob} + e_{S_{\Delta,0}^p}^{glob} \tag{4.3.55a}$$

and proved that

$$\|e_{S_{\Delta,0}^p}^{glob}\|_{\mathcal{U}} \le \|e_{S_\Delta^1}^{glob}\|_{\mathcal{U}} \le \mathscr{C} \sum_{j=1}^{N(\Delta)} \sqrt{h_{I_j^\Delta}} \sqrt{(e_j^\Delta + e_{j-1}^\Delta)^2 + \left(\frac{e_j^\Delta - e_{j-1}^\Delta}{h_{I_j^\Delta}}\right)^2}. \tag{4.3.55b}$$

Further, in the case that we have

$$\left\| \frac{d}{dx}(AE) \right\|_{L^\infty(I)} \le K \tag{4.3.56a}$$

we also have

$$\|e_{S_{\Delta,0}^p}^{glob}\|_{\mathcal{U}} \le \mathscr{C} h^\Delta \|e_{S_\Delta^1}^{glob}\|_{\mathcal{U}} \tag{4.3.56b}$$

and

$$\|\mathcal{A}_\Delta^p u_{EX} - \mathcal{B}_\Delta^p u_{EX}\|_{\mathcal{U}} \le \mathscr{C} \sqrt{\sum_{j=1}^{N(\Delta)} h_{I_j^\Delta}^2 \|\tilde{e}_{S^p(I_j^\Delta)}^{loc}\|_{\mathcal{U}(I_j^\Delta)}^2} \tag{4.3.56c}$$

where, in both cases, $\mathscr{C}$ depends only on K and γ. Further, under the assumption

$$\left\| \frac{d^r}{dx^r}(u_{EX}) \right\|_{L^\infty(I_j^\Delta)} < \infty \tag{4.3.57a}$$

we proved the L^∞ estimates

$$\left\| \frac{d}{dx}(u_{EX} - \mathcal{A}_{I_j^\Delta}^p u_{EX}) \right\|_{L^\infty(I_j^\Delta)} \le \mathscr{C}(h_{I_j^\Delta})^\mu \left\| \frac{d^r}{dx^r}(u_{EX}) \right\|_{L^\infty(I_j^\Delta)} \tag{4.3.57b}$$

and

$$\left\|\frac{d}{dx}(e^{glob}_{S^p(I_j^\Delta)})\right\|_{L^\infty(I_j^\Delta)} \leq \mathscr{C}\sqrt{(e_j^\Delta + e_{j-1}^\Delta)^2 + \left(\frac{e_j^\Delta - e_{j-1}^\Delta}{h_{I_j^\Delta}}\right)^2} \tag{4.3.57c}$$

and under the additional assumption

$$\left\|\frac{d}{dx}(AE)\right\|_{L^\infty(I_j^\Delta)} \leq K < \infty \tag{4.3.57d}$$

we showed that

$$\left\|\frac{d}{dx}(\mathcal{B}^p_{I_j^\Delta}u_{EX} - \mathcal{A}^p_{I_j^\Delta}u_{EX})\right\|_{L^\infty(I_j^\Delta)} \leq \mathscr{C}\,(h_{I_j^\Delta})^{\mu+1}\left\|\frac{d^r}{dx^r}(u_{EX})\right\|_{L^\infty(I_j^\Delta)}. \tag{4.3.57e}$$

Finally, we showed that in the case of the singular model problem which is a one-dimensional analogue of two-dimensional problems in domains with corner points, the pollution cannot be neglected, in general.

Problems

P4.3.1. Consider the regular model problem with input data $AE \equiv 1+ax$, $c \equiv b$, $\alpha_2 = 0$, and exact solutions $u_{EX} \equiv \sin x$, and $u_{EX}(x) \equiv \cos x$. Let $I_j^\Delta = (0, h)$ and determine the constant $\mathscr{C}(p)$ in the estimate

$$\mathscr{C}(p)\|u_{EX} - \mathcal{I}^\Delta_{I_j^\Delta}u_{EX}\|_{\mathcal{U}(I_j^\Delta)} \leq \|u_{EX} - \mathcal{A}^p_{I_j^\Delta}u_{EX}\|_{\mathcal{U}(I_j^\Delta)} \tag{4.3.58}$$

for $p = 2$, 3 and for various values of h, a, b.

P4.3.2. Consider the singular model problem with $AE \equiv x^\vartheta$, $0 < \vartheta < 1$, $c \equiv 0$, and the exact solutions considered in P4.3.1. Determine the constant $\mathscr{C}(p)$, defined in P4.3.1, for various values of ϑ and for $p = 2$, 3.

P4.3.3. Determine the constants $\mathscr{C}$ in the inequalities in Corollary 4.3.1 for the input data and the exact solutions given in P4.3.1 and P4.3.2.

P4.3.4. Determine the constant $\mathscr{C}$ in Theorem 4.3.3 for the input data and the exact solutions given in P4.3.1 and P4.3.2.

P4.3.5. Prove (4.3.52a) and (4.3.52b).

P4.3.6. Let $u_{EX} = x^{1-\vartheta}$, and determine $\|u_{EX} - \mathcal{O}^p_{I_1^\Delta}u_{EX}\|_{S_1(I_1^\Delta)}$ and $\|(d/dx)(u_{EX} - \mathcal{O}^p_{I_1^\Delta}u_{EX})\|_{S(I_1^\Delta)}$ for $\mathcal{O} \equiv \mathcal{A}$, $\mathcal{I}$, and $\mathcal{B}$, and for $S_1 = L^2$, $\mathcal{U}^\vartheta$, and $S = L^2$, and $\|\mathcal{B}^p_{I_1^\Delta}u_{EX} - \mathcal{A}^p_{I_1^\Delta}u_{EX}\|_{\mathcal{U}^\vartheta(I_1^\Delta)}$ and $\|(d/dx)(\mathcal{B}^p_{I_1^\Delta}u_{EX} - \mathcal{A}^p_{I_1^\Delta}u_{EX})\|_{\mathcal{U}^\vartheta(I_1^\Delta)}$.

P4.3.7. Let $\bar{x} \in I_j^\Delta$, and

$$u_{EX}(x) = (x - \bar{x})^+ = \begin{cases} x - \bar{x} & \text{for } x \geq \bar{x} \\ 0 & \text{for } x < \bar{x}. \end{cases} \tag{4.3.59}$$

In the case $AE \equiv 1$, $c \equiv 0$, compute $\|u_{EX} - \mathcal{A}^p_{I_j^\Delta}u_{EX}\|_{\mathcal{U}(I_j^\Delta)}$, $\|u_{EX} - \mathcal{I}^p_{I_j^\Delta}u_{EX}\|_{\mathcal{U}(I_j^\Delta)}$, $\|u_{EX} - \mathcal{B}^p_{I_j^\Delta}u_{EX}\|_{\mathcal{U}(I_j^\Delta)}$, and $\|\mathcal{A}^p_{I_j^\Delta}u_{EX} - \mathcal{B}^p_{I_j^\Delta}u_{EX}\|_{\mathcal{U}(I_j^\Delta)}$, for $p = 1, 2$, as functions of $\bar{x}$.

P4.3.8. Let $x_i = i\,h$, $i = 0, \ldots, 3$, $I^{(h)} = (x_0^{\Delta}, x_3^{\Delta})$, $I_j^{(h)} = (x_{j-1}^h, x_j^h)$. Assuming that $u_{EX} \in H^r(I^{(h)})$, prove that

$$\|\mathcal{B}^p_{I^{(h)}} u_{EX} - u_{EX}\|_{\mathcal{U}(I^{(h)})} = \min_{\chi \in S^p_{\Delta_h}} \|\chi - u_{EX}\|_{\mathcal{U}(I^{(h)})}$$

$$\leq \mathcal{C} h^{\mu} \left\| \frac{d^r}{dx^r}(u_{EX}) \right\|_{L^2(I^{(h)})} \tag{4.3.60a}$$

where

$$\mu = \min(p, r-1), \quad 0 < p \leq r+1 \tag{4.3.60b}$$

and that

$$\|\mathcal{B}^p_{I^{(h)}} u_{EX} - u_{EX}\|_{\mathcal{U}(I^{(h)})} \leq \|\mathcal{I}^p_{I^{(h)}} u_{EX} - u_{EX}\|_{\mathcal{U}(I^{(h)})}$$

$$\leq \mathcal{C} \|\mathcal{B}^p_{I^{(h)}} u_{EX} - u_{EX}\|_{\mathcal{U}(I^{(h)})}. \tag{4.3.60c}$$

4.3* The splitting of the error in the two-dimensional case

In the one-dimensional case, we introduced the local error in the element I_j^{Δ}

$$e^{loc}_{S^p(I_j^{\Delta})} = u_{EX} - \mathcal{A}^p_{I_j^{\Delta}} u_{EX} \approx u_{EX} - \mathcal{B}^p_{I_j^{\Delta}} u_{EX} \tag{4.3*.1a}$$

by employing either $\mathcal{A}^p_{I_j^{\Delta}} u_{EX}$ or $\mathcal{B}^p_{I_j^{\Delta}} u_{EX}$ as the local best approximation of u_{EX}, where the symbol $\approx$ means that the two definitions are practically the same, asymptotically, and the equivalence holds for the regular model problem under the additional assumption that $(d/dx)(AE)$ is bounded. We also defined the pollution error as

$$e^{glob}_{S^p(I_j^{\Delta})} = \mathcal{A}^p_{I_j^{\Delta}} u_{EX} - u_{S^p_{\Delta}}|_{I_j^{\Delta}} \approx \mathcal{B}^p_{I_j^{\Delta}} u_{EX} - u_{S^p_{\Delta}}|_{I_j^{\Delta}}. \tag{4.3*.1b}$$

In the cases that the pollution error $e^{glob}_{S^p(I_j^{\Delta})}$ is negligible, we have

$$u_{S^p(I_j^{\Delta})} \approx \mathcal{A}^p_{I_j^{\Delta}} u_{EX} \approx \mathcal{B}^p_{I_j^{\Delta}} u_{EX} \tag{4.3*.1c}$$

which means that both $\mathcal{A}^p_{I_j^{\Delta}} u_{EX}$ and $\mathcal{B}^p_{I_j^{\Delta}} u_{EX}$ have the meaning of local finite element solution.

In the higher-dimensional case we have an analogous situation, but here only the analogue of $\mathcal{B}^p_{I_j^{\Delta}} u_{EX}$ can be used. More precisely, $\mathcal{B}^p_{\omega_\tau} u_{EX}$, the unconstrained best approximation over ω_τ, a neighborhood of the element τ, will be employed, and we will let

$$e^{loc,\omega_\tau}_{S^p(\tau)} \stackrel{\text{def}}{=} u_{EX}|_\tau - \mathcal{B}^p_{\omega_\tau} u_{EX}|_\tau \tag{4.3*.2a}$$

for the local error, and

$$e^{glob,\omega_\tau}_{S^p(\tau)} \stackrel{\text{def}}{=} \mathcal{B}^p_{\omega_\tau} u_{EX}|_\tau - u_{S^p_{\Delta}}|_\tau \tag{4.3*.2b}$$

for the pollution error, and in the cases that $e^{glob}_{S^p(\tau)}$ is negligible, we have

$$u_{S^p_{\Delta}}|_\tau \approx \mathcal{B}^p_{\omega_\tau} u_{EX}|_\tau \tag{4.3*.2c}$$

and $\mathcal{B}^p_{\omega_\tau} u_{EX}|_\tau$ has the meaning of local finite element solution. Below we will suppress the dependence of $e^{loc,\omega_\tau}_{S^p(\tau)}$ and $e^{glob,\omega_\tau}_{S^p(\tau)}$ on ω_τ.

Remark 4.3*.1 The local best approximation $\mathcal{A}^p_{I^\Delta_j} u_{EX}$, cannot be extended to higher dimensions because the requirement $u_{EX} \in \mathcal{U}$ does not guarantee that u_{EX} has well-defined values at every point.

Let us now consider a case which reflects the character of two-dimensional engineering computations, which most often have the following characteristics:

1. The domain Ω is polygonal with straight or curved edges.
2. The right-hand side f is most often equal to zero, and the functions $\bar{u}$ and g which, respectively, prescribe the nonhomogeneous essential and the flux boundary conditions are smooth on each edge.
3. The finite element solution is computed using almost uniform meshes and elements of low degree, e.g., linear ($p = 1$) or quadratic ($p = 2$) elements.

We will now discuss the structure of the error for finite element solutions of problems with the above characteristics.

Let us, for simplicity, consider the Laplace equation

$$-\Delta u|_\Omega \equiv 0, \qquad u|_{\Gamma_\mathrm{D}} \equiv 0, \qquad \left.\frac{\partial u}{\partial n}\right|_{\Gamma_\mathrm{N}} = g \tag{4.3*.3a}$$

where Ω is a polygonal domain (e.g., the domain shown in Fig. 4.3*.1), and $\Gamma_\mathrm{D} \not\equiv \emptyset$, consists of one or several edges. The variational formulation of (4.3*.3a) is

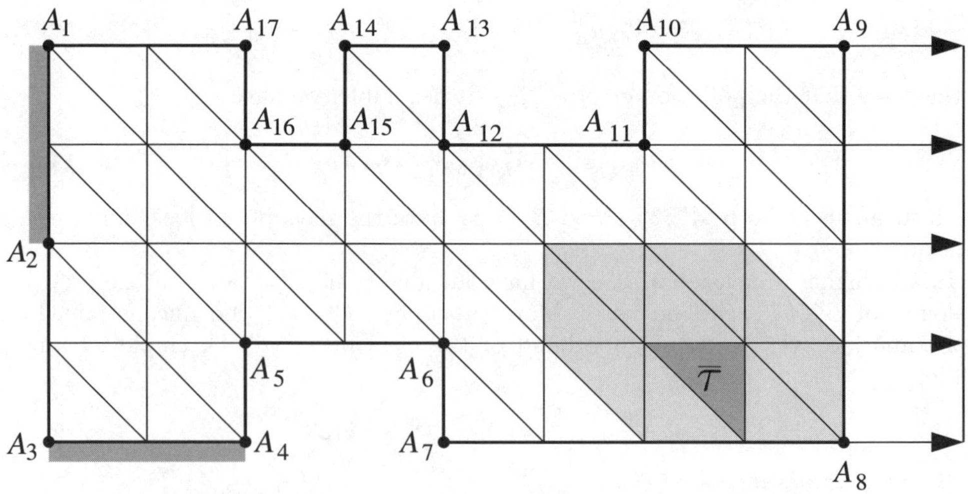

FIG. 4.3*.1. An example of a polygonal domain. A domain with straight edges discretized by a mesh of triangles with straight edges. The points $A_1, \ldots, A_{17}$ denote the corner points where the solution has algebraic and/or logarithmic singularities. Note that A_2 is also referred to as a corner point because it is a singular point (the boundary condition type changes from homogeneous Dirichlet to homogeneous Neumann)

given by:

Find $u_{EX} \in \mathcal{U} \equiv H^1_{\Gamma_D}(\Omega)$ such that

$$\mathcal{B}_\Omega(u_{EX}, v) = \int_\Omega \nabla u_{EX} \cdot \nabla v = \int_{\Gamma_N} gv \quad \forall v \in H^1_{\Gamma_D}(\Omega). \tag{4.3*.3b}$$

Let now $u_{S^1_\Delta}$ be the finite element approximation of u_{EX} using the mesh of linear ($p = 1$) triangles shown in Fig. 4.3*.1. Letting

$$u_{EX} = u_{S^1_\Delta} + e_{S^1_\Delta} \tag{4.3*.3c}$$

into (4.3*.3b) we get that the error $e_{S^1_\Delta} \in \mathcal{U}$ satisfies the variational equation

$$\begin{aligned}\mathcal{B}_\Omega(e_{S^1_\Delta}, v) &= \int_\Omega \nabla e_{S^1_\Delta} \cdot \nabla v = \mathcal{R}(v)\\ &\stackrel{\text{def}}{=} \int_{\Gamma_N} gv - \int_\Omega \nabla u_{S^1_\Delta} \cdot \nabla v \quad \forall v \in H^1_{\Gamma_D}(\Omega).\end{aligned} \tag{4.3*.3d}$$

The linear functional $\mathcal{R} : H^1_{\Gamma_D}(\Omega) \mapsto \mathbb{R}$ is known as the *residuum*, and the variational equation (4.3*.3d) as the *residual equation*; both will be used extensively in Chapter 5.

Let us now express the residuum, $\mathcal{R}$, in terms of the residual loads corresponding to the finite element solution $u_{S^1_\Delta}$. These loads are the *interior element residuals*

$$r_\tau = \Delta u_{S^1_\Delta} \tag{4.3*.4a}$$

in each $\tau \in \Delta$, and the *jumps* of the normal derivative on the edges

$$J^\Delta_\epsilon \stackrel{\text{def}}{=} \begin{cases} \left(\nabla u_{S_{\Delta^1}}\big|_{\tau^*_\epsilon} - \nabla u_{S^1_\Delta}\big|_{\tau_\epsilon}\right) \cdot \mathbf{n}_\epsilon, & \epsilon \in \mathcal{E}_{int}(\Delta)\\ g - \nabla u_{S^1_\Delta} \cdot \mathbf{n}_{\Gamma_N}, & \epsilon \subseteq \partial\tau_\epsilon \subseteq \Gamma_N\\ 0, & \epsilon \subseteq \Gamma_D. \end{cases} \tag{4.3*.4b}$$

Here ϵ denotes an edge, $\mathcal{E}_{int}(\Delta)$ is the set of interior edges for the mesh Δ, τ_ϵ, τ^*_ϵ are the two elements sharing the interior edge ϵ, and $\mathbf{n}_\epsilon$ is the exterior normal of τ_ϵ on ϵ.

Employing the integration by parts

$$\mathcal{B}_\tau(u_{S^1_\Delta}, v) = \int_\tau \nabla u_{S^1_\Delta} \cdot \nabla v = \int_\tau \Delta(u_{S^1_\Delta}|_\tau)\, v - \oint_{\partial\tau} \frac{\partial}{\partial n}(u_{S^1_\Delta}|_\tau)\, v \tag{4.3*.5a}$$

taking now into account that

$$\Delta(u_{S^1_\Delta}|_\tau) \equiv 0, \qquad \nabla(u_{S^1_\Delta}|_\tau) \equiv \mathit{constant} \tag{4.3*.5b}$$

and summing over the elements we get

$$\sum_{\tau\in\Delta} \int_\tau \nabla u_{S^1_\Delta} \cdot \nabla v = - \sum_{\epsilon\in\mathcal{E}(\Delta)} J^\Delta_\epsilon \int_\epsilon v \tag{4.3*.5c}$$

where $\mathcal{E}(\Delta)$ is the set of edges of Δ, and we assumed that, as it often occurs in practice, that g is edgewise constant.

Hence the residual equation can be written as:

Find $e_{S^1_\Delta} \in \mathcal{U}$ such that

$$\sum_{\tau\in\Delta} \int_\tau \nabla e_{S^1_\Delta} \cdot \nabla v = \mathcal{R}(v) = \sum_{\epsilon\in\mathcal{E}(\Delta)} J^\Delta_\epsilon \int_\epsilon v \quad \forall v \in \mathcal{U}. \tag{4.3*.5d}$$

Let now G_ϵ, be the solution of the variational problem:

Find $G_\epsilon \in \mathcal{U}$ such that

$$\sum_{\tau\in\Delta} \int_\tau \nabla G_\epsilon \cdot \nabla v = \int_\epsilon v \quad \forall v \in \mathcal{U}. \tag{4.3*.6a}$$

The function G_ϵ is an *influence function* corresponding to the edge ϵ. It exists because *we assumed that* Γ_{D} *has finite length.* Letting

$$e^\epsilon_{S^p_\Delta} = J^\Delta_\epsilon G_\epsilon \tag{4.3*.6b}$$

by superposition, we have

$$e_{S^1_\Delta} = \sum_{\epsilon\in\mathcal{E}(\Delta)} e^\epsilon_{S^p_\Delta} = \sum_{\epsilon\in\mathcal{E}(\Delta)} J^\Delta_\epsilon G_\epsilon. \tag{4.3*.6c}$$

Note that $e^\epsilon_{S^p_\Delta}$ is the contribution of the jump J^Δ_ϵ to the error $e_{S^p_\Delta}$. For any element $\bar\tau$, let

$$e^{\omega_{\bar\tau}}_{S^1_\Delta} = \sum_{\epsilon\in\mathcal{E}(\omega_{\bar\tau})} J^\Delta_\epsilon G_\epsilon \tag{4.3*.7a}$$

where $\mathcal{E}(\omega_{\bar\tau})$ is the set of edges in the mesh patch $\omega_{\bar\tau}$, and

$$e^{\Omega-\omega_{\bar\tau}}_{S^1_\Delta} = e_{S^1_\Delta} - e^{\omega_{\bar\tau}}_{S^1_\Delta} = \sum_{\epsilon\in\mathcal{E}(\Delta)-\mathcal{E}(\omega_{\bar\tau})} J^\Delta_\epsilon G_\epsilon. \tag{4.3*.7b}$$

A two-dimensional analogue of the local error may be defined by

$$e^{loc}_{S^p(\bar\tau)} \stackrel{\text{def}}{=} e^{\omega_{\bar\tau}}_{S^1_\Delta}\big|_{\bar\tau}. \tag{4.3*.7c}$$

It is the error in the element $\bar\tau$ due to the jumps on the edges in $\omega_{\bar\tau}$. Similarly, the pollution error may be defined by

$$e^{glob}_{S^p(\bar\tau)} \stackrel{\text{def}}{=} e^{\Omega-\omega_{\bar\tau}}_{S^1_\Delta}\big|_{\bar\tau} \tag{4.3*.7d}$$

namely, as the error due to the jumps in all the edges outside $\omega_{\bar\tau}$. The local error $e^{loc}_{S^p(\bar\tau)}$ is the *near-field component* of the error in $\bar\tau$, and $e^{glob}_{S^p(\bar\tau)}$ is the *far-field component* of the error in $\bar\tau$.

The energy norm $\|e_{S^1_\Delta}\|_{\mathcal{U}(\tilde{\tau})}$ can be written as

$$\|e_{S^1_\Delta}\|_{\mathcal{U}(\tilde{\tau})} = \sqrt{\sum_{\epsilon,\epsilon' \in \mathcal{E}(\Delta)} J^\Delta_\epsilon J^\Delta_{\epsilon'} \mathcal{B}_{\tilde{\tau}}(G_\epsilon, G_{\epsilon'})} \tag{4.3*.8a}$$

and similarly for the local error and the pollution error, where the sum has to be taken over the edges in $\mathcal{E}(\omega_{\tilde{\tau}})$, and $\mathcal{E}(\Delta) - \mathcal{E}(\omega_{\tilde{\tau}})$, respectively. Further,

$$\|e_{S^1_\Delta}\|_{\mathcal{U}(\tilde{\tau})} = \sqrt{\|e^{loc}_{S^1(\tilde{\tau})}\|^2_{\mathcal{U}(\tilde{\tau})} + \|e^{glob}_{S^1(\tilde{\tau})}\|^2_{\mathcal{U}(\tilde{\tau})} + 2\,\mathcal{B}_{\tilde{\tau}}(e^{loc}_{S^1(\tilde{\tau})}, e^{glob}_{S^1(\tilde{\tau})})}. \tag{4.3*.8b}$$

Under the assumption that the mesh Δ is nearly uniform in $\omega_{\tilde{\tau}}$, we have

$$\left|\mathcal{B}_{\tilde{\tau}}(e^{loc}_{S^1(\tilde{\tau})}, e^{glob}_{S^1(\tilde{\tau})})\right| \ll \|e^{loc}_{S^1(\tilde{\tau})}\|_{\mathcal{U}(\tilde{\tau})} \|e^{glob}_{S^1(\tilde{\tau})}\|_{\mathcal{U}(\tilde{\tau})} \tag{4.3*.8c}$$

i.e., $e^{loc}_{S^1(\tilde{\tau})}$ and $e^{glob}_{S^1(\tilde{\tau})}$ are *almost orthogonal*, and hence

$$\|e_{S^1_\Delta}\|_{\mathcal{U}(\tilde{\tau})} \approx \sqrt{\|e^{loc}_{S^1(\tilde{\tau})}\|^2_{\mathcal{U}(\tilde{\tau})} + \|e^{glob}_{S^1(\tilde{\tau})}\|^2_{\mathcal{U}(\tilde{\tau})}}. \tag{4.3*.8d}$$

Remark 4.3*.2 The above observations also hold in the one-dimensional case.

Remark 4.3*.3 In two-dimensional computations in polygonal domains we often have $\|e^{loc}_{S^p(\tilde{\tau})}\|_{\mathcal{U}(\tilde{\tau})} \ll \|e^{glob}_{S^p(\tilde{\tau})}\|_{\mathcal{U}(\tilde{\tau})}$, i.e., the pollution is the major component of the error in many elements.

Remark 4.3*.4 We can also define $e^{loc}_{S^p(\tilde{\omega})}$ and $e^{glob}_{S^p(\tilde{\omega})}$, the local and pollution error with respect to an entire patch of elements, $\tilde{\omega}$, by replacing $\tilde{\tau}$ by $\tilde{\omega}$ in the above definitions.

Let us now give an example which illustrates the local and pollution errors introduced above in a typical case.

Example 4.3*.1 Consider the mixed boundary value

$$\Delta u|_\Omega \equiv 0, \qquad u|_{A_1A_2} \equiv 0, \qquad \left.\frac{\partial u}{\partial n}\right|_{A_1A_6} \equiv 0, \qquad \left.\frac{\partial u}{\partial n}\right|_{\Gamma_N - A_1A_6} = g \tag{4.3*.9a}$$

where Ω is the L-shaped domain which is shown in Fig. 4.3*.2a, where $g = \nabla u_{EX}|_{\Gamma_N} \cdot \mathbf{n}_{\Gamma_N}$ is the normal derivative corresponding to the exact solution

$$u_{EX}(r,\theta) = r^{1/3} \sin\left(\frac{\theta}{3}\right) \tag{4.3*.9b}$$

where (r,θ) are the polar coordinates with respect to the vertex A_1. Let Δ be the uniform mesh of triangles shown in Fig. 4.3*.2b, and assume that we are interested in $e^{loc}_{S^1(\omega_0)}$ and $e^{glob}_{S^1(\omega_0)}$, where ω_0 is the shaded patch of elements in Fig. 4.3*.2b. Let τ_i, $i = 1,\ldots,6$, be the elements with a vertex at the singular point, as shown in Fig. 4.3*.2c, and $e^{glob}_{S^1(\omega_0),\{\tau_i\}_{i=1}^k}$, be the part of the error due to the edge residuals in the elements τ_i, $i = 1,\ldots,k$; details about the employed splitting of the jumps are given in Chapter 5. Figure 4.3*.2d gives the graphs of $((\partial/\partial x_1)(e_{S^1_\Delta})/\|(\partial/\partial x_1)(u_{EX})\|_{L^\infty(\omega_0)})100\%$, and

$((\partial/\partial x_1)(e^{glob}_{S^1(\omega_0),\{\tau_i\}_{i=1}^k})/\|(\partial/\partial x_1)(u_{EX})\|_{L^\infty(\omega_0)})100\%$, $k = 1,\ldots,6$, along the line aa which passes through the mid points of the edges of the elements that it intersects (the line aa is shown in Fig. 4.3*.2b). Note that $(\partial/\partial x_1)(e^{glob}_{S^1(\omega_0),\{\tau_i\}_{i=1}^6})$ is practically 70% of the derivative of the total error $(\partial/\partial x_1)(e_{S^1_\Delta})|_{\omega_0}$. Figure 4.3*.2e gives the graphs of $((\partial/\partial x_1)(e^{loc}_{S^1(\omega_0)})/\|(\partial/\partial x_1)(u_{EX})\|_{L^\infty(\omega_0)})$ 100%. Figures 4.3*.3a,b give

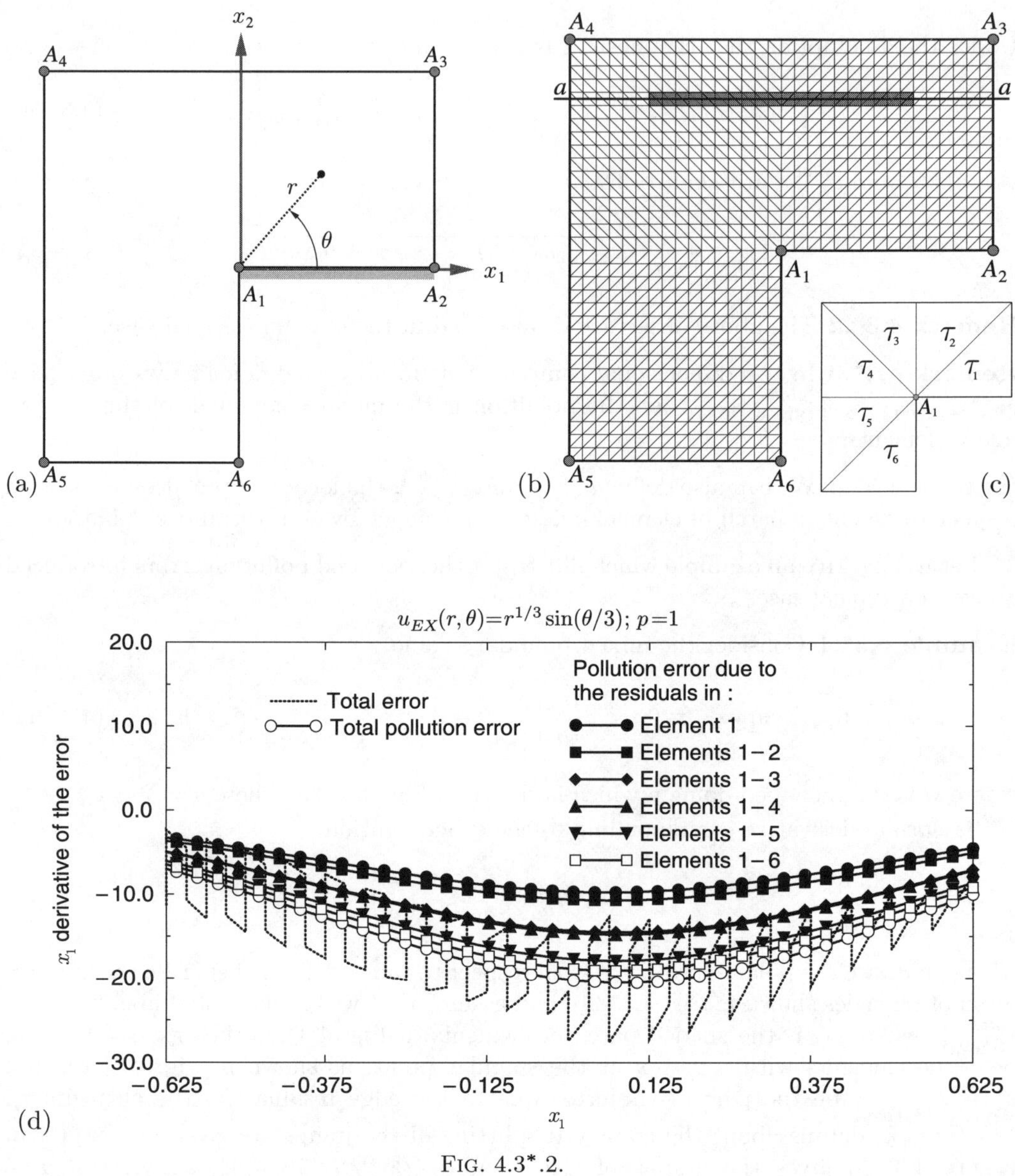

FIG. 4.3*.2.

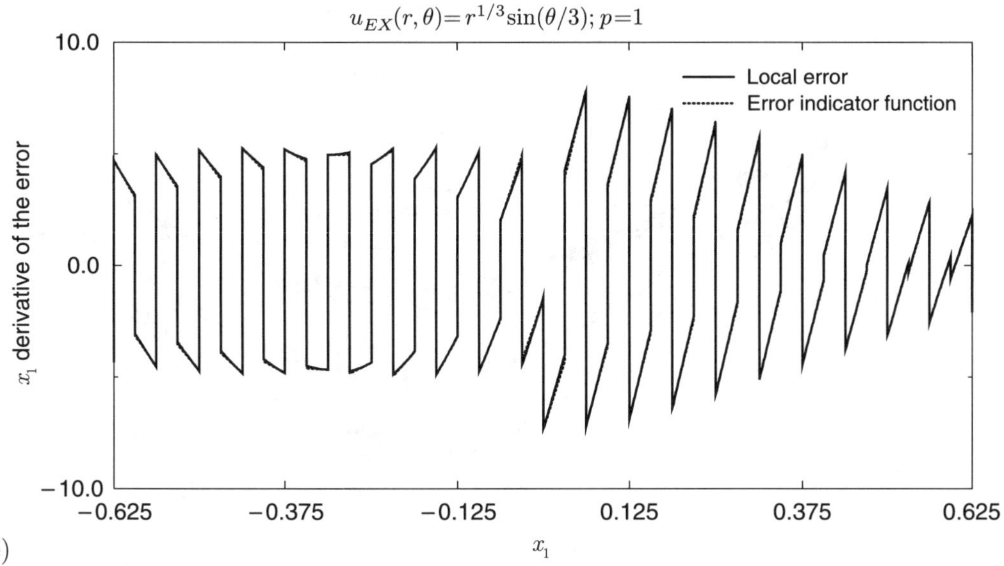

(e)

FIG. 4.3*.2. An example of the local and global component of the error in the finite element solution of the Laplace equation in an L-shaped domain. Input data: Laplace's equation with mixed boundary conditions, $u_{EX}(r,\theta) = r^{1/3}\sin(\theta/3)$. Mesh data: Uniform mesh of linear triangles ($p = 1$). (a) The system of polar coordinates and the notations employed in the definition of the boundary conditions ($A_1A_2 \equiv \Gamma_D$, $\Gamma_N = \partial\Omega - \Gamma_D$). (b) The finite element mesh with the subdomain ω_0 shown shaded gray and the line aa along which the graphs of the error are shown. (c) Enumeration of the elements with a vertex at the singular point A_1. (d) Graphs of the relative errors $((\partial/\partial x_1)(e_{S^1_\Delta})/\|(\partial/\partial x_1)(u_{EX})\|_{L^\infty(\omega_0)})100\%$, $((\partial/\partial x_1)(e^{glob}_{S^1(\omega_0),\{\tau_i\}_{i=1}^k})/\|(\partial/\partial x_1)(u_{EX})\|_{L^\infty(\omega_0)})100\%$, $k = 1,\ldots,6$ and $((\partial/\partial x_1)(e^{glob}_{S^1(\omega_0)})/\|(\partial/\partial x_1)(u_{EX})\|_{L^\infty(\omega_0)})100\%$ along the line aa shown in (b). (e) Graphs of $((\partial/\partial x_1)(e^{loc}_{S^1(\omega_0)})/\|(\partial/\partial x_1)(u_{EX})\|_{L^\infty(\omega_0)})100\%$

the corresponding graphs in the case of quadratic elements ($p = 2$). By comparing Figs 4.3*.2d and 4.3*.3a we see that *the significance of the pollution increases with p.*

Remark 4.3*.5 The effect of the pollution for the approximations of the solutions of the Laplacian in the L-shaped domain is similar to the effect of the pollution in the approximation of the solutions of the singular model problem; see also P4.3*.2.

Remark 4.3*.6 The pollution is due to the singular character of the solution u_{EX}, and the Green's function G_ϵ. [Blum and Rannacher, 1988] attempted to take into account this fact, and to construct extrapolation methods to reduce the pollution.

In this section we introduced the local and the pollution error in the two-dimensional case. We gave an example which illustrates that, for the polygonal domains which occur in practical computations, the pollution is not negligible and becomes more significant as the degree p of the elements increases.

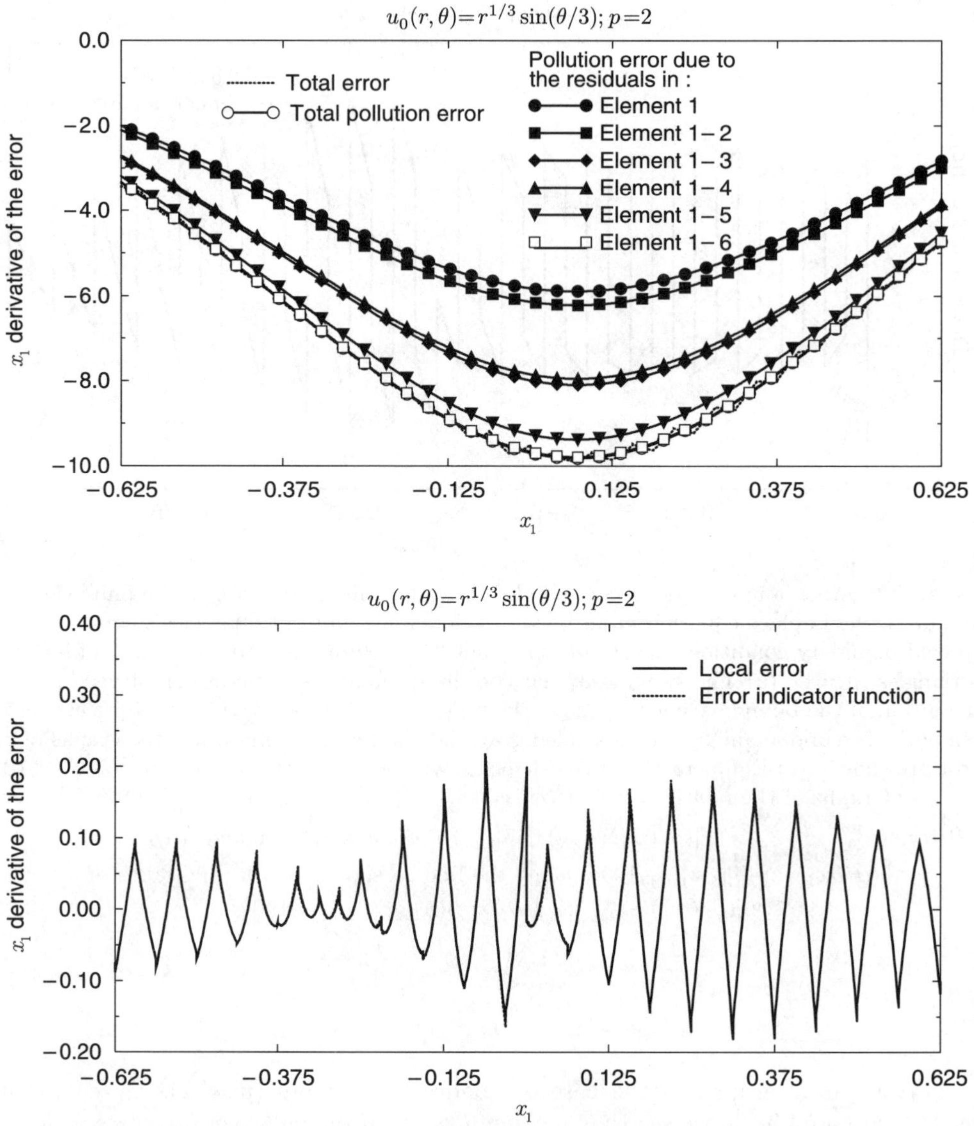

FIG. 4.3*.3. An example of the local and global component of the error in the finite element solution of the Laplace equation in an L-shaped domain. Input data: Laplace's equation with mixed boundary conditions, $u_{EX}(r,\theta) = r^{1/3}\sin(\theta/3)$. Mesh data: Uniform mesh of quadratic triangles ($p = 2$). (a) Graphs of the relative errors $((\partial/\partial x_1)(e_{S^2_\Delta})/\|(\partial/\partial x_1)(u_{EX})\|_{L^\infty(\omega_0)})100\%$, $((\partial/\partial x_1)(e^{glob}_{S^2(\omega_0),\{\tau_i\}_{i=1}^k})/\|(\partial/\partial x_1)(u_{EX})\|_{L^\infty(\omega_0)})100\%$, $k = 1,\ldots,6$ and $((\partial/\partial x_1)(e^{glob}_{S^2(\omega_0)})/\|(\partial/\partial x_1)(u_{EX})\|_{L^\infty(\omega_0)})100\%$ along the line aa shown in Fig. 4.3*.2b. (b) Graphs of $((\partial/\partial x_1)(e^{loc}_{S^2(\omega_0)})/\|(\partial/\partial x_1)(u_{EX})\|_{L^\infty(\omega_0)})100\%$. By comparing Figs 4.3*.2d and 4.3*.3a we see that, *in the case of quadratic elements, the effect of the pollution is much more significant than in the case of linear elements*

4.4 A-priori estimates for the pollution error

In Section 4.3, we introduced the splitting of the error

$$e_{S^p_\Delta} = u_{EX} - u_{S^p_\Delta} = \underbrace{u_{EX} - \mathcal{A}^p_{I^\Delta_j} u_{EX}}_{e^{loc}_{S^p_\Delta}} + \underbrace{\mathcal{A}^p_{I^\Delta_j} u_{EX} - u_{S^p_\Delta}}_{e^{glob}_{S^p_\Delta}}. \tag{4.4.1a}$$

In the special case

$$\left.\frac{1}{AE}\right|_{I^\Delta_k} \in \mathcal{P}^{p-1}(I^\Delta_k) \quad k = 1, \dots, N(\Delta), \qquad c \equiv 0 \tag{4.4.1b}$$

we have

$$u_{S^p_\Delta} \equiv \mathcal{A}^p_\Delta u_{EX} \tag{4.4.1c}$$

and

$$e^{glob}_{S^p_\Delta} \equiv 0. \tag{4.4.1d}$$

In general, however, this is not the case, and to assess the importance of $e^{glob}_{S^p_\Delta}$ relative to $e^{loc}_{S^p_\Delta}$ we need estimates, for classes of input data, which are more general than (4.4.1b). This is the topic of this section.

We will analyze the pollution error relative to the local error for sequences of meshes $\{\Delta_n\}_{n=1}^\infty$ with the property that the number of elements $N(\Delta_n)$ diverges to infinity with n. In particular: We will be mostly interested in sequences of *almost-uniform meshes*, which can be characterized as follows:

We will say that the meshes Δ_n, $n = 1, 2, \dots$, are *κ-uniform* or *κ-quasi-uniform* if there exists a fixed κ, $0 < \kappa \le 1$, such that for every mesh Δ_n

$$0 < \kappa\, h^{\Delta_n} \le h_{I^{\Delta_n}_j} \le h^{\Delta_n}, \quad j = 1, \dots, N(\Delta_n). \tag{4.4.2}$$

For κ-quasi-uniform meshes we have

$$\frac{l}{N} \le h^{\Delta_n} \le \frac{l}{(N-1)\kappa + 1} \tag{4.4.3}$$

and thus the maximal element-size h^{Δ_n} tends to zero as the number of elements $N(\Delta_n)$ diverges to infinity. Below we will always employ such meshes, unless it is otherwise stated.

We will say that the pollution at a point $\bar{x}$ is negligible, if there exists an $\alpha > 0$ such that

$$\|e^{glob}_{S^p(I^{\Delta_n}_{j(\bar{x})})}\|_{\mathcal{U}(I^{\Delta_n}_{j(\bar{x})})} \le \mathscr{C}(h^{\Delta_n})^\alpha \|e^{loc}_{S^p(I^{\Delta_n}_{j(\bar{x})})}\|_{\mathcal{U}(I^{\Delta_n}_{j(\bar{x})})} \tag{4.4.4}$$

where $I^{\Delta_n}_{j(\bar{x})}$ an element which includes $\bar{x}$ in its interior, or as one of its end points, i.e., $\bar{x} \in I^{\Delta_n}_{j(\bar{x})}$; note that there can be at most two of such elements.

Let us now use the results of Theorem 4.3.2 to state a sufficient condition for the pollution to be negligible relative to the local error.

Lemma 4.4.1 *If there exists an $\alpha > 0$ such that*

$$\max\left(\left|e^{\Delta_n}_{j(\bar{x})} + e^{\Delta_n}_{j(\bar{x})-1}\right|, \frac{\left|e^{\Delta_n}_{j(\bar{x})} - e^{\Delta_n}_{j(\bar{x})-1}\right|}{h^{\Delta_n}_{j(\bar{x})}}\right) \leq \mathscr{C}\,(h^{\Delta_n})^{\alpha-1/2}\|e^{loc}_{S^p(I^{\Delta_n}_{j(\bar{x})})}\|_{\mathscr{U}(I^{\Delta_n}_{j(\bar{x})})} \tag{4.4.5}$$

then (4.4.4) holds.

Proof Employing the splitting

$$e^{glob}_{S^p(I^{\Delta_n}_{j(\bar{x})})} = e^{glob}_{S^1(I^{\Delta_n}_{j(\bar{x})})} + e^{glob}_{S^p_0(I^{\Delta_n}_{j(\bar{x})})} \tag{4.4.6a}$$

we get

$$\begin{aligned}\|e^{glob}_{S^p(I^{\Delta_n}_{j(\bar{x})})}\|_{\mathscr{U}(I^{\Delta_n}_{j(\bar{x})})} &\leq \|e^{glob}_{S^1(I^{\Delta_n}_{j(\bar{x})})}\|_{\mathscr{U}(I^{\Delta_n}_{j(\bar{x})})} + \|e^{glob}_{S^p_0(I^{\Delta_n}_{j(\bar{x})})}\|_{\mathscr{U}(I^{\Delta_n}_{j(\bar{x})})} \\ &\leq 2\|e^{glob}_{S^1(I^{\Delta_n}_{j(\bar{x})})}\|_{\mathscr{U}(I^{\Delta_n}_{j(\bar{x})})}\end{aligned} \tag{4.4.6b}$$

and hence

$$\|e^{glob}_{S^p(I^{\Delta_n}_{j(\bar{x})})}\|_{\mathscr{U}(I^{\Delta_n}_{j(\bar{x})})} \leq 2\|e^{glob}_{S^1(I^{\Delta_n}_{j(\bar{x})})}\|_{\mathscr{U}(I^{\Delta_n}_{j(\bar{x})})} \leq \mathscr{C}\sqrt{h_{I^{\Delta_n}_{j(\bar{x})}}}\max\left(|\rho_1^{I^{\Delta_n}_{j(\bar{x})}}|, |\rho_2^{I^{\Delta_n}_{j(\bar{x})}}|\right) \tag{4.4.6c}$$

where

$$\rho_1^{I^{\Delta_n}_{j(\bar{x})}} = \left|e^{\Delta_n}_{j(\bar{x})} + e^{\Delta_n}_{j(\bar{x})-1}\right|, \qquad \rho_2^{I^{\Delta_n}_{j(\bar{x})}} = \frac{\left|e^{\Delta_n}_{j(\bar{x})} - e^{\Delta_n}_{j(\bar{x})-1}\right|}{h^{\Delta_n}_{j(\bar{x})}} \tag{4.4.6d}$$

and using assumption (4.4.5), we get

$$\|e^{glob}_{S^p(I^{\Delta_n}_{j(\bar{x})})}\|_{\mathscr{U}(I^{\Delta_n}_{j(\bar{x})})} \leq \mathscr{C}\,(h^{\Delta_n})^{\alpha}\|e^{loc}_{S^p(I^{\Delta_n}_{j(\bar{x})})}\|_{\mathscr{U}(I^{\Delta_n}_{j(\bar{x})})} \tag{4.4.6e}$$

with $\alpha > 0$. This proves that the pollution in $I^{\Delta_n}_{j(\bar{x})}$ is negligible, asymptotically. □

Below we will use condition (4.4.5) to analyze the pollution in finite element solutions of the regular model problem under various assumptions about the smoothness of the input data. We will often omit $\bar{x}$ from $j(\bar{x})$ (i.e., we will write $I^{\Delta_n}_j$ instead of $I^{\Delta_n}_{j(\bar{x})}$, etc.), except when it is needed for clarity.

Let us now give the main result, which states that for sufficiently smooth AE, c, and f, the pollution is negligible.

Theorem 4.4.1 *Consider the regular model problem and assume that*

$$\left\|\frac{d^{r'}}{dx^{r'}}(AE)\right\|_{L^\infty(\bar{I})} \leq K, \qquad \left\|\frac{d^{q'}c}{dx^{q'}}\right\|_{L^\infty(\bar{I})} \leq \gamma, \qquad \left\|\frac{d^{t'}f}{dx^{t'}}\right\|_{L^\infty(\bar{I})} \leq C \tag{4.4.7a}$$

for all integers r', q', t', which are such that $0 \le r' \le r$, $0 \le q' \le q$, $0 \le t' \le t$. Let $\{\Delta_n\}_{n=1}^{\infty}$ be a sequence of κ-quasi-uniform meshes, and $\bar{x} \in I$, be such that

$$\mathscr{C}\, h^{\tilde{\mu}+1/2} \le \left\| e^{loc}_{S^p(I^{\Delta_n}_{j(\bar{x})})} \right\|_{\mathcal{U}(I^{\Delta}_{j(\bar{x})})} \tag{4.4.7b}$$

where

$$\tilde{\mu} = \min(\mu, t+1), \qquad \mu = \min(p, r, q+1) > \tfrac{1}{2}. \tag{4.4.7c}$$

Then, for $r \ge 1$, we have

$$\left\| e^{glob}_{S^p(I^{\Delta_n}_{j(\bar{x})})} \right\|_{\mathcal{U}(I^{\Delta_n}_{j(\bar{x})})} \le \mathscr{C}\,(h^{\Delta_n})^{\mu} \left\| e^{loc}_{S^p(I^{\Delta_n}_{j(\bar{x})})} \right\|_{\mathcal{U}(I^{\Delta_n}_{j(\bar{x})})}. \tag{4.4.7d}$$

Proof Below we will suppress $\bar{x}$ from $j(\bar{x})$, and n from Δ_n.

From Theorem 4.1.1 we have

$$\max\left(|e^{\Delta}_{j-1}|, |e^{\Delta}_{j}|\right) \le \mathscr{C}\, h^{\mu} \|e_{S^p_{\Delta}}\|_{\mathcal{U}} \tag{4.4.8a}$$

where $\mu = \min(p, r, q+1)$, and we get

$$|\rho_1^{I_j^{\Delta}}| = |e^{\Delta}_{j} + e^{\Delta}_{j-1}| \le \mathscr{C}\, h^{\mu} \|e_{S^p_{\Delta}}\|_{\mathcal{U}} \le \mathscr{C} h^{\mu+\tilde{\mu}}. \tag{4.4.8b}$$

We will estimate $|\rho_2^{I_j^{\Delta}}| = |e^{\Delta}_{j} - e^{\Delta}_{j-1}|/h_{I_j^{\Delta}}$, by employing the function

$$G'_j(\cdot) \stackrel{\text{def}}{=} \frac{1}{h_{I_j^{\Delta}}}(G(\cdot; x^{\Delta}_j) - G(\cdot; x^{\Delta}_{j-1})) \tag{4.4.9a}$$

where $G(\cdot; x^{\Delta}_j)$ is the Green's function. Using similar arguments, as in the proof of Theorem 2.3.1, from assumption (4.4.7a) it follows that G'_j has $s = \min(q+2, r+1)$ bounded derivatives outside I^{Δ}_j. Employing Theorem 3.3.1, we get

$$\left\|G'_j - \mathcal{A}^p_{I_i^{\Delta}} G'_j\right\|_{\mathcal{U}(I_i^{\Delta})} \le \left\|G'_j - \mathcal{I}^p_{I_i^{\Delta}} G'_j\right\|_{\mathcal{U}(I_i^{\Delta})} \le \mathscr{C}\, h^{\mu}_{I_i^{\Delta}} \sqrt{h_{I_i^{\Delta}}} \tag{4.4.9b}$$

for $i = 1, \dots, j-1, j+1, \dots, N(\Delta)$, and taking into account that inside I^{Δ}_j the sth derivative of G'_j with respect to its first argument is bounded by $\mathscr{C}/h_{I_j^{\Delta}}$, we get

$$\left\|G'_j - \mathcal{A}^p_{I_j^{\Delta}} G'_j\right\|_{\mathcal{U}(I_j^{\Delta})} \le \left\|G'_j - \mathcal{I}^p_{I_j^{\Delta}} G'_j\right\|_{\mathcal{U}(I_j^{\Delta})} \le \mathscr{C}\, h^{\mu-1}_{I_j^{\Delta}} \sqrt{h_{I_j^{\Delta}}}. \tag{4.4.9c}$$

Now, from the construction of G'_j, we have

$$|\rho_2^{I_j^{\Delta}}| = \frac{|e^{\Delta}_j - e^{\Delta}_{j-1}|}{h_{I_j^{\Delta}}} = |\mathscr{B}(e_{S^p_{\Delta}}, G'_j)| = \left| \sum_{i=1}^{N(\Delta)} \mathscr{B}_{I_i^{\Delta}}(e_{S^p_{\Delta}}, G'_j - \mathcal{A}^p_i G'_j)\right| \tag{4.4.10a}$$

and, using the Cauchy–Schwarz inequality for the energy inner product over each element, we get

$$|\rho_2^{I_j^{\Delta}}| \le \sum_{i=1}^{N(\Delta)} \|e_{S^p_{\Delta}}\|_{\mathcal{U}(I_i^{\Delta})} \left\|G'_j - \mathcal{A}^p_{I_i^{\Delta}} G'_j\right\|_{\mathcal{U}(I_i^{\Delta})}. \tag{4.4.10b}$$

Employing (4.4.9b–c), and using the Cauchy–Schwarz inequality, for the $\mathbb{R}^{N(\Delta)}$ inner product, we get

$$|\rho_2^{I_j^\Delta}| \leq \sqrt{\sum_{\substack{i=1\\ i\neq j}}^{N(\Delta)} \mathscr{C} h_{I_i^\Delta}^{2\mu} h_{I_i^\Delta}} \sqrt{\sum_{\substack{i=1\\ i\neq j}}^{N(\Delta)} \|e_{S_\Delta^p}\|_{\mathcal{U}(I_i)}^2} + \mathscr{C}\, h_{I_j^\Delta}^{\mu-1+1/2} \|e_{S_\Delta^p}\|_{\mathcal{U}(I_j^\Delta)}. \tag{4.4.10c}$$

Recalling now the assumption of quasi-uniformity of the meshes

$$\kappa\, h^\Delta \leq h_{I_j^\Delta} \leq h^\Delta, \quad 0 < \kappa \leq 1 \tag{4.4.10d}$$

we obtain

$$|\rho_2^{I_j^\Delta}| \leq \mathscr{C}\, h^\mu \, \|e_{S_\Delta^p}\|_{\mathcal{U}} + \mathscr{C}\, h^{\mu-1/2} \, \|e_{S_\Delta^p}\|_{\mathcal{U}(I_j^\Delta)}. \tag{4.4.10e}$$

Further, noting that

$$\|e_{S_\Delta^p}\|_{\mathcal{U}(I_j^\Delta)} \leq \|e_{S_\Delta^p}\|_{\mathcal{U}} \leq \mathscr{C}\, h^{\tilde{\mu}} \tag{4.4.10f}$$

with $\tilde{\mu} = \min(\mu, t+1)$, we get

$$|\rho_2^{I_j^\Delta}| \leq \mathscr{C} h^{\mu+\tilde{\mu}-1/2}. \tag{4.4.10g}$$

Using that

$$\left\|e_{S^p(I_j^\Delta)}^{glob}\right\|_{\mathcal{U}(I_j^\Delta)} \leq \mathscr{C} \sqrt{h_{I_j^\Delta}} \sqrt{(e_j^\Delta + e_{j-1}^\Delta)^2 + \left(\frac{e_j^\Delta - e_{j-1}^\Delta}{h_{I_j^\Delta}}\right)} \tag{4.4.11a}$$

we get

$$\left\|e_{S^p(I_j^\Delta)}^{glob}\right\|_{\mathcal{U}(I_j^\Delta)} \leq \mathscr{C}\sqrt{h}(h^{\mu+\tilde{\mu}} + h^{\mu+\tilde{\mu}-1/2}) \leq \mathscr{C} h^{\mu+\tilde{\mu}} \tag{4.4.11b}$$

and using assumption (4.4.7b) we obtain

$$\left\|e_{S^p(I_{j(\bar{x})}^{\Delta_n})}^{glob}\right\|_{\mathcal{U}(I_{j(\bar{x})}^{\Delta_n})} \leq \mathscr{C}(h^{\Delta_n})^{\mu-1/2} \left\|e_{S^p(I_{j(\bar{x})}^{\Delta_n})}^{loc}\right\|_{\mathcal{U}(I_{j(\bar{x})}^{\Delta_n})}. \tag{4.4.11c}$$

Further, noting that

$$\left\|e_{S^p(I_j^\Delta)}^{loc}\right\|_{\mathcal{U}(I_j^\Delta)} \leq \mathscr{C}\sqrt{h} h^{\tilde{\mu}} \tag{4.4.11d}$$

we get

$$\|e_{S_\Delta^p}\|_{\mathcal{U}(I_j^\Delta)} \leq \mathscr{C}\, h^{\tilde{\mu}+1/2} \, (1 + h^{\mu-1/2}) \leq \mathscr{C} h^{\tilde{\mu}+1/2} \tag{4.4.11e}$$

for $\mu > \frac{1}{2}$. Using this to estimate the second term in the right-hand side of (4.4.10c) we get (4.4.7d). □

Remark 4.4.1 In the cases that $0 < \mathscr{C} \le \min_{x\in I^{\Delta}_{j(\bar{x})}} |(d^{p+1}/dx^{p+1})(u_{EX})|(x)$, assumption (4.4.7b) is satisfied with $\mu = p$.

It immediately follows that the pollution in finite element solutions of the regular model problem with smooth input data is negligible.

Corollary 4.4.1 *Consider the setting of the regular model problem, and assume that AE, c, and f are sufficiently smooth, such that*

$$\|e^{loc}_{S^p(I^{\Delta}_{j(\bar{x})})}\|_{\mathscr{U}(I^{\Delta}_{j(\bar{x})})} \approx \mathscr{C}(h_{I^{\Delta}_{j(\bar{x})}})^{p+1/2}. \tag{4.4.12a}$$

Then

$$\|e_{S^p_{\Delta}}\|_{\mathscr{U}(I^{\Delta}_{j(\bar{x})})} = \|e^{loc}_{S^p(I^{\Delta}_{j(\bar{x})})}\|_{\mathscr{U}(I^{\Delta}_{j(\bar{x})})}(1 + \mathscr{C}\, h^p). \tag{4.4.12b}$$

Proof We have

$$\min_{\chi\in S^p(I^{\Delta}_{j(\bar{x})})} \|u_{EX} - \chi\|_{\mathscr{U}(I^{\Delta}_{j(\bar{x})})} \le \|e_{S^p_{\Delta}}\|_{\mathscr{U}(I^{\Delta}_{j(\bar{x})})} \le \|e^{loc}_{S^p(I^{\Delta}_{j(\bar{x})})}\|_{\mathscr{U}(I^{\Delta}_{j})} + \|e^{glob}_{S^p(I^{\Delta}_{j(\bar{x})})}\|_{\mathscr{U}(I^{\Delta}_{j})}. \tag{4.4.13a}$$

Now employing that

$$\min_{\chi\in S^p(I^{\Delta}_{j(\bar{x})})} \|u_{EX} - \chi\|_{\mathscr{U}(I^{\Delta}_{j(\bar{x})})} \le \|e^{loc}_{S^p(I^{\Delta}_{j(\bar{x})})}\|_{\mathscr{U}(I^{\Delta}_{j(\bar{x})})} \le \mathscr{C}\,(h_{I^{\Delta}_{j(\bar{x})}})^{p+1/2} \tag{4.4.13b}$$

and using assumption (4.4.7d), which, in this case, reads

$$\|e^{glob}_{S^p(I^{\Delta}_{j(\bar{x})})}\|_{\mathscr{U}(I^{\Delta}_{j(\bar{x})})} \le \mathscr{C}\,(h^{\Delta})^p \|e^{loc}_{S^p(I^{\Delta}_{j(\bar{x})})}\|_{\mathscr{U}(I^{\Delta}_{j(\bar{x})})} \tag{4.4.13c}$$

(4.4.12b) follows. □

Remark 4.4.2 For the type of solutions which occur in practical applications, the $(p+1)$-derivative of the solution is piecewise smooth and has only isolated zeros, and we may assume that the point $\bar{x}$ is at a fixed distance from these zeros. Hence assumption (4.4.12a) is satisfied in most of the cases of practical interest.

Let us now examine the effect of unsmoothness of the exact solution.

Theorem 4.4.2 *Consider the regular model problem, and assume that*

$$\left\|\frac{d^{\mu+1}}{dx^{\mu+1}}(u_{EX})\right\|_{L^\infty(I - I^{\Delta}_{j(\hat{x})})} \le D_1 \tag{4.4.14a}$$

for $\mu \ge 2$. *Let* Δ *be quasi-uniform meshes, and assume that*

$$\mathscr{C}_1 h^{t-1/2} \le \|e^{loc}_{S^p(I^{\Delta}_{j(\hat{x})})}\|_{\mathscr{U}(I^{\Delta}_{j(\hat{x})})} \le \mathscr{C}_2 h^{t-1/2} \tag{4.4.14b}$$

with $1 < t \le \mu + 1$, *and*

$$\mathscr{C}\, h^{\mu+1/2} \le \|e^{loc}_{S^p(I^{\Delta}_{j(\bar{x})})}\|_{\mathscr{U}(I^{\Delta}_{j(\bar{x})})}, \quad \bar{x} \ne \hat{x}. \tag{4.4.14c}$$

Then we have

$$\|e^{glob}_{S^p(I^\Delta_{j(\bar{x})})}\|_{\mathcal{U}(I^\Delta_{j(\bar{x})})} \le \begin{cases} \mathscr{C}_1\, h^\sigma \|e^{loc}_{S^p(I^\Delta_{j(\bar{x})})}\|_{\mathcal{U}(I^\Delta_{j(\bar{x})})} & \text{for } \bar{x} \neq \hat{x} \\ \mathscr{C}_2\, h^\mu \|e^{loc}_{S^p(I^\Delta_{j(\bar{x})})}\|_{\mathcal{U}(I^\Delta_{j(\bar{x})})} & \text{for } \bar{x} = \hat{x} \end{cases} \tag{4.4.14d}$$

where

$$\sigma = \min\left(\mu, t - \tfrac{1}{2}\right) \tag{4.4.14e}$$

and $\mathscr{C}_1$ *depends on the distance* $|\bar{x} - \hat{x}|$.

Proof From the best-approximation property of $u_{S^p_\Delta}$, we have

$$\|e_{S^p_\Delta}\|_{\mathcal{U}} \le \|u_{EX} - \mathcal{I}^p_\Delta u_{EX}\|_{\mathcal{U}} \tag{4.4.15a}$$

and hence

$$\|e_{S^p_\Delta}\|_{\mathcal{U}} \le \sqrt{\sum_{\substack{i=1 \\ i \neq j(\hat{x})}}^{N} \|u_{EX} - \mathcal{I}^p_{I^\Delta_i} u_{EX}\|^2_{\mathcal{U}(I^\Delta_i)} + \|u_{EX} - \mathcal{I}^p_{I^\Delta_{j(\hat{x})}} u_{EX}\|^2_{\mathcal{U}(I^\Delta_{j(\hat{x})})}}. \tag{4.4.15b}$$

Using now that

$$\mathscr{C}(h_{I^\Delta_i})^{\mu+1/2} \le \|e^{loc}_{S^p(I^\Delta_i)}\|_{\mathcal{U}(I^\Delta_i)} \le \|u_{EX} - \mathcal{I}^p_{I^\Delta_i} u_{EX}\|_{\mathcal{U}(I^\Delta_i)} \le \mathscr{C}(h_{I^\Delta_i})^{\mu+1/2} \tag{4.4.15c}$$

and

$$\|u_{EX} - \mathcal{I}^p_{I^\Delta_{j(\hat{x})}} u_{EX}\|_{\mathcal{U}(I^\Delta_{j(\hat{x})})} \le \|e^{loc}_{S^p(I^\Delta_{j(\hat{x})})}\|_{\mathcal{U}(I^\Delta_{j(\hat{x})})} \approx \mathscr{C} h^{t-1/2} \tag{4.4.15d}$$

from (4.4.15) it follows that

$$\|e_{S^p_\Delta}\|_{\mathcal{U}} \le \sqrt{\sum_{\substack{i=1 \\ i \neq j(\hat{x})}}^{N} \mathscr{C} h^{2\mu+1}_{I^\Delta_i} + \mathscr{C} h^{2t-1}_{I^\Delta_{j(\hat{x})}}} \le \mathscr{C} h^\sigma. \tag{4.4.15e}$$

Using this result we get

$$|\rho_1^{I^\Delta_j}| = |e^\Delta_j + e^\Delta_{j-1}| \le \mathscr{C} h^\mu \|e_{S^p_\Delta}\|_{\mathcal{U}} \le \mathscr{C} h^{\mu+\sigma} \tag{4.4.16a}$$

and

$$|\rho_2^{I^\Delta_j}| \le \mathscr{C} h^\mu \|e_{S^p_\Delta}\|_{\mathcal{U}} + \mathscr{C} h^{\mu-1/2} \|e_{S^p_\Delta}\|_{\mathcal{U}(I^\Delta_j)} \le \mathscr{C} h^{\mu-1/2} h^\sigma \tag{4.4.16b}$$

where, once more, we used that

$$\|e_{S^p_\Delta}\|_{\mathcal{U}(I^\Delta_j)} \le \|e_{S^p_\Delta}\|_{\mathcal{U}} \le \mathscr{C} h^\sigma. \tag{4.4.16c}$$

The estimate (4.4.14d) follows from

$$\left\|e^{glob}_{S^p(I^{\Delta}_{j(\bar{x})})}\right\|_{\mathcal{U}(I^{\Delta}_{j(\bar{x})})} \leq \mathcal{C}\sqrt{h}\sqrt{(\rho_1^{I^{\Delta}_{j(\bar{x})}})^2 + (\rho_2^{I^{\Delta}_{j(\bar{x})}})^2}. \tag{4.4.16d}$$

□

Remark 4.4.3 In Theorem 4.4.2 we assumed that u_{EX} has $\mu+1$ derivatives everywhere, with the exception of the point $\hat{x}$, where u_{EX} has $t < \mu + 1$ bounded derivatives. The theorem can also be extended to allow for any finite number of points $\hat{x}$.

Remark 4.4.4 From Theorem 4.4.2 it follows that if $t > 1$, the pollution is negligible.

Remark 4.4.5 From Theorems 4.4.1–4.4.2 we conclude that the pollution error in finite element approximations of the solution of the regular model problem, computed using quasi-uniform meshes, is negligible, under relatively weak assumptions about the smoothness of the input data.

Remark 4.4.6 Using the spaces of fractional derivatives introduced in Section 3.3, Theorems 4.4.1–4.4.2 can be extended to non-integer values of t.

Remark 4.4.7 Theorem 4.4.2 shows that the pollution caused by local unsmoothness of f is small. In constrast, as we have already seen in Example 4.3.3, local unsmoothness in AE can cause severe pollution.

Remark 4.4.8 In the two-dimensional case, the smoothness of the boundary must also be considered. If all input data including the boundary are smooth, and the meshes are quasi-uniform, the pollution is negligible. The pollution due to local unsmoothness of f is confined to a neighborhood of the points where f is unsmooth. The pollution due to unsmoothness of the coefficients of the differential equation, or due to corner points at the boundary spreads over the entire domain.

We will now prove that *the quasi-uniformity of the meshes is a necessary condition for the pollution to be negligible.* We will do this by constructing an example where the assumption of quasi-uniformity is violated, and the pollution is not negligible.

Example 4.4.1 Consider the regular model problem in the domain $I = (0, 1)$, with input data

$$AE \equiv 1 + x, \qquad c \equiv 0, \qquad f \equiv 0, \qquad a_2 = 1, \qquad \alpha_2 = 1 \tag{4.4.17a}$$

namely, the problem

$$\frac{d}{dx}\left(AE\frac{du}{dx}\right) \equiv 0, \qquad u(0) = 0, \qquad 2\frac{du}{dx}(1) + u(1) = 1 \tag{4.4.17b}$$

with exact solution given by

$$u_{EX}(x) = C\ln(1 + x), \qquad C = \frac{1}{1 + \ln(2)}. \tag{4.4.17c}$$

We will employ meshes Δ with nodes at

$$x_0^{\Delta} = 0, \qquad x_1^{\Delta} = h_1, \qquad x_j^{\Delta} = h_1 + (j-1)h, \quad j = 1, \ldots, N(\Delta) \tag{4.4.18}$$

and elements of degree $p = 1$, and we will analyze the pollution in $u_{S^1_\Delta}$ for cases that h_1 and h tend to zero with very different rates.

Let us modify the problem (4.4.17) by replacing $AE \equiv 1 + x$, by

$$\overline{AE}(x) = \begin{cases} a, & 0 < x < h_1 \\ 1 + x, & h_1 < x < 1 \end{cases} \tag{4.4.19a}$$

where a will be specified in (4.4.20a) below. The exact solution $\bar{u}_{EX}$ of the problem (4.4.17) with AE replaced by $\overline{AE}$ is

$$\bar{u}_{EX}(x) = \begin{cases} \bar{C}(h_1)x/a, & 0 < x < h_1 \\ \bar{C}(h_1)(\ln(1+x) - \ln(1+h_1) + h_1/a), & h_1 \leq x < 1 \end{cases} \tag{4.4.19b}$$

with

$$\bar{C}(h_1) = (h_1/a - \ln(1 + h_1) + (1 + \ln(2)))^{-1}. \tag{4.4.19c}$$

We will call the boundary value problem (4.4.17b) with $AE \equiv 1+x$, the original problem, and the same problem with AE replaced by $\overline{AE}$, the modified problem. Note that the solution u_{EX} of the original problem is recovered from the solution $\bar{u}_{EX}$ of the modified problem by letting $h_1 = 0$.

Below we will denote by $\bar{u}_{S^1_\Delta}$ (respectively $u_{S^1_\Delta}$) the finite element solution of the modified (respectively original) problem. For the choice

$$a = \frac{1}{h_1} \int_0^{h_1} (1 + x) \tag{4.4.20a}$$

the discrete versions of the original and the modified problems have identical stiffness matrices and load vectors, and we have

$$u_{S^1_\Delta} \equiv \bar{u}_{S^1_\Delta} \tag{4.4.20b}$$

namely, the finite element solutions of the original and the modified problem coincide. Letting

$$\bar{e}_{S^1_\Delta} \stackrel{\text{def}}{=} \bar{u}_{EX} - \bar{u}_{S^1_\Delta} \tag{4.4.21a}$$

from the best-approximation property, we have

$$\|\bar{e}_{S^1_\Delta}\|_{\mathcal{U}} \leq \sqrt{\sum_{j=2}^{N(\Delta)} \|\bar{u}_{EX} - \mathcal{I}^1_{I_j^\Delta} \bar{u}_{EX}\|^2_{\mathcal{U}(I_j^\Delta)}} \leq \mathscr{C}h \tag{4.4.21b}$$

where we utilized that

$$\bar{u}_{EX}|_{I_1^\Delta} \in \mathcal{P}^1(I_1^\Delta) \tag{4.4.21c}$$

and hence

$$\|\bar{u}_{EX} - \mathcal{I}^1_{I_1^\Delta}\bar{u}_{EX}\|_{\mathcal{U}(I_1^\Delta)} = 0. \tag{4.4.21d}$$

Denoting now by $\bar{G}(\cdot;h_1)$ (*respectively* $\bar{G}_{S^1_\Delta}(\cdot;h_1)$) the exact (respectively the finite element approximation of the) Green's function for the modified problem, we also have

$$\|\bar{G}(\cdot;h_1) - \bar{G}_{S^1_\Delta}(\cdot;h_1)\|_{\mathcal{U}} \le \sqrt{\sum_{j=2}^{N(\Delta)} \|\bar{G}(\cdot;h_1) - \mathcal{I}^p_{I_j^\Delta}\bar{G}(\cdot;h_1)\|^2_{\mathcal{U}(I_j^\Delta)}} \le \mathcal{C}h. \tag{4.4.22}$$

It follows that

$$\left|\bar{e}_{S^1_\Delta}(h_1)\right| \le \left\|\bar{e}_{S^1_\Delta}\right\|_{\mathcal{U}}\|\bar{G}(\cdot;h_1) - \bar{G}_{S^1_\Delta}(\cdot;h_1)\|_{\mathcal{U}} \le \mathcal{C}h^2 \tag{4.4.23a}$$

and because

$$u_{S^1_\Delta}(h_1) = \bar{u}_{S^1_\Delta}(h_1) \tag{4.4.23b}$$

we also have

$$\left|u_{S^1_\Delta}(h_1) - \bar{u}_{EX}(h_1)\right| = \left|\bar{e}_{S^1_\Delta}(h_1)\right| \le \mathcal{C}h^2 \tag{4.4.23c}$$

and from this result it follows that

$$\left\|u_{S^1_\Delta} - \bar{u}_{EX}\right\|_{\mathcal{U}(I-I_1^\Delta)} \le \mathcal{C}h. \tag{4.4.23d}$$

This will be proven in Remark 4.4.9 which follows this example.

We also have

$$\left\|u_{EX} - \bar{u}_{EX}\right\|_{\mathcal{U}(I-I_1^\Delta)} = \sqrt{\sum_{j=2}^{N(\Delta)} \left\|u_{EX} - \bar{u}_{EX}\right\|^2_{\mathcal{U}(I_j^\Delta)}} \approx Q'h_1^3 \tag{4.4.24a}$$

with $Q' = \frac{1}{12}\sqrt{\ln(2)}/(1+\ln(2))^2$. This follows from the observation

$$|C - \bar{C}(h_1)| \approx Qh_1^3, \qquad Q = \frac{1}{12(1+\ln(2))^2}. \tag{4.4.24b}$$

The symbol $\approx$ means an asymptotic equality, where terms of order higher than h_1^3 have been omitted.

Using now the triangle inequality

$$\left\|u_{EX} - \bar{u}_{EX}\right\|_{\mathcal{U}(I-I_1^\Delta)} \le \left\|u_{EX} - u_{S^1_\Delta}\right\|_{\mathcal{U}(I-I_1^\Delta)} + \left\|u_{S^1_\Delta} - \bar{u}_{EX}\right\|_{\mathcal{U}(I-I_1^\Delta)} \tag{4.4.25a}$$

and employing (4.4.23d) and (4.4.24a), we obtain

$$Q'h_1^3 \le \left\|u_{EX} - u_{S^1_\Delta}\right\|_{\mathcal{U}(I-I_1^\Delta)} + \mathcal{C}h \tag{4.4.25b}$$

asymptotically, as h_1 tends to zero. Letting

$$h_1 = \mathscr{C}h^\beta \tag{4.4.25c}$$

we obtain

$$\mathscr{C}h^{3\beta} \leq \|u_{EX} - u_{S^1_\Delta}\|_{\mathscr{U}(I-I_1^\Delta)} + \mathscr{C}\,h. \tag{4.4.25d}$$

We will use (4.4.25d) to find the range of β for which the pollution is negligible. In the cases that there is no pollution, we have

$$\|u_{EX} - u_{S^1_\Delta}\|_{\mathscr{U}(I_j^\Delta)} \leq \mathscr{C}h\sqrt{h}, \quad j = 2, \dots, N(\Delta) \tag{4.4.26a}$$

and

$$\|u_{EX} - u_{S^1_\Delta}\|_{\mathscr{U}(I-I_1^\Delta)} \leq \mathscr{C}h\sqrt{\sum_{j=2}^{N(\Delta)} h} = \mathscr{C}h\sqrt{1-h_1} = \mathscr{C}h\sqrt{1-\mathscr{C}h^\beta} \leq \mathscr{C}h. \tag{4.4.26b}$$

Substituting into (4.4.25d) we get

$$\mathscr{C}h^{3\beta} \leq \|u_{EX} - u_{S^1_\Delta}\|_{\mathscr{U}(I-I_1^\Delta)} + \mathscr{C}h \leq \mathscr{C}h \tag{4.4.27a}$$

which holds only if

$$\beta > \tfrac{1}{3}. \tag{4.4.27b}$$

Hence we have shown that for

$$h_1 \leq \mathscr{C}h^\beta, \quad \beta > \tfrac{1}{3} \tag{4.4.27c}$$

the pollution in $u_{S^1_\Delta}$ is negligible.

Remark 4.4.9 We will now prove that

$$|u_{S^1_\Delta} - \bar{u}_{EX}(h_1)| \leq \mathscr{C}h^2 \tag{4.4.28a}$$

implies that

$$\|u_{S^1_\Delta} - \bar{u}_{EX}\|_{\mathscr{U}(I-I_1^\Delta)} \leq \mathscr{C}h. \tag{4.4.28b}$$

Let $\hat{u}_{EX}$ be the solution of the problem:

$$\begin{aligned} &\frac{d}{dx}\left(AE(x)\frac{du}{dx}(x)\right) = 0, \quad h_1 < x < 1, \\ &u(h_1) = u_{S^p_\Delta}(h_1), \qquad 2\frac{du}{dx}(1) + u(1) = 1. \end{aligned} \tag{4.4.29a}$$

Then, from (4.4.28a) we have

$$|\bar{u}_{EX}(h_1) - \hat{u}_{EX}(h_1)| \leq \mathscr{C}h^2 \tag{4.4.29b}$$

and by employing the same procedure used in the derivation of (4.4.24a), we get

$$\|\bar{u}_{EX} - \hat{u}_{EX}\|_{\mathscr{U}(I-I_1^\Delta)} \leq \mathscr{C}h^2 \tag{4.4.29c}$$

with $\mathscr{C}$ independent of h. Noting that $u_{S^1_\Delta}$ is identical to $\hat{u}_{S^1_\Delta}$, we get

$$\|\hat{u}_{EX} - u_{S^1_\Delta}\|_{\mathscr{U}(I-I_1^\Delta)} = \|\hat{u}_{EX} - \hat{u}_{S^1_\Delta}\|_{\mathscr{U}(I-I_1^\Delta)} \leq \mathscr{C}h \tag{4.4.29d}$$

and employing now the triangle inequality, we obtain

$$\|\bar{u}_{EX} - u_{S^1_\Delta}\|_{\mathscr{U}(I-I_1^\Delta)} \leq \|\bar{u}_{EX} - \hat{u}_{EX}\|_{\mathscr{U}(I-I_1^\Delta)} + \|\hat{u}_{EX} - u_{S^1_\Delta}\|_{\mathscr{U}(I-I_1^\Delta)} \leq \mathscr{C}h \tag{4.4.29e}$$

and (4.4.29b) has been proven.

Let us now illustrate the results of Example 4.4.1 using numerical computations.

Example 4.4.2 Consider Example 4.4.1 with meshes Δ_n, obtained by letting $\beta = 0.15$, $N(\Delta_n) = 10\,n + 1$, and $h = h_n$, obtained from the condition

$$h_1 + (N(\Delta_n) - 1)h = h_n^{0.15} + 10nh_n = 1. \tag{4.4.30}$$

Table 4.4.1 gives the values of $\|e_{S^1_{\Delta_n}}\|_{\mathscr{U}(I-I_1^{\Delta_n})}$, $h_n^{-0.45}\|e_{S^1_{\Delta_n}}\|_{\mathscr{U}(I-I_1^{\Delta_n})}$, $\|\check{e}_{S^1_{\Delta_n}}\|_{\mathscr{U}(I-I_1^{\Delta_n})}$, and $h_n^{-1}\|\check{e}_{S^1_{\Delta_n}}\|_{\mathscr{U}(I-I_1^{\Delta_n})}$. Here

$$\check{e}_{S^1_{\Delta_n}} = u_{EX} - \check{u}_{S^1_{\Delta_n}} \tag{4.4.31a}$$

where $\check{u}_{S^1_{\Delta_n}}$ is the *pollution-free finite element solution* which is constrained to be exact at node $x_1^{\Delta_n} = h_n^{0.15}$, namely,

$$\check{u}_{S^1_{\Delta_n}}(h_n^{0.15}) = u_{EX}(h_n^{0.15}). \tag{4.4.31b}$$

TABLE 4.4.1. An example which shows that the pollution in finite element solutions of the regular model problem may be arbitrarily large, asymptotically, for very non-uniform meshes. Input data: $AE(x) = 1 + x$, $0 < x < 1$, $c \equiv 0$, $a_2 = 1$, $\alpha_2 = 10$. Mesh data: $I_1^{\Delta_n} = (0, h_1)$, $I_j^{\Delta_n} = (h_1 + (j-2)\,h_n, h_1 + (j-1)\,h_n)$, $j = 2, \ldots, N(\Delta_n)$, $h_n = (1 - h_1)/(N-1)$, $h_1 = h_n^{0.15}$. The values of the energy norm of the error for $u_{S^1_{\Delta_n}}$ and $\check{u}_{S^1_{\Delta_n}}$ in $I' = I - I_1^{\Delta_n} = (h_1, 1)$. Note that we have $\|e_{S^1_{\Delta_n}}\|_{\mathscr{U}(I')} \approx \mathscr{C}\,h_n^{0.45} \gg h_n$, which means that the pollution in I' is unbounded, relative to the local error, as h tends to zero. Note that the ratio $\|e_{S^1_{\Delta_n}}\|_{\mathscr{U}(I')}/\|\check{e}_{S^1_{\Delta_n}}\|_{\mathscr{U}(I')}$ is less than two, and hence, the pollution in this particular example is not very severe for low accuracies

$N(\Delta_n)$	h_n	$\|e_{S^1_{\Delta_n}}\|_{\mathscr{U}(I')}$	$h_n^{-0.45}\|e_{S^1_{\Delta_n}}\|_{\mathscr{U}(I')}$	$\|\check{e}_{S^1_{\Delta_n}}\|_{\mathscr{U}(I')}$	$h_n^{-1}\|\check{e}_{S^1_{\Delta_n}}\|_{\mathscr{U}(I')}$
11	0.0386	0.2239×10^{-2}	0.9681×10^{-2}	0.1704×10^{-2}	0.4412×10^{-1}
21	0.0218	0.1646×10^{-2}	0.9203×10^{-2}	0.1050×10^{-2}	0.4809×10^{-1}
31	0.0155	0.1399×10^{-2}	0.9123×10^{-2}	0.7798×10^{-3}	0.5033×10^{-1}
41	0.0121	0.1255×10^{-2}	0.9147×10^{-2}	0.6282×10^{-3}	0.5189×10^{-1}
51	0.0100	0.1158×10^{-2}	0.9208×10^{-2}	0.5297×10^{-3}	0.5308×10^{-1}
61	0.0085	0.1087×10^{-2}	0.9282×10^{-2}	0.4600×10^{-3}	0.5404×10^{-1}
71	0.0074	0.1031×10^{-2}	0.9359×10^{-2}	0.4078×10^{-3}	0.5484×10^{-1}
81	0.0066	0.9863×10^{-3}	0.9437×10^{-2}	0.3671×10^{-3}	0.5553×10^{-1}

From the numerical results in Table 4.4.1, we see that

$$\|e_{S^1_{\Delta_n}}\|_{\mathcal{U}(I-I_1^{\Delta_n})} \approx \mathcal{C} h_n^{0.45} \tag{4.4.32a}$$

while

$$\|\check{e}_{S^1_{\Delta_n}}\|_{\mathcal{U}(I-I_1^{\Delta_n})} \approx \mathcal{C} h_n \tag{4.4.32b}$$

and thus

$$\|e_{S^1_{\Delta_n}}\|_{\mathcal{U}(I-I_1^{\Delta_n})} \approx \mathcal{C} h_n^{-0.55} \|\check{e}_{S^1_{\Delta_n}}\|_{\mathcal{U}(I-I_1^{\Delta_n})} \tag{4.4.32c}$$

and, as expected, the pollution error dominates the local error, as h_n tends to zero. From Table 4.4.1 we can also see that, for relatively coarse meshes, with number of elements $N(\Delta_n) \leq 31$, we have

$$\|e_{S^1_{\Delta_n}}\|_{\mathcal{U}(I-I_1^{\Delta_n})} \leq 2\|\check{e}_{S^1_{\Delta_n}}\|_{\mathcal{U}(I-I_1^{\Delta_n})} \tag{4.4.32d}$$

and the pollution could be still acceptable for engineering accuracies, if it is measured in the energy norm. Figures 4.4.1a,b give the graphs of the values of the errors $e_{S^1_{\Delta_5}}$, $\check{e}_{S^1_{\Delta_5}}$ and their derivatives in the subdomain $I - I_1^{\Delta_n}$. From these graphs, it can be clearly seen that the pollution in the value is much more significant than the pollution in the derivative.

Remark 4.4.10 Let us emphasize that the pollution error depends both on the error in the finite element solution, and the Green's function. If the Green's function is smooth, as it is the case for the regular model problem, the lack of smoothness of the exact solution in a subregion does not influence the error outside the region of unsmoothness of the solution, if quasi-uniform meshes are employed. As we will see immediately below, the situation is different if the Green's function is unsmooth. Let us underline that the majority of engineering problems in two and three dimensions are set in domains with re-entrant corners, and/or points where there is a change in the type of boundary condition and the Green's function has algebraic and logarithmic singularities at these points.

We will now give an illustrative example (see also Example 4.3.2) which shows that, *when the Green's function is unsmooth, we should, in general, expect significant pollution when quasi-uniform meshes are employed.*

Example 4.4.3 Let us consider the singular model problem with $f \equiv 0$, $a_2 = 1$, $\alpha_2 \neq 0$, namely, the boundary value problem

$$\frac{d}{dx}\left(x^{\vartheta}\frac{du}{dx}\right) = 0, \qquad u(0) = 0, \qquad \frac{du}{dx}(1) + \alpha_2 u(1) = 1 \tag{4.4.33a}$$

with the exact solution

$$u_{EX}(x) = \frac{1}{1+\alpha_2-\vartheta}x^{1-\vartheta}, \quad 0 \leq x \leq 1. \tag{4.4.33b}$$

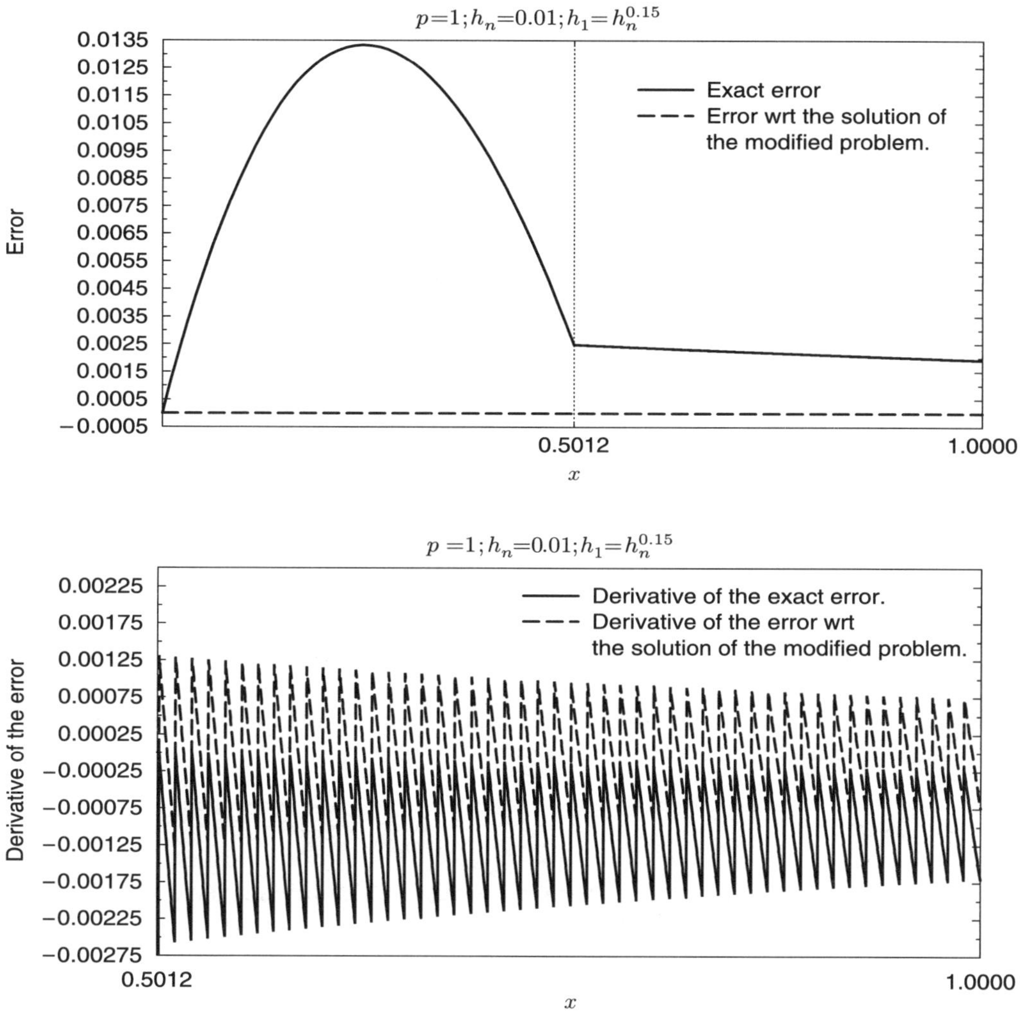

FIG. 4.4.1. An example which shows that the pollution in finite element solutions of the regular model problem may be arbitrarily large, asymptotically, for very non-uniform meshes. Input data: $AE(x) = 1 + x$, $0 < x < 1$, $c \equiv 0$, $a_2 = 1$, $\alpha_2 = 10$. Mesh data: $I_1^{\Delta_n} = (0, h_1)$, $I_j^{\Delta_n} = (h_1 + (j-2)\,h_n, h_1 + (j-1)\,h_n)$, $j = 2, \ldots, N(\Delta_n)$, $h_n = (1 - h_1)/(N-1)$, $h_1 = h_n^{0.15}$. Graphs of. (a) The error $e_{S^1_{\Delta_n}}(x) = (u_{EX} - u_{S^1_{\Delta_n}})(x)$ (solid line) and $\check{e}_{S^1_{\Delta_n}}(x) = (u_{EX} - \check{u}_{S^1_{\Delta_n}})(x)$ (dashed line). (b) Derivative of the error $(d/dx)(e_{S^1_{\Delta_n}})(x)$ (solid line) and $(d/dx)(\check{e}_{S^1_{\Delta_n}})(x)$ (dashed line), for $h_1 = 0.5012 < x < 1$

The Green's function for this problem has the general form

$$G(x; \hat{x}) = \begin{cases} \mathscr{C}_1 x^{1-\vartheta}, & 0 < x < \hat{x} \\ \mathscr{C}_2 x^{1-\vartheta} + \mathscr{C}_3, & \hat{x} < x < 1. \end{cases} \tag{4.4.34a}$$

Employing the continuity of the value

$$G(\hat{x}+0;\hat{x}) = G(\hat{x}-0;\hat{x}) \tag{4.4.34b}$$

and the jump condition for the derivative of the Green's function at $x = \hat{x}$

$$\hat{x}^{\vartheta}\Big(\frac{dG}{dx}(\hat{x}+0;\hat{x}) - \frac{dG}{dx}(\hat{x}-0;\hat{x})\Big) = -1 \tag{4.4.34c}$$

and the boundary condition at $x = 1$,

$$\frac{dG}{dx}(1;\hat{x}) + \alpha_2 G(1;\hat{x}) = 0 \tag{4.4.34d}$$

we get

$$G(x;\hat{x}) = \begin{cases} \dfrac{x^{1-\vartheta}}{1-\vartheta}\Big(1 - \dfrac{\alpha_2\,\hat{x}^{1-\vartheta}}{1-\vartheta+\alpha_2}\Big), & 0 < x < \hat{x} \\[2ex] \dfrac{\hat{x}^{1-\vartheta}}{1-\vartheta}\Big(1 - \dfrac{\alpha_2}{1-\vartheta+\alpha_2}\Big)x^{1-\vartheta}, & \hat{x} < x < 1. \end{cases} \tag{4.4.34e}$$

Note that the Green's function is proportional to $x^{1-\vartheta}$ in $(0,\hat{x})$, and its derivatives have an algebraic singularity at $x = 0$ for $0 < \vartheta < 1$.

We will now show that under the assumption $\alpha_2 \neq 0$, and for uniform meshes Δ_n, the pollution error dominates the local error at $\bar{x} = l$. Noting that

$$\min_{\chi \in S^p(I_1^{\Delta_n})} \|u_{EX} - \chi\|^2_{\mathcal{U}^{\vartheta}(I_1^{\Delta_n})} = \mathscr{C}(p)h^{1-\vartheta} \leq \|e_{S^p_{\Delta_n}}\|^2_{\mathcal{U}^{\vartheta}(I_1^{\Delta_n})} \leq \|e_{S^p_{\Delta_n}}\|^2_{\mathcal{U}^{\vartheta}} \tag{4.4.35a}$$

where $h = h^{\Delta_n}$ is the element size for the mesh Δ_n, and employing the orthogonality

$$\mathscr{B}(e_{S^p_{\Delta_n}}, u_{S^p_{\Delta_n}}) = 0 \tag{4.4.35b}$$

we get

$$\mathscr{C}(p)h^{1-\vartheta} \leq \|e_{S^p_{\Delta_n}}\|^2_{\mathcal{U}^{\vartheta}} = \mathscr{B}(u_{EX} - u_{S^p_{\Delta_n}}, e_{S^p_{\Delta_n}}) = \mathscr{B}(u_{EX}, e_{S^p_{\Delta_n}}) = e_{S^p_{\Delta_n}}(1) \tag{4.4.35c}$$

where we employed that

$$\mathscr{B}(u_{EX}, e_{S^p_{\Delta_n}}) = e_{S^p_{\Delta_n}}(1). \tag{4.4.35d}$$

Hence we have proven that

$$\mathscr{C}(p)h^{1-\vartheta} \leq e_{S^p_{\Delta_n}}(1). \tag{4.4.35e}$$

Further, employing the orthogonality condition

$$\alpha_2 e_{S^p_{\Delta_n}}(1)v(1) + \int_{I_N^{\Delta_n}} x^{\vartheta}\frac{d}{dx}(e_{S^p_{\Delta_n}})\frac{dv}{dx} = \mathscr{B}(e_{S^p_{\Delta_n}}, v) = 0 \tag{4.4.36a}$$

with $v = \varphi_N^{\Delta_n}$, where $\varphi_N^{\Delta_n}$ is the hat function for the Nth node, and using that

$$\varphi_N^{\Delta_n}(1) = 1, \qquad \frac{d}{dx}(\varphi_N^{\Delta_n})\Big|_{I_N^{\Delta_n}} = \frac{1}{h} \tag{4.4.36b}$$

we get

$$\alpha_2 e_{S^p_{\Delta_n}}(1) = -\frac{1}{h}\int_{I_N^{\Delta_n}} x^\vartheta \frac{d}{dx}(e_{S^p_{\Delta_n}}) \leq \frac{1}{h}\sqrt{\int_{I_N^{\Delta_n}} x^\vartheta}\, \|e_{S^p_{\Delta_n}}\|_{\mathcal{U}^\vartheta(I_N^{\Delta_n})}. \tag{4.4.36c}$$

Employing now (4.4.35e), we get

$$\mathcal{C}(p)\sqrt{h}\alpha_2 h^{1-\vartheta} \leq \sqrt{h}\alpha_2 |e_{S^p_{\Delta_n}}(1)| \leq \|e_{S^p_{\Delta_n}}\|_{\mathcal{U}^\vartheta(I_N^{\Delta_n})} \tag{4.4.36d}$$

where we have used that $\sqrt{\int_{I_N^{\Delta_n}} x^\vartheta} \approx \sqrt{h}$. Employing now the triangle inequality

$$\|e_{S^p_{\Delta_n}}\|_{\mathcal{U}^\vartheta(I_N^{\Delta_n})} \leq \|e^{loc}_{S^p(I_N^{\Delta_n})}\|_{\mathcal{U}^\vartheta(I_N^{\Delta_n})} + \|e^{glob}_{S^p(I_N^{\Delta_n})}\|_{\mathcal{U}^\vartheta(I_N^{\Delta_n})} \tag{4.4.37a}$$

we get

$$\mathcal{C}(p)\alpha_2 h^{1-\vartheta}\sqrt{h} \leq \|e_{S^p_{\Delta_n}}\|_{\mathcal{U}^\vartheta(I_N^{\Delta_n})} - \underbrace{\|e^{loc}_{S^p(I_N^{\Delta_n})}\|_{\mathcal{U}^\vartheta(I_N^{\Delta_n})}}_{\approx \mathcal{C}h^p\sqrt{h}} \leq \|e^{glob}_{S^p(I_N^{\Delta_n})}\|_{\mathcal{U}(I_N^{\Delta})} \tag{4.4.37b}$$

where we neglected the term $\|e^{loc}_{S^p(I_N^{\Delta_n})}\|_{\mathcal{U}^\vartheta(I_N^{\Delta_n})}$, as higher order, and hence we have proven that

$$\mathcal{C}(p)\alpha_2\, h^{1-p-\vartheta}\|e^{loc}_{S^p(I_N^{\Delta_n})}\|_{\mathcal{U}(I_N^{\Delta})} \leq \|e^{glob}_{S^p(I_N^{\Delta_n})}\|_{\mathcal{U}(I_N^{\Delta})}. \tag{4.4.37c}$$

Note that for $\alpha_2 \neq 0$, any $\vartheta > 0$, and any $p = 1, 2, \ldots$, we have

$$1 - p - \vartheta < 0. \tag{4.4.37d}$$

We conclude that:

- the pollution error dominates the local error in the last element, as h tends to zero
- for fixed ϑ, *the pollution becomes more severe as the degree p of the elements is increased.*

Recalling (4.4.35e), we have

$$\mathcal{C}h^{1-\vartheta} \leq |e_{S^p_{\Delta_n}}(1)| \leq \|e_{S^p_{\Delta_n}}\|_{\mathcal{U}^\vartheta}\|G(\cdot;1) - \mathcal{I}^p_\Delta G(\cdot;1)\|_{\mathcal{U}^\vartheta} \tag{4.4.38a}$$

and recalling that

$$\|e_{S^p_{\Delta_n}}\|_{\mathcal{U}^\vartheta} \leq \mathcal{C}h^{(1-\vartheta)/2} \tag{4.4.38b}$$

and

$$\|G(\cdot;1) - \mathcal{I}^p_\Delta G(\cdot;1)\|_{\mathcal{U}^\vartheta} \leq \mathcal{C}h^{1-\vartheta} \tag{4.4.38c}$$

we get

$$\mathscr{C}\, h^{(1-\vartheta)/2} \leq \left|e_{S^p_{\Delta_n}}(1)\right| \leq \mathscr{C}\, h^{1-\vartheta} \tag{4.4.38d}$$

which proves that the estimate, based on the duality principle, is optimal. We thus have

$$\left|e_{S^p_{\Delta_n}}(1)\right| \approx \left|\mathscr{B}_{I_1^{\Delta_n}}\left(e_{S^p_{\Delta_n}}, G(\cdot;1) - \mathcal{I}^p_{\Delta} G(\cdot;1)\right)\right| \approx \mathscr{C} h^{1-\vartheta} \tag{4.4.38e}$$

and we can see that the pollution is due to the unsmoothness of both u_{EX} and $G(\cdot;1)$ in the first element $I_1^{\Delta_n}$.

Let us now illustrate the results of Example 4.4.3 using numerical computations.

Example 4.4.4 Above, we have shown that for the singular model problem with input data $AE \equiv x^{\vartheta}$, $0 < \vartheta < 1$, $c \equiv 0$, $f \equiv 0$, $a_2 = 1$, $\alpha_2 > 0$, we have

$$\mathscr{C}\, \alpha_2\, h^{1-p-\vartheta} \|e^{loc}_{S^p(I_N^{\Delta_n})}\|_{\mathscr{U}(I_N^{\Delta})} \leq \|e^{glob}_{S^p(I_N^{\Delta_n})}\|_{\mathscr{U}(I_N^{\Delta})} \tag{4.4.39a}$$

where Δ is a uniform mesh with mesh size h. Let us now consider the more general case of γ-radical meshes, Δ_n^{γ}, and examine numerically the dependence of the pollution error on the value of the exponent ϑ, the ratio $\alpha_2/(1-\vartheta)$, the parameter γ, and the degree p of the elements. Note that the exponent ϑ describes the unsmoothness of AE, and the ratio $\alpha_2/(1-\vartheta)$, the stiffness of the end-support at $x = 1$ relative to the stiffness of the bar. It can be proven that

$$\alpha_2 \gamma^{-p} N^{-\gamma(1-\vartheta)+p} \|e^{loc}_{S^p(I_N^{\Delta_n})}\|_{\mathscr{U}(I_N^{\Delta_n^{\gamma}})} \leq \|e^{glob}_{S^p(I_N^{\Delta_n})}\|_{\mathscr{U}(I_N^{\Delta_n^{\gamma}})} \tag{4.4.39b}$$

and hence if

$$-\gamma(1-\vartheta) + p > 0 \quad \Longrightarrow \quad \gamma < \bar{\gamma}_{cr} \stackrel{\text{def}}{=} \frac{p}{1-\vartheta} \tag{4.4.39c}$$

the pollution is not negligible. For example, for $\vartheta = \frac{1}{4}$ we have $\bar{\gamma}_{cr} = 4p/3$, and for $\vartheta = \frac{3}{4}$, $\bar{\gamma}_{cr} = 4p$.

Tables 4.4.2–4.4.5 report the values of the ratio $\|e_{S^p_{\Delta}}\|_{\mathscr{U}(I_N^{\Delta_n^{\gamma}})} / \|e^{loc}_{S^p(I_N^{\Delta_n})}\|_{\mathscr{U}(I_N^{\Delta_n^{\gamma}})}$ for radical meshes ($x_j^{\Delta_n^{\gamma}} = N^{-\gamma} j^{\gamma}$, $j = 0, \ldots, n = N(\Delta_n^{\gamma})$) with $N(\Delta_n^{\gamma}) = 2, 4, 8$, elements of degree $p = 1, 2, 3, 4$, for several values of γ. Recalling that $e_{S^p(I_N^{\Delta_n})} = e^{loc}_{S^p(I_N^{\Delta_n})} + e^{glob}_{S^p(I_N^{\Delta_n})}$ and employing the triangle inequality, we see that

$$\max\left(0, 1 - \frac{\left\|e^{glob}_{S^p(I_N^{\Delta_n})}\right\|_{\mathscr{U}(I_N^{\Delta_n^{\gamma}})}}{\left\|e^{loc}_{S^p(I_N^{\Delta_n})}\right\|_{\mathscr{U}(I_N^{\Delta_n^{\gamma}})}}\right) \leq \frac{\|e_{S^p_{\Delta_n^{\gamma}}}\|_{\mathscr{U}(I_N^{\Delta_n^{\gamma}})}}{\left\|e^{loc}_{S^p(I_N^{\Delta_n})}\right\|_{\mathscr{U}(I_N^{\Delta_n^{\gamma}})}} \leq 1 + \frac{\left\|e^{glob}_{S^p(I_N^{\Delta_n})}\right\|_{\mathscr{U}(I_N^{\Delta_n^{\gamma}})}}{\left\|e^{loc}_{S^p(I_N^{\Delta_n})}\right\|_{\mathscr{U}(I_N^{\Delta_n^{\gamma}})}}. \tag{4.4.40a}$$

Hence we expect that

$$\frac{\|e_{S^p_{\Delta_n^{\gamma}}}\|_{\mathscr{U}(I_N^{\Delta_n^{\gamma}})}}{\left\|e^{loc}_{S^p(I_N^{\Delta_n})}\right\|_{\mathscr{U}(I_N^{\Delta_n^{\gamma}})}} \approx 1 \tag{4.4.40b}$$

TABLE 4.4.2. Pollution error for the singular model problem ($AE \equiv x^{\vartheta}$, $c \equiv 0$, $a_2 = 1$, $f \equiv 0$). The values of the ratio between the energy norm of the error in the finite element solution and the energy norm of the local error in the Nth element for $\vartheta = \frac{1}{4}$, $\alpha_2/(1-\vartheta) = 10$ (stiff elastic support at $x = 1$) are given for various radical meshes Δ_n^γ, $n = N(\Delta_n^\gamma)$, and elements of degree p. Note that $\bar{\gamma}_{cr} = 4p/3$ ($\bar{\gamma}_{cr} = \frac{4}{3} \simeq 1.333$ for $p = 1$, $\bar{\gamma}_{cr} = \frac{8}{3} \simeq 2.666$ for $p = 2$, $\bar{\gamma}_{cr} = \frac{12}{3} = 4$ for $p = 3$, $\bar{\gamma}_{cr} = \frac{16}{3} \simeq 5.333$ for $p = 4$) and the pollution is negligible asymptotically only if $\gamma > \bar{\gamma}_{cr}$

$N(\Delta_n^\gamma)$	$\Vert e_{S^p_{\Delta_n^\gamma}}\Vert_{\mathcal{U}(I_N^\Delta)} / \Vert e^{loc}_{S^p(I_N^{\Delta_n})}\Vert_{\mathcal{U}(I_N^\Delta)}$					
	$p = 1$			$p = 2$		
	$\gamma = \frac{3}{4}$	$\gamma = 1.0$	$\gamma = \frac{5}{2}$	$\gamma = \frac{3}{4}$	$\gamma = 1.0$	$\gamma = \frac{5}{2}$
2	1.4764	1.2354	1.0158	5.0303	2.6508	1.0233
4	2.0149	1.4356	1.0145	19.1187	8.3414	1.0498
8	2.7220	1.6504	1.0080	59.7170	22.7379	1.0767
16	3.6664	1.8952	1.0036	172.6308	57.6842	1.1026
64	6.6611	2.5301	1.0006	1325.7778	341.7870	1.1557
	$p = 3$			$p = 4$		
	$\gamma = 1.0$	$\gamma = \frac{5}{2}$	$\gamma = 4.0$	$\gamma = \frac{5}{2}$	$\gamma = 4.0$	$\gamma = 5.0$
2	10.5280	1.0643	1.0024	1.2247	1.0032	1.0007
4	84.7189	1.5548	1.0049	5.3969	1.0228	1.0019
8	500.6866	3.3148	1.0069	30.1072	1.1321	1.0038
16	2631.7119	7.6537	1.0082	149.2361	1.5570	1.0066
32	13133.5107	17.3829	1.0090	693.5291	2.7263	1.0105

TABLE 4.4.3. Pollution error for the singular model problem ($AE \equiv x^{\vartheta}$, $c \equiv 0$, $a_2 = 1$, $f \equiv 0$). The values of the ratio between the energy norm of the error in the finite element solution and the energy norm of the local error in the Nth element for $\vartheta = \frac{1}{4}$, $\alpha_2/(1-\vartheta) = \frac{1}{100}$ (soft elastic support at $x = 1$) are given for various radical meshes Δ_n^γ, $n = N(\Delta_n^\gamma)$, and elements of degree p

$N(\Delta_n^\gamma)$	$\Vert e_{S^p_{\Delta_n^\gamma}}\Vert_{\mathcal{U}(I_N^\Delta)} / \Vert e^{loc}_{S^p(I_N^{\Delta_n})}\Vert_{\mathcal{U}(I_N^\Delta)}$					
	$p = 1$			$p = 2$		
	$\gamma = \frac{3}{4}$	$\gamma = 1.0$	$\gamma = \frac{5}{2}$	$\gamma = \frac{3}{4}$	$\gamma = 1.0$	$\gamma = \frac{5}{2}$
2	0.9994	0.9988	0.9933	1.0014	1.0003	0.9995
4	1.0001	0.9998	0.9987	1.0210	1.0040	1.0000
8	1.0003	1.0001	0.9997	1.1904	1.0299	1.0000
16	1.0007	1.0001	0.9999	2.1218	1.1800	1.0000
32	1.0014	1.0002	1.0000	5.3304	1.8353	1.0000
64	1.0025	1.0003	1.0000	14.4431	3.8509	1.0000
	$p = 3$			$p = 4$		
	$\gamma = \frac{3}{4}$	$\gamma = 1.0$	$\gamma = \frac{5}{2}$	$\gamma = 1.0$	$\gamma = \frac{5}{2}$	$\gamma = 4.0$
2	1.0445	1.0064	1.0000	1.1327	1.0000	0.9999
4	2.9893	1.3579	1.0001	10.3578	1.0017	1.0000
8	19.0558	5.5300	1.0006	131.3400	1.0523	1.0000
16	114.0740	28.6353	1.0034	1429.4303	1.9082	1.0001
32	648.7337	142.9102	1.0177	14509.2334	7.6191	1.0004

TABLE 4.4.4. Pollution error for the singular model problem ($AE \equiv x^{\vartheta}$, $c \equiv 0$, $a_2 = 1$, $f \equiv 0$). The values of the ratio between the energy norm of the error in the finite element solution and the energy norm of the local error in the Nth element for $\vartheta = \frac{3}{4}$, $\alpha_2/(1-\vartheta) = 10$ (stiff elastic support at $x = 1$) are given for various radical meshes Δ^{γ}_{n}, $n = N(\Delta^{\gamma}_{n})$, and elements of degree p

N(Δ^{γ}_{n})	$\\|e_{S^p_{\Delta^{\gamma}_{n}}}\\|_{\mathcal{U}(I^{\Delta}_{N})} / \\|e^{loc}_{S^p(I^{\Delta_n}_{N})}\\|_{\mathcal{U}(I^{\Delta}_{N})}$					
	$p = 1$			$p = 2$		
	$\gamma = \frac{3}{4}$	$\gamma = 1.0$	$\gamma = \frac{5}{2}$	$\gamma = \frac{3}{4}$	$\gamma = 1.0$	$\gamma = \frac{5}{2}$
2	7.3345	5.1694	1.7948	40.2048	21.4322	2.9024
4	14.0596	9.2803	2.3473	190.4066	95.0177	8.2762
8	24.6068	15.3702	2.9053	735.8015	348.3686	22.6456
16	41.9641	24.9561	3.5942	2648.4038	1195.4095	59.9485
32	71.0883	40.4027	4.4916	9266.4707	3998.2200	156.2575
	$p = 3$			$p = 4$		
	$\gamma = \frac{5}{2}$	$\gamma = 4.0$	$\gamma = 5.0$	$\gamma = 4.0$	$\gamma = 5.0$	$\gamma = 6.0$
2	5.7307	1.7197	1.2837	2.2791	1.4087	1.1724
4	39.5485	6.1577	2.7839	19.1141	6.3205	2.8779
8	235.7891	27.0006	9.4302	180.3137	47.4057	16.3142
16	1300.7651	115.0359	33.5112	1589.3722	349.2124	99.8579
32	6924.1006	476.0811	116.7879	13381.6592	2474.8250	594.3 759

TABLE 4.4.5. Pollution error for the singular model problem ($AE \equiv x^{\vartheta}$, $c \equiv 0$, $a_2 = 1$, $f \equiv 0$). The values of the ratio between the energy norm of the error in the finite element solution and the energy norm of the local error in the Nth element for $\vartheta = \frac{3}{4}$, $\alpha_2/(1-\vartheta) = \frac{1}{100}$ (soft elastic support at $x = 1$) are given for various radical meshes Δ^{γ}_{n}, $n = N(\Delta^{\gamma}_{n})$, and elements of degree p

N(Δ^{γ}_{n})	$\\|e_{S^p_{\Delta^{\gamma}_{n}}}\\|_{\mathcal{U}(I^{\Delta}_{N})} / \\|e^{loc}_{S^p(I^{\Delta_n}_{N})}\\|_{\mathcal{U}(I^{\Delta}_{N})}$					
	$p = 1$			$p = 2$		
	$\gamma = \frac{3}{4}$	$\gamma = 1.0$	$\gamma = \frac{5}{2}$	$\gamma = \frac{3}{4}$	$\gamma = 1.0$	$\gamma = \frac{5}{2}$
2	0.9947	0.9896	0.9413	1.0406	1.0120	0.9923
4	1.0032	1.0001	0.9884	1.7633	1.2503	1.0025
8	1.0150	1.0063	0.9977	5.9901	3.0846	1.0238
16	1.0494	1.0201	1.0000	22.2073	10.5398	1.1698
32	1.1501	1.0579	1.0008	80.4742	36.4539	1.9076
	$p = 3$			$p = 4$		
	$\gamma = 1.0$	$\gamma = \frac{5}{2}$	$\gamma = 4.0$	$\gamma = \frac{5}{2}$	$\gamma = 4.0$	$\gamma = 5.0$
2	1.2969	1.0000	0.9892	1.0058	0.9966	0.9887
4	9.4936	1.0667	1.0017	2.1782	1.0184	1.0020
8	78.9963	2.5439	1.0388	26.0711	2.1399	1.1199
16	587.5448	13.3856	1.5819	307.0932	17.0227	3.8963
32	4148.0225	72.5932	5.2254	3393.4016	144.4111	26.8604

(respectively >1) if the pollution is negligible (respectively is the dominant component of the error). From the values reported in the tables we see that even for coarse meshes the pollution can be very severe, as the unsmoothness of the data is increased (as $\vartheta \to 1$), the spring support becomes stiffer (as $\alpha_2/(1-\vartheta) \to \infty\big)$, and the degree p of the elements is increased. We also see that it is possible to reduce, and hence control, the effect of the pollution error in an element of interest, by choosing a mesh which is sufficiently refined in the neighborhood of zero (which is the singular point of the Green's function); see, for example, the results in Table 4.4.2 for $p = 1$, $\gamma = \frac{5}{2}$, $\bar{\gamma}_{cr} = \frac{4}{3}$. Note however that for $p > 2$ and for strong singularities (ϑ close to one) it may be practically impossible to employ the strong grading of the mesh which is required in order to control the pollution, because of difficulties with the roundoff error.

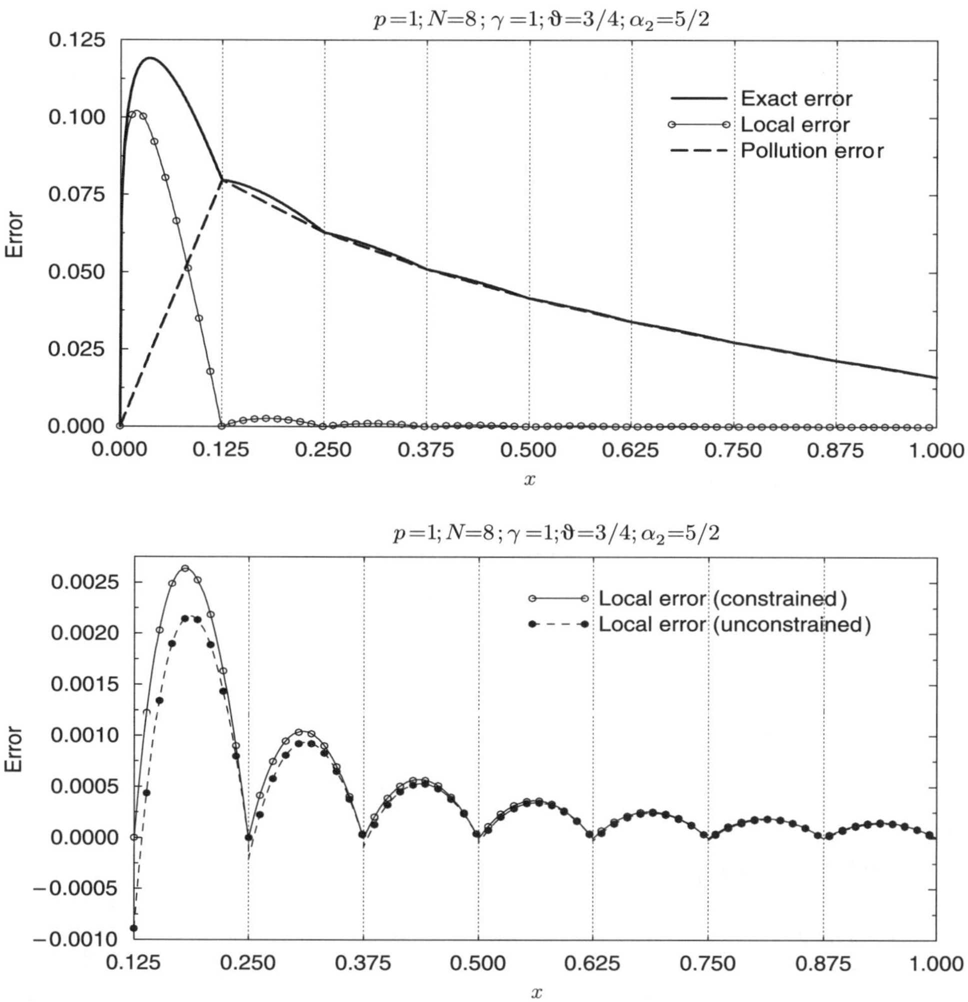

(a)

(b)

FIG. 4.4.2.

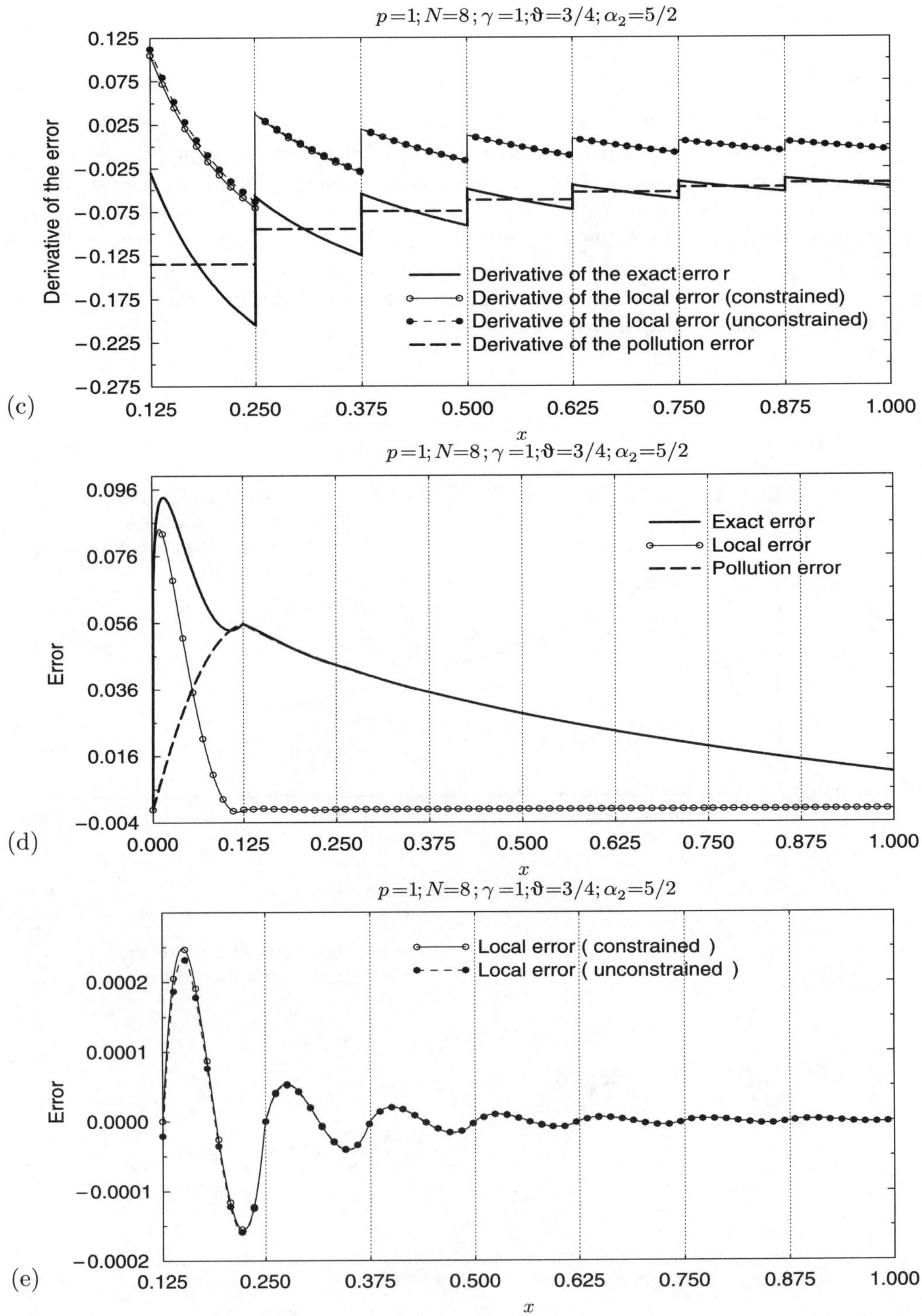

(c)
$p=1; N=8; \gamma=1; \vartheta=3/4; \alpha_2=5/2$
Derivative of the error
Derivative of the exact error
Derivative of the local error (constrained)
Derivative of the local error (unconstrained)
Derivative of the pollution error
x
(d)
$p=1; N=8; \gamma=1; \vartheta=3/4; \alpha_2=5/2$
Error
Exact error
Local error
Pollution error
x
(e)
$p=1; N=8; \gamma=1; \vartheta=3/4; \alpha_2=5/2$
Error
Local error (constrained)
Local error (unconstrained)
x

FIG. 4.4.2.

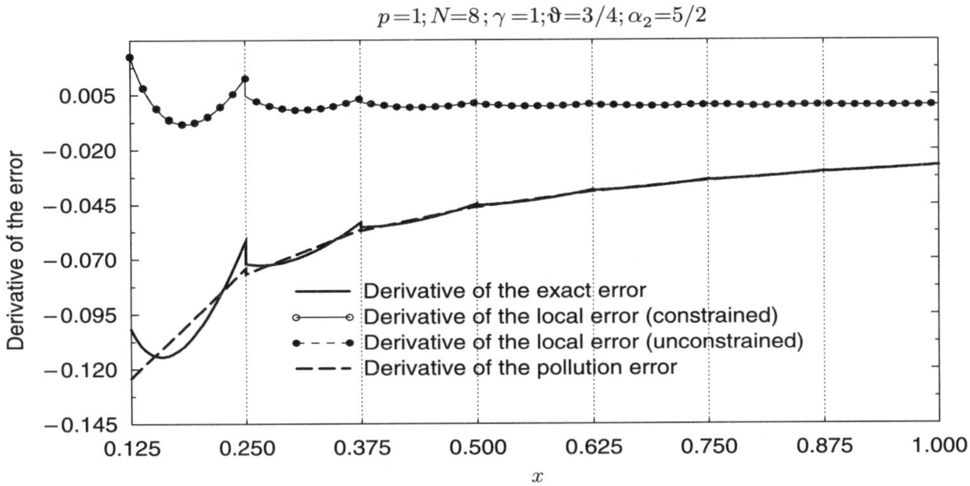

(f)

FIG. 4.4.2. An example which illustrates the relative magnitudes of the local and the pollution error for the singular model problem. Input data: $AE(x) = x^\vartheta$, $0 \le x \le 1$; $\vartheta = \frac{3}{4}$; $c \equiv 0$; $f \equiv 0$; $\alpha_2/(1-\vartheta) = 10$; $a_2 = 1$. Uniform mesh Δ of $N = 8$ elements. (a) Graphs of the error $e_{S^1_\Delta}$, the local error $e^{loc}_{S^p_\Delta} = u_{EX} - \mathcal{A}^p_\Delta u_{EX}$, and the pollution error $e^{glob}_{S^p_\Delta} = \mathcal{A}^p_\Delta u_{EX} - u_{S^p_\Delta}$ for linear elements ($p = 1$). The graphs are shown over the entire domain. (b) Graphs of the local errors $e^{loc}_{S^p_\Delta} = u_{EX} - \mathcal{A}^p_\Delta u_{EX}$ and $\tilde{e}^{loc}_{S^p_\Delta} = u_{EX} - \mathcal{B}^p_\Delta u_{EX}$ which are based on two different definitions of local best approximation for linear elements ($p = 1$). The graphs are shown over the subdomain $I - I^\Delta_1$. (c) Graphs of $(d/dx)(e_{S^p_\Delta})$, $(d/dx)(e^{loc}_{S^p_\Delta})$, $(d/dx)(\tilde{e}^{loc}_{S^p_\Delta})$, and $(d/dx)(e^{glob}_{S^p_\Delta}) = (d/dx)(e_{S^p_\Delta}) - (d/dx)(e^{loc}_{S^p_\Delta})$ for linear elements ($p = 1$). The graphs are shown over the subdomain $I - I^\Delta_1$. The graphs in (d), (e), and (f) are the graphs corresponding to (a), (b), and (c), respectively, in the case of quadratic elements ($p = 2$). Note that the graphs of the local errors $e^{loc}_{S^p_\Delta}$ and $\tilde{e}^{loc}_{S^p_\Delta}$ are practically identical in $I - I^\Delta_1$ and that the pollution error is the major part of the error in all the elements except the first element. By comparing (a), (b), (c) with (d), (e), (f) we see that the magnitude of the pollution error relative to the local error increases with the degree p of the elements

In order to get an idea about the distribution of the pollution and the local error over the elements of the mesh, we constructed graphs of the error in the finite element solution, the pollution error $e^{glob}_{S^p_\Delta} = \mathcal{A}^p_\Delta u_{EX} - u_{S^p_\Delta}$, and the local error $e^{loc}_{S^p_\Delta} = u_{EX} - \mathcal{A}^p_\Delta u_{EX}$ and $\tilde{e}^{loc}_{S^p_\Delta} = u_{EX} - \mathcal{B}^p_\Delta u_{EX}$. The graphs are given for the case $\vartheta = \frac{3}{4}$, $\alpha_2/(1-\vartheta) = 10$, and a uniform mesh of $N = 8$ elements. Figure 4.4.2 shows the graphs for $p = 1$ and $p = 2$. Note that the pollution error is much more significant than the local error in all the elements except the first one, where the local error dominates the pollution error. Note also that the severity of the pollution increases with the degree p of the elements. Figures 4.4.2c,f are very typical and have a direct analogues to the two-dimensional case. Note that the two definitions of the local error $e^{loc}_{S^p_\Delta}$ and $\tilde{e}^{loc}_{S^p_\Delta}$, yield practically identical results.

Remark 4.4.12 The pollution can be understood as a shift in the finite element solution. For example, from Figs 4.4.2c,f it can be seen that $u_{S^p_\Delta}$ is close to the best approximation of the shifted exact solution $\bar{u}_{EX}^{\Delta,p}$ in the subdomain $\hat{I} = (\frac{1}{2}, 1) \subseteq I$, namely

$$\left\|\bar{u}_{EX}^{\Delta,p} - u_{S_{\Delta}^{p}}\right\|_{H^{l}(\tilde{I})} \leq \mathscr{C} h^{p+1-l}, \quad -p \leq l \leq 1. \tag{4.4.41}$$

For example, Figs 4.4.2c,f show that the derivative of the finite element solution $(d/dx)(u_{S_{\Delta}^{p}})$ oscillates about the shifted exact solution $(d/dx)(\bar{u}_{EX}^{\Delta,p})$. The difference $u_{EX} - \bar{u}_{EX}^{\Delta,p}$ is the pollution and converges to zero with h^{Δ}, but possibly with much slower rate than the local error.

Remark 4.4.13 Note that the derivative of the pollution $(d/dx)(\pi_{\Delta})$ is a smooth function which may be approximated as constant or affine over each element. In Section 4.5 we will see that the magnitude of the pollution in the $H^{1}(\tilde{I})$ norm is of the same order as the magnitude of the error measured in the $H^{-l}(I)$ norm. In contrast, the local error, $\bar{u}_{EX}^{\Delta,p} - u_{S_{\Delta}^{p}}$, is oscillatory over the patch with period equal to the element size.

Remark 4.4.14 The above observations have direct analogues in higher dimensions. The main cause for pollution in finite element computations in two dimensions is the singular behavior of the Green's function in the neighborhood of corner points, and it can be controlled by refining the mesh in the neighborhood of these points.

Remark 4.4.15 The pollution error could be especially significant when special approaches, which are designed for computing solutions in complex domains and are known as *global/local* or *zooming* methods, are used. These methods compute an approximate solution over the entire domain using a rather coarse mesh, and then attempt to enhance it locally by solving local boundary value problems with boundary conditions extracted from the coarse mesh solution. Nevertheless any local refinements of the mesh in an area of interest reduce only the local error, and the magnitude of the pollution error is determined by the density of the mesh in the area of interest compared with the density of the mesh in the rest of the domain. Thus, *unless special care is taken to design the global mesh and the local meshes to balance the pollution and local error in the areas of interest, the pollution error may be the dominant component of the error and, if this is the case, only minimal gains in accuracy may be achieved by local refinements.* The justification of the global/local or zooming approaches used in engineering is based on an intuitive understanding of Saint Venant's principle for the error. The error is the exact solution of the model problem loaded by the residuals in the finite element solution (see Chapter 5), and since the residuals are oscillatory, it is argued that they influence the error only locally and hence the pollution error cannot significant. This intuitive understanding of Saint Venant's principle could be misleading, and a quantitative analysis is needed; see Chapter 6. This can be easily seen from Examples 4.4.1 and 4.4.2 where the residuals in the first element are responsible for the pollution error over the entire mesh.

In this section, we analyzed the pollution error in the h version of the finite element method for the model problem. We have shown that there are three types of pollution:

(a) Pollution due to local unsmoothness of the right-hand side f. As shown in Theorems 4.4.1–4.4.3, this type of pollution is essentially limited in the neighborhood of the points of unsmoothness of f.

(b) Pollution due to non-uniformity of the mesh.

(c) Pollution due to unsmoothness in the Green's function, which is related also to the unsmoothness of the coefficients in the differential equation.

As illustrated by Examples 4.4.1 and 4.4.2, pollution of type (b), and (c) spreads over the entire mesh, and the pollution in an element may be due to the behavior of the approximation in a very distant element, e.g., the element next to the singular point for the Green's function. The pollution may be controlled by employing a mesh which is sufficiently refined in the appropriate places, e.g., in the neighborhood of the singular points for the Green's function.

Problems

P4.4.1. Consider the singular model problem for the input data and the radical meshes employed in Example 4.4.2. Obtain a lower bound of the ratio $\|e_{S^p_{\Delta^\gamma_n}}\|_{\mathcal{U}(I_N^{\Delta^\gamma_n})}/\|e^{loc}_{S^p(I_N^{\Delta^\gamma_n})}\|_{\mathcal{U}(I_N^{\Delta^\gamma_n})}$ as a function of γ, p, and N.
Hint: Neglect the errors in all elements, except the first and the last element, and use the Green's function given in Example 4.4.2 and the estimates for the relative size of the elements in the radical meshes given in Lemma 3.4.1.

P4.4.2. Estimate the global energy norm of the error, $\|e_{S^p_\Delta}\|_{\mathcal{U}}$, for the problem considered in Example 3.3.1.

P4.4.3. Derive the optimal density function for the norm $\|\cdot\|_{\mathcal{U}^\vartheta}$ and exact solution considered in Example 4.4.2.
Hint: In Section 3.5 the optimal density for the energy norm was derived for $u_{EX} \equiv x^\alpha$, $AE \equiv 1$, $c \equiv 0$. Here we have $AE \equiv x^\vartheta$, $0 < \vartheta < 1$, $c \equiv 0$.

P4.4.4. Consider the model problem of Example 4.4.1, and assume that the mesh is constructed to correspond to a given density function. Construct an estimate of the error in $I_N^{\Delta^n}$ in terms of the density function.
Hint: Employ estimates of the interpolation error for the solution and the Green's function in terms of the density function.

P4.4.5. Based on the analysis of P4.4.4, find the minimal value for γ as a function of p, such that

$$\frac{\|e_{S^p_{\Delta^\gamma_n}}\|_{\mathcal{U}(I_N^{\Delta^\gamma_n})}}{\|e_{S^p(I_N^{\Delta^\gamma_n})}\|_{\mathcal{U}(I_N^{\Delta^\gamma_n})}} \le 2 \quad \text{as } N \to \infty. \tag{4.4.42}$$

4.4* A-priori estimates for the pollution error in two dimensions

In Section 4.4, we analyzed the pollution error in finite element solutions of the one-dimensional model problem. A similar analysis can also be done in the higher-dimensional case. The main characteristic of problems in higher dimensions is that they are set in domains with corners, and points of abrupt change in the type of boundary conditions from Dirichlet to Neumann, and the Green's function has algebraic singularities at these points, and as it was illustrated in Example 4.4.2 the pollution may be the dominant component of the error in the majority of the elements. As in the one-dimensional case, local unsmoothness in u_{EX} caused by local unsmoothness in the load f does not lead to significant pollution of the finite element solution $u_{S^p_\Delta}$ in elements far from the points where f

is unsmooth, and the pollution error decays with the number of elements which separate the element of interest from the points of unsmoothness of f; these results are proven in [Wahlbin, 1991]. Highly non-uniform meshes will cause pollution in higher dimensions similarly as in the one-dimensional case which was illustrated by Example 4.4.1.

Let us now give some estimates which underline the significance of the pollution in two-dimensional computations.

As we have seen in Section 4.3*, the error $e_{S^1_\Delta}$ is the solution of the residual equation which, in the case of the Laplacian with piecewise constant boundary conditions, and meshes of linear triangles, reads:

Find $e_{S^1_\Delta} \in \mathcal{U}$ such that

$$\mathcal{B}(e_{S^1_\Delta}, v) = \int_\Omega \nabla e_{S^1_\Delta} \cdot \nabla v = \mathcal{R}(v) = \sum_{\epsilon \in \mathcal{E}(\Delta)} J^\Delta_\epsilon \int_\epsilon v \quad \forall v \in \mathcal{U}. \tag{4.4*.1}$$

In the more general case, the residual equation can be written as:

Find $e_{S^p_\Delta} \in \mathcal{U}$ such that

$$\mathcal{B}(e_{S^p_\Delta}, v) = \mathcal{R}(v) \stackrel{\text{def}}{=} \mathcal{L}(v) - \mathcal{B}(u_{S^p_\Delta}, v). \tag{4.4*.2}$$

As we will see in Chapter 5, the residuum $\mathcal{R}$ can be split into element contributions

$$\mathcal{R}(v) = \sum_{\tau \in \Delta} \mathcal{R}^{EQ,p}_\tau (v|_\tau) \quad \forall v \in \mathcal{U} \tag{4.4*.3a}$$

where $\mathcal{R}^{EQ,p}_\tau$ is the *equilibrated element residuum* in the element τ, given by

$$\mathcal{R}^{EQ,p}_\tau (v) \stackrel{\text{def}}{=} \int_\tau r_\tau v + \sum_{\epsilon \subseteq \partial\tau} \int_\epsilon \Big(\frac{1}{2} J_\epsilon + \text{sg}(\epsilon, \tau)\, \theta_\epsilon\Big) v, \quad v \in \mathcal{U}(\tau) \tag{4.4*.3b}$$

where

$$r_\tau = f|_\tau + \mathfrak{D} u_{S^p_\tau} \tag{4.4*.3c}$$

is the *interior residual* in the element τ, J_ϵ is the *jump* of the normal flux on the edge ϵ,

$$\text{sg}(\epsilon, \tau) = +1 \text{ or } -1, \qquad \text{sg}(\epsilon, \tau) + \text{sg}(\epsilon, \tau^*) = 0, \qquad \epsilon = \partial\tau \cap \partial\tau^* \tag{4.4*.3d}$$

for each interior edge, and $\text{sg}(\epsilon, \tau) = 0$, if $\epsilon = \partial\tau \cap \Gamma_N$. Every θ_ϵ can be constructed such that each $\mathcal{R}^{EQ,p}_\tau$ satisfies the orthogonality condition

$$\mathcal{R}^{EQ,p}_\tau (v) = 0 \quad \forall v \in S^p(\tau). \tag{4.4*.3e}$$

For particular constructions of θ_ϵ, see Section 5.3*.

We will now split the error into components corresponding to the splitting (4.4*.3) of $\mathcal{R}$. We will denote by $e^\tau_{S^p_\Delta}$ the component of the error due only to $\mathcal{R}^{EQ,p}_\tau$, namely, the solution of the variational problem:

Find $e^\tau_{S^p_\Delta} \in \mathcal{U}$ such that

$$\mathcal{B}(e^\tau_{S^p_\Delta}, v) = \mathcal{R}^{EQ,p}_\tau (v|_\tau) \quad \forall v \in \mathcal{U}. \tag{4.4*.4a}$$

We then have by superposition

$$e_{S^p_\Delta} = \sum_{\tau\in\Delta} e^\tau_{S^p_\Delta}. \tag{4.4*.4b}$$

Let us underline that

$$e^\tau_{S^p_\Delta} \neq e_{S^p_\Delta}|_\tau. \tag{4.4*.4c}$$

Given an element $\bar{\tau}$, we will let $\omega_{\bar{\tau}}$ be the patch of elements consisting of $\bar{\tau}$ and a few (e.g. two) layers of elements around $\bar{\tau}$, as is, for example, shown in Fig. 4.4*.1 and we will partition the error into two components with respect to $\bar{\tau}$, namely,

$$e^{\omega_{\bar{\tau}}}_{S^p_\Delta} \stackrel{\text{def}}{=} \sum_{\substack{\tau\in\Delta\\ \tau\subseteq\omega_{\bar{\tau}}}} e^\tau_{S^p_\Delta} \tag{4.4*.5a}$$

and

$$e^{\Omega-\omega_{\bar{\tau}}}_{S^p_\Delta} = e_{S^p_\Delta} - e^{\omega_{\bar{\tau}}}_{S^p_\Delta} \stackrel{\text{def}}{=} \sum_{\substack{\tau\in\Delta\\ \tau\subseteq\Omega-\omega_{\bar{\tau}}}} e^\tau_{S^p_\Delta}. \tag{4.4*.5b}$$

The restriction of $e^{\omega_{\bar{\tau}}}_{S^p_\Delta}$ in $\bar{\tau}$

$$e^{loc}_{S^p(\bar{\tau})} = e^{\omega_{\bar{\tau}}}_{S^p_\Delta}\Big|_{\bar{\tau}} \tag{4.4*.5c}$$

is the error in $\bar{\tau}$ due to the residuals in $\bar{\tau}$ and its neighborhood, and hence it has the meaning of local error, and the restriction of $e^{\Omega-\omega_{\bar{\tau}}}_{S^p_\Delta}$ in $\bar{\tau}$

$$e^{glob}_{S^p(\bar{\tau})} = e^{\Omega-\omega_{\bar{\tau}}}_{S^p_\Delta}\Big|_{\bar{\tau}} \tag{4.4*.5d}$$

is the corresponding pollution error.

Using the above constructions, and letting

$$G_i(\mathbf{x};\hat{\mathbf{x}}) = \frac{\partial}{\partial x_i}(G(\mathbf{x};\hat{\mathbf{x}})) \tag{4.4*.6a}$$

denote the Green's function and its first derivative in the ith coordinate direction for any $\hat{\mathbf{x}} \in \Omega$, we have

$$\frac{\partial}{\partial x_i}(e^{\omega_{\bar{\tau}}}_{S^p_\Delta})(\hat{\mathbf{x}}) = \sum_{\substack{\tau\in\Delta\\ \tau\subseteq\omega_{\bar{\tau}}}} \mathcal{R}^{EQ,p}_\tau(G_i(\cdot;\hat{\mathbf{x}})) = \sum_{\substack{\tau\in\Delta\\ \tau\subseteq\omega_{\bar{\tau}}}} \mathcal{R}^{EQ,p}_\tau(G_i(\cdot;\hat{\mathbf{x}}) - \chi_\tau) \tag{4.4*.6b}$$

and similarly

$$\frac{\partial}{\partial x_i}(e^{\Omega-\omega_{\bar{\tau}}}_{S^p_\Delta})(\hat{\mathbf{x}}) = \sum_{\substack{\tau\in\Delta\\ \tau\not\subseteq\omega_{\bar{\tau}}}} \mathcal{R}^{EQ,p}_\tau(G_i(\cdot;\hat{\mathbf{x}})) = \sum_{\substack{\tau\in\Delta\\ \tau\not\subseteq\omega_{\bar{\tau}}}} \mathcal{R}^{EQ,p}_\tau(G_i(\cdot;\hat{\mathbf{x}}) - \chi_\tau) \tag{4.4*.6c}$$

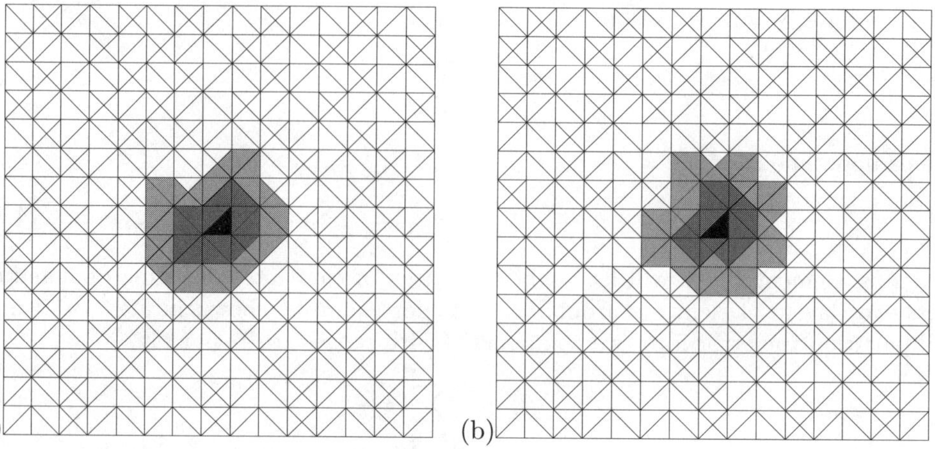

FIG. 4.4*.1. Examples of one- and two-layer neighborhoods $\omega_{\bar{\tau}}$

for every $\chi_\tau \in S^p(\tau)$, where we employed that

$$\mathcal{R}_\tau^{EQ,p}(\chi_\tau) = 0 \quad \forall \chi_\tau \in S^p(\tau) \tag{4.4*.6d}$$

and the integrals have to be understood in the proper sense. It follows that

$$\left| \frac{\partial}{\partial x_i} \left(e_{S_\Delta^p}^{\Omega - \omega_\tau} \right)(\hat{\mathbf{x}}) \right| \le \sum_{\substack{\tau \in \Delta \\ \tau \not\subseteq \omega_\tau}} \|\mathcal{R}_\tau^{EQ,p}\|_{L(\mathcal{U}(\tau),\mathbb{R})} \|G_i(\cdot;\hat{\mathbf{x}}) - \chi_\tau\|_{\mathcal{U}(\tau)}. \tag{4.4*.8a}$$

Employing now that

$$\|e_{S_\Delta^p}\|_{\mathcal{U}} = \|\mathcal{R}\|_{L(\mathcal{U},\mathbb{R})} \approx \sqrt{\sum_{\tau \in \Delta} \|\mathcal{R}_\tau^{EQ,p}\|^2_{L(\mathcal{U}(\tau),\mathbb{R})}} \tag{4.4*.8b}$$

which will be proven in Chapter 5, we get

$$\left| \frac{\partial}{\partial x_i} (e_{S_\Delta^p}^{\Omega - \omega_{\bar{\tau}}})(\hat{\mathbf{x}}) \right| \le \|e_{S_\Delta^p}\|_{\mathcal{U}} \sqrt{\sum_{\substack{\tau \in \Delta \\ \tau \not\subseteq \omega_{\bar{\tau}}}} \|G_i(\cdot;\hat{\mathbf{x}}) - \chi_\tau\|^2_{\mathcal{U}(\tau)}}. \tag{4.4*.8c}$$

Under relatively general assumptions on the input data, we have

$$|D^m G_i(\mathbf{x};\hat{\mathbf{x}})| \le \frac{\mathcal{C}(m)}{R^{m+1}}, \quad \mathbf{x} \in \Omega - B(\hat{\mathbf{x}}; R) \tag{4.4*.9a}$$

with $m > 0$, where $B(\hat{\mathbf{x}}; R)$ is the disk of radius R centered at $\hat{\mathbf{x}}$. For general polygonal domains (4.4*.9a) holds for $m < \hat{m}$, where $\hat{m}$ depends on the angles and the type of boundary conditions. Assuming that $B(\hat{\mathbf{x}}, R) \subset \overline{\omega_{\bar{\tau}}}$, and letting χ_τ be the best approximation of $G_i(\cdot;\hat{\mathbf{x}})$ from $S^p(\tau)$ over the element τ, we have

$$\|G_i(\cdot;\hat{\mathbf{x}}) - \chi_\tau\|_{\mathcal{U}(\tau)} \le \mathcal{C} \frac{h^{p+1}}{R^{p+2}} \tag{4.4*.9b}$$

where χ_τ is fixed by imposing the condition

$$\int_\tau (G_i(\cdot;\hat{\mathbf{x}}) - \chi_\tau) = 0. \qquad (4.4^*.9c)$$

For u_{EX} sufficiently smooth, we get

$$\|e_{S^p_\Delta}\|_{\mathcal{U}} \approx \mathscr{C} h^p \qquad (4.4^*.10a)$$

and we have

$$\left\| \frac{\partial}{\partial x_i} \left(e^{\Omega-\omega_{\tilde{\tau}}}_{S^p_\Delta}\right) \right\|_{L^2(\tilde{\tau})} \leq \mathscr{C} \frac{h^{2p+2}}{R^{p+2}}. \qquad (4.4^*.10b)$$

Recalling that

$$\mathscr{C} h^{p+1} \leq \|e^{\omega_{\tilde{\tau}}}_{S^p_\Delta}\|_{\mathcal{U}(\tilde{\tau})} \qquad (4.4^*.10c)$$

we see that if

$$R \geq h^{(p-\epsilon)/(p+2)} \qquad (4.4^*.10d)$$

the pollution is negligible, asymptotically. This result indicates that for higher p, a smaller number of layers may be used in the definition of the local and hence the pollution error. This observation is also related to the theoretical results in [Schatz and Wahlbin, 1997] and [Schatz, 1998, 2001], and can also be seen in numerical computations.

Let us now assume that $\partial\Omega$ has corner points, and let $r^\alpha \psi(\theta)$ be the leading singular term of the asymptotic expansions at these points. Then it can be proven (see [Wahlbin, 1991]) that

$$\frac{\mathscr{C} h^{2\alpha+1}}{\rho^2} \leq \|e^{\Omega-\omega_{\tilde{\tau}}}_{S^p_\Delta}\|_{\mathcal{U}(\tilde{\tau})} \qquad (4.4^*.11a)$$

where $\rho > \rho_0 > 0$ is the distance of $\omega_{\tilde{\tau}}$ to the vertices, and

$$\|e^{\omega_{\tilde{\tau}}}_{S^p_\Delta}\|_{\mathcal{U}(\tilde{\tau})} \approx \mathscr{C} \rho_0 h^{p+1}. \qquad (4.4^*.11b)$$

It follows that:

(1) For $\alpha < p/2$, we have

$$\|e^{\omega_{\tilde{\tau}}}_{S^p_\Delta}\|_{\mathcal{U}(\tilde{\tau})} \approx \mathscr{C} \rho_0 h^{p+1} \ll \frac{\mathscr{C} h^{2\alpha+1}}{\rho_0^2} \leq \|e^{\Omega-\omega_{\tilde{\tau}}}_{S^p_\Delta}\|_{\mathcal{U}(\tilde{\tau})} \qquad (4.4^*.12a)$$

and the pollution is the dominant part of the error in $\tilde{\tau}$.

(2) For $\alpha > p/2$, we have

$$\|e^{\Omega-\omega_{\tilde{\tau}}}_{S^p_\Delta}\|_{\mathcal{U}(\tilde{\tau})} \leq \frac{\mathscr{C} h^{2\alpha+1}}{\rho_0^2} \ll \mathscr{C} \rho_0 h^{p+1} \approx \|e^{\omega_{\tilde{\tau}}}_{S^p_\Delta}\|_{\mathcal{U}(\tilde{\tau})} \qquad (4.4^*.12b)$$

and there is no pollution.

Let us now show that the above asymptotic results also reflect the behavior for relatively coarse grids.

Example 4.4*.1 (*The case of significant pollution,* $2\alpha < p$) Consider the mixed boundary value problem for the Laplacian in the L-shaped domain, the exact solution $u_{EX} = r^{1/3}\sin(\theta/3)$, shown in Fig. 4.4*.3a, where the boundary conditions are given in Example 4.3*.1, and let $u_{S^p_\Delta}$ for the uniform meshes, as the mesh shown in Fig. 4.4*.3a. Figure 4.4*.2 gives the graphs of $\|e^{\tilde{\omega}^h}_{S^2_\Delta}\|_{\mathcal{U}(\Omega_0)}$, $\|e^{\Omega-\tilde{\omega}^h}_{S^2_\Delta}\|_{\mathcal{U}(\Omega_0)}$, and $\|e_{S^p_\Delta}\|_{\mathcal{U}}$ as functions of the mesh size h, for the mesh patches ω_h which coincide with the subdomain $\Omega_0 = (0.375, 0.625)^2$, which is shown with dashed perigram in Fig. 4.4*.3a. Note that for the pollution error in Ω_0, we have

$$\|e^{\Omega-\tilde{\omega}^h}_{S^p_\Delta}\|_{\mathcal{U}(\Omega_0)} \approx \mathcal{C}h^{2\alpha} = \mathcal{C}h^{2/3} \tag{4.4*.13a}$$

while the local error in Ω_0 satisfies

$$\|e^{\tilde{\omega}^h}_{S^p_\Delta}\|_{\mathcal{U}(\Omega_0)} \approx \mathcal{C}h^2 \tag{4.4*.13b}$$

and the global energy norm of the error

$$\|e_{S^p_\Delta}\|_{\mathcal{U}} \approx \mathcal{C}h^{\alpha} = \mathcal{C}h^{1/3}. \tag{4.4*.13c}$$

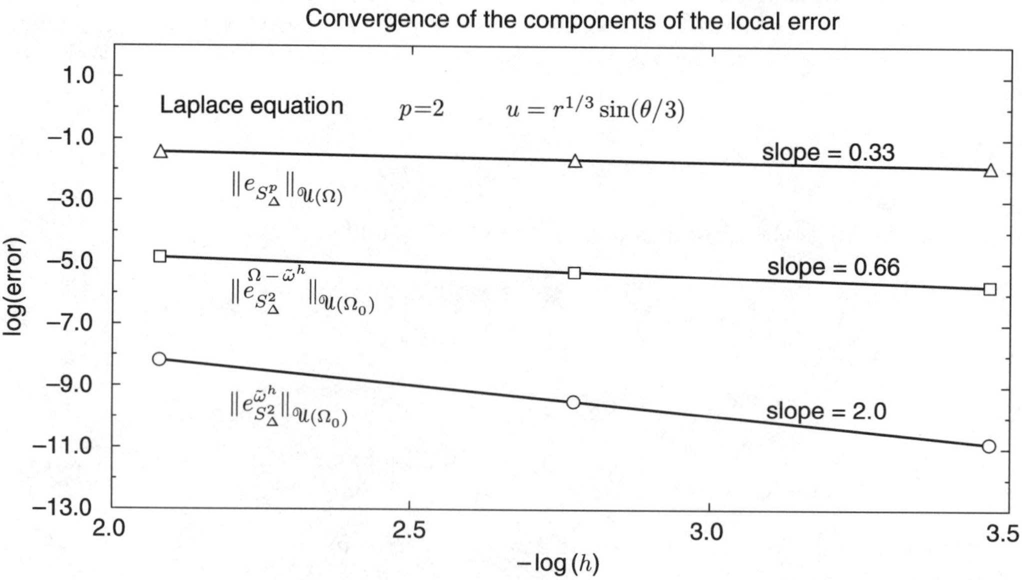

FIG. 4.4*.2. An example which illustrates the rate of convergence of the various components of the error: Laplace equation in the L-shaped domain, $u_{EX} = r^{1/3}\sin(\theta/3)$, uniform mesh of quadratic elements ($p = 2$). The convergence of $\|e^{\tilde{\omega}^h}_{S^p_\Delta}\|_{\mathcal{U}(\Omega_0)}$, $\|e^{\Omega-\tilde{\omega}^h}_{S^p_\Delta}\|_{\mathcal{U}(\Omega_0)}$, $\|e_{S^p_\Delta}\|_{\mathcal{U}(\Omega)}$ as a function of h, where $\Omega_0 = (0.375, 0.625)^2$. Note that $\|e^{\Omega-\tilde{\omega}^h}_{S^p_\Delta}\|_{\mathcal{U}(\Omega_0)}$ converges as $\mathcal{C}\,h^{2\alpha}$, as predicted

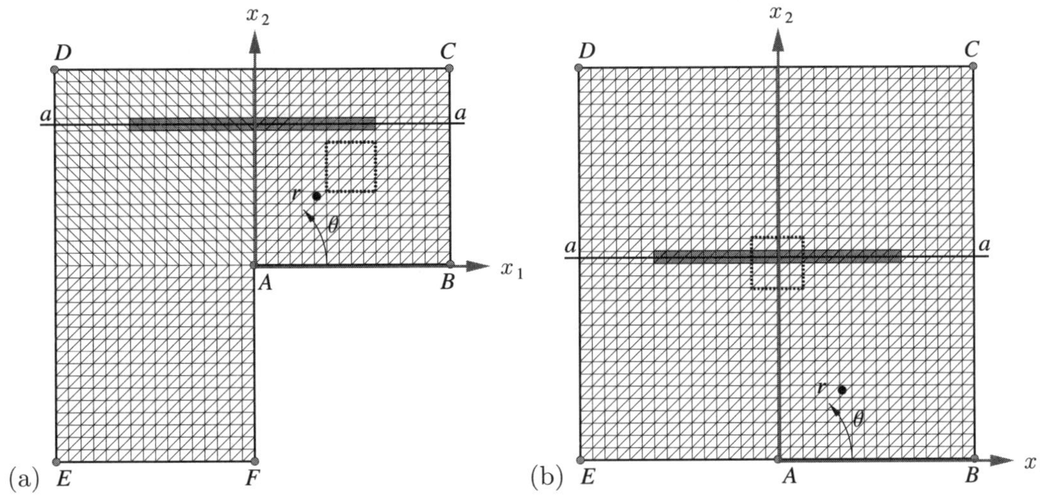

FIG. 4.4*.3. The domains and meshes employed in Example 4.4*.1: (a) The L-shaped domain with $A = (0, 0)$, $B = (1, 0)$, $C = (1, 1)$, $D = (-1, 1)$, $E = (-1, -1)$, $F = (0, -1)$, and the line aa given by $x_2 = 0.71875$. (b) The square domain with $A = (0, 0)$, $B = (1, 0)$, $C = (1, 1)$, $D = (-1, 1)$, $E = (-1, 0)$, and the line aa given by $x_2 = 1.03125$. In both cases the singular vertex (and the origin of the polar coordinate system) is located at A

Let us now illustrate the differences between the cases $2\alpha > p$, $2\alpha = p$, and $2\alpha < p$.

Example 4.4*.2 (The dependence of the relative magnitude of the pollution and the local error on the relation between $2\,\alpha$ and p)

(a) *The case* $2\,\alpha > p$ (*negligible pollution*): Let $u_{S^1_\Delta}$ be the finite element solution of the Laplacian in the L-shaped domain of Fig. 4.4*.3a with homogeneous Dirichlet boundary conditions on the edges AB and FA and Neumann boundary conditions corresponding to the exact solution $u_{EX} = r^{2/3}\,\sin(2\theta/3)$ on the remaining edges, obtained using the mesh shown in Fig. 4.4*.3a and linear elements. Fig. 4.4*.4 shows the graphs of $(\partial/\partial x_1)(e_{S^1_\Delta})$, $(\partial/\partial x_1)(e^{loc}_{S^1_\Delta})$, and $(\partial/\partial x_1)(e^{glob}_{S^1_\Delta})$, along the line aa given by $x_2 = 0.71875$. From this graph, it can be seen that the pollution error is negligible, and $e_{S^p_\Delta} \approx e^{loc}_{S^p_\Delta}$.

(b) *The case* $2\alpha = p$ (*the pollution error and the local error have comparable magnitudes*): Let $u_{S^1_\Delta}$ be the finite element solution of the Laplacian in the domain of Fig. 4.4*.3b with homogeneous Dirichlet boundary conditions on the edge AB, homogeneous Neumann boundary conditions on the edge EA and Neumann boundary conditions corresponding to the exact solution $u_{EX} = r^{1/2}\,\sin(\theta/2)$ on the remaining edges, for the mesh shown in Fig. 4.4.3b. Figure 4.4*.5 gives the graphs of $(\partial/\partial x_1)(e_{S^1_\Delta})$, $(\partial/\partial x_1)(e^{loc}_{S^1_\Delta})$, and $(\partial/\partial x_1)(e^{glob}_{S^1_\Delta})$ along the line $x_2 = 1.03125$, for the mesh of Fig. 4.4*.3b. From these graphs, it can be seen that the error $(\partial/\partial x_1)(e_{S^1_\Delta})$ oscillates, taking greater or smaller values than the local error $(\partial/\partial x_1)(e^{loc}_{S^1_\Delta})$. *This oscillation is due to cancellation between the local and the pollution error* (see [Babuška, Plank, and Rodriguez, 1992]).

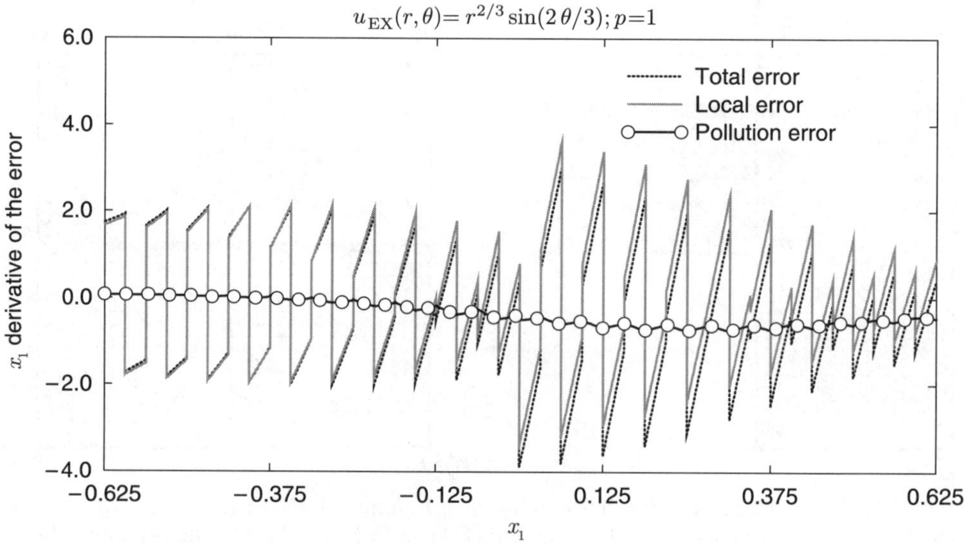

FIG. 4.4*.4. An example of the case $2\alpha > p$ where the pollution is negligible asymptotically: L-shaped domain (given in Fig. 4.4*.3a); exact solution $u_{EX}(r,\theta) = r^{2/3}\sin(2\theta/3)$; linear elements ($p = 1$). The x_1 derivative of the total error, the local error and the pollution error (relative to the maximum x_1 derivative of the solution along aa), along the line aa (given by $x_2 = 0.71875$)

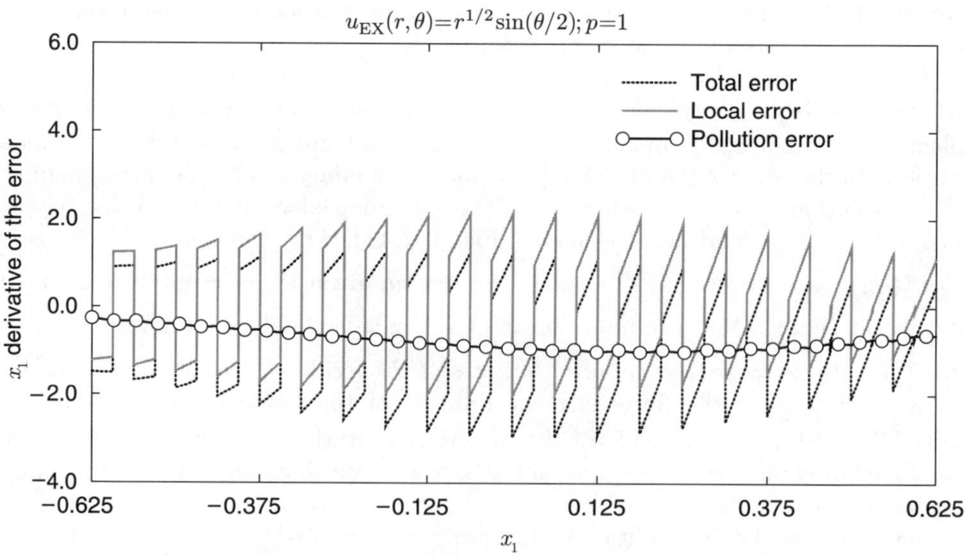

FIG. 4.4*.5. An example of the case $2\alpha = p$ where the pollution error is comparable to the local error: Square domain (given in Fig. 4.4*.3a); exact solution $u_{EX}(r,\theta) = r^{1/2}\sin(\theta/2)$; linear elements ($p = 1$). The x_1 derivative of the total error, the local error and the pollution error (relative to the maximum x_1 derivative of the solution along aa), along the line aa (given by $x_2 = 1.03125$)

(c) *The case $2\alpha < p$ (the pollution error is the dominant part of the total error)*: Once more, let $u_{S^p_\Delta}$ be the finite element solution of the Laplacian in the L-shaped domain of Fig. 4.4*.3a, with boundary conditions corresponding to the exact solution $u_{EX} = r^{1/3}\sin(\theta/3)$, obtained using the mesh shown in Fig. 4.4*.3a and elements of

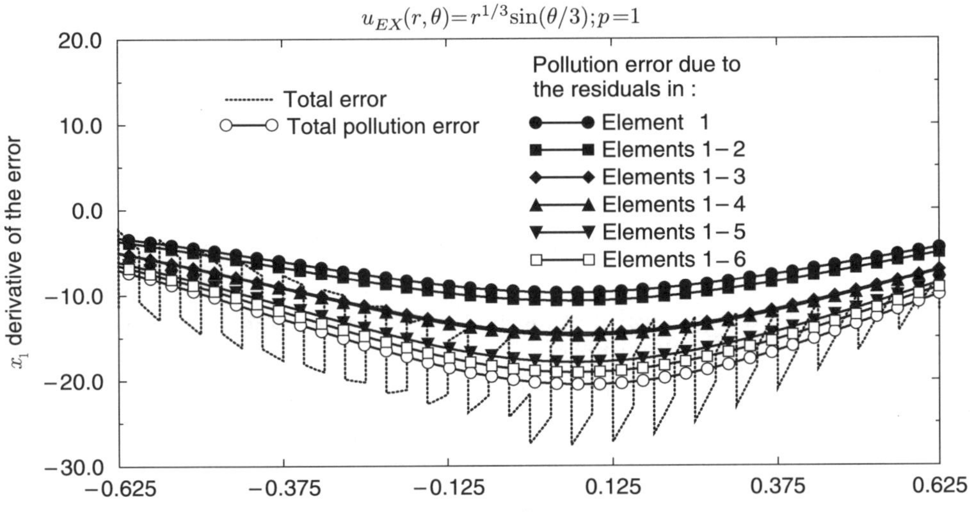

(a)

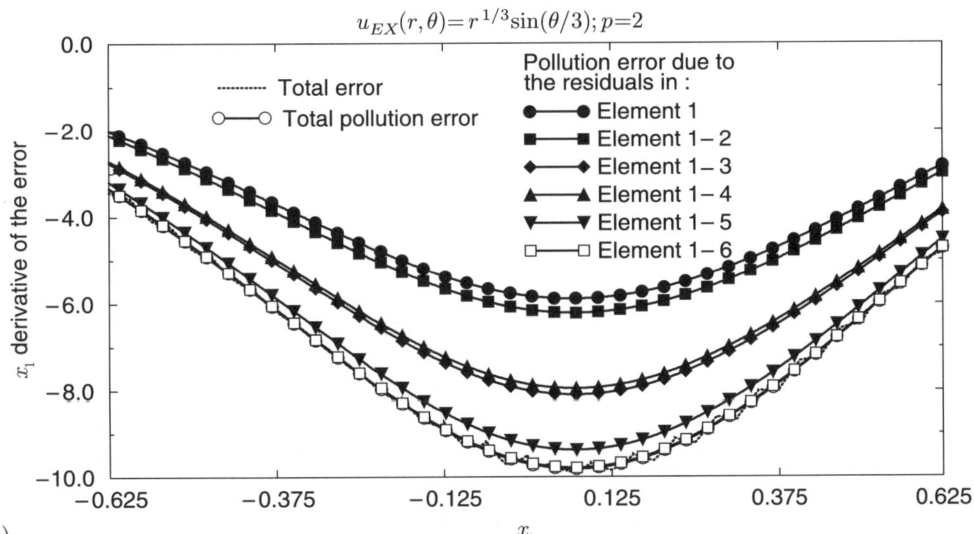

(b)

FIG. 4.4*.6. An example of the case $2\alpha < p$ where the pollution error dominates the local error: L-shaped domain (given in Fig. 4.4*.3a); exact solution $u_{EX}(r,\theta) = r^{1/3}\sin(\theta/3)$. (a) The x_1 derivative of the total error, and the pollution error (relative to the maximum x_1 derivative of the solution along aa) in the case of linear elements ($p = 1$). (b) The x_1 derivative of the total error, and the pollution error (relative to the maximum x_1 derivative of the solution along aa) in the case of quadratic elements ($p = 2$)

degree p. Figure 4.4*.6a shows the graphs of $(\partial/\partial x_1)(e_{S^1_\Delta})$ and $(\partial/\partial x_1)(e^{loc}_{S^1_\Delta})$ along the line $x_2 = 0.71875$. In this case we also give the contributions to the pollution error by the elements connected to the singular point. It can be seen that all the pollution error is essentially due to the six elements at the singularity, and that the pollution error is almost 20% of the maximum value of $\partial u/\partial x_1$ along the line. Note that the local error is small compared to the pollution error. Figure 4.4*.6b shows the analogous results for the case of quadratic elements ($p = 2$); in this case pollution error is almost 100% of the total error.

In this section we discussed the two-dimensional analogues of the local and the pollution error. In particular we gave another definition of the pollution error and the local error based on a splitting of the residuum. This definition has the same characteristics as the one-dimensional definitions given in Section 4.6, and has good meaning in the higher-dimensional case. Analogously as in the one-dimensional case, in higher dimensions:

1. The pollution due to local unsmoothness of the right-hand side is negligible away from the points of unsmoothness.
2. The pollution due to non-uniformity of the mesh spreads over the entire domain.
3. For the quasi-uniform meshes, pollution stems from the corners of the domain where the Green's function has singular behavior.

Here we focused in the case of domains with corner points and quasi-uniform meshes. We showed that the significance of the pollution depends on the relation between 2α and p, where α is the smallest exponent of the asymptotic expansions at the singular points, and p is the degree of the elements, and that the asymptotic results are visible for rather coarse meshes.

Problems

P4.4*.1. Let Ω be a triangular domain and consider the Dirichlet problem for the Poisson equation

$$-\Delta u|_\Omega = f, \qquad u|_{\partial\Omega} = 0 \tag{4.4*.14}$$

where f is such that u_{EX} is smooth. Let $\{\Delta_n\}_{n=1}^{\infty}$ be the sequence of uniform meshes obtained by employing uniform nested subdivisions of Ω, and let $\Omega_0 \subset \Omega$ and interior subdomain of Ω. What are the degrees p for which the pollution in $u_{S^p_\Delta}$ in Ω_0 is negligible (respectively not negligible)?

P4.4*.2. Let $\Omega_\epsilon = \Omega - B(\bar{\mathbf{x}}, \epsilon)$ be the L-shaped domain which is shown in Fig. 4.4*.3a, with an ϵ neighborhood of the re-entrant corner point at $\bar{\mathbf{x}}$ removed, and consider the mixed boundary value for the Laplacian with exact solution $u_{EX} = r^{1/3}\sin(\theta/3)$, with mixed boundary conditions as in Example 4.4*.1 on the edges of the original L-shaped domain, and Neumann boundary condition on the boundary of the cut out. Prove that, for quasi-uniform meshes, there is, asymptotically, no pollution in any interior domain $\Omega_0 \subset \Omega_\epsilon$.

4.5 Further analysis of the pollution error: Interior estimates

Let us now analyze further the problem of the pollution. The best approximation, $\mathcal{B}^p_{I^\Delta_j} u_{EX}$, oscillates around the exact solution u_{EX} with maximum error of order $(h_{I^\Delta_j})^{p+1}$, namely,

$$\|u_{EX} - \mathcal{B}^p_{I^\Delta_j} u_{EX}\|_{L^\infty(I^\Delta_j)} \leq \mathscr{C}(h_{I^\Delta_j})^{p+1}. \tag{4.5.1a}$$

Similarly, the finite element solution, $u_{S^p_\Delta}$, oscillates with maximum error $(h_{I^\Delta_j})^{p+1}$ around a *shifted* exact solution $\bar{u}^{\Delta,p}_{EX}$, and we have

$$\left\|\bar{u}^{\Delta,p}_{EX} - u_{S^p_\Delta}\right\|_{L^\infty(I^\Delta_j)} \approx \mathscr{C}(h_{I^\Delta_j})^{p+1}. \tag{4.5.1b}$$

The difference $\bar{u}^{\Delta,p}_{EX} - u_{EX}$ is the pollution. In Theorem 4.4.1, we have proven that *for sufficiently smooth input data*, we have

$$\|e^{glob}_{S^p_\Delta}\|_{\mathcal{U}(I^\Delta_j)} \approx \|u_{EX} - \bar{u}^{\Delta,p}_{EX}\|_{\mathcal{U}(I^\Delta_j)} \approx \mathscr{C}(h_{I^\Delta_j})^{2p+1/2} \approx \sqrt{h^\Delta}\|e_{S^p_\Delta}\|_{H^{-p+1}(I)} \tag{4.5.1c}$$

and hence the pollution on an interval $I'' \subset I$ measured in the energy norm is of order h^{2p}, which is the same order as the error, $e_{S^p_\Delta}$, measured in the $\|\cdot\|_{H^{-p+1}(I)}$ norm. In this section we will show a more general result, namely, for

$$I'' \subset \bar{I}'' \subset I' \subset I \tag{4.5.2a}$$

we have

$$\|e_{S^p_\Delta}\|_{\mathcal{U}(I'')} \leq \mathscr{C}_1 \min_{\chi \in S^p_\Delta(I')} \|u_{EX} - \chi\|_{\mathcal{U}(I')} + \mathscr{C}_2\|e_{S^p_\Delta}\|_{H^{-p+1}(I)}. \tag{4.5.2b}$$

This means that $\|e_{S^p_\Delta}\|_{\mathcal{U}(I'')}$, the energy norm of the error of the finite element solution in I'' can be estimated in terms of $\min_{\chi\in S^p_\Delta(I')}\|u_{EX} - \chi\|_{\mathcal{U}(I')}$, the energy norm of the error in the best approximation over I', plus $\|e_{S^p_\Delta}\|_{H^{-p+1}(I)}$, the error measured in the negative norm $\|\cdot\|_{H^{-p+1}(I)}$. Estimates of this type are called *interior estimates*, and can be constructed in the one and the higher-dimensional cases by employing analogous arguments. Below we will give a discussion of interior estimates for the one-dimensional model problem, which follows closely the general proof given in [Nitsche and Schatz, 1974] for the higher-dimensional case. See also [Wahlbin, 1991] and the references therein.

First let us give an example to motivate the idea of the interior estimates.

Example 4.5.1 Let us recall the setting of Example 4.4.1. Let $I = (0,1)$, and u_{EX} be the exact solution of the problem

$$\frac{d}{dx}\left((1+x)\frac{du}{dx}(x)\right) = 0, \quad x \in I, \qquad u(0) = 0, \qquad 2\frac{du}{dx}(1) + u(1) = 1 \tag{4.5.3a}$$

and let $\bar{u}_{EX}$ be the exact solution of the problem

$$\frac{d}{dx}\left(\overline{AE}(x)\frac{du}{dx}(x)\right) = 0, \quad x \in I, \qquad u(0) = 0, \qquad 2\frac{du}{dx}(1) + u(1) = 1 \tag{4.5.3b}$$

where

$$\overline{AE}(x) = \begin{cases} (1/h_1)\int_0^{h_1}(1+x), & 0 < x < h_1 \\ 1+x, & h_1 < x < 1. \end{cases} \tag{4.5.3c}$$

Choosing the mesh Δ such that $h_{I_1^\Delta} \equiv h_1$ and $p = 1$, we have

$$u_{S_\Delta^1} \equiv \bar{u}_{S_\Delta^1} \tag{4.5.3d}$$

where $u_{S_\Delta^1}$ (resp. $\bar{u}_{S_\Delta^1}$) is the finite element approximation of u_{EX} (respectively $\bar{u}_{EX}$).

Let now

$$I'' = (a'', b'') \subset I' = (a', b') \subset I, \quad 0 < a' < a'' < b'' < b' < 1 \tag{4.5.4}$$

and for a given mesh Δ, let $x_m^\Delta \in I'$ (respectively $x_n^\Delta \in I'$) be the nodal point of the mesh which is closest to a' (respectively b'), let $I'_\Delta = (x_m^\Delta, x_n^\Delta)$, and assume that $h_{I_1^\Delta} < a'$, which implies $m > 1$. Employing the L^2 norm estimate given in Corollary 4.2.2, we get

$$\|\bar{e}_{S_\Delta^1}\|_{L^2(I'_\Delta)} \leq \|\bar{e}_{S_\Delta^1}\|_{L^2(I-I_1^\Delta)} \leq \mathscr{C}h\|\bar{e}_{S_\Delta^1}\|_{\mathscr{U}(I-I_1^\Delta)} \leq \mathscr{C}h^2 \tag{4.5.5a}$$

where we utilized that

$$\|\bar{e}_{S_\Delta^1}\|_{\mathscr{U}(I-I_1^\Delta)} = \|\bar{u}_{EX} - \bar{u}_{S_\Delta^1}\|_{\mathscr{U}(I-I_1^\Delta)} \approx \mathscr{C}h \tag{4.5.5b}$$

which was proven in Example 4.4.1. Further, using that

$$e_{S_\Delta^1} = u_{EX} - u_{S_\Delta^1} = (u_{EX} - \bar{u}_{EX}) + (\bar{u}_{EX} - u_{S_\Delta^1}) = (u_{EX} - \bar{u}_{EX}) + \bar{e}_{S_\Delta^1} \tag{4.5.6a}$$

where we used that $u_{S_\Delta^1} \equiv \bar{u}_{S_\Delta^1}$, and the triangle inequality, we obtain

$$\|e_{S_\Delta^1}\|_{L^2(I'_\Delta)} \leq \|u_{EX} - \bar{u}_{EX}\|_{L^2(I'_\Delta)} + \|\bar{e}_{S_\Delta^1}\|_{L^2(I'_\Delta)} \approx \mathscr{C}(h^{3\beta} + h^2). \tag{4.5.6b}$$

Here we utilized (4.5.5a) and that

$$\|u_{EX} - \bar{u}_{EX}\|_{L^2(I'_\Delta)} \approx \mathscr{C}h^{3\beta}. \tag{4.5.7a}$$

This follows from the observation that

$$\|u_{EX} - \bar{u}_{EX}\|_{L^2(I'_\Delta)} \leq \|u_{EX} - \bar{u}_{EX}\|_{L^2(I-I_1^\Delta)} \leq \mathscr{C}\|u_{EX} - \bar{u}_{EX}\|_{\mathscr{U}(I-I_1^\Delta)} \tag{4.5.7b}$$

and from

$$\|u_{EX} - \bar{u}_{EX}\|_{\mathscr{U}(I-I_1^\Delta)} \approx \mathscr{C}h_1^3 \tag{4.5.7c}$$

which was proven in Example 4.4.1, and the choice $h_1 = h^\beta$. Using now that

$$\|e_{S_\Delta^1}\|_{\mathscr{U}(I'')} \leq \|u_{EX} - \bar{u}_{EX}\|_{\mathscr{U}(I'')} + \|\bar{e}_{S_\Delta^1}\|_{\mathscr{U}(I'')} \leq \mathscr{C}(h^{3\beta} + h) \tag{4.5.8a}$$

and that

$$\inf_{\chi \in S^1_\Delta(I'_\Delta)} \|u_{EX} - \chi\|_{\mathcal{U}(I'_\Delta)} \approx \mathscr{C}h \tag{4.5.8b}$$

where $S^1_\Delta(I'_\Delta)$ is the space of continuous, piecewise linear functions defined over the submesh in I'_Δ, and (4.5.6b), we get

$$\|e_{S^1_\Delta}\|_{\mathcal{U}(I'')} \leq \mathscr{C}\Big(\inf_{\chi \in S^1_\Delta(I'_\Delta)} \|u_{EX} - \chi\|_{\mathcal{U}(I'_\Delta)} + \|e_{S^1_\Delta}\|_{L^2(I'_\Delta)}\Big). \tag{4.5.8c}$$

This is a typical interior estimate. The first term, $\inf_{\chi \in S^1_\Delta(I'_\Delta)} \|u_{EX} - \chi\|_{\mathcal{U}(I'_\Delta)}$, reflects the local error, while the second term, $\|e_{S^1_\Delta}\|_{L^2(I'_\Delta)}$, accounts for the pollution error.

Let us explain now the main idea in the proof of the interior estimate in the general case.

Example 4.5.2 We will employ a slightly modified version of Example 4.4.1, namely, we will let u_{EX} be the exact solution of the problem

$$\frac{d}{dx}\Big(AE(x)\frac{du}{dx}(x)\Big) = 0, \quad x \in I, \qquad u(0) = 0, \qquad 2\frac{du}{dx}(1) + u(1) = 1 \tag{4.5.9a}$$

where

$$AE(x) = \begin{cases} 1 + x, & 0 < x < \frac{1}{4} \\ 2, & \frac{1}{4} < x < 1. \end{cases} \tag{4.5.9b}$$

We will let

$$I'_\Delta = (x^\Delta_m, x^\Delta_n) \subset \big(\tfrac{1}{2}, 1\big) \tag{4.5.10a}$$

and we will be interested in $e_{S^p_\Delta}|_{I'_\Delta}$, the restriction of the error in I'_Δ, which satisfies the homogeneous differential equation

$$\frac{d}{dx}\Big(AE(x)\frac{d}{dx}(e_{S^p_\Delta}|_{I'_\Delta})(x)\Big) = 0, \quad x \in I'_\Delta \tag{4.5.10b}$$

or, equivalently, the orthogonality condition

$$\mathscr{B}_{I'_\Delta}(e_{S^p_\Delta}, v) = 0 \quad \forall v \in S^p_{\Delta,0}(I'_\Delta) \stackrel{\text{def}}{=} \big\{v \in S^p_\Delta(I'_\Delta) \,\big|\, v(x^\Delta_m) = v(x^\Delta_n) = 0\big\}. \tag{4.5.10c}$$

Noting that $x^\Delta_m > \frac{1}{2}$, and $AE|_{I'_\Delta} \equiv 2$, from (4.5.10b) we obtain

$$e_{S^p_\Delta}|_{I'_\Delta} \in \mathscr{P}^1(I'_\Delta) \tag{4.5.11a}$$

and we have

$$\|e_{S^p_\Delta}\|_{\mathcal{U}(I'')} \leq \mathscr{C}\|e_{S^p_\Delta}\|_{L^2(I'_\Delta)} \tag{4.5.11b}$$

with $\mathscr{C}$ depending on a', a'', b', b''; this will be proven below. Note also that

$$\inf_{\chi \in S^1_\Delta(I'_\Delta)} \|u_{EX} - \chi\|_{\mathscr{U}(I'_\Delta)} = 0 \tag{4.5.11c}$$

and there is no contribution from the local error, and thus (4.5.11b) is the interior estimate.

As we have seen in the above example, the interior estimate (4.5.11b) in $I'' \subset I'_\Delta$, follows by employing that the error $e_{S^1_\Delta}$ is harmonic in I'_Δ. In the particular example, this also implied that $e_{S^1_\Delta}|_{I'_\Delta} \in \mathscr{P}^1(I'_\Delta)$, which is a special case. The general case can be addressed by employing the following lemma:

Lemma 4.5.1 *For any $\psi \in S^p_\Delta$ which satisfies the orthogonality*

$$\mathscr{B}_{I'_\Delta}(\psi, v) = 0 \quad \forall v \in S^p_{\Delta,0}(I'_\Delta) \tag{4.5.12a}$$

and $I'' \subset I'_\Delta$, we have

$$\|\psi\|_{\mathscr{U}(I'')} \leq \mathscr{C}\|\psi\|_{L^2(I'_\Delta)} \tag{4.5.12b}$$

where $I'' \subset I'_\Delta$.

Remark 4.5.1 In the limiting case as h^Δ tends to zero, and/or p diverges to infinity, we get the continuous version of the lemma:

For any $\psi \in \mathscr{U}$ which satisfies the orthogonality

$$\mathscr{B}_{I'_\Delta}(\psi, v) = 0 \quad \forall v \in \mathscr{U}_0(I'_\Delta) \tag{4.5.13a}$$

also satisfies (4.5.12b). This is the classical Harnack's lemma which states that:

For every harmonic function in I'_Δ, the magnitude of the derivatives of the function in $I'' \subset I'_\Delta$, may be bounded by the values of the function.

Lemma 4.5.1 is a discrete version of the classical Harnack's lemma.

Let us prove Harnack's lemma in the simplest one-dimensional setting.

Example 4.5.3 (Harnack's lemma in the simplest one-dimensional setting) Consider the simplest possible case:

$$AE \equiv 1, \qquad c \equiv 0. \tag{4.5.14a}$$

In this special case we have

$$\frac{d^2\psi}{dx^2} \equiv 0 \tag{4.5.14b}$$

from which it follows that

$$\psi \equiv a + bx \tag{4.5.14c}$$

and using this, we get

$$\|\psi\|^2_{L^2(I'_\Delta)} = a^2 \int_{I'_\Delta} 1 + b^2 \int_{I'_\Delta} x^2 + 2ab \int_{I'_\Delta} x \tag{4.5.14d}$$

and

$$\|\psi\|^2_{\mathcal{U}(I'')} = b^2|I''| \tag{4.5.14e}$$

and we obtain

$$\|\psi\|_{\mathcal{U}(I'')} \leq \sqrt{\max_{a,b} \frac{b^2|I''|}{a^2 \int_{I'_\Delta} 1 + b^2 \int_{I'_\Delta} x^2 + 2ab \int_{I'_\Delta} x}} \|\psi\|_{L^2(I'_\Delta)}. \tag{4.5.14f}$$

A similar argument can be used to prove the inequality $\|\psi\|_{\mathcal{U}(I'')} \leq \mathscr{C}\|\psi\|_{L^2(I'_\Delta)}$ in the more general case, where $AE \not\equiv$ *constant*, and $c \not\equiv 0$.

Remark 4.5.2 Let us also elaborate on the two-dimensional version of Harnack's Lemma. Let $\Omega = \{\mathbf{x} \mid |\mathbf{x}| < 1\}$ be an analogue of I'_Δ, and $\Omega_R = \{\mathbf{x} \mid |\mathbf{x}| < R\}$, with $R < 1$, an analogue of I'', and let ψ be harmonic in Ω, namely,

$$\Delta\psi|_\Omega = 0. \tag{4.5.15a}$$

Then ψ can be written as the convergent harmonic series

$$\psi = \sum_{k=0}^{\infty} r^k(a_k \sin k\theta + b_k \cos k\theta), \quad a_0 = 0 \tag{4.5.15b}$$

where r, θ are the polar coordinates centered at the origin. It follows that

$$\|\psi\|^2_{L^2(\Omega)} = \pi \sum_{k=1}^{\infty} \frac{1}{2k+2}(a_k^2 + b_k^2) + b_0^2\pi \tag{4.5.15c}$$

and

$$\|\psi\|^2_{\mathcal{U}(\Omega_R)} = 2\pi \sum_{k=1}^{\infty} \frac{k^2}{2k}(a_k^2 + b_k^2)R^{2k} \leq 2\pi \sum_{k=1}^{\infty} \frac{a_k^2 + b_k^2}{2k+2}\left(\frac{2k+2}{2k}\right)k^2 R^{2k} \tag{4.5.15d}$$

which proves that

$$\|\psi\|^2_{\mathcal{U}(\Omega_R)} \leq 2 \max_{k \geq 1}\left[\left(\frac{2k+2}{2k}\right)k^2 R^{2k}\right]\|\psi\|^2_{L^2(\Omega)} \leq \frac{\mathscr{C}}{(1-R)^2}\|\psi\|^2_{L^2(\Omega)}. \tag{4.5.15e}$$

Following similar steps, we can also obtain

$$\left\|\frac{\partial\psi}{\partial x_1}\right\|^2_{L^\infty(\Omega_R)} + \left\|\frac{\partial\psi}{\partial x_2}\right\|^2_{L^\infty(\Omega_R)} \leq \frac{\mathscr{C}}{(1-R)^2}\|\psi\|^2_{L^2(\Omega)}. \tag{4.5.15f}$$

We will now prove a more general discrete version of Harnack's lemma.

Lemma 4.5.2 *Consider the regular model problem, with the additional assumption*

$$\left\|\frac{d}{dx}(AE)\right\|_{L^\infty(I)} \leq K \tag{4.5.16}$$

or the singular model problem, and let

$$0 < a' < a'' < b'' < b' < 1, \quad d \leq \min(a'' - a', b' - b'') \tag{4.5.17a}$$

and

$$I'' = (a'', b'') \subset I'_\Delta = (x_m^\Delta, x_n^\Delta) \subset I' = (a', b') \quad I'' \neq I'_\Delta \tag{4.5.17b}$$

and let Δ *be quasi-uniform on* I'_Δ. *Then, for any* $\psi \in S_\Delta^p$, *which satisfies the orthogonality*

$$\mathcal{B}_{I'_\Delta}(\psi, v) = 0 \quad \forall v \in S_{\Delta,0}^p(I'_\Delta) \tag{4.5.18a}$$

we have

$$\|\psi\|_{\mathcal{U}(I'')} \leq \frac{\mathcal{C}}{d} \|\psi\|_{L^2(I'_\Delta)} \tag{4.5.18b}$$

with $\mathcal{C}$ *independent of* ψ *and* Δ, *and where*

$$d \leq \min(a' - a'', b'' - b'). \tag{4.5.18c}$$

Proof Note that

$$I'' \subset \hat{I} \stackrel{\text{def}}{=} \left(\frac{a' + a''}{2}, \frac{b' + b''}{2} \right) \subset I'_\Delta \tag{4.5.19a}$$

where the inclusion holds for sufficiently small h, and let ω be a smooth function, $\omega \in C^\infty(I)$, such that

$$\omega|_{I''} \equiv 1, \qquad \omega|_{I - \hat{I}} \equiv 0. \tag{4.5.19b}$$

Then we have

$$\|\psi\|^2_{\mathcal{U}(I'')} \leq \int_{I'_\Delta} \left(AE \left(\frac{d\psi}{dx} \right)^2 \omega + c\psi^2 \omega \right) \tag{4.5.20a}$$

which can also be written as

$$\|\psi\|^2_{\mathcal{U}(I'')} \leq \int_{I'_\Delta} \left(AE \frac{d\psi}{dx} \frac{d}{dx}(\omega\psi) + c\psi(\omega\psi) \right) - \int_{I'_\Delta} AE \frac{d\psi}{dx} \frac{d\omega}{dx} \psi \tag{4.5.20b}$$

and

$$\|\psi\|^2_{\mathcal{U}(I'')} \leq \mathcal{B}_{I'_\Delta}(\psi, \omega\psi) - \int_{I'_\Delta} AE \frac{d\psi}{dx} \frac{d\omega}{dx} \psi. \tag{4.5.20c}$$

Recalling now that ψ satisfies the orthogonality

$$\mathcal{B}_{I'_\Delta}(\psi, \chi) = 0, \quad \forall \chi \in S_{\Delta,0}^p(I'_\Delta) \tag{4.5.20d}$$

we get

$$\|\psi\|^2_{\mathcal{U}(I'')} \leq \mathcal{B}_{I'_\Delta}(\psi, \omega\psi - \chi) - \int_{I'_\Delta} AE \frac{d\psi}{dx} \frac{d\omega}{dx} \psi \quad \forall \chi \in S_{\Delta,0}^p(I'_\Delta) \tag{4.5.20e}$$

Using now that $\omega|_{I-\hat{I}} \equiv 0$, we obtain

$$\int_{I'_\Delta} \frac{d}{dx}\Big(AE\frac{d\omega}{dx}\psi^2\Big) = \Big(AE\frac{d\omega}{dx}\psi^2\Big)(x_n^\Delta) - \Big(AE\frac{d\omega}{dx}\psi^2\Big)(x_m^\Delta) = 0 \tag{4.5.21a}$$

and expanding the left-hand side, we get

$$2\int_{I'_\Delta} AE\frac{d\psi}{dx}\frac{d\omega}{dx}\psi + \int_{I'_\Delta} \frac{d}{dx}\Big(AE\frac{d\omega}{dx}\Big)\psi^2 = 0. \tag{4.5.21b}$$

Using this result in (4.5.20e), we get

$$\|\psi\|^2_{\mathcal{U}(I'')} \leq \mathcal{B}_{I'_\Delta}(\psi, \omega\psi - \chi) + \frac{1}{2}\int_{I'_\Delta} \frac{d}{dx}\Big(AE\frac{d\omega}{dx}\Big)\psi^2 \quad \forall\chi \in S^p_{\Delta,0}(I'_\Delta) \tag{4.5.22a}$$

and employing the Cauchy–Schwarz inequality, we obtain

$$\|\psi\|^2_{\mathcal{U}(I'')} \leq \|\psi\|_{\mathcal{U}(I'_\Delta)}\|\omega\psi - \chi\|_{\mathcal{U}(I'_\Delta)} + \frac{1}{2}\underbrace{\Big\|\frac{d}{dx}\Big(AE\frac{d\omega}{dx}\Big)\Big\|_{L^\infty(I'_\Delta)}}_{\leq K}\|\psi\|^2_{L^2(I'_\Delta)}. \tag{4.5.22b}$$

In Lemma 4.5.2 below we will prove that $\chi \in S^p_{\Delta,0}(I'_\Delta)$ can be chosen such that

$$\|\omega\psi - \chi\|_{\mathcal{U}(I'_\Delta)} \leq \mathcal{C}h\big(\|\psi\|_{\mathcal{U}(I'_\Delta)} + \|\psi\|_{L^2(I'_\Delta)}\big) \tag{4.5.23}$$

with $\mathcal{C}$ independent of ψ and h. Using this result in (4.5.22b), we get

$$\|\psi\|^2_{\mathcal{U}(I'')} \leq \mathcal{C}h\big(\|\psi\|_{\mathcal{U}(I'_\Delta)} + \|\psi\|_{L^2(I'_\Delta)}\big)\|\psi\|_{\mathcal{U}(I'_\Delta)} + \tfrac{1}{2}K\|\psi\|^2_{L^2(I'_\Delta)}. \tag{4.5.24a}$$

It follows that, for sufficiently small h, we get

$$\|\psi\|_{\mathcal{U}(I'')} \leq \mathcal{C}\big(\sqrt{h}\|\psi\|_{H^1(I'_\Delta)} + \|\psi\|_{L^2(I'_\Delta)}\big). \tag{4.5.24b}$$

Replacing I'' by I'_Δ, and I'_Δ by $\check{I}_\Delta \supset I'_\Delta$, we also get

$$\|\psi\|_{H^1(I'_\Delta)} \leq \mathcal{C}\big(\sqrt{h}\|\psi\|_{H^1(\check{I}_\Delta)} + \|\psi\|_{L^2(\check{I}_\Delta)}\big). \tag{4.5.24c}$$

Combining the last two inequalities we get

$$\|\psi\|_{H^1(I'')} \leq \mathcal{C}\big(h\|\psi\|_{H^1(\check{I}_\Delta)} + \sqrt{h}\|\psi\|_{L^2(\check{I}_\Delta)} + \|\psi\|_{L^2(I'_\Delta)}\big) \tag{4.5.25a}$$

and renaming $\check{I}_\Delta$ to I'_Δ we get

$$\|\psi\|_{H^1(I'')} \leq \mathcal{C}\big(h\|\psi\|_{H^1(I'_\Delta)} + \|\psi\|_{L^2(I'_\Delta)}\big). \tag{4.5.25b}$$

Further, because $\psi \in S^p_\Delta$, we have

$$\|\psi\|_{H^1(I'_\Delta)} \leq \frac{\mathcal{C}}{h}\|\psi\|_{L^2(I'_\Delta)}. \tag{4.5.26a}$$

Using this result in (4.5.25b), and recalling that $\|\psi\|_{\mathcal{U}(I'')} \le \mathcal{C}\|\psi\|_{H^1(I'')}$, we finally get

$$\|\psi\|_{\mathcal{U}(I'')} \le \mathcal{C}\|\psi\|_{H^1(I'')} \le \mathcal{C}\|\psi\|_{L^2(I'_\Delta)}. \tag{4.5.26b}$$

Let us now determine the explicit dependence of $\mathcal{C}$ on d, where

$$d \le \min(a' - a'', b'' - b'). \tag{4.5.27a}$$

Employing the linear transformation, $\tilde{x} = x/d$, and denoting by $\tilde{I}'_{\tilde{\Delta}}$, $\tilde{a}'$, $\tilde{a}''$, $\tilde{b}'$, $\tilde{b}''$, the images of I'_Δ, a', a'', b', b'', respectively, we get

$$1 \le \min(\tilde{a}' - \tilde{a}'', \tilde{b}'' - \tilde{b}') \tag{4.5.27b}$$

and

$$\|\tilde{\psi}\|_{H^1(\tilde{I}'')} \le \mathcal{C}\|\tilde{\psi}\|_{L^2(\tilde{I}'_\Delta)} \tag{4.5.27c}$$

with $\mathcal{C}$ independent of d, where (4.5.27c) can be proven following similar steps as above in the mapped domain. Using now that

$$\|\tilde{\psi}\|_{L^2(\tilde{I}'_\Delta)} = \sqrt{\frac{d\tilde{x}}{dx}}\|\psi\|_{L^2(I'_\Delta)} = \frac{1}{\sqrt{d}}\|\psi\|_{L^2(I'_\Delta)} \tag{4.5.28a}$$

and

$$\left\|\frac{d\tilde{\psi}}{d\tilde{x}}\right\|_{L^2(\tilde{I}'_\Delta)} = \sqrt{\frac{dx}{d\tilde{x}}}\left\|\frac{d\psi}{dx}\right\|_{L^2(I_\Delta)} = \sqrt{d}\left\|\frac{d\psi}{dx}\right\|_{L^2(I_\Delta)} \tag{4.5.28b}$$

the estimate (4.5.18b) follows. □

Remark 4.5.3 The proof of Lemma 4.5.1 given above, does not exploit the one-dimensional setting. The same proof can also be used in higher dimensions.

Let us now prove the approximation result used in the above proof.

Lemma 4.5.3 *Consider the setting of Lemma 4.5.2, and let $\omega \in C_0^\infty(I)$, with* supp$(\omega) \subseteq I'_\Delta$*, and $\omega|_{I''} \equiv 1$. Then, for any $\psi \in S^p_\Delta$, there exists a $\chi \in S^p_{\Delta,0}(I'_\Delta)$, such that*

$$\|\omega\psi - \chi\|_{\mathcal{U}(I'_\Delta)} \le \mathcal{C}h\big(\|\psi\|_{\mathcal{U}(I'_\Delta)} + \|\psi\|_{L^2(I'_\Delta)}\big). \tag{4.5.29}$$

Proof Using the chain rule we get

$$\frac{d^{p+1}}{dx^{p+1}}(\omega\psi) = \sum_{s=0}^{p+1}\binom{p+1}{s}\frac{d^s\omega}{dx^s}\frac{d^{p+1-s}\psi}{dx^{p+1-s}}. \tag{4.5.30a}$$

Now, because $\psi|_{I_j^\Delta} \in \mathcal{P}^p(I_j^\Delta)$, we have

$$\left.\frac{d^{p+1}\psi}{dx^{p+1}}\right|_{I_j^\Delta} \equiv 0 \tag{4.5.30b}$$

and for $I_j^\Delta \subset I'_\Delta$, we have

$$\left\| \frac{d^{p+1}}{dx^{p+1}}(\omega\psi) \right\|_{L^2(I_j^\Delta)} \leq \mathscr{C} \sum_{s=0}^{p} \left\| \frac{d^s \psi}{dx^s} \right\|_{L^2(I_j^\Delta)} \tag{4.5.30c}$$

and employing the standard interpolation estimate, we get

$$\|\omega\psi - \mathcal{I}_\Delta^p(\omega\psi)\|_{\mathscr{U}(I_j^\Delta)} \leq \mathscr{C}h^p \left\| \frac{d^{p+1}}{dx^{p+1}}(\omega\psi) \right\|_{L^2(I_j^\Delta)} \leq \mathscr{C}h^p \sum_{s=0}^{p} \left\| \frac{d^s \psi}{dx^s} \right\|_{L^2(I_j^\Delta)}. \tag{4.5.31a}$$

Now, because $\psi|_{I_j^\Delta} \in \mathscr{P}^p(I_j^\Delta)$, for $s \geq 1$, we also have

$$\left\| \frac{d^s \psi}{dx^s} \right\|_{L^2(I_j^\Delta)} \leq \mathscr{C}h^{-(s-1)} \left(\left\| \frac{d\psi}{dx} \right\|_{L^2(I_j^\Delta)} + \|\psi\|_{L^2(I_j^\Delta)} \right) \tag{4.5.31b}$$

and employing this result in the interpolation estimate (4.5.31a), we obtain

$$\|\omega\psi - \mathcal{I}_\Delta^p(\omega\psi)\|^2_{\mathscr{U}(I_j^\Delta)} \leq \mathscr{C}h^2 \left(\|\psi\|^2_{\mathscr{U}(I_j^\Delta)} + \|\psi\|^2_{L^2(I_j^\Delta)} \right). \tag{4.5.31c}$$

Summing over all $I_j^\Delta \subset I'_\Delta$, we obtain (4.5.29) with $\chi \equiv \mathcal{I}_\Delta^p(\omega\psi)$. □

We can also prove a sharper version of Lemma 4.5.2.

Lemma 4.5.4 *Consider the setting of Lemma 4.5.1. We then have*

$$\left\| \frac{d\psi}{dx} \right\|_{L^\infty(I'')} \leq \frac{\mathscr{C}}{d^{3/2}} \|\psi\|_{L^2(I'_\Delta)}. \tag{4.5.32}$$

Proof Let us, without loss of generality, assume that a'', b'' are nodal points of the mesh, and let $z^{I''}$ be the linear interpolant of ψ through these points, namely

$$z^{I''} = \frac{b'' - x}{b'' - a''}\psi(a'') + \frac{x - a''}{b'' - a''}\psi(b''). \tag{4.5.33a}$$

We then have

$$\|z^{I''}\|_{\mathscr{U}(I'')} \leq \mathscr{C}\|\psi\|_{H^1(I'')}. \tag{4.5.33b}$$

Let $\xi^{I''}$ be the energy projection of $z^{I''}$ into $\mathscr{U}_0(I'')$, and $\xi^{I''}_{S^p_{\Delta,0}(I'')}$ its finite element approximation. We then have

$$\mathscr{B}_{I''}(\xi^{I''}, v) = \mathscr{B}_{I''}(z^{I''}, v), \quad \forall v \in \mathscr{U}_0(I'') \tag{4.5.33c}$$

and

$$\mathscr{B}_{I''}(\xi^{I''}_{S^p_\Delta}, v) = \mathscr{B}_{I''}(\xi^{I''}, v) = \mathscr{B}_{I''}(z^{I''}, v), \quad \forall v \in S^p_{\Delta,0}(I''). \tag{4.5.33d}$$

Using now that $z^{I''} \in \mathscr{P}^1(I'')$, and Theorem 2.3.1, we get

$$\left\| \frac{d^2}{dx^2}(\xi^{I''}) \right\|_{L^\infty(I'')} \leq \mathscr{C}\|\psi\|_{H^1(I'')} \tag{4.5.34a}$$

and employing Theorems 3.3.1 and 3.1.2, we obtain

$$\|\xi^{I''} - \xi^{I''}_{S^p_{\Delta,0}(I'')}\|_{\mathcal{U}(I'')} \leq \mathcal{C}h\|\psi\|_{H^1(I'')}. \tag{4.5.34b}$$

Noting that

$$\psi|_{I''} - z^{I''} + \xi^{I''}_{S^p_{\Delta,0}(I'')} \in S^p_{\Delta,0}(I'') \tag{4.5.35a}$$

and

$$\mathcal{B}_{I''}(\psi - z^{I''} + \xi^{I''}_{S^p_{\Delta,0}(I'')}, v) = 0 \quad \forall v \in S^p_{\Delta,0}(I'') \tag{4.5.35b}$$

we get

$$\psi|_{I''} = z^{I''} - \xi^{I''}_{S^p_{\Delta,0}(I'')} = z^{I''} - \xi^{I''} + \xi^{I''} - \xi^{I''}_{S^p_{\Delta,0}(I'')} \tag{4.5.35c}$$

and hence

$$\|\psi\|_{\mathcal{U}(I_j^\Delta)} \leq \|z^{I''} - \xi^{I''}\|_{\mathcal{U}(I_j^\Delta)} + \|\xi^{I''} - \xi^{I''}_{S^p_{\Delta,0}(I'')}\|_{\mathcal{U}(I_j^\Delta)} \tag{4.5.35d}$$

from which we obtain

$$\|\psi\|_{\mathcal{U}(I_j^\Delta)} \leq \mathcal{C}\big(\sqrt{h}\|\psi\|_{\mathcal{U}(I'')} + h\|\psi\|_{\mathcal{U}(I'')}\big) \leq \mathcal{C}\sqrt{h}\|\psi\|_{\mathcal{U}(I'')}. \tag{4.5.35e}$$

Moreover, because $\psi \in \mathcal{P}^p(I_j^\Delta)$, we have

$$\left\|\frac{d\psi}{dx}\right\|_{L^\infty(I_j^\Delta)} \leq \mathcal{C}h^{-1/2}\|\psi\|_{\mathcal{U}(I_j^\Delta)} \leq \mathcal{C}\|\psi\|_{\mathcal{U}(I'')} \tag{4.5.36a}$$

and using (4.5.26b) we get

$$\left\|\frac{d\psi}{dx}\right\|_{L^\infty(I'')} \leq \mathcal{C}\|\psi\|_{L^2(I'_\Delta)}. \tag{4.5.36b}$$

Inequality (4.5.32) is now obtained by employing the same scaling argument as in Lemma 4.5.1. □

Remark 4.5.4 For simplicity in the above proof, we employed one-dimensional arguments, e.g. we employed $z^{I''}$, the interpolant of ψ, which may not be well defined in higher dimensions. A more general proof, which does not exploit the one-dimensional setting, is given in [Nitsche and Schatz, 1974] and [Wahlbin, 1991].

We are now ready to prove the interior estimate.

Theorem 4.5.1 *Consider the regular model problem with* $\|(d/dx)(AE)\|_{L^\infty(I)} \leq K$, *or the singular model problem. Let* I', I'_Δ, I'', *and* d *be defined as above, and assume that the meshes* Δ *are quasi-uniform on* I'_Δ. *Then, for sufficiently small* h, *we have*

$$\|e_{S^p_\Delta}\|_{\mathcal{U}(I'')} \leq \mathcal{C}\Big(\inf_{\chi \in S^p_\Delta}\Big(\|u_{EX} - \chi\|_{\mathcal{U}(I'_\Delta)} + \frac{1}{d}\|u_{EX} - \chi\|_{L^2(I'_\Delta)}\Big) + \frac{1}{d}\|e_{S^p_\Delta}\|_{L^2(I'_\Delta)}\Big) \tag{4.5.37}$$

with $\mathcal{C}$ *independent of* u_{EX}, $u_{S^p_\Delta}$, d, *and* Δ.

Proof Let $\mathcal{B}^p_\Delta u_{EX} \in S^p_\Delta(I'_\Delta)$ be the best approximation of u_{EX} from $S^p_\Delta(I'_\Delta)$; then we have

$$\mathscr{B}_{I'_\Delta}(u_{EX} - \mathcal{B}^p_\Delta u_{EX}, v) = 0 \quad \forall v \in S^p_{\Delta,0}(I'_\Delta) \tag{4.5.38a}$$

and recalling that the error $e_{S^p_\Delta} = u_{EX} - u_{S^p_\Delta}$ satisfies the orthogonality condition

$$\mathscr{B}_{I'_\Delta}\big(u_{EX} - u_{S^p_\Delta}, v\big) = 0 \quad \forall v \in S^p_{\Delta,0}(I'_\Delta) \tag{4.5.38b}$$

we obtain that $u_{S^p_\Delta}|_{I'_\Delta} - \mathcal{B}^p_\Delta u_{EX} \in S^p_\Delta(I'_\Delta)$ satisfies

$$\mathscr{B}_{I'_\Delta}\big(u_{S^p_\Delta} - \mathcal{B}^p_\Delta u_{EX}, v\big) = 0 \quad \forall v \in S^p_{\Delta,0}(I'_\Delta) \tag{4.5.38c}$$

and applying Lemma 4.5.1, we obtain

$$\big\|u_{S^p_\Delta} - \mathcal{B}^p_\Delta u_{EX}\big\|_{\mathscr{U}(I'')} \leq \frac{\mathscr{C}}{d}\|u_{S^p_\Delta} - \mathcal{B}^p_\Delta u_{EX}\|_{L^2(I')}. \tag{4.5.38d}$$

Now, partitioning the error

$$e_{S^p_\Delta} = (u_{EX} - \mathcal{B}^p_\Delta u_{EX}) + (\mathcal{B}^p_\Delta u_{EX} - u_{S^p_\Delta}) \tag{4.5.39a}$$

and using the triangle inequality we obtain

$$\|e_{S^p_\Delta}\|_{\mathscr{U}(I'')} \leq \|u_{EX} - \mathcal{B}^p_\Delta u_{EX}\|_{\mathscr{U}(I'')} + \|\mathcal{B}^p_\Delta u_{EX} - u_{S^p_\Delta}\|_{\mathscr{U}(I'')} \tag{4.5.39b}$$

and employing (4.5.38d) we get

$$\|e_{S^p_\Delta}\|_{\mathscr{U}(I'')} \leq \|u_{EX} - \mathcal{B}^p_\Delta u_{EX}\|_{\mathscr{U}(I'')} + \frac{\mathscr{C}}{d}\big\|u_{S^p_\Delta} - \mathcal{B}^p_\Delta u_{EX}\big\|_{L^2(I'_\Delta)}. \tag{4.5.39c}$$

Using now the triangle inequality for the second term of the right-hand side, we obtain

$$\begin{aligned}\|e_{S^p_\Delta}\|_{\mathscr{U}(I'')} &\leq \|u_{EX} - \mathcal{B}^p_\Delta u_{EX}\|_{\mathscr{U}(I'')} \\ &\quad + \frac{\mathscr{C}}{d}\big(\|u_{S^p_\Delta} - u_{EX} + u_{EX} - \mathcal{B}^p_\Delta u_{EX}\|_{L^2(I'_\Delta)}\big) \\ &\leq \|u_{EX} - \mathcal{B}^p_\Delta u_{EX}\|_{\mathscr{U}(I'')} + \frac{\mathscr{C}}{d}\big(\|e_{S^p_\Delta}\|_{L^2(I'_\Delta)} \\ &\quad + \|u_{EX} - \mathcal{B}^p_\Delta u_{EX}\|_{L^2(I'_\Delta)}\big),\end{aligned} \tag{4.5.39d}$$

and hence

$$\begin{aligned}\|e_{S^p_\Delta}\|_{\mathscr{U}(I'')} &\leq \Big(\|u_{EX} - \mathcal{B}^p_\Delta u_{EX}\|_{\mathscr{U}(I'_\Delta)} + \frac{\mathscr{C}}{d}\|u_{EX} - \mathcal{B}^p_\Delta u_{EX}\|_{L^2(I'_\Delta)}\Big) \\ &\quad + \frac{\mathscr{C}}{d}\|e_{S^p_\Delta}\|_{L^2(I'_\Delta)}\end{aligned} \tag{4.5.39e}$$

and (4.5.37) has been proven. □

Remark 4.5.5 The term $(1/d)\|e_{S^p_\Delta}\|_{L^2(I'_\Delta)}$ may be replaced by $(1/d^{l+1})\|e_{S^p_\Delta}\|_{H^{-l}_{EX}(I'_\Delta)}$.

Remark 4.5.6 Recalling that $\mathcal{B}^p_\Delta u_{EX}$ is the solution of the minimization problem

$$\|u_{EX} - \mathcal{B}^p_\Delta u_{EX}\|_{\mathcal{U}(I'_\Delta)} = \min_{\chi \in S^p_\Delta(I'_\Delta)} \|u_{EX} - \chi\|_{\mathcal{U}(I'_\Delta)} \tag{4.5.40a}$$

and the L^2 norm estimate

$$\|u_{EX} - \mathcal{B}^p_\Delta u_{EX}\|_{L^2(I'_\Delta)} \leq \mathcal{C}h\|u_{EX} - \mathcal{B}^p_\Delta u_{EX}\|_{\mathcal{U}(I'_\Delta)} \tag{4.5.40b}$$

and using these estimates in the last step of the above proof, we get

$$\|e_{S^p_\Delta}\|_{\mathcal{U}(I'')} \leq \mathcal{C}\Big(1 + \frac{h}{d}\Big) \min_{\chi \in S^p_\Delta(I'_\Delta)} \|u_{EX} - \chi\|_{\mathcal{U}(I'_\Delta)} + \frac{\mathcal{C}}{d}\|e_{S^p_\Delta}\|_{L^2(I'_\Delta)}. \tag{4.5.40c}$$

Using Lemma 4.5.3 instead of Lemma 4.5.1 in the proof of Theorem 4.5.1, we obtain:

Theorem 4.5.2 *Let the assumptions of Theorem 4.5.1 hold. Then*

$$\begin{aligned}\Big\|\frac{d}{dx}(e_{S^p_\Delta})\Big\|_{L^\infty(I'')} \leq \mathcal{C} \inf_{\chi \in S^p_\Delta(I'_\Delta)} \Big(\Big\|\frac{d}{dx}(u_{EX} - \chi)\Big\|_{L^\infty(I'_\Delta)} + \frac{\mathcal{C}}{d^{3/2}}\|u_{EX} - \chi\|_{L^2(I'_\Delta)}\Big) \\ + \frac{\mathcal{C}}{d^{3/2}}\|e_{S^p_\Delta}\|_{L^2(I'_\Delta)}\end{aligned} \tag{4.5.41}$$

with $\mathcal{C}$ *independent of* u_{EX}, $u_{S^p_\Delta}$, d, Δ.

Remark 4.5.7 Theorems 4.5.1 and 4.5.2 hold also in the special case that $\bar{I}''$ ends at the boundary; see [Wahlbin, 1991], [Wahlbin, 1995].

Remark 4.5.8 The above estimates are also valid in the case of the pure Neumann problem also in the cases that the error is only unique up to a constant. In these cases, the condition $\int_I e_{S^p_\Delta} = 0$ must be imposed.

The main result of this section is that the error of the finite element solution in the domain I'' is the error of the local best approximation on a slightly larger domain I' and the error of the finite element solution in a *lower norm*. Estimates of this type are called interior estimates, and they are of interest because they show the interplay between the local and the pollution errors. The error in the local best approximation reflects the local error, while the pollution error is accounted for by measuring the error in a lower norm.

Problems

P4.5.1. Consider the setting of Example 4.5.1, and prove the interior estimate

$$\|e_{S^p_\Delta}\|_{\mathcal{U}(I'')} \leq \mathcal{C}(l)\Big(\inf_{\chi \in S^p_\Delta(I'_\Delta)} \|u_{EX} - \chi\|_{\mathcal{U}(I'_\Delta)} + \|e_{S^p_\Delta}\|_{H^{-l}(I')}\Big). \tag{4.5.42}$$

P4.5.2. Estimate the value of $\|e_{S^p_\Delta}\|_{H^{-l}(I)}$ as a function of s and p in the case of the regular model problem with sufficiently smooth AE, c, and f.

4.5* The interior estimates in two dimensions

In this section we will state without proof the two-dimensional versions of the interior estimates which were derived in Section 4.5.1 for the one-dimensional case.

Let $\Omega^1 \subset \Omega^0 \subset \Omega$ denote interior subdomains, and assume, for simplicity, that Ω^1, Ω^0, always coincide with submeshes Δ^1, Δ^0, respectively, of the original mesh Δ. We then have

$$\|\nabla e_{S^p_\Delta}\|_{L^2(\Omega^1)} \leq \mathscr{C} \min_{\chi\in S^p_\Delta(\Omega^0)} \Big(\|\nabla(u_{EX}-\chi)\|_{L^2(\Omega^0)} + \frac{1}{d}\|u_{EX}-\chi\|_{L^2(\Omega^0)}\Big) + \frac{\mathscr{C}}{d}\|e_{S^p_\Delta}\|_{L^2(\Omega^0)} \tag{4.5*.1}$$

and

$$\|\nabla e_{S^p_\Delta}\|_{L^\infty(\Omega^1)} \leq \mathscr{C} \min_{\chi\in S^p_\Delta(\Omega^0)} \Big(\|\nabla(u_{EX}-\chi)\|_{L^\infty(\Omega^0)} + \frac{1}{d}\|u_{EX}-\chi\|_{L^\infty(\Omega^0)}\Big) + \frac{\mathscr{C}}{d^2}\|e_{S^p_\Delta}\|_{L^2(\Omega^0)}. \tag{4.5*.2}$$

The second term in (4.5*.1) can also be written in terms of $\|e_{S^p_\Delta}\|_{H^{-s}(\Omega_0)}$, to obtain

$$\|\nabla e_{S^p_\Delta}\|_{L^2(\Omega^1)} \leq \mathscr{C} \min_{\chi\in S^p_\Delta(\Omega^0)} \Big(\|\nabla(u_{EX}-\chi)\|_{L^2(\Omega^0)} + \frac{1}{d}\|u_{EX}-\chi\|_{L^2(\Omega^0)}\Big) + \frac{\mathscr{C}}{d^{s+1}}\|e_{S^p_\Delta}\|_{H^{-s}(\Omega^0)} \tag{4.5*.3}$$

and similarly for (4.5*.2). The above estimates are analogous to the estimates obtained in the one-dimensional case, and also hold in the case of the pure Neumann problem with no zeroth-order term, in which the error is unique up to an additive constant; in this case the condition $\int_\Omega e_{S^p_\Delta} = 0$ must be imposed. The idea of the splitting the error into a local part and a global part, measured in the lower norm, was first introduced in [Nitsche and Schatz, 1974]; see also [Wahlbin, 1991, 1995] and the references therein.

In this section we have stated without proof some interior estimates in the two-dimensional case. These estimates are analogous to the estimates proven in the one-dimensional case.

Problems

P4.5*.1. Let Ω be the L-shaped domain $A\,B\,C\,D\,E\,F\,A$ with internal angle $4\pi/3$ at A, and consider the mixed boundary value problem

$$\Delta u|_\Omega = 0 \qquad \frac{\partial u}{\partial n}\Big|_{BCDEF} = g_i \qquad u|_{FABE} = 0 \tag{4.5*.4a}$$

for $i = 1, 2$, where g_1 and g_2 are such that $u^{(1)}_{EX} = r^{3/4}\sin(3\theta/4)$, and $u^{(2)}_{EX} = r^{6/4}\sin(6\theta/4)$, respectively.

(a) Prove that $u^{(1)}_{EX} \in B^{7/4}(\Omega)$ and $u^{(2)}_{EX} \in B^{10/4}(\Omega)$, where $B^\alpha(\Omega)$ is the Besov space introduced in Chapter 3.

(b) Let $e^{(i)}_{S^p_\Delta}$ be the error in the finite element approximation of $u^{(i)}_{EX}$ and show that

$$\|e^{(1)}_{S^p_\Delta}\|_{\mathcal{U}} \leq \mathscr{C}h^{3/4} \tag{4.5*.4b}$$

and

$$\|e^{(2)}_{S^p_\Delta}\|_{\mathcal{U}} \leq \mathscr{C}h^{6/4}. \tag{4.5*.4c}$$

(c) Let $\omega \subset \Omega$ be an interior subdomain of Ω. Use the interior estimates to show that $u^{(1)}_{S^p_\Delta}$ for $p = 2$ has pollution in ω but not $u^{(2)}_{S^p_\Delta}$, while for $p = 1$ there is no pollution in both cases.

(d) Let now u_{EX} be the solution of the boundary value problem

$$-\Delta u|_\Omega = f \qquad \frac{\partial u}{\partial n}\bigg|_{BCDEF} = 0 \qquad u|_{FAB} = 0 \tag{4.5*.4d}$$

and prove that

$$\|u_{EX}\|_{B^{3/4}(\Omega)} \leq \mathscr{C}\|f\|_{L^2(\Omega)}. \tag{4.5*.4e}$$

4.6 A-priori estimates for the local error: superconvergence

In Section 4.3, we partitioned the error in the finite element solution in any element I^Δ_j into $e^{loc}_{S^p(I^\Delta_j)}$, the local error, and $e^{glob}_{S^p(I^\Delta_j)}$, the pollution error. In Sections 4.4 and 4.5, we constructed asymptotic estimates for the pollution error and showed that, in some cases, e.g., the regular model problem with smooth input data and for finite element solutions computed using quasi-uniform meshes, it is negligible. We will now *assume that the pollution error is negligible* and we will analyze the asymptotic behavior of the error $e_{S^p_\Delta}|_{I^\Delta_j}$ by analyzing the local error $e^{loc}_{S^p(I^\Delta_j)}$.

We will analyze the local error $e^{loc}_{S^p(I^\Delta_j)}$ by splitting it into two components, $e^{loc}_{S^1(I^\Delta_j)}$, the error in the linear interpolant,

$$e^{loc}_{S^1(I^\Delta_j)} \stackrel{\text{def}}{=} u_{EX}|_{I^\Delta_j} - \mathcal{A}^1_{I^\Delta_j} u_{EX} \in S^1(I^\Delta_j) \tag{4.6.1a}$$

and $\mathcal{A}^p_{I^\Delta_j,0} u_{EX}$, the bubble part of the local finite element solution

$$\mathcal{A}^p_{I^\Delta_j,0} u_{EX} \stackrel{\text{def}}{=} \mathcal{A}^p_{I^\Delta_j} u_{EX} - \mathcal{A}^1_{I^\Delta_j} u_{EX} \in S^p_0(I^\Delta_j). \tag{4.6.1b}$$

We have

$$\begin{aligned} e^{loc}_{S^p(I^\Delta_j)} &= \left(u_{EX}|_{I^\Delta_j} - \mathcal{A}^1_{I^\Delta_j} u_{EX}\right) - \left(\mathcal{A}^p_{I^\Delta_j} u_{EX} - \mathcal{A}^1_{I^\Delta_j} u_{EX}\right) \\ &= e^{loc}_{S^1(I^\Delta_j)} - \mathcal{A}^p_{I^\Delta_j,0} u_{EX}. \end{aligned} \tag{4.6.1c}$$

From the orthogonality condition

$$\mathscr{B}_{I^\Delta_j}\left(e_{S^p(I^\Delta_j)}, v\right) = \mathscr{B}_{I^\Delta_j}\left(e^{loc}_{S^1(I^\Delta_j)} - \mathcal{A}^p_{I^\Delta_j,0} u_{EX}, v\right) = 0 \quad \forall v \in S^p_0(I^\Delta_j) \tag{4.6.1d}$$

it follows that

$$\mathscr{B}_{I_j^\Delta}\big(\mathcal{A}^p_{I_j^\Delta,0}u_{EX},v\big)=\mathscr{B}_{I_j^\Delta}\big(e^{loc}_{S^1(I_j^\Delta)},v\big)\quad \forall v\in S^p_0(I_j^\Delta). \tag{4.6.1e}$$

We will use this property in the derivations that follow.

Below we will employ the master interval $\tilde{I}\overset{\text{def}}{=}(-1,1)$ and the affine transformations $F_{I_j^\Delta}:\tilde{I}\mapsto I_j^\Delta$, given by

$$F_{I_j^\Delta}(\tilde{x})\overset{\text{def}}{=}x^\Delta_{j-1/2}+\frac{h_{I_j^\Delta}}{2}\tilde{x},\quad \tilde{x}\in\tilde{I}. \tag{4.6.2a}$$

Given a function $v\in C^0(I_j^\Delta)$, we will construct a corresponding function over the master element $\tilde{I}$ by

$$\tilde{v}_j(\tilde{x})\overset{\text{def}}{=}v\big(F_{I_j^\Delta}(\tilde{x})\big),\quad \tilde{x}\in I. \tag{4.6.2b}$$

For example, $\tilde{u}_{EX,I_j^\Delta}$, $\widetilde{AE}_{I_j^\Delta}$, and $\tilde{c}_{I_j^\Delta}$, are the functions over $\tilde{I}$ which correspond to $u_{EX}|_{I_j^\Delta}$, $AE|_{I_j^\Delta}$, and $c|_{I_j^\Delta}$, respectively. We also let $\tilde{\mathscr{B}}_{I_j^\Delta}:\mathscr{U}(\tilde{I})\times\mathscr{U}(\tilde{I})\mapsto\mathbb{R}$ be the bilinear form

$$\tilde{\mathscr{B}}_{I_j^\Delta}\big(\tilde{u}_{I_j^\Delta},\tilde{v}_{I_j^\Delta}\big)\overset{\text{def}}{=}\int_{-1}^{1}\Big(\widetilde{AE}_{I_j^\Delta}\frac{d}{d\tilde{x}}(\tilde{u}_{I_j^\Delta})\frac{d}{d\tilde{x}}(\tilde{v}_{I_j^\Delta})\frac{2}{h_{I_j^\Delta}}+\frac{h_{I_j^\Delta}}{2}\tilde{c}_{I_j^\Delta}\tilde{u}_{I_j^\Delta}\tilde{v}_{I_j^\Delta}\Big) \tag{4.6.2c}$$

for $\tilde{u}_{I_j^\Delta},\tilde{v}_{I_j^\Delta}\in\mathscr{U}(\tilde{I})$. Below we will suppress the subscript I_j^Δ from the functions $\tilde{u}_{I_j^\Delta}$ and their errors, keeping in mind that any function defined over the master element corresponds to a function defined over an element I_j^Δ. Employing the notation

$$\tilde{e}^{loc}_{\tilde{S}^1}\overset{\text{def}}{=}e^{loc}_{S^1(I_j^\Delta)}\circ F_{I_j^\Delta},\qquad \tilde{\mathcal{A}}^p_0\tilde{u}_{EX}=\mathcal{A}^p_{I_j^\Delta,0}u_{EX}\circ F_{I_j^\Delta} \tag{4.6.2d}$$

$\tilde{S}^p\overset{\text{def}}{=}S^p(\tilde{I})$ and $\tilde{S}^p_0=S^p_0(\tilde{I})$, (4.6.1e) can be rewritten as

$$\tilde{\mathscr{B}}_{I_j^\Delta}(\tilde{\mathcal{A}}^p_0\tilde{u}_{EX},v)=\tilde{\mathscr{B}}_{I_j^\Delta}(\tilde{e}^{loc}_{\tilde{S}^p},v)\quad\forall v\in\tilde{S}^p_0. \tag{4.6.2e}$$

We will first analyze the local error, $\tilde{e}^{loc}_{\tilde{S}^p}$, in the special case that

$$AE\equiv constant,\qquad c\equiv 0 \tag{4.6.3}$$

and, without loss of generality, we will assume $AE=1$. We will prove that, in this special case, the error in the derivative of the finite element solution coincides with the error in the $(p-1)$-degree Legendre expansion of the derivative of the exact solution, namely

$$\frac{d}{d\tilde{x}}(\tilde{e}^{loc}_{\tilde{S}^p})=\sum_{l=p}^{\infty}b_l\tilde{P}_l \tag{4.6.4a}$$

where $\tilde{P}_l$ is the lth Legendre polynomial over $\tilde{I}$, and

$$\frac{d}{d\tilde{x}}(\tilde{u}_{EX})=\sum_{l=0}^{\infty}b_l\tilde{P}_l \tag{4.6.4b}$$

with

$$b_l = \frac{2l+1}{2}\int_{-1}^{1}\frac{d}{d\tilde{x}}(\tilde{u}_{EX})\tilde{P}_l \tag{4.6.4c}$$

is the Legendre expansion of $(d/d\tilde{x})(\tilde{u}_{EX})$.

For this we will need the following estimates for the error in the Legendre expansion.

Lemma 4.6.1 *Given any $\tilde{u} \in L^2(\tilde{I})$, let*

$$\tilde{u}_n \stackrel{\text{def}}{=} \sum_{l=0}^{n} a_l\tilde{P}_l, \qquad a_l \stackrel{\text{def}}{=} \int_{-1}^{1}\frac{\tilde{u}\tilde{P}_l}{\int_{-1}^{1}\tilde{P}_l^2} \tag{4.6.5a}$$

be the nth partial sum of its Legendre expansion, and let

$$\tilde{e}_n^{\text{Leg}}(\tilde{u})(\tilde{x}) \stackrel{\text{def}}{=} \tilde{u}(\tilde{x}) - \tilde{u}_{n-1}(\tilde{x}) = \sum_{l=n}^{\infty} a_l\tilde{P}_l(\tilde{x}), \quad \tilde{x} \in \tilde{I}. \tag{4.6.5b}$$

Then, under the assumption that

$$\frac{d^r\tilde{u}}{d\tilde{x}^r} \in L^q(\tilde{I}), \quad 0 \le r \le n \tag{4.6.5c}$$

we have

$$\|\tilde{e}_n^{\text{Leg}}(\tilde{u})\|_{L^q(\tilde{I})} \le \mathscr{C}\left\|\frac{d^r\tilde{u}}{d\tilde{x}^r}\right\|_{L^q(\tilde{I})} \tag{4.6.5d}$$

for $q=2$ and $q=\infty$, with $\mathscr{C}$ depending on r and n.

Proof The term $\tilde{e}_n^{\text{Leg}}(\tilde{u})$ is the error in $\tilde{u}_{n-1}$, and satisfies the orthogonality condition

$$\int_{\tilde{I}}\tilde{e}_n^{\text{Leg}}(\tilde{u})\,\tilde{P}_l = \sum_{k=n}^{\infty} a_k\int_{\tilde{I}}\tilde{P}_k\,\tilde{P}_l = 0, \quad l = 0,\ldots,(n-1). \tag{4.6.6a}$$

It follows that

$$\|\tilde{e}_n^{\text{Leg}}(\tilde{u})\|_{L^2(\tilde{I})} = \min_{\tilde{\chi}\in\mathscr{P}^{n-1}(\tilde{I})}\|\tilde{u}-\tilde{\chi}\|_{L^2(\tilde{I})} \tag{4.6.6b}$$

and hence

$$\|\tilde{e}_n^{\text{Leg}}(\tilde{u})\|_{L^2(\tilde{I})} \le \|\tilde{u}-\tilde{\mathcal{I}}^{(n-1)}\tilde{u}\|_{L^2(\tilde{I})} \le \mathscr{C}(n,r)\left\|\frac{d^r\tilde{u}}{d\tilde{x}^r}\right\|_{L^2(\tilde{I})} \tag{4.6.6c}$$

where $\tilde{\mathcal{I}}^{(n-1)}\tilde{u}$ is the $(n-1)$th degree Lagrangian interpolant of $\tilde{u}$ over $\tilde{I}$. Hence (4.6.5d) for $q=2$, follows from the L^2 estimate for the interpolation error.

The proof of (4.6.5d) for $q=\infty$ can be obtained by applying Lemma 4.3.1. □

We will now show that, in the special case $AE \equiv constant \equiv 1$, $c \equiv 0$, we have

$$\frac{d}{d\tilde{x}}(\tilde{e}_{\tilde{S}^p}^{loc}) = \tilde{e}_p^{\text{Leg}}\left(\frac{d}{d\tilde{x}}(\tilde{u}_{EX})\right) \tag{4.6.7}$$

and hence we can obtain L^2 and L^∞ estimates for $(d/d\tilde{x})(\tilde{e}_{\tilde{S}^p}^{loc})$ by applying Lemma 4.6.1.

Theorem 4.6.1 *Consider the regular model problem with $AE \equiv$ constant, $c \equiv 0$. Then*

$$\left\| \frac{d}{dx}(e^{loc}_{S^p(I_j^\Delta)}) \right\|_{L^q(I_j^\Delta)} \leq \mathscr{C}(h_{I_j^\Delta})^\mu \left\| \frac{d^r}{dx^r}(u_{EX}) \right\|_{L^q(I_j^\Delta)} \tag{4.6.8a}$$

for $q = 2$, and $q = \infty$, where

$$\mu = \min(p, r-1), \quad 1 \leq r \leq p+1 \tag{4.6.8b}$$

and $\mathscr{C}$ is independent of $h_{I_j^\Delta}$, u_{EX}, and q.

Proof Employing the Legendre expansion

$$\frac{d}{d\tilde{x}}(\tilde{e}^{loc}_{\tilde{S}^1}) = \sum_{l=0}^{\infty} b_l \tilde{P}_l \tag{4.6.9a}$$

and noting that

$$b_0 = \int_{-1}^{1} \frac{d}{d\tilde{x}}(\tilde{e}^{loc}_{\tilde{S}^1})\tilde{P}_0 = \int_{-1}^{1} \frac{d}{d\tilde{x}}(\tilde{e}^{loc}_{\tilde{S}^1}) = \tilde{e}^{loc}_{\tilde{S}^1}(1) - \tilde{e}^{loc}_{\tilde{S}^1}(-1) = 0 \tag{4.6.9b}$$

and integrating (4.6.9a) we get

$$\tilde{e}^{loc}_{\tilde{S}^1}(\tilde{x}) = \sum_{l=1}^{\infty} b_l \int_{-1}^{\tilde{x}} \tilde{P}_l = \sum_{l=1}^{\infty} b_l \tilde{\psi}_l(\tilde{x}), \quad x \in \tilde{I} \tag{4.6.9c}$$

where

$$\tilde{\psi}_l(\tilde{x}) \stackrel{\text{def}}{=} \int_{-1}^{\tilde{x}} \tilde{P}_l \tag{4.6.9d}$$

is the primitive of the lth Legendre polynomial, $l = 1, 2, \ldots$. Note that

$$\tilde{\psi}_l(\pm 1) = 0 \tag{4.6.9e}$$

and hence $\tilde{\psi}_l \in \mathscr{P}_0^{l+1}(\tilde{I})$, and $\{\tilde{\psi}_l\}_{l=1}^{p-1}$ is a basis for the bubble space $\mathscr{P}_0^p(\tilde{I})$.

Let us now write down (4.6.1e) for the case that $AE \equiv 1$, $c \equiv 0$; we have

$$\int_{-1}^{1} \frac{d\tilde{v}}{d\tilde{x}} \frac{d}{d\tilde{x}}(\tilde{\mathcal{A}}_0^p \tilde{u}_{EX}) = \int_{-1}^{1} \frac{d\tilde{v}}{d\tilde{x}} \frac{d}{d\tilde{x}}(\tilde{e}^{loc}_{\tilde{S}^1}) \quad \forall \tilde{v} \in \mathscr{P}_0^p(\tilde{I}). \tag{4.6.10a}$$

Letting $\tilde{v} = \tilde{\psi}_l$, $l = 1, \ldots, p-1$, and recalling that $\tilde{P}_l = (d/d\tilde{x})(\tilde{\psi}_l)$, we obtain

$$\int_{-1}^{1} \tilde{P}_l \frac{d}{d\tilde{x}}(\tilde{\mathcal{A}}_0^p \tilde{u}_{EX}) = \int_{-1}^{1} \tilde{P}_l \frac{d}{d\tilde{x}}(\tilde{e}^{loc}_{\tilde{S}^1}), \quad l = 1, \ldots, p-1. \tag{4.6.10b}$$

This proves that $(d/d\tilde{x})(\tilde{\mathcal{A}}_0^p \tilde{u}_{EX})$ is the best L^2 approximation of $(d/dx)(\tilde{e}^{loc}_{\tilde{S}^1})$ from $\mathcal{P}^{p-1}(\tilde{I})$, and hence

$$\frac{d}{d\tilde{x}}(\tilde{\mathcal{A}}_0^p \tilde{u}_{EX}) = \sum_{l=1}^{p-1} b_l \tilde{P}_l \tag{4.6.11a}$$

and integrating we get

$$\tilde{\mathcal{A}}_0^p \tilde{u}_{EX} = \sum_{l=1}^{p-1} b_l \tilde{\psi}_l. \tag{4.6.11b}$$

Recalling (4.6.1c) we have

$$\tilde{e}^{loc}_{\tilde{S}^p} = \tilde{e}^{loc}_{\tilde{S}^1} - \tilde{\mathcal{A}}_0^p \tilde{u}_{EX} = \sum_{l=p}^{\infty} b_l \tilde{\psi}_l \tag{4.6.12a}$$

and differentiating, we get

$$\frac{d}{d\tilde{x}}(\tilde{e}^{loc}_{\tilde{S}^p}) = \sum_{l=p}^{\infty} b_l \tilde{P}_l = \tilde{e}_p^{\text{Leg}}\Big(\frac{d}{d\tilde{x}}(\tilde{e}^{loc}_{\tilde{S}^1})\Big) \tag{4.6.12b}$$

and applying Lemma 4.6.1, we obtain

$$\begin{aligned}\Big\|\frac{d}{d\tilde{x}}(\tilde{e}^{loc}_{\tilde{S}^p})\Big\|_{L^q(\tilde{I})} &= \Big\|\tilde{e}_p^{\text{Leg}}\Big(\frac{d}{d\tilde{x}}(\tilde{e}^{loc}_{\tilde{S}^1})\Big)\Big\|_{L^q(\tilde{I})} \le \mathscr{C}\Big\|\frac{d^{k+1}}{d\tilde{x}^{k+1}}(\tilde{e}^{loc}_{\tilde{S}^1})\Big\|_{L^q(\tilde{I})} \\ &= \mathscr{C}\Big\|\frac{d^{k+1}}{d\tilde{x}^{k+1}}(\tilde{u}_{EX})\Big\|_{L^q(\tilde{I})}\end{aligned} \tag{4.6.13a}$$

for $q = 2$ and $q = \infty$, with $1 \le k \le p$, where we used that for $k \ge 1$, we have $(d^{k+1}/d\tilde{x}^{k+1})(\tilde{\mathcal{I}}^1 \tilde{u}_{EX}) \equiv 0$, and hence

$$\frac{d^{k+1}}{d\tilde{x}^{k+1}}(\tilde{e}^{loc}_{\tilde{S}^1}) = \frac{d^{k+1}}{d\tilde{x}^{k+1}}(\tilde{u}_{EX}), \quad k \ge 1. \tag{4.6.13b}$$

Transforming (4.6.13a) to I_j^Δ, we get

$$\frac{h_{I_j^\Delta}}{2}\Big\|\frac{d}{dx}(e^{loc}_{S^p(I_j^\Delta)})\Big\|_{L^q(I_j^\Delta)} \le \mathscr{C}\Big(\frac{h_{I_j^\Delta}}{2}\Big)^{k+1}\Big\|\frac{d^k}{dx^k}(u_{EX})\Big\|_{L^q(I_j^\Delta)} \tag{4.6.13c}$$

which is (4.6.8a). □

Using Theorem 4.6.1, we can now derive estimates for the L^2 and the L^∞ norm of the local error.

Corollary 4.6.1 *Consider the regular model problem with $AE \equiv$ constant, $c \equiv 0$. Then*

$$\|e^{loc}_{S^p(I_j^\Delta)}\|_{L^q(I_j^\Delta)} \le \mathscr{C}(h_{I_j^\Delta})^{\mu+1}\Big\|\frac{d^r}{dx^r}(u_{EX})\Big\|_{L^q(I_j^\Delta)} \tag{4.6.14a}$$

with $q = 2, \infty$, *and*

$$\mu = \min(p, r-1), \quad 1 \le r \le p+1. \tag{4.6.14b}$$

Proof The proof follows from the Poincaré inequalities. Noting that

$$e^{loc}_{S^p(I_j^\Delta)}(x_j^\Delta) = 0 \tag{4.6.15a}$$

we get

$$\|e^{loc}_{S^p(I_j^\Delta)}\|_{L^2(I_j^\Delta)} \le \mathscr{C} h_{I_j^\Delta} \left\| \frac{d}{dx}(e^{loc}_{S^p(I_j^\Delta)}) \right\|_{L^2(I_j^\Delta)} \le \mathscr{C} h^{\mu+1}_{I_j^\Delta} \left\| \frac{d^r}{dx^r}(u_{EX}) \right\|_{L^2(I_j^\Delta)} \tag{4.6.15b}$$

which proves (4.6.14) for $q = 2$. The case $q = \infty$ follows from the inequality

$$|e^{loc}_{S^p(I_j^\Delta)}(x)| = \left| \int_{x_{j-1}^\Delta}^{x} \frac{d}{dx}(e^{loc}_{S^p(I_j^\Delta)}) \right| \le h_{I_j^\Delta} \left\| \frac{d}{dx}(e^{loc}_{S^p(I_j^\Delta)}) \right\|_{L^\infty(I_j^\Delta)}. \tag{4.6.15c}$$

□

Remark 4.6.1 The explicit dependence of the constants $\mathscr{C}$ in Theorem 4.6.1 and Corollary 4.6.1, on the degree p of the elements can be obtained following similar steps as in Section 3.4, and is essential in the study of the p and h–p versions of the finite element method.

Above we derived a-priori estimates for the local error for the special case that $AE \equiv$ *constant* and $c \equiv 0$. Let us now consider the general case.

Theorem 4.6.2 *Consider the regular model problem, under the additional assumption that*

$$\left\| \frac{d}{dx}(AE) \right\|_{L^\infty(I)} \le K < \infty. \tag{4.6.16}$$

Then

$$\|e^{loc}_{S^p(I_j^\Delta)}\|_{\mathscr{U}(I_j^\Delta)} = (1 + \mathscr{C} h_{I_j^\Delta} \theta) \|\bar{e}^{loc}_{S^p(I_j^\Delta)}\|_{\mathscr{U}(I_j^\Delta)} \tag{4.6.17a}$$

and

$$\left\| \frac{d}{dx}(e^{loc}_{S^p(I_j^\Delta)}) \right\|_{L^\infty(I_j^\Delta)} = (1 + \mathscr{C} h_{I_j^\Delta} \theta) \left\| \frac{d}{dx}(\bar{e}^{loc}_{S^p(I_j^\Delta)}) \right\|_{L^\infty(I_j^\Delta)} \tag{4.6.17b}$$

with $-1 < \theta < 1$, *where*

$$\bar{e}^{loc}_{S^p(I_j^\Delta)} \overset{\text{def}}{=} u_{EX}|_{I_j^\Delta} - \bar{\mathscr{A}}^p_{I_j^\Delta} u_{EX} \tag{4.6.18a}$$

and $\bar{\mathscr{A}}^p_{I_j^\Delta} u_{EX}$ *is the local finite element solution based on the simplified bilinear form,*

$$\mathscr{B}^0_{I_j^\Delta}(u, v) \overset{\text{def}}{=} AE(x^\Delta_{j-1/2}) \int_{I_j^\Delta} \frac{du}{dx} \frac{dv}{dx} \tag{4.6.18b}$$

namely, the solution of the problem:

Find $\bar{\mathcal{A}}^p_{I^\Delta_j} u_{EX} \in S^p_{u_{EX}}(I^\Delta_j) \overset{\text{def}}{=} \{w \in S^p(I^\Delta_j) \mid w(x^\Delta_k) = u_{EX}(x^\Delta_k),\ k = j-1, j\}$ *such that*

$$\mathcal{B}^0_{I^\Delta_j}\big(\bar{\mathcal{A}}^p_{I^\Delta_j} u_{EX}, v\big) = \mathcal{B}^0_{I^\Delta_j}(u_{EX}, v) \quad \forall v \in S^p_0(I^\Delta_j). \tag{4.6.18c}$$

Proof Letting

$$\delta e^{loc}_{S^p(I^\Delta_j)} \overset{\text{def}}{=} e^{loc}_{S^p(I^\Delta_j)} - \bar{e}^{loc}_{S^p(I^\Delta_j)} \tag{4.6.19a}$$

we have

$$\begin{aligned} 0 &= \mathcal{B}_{I^\Delta_j}\big(e^{loc}_{S^p(I^\Delta_j)}, v\big) \\ &= \mathcal{B}^0_{I^\Delta_j}\big(\bar{e}^{loc}_{S^p(I^\Delta_j)}, v\big) + \mathcal{B}^1_{I^\Delta_j}\big(\bar{e}^{loc}_{S^p(I^\Delta_j)}, v\big) + \mathcal{B}_{I^\Delta_j}\big(\delta e^{loc}_{S^p(I^\Delta_j)}, v\big) \end{aligned} \tag{4.6.19b}$$

for any $v \in S^p_0(I^\Delta_j)$, where

$$\begin{aligned} \mathcal{B}^1_{I^\Delta_j}(u, v) &\overset{\text{def}}{=} \mathcal{B}_{I^\Delta_j}(u, v) - \mathcal{B}^0_{I^\Delta_j}(u, v) \\ &= \int_{I^\Delta_j} \Big((AE - AE(x^\Delta_{j-1/2})) \frac{du}{dx}\frac{dv}{dx} + cuv \Big) \end{aligned} \tag{4.6.19c}$$

is the difference between the original and the simplified variational forms. From (4.6.19b) it follows that $\delta e^{loc}_{S^p(I^\Delta_j)}$ is the solution of the variational problem:

Find $\delta e^{loc}_{S^p(I^\Delta_j)} \in S^p_0(I^\Delta_j)$ such that

$$\begin{aligned} -\mathcal{B}_{I^\Delta_j}\big(\delta e^{loc}_{S^p(I^\Delta_j)}, v\big) &= \mathcal{B}^1_{I^\Delta_j}\big(\bar{e}^{loc}_{S^p(I^\Delta_j)}, v\big) \\ &= \int_{I^\Delta_j} \Big((AE - AE(x^\Delta_{j-1/2})) \frac{d}{dx}\big(\bar{e}^{loc}_{S^p(I^\Delta_j)}\big) \frac{dv}{dx} + c\bar{e}^{loc}_{S^p(I^\Delta_j)} v \Big) \end{aligned} \tag{4.6.20a}$$

for all $v \in S^p_0(I^\Delta_j)$.

Employing the Cauchy–Schwarz inequality we get

$$\begin{aligned} &\big|\mathcal{B}^1_{I^\Delta_j}\big(\bar{e}^{loc}_{S^p(I^\Delta_j)}, v\big)\big| \\ &\quad \le \|AE - AE(x^\Delta_{j-1/2})\|_{L^\infty(I^\Delta_j)} \Big\| \frac{d}{dx}\big(\bar{e}^{loc}_{S^p(I^\Delta_j)}\big)\Big\|_{L^2(I^\Delta_j)} \Big\|\frac{dv}{dx}\Big\|_{L^2(I^\Delta_j)} \\ &\qquad + \|c\|_{L^\infty(I^\Delta_j)} \big\|\bar{e}^{loc}_{S^p(I^\Delta_j)}\big\|_{L^2(I^\Delta_j)} \|v\|_{L^2(I^\Delta_j)} \end{aligned} \tag{4.6.20b}$$

and noting that from assumption (4.6.16) we have

$$\|AE - AE(x^\Delta_{j-1/2})\|_{L^\infty(I^\Delta_j)} \le \mathcal{C} h_{I^\Delta_j} \Big\| \frac{d}{dx}(AE) \Big\|_{L^\infty(I^\Delta_j)} \le \mathcal{C} h_{I^\Delta_j} K \tag{4.6.20c}$$

we obtain

$$\left|\mathcal{B}_{I_j^\triangle}\left(\delta e^{loc}_{S^p(I_j^\triangle)}, v\right)\right| = \left|\mathcal{B}^1_{I_j^\triangle}\left(\bar{e}^{loc}_{S^p(I_j^\triangle)}, v\right)\right| \le \mathcal{C} h_{I_j^\triangle} \left\|\bar{e}^{loc}_{S^p(I_j^\triangle)}\right\|_{\mathcal{U}(I_j^\triangle)} \|v\|_{\mathcal{U}(I_j^\triangle)} \tag{4.6.20d}$$

for any $v \in S^p_0(I_j^\triangle)$, where $\mathcal{C} = \mathcal{C}'(K + \gamma h_{I_j^\triangle})$, and we employed that

$$\left\|\bar{e}^{loc}_{S^p(I_j^\triangle)}\right\|_{L^2(I_j^\triangle)} \le \mathcal{C} h_{I_j^\triangle} \left\|\frac{d}{dx}\left(\bar{e}^{loc}_{S^p(I_j^\triangle)}\right)\right\|_{L^2(I_j^\triangle)}. \tag{4.6.20e}$$

Letting $v = \delta e^{loc}_{S^p(I_j^\triangle)}$ we get

$$\left\|\delta e^{loc}_{S^p(I_j^\triangle)}\right\|_{\mathcal{U}(I_j^\triangle)} \le \mathcal{C} h_{I_j^\triangle} \left\|\bar{e}^{loc}_{S^p(I_j^\triangle)}\right\|_{\mathcal{U}(I_j^\triangle)}. \tag{4.6.21a}$$

Employing now the triangle inequality, we obtain

$$\begin{aligned}\left\|e^{loc}_{S^p(I_j^\triangle)}\right\|_{\mathcal{U}(I_j^\triangle)} &\le \left\|\bar{e}^{loc}_{S^p(I_j^\triangle)}\right\|_{\mathcal{U}(I_j^\triangle)} + \left\|\delta e^{loc}_{S^p(I_j^\triangle)}\right\|_{\mathcal{U}(I_j^\triangle)} \\ &\le \left\|\bar{e}^{loc}_{S^p(I_j^\triangle)}\right\|_{\mathcal{U}(I_j^\triangle)} + \mathcal{C} h_{I_j^\triangle} \left\|\bar{e}^{loc}_{S^p(I_j^\triangle)}\right\|_{\mathcal{U}(I_j^\triangle)}\end{aligned} \tag{4.6.21b}$$

which proves (4.6.17a).

From (4.6.21a) it follows that

$$\left\|\delta e^{loc}_{S^p(I_j^\triangle)}\right\|_{\mathcal{U}(I_j^\triangle)} \le \mathcal{C} h_{I_j^\triangle} \sqrt{h_{I_j^\triangle}} \left\|\frac{d}{dx}\left(\bar{e}^{loc}_{S^p(I_j^\triangle)}\right)\right\|_{L^\infty(I_j^\triangle)} \tag{4.6.22}$$

and noting that $\delta e^{loc}_{S^p(I_j^\triangle)} \in \mathcal{P}^p_0(I_j^\triangle)$, we get

$$\left\|\frac{d}{dx}\left(\delta e^{loc}_{S^p(I_j^\triangle)}\right)\right\|_{L^\infty(I_j^\triangle)} \le \frac{\mathcal{C}(p)}{\sqrt{h_{I_j^\triangle}}} \left\|\delta e^{loc}_{S^p(I_j^\triangle)}\right\|_{\mathcal{U}(I_j^\triangle)} \tag{4.6.23}$$

and hence

$$\left\|\frac{d}{dx}\left(\delta e^{loc}_{S^p(I_j^\triangle)}\right)\right\|_{L^\infty(I_j^\triangle)} \le \mathcal{C} h_{I_j^\triangle} \left\|\frac{d}{dx}\left(\bar{e}^{loc}_{S^p(I_j^\triangle)}\right)\right\|_{L^\infty(I_j^\triangle)} \tag{4.6.24}$$

which proves (4.6.17b). □

Let us also state the estimates in the L^2 and the L^∞ norm.

Corollary 4.6.3 *Consider the regular model problem under the additional assumption (4.6.16). Then*

$$\left\|e^{loc}_{S^p(I_j^\triangle)}\right\|_{L^q(I_j^\triangle)} \le \left(1 + \mathcal{C} h_{I_j^\triangle}\theta\right) \left\|\bar{e}^{loc}_{S^p(I_j^\triangle)}\right\|_{L^q(I_j^\triangle)} \tag{4.6.25}$$

for $q = 2$ and $q = \infty$, with $-1 < \theta < 1$.

Proof The proof follows from the Poincaré inequalities, and (4.6.17a), (4.6.17b). □

We are now ready to prove the *superconvergence* of the derivatives at the Gauss points.

Theorem 4.6.3 *Consider the regular model problem under the additional assumption (4.6.16). Further, assume that:*

Assumption I. *The exact solution u_{EX} satisfies*

$$\left\| \frac{d^{p+2}}{dx^{p+2}}(u_{EX}) \right\|_{L^\infty(I_j^\Delta)} \leq \mathscr{C} < \infty \tag{4.6.26a}$$

and

$$0 < \mathscr{C} \leq \min_{x \in I_j^\Delta} \left| \frac{d^{p+1}}{dx^{p+1}}(u_{EX})(x) \right|. \tag{4.6.26b}$$

Assumption II. *The sequence of meshes $\{\Delta_n\}_{n=1}^\infty$ is such that*

$$\max\left(|e_j^{\Delta_n} + e_{j-1}^{\Delta_n}|, \frac{|e_j^{\Delta_n} - e_{j-1}^{\Delta_n}|}{h_{I_j^{\Delta_n}}} \right) \leq \mathscr{C}(h_{I_j^{\Delta_n}})^{p+1} \left\| \frac{d^{p+2}}{dx^{p+2}}(u_{EX}) \right\|_{L^\infty(I_j^{\Delta_n})} \tag{4.6.26c}$$

as n diverges to infinity.

Under these assumptions, we have

$$\left| \frac{d}{dx}(e_{S^p_{\Delta_n}})(x_l^{G,I_j^{\Delta_n}}) \right| \leq \mathscr{C}(h_{I_j^{\Delta_n}})^{p+1} \left\| \frac{d^{p+2}}{dx^{p+2}}(u_{EX}) \right\|_{L^\infty(I_j^{\Delta_n})} \tag{4.6.27a}$$

for $l = 1, \ldots, p$, where $x_l^{G,I_j^{\Delta_n}}$, are the mapped Gauss–Legendre points of order p in $I_j^{\Delta_n}$ (the mapped zeros of the Legendre polynomial of degree p), namely,

$$x_l^{G,I_j^{\Delta_n}} \stackrel{\text{def}}{=} F_{I_j^{\Delta_n}}(\tilde{x}_l^G), \quad \tilde{P}_p(\tilde{x}_l^G) = 0, \qquad 1 \leq l \leq p \tag{4.6.27b}$$

while

$$\mathscr{C}(h_{I_j^{\Delta_n}})^p \left\| \frac{d^{p+1}}{dx^{p+1}}(u_{EX}) \right\|_{L^\infty(I_j^{\Delta_n})} \leq \left\| \frac{d}{dx}(e_{S^p_{\Delta_n}}) \right\|_{L^\infty(I_j^{\Delta_n})}. \tag{4.6.27c}$$

We also have

$$\left\| \frac{d}{dx}(e_{S^p_{\Delta_n}}) - b_p P_p \right\|_{L^q(I_j^{\Delta_n})} \leq \mathscr{C}(h_{I_j^{\Delta_n}})^{p+1} \left\| \frac{d^{p+2}}{dx^{p+2}}(u_{EX}) \right\|_{L^q(I_j^{\Delta_n})} \tag{4.6.28a}$$

for $q = 2, \infty$, where

$$b_p = \frac{2l+1}{2} \int_{-1}^{1} \frac{d}{d\tilde{x}}(\tilde{u}_{EX}) \tilde{P}_p \tag{4.6.28b}$$

while

$$\mathscr{C}\big(h_{I_j^{\Delta_n}}\big)^p \Big\|\frac{d^{p+1}}{dx^{p+1}}(u_{EX})\Big\|_{L^q(I_j^{\Delta_n})} \leq \Big\|\frac{d}{dx}(e_{S^p_{\Delta_n}})\Big\|_{L^q(I_j^{\Delta_n})}. \tag{4.6.28c}$$

Further, for any $x \in I_j^{\Delta_n}$

$$b_p = \mathscr{C}(p)\frac{d^{p+1}}{dx^{p+1}}(u_{EX})(x)\big(h_{I_j^{\Delta_n}}\big)^p + \epsilon_p \tag{4.6.29a}$$

with

$$|\epsilon_p| \leq \mathscr{C}\big(h_{I_j^{\Delta_n}}\big)^{p+1} \tag{4.6.29b}$$

and $\mathscr{C}(p)$ *depending only on the degree* p.

Proof Let us first consider the special case $AE \equiv 1$ and $c \equiv 0$. Recalling(4.6.12b) we have

$$\frac{d}{d\tilde{x}}(\tilde{e}^{loc}_{\tilde{S}^p}) = \sum_{l=p}^{\infty} b_l\tilde{P}_l = b_p\tilde{P}_p + \sum_{l=p+1}^{\infty} b_l\tilde{P}_l = b_p\tilde{P}_p + \frac{d}{d\tilde{x}}(\tilde{e}^{loc}_{\tilde{S}^{p+1}}). \tag{4.6.30a}$$

Evaluating this equality at $\tilde{x}^G_m$ the zeros of $\tilde{P}_p$ ($\tilde{P}_p(\tilde{x}^G_m) = 0, m = 1, \ldots, p$), we get

$$\frac{d}{d\tilde{x}}(\tilde{e}^{loc}_{\tilde{S}^p})(\tilde{x}^G_m) = \frac{d}{d\tilde{x}}(\tilde{e}^{loc}_{\tilde{S}^{p+1}})(\tilde{x}^G_m), \quad m = 1, \ldots, p \tag{4.6.30b}$$

and transforming to $I_j^{\Delta_n}$, we obtain

$$\begin{aligned}\Big|\frac{d}{dx}(e^{loc}_{S^p(I_j^{\Delta})})(x_l^{G,I_j^{\Delta_n}})\Big| &= \Big|\frac{d}{dx}(e^{loc}_{S^{p+1}(I_j^{\Delta})})(x_l^{G,\,I_j^{\Delta_n}})\Big| \\ &\leq \Big\|\frac{d}{dx}(e^{loc}_{S^{p+1}(I_j^{\Delta})})\Big\|_{L^\infty(I_j^{\Delta_n})}\end{aligned} \tag{4.6.30c}$$

and employing (4.6.8) we get

$$\Big|\frac{d}{dx}(e_{S^p(I_j^{\Delta})})(x_l^{G,I_j^{\Delta_n}})\Big| \leq \mathscr{C}\big(h_{I_j^{\Delta_n}}\big)^{p+1}\Big\|\frac{d^{p+2}}{dx^{p+2}}(u_{EX})\Big\|_{L^\infty(I_j^{\Delta_n})}. \tag{4.6.30d}$$

The proof of (4.6.30d) in the general case follows from (4.6.17b).

Further, from assumption (4.6.26b), it follows that

$$\mathscr{C}\big(h_{I_j^{\Delta_n}}\big)^p \leq \Big\|\frac{d}{dx}(e^{loc}_{S^p(I_j^{\Delta})})\Big\|_{L^\infty(I_j^{\Delta_n})} \tag{4.6.31}$$

which is (4.6.27c). Hence we have proven that

$$\Big|\frac{d}{dx}\Big(e^{loc}_{S^p(I_j^{\Delta})}(x_l^{G,I_j^{\Delta_n}})\Big)\Big| \leq \mathscr{C}h_{I_j^{\Delta_n}}\Big\|\frac{d}{dx}(e^{loc}_{S^p(I_j^{\Delta})})\Big\|_{L^\infty(I_j^{\Delta_n})}. \tag{4.6.32}$$

Thus far we have addressed only the local error. Recalling now that

$$\left\| \frac{d}{dx}(e^{glob}_{S^p(I_j^{\Delta})}) \right\|_{L^2(I_j^{\Delta_n})} \leq \mathscr{C}\sqrt{h_{I_j^{\Delta_n}}}\sqrt{(e_j^{\Delta_n} + e_{j-1}^{\Delta_n})^2 + \left(\frac{e_j^{\Delta_n} - e_{j-1}^{\Delta_n}}{h_{I_j^{\Delta_n}}}\right)^2} \tag{4.6.33a}$$

and employing that

$$\left\| \frac{d}{dx}(e^{glob}_{S^p(I_j^{\Delta})}) \right\|_{L^\infty(I_j^{\Delta_n})} \leq \frac{\mathscr{C}}{\sqrt{h_{I_j^{\Delta_n}}}} \left\| \frac{d}{dx}(e^{glob}_{S^p(I_j^{\Delta})}) \right\|_{L^2(I_j^{\Delta_n})} \tag{4.6.33b}$$

we get

$$\left\| \frac{d}{dx}(e^{glob}_{S^p(I_j^{\Delta})}) \right\|_{L^\infty(I_j^{\Delta_n})} \leq \mathscr{C}\sqrt{(e_j^{\Delta_n} + e_{j-1}^{\Delta_n})^2 + \left(\frac{e_j^{\Delta_n} - e_{j-1}^{\Delta_n}}{h_{I_j^{\Delta_n}}}\right)^2} \tag{4.6.33c}$$

and using Assumption II, we obtain

$$\begin{aligned} \left| \frac{d}{dx}(e^{glob}_{S^p(I_j^{\Delta})})(x_l^{G,I_j^{\Delta_n}}) \right| &\leq \left\| \frac{d}{dx}(e^{glob}_{S^p(I_j^{\Delta})}) \right\|_{L^\infty(I_j^{\Delta_n})} \\ &\leq \mathscr{C}(h_{I_j^{\Delta_n}})^{p+1} \left\| \frac{d^{p+2}}{dx^{p+2}}(u_{EX}) \right\|_{L^\infty(I_j^{\Delta_n})} \end{aligned} \tag{4.6.33d}$$

for $l = 1, \ldots, p$, and (4.6.27a) follows from the triangle inequality

$$\left| \frac{d}{dx}(e_{S^p_{\Delta_n}}) \right| \leq \left| \frac{d}{dx}(e^{loc}_{S^p_{\Delta_n}}) \right| + \left| \frac{d}{dx}(e^{glob}_{S^p_{\Delta_n}}) \right|. \tag{4.6.33e}$$

□

Remark 4.6.2 The superconvergence points for a solution quantity (e.g., the derivative of the solution, the axial force, the value of the solution, etc.) are defined as the points where the computed value from the finite element solution converges with a rate which is higher than the rate of convergence of the maximum error in the solution quantity in the element. Above we gave an example of superconvergence for the derivative where a full power of h is gained at the superconvergence points, namely

$$\left| \frac{d}{dx}(e_{S^p_{\Delta_n}}(x_l^{G,I_j^{\Delta_n}})) \right| \leq \mathscr{C}h_{I_j^{\Delta_n}} \left\| \frac{d}{dx}(e_{S^p_{\Delta_n}}) \right\|_{L^\infty(I_j^{\Delta_n})}. \tag{4.6.34}$$

A more general definition is that a point $\bar{x}^{I_j^{\Delta_n}}$ is a superconvergence point for the derivative (and similarly for any other solution quantity) if

$$\left| \frac{d}{dx}(e_{S^p_{\Delta_n}})(\bar{x}^{I_j^{\Delta_n}}) \right| \leq \mathscr{C}(h_{I_j^{\Delta_n}})^{\alpha} \left\| \frac{d}{dx}(e_{S^p_{\Delta_n}}) \right\|_{L^\infty(I_j^{\Delta_n})} \tag{4.6.35}$$

for some $\alpha > 0$. It can be proven that the mapped Gauss–Legendre points are superconvergence points for the derivative in the sense (4.6.35), under a weaker version of Assumption I.

Remark 4.6.3 We are interested in the superconvergence points because *we want to sample the solution quantities from the computed finite element solution with the best possible accuracy.*

Remark 4.6.4 As we will see below, in the higher-dimensional case the superconvergence points may not exist. In that case we will be interested in the *points of least error* in the element which, as we will see below, always exist.

Remark 4.6.5 A more general definition of superconvergence which has meaning for any mesh in one or higher dimension is the $\eta\%$ superconvergence. We will say that $\bar{x}_{\eta\%}^{I_j^{\Delta_n}}$ is an *$\eta\%$-superconvergence point for the derivative* if

$$\left|\frac{d}{dx}(e_{S^p_{\Delta_n}})(\bar{x}_{\eta\%}^{I_j^{\Delta_n}})\right| \leq \left(1 - \frac{\eta}{100}\right)\left\|\frac{d}{dx}(e_{S^p_{\Delta_n}})\right\|_{L^\infty(I_j^{\Delta_n})} \tag{4.6.36}$$

asymptotically as n diverges to infinity. Note that $\bar{x}_{100\%}^{I_j^{\Delta_n}}$ is a superconvergence point in the classical sense, if it exists. The usefulness of $\eta\%$ superconvergence is illustrated in Section 4.6* below, where it is used to find the points of least error. The notion of $\eta\%$ superconvergence was introduced in [Babuška, Strouboulis *et al.*, 1994].

Let us now discuss how the Assumptions Ia, Ib, and II can be verified. We have:

1. Assumption Ia (inequality (4.6.26a)) states that the exact solution, u_{EX}, is sufficiently smooth in the neighborhood of the element of interest. This assumption can be verified a-priori by employing Theorem 2.3.1 to characterize the smoothness of u_{EX} from the known smoothness of the input data AE, c, and f, of interest. For example, in the case of the Laplace equation in two or three dimensions this assumption holds everywhere with the exception of small neighborhoods of corner points, edges, and vertices.
2. Assumption Ib (inequality (4.6.26b)) states that the $(p+1)$ derivative of the exact solution does not change sign in the neighborhood of the element. This assumption is satisfied for the type of solutions which occur in practical applications, with the exception of a few elements.
3. Assumption II states that the employed meshes must be such that the pollution is negligible. This assumption can be verified a-priori in some cases; e.g., for AE, c smooth and quasi-uniform meshes. However, in the majority of practical cases of interest in the multi-dimensional case, it is not easy to design a-priori meshes for which the pollution is negligible. As we will see in Chapter 6, the pollution error can be estimated using a-posteriori computations, and Assumption II can be checked for the concrete case of interest, and can be enforced by refining the elements which are responsible for the pollution error in the patch of elements of interest.

Let us now give an example which illustrates the meaning of Theorem 4.6.3.

Example 4.6.1 Let us consider the singular model problem with

$$AE \equiv x^{3/4}, \qquad c \equiv 0, \qquad f \equiv 0, \qquad \frac{\alpha_2}{1-\vartheta} = 10, \qquad a_2 = 1 \tag{4.6.37a}$$

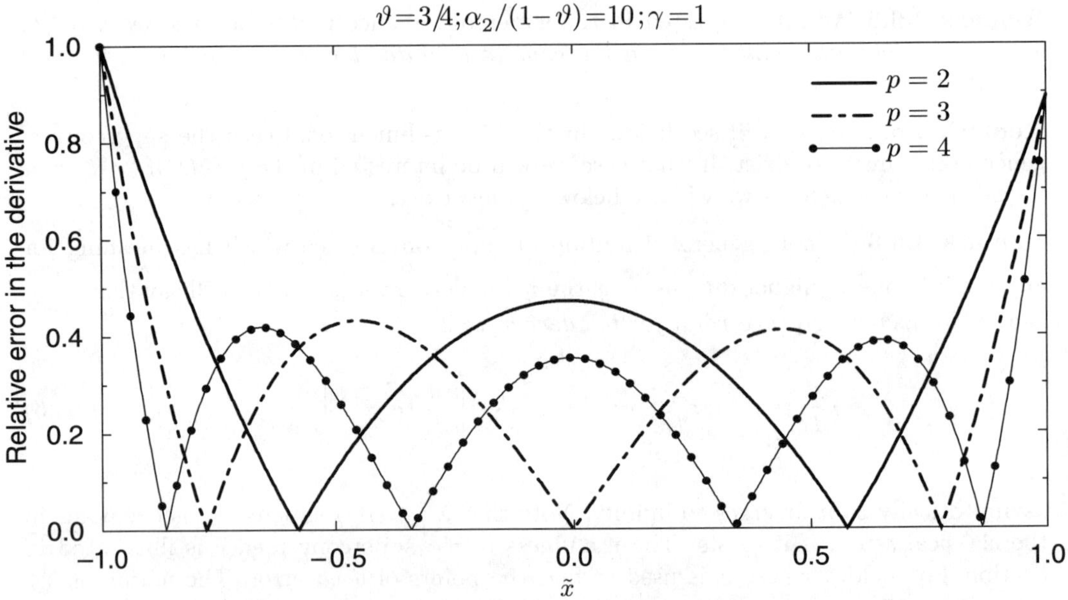

FIG. 4.6.1. An example which illustrates the superconvergence of the derivative of the local error. Input data: $AE \equiv x^{3/4}$, $\alpha_2/(1-\frac{3}{4}) = 10$. Graphs of the absolute value of the relative local error in the derivative, $|(d/dx)(e^{loc}_{S^p(I^\Delta_N)})|/\|(d/dx)(e^{loc}_{S^p(I^\Delta_N)})\|_{L^\infty(I^\Delta_N)}$, over the master element, for $p = 2, 3, 4$ and $h = \frac{1}{32}$. The zero points for each graph occur very close to the Gauss–Legendre points of order p and converge to these points as h tends to zero

and compute the local finite element solution $\mathcal{A}^p_{\Delta_k} u_{EX}$, and the finite element solution $u_{S^p_{\Delta_k}}$, by employing uniform meshes Δ_k, elements of degree $p = 2, 3, 4$, and mesh sizes $h^{\Delta_k} = 2^{-k}$, $k = 1, 2, \ldots, 5$. Figure 4.6.1 shows the graph of the absolute value of the relative local error in the derivative, $|(d/dx)(e^{loc}_{S^p(I^\Delta_N)})|/\|(d/dx)(e^{loc}_{S^p(I^\Delta_N)})\|_{L^\infty(I^\Delta_N)}$, over the master element for $p = 2, 3, 4$. Note that the relative local error is practically equal to zero at the mapped Gauss–Legendre points of order p. The corresponding graph for the total error, namely $|(d/dx)(e_{S^p(I^\Delta_N)})|/\|(d/dx)(e_{S^p(I^\Delta_N)})\|_{L^\infty(I^\Delta_N)}$, is practically constant over I^Δ_N, because of the pollution, and is not shown. Table 4.6.1 gives the maximum values of the absolute value of the relative local error $E^{du/dx}_{loc}$, and the relative error $E^{du/dx}$, over the Gauss–Legendre points in the last element I^Δ_N, divided by the mesh size h, namely,

$$E^{du/dx}_{loc}(I^\Delta_N) \stackrel{\text{def}}{=} \max_{m=1,\ldots,p} \left| \frac{d}{dx}(e^{loc}_{S^p(I^\Delta_N)})(x^G_m) \right| \Big/ \left(h \left\| \frac{d}{dx}(e^{loc}_{S^p(I^\Delta_N)}) \right\|_{L^\infty(I^\Delta_N)} \right) \tag{4.6.37b}$$

where $E^{du/dx}(I^\Delta_N)$ is obtained by replacing $e^{loc}_{S^p(I^\Delta_N)}$ by $e_{S^p_\Delta}|_{I^\Delta_N}$. Note that, as h tends to zero, $E^{du/dx}_{loc}(I^\Delta_N)$ converges to a constant, which means that

$$\max_{m=1,\ldots,p} \left| \frac{d}{dx}(e^{loc}_{S^p(I^\Delta_N)})(x^G_m) \right| \approx \mathscr{C} h \left\| \frac{d}{dx}(e^{loc}_{S^p(I^\Delta_j)}) \right\|_{L^\infty(I^\Delta_j)} \tag{4.6.37c}$$

TABLE 4.6.1. Values of the maximum derivative of the error at the Gauss–Legendre points, normalized by the maximum derivative of the error in the element times h. Input data: $AE \equiv x^{3/4}$, $c \equiv 0$, $\alpha_2/(1-\frac{3}{4}) = 10$. Uniform mesh $\Delta \equiv \Delta_k$ of mesh size $h^{\Delta_k} = 1/2^k$, $k = 1, \ldots, 5$. Note that $E^{du/dx}_{loc}(I^{\Delta}_N)$ converges to a constant value while $E^{du/dx}(I^{\Delta}_N)$ diverges to infinity as the mesh size h tends to zero. This means that the derivative of the local error at the Gauss–Legendre points converges with one order higher in h than the maximum error in the element, while this is not the case for the derivative of the error in the finite element solution. This is because, as we have seen in Example 4.4.2, the pollution error is much more significant than the local error, and hence the derivative of the finite element solution at the Gauss–Legendre points is not superconvergent

h	$p = 2$		$p = 3$		$p = 4$	
	$E^{du/dx}_{loc}(I^{\Delta}_N)$	$E^{du/dx}(I^{\Delta}_N)$	$E^{du/dx}_{loc}(I^{\Delta}_N)$	$E^{du/dx}(I^{\Delta}_N)$	$E^{du/dx}_{loc}(I^{\Delta}_N)$	$E^{du/dx}(I^{\Delta}_N)$
1/2	0.3490×10^{-1}	0.1997×10^{1}	0.2394×10^{-1}	0.1916×10^{1}	0.2030×10^{-1}	0.1916×10^{1}
1/4	0.2429×10^{-1}	0.3931×10^{1}	0.2025×10^{-1}	0.3903×10^{1}	0.1957×10^{-1}	0.3933×10^{1}
1/8	0.2107×10^{-1}	0.7890×10^{1}	0.2020×10^{-1}	0.7908×10^{1}	0.1895×10^{-1}	0.7941×10^{1}
1/16	0.1997×10^{-1}	0.1587×10^{2}	0.2016×10^{-1}	0.1591×10^{2}	0.1863×10^{-1}	0.1594×10^{2}
1/32	0.1956×10^{-1}	0.3186×10^{2}	0.2013×10^{-1}	0.3191×10^{2}	0.1846×10^{-1}	0.3195×10^{2}

and hence, the values of $(d/dx)(e^{loc}_{S^p(I^{\Delta}_N)})(x^G_m)$, $m = 1, \ldots, p$, are superconvergent. On the other hand, $E^{du/dx}(I^{\Delta}_N)$ diverges to infinity as $1/h$, and thus the values of $(d/dx)(e_{S^p(I^{\Delta}_N)})(x^G_m)$ are not superconvergent. This is because for the singular model problem and the uniform meshes employed, the pollution error is the most significant component of the error in the element I^{Δ}_N (see also Example 4.4.2), and Assumption II does not hold. It follows that, *it makes no sense to talk about superconvergence in the finite element solution unless it can be verified that the pollution is negligible.*

Above we have considered the superconvergence of the derivative. Let us now consider the superconvergence of the value.

Theorem 4.6.4 *Consider the setting of Theorem 4.6.3, and replace (4.6.26c) in Assumption II by*

$$\max\left(|e^{\Delta_n}_j + e^{\Delta_n}_{j-1}|, |e^{\Delta_n}_j - e^{\Delta_n}_{j-1}|\right) \leq \mathscr{C}\left(h_{I^{\Delta_n}_j}\right)^{\mu} \left\| \frac{d^{p+2}}{dx^{p+2}}(u_{EX}) \right\|_{L^\infty(I^{\Delta_n}_j)} \tag{4.6.38a}$$

with

$$\mu = \min(2p, p+2). \tag{4.6.38b}$$

Then we have

$$\left|e_{S^p_{\Delta_n}}(x^{L,I^{\Delta_n}_j}_l)\right| \leq \mathscr{C}\left(h_{I^{\Delta_n}_j}\right)^{\mu} \left\| \frac{d^{p+2}}{dx^{p+2}}(u_{EX}) \right\|_{L^\infty(I^{\Delta_n}_j)}, \quad 1 \leq l \leq p+1 \tag{4.6.39a}$$

where

$$x^{L,I^{\Delta_n}_j}_l \stackrel{\text{def}}{=} F_{I^{\Delta_n}_j}(\tilde{x}^L_l), \quad \tilde{\psi}_p(\tilde{x}^L_l) = \int_{-1}^{\tilde{x}^L_l} \tilde{P}_p = 0, \qquad l = 1, \ldots, p+1 \tag{4.6.39b}$$

are the mapped Lobatto–Legendre points in $I_j^{\Delta_n}$*, while we have*

$$\mathscr{C}\big(h_{I_j^{\Delta_n}}\big)^{p+1}\Big\|\frac{d^{p+1}}{dx^{p+1}}(u_{EX})\Big\|_{L^\infty(I_j^{\Delta_n})} \le \|e_{S^p_{\Delta_n}}\|_{L^\infty(I_j^{\Delta_n})}. \tag{4.6.39c}$$

Further

$$\|e_{S^p_{\Delta_n}} - b_p\psi_p\|_{L^2(I_j^{\Delta_n})} \le \mathscr{C}\big(h_{I_j^{\Delta_n}}\big)^{p+2}\Big\|\frac{d^{p+2}}{dx^{p+2}}(u_{EX})\Big\|_{L^2(I_j^{\Delta_n})} \tag{4.6.40a}$$

while

$$\mathscr{C}\big(h_{I_j^{\Delta_n}}\big)^{p+1}\Big\|\frac{d^{p+1}}{dx^{p+1}}(u_{EX})\Big\|_{L^2(I_j^{\Delta_n})} \le \|e_{S^p_{\Delta_n}}\|_{L^2(I_j^{\Delta_n})}. \tag{4.6.40b}$$

Proof As in the proof of Theorem 4.6.3, we will first consider the special case $AE \equiv 1$, $c \equiv 0$. Integrating (4.6.30a) we get

$$\tilde{e}^{loc}_{\tilde{S}^p} = \sum_{l=p}^{\infty} b_l\tilde{\psi}_l = b_p\tilde{\psi}_p + \sum_{l=p+1}^{\infty} b_l\tilde{\psi}_l = b_p\tilde{\psi}_p + \tilde{e}^{loc}_{\tilde{S}^{p+1}} \tag{4.6.41a}$$

and evaluating this equality at the zeros of $\tilde{\psi}_p$ ($\tilde{\psi}_p(\tilde{x}^L_m) = 0, m = 1, \dots, p$) we get

$$\tilde{e}^{loc}_{\tilde{S}^p}(\tilde{x}^L_m) = \tilde{e}^{loc}_{\tilde{S}^{p+1}}(\tilde{x}^L_m), \quad m = 1, \dots, p. \tag{4.6.41b}$$

The zeros of $\tilde{\psi}_p$, which was defined in (4.6.9d), include the two end points of the interval and, for $p > 2$, $(p-1)$ interior points, and are known as the Lobatto–Legendre points of order p. It follows that

$$|e^{loc}_{S^p(I_j^\Delta)}(x_l^{L,I_j^\Delta})| = |e^{loc}_{S^{p+1}(I_j^\Delta)}(x_l^{L,I_j^\Delta})| \le \mathscr{C}h^{p+2}_{I_j^\Delta}\Big\|\frac{d^{p+2}}{dx^{p+2}}(u_{EX})\Big\|_{L^\infty(I_j^\Delta)}. \tag{4.6.41c}$$

In the general case where $AE \not\equiv$ *constant*, and/or $c \not\equiv 0$, the proof follows from (4.6.25) with $q = \infty$. □

Remark 4.6.6 The Lobatto points are superconvergence points for the value for $p > 1$. In the case $p = 1$, the rate of convergence of the values at the nodal points, which are the Lobatto points for $p = 1$, is $\mu = 2$, and is the same as the rate of convergence of the maximum error over the element.

Let us now illustrate the above results.

Example 4.6.2 Consider the singular model problem and the meshes of Example 4.6.1. Figure 4.6.2 shows the graph of $|e^{loc}_{S^p(I_N^\Delta)}|/\|e^{loc}_{S^p(I_N^\Delta)}\|_{L^\infty(I_N^\Delta)}$ over the element, for $p = 2, 3, 4$ for $h = \frac{1}{32}$. From this graph it can be seen that the zeros of the relative local error occur very close to the mapped Lobatto–Legendre points of order p, and converge to these points as h tends to zero. Table 4.6.2 gives the values of

$$E^u_{loc}(I_N^\Delta) \overset{\text{def}}{=} \frac{\max_{m=1,\dots,p+1} |e^{loc}_{S^p(I_N^\Delta)}(x^L_m)|}{h\|e^{loc}_{S^p(I_N^\Delta)}\|_{L^\infty(I_N^\Delta)}} \tag{4.6.42a}$$

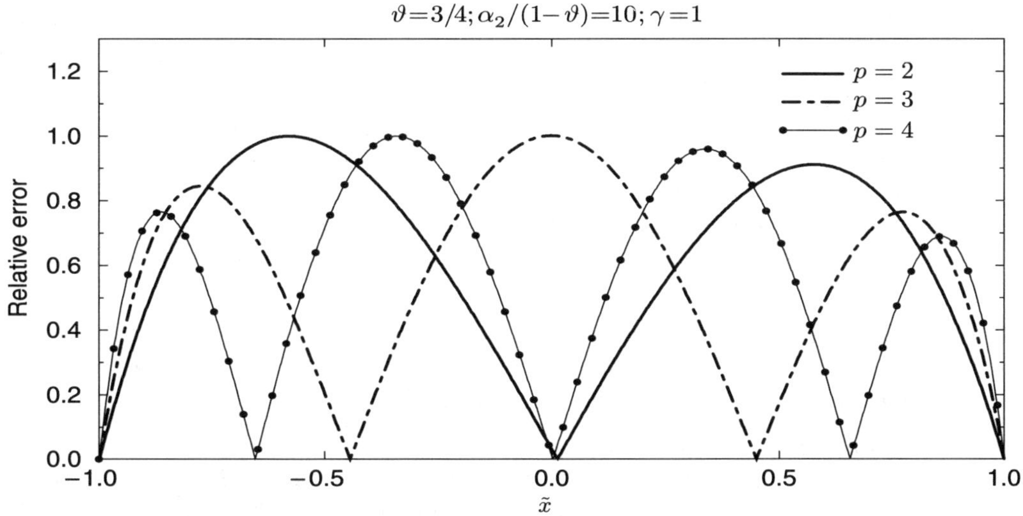

FIG. 4.6.2. An example which illustrates the superconvergence of local error. Input data: $AE \equiv x^{3/4}$, $\alpha_2/(1-\frac{3}{4}) = 10$, $\gamma = 1$. Graphs of the absolute value of the relative local error $|e^{loc}_{S^p(I^{\Delta}_N)}|/\|e^{loc}_{S^p(I^{\Delta}_N)}\|_{L^\infty(I^{\Delta}_N)}$ over the master element, for $p = 2, 3, 4$ and $h = \frac{1}{32}$. The zero points for each graph occur very close to the Lobatto–Legendre points of order p and they converge to these points as h tends to zero

TABLE 4.6.2. Values of the maximum error at the Lobatto–Legendre superconvergence points, normalized by the maximum error in the element and times h. Input data: $AE \equiv x^{3/4}$, $c \equiv 0$, $\alpha_2/(1-\frac{3}{4}) = 10$. Uniform mesh of mesh size h. Note that $E^u_{loc}(I^{\Delta}_N)$ converges to a constant value which means that the local error at the Lobatto–Legendre points converges with one order higher in h than the maximum error in the element. On the other hand, $E^u(I^{\Delta}_N)$ diverges to infinity as h tends to zero, because, as we have seen in Example 4.4.2, the pollution error is much more significant than the local error, and hence, the values of the finite element solution at the Lobatto–Legendre points are not superconvergent

h	$p=2$		$p=3$		$p=4$	
	$E^u_{loc}(I^{\Delta}_N)$	$E^u(I^{\Delta}_N)$	$E^u_{loc}(I^{\Delta}_N)$	$E^u(I^{\Delta}_N)$	$E^u_{loc}(I^{\Delta}_N)$	$E^u(I^{\Delta}_N)$
1/2	0.2693×10^{0}	0.1311×10^{1}	0.1679×10^{0}	0.1592×10^{1}	0.8884×10^{-1}	0.1736×10^{1}
1/4	0.2416×10^{0}	0.3137×10^{1}	0.1119×10^{0}	0.3510×10^{1}	0.5607×10^{-1}	0.3690×10^{1}
1/8	0.2276×10^{0}	0.6987×10^{1}	0.9161×10^{-1}	0.7433×10^{1}	0.5479×10^{-1}	0.7644×10^{1}
1/16	0.2199×10^{0}	0.1488×10^{2}	0.8368×10^{-1}	0.1538×10^{2}	0.5383×10^{-1}	0.1561×10^{2}
1/32	0.2157×10^{0}	0.3082×10^{2}	0.8014×10^{-1}	0.3135×10^{2}	0.5329×10^{-1}	0.3159×10^{2}

and $E^u(I^{\Delta}_N)$ which is obtained from (4.6.41a) by replacing $e^{loc}_{S^p(I^{\Delta}_N)}$ by $e_{S^p_{\Delta}}|_{I^{\Delta}_N}$. It can be seen that, as h tends to zero, the values of $E^u_{loc}(I^{\Delta}_N)$ converge to a constant value, while the values of $E^u(I^{\Delta}_N)$ diverge to infinity. This means that

$$\max_{m=1,\ldots,p+1} \left|e^{loc}_{S^p(I^{\Delta}_N)}(x^L_m)\right| \approx \mathscr{C}h\left\|e^{loc}_{S^p(I^{\Delta}_N)}\right\|_{L^\infty(I^{\Delta}_N)} \tag{4.6.42b}$$

and hence the values of the local finite element solution at the Lobatto–Legendre points of order p are superconvergent with one order higher in h than the maximum error in the element. On the other hand the values of the finite element solution at the same points are not superconvergent because of the pollution.

Remark 4.6.7 In the case of the regular model problem, Assumption I is satisfied in all the elements, and if the finite element solution is computed using quasi-uniform meshes, the pollution is asymptotically negligible and Assumption II is also satisfied. Thus, in each element there exist p superconvergence points for the derivative of the finite element solution, while, for $p \geq 2$, there are $p + 1$ superconvergence points for the value of the finite element solution.

Remark 4.6.8 In the case of the singular problem we can have superconvergence in elements away from the singular point at $x = 0$ only if the mesh is sufficiently refined near the singular point so that the pollution in the elements of interest is negligible.

Remark 4.6.9 Let us emphasize that *superconvergence is strongly related with the pollution.* Therefore *to speak about superconvergence without addressing the pollution is meaningless.* As we have seen in Examples 4.4.1 and 4.4.2, significant pollution can be caused, throughout almost the entire mesh, by a local non-uniformity of the mesh, or by local unsmoothness of the function AE.

Remark 4.6.10 The superconvergence has asymptotic character, which means that it can be realized only if the mesh size is sufficiently small for the input data of interest. In practical computations with meshes of element of higher degree (e.g. $p \geq 3$), the required accuracy may be achieved for a relatively coarse mesh before the superconvergence effects become significant.

Remark 4.6.11 It is possible to construct post-processing schemes which can be used to extract solution quantities (e.g. the values, the derivative, second derivative, etc.) with accuracy of order h^{2p}, in contrast with the h^{p+1} accuracy achieved by direct sampling at the superconvergence points. Such schemes can be employed to realize significant gains in accuracy even for very coarse meshes. See also Section 6.2.

Remark 4.6.12 We addressed here only the problem of superconvergence of direct sampling of the computed solution which is the most simple case of the recovery technique. As we will see in Section 4.8, various averaging techniques can be used to obtain superconvergence at any point.

In this section we have shown that there exist superconvergent points in the element where the accuracy is higher than in all other points. This relates to the fact that the finite element solution 'oscillates' around the exact solution. The major assumption for realizing the superconvergence is the absence of pollution. The arguments employed in this section have one-dimensional character and cannot be used in higher dimensions. In Section 4.7 we present another approach which has a direct analogue in higher dimensions.

4.6* Superconvergence in two dimensions

In the one-dimensional case the analysis of superconvergence is based on the analysis of the error for the regular model problem with $AE \equiv$ *constant*, $c \equiv 0$. Analogously, in the two-dimensional case the analysis of superconvergence is based on the analysis of the error for the Poisson equation, namely $-\Delta u = f$. The main differences between the one and the two-dimensional cases are:

1. In the case of the one-dimensional model problem with $AE \equiv constant$, $c \equiv 0$, *the nodal errors are zero*, and we have

$$e_{S^p_\Delta}|_{I_j^\Delta} \equiv e_{S_j^p}^{loc} = \sum_{l=p}^{\infty} b_l \psi_l, \qquad b_l = \frac{2l+1}{2} \int_{I_j^\Delta} \frac{du_{EX}}{dx} P_l \tag{4.6*.1a}$$

where P_l denotes the lth Legendre polynomial mapped into I_j^Δ and

$$\psi_l(x) = \int_{x_{j-1}^\Delta}^{x} P_l \tag{4.6*.1b}$$

is its primitive (with $\psi_l(x_{j-1}^\Delta) = 0$). Hence the error *localizes* in each element (the error in the element I_j^Δ is not influenced by the exact solution and the mesh outside I_j^Δ), and the superconvergence points for the value, for an element of degree p, are the roots of ψ_p (respectively the superconvergence points for the derivative are the roots of $P_p = (d/dx)(\psi_p)$), where ψ_p and P_p are defined in each element independently of its neighbors. In contrast, in the two-dimensional case, the error in finite element approximations of the Poisson equation in an element τ (e.g., the element τ in Fig. 4.6*.1b) depends on the mesh in the elements connected to τ and their neighbors (e.g., the neighborhoods of τ shown in Fig. 4.6*.1b). Hence *in two-dimensional computations it is not, in general, possible to determine the superconvergence in an element by employing a function defined only from the element.*

2. *In one dimension the $(p+1)$-degree term of the local Taylor series expansion consists of only one monomial*, namely $(x-\bar{x})^{p+1}$, while *in two-dimensions it is a linear*

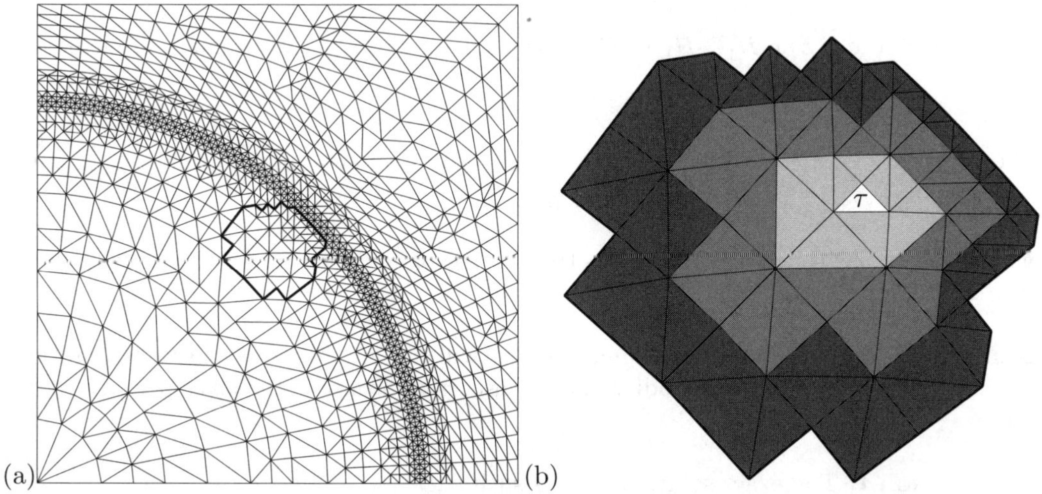

FIG. 4.6*.1. An example of a mesh of triangles constructed using a commercial mesh generator. (a) The entire mesh. (b) A detail of the mesh which shows an element τ and its neighborhoods which practically influence the local error in τ

combination of $(p+2)$ *monomials*, namely $(x_1 - \bar{x}_1)^i(x_2 - \bar{x}_2)^{p+1-i}$, $i = 0, 1, \ldots, p+1$. This means that while in the one-dimensional case we can study the superconvergence by analyzing one function, the error in the finite element approximation of $(x-\bar{x})^{p+1}$, in the two-dimensional case we must analyze simultaneously $(p+2)$ functions, the errors in the finite element approximations of the monomials $(x_1-\bar{x}_1)^i(x_2-\bar{x}_2)^{p+1-i}$, $i = 0, \ldots, p+2$.

4.7 A general approach for the analysis of local behavior of finite element methods for locally periodic meshes and its applications

As we have seen in Section 4.6*, the approach employed for analyzing the superconvergence of the finite element solution in Section 4.6, has essentially one-dimensional character and cannot be used in higher dimensions, except in some very special cases, like for example, the finite element approximation of Poisson's equation on meshes of bi-p squares. We will now present a general approach for analyzing the local behavior of the finite element solution. This approach was first presented in the two-dimensional setting in [Babuška, Strouboulis *et al.*, 1996], and can be used to analyze the superconvergence of finite element methods in any dimension.

Below, we will give the assumptions on which the analysis is based. We will denote the interval (subdomain) of size H centered at the point $\bar{x}$ by

$$S(\bar{x}, H) \stackrel{\text{def}}{=} \left(\bar{x} - \tfrac{1}{2}H, \bar{x} + \tfrac{1}{2}H\right) \tag{4.7.1}$$

and we will assume that:

Assumption 0. The meshes Δ are such that:

(a) The restriction of each mesh Δ in $S(\bar{x}, H)$ is uniform, namely,

$$h_{I_j^\Delta} = h, \quad \forall I_j^\Delta \subseteq S(\bar{x}, H). \tag{4.7.2a}$$

(b) $S(\bar{x}, H_0)$ and $S(\bar{x}, H_1)$, $H_1 < H_0 \leq H$ coincide exactly with a patch of elements, namely

$$\bar{S}(\bar{x}, H_i) = \bigcup_{j=n(\bar{x},H_i;\Delta)}^{m(\bar{x},H_i;\Delta)} \bar{I}_j^\Delta, \quad i = 0, 1 \tag{4.7.2b}$$

where $n(\bar{x}, H_i; \Delta)$ (resp. $m(\bar{x}, H_i; \Delta)$) is the first (resp. last) element from Δ which belong to $S(\bar{x}, H_i)$.

Remark 4.7.1 In the two-dimensional case, we will employ the more general assumption that the mesh in $S(\bar{x}, H)$ is periodic. In that case, h is the size of the basic periodic mesh cell.

Assumption I. The exact solution u_{EX} is such that

$$\left\| \frac{d^{p+2}}{dx^{p+2}}(u_{EX}) \right\|_{L^\infty(S(\bar{x},H))} \leq K < \infty \tag{4.7.3a}$$

and

$$0 < \mathscr{C} \leq \left| \frac{d^{p+1}}{dx^{p+1}}(u_{EX})(\bar{x}) \right|. \tag{4.7.3b}$$

Below we will refer to (4.7.3a) as Assumption Ia, and to (4.7.3b) as Assumption Ib.

Remark 4.7.2 Assumption I is the same as Assumption I in Section 4.6.

Assumption II. The meshes Δ are such that

$$\|e_{S^p_\Delta}\|_{L^2(S(\bar{x},H))} \leq \mathscr{C} h^\beta \sqrt{H} \tag{4.7.4a}$$

with

$$\beta \geq (p+1) - \varepsilon, \quad 0 < \varepsilon < \tfrac{3}{4} \tag{4.7.4b}$$

and $\mathscr{C}$ depending only on K.

Remark 4.7.3 Assumption II is a more general characterization of the case of negligible pollution which has meaning for general meshes in higher dimensions, as well as in the simplest one-dimensional setting.

In the following, we will present a series of lemmas which lead to the major result

$$\frac{d}{dx}(e_{S^p_\Delta})(x) \approx \frac{d}{dx}(\Psi^{\bar{x},H}_{S^p_\Delta})(x) \quad x \in S(\bar{x}, \gamma H) \tag{4.7.5}$$

where $\Psi^{\bar{x},H}_{S^p_\Delta}$ is a periodic function which can be constructed in the master element, $0 < \gamma < 1$ is sufficiently small, H converges to zero with a rate slower than h and is sufficiently small, depending on K and $\mathscr{C}$, the constants in Assumptions Ia and Ib.

Let us first recall that Assumption Ib implies that the L^∞ norm of the derivative of the error in $S(\bar{x}, H)$ cannot converge to zero faster than h^p.

Lemma 4.7.1 *Under Assumption I we have*

$$\mathscr{C} h^{p+1} \leq \|e_{S^p_\Delta}\|_{L^\infty(I^\Delta_j)}, \quad I^\Delta_j \subseteq S(\bar{x}, H) \tag{4.7.6a}$$

and

$$\mathscr{C} h^p \leq \left\| \frac{d}{dx}(e_{S^p_\Delta}) \right\|_{L^\infty(S(\bar{x},H_i))} \tag{4.7.6b}$$

with $\mathscr{C}$ depending only on $\mathscr{C}$, the constant in Assumption Ib, H, γ, and p.

Proof Inequality (4.7.6a) has already been proven in Theorem 4.1.2, and (4.7.6b) follows from (4.7.6a). □

We will now employ Assumption Ia to derive estimates for the error in the $(p+1)$-degree Taylor series expansion of u_{EX} in $S(\bar{x}, H)$.

Lemma 4.7.2 *Let u_{EX} satisfy Assumption Ia, and let $U_{EX}^{\bar{x},(p+1)}$ be the $(p+1)$ degree Taylor series expansion of u_{EX} centered at $\bar{x}$, namely,*

$$U_{EX}^{\bar{x},(p+1)}(x) \stackrel{\text{def}}{=} \sum_{k=0}^{p+1} \frac{1}{k!} \frac{d^k}{dx^k}(u_{EX})(\bar{x})(x-\bar{x})^k. \tag{4.7.7a}$$

Then we have

$$\left\| \frac{d^n}{dx^n}(u_{EX} - U_{EX}^{\bar{x},(p+1)}) \right\|_{L^\infty(S(\bar{x},H_i))} \leq \mathscr{C} H_i^{p+2-n} \tag{4.7.7b}$$

for $n = 0, 1, 2$, with constants $\mathscr{C}$ depending only on K and p.

Proof We have

$$u_{EX}(x) = U_{EX}^{\bar{x},(p+1)}(x) + R_{EX}^{\bar{x},(p+1)}(x) \tag{4.7.8a}$$

where

$$R_{EX}^{\bar{x},(p+1)}(x) \stackrel{\text{def}}{=} \underbrace{\int_{\bar{x}}^{x} \cdots \int_{\bar{x}}^{x'}}_{(p+2)\text{ times}} \frac{d^{p+2}}{dx^{p+2}}(u_{EX}) \tag{4.7.8b}$$

is the *remainder*. Noting that

$$\left| \underbrace{\int_{\bar{x}}^{x} \cdots \int_{\bar{x}}^{x''} \int_{\bar{x}}^{x'}}_{(p+2)\text{ times}} \frac{d^{p+2}}{dx^{p+2}}(u_{EX}) \right| \leq \frac{1}{(p+2)!} |x-\bar{x}|^{p+2} \left\| \frac{d^{p+2}}{dx^{p+2}}(u_{EX}) \right\|_{L^\infty(S(\bar{x},H))} \tag{4.7.8c}$$

we get the estimate

$$\|R_{EX}^{\bar{x},(p+1)}\|_{L^\infty(S(\bar{x},H_i))} \leq \frac{1}{(p+2)!} H_i^{p+2} \left\| \frac{d^{p+2}}{dx^{p+2}}(u_{EX}) \right\|_{L^\infty(S(\bar{x},H))} \tag{4.7.8d}$$

which can also be written as

$$\|u_{EX} - U_{EX}^{\bar{x},(p+1)}\|_{L^\infty(S(\bar{x},H_i))} \leq \frac{1}{(p+2)!} H_i^{p+2} \left\| \frac{d^{p+2}}{dx^{p+2}}(u_{EX}) \right\|_{L^\infty(S(\bar{x},H))} \tag{4.7.8e}$$

and using Assumption Ia we obtain

$$\|u_{EX} - U_{EX}^{\bar{x},(p+1)}\|_{L^\infty(S(\bar{x},H_i))} \leq \frac{K}{(p+2)!} H_i^{p+2}. \tag{4.7.8f}$$

Following similar steps we can also get the L^∞ estimates (4.7.7b) for $(d^n/dx^n)(u_{EX} - U_{EX}^{\bar{x},(p+1)})$, for $n = 1$, and 2. □

We will denote by $U_{S_\Delta^p}^{\bar{x},H}$, the best approximation of $U_{EX}^{\bar{x},(p+1)}$, from $S_\Delta^p(S(\bar{x},H))$ the restriction of the finite element space S_Δ^p in the patch of elements which belong to $S(\bar{x},H)$, and we will prove that asymptotically

$$\frac{d}{dx}(e_{S_\Delta^p})(x) \approx \frac{d}{dx}\left(U_{EX}^{\bar{x},(p+1)} - U_{S_\Delta^p}^{\bar{x},H}\right)(x), \quad x \in S(\bar{x},\gamma H) \tag{4.7.9}$$

where $0 < \gamma < 1$, is sufficiently small.

Let

$$S^p_\Delta(S(\bar{x},H)) \stackrel{\text{def}}{=} \left\{v \in C^0(S(\bar{x},H)) \mid \exists w \in S^p_\Delta : \; v \equiv w|_{S(\bar{x},H)}\right\} \tag{4.7.10a}$$

be the restriction of S^p_Δ in $S(\bar{x},H)$, and $U^{\bar{x},H}_{S^p_\Delta}$ be the $\mathscr{B}^0_{S(\bar{x},H)}$-*projection* of $U^{\bar{x},(p+1)}_{EX}$ into $S^p_\Delta(S(\bar{x},H))$, namely, the solution of the discrete problem:

Find $U^{\bar{x},H}_{S^p_\Delta} \in S^p_\Delta(S(\bar{x},H))$ such that

$$\mathscr{B}^0_{S(\bar{x},H)}\big(U^{\bar{x},H}_{S^p_\Delta}, v\big) = \mathscr{B}^0_{S(\bar{x},H)}\big(U^{\bar{x},(p+1)}_{EX}, v\big) \quad \forall v \in S^p_\Delta(S(\bar{x},H)) \tag{4.7.10b}$$

and

$$\int_{S(\bar{x},H)} U^{\bar{x},H}_{S^p_\Delta} = \int_{S(\bar{x},H)} U^{\bar{x},(p+1)}_{EX} \tag{4.7.10c}$$

where

$$\mathscr{B}^0_{S(\bar{x},H)}(u,v) = AE(\bar{x}) \int_{S(\bar{x},H)} \frac{du}{dx}\frac{dv}{dx}. \tag{4.7.10d}$$

Note that $U^{\bar{x},H}_{S^p_\Delta}$ exists, and is uniquely determined.

We will also let

$$Q^{\bar{x},(p+1)}_{EX}(x) = \frac{1}{(p+1)!}\frac{d^{p+1}}{dx^{p+1}}(u_{EX})(\bar{x})(x-\bar{x})^{p+1} \tag{4.7.10e}$$

and we will denote by $Q^{\bar{x},H}_{S^p_\Delta} \in S^p_\Delta(S(\bar{x},H))$ its $\mathscr{B}^0_{S(\bar{x},H)}$ projection.

Let us give some estimates for the errors in the above projections.

Lemma 4.7.3 *Under Assumption Ia, we have*

$$\begin{aligned}\left\|\frac{d}{dx}\big(U^{\bar{x},(p+1)}_{EX} - U^{\bar{x},H}_{S^p_\Delta}\big)\right\|_{L^2(S(\bar{x},H))} &= \left\|\frac{d}{dx}\big(Q^{\bar{x},(p+1)}_{EX} - Q^{\bar{x},H}_{S^p_\Delta}\big)\right\|_{L^2(S(\bar{x},H))} \\ &\le \mathscr{C} h^p \sqrt{H}\end{aligned} \tag{4.7.11a}$$

and

$$\left\|U^{\bar{x},(p+1)}_{EX} - U^{\bar{x},H}_{S^p_\Delta}\right\|_{L^2(S(\bar{x},H))} = \left\|Q^{\bar{x},(p+1)}_{EX} - Q^{\bar{x},H}_{S^p_\Delta}\right\|_{L^2(S(\bar{x},H))} \le \mathscr{C} h^{p+1}\sqrt{H}. \tag{4.7.11b}$$

Proof By its construction, $U^{\bar{x},H}_{S^p_\Delta}$ satisfies the orthogonality

$$\int_{S(\bar{x},H)} \left(\frac{d}{dx}\big(U^{\bar{x},H}_{S^p_\Delta}\big) - \frac{d}{dx}\big(U^{\bar{x},(p+1)}_{EX}\big)\right)\frac{dv}{dx} = 0 \quad \forall v \in S^p_\Delta(S(\bar{x},H)). \tag{4.7.12a}$$

It follows that $(d/dx)(U^{\bar{x},H}_{S^p_\Delta})$ is the best approximation of $(d/dx)(U^{\bar{x},(p+1)}_{EX})$ from $S^{p-1}_\Delta(S(\bar{x},H))$ in the L^2 norm, and hence

$$\left\|\frac{d}{dx}\big(U^{\bar{x},(p+1)}_{EX} - U^{\bar{x},H}_{S^p_\Delta}\big)\right\|_{L^2(S(\bar{x},H))} \le \mathscr{C} h^p \left\|\frac{d^{p+1}}{dx^{p+1}}\big(U^{\bar{x},(p+1)}_{EX}\big)\right\|_{L^2(S(\bar{x},H))}. \tag{4.7.12b}$$

Noting that

$$U_{EX}^{\bar{x},p} \in \mathcal{P}^p(I) \subseteq S_\Delta^p \tag{4.7.13a}$$

we get

$$U_{EX}^{\bar{x},(p+1)} - U_{S_\Delta^p}^{\bar{x},H} \equiv Q_{EX}^{\bar{x},(p+1)} - Q_{S_\Delta^p}^{\bar{x},H} \tag{4.7.13b}$$

and employing that

$$\left\| \frac{d^{p+1}}{dx^{p+1}}(U_{EX}^{\bar{x},(p+1)}) \right\|_{L^2(S(\bar{x},H))} \leq \sqrt{H} \left\| \frac{d^{p+1}}{dx^{p+1}}(U_{EX}^{\bar{x},(p+1)}) \right\|_{L^\infty(S(\bar{x},H))} \tag{4.7.13c}$$

(4.7.11a) follows. From Corollary 4.2.2 we have

$$\|U_{EX}^{\bar{x},(p+1)} - U_{S_\Delta^p}^{\bar{x},H}\|_{L^2(S(\bar{x},H))} \leq \mathscr{C}h \left\| \frac{d}{dx}(U_{EX}^{\bar{x},(p+1)} - U_{S_\Delta^p}^{\bar{x},H}) \right\|_{L^2(S(\bar{x},H))} \tag{4.7.13d}$$

and (4.7.11b) follows. □

Next we will show that the interpolation error, $U_{EX}^{\bar{x},(p+1)} - \mathcal{I}_\Delta^p U_{EX}^{\bar{x},(p+1)}$, is periodic in $S(\bar{x}, H)$ with period equal to h.

Lemma 4.7.4 *The interpolation error*

$$\rho_{S_\Delta^p}^{\bar{x},H} \stackrel{\text{def}}{=} U_{EX}^{\bar{x},(p+1)} - \mathcal{I}_\Delta^p U_{EX}^{\bar{x},(p+1)} = Q_{EX}^{\bar{x},(p+1)} - \mathcal{I}_\Delta^p Q_{EX}^{\bar{x},(p+1)} \tag{4.7.14a}$$

satisfies the periodicity condition

$$\rho_{S_\Delta^p}^{\bar{x},H}(x) = \rho_{S_\Delta^p}^{\bar{x},H}(x \pm kh), \quad x \in I_j^\Delta \tag{4.7.14b}$$

with j, k *such that*

$$n(\bar{x}, H; \Delta) \leq j \pm k \leq m(\bar{x}, H; \Delta). \tag{4.7.14c}$$

Proof Expanding $U_{EX}^{\bar{x},(p+1)} \in \mathcal{P}^{p+1}$ in a Taylor series about $x_{j-1/2}^\Delta$, we obtain

$$\begin{aligned} U_{EX}^{\bar{x},(p+1)}(x) &= U^{x_{j-1/2}^\Delta,p}(x) \\ &\quad + \frac{1}{(p+1)!}\frac{d^{p+1}}{dx^{p+1}}(U_{EX}^{\bar{x},(p+1)})(x_{j-1/2}^\Delta)(x - x_{j-1/2}^\Delta)^{p+1} \end{aligned} \tag{4.7.15a}$$

where we let

$$U^{x_{j-1/2}^\Delta,p}(x) = \sum_{k=0}^{p} \frac{1}{k!}\frac{d^k}{dx^k}(U_{EX}^{\bar{x},(p+1)})(x_{j-1/2}^\Delta)(x - x_{j-1/2}^\Delta)^k. \tag{4.7.15b}$$

Now because

$$U^{x_{j-1/2}^\Delta,p} \in \mathcal{P}^p(I) \tag{4.7.15c}$$

we have

$$U^{x^{\Delta}_{j-1/2},p} - \mathcal{I}^p_{\Delta} U^{x^{\Delta}_{j-1/2},p} \equiv 0 \tag{4.7.15d}$$

and it follows that

$$\rho^{\bar{x},H}_{S^p_{\Delta}}|_{I^{\Delta}_j} = \frac{1}{(p+1)!}\frac{d^{p+1}}{dx^{p+1}}(U^{\bar{x},(p+1)}_{EX})(x^{\Delta}_{j-1/2}) \times \Big((x - x^{\Delta}_{j-1/2})^{p+1} - \mathcal{I}^p_{I^{\Delta}_j}\big((x - x^{\Delta}_{j-1/2})^{p+1}\big)\Big). \tag{4.7.15e}$$

Noting that

$$\frac{d^{p+1}}{dx^{p+1}}(U^{\bar{x},(p+1)}_{EX})(x^{\Delta}_{j-1/2}) \equiv \mathit{constant} \tag{4.7.15f}$$

independent of $x^{\Delta}_{j-1/2}$, it follows that $\rho^{\bar{x},H}_{S^p_{\Delta}}|_{I^{\Delta}_j}$ is the same on every element $I^{\Delta}_j \subseteq S(\bar{x}, H)$. □

Below we will employ $\pi^0_{S^p_{PER}(I^{\Delta}_j)}(\rho^{\bar{x},H}_{S^p_{\Delta}})$, the periodic best approximation of $\rho^{\bar{x},H}_{S^p_{\Delta}}$. Let us first define the periodic projection $\pi^0_{S^p_{PER}(I^{\Delta}_j)}$.

Let

$$H^1_{\mathrm{PER}}(I^{\Delta}_j) \stackrel{\mathrm{def}}{=} \Big\{v \in H^1(I^{\Delta}_j) \,\Big|\, v\Big(x^{\Delta}_{j-1/2} + \frac{h}{2}\Big) = v\Big(x^{\Delta}_{j-1/2} - \frac{h}{2}\Big)\Big\} \tag{4.7.16a}$$

denote the space of periodic functions with square-integrable derivatives in I^{Δ}_j, and

$$S^p_{PER}(I^{\Delta}_j) = S^p(I^{\Delta}_j) \cap H^1_{PER}(I^{\Delta}_j) \tag{4.7.16b}$$

denote the periodic element space. Given a $\varphi \in H^1_{PER}(I^{\Delta}_j)$, we will define $\pi^0_{S^p_{PER}(I^{\Delta}_j)}(\varphi)$, the periodic $\mathcal{B}^0_{I^{\Delta}_j}$ projection of φ, as the solution of the discrete problem:

Find $\pi^0_{S^p_{PER}(I^{\Delta}_j)}(\varphi) \in S^p_{PER}(I^{\Delta}_j)$ such that

$$\mathcal{B}^0_{I^{\Delta}_j}\big(\pi^0_{S^p_{PER}(I^{\Delta}_j)}(\varphi), v\big) = \mathcal{B}^0_{I^{\Delta}_j}(\varphi, v) \quad \forall v \in S^p_{PER}(I^{\Delta}_j) \tag{4.7.16c}$$

and

$$\int_{I^{\Delta}_j} \pi^0_{S^p_{PER}(I^{\Delta}_j)}(\varphi) = \int_{I^{\Delta}_j} \varphi. \tag{4.7.16d}$$

Note that $\pi^0_{S^p_{PER}(I^{\Delta}_j)}(\varphi)$ exists and is uniquely determined.

Given a periodic function $\varphi \in H^1_{PER}(I^{\Delta}_j)$, we will denote by $\bar{\varphi}$ the *periodic extension* of φ over $S(\bar{x}, H)$, defined by

$$\bar{\varphi}(x) = \varphi\Big(x^{\Delta}_{j-1/2} + (x - x^{\Delta}_{k-1/2})\Big), \quad x \in I^{\Delta}_k \subseteq S(\bar{x}, H). \tag{4.7.17}$$

We will now prove that the periodic extension of $\bar{\varphi} - \pi^0_{S^p_{PER}(I^{\Delta}_j)}(\varphi)$ is orthogonal to $S^p_{\Delta,0}(S(\bar{x}, H))$.

Lemma 4.7.5 *For any $\varphi \in H^1_{PER}(I_j^\Delta)$, we have*

$$\mathcal{B}^0_{S(\bar{x},H)}\left(\bar{\pi}^0_{S^p_{PER}(I_j^\Delta)}(\varphi) - \bar{\varphi}, v\right) = 0 \quad \forall v \in S^p_{\Delta,0}(S(\bar{x},H)) \tag{4.7.18a}$$

where

$$S^p_{\Delta,0}(S(\bar{x},H)) \stackrel{\text{def}}{=} S^p_\Delta(S(\bar{x},H)) \cap H^1_0(S(\bar{x},H)). \tag{4.7.18a}$$

Proof Given $v \in S^p_{\Delta,0}(S(\bar{x},H))$ and $I_k^\Delta, I_j^\Delta \subseteq S(\bar{x},H)$, let $v_{I_k^\Delta \to I_j^\Delta} \in H^1(I_j^\Delta)$ denote the *translation*

$$v_{I_k^\Delta \to I_j^\Delta}(x) = v|_{I_k^\Delta}(x^\Delta_{k-1/2} + (x - x^\Delta_{j-1/2})), \quad x \in I_j^\Delta. \tag{4.7.19a}$$

Then we have

$$v \in S^p_{\Delta,0}(S(\bar{x},H)) \implies \sum_{I_k^\Delta \subseteq S(\bar{x},H)} v_{I_k^\Delta \to I_j^\Delta} \in S^p_{PER}(I_j^\Delta) \tag{4.7.19b}$$

and employing this result for any $v \in S^p_{\Delta,0}(I_j^\Delta)$, we obtain

$$\begin{aligned}
\sum_{I_k^\Delta \subseteq S(\bar{x},H)} \mathcal{B}^0_{I_k^\Delta}\left(\bar{\pi}^0_{S^p_{PER}(I_j^\Delta)}(\varphi), v\right) &= \mathcal{B}^0_{I_j^\Delta}\left(\pi^0_{S^p_{PER}(I_j^\Delta)}(\varphi), \sum_{I_k^\Delta \subseteq S(\bar{x},H)} v_{I_k^\Delta \to I_j^\Delta}\right) \\
&= \mathcal{B}^0_{I_j^\Delta}\left(\varphi, \sum_{I_k^\Delta \subseteq S(\bar{x},H)} v_{I_k^\Delta \to I_j^\Delta}\right) \\
&= \sum_{I_k^\Delta \subseteq S(\bar{x},H)} \mathcal{B}^0_{I_k^\Delta}(\bar{\varphi}, v) = \mathcal{B}^0_{S(\bar{x},H)}(\bar{\varphi}, v)
\end{aligned} \tag{4.7.19c}$$

which proves (4.7.18). □

We will now let $\Psi^{\bar{x},H}_{S^p_\Delta}$ denote the error in the projection of the interpolation error $\rho^{\bar{x},H}_{S^p(I_j^\Delta)}$, namely,

$$\Psi^{\bar{x},H}_{S^p_\Delta} = \overline{\rho^{\bar{x},H}_{S^p(I_j^\Delta)} - \pi^0_{S^p_{PER}(I_j^\Delta)}(\rho^{\bar{x},H}_{S^p(I_j^\Delta)})} \tag{4.7.20a}$$

and we will prove that

$$\frac{d}{dx}(U^{\bar{x},(p+1)}_{EX} - U^{\bar{x},H}_{S^p_\Delta}) \equiv \frac{d}{dx}(Q^{\bar{x},(p+1)}_{EX} - Q^{\bar{x},H}_{S^p_\Delta}) \approx \frac{d}{dx}(\Psi^{\bar{x},H}_{S^p_\Delta}) \tag{4.7.20b}$$

on $S(\bar{x},\gamma H)$, for γ, $0 < \gamma < 1$, sufficiently small.

Lemma 4.7.6 *Under the Assumption Ia,*

$$\delta\Psi \stackrel{\text{def}}{=} (U^{\bar{x},(p+1)}_{EX} - U^{\bar{x},H}_{S^p_\Delta}) - \Psi^{\bar{x},H}_{S^p_\Delta} \tag{4.7.21a}$$

satisfies the estimate

$$\left\| \frac{d}{dx}(\delta\Psi) \right\|_{L^\infty(S(\bar{x},\gamma H))} \leq \mathscr{C} h^{p+1}((1-\gamma)\,H)^{-3/2}\sqrt{H}. \tag{4.7.21b}$$

Proof We will prove that $\delta\Psi$ is discrete harmonic and we will apply Lemma 4.5.4. Recalling the definition of $\Psi^{\bar{x},H}_{S^p_\Delta}$, we have

$$\begin{aligned} \delta\Psi &= U^{\bar{x},(p+1)}_{EX} - U^{\bar{x},H}_{S^p_\Delta} - \Psi^{\bar{x},H}_{S^p_\Delta} \\ &= U^{\bar{x},(p+1)}_{EX} - U^{\bar{x},H}_{S^p_\Delta} - \left(\rho^{\bar{x},H}_{S^p_\Delta} - \bar{\pi}^0_{S^p_{PER}(I^\Delta_j)}\left(\rho^{\bar{x},H}_{S^p_\Delta}\right)\right) \end{aligned} \tag{4.7.22a}$$

and employing now the orthogonality conditions

$$\mathscr{B}^0_{S(\bar{x},H)}\left(U^{\bar{x},(p+1)}_{EX} - U^{\bar{x},H}_{S^p_\Delta}, v\right) = 0 \quad \forall v \in S^p_{\Delta,0}(S(\bar{x},H)) \tag{4.7.22b}$$

and

$$\mathscr{B}^0_{S(\bar{x},H)}\left(\rho^{\bar{x},H}_{S^p_\Delta} - \bar{\pi}^0_{S^p_{PER}(I^\Delta_j)}\left(\rho^{\bar{x},H}_{S^p_\Delta}\right), v\right) = 0 \quad \forall v \in S^p_{\Delta,0}(S(\bar{x},H)) \tag{4.7.22c}$$

we obtain

$$\mathscr{B}^0_{S(\bar{x},H)}(\delta\Psi, v) = \mathscr{B}^0_{S(\bar{x},H)}\left(U^{\bar{x},(p+1)}_{EX} - U^{\bar{x},H}_{S^p_\Delta} - \Psi^{\bar{x},H}_{S^p_\Delta}, v\right) = 0 \tag{4.7.22d}$$

for all $v \in S^p_{\Delta,0}(S(\bar{x},H))$. Hence $\delta\Psi \in \mathscr{P}^{p+1}(S(\bar{x},H))$ is discrete harmonic, and applying Lemma 4.5.2 we get

$$\left\| \frac{d}{dx}(\delta\Psi) \right\|_{L^\infty(S(\bar{x},\gamma\, H))} \leq \mathscr{C}((1-\gamma), H)^{-3/2}\|\delta\Psi\|_{L^2(S(\bar{x},H))} \tag{4.7.23}$$

with $0 < \gamma < 1$. Employing now the triangle inequality

$$\|\delta\Psi\|_{L^2(S(\bar{x},H))} \leq \left\|U^{\bar{x},(p+1)}_{EX} - U^{\bar{x},H}_{S^p_\Delta}\right\|_{L^2(S(\bar{x},H))} + \left\|\bar{\Psi}^{\bar{x},H}_{S^p_\Delta}\right\|_{L^2(S(\bar{x},H))} \tag{4.7.24a}$$

and using that

$$\left\|\bar{\Psi}^{\bar{x},H}_{S^p_\Delta}\right\|_{L^2(S(\bar{x},H))} \leq \mathscr{C}h\left\|\frac{d}{dx}\left(\bar{\Psi}^{\bar{x},H}_{S^p_\Delta}\right)\right\|_{L^2(S(\bar{x},H))} \leq \mathscr{C}hh^p\sqrt{H} \tag{4.7.24b}$$

and that

$$\left\|U^{\bar{x},(p+1)}_{EX} - U^{\bar{x},H}_{S^p_\Delta}\right\|_{L^2(S(\bar{x},H))} \leq \mathscr{C}h^{p+1}\sqrt{H} \tag{4.7.24c}$$

which was proven in Lemma 4.7.3, we obtain

$$\|\delta\Psi\|_{L^2(S(\bar{x},H))} \leq \mathscr{C}h^{p+1}\sqrt{H} + \mathscr{C}h^{p+1}\sqrt{H}. \tag{4.7.24d}$$

Using this result in (4.7.23), we obtain (4.7.21b). □

We will now prove the asymptotic equality

$$\frac{d}{dx}(e_{S^p_\Delta})\Big|_{S(\bar{x},\gamma H)} \approx \frac{d}{dx}(\Psi^{\bar{x},H}_{S^p_\Delta})\Big|_{S(\bar{x},\gamma H)} \tag{4.7.25a}$$

from which it follows that the points $\bar{x}^{sup}$ which satisfy

$$\frac{d}{dx}\left(\Psi^{\bar{x},H}_{S^p_\Delta}\right)(\bar{x}^{sup}) = 0 \tag{4.7.25b}$$

are superconvergence points for $(d/dx)(e_{S^p_\Delta})$ in $S(\bar{x},\gamma H)$.

Theorem 4.7.1 *Consider the regular model problem under the additional assumption*

$$\left\|\frac{d}{dx}(AE)\right\|_{L^\infty(I)} \le K < \infty. \tag{4.7.26a}$$

Then, under Assumptions 0, I, and II, there exists an α, $0 < \alpha < 1$ such that

$$\mathscr{C}_1 h^\alpha \le H \le \mathscr{C}_2 h^\alpha \tag{4.7.26b}$$

and

$$\left\|\frac{d}{dx}(e_{S^p_\Delta} - \Psi^{\bar{x},H}_{S^p_\Delta})\right\|_{L^\infty(S(\bar{x},H/2))} \le \mathscr{C} h^{\bar{\nu}} \left\|\frac{d}{dx}(e_{S^p_\Delta})\right\|_{L^\infty(S(\bar{x},H))} \tag{4.7.27}$$

with $\bar{\nu} > 0$.

Proof Let $\xi_{S^p_\Delta} \in S^p_\Delta(S(\bar{x},H))$ be such that

$$\mathscr{B}_{S(\bar{x},H)}(\xi_{S^p_\Delta}, v) = \mathscr{B}^1_{S(\bar{x},H)}(U^{\bar{x},(p+1)}_{EX} - U^{\bar{x},H}_{S^p_\Delta}, v) \quad \forall v \in S^p_\Delta(S(\bar{x},H)) \tag{4.7.28a}$$

where

$$\mathscr{B}^1_{S(\bar{x},H)}(u,v) \stackrel{\text{def}}{=} \int_{S(\bar{x},H)} (AE - AE(\bar{x}))\frac{du}{dx}\frac{dv}{dx} + cuv \tag{4.7.28b}$$

and in the case $c \equiv 0$, $\xi_{S^p_\Delta}$ is uniquely determined by imposing the additional condition

$$\int_{S(\bar{x},H)} \xi_{S^p_\Delta} = 0. \tag{4.7.28c}$$

Utilizing assumption (4.7.26a), we get

$$|AE(x) - AE(\bar{x})| \le \mathscr{C} H \left\|\frac{d}{dx}(AE)\right\|_{L^\infty(S(\bar{x},H))} \le \mathscr{C} HK. \tag{4.7.29a}$$

Working now as in the proof of Theorem 4.6.2, we get

$$\begin{aligned} &\left|\mathscr{B}^1_{S(\bar{x},H)}(U^{\bar{x},(p+1)}_{EX} - U^{\bar{x},H}_{S^p_\Delta}, v)\right| \\ &\quad \le \|AE(x) - AE(\bar{x})\|_{L^\infty(S(\bar{x},H))} \left\|\frac{d}{dx}(U^{\bar{x},(p+1)}_{EX} - U^{\bar{x},H}_{S^p_\Delta})\right\|_{L^2(S(\bar{x},H))} \\ &\quad \times \left\|\frac{dv}{dx}\right\|_{L^2(S(\bar{x},H))} + \|c\|_{L^\infty(S(\bar{x},H))} \|U^{\bar{x},(p+1)}_{EX} - U^{\bar{x},H}_{S^p_\Delta}\|_{L^2(S(\bar{x},H))} \\ &\quad \times \|v\|_{L^2(S(\bar{x},H))} \end{aligned} \tag{4.7.29b}$$

and utilizing (4.7.29a), and the estimates given in Lemma 4.7.3, we get

$$\left|\mathcal{B}^1_{S(\bar{x},H)}\big(U^{\bar{x},(p+1)}_{EX} - U^{\bar{x},H}_{S^p_\Delta}, v\big)\right| \le \mathcal{C} h^p H^{3/2} \|v\|_{\mathcal{U}(S(\bar{x},H))} \tag{4.7.29c}$$

which proves that

$$\|\xi_{S^p_\Delta}\|_{\mathcal{U}(S(\bar{x},H))} \le \mathcal{C} h^p H^{3/2}. \tag{4.7.30a}$$

Using this estimate, we get

$$\|\xi_{S^p_\Delta}\|_{L^2(S(\bar{x},H))} \le \mathcal{C}\sqrt{H} \|\xi_{S^p_\Delta}\|_{\mathcal{U}(S(\bar{x},H))} \le \mathcal{C} h^p H^2 \tag{4.7.30b}$$

and

$$\left\|\frac{d}{dx}(\xi_{S^p_\Delta})\right\|_{L^\infty(S(\bar{x},H))} \le \frac{\mathcal{C}}{\sqrt{h}} \|\xi_{S^p_\Delta}\|_{\mathcal{U}(S(\bar{x},H))} \le \mathcal{C} h^{p-1/2} H^2. \tag{4.7.30c}$$

We will estimate the difference between $e_{S^p_\Delta}$ and $\Psi^{\bar{x},H}_{S^p_\Delta}$

$$e_{S^p_\Delta} - \Psi^{\bar{x},H}_{S^p_\Delta} = u_{EX} - u_{S^p_\Delta} - \bar{\Psi}^{\bar{x},H}_{S^p_\Delta} = -\xi_{S^p_\Delta} + \delta\Psi - w \tag{4.7.31a}$$

where we let

$$w \stackrel{\text{def}}{=} u_{S^p_\Delta} - U^{\bar{x},H}_{S^p_\Delta} - \big(u_{EX} - U^{\bar{x},(p+1)}_{EX}\big) + \xi_{S^p_\Delta}. \tag{4.7.31b}$$

We have already constructed asymptotic estimates for the L^∞ norm of $(d/dx)(\xi_{S^p_\Delta})$ and $(d/dx)(\delta\Psi)$. To complete the proof we need to estimate the L^∞ norm of (dw/dx). We will do this by applying Lemma 4.5.4, and for this we need to first prove that w is discrete harmonic. We have

$$\begin{aligned}
&\mathcal{B}_{S(\bar{x},H)}\big(U^{\bar{x},(p+1)}_{EX} - U^{\bar{x},H}_{S^p_\Delta} - \xi_{S^p_\Delta}, v\big)\\
&\quad= \mathcal{B}^0_{S(\bar{x},H)}\big(U^{\bar{x},(p+1)}_{EX} - U^{\bar{x},H}_{S^p_\Delta}, v\big)\\
&\qquad+ \underbrace{\mathcal{B}^1_{S(\bar{x},H)}\big(U^{\bar{x},(p+1)}_{EX} - U^{\bar{x},H}_{S^p_\Delta}, v\big) + \mathcal{B}_{S(\bar{x},H)}\big(\xi_{S^p_\Delta}, v\big)}_{0} = 0
\end{aligned} \tag{4.7.31c}$$

for all $v \in S^p_{\Delta,0}(S(\bar{x},H))$, and hence

$$\mathcal{B}_{S(\bar{x},H)}\big(U^{\bar{x},(p+1)}_{EX} - U^{\bar{x},H}_{S^p_\Delta} - \xi_{S^p_\Delta}, v\big) = \mathcal{B}^0_{S(\bar{x},H)}\big(U^{\bar{x},(p+1)}_{EX} - U^{\bar{x},H}_{S^p_\Delta}, v\big) = 0 \tag{4.7.31d}$$

for all $v \in S^p_{\Delta,0}(S(\bar{x},H))$, and recalling the orthogonality condition

$$\mathcal{B}_{S(\bar{x},H)}\big(u_{EX} - u_{S^p_\Delta}, v\big) = 0 \quad \forall v \in S^p_{\Delta,0}(S(\bar{x},H)) \tag{4.7.31e}$$

we get the orthogonality condition

$$\begin{aligned}
&\mathcal{B}_{S(\bar{x},H)}\big((u_{S^p_\Delta} - u_{EX}) + \big(U^{\bar{x},(p+1)}_{EX} - U^{\bar{x},H}_{S^p_\Delta} - \xi_{S^p_\Delta}\big), v\big) = 0\\
&\quad \forall v \in S^p_{\Delta,0}(S(\bar{x},H)).
\end{aligned} \tag{4.7.32a}$$

Hence we have proven that $w = u_{S^p_\Delta} - U^{\bar{x},H}_{S^p_\Delta} - (u_{EX} - U^{\bar{x},(p+1)}_{EX}) + \xi_{S^p_\Delta}$ is discrete harmonic in $S(\bar{x}, H)$, and applying Lemma 4.5.3 we get

$$\left\|\frac{dw}{dx}\right\|_{L^\infty(S(\bar{x},\gamma H))} \leq \mathscr{C}((1-\gamma)H)^{-3/2}\|w\|_{L^2(S(\bar{x},H))}. \tag{4.7.32b}$$

Employing now the triangle inequality

$$\begin{aligned}\left\|\frac{d}{dx}(e_{S^p_\Delta} - \Psi^{\bar{x},H}_{S^p_\Delta})\right\|_{L^\infty} \leq &\left\|\frac{d}{dx}(U^{\bar{x},(p+1)}_{EX} - U^{\bar{x},H}_{S^p_\Delta} - \Psi^{\bar{x},H}_{S^p_\Delta})\right\|_{L^\infty} \\ &+ \left\|\frac{d}{dx}(\xi_{S^p_\Delta})\right\|_{L^\infty} + \left\|\frac{dw}{dx}\right\|_{L^\infty}\end{aligned} \tag{4.7.33a}$$

where we wrote $\|\cdot\|_{L^\infty}$ instead of $\|\cdot\|_{L^\infty(S(\bar{x},\gamma H))}$, and using the estimates derived above for the terms in the right-hand side, we get

$$\begin{aligned}\left\|\frac{d}{dx}(e_{S^p_\Delta} - \Psi^{\bar{x},H}_{S^p_\Delta})\right\|_{L^\infty} \leq &\mathscr{C}(h^{p-1/2}H^2 \\ &+ ((1-\gamma)H)^{-3/2}(h^\beta H^{1/2} + h^{p+1}H^{1/2} + h^p H^2)).\end{aligned} \tag{4.7.33b}$$

Letting now $H = \mathscr{C}h^\alpha$ we get

$$\left\|\frac{d}{dx}(e_{S^p_\Delta} - \bar{\Psi}^{\bar{x},H}_{S^p_\Delta})\right\|_{L^\infty(S(\bar{x},\gamma H))} \leq \mathscr{C}h^{\bar{\nu}+p} \leq \mathscr{C}h^{\bar{\nu}}\left\|\frac{d}{dx}(e_{S^p_\Delta})\right\|_{L^\infty(S(\bar{x},H))} \tag{4.7.34a}$$

where

$$\bar{\nu} = \min\left(p - \frac{1}{2} + 2\,\alpha, -\alpha + \beta, p + 1 - \alpha, p + \frac{\alpha}{2}\right) - p. \tag{4.7.34b}$$

Letting

$$\beta = p + 1 - \epsilon \tag{4.7.34c}$$

and using Assumption II, there exists $0 < \alpha < 1$ such that $\bar{\nu} > 0$; this completes the proof. □

Remark 4.7.4 The above approach is a *constructive approach* for the local $(p+1)$ asymptotic analysis of finite element methods. It is based on the *construction* of the function $\Psi^{\bar{x},H}_{S^p_{\Delta_n}}$ which is the error in the periodic $\mathscr{B}^0_{S(\bar{x},H)}$ projection of the $(p+1)$-degree monomial

$$Q^{\bar{x},(p+1)}_{EX}(x) = \frac{1}{(p+1)!}\frac{d^{p+1}}{dx^{p+1}}(u_{EX})(\bar{x})(x-\bar{x})^{p+1}. \tag{4.7.35}$$

Remark 4.7.5 For $p \geq 1$ we have

$$\|e_{S^p_\Delta} - \bar{\Psi}^{\bar{x},H}_{S^p_\Delta}\|_{\mathscr{U}(S(\bar{x},\gamma H))} \leq \mathscr{C}h^{\bar{\nu}}\|e_{S^p_\Delta}\|_{\mathscr{U}(S(\bar{x},H))} \tag{4.7.36a}$$

and for $p \geq 2$ we have

$$\|e_{S^p_\Delta} - \bar{\Psi}^{\bar{x},H}_{S^p_\Delta}\|_{L^q(S(\bar{x},\gamma H))} \leq \mathscr{C}h^{\bar{\nu}}\|e_{S^p_\Delta}\|_{L^q(S(\bar{x},H))} \tag{4.7.36b}$$

for $q = 2, \infty$.

Remark 4.7.6 The superconvergence points for the value (respectively the derivative) of the finite element solution are the zeros of $\Psi^{\bar{x},H}_{S^p_{\Delta_n}}$ (resp. $(d/dx)(\Psi^{\bar{x},H}_{S^p_{\Delta_n}})$). Note that

$$\Psi^{\bar{x},H}_{S^p_{\Delta_n}}(x) \equiv \frac{1}{(p+1)!}\frac{d^{p+1}}{dx^{p+1}}(u_{EX})(\bar{x})\tilde{\psi}_p\big(F^{-1}_{I^{\Delta_n}_j}(x)\big), \quad x \in I^{\Delta_n}_j \tag{4.7.37}$$

where $\tilde{\psi}_p$ is the primitive of the Legendre polynomial of degree p in $\tilde{I}$, and the above construction does not give anything different than the analysis given in Section 4.6. *The merit of the above approach will become clear from its application to the analysis of the two-dimensional case.*

Remark 4.7.7 The above approach can be extended to analyze higher-order asymptotics.

Remark 4.7.8 The function $\Psi^{\bar{x},H}_{S^p_{\Delta_n}}$ can be constructed in the master element. Let

$$\tilde{Q}^{(p+1)}_{EX}(\tilde{x}) = \frac{1}{(p+1)!}\frac{d^{p+1}u_{EX}}{dx^{p+1}}(\bar{x})\tilde{x}^{p+1}, \quad \tilde{x} \in \tilde{I} = (-1,1) \tag{4.7.38a}$$

and

$$\tilde{\rho} = \tilde{Q}^{(p+1)}_{EX} - \tilde{\mathcal{I}}^p\tilde{Q}^{(p+1)}_{EX} \tag{4.7.38b}$$

and let $\pi^0_{\tilde{S}^p_{PER}}(\tilde{\rho})$ be the solution of the problem:

Find $\pi^0_{\tilde{S}^p_{PER}}(\tilde{\rho}) \in \tilde{S}^p_{PER}$, such that

$$\mathcal{B}^0_{\tilde{I}}\big(\pi^0_{\tilde{S}^p_{PER}}(\tilde{\rho}), \tilde{v}\big) = \mathcal{B}^0_{\tilde{I}}(\tilde{\rho}, \tilde{v}) \quad \forall \tilde{v} \in \tilde{S}^p_{PER}(\tilde{I}) = \tilde{S}^p(\tilde{I}) \cap H^1_{PER}(\tilde{I}) \tag{4.7.38c}$$

and

$$\int_{\tilde{I}} \pi^0_{\tilde{S}^p_{PER}}(\tilde{\rho}) = \int_{\tilde{I}} \tilde{\rho}. \tag{4.7.38d}$$

Then let

$$\tilde{\Psi}_{\tilde{S}^p_{PER}} \stackrel{\text{def}}{=} \tilde{\rho} - \pi^0_{\tilde{S}^p_{PER}}(\tilde{\rho}) \tag{4.7.38e}$$

and note that

$$\Psi^{\bar{x},H}_{S^p_{\Delta_n}} = h^{p+1}\tilde{\Psi}_{\tilde{S}^p_{PER}} \circ F^{-1}_{I^{\Delta}_j} \tag{4.7.38f}$$

and

$$\frac{d}{dx}\big(\Psi^{\bar{x},H}_{S^p_{\Delta_n}}\big) = h^p\frac{d}{d\tilde{x}}\big(\tilde{\Psi}_{\tilde{S}^p_{PER}}\big) \circ F^{-1}_{I^{\Delta}_j} \tag{4.7.38g}$$

where $F^{-1}_{I^{\Delta}_j}$ is the inverse of the affine transformation $F_{I^{\Delta}_j}$ which maps $\tilde{I}$ onto I^{Δ}_j.

Remark 4.7.9 The superconvergence of the finite element solution and its derivative can be determined by analyzing $\tilde{\Psi}_{\tilde{S}^p_{PER}}$ and $(d/d\tilde{x})(\tilde{\Psi}_{\tilde{S}^p_{PER}})$.

Remark 4.7.10 Theorem 4.7.1 shows that the local error in the finite element solution can be constructed by employing a local analysis which takes into account the character of the mesh and the solution in the neighborhood of the point of interest.

Remark 4.7.11 Theorem 4.7.1 was proven above under the assumption that the mesh is *locally uniform.* As we will see in Section 4.7*, the extension of the proof to the case that the mesh is *locally periodic* is straightforward.

Following analogous steps as in the proof of Theorem 4.7.1, we can prove that:

Theorem 4.7.2 *Let the assumptions of Theorem 4.7.1 hold, and let $p \geq 2$. Then, we have*

$$u_{S^p_\Delta} = \mathcal{I}^p_\Delta u_{EX} + \Psi^{\bar{x},H}_{S^p_\Delta} + \mathscr{C}h^{p+1+\nu}. \tag{4.7.39}$$

Below, we will replace the notation $\Psi^{\bar{x},H}_{S^p_\Delta}$ by $\Psi^{\bar{x},PER(h)}_{S^p_\Delta}$, in which no distinction is made about whether Ψ is defined over one element or an entire patch of elements. We will also introduce $U^{PER(h)}_{S^p_\Delta}$, the *asymptotic finite element solution* in the interior of the mesh by

$$U^{PER(h)}_{S^p_\Delta} = \mathcal{I}^p_\Delta U^{\bar{x},(p+1)}_{EX} + \Psi^{\bar{x},PER(h)}_{S^p_\Delta}. \tag{4.7.40a}$$

Note that we have

$$\mathscr{B}^0_{S(\bar{x},H)}\big(U^{\bar{x},(p+1)}_{EX} - U^{PER(h)}_{S^p_\Delta}, v\big) = 0 \quad \forall v \in S^p_{\Delta,0}(S(\bar{x},H)). \tag{4.7.40b}$$

Note that $U^{PER(h)}_{S^p_\Delta}$ is not periodic.

Theorems 4.7.1 and 4.7.2 were based on the assumption that $S(\bar{x},H)$ is an interior patch; we will now generalize them for patches $S(\bar{x},H)$ which extend up to the boundary. We have:

Theorem 4.7.3 *Let the assumptions of Theorem 4.7.1 hold for $\bar{x} = 0$. We then have*

$$u_{S^p_\Delta} = \mathcal{I}^p_\Delta u_{EX} + \Psi^{0,PER(h)}_{S^p_\Delta} + \Phi^{0,h} + \mathscr{C}h^{p+1+\nu} \tag{4.7.41a}$$

where $\Phi^{0,h} \in S^p_\Delta(S(0,H))$ is such that $\bar{u}_{S^p_\Delta} + \Phi^{0,h}$ satisfies the imposed boundary condition at $x = 0$, and

$$\mathscr{B}^0_{S(0,\infty)}(\Phi^{0,h}, v_h) = 0 \quad \forall v_h \in S^p_\Delta(S(0,\infty)) \tag{4.7.41b}$$

and

$$\lim_{x\to\infty} \frac{d}{dx}(\Phi^{0,h}(x)) = 0. \tag{4.7.41c}$$

We will refer to $\Phi^{0,h}$ as the *boundary-layer correction* and (4.7.41c) as the *decay condition.* In the case of the homogeneous Dirichlet boundary condition, $u(0) = 0$, $\Phi^{0,h}$ must satisfy the Dirichlet boundary condition

$$\Phi^{0,h}(0) = -\Psi^{0,PER(h)}_{S^p_\Delta}(0). \tag{4.7.42a}$$

In the case of the homogeneous Neumann boundary condition, $AE(0)(du/dx)(0) = 0$, $\Phi^{0,h}$ must satisfy

$$\mathscr{B}^0_{S(0,\infty)}(\Phi^{0,h}, \varphi_0^\Delta) = -\mathscr{B}^0_{S(0,\infty)}(\Psi^{0,PER(h)}_{S^p_\Delta}, \varphi_0^\Delta) = 0 \tag{4.7.42b}$$

where φ_0^Δ is the hat function for the node x_0^Δ. The right-hand side in (4.7.42b) vanishes identically, because $\Psi^{0,PER(h)}_{S^p_\Delta}$ is an h-periodic function. In both cases we have

$$\Phi^{0,h} \equiv constant \tag{4.7.42c}$$

and hence there is no boundary-layer correction in the case of the one-dimensional model problem.

Remark 4.7.12 In two dimensions, $\Phi^{0,h}$ is non-zero, in general, and is the solution of an infinite dimensional system of linear equations.

Remark 4.7.13 The superconvergence of the finite element solution in the neighborhood of the boundary can be determined by analyzing the function $\Psi^{0,PER(h)}_{S^p_\Delta} + \Phi^{0,h}$ and its derivative.

Remark 4.7.14 As we have seen above, in the considered one-dimensional model problem $\Phi^{0,h}$ is constant, and the superconvergence points for interior elements are valid up to the boundary. However, this is not the case, in general, for higher dimensions.

Above we analyzed the asymptotic behavior of the error in the finite element solution under the assumption that the exact solution u_{EX} is smooth. We will now consider the case that u_{EX} has an algebraic singularity of the type x^α. This is done in order to construct a one-dimensional analogue of the asymptotic behavior of the finite element solution in the neighborhood of corner points, or points where there is a change in the type of boundary condition.

We will replace Assumption I by

Assumption I′. We will assume that u_{EX} is given by

$$u_{EX}(x) = a_1 x^{\alpha_1} + a_2 x^{\alpha_2} + \hat{u}_{EX}(x) \tag{4.7.43a}$$

where

$$\tfrac{1}{2} < \alpha_1 < \tfrac{3}{2}, \quad \alpha_1 \neq \text{integer}, \qquad \alpha_2 > \alpha_1. \tag{4.7.43b}$$

$\hat{u}_{EX}$ is smoother than the first two terms, $a_1 \neq 0$ and, without loss of generality, we will let $a_2 = 1$.

We will also replace Assumption II by:

Assumption II′. Let $H_1 > 0$, $d > 0$, with $H_1 + d \leq H$, and let

$$\omega_1(H_1, d) \stackrel{\text{def}}{=} (H_1, H_1 + d), \qquad \omega_2(H_1, d) \stackrel{\text{def}}{=} (H_1 + \tfrac{1}{4}d, H_1 + \tfrac{3}{4}d) \tag{4.7.44a}$$

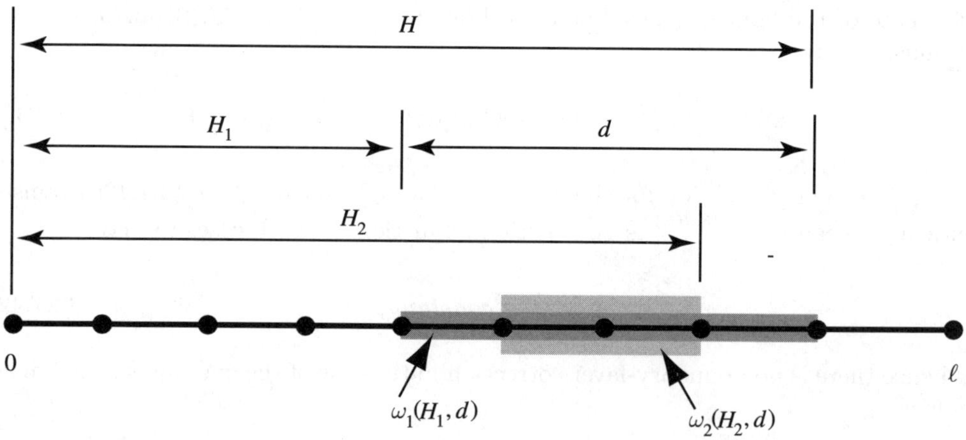

FIG. 4.7.1. The notation employed in the asymptotic analysis

where $H_2 \stackrel{\text{def}}{=} H_1 + \frac{3}{4}d$, and let $\omega_1(H_1, d)$, $\omega_2(H_2, d)$ coincide with an entire patches of elements, as shown in Fig. 4.7.1. We will assume that

$$\|e_{S^p_\Delta}\|_{L^2(\omega_1(H_1,d))} \leq \mathscr{C} h^{2(\alpha_1-1/2)-\varepsilon}\, d^{1/2-(\alpha_1-1/2)-\varepsilon} \tag{4.7.44b}$$

with $\varepsilon > 0$ sufficiently small, and $\mathscr{C}$ depending on ε, but not on Δ, h, or d.

Remark 4.7.15 Assumption (4.7.44b) is an assumption about the pollution, and is a one-dimensional analogue of a corresponding assumption employed in two dimensions by [Wahlbin, 1991].

Below we will employ the shorthand notation $\pi_{S^p_\Delta(H_2)} \stackrel{\text{def}}{=} \pi_{S^p_\Delta(S(0,H_2))}$ to denote the energy projection into $S^p_\Delta(S(0, H_2))$.

Let us now give the main asymptotic result:

Theorem 4.7.4 *Under Assumptions 0, I′, and II′, there exist* $0 < \gamma_1, \gamma_2 < 1$ *such that, for* $H_1 = h^{\gamma_2}$ *and* $d = h^{\gamma_1}$, *and for sufficiently small* h *we have*

$$\|\pi_{S^p_\Delta(H_2)}(u_{EX}) - u_{S^p_\Delta}\|_{\mathscr{U}(S(0,H_2))} \leq \mathscr{C} h^\nu \|e_{S^p_\Delta}\|_{\mathscr{U}(S(0,H_2))} \tag{4.7.45}$$

for some $\nu > 0$.

Proof We will first prove (4.7.45) in the special case that $u_{EX} \equiv x^{\alpha_1}$, and we will let $x^{\alpha_1}_{S^p_\Delta}$ denote the corresponding finite element solution. Adding the orthogonalities of the error for the finite element solution and the energy projection,

$$\mathscr{B}_{S(0,H_2)}\left(x^{\alpha_1} - x^{\alpha_1}_{S^p_\Delta}, v\right) = 0 \quad \forall v \in S^p_{\Delta,0}(S(0, H_2)) \tag{4.7.46a}$$

and

$$\mathscr{B}_{S(0,H_2)}\left(\pi_{S^p_\Delta(H_2)}(x^{\alpha_1}) - x^{\alpha_1}, v\right) = 0 \quad \forall v \in S^p_\Delta(S(0, H_2)) \tag{4.7.46b}$$

we get

$$\mathscr{B}_{S(0,H_2)}\big(\pi_{S^p_\Delta(H_2)}(x^{\alpha_1}) - x^{\alpha_1}_{S^p_\Delta}, v\big) = 0 \quad \forall v \in S^p_{\Delta,0}(S(0,H_2)). \tag{4.7.46c}$$

It follows that

$$\begin{aligned}\|\pi_{S^p_\Delta(H_2)}(x^{\alpha_1}) - x^{\alpha_1}_{S^p_\Delta}\|^2_{\mathscr{U}(S(0,H_2))}\\ = \mathscr{B}_{S(0,H_2)}\big(x^{\alpha_1} - x^{\alpha_1}_{S^p_\Delta}, \pi_{S^p_\Delta(H_2)}(x^{\alpha_1}) - x^{\alpha_1}_{S^p_\Delta} - v\big)\end{aligned} \tag{4.7.47a}$$

for all $v \in S^p_{\Delta,0}(S(0,H_2))$, and employing the Cauchy–Schwarz inequality we obtain

$$\begin{aligned}&\|\pi_{S^p_\Delta(H_2)}(x^{\alpha_1}) - x^{\alpha_1}_{S^p_\Delta}\|^2_{\mathscr{U}(S(0,H_2))}\\ &\quad\le \|x^{\alpha_1} - x^{\alpha_1}_{S^p_\Delta}\|_{\mathscr{U}(\omega_2(H_2,d))}\|\pi_{S^p_\Delta(H_2)}(x^{\alpha_1}) - x^{\alpha_1}_{S^p_\Delta} - v\|_{\mathscr{U}(S(0,H_2))}.\end{aligned} \tag{4.7.47b}$$

Constructing now a $v \in S^p_{\Delta,0}(S(0,H_2))$ such that

$$\big(\pi_{S^p_\Delta(H_2)}(x^{\alpha_1}) - x^{\alpha_1}_{S^p_\Delta} - v\big)\big|_{(0,H_1+d/4)} \equiv 0 \tag{4.7.48a}$$

and employing Lemma 4.5.2 we obtain

$$\begin{aligned}&\|\pi_{S^p_\Delta(H_2)}(x^{\alpha_1}) - x^{\alpha_1}_{S^p_\Delta} - v\|_{\mathscr{U}(S(0,H_2))}\\ &\quad\le \mathscr{C}\Big(1+\frac{h}{d}\Big)\|\pi_{S^p_\Delta(H_2)}(x^{\alpha_1}) - x^{\alpha_1}_{S^p_\Delta}\|_{\mathscr{U}(S(0,H_2))}.\end{aligned} \tag{4.7.48b}$$

Using the same v in (4.7.47a), we get

$$\|\pi_{S^p_\Delta(H_2)}(x^{\alpha_1}) - x^{\alpha_1}_{S^p_\Delta}\|_{\mathscr{U}(S(0,H_2))} \le \mathscr{C}\Big(1+\frac{h}{d}\Big)\|x^{\alpha_1}_{S^p_\Delta} - x^{\alpha_1}\|_{\mathscr{U}(\omega_2(H_2,d))} \tag{4.7.49a}$$

and using Theorem 4.5.1 and Remark 4.5.6, we get

$$\begin{aligned}\|\pi_{S^p_\Delta(H_2)}(x^{\alpha_1}) - x^{\alpha_1}_{S^p_\Delta}\|_{\mathscr{U}(S(0,H_2))} &\le \mathscr{C}\Big[h^p\, H_1^{\alpha_1-p-1}\sqrt{d} + \frac{1}{d}h^{p+1}H_1^{\alpha_1-p-1}\sqrt{d}\\ &\quad + h^{2(\alpha_1-1/2)-\varepsilon}d^{-1/2-(\alpha_1-1/2)-\varepsilon}\Big]\Big(1+\frac{h}{d}\Big),\end{aligned} \tag{4.7.49b}$$

where we employed that

$$\Big|\frac{d^{p+1}}{dx^{p+1}}(x^{\alpha_1})\Big| \le \mathscr{C}H_1^{\alpha_1-p-1}. \tag{4.7.49c}$$

Letting

$$\gamma_1 < \gamma_2 < 1, \qquad d = h^{\gamma_2}, \qquad H_1 = h^{\gamma_1} \tag{4.7.50a}$$

we can choose such γ_1 and γ_2 that

$$\|\pi_{S^p_\Delta(H_2)}(x^{\alpha_1}) - x^{\alpha_1}_{S^p_\Delta}\|_{\mathscr{U}(S(0,H_2))} \le \mathscr{C}h^{(\alpha_1-1/2)(1+\bar{\nu})} \tag{4.7.50b}$$

with $0 < \bar{\nu} < 1$. Noting that

$$\inf_{\chi \in S_\Delta^p(I_1^\Delta)} \|\chi - x^{\alpha_1}\|_{\mathcal{U}(I_1^\Delta)} \leq \|x_{S_\Delta^p}^{\alpha_1} - x^{\alpha_1}\|_{\mathcal{U}(S(0,H_2))} \tag{4.7.51}$$

we obtain (4.7.45) in the special case that $u_{EX} \equiv x^{\alpha_1}$.

In the general case where

$$u_{EX} = x^{\alpha_1} + \bar{u}_{EX} \tag{4.7.52a}$$

we have

$$\begin{aligned} \|\pi_{S_\Delta^p(H_2)}(x^{\alpha_1}) - \pi_{S_\Delta^p(H_2)}(u_{EX})\|_{\mathcal{U}(S(0,H_2))} &\leq \mathscr{C} H_2^{\alpha_2+1/2} \\ &\leq \mathscr{C} h^{(\alpha_2+1/2)\min(\gamma_1,\gamma_2)}. \end{aligned} \tag{4.7.52b}$$

Noting that

$$\|\pi_{S_\Delta^p(H_2)}(u_{EX}) - u_{S_\Delta^p}\|_{\mathcal{U}(S(0,H_2))} \leq \mathscr{C} h^{(\alpha_1-1/2)(1+\bar{\nu})} \tag{4.7.52c}$$

and that

$$\mathscr{C} h^{\alpha_1-1/2} \leq \|e_{S_\Delta^p}\|_{\mathcal{U}(S(0,H_2))} \tag{4.7.52d}$$

and employing the triangle inequality, and selecting h sufficiently small, (4.7.45) follows. □

Remark 4.7.16 From the above theorem, it follows that the finite element solution $u_{S_\Delta^p}$ may be replaced by $a_1\pi_{S_\Delta^p(H_2)}(x^{\alpha_1})$ in $S(0, H_2)$, with negligible error as h tends to zero.

The function $\pi_{S_\Delta^p(H_2)}(x^{\alpha_1})$ plays the same role as the function $\Psi_{S_\Delta^p}^{\bar{x},PER(h)}$ for interior mesh patches, and it can be utilized for finding the superconvergence points in the first k elements elements in $S(0, H_2)$. We will now employ $\pi_{S_\Delta^p(H_2)}(x^{\alpha_1})$, to analyze the superconvergence in the case of linear elements ($p = 1$); the case of higher order elements ($p > 1$) can be addressed analogously.

Let $\bar{x} \in I_j^\Delta$, $1 \leq j \leq k$, denote the point of zero error for $(d/dx)(u_{S_\Delta^1})$, namely,

$$\frac{d}{dx}(e_{S_\Delta^1})(\tilde{x}) = \frac{d}{dx}(u_{EX} - u_{S_\Delta^1})(\bar{x}) = 0 \tag{4.7.53a}$$

and let $\bar{\bar{x}} \in I_j^\Delta$ be the point of zero error for $(d/dx)(\pi_{S_\Delta^p(H_2)}(x^{\alpha_1}))$, namely

$$\frac{d}{dx}(\pi_{S_\Delta^p(H_2)}(x^{\alpha_1}) - x^{\alpha_1})(\bar{\bar{x}}) = 0. \tag{4.7.53b}$$

Noting that $(d/dx)(u_{S_\Delta^1})$ and $(d/dx)(\pi_{S_\Delta^p(H_2)}(x^{\alpha_1}))$ are constant in each element, and that

$$\left\|\frac{d}{dx}(u_{S_\Delta^1} - \pi_{S_\Delta^p(H_2)}(x^{\alpha_1}))\right\|_{L^2(I_j^\Delta)} \leq \mathscr{C} h^{(\alpha-1/2)(1+\bar{\nu})} \tag{4.7.54a}$$

it follows that

$$\left\| \frac{d}{dx}\left(u_{S^1_\Delta} - \pi_{S^p_\Delta(H_2)}(x^{\alpha_1})\right) \right\|_{L^\infty(I^\Delta_j)} \leq \mathscr{C} h^{(\alpha-1/2)(1+\bar{\nu})-1/2}. \tag{4.7.54b}$$

Realizing that

$$\mathscr{C} h^{\alpha-2} \leq \left| \frac{d^2}{dx^2}(u_{EX}) \right| \tag{4.7.55a}$$

we get

$$|\bar{x} - \bar{\bar{x}}| \leq \mathscr{C} h^{(\alpha-1/2)(1+\bar{\nu})-1/2} h^{2-\alpha} \leq \mathscr{C} h^{1+\tilde{\nu}}. \tag{4.7.55b}$$

Noting that $\bar{\nu} > 0$ and $\alpha - \frac{1}{2} > 0$, it follows that $\tilde{\nu} > 0$, and the point $\bar{\bar{x}}$ is a superconvergence point. The above analysis has an analogue in the two-dimensional case, where we have superconvergence lines for each component of the derivative.

Remark 4.7.17 The asymptotic analysis given above is the one-dimensional version of the analysis presented in [Babuška, Strouboulis *et al.*, 1996] and [Babuška, Strouboulis *et al.*, 1997].

Remark 4.7.18 Above we analyzed the case where the mesh is uniform in the neighborhood of the singular point. The above results are also valid in the case that the mesh is only quasi-uniform. The case of strongly refined meshes can be also analyzed following similar lines, however, in this case a more careful analysis of the pollution is necessary.

In this section we have presented a general approach for analyzing the local asymptotics of finite element methods. The asymptotics of finite element solutions in the interior of the mesh can be analyzed by employing the function $\Psi^{\bar{x},PER(h)}_{S^p_\Delta}$, which, in the one-dimensional case, can be determined from one element. The analysis of the asymptotics in the neighborhood of the boundary and corner points requires the solution of an infinite-dimensional system of equations for the determination of the boundary-layer correction, $\Phi^{0,h}$, and the projection $\pi_{S^p_\Delta(H_2)}(x^{\alpha_1})$, respectively. In both cases the functions can be approximated by using a sufficiently large finite-dimensional system.

Problems

P4.7.1. Explain the constructive use of Theorem 4.7.1 for determining the superconvergence points for the derivative.

P4.7.2. Show that in the one-dimensional case the superconvergence points in an element are independent of the size of its neighbors.

P4.7.3. Determine the superconvergence points for the element of degree p, adjacent to $x = 0$, when $u_{EX} = x^{\alpha}$. Are these points close to the Gauss–Legendre points of degree p?

4.7* A general approach for local asymptotic analysis and analysis of the superconvergence in two dimensions

In this section we will show give a general approach for analyzing the local asymptotics and the superconvergence of finite elements approximations in the interior, at smooth boundaries, and in the neighborhood of corner points for meshes of triangles and quadrilaterals. This section is based on [Babuška, Strouboulis *et al.*, 1996] and [Babuška, Strouboulis *et al.*, 1997].

We will first consider the case of interior mesh cells.

As in the one-dimensional case, we will make three assumptions:

Assumption 0. The mesh is uniform in the region of interest, and can be constructed by the periodic repetition of a square basic cell, as shown in Fig. 4.7*.1. More precisely, let $S(\bar{\mathbf{x}}, H) \subseteq \Omega$ denote a square domain of size H centered at $\bar{\mathbf{x}}$, as shown in Fig. 4.7*.1. We will assume that the meshes are such that $S(\bar{\mathbf{x}}, H)$ and $S(\bar{\mathbf{x}}, \frac{1}{2}H)$ are exactly covered by a uniform patch of basic cells $c^h_{(k,l)}$, namely,

$$\bar{S}(\bar{\mathbf{x}}, H) = \bigcup_{(k,l)\in I} \bar{c}^h_{(k,l)}, \qquad c^h_{(k,l)} = S\big(\mathbf{x}(k,l), \tfrac{1}{2}h\big) \tag{4.7*.1}$$

and similarly for $S(\bar{\mathbf{x}}, \frac{1}{2}H)$. Here I is a set of indices which enumerates the cells which belong in $S(\bar{\mathbf{x}}, H)$.

Remark 4.7*.1 Let us underline that h *is the uniform size of the basic cells in* $S(\bar{\mathbf{x}}, H)$. The size of the elements outside $S(\bar{\mathbf{x}}, H)$ must be such that Assumption II is satisfied, and could be very different than h.

Remark 4.7*.2 The case of general meshes, can be addressed by employing a local *periodization* of the mesh in the region of interest, as shown in Fig. 4.7*.2.

(a) 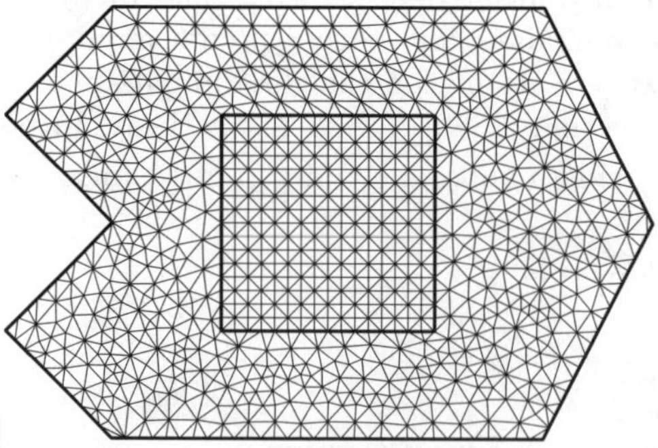

(b) 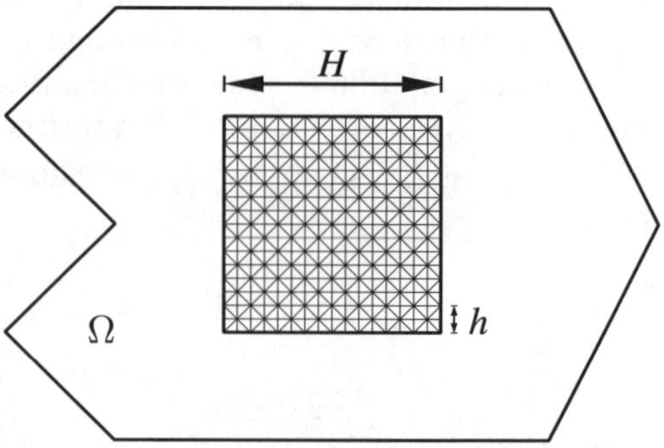

FIG. 4.7*.1. A finite element mesh of triangles which satisfies the assumptions of the analysis. (a) The entire mesh with the subdomain $S(\bar{\mathbf{x}}, H)$ shown with thick perigram; (b) The mesh in the subdomain $S(\bar{\mathbf{x}}, H)$. The mesh in $S(\bar{\mathbf{x}}, H)$ is uniform and is obtained by the periodic repetition of the basic cell of the Union-Jack pattern shown in Fig. 4.7*.3c

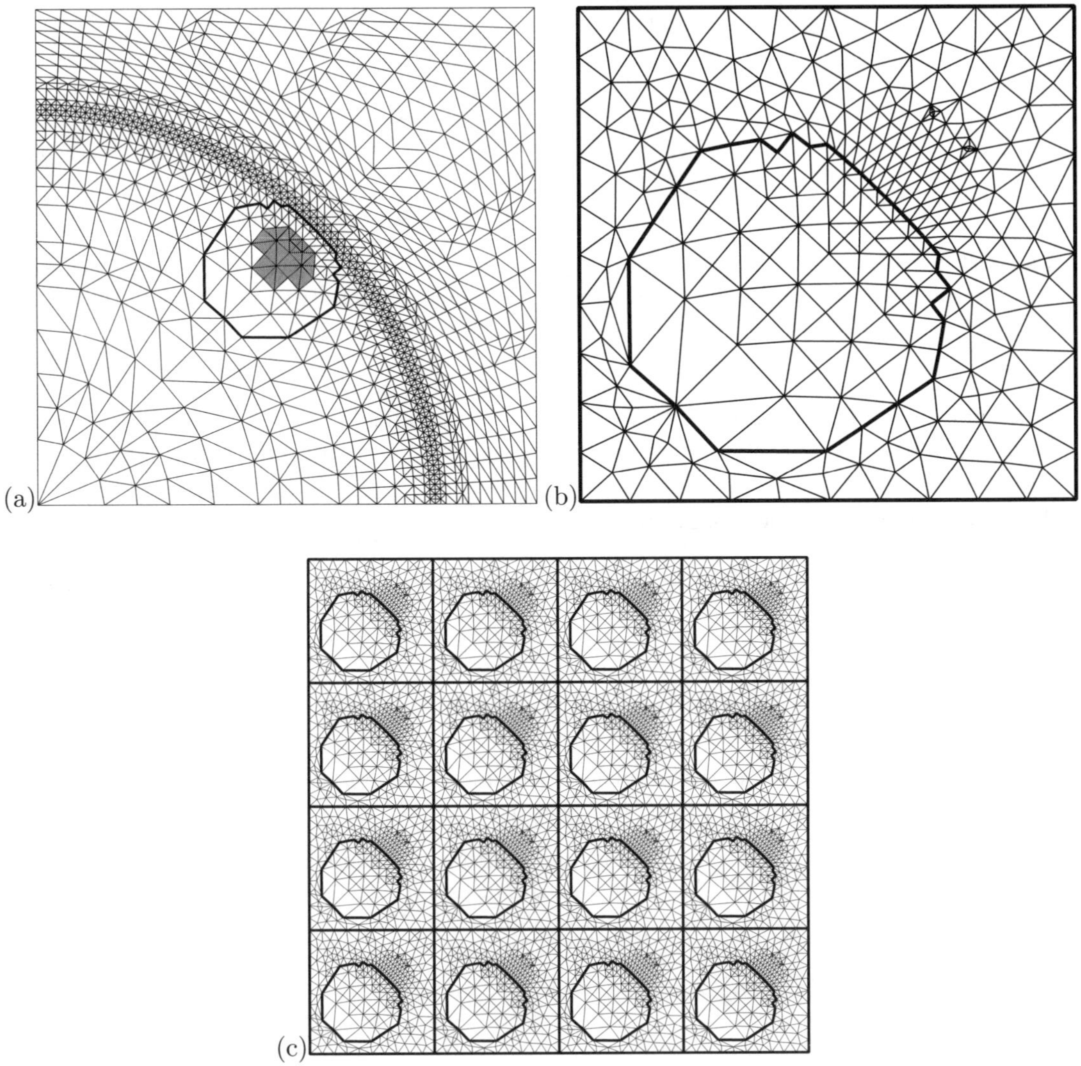

FIG. 4.7*.2. Periodization of the mesh around a mesh cell of interest: The mesh cell of interest, shown shaded gray, with the two neighborhoods of elements embedded in (a) the actual mesh, (b) the periodicized mesh cell, and (c) the mesh assumed by the periodization. This figure shows how Assumption I is understood for general meshes

Assumption I. The exact solution u_{EX} is such that

$$\max_{\substack{0\le i,j\le p+2\\ i+j=p+2}} \left\| \frac{\partial^{p+2} u_{EX}}{\partial x_1^i \partial x_2^j} \right\|_{L^\infty(S(\bar{x},H))} \le K < \infty \tag{4.7*.2a}$$

and

$$\sum_{\substack{0\le i,j\le p+1\\ i+j=p+1}} \left| \frac{\partial^{p+1} u_{EX}}{\partial x_1^i \partial x_2^j} \right| (\bar{\mathbf{x}}) \neq 0. \tag{4.7*.2b}$$

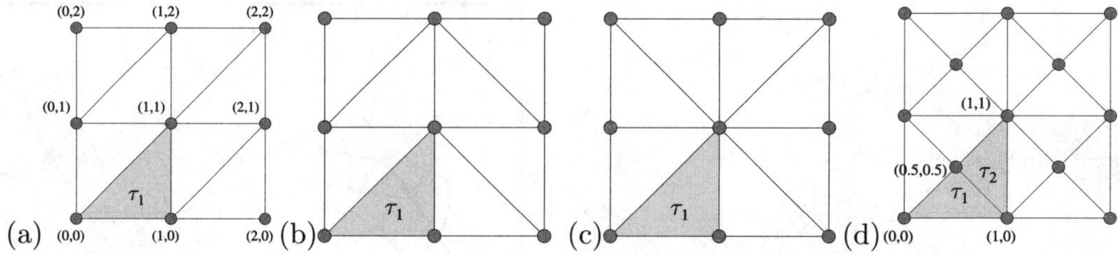

FIG. 4.7*.3. Examples of some typical mesh patterns encountered in meshes of triangles which are employed in engineering computations: (a) regular pattern; (b) chevron pattern; (c) Union-Jack pattern; (d) criss-cross pattern

Assumptions (4.7*.2a) and (4.7*.2b) are direct generalizations of Assumptions Ia and Ib of Section 4.7, to the two-dimensional case. They govern the smoothness of u_{EX} in $S(\bar{\mathbf{x}}, H)$.

Assumption II. The meshes Δ are such that

$$\|e_{S^p_\Delta}\|_{L^2(S(\bar{x},H))} \leq \mathscr{C} h^\beta H \tag{4.7*.3a}$$

with

$$(p+1) - \varepsilon < \beta, \quad 0 < \varepsilon < 1 - \left(\frac{p}{p+1}\right). \tag{4.7*.3b}$$

Assumption II states that the pollution error in $S(\bar{x}, H)$ is negligible, and is the two-dimensional version of Assumption II of Section 4.7.

Let us now describe the two-dimensional analogue of the constructive approach for the local asymptotic analysis. As in the one-dimensional case, we will employ the $(p+1)$ degree Taylor series expansion of u_{EX} at $\bar{\mathbf{x}}$, namely,

$$u_{EX} = U^{\bar{\mathbf{x}},p}_{EX} + Q^{\bar{\mathbf{x}},(p+1)}_{EX} + R^{\bar{\mathbf{x}},(p+1)}_{EX} \tag{4.7*.4a}$$

where

$$Q^{\bar{\mathbf{x}},k}_{EX}(\mathbf{x}) \stackrel{\text{def}}{=} \sum_{\substack{0\leq i,j\leq k\\ i+j=k}} \frac{1}{i!j!}\frac{\partial^k u_{EX}}{\partial x_1^i \partial x_2^j}(\bar{\mathbf{x}})(x_1-\bar{x}_1)^i(x_2-\bar{x}_2)^j \tag{4.7*.4b}$$

and

$$U^{\bar{\mathbf{x}},p}_{EX}(\mathbf{x}) \stackrel{\text{def}}{=} \sum_{k=0}^{p} Q^{\bar{\mathbf{x}},k}_{EX}(\mathbf{x}) = \sum_{k=0}^{p} \sum_{\substack{0\leq i,j\leq k\\ i+j=k}} \frac{1}{i!j!}\frac{\partial^k u_{EX}}{\partial x_1^i \partial x_2^j}(\bar{\mathbf{x}})(x_1-\bar{x}_1)^i(x_2-\bar{x}_2)^j \tag{4.7*.4c}$$

and $R^{\bar{\mathbf{x}},(p+1)}_{EX}$ is the remainder. We will also let

$$Q^{\bar{\mathbf{x}},(p+1)}_{(i,j)}(x_1,x_2) \stackrel{\text{def}}{=} (x_1-\bar{x}_1)^i(x_2-\bar{x}_2)^j \tag{4.7*.4d}$$

where i, j are positive integers which satisfy $0 \le i, j \le p+1$, $i+j = p+1$, and

$$\rho^{\bar{\mathbf{x}},H}_{S^p_\Delta,(i,j)} \stackrel{\text{def}}{=} Q^{\bar{\mathbf{x}},(p+1)}_{(i,j)} - \mathcal{I}^p_\Delta Q^{\bar{\mathbf{x}},(p+1)}_{(i,j)} \tag{4.7*.4e}$$

and

$$\Psi^{\bar{\mathbf{x}},H}_{S^p_\Delta,(i,j)} \stackrel{\text{def}}{=} \rho^{\bar{\mathbf{x}},H}_{S^p_\Delta,(i,j)} - \pi^0_{S^p_\Delta,PER}\left(\rho^{\bar{\mathbf{x}},H}_{S^p_\Delta,(i,j)}\right). \tag{4.7*.4f}$$

Then

$$\Psi^{\bar{\mathbf{x}},H}_{S^p_\Delta} = \sum_{\substack{0 \le i,j \le p+1 \\ i+j=p+1}} \frac{1}{i!j!} \frac{\partial^{p+1} u_{EX}}{\partial x_1^i \partial x_2^j}(\bar{\mathbf{x}}) \Psi^{\bar{\mathbf{x}},H}_{S^p_\Delta,(i,j)}. \tag{4.7*.4g}$$

The two-dimensional analogue of Theorem 4.7.1 can now be stated.

Theorem 4.7*.1 *Under Assumptions 0, I, and II, there exists an α, $0 < \alpha < 1$, such that*

$$\mathcal{C}_1 h^\alpha \le H \le \mathcal{C}_2 h^\alpha \tag{4.7*.5a}$$

and

$$\max_{i=1,2} \left\| \frac{\partial}{\partial x_i}(e_{S^p_\Delta}) - \frac{\partial}{\partial x_i}\left(\Psi^{\bar{\mathbf{x}},H}_{S^p_\Delta}\right) \right\|_{L^\infty(S(\bar{\mathbf{x}},\gamma H))} \le \mathcal{C} h^{p+\bar{\nu}} \tag{4.7*.5b}$$

with $\bar{\nu} > 0$.

The proof follows analogous steps as in the one-dimensional case.

Theorem 4.7*.1 shows that $\boldsymbol{\nabla} e_{S^p_\Delta} \approx \boldsymbol{\nabla} \Psi^{\bar{\mathbf{x}},H}_{S^p_\Delta}$.

We note that using results in [Schatz, 2001], H can be chosen so that $\tilde{\nu}$ is arbitrarily close to 1.

Analogously as in the one-dimensional case, the function $\Psi^{\bar{\mathbf{x}},H}_{S^p_\Delta,(i,j)}$ can be constructed from the error in the periodic $\tilde{\mathcal{B}}^0_{\tilde{c}}$ projection of $\tilde{x}_1^i\, \tilde{x}_2^j$ in the master cell $\tilde{c} = (-1,1) \times (-1,1)$. Let

$$\tilde{Q}^{(p+1)}_{(i,j)}(\tilde{x}_1, \tilde{x}_2) \stackrel{\text{def}}{=} \tilde{x}_1^i\, \tilde{x}_2^j \tag{4.7*.6a}$$

and

$$\tilde{\rho}_{(i,j)} \stackrel{\text{def}}{=} \tilde{Q}^{(p+1)}_{(i,j)} - \tilde{\mathcal{I}}^p \tilde{Q}^{(p+1)}_{(i,j)} \tag{4.7*.6b}$$

where $0 \le i, j \le p+1$, $i+j = p+1$, and construct $\tilde{Z}^{(i,j)}_{\tilde{S}^p_{\tilde{\Delta},PER}}$ as the periodic energy projection of $\tilde{\rho}_{(i,j)}$, namely:

Find $\tilde{Z}^{(i,j)}_{\tilde{S}^p_{\tilde{\Delta},PER}} \stackrel{\text{def}}{=} \pi^0_{\tilde{S}^p_{\tilde{\Delta},PER}}(\tilde{\rho}_{(i,j)}) \in \tilde{S}^p_{\tilde{\Delta},PER}$ such that

$$\tilde{\mathcal{B}}^0_{\tilde{c}}\left(\tilde{Z}^{(i,j)}_{\tilde{S}^p_{\tilde{\Delta},PER}}, \tilde{v}\right) = \tilde{\mathcal{B}}^0_{\tilde{c}}(\tilde{\rho}_{(i,j)}, \tilde{v}) \quad \forall \tilde{v} \in \tilde{S}^p_{\tilde{\Delta},PER} \tag{4.7*.7a}$$

and

$$\int_{\tilde{c}} \tilde{Z}^{(i,j)}_{\tilde{S}^p_{\tilde{\Delta},PER}} = \int_{\tilde{c}} \tilde{\rho}_{(i,j)}. \tag{4.7*.7b}$$

Here

$$\tilde{\mathcal{B}}^0_{\tilde{c}}(\tilde{u},\tilde{v}) = \int_{\tilde{c}} \sum_{ij=1,2} a_{ij} \frac{\partial \tilde{u}}{\partial \tilde{x}_i} \frac{\partial \tilde{v}}{\partial \tilde{x}_j} \tag{4.7*.7c}$$

and $\tilde{\Delta}$ is the master mesh in $\tilde{c}$ which has the same geometry as the mesh in the cells $c^h_{(k,l)}$ in the subdomain $S(\bar{x}, H)$; and $\tilde{S}^p_{\tilde{\Delta},PER} = H^1_{PER}(\tilde{c}) \cap \tilde{S}^p_{\tilde{\Delta}}$ is the space of periodic functions in $\tilde{c}$ which are continuous and are polynomials of degree p when restricted in any element of the master mesh $\tilde{\Delta}$, where

$$H^1_{PER}(\tilde{c}) = \{v \in H^1(\tilde{c}) \mid v(\tilde{x}_1, 1) = v(\tilde{x}_1, -1),\ v(1, \tilde{x}_2) = v(-1, \tilde{x}_2)\}. \tag{4.7*.7d}$$

Then, letting

$$\tilde{\Psi}^{PER}_{\tilde{S}^p_{\tilde{\Delta}},(i,j)} \stackrel{\text{def}}{=} \tilde{\rho}_{(i,j)} - \tilde{Z}^{(i,j)}_{\tilde{S}^p_{\tilde{\Delta},PER}} \tag{4.7*.8a}$$

we have

$$\Psi^{\bar{\mathbf{x}},H}_{S^p_{\Delta},(i,j)} = h^{p+1} \tilde{\Psi}^{PER}_{\tilde{S}^p_{\tilde{\Delta}},(i,j)} \circ \tilde{\mathbf{F}}^{-1}_{c^h_{(k,l)}} \tag{4.7*.8b}$$

and

$$\frac{\partial}{\partial x_k}\left(\Psi^{\bar{\mathbf{x}},H}_{S^p_{\Delta},(i,j)}\right) = h^p \frac{\partial}{\partial \tilde{x}_k}\left(\tilde{\Psi}^{PER}_{\tilde{S}^p_{\tilde{\Delta}},(i,j)}\right) \circ \tilde{\mathbf{F}}^{-1}_{c^h_{(k,l)}} \tag{4.7*.8c}$$

where $k = 1, 2$, and $\tilde{\mathbf{F}}^{-1}_{c^h_{(k,l)}}$ is the inverse of the affine transformation which maps $\tilde{c}$ onto $c^h_{(k,l)}$.

Remark 4.7*.3 Note that $\int_c \Psi^{\bar{\mathbf{x}},H}_{S^p_{\Delta},(i,j)} = 0$, and hence, for negligible pollution, $\int_c e_{S^p_{\Delta}} \approx 0$ asymptotically. The same is true for the average of the derivatives. Thus, non-zero averages of the error and its derivatives indicate pollution.

Remark 4.7*.4 The error $\Psi^{\bar{\mathbf{x}},H}_{S^p_{\Delta},(i,j)}$ is due to the residuals in all elements in the infinite periodic mesh. However, the effect of the residuals outside the first two mesh layers around each element is negligible, for all practical purposes.

Remark 4.7*.5 Above we addressed the case of cell-periodic meshes, and in Theorem 4.7*.1 we have proven the existence of an asymptotic finite element solution. One of the main assumptions was that every h cell can be scaled to the basic cell. This assumption is also satisfied for meshes of general quadrilaterals; it suffices to employ a master cell of general quadrilaterals. The other main assumption of the above theory was that $\mathcal{P}^p(c) \subseteq S^p_{\Delta}(c)$, namely, that the finite element approximation is able to reproduce exactly all polynomials of degree p. This does not hold for cells of general quadrilaterals. Even in the case of quadrilaterals with straight edges which are not parallelograms, the assumption $\mathcal{P}^p(c) \subseteq S^p_{\Delta}(c)$, does not hold if the serendipity element space is used but holds for the tensor product (bi-p) elements. Nevertheless, the above theory can be extended to also address the more general case. For a discussion of the serendipity quadrilateral elements, see [Arnold, Boffi, and Falk, 2000].

Let us now give some examples of the determination of the asymptotic error in the finite element solution. We will address the case of the isotropic, or orthotropic, Laplace operator.

Example 4.7*.1 (*Construction of the asymptotic finite element solution in the interior of the mesh*) Consider the criss-cross pattern shown in Fig. 4.7*.4a, and let

$$\tilde{Z}^{(2,0)}_{S^1_{\tilde{\Delta}},PER} = \sum_{i=1}^{5} z_i \tilde{\varphi}_i \tag{4.7*.9a}$$

where z_i are the nodal degrees of freedom, $\tilde{\varphi}_i$ are the nodal hat functions, and the enumeration of the vertices is shown in Fig. 4.7*.4a. From the periodicity it

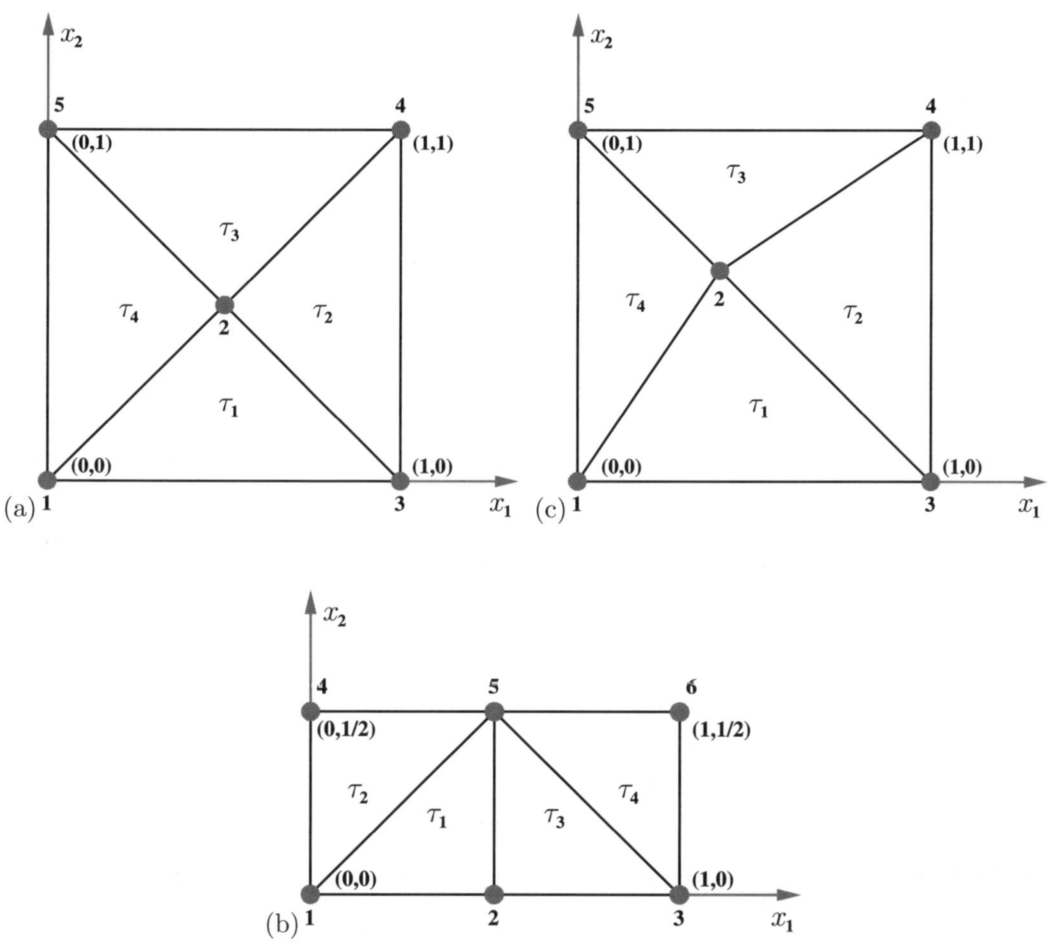

FIG. 4.7*.4. The basic cell for the (a) criss-cross pattern, (b) chevron pattern, (c) distorted criss-cross pattern (with the central node at (0.4, 0.6))

follows that

$$z_3 = z_4 = z_5 = z_1 \tag{4.7*.9b}$$

and hence z_1 and z_2 are the only unknowns to be determined from (4.7*.7a) and (4.7*.7b). Here we will consider the case of the isotropic Laplacian, and we will let

$$k_{lm}^{\tau^i} = \int_{\tau_i} \nabla\psi_l^{\tau_i} \cdot \nabla\psi_m^{\tau_i}, \qquad f_l^{\tau_i} = \int_{\tau_i} \nabla(\tilde{x}_1^2 - \mathcal{I}_{\tau_i}^1(\tilde{x}_1^2)) \cdot \nabla\psi_l^{\tau_i} \tag{4.7*.9c}$$

$l, m = 1, 2, 3$, where $\psi_l^{\tau_i}$ is the lth nodal shape function of τ_i, and $\mathcal{I}_{\tau_i}^1$ is the linear interpolation operator in τ_i. We have

$$\mathbf{k}^{\tau_1} = \begin{bmatrix} 1 & -\frac{1}{2} & -\frac{1}{2} \\ -\frac{1}{2} & \frac{1}{2} & 0 \\ -\frac{1}{2} & 0 & \frac{1}{2} \end{bmatrix}, \qquad \mathbf{f}^{\tau_1} = \begin{Bmatrix} \frac{1}{4} \\ -\frac{1}{8} \\ -\frac{1}{8} \end{Bmatrix} \tag{4.7*.9d}$$

where the first row corresponds to z_2, the second row corresponds to z_1 and the third row corresponds to z_3. Condensing (see Section 3.2) the rows and columns of $\mathbf{k}^{\tau_i}$, $\mathbf{f}^{\tau_i}$, which correspond to the same degrees of freedom due to the periodicity, we get

$$\mathbf{k}_{\text{cond}}^{\tau_1} = \begin{bmatrix} 1 & -1 \\ -1 & 1 \end{bmatrix}, \qquad \mathbf{f}_{\text{cond}}^{\tau_1} = \frac{1}{4}\begin{Bmatrix} 1 \\ -1 \end{Bmatrix} \tag{4.7*.9e}$$

where the first row corresponds to z_2 and the second row corresponds to z_1. Assembling the element arrays in the global arrays, we get the global equations

$$\begin{bmatrix} 4 & -4 \\ -4 & 4 \end{bmatrix} \begin{Bmatrix} z_1 \\ z_2 \end{Bmatrix} = \begin{Bmatrix} -\frac{1}{3} \\ \frac{1}{3} \end{Bmatrix} \tag{4.7*.9f}$$

and letting $z_1 = \tilde{\mathscr{C}}$, we obtain solution $z_2 = \frac{1}{12} + \tilde{\mathscr{C}}$. Employing (4.7*.7b) we get

$$\sum_{i=1}^{4} \int_{\tau_i} \left(\tilde{Z}^{(2,0)}_{S^1_{\tilde{\Delta}},PER} + \tilde{\mathscr{C}} \right) = \sum_{i=1}^{4} \int_{\tau_i} (\tilde{x}_1^2 - \tilde{\mathcal{I}}_{\tau_i}^1(\tilde{x}_1^2)) \tag{4.7*.9g}$$

which gives $\tilde{\mathscr{C}} = -\frac{1}{36}$. Following similar steps, we can also construct the asymptotic solution for the cells shown in Figs 4.7*.4b,c.

In the one-dimensional case the superconvergence points for the derivative, $\tilde{x}^{sup}_{du/dx}$, are determined from the condition

$$\frac{d}{d\tilde{x}}(\tilde{\Psi}^{PER}_{\tilde{S}^p_{\tilde{\Delta}}})(\tilde{x}^{sup}_{du/dx}) = 0. \tag{4.7*.10a}$$

In the two-dimensional case the superconvergence points for the gradient, $\tilde{\mathbf{x}}^{sup}_{\nabla u}$, must satisfy the $2(p+2)$ conditions

$$\frac{\partial}{\partial \tilde{x}_k}(\tilde{\Psi}^{PER}_{\tilde{S}^p_{\tilde{\Delta}},(i,j)})(\tilde{\mathbf{x}}^{sup}_{\nabla u}) = 0, \quad 0 \le i, j \le p+1, \;\; i+j = p+1, \;\; k = 1, 2. \tag{4.7*.10b}$$

For example, in the case of quadratic elements ($p = 2$) the point $\tilde{\mathbf{x}}^{sup}_{\nabla u}$ must satisfy eight such conditions. It follows that the points $\tilde{\mathbf{x}}^{sup}_{\nabla u}$ do not exist for general meshes. Of course, we can seek superconvergence points for each component of the gradient separately. For example, the point $\tilde{\mathbf{x}}^{sup}_{\partial u/\partial x_k}$, which must satisfy the $p+2$ conditions

$$\frac{\partial}{\partial \tilde{x}_k}\big(\tilde{\Psi}^{PER}_{\tilde{S}^p_{\tilde{\Delta}},(i,j)}\big)(\tilde{\mathbf{x}}^{sup}_{\partial u/\partial x_k}) = 0, \quad 0 \le i, j, \;\; i + j = p = 1. \tag{4.7*.10c}$$

Nevertheless, these conditions are still too many for elements of higher degree, and the superconvergence points $\tilde{\mathbf{x}}^{sup}_{\partial u/\partial x_k}$, may also not exist.

The number of conditions which must be satisfied by the superconvergence points can be drastically reduced, in the cases that $f \equiv 0$, where it is known a-priori that u_{EX}, satisfies the homogeneous differential equation $\mathfrak{D}u_{EX} \equiv 0$. Let us, for example, consider the case of the isotropic Laplacian

$$\Delta u_{EX} = \frac{\partial^2}{\partial x_1^2}(u_{EX}) + \frac{\partial^2}{\partial x_2^2}(u_{EX}) = 0. \tag{4.7*.11a}$$

In this case we have

$$\begin{aligned} Q^{\bar{\mathbf{x}},(p+1)}_{EX}(\mathbf{x}) = {} & \frac{1}{(p+1)!}\frac{\partial^{p+1}}{\partial x_1^{p+1}}(u_{EX})(\bar{\mathbf{x}})Q^{\bar{\mathbf{x}},(p+1)}_{1,hom}(\mathbf{x}) \\ & + B(u_{EX})(\bar{\mathbf{x}})Q^{\bar{\mathbf{x}},(p+1)}_{2,hom}(\mathbf{x}), \end{aligned} \tag{4.7*.11b}$$

where

$$B(u_{EX})(\bar{\mathbf{x}}) = \begin{cases} \dfrac{1}{(p+1)!}\dfrac{\partial^{p+1}u_{EX}}{\partial x_1^p \partial x_2}(\bar{\mathbf{x}}), & p \text{ even} \\ (-1)^{(p-1)/2}\dfrac{1}{(p+1)!}\dfrac{\partial^{p+1}}{\partial x_2^{p+1}}(u_{EX})(\bar{\mathbf{x}}), & p \text{ odd} \end{cases} \tag{4.7*.11c}$$

and

$$Q^{\bar{\mathbf{x}},(p+1)}_{1,hom}(x_1, x_2) = \mathfrak{Re}((z-\bar{z})^k); \qquad Q^{\bar{\mathbf{x}},(p+1)}_{2,hom}(x_1, x_2) = \mathfrak{Im}((z-\bar{z})^k) \tag{4.7*.11d}$$

are the *harmonic monomials* of degree $(p+1)$ centered at $\bar{\mathbf{x}}$, where $z = x_1 + ix_2$ and $\bar{z} = \bar{x}_1 + i\bar{x}_2$. It follows that

$$\Psi^{\bar{\mathbf{x}},H}_{S^p_\Delta} = \frac{1}{(p+1)!}\frac{\partial^{p+1}}{\partial x_1^{p+1}}(u_{EX})(\bar{\mathbf{x}})\Psi^{\bar{\mathbf{x}},H}_{S^p_\Delta,1,hom} + B(u_{EX})(\bar{\mathbf{x}})\Psi^{\bar{\mathbf{x}},H}_{S^p_\Delta,2,hom} \tag{4.7*.11e}$$

where $\Psi^{\bar{\mathbf{x}},H}_{S^p_\Delta,i,hom}$ is the error corresponding to $Q^{\bar{\mathbf{x}},(p+1)}_{i,hom}$.

Using this result we obtain:

A point $\mathbf{x}^{sup}_{\nabla u}$ is a point of superconvergence for the gradient for finite element solutions of the Laplacian if and only if

$$\frac{\partial}{\partial \tilde{x}_k}\big(\tilde{\Psi}^{PER}_{\tilde{S}^p_{\tilde{\Delta}},i,hom}\big)(\mathbf{x}^{sup}_{\nabla u}) = 0, \quad i = 1, 2, \;\; k = 1, 2 \tag{4.7*.12a}$$

where $\tilde{\Psi}^{PER}_{\tilde{S}^p_{\tilde{\Delta}},i,hom}$ is defined as in (4.7*.8a), by replacing $\tilde{\rho}_{(i,j)}$ by $\tilde{\rho}_{i,hom} = \tilde{Q}^{(p+1)}_{i,hom} - \tilde{\mathcal{I}}^p\tilde{Q}^{(p+1)}_{i,hom}$.

Thus, in the case of the Laplacian *a superconvergence point for the gradient,* $\mathbf{x}^{sup}_{\boldsymbol{\nabla}u}$, *must satisfy only four conditions, independently of the degree* p *of the elements.* Nevertheless, these four conditions may be incompatible, and $\mathbf{x}^{sup}_{\boldsymbol{\nabla}u}$ may not exist even in this case. However, *a superconvergence point for the* k*th component of the gradient,* $\mathbf{x}^{sup}_{\partial u/\partial x_k}$, *must satisfy only the two conditions and always exists.*

$$\frac{\partial}{\partial \tilde{x}_k}(\tilde{\Psi}^{PER}_{\tilde{S}^p_{\tilde{\Delta}},i,hom})(\mathbf{x}^{sup}_{\partial u/\partial x_k}) = 0, \quad i = 1,2 \tag{4.7*.12b}$$

More generally, let $\mathcal{F}(u)$ denote the quantity of interest where $\mathcal{F}$ is *linear*, namely

$$\mathcal{F}(\alpha u_1 + \beta u_2) = \alpha\mathcal{F}(u_1) + \beta\mathcal{F}(u_2). \tag{4.7*.13a}$$

For example, $\mathcal{F}(u) = u$, or $\mathcal{F}(u) = \boldsymbol{\nabla}u$, or $\mathcal{F}(u) = \partial u/\partial x_1$, etc. We then have:

The point $\tilde{\mathbf{x}}^{sup}_{\mathcal{F}(u)}$ *is a superconvergence point for* $\mathcal{F}(u)$ *if and only if*

$$\mathcal{F}(\tilde{\Psi}^{PER}_{\tilde{S}^p_{\tilde{\Delta}},l})(\tilde{\mathbf{x}}^{sup}_{\mathcal{F}(u)}) = 0 \tag{4.7*.13b}$$

where in the nonhomogeneous case (e.g., the case of Poisson's equation $\Delta u = f \not\equiv 0$) l is the multi-index (i,j) with $nd = p+2$ and $i+j = p+1$, $0 \le i,j \le p+1$; and in the homogeneous case (e.g., the case of the Laplace equation $\Delta u \equiv 0$) $l = 1,2$, $nd = 2$, and $\Psi^{PER}_{\tilde{S}^p_{\tilde{\Delta}},l} = \Psi^{PER}_{\tilde{S}^p_{\tilde{\Delta}},l,hom}$.

It follows that the superconvergence points $\tilde{\mathbf{x}}^{sup}_{\mathcal{F}(u)}$ for the class $\mathfrak{Q}^{(p+1)}$ are the points of common intersection of the contours

$$C_{\mathcal{F}(u)}(\tilde{Q}^{(p+1)}_l; \tilde{\tau}, \tilde{\Delta}) \stackrel{\text{def}}{=} \{\mathbf{x} \in \tilde{\tau} \mid \mathcal{F}(\tilde{\Psi}^{PER}_{\tilde{S}^p_{\tilde{\Delta}},l}) = 0\} \tag{4.7*.14a}$$

where $\tilde{Q}^{(p+1)}_l$ denotes the lth basis monomial of the class $\mathfrak{Q}^{(p+1)}$, and hence

$$\mathcal{X}^{sup}_{\mathcal{F}(u)}(\mathfrak{Q}^{(p+1)}; \tilde{\tau}, \tilde{\Delta}) \stackrel{\text{def}}{=} \bigcap_{l=1}^{nd} C_{\mathcal{F}(u)}(\tilde{Q}^{(p+1)}_l; \tilde{\tau}, \tilde{\Delta}) \tag{4.7*.14b}$$

where $\mathcal{X}^{sup}_{\mathcal{F}(u)}(\mathfrak{Q}^{(p+1)}; \tilde{\tau}, \tilde{\Delta})$ is the set of superconvergence points in the element $\tilde{\tau}$ for the class $\mathfrak{Q}^{(p+1)}$.

In the nonhomogeneous case (e.g., the Poisson equation), we employ $\mathfrak{Q}^{(p+1)} = \mathfrak{Q}^{(p+1)}_{gen}$, where

$$\mathfrak{Q}^{(p+1)}_{gen} \stackrel{\text{def}}{=} \Big\{\tilde{Q} \mid \tilde{Q} = \sum_{\substack{0\le i,j\le p+1\\ i+j=p+1}} \alpha_{ij}\tilde{Q}^{(p+1)}_{(i,j)},\ \tilde{Q}^{(p+1)}_{(i,j)}(\tilde{x}_1,\tilde{x}_2) = \tilde{x}_1^i\tilde{x}_2^j\Big\} \tag{4.7*.15a}$$

while in the homogeneous case (e.g., the Laplace equation), we let $\mathfrak{Q}^{(p+1)} = \mathfrak{Q}^{(p+1)}_{hom}$, where

$$\mathfrak{Q}^{(p+1)}_{hom} \stackrel{\text{def}}{=} \Big\{\tilde{Q} \mid \tilde{Q} = \sum_{l=1}^{2} \alpha_l\tilde{Q}^{(p+1)}_{l,hom}\Big\}. \tag{4.7*.15b}$$

Let us underline that the presented approach leads to the complete characterization of the set of superconvergent points.

Let us now give an illustrative example.

Example 4.7*.2 (*Superconvergence points for the derivative in the four patterns of triangles*) Let us be interested in the points $\tilde{\mathbf{x}}^{sup}_{\partial u/\partial x_k}$ for the Poisson equation and the basic mesh cells shown in Fig. 4.7*.3, in the case of linear elements ($p = 1$). Letting

$$\tilde{Q}_1^{(2)} = \tilde{x}_1^2, \qquad \tilde{Q}_2^{(2)} = \tilde{x}_1\tilde{x}_2, \qquad \tilde{Q}_3^{(2)} = \tilde{x}_2^2 \tag{4.7*.16a}$$

we obtain for the element τ_1 of the criss-cross pattern, shown in Fig. 4.7*.3d

$$\tilde{\Psi}^{PER}_{S^1_{\tilde{\Delta}},1} = \tilde{x}_1^2 - \left((\tilde{x}_1 - \tilde{x}_2) + \tfrac{2}{3}\,\tilde{x}_2\right), \qquad \tilde{\Psi}^{PER}_{S^1_{\tilde{\Delta}},2} = \tilde{x}_1\tilde{x}_2 - \tfrac{1}{2}\tilde{x}_2 \tag{4.7*.16b}$$

and

$$\tilde{\Psi}^{PER}_{S^1_{\tilde{\Delta}},3} = \tilde{x}_1^2 - \tfrac{2}{3}\tilde{x}_2 \tag{4.7*.16c}$$

and hence

$$\frac{\partial}{\partial \tilde{x}_1}(\tilde{\Psi}^{PER}_{S^1_{\tilde{\Delta}},1}) = 2\tilde{x}_1 - 1, \qquad \frac{\partial}{\partial \tilde{x}_1}(\tilde{\Psi}^{PER}_{S^1_{\tilde{\Delta}},2}) = \tilde{x}_2, \qquad \frac{\partial}{\partial \tilde{x}_1}(\tilde{\Psi}^{PER}_{S^1_{\tilde{\Delta}},3}) = 0. \tag{4.7*.16d}$$

It follows that the contours of zero error in the x_1 derivative are

$$\tilde{C}_{\partial u/\partial \tilde{x}_1}(\tilde{Q}_1^{(2)}; \tau_1, \tilde{\Delta}) = \{(\tfrac{1}{2}, \tilde{x}_2) \in \tau_1\} \tag{4.7*.16e}$$

$$\tilde{C}_{\partial u/\partial \tilde{x}_1}(\tilde{Q}_2^{(2)}; \tau_1, \tilde{\Delta}) = \{(\tilde{x}_1, 0) \in \tau_1\}, \qquad \tilde{C}_{\partial u/\partial \tilde{x}_1}(\tilde{Q}_3^{(2)}; \tau_1, \tilde{\Delta}) \equiv \tau_1$$

and the set of superconvergence points is

$$\tilde{\mathcal{X}}^{\text{sup}}_{\partial u/\partial \tilde{x}_1}(\mathfrak{Q}^{(2)}_{gen}; \tau_1, \tilde{\Delta}) \stackrel{\text{def}}{=} \bigcap_{i=1,2,3} \tilde{C}_{\partial u/\partial \tilde{x}_1}(\tilde{Q}_i^{(2)}; \tau_1, \tilde{\Delta}) = \{(\tfrac{1}{2}, 0)\}. \tag{4.7*.16f}$$

The contours $\tilde{C}_{\partial u/\partial \tilde{x}_1}(\tilde{Q}_1^{(2)}; \tau_1, \tilde{\Delta})$ and $\tilde{C}_{\partial u/\partial \tilde{x}_1}(\tilde{Q}_2^{(2)}; \tau_1, \tilde{\Delta})$ are shown in Fig. 4.7*.5a.

Similarly, for the x_2 derivative we have

$$\tilde{C}_{\partial u/\partial \tilde{x}_2}(\tilde{Q}_1^{(2)}; \tau_1, \tilde{\Delta}) = \emptyset, \qquad \tilde{C}_{\partial u/\partial \tilde{x}_2}(\tilde{Q}_2^{(2)}; \tau_1, \tilde{\Delta}) = \{(\tfrac{1}{2}, \tilde{x}_2) \in \tau_1\} \tag{4.7*.16g}$$

and

$$\tilde{C}_{\partial u/\partial \tilde{x}_2}(\tilde{Q}_3^{(2)}; \tau_1, \tilde{\Delta}) = \emptyset \tag{4.7*.16h}$$

and hence

$$\tilde{\mathcal{X}}^{\text{sup}}_{\partial u/\partial \tilde{x}_2}(\mathfrak{Q}^{(2)}_{gen}; \tau_1, \tilde{\Delta}) = \emptyset. \tag{4.7*.16i}$$

For the chevron pattern, shown in Fig. 4.7*.3b, and the x_1 derivative, we have

$$\begin{aligned} &\tilde{C}_{\partial u/\partial \tilde{x}_1}(\tilde{Q}_1^{(2)}; \tau_1, \tilde{\Delta}) = \{(\tfrac{1}{4}, \tilde{x}_2) \in \tau_1\}, \\ &\tilde{C}_{\partial u/\partial \tilde{x}_1}(\tilde{Q}_2^{(2)}; \tau_1, \tilde{\Delta}) = \{(\tilde{x}_1, 0) \in \tau_1\} \end{aligned} \tag{4.7*.17a}$$

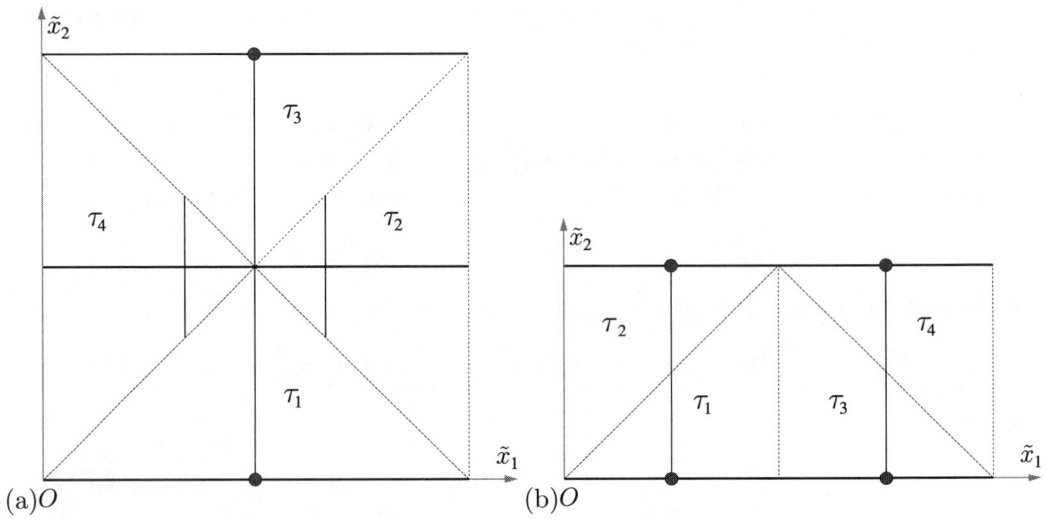

FIG. 4.7*.5. An example of the determination of the superconvergence points at the intersections of the zero contours of the basis monomials: Poisson equation ($-\Delta u = f$), quadratic polynomial solution, linear elements ($p = 1$). The contours of zero error in the x_1 derivative of the solution, $\tilde{C}_{\partial u/\partial \tilde{x}_1}(\tilde{Q}_1^{(2)}; \tau_j, \tilde{\Delta})$ and $\tilde{C}_{\partial u/\partial \tilde{x}_1}(\tilde{Q}_2^{(2)}; \tau_j, \tilde{\Delta})$, for the elements τ_j ($j = 1, \ldots, 4$) are shown with the thin and thick lines, respectively, for the (a) criss-cross pattern, (b) chevron pattern. The superconvergence points are shown with gray circles. Note that *the number and location of the superconvergence points depend on the element* τ *and the geometry of the mesh*

and

$$\tilde{C}_{\partial u/\partial \tilde{x}_1}(\tilde{Q}_3^{(2)}; \tau_1, \tilde{\Delta}) \equiv \tau_1 \tag{4.7*.17b}$$

and

$$\tilde{\mathcal{X}}^{\text{sup}}_{\partial u/\partial \tilde{x}_1}(\mathcal{Q}^{(2)}_{gen}; \tau_1, \tilde{\Delta}) = \{(\tfrac{1}{4}, 0)\}. \tag{4.7*.17c}$$

For the x_2 derivative we have

$$\begin{aligned} \tilde{C}_{\partial u/\partial \tilde{x}_2}(\tilde{Q}_1^{(2)}; \tau_1, \tilde{\Delta}) &= \{(\tilde{x}_1, \tilde{x}_2) \in \tau_1\}, \\ \tilde{C}_{\partial u/\partial \tilde{x}_2}(\tilde{Q}_2^{(2)}; \tau_1, \tilde{\Delta}) &= \{(\tfrac{1}{2}, \tilde{x}_2) \in \tau_1\} \end{aligned} \tag{4.7*.17d}$$

and

$$\tilde{C}_{\partial u/\partial \tilde{x}_2}(\tilde{Q}_3^{(2)}; \tau_1, \tilde{\Delta}) = \{(\tilde{x}_1, \tfrac{1}{4}) \in \tau_1\} \tag{4.7*.17e}$$

and hence

$$\tilde{\mathcal{X}}^{\text{sup}}_{\partial u/\partial \tilde{x}_2}(\mathcal{Q}^{(2)}_{gen}; \tau_1, \tilde{\Delta}) = \{(\tfrac{1}{2}, \tfrac{1}{4})\}. \tag{4.7*.17f}$$

In the case of the Laplacian ($\Delta u = 0$), letting

$$\tilde{Q}_1^{(2)} = \tilde{x}_1^2 - \tilde{x}_2^2, \qquad \tilde{Q}_2^{(2)} = \tilde{x}_1\tilde{x}_2 \tag{4.7*.18a}$$

for the element τ_1 of the criss-cross pattern we have

$$\tilde{\Psi}^{PER}_{S^1_{\tilde{\Delta}},1} = (\tilde{x}_1^2 - \tilde{x}_2^2) - (\tilde{x}_1 - \tilde{x}_2), \qquad \tilde{\Psi}^{PER}_{S^1_{\tilde{\Delta}},2} = \tilde{x}_1\tilde{x}_2 - \tfrac{1}{2}\tilde{x}_2 \tag{4.7*.18b}$$

and for the x_1 derivative we have

$$\begin{aligned} \tilde{C}_{\partial u/\partial \tilde{x}_1}(\tilde{Q}_1^{(2)}; \tau_1, \tilde{\Delta}) &= \{(\tfrac{1}{2}, \tilde{x}_2) \in \tau_1\}, \\ \tilde{C}_{\partial u/\partial \tilde{x}_1}(\tilde{Q}_2^{(2)}; \tau_1, \tilde{\Delta}) &= \{(\tilde{x}_1, 0) \in \tau_1\} \end{aligned} \tag{4.7*.18c}$$

and hence

$$\tilde{\mathcal{X}}^{\text{sup}}_{\partial u/\partial \tilde{x}_1}(\mathfrak{Q}^{(2)}_{hom}; \tau_1, \tilde{\Delta}) = \{(\tfrac{1}{2}, 0)\}. \tag{4.7*.18d}$$

For the x_2 derivative

$$\begin{aligned} \tilde{C}_{\partial u/\partial \tilde{x}_2}(\tilde{Q}_1^{(2)}; \tau_1, \tilde{\Delta}) &= \{(\tilde{x}_1, \tfrac{1}{2}) \in \tau_1\}, \\ \tilde{C}_{\partial u/\partial \tilde{x}_2}(\tilde{Q}_2^{(2)}; \tau_1, \tilde{\Delta}) &= \{(\tfrac{1}{2}, \tilde{x}_2) \in \tau_1\} \end{aligned} \tag{4.7*.18e}$$

and hence

$$\tilde{\mathcal{X}}^{\text{sup}}_{\partial u/\partial \tilde{x}_2}(\mathfrak{Q}^{(2)}_{hom}; \tau_1, \tilde{\Delta}) = \{(\tfrac{1}{2}, \tfrac{1}{2})\}. \tag{4.7*.18f}$$

Thus the midpoints of the vertical sides of the chevron mesh are superconvergence points for the y derivatives for the finite element solution of the Laplacian, but not for the Poisson equation.

Hence *the existence and location of the superconvergence points depends on the local geometry of the mesh, the class of solutions and the solution quantity under consideration.*

In practice, the superconvergence points $\mathbf{x}^{sup}_{\mathcal{F}(u)}$, may be determined by employing the following steps:

(1) A *uniform background mesh* over the element τ (e.g., for the element τ_1 of the criss-cross pattern as shown in Fig. 4.7*.6a), is created.

(2) For each monomial $\tilde{Q}_i^{(p+1)}$, the values of $\mathcal{F}(\tilde{\Psi}_i^{(p+1)})$, are determined at the vertices of the background mesh, and an approximate contour $\tilde{C}^{\tilde{\Delta}^{back}}_{\mathcal{F}(u)}(\tilde{Q}_i^{(p+1)}; \tau, \tilde{\Delta})$ is constructed by connecting all the points where the interpolated value of $\mathcal{F}(\tilde{Q}_i^{(p+1)})$ is zero.

(3) The approximate superconvergence points $\tilde{\mathbf{x}}^{sup,\tilde{\Delta}^{back}}_{\mathcal{F}(u)}$, are determined as the points of intersection of the approximate contours for the basis monomials $\tilde{Q}_i^{(p+1)}$.

(4) The superconvergence points $\tilde{\mathbf{x}}^{sup}_{\mathcal{F}(u)}$, are determined by employing the points $\tilde{\mathbf{x}}^{sup,\tilde{\Delta}^{back}}_{\mathcal{F}(u)}$ as an initial guess and by minimizing $\max_{i=1,\ldots,nd} |\mathcal{F}(\tilde{Q}_i^{(p+1)})|$, to a desired level of accuracy.

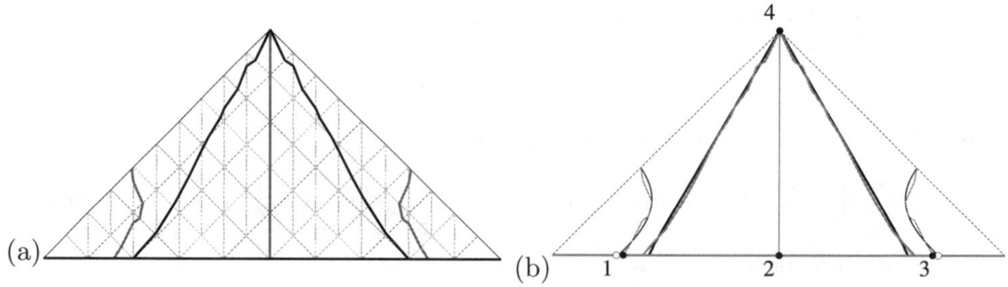

FIG. 4.7*.6. Numerical construction of the contours of zero error and the points of superconvergence: Laplace equation ($\Delta u = 0$), cubic elements ($p = 3$), element τ_1 of the criss-cross pattern. (a) The approximate contours $\tilde{C}^{\tilde{\Delta}_1^{back}}_{\partial u/\partial \tilde{x}_1}(\tilde{Q}_1^{(4)}; \tau_j, \tilde{\Delta})$ and $\tilde{C}^{\tilde{\Delta}_1^{back}}_{\partial u/\partial \tilde{x}_1}(\tilde{Q}_2^{(4)}; \tau_j, \tilde{\Delta})$ shown with the gray and dark lines, constructed by employing the background mesh shown in the figure ($np = 66$). (b) The contours $\tilde{C}^{\tilde{\Delta}_2^{back}}_{\partial u/\partial \tilde{x}_1}(\tilde{Q}_1^{(4)}; \tau_j, \tilde{\Delta})$ and $\tilde{C}^{\tilde{\Delta}_2^{back}}_{\partial u/\partial \tilde{x}_1}(\tilde{Q}_2^{(4)}; \tau_j, \tilde{\Delta})$ obtained by employing a refined background mesh with $np = 3321$ vertices shown with the thin and thick dark lines respectively. The exact superconvergence points $\tilde{\mathbf{x}}^{\mathrm{sup}}_{\partial u/\partial \tilde{x}_1}$ (numbered 1 to 4) are shown with empty circles. The approximate contours from (a), determined using the background mesh with $np = 66$ vertices, are shown with the gray lines and the corresponding approximate superconvergence points are shown with black circles. Note that the approximate contours and superconvergence points which are determined from the coarse background mesh are rather close to the ones determined from the refined background mesh

Note that in the case of linear triangles (e.g., Example 4.7*.2) the contours are straight lines in each triangle of the mesh pattern and the points determined in step (ii) are the exact superconvergence points. For $p \geq 2$, we should expect an error $|\tilde{\mathbf{x}}^{sup}_{\mathcal{F}(u)} - \tilde{\mathbf{x}}^{sup,\tilde{\Delta}^{back}}_{\mathcal{F}(u)}| \neq 0$, which can be made sufficiently small by employing a refined background mesh. The superconvergence points can then be determined with the desired number of significant digits by employing the minimization of $\max_{i=1,\ldots,nd} |\mathcal{F}(\tilde{Q}_i^{(p+1)})|$, and *interval arithmetic*.

Remark 4.7*.6 The problem of constructing the level lines of a function occurs frequently in the theory of visualization. In our case, the functions are polynomials, and the contours are algebraic curves. For analysis and construction of such lines, we refer to [Bajaj and Xu, 1999], [Bajaj and Xu, 1997].

Let us now use the above approach to study the superconvergence of the derivatives of the solution for the four mesh patterns shown in Fig. 4.7*.3, for the Laplace and the Poisson equations.

Example 4.7*.3 Let us first consider the case of the Laplace equation and $\mathcal{F}(u) = \partial u/\partial x_1$. Figure 4.7*.7 shows the contours $C_{\partial u/\partial x_1}(\tilde{Q}^{(4)}_{1,hom}; \tilde{\Delta})$, $\tilde{Q}^{(4)}_{1,hom}(\tilde{x}_1, \tilde{x}_2) = \mathfrak{Re}(\tilde{z}^4) = \tilde{x}_1^4 + \tilde{x}_2^4 - 6\,\tilde{x}_1^2\tilde{x}_2^2$, with thin line, and $C_{\partial u/\partial x_1}(\tilde{Q}^{(4)}_{2,hom}; \tilde{\Delta})$, for $\tilde{Q}^{(4)}_{2,hom}(\tilde{x}_1, \tilde{x}_2) = \mathfrak{Im}(\tilde{z}^4) = \tilde{x}_1^3\,\tilde{x}_2 - \tilde{x}_1\,\tilde{x}_2^3$, with thick line, with the superconvergence points $\tilde{\mathbf{x}}^{sup}_{\partial u/\partial x_1,k}$ at their intersections. Note that in the regular, chevron and Union-Jack patterns, it is sufficient to determine the superconvergence points for the element τ_1, shown in Figs 4.7*.3a–c, and the superconvergence points for the other elements in the basic cell

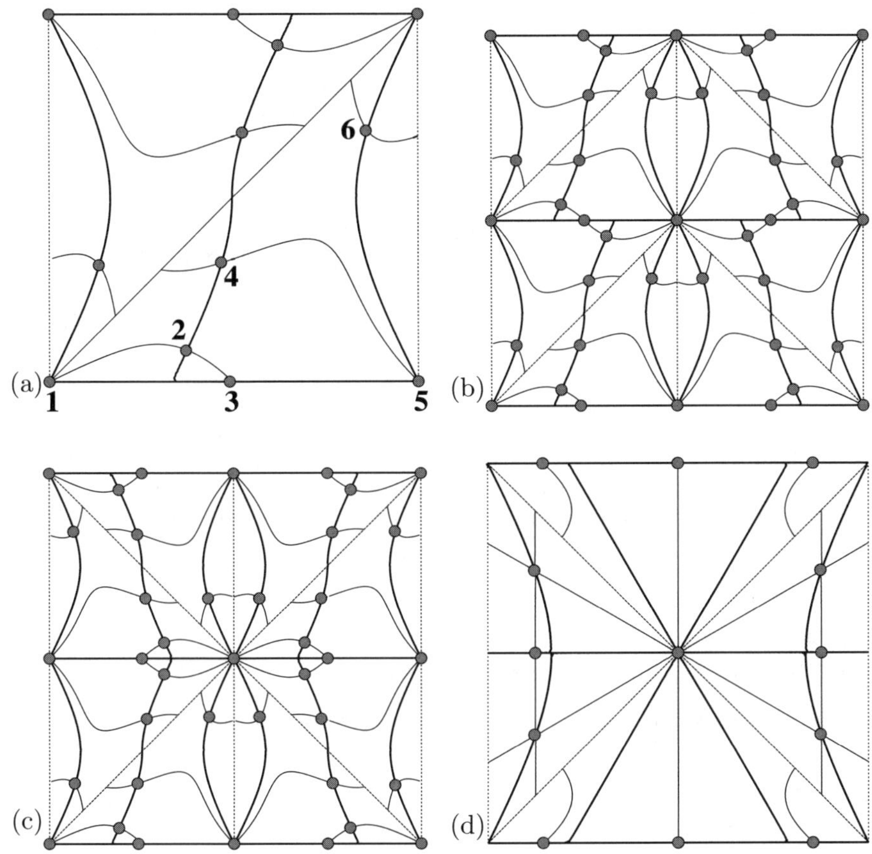

FIG. 4.7*.7. Superconvergence points for finite element solutions of the Laplace equation. Superconvergence points for $\mathcal{F}(u) = \partial u/\partial x_1$ and cubic triangular elements for the cell mesh geometries shown in Fig. 4.7*.3. (a) regular pattern; (b) chevron pattern; (c) Union-Jack pattern; (d) criss-cross pattern. The superconvergence points in the element $\tau \in \tilde{\Delta}$ are obtained as the points of intersection of the zero contours $C_{\partial u/\partial x_1}(\tilde{Q}_l^{(4)}; \tau, \tilde{\Delta})$, $l = 1, 2$, for $\tilde{Q}_{1,hom}^{(4)}(\tilde{\mathbf{x}}) = \tilde{x}_1^4 + \tilde{x}_2^4 - 6\tilde{x}_1^2\tilde{x}_2^2$ (thin line) and $\tilde{Q}_{2,hom}^{(4)}(\tilde{\mathbf{x}}) = \tilde{x}_1^3\tilde{x}_2 - \tilde{x}_1\tilde{x}_2^3$ (thick line). Note that the edges of the elements are shown with dashed line

can be easily deduced. The coordinates of the superconvergence points are:

$$\tilde{\mathbf{x}}^{sup}_{\partial u/\partial x_1,1} = (-1,-1), \qquad \tilde{\mathbf{x}}^{sup}_{\partial u/\partial x_1,2} = (-0.6247818, -0.9141175) \qquad (4.7^*.19a)$$

$$\tilde{\mathbf{x}}^{sup}_{\partial u/\partial x_1,3} = (-\tfrac{1}{2},-1), \qquad \tilde{\mathbf{x}}^{sup}_{\partial u/\partial x_1,4} = (-0.5284881, -0.6758267) \qquad (4.7^*.19b)$$

$$\tilde{\mathbf{x}}^{sup}_{\partial u/\partial x_1,5} = (0,-1), \qquad \tilde{\mathbf{x}}^{sup}_{\partial u/\partial x_1,6} = (-0.1361824, -0.3175934) \qquad (4.7^*.19c)$$

The point $\tilde{\mathbf{x}}^{sup}_{\partial u/\partial x_1,3}$ is the only point which is also a superconvergence point for finite element solutions of the Poisson equation and only for the regular (see Fig. 4.7*.8) and the chevron pattern. There are no superconvergence points for $\partial u/\partial x_1$ in the Union-Jack pattern for finite element solutions of the Poisson equation. The point $\tilde{\mathbf{x}}^{sup}_{\partial u/\partial x_1,5}$ is the

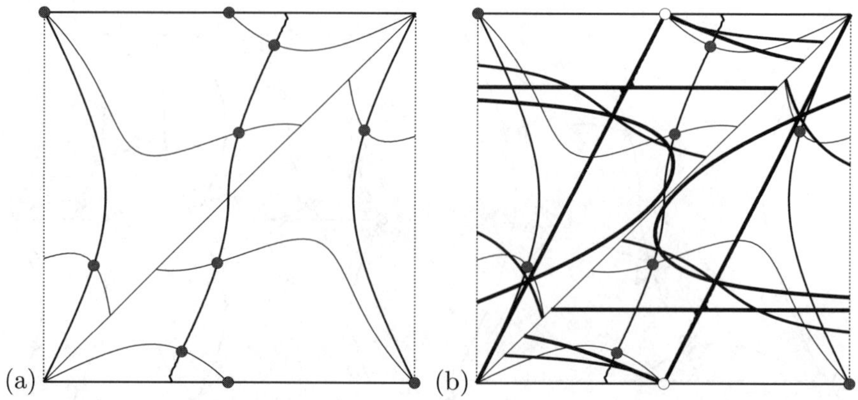

FIG. 4.7*.8. Superconvergence points for finite element solutions of the Laplace and Poisson equations. Superconvergence points for $\mathcal{F}(u) = \partial u/\partial x_1$ and cubic triangular elements arranged in the regular pattern: (a) Laplace equation; (b) Poisson equation. Note that the mid-points of the sides which are parallel to the x_1 axis are the only superconvergence points in the case of the Poisson equation

only point of simultaneous superconvergence for $\partial u/\partial x_1$ and $\partial u/\partial x_2$ for finite element solutions of the Laplace equation; this can be seen from Fig. 4.7*.8.

Above we gave examples of the determination of superconvergence points for meshes of triangles. Let us now give an example of serendipity squares.

Example 4.7*.4 (*Superconvergence points for the derivatives in serendipity squares*) Let the master mesh $\tilde{\Delta}$ consist of one square element which coincides with the master cell $\tilde{c} = (-1, 1)^2$. Figures 4.7*.9a–d give the points $\mathbf{x}^{sup}_{\partial u/\partial x_1}$, for $p = 2, 3, 4, 5$. Note that for $p = 2$ only the monomial $\tilde{x}_k^3$ gives non-zero error for $\partial u/\partial x_k$, $k = 1, 2$, and there are superconvergence lines for $\partial u/\partial x_k$, $k = 1, 2$, which are perpendicular to the x_k axis through the Gauss points of order two. It follows that in the case $p = 2$ the two by two Gauss–Legendre points are points of simultaneous superconvergence for $\partial u/\partial x_1$ and $\partial u/\partial x_2$. In the case of $p = 3$ there are four monomials of degree $p = 4$ which do not belong to the span of the basis of the cubic serendipity square but only two of these monomials have non-zero x_1 derivative of the error, namely $\tilde{Q}_1^{(4)}(\tilde{x}_1, \tilde{x}_2) = \tilde{x}_1^4$ and $\tilde{Q}_2^{(4)}(\tilde{x}_1, \tilde{x}_2) = \tilde{x}_1^2\, \tilde{x}_2^2$. Figure 4.7*.9b shows the contours $C_{\partial u/\partial x_1}(\tilde{Q}_1^{(4)}; \tau, \tilde{\Delta})$, by thin line, and $C_{\partial u/\partial x_1}(\tilde{Q}_2^{(4)}; \tau, \tilde{\Delta})$, by thick line. The intersections of these contours are four superconvergence points, namely

$$\mathbf{x}_2 = (0.77459667, 0.57735027), \qquad \mathbf{x}_3 = (-0.77459677, 0.57735027) \qquad (4.7^*.20a)$$

$$\mathbf{x}_4 = (-0.77459667, -0.57735027), \qquad \mathbf{x}_5 = (0.77459667, -0.57735027) \qquad (4.7^*.20b)$$

and the superconvergence line $\tilde{x}_1 = 0$. Figures 4.7*.9c,d shows the superconvergence lines $C_{\partial u/\partial x_1}(\Re\mathfrak{e}(z^{p+1}), \tau, \tilde{\Delta})$, by thin line, and $C_{\partial u/\partial x_1}(\Im\mathfrak{m}(z^{p+1}), \tau, \tilde{\Delta})$, by thick line, with the superconvergence points $\tilde{\mathbf{x}}^{sup}_{\partial u/\partial x_1}$ at their intersections. Note that for $p = 3, 5, 7$ the

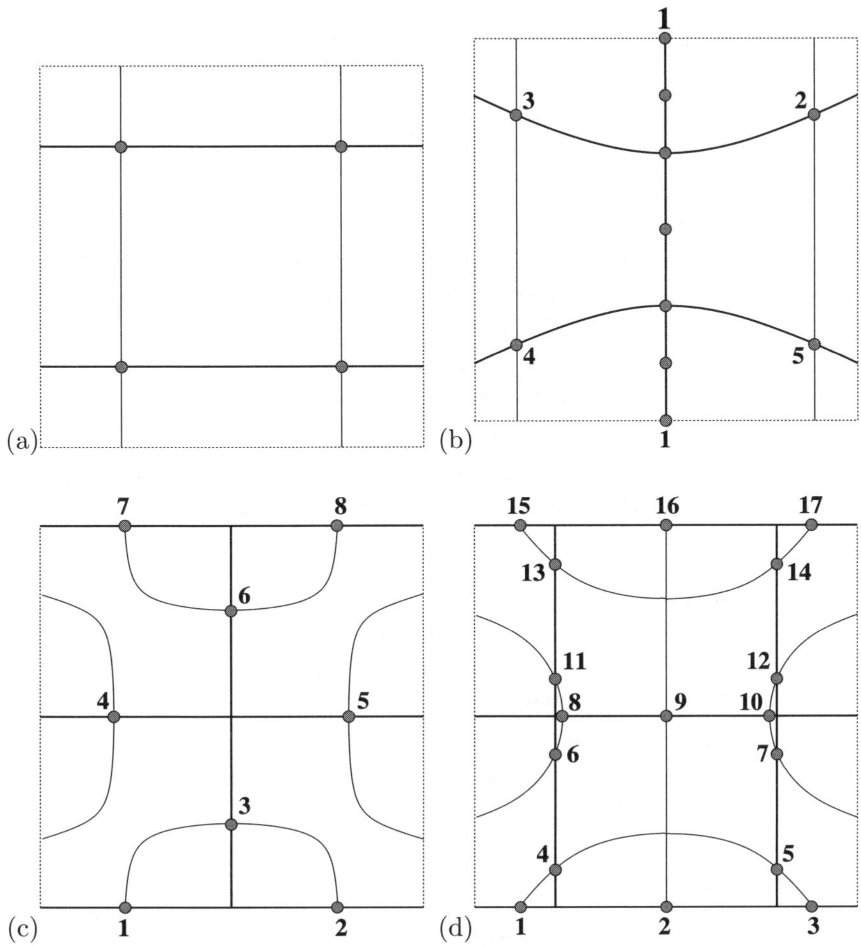

FIG. 4.7*.9. Superconvergence points for finite element solutions of the Laplace and the Poisson equations for serendipity squares. Superconvergence points for $\mathcal{F}(u) = \partial u/\partial x_1$ and serendipity squares of degree p. (a) $p = 2$, Poisson equation; $\mathcal{F}(u) = \partial u/\partial x_1$ and $\partial u/\partial x_2$ (b) $p = 3$, Poisson equation; (c) $p = 4$, Laplace equation; (d) $p = 5$, Laplace equation. Note that for $p = 2$ there exist superconvergence lines for $\partial u/\partial x_1$ and $\partial u/\partial x_2$; these lines intersect at the two by two Gauss–Legendre points which are points of simultaneous superconvergence for $\partial u/\partial x_1$ and $\partial u/\partial x_2$. For $p = 3$ and the Poisson equation there exist four superconvergence points and a superconvergence line which coincides with the x_2 axis through the center of the element. *Note that for $p \geq 3$ the p by p Gauss–Legendre points are not superconvergence points for the derivatives with the exception of the center of the element for p odd.* Further, for $p > 2$ even, there are no points of simultaneous superconvergence for both components of the derivative

center of the element is the only point of simultaneous superconvergence for $\partial u/\partial x_1$ and $\partial u/\partial x_2$ for both the classes of harmonic and general solutions. For p even, $p > 2$, there are no points of simultaneous superconvergence for both components of the gradient.

Remark 4.7*.7 Let us give a summary of results of the analysis of the superconvergence of the x_1 derivative for the case of the Laplace and the Poisson equation: For meshes of triangles:

(1) For the class of solutions of the Laplace equation there exist points which can be determined as in Example 4.7*.3 above.

(2) For the class of solutions of the Poisson equation there are the following points:

 (i) *Regular pattern*: For odd p, the midpoint of the edge which is parallel to the x_1 axis. For $p = 2$, the two Gauss–Legendre points on the edge which is parallel to the x_1 axis. For $p > 2$ and even, there are no superconvergence points.

 (ii) *Chevron pattern*: For odd p, the same points as in the regular pattern. For even p, there are no superconvergence points.

 (iii) *Union-Jack pattern*: There are no superconvergence points.

 (iv) *Criss-cross pattern*: For element τ_1, the same points as in the regular pattern. For element τ_2, there are no superconvergence points.

For meshes of squares:

(1) For the elements of the *tensor product* and the *intermediate family*, superconvergence of the x_1 derivative occurs along the constant x_1 Gauss lines which pass through the Gauss–Legendre points of degree p along the x_1 axis, for both the Laplace and the Poisson equations.

(2) For the *serendipity elements* we have:

 (i) For $p = 1, 2$, the points are exactly the same as for the tensor product and the intermediate family.

 (ii) For $p = 3$ and the Laplace equation there are three superconvergence lines which are shown with thick line in Fig. 4.7*.9b. For the class of general solution there are four superconvergence points and one superconvergence line (the line $\tilde{x}_1 = 0$) as shown in Fig. 4.7*.9b.

 (iii) For $p \geq 4$: For the Laplace equation there exist several superconvergence points, as shown in Fig. 4.7*.9 for $p = 4, 5$. For the Poisson equation and even p ($p = 4, 6, \dots$) there are no superconvergence points while for odd p ($p = 5, 7, 9, \dots$) there are three points, the point at the center of the element and the midpoints of the edges which are parallel to the $\tilde{x}_1$ axis. The coordinates of these points can be found in [Babuška, Strouboulis *et al.*, 1996].

We have been interested in the existence of the 'direct' superconvergence points. Although in the next section we will create superconvergence by averaging, let us comment here on the 'simple' averaging of the derivatives from both sides of the edge. For example, there is no 'direct' superconvergence point for $\mathcal{F}(u) = \partial u/\partial x_1$ on the vertical sides and the diagonal of the cell shown in Fig. 4.7*.8. Using simple averaging we will obtain the superconvergent points in the middle of the vertical line and the diagonal.

Let us now present the idea of the symmetry theory from [Schatz, Sloan, and Wahlbin, 1996] to find the superconvergent points.

A point x_{symm} is a symmetry point if the finite element space S^p_Δ is locally invariant under the reflection through x_{symm} (i.e., it is polary symmetric). For example, for the criss-cross pattern shown in Fig. 4.7*.1 the symmetry points are the vertices as well as the midpoints of the vertical and horizontal sides of the cell. Then it is easy to see that for p odd the function $\tilde{\Psi}^{PER}_{\tilde{S}^p_\Delta(i,j)}$ is symmetric for $i+j=p$ because all monomials of degree $p+1$ are polary symmetric. Hence $\partial/\partial x_i \tilde{\Psi}_{S^p_\Delta(i,j)}$ is odd at the symmetry points. From this immediately follows that the simple averages are superconvergent (see problem P4.7*.3). The results in [Schatz, Sloan, and Wahlbin, 1996] deal with general meshes which are locally polary symmetric.

The symmetry principle leads to the simple description of the superconvergent points (with simple averaging). Nevertheless, it does not characterize all points, e.g., when elements are quadrilateral or we analyze the superconvergence for the solution satisfying the partial differential equation, e.g., $\Delta u = 0$. On the other hand, our approach describes completely the set of superconvergence points, either direct ones or ones based on simple averaging. It is interesting to note that for the patterns shown in Fig. 4.7*.3, *all* the superconvergence points for the Poisson equation are symmetry points.

Another idea to obtain the superconvergence points is to use difference quotients on translation invariant meshes [Nitsche and Schatz, 1974] and [Schatz and Wahlbin, 1977]. Here if the mesh is translation invariant and when the difference quotient approximates the derivative with the accuracy h^{p+1} for any smooth function, then the difference quotient leads to the superconvergence point. Intuitively we assume that the finite element solution is exact at the nodal points.

Remark 4.7*.8 The theory presented here can be interpreted as the computer proof of the existence of the superconvergent points which are (philosophically) similar to the four-color problem [Appel and Haken, 1978].

As we have seen above, superconvergence points do not exist in general. For example, in the case of the Union-Jack mesh pattern shown in Fig. 4.7*.3c the superconvergence points $\tilde{\mathbf{x}}^{sup}_{\partial u/\partial x_k}$, for either $k=1$ or $k=2$, do not exist even in the case of linear elements ($p=1$). We will now give a more general definition of superconvergence points which has meaning even in the cases that no classical superconvergence points exist.

A point $\mathbf{x}^{sup,\eta\%}_{\mathcal{F}(u)}$ *is a point of* $\eta\%$ *superconvergence for* $\mathcal{F}(u)$ *in the element* τ, *if*

$$\left|\mathcal{F}(e_{S^p_\Delta})\left(\mathbf{x}^{sup,\eta\%}_{\mathcal{F}(u)}\right)\right| \le \left(1-\frac{\eta}{100}\right)\|\mathcal{F}(e_{S^p_\Delta})\|_{L^\infty(\tau)}. \tag{4.7*.21}$$

Note that *a 100%-superconvergence point is a superconvergence point in the classical sense.*

We will also introduce the $\eta\%$-superconvergence regions

$$\tilde{\mathscr{G}}^{\eta\%}_{\mathcal{F}(u)}(\mathcal{Q}^{(p+1)};\tilde{\tau},\tilde{\Delta}) \stackrel{\text{def}}{=} \left\{\tilde{\mathbf{x}}\in\tilde{\tau} \mid \mathcal{H}_{\mathcal{F}(u)}(\mathcal{Q}^{(p+1)};\tilde{\mathbf{x}},\tilde{\tau},\tilde{\Delta}) < 1-\frac{\eta}{100}\right\} \tag{4.7*.22a}$$

where

$$\mathcal{H}_{\mathcal{F}(u)}(\mathcal{Q}^{(p+1)};\bar{\mathbf{x}},\tilde{\tau},\tilde{\Delta}) \stackrel{\text{def}}{=} \max_{\alpha_l}\left(\frac{\left|\sum_{l=1}^{nd}\alpha_l\mathcal{F}(\Psi^{PER}_{\tilde{S}^p_\Delta,l})(\bar{\mathbf{x}})\right|}{\left\|\sum_{l=1}^{nd}\alpha_l\mathcal{F}(\Psi^{PER}_{\tilde{S}^p_\Delta,l})\right\|_{L^\infty(\tilde{\tau})}}\right). \tag{4.7*.22b}$$

The superconvergence regions $\tilde{\mathcal{G}}^{\eta\%}_{\mathcal{F}(u)}(\mathcal{Q}^{(p+1)};\tilde{\tau},\tilde{\Delta})$ can be approximated by

$$\tilde{\mathcal{G}}^{\eta\%,\Xi}_{\mathcal{F}(u)}(\mathcal{Q}^{(p+1)};\tilde{\tau},\tilde{\Delta}) = \left\{\tilde{\mathbf{x}} \in \tilde{\tau} \;\middle|\; \tilde{\mathcal{H}}^{\Xi}_{\mathcal{F}(u)}(\mathcal{Q}^{(p+1)};\tilde{\mathbf{x}},\tilde{\tau},\tilde{\Delta}) \le 1 - \frac{\eta}{100}\right\} \qquad (4.7^*.23a)$$

where

$$\tilde{\mathcal{H}}^{\Xi}_{\mathcal{F}(u)}(\mathcal{Q}^{(p+1)};\tilde{\boldsymbol{\xi}}_k,\tilde{\tau},\tilde{\Delta}) = \max_{\alpha_i}\left(\frac{\left|\sum_{i=1}^{nd}\alpha_i\mathcal{F}(\tilde{\Psi}^{PER}_{\tilde{S}^p_{\tilde{\Delta}},i})(\tilde{\boldsymbol{\xi}}_k)\right|}{\max_{\tilde{\xi}_l\in\Xi}\left|\sum_{i=1}^{nd}\alpha_i\mathcal{F}(\tilde{\Psi}^{PER}_{\tilde{S}^p_{\tilde{\Delta}},i})(\tilde{\boldsymbol{\xi}}_l)\right|}\right) \qquad (4.7^*.23b)$$

and $\Xi = \{\tilde{\boldsymbol{\xi}}_k, k = 1,\ldots,nv\}$ is the set of vertices of the uniform background mesh in the element $\tilde{\tau}$. The vertex values $\tilde{\mathcal{H}}^{\Xi}_{\mathcal{F}(u)}(\mathcal{Q},\tilde{\boldsymbol{\xi}}_k,\tilde{\tau},\tilde{\Delta})$ are determined using numerical optimization. The set $\mathcal{Q}^{(p+1)}$ describes the set of solutions under consideration, e.g., only the set of harmonic polynomials of degree $p+1$, and nd is the dimension of $\mathcal{Q}^{(p+1)}$. The term 'η% superconvergence regions' thus refers to regions for the *worst* possible solution in the set $\mathcal{Q}^{(p+1)}$. Note that each evaluation of the ratio in (4.7*.23b) requires the determination of the maximum of the denominator over all points in Ξ, and is rather time consuming, especially in the case of the general class of solutions and for large p. We will call the above the *direct approach* for approximating the superconvergence regions.

A simpler approach is to employ the regions $\tilde{\tilde{\mathcal{G}}}^{\eta\%,\Xi}(\mathcal{Q}^{(p+1)};\tilde{\tau},\tilde{\Delta})$, corresponding to

$$\tilde{\tilde{\mathcal{H}}}^{\Xi}_{\mathcal{F}(u)}(\mathcal{Q}^{(p+1)};\tilde{\mathbf{x}},\tilde{\tau},\tilde{\Delta}) \stackrel{\text{def}}{=} \frac{\sqrt{\sum_{i=1}^{nd}\left(\mathcal{F}(\tilde{\Psi}^{PER}_{\tilde{S}^p_{\tilde{\Delta}},i})(\tilde{\mathbf{x}})\right)^2}}{V(\mathcal{Q}^{(p+1)};\tilde{\tau},\tilde{\Delta})} \qquad (4.7^*.24a)$$

where

$$V(\mathcal{Q}^{(p+1)};\tilde{\tau},\tilde{\Delta}) \stackrel{\text{def}}{=} \min_{\alpha_i}\left(\frac{\max_{\tilde{\xi}_l\in\Xi}\left|\sum_{i=1}^{nd}\alpha_i\mathcal{F}(\tilde{\Psi}^{PER}_{\tilde{S}^p_{\tilde{\Delta}},i})(\tilde{\boldsymbol{\xi}}_l)\right|}{\sqrt{\sum_{i=1}^{nd}\alpha_i^2}}\right). \qquad (4.7^*.24b)$$

Note that $V(\mathcal{Q}^{(p+1)};\tilde{\tau},\tilde{\Delta})$ needs to be determined only once. We will call this approach the *simplified approach* for determining the superconvergence regions.

Note that

$$\tilde{\mathcal{H}}^{\Xi}_{\mathcal{F}(u)}(\mathcal{Q}^{(p+1)};\tilde{\mathbf{x}},\tilde{\tau},\tilde{\Delta}) \le \tilde{\tilde{\mathcal{H}}}^{\Xi}_{\mathcal{F}(u)}(\mathcal{Q}^{(p+1)};\tilde{\mathbf{x}},\tilde{\tau},\tilde{\Delta}) \qquad (4.7^*.25a)$$

for any $\tilde{\mathbf{x}} \in \tilde{\tau}$. This can be proven for each vertex $\tilde{\boldsymbol{\xi}}_k \in \Xi$ by using the Cauchy–Schwarz inequality; assuming that $\tilde{\mathcal{H}}^{\Xi}_{\mathcal{F}(u)}$ and $\tilde{\tilde{\mathcal{H}}}^{\Xi}_{\mathcal{F}(u)}$ are constructed by employing linear interpolation in each subtriangle of the background mesh, it also holds for any point. It follows that

$$\tilde{\tilde{\mathcal{G}}}^{\eta\%}_{\mathcal{F}(u)}(\mathcal{Q}^{(p+1)};\tilde{\tau},\tilde{\Delta}) \subseteq \tilde{\mathcal{G}}^{\eta\%}_{\mathcal{F}(u)}(\mathcal{Q}^{(p+1)};\tilde{\tau},\tilde{\Delta}) \qquad (4.7^*.25b)$$

and hence the regions $\tilde{\tilde{\mathcal{G}}}^{\eta\%}_{\mathcal{F}(u)}(\mathcal{Q}^{(p+1)};\tilde{\tau},\tilde{\Delta})$ determined by the simplified approach are smaller than the regions $\tilde{\mathcal{G}}^{\eta\%}_{\mathcal{F}(u)}(\mathcal{Q}^{(p+1)};\tilde{\tau},\tilde{\Delta})$ determined by the direct approach; this can be seen in the example shown in Fig. 4.7*.10.

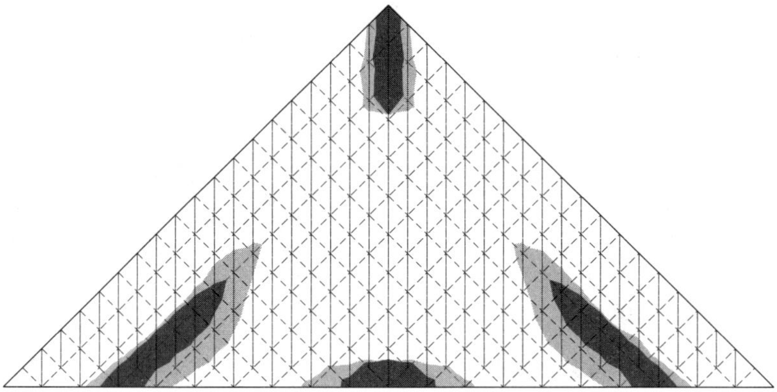

FIG. 4.7*.10. An example of the approximate determination of the $\eta\%$-superconvergence regions: Cubic elements ($p = 3$), element τ_1 of the criss-cross pattern, background mesh with $np = 66$ nodes. The regions $\tilde{\mathcal{G}}^{80\%}_{\partial u/\partial x_1}(\mathfrak{Q}^{(4)}_{hom}; \tilde{\tau}_1, \tilde{T})$ obtained by the direct approach, are shown shaded gray, and the regions $\tilde{\tilde{\mathcal{G}}}^{80\%}_{\partial u/\partial x_1}(\mathfrak{Q}^{(4)}_{hom}; \tilde{\tau}_1, \tilde{T})$ obtained by the simplified approach are shown shaded dark gray. Note that the regions determined by the simplified approach are smaller than the regions determined by the direct approach

Remark 4.7*.9 Note that while the $\eta\%$ regions obtained from both the simplified and direct approaches are rather close for the Laplacian, this is not the case for the Poisson equation where the regions determined by the simplified approach are very small or nonexistent.

Let us now give examples of $\eta\%$-superconvergence regions for the gradient for the four basic patterns of triangles; see also [Babuška, Strouboulis *et al.*, 1994].

Example 4.7*.5 ($\eta\%$-*superconvergence regions for triangles*) Figure 4.7*.11 gives the regions of 90%, 80%, and 70% superconvergence for the x_1 derivative of finite element solutions of the Laplace equation and cubic triangles in the regular, criss-cross, Union-Jack, and the perturbed Union-Jack patterns with the central node, placed at (0.2, 0.2) instead of at (0, 0). It can be seen that the 90%-superconvergence regions are of significant size for all the mesh patterns. Comparing Figs 4.7*.11c,d, it can be seen that the $\eta\%$-superconvergence regions are not very sensitive to perturbations of the mesh.

Figure 4.7*.12 gives the regions of simultaneous superconvergence for the x_1 and x_2 components of the gradient of finite element solutions of the Laplacian for the regular pattern and $p = 1, 2, 3$, and 4. Identical regions can also be obtained for the chevron and Union-Jack pattern; see, for example, Figs 4.7*.11a–c where the correspondence of the regions for the three patterns is clearly seen. Note that for $p = 1$ the centroid is a 66.66%-superconvergence point for both components of the gradient. It can be seen that the size of the $\eta\%$-superconvergence regions for the Laplacian, for fixed η, increases with the degree p of the elements. In contrast, in the case of the Poisson equation, the $\eta\%$-superconvergence regions, for fixed η, disappear as p is increased. Figure 4.7*.13 compares the 90%, 80%, 70% superconvergence regions for $\mathcal{F}(u) = \partial u/\partial x_1$ in the basic cell of the regular pattern for the class of harmonic and the class of general solutions for elements of degree $p = 1, 2, 3, 4$. Note that as p is increased the size of the fixed $\eta\%$

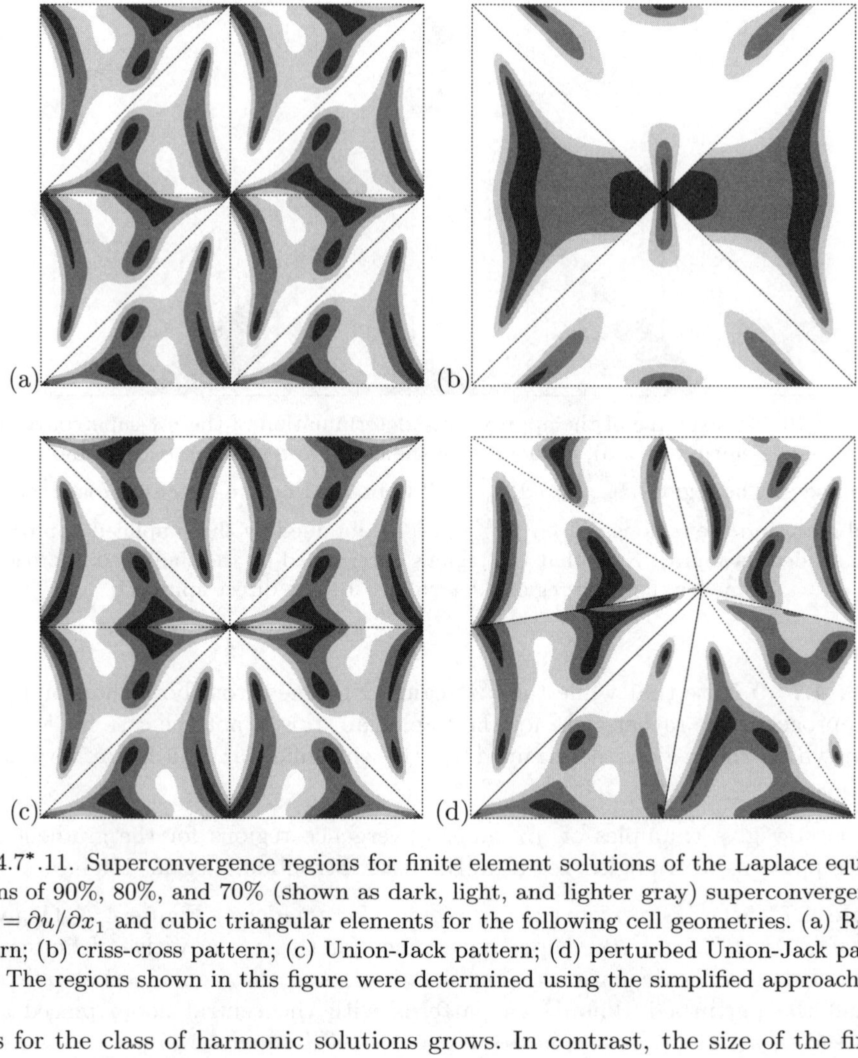

FIG. 4.7*.11. Superconvergence regions for finite element solutions of the Laplace equation. Regions of 90%, 80%, and 70% (shown as dark, light, and lighter gray) superconvergence for $\mathcal{F}(u) = \partial u/\partial x_1$ and cubic triangular elements for the following cell geometries. (a) Regular pattern; (b) criss-cross pattern; (c) Union-Jack pattern; (d) perturbed Union-Jack pattern. The regions shown in this figure were determined using the simplified approach

regions for the class of harmonic solutions grows. In contrast, the size of the fixed $\eta\%$ regions for the class of general solutions shrinks as p is increased.

Let us also consider the case of serendipity squares.

Example 4.7*.6 (*$\eta\%$-superconvergence regions for the serendipity squares*) Figure 4.7*.14 gives the 90%, 80%, and 70% superconvergence regions for the x_1 derivative of finite element solutions of the Laplace and the Poisson equation for serendipity squares of degree p, $2 \le p \le 5$. Figures 4.7*.14a,b give the regions for the case of the Poisson equation, for $p = 2$ and $p = 3$. Figures 4.7*.14c,d (respectively Figs 4.7*.14e,f) give the regions of $\eta\%$ superconvergence for $\mathcal{F}(u) = \partial u/\partial x_1$ for the Laplace and the Poisson equations, for $p = 4$ (respectively for $p = 5$), respectively. Note that the size of the regions for the Poisson equation shrinks as p is increased. Figure 4.7*.15 shows the 90%, 85%, and 75% regions of simultaneous superconvergence for both components of the gradient for finite element solutions of the Laplace equation and the serendipity squares of degree $p = 2, 3, 4, 5$. Note that, with exception of the case $p = 2$ (and the case $p = 1$, not

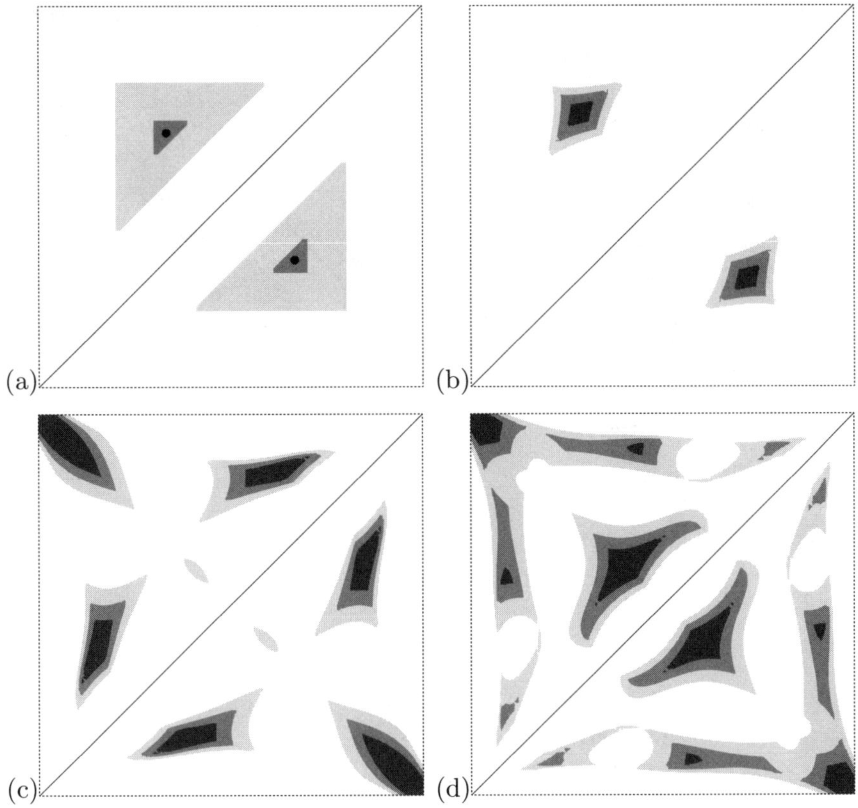

FIG. 4.7*.12. Superconvergence regions for finite element solutions of the Laplace equation. Regions of 85%, 80%, and 75% (shown as dark, light, and lighter gray) simultaneous superconvergence for $\mathcal{F}(u) = \partial u/\partial x_1$ and $\partial u/\partial x_2$ for meshes of triangular elements in the regular pattern. (a) Linear elements ($p = 1$) (in this case we give the 60% and 40% superconvergence regions)—note that the centroid is a 66.66%-superconvergence point; (b) quadratic elements ($p = 2$); (c) cubic elements ($p = 3$); (d) quartic elements ($p = 4$). The regions shown in this figure were determined using the simplified approach. Note that as the degree p of the elements increases, the size of the regions of η% superconvergence, for fixed η, also increases

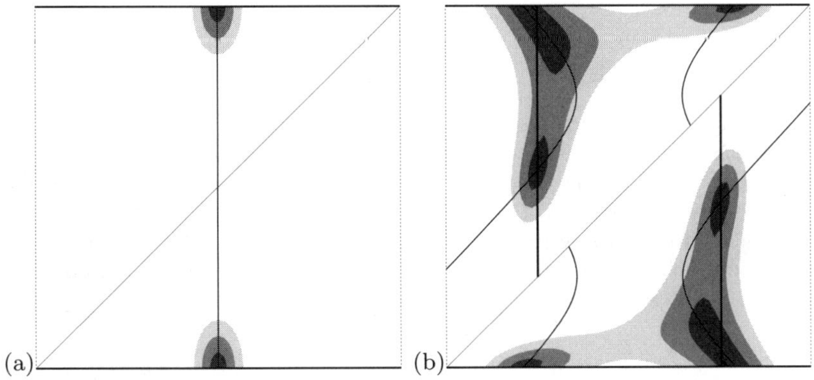

FIG. 4.7*.13.

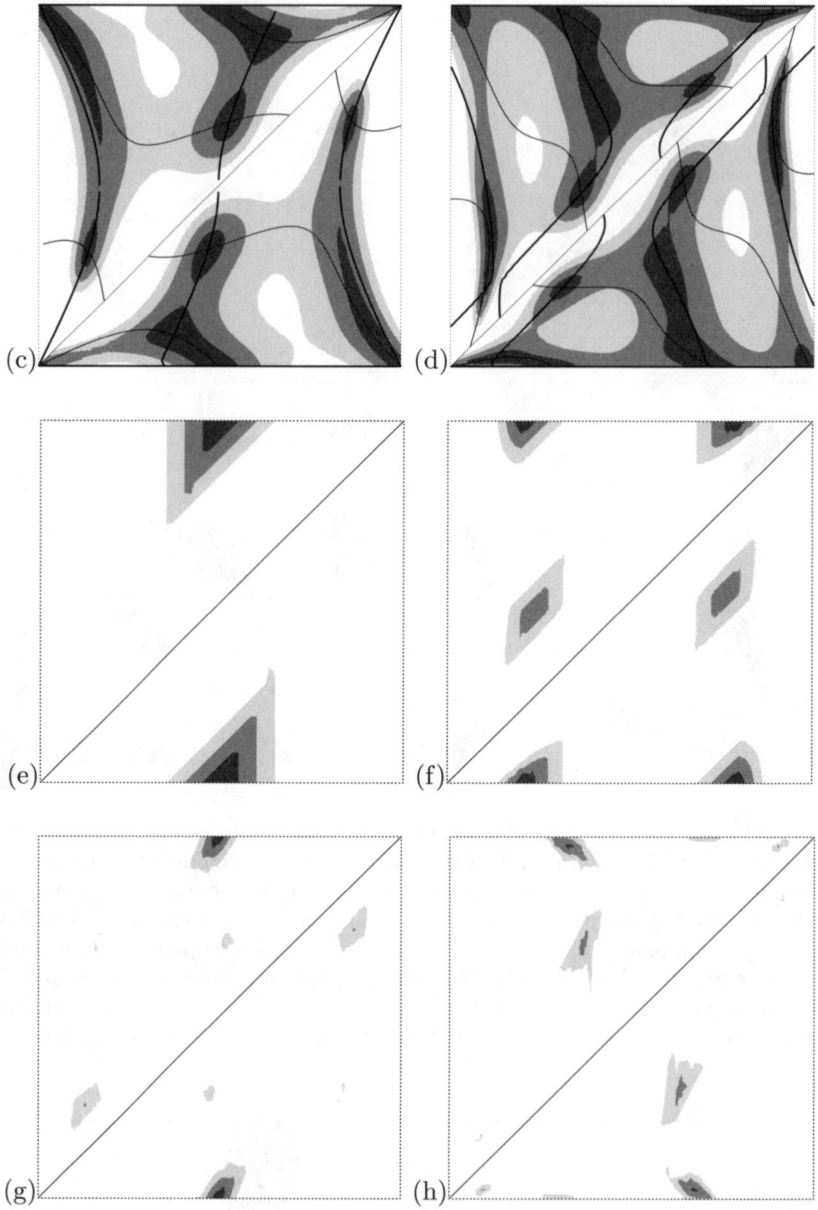

FIG. 4.7*.13. Superconvergence regions for finite element solutions of the Laplace and the Poisson equations. Regions of 90%, 80%, and 70% (shown as dark, light, and lighter gray) superconvergence for $\mathcal{F}(u) = \partial u/\partial x_1$ in the regular pattern for the class of harmonic solutions and element of degree (a) $p = 1$, (b) $p = 2$, (c) $p = 3$, (d) $p = 4$; and the class of general solutions and elements of degree (e) $p = 1$, (f) $p = 2$, (g) $p = 3$, (h) $p = 4$. In the case of the class of harmonic solutions the superconvergence lines for the two basic harmonic monomials are also shown

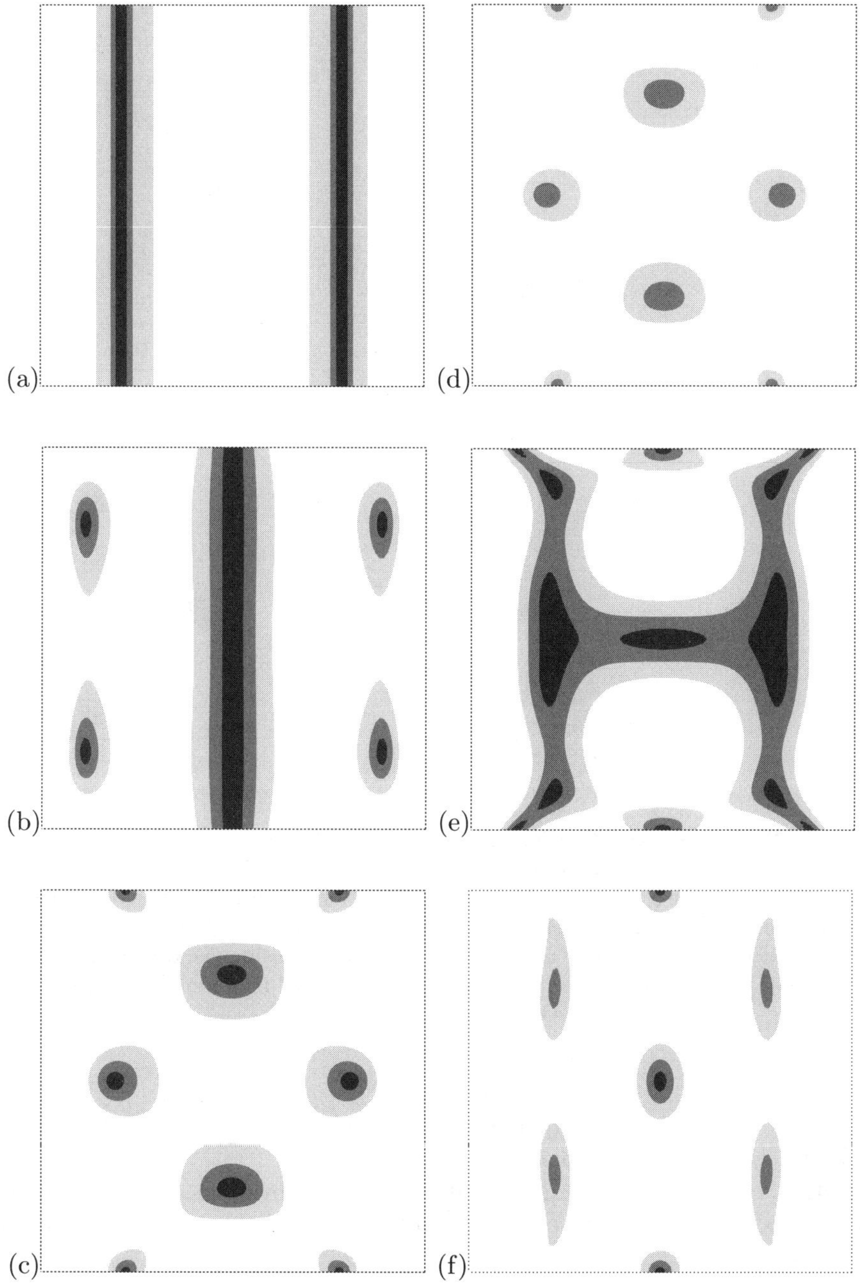

FIG. 4.7*.14. Superconvergence regions for finite element solutions of the Laplace and the Poisson equation. Regions of 90%, 80%, and 70% (shown as dark, light, and lighter gray) superconvergence for $\mathcal{F}(u) = \partial u/\partial x_1$ for square serendipity elements of degree p: (a) $p = 2$, Poisson equation; (b) $p = 3$, Poisson equation; (c) $p = 4$, Laplace equation; (d) $p = 4$, Poisson equation; (e) $p = 5$, Laplace equation; (f) $p = 5$, Poisson equation. The regions shown in this figure were determined using the direct approach

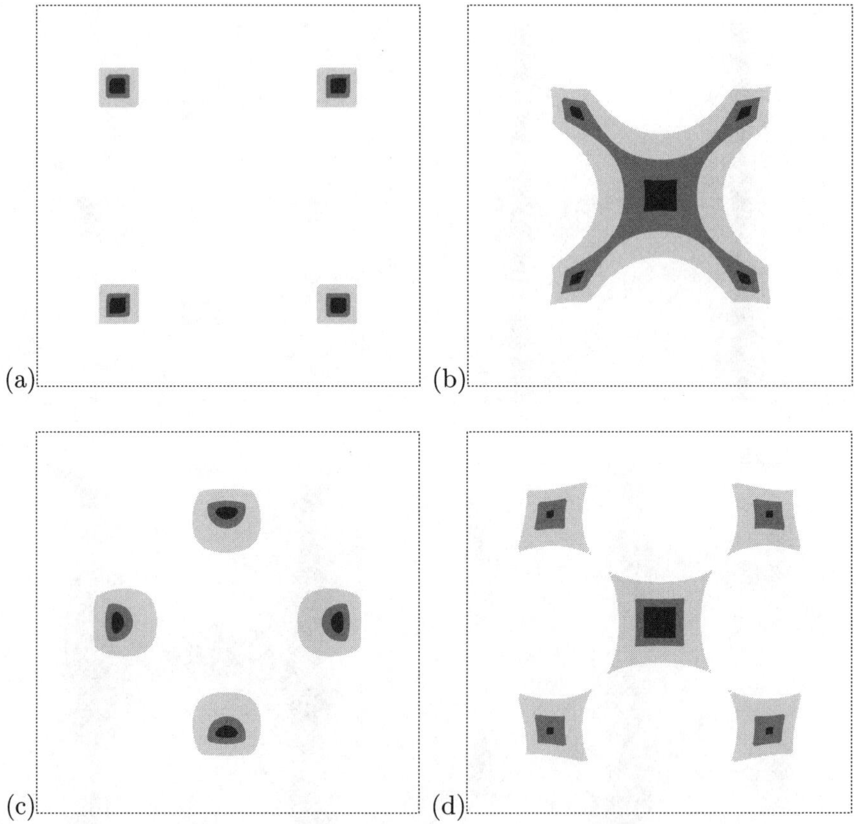

FIG. 4.7*.15. Superconvergence regions for finite element solutions of the Laplace equation. Regions of 90%, 85%, and 75% (shown as dark, light, and lighter gray) simultaneous superconvergence for $\mathcal{F}(u) = \partial u/\partial x_1$ and $\partial u/\partial x_2$ for serendipity squares of degree p. (a) Quadratic elements ($p = 2$); (b) cubic elements ($p = 3$); (c) quartic elements ($p = 4$); (d) quintic elements ($p = 5$). The regions were determined using the direct approach. Note that for $p \geq 3$ the 90%-superconvergence regions are not neighborhoods of the $p \times p$ Gauss–Legendre points

shown here) the regions of 90% superconvergence for the gradient are not neighborhoods of the p by p Gauss–Legendre points.

Remark 4.7*.10 Above we discussed only the superconvergence of the gradient. The superconvergence of any other solution quantity, e.g., the value, can be analyzed in exactly the same way.

We have given some examples of the superconvergence contours and the η%-superconvergence regions for the triangular and quadrilateral elements. We will now give the complete set of the results about the superconvergence contours and the η%-superconvergence regions for the linear and quadratic triangular elements.

Example 4.7*.7 (*Illustration of the superconvergence results for the four patterns of meshes of triangles, and elements of degree* $p = 1$ *and* $p = 2$)

Figure 4.7*.16 through Fig. 4.7*.19 show the superconvergence lines and the η%-superconvergence regions for the class of harmonic solutions. In the figures, the thin lines and the thick lines are the superconvergence contours $C_{\mathcal{F}(u)}(\tilde{Q}_l^{(p+1)}; \tilde{\tau}, \tilde{\Delta})$, which correspond to the two harmonic exact solutions $\tilde{Q}_1^{(p+1)} = \mathrm{Re}(z^{(p+1)})$, $\tilde{Q}_2^{(p+1)} = \mathrm{Im}(z^{(p+1)})$, respectively.

Figures 4.7*.20 and 4.7*.21 present the superconvergence results for the general polynomial exact solutions. In the figures, the lines (from the thinnest to the thickest) are

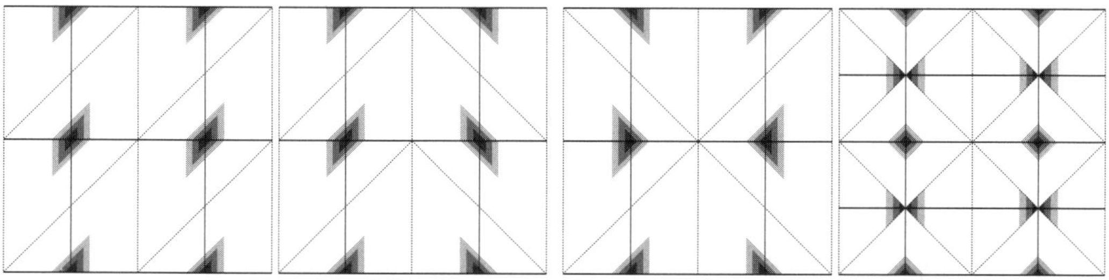

FIG. 4.7*.16. Harmonic exact solution, linear element ($p = 1$). Superconvergence contours and η%-superconvergence regions (90%-dark, 80%-gray, 70%-light gray) for $\mathcal{F}(u) = \partial u/\partial x_1$ for the four patterns of the mesh

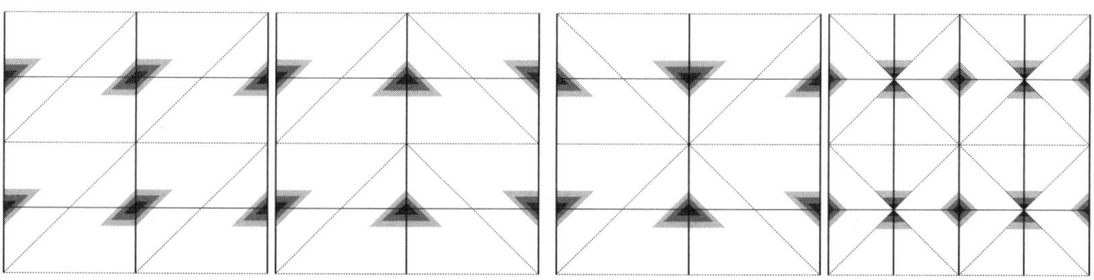

FIG. 4.7*.17. Harmonic exact solution, linear element ($p = 1$). Superconvergence contours and η%-superconvergence regions (90%-dark, 80%-gray, 70%-light gray) for $\mathcal{F}(u) = \partial u/\partial x_2$ for the four patterns of the mesh

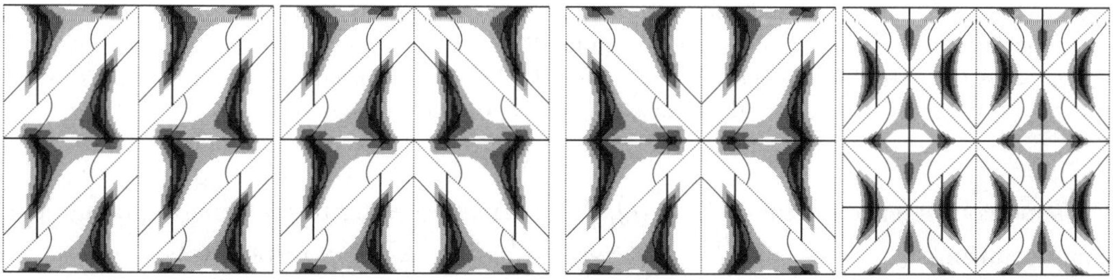

FIG. 4.7*.18. Harmonic exact solution, quadratic element ($p = 2$). Superconvergence contours and η%-superconvergence regions (90%-dark, 80%-gray, 70%-light gray) for $\mathcal{F}(u) = \partial u/\partial x_1$ for the four patterns of the mesh

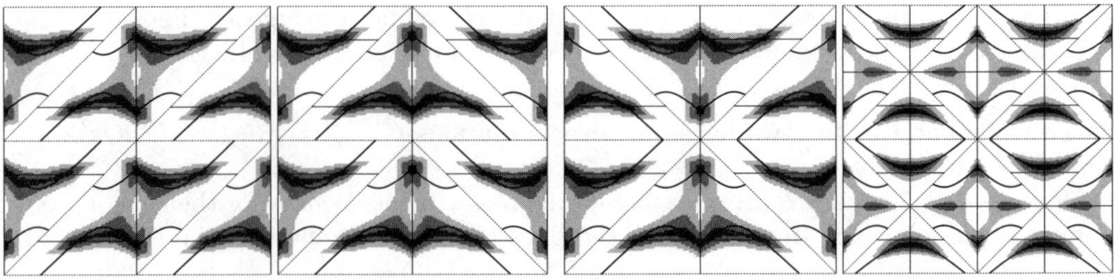

FIG. 4.7*.19. Harmonic exact solution, quadratic element ($p = 2$). Superconvergence contours and η%-superconvergence regions (90%-dark, 80%-gray, 70%-light gray) for $\mathcal{F}(u) = \partial u/\partial x_2$ for the four patterns of the mesh

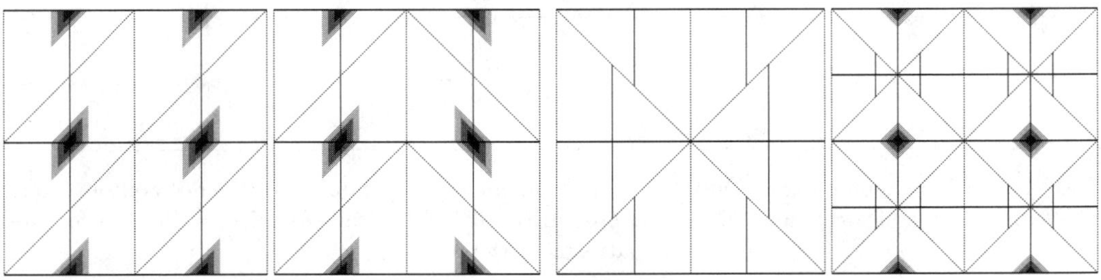

FIG. 4.7*.20. General polynomial exact solution, linear element ($p = 1$). Superconvergence contours and η%-superconvergence regions (90%-dark, 80%-gray, 70%-light gray) for $\mathcal{F}(u) = \partial u/\partial x_1$ for the four patterns of the mesh. The zero contours $C_{\partial u/\partial x_1}(\tilde{Q}_3^{(2)}; \tilde{\tau}, \tilde{\Delta})$ are not plotted because the corresponding error is zero

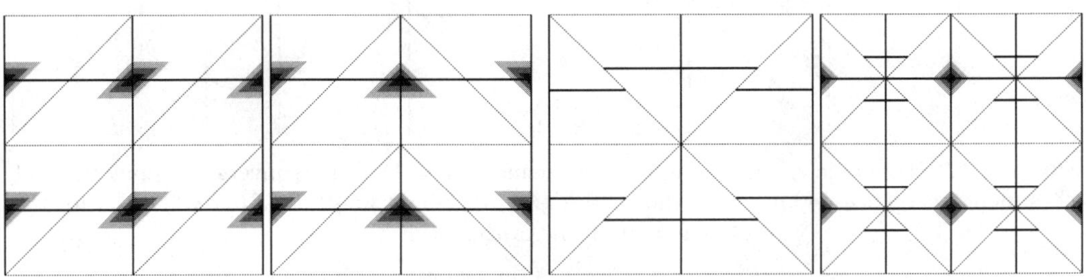

FIG. 4.7*.21. General polynomial exact solution, linear element ($p = 1$). Superconvergence contours and η%-superconvergence regions (90%-dark, 80%-gray, 70%-light gray) for $\mathcal{F}(u) = \partial u/\partial x_2$ for the four patterns of the mesh. The zero contours $C_{\partial u/\partial x_2}(\tilde{Q}_1^{(2)}; \tilde{\tau}, \tilde{\Delta})$ are not plotted because the corresponding error is zero

the superconvergence contours $C_{\mathcal{F}(u)}(\tilde{Q}_l^{(p+1)}; \tilde{\tau}, \tilde{\Delta})$, according to the general polynomial exact solutions $\tilde{Q}_l^{(p+1)}$ (from $x_1^{(p+1)}$, $x_1^p x_2$, $x_1^{(p-1)} x_2^2$, to $x_2^{(p+1)}$), respectively.

Let us also give examples which illustrate the robustness of the η%-superconvergence regions with respect to the class of solutions, the distortion and the local geometry of the mesh.

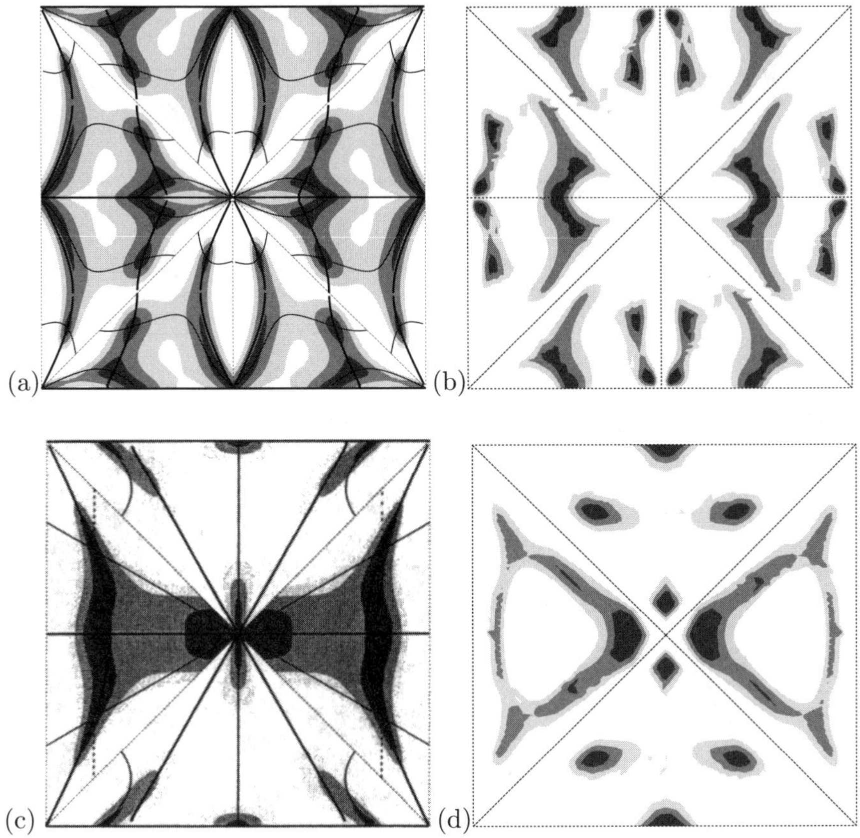

FIG. 4.7*.22. Robustness of the $\eta\%$-superconvergence regions with respect to the class of solutions. The $\eta\%$ regions for the Union-Jack and the criss-cross pattern and cubic elements ($p = 3$) for (a) and (c) the class of harmonic solutions with $\eta = 90, 80, 70$ (the regions were determined by the simplified approach); (b) and (d) the class of general solutions with $\eta = 70, 60, 50$ (the regions were determined by the direct approach)

Example 4.7*.8 (*Robustness of the $\eta\%$-superconvergence regions*)

(a) *Effect of the class of solutions*: As we have seen in Fig. 4.7*.13 the $\eta\%$-superconvergence regions for the Poisson equation are smaller than the corresponding regions for the Laplace equation. Nevertheless, for fixed p the $\eta\%$-superconvergence regions always exist for a sufficiently small η, as it can be seen from Fig. 4.7*.22.

(b) *Effect of mesh distortion*: Figures 4.7*.23a,b show the $\eta\%$-superconvergence regions $\tilde{\tilde{\mathscr{G}}}^{\eta\%}_{\partial u/\partial x_1}(\mathcal{Q}^{(p+1)}_{hom};\tilde{\tau},\tilde{\Delta})$ (η = 90, 80, 70) for the regular pattern with the central node placed, respectively, at (0.5, 0.5) and at (0.4, 0.4). It can be seen that for small distortions of the pattern the regions for $\eta\%$ error (η = 10, 20, 30) are not significantly different from the corresponding regions in the undistorted pattern.

(c) *Effect of material orthotropy*: Given a mesh and an orthotropy matrix $\mathbf{K}$, we will denote the $\eta\%$-superconvergence region $\tilde{\tilde{\mathscr{G}}}^{\eta\%}_{\mathcal{F}(u)}(\mathcal{Q}^{(p+1)}_{\text{“}hom\text{”}};\mathbf{K};\tilde{\tau},\tilde{\Delta})$. By fixing the ratio K_1/K_2 and varying the angle θ of orientation of the principal plane of orthotropy

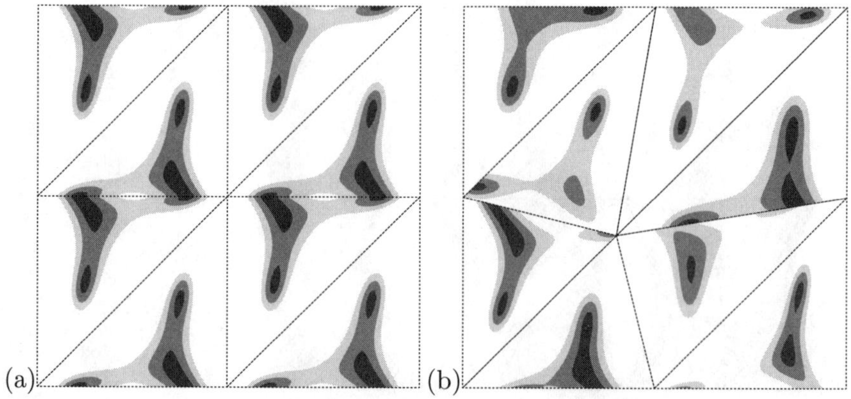

FIG. 4.7*.23. Effect of mesh distortion on the $\eta\%$ regions: Laplace equation, quadratic elements ($p = 2$). The $\eta\% = 90\%, 80\%, 70\%$ regions for (a) regular pattern (with the central node at (0.5,0.5)), (b) distorted regular pattern (with the central node at (0.4,0.4)). The $\eta\%$ regions were determined by the simplified approach

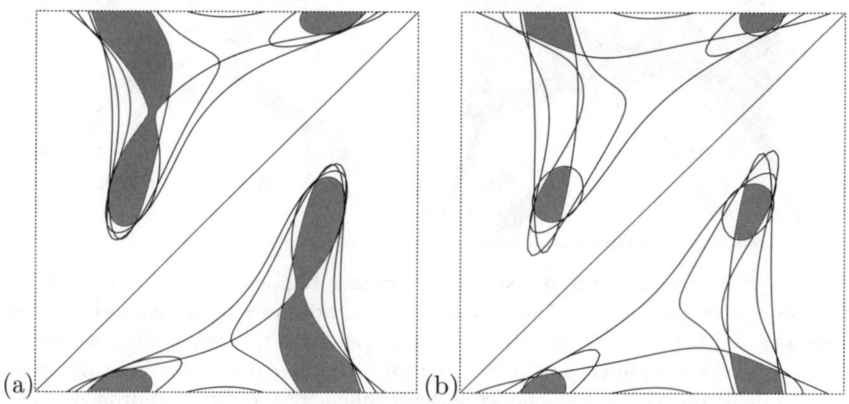

FIG. 4.7*.24. Sensitivity of the $\eta\%$-superconvergence regions to the material orthotropy: Orthotropic heat conduction, cubic 'harmonic' solution, quadratic elements ($p = 2$), regular pattern. The common 75% regions for the x_1 derivative of the error for $\theta = 0°$, $30°$, $60°$, $90°$ when (a) $K_1/K_2 = 2$; (b) $K_1/K_2 = 4$. The $\eta\%$ regions were determined by the simplified approach

with respect to the mesh from 0° to 90°, and computing the intersection of the $\eta\%$-superconvergence regions for all the θs we obtain the common region $\bigcap_{\substack{\theta\in[0°,90°]\\ K_1/K_2=r}} \tilde{\mathscr{G}}^{75\%}_{\partial u/\partial x_1}(\mathfrak{Q}^{(p+1)}_{\text{“}hom\text{”}}; \mathbf{K}; \tilde{\tau}, \tilde{\Delta})$. This is shown in Fig. 4.7*.24a for $K_1/K_2 = 2$, and the regular pattern and it was determined by taking the intersections of the 75%-superconvergence regions for $\theta = 0°$, $30°$, $60°$, $90°$. Fig. 4.7*.24b shows the common 75%-superconvergence region for $K_1/K_2 = 4$. It can be seen that the regions shrink as K_1/K_2 grows.

(d) *Effect of the geometry of the surrounding mesh*: Let us now show that the $\eta\%$-superconvergence regions in an element τ are practically affected by the geometry of the mesh in one or two surrounding layers of elements. We considered the four

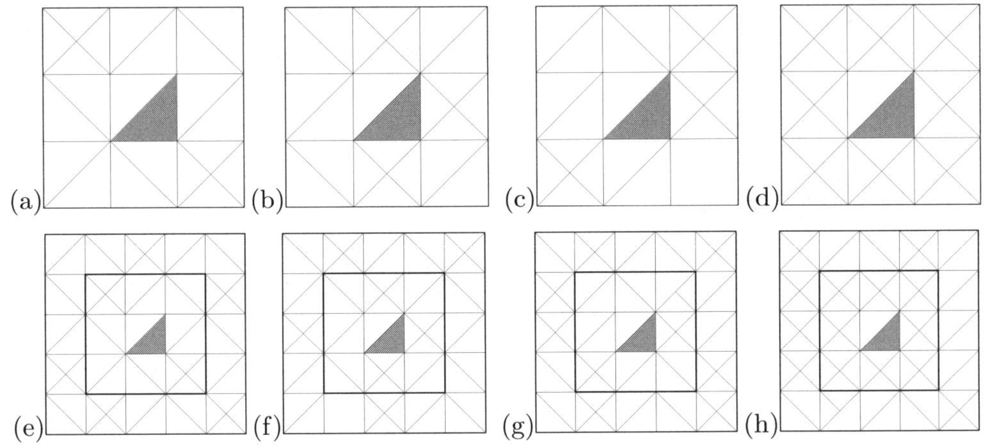

FIG. 4.7*.25. Effect of the surrounding mesh on the $\eta\%$-superconvergence regions: The element τ, shown shaded gray, embedded in the mesh (a) $\hat{\Delta}_1$, (b) $\hat{\Delta}_2$, (c) $\hat{\Delta}_3$, (d) $\hat{\Delta}_4$, (e) $\hat{\Delta}'_1$, (f) $\hat{\Delta}'_2$, (g) $\hat{\Delta}'_3$, (h) $\hat{\Delta}'_4$. Note that the mesh $\hat{\Delta}'_i$ contains the mesh $\hat{\Delta}_i$, $i = 1, 2, 3, 4$

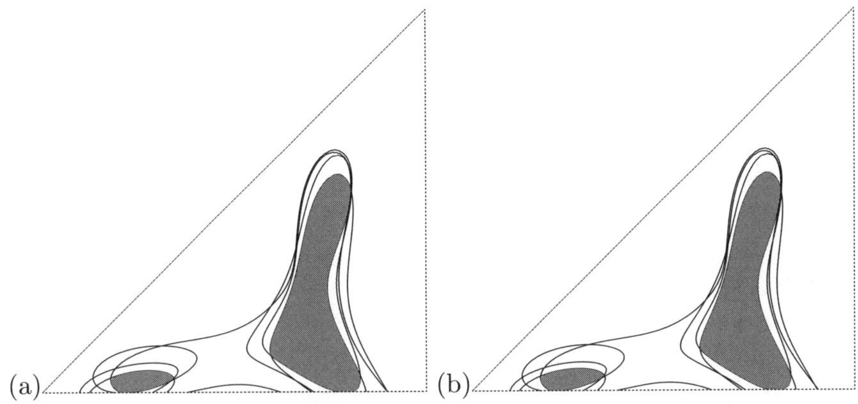

FIG. 4.7*.26. Effect of the surrounding mesh: Laplace equation ($\Delta u = 0$), harmonic solutions, quadratic elements ($p = 2$). The common 75%-superconvergence regions, for $\partial u/\partial x_1$, in the element τ (shown shaded gray in Figs 4.7*.25a–h) for the meshes (a) $\Delta_1 - \Delta_4$, (b) $\hat{\Delta}_1 - \hat{\Delta}_4$. Note that the regions are shown in (a) and (b) are practically identical. The $\eta\%$ regions shown here were determined by the simplified approach

3×3-cell periodic meshes, $\hat{\Delta}_1, \ldots, \hat{\Delta}_4$, shown in Figs 4.7*.25a–d, and determined the 75%-superconvergence regions common to the four meshes, namely, $\cap_{i=1}^{4} \tilde{\tilde{\mathscr{G}}}^{75\%}_{\partial u/\partial x_1}(\mathscr{Q}^{(p+1)}_{hom};$ $\tau, \hat{\Delta}_i)$, for the element τ shown shaded gray in Figs 4.7*.25a–d, and quadratic elements ($p = 2$). The common region is given in Fig. 4.7*.26a; it can be seen that there are large common regions of 75% superconvergence. $\hat{\Delta}'_1, \ldots, \hat{\Delta}'_4$ of Figs 4.7*.25e–h, which include $\hat{\Delta}_1, \ldots, \hat{\Delta}_4$. We also considered the meshes in their interior and determined the common region $\bigcap_{i=1}^{4} \tilde{\tilde{\mathscr{G}}}\hat{\mathscr{G}}^{75\%}_{\partial u/\partial x_1}(\mathscr{Q}^{(p+1)}_{hom}; \tau,\ \hat{\Delta}'_i)$, which is shown in Fig. 4.7*.26b. Comparing Figs 4.7*.26a,b, we see that the regions are essentially the same. Thus, the

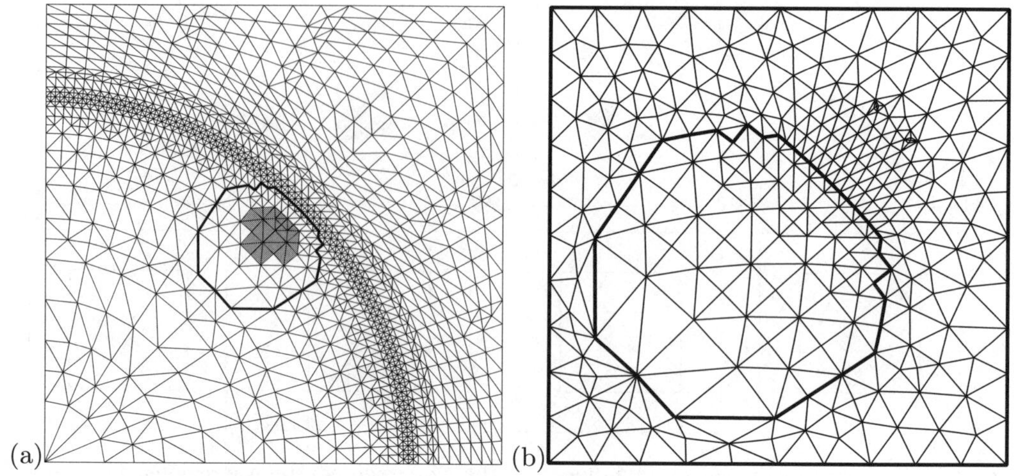

FIG. 4.7*.27. Periodization of the mesh around a mesh cell of interest. The mesh cell of interest, shown shaded gray, with the two neighborhoods of elements embedded in (a) the actual mesh, (b) the periodicized mesh cell used in the determination of the superconvergence regions shown in Fig. 4.7*.28

η%-superconvergence regions in an element are, for all practical purposes, affected only by the geometry of the mesh in the first layer of elements around the element.

Let us now show that the η% regions can be utilized to extract *robust sampling points* in practical meshes.

Example 4.7*.9 (η% *superconvergence in meshes occuring in engineering practice*) Let us be interested in the superconvergence in the mesh cell ω_0^h, shown shaded gray in Fig. 4.7*.27. We determined the η%-superconvergence regions $\tilde{\mathscr{G}}^{\eta\%}_{\partial u/\partial x_1}(\mathfrak{Q}^{(p+1)}_{hom}; \tilde{\tau}, \Delta_h)$ for $\eta = 90, 80, 70$, shown in Fig. 4.7*.28a for $p = 2$, by employing the periodicized mesh cell shown in Fig. 4.7*.27b. To illustrate the usefulness of the η%-superconvergence regions for extracting robust sampling points in practical meshes, we selected several sampling points within the 90% regions, and used them to sample the finite element solution of the Neumann boundary value problem over the mesh shown in Fig. 4.7*.27a, with boundary conditions corresponding to the exact solutions

$$u_1(z) = \frac{1}{1.5^2 - (z+z_0)^2} + \frac{1}{1.5^2 + (z+z_0)^2} - \frac{1}{1.5^2 - 1.0} + \frac{1}{1.5^2 + 1.0} \qquad (4.7^*.26a)$$

where $z_0 = 0.5(1+i), \Omega^1 = (-0.5, 0.5)^2$, and

$$u_2(z) = \Re\mathfrak{e}(\sqrt{z - z_0}) \qquad (4.7^*.26b)$$

where $z_0 = 0.5 + 2.0\,i$, and domain $\Omega^2 = (0,1)^2$. Computing the relative error in the x_1 derivative of the solution at the sampling points, we see that, with the exception of one point, all the points are 90%-superconvergence points in the actual mesh. Hence the superconvergence regions determined from the periodicized mesh cell practically hold for the mesh cell embedded in the actual mesh.

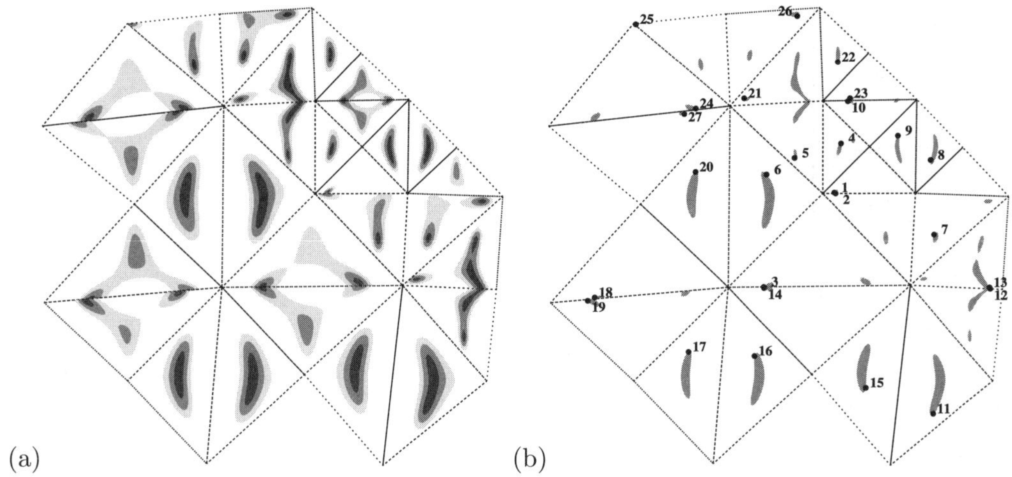

Point	% accuracy		Point	% accuracy		Point	% accuracy		Point	% accuracy	
	u_1	u_2		u_1	u_2		u_1	u_2		u_1	u_2
1	88.06	97.04	8	98.06	93.08	15	93.19	97.18	22	94.58	98.42
2	91.26	99.91	9	92.28	99.12	16	93.68	96.44	23	91.63	99.57
3	95.63	93.53	10	94.97	93.15	17	93.04	99.95	24	92.93	94.81
4	99.42	94.44	11	94.24	93.15	18	91.90	97.86	25	98.41	94.53
5	95.41	98.65	12	94.67	99.29	19	92.44	95.23	26	95.84	98.69
6	99.95	93.96	13	95.69	99.15	20	93.36	99.43	27	92.31	99.51
7	97.91	93.28	14	98.55	92.01	21	96.43	98.92			

(c)

FIG. 4.7*.28. Robustness of the sampling points which are extracted from the η%-superconvergence regions for the meshes employed in practical computations: Laplace equation, quadratic elements ($p = 2$). (a) The η%-superconvergence regions ($\eta = 90$, 80, 70) determined using the periodicized mesh shown in Fig. 4.7*.27b, and (b) sampling points arbitrarily selected from the 90%-superconvergence region for the x_1 derivative of the solution. The η% regions were determined by the simplified approach. (c) The value of the relative error computed using the mesh of Fig. 4.7*.27a at the sampling points. Note that with the exception of point 1 for solution u_1, all other points are 90%-superconvergence points in the actual mesh for both u_1 and u_2

Let us now address the asymptotic analysis for mesh cells adjacent to the boundary (see also [Babuška, Strouboulis *et al.*, 1997]).

Let $\tilde{S} = (0,1) \times (0,\infty)$ be the *master strip* with boundary $\Gamma = \{(\tilde{x}_1, 0) \mid 0 \le \tilde{x}_1 \le 1\}$, as shown in Fig. 4.7*.29, and let the cells be denoted by $\tilde{c}_{0,j}$, $j = 1, 2, 3, \ldots$. The finite element solution $\tilde{u}_{S^p_\Delta}$ in the cell $\tilde{c}_{0,j}$ can be written asymptotically in the form

$$\tilde{u}_{S^p_\Delta} \approx \tilde{U}^{bnd}_{S^p_\Delta,j} = \mathcal{I}^p_\Delta \tilde{Q}^{\mathbf{0},(p+1)}_{EX,j} + \tilde{Z}^{\mathbf{0},p,bnd}_{\tilde{\Delta},PER} \qquad (4.7^*.27\text{a})$$

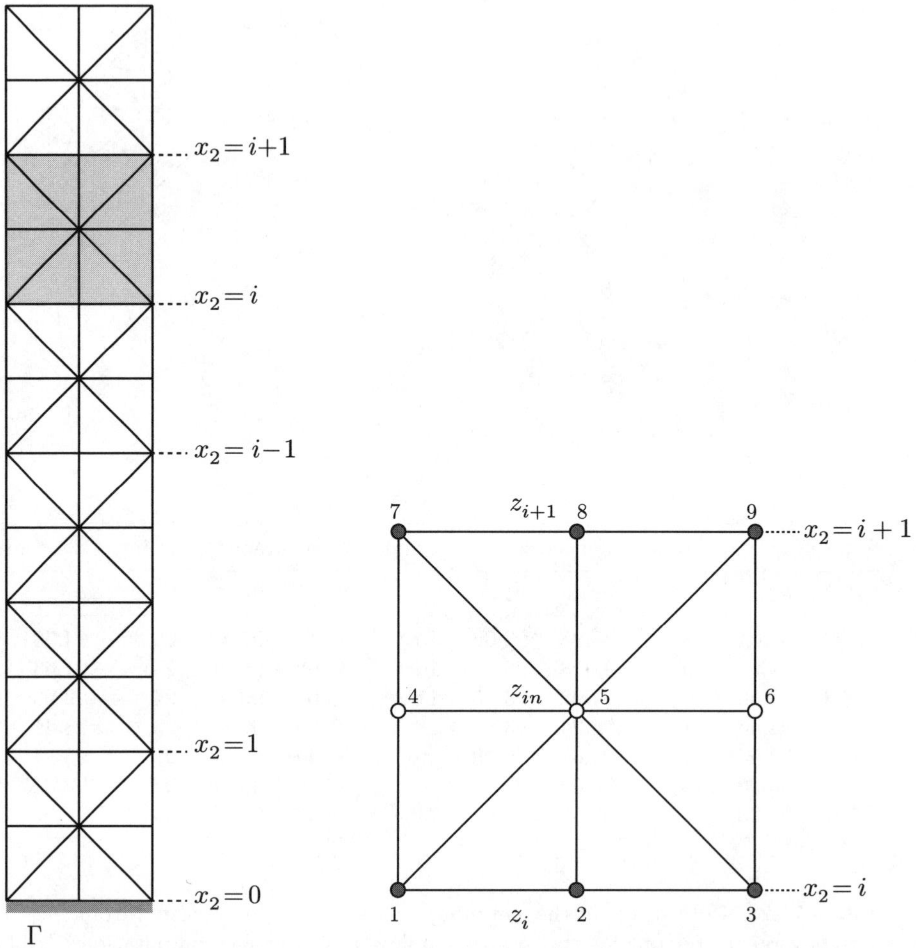

FIG. 4.7*.29. The periodic problem at the boundary. (a) A typical periodic strip formed by the periodic repetition the Union-Jack pattern. (b) The ith cell with the enumeration of the degrees of freedom

where

$$\tilde{Z}^{\mathbf{0},p,bnd}_{\tilde{\Delta},PER} \stackrel{\text{def}}{=} \tilde{Z}^{\mathbf{0},p,interior}_{\tilde{\Delta},PER} + \tilde{Z}^{\mathbf{0},p,BL}_{\tilde{\Delta},PER} \tag{4.7*.27b}$$

and $\tilde{Q}^{\mathbf{0},(p+1)}$ is the polynomial from the Taylor series expansion at the point $(\frac{1}{2}, 0)$ located at the boundary. Here $\tilde{Z}^{\mathbf{0},p,interior}_{\tilde{\Delta},PER}$ satisfies

$$\mathscr{B}_{\tilde{c}}(\tilde{Z}^{\mathbf{0},p,interior}_{\tilde{\Delta},PER}, \tilde{v}) = \mathscr{B}_{\tilde{c}}(\tilde{Q}^{\mathbf{0},(p+1)}_{EX} - \mathcal{I}^p_{\tilde{\Delta}} \tilde{Q}^{\mathbf{0},(p+1)}_{EX}, \tilde{v}) \quad \forall \tilde{v} \in \tilde{S}^p_{\tilde{\Delta},PER}(\tilde{c}) \tag{4.7*.28a}$$

with

$$\int_{\tilde{c}} \tilde{Z}^{\mathbf{0},p,interior}_{\tilde{\Delta},PER} = \int_{\tilde{c}} (\tilde{Q}^{\mathbf{0},(p+1)}_{EX} - \mathcal{I}^p_{\tilde{\Delta}} \tilde{Q}^{\mathbf{0},(p+1)}_{EX}). \tag{4.7*.28b}$$

Note that $\tilde{Z}^{\mathbf{0},p,interior}_{\tilde{\Delta},PER}$ is the periodic extension of the function $\tilde{Z}^{\mathbf{0},p}_{\tilde{\Delta},PER}$ obtained over a master cell in the interior of the mesh, as described above. Also note that $\tilde{Q}^{\mathbf{0},(p+1)}_{EX,j} - \mathcal{I}^p_{\tilde{\Delta}}\tilde{Q}^{\mathbf{0},(p+1)}_{EX,j}$ is independent of j.

The function $\tilde{Z}^{\mathbf{0},p,BL}_{\tilde{\Delta},PER}$ is the correction which must be added to $\tilde{Z}^{\mathbf{0},p,interior}_{\tilde{\Delta},PER}$ in order to ensure that $\tilde{Z}^{\mathbf{0},p,bnd}_{\tilde{\Delta},PER}$ satisfies the boundary condition (Dirichlet or Neumann) along the edge Γ. In the case of the homogeneous Dirichlet boundary condition $\Gamma \equiv \Gamma_{\mathrm{D}}$, $\tilde{Z}^{\mathbf{0},p,BL}_{\tilde{\Delta},PER}$ is the solution of the discrete problem:

Find $\tilde{Z}^{\mathbf{0},p,BL}_{\tilde{\Delta},PER} \in \tilde{S}^{bnd}_{\tilde{\Delta},\Gamma_{\mathrm{D}},PER}(\tilde{S})$ such that $\tilde{Z}^{\mathbf{0},p,BL}_{\tilde{\Delta},PER} = -\tilde{Z}^{\mathbf{0},p,interior}_{\tilde{\Delta},PER}$ on Γ_{D}, and

$$\mathscr{B}_{\tilde{\Omega}}\big(\tilde{Z}^{\mathbf{0},p,BL}_{\tilde{\Delta},PER}, \tilde{v}\big) = \mathscr{B}_{\tilde{\Omega}}\big(\tilde{Q}^{\mathbf{0},(p+1)}_{EX} - \mathcal{I}^p_{\tilde{\Delta}}\tilde{Q}^{\mathbf{0},(p+1)}_{EX} - \tilde{Z}^{\mathbf{0},p,interior}_{PER}, \tilde{v}\big)$$
$$\forall \tilde{v} \in \tilde{S}^{p,bnd}_{\tilde{\Delta},\Gamma_{\mathrm{D}},PER}(\tilde{S}) \qquad (4.7^*.29\text{a})$$

where $\tilde{S}_{\tilde{\Delta},\Gamma_{\mathrm{D}},PER}(\tilde{S})$ is the finite element space of functions which are periodic in $\tilde{x}_1$ and zero at $\tilde{x}_2 = 0$, and

$$\lim_{\tilde{x}_2\to\infty} \boldsymbol{\nabla}\tilde{Z}^{\mathbf{0},p,BL}_{\tilde{\Delta},PER}(\tilde{x}_1,\tilde{x}_2) = \mathbf{0}. \qquad (4.7^*.29\text{b})$$

Recalling (4.7*.28a) we get that

$$\mathscr{B}_{\tilde{\Omega}}\big(\tilde{Z}^{\mathbf{0},p,BL}_{\tilde{\Delta},PER}, \tilde{v}\big) = 0 \quad \forall \tilde{v} \in \tilde{S}^{p,bnd}_{\tilde{\Delta},\Gamma_{\mathrm{D}},PER}(\tilde{S}). \qquad (4.7^*.30)$$

Therefore, $\tilde{Z}^{\mathbf{0},p,BL}_{\tilde{\Delta},PER}$ *is a solution of the homogeneous discrete partial differential equation in* $\tilde{S}$.

Let us now give an example of the above constructions.

Example 4.7*.10 (*Construction of the asymptotic finite element solution for mesh cells adjacent to the boundary*) We will consider the case of the Union-Jack pattern shown in Fig. 4.7*.29, linear elements ($p = 1$), and $\tilde{Q}^{\mathbf{0},(p+1)}_{EX}(\tilde{x}_1,\tilde{x}_2) = \tilde{x}_2^2$. As we have seen above, the asymptotic finite element solution is obtained in two steps:

(i) Construction of $\tilde{Z}^{interior}_{\tilde{\Delta},PER}$

For any unit cell $\tilde{c}^i$, we have

$$\tilde{Z}^{interior}_{\tilde{\Delta},PER} = \tilde{Z}_{\tilde{\Delta},PER} = \sum_{j=1}^{9} z^{int}_j \tilde{\varphi}_j \qquad (4.7^*.31\text{a})$$

where $\tilde{\varphi}_j$ are the linear hat function in over the cell $\tilde{c}^i$, and, in our case, we have

$$z^{int}_1 = z^{int}_3 = z^{int}_7 = z^{int}_9 = -0.0555,\ z^{int}_2 = z^{int}_8 = -0.01388$$
$$z^{int}_4 = z^{int}_6 = -0.01388,\ z^{int}_5 = -0.0555 \qquad (4.7^*.31\text{b})$$

where the enumeration is shown in Fig. 4.7*.29b.

(ii) Construction of $\tilde{Z}^{BL}_{PER}$

(a) For the ith unit cell $\tilde{c}^i$ obtain the stiffness matrix $\mathbf{K}^i$ corresponding to the degrees of freedom $z^{BL}_{i,j}$, $j = 1, \ldots, 9$, for the boundary-layer solution where $z^{BL}_{i,j}$ denotes the jth degree of freedom in the ith layer, and impose the periodicity of the degrees of freedom

$$z^{BL}_{i,1} = z^{BL}_{i,3}, \qquad z^{BL}_{i,4} = z^{BL}_{i,6}, \qquad z^{BL}_{i,7} = z^{BL}_{i,9} \tag{4.7*.32a}$$

to get the modified stiffness matrix

$$\mathbf{K}^i = \begin{bmatrix} 2 & -1 & -1 & 0 & 0 & 0 \\ -1 & 2 & 0 & -1 & 0 & 0 \\ -1 & 0 & 4 & -2 & -1 & 0 \\ 0 & -1 & -2 & 4 & 0 & -1 \\ 0 & 0 & -1 & 0 & 2 & -1 \\ 0 & 0 & 0 & -1 & -1 & 2 \end{bmatrix} \begin{matrix} z^{BL}_{i,1} \\ z^{BL}_{i,2} \\ z^{BL}_{i,4} \\ z^{BL}_{i,5} \\ z^{BL}_{i,7} \\ z^{BL}_{i,8} \end{matrix} . \tag{4.7*.32b}$$

(b) Eliminate the interior degrees of freedom $\mathbf{z}_{in}$, where $\mathbf{z}_{in} = [z^{BL}_{i,4}, z^{BL}_{i,5}]^T$,

$$\begin{aligned} z^{BL}_{i,4} &= \tfrac{1}{6}(2z^{BL}_{i,1} + z^{BL}_{i,2} + 2z^{BL}_{i,7} + z^{BL}_{i,8}) \\ z^{BL}_{i,5} &= \tfrac{1}{6}(z^{BL}_{i,1} + 2z^{BL}_{i,2} + z^{BL}_{i,7} + 2z^{BL}_{i,8}) \end{aligned} \tag{4.7*.33a}$$

to get the reduced stiffness matrix $\tilde{\mathbf{K}}^i$ corresponding to the vector $[\mathbf{z}_i{}^T, \mathbf{z}_{i+1}{}^T]^T$ with $\mathbf{z}_i = [z^{BL}_{i,1}, z^{BL}_{i,2}]^T$ and $\mathbf{z}_{i+1} = [z^{BL}_{i,7}, z^{BL}_{i,8}]^T$

$$\tilde{\mathbf{K}}^i = \frac{1}{6} \begin{bmatrix} 10 & -7 & -2 & -1 \\ -7 & 10 & -1 & -2 \\ -2 & -1 & 10 & -7 \\ -1 & -2 & -7 & 10 \end{bmatrix} \begin{matrix} z^{BL}_{i,1} \\ z^{BL}_{i,2} \\ z^{BL}_{i+1,1} \\ z^{BL}_{i+1,2} \end{matrix} . \tag{4.7*.33b}$$

(c) Assemble $\tilde{\mathbf{K}}^{i-1}$ and $\tilde{\mathbf{K}}^i$, to obtain

$$[\mathbf{A} \;\; \mathbf{B} \;\; \mathbf{C}] \begin{Bmatrix} \mathbf{z}_{i+1} \\ \mathbf{z}_i \\ \mathbf{z}_{i-1} \end{Bmatrix} = \{\mathbf{0}\} . \tag{4.7*.34a}$$

In our case, we have

$$\mathbf{A} = \mathbf{C} = \begin{bmatrix} -\frac{1}{3} & -\frac{1}{6} \\ -\frac{1}{6} & -\frac{1}{3} \end{bmatrix} \qquad \mathbf{B} = \begin{bmatrix} \frac{10}{3} & -\frac{7}{3} \\ -\frac{7}{3} & \frac{10}{3} \end{bmatrix} \tag{4.7*.34b}$$

with

$$\mathbf{z}_{i+1} = [z^{BL}_{i+1,1}, z^{BL}_{i+1,2}]^T = [z^{BL}_{i,7}, z^{BL}_{i,8}]^T, \qquad \mathbf{z}_{i-1} = [z^{BL}_{i-1,1}, z^{BL}_{i-1,2}]^T . \tag{4.7*.34c}$$

(d) Rewriting (4.7*.34a) to express $[\mathbf{z}_{i+1}{}^T, \mathbf{z}_i{}^T]^T$ in terms of $[\mathbf{z}_i{}^T, \mathbf{z}_{i-1}{}^T]^T$, we get

$$\begin{bmatrix} \mathbf{A} & \mathbf{B} \\ \mathbf{0} & \mathbf{I} \end{bmatrix} \begin{Bmatrix} \mathbf{z}_{i+1} \\ \mathbf{z}_i \end{Bmatrix} = \begin{bmatrix} \mathbf{0} & -\mathbf{C} \\ \mathbf{I} & \mathbf{0} \end{bmatrix} \begin{Bmatrix} \mathbf{z}_i \\ \mathbf{z}_{i-1} \end{Bmatrix}. \tag{4.7*.35}$$

(e) The solution of (4.7*.35) can be written in the form

$$\mathbf{z}_i = \lambda^i \mathbf{v} \tag{4.7*.36a}$$

and we have

$$\lambda \begin{bmatrix} \mathbf{A} & \mathbf{B} \\ \mathbf{0} & \mathbf{I} \end{bmatrix} \begin{Bmatrix} \lambda\mathbf{v} \\ \mathbf{v} \end{Bmatrix} = \begin{bmatrix} \mathbf{0} & -\mathbf{C} \\ \mathbf{I} & \mathbf{0} \end{bmatrix} \begin{Bmatrix} \lambda\mathbf{v} \\ \mathbf{v} \end{Bmatrix}. \tag{4.7*.36b}$$

It follows that λ and $\mathbf{v}$ are obtained by solving the above generalized eigenvalue problem. For our example, the eigenvalues are

$$\lambda_1 = 17 + 12\sqrt{2}, \qquad \lambda_2 = 1, \qquad \lambda_3 = 1, \qquad \lambda_4 = 17 - 12\sqrt{2} \tag{4.7*.37a}$$

with the corresponding eigenvectors

$$\mathbf{v}_1 = \begin{Bmatrix} 1 \\ -1 \end{Bmatrix}, \qquad \mathbf{v}_2 = \begin{Bmatrix} 1 \\ 1 \end{Bmatrix}, \qquad \mathbf{v}_3 = \begin{Bmatrix} 1 \\ 1 \end{Bmatrix}, \qquad \mathbf{v}_4 = \begin{Bmatrix} 1 \\ -1 \end{Bmatrix}. \tag{4.7*.37b}$$

Taking into account that $\mathbf{A} \equiv \mathbf{C}$, from (4.7*.36b) we get

$$-\mathbf{C}\mathbf{v} = \lambda^2 \mathbf{A}\mathbf{v} + \lambda \mathbf{B}\mathbf{v} = \lambda^2 \mathbf{C}\mathbf{v} + \lambda \mathbf{B}\mathbf{v} \tag{4.7*.37c}$$

and letting $\tilde{\lambda} = 1/\lambda$ we obtain

$$-\mathbf{C}\mathbf{v} = \tilde{\lambda}^2 \mathbf{A}\mathbf{v} + \tilde{\lambda} \mathbf{B}\mathbf{v}. \tag{4.7*.37d}$$

Thus if λ is an eigenvalue, then $\tilde{\lambda} = 1/\lambda$ is also an eigenvalue with the same eigenvector $\mathbf{v}$.

(f) Rejecting λ_1 because it would not lead to a solution with finite energy, and noting that $\lambda_2 = \lambda_3 = 1$ is a multiple eigenvalue with $\mathbf{v}_2 = \mathbf{v}_3$, we get

$$\mathbf{z}_i = \begin{Bmatrix} z_{i,1}^{BL} \\ z_{i,2}^{BL} \end{Bmatrix} = \mathscr{C}_1 \lambda_3^i \mathbf{v}_3 + \mathscr{C}_2 \lambda_4^i \mathbf{v}_4, \quad i = 1, 2, \ldots \tag{4.7*.38}$$

where $\mathscr{C}_1$ and $\mathscr{C}_2$ are to be determined from the boundary conditions at $x_2 = 0$.

(g) In the case of homogeneous Dirichlet boundary conditions we have

$$\mathbf{z}_0 = \begin{Bmatrix} z_{0,1}^{BL} \\ z_{0,2}^{BL} \end{Bmatrix} = -\begin{Bmatrix} z_1^{int} \\ z_2^{int} \end{Bmatrix} = \begin{Bmatrix} 0.05555 \\ 0.01388 \end{Bmatrix}. \tag{4.7*.39a}$$

Thus, from (4.7*.35) for $i = 1$, we obtain

$$\begin{bmatrix} \mathbf{A} & \mathbf{B} \end{bmatrix} \begin{Bmatrix} \mathbf{z}_2 \\ \mathbf{z}_1 \end{Bmatrix} = -\mathbf{C}\mathbf{z}_0 = -\mathbf{C} \begin{Bmatrix} 0.05555 \\ 0.01388 \end{Bmatrix} \tag{4.7*.39b}$$

or

$$[(\lambda_3^2\mathbf{A} + \lambda_3\mathbf{B})\mathbf{v}_3 \quad (\lambda_4^2\mathbf{A} + \lambda_4\mathbf{B})\mathbf{v}_4] \begin{Bmatrix} \mathscr{C}_1 \\ \mathscr{C}_2 \end{Bmatrix} = -\mathbf{C}\mathbf{z}_0 = -\mathbf{C} \begin{Bmatrix} 0.05555 \\ 0.01388 \end{Bmatrix} \qquad (4.7^*.39c)$$

and we get

$$\mathscr{C}_1 = 0.0347, \qquad \mathscr{C}_2 = -0.0208. \qquad (4.7^*.39d)$$

Remark 4.7*.11 The value of λ indicates the strength of the boundary layer. In the above example $\lambda = 0.0294$, and the boundary layer of the derivatives is negligible in all the elements except the element adjacent to the boundary. Note that the term $\lambda_3^i \mathbf{v}_3$ is constant, and it does not influence the derivatives.

Let us now give another example where the boundary layer in the finite element solution is significant.

Example 4.7*.11 (*The effect of the boundary layer in the superconvergence in the mesh cells adjacent to the boundary*) Let us consider the finite element approximation of the orthotropic heat conduction equation

$$100\frac{\partial^2 u_{EX}}{\partial x_1^2} + \frac{\partial^2 u_{EX}}{\partial x_2^2} = f \qquad (4.7^*.40a)$$

for the mesh formed by the periodic repetition of the chevron pattern (shown in Fig. 4.7*.3b), let

$$u_{EX}(x_1, x_2) = x_1^2 + x_1x_2 + x_2^2 \qquad (4.7^*.40b)$$

and linear elements ($p = 1$), in the case that Neumann boundary conditions are imposed on the edge $x_2 = 0$. Figure 4.7*.30 gives the graph of the x_2 derivative of the boundary layer in the finite element solution and the x_2 derivative of the exact asymptotic error along the line $x_1 = 0.75$. It can be seen that the magnitude of the x_2 derivative of the boundary layer in the finite element solution $(\partial/\partial x_2)(Z^{1,BL}_{\tilde{\Delta},PER})$, is about 25% of the magnitude of the x_2 derivative of the exact error $(\partial/\partial x_2)(e_{S^1_{\tilde{\Delta}}})$, in the first element, and hence the boundary layer is not negligible. It can also be seen that the x_2 derivative of the boundary layer decays to zero (although slowly) as we move away from the boundary.

Let us now utilize the above constructions to analyze the superconvergence up to the boundary.

Example 4.7*.12 (*η% superconvergence up to the boundary*) Consider the Laplace equation with Neumann boundary condition imposed at the boundary $x_2 = 0$ (the bottom edge of the strip). Figure 4.7*.31 gives the contours of zero error and the superconvergence regions $\tilde{\mathscr{G}}^{\eta\%}_{\partial u/\partial x_1}(\mathscr{Q}^{(p+1)}_{har}; \tau, \Delta_h)$, for $\eta = 90, 80, 70$, for a strip of elements formed with the regular, chevron, Union-Jack, and criss-cross patterns, when quadratic ($p = 2$) elements are employed. The superconvergence points are located at the intersection of the zero-level lines which are also shown in the figure. Note that the influence of the boundary on the location of the superconvergence points near the boundary depends on the cell.

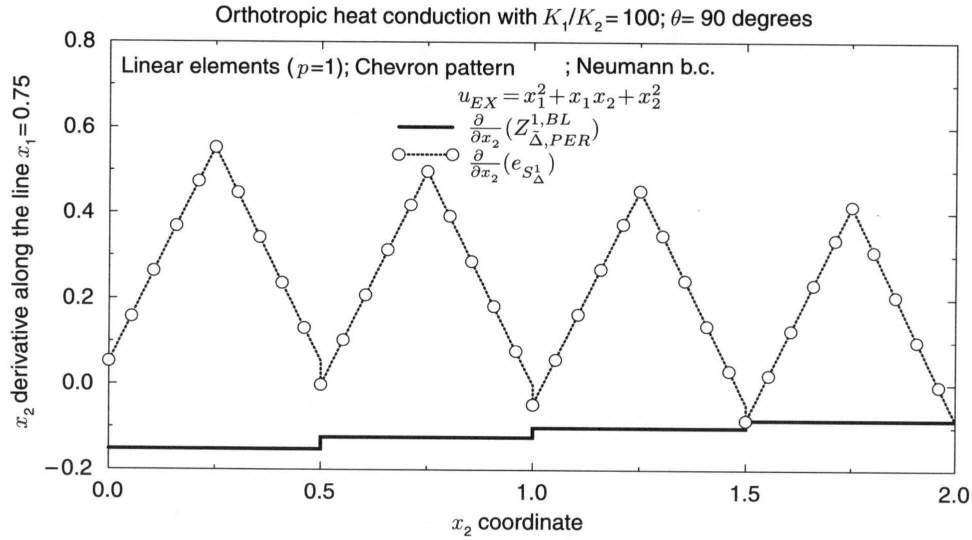

FIG. 4.7*.30. An example which illustrates the boundary layer in the finite element solution: orthotropic heat conduction with $K_1/K_2 = 100$, $\theta = 90°$; $u_{EX}(x_1, x_2) = x_1^2 + x_1x_2 + x_2^2$, Neumann boundary condition on the edge $x_2 = 0$, linear elements ($p = 1$). The graph of the x_2 derivative of the exact asymptotic error, $(\partial/\partial x_2)(e_{S^1_\Delta})$, and the x_2 derivative of the boundary layer in the finite element solution, $(\partial/\partial x_2)(Z^{1,BL}_{\Delta,PER})$, along the line $x_1 = 0.75$, for the uniform mesh formed by the chevron pattern

Example 4.7*.13 (*η% superconvergence up to the the boundary*) Consider the Poisson equation with Neumann boundary condition imposed at the boundary $x_2 = 0$ (the bottom edge of the strip). Figure 4.7*.32 gives the superconvergence regions $\tilde{\mathcal{G}}^{\eta\%}_{\partial u/\partial x_1}(\mathcal{Q}^{(p+1)}_{gen}; \tau, \Delta_h)$, for $\eta = 90, 80, 70$, for a strip of elements formed with the regular, chevron, and criss-cross patterns, when quadratic ($p = 2$) elements are employed.

From Figs 4.7*.31 and 4.7*.32, we note that

(1) The effect of the boundary layer in the finite element solution may be significant. The contours of zero error, the superconvergence points, and the η%-superconvergence regions for the elements near the boundary could be markedly different from those in elements one or two layers away from the boundary.

(2) The contours of zero error and the η%-superconvergence regions converge rather quickly to the corresponding contours and η%-superconvergence regions for an interior cell, as the mesh cell moves away from the boundary.

Remark 4.7*.12 Sometimes in the literature superconvergence at the boundary is discussed only for the energy norm and the Poisson problem, in which case the boundary layer may not be visible. Also the regular pattern of triangles and square elements with bi-p basis are almost exclusively addressed.

Finally, we will address the asymptotics and the superconvergence in the neighborhood of the corner points (see also [Babuška, Strouboulis *et al.*, 2001]).

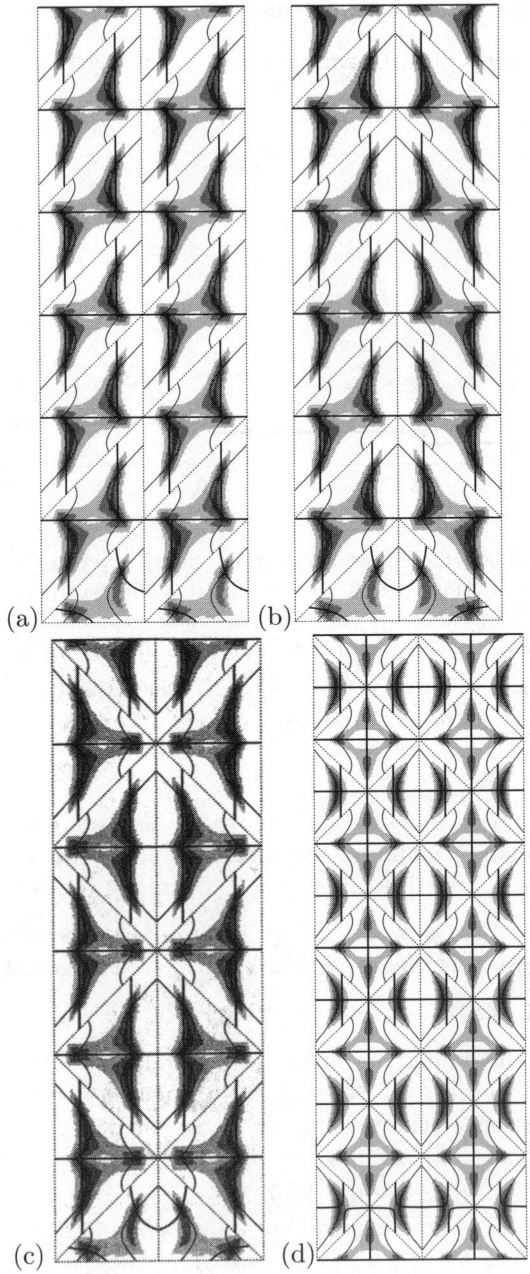

FIG. 4.7*.31. Superconvergence up to the boundary: Laplace equation, harmonic polynomial solution, Neumann boundary conditions along the edge $x_2 = 0$ (the bottom edge of the strip). Quadratic elements ($p = 2$). The contours of zero error and the regions of $\eta\%$ superconvergence ($\eta = 90$ (dark), 80 (gray), 70 (light gray)) corresponding to the x_1 derivative of the solution for (a) regular pattern, (b) chevron pattern, (c) Union-Jack pattern, (d) criss-cross pattern. The $\eta\%$ regions were determined by the direct approach

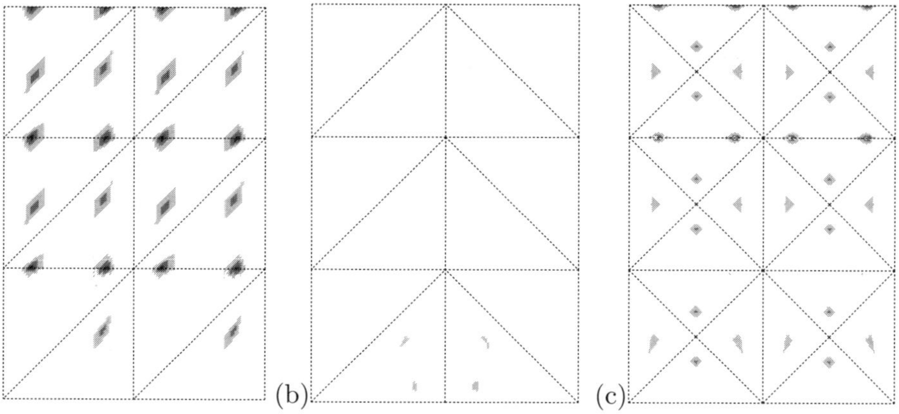

FIG. 4.7*.32. Superconvergence up to the boundary: Poisson equation, general polynomial solution, Neumann boundary conditions along the edge $x_2 = 0$ (the bottom edge of the strip). Quadratic elements ($p = 2$). The regions of $\eta\%$ superconvergence ($\eta = 90$ (dark), 80 (gray), 70 (light gray)) corresponding to the x_1 derivative of the solution for (a) regular pattern, (b) chevron pattern, and (c) criss-cross pattern. The $\eta\%$ regions were determined by the direct approach

Let us consider the case of the Laplacian in an L-shaped or a cracked domain, as shown in Figs 4.7*.33a,b, respectively. The solution in the neighborhood of the corner point can be written in the form

$$u_{EX} = Ar^{\alpha} \sin(\alpha\theta) + \bar{u}_{EX} \tag{4.7*.41}$$

where r, θ are polar coordinates centered at the corner point, α is the exponent of the leading algebraic singularity at the corner, and $\bar{u}_{EX}$ is a function which is smoother than φ in the neighborhood of the corner point. For the L-shaped (respectively the cracked) domain $\alpha = \frac{2}{3}$ (respectively $\alpha = \frac{1}{2}$) in the case that either homogeneous Dirichlet or Neumann boundary conditions are imposed on both the edges emanating from the corner point, and $\alpha = \frac{1}{3}$ (respectively $\alpha = \frac{1}{4}$) in the case that homogeneous Dirichlet boundary conditions are imposed on one edge, and homogeneous Neumann boundary conditions on the other edge.

Analogously as in the one-dimensional case, it can be shown that for sufficiently small h, we have

$$\left\| \frac{\partial}{\partial x_i} \left(u_{S^p_\Delta} - \pi_{S^p_\Delta(S_L(H))}(r^{\alpha} \sin(\alpha\theta)) \right) \right\|_{L^\infty(S_L(H))} \leq \mathscr{C} h^{\alpha(1+\varepsilon)-1} \tag{4.7*.42a}$$

with $\varepsilon > 0$, where $H = h^{\gamma}$, and γ, $0 < \gamma < 1$ is properly selected. Here $\pi_{S^p_\Delta(S_L(H))}(r^{\alpha} \sin(\alpha\theta))$ is the energy projection of $r^{\alpha} \sin(\alpha\,\theta)$ into $S^p_\Delta(S_L(H))$, namely, the solution of the discrete problem:

Find $\pi_{S^p_\Delta(S_L(H))}(r^{\alpha} \sin(\alpha\theta)) \in S^p_\Delta(S_L(H))$ such that

$$\mathscr{B}_{S_L(H)}\left(\pi_{S^p_\Delta(S_L(H))}(r^{\alpha} \sin(\alpha\theta)), v\right) = \mathscr{B}_{S_L(H)}(r^{\alpha} \sin(\alpha\theta), v) \tag{4.7*.42b}$$
$$\forall v \in S^p_\Delta(S_L(H)).$$

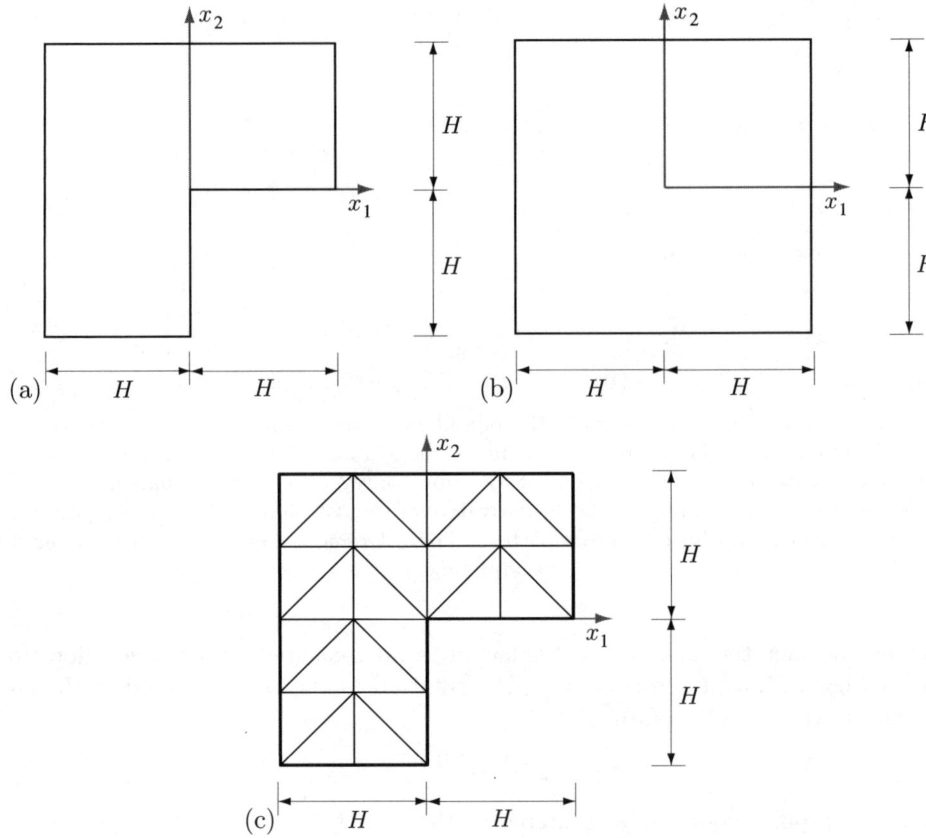

FIG. 4.7*.33. Superconvergence in the neighborhood of corner points. (a) The L-shaped domain $S_L(H)$. (b) The cracked domain $S_c(H)$. The squares represent mesh cells like the ones shown in Fig. 4.7*.3. (c) Detail for the chevron pattern on the L-shaped domain

In the case of the cracked domain the same results hold by replacing $S_L(H)$ by $S_c(H)$. Note that in the elements with a vertex at the corner point

$$\mathscr{C}h^{\alpha-1} \leq |\boldsymbol{\nabla} r^\alpha \sin(\alpha\theta)| \tag{4.7*.43a}$$

while

$$\left|\boldsymbol{\nabla}\big(u_{S^p_\Delta} - \pi_{S^p_\Delta(S_L(H))}(r^\alpha \sin(\alpha\theta))\big)\right| \leq \mathscr{C}h^{\alpha-1+\varepsilon}. \tag{4.7*.43b}$$

Remark 4.7*.13 Computational experience indicates that the asymptotic behavior of the finite element solution in the first mesh layer around the corner point can be realized, for all practical purposes, by employing $H \approx 10\,h$.

From (4.7*.42), it follows that the superconvergence points for $\mathcal{F}(u) = \partial u/\partial x_i$ can be obtained for the condition

$$\frac{\partial}{\partial x_i}\big(r^\alpha \sin(\alpha\theta) - \pi_{S^p_\Delta(S_L(H))}(r^\alpha \sin(\alpha\theta))\big)(\mathbf{x}^{sup}_{\partial u/\partial x_i}) = 0. \tag{4.7*.44}$$

It follows that there are superconvergence lines for each component of the derivative, and hence there exist points of simultaneous superconvergence for both components of the derivative located at the intersection of these lines.

Let us now give some results.

Example 4.7*.14 (*Superconvergence in the neighborhood of the corner points*) Consider the Laplace equation in the L-shaped and the cracked domain. Figure 4.7*.34 shows the superconvergence lines for the components of the gradient in the element with a vertex at the corner point for the various cell patterns. Here only the elements with a vertex at the corner point are shown, and not the entire cells. Note that in some elements the superconvergence lines intersect outside the element, and these points may not be admissible because usually we are interested in superconvergence points in the interior, or at the boundary of the element.

Remark 4.7*.14 Analogous results can be also obtained for corner points with any internal angle. We need only to assume that the character of the mesh does not change in the neighborhood of the corner as the mesh size tends to zero.

Remark 4.7*.15 The above asymptotic analysis does not hold in the case of strongly refined meshes because, in this case, the pollution effects due to the non-uniformity of the mesh can be rather significant.

Remark 4.7*.16 Above we discussed only the case of the scalar model problem. The same approach can also be used for the elasticity problem, in which case we must take into account that there are, in general, two leading singular terms and hence there exist only superconvergence points for each component of the derivative.

Remark 4.7*.17 We have used the function $\Psi^{\bar{\mathbf{x}},H}$ for the analyses of the superconvergence for the derivatives $\partial u/\partial x_1$. For elements of degree $p \geq 2$, $\Psi^{\bar{\mathbf{x}},H}$, can also be used for the analysis of the superconvergence points for the value.

Remark 4.7*.18 The superconvergence effect in finite elements has been analyzed by many investigators in the last 30 years. We refer to [Wahlbin, 1995] for a thorough analysis of a 'classical' character, in contrast to the analysis presented here, and for a long list of references. An additional extensive list of references on superconvergence may be found in [Krizek and Neittaanmaki, 1998].

In this section we analyzed the local asymptotics, and the superconvergence of finite element solutions in the interior of the mesh, up to a smooth boundary, and the neighborhood of corner points. We gave a constructive approach for analyzing the local asymptotics and superconvergence. Using this approach, we gave examples of analysis of classical and $\eta\%$ superconvergence in the interior, up to a smooth boundary, and in the neighborhood of corner points.

Problems

P4.7*.1. Consider the finite element approximation of the solution of the Laplace equation and the Poisson equation on a locally uniform mesh made from the regular pattern and the Union-Jack pattern shown, respectively, in Figs 4.7*.35a,b. Find the superconvergence points for the quantity $\mathcal{F}(u) = \alpha(\partial u/\partial x_1) + \beta(\partial u/\partial x_2)$. Check if the values $(\partial/\partial x_1)(u_{S^1_\Delta})(A)$ and $\frac{1}{2}[(\partial/\partial x_2)(u_{S^1_\tau})(A)+(\partial/\partial x_2)(u_{S^1_{\tau*}})(A)]$ are superconvergent

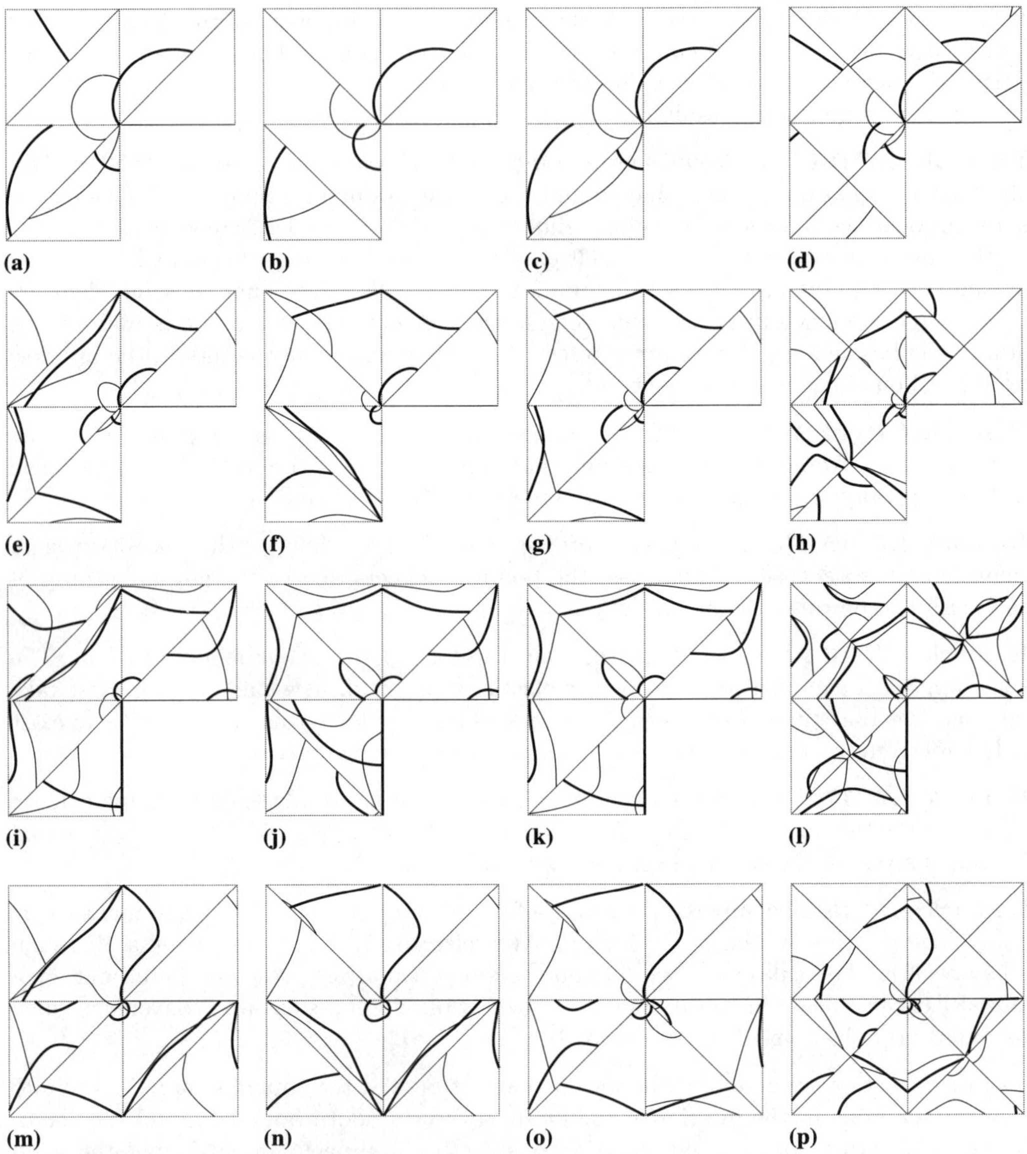

FIG. 4.7*.34. Superconvergence in the elements with a vertex at a corner point. Lines of superconvergence in the elements at the re-entrant corner of the L-shaped domain for $\alpha = \frac{1}{3}$ and linear elements ($p = 1$): (a) regular pattern, (b) chevron pattern, (c) Union-Jack pattern, (d) criss-cross pattern; and for the case of $\alpha = \frac{1}{3}$ and quadratic elements ($p = 2$): (e) regular pattern, (f) chevron pattern, (g) Union-Jack pattern, (h) criss-cross pattern; and for the case of $\alpha = \frac{2}{3}$ and quadratic elements ($p = 2$): (i) regular pattern, (j) chevron pattern, (k) Union-Jack pattern, (l) criss-cross pattern. Superconvergence lines in the element at the crack-tip of the cracked domain for $\alpha = \frac{1}{2}$ and quadratic element ($p = 2$): (m) regular pattern, (n) chevron pattern, (o) Union-Jack pattern, (p) criss-cross pattern. The thin lines correspond to $\partial u/\partial x_1$ and the thick lines to $\partial u/\partial x_2$

approximations of $(\partial/\partial x_1)(u_{EX})(A)$ and $(\partial/\partial x_2)(u_{EX})(A)$, where A is the midpoint of the horizontal edges in each pattern.

P4.7*.2. Construct the asymptotic approach, which was given in Section 4.7*, for the case of the elasticity problem. Repeat the examples given in Section 4.7* in the case of elasticity.

P4.7*.3. Use Example 4.7*.1 to show that the function $\tilde{\Psi}^{PER}_{\tilde{S}^p_\Delta}$ for $p = 1$ and the criss-cross pattern (Fig. 4.7*.4a) is invariant under reflection (i.e., it is symmetric) with respect to the center of the cell.

4.8 Superconvergence via local averaging

In Section 4.6 we have seen that, under the assumptions that u_{EX} satisfies

$$\left\| \frac{d^{p+2}}{dx^{p+2}}(u_{EX}) \right\|_{L^\infty(I_j^\Delta)} \le \mathscr{C} < \infty \tag{4.8.1a}$$

and

$$0 < \mathscr{C} \le \min_{x \in I_j^\Delta} \left| \frac{d^{p+1}}{dx^{p+1}}(u_{EX})(x) \right| \tag{4.8.1b}$$

and for meshes Δ which are such that

$$\max\left(|e_j^\Delta + e_{j-1}^\Delta|, \frac{|e_j^\Delta - e_{j-1}^\Delta|}{h_{I_j^\Delta}} \right) \le \mathscr{C}(h_{I_j^\Delta})^{p+1} \left\| \frac{d^{p+2}}{dx^{p+2}}(u_{EX}) \right\|_{L^\infty(I_j^\Delta)} \tag{4.8.1c}$$

we have

$$\left| \frac{d}{dx}(e_{S_\Delta^p})(x_k^{G,I_j^\Delta}) \right| \le \mathscr{C}(h_{I_j^\Delta})^{p+1} \left\| \frac{d^{p+2}}{dx^{p+2}}(u_{EX}) \right\|_{L^\infty(I_j^\Delta)} \tag{4.8.2a}$$

while

$$\left\| \frac{d}{dx}(e_{S_\Delta^p}) \right\|_{L^\infty(I_j^\Delta)} \approx \mathscr{C}(h_{I_j^\Delta})^p \left\| \frac{d^{p+1}}{dx^{p+1}}(u_{EX}) \right\|_{L^\infty(I_j^\Delta)}. \tag{4.8.2b}$$

(a) 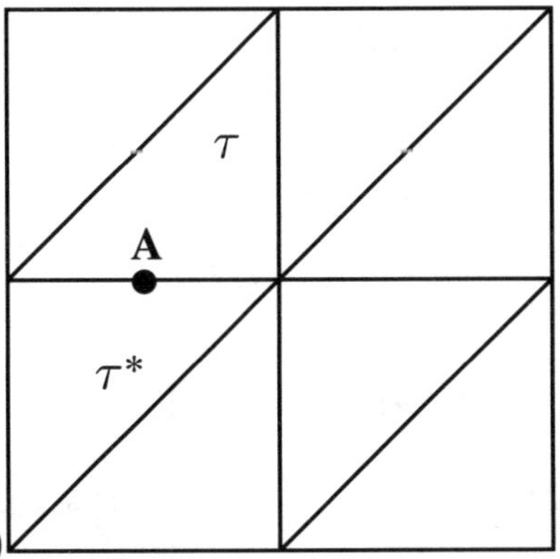

(b) 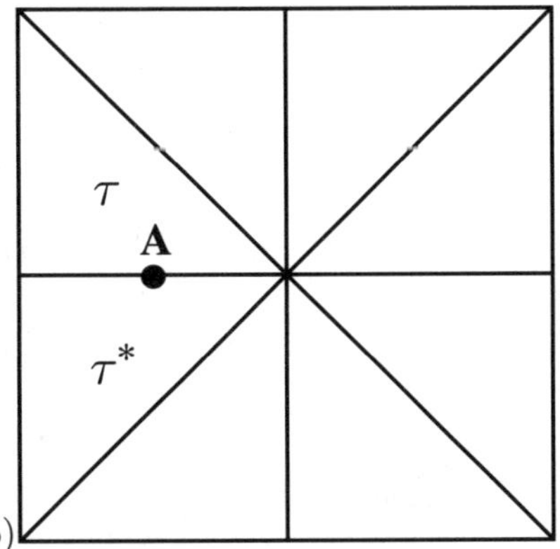

FIG. 4.7*.35. The mesh cell for (a) the regular pattern and (b) the Union-Jack pattern

Thus the value of $(d/dx)(u_{S^p_\Delta})$ at x_k^{G,I_j^Δ} converges with rate $p+1$, while everywhere else it converges only with rate p. This observation suggests the construction of a *recovered derivative* $\mathcal{A}^*_{du/dx}(u_{S^p_\Delta})$ by using a least-squares approximation of the superconvergent values $(d/dx)(u_{S^p_\Delta})(x_k^{G,I_j^\Delta})$ in the element I_j^Δ and its neighbors. For this purpose we will introduce

$$\mathcal{A}^*_{du/dx,\omega^{(m)}_{I_j^\Delta}}: \prod_{I_k^\Delta \subseteq \omega^{(m)}_{I_j^\Delta}} C^\infty(I_k^\Delta) \longmapsto C^\infty\left(\omega^{(m)}_{I_j^\Delta}\right) \tag{4.8.3}$$

the *local recovery operator* for the derivative in the patch $\omega^{(m)}_{I_j^\Delta} \stackrel{\text{def}}{=} \bigcup_{k=j-m}^{j+m} I_k^\Delta$ which includes the element I_j^Δ and m mesh layers around it. $\mathcal{A}^*_{du/dx,\omega^{(m)}_{I_j^\Delta}}$ will be constructed by employing a least-squares fitting of the superconvergent values in $\omega^{(m)}_{I_j^\Delta}$. We can then define $\mathcal{A}^*_{du/dx}$, the global recovery operator, by employing the simple restriction

$$\mathcal{A}^*_{du/dx}(u_{S^p_\Delta})(x) = \mathcal{A}^*_{du/dx,\omega^{(m)}_{I_j^\Delta}}(u_{S^p_\Delta})\Big|_{I_j^\Delta}(x), \quad x \in I_j^\Delta. \tag{4.8.4}$$

There exist various other possibilities, as shown in the following example.

Example 4.8.1 Consider the case of linear elements ($p = 1$), and, for $2 \le j \le N(\Delta) - 1$, let

$$\mathcal{A}^*_{du/dx}(u_{S^1_\Delta}) = \frac{1}{2}\left(\mathcal{A}^{*,L}_{du/dx,\omega^{(1)}_{I_j^\Delta}}(u_{S^1_\Delta}) + \mathcal{A}^{*,R}_{du/dx,\omega^{(1)}_{I_j^\Delta}}(u_{S^1_\Delta})\right) \tag{4.8.5a}$$

where

$$\begin{aligned}\mathcal{A}^{*,L}_{du/dx,\omega^{(1)}_{I_j^\Delta}}(v)(x) &= \frac{dv}{dx}(x^\Delta_{j-1/2})\frac{x - x^\Delta_{j-3/2}}{\frac{1}{2}(h_{I_j^\Delta} + h_{I_{j-1}^\Delta})} \\ &\quad + \frac{dv}{dx}(x^\Delta_{j-3/2})\frac{x^\Delta_{j-1/2} - x}{\frac{1}{2}(h_{I_j^\Delta} + h_{I_{j-1}^\Delta})}\end{aligned} \tag{4.8.5b}$$

and

$$\begin{aligned}\mathcal{A}^{*,R}_{du/dx,\omega^{(1)}_{I_j^\Delta}}(v)(x) &= \frac{dv}{dx}(x^\Delta_{j+1/2})\frac{x - x^\Delta_{j-1/2}}{\frac{1}{2}(h_{I_{j+1}^\Delta} + h_{I_j^\Delta})} \\ &\quad + \frac{dv}{dx}(x^\Delta_{j-1/2})\frac{x^\Delta_{j+1/2} - x}{\frac{1}{2}(h_{I_{j+1}^\Delta} + h_{I_j^\Delta})}.\end{aligned} \tag{4.8.5c}$$

In the first element let $\mathcal{A}^*_{du/dx}(u_{S^1_\Delta}) = \mathcal{A}^{*,R}_{du/dx,\omega^{(1)}_{I_1^\Delta}}(u_{S^1_\Delta})$, and in the last element $\mathcal{A}^*_{du/dx}(u_{S^1_\Delta}) = \mathcal{A}^{*,L}_{du/dx,\omega^{(1)}_{I_4^\Delta}}(u_{S^1_\Delta})$. Let us now consider the regular model problem with $AE \equiv 1$, $c \equiv 0$, with exact solution $u_{EX} \equiv (4/\pi^2)\sin(\pi x/2)$, and compare the accuracy of $\mathcal{A}^*_{du/dx}(u_{S^1_\Delta})$ with the accuracy of $(d/dx)(u_{S^1_\Delta})$. Figure 4.8.1b compares the graphs of the absolute value of the errors $|(d/dx)(u_{EX}) - \mathcal{A}^*_{du/dx}(u_{S^1_\Delta})|$ and $|(d/dx)(u_{EX}) - (d/dx)(u_{S^1_\Delta})|$. Note that, with the exception of the first element, the recovered derivative $\mathcal{A}^*_{du/dx}(u_{S^1_\Delta})$ is overall much more accurate than $(d/dx)(u_{S^1_\Delta})$.

Remark 4.8.1 The recovery (4.8.5) combines linear interpolants of the superconvergent values in the element and its neighbors. Various other constructions are possible; for example, we could have used the quadratic interpolant of the values $(d/dx)(u_{S^1_\Delta})(x^\Delta_{j-3/2+m})$, through the midpoints $x^\Delta_{j-3/2+m}$ of I^Δ_{j-1+m} for $m = 0, 1, 2$.

Following similar steps, we will construct $\mathcal{A}^*_{\mathcal{F}(u),\omega^{(m)}_{I^\Delta_j}}$, a recovery for $\mathcal{F}(u)$ in $\omega^{(m)}_{I^\Delta_j}$, and $\mathcal{A}^*_{\mathcal{F}(u)}$ the corresponding recovery over the entire mesh, where $\mathcal{F}(u)$ is the solution quantity of interest (e.g., $\mathcal{F}(u) = u$ or du/dx, etc.). Let us now give some possible constructions of $\mathcal{A}^*_{\mathcal{F}(u),\omega^{(m)}_{I^\Delta_j}}$.

(a) *Lagrangian interpolation of superconvergent values*: Given a $v \in \prod_{I^\Delta_k \subseteq \omega^{(m)}_{I^\Delta_j}} C^\infty(I^\Delta_k)$, let

$$\mathcal{A}^{*,int}_{\mathcal{F}(u),\omega^{(m)}_{I^\Delta_j}}(v) \stackrel{\text{def}}{=} \sum_{l=1}^{nsup} \mathcal{F}(v)(x^{sup}_{\mathcal{F}(u),k})\mathcal{L}^{sup}_l(\cdot;\chi^{sup}_{\mathcal{F}(u),\omega^{(m)}_{I^\Delta_j}}). \tag{4.8.6a}$$

Here $\mathcal{L}^{sup}_l(\cdot;\chi^{sup}_{\mathcal{F}(u),\omega^{(m)}_{I^\Delta_j}})$ is the lth Lagrangian polynomial of degree $q = (nsup - 1)$ corresponding to the set of interpolation points

$$\chi^{sup}_{\mathcal{F}(u),\omega^{(m)}_{I^\Delta_j}} \stackrel{\text{def}}{=} \{x^{sup}_{\mathcal{F}(u),k}\}_{k=1}^{nsup} \tag{4.8.6b}$$

which is also the set of $nsup$ superconvergence points for $\mathcal{F}(u)$ in $\omega^{(m)}_{I^\Delta_j}$. In order for $\mathcal{A}^{*,int}_{\mathcal{F}(u),\omega^{(m)}_{I^\Delta_j}}(u_{S^p_\Delta})$ to be superconvergent, $nsup$ must be selected sufficiently large. For example, for $\mathcal{F}(u) = du/dx$ (respectively $\mathcal{F}(u) = u$) it is sufficient to employ $nsup \geq p+1$ (respectively $nsup \geq p + 2$) distinct superconvergence points for the derivative (respectively for the value). It is not necessary to include all the available superconvergence points in $\omega^{(m)}_{I^\Delta_j}$. For example, in the case of $\mathcal{F}(u) = du/dx$, it is sufficient to employ the Gauss–Legendre points of order p in the element I^Δ_j, and one from one of its neighbors, I^Δ_{j-1} or I^Δ_{j+1}. We may also employ $m > 1$, and include in $\chi^{sup}_{\mathcal{F}(u),\omega^{(m)}_{I^\Delta_j}}$ additional superconvergence points from the neighbors of the neighbors of I^Δ_j.

Let us now give an example which illustrates the accuracy of $\mathcal{A}^{*,int}_{du/dx}$.

Example 4.8.2 Let us employ the model problem of Example 4.8.1, a uniform mesh of four quadratic elements ($p = 2$, $N = 4$), and $\mathcal{A}^{*,int,L}_{du/dx}$ (respectively $\mathcal{A}^{*,int,R}_{du/dx}$), the recovery given in (4.8.6) with $q = p = 2$ and $nsup = 3$, which uses two superconvergence points

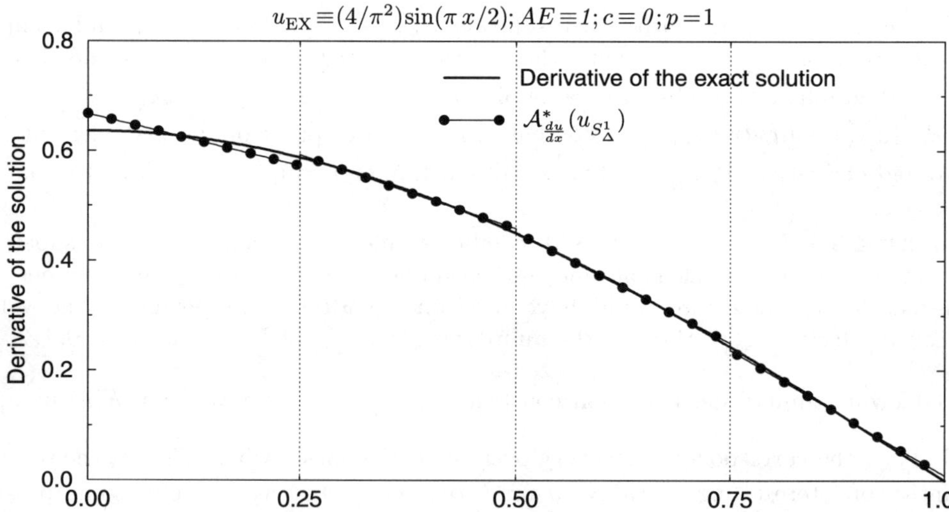

(a)

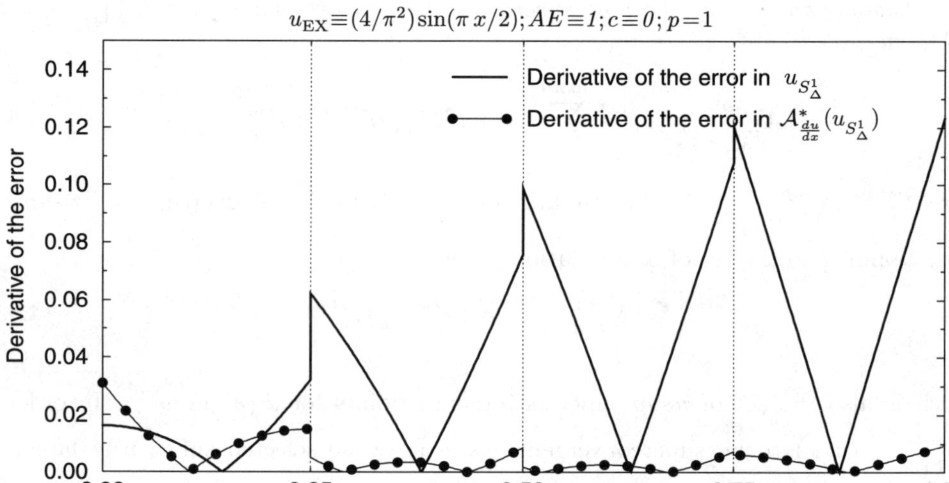

(b)

FIG. 4.8.1. Recovery of the derivative by interpolation of its superconvergent values at the Gauss–Legendre points of order p. Input data: $AE \equiv 1$, $c \equiv 0$, $u_{EX} \equiv (4/\pi^2)\sin(\pi\,x/2)$. Mesh data: Uniform mesh of four ($N = 4$) linear elements ($p = 1$). (a) Graphs of the derivative of the exact solution, $(d/dx)(u_{EX})$, and the recovered derivative $\mathcal{A}^*_{du/dx}(u_{S^1_\Delta})$, given by $\mathcal{A}^*_{du/dx}(u_{S^1_\Delta})|_{I^\Delta_j} = \frac{1}{2}(\mathcal{A}^{*,L}_{du/dx,\omega^{(1)}_{I^\Delta_j}}(u_{S^1_\Delta}) + \mathcal{A}^{*,R}_{du/dx,\omega^{(1)}_{I^\Delta_j}}(u_{S^1_\Delta}))$. (b) Graphs of the absolute value of the error in the recovered derivative $|(d/dx)(u_{EX}) - \mathcal{A}^*_{du/dx}(u_{S^1_\Delta})|$ and the absolute value of the error in the derivative of the finite element solution $|(d/dx)(u_{EX}) - (d/dx)(u_{S^1_\Delta})|$

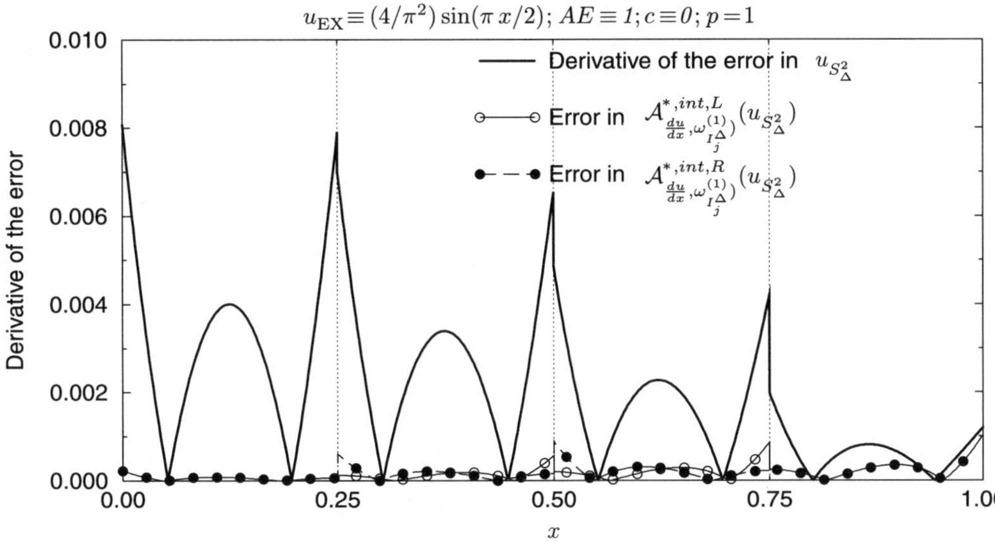

FIG. 4.8.2. Recovery of the derivative by interpolation of its superconvergent values at the Gauss–Legendre points of order p. Input data: $AE \equiv 1$, $c \equiv 0$, $u_{EX} \equiv (4/\pi^2)\sin(\pi x/2)$. Mesh data: Uniform mesh of four ($N = 4$) quadratic elements ($p = 2$). Graphs of the absolute value of the error in the recovered derivative $|(d/dx)(u_{EX}) - \mathcal{A}^{*,int}_{du/dx}(u_{S^2_\Delta})|$ and the absolute value of the error in the derivative of the finite element solution $|(d/dx)(u_{EX}) - (d/dx)(u_{S^2_\Delta})|$. Here we let $\mathcal{A}^{*,int}_{du/dx}(u_{S^2_\Delta}) = \frac{1}{2}(\mathcal{A}^{*,int,L}_{du/dx,\omega^{(1)}_{I^\Delta_j}}(u_{S^2_\Delta}) + \mathcal{A}^{*,int,R}_{du/dx,\omega^{(1)}_{I^\Delta_j}}(u_{S^2_\Delta}))|_{I^\Delta_j}$, $j = 2, 3$, for the interior elements, while for the first element we let $\mathcal{A}^{*,int}_{du/dx}(u_{S^2_\Delta}) = \mathcal{A}^{*,int,R}_{du/dx,\omega^{(1)}_{I^\Delta_1}}(u_{S^2_\Delta})|_{I^\Delta_j}$, and for the last element $\mathcal{A}^{*,int}_{du/dx}(u_{S^2_\Delta}) = \mathcal{A}^{*,int,L}_{du/dx,\omega^{(1)}_{I^\Delta_4}}(u_{S^2_\Delta})|_{I^\Delta_j}$. The recovered derivative $\mathcal{A}^{*,int,L}_{du/dx,\omega^{(1)}_{I^\Delta_j}}$ (respectively $\mathcal{A}^{*,int,R}_{du/dx,\omega^{(1)}_{I^\Delta_j}}$) was constructed using the two superconvergence points from the element and the point from I^Δ_{j-1} closest to x^Δ_{j-1} (respectively the point from I^Δ_{j+1} closest to x^Δ_j)

from the element and one point from the left (respectively right) neighbor. Then for the interior elements we let

$$\mathcal{A}^{*,int}_{du/dx}(v) = \frac{1}{2}\Big(\mathcal{A}^{*,int,L}_{du/dx,\omega^{(1)}_{I^\Delta_j}}(v) + \mathcal{A}^{*,int,R}_{du/dx,\omega^{(1)}_{I^\Delta_j}}(v)\Big). \tag{4.8.7}$$

From Fig. 4.8.2 it can be seen that, with the exception of small neighborhoods of the superconvergence points, we have

$$\Big|\frac{d}{dx}(u_{EX}) - \mathcal{A}^{*,int}_{du/dx}(u_{S^2_\Delta})\Big|(x) \ll \Big|\frac{d}{dx}(u_{EX}) - \frac{d}{dx}(u_{S^2_\Delta})\Big|(x) \tag{4.8.8}$$

i.e., the recovered derivative $\mathcal{A}^{*,int}_{du/dx}(u_{S^2_\Delta})$, is much more accurate than the derivative of the finite element solution $(d/dx)(u_{S^2_\Delta})$.

(b) *Least-squares fit of superconvergent values*: Given $v \in \prod_{I_k^\Delta \subseteq \omega_{I_j^\Delta}^{(m)}} C^\infty(I_k^\Delta)$, we will construct $\mathcal{A}^{*,ls}_{\mathcal{F}(u),\omega_{I_j^\Delta}^{(m)}}(v) \in \mathcal{P}^q(\omega_{I_j^\Delta}^{(m)})$, as the solution of the minimization problem

$$\|\mathcal{A}^{*,ls}_{\mathcal{F}(u),\omega_{I_j^\Delta}^{(m)}}(v) - \mathcal{F}(v)\|_{*,sup,\omega_{I_j^\Delta}^{(m)}} = \min_{\sigma \in \mathcal{P}^q(\omega_{I_j^\Delta}^{(m)})} \|\sigma - \mathcal{F}(v)\|_{*,sup,\omega_{I_j^\Delta}^{(m)}} \tag{4.8.9a}$$

where

$$\|\sigma\|_{*,sup,\omega_{I_j^\Delta}^{(m)}} = \|\sigma\|_{l^2(\omega_{I_j^\Delta}^{(m)}),\chi^{sup}_{\mathcal{F}(u),\omega_{I_j^\Delta}^{(m)}},W^{sup}_{\mathcal{F}(u),\omega_{I_j^\Delta}^{(m)}}} \stackrel{\text{def}}{=} \sqrt{\sum_{l=1}^{nsup} w_l(\sigma(x_l^{sup}))^2} \tag{4.8.9b}$$

is a weighted l^2 norm, $\chi^{sup}_{\mathcal{F}(u),\omega_{I_j^\Delta}^{(1)}}$ is the set of superconvergence points employed in the least-squares fit, $W^{sup}_{\mathcal{F}(u),\omega_{I_j^\Delta}^{(1)}} = \{w_l^{sup}\}_{l=1}^{nsup}$ is a set of weights assigned to the superconvergence points in $\chi^{sup}_{\mathcal{F}(u),\omega_{I_j^\Delta}^{(1)}}$, and q is the polynomial order of the least-squares fit. Note that we must have $nsup - 1 \geq q$, where, as before, q must be selected sufficiently high depending on the quantity $\mathcal{F}(u)$.

Let us give an example of a least-squares recovery $\mathcal{A}^{*,ls}_{du/dx,\omega_{I_j^\Delta}^{(1)}}$:

Example 4.8.3 Consider the setting of Example 4.8.2, and let $\mathcal{A}^{*,ls}_{du/dx,\omega_{I_j^\Delta}^{(1)}}$ be the quadratic least-squares fit through all six superconvergence points for the derivative in $\omega_{I_j^\Delta}^{(1)}$. Figure 4.8.3 compares the graphs of the error $|(d/dx)(u_{EX}) - \mathcal{A}^{*,ls}_{du/dx}(u_{S^2_\Delta})|$, with the error $|(d/dx)(u_{EX}) - \mathcal{A}^{*,int}_{du/dx}(u_{S^2_\Delta})|$, where $\mathcal{A}^{*,int}_{du/dx}(u_{S^2_\Delta})$ was constructed as in Example 4.8.2. Note that, although the error in both recoveries is of order h^3, asymptotically as h tends to zero, for the fixed h employed here, $\mathcal{A}^{*,int}_{du/dx}(u_{S^2_\Delta})$ is, overall, more accurate than $\mathcal{A}^{*,ls}_{du/dx}(u_{S^2_\Delta})$ in the interior elements.

Remark 4.8.2 $\mathcal{A}^{*,ls}_{du/dx}$ defined as in (4.8.9) with weights $w_l = (AE(x_l^{sup}))^{-1}$ is a one-dimensional analogue of a recovery often used in practice. For this choice of weights, the norm $\|\cdot\|_{*,sup,\omega_{I_j^\Delta}^{(m)}}$ has the meaning of a discrete version of the energy norm in the patch.

In the above examples, the recovered quantity $\mathcal{A}^*_{\mathcal{F}(u)}(u_{S^p_\Delta})$ was obtained by employing the *restriction*

$$\mathcal{A}^*_{\mathcal{F}(u)}(u_{S^p_\Delta})(x) \stackrel{\text{def}}{=} \mathcal{A}^*_{\mathcal{F}(u),\omega_{I_j^\Delta}^{(m)}}(u_{S^p_\Delta})|_{I_j^\Delta}(x), \quad x \in I_j^\Delta. \tag{4.8.10}$$

It follows that $\mathcal{A}^*_{\mathcal{F}(u)}(u_{S^p_\Delta})$ is discontinuous at the nodes, and we will underline this by calling $\mathcal{A}^*_{\mathcal{F}(u)}$ a C^{-1} recovery.

In practice, it is often desirable to construct $\mathcal{A}^*_{\mathcal{F}(u)}(u_{S^p_\Delta})$ to be continuous over the entire domain, with the exception of points where it is known a-priori that $\mathcal{F}(u)$ is

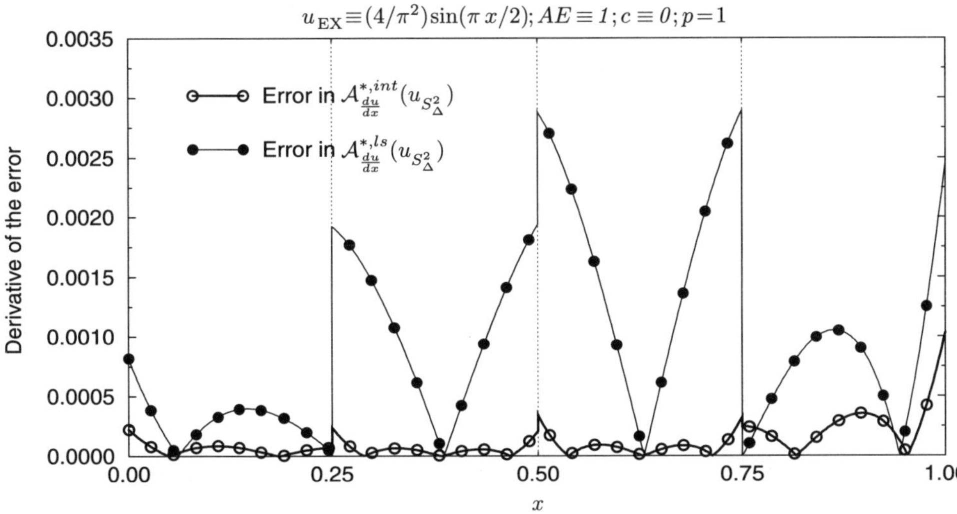

FIG. 4.8.3. Recovery of the derivative by a least-squares fit of its superconvergent values at the Gauss–Legendre points of order p. Input data: $AE \equiv 1$, $c \equiv 0$, $u_{EX} \equiv (4/\pi^2)\sin(\pi x/2)$. Mesh data: Uniform mesh of four ($N = 4$) quadratic elements ($p = 2$). Graphs of the absolute value of the error in the recovered derivative, $|(d/dx)(u_{EX}) - \mathcal{A}^{*,ls}_{du/dx}(u_{S^2_\Delta})|$ and $|(d/dx)(u_{EX}) - \mathcal{A}^{*,int}_{du/dx}(u_{S^2_\Delta})|$, where $\mathcal{A}^{*,int}_{du/dx}$ is the interpolation recovery which was defined in the caption of Fig. 4.8.2. The recovered derivative $\mathcal{A}^{*,ls}_{du/dx}(u_{S^2_\Delta})$ for the interior elements (respectively end elements) was obtained by fitting a quadratic polynomial through all six (respectively four) superconvergence points from the element and its two neighbors (respectively neighbor) with weights equal to one. Note that, with the exception of the neighborhood of a few points, $\mathcal{A}^{*,int}_{du/dx}(u_{S^2_\Delta})$ is much more accurate than $\mathcal{A}^{*,ls}_{du/dx}(u_{S^2_\Delta})$ in the interior elements

discontinuous. We will call a recovery $\mathcal{A}^*_{\mathcal{F}(u)}$, with this property a C^0 recovery, and we will denote it by $\mathcal{A}^{C^0}_{\mathcal{F}(u)}$. Let us now give some possible constructions of C^0 recoveries.

(c) C^0 *recovery based either on interpolation or on least-squares fit.* We will construct $\mathcal{A}^{C^0}_{\mathcal{F}(u)}$ in two steps:

(i) *Recovery over the nodal mesh patches associated with the element* I^Δ_j: Let

$$\omega^{(1),L}_{I^\Delta_j} = I^\Delta_{j-1} \cup I^\Delta_j, \qquad \omega^{(1),R}_{I^\Delta_j} = I^\Delta_j \cup I^\Delta_{j+1} \tag{4.8.11a}$$

be the left and the right nodal patch associated with I^Δ_j, and denote by $\mathcal{A}^*_{\mathcal{F}(u),\omega^{(1),M}_{I^\Delta_j}}$, with $M = L$ and $M = R$, respectively, the recovery which is constructed as in (4.8.6a) or (4.8.9), by replacing $\omega^{(1)}_{I^\Delta_j}$ by $\omega^{(1),L}_{I^\Delta_j}$ and $\omega^{(1),R}_{I^\Delta_j}$, respectively.

(ii) *Patching of the left and right recoveries into a continuous recovered function*: Given a $v \in \prod_{k=1}^{N(\Delta)} C^\infty(I_k^\Delta)$, we will construct

$$\mathcal{A}^*_{\mathcal{F}(u)}(v) \mid_{I_j^\Delta} \in \mathcal{P}^{q'}(I_j^\Delta) \tag{4.8.11b}$$

such that

$$\begin{aligned} \mathcal{A}^{C^0}_{\mathcal{F}(u)}(v)(x_{j-1}^\Delta) &= \mathcal{A}^*_{\mathcal{F}(u),\omega^{(1),L}_{I_j^\Delta}}(v)(x_{j-1}^\Delta), \\ \mathcal{A}^{C^0}_{\mathcal{F}(u)}(v)(x_j^\Delta) &= \mathcal{A}^*_{\mathcal{F}(u),\omega^{(1),R}_{I_j^\Delta}}(v)(x_j^\Delta) \end{aligned} \tag{4.8.11c}$$

and

$$\|\mathcal{A}^{C^0}_{\mathcal{F}(u)}(v) - \mathcal{A}^{*,LR}_{\mathcal{F}(u),I_j^\Delta}(v)\|_{**,I_j^\Delta} = \min_{\sigma \in \mathcal{P}^{q'}_{\mathcal{A}^*_{\mathcal{F}(u)}}(I_j^\Delta)} \|\sigma - \mathcal{A}^{*,LR}_{\mathcal{F}(u),j}(v)\|_{**,I_j^\Delta} \tag{4.8.11d}$$

where

$$\mathcal{A}^{*,LR}_{\mathcal{F}(u),I_j^\Delta}(v) = \tfrac{1}{2}\big(\mathcal{A}^*_{\mathcal{F}(u),\omega^{(1),L}_{I_j^\Delta}}(v) + \mathcal{A}^*_{\mathcal{F}(u),\omega^{(1),R}_{I_j^\Delta}}(v)\big) \tag{4.8.11e}$$

and

$$\mathcal{P}^{q'}_{\mathcal{A}^*_{\mathcal{F}(u)}}(I_j^\Delta) \stackrel{\text{def}}{=} \{w \in \mathcal{P}^{q'}(I_j^\Delta) \mid w(x_k^\Delta) = \mathcal{A}^*_{\mathcal{F}(u),\omega^{(1),M(k)}_{I_j^\Delta}}(v)(x_k^\Delta), \quad k = j-1, j\} \tag{4.8.11f}$$

where $M(j-1) = L$, $M(j) = R$, q' is the polynomial order of $\mathcal{A}^{C^0}_{\mathcal{F}(u)}(v)$ (typically $q = q' = p$ for $\mathcal{F}(u) = du/dx$), and $\|\cdot\|_{**,I_j^\Delta}$ denotes the norm employed in the construction. For example, choosing $\|\cdot\|_{**,I_j^\Delta}$ as the weighted l^∞ norm corresponding to the Lagrangian interpolation points in the interior of the element, we obtain

$$\mathcal{A}^{C^0}_{\mathcal{F}(u)}(v) = \sum_{l=0}^{p} \mathcal{A}^{*,LR}_{\mathcal{F}(u),I_j^\Delta}(v)(\bar{x}_l)\mathcal{L}^j_{l+1} \tag{4.8.12a}$$

where we let

$$\mathcal{A}^{*,LR}_{\mathcal{F}(u),I_j^\Delta}(v)(\bar{x}_0) = \mathcal{A}^*_{\mathcal{F}(u),\omega^{(1),L}_{I_j^\Delta}}(v)(x_{j-1}^\Delta) \tag{4.8.12b}$$

and

$$\mathcal{A}^{*,LR}_{\mathcal{F}(u),I_j^\Delta}(v)(\bar{x}_p) = \mathcal{A}^*_{\mathcal{F}(u),\omega^{(1),R}_{I_j^\Delta}}(v)(x_j^\Delta) \tag{4.8.12c}$$

where $\mathcal{L}^j_l$ is the lth Lagrangian polynomial of degree p, corresponding to the set of uniformly distributed points $\bar{x}_l = x_{j-1}^\Delta + (l/p)h_{I_j^\Delta}$, $l = 1, \ldots, p-1$.

There are, of course, many other possibilities for construction C^0 recoveries, like

$$\mathcal{A}^{C^0}_{\mathcal{F}(u)}(v)(x) = \mathcal{A}^*_{\mathcal{F}(u),\omega^{(1),L}_{I^\Delta_j}}(v)(x)\,\psi^j_L(x) + \mathcal{A}^*_{\mathcal{F}(u),\omega^{(1),R}_{I^\Delta_j}}(v)(x)\,\psi^j_R(x) \tag{4.8.13a}$$

where ψ^j_L, $\psi^j_R \geq 0$, are smooth functions such that

$$\psi^j_L + \psi^j_R \equiv 1, \qquad \psi^j_L(x^\Delta_{j-1}) = 1, \qquad \psi^j_R(x^\Delta_j) = 1. \tag{4.8.13b}$$

For example, ψ^j_L, ψ^j_R could be chosen as the linear shape-functions in I^Δ_j. Another possibility is to modify the construction given in (4.8.11) by employing

$$\mathcal{A}^{*,LR}_{\mathcal{F}(u),I^\Delta_j}(v) \stackrel{\text{def}}{=} \mathcal{A}^*_{\mathcal{F}(u),\omega^{(1),L}_{I^\Delta_j}}(v)\psi^j_L + \mathcal{A}^*_{\mathcal{F}(u),\omega^{(1),R}_{I^\Delta_j}}(v)\psi^j_R \tag{4.8.14}$$

instead of the simpler one-half averaging used in (4.8.11e).

Remark 4.8.3 We have shown that the recovery leads to a convergence rate which is better than the rate of the finite element solution by the factor h. For elements of even degree and for a uniform mesh, the recovery can lead to h^2 improvement in the accuracy. This is called *ultraconvergence*. This is valid in one dimension and in two dimensions on rectangular elements; see [Zhang, 1996], [Zhang, 1999].

Let us give an example of a continuous recovery.

Example 4.8.4 Consider the model problem, exact solution, and mesh employed in Examples 4.8.2 and 4.8.3. Figure 4.8.4 shows the absolute values of the error $|(d/dx)(u_{EX}) - \mathcal{A}^{C^0}_{du/dx}(u_{S^2_\Delta})|$ where $\mathcal{A}^{C^0}_{du/dx}$ is constructed as in (4.8.12); this is compared with the error $|(d/dx)(u_{EX}) - \mathcal{A}^{*,int}_{du/dx}(u_{S^2_\Delta})|$. As before, we note that there may be significant differences in the accuracy of various recoveries non-asymptotically.

Another approach for constructing $\mathcal{A}^*_{\mathcal{F}(u)}$ is by employing a recovery for a related solution quantity. For example, we may let

$$\mathcal{A}^*_{du/dx}(u_{S^p_\Delta})(x) = \frac{d}{dx}(\mathcal{A}^*_u(u_{S^p_\Delta})(x)), \quad x \in I^\Delta_j \tag{4.8.15}$$

where $\mathcal{A}^*_u(u_{S^p_\Delta})$ is a recovery for $\mathcal{F}(u) = u$, constructed by employing a least-squares fit of the superconvergent values at the Lobatto–Legendre points of order p.

Let us now give an example where (4.8.15) is used.

Example 4.8.5 Once more, consider the case employed in Examples 4.8.2–4.8.4. Figure 4.8.5 compares the error $|(d/dx)(u_{EX}) - (d/dx)(\mathcal{A}^{*,ls}_u(u_{S^2_\Delta}))|$, where $\mathcal{A}^{*,ls}_u(u_{S^2_\Delta})$ was recovered in each element by employing a least-squares fit of the values of $u_{S^2_\Delta}$ at the Lobatto–Legendre points in the element and its neighbors, with $|(d/dx)(u_{EX}) - \mathcal{A}^{C^0}_{du/dx}(u_{S^2_\Delta})|$, where $\mathcal{A}^{C^0}_{du/dx}$ is defined by (4.8.12). Note that $\mathcal{A}^{C^0}_{du/dx}(u_{S^2_\Delta})$ is more accurate than $(d/dx)(\mathcal{A}^{*,ls}_u(u_{S^2_\Delta}))$ in the interior elements.

Let us now summarize the properties of the recovery operators $\mathcal{A}^*_{\mathcal{F}(u)}$. For concreteness we will let $\mathcal{F}(v) = dv/dx$.

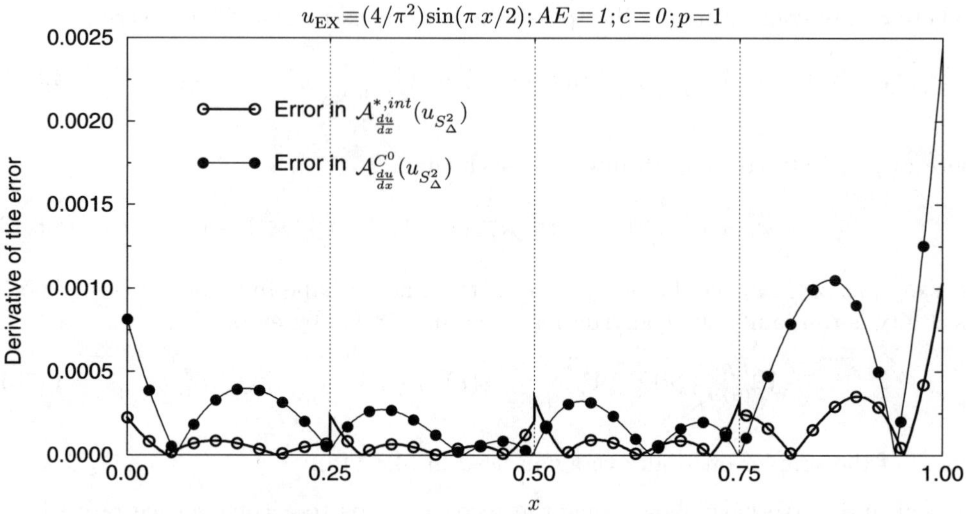

FIG. 4.8.4. C^0 least-squares recovery of the derivative by a least-squares fit of its superconvergent values at the Gauss–Legendre points of order p. Input data: $AE \equiv 1$, $c \equiv 0$, $u_{EX} \equiv (4/\pi^2)\sin(\pi x/2)$. Mesh data: Uniform mesh of four $N = 4$ quadratic elements ($p = 2$). Graph of the absolute value of the error in the recovered derivative $|(d/dx)(u_{EX}) - \mathcal{A}^{C^0}_{du/dx}(u_{S^2_\Delta})|$ where $\mathcal{A}^{C^0}_{du/dx}$ is given in (4.8.12), compared with $|(d/dx)(u_{EX}) - \mathcal{A}^{*,int}_{du/dx}(u_{S^2_\Delta})|$, which was also given in Fig. 4.8.2. Note that $\mathcal{A}^{*,int}_{du/dx}(u_{S^2_\Delta})$ is overall more accurate than $\mathcal{A}^{C^0}_{du/dx}(u_{S^2_\Delta})$

(i) The operator $\mathcal{A}^*_{du/dx,\omega^{(m)}_{I^\Delta_j}}$ is a *linear operator*

$$v \in L^\infty(\omega^{(m)}_{I^\Delta_j}) \longrightarrow \mathcal{A}^*_{du/dx,\omega^{(m)}_{I^\Delta_j}}(v) \in L^\infty(I^\Delta_j) \tag{4.8.16a}$$

and

$$\mathcal{A}^*_{du/dx,\omega^{(m)}_{I^\Delta_j}}(\alpha v + \beta w) = \alpha \mathcal{A}^*_{du/dx,\omega^{(m)}_{I^\Delta_j}}(v) + \beta \mathcal{A}^*_{du/dx,\omega^{(m)}_{I^\Delta_j}}(w). \tag{4.8.16b}$$

(ii) The operator $\mathcal{A}^*_{du/dx,\omega^{(m)}_{I^\Delta_j}}(v)$ depends only on the values of dv/dx at the superconvergence points $x^{sup}_{du/dx,k}$, $k = 1, \ldots, nsp$, in the elements $I^\Delta_l \subseteq \omega^{(m)}_{I^\Delta_j}$.

(iii) The operator $\mathcal{A}^*_{du/dx,\omega^{(m)}_{I^\Delta_j}}$ reproduces any polynomial of degree $(p+1)$ on $\omega^{(m)}_{I^\Delta_j}$, namely

$$\mathcal{A}^*_{du/dx,\omega^{(m)}_{I^\Delta_j}}(v) \equiv \frac{dv}{dx} \quad \forall v \in \mathcal{P}^{p+1}(\omega^{(m)}_{I^\Delta_j}). \tag{4.8.16c}$$

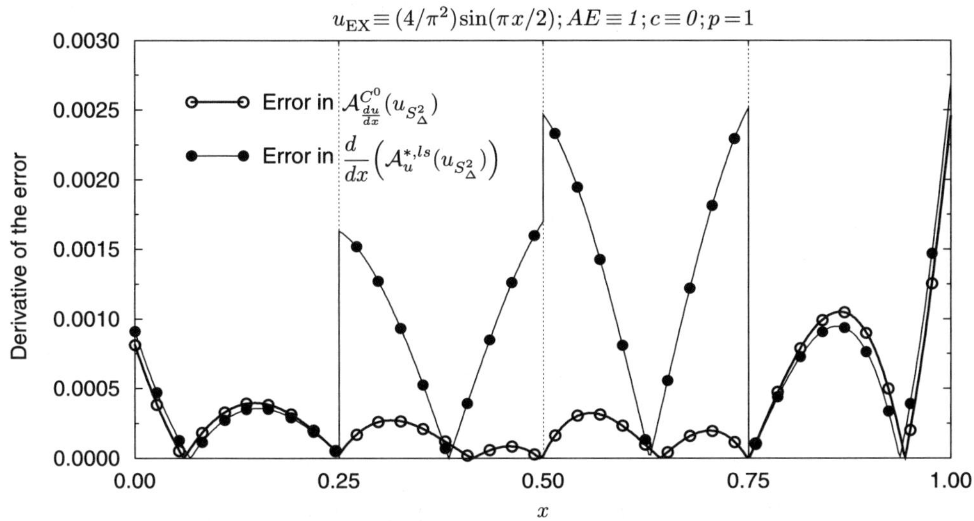

FIG. 4.8.5. Recovery of the derivative by differentiating a least-squares fit of its superconvergent values at the Lobatto–Legendre points of order p. Input data: $AE \equiv 1$, $c \equiv 0$, $u_{EX} \equiv (4/\pi^2)\sin(\pi x/2)$. Mesh data: Uniform mesh of four ($N = 4$) quadratic elements ($p = 2$). Graphs of the absolute value of the errors $|(d/dx)(u_{EX}) - (d/dx)(\mathcal{A}^{*,ls}_u(u_{S^2_\Delta}))|$ and $|(d/dx)(u_{EX}) - \mathcal{A}^{C^0}_{du/dx}(u_{S^2_\Delta})|$. Here $\mathcal{A}^{C^0}_{du/dx}$ is the combined recovery also employed in Fig. 4.8.4. The recovery $\mathcal{A}^{*,ls}_u$ was constructed in the interior element by employing a cubic least-squares fit of the values at all seven superconvergence points (Lobatto–Legendre points of order two) from the element and its two neighbors, and, for an end element, a least-squares fit of the values at the five superconvergence points in the element and its neighbor (weights equal to one were employed). Note that $\mathcal{A}^{C^0}_{du/dx}(u_{S^2_\Delta})$ is more accurate than $(d/dx)(\mathcal{A}^{*,ls}_u(u_{S^2_\Delta}))$ in the interior elements

(iv) Under the assumption that $1/\kappa \leq h_{I^\Delta_j}/h_{I^\Delta_{j-1}}, h_{I^\Delta_j}/h_{I^\Delta_{j+1}} \leq \kappa$, for a fixed $\kappa \geq 1$ we have

$$\|\mathcal{A}^*_{du/dx,\omega^{(m)}_{I^\Delta_j}}(v)\|_{L^\infty(I^\Delta_j)} \leq \mathscr{C} \max_{k=1,\dots,nsp} \left|\frac{dv}{dx}(x^{sup}_{du/dx,k})\right| \tag{4.8.16d}$$

where $\mathscr{C}$ depends only on $x^{sup}_{du/dx,k}$, $k = 1, \dots, nsp$, and on κ.

Let us now prove that given any recovery operator $\mathcal{A}^*_{du/dx,\omega^{(m)}_{I^\Delta_j}}$, with the above properties, $\mathcal{A}^*_{du/dx}(u_{S^p_\Delta})$ is superconvergent.

Theorem 4.8.1 *Let Assumptions I and II of Theorem 4.6.3 hold, and assume that* $\mathcal{A}^*_{du/dx,\omega^{(m)}_{I^\Delta_j}}$ *satisfies conditions (i)–(iv) given above, and let* $\mathcal{A}^*_{du/dx}$ *be defined by the simple restriction (4.8.4). Then we have*

$$\left\|\frac{d}{dx}(u_{EX}) - \mathcal{A}^*_{du/dx}(u_{S^p_\Delta})\right\|_{L^\infty(I^\Delta_j)} \leq \mathscr{C}(h_{I^\Delta_j})^{p+1} \leq \mathscr{C}h\left\|\frac{d}{dx}(e_{S^p_\Delta})\right\|_{L^\infty(I^\Delta_j)}. \tag{4.8.17}$$

Proof From the linearity of $\mathcal{A}^*_{du/dx}$, we have

$$\mathcal{A}^*_{du/dx}(e_{S^p_\Delta}) = \mathcal{A}^*_{du/dx}(u_{EX} - u_{S^p_\Delta}) = \mathcal{A}^*_{du/dx}(u_{EX}) - \mathcal{A}^*_{du/dx}(u_{S^p_\Delta}) \tag{4.8.18a}$$

and hence

$$\frac{d}{dx}(u_{EX}) - \mathcal{A}^*_{du/dx}(u_{S^p_\Delta}) = \frac{d}{dx}(u_{EX}) - \mathcal{A}^*_{du/dx}(u_{EX}) + \mathcal{A}^*_{du/dx}(e_{S^p_\Delta}). \tag{4.8.18b}$$

From (4.8.16d), we have

$$\|\mathcal{A}^*_{du/dx}(e_{S^p_\Delta})\|_{L^\infty(I^\Delta_j)} \leq \mathscr{C} \max_{k=1,\ldots,nsp} \left|\frac{d}{dx}(e_{S^p_\Delta})(x^{sup}_{du/dx,k})\right|. \tag{4.8.19a}$$

Recalling that from Theorem 4.6.3, we have

$$\max_{k=1,\ldots,nsp} \left|\frac{d}{dx}(e_{S^p_\Delta})(x^{sup}_{du/dx,k})\right| \leq \mathscr{C}(h_{I^\Delta_j})^{p+1} \left\|\frac{d^{p+2}}{dx^{p+2}}(u_{EX})\right\|_{L^\infty(\omega^{(m)}_{I^\Delta_j})} \tag{4.8.19b}$$

we obtain

$$\|\mathcal{A}^*_{du/dx}(e_{S^p_\Delta})\|_{L^\infty(I^\Delta_j)} \leq \mathscr{C}(h_{I^\Delta_j})^{p+1} \left\|\frac{d^{p+2}}{dx^{p+2}}(u_{EX})\right\|_{L^\infty(\omega^{(m)}_{I^\Delta_j})}. \tag{4.8.19c}$$

Further, because $\mathcal{A}^*_{du/dx}$ reproduces all polynomials of degree $(p+1)$, we have

$$\left\|\frac{d}{dx}(u_{EX}) - \mathcal{A}^*_{du/dx}(u_{EX})\right\|_{L^\infty(I^\Delta_j)} \leq \mathscr{C}(h_{I^\Delta_j})^{p+1} \left\|\frac{d^{p+2}}{dx^{p+2}}(u_{EX})\right\|_{L^\infty(\omega^{(m)}_{I^\Delta_j})}. \tag{4.8.19d}$$

Employing (4.8.18b), and the triangle inequality, (4.8.17) follows. □

The recoveries given above were based on a local interpolation or least-squares approximation of superconvergent values. Let us now give an example of a recovery which is not based on superconvergence.

Given a $v \in \prod_{I^\Delta_k \subseteq \omega^{(1)}_{I^\Delta_j}} C^\infty(I^\Delta_k)$, we will let $\mathcal{A}^{*,cls}_{\mathcal{F}(u),\omega^{(1)}_{I^\Delta_j}}(v) \in \mathscr{P}^q(\omega^{(1)}_{I^\Delta_j})$ be the solution of the minimization problem

$$\|\mathcal{A}^{*,cls}_{\mathcal{F}(u),\omega^{(1)}_{I^\Delta_j}}(v) - \mathcal{F}(v)\|_{L^2(\omega^{(1)}_{I^\Delta_j}),w} = \min_{\sigma \in \mathscr{P}^q(\omega^{(1)}_{I^\Delta_j})} \|\sigma - \mathcal{F}(v)\|_{L^2(\omega^{(1)}_{I^\Delta_j}),w} \tag{4.8.20}$$

where $\|\sigma\|_{L^2(\omega^{(1)}_{I^\Delta_j}),w} \stackrel{\text{def}}{=} \|\sqrt{w}\sigma\|_{L^2(\omega^{(1)}_{I^\Delta_j})}$ is the weighted L^2 norm in $\omega^{(1)}_{I^\Delta_j}$ and w is a positive weight function. A C^0 version of the above recovery, can also be constructed as in (4.8.11)–(4.8.12).

Although $\mathcal{A}^{*,cls}_{\mathcal{F}(u)}$ is not based on superconvergent values, in certain cases it also results in the recovery of superconvergent values for $\mathcal{F}(u)$. This can be seen from the following example.

Example 4.8.6 Consider the regular model problem with $AE \equiv 1$, $c \equiv 0$, linear elements ($p = 1$), and let

$$u_{EX} \equiv \left(x - \tfrac{1}{2}\right)^2 \tag{4.8.21a}$$

and

$$I^{\Delta}_{j(1/2)} = \left(\frac{1}{2} - \frac{h}{2}, \frac{1}{2} + \frac{h}{2}\right) \tag{4.8.21b}$$

and

$$I^{\Delta}_{j(1/2)-1} = \left(\frac{1}{2} - \frac{h}{2} - \beta\, h, \frac{1}{2} - \frac{h}{2}\right), \qquad I^{\Delta}_{j(1/2)+1} = \left(\frac{1}{2} + \frac{h}{2}, \frac{1}{2} + \frac{h}{2} + \frac{h}{\beta}\right). \tag{4.8.21c}$$

Constructing $\mathcal{A}^{*,cls}_{du/dx}$ for $q = 1$, $w \equiv 1$, we get

$$\begin{aligned}\left\| \frac{d}{dx}(u_{EX}) - \mathcal{A}^{*,cls}_{du/dx}(u_{S^1_\Delta}) \right\|_{L^\infty(I^{\Delta}_{j(1/2)})} \\ = \frac{|\beta^8 - \beta^7 + \beta^6 - \beta^5 + \beta^4 - \beta^3 + \beta^2 - 1|}{\beta^7 + 3\,\beta^6 + 6\,\beta^5 + 7\,\beta^4 + 6\,\beta^3 + 3\,\beta^2 + \beta} h\end{aligned} \tag{4.8.21d}$$

while we have

$$\left\| \frac{d}{dx}(e_{S^1_\Delta}) \right\|_{L^\infty(I^{\Delta}_{j(1/2)})} = h. \tag{4.8.21e}$$

It follows that $\mathcal{A}^{*,cls}_{du/dx}(u_{S^1_\Delta})$ is superconvergent only for the values of β for which the right-hand side of (4.8.21d) is zero, and, as it can be also seen from the graph given in Fig. 4.8.6, this occurs only for the uniform mesh ($\beta = 1$). Note that for sufficiently small or sufficiently large β the error in $\mathcal{A}^{*,cls}_{du/dx}$ is bigger than the error in $(d/dx)(u_{S^1_\Delta})$.

Let us also analyze $\mathcal{A}^{*,ls}_{du/dx}(u_{S^1_\Delta})$, which is constructed as in (4.8.9) or (4.8.11), by employing least-squares fits of the values of the derivative at the Gauss–Legendre points of order two in each element and its neighbors (these points are not superconvergence points for the derivative for $p = 1$). In this case we have

$$\left\| (d/dx)(u_{EX}) - \mathcal{A}^{*,ls}_{du/dx}(u_{S^1_\Delta}) \right\|_{L^\infty(I^{\Delta}_{j(1/2)})} = \frac{|\sqrt{3}\beta^3 - 3\,\beta^2 + 2\,\beta - \sqrt{3} + 1|}{\sqrt{3}\beta(\beta^2 + 4\,\beta + 7)} h. \tag{4.8.21f}$$

Once more, we see that, with the exception of the uniform meshes, $\mathcal{A}^{*,ls}_{du/dx}(u_{S^1_\Delta})$ is not superconvergent.

Let us now give a general approach for verifying the superconvergence of $\mathcal{A}^*_{\mathcal{F}(u)}$ which is based on the results of Section 4.7. We have:

Given a $\mathcal{A}^*_{du/dx,\omega^{(m)}_{I^\Delta_j}}$ which satisfies (i)–(iv), the property

$$\mathcal{A}^*_{\mathcal{F}(u)}(U^{PER(h)}_{S^p_\Delta}) = \mathcal{F}(U^{\bar{x},(p+1)}_{EX}) \quad U^{\bar{x},(p+1)}_{EX} \in \mathcal{P}^{p+1}(I) \tag{4.8.22}$$

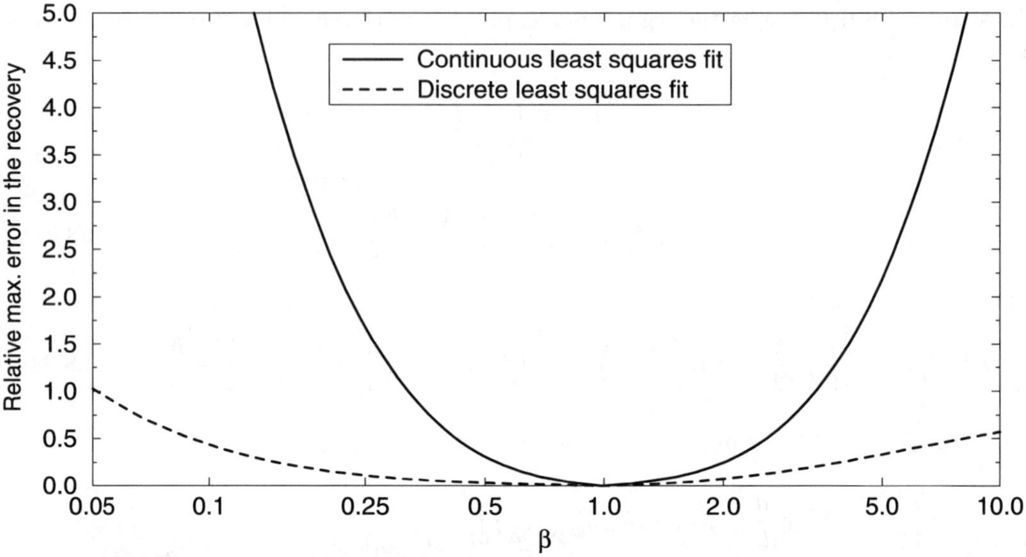

FIG. 4.8.6. The effect of the non-uniformity of the mesh on the asymptotic accuracy of the derivative recovered by the continuous least-squares fit. Input data: $AE \equiv 1$, $c \equiv 0$ and $u_{EX} \equiv (x - \frac{1}{2})^2$. Mesh data: mesh patch of three elements, $I^{\Delta}_{j(1/2)} = (\frac{1}{2} - h/2, \frac{1}{2} + h/2)$, $I^{\Delta}_{j(1/2)-1} = (\frac{1}{2} - h/2 - \beta h, \frac{1}{2} - h/2)$ and $I^{\Delta}_{j(1/2)+1} = (\frac{1}{2} + h/2, \frac{1}{2} + h/2 + h/\beta)$, linear elements ($p = 1$). The relative maximum error in the recovery $\|(d/dx)(u_{EX}) - \mathcal{A}^*_{du/dx}(u_{S^1_\Delta})\|_{L^\infty(I^{\Delta}_{j(1/2)})} / \|(d/dx)(e_{S^1_\Delta})\|_{L^\infty(I^{\Delta}_{j(1/2)})}$ for $\mathcal{A}^*_{du/dx} \equiv \mathcal{A}^{*,cls}_{du/dx}$, and for $\mathcal{A}^*_{du/dx} \equiv \mathcal{A}^{*,ls}_{du/dx}$ (defined in Example 4.8.2), as a function of β. Note that for sufficiently small or sufficiently large β the ratio is greater than one and hence the maximum error in the recovered derivative is greater than the maximum error in the derivative of the finite element solution

where $\Psi^{PER(h)}_{S^p_\Delta}$ is the cell-periodic asymptotic error function given in Theorem 4.7.1, and hence $U^{PER(h)}_{S^p_\Delta}$ is the finite element approximation of $U^{\bar{x},(p+1)}_{S^p_\Delta}$. Equation (4.8.22) is a necessary and sufficient condition for superconvergence.

Remark 4.8.4 Note that (4.8.22) is satisfied if a sufficient number of the superconvergence points $x^{sup}_{du/dx,\omega^{(m)}_{I^\Delta_j}}$, are used as the sampling points, but this is not necessary.

When some or all of the sampling points are chosen to be different than the superconvergence points, superconvergence can only be achieved by *cancellation* of the errors at the sampling points. Let us illustrate this by an example.

Example 4.8.7 We will let

$$\mathcal{A}^*_{du/dx}(v)(\hat{x}) = \sum_{i=1}^{nsp} \gamma_i(\hat{x}) \frac{dv}{dx}(\bar{\bar{x}}_i), \quad \hat{x} \in I^{\Delta}_0 \tag{4.8.23}$$

and we will show that the coefficients γ_i, $i = 1, \ldots, nsp$, can be determined such that $\mathcal{A}^*_{du/dx}(\cdot)(\hat{x})$ is superconvergent. It is sufficient to have

$$\mathcal{A}^*_{du/dx}(U^{PER(h)}_{S^1_\Delta})(\hat{x}) = \frac{d}{dx}(U^{\bar{x},2}_{EX})(\hat{x}) \quad \forall U^{\bar{x},2}_{EX}(x) = \sum_{k=0}^{2} a_k (x - \bar{x})^k \tag{4.8.24}$$

and we obtain

$$\begin{bmatrix} 1 & 1 & \cdots & 1 \\ 2(\hat{x} - \bar{\bar{x}}_1) & 2(\hat{x} - \bar{\bar{x}}_2) & \cdots & 2(\hat{x} - \bar{\bar{x}}_6) \end{bmatrix} \begin{Bmatrix} \gamma_1(\hat{x}) \\ \gamma_2(\hat{x}) \\ \gamma_3(\hat{x}) \\ \gamma_4(\hat{x}) \\ \gamma_5(\hat{x}) \\ \gamma_6(\hat{x}) \end{Bmatrix} = \begin{Bmatrix} 1 \\ 2(\hat{x} - \bar{x}) \end{Bmatrix}. \tag{4.8.25}$$

Using the Gauss–Legendre points of order two in each element as our sampling points $\bar{\bar{x}}_i$ (these points are not superconvergence points for the derivative for $p = 1$), and choosing the solution of (4.8.25) with minimal l^2 norm, for $\hat{x} = \bar{x}$ and $h_{I^\Delta_{j-1}}/h_{I^\Delta_j} = 5$, $h_{I^\Delta_{j+1}}/h_{I^\Delta_j} = \frac{1}{6}$, we get

$$\boldsymbol{\gamma} = \frac{1}{12393}\left[920 - 5\,A, 920 + 5\,A, 2486 - A, 2486 + A, 5581 - \frac{A}{6}, 5581 + \frac{A}{6}\right]^{\mathrm{T}} \tag{4.8.26}$$

with $A = 87\sqrt{3}$.

We can also employ a superconvergent recovery which employs the values

$$\mathcal{A}^{**}_{du/dx}(v)(\hat{x}) = \sum_{i=1}^{nsp} \beta_i(\hat{x}) v(\bar{\bar{x}}_i). \tag{4.8.27a}$$

In this case, condition (4.8.24) gives

$$\begin{bmatrix} 1 & 1 & \cdots & 1 \\ (\hat{x} - \bar{\bar{x}}_1) & (\hat{x} - \bar{\bar{x}}_2) & \cdots & (\hat{x} - \bar{\bar{x}}_6) \\ (\hat{x} - \bar{\bar{x}}_1)^2 & (\hat{x} - \bar{\bar{x}}_2)^2 & \cdots & (\hat{x} - \bar{\bar{x}}_6)^2 \end{bmatrix} \begin{Bmatrix} \beta_1(\hat{x}) \\ \beta_2(\hat{x}) \\ \beta_3(\hat{x}) \\ \beta_4(\hat{x}) \\ \beta_5(\hat{x}) \\ \beta_6(\hat{x}) \end{Bmatrix} = \begin{Bmatrix} 0 \\ 1 \\ 2(\hat{x} - \bar{x}) \end{Bmatrix} \tag{4.8.27b}$$

and, the solution with minimal l^2 norm, for $\hat{x} = \bar{x}$, $h_{I^\Delta_{j-1}}/h_{I^\Delta_j} = 5$, $h_{I^\Delta_{j+1}}/h_{I^\Delta_j} = \frac{1}{6}$ is

$$\begin{aligned} \boldsymbol{\beta} = \frac{1}{4682649277}[&B_{1,1} - B_{1,2}, B_{1,1} + B_{1,2}, B_{2,1} - B_{2,2}, B_{2,1} + B_{2,2}, B_{3,1} \\ &- B_{3,2}, B_{3,1} + B_{3,2}]^{\mathrm{T}} \end{aligned} \tag{4.8.27c}$$

where $B_{1,1} = -1128376557$, $B_{1,2} = -804240990\sqrt{3}$, $B_{2,1} = -185058657$, $B_{2,2} = -387521154\sqrt{3}$, $B_{3,1} = 1313435214$, $B_{3,2} = 82358088\sqrt{3}$.

Remark 4.8.5 Starting in the late 1960s, various averagings have been implemented in the engineering codes, like for example, the continuous least-squares fitting. The idea of superconvergent recovery gained significant popularity with the appearance of the papers [Zienkiewicz and Zhu, 1992a,b, 1994] and [Zienkiewicz, Zhu, and Wu, 1993].

Finally, let us give an example of another type of averaging based on Harnack's Lemma, which was discussed in Section 4.5.

Example 4.8.8 Consider the regular model problem with $AE \equiv 1$, $c \equiv 1$, and $f \equiv 1$, and let $I' = (a, b) \subset I$, $b - a = H$ and $\bar{a} = (a + b)/2$. We then have

$$u_{EX} = \mathscr{C}_1 e^{x-\bar{a}} + \mathscr{C}_2 e^{-(x-\bar{a})} + \mathscr{C}_3 \tag{4.8.28}$$

and

$$\|e_{S^p_\Delta}\|_{L^2(I')} \leq \mathscr{C} h^{p+1} \tag{4.8.29}$$

with $\mathscr{C}$ independent of H. Now, let

$$\mathcal{A}^{*,harm}_{u,I'}(v) \stackrel{\text{def}}{=} \mathscr{C}_1(v) e^{x-\bar{a}} + \mathscr{C}_2(v) e^{-(x-\bar{a})} + \mathscr{C}_3(v) \tag{4.8.30a}$$

where the $\mathscr{C}_i(v)$, $i = 1, 2, 3$, are obtained by solving the minimization problem

$$\|v - \mathcal{A}^{*,harm}_{u,I'}(v)\|_{L^2(I')} = \min_{\alpha,\beta,\gamma} \|v - (\alpha\, e^{x-\bar{a}} + \beta\, e^{-(x-\bar{a})} + \gamma)\|_{L^2(I')}. \tag{4.8.30b}$$

Using (4.8.29), we get

$$|\mathscr{C}_i(u_{EX}) - \mathscr{C}_i(u_{S^p_\Delta})| \leq \mathscr{C} \frac{h^{p+1}}{H} \tag{4.8.31a}$$

for $i = 1, 2, 3$, and hence

$$\left\| \frac{d}{dx}(u_{EX}) - \frac{d}{dx}(\mathcal{A}^{*,harm}_{u,I'}) \right\|_{L^2(I')} \leq \mathscr{C} \frac{h^{p+1}}{H}. \tag{4.8.31b}$$

Selecting now $H = h^{\alpha}$, $\alpha < \frac{1}{2}$, we get superconvergence.

Remark 4.8.6 We call the above recovery *harmonic* because it employs the form of the exact solution of the homogeneous differential equation. It can also be generalized to the cases of non-constant AE, c, and f.

Remark 4.8.7 In Chapter 5 we will utilize the recoveries for a-posteriori error estimation. Using the triangle inequality

$$\begin{aligned} \left\| \frac{d}{dx}(e_{S^p_\Delta}) \right\|_{L^2(I_j)} \leq & \left\| \mathcal{A}^*_{du/dx}(u_{S^p_\Delta}) - \frac{d}{dx}(u_{S^p_\Delta}) \right\|_{L^2(I_j)} \\ & + \left\| \frac{d}{dx}(u_{EX}) - \mathcal{A}^*_{du/dx}(u_{S^p_\Delta}) \right\|_{L^2(I_j)} \end{aligned} \tag{4.8.32a}$$

and noting that

$$\left\| \frac{d}{dx}(u_{EX}) - \mathcal{A}^*_{du/dx}(u_{S^p_\Delta}) \right\|_{L^2(I_j)} \ll \left\| \frac{d}{dx}(e_{S^p_\Delta}) \right\|_{L^2(I_j)} \tag{4.8.32b}$$

we get

$$\left\| \frac{d}{dx}(e_{S^p_\Delta}) \right\|_{L^2(I_j)} \approx \left\| \mathcal{A}^*_{du/dx}(u_{S^p_\Delta}) - \frac{d}{dx}(u_{S^p_\Delta}) \right\|_{L^2(I_j)}. \tag{4.8.32c}$$

The function

$$\hat{e}_{du/dx}(u_{S^p_\Delta}) \stackrel{\text{def}}{=} \mathcal{A}^*_{du/dx}(u_{S^p_\Delta}) - \frac{d}{dx}(u_{S^p_\Delta}) \tag{4.8.33a}$$

is an example of an error indicator function for $(d/dx)(e_{S^p_\Delta})$. In the cases that $\mathcal{A}^*_{du/dx}$ is superconvergent, we have

$$\frac{d}{dx}(u_{EX}) \approx \frac{d}{dx}(u_{S^p_\Delta}) + \hat{e}_{du/dx}(u_{S^p_\Delta}). \tag{4.8.33b}$$

In this section we discussed the construction of superconvergent recoveries based on local least-squares averaging of the finite element solution. We have shown that some approaches which are intuitively very natural, e.g., the continuous least-squares fitting, could lead in many cases to accuracies which are inferior to the accuracy of the original finite element solution. Once more, it is necessary to underline that for the superconvergence to be realized, the pollution must be negligible.

Problems

P4.8.1. Formulate and prove the analogue of Theorem 4.8.1, for the case $\mathcal{F}(u) = u$.

P4.8.2. Given a recovery $\mathcal{A}^*_u$ which is such that

$$\|u_{EX} - \mathcal{A}^*_u(u_{S^p_\Delta})\|_{L^\infty(I^\Delta_j)} \leq \mathscr{C}(h_{I^\Delta_j})^{p+2} \tag{4.8.33a}$$

prove that

$$\left\| \frac{d}{dx}(u_{EX} - \mathcal{A}^*_u u_{S^p_\Delta}) \right\|_{L^\infty(I^\Delta_j)} \leq \mathscr{C}(h_{I^\Delta_j})^{p+1}. \tag{4.8.33b}$$

Hence, if $\mathcal{A}^*_u$ is a superconvergent recovery for the value, then $(d/dx)(\mathcal{A}^*_u)$ is a superconvergent recovery for the derivative.

P4.8.3. Discuss the dependence of the error $\|\mathcal{F}(u) - \mathcal{A}^*_{\mathcal{F}(u)}(u_{S^p_\Delta})\|_{L^\infty(I^\Delta_j)}$ on the parameter κ which controls the local uniformity of the mesh. Consider, for example, $\mathcal{F}(u) = u$ and $\mathcal{F}(u) = du/dx$.

P4.8.4. Use a counterexample to show that in the cases that Assumption II is violated, there is no local recovery $\mathcal{A}^*_{\mathcal{F}(u)}$ such that $\mathcal{A}^*_{\mathcal{F}(u)}(u_{S^p_\Delta})$ has better accuracy than $\mathcal{F}(u_{S^p_\Delta})$.

P4.8.5. Show that, although the pollution in $(d/dx)(u_{S^p_\Delta})$ may be negligible, this may not be the case for $\mathcal{A}^*_{du/dx}(u_{S^p_\Delta})$.

Hint: The derivative $(d/dx)(u_{S^p_\Delta})$ has accuracy of order h^p, and hence, pollution of order h^{p+1} is negligible. On the other hand, $\mathcal{A}^*_{du/dx}(u_{S^1_\Delta})$ has accuracy of h^{p+1}, and hence, the pollution of order h^{p+1} cannot be neglected.

4.8* Superconvergence based on local recovery: the two-dimensional case

The construction of superconvergent recoveries in higher dimensions is primarily based on the principle of cancellation of error. This is because in higher dimensions:

1. *There may not exist any superconvergence points or there may exist too few points.* For example, there are no superconvergence points for the derivative for finite element solutions of the Poisson equation in meshes of quadratic triangles arranged in the Union-Jack pattern.
2. *The superconvergence points may be cumbersome to use.* For example, the superconvergence points for the components of the derivative for finite element solutions of the Laplace equation are different for each component, and depend on the mesh pattern and the orientation of the axes of coordinates with respect to the mesh pattern.

For these reasons, the construction of superconvergent recoveries in the two-dimensional case is mainly based on *cancellation of error*.

As in the one-dimensional case, we will denote by $\mathcal{A}^*_{\mathcal{F}(u),\omega_\tau}$, a recovery which employs a least-squares fit of $\mathcal{F}(u_{S^p_\Delta})$ in the mesh patch ω_τ, which is the patch of elements which are connected to the vertices of τ. Let us give an example of the two-dimensional analogue of $\mathcal{A}^{C^0}_{\mathcal{F}(u)}$, which is known as *Zienkiewicz–Zhu Superconvergent Patch Recovery* (ZZ-SPR) (see [Zienkiewicz and Zhu, 1992a,b,c, 1993, and 1994]). Note that *for general meshes ZZ-SPR recovery is not superconvergent.*

As in the one-dimensional case, the construction of $\mathcal{A}^{C^0}_{\mathcal{F}(u)}$ is done in two steps:

(i) *Least-squares fits over the nodal patches*: Let $\omega_{X_i} = \bigcup_{X_i \in \mathcal{N}(\tau')} \tau$ be the patch of elements connected to the vertex X_i^τ, where X_i^τ is the ith vertex of the element τ.

Given $v \in \prod_{\tau' \in \omega_{X_i}} C^\infty(\tau)$, we will let $\mathcal{A}^*_{\mathcal{F}(u),\omega_{X_i^\tau}}(v) \in \mathcal{P}^q(\omega_{X_i})$ be the solution of the minimization problem

$$\|\mathcal{A}^*_{\mathcal{F}(u),\omega_{X_i}}(v) - \mathcal{F}(v)\|_{*,\omega_{X_i}} = \min_{\sigma^* \in \mathcal{P}^q(\omega_{X_i})} \|\sigma^* - \mathcal{F}(v)\|_{*,\omega_{X_i}} \qquad (4.8^*.1a)$$

where $\|\cdot\|_{*,\omega_{X_i}}$ is the norm which is employed in the recovery. This recovery was first introduced in the case $\mathcal{F}(u) = du/dx$, with $q = p$, and

$$\|v\|_{*,\omega_{X_i}} = \sqrt{\sum_{\tau' \in \omega_{X_i}} \sum_{\bar{x}_k^{\tau'} \in \tau'} w_k (v(\bar{x}_k^{\tau'}))^2} \qquad (4.8^*.1b)$$

where $\bar{x}_k^{\tau'}$ denotes the kth *sampling point* in the element τ'. The sampling points $\bar{x}_k^{\tau'}$ are often taken to be the mapped points of a quadrature rule defined in the master element. Note that the points $\bar{x}_k^{\tau'}$ are referred to as *sampling points because, in general, they are not superconvergence points for* $\mathcal{F}(u)$. For example, in the case of linear triangles the centroid is chosen as the sampling point for both components of the gradient.

(ii) *Patching of the recovered values into a continuous function*: The recovered quantity $\mathcal{A}^{C^0}_{\mathcal{F}(u)}(v)$ is constructed by averaging the $\mathcal{A}^*_{\mathcal{F}(u),\omega_{X_i^\tau}}(v)$ as follows:

(a) Nodal degrees of freedom

$$\mathcal{A}^{C^0}_{\mathcal{F}(u)}(v)(X^{\tau}_i) = \mathcal{A}^*_{\mathcal{F}(u),\omega_{X^{\tau}_i}}(v)(X^{\tau}_i). \tag{4.8*.2a}$$

(b) Edge degrees of freedom

$$\mathcal{A}^{C^0}_{\mathcal{F}(u)}(v)(\bar{X}^{\epsilon}_k) = \frac{1}{2}\Big(\mathcal{A}^*_{\mathcal{F}(u),\omega_{X^{\epsilon}_1}}(v)(\bar{X}^{\epsilon}_k) + \mathcal{A}^*_{\mathcal{F}(u),\omega_{X^{\epsilon}_2}}(v)(\bar{X}^{\epsilon}_k)\Big) \tag{4.8*.2b}$$

where X^{ϵ}_1, X^{ϵ}_2 are the end points of the edge ϵ and $\bar{X}^{\epsilon}_k$, $k = 1, \ldots, (p-1)$ are the interior nodal points on the edge ϵ.

(c) Interior degrees of freedom

$$\mathcal{A}^{C^0}_{\mathcal{F}(u)}(v)(\bar{X}^{\tau}_k) = \frac{1}{nv}\sum_{i=1}^{nv} \mathcal{A}^*_{\mathcal{F}(u),\omega_{X_i}}(v)(\bar{X}^{\tau}_k) \tag{4.8*.2c}$$

where $\bar{X}^{\tau}_k$, $k = 1, 2, \ldots,$ are the interior nodes of the element τ, and $nv = 3$ when τ is a triangular element (respectively $nv = 4$ when τ is a quadrilateral element).

Note that above we assumed that a Lagrangian basis is employed. The above construction may also be employed for elements with one or more vertices on the boundary, or another form which employs extrapolation from the interior patches may be used [Zienkiewicz and Zhu, 1994].

Remark 4.8*.1 The above recovery is superconvergent only for the special cases, e.g., for a mesh of triangles arranged in the regular pattern and $p = 1$.

Remark 4.8*.2 In the more general case, when a hierarchic basis is employed the edge degrees of freedom can be obtained by fixing the vertex degrees of freedom and by solving a one-dimensional minimization problem over each edge, and the interior degrees of freedom can be determined by fixing the vertex and the edge degrees of freedom and by solving a minimization problem in the interior of the element.

From the asymptotic analysis of Section 4.7*, it follows that a necessary and sufficient condition for superconvergence is

$$\mathcal{A}^*_{\mathcal{F}(u)}\big(U^{PER(h)}_{S^p_\Delta}\big) = \mathcal{F}(U^{\bar{\mathbf{x}},(p+1)}_{EX}) \quad \forall U^{\bar{\mathbf{x}},(p+1)}_{EX} \in \mathcal{P}^{p+1} \tag{4.8*.3}$$

where we employed the notations introduced in Section 4.7*.

A recovery $\mathcal{A}^*_{\mathcal{F}(u)}$ may be analyzed by analyzing the relative error

$$\Theta_{\mathcal{A}^*_{\mathcal{F}(u)}}(\bar{\mathbf{x}}; u_{EX}, u_{S^p_\Delta}, \tau) \stackrel{\text{def}}{=} \frac{\Big|\big(\mathcal{F}(u_{EX}) - \mathcal{A}^*_{\mathcal{F}(u)}(u_{S^p_\Delta})\big)(\bar{\mathbf{x}})\Big|}{\|\mathcal{F}(u_{EX}) - \mathcal{F}(u_{S^p_\Delta})\|_{L^\infty(\tau)}} \tag{4.8*.4}$$

where we assumed that $\|\mathcal{F}(u_{EX}) - \mathcal{F}(u_{S^p_\Delta})\|_{L^\infty(\tau)} \neq 0$. Note that, in many cases we could have $\Theta_{\mathcal{A}^*_{\mathcal{F}(u)}}(\bar{\mathbf{x}}; u_{EX}, u_{S^p_\Delta}, \tau) \gg 1$.

As in the case of direct superconvergence which was discussed in Section 4.7*, we will introduce the $\eta\%$-superconvergence regions for $\mathcal{A}^*_{\mathcal{F}(u)}$ by

$$\mathcal{G}^{\eta\%}_{\mathcal{A}^*_{\mathcal{F}(u)}}(\mathfrak{Q};\tau,\Delta) \stackrel{\text{def}}{=} \left\{ \mathbf{x}\in\tau \,\Big|\, \sup_{u_{EX}\in\mathfrak{Q}} \Theta_{\mathcal{A}^*_{\mathcal{F}(u)}}(\bar{\mathbf{x}};u_{EX},u_{S^p_\Delta},\tau) \le 1-\frac{\eta}{100} \right\}. \qquad (4.8^*.5)$$

The relative quality of two recoveries $\mathcal{A}^{*,1}_{\mathcal{F}(u)}$ and $\mathcal{A}^{*,2}_{\mathcal{F}(u)}$ at a point $\bar{\mathbf{x}}$ can be measured by computing the *index of relative robustness of* $\mathcal{A}^{*,1}_{\mathcal{F}(u)}$ *with respect to* $\mathcal{A}^{*,2}_{\mathcal{F}(u)}$, namely

$$r\left(\bar{\mathbf{x}};\mathfrak{Q};\frac{\mathcal{A}^{*,1}_{\mathcal{F}(u)}}{\mathcal{A}^{*,2}_{\mathcal{F}(u)}};(\Delta,p),\tau\right) \stackrel{\text{def}}{=} \frac{\sup_{u_{EX}\in\mathfrak{Q}} \Theta_{\mathcal{A}^{*,1}_{\mathcal{F}(u)}}(\bar{\mathbf{x}};u_{EX},u_{S^p_\Delta},\tau)}{\sup_{u_{EX}\in\mathfrak{Q}} \Theta_{\mathcal{A}^{*,2}_{\mathcal{F}(u)}}(\bar{\mathbf{x}};u_{EX},u_{S^p_\Delta},\tau)} \qquad (4.8^*.6)$$

where $\mathfrak{Q}$ is the class of solutions under consideration. In the cases that the denominator is equal to zero, the ratio will be taken to be ∞.

The regions where the recovery $\mathcal{A}^{*,1}_{\mathcal{F}(u)}$ is more robust than the recovery $\mathcal{A}^{*,2}_{\mathcal{F}(u)}$ are identified by

$$\mathcal{G}_{\mathcal{A}^{*,1}_{\mathcal{F}(u)}>\mathcal{A}^{*,2}_{\mathcal{F}(u)}}(\mathfrak{Q};\tau,\Delta) \stackrel{\text{def}}{=} \left\{ \mathbf{x}\in\tau \,\Big|\, r\left(\bar{\mathbf{x}};\mathfrak{Q};\frac{\mathcal{A}^{*,1}_{\mathcal{F}(u)}}{\mathcal{A}^{*,2}_{\mathcal{F}(u)}};(\Delta,p),\tau\right) < 1 \right\}. \qquad (4.8^*.7)$$

We will say that $\mathcal{A}^*_{\mathcal{F}(u)}$ is effective for the class $\mathfrak{Q}$ if the regions $\mathcal{G}_{\mathcal{A}^*_{\mathcal{F}(u)}>\mathcal{F}(u)}(\mathfrak{Q};\tau,\Delta)$ are of significant size.

Remark 4.8*.3 In the engineering literature comparisons between various recoveries are often based on the study of their accuracy on one or two example problems. Our approach which is based on (4.8*.3)–(4.8*.5) removes any bias related to the selection of the example problems.

Let us now give some examples of how the *index of relative robustness* can be used to compare various recoveries.

Example 4.8*.1 (*Comparison of the pointwise accuracy of local recoveries for periodic meshes*) Let us be interested in the relative accuracy of the recoveries at the Neumann boundary (the bottom edge of the boundary strip) for the four patterns of mesh. Figure 4.8*.1 (respectively Figs 4.8*.2–4.8*.4) compares the regions $\hat{\mathcal{G}}^{\eta\%}_{\mathcal{A}^*_{\mathcal{F}(u)}}(\mathfrak{Q}^{(3)}_{hom};\tau,\Delta)$ for the *ERpB*, *ERpPp*+1 recoveries, which will be defined in Section 5.3*, and *ZZ-SPR*, for the regular pattern (respectively chevron pattern, Union-Jack pattern, and criss-cross pattern) when quadratic elements ($p=2$) are employed. Here $\mathfrak{Q}^{(3)}_{hom}$ is the set of polynomials of degree 3 satisfying the Neumann homogeneous condition. Figures 4.8*.1d, 4.8*.2d, 4.8*.3d, and 4.8*.4d show the regions $\mathcal{G}_{\mathcal{A}^{ERpB}_{\nabla u}>\mathcal{A}^{ZZ}_{\nabla u}}(\mathfrak{Q}^{(3)}_{hom};\tau,\Delta)$ for the regular, chevron, Union-Jack, and criss-cross patterns, respectively. We note the following:

(a) The recovered gradient $\mathcal{A}^{ERpB}_{\nabla u}$ is a marked improvement over $\nabla u_{S^p_\Delta}$, the gradient of the finite element solution, for all the four patterns.

(b) When *ZZ-SPR* recovery is employed, the more than 50% regions are significant for the regular and chevron patterns.

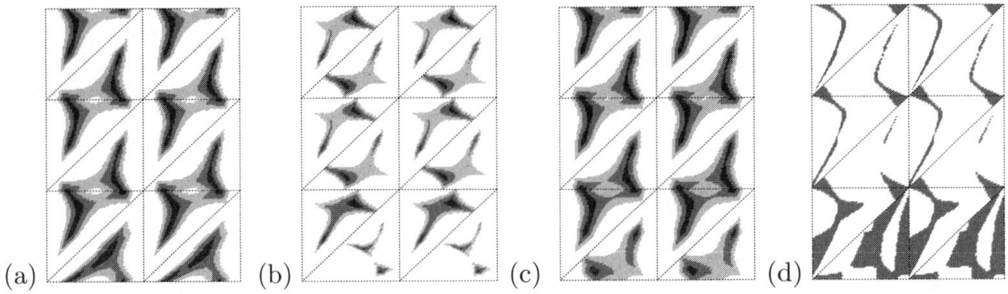

FIG. 4.8*.1. Pointwise accuracy of local recoveries: Laplace equation, quadratic elements ($p = 2$), a strip of elements at the Neumann boundary (the bottom edge of the strip) made by repeating the regular pattern. The regions of superconvergence ($\eta = 90, 80, 70$) for the x_1 derivative, $\mathcal{G}^{\eta\%}_{\mathcal{A}^*_{\nabla u}}(\mathfrak{Q}^{(3)}_{hom}; \tau, \Delta)$, when the (a) $ERpB$, (b) $ERpPp+1$, (c) ZZ, are employed. (d) The region $\mathcal{G}_{\mathcal{A}^{ERpB}_{\nabla u} > \mathcal{A}^{ZZ}_{\nabla u}}(\mathfrak{Q}^{(3)}_{hom}; \tau, \Delta)$ of robustness of the ERpB recovery, relative to ZZ recovery is shown shaded gray

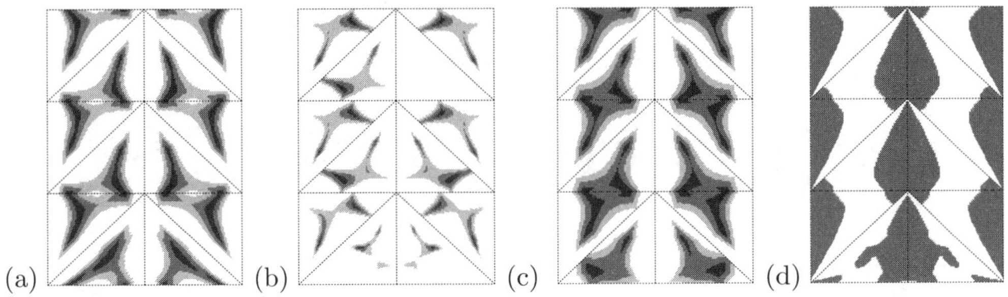

FIG. 4.8*.2. Pointwise accuracy of local recoveries: Laplace equation, quadratic elements ($p = 2$), a strip of elements at the Neumann boundary (the bottom edge of the strip) made by repeating the chevron pattern. The regions of superconvergence ($\eta = 90, 80, 70$) for the x_1 derivative, $\mathcal{G}^{\eta\%}_{\mathcal{A}^*_{\nabla u}}(\mathfrak{Q}^{(3)}_{hom}; \tau, \Delta)$, when the (a) $ERpB$, (b) $ERpPp+1$, (c) ZZ, are employed. (d) The region $\mathcal{G}_{\mathcal{A}^{ERpB}_{\nabla u} > \mathcal{A}^{ZZ}_{\nabla u}}(\mathfrak{Q}^{(3)}_{hom}; \tau, \Delta)$ of robustness of the ERpB recovery, relative to ZZ recovery is shown shaded gray

(c) The recovered gradient $\mathcal{A}^{ERpPp+1}_{\boldsymbol{\nabla} u}(u_{S^p_\Delta}) = \boldsymbol{\nabla} u_{S^p_\Delta} + \boldsymbol{\nabla} \hat{e}^{ERpPp+1}_\tau$, which corresponds to the error indicator function of the $ERpPp$+1 estimator, has very small regions of 75% superconvergence, and hence its pointwise accuracy is inferior to that for $\mathcal{A}^{ERpB}_{\boldsymbol{\nabla} u}(u_{S^p_\Delta})$ and $\mathcal{A}^{ZZ}_{\boldsymbol{\nabla} u}(u_{S^p_\Delta})$.

(d) For the elements at the Neumann boundary, the quality of the recoveries is rather different from that in the interior of the domain.

It is also interesting to consider the case of the orthotropic Laplacian, $\boldsymbol{\nabla} \cdot (\mathbf{K} \boldsymbol{\nabla} u) = 0$, in order to see the effect of the orthortopy.

Example 4.8*.2 (*Effect of the material orthotropy on the robustness of the recoveries*) Let us consider the orthotropic Laplacian ($\boldsymbol{\nabla} \cdot (\mathbf{K} \boldsymbol{\nabla} u) = 0$) with ratio of the principal values of orthotropy $K_1/K_2 = 4$, and determine the common $\eta\%$ for the components of the gradient, by taking the intersection of the $\eta\%$ regions for $\theta = 0°, 30°, 60°, 90°, 120°, 150°$,

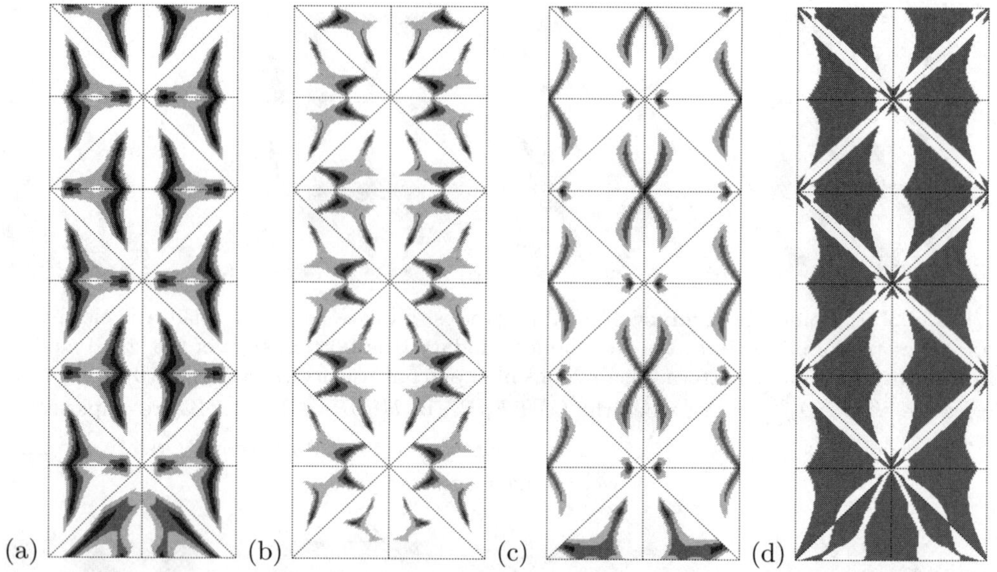

FIG. 4.8*.3. Union-Jack pattern. All other settings are the same as in Fig. 4.8*.1

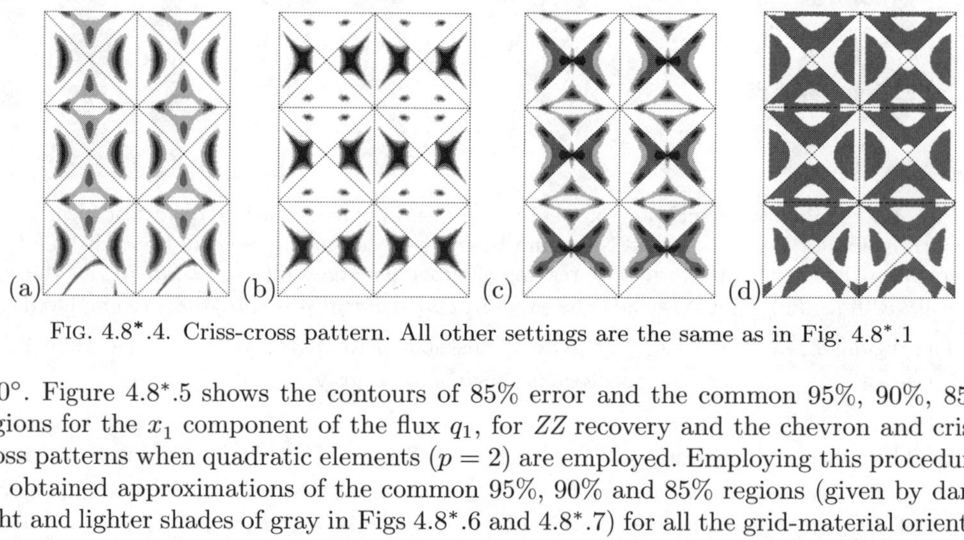

FIG. 4.8*.4. Criss-cross pattern. All other settings are the same as in Fig. 4.8*.1

180°. Figure 4.8*.5 shows the contours of 85% error and the common 95%, 90%, 85% regions for the x_1 component of the flux q_1, for *ZZ* recovery and the chevron and criss-cross patterns when quadratic elements ($p = 2$) are employed. Employing this procedure, we obtained approximations of the common 95%, 90% and 85% regions (given by dark, light and lighter shades of gray in Figs 4.8*.6 and 4.8*.7) for all the grid-material orientations $0 \leq \theta \leq 180°$ for the components of the flux, namely the regions $\mathscr{G}^{\eta\%}_{q_1^*}(\mathscr{Q}^{(3)}_{hom}; \mathbf{K}; \tau, \Delta)$, $\mathscr{G}^{\eta\%}_{q_2^*}(\mathscr{Q}^{(3)}_{hom}; \mathbf{K}; \tau, \Delta)$ and the common regions of simultaneous superconvergence for both components of the flux, $\mathscr{G}^{\eta\%}_{\mathbf{q}^*}(\mathscr{Q}^{(3)}_{hom}; \mathbf{K}; \tau, \Delta)$, for the *ZZ* and *ERpB* recoveries and the chevron and criss-cross patterns when quadratic element ($p = 2$) are employed, as shown in Figs 4.8*.6 and 4.8*.7. Note that for the criss-cross pattern, there are no common 85% simultaneous superconvergence regions for *ERpB* recovery, and for this reason in Fig. 4.8*.7e we employed $\eta = 80$, 75, 70. We note that:

(a) Both the *ZZ* and *ERpB* recoveries have significant 85%-superconvergence regions, for q_1 and q_2, for both the mesh patterns considered, while there are no common

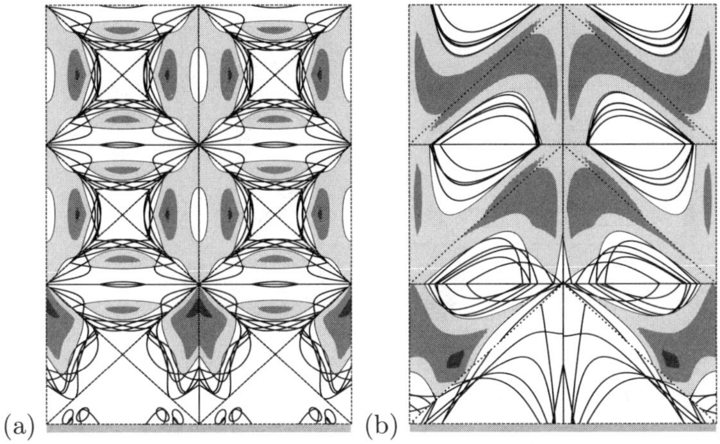

FIG. 4.8*.5. Construction of the common $\eta\%$ regions for all grid-material orientations: Orthotropic heat conduction ($\nabla \cdot (\mathbf{K}\nabla u) = 0$) with $K_1/K_2 = 4$, quadratic elements ($p = 2$), a strip of elements at the Neumann boundary (the edge $x_2 = 0$ shown shaded gray). The contours of 85% error for the x_1 component of the flux, $q_1 = K_{11}(\partial u/\partial x_1) + K_{12}(\partial u/\partial x_2)$, for grid-material orientations $\theta = 0°, 30°, 60°, 90°, 120°, 150°, 180°$ are shown with thick lines. The common 95%, 90%, 85% regions are shown with dark, light and lighter shades of gray for the (a) chevron pattern, (b) criss-cross pattern for *ZZ* recovery

regions of 85% superconvergence for the fluxes computed directly from the finite element solution.

(b) The common regions of 85% superconvergence for *ZZ* recovery are greater than those for *ERpB* recovery for both the mesh patterns, and we can conclude that *ZZ recovery is superior to ERpB recovery.*

(c) The common 85% simultaneous superconvergence regions for the components of the flux are significant for *ZZ* recovery, while for *ERpB* recovery the corresponding regions are small. For the criss-cross pattern *ERpB* recovery has no 80% simultaneous superconvergence regions for the components of the flux.

The above examples demonstrate that *ZZ* and *ERpB* recoveries are not superconvergent in the classical sense, and their accuracy can deteriorate significantly with increasing orthotropy.

Let us also address briefly the case of elements with a vertex at a corner point. We will distinguish two cases, namely, $\alpha > p + 1$ and $\alpha \leq p + 1$, where the solution in the neighborhood of the corner is given by $u \approx r^\alpha \varphi(\theta)$. In the first case the singular behavior of the exact solution is insignificant and the analysis can proceed similarly as above. In the case that the terms of the singular asymptotic expansion in the neighborhood of the corner point is known, it can be used in the recovery, where the goal is to recover the coefficients (*intensity factors*) for the leading terms; see [Babuška and Miller, 1984], [Wahlbin, 1991], and [Babuška, Strouboulis *et al.*, 2001].

Above, we constructed the recoveries without taking into account what we already know about the solution, namely, that it satisfies the differential equation (the Laplacian, orthotropic Laplacian, etc.), the boundary conditions, etc. Let us now show that by

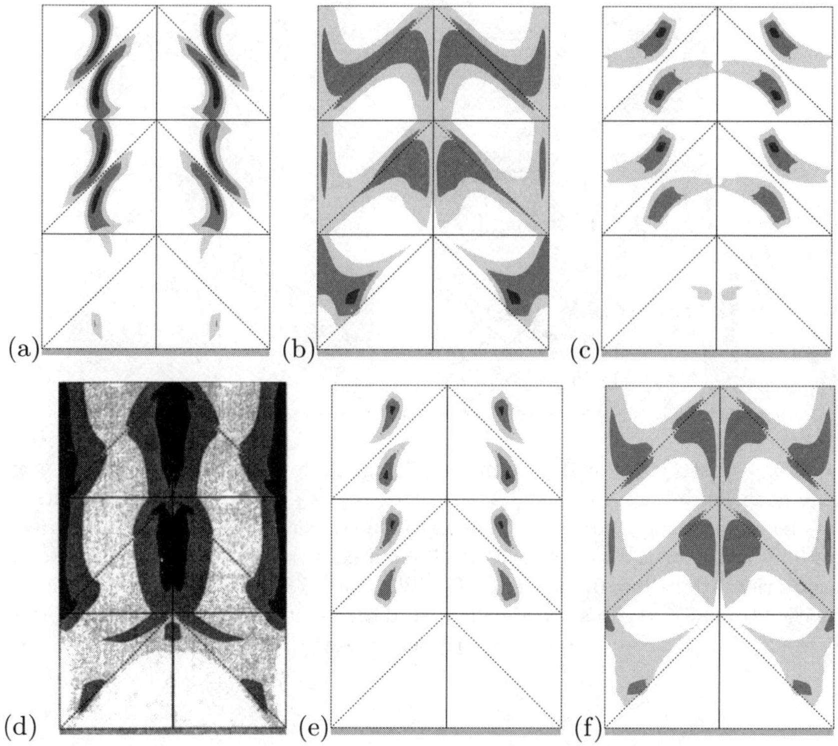

FIG. 4.8*.6. Pointwise accuracy of local recoveries: Orthotropic heat conduction ($\boldsymbol{\nabla} \cdot (\mathbf{K}\boldsymbol{\nabla} u) = 0$) with $K_1/K_2 = 4$, quadratic elements ($p = 2$), a strip of elements at the Neumann boundary (the edge $x_2 = 0$ shown shaded gray) made by repeating the chevron pattern. The common regions of η%-superconvergence for all grid-material orientations ($0 \le \theta \le 180°$) for (a) $\mathcal{G}^{\eta\%}_{q_1^{ERpB}}(\mathfrak{Q}^{(3)}_{hom}; \mathbf{K}; \tau, \Delta)$; (b) $\mathcal{G}^{\eta\%}_{q_1^{ZZ}}(\mathfrak{Q}^{(3)}_{hom}; \mathbf{K}; \tau, \Delta)$; (c) $\mathcal{G}^{\eta\%}_{q_2^{ERpB}}(\mathfrak{Q}^{(3)}_{hom}; \mathbf{K}; \tau, \Delta)$; (d) $\mathcal{G}^{\eta\%}_{q_2^{ZZ}}(\mathfrak{Q}^{(3)}_{hom}; \mathbf{K}; \tau, \Delta)$; (e) $\mathcal{G}^{\eta\%}_{\mathbf{q}^{ERpB}}(\mathfrak{Q}^{(3)}_{hom}; \mathbf{K}; \tau, \Delta)$; (f) $\mathcal{G}^{\eta\%}_{\mathbf{q}^{ZZ}}(\mathfrak{Q}^{(3)}_{hom}; \mathbf{K}; \tau, \Delta)$. The shades indicate the regions of 95, 90, and 85% superconvergence (dark, lighter, light gray)

including this information in the construction of a recovery we can improve its accuracy. Here we will consider the case where the recovery uses the information about the differential equation.

Consider the problem of the orthotropic Laplacian $\boldsymbol{\nabla} \cdot (\mathbf{K}\,\boldsymbol{\nabla} u) = 0$. There are only two independent monomials of degree l which satisfy the differential equation, namely,

$$Q_1^{(l)}(\tilde{x}_1, \tilde{x}_2) = \mathrm{Re}(\tilde{z}^l), \qquad Q_2^{(l)}(\tilde{x}_1, \tilde{x}_2) = \mathrm{Im}(\tilde{z}^l) \tag{4.8*.8a}$$

with $\tilde{z} = \tilde{x}_1 + i\tilde{x}_2$ and

$$\tilde{x}_1 = \frac{1}{\sqrt{K_{11}}}(x_1 \cos\theta + x_2 \sin\theta), \qquad \tilde{x}_2 = \frac{1}{\sqrt{K_{22}}}(-x_1 \sin\theta + x_2 \cos\theta) \tag{4.8*.8b}$$

where θ is the angle of the grid-material orientation.

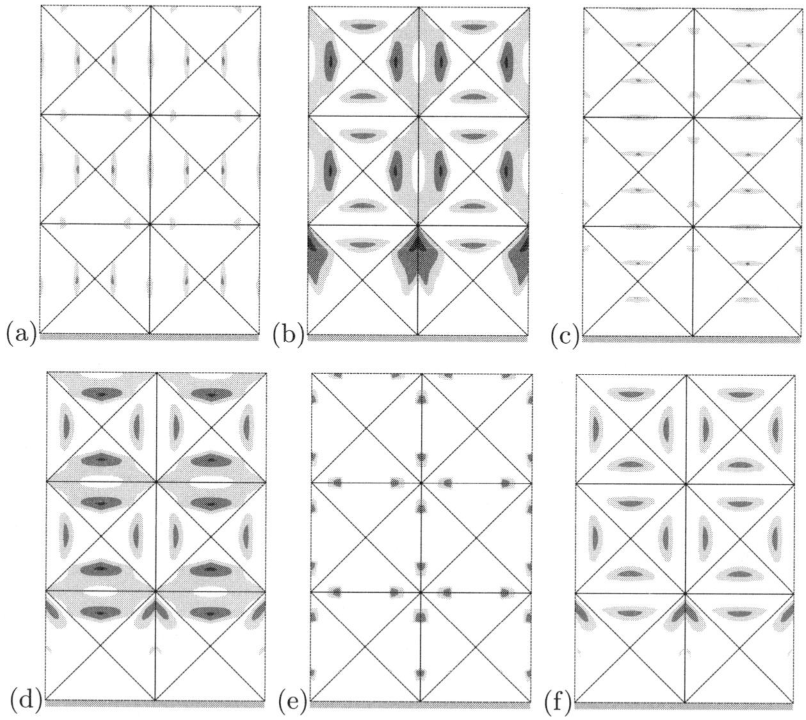

FIG. 4.8*.7. Pointwise accuracy of local recoveries: Orthotropic heat conduction ($\boldsymbol{\nabla} \cdot (\mathbf{K}\boldsymbol{\nabla} u) = 0$) with $K_1/K_2 = 4$, quadratic elements ($p = 2$), a strip of elements at the Neumann boundary (the edge $x_2 = 0$ shown shaded gray) made by repeating the criss-cross pattern. The common regions of η%-superconvergence for all grid-material orientations ($0 \leq \theta \leq 180°$) for (a) $\mathcal{G}^{\eta\%}_{q_1^{ERpB}}(\mathcal{Q}^{(3)}_{hom}; \mathbf{K}; \tau, \Delta)$; (b) $\mathcal{G}^{\eta\%}_{q_1^{ZZ}}(\mathcal{Q}^{(3)}_{hom}; \mathbf{K}; \tau, \Delta)$; (c) $\mathcal{G}^{\eta\%}_{q_2^{ERpB}}(\mathcal{Q}^{(3)}_{hom}; \mathbf{K}; \tau, \Delta)$; (d) $\mathcal{G}^{\eta\%}_{q_2^{ZZ}}(\mathcal{Q}^{(3)}_{hom}; \mathbf{K}; \tau, \Delta)$; (e) $\mathcal{G}^{\eta\%}_{\mathbf{q}^{ERpB}}(\mathcal{Q}^{(3)}_{hom}; \mathbf{K}; \tau, \Delta)$; (f) $\mathcal{G}^{\eta\%}_{\mathbf{q}^{ZZ}}(\mathcal{Q}^{(3)}_{hom}; \mathbf{K}; \tau, \Delta)$. The shades indicate the regions of 95, 90, and 85% superconvergence (dark, lighter, light gray) except for (e), where the shades indicate 80, 75, 70% superconvergence

Let ω_h^τ be the patch of elements which consists of the element τ and one layer of elements around it. We have, among others, the following possibilities:

(a) *'Harmonic' recovery of the gradient* ($RecHG$).
Let $\mathbf{q}_\tau^{q,RecHG} = \mathbf{q}(u_{\omega_h^\tau}^{q,RecHG} \mid_\tau)$ where

$$u_{\omega_h^\tau}^{q,RecHG} \in \mathcal{P}^q_{hom}(\omega_h^\tau) \stackrel{\text{def}}{=} \{v \in \mathcal{P}^q(\omega_h^\tau) \mid \boldsymbol{\nabla} \cdot (\mathbf{K}\boldsymbol{\nabla} v) = 0\} \tag{4.8*.9a}$$

is determined from

$$|||u_{\omega_h^\tau}^{q,RecHG} - u_h|||_{\omega_h^\tau} = \min_{v \in \mathcal{P}^q_{hom}(\omega_h^\tau)} |||v - u_h|||_{\omega_h^\tau}. \tag{4.8*.9b}$$

The condition $\boldsymbol{\nabla} \cdot (\mathbf{K}\boldsymbol{\nabla} u^{q,RecHG}_{\omega^\tau_h}) = 0$ is enforced by letting

$$u^{q,RecHG}_{\omega^\tau_h} = \alpha_1 + \sum_{l=1}^{q}\sum_{i=1}^{2} \alpha_{(l-1)2+1+i} Q^{(l)}_i. \tag{4.8*.9c}$$

Note that $\mathbf{q}^{q,RecHG}_\tau$ is not continuous from element to element.

(b) *Patch recovery based on the value* (*RecHV*):
Let $\mathbf{q}^{q,RecHV}_\tau = \mathbf{q}(u^{q,RecHV}_{\omega^\tau_h}|_\tau)$, where

$$u^{q,RecHV}_{\omega^\tau_h} = \alpha_1 + \sum_{l=1}^{q}\sum_{i=1}^{2} \alpha_{(l-1)2+1+i} Q^{(l)}_i \in \mathcal{P}^q_{hom}(\omega^\tau_h) \tag{4.8*.10a}$$

is determined from

$$\left\| u^{q,RecHV}_{\omega^\tau_h} - u_h \right\|_{L^2(\omega^\tau_h)} = \min_{v \in \mathcal{P}^{(q)}_{hom}(\omega^\tau_h)} \|v - u_h\|_{L^2(\omega^\tau_h)}. \tag{4.8*.10b}$$

Note that we can also construct C^0-continuous versions of *RecHG* and *RecHV* recoveries, analogously as ZZ-SPR.

Let us now analyze the accuracy of *RecHG* and *RecHV*.

Example 4.8*.3 (*Robustness of recoveries which employ the known local character of the exact solution*) Consider the problem of orthotropic heat conduction ($\boldsymbol{\nabla} \cdot (\mathbf{K}\boldsymbol{\nabla} u) = 0$) with $K_1/K_2 = 4$. Figure 4.8*.8 shows the common 95%, 90%, and 85% regions for all grid-material orientations $0 \leq \theta \leq 180°$ for the components of the flux, namely $\mathcal{G}^{\eta\%}_{q^*_1}(\mathcal{Q}^{(3)}_{hom}; \mathbf{K}; \tau, \Delta)$, $\mathcal{G}^{\eta\%}_{q^*_2}(\mathcal{Q}^{(3)}_{hom}; \mathbf{K}; \tau, \Delta)$, and the regions of simultaneous superconvergence $\mathcal{G}^{\eta\%}_{\mathbf{q}^*}(\mathcal{Q}^{(3)}_{hom}; \mathbf{K}; \tau, \Delta)$, for the *RecHG* and *RecHV* recoveries and the criss-cross pattern when quadratic element ($p = 2$) are employed. Note that the common regions of 85% superconvergence are significant for all the elements (including the elements at the boundary) for both *RecHG* and *RecHV* recoveries. Further, for the elements in the interior, the 90% regions cover all the elements completely for both recoveries. From a comparison of Fig. 4.8*.8 with Fig. 4.8*.6, it can be clearly seen that *both RecHG and RecHV recoveries are much superior pointwise to ZZ and ERpB recoveries.* As an improvement of the *ERpPp+1* recovery one can employ the patch residual recovery, *PRpPp+1* (defined in Chapter 5). For this recovery note, from Figs 4.8*.8c, 4.8*.8f, and 4.8*.8i that (for the criss-cross pattern) the common regions of 90% simultaneous superconvergence, for the flux $\mathbf{q}$, for the orthotropic material with $K_1/K_2 = 4$ are significant, even in the elements adjacent to the boundary. Note that while for *ERpPp+1* and *ERpB* recoveries there are no regions of 90% superconvergence, *PRpPp+1* has significant 90% regions and thus it is more robust pointwise.

As in the one-dimensional case, given a sufficient number of arbitrary sampling points we can construct the weights so that we obtain a superconvergent recovery at a given point. Let $\hat{\mathbf{x}}$ be the point where we would like to obtain a superconvergent value for

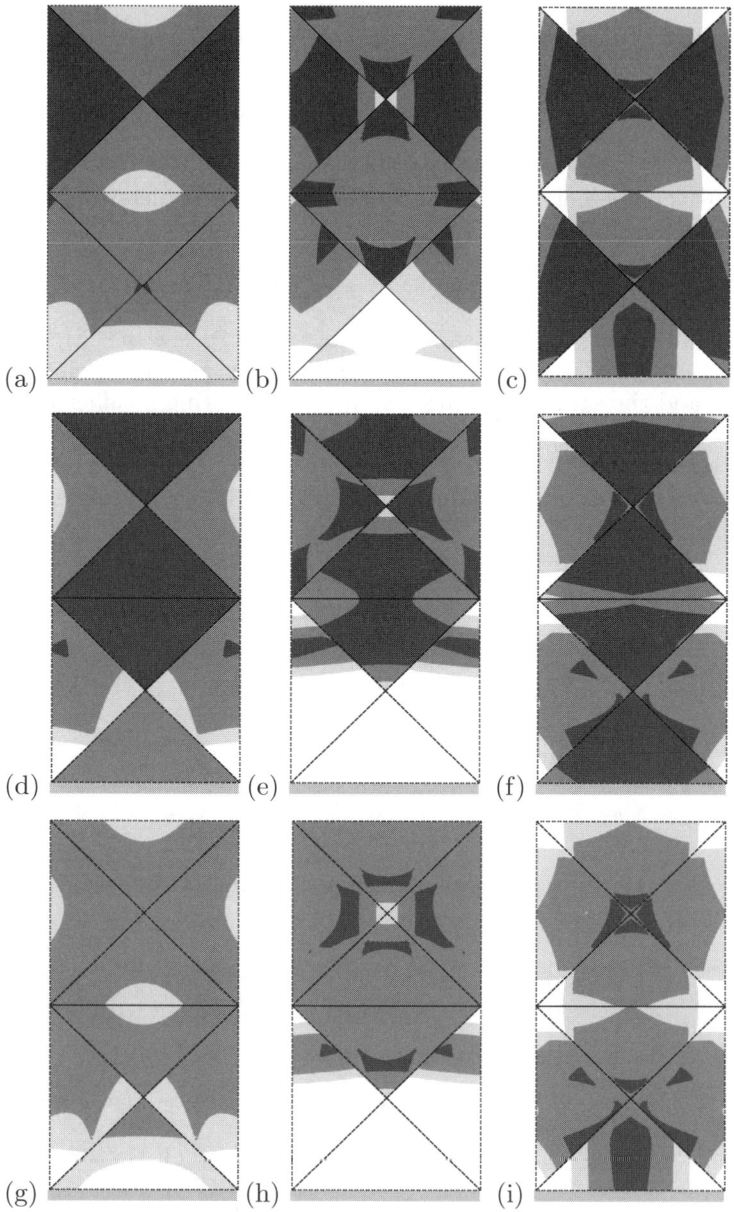

FIG. 4.8*.8. Pointwise accuracy of recoveries which employ the 'harmonic' character of the exact solution: Orthotropic heat conduction ($\boldsymbol{\nabla}\cdot(\mathbf{K}\boldsymbol{\nabla}u)=0$) with $K_1/K_2=4$, quadratic elements ($p=2$), a strip of eight elements at the Neumann boundary (the edge $x_2=0$ shown shaded gray) made by repeating the criss-cross pattern. The common regions of $\eta\%$ superconvergence for all grid-material orientations ($0\leq\theta\leq180°$) for (a) $\mathcal{G}^{\eta\%}_{q_1^{RecHG}}(\mathfrak{Q}^{(3)}_{hom};\mathbf{K};\tau,\Delta)$; (b) $\mathcal{G}^{\eta\%}_{q_1^{RecHV}}(\mathfrak{Q}^{(3)}_{hom};\mathbf{K};\tau,\Delta)$; (c) $\mathcal{G}^{\eta\%}_{q_1^{PRpPp+1}}(\mathfrak{Q}^{(3)}_{hom};\mathbf{K};\tau,\Delta)$; (d) $\mathcal{G}^{\eta\%}_{q_2^{RecHG}}(\mathfrak{Q}^{(3)}_{hom};\mathbf{K};\tau,\Delta)$; (e) $\mathcal{G}^{\eta\%}_{q_2^{RecHV}}(\mathfrak{Q}^{(3)}_{hom};\mathbf{K};\tau,\Delta)$; (f) $\mathcal{G}^{\eta\%}_{q_2^{PRpPp+1}}(\mathfrak{Q}^{(3)}_{hom};\mathbf{K};\tau,\Delta)$. (g) $\mathcal{G}^{\eta\%}_{\mathbf{q}^{RecHG}}(\mathfrak{Q}^{(3)}_{hom};\mathbf{K};\tau,\Delta)$; (h) $\mathcal{G}^{\eta\%}_{\mathbf{q}^{RecHV}}(\mathfrak{Q}^{(3)}_{hom};\mathbf{K};\tau,\Delta)$; (i) $\mathcal{G}^{\eta\%}_{\mathbf{q}^{PRpPp+1}}(\mathfrak{Q}^{(3)}_{hom};\mathbf{K};\tau,\Delta)$. The shades indicate the regions of 95, 90, and 85% superconvergence (dark, lighter, light gray)

$\mathcal{F}(u)$, and let

$$\mathcal{A}^*_{\mathcal{F}(u)}(u_{S^p_\Delta})(\hat{\mathbf{x}}) = \sum_{k=1}^{nsp} \beta_k u_{S^p_\Delta}(\bar{\bar{\mathbf{x}}}_k) \qquad (4.8^*.11a)$$

where $\bar{\bar{\mathbf{x}}}_k$, $k = 1, \ldots, nsp$, are the sampling points, and β_k, $k = 1, \ldots, nsp$, are the weight. To achieve superconvergence at $\hat{\mathbf{x}}$, it is sufficient to determine a set of weights β_k, $k = 1, \ldots, nsp$, such that

$$\mathcal{A}^*_{\mathcal{F}(u)}(U^{PER(h)}_{S^p_\Delta}) = \mathcal{F}(U_{EX}) \quad \forall U_{EX} \in \mathcal{P}^{p+1}(S(\bar{x}, H)). \qquad (4.8^*.11b)$$

Condition (4.8*.11b) leads to a linear system of $neq = (p+2)(p+3)/2$ equations with nsp unknowns, $\beta_k = \beta_k(\hat{\mathbf{x}}; \mathcal{F}(u); \{\bar{\bar{\mathbf{x}}}_l\}_{l=1}^{nsp})$, which depend on the point $\hat{\mathbf{x}}$, the solution quantity $\mathcal{F}(u)$, and the sampling points $\{\bar{\bar{\mathbf{x}}}_l\}_{l=1}^{nsp}$. By selecting a sufficiently high number of sampling points, $nsp > neq$, in an h neighborhood of the point X the existence of solutions can be guaranteed.

Let us now give an example of the construction of an asymptotically exact formula $\mathcal{A}^*_{du/dx}(u_{S^p_\Delta})(\hat{\mathbf{x}})$.

Example 4.8*.4 Consider a mesh of linear triangles arranged in the chevron pattern. Let us be interested in determining the coefficients in the recovery formula

$$\mathcal{A}^*_{\partial u/\partial x_l}(u_{S^1_\Delta})(0,0) = \sum_{k=1}^{9} \beta_k^{(l)} u_{S^1_\Delta}(\bar{\bar{\mathbf{x}}}_k) \qquad (4.8^*.12a)$$

where we employed the abbreviation

$$\beta_k^{(l)} = \beta_k\left((0,0); \frac{\partial u}{\partial x_l}; \{\bar{\bar{\mathbf{x}}}_k\}_{k=1}^{9}\right) \qquad (4.8^*.12b)$$

where $\{\bar{\bar{\mathbf{x}}}_k\}_{k=1}^{9}$ are the sampling points which are located at the vertices of the elements namely

$$\begin{aligned} &\bar{\bar{\mathbf{x}}}_1 = (-1,-1), && \bar{\bar{\mathbf{x}}}_2 = (0,-1), && \bar{\bar{\mathbf{x}}}_3 = (1,-1),\\ &\bar{\bar{\mathbf{x}}}_4 = (-1,0), && \bar{\bar{\mathbf{x}}}_5 = (0,0), && \bar{\bar{\mathbf{x}}}_6 = (1,0),\\ &\bar{\bar{\mathbf{x}}}_7 = (-1,1), && \bar{\bar{\mathbf{x}}}_8 = (0,1), && \bar{\bar{\mathbf{x}}}_9 = (1,1). \end{aligned}$$

Employing (4.8*.11b) for all the basis monomials up to degree $p = 2$ we get

$$\begin{matrix} 1 \\ x_1 \\ x_2 \\ x_1^2 \\ x_1 x_2 \\ x_2^2 \end{matrix} \begin{bmatrix} 1 & 1 & 1 & 1 & 1 & 1 & 1 & 1 & 1 \\ -1 & 0 & 1 & -1 & 0 & 1 & -1 & 0 & 1 \\ -1 & -1 & -1 & 0 & 0 & 0 & 1 & 1 & 1 \\ 1 & 0 & 1 & 1 & 0 & 1 & 1 & 0 & 1 \\ 1 & 0 & -1 & 0 & 0 & 0 & -1 & 0 & 1 \\ 1 & 1 & 1 & 0 & 0 & 0 & 1 & 1 & 1 \end{bmatrix} \begin{Bmatrix} \beta_1^{(1)} \\ \beta_2^{(1)} \\ \beta_3^{(1)} \\ \beta_4^{(1)} \\ \beta_5^{(1)} \\ \beta_6^{(1)} \\ \beta_7^{(1)} \\ \beta_8^{(1)} \\ \beta_9^{(1)} \end{Bmatrix} = \begin{Bmatrix} 0 \\ 1 \\ 0 \\ 0 \\ 0 \\ 0 \end{Bmatrix} \qquad (4.8^*.13a)$$

TABLE 4.8*.1. Construction of the asymptotically exact formula. The coefficients β_i of the formula (4.8*.6) in the case that $\hat{\mathbf{x}} = (0,0)$ (in master cell coordinates) and $\mathcal{F}(u) = \partial u/\partial x_1$ or $\partial u/\partial x_2$ for the chevron and the Union-Jack pattern and linear elements ($p = 1$). The coordinates of the sampling points in the master cell are $\bar{\bar{\mathbf{x}}}_1 = (-1,-1)$, $\bar{\bar{\mathbf{x}}}_2 = (0,-1)$, $\bar{\bar{\mathbf{x}}}_3 = (1,-1)$, $\bar{\bar{\mathbf{x}}}_4 = (-1,0)$, $\bar{\bar{\mathbf{x}}}_5 = (0,0)$, $\bar{\bar{\mathbf{x}}}_6 = (1,0)$, $\bar{\bar{\mathbf{x}}}_7 = (-1,1)$, $\bar{\bar{\mathbf{x}}}_8 = (0,1)$, $\bar{\bar{\mathbf{x}}}_9 = (1,1)$

$\mathcal{F}(u)$	Mesh paterrn	Values of the coefficients								
		β_1	β_2	β_3	β_4	β_5	β_6	β_7	β_8	β_9
$\dfrac{\partial u}{\partial x_1}$	chevron	0	0	1/2	0	1/2	0	0		
	Union-Jack	0	1/6	1/6	1/6	0	−1/6	−1/6	−1/6	0
$\dfrac{\partial u}{\partial x_2}$	chevron	−1/3	−1/12	1/12	1/2	1/12	−1/12	−1/6		
	Union-Jack	−1/6	−1/6	0	1/6	1/6	1/6	0	−1/6	0

or in abbreviated form $\mathbf{M}\boldsymbol{\beta}_1 = \mathbf{F}_1$, and similarly for the x_2 derivative we have $\mathbf{M}\boldsymbol{\beta}_2 = \mathbf{F}_2$ with $\mathbf{F}_2 = [0,0,1,0,0,0]^{\mathrm{T}}$. Letting

$$L(\boldsymbol{\beta},\boldsymbol{\Lambda}) = \tfrac{1}{2}\boldsymbol{\beta}^{\mathrm{T}}\boldsymbol{\beta} + (\mathbf{M}\boldsymbol{\beta} - \mathbf{F})^{\mathrm{T}}\boldsymbol{\Lambda} \qquad (4.8^*.13\mathrm{b})$$

and from the condition of stationary value of L, $\dfrac{\partial L}{\partial \boldsymbol{\beta}} = \mathbf{0}$, we get $\boldsymbol{\beta} = -\mathbf{M}^{\mathrm{T}}\boldsymbol{\Lambda}$ and employing the above into (4.8*.13b) we get

$$\mathbf{M}\mathbf{M}^{\mathrm{T}}\boldsymbol{\Lambda} = -\mathbf{F}. \qquad (4.8^*.13\mathrm{c})$$

Letting $\mathbf{F} = \mathbf{F}_1$ into (4.8*.13c) we get

$$\boldsymbol{\beta}_1 = \tfrac{1}{6}[-1,0,1,-1,0,1,-1,0,1]^{\mathrm{T}} \qquad (4.8^*.13\mathrm{d})$$

and we have

$$\begin{aligned} \mathcal{A}^*_{\partial u/\partial x_1}(u_{S^1_\Delta})(\hat{\mathbf{x}}) = \tfrac{1}{6}\Big(&\big(u_{S^1_\Delta}(\bar{\bar{\mathbf{x}}}_3) - u_{S^1_\Delta}(\bar{\bar{\mathbf{x}}}_1)\big) + \big(u_{S^1_\Delta}(\bar{\bar{\mathbf{x}}}_6) - u_{S^1_\Delta}(\bar{\bar{\mathbf{x}}}_4)\big) \\ &+ \big(u_{S^1_\Delta}(\bar{\bar{\mathbf{x}}}_9) - u_{S^1_\Delta}(\bar{\bar{\mathbf{x}}}_7)\big)\Big) \end{aligned} \qquad (4.8^*.14)$$

where $\hat{\mathbf{x}} = (0,0) = \bar{\bar{\mathbf{x}}}_5$, and similarly for the x_2 derivative. The above formula was derived in the master cell which has size 2. For the cell of size $2h$ the entries of $\boldsymbol{\beta}$ must be divided by h. Table 4.8*.1 gives the values of the coefficients $\{\beta_k\}_{k=1}^{nsp}$ which were determined by the above procedure for the recovery of the x_1 and the x_2 derivative at the central node of the chevron and the Union-Jack pattern for linear elements ($p = 1$).

Let us now consider another example where the asymptotically exact formula $\mathcal{A}^*_{du/dx}(u_{S^p_\Delta})(\hat{\mathbf{x}})$ is far from being obvious.

Example 4.8*.5 Consider a mesh of quadratic triangles ($p = 2$) arranged in the chevron pattern with aspect ratio $1/a$. Following similar steps as above, construct an asymptotically exact formula for the components of the gradient at $\hat{\mathbf{x}} = (0,0)$ for the class of solutions of the homogeneous differential equation.

We have $\mathbf{M}\boldsymbol{\beta}_1 = \mathbf{F}_1$, where

$$\mathbf{M} = \begin{bmatrix} 1 & 1 & 1 & 1 & 1 & 1 & 1 & 1 & 1 \\ -1 & 0 & 1 & -1 & 0 & 1 & -1 & 0 & 1 \\ -a & -a & -a & 0 & 0 & 0 & a & a & a \\ (1-a^2) & -a^2 & (1-a^2) & 1 & 0 & 1 & (1-a^2) & -a^2 & (1-a^2) \\ a & 0 & -a & 0 & 0 & 0 & -a & 0 & a \\ (-1+3a^2) & 0 & (1-3a^2) & -1 & 0 & 1 & (-1+3a^2) & 0 & (1-3a^2) \\ (-3a+a^3) & a^3 & (-3a+a^3) & 0 & 0 & 0 & (3a-a^3) & -a^3 & (3a-a^3) \end{bmatrix} \begin{array}{r} 1 \\ x_1 \\ x_2 \\ x_1^2 - x_2^2 \\ x_1 x_2 \\ x_1^3 - 3x_1 x_2^2 \\ 3x_1^2 x_2 - x_2^3 \end{array} \tag{4.8*.15a}$$

and $\mathbf{F}_1 = [0, 1, 0, 0, 0, 0, 0]^{\mathrm{T}}$, and we obtain

$$\boldsymbol{\beta}_1 = \frac{1}{12a^2}[-1, 0, 1, 2(1-3a^2), 0, 2(-1+3a^2), -1, 0, 1\ ]^{\mathrm{T}} \tag{4.8*.15b}$$

and similarly

$$\boldsymbol{\beta}_2 = \frac{1}{12a}[\ -1, 2(-3+a^2), -1, 0, 0, 0, 1, 2(3-a^2), 1]^{\mathrm{T}}. \tag{4.8*.15c}$$

Remark 4.8*.4 Note that the asymptotically exact formulae derived above employ the values of the finite element solution in the h cell, and are valid for all aspect ratios. It is conceivable that *a sophisticated finite element code could include such formulae for all the mesh geometries which can be constructed by the mesh generator, solution quantities and classes of solutions of interest.*

Remark 4.8*.5 The above technique for constructing superconvergent recoveries can also be used in the case of the elasticity problem.

In this section we addressed the local post-processing techniques of the finite element solution $u_{S^p_\Delta}$ which leads to the approximation of the exact solution u_{EX} which has smaller error than $u_{S^p_\Delta}$. The main assumption is that *pollution is negligible.* Sometimes the difference between the post-processed solution and the finite element solution is used as indication of the local accuracy of the finite element solution. This approach can be very misleading when the pollution is not negligible.

Problems

P4.8*.1. Consider the Poisson equation $-\Delta u = f$, in a square domain meshed by the regular pattern and elements of degree $p = 1$. Assume that there is no pollution. Show that the centroids of the elements are not superconvergence points for $\partial u/\partial x_1$ or $\partial u/\partial x_2$. Nevertheless, the recovery $\mathcal{A}^*_{\partial u/\partial x_i}$ of the type given in (4.8*.1) and (4.8*.2) (e.g., *ZZ-SPR* recovery) is superconvergent. Explain how the averaging results in cancellation of the error in this particular case.

P4.8*.2. Repeat P4.8*.1 for the chevron pattern. Show that in this case, we can obtain superconvergent values by employing a weighted recovery.

P4.8*.3. Show that for linear elements ($p = 1$), arranged in the regular pattern, the superconvergent region for *ZZ-SPR* is the entire cell.

4.9 Historical and bibliographical comments

The duality principles for analyzing L^2 and lower norms were used independently by Aubin in 1967 [Aubin, 1967], Nitsche in 1968 [Nitsche, 1968], and Oganesjan and Rukhovic in 1969 [Oganesjan and Rukhovec, 1969]. Duality principles were utilized for computation of data of interest in 1984 [Babuška and Miller, 1984]. The basic interior estimate was proven first in 1974 in [Nitsche and Schatz, 1974], and then further important results were published in [Schatz and Wahlbin, 1977] and [Schatz and Wahlbin, 1995].

Superconvergence in finite element methods was investigated for a long time in the mathematical and engineering literature. As early papers, we refer to [Striclin, 1966], [Filho, 1968], [Oganesjan and Rukhovec, 1969], [Oganesjan and Rukhovec, 1979]. An early paper on superconvergence in one dimension is [Douglas and Dupont, 1973]. Early papers concentrated on superconvergence at isolated points. Later various types of superconvergent recoveries were established, either in the strict sense or in an approximate way. The $\eta\%$ superconvergence, describing such approximate superconvergence, is addressed in [Babuška, Strouboulis *et al.*, 1994, 1995]. For the translation-invariant meshes, [Bramble and Schatz, 1977] is an early paper on improvement of the accuracy by averaging. In [Babuška and Miller, 1984] a post-processing technique achieving higher accuracy is addressed. Today there are many results on superconvergence. For example, the survey [Krizek and Neittaanmäki, 1998] contains 600 references. There are already a few monographs on superconvergence, e.g., [Wahlbin, 1995], [Chen and Huang, 1995], [Lin and Yan, 1996]. In China, research on superconvergence for meshes of triangles arranged in the regular pattern has been under way for many years. The main idea is to use orthogonal expansion in the special shape functions (which are identical to the hierarchical shape functions used in the p version) [see, e.g., Szabo and Babuška, 1991]. For the latest results, see [Chen, 1999]. For a survey of the ideas and principles in the area of superconvergence, we refer to [Wahlbin, 1998a], [Zhang, 1996, 1998, 1999, 2000], [Zhu, 1998, 2000]. All the above results are of analytic type. [Schatz *et al.*, 1996] presented very powerfull ideas about local symmetry leading to superconvergence. A majority of the mentioned results are for the scalar equation. In the lecture notes [Wahlbin, 1998b], some interesting results related to the topics addressed in this chapter are given and additional references cited. In [Babuška, Strouboulis *et al.*, 1996] a novel idea of investigation of superconvergence based on computer proof for general 'cell' translation-invariant meshes is given. This allows one to prove the existence of superconvergent points and superconvergent recovery, not only in the classical way for all solutions, but also for solutions satisfying a differential equation (e.g. a homogeneous one). The approach allows one to analyze general elliptic equations, e.g., the elasticity equations for the Cosserat medium, superconvergence at the boundary elements, in the neighborhood of corners for arbitrary type of elements.

References

K. Appel and W. Hakin (1978) The four-color problem. In: L.A. Steen (ed.) Mathematics Today, Springer, Berlin, pp. 153–180.

D. Arnold, D. Boffi, and R. Falk (2000) On the serendipity elements, Penn State University, preprint.

K.E. Atkinson (1989) *An Introduction to numerical analysis.* John Wiley, New York.

J.P. Aubin (1967) Behavior of the error of the approximate solutions of boundary value problems for linear elliptic operators by Galerkin and finite difference methods. *Annali della Scuola Normale Superiore di Pisa. Classe di Scienz*, **21**, 599–637.

I. Babuška and A. Miller (1984) The post-processing approach in the finite element method. Part 1: Calculation of displacements, stresses, and other higher derivatives of the displacements; Part 2: The calculations of stress intensity factors; Part 3: A-posteriori error estimates and adaptive mesh selection. *International Journal for Numerical Methods in Engineering*, **20**, 1085–1109, 1111–1129, 2311–2325.

I. Babuška, L. Plank, and R. Rodriguez (1992) Quality assessment of the a-posteriori error estimation in finite elements. *Finite Elements in Analysis and Design*, **11**, 285–306.

I. Babuška, T. Strouboulis, S.K. Gangaraj, and C.S. Upadhyay (1994) $\eta\%$ superconvergence in the interior of locally refined meshes of quadrilaterals: superconvergence of the gradient in finite element solutions of Laplace's and Poisson's equations. *Applied Numerical Mathematics*, **122**, 273–305.

I. Babuška, T. Strouboulis, and C.S. Upadhyay (1994) $\eta\%$ superconvergence of finite element approximations in the interior of meshes of triangles. *Computer Methods in Applied Mechanics and Engineering*, **16**, 3–49.

I. Babuška, T. Strouboulis, C.S. Upadhyay, and S.K. Gangaraj (1996) Computer-based proof of the existence of superconvergence points in the finite element method. Superconvergence of the derivatives in finite element solutions of Laplace's, Poisson's and the elasticity equations. *Numerical Methods for PDEs*, **12**, 347–392.

I. Babuška, T. Strouboulis, and C.S. Upadhyay (1997) A model study of the quality of a-posteriori error estimators for finite element solutions of linear elliptic problems, with particular reference to the behavior near the boundary. *International Journal for Numerical Methods in Engineering*, **40**, 2521–2577.

I. Babuška, T. Strouboulis, and C.S. Upadhyay (2001) A model study of the quality of a-posteriori error estimators and superconvergence in the corner region of a domain (in preparation).

C.L. Bajaj and G.L. Xu (1997) Piecewise rational approximations of real algebraic curves. *Journal of Computational Mathematics*, **15**, 55–71.

C.L. Bajaj and G.L. Xu (1999) A-splines: local interpolation and approximation using G^k-continuous piecewise real algebraic curves. *Computer Aided Geometric Design*, **16**, 39–59.

J.H. Bramble and A.H. Schatz (1977) Higher-order local accuracy by averaging in the finite element method. *Mathematics of Computation*, **31**, 94–111.

C.M. Chen and Y.Q. Huang (1995) *High accuracy theory of finite element methods.* Hunan Science and Technology Press, Changsha (in Chinese).

C. Chen (1999) Superconvergence for triangular finite elements. *Science in China* (Ser A), **42**, 917–934.

P.G. Ciarlet (1991) Basic error estimates for elliptic problems. In: P.G. Ciarlet and J.L. Lions (eds) *Handbook of numerical analysis, Vol. II; Finite Element Methods* (Part 1). Elsevier Science Publishers B. V., North-Holland, Amsterdam, pp. 19–351.

J. Douglas and T. Dupont (1973) Some superconvergence results for Galerkin methods for the approximate solution of two-point boundary value problems. *Topics in Numerical Analysis*, Academic Press, pp. 89–92.

F.V. Filho (1968) Comment on 'computation of stress resultants from the element stiffness matrices'. *AIAA Journal*, **6**, 571–572.

I.M. Gel'fand and I.M. Shilov (1964) *Generalized functions*. Academic Press, New York.

M. Krizek and P. Neittaanmäki (1998) Biography on superconvergence. In: M. Krizek, P. Neittaanmäki, and R. Sternberg (eds) *Finite Element Methods, superconvergence post-processing, and a-posteriori estimates*, Marcel Dekker, pp. 315–348.

Q. Lin and N.N. Yan (1996) *The construction and analysis for efficient finite elements.* Hibei Univ. Pabe H (in Chinese).

J.A. Nitsche (1968) Ein Kriterium für die Quasioptimalität des Ritzchen Verfahrens. *Numerische Mathematik*, **11**, 346–358.

J. Nitsche and A.H. Schatz (1974) Interior estimates for Ritz–Galerkin methods. *Mathematics of Computation*, **28**, 937–958.

L.A. Oganesjan and L.A. Rukhovec (1969) Investigation of the convergence rate of variation-difference schemes for elliptic second order equations in two dimensional domain with a smooth boundary. *Zhurnal Vychislitel'noi Matematiki i Matematicheskoĭ Fiziki*, **9**, 1102–1120 (in Russian).

L.A. Oganesjan and L.A. Rukhovec (1979) Variational-difference methods for the solution of elliptic equations. *Akademii Nauk Armjanskoj SSR* (in Russian).

A.H. Schatz and L.B. Wahlbin (1977) Interior maximum norm estimates for finite element methods. *Mathematics of Computation*, **31**, 414–442.

A.H. Schatz and L.B. Wahlbin (1995) Interior maximum norm estimates for finite element methods, Part II. *Mathematics of Computation*, **64**, 907–928.

A.H. Schatz, I.H. Sloan, and L.B. Wahlbin (1996) Superconvergence in finite element methods and meshes which are locally symmetric with respect to a point. *SIAM Journal on Numerical Analysis*, **33**, 505–521.

A.H. Schatz (1998) Pointwise error estimates and asymptotic expansion error inequalities for the finite element method on irregular grids, Part I: global estimates. *Mathematics of Computation*, **67**, 877–899.

A.H. Schatz (2001) Pointwise error estimates and asymptotic expansion error inequalities for the finite element method on irregular grids, Part II: interior estimates. *SIAM Journal on Numerical Analysis* (to appear).

R. Scott (1976) Optimal L^∞ estimates for the finite element method on irregular meshes. *Mathematics of Computation*, **30**, 681–697.

J.A. Striclin (1966) Computation of stress resultants from the element stiffness matrices. *AIAA Journal*, **4**, 1095–1096.

B.A. Szabo and I. Babuška (1991) *Finite element analysis.* John Wiley & Sons, Inc., New York.

L.B. Wahlbin (1991) Local behaviour in finite element methods. In: P.G. Ciarlet and J.L. Lions (eds) *Handbook of numerical analysis, Vol. II; Finite Element Methods* (Part 1), Nord Holland, pp. 409–416.

L.B. Wahlbin (1995) *Superconvergence in Galerkin Finite Element Methods.* Lecture Notes in Mathematics, **1605**, Springer, Berlin.

L.B. Wahlbin (1998a) General principles of superconvergence in Galerkin Finite Element Methods. In: M. Krizek, P. Neittaanmäki, and R. Sternberg (eds) *Finite Element Methods, superconvergence post-processing, and a-posteriori estimates*, Marcel Dekker, pp. 269–286.

Z. Zhang (1996) Ultraconvergence of patch recovery technique. *Mathematics of Computation*, **65**, 1431–1437.

Z. Zhang (1998) Derivative superconvergence points in finite element solutions of Poisson's equation for the serendipity and intermediate families; a theoretical justification. *Mathematics of Computation*, **67**, 541–552.

Z. Zhang (2000) Ultraconvergence of patch recovery technique II. *Mathematics of Computation*, **69**, 141–168.

Z. Zhang (2000) Derivative superconvergent points in Finite Element solutions of Harmonic Functions (preprint).

Qiding Zhu (1998) A survey of superconvergence techniques in finite element methods. In: M. Krizek, P. Neittaanmäki, and R. Sternberg (eds) *Finite Element Methods, superconvergence post-processing, and a-posteriori estimates*, Marcel Dekker, pp. 287–302.

Qiding Zhu (2000) Superconvergence analysis of cubic triangular elements in the finite element method. *Journal of Computational Mathematics*, **18**, 545–550.

O.C. Zienkiewicz and J.Z. Zhu (1992a) The superconvergent patch recovery and a posteriori error estimates. Part 1: The recovery technique. *International Journal for Numerical Methods in Engineering*, **33**, 1331–1364.

O.C. Zienkiewicz and J.Z. Zhu (1992b) The superconvergent patch recovery and a posteriori error estimates. Part 2: Error estimates and adaptivity. *International Journal for Numerical Methods in Engineering*, **33**, 1365–1382.

O.C. Zienkiewicz and J.Z. Zhu (1992c) The superconvergent patch recovery (SPR) and adaptive finite element refinement. *Computer Methods in Applied Mechanics and Engineering*, **101**, 207–224.

O.C. Zienkiewicz, J.Z. Zhu and J. Wu (1993) Superconvergent recovery techniques, some further tests. *Computational Applications in Science and Engineering*, **9**, 251–258.

O.C. Zienkiewicz and J.Z. Zhu (1994) *Superconvergence and superconvergent patch recovery, or the obvious which escaped attention*, CR/823/94, Department of Civil Engineering., University of Wales Swansea, U.K., March 1994.

5

A-POSTERIORI ESTIMATION OF THE ERROR

5.0 Introduction

In Chapters 3 and 4, we analyzed the properties of the Finite Element Method. We have shown that the finite element solution converges to the exact solution by establishing asymptotic estimates for the energy norm, the L^2 norm, and various other norms of the error. For example, under the assumption that the input data are such that

$$\left\| \frac{d^r}{dx^r}(u_{EX}) \right\|_{L^q(I)} < \infty \tag{5.0.1a}$$

for $0 \leq r$, integer, and $q = 2$, or ∞, we have proven that

$$\|e_{S^p_\Delta}\|_{\mathcal{U}} \leq \mathscr{C}(h^\Delta)^\mu \left\| \frac{d^r}{dx^r}(u_{EX}) \right\|_{L^q(I)} \tag{5.0.1b}$$

where

$$\mu = \min(p, r-1) \tag{5.0.1c}$$

is the rate of convergence, and $\mathscr{C}$ depends only on the class of the employed meshes and the degree p of the elements. In higher dimensions for linear elements ($p = 1$) and $q = \infty$, h^Δ is replaced by $h^\Delta \mid \ln h^\Delta \mid$. Although in some cases, like, for example $u_{EX} = x^\alpha$, and for the class of radical meshes Δ^γ_n, we analyzed the constants $\mathscr{C}$, we were mostly interested in the rate of convergence μ. The estimates of the type (5.0.1) are called *a-priori* because they utilize information about the smoothness of the exact solution, the input data, the span of the finite element basis functions, and the geometry of the elements, which are all known prior to the computation of the finite element solution for the problem of interest. The a-priori estimates are important because they guarantee that, in principle, the finite element solution $u_{S^p_\Delta}$, can approximate the exact solution u_{EX}, with arbitrarily high accuracy, and because they characterize the factors on which the accuracy depends on. A priori estimates however cannot be used effectively, for the quantitative estimation of the error. For example, an estimate of the form

$$\mathscr{E} = \mathscr{C}(h^\Delta)^\mu \| \mathcal{A}^*_{d^r u/dx^r}(u_{S^p_\Delta}) \|_{L^q(I)} \tag{5.0.2}$$

where $\mathcal{A}^*_{d^r u/dx^r}$, a recovery for the derivative of order r, which follows from (5.0.1), cannot, in general, give an accurate estimate for $\|e_{S^p_\Delta}\|_{\mathcal{U}}$, because of the uncertainties in the values of $\mathscr{C}$, μ, r, and the error in $\mathcal{A}^*_{d^r u/dx^r}(u_{S^p_\Delta})$.

In this and the next chapter, we will address another class of error estimates which we will call *a-posteriori* because they are constructed after the finite element solution has been computed, and they utilize the finite element solution and the input data for the concrete case of interest. We will say that $\mathscr{E} = \mathscr{E}(u_{S^p_\Delta}; (f, a_2))$ is an estimator for the $|||\cdot|||$ norm of the error, if there exist positive constants $\mathscr{C}_L$, $\mathscr{C}_U$, which depend only on the classes of input data, and the classes of meshes of interest, such that

$$\mathscr{C}_L \mathscr{E} \leq |||e_{S^p_\Delta}||| \leq \mathscr{C}_U \mathscr{E}. \tag{5.0.3}$$

In this chapter we will focus in the construction of various estimators $\mathscr{E}$, for the energy norm of the error $\|e_{S^p_\Delta}\|_{\mathscr{U}}$, and the analysis of their asymptotic properties. In Chapter 6 we will also address the construction of a-posteriori estimates for the error in the quantity of interest, namely, the construction of $\mathscr{E}$, such that we have

$$\mathscr{C}_L \mathscr{E} \leq \left|\mathcal{F}(u_{EX}) - \mathcal{F}\left(u_{S^p_\Delta}\right)\right| \leq \mathscr{C}_U \mathscr{E} \tag{5.0.4}$$

where $\mathcal{F}$ is a functional which expresses the output of interest, and the construction of *guaranteed upper and lower computable estimates* ($\mathscr{E}^U$ and $\mathscr{E}^L$) for the energy norm, namely

$$\mathscr{E}^L \leq \|e_{S^p_\Delta}\|_{\mathscr{U}} \leq \mathscr{E}^U \tag{5.0.5}$$

and the error in the quantity of interest, and the a-posteriori estimation of the pollution error.

The state of the art of a-posteriori estimation was first developed for the *global energy norm* of the error; for surveys of the state of the art, see [Verfürth, 1996], [Ainsworth and Oden, 1997], and [Ainsworth and Oden, 2000]. Based on the way that $\mathscr{E}$ is constructed, we will call it either a *residual estimator*, or a *recovery estimator*. Further, the residual estimators are classified as *implicit* or *explicit*. An implicit residual estimator $\mathscr{E}$ is given in the form

$$\mathscr{E} = \sqrt{\sum_{j=1}^{M} \|\hat{e}_{\omega_j}\|^2_{\mathscr{U}(\omega_j)}} \tag{5.0.6a}$$

where $\{\omega_j\}_{j=1}^M$ is a partition of the domain Ω into subdomains ω_j, which may overlap, and $\hat{e}_{\omega_j}$ is the solution of a boundary value problem of the form:

Find $\hat{e}_{\omega_j} \in S(\omega_j)$ such that

$$\mathscr{B}_{\omega_j}(\hat{e}_{\omega_j}, v) = \mathscr{R}_{\omega_j}(v) \quad \forall v \in S(\omega_j) \tag{5.0.6b}$$

where $S(\omega_j)$ is the employed solution space, and $\mathscr{R}_{\omega_j}$ is defined in terms of the residuals of the finite element solution $u_{S^p_\Delta}$. Explicit estimators are computed directly from the residual in the interior and at the boundaries of the elements, such as, for example,

$$\mathscr{E} = \sqrt{\sum_{j=1}^{N(\Delta)} (h_{I_j^\Delta})^2 \|f + \mathfrak{D} u_{S^p_\Delta}\|^2_{L^2(I_j^\Delta)} + \sum_{j=1}^{N(\Delta)} (h_{I_{j-1}^\Delta} + h_{I_j^\Delta}) \left(\left[\!\left[AE\frac{d}{dx}(u_{S^p_\Delta})\right]\!\right](x_j^\Delta)\right)^2}. \tag{5.0.7}$$

where $[\![\cdot]\!]$ denotes the jump.

A recovery estimator is given in the form

$$\mathscr{E} = \left|\left|\left| \mathcal{A}^*(u_{S^p_\Delta}) - q(u_{S^p_\Delta}) \right|\right|\right|_* \qquad (5.0.8)$$

where $\mathcal{A}^*$ is a recovery operator, $q(u)$ is a solution quantity, e.g., u or du/dx, and $|||\cdot|||_*$ is an appropriate norm.

In engineering practice it is not sufficient to estimate only the energy norm of the the error because *a small value of the global energy norm of the error does not necessarily imply that the error in the outputs of interest is also small* (e.g. a 5% relative error in the global energy norm does not imply 5% relative error in the maximum stress in a region of interest). The a-posteriori estimation of the error in the solution quantity of interest (for example, the error in the principal stresses or the modulus of the stress in a region, the stress intensity factors, etc.) will be addressed in Chapter 6. An essential requirement is that the quantity of interest has to be well defined; for example, it is meaningless to ask for an estimate of the maximum error in the derivative, flux, or stress for a problem set in a polygonal non-convex domain, because the exact value does not exist (the derivative, flux, or stress in the neighborhood of a corner point is usually unbounded). The a-posteriori estimation can be *asymptotic* or *guaranteed*. In this chapter we will mainly address the asymptotic approach, while in Chapter 6 we will address the guaranteed estimates.

Below we will often speak about an error *estimator* and an element error *indicator*. Let us explain the difference. The error estimator gives an estimate for the error in a solution quantity, and serves as criterion for the acceptance of the results, whether the desired accuracy in the outputs of interest has been achieved. Thus the estimator is used as a *stopping* criterion. The error indicator is a tool for the adaptive construction of the approximation. In many cases an error indicator is the contribution of the element in an estimate for a quantity of interest, e.g., the global energy norm of the error, but often the indicator is based on heuristics, like, for example, the indicator which employs the gradient of the computed solution, or some function of the residual, and may have no direct relation with the error.

A major requirement is that the computation of the a-posteriori estimates should have reasonable cost. The admissible cost of a reliable a-posteriori error estimation depends on various factors, like, for example, on its importance in concrete cases. As an example, let us mention the Sleipner A Condeep platform accident (see Chapter 1), in which an error in the computation led to a faulty design and to a loss of capital of almost one billion dollars. The fault in the design was traced to a 40% underestimation of the shear forces by the employed finite element approximation which was not fully compensated by the safety factor which was used, and could have been avoided if a reliable estimate of the error in the shear stresses had been used. This is an example where the cost of unreliable finite element computation was incomparable with any cost of a-posteriori error estimation which could have prevented the accident. In contrast, the error indicator is used in the context of adaptive procedures, and its cost should be as small as possible.

Another important requirement is that the a-posteriori estimates must be *accurate* or *effective*. The accuracy, or effectiveness, of an error estimate is measured by its *effectivity index* κ, which is the ratio of the estimated error to the exact error. Not all estimates are effective; for example, as we will see below various *explicit residual error estimators* involve several constants the values of which are often difficult to obtain, and are pessimistic, because they are obtained from the worst possible case, so that their use leads to ineffective estimates.

The cost of the error estimator can be divided into two parts, the *direct*, and the *indirect cost*. The *direct cost* is the cost of computing the estimator; it depends on the number of required arithmetic operations and the implementation. Most often in the literature the cost of an estimator is understood as the direct cost. The *indirect cost* of the estimator is the cost of additional computations which are due to the fact that the estimator is unreliable. To illustrate this, let us assume that the goal of the computation is to achieve 3% accuracy in an error measure, and that the employed estimator indicates that 3% κ accuracy has been achieved, where κ is the effectivity index of the estimator, the ratio between the estimator and the exact error. If $\kappa > 1$ the estimator overestimates the error, and if it used as the stopping criterion it will lead to overrefinement of the mesh, because the mesh will be adapted to achieve the accuracy of $(3/\kappa)$% which is higher than the 3% accuracy requested. This could lead to the increase of the degrees of freedom by a factor of κ^2 in two dimensions, and κ^3 in three dimensions, and a corresponding increase of the computational cost by κ^3 in two dimensions, and $\kappa^{9/2}$ in three dimensions; here we assumed that the computational cost is proportional to $N^{3/2}$, where N is the number of degrees of freedom, as it is the case in most engineering computations. For example, if $\kappa = 2$, the cost of unnecessary computation due to overestimation, or the *indirect cost* of the estimators, is, in two dimensions, 8 times, and in three dimensions 22.6 times the cost of the entire computation. Hence an estimator may have low direct cost but, if it leads to gross overestimation, very high indirect cost, especially in three dimensions. If the effectivity index κ is less than one ($\kappa < 1$), the estimator underestimates the error and this can be dangerous, and the indirect cost could depend on the circumstances. For example, if it leads to a costly accident, like for example the Sleipner accident (see Chapter 1), the indirect cost is the cost of the accident. Hence in many cases the indirect cost is the main cost of the error estimation and *to focus only on the direct cost it could be rather misleading and not appropriate to the overall goal of engineering computation.* Because the estimator is intended as a stopping criterion, for safety we should have a guaranteed upper estimator. At the same time we also need a guaranteed lower estimator because the difference between the upper and lower estimators indicates their effectivity, and hence the indirect cost. The construction of guaranteed upper and lower estimators will be addressed in Chapter 6.

5.1 The residuum and the construction of lower and upper estimators for the energy norm of the error

The main idea employed in the construction of the a-posteriori error estimators is that the error

$$e_{S^p_\Delta} = u_{EX} - u_{S^p_\Delta} \tag{5.1.1a}$$

is the exact solution of the following variational problem:

Find $e_{S^p_\Delta} \in \mathcal{U}$ such that

$$\mathcal{B}\big(e_{S^p_\Delta}, v\big) = \mathcal{R}(v) \stackrel{\text{def}}{=} \mathcal{L}(v) - \mathcal{B}\big(u_{S^p_\Delta}, v\big) \quad \forall v \in \mathcal{U}. \tag{5.1.1b}$$

This variational problem is known as the *residual equation*, and the linear functional $\mathcal{R}$ as the *global residuum*, or the *residuum*.

The proof of (5.1.1b) follows directly from the observation that

$$\mathcal{B}(e_{S^p_\Delta}, v) = \mathcal{B}(u_{EX} - u_{S^p_\Delta}, v) = \mathcal{B}(u_{EX}, v) - \mathcal{B}(u_{S^p_\Delta}, v) = \mathcal{L}(v) - \mathcal{B}(u_{S^p_\Delta}, v) \quad (5.1.1c)$$

for any $v \in \mathcal{U}$, where we recalled that

$$\mathcal{L}(v) = \mathcal{B}(u_{EX}, v) \quad \forall v \in \mathcal{U}. \quad (5.1.1d)$$

Employing the definition for the energy norm via duality, we immediately get

$$\|e_{S^p_\Delta}\|_{\mathcal{U}} = \sup_{v\in\mathcal{U}} \frac{\mathcal{B}(e_{S^p_\Delta}, v)}{\|v\|_{\mathcal{U}}} = \sup_{v\in\mathcal{U}} \frac{\mathcal{R}(v)}{\|v\|_{\mathcal{U}}} = \|\mathcal{R}\|_{L(\mathcal{U},\mathbb{R})}. \quad (5.1.2)$$

Hence $\|e_{S^p_\Delta}\|_{\mathcal{U}} = \|\mathcal{R}\|_{L(\mathcal{U},\mathbb{R})}$ and *the problem of the a-posteriori estimation of* $\|e_{S^p_\Delta}\|_{\mathcal{U}}$ *is the same as the problem of estimating* $\|\mathcal{R}\|_{L(\mathcal{U},\mathbb{R})}$.

Let us now derive the loads which correspond to $\mathcal{R}$. We have

$$\mathcal{R}(v) \overset{\text{def}}{=} \mathcal{L}(v) - \mathcal{B}(u_{S^p_\Delta}, v) = \sum_{j=1}^{N(\Delta)} \int_{I_j^\Delta} fv + a_2 v(l) - \sum_{j=1}^{N(\Delta)} \mathcal{B}_{I_j^\Delta}(u_{S^p_\Delta}, v) \quad (5.1.3a)$$

where

$$\mathcal{B}_{I_j^\Delta}(u, v) = \int_{I_j^\Delta} \left(AE\frac{du}{dx}\frac{dv}{dx} + cuv\right) + \delta_{jN}\alpha_2 u(l)v(l) \quad (5.1.3b)$$

and δ_{jN} is Kronecker's delta. Letting

$$P_{S^p_\Delta} \overset{\text{def}}{=} AE\frac{d}{dx}(u_{S^p_\Delta}) \quad (5.1.4a)$$

and employing integration by parts we get

$$\int_{I_j^\Delta} P_{S^p_\Delta}\frac{dv}{dx} = P_{S^p_\Delta}(x_j^{\Delta,-})v(x_j^\Delta) - P_{S^p_\Delta}(x_{j-1}^{\Delta,+})v(x_{j-1}^\Delta) - \int_{I_j^\Delta} \frac{d}{dx}(P_{S^p_\Delta})v \quad (5.1.4b)$$

where we let

$$P_{S^p_\Delta}(x_j^{\Delta,\pm}) = \lim_{\substack{\delta\to 0\\ \delta>0}} P_{S^p_\Delta}(x_j^\Delta \pm \delta). \quad (5.1.4c)$$

Using this result in (5.1.3a) we obtain

$$\mathcal{R}(v) = \sum_{j=1}^{N(\Delta)} \int_{I_j^\Delta} r_{I_j^\Delta} v + \sum_{k=1}^{N(\Delta)} J_k^\Delta v(x_k^\Delta) \quad (5.1.4d)$$

where $r_{I_j^\Delta}$ is the *distributed residual load* in I_j^Δ, given by

$$r_{I_j^\Delta} \overset{\text{def}}{=} \left(f + \frac{d}{dx}(P_{S^p_\Delta}) - cu_{S^p_\Delta}\right)\Big|_{I_j^\Delta} = \left(f + \frac{d}{dx}\left(AE\frac{d}{dx}(u_{S^p_\Delta})\right) - cu_{S^p_\Delta}\right)\Big|_{I_j^\Delta} \quad (5.1.5a)$$

and J_k^Δ is the *concentrated residual load* or *jump* at the kth node, given by

$$\begin{aligned} J_k^\Delta &\stackrel{\text{def}}{=} P_{S_\Delta^p}(x_k^{\Delta,+}) - P_{S_\Delta^p}(x_k^{\Delta,-}) \\ &= \Big(AE\frac{d}{dx}\big(u_{S_\Delta^p}\big)\Big)(x_k^{\Delta,+}) - \Big(AE\frac{d}{dx}\big(u_{S_\Delta^p}\big)\Big)(x_k^{\Delta,-}) \end{aligned} \tag{5.1.5b}$$

for $k = 1, \ldots, N(\Delta) - 1$, and

$$J_N^\Delta = a_2 - \alpha_2 u_{S_\Delta^p}(l) - P_{S_\Delta^p}(l) = a_2 - \alpha_2 u_{S_\Delta^p}(l) - \Big(AE\frac{d}{dx}\big(u_{S_\Delta^p}\big)\Big)(l). \tag{5.1.5c}$$

It follows that $\mathscr{R}$ is the functional corresponding to the load

$$\mathfrak{R} \stackrel{\text{def}}{=} r^\Delta + \sum_{k=1}^{N(\Delta)} J_k^\Delta \delta(\cdot\,; x_k^\Delta) \tag{5.1.5d}$$

where $\delta(\cdot; x_k^\Delta)$ is the Dirac load at x_k^Δ, and r^Δ is defined elementwise by $r^\Delta|_{I_j^\Delta} = r_{I_j^\Delta}$.

From the definition of the residuum $\mathscr{R}$

$$\mathscr{R}(v) \stackrel{\text{def}}{=} \mathscr{L}(v) - \mathscr{B}\big(u_{S_\Delta^p}, v\big) \quad \forall v \in \mathscr{U} \tag{5.1.6a}$$

it follows that the finite element solution, $u_{S_\Delta^p}$, is the *exact solution* of the variational problem:

Find $u_{S_\Delta^p} \in \mathscr{U}$ such that

$$\mathscr{B}\big(u_{S_\Delta^p}, v\big) = \mathscr{L}(v) - \mathscr{R}(v) \quad \forall v \in \mathscr{U}. \tag{5.1.6b}$$

Hence $u_{S_\Delta^p}$ is the exact solution corresponding to $\mathscr{L} - \mathscr{R}$.

Let us illustrate the definition of the residuum $\mathscr{R}$ in a concrete case.

Example 5.1.1 Consider the model problem with input data

$$AE \equiv 1, \qquad c \equiv 0, \qquad a_2 = 0, \qquad \alpha_2 = 0, \qquad f(x) = \sin\Big(\frac{\pi}{2}x\Big), \qquad l = 1 \tag{5.1.7a}$$

namely, the boundary value problem

$$-\frac{d^2u}{dx^2}(x) = \sin\Big(\frac{\pi}{2}x\Big), \quad x \in (0,1), \qquad u(0) = 0, \qquad \frac{du}{dx}(1) = 0 \tag{5.1.7b}$$

with exact solution

$$u_{EX}(x) = \frac{4}{\pi^2}\sin\Big(\frac{\pi}{2}x\Big). \tag{5.1.7c}$$

Let $u_{S_\Delta^1} \in S_\Delta^1$ be the piecewise linear finite element approximation of u_{EX} computed using the mesh Δ. From Corollary 4.1.1, it follows that $u_{S_\Delta^1}$ is exact at the nodes, namely,

$$u_j^\Delta = u_{S_\Delta^1}(x_j^\Delta) = u_{EX}(x_j^\Delta) = \frac{4}{\pi^2}\sin\Big(\frac{\pi}{2}x_j^\Delta\Big) \tag{5.1.8a}$$

for $j = 0, \dots, N(\Delta)$. The computed axial force in the element I_j^Δ is

$$P_{S_\Delta^1}|_{I_j^\Delta} \stackrel{\text{def}}{=} \frac{d}{dx}(u_{S_\Delta^1})|_{I_j^\Delta} = \frac{u_j^\Delta - u_{j-1}^\Delta}{h_{I_j^\Delta}} = \frac{4}{\pi^2}\frac{1}{h_{I_j^\Delta}}\left(\sin\left(\frac{\pi}{2}x_j^\Delta\right) - \sin\left(\frac{\pi}{2}x_{j-1}^\Delta\right)\right)$$
$$\equiv \textit{constant} \tag{5.1.8b}$$

while the exact axial force is

$$P_{EX}(x) = \frac{d}{dx}(u_{EX}) = \frac{2}{\pi}\cos\left(\frac{\pi}{2}x\right). \tag{5.1.8c}$$

It follows that

$$r_{I_j^\Delta} = f + \underbrace{\frac{d}{dx}(P_{S_\Delta^1}|_{I_j^\Delta})}_{0} = f \tag{5.1.8e}$$

and

$$J_j^\Delta \stackrel{\text{def}}{=} P_{S_\Delta^1}(x_j^{\Delta,+}) - P_{S_\Delta^1}(x_j^{\Delta,-}) = \frac{u_{j+1}^\Delta - u_j^\Delta}{h_{I_{j+1}^\Delta}} - \frac{u_j^\Delta - u_{j-1}^\Delta}{h_{I_j^\Delta}} \tag{5.1.8f}$$

for $j = 1, \dots, N(\Delta) - 1$, and

$$J_N^\Delta \stackrel{\text{def}}{=} a_2 - P_{S_\Delta^1}(x_N^{\Delta,-}) = -\frac{u_N^\Delta - u_{N-1}^\Delta}{h_{I_N^\Delta}} \tag{5.1.8g}$$

where we recalled that $a_2 = 0$. Recalling that $P_{S_\Delta^1}|_{I_j^\Delta} \equiv \textit{constant}$, we get the integration by parts formula

$$\mathscr{B}_{I_j^\Delta}(u_{S_\Delta^1}, v) = \int_{I_j^\Delta} P_{S_\Delta^1}\frac{dv}{dx} = P_{S_\Delta^1}|_{I_j^\Delta}\int_{I_j^\Delta}\frac{dv}{dx} = P_{S_\Delta^1}|_{I_j^\Delta}(v(x_j^\Delta) - v(x_{j-1}^\Delta)) \tag{5.1.9a}$$

and hence we have

$$\mathscr{R}(v) = \mathscr{L}(v) - \sum_{j=1}^{N(\Delta)}\mathscr{B}_{I_j^\Delta}(u_{S_\Delta^1}, v) = \mathscr{L}(v) - \sum_{j=1}^{N(\Delta)} P_{S_\Delta^1}|_{I_j^\Delta}(v(x_j^\Delta) - v(x_{j-1}^\Delta)) \tag{5.1.9b}$$

for any $v \in \mathscr{U}$. Rearranging the sum, we obtain that

$$\mathscr{R}(v) = \mathscr{L}(v) - \sum_{k=1}^{N(\Delta)-1} v(x_k^\Delta)(P_{S_\Delta^1}|_{I_k^\Delta} - P_{S_\Delta^1}|_{I_{k+1}^\Delta}) - v(x_N^\Delta)P_{S_\Delta^1}|_{I_N^\Delta}$$
$$= \int_I fv + \sum_{k=1}^{N(\Delta)} J_k^\Delta v(x_k^\Delta) \tag{5.1.9c}$$

for any $v \in \mathcal{U}$. It follows that $u_{S^1_\Delta}$ is the solution of the variational problem:
Find $u_{S^1_\Delta} \in \mathcal{U}$ such that

$$\mathcal{B}(u_{S^1_\Delta}, v) = \mathcal{L}(v) - \mathcal{R}(v) = -\sum_{j=1}^{N(\Delta)} J_j^\Delta v(x_j^\Delta) \quad \forall v \in \mathcal{U}. \tag{5.1.9d}$$

Hence $e_{S^1_\Delta}$ is the exact solution corresponding to the load $\sum_{k=1}^{N(\Delta)} J_k \delta(\cdot; x_k^\Delta)$.

As we have seen above, the error $e_{S^p_\Delta}$ is the exact solution of the original problem with load

$$\mathfrak{R} = r^\Delta + \sum_{k=1}^{N(\Delta)} J_k^\Delta \delta(\cdot; x_k^\Delta). \tag{5.1.10a}$$

From the linearity of the model problem, it follows that

$$e_{S^p_\Delta}(x) = e^{r^\Delta}_{S^p_\Delta}(x) + e^{J^\Delta}_{S^p_\Delta}(x) = \sum_{j=1}^{N(\Delta)} \int_{I_j^\Delta} r_{I_j^\Delta}(t) G(x;t)\, dt + \sum_{k=1}^{N(\Delta)} J_k^\Delta G(x; x_k^\Delta) \tag{5.1.10b}$$

where

$$e^{r^\Delta}_{S^p_\Delta}(x) \stackrel{\text{def}}{=} \int_I r^\Delta(t) G(x;t)\, dt = \sum_{j=1}^{N(\Delta)} \int_{I_j^\Delta} r_{I_j^\Delta}(t) G(x;t)\, dt \tag{5.1.10c}$$

is the component of the error, $e_{S^p_\Delta}$, due to the distributed residual r^Δ, and

$$e^{J^\Delta}_{S^p_\Delta}(x) \stackrel{\text{def}}{=} \sum_{k=1}^{N(\Delta)} J_k^\Delta G(x; x_k^\Delta) \tag{5.1.10d}$$

is the component of the error $e_{S^p_\Delta}$ due to the nodal jumps $\sum_{k=1}^{N(\Delta)} J_k^\Delta \delta(\cdot; x_k^\Delta)$, and $G(\cdot; \bar{x})$, the Green's function for the problem. It can be proven that:

(a) For *p even*, $\|e^{J^\Delta}_{S^p_\Delta}\|_{\mathcal{U}}$ is much smaller than $\|e^{r^\Delta}_{S^p_\Delta}\|_{\mathcal{U}}$, asymptotically as h^Δ tends to zero.

(b) For odd p, the components $e^{r^\Delta}_{S^p_\Delta}$ and $e^{J^\Delta}_{S^p_\Delta}$ could have comparable magnitudes.

Let us now give an example which illustrates the relative magnitude of $e^{r^\Delta}_{S^p_\Delta}$ and $e^{J^\Delta}_{S^p_\Delta}$.

Example 5.1.2 Let us consider the model problem, employed in Example 5.1.1, and let us determine the jumps J_k^Δ, and the interior residuals $r_{I_j^\Delta}$, in the case of a uniform mesh of size h, for $p = 1$, and $p = 2$.

In the case of linear elements ($p = 1$), the jumps at the interior nodes are given by

$$J_k^\Delta = \frac{4}{\pi^2}\left(\frac{\sin((\pi/2)x_{k+1}^\Delta) - \sin((\pi/2)x_k^\Delta)}{h} - \frac{\sin((\pi/2)x_k^\Delta) - \sin((\pi/2)x_{k-1}^\Delta)}{h}\right) \tag{5.1.11a}$$

and using well known trigonometric identities, we get

$$J_k^\Delta = \frac{2}{\pi}\Big(\cos\left(\frac{\pi}{2}x_{k+1/2}^\Delta\right) - \cos\left(\frac{\pi}{2}x_{k-1/2}^\Delta\right)\Big)\frac{\sin((\pi/4)h)}{(\pi/4)h}$$

$$= -\frac{4}{\pi}\sin\left(\frac{\pi}{2}x_k^\Delta\right)\frac{\sin^2((\pi/4)h)}{(\pi/4)h} \tag{5.1.11b}$$

for $k = 1, \dots, N(\Delta) - 1$, where we employed that $x_{j+1}^\Delta - x_j^\Delta = x_j^\Delta - x_{j-1}^\Delta = h$, and we used the notations $x_{j+1/2}^\Delta = \frac{1}{2}(x_{j+1}^\Delta + x_j^\Delta)$ and $\sin^2(\cdot) = (\sin(\cdot))^2$. Expanding the sine function in Taylor series we obtain

$$J_k^\Delta = -\sin\left(\frac{\pi}{2}x_k^\Delta\right)h(1 + O(h^2)), \quad k = 1, \dots, N(\Delta) - 1. \tag{5.1.11c}$$

The jump at the Nth node is given by

$$J_N^\Delta = -\frac{4}{\pi^2}\frac{1 - \sin((\pi/2)(1-h))}{h} = -\frac{2}{\pi}\frac{\sin^2((\pi/4)h)}{(\pi/4)h} = -\frac{h}{2}(1 + O(h^2)). \tag{5.1.11d}$$

The interior residuals are given by

$$r_{I_j^\Delta}(x) = f(x) + \frac{d^2}{dx^2}(u_{S_\Delta^1})|_{I_j^\Delta} = \sin\left(\frac{\pi}{2}x\right). \tag{5.1.11e}$$

In the case of quadratic elements ($p = 2$), we have

$$P_{S_\Delta^2}(x_k^{\Delta,-}) = \frac{d}{dx}(u_{S_\Delta^2})(x_k^{\Delta,-})$$

$$= \frac{1}{h_{I_k^\Delta}}(u_{EX}(x_{k-1}^\Delta) - 4u_{EX}(x_{k-1/2}^\Delta) + 3u_{EX}(x_k^\Delta)) \tag{5.1.12a}$$

and

$$P_{S_\Delta^2}(x_k^{\Delta,+}) = \frac{d}{dx}(u_{S_\Delta^2})(x_k^{\Delta,+})$$

$$= \frac{1}{h_{I_{k+1}^\Delta}}(-3u_{EX}(x_k^\Delta) + 4u_{EX}(x_{k+1/2}^\Delta) - u_{EX}(x_{k+1}^\Delta)) \tag{5.1.12b}$$

for $k = 1, \dots, N(\Delta) - 1$. The jumps at the interior nodes are given by

$$J_k^\Delta = P_{S_\Delta^2}(x_k^{\Delta,+}) - P_{S_\Delta^2}(x_k^{\Delta,-}) \tag{5.1.12c}$$

which can be written as

$$J_k^\Delta = \frac{1}{h}\Big(- u_{EX}((k-1)h) + 4u_{EX}\Big(\Big(k - \frac{1}{2}\Big)h\Big) - 6u_{EX}(kh)$$

$$+ 4u_{EX}\Big(\Big(k + \frac{1}{2}\Big)h\Big) - u_{EX}((k+1)h)\Big), \tag{5.1.12d}$$

from which we get

$$J_k^\Delta = h\sin\left(\frac{\pi}{2}kh\right)\left(-\left(\frac{\pi}{8}h\right)^2\right)(1+O(h^2)) \tag{5.1.12e}$$

for $k = 1, \ldots, N(\Delta) - 1$. For the jump at the Nth node we have

$$J_N^\Delta = a_2 - P_{S^2_{I_N^\Delta}}(x_N^\Delta - 0) = -\frac{1}{h}(u_{EX}(x_{N-1}^\Delta) - 4u_{EX}(x_{N-1/2}^\Delta) + 3u_{EX}(x_N^\Delta)) \tag{5.1.12f}$$

and we get

$$J_N^\Delta = \frac{3h}{8}\left(-\frac{13\pi^2h^2}{1536} + \frac{28\pi^4h^4}{262180}\right) + O(h^5). \tag{5.1.12g}$$

The interior residuals are given by

$$\begin{aligned} r_{I_j^\Delta}(x) &= f(x) + \frac{d^2}{dx^2}(u_{S^2_\Delta}|_{I_j^\Delta})(x) \\ &= \sin\left(\frac{\pi}{2}x\right) + \frac{4}{h^2}(u_{EX}(x_{j-1}^\Delta) - 2u_{EX}(x_{j-1/2}^\Delta) + u_{EX}(x_j^\Delta)) \end{aligned} \tag{5.1.12h}$$

and taking into account that the mesh is uniform, and expanding into Taylor series, we get

$$r_{I_j^\Delta}(x) = \frac{\pi}{2}\cos\left(\frac{\pi}{4}\left(x + \left(j - \frac{1}{2}\right)h\right)\right)h(1+O(h^2)) \tag{5.1.12i}$$

with $(j-1)h < x < jh$.

Let us now compare the cases of linear ($p = 1$) and quadratic ($p = 2$) elements. In the case $p = 1$, we have

$$\int_{x_{j-1/2}^\Delta}^{x_{j+1/2}^\Delta} r^\Delta \approx -J_j^\Delta \tag{5.1.13}$$

which indicates that $e^{r^\Delta}_{S^1_\Delta}$ and $e^{J^\Delta}_{S^1_\Delta}$ are of the same magnitude. Hence instead of estimating $e_{S^1_\Delta}$, it is sufficient to estimate either $e^{r^\Delta}_{S^1_\Delta}$ or $e^{J^\Delta}_{S^1_\Delta}$ by employing only the magnitudes of r^Δ or J^Δ, respectively. This is strictly a one-dimensional result. In two dimensions, the interior residual r^Δ could be negligible in comparison with the jumps J^Δ, and hence the construction of an estimator must include the jumps J^Δ, or both the jumps J^Δ and the interior residual r^Δ; this has been shown in [Babuška and Miller, 1981], [Babuška and Miller, 1987], [Babuška and Yu, 1987], and in [Yu, 1991a,b] for meshes of quadrilaterals, while in [Carstensen and Verfürth, 1999] and [Kunert and Verfürth, 1999] it was proven that, for general meshes of triangles of degree $p = 1$, the energy norm of the error may be estimated by employing only the jumps J^Δ. It does not mean that the jump estimate necessarily dominates the interior residual, but still leads to an estimate of the total error.

In the case of quadratic elements ($p = 2$), we have

$$\int_{I_j^\Delta} |r^\Delta| \approx O(h^3), \qquad J_j^\Delta \approx O(h^4) \tag{5.1.14}$$

and hence $|e^{J^\Delta}_{S^2_\Delta}|$ is much smaller than $|e^{r^\Delta}_{S^2_\Delta}|$, and there is no cancellation between $e^{J^\Delta}_{S^2_\Delta}$ and $e^{r^\Delta}_{S^2_\Delta}$. In this case the estimator may be based on the interior residual r^Δ only; see, for example, [Babuška and Yu, 1987] and [Yu, 1991b], where this is proven for special meshes of quadrilaterals. The above observations for $p = 1$, can be extended to any odd p, and for $p = 2$, to any even p.

Remark 5.1.1 The different character of the error for odd and even p, is known as the *dichotomy* and was first addressed in [Babuška and Yu, 1987] and [Yu, 1991a,b].

As we have already seen in (5.1.2) above, we have

$$\|e_{S^p_\Delta}\|_{\mathcal{U}} = \|\mathcal{R}\|_{L(\mathcal{U},\mathbb{R})} \tag{5.1.15}$$

which means that $\|e_{S^p_\Delta}\|_{\mathcal{U}}$ can be estimated by estimating $\|\mathcal{R}\|_{L(\mathcal{U},\mathbb{R})}$. Let us now introduce the definition of the a-posteriori estimator.

We will say that $\mathcal{E}$ is an estimator for $\|e_{S^p_\Delta}\|_{\mathcal{U}}$ if there exist constants $\mathcal{C}_L$, $\mathcal{C}_U$, independent of the mesh size h and the computed solution $u_{S^p_\Delta}$, such that

$$\underline{\mathcal{E}} \overset{\text{def}}{=} \mathcal{C}_L\mathcal{E} \le \|e_{S^p_\Delta}\|_{\mathcal{U}} \le \bar{\mathcal{E}} \overset{\text{def}}{=} \mathcal{C}_U\mathcal{E}. \tag{5.1.16}$$

The estimator $\bar{\mathcal{E}}$ (respectively $\underline{\mathcal{E}}$) is a *guaranteed upper* (respectively *lower*) *estimator* for $\|e_{S^p_\Delta}\|_{\mathcal{U}}$.

Remark 5.1.2 An upper estimator $\bar{\mathcal{E}}$ is essential for a *safe stopping criterion.* Let us assume that the goal is

$$\frac{\|e_{S^p_\Delta}\|_{\mathcal{U}}}{\|u_{EX}\|_{\mathcal{U}}} \le \delta. \tag{5.1.17a}$$

Recalling that, in the case that the Dirichlet conditions are homogeneous, we have

$$\left\|u_{S^p_\Delta}\right\|_{\mathcal{U}} \le \|u_{EX}\|_{\mathcal{U}} \tag{5.1.17b}$$

we get

$$\frac{\|e_{S^p_\Delta}\|_{\mathcal{U}}}{\|u_{EX}\|_{\mathcal{U}}} \le \frac{\bar{\mathcal{E}}}{\left\|u_{S^p_\Delta}\right\|_{\mathcal{U}}}. \tag{5.1.17c}$$

It follows that, by employing

$$\bar{\mathcal{E}} \le \delta \left\|u_{S^p_\Delta}\right\|_{\mathcal{U}} \tag{5.1.17d}$$

as the stopping criterion we are certain that the goal (5.1.17a) will be achieved.

Remark 5.1.3 A lower estimator $\underline{\mathcal{E}}$ is needed to check that the use of $\bar{\mathcal{E}}$ in the stopping criterion will not lead to overkill.

We will now show that upper and lower estimators, $\underline{\mathcal{E}}$ and $\bar{\mathcal{E}}$, can be constructed by employing the solutions of local boundary value problems loaded by the residuals. We will first give the construction of $\underline{\mathcal{E}}$.

Let

$$\mathcal{U}_0(I_j^\Delta) \stackrel{\text{def}}{=} \{v \in \mathcal{U}(I_j^\Delta) \mid v(x_{j-1}^\Delta) = v(x_j^\Delta) = 0\} \tag{5.1.18a}$$

and

$$\mathcal{U}_{0,\Delta} \stackrel{\text{def}}{=} \prod_{j=1}^{N(\Delta)} \mathcal{U}_0(I_j^\Delta) \stackrel{\text{def}}{=} \{v \in \mathcal{U} \mid v(x_k^\Delta) = 0,\ k = 0, \dots, N(\Delta)\}. \tag{5.1.18b}$$

Let $\hat{e}^{\text{Dir}}$ be the solution of the variational problem:

Find $\hat{e}^{\text{Dir}} \in \mathcal{U}_{0,\Delta}$ such that

$$\mathcal{B}(\hat{e}^{\text{Dir}}, v) = \sum_{j=1}^{N(\Delta)} \int_{I_j^\Delta} r_{I_j^\Delta} v \quad \forall v \in \mathcal{U}_{0,\Delta}. \tag{5.1.18c}$$

We will call $\hat{e}^{\text{Dir}}$ the *Dirichlet indicator function.* We have

$$\mathcal{B}(\hat{e}^{\text{Dir}}, v) = \sum_{j=1}^{N(\Delta)} \int_{I_j^\Delta} r_{I_j^\Delta} v + \underbrace{\sum_{j=1}^{N(\Delta)} J_k^\Delta v(x_k^\Delta)}_{0} = \mathcal{R}(v) = \mathcal{B}\big(e_{S_\Delta^p}, v\big) \quad \forall v \in \mathcal{U}_{0,\Delta}. \tag{5.1.19a}$$

This means that $\hat{e}^{\text{Dir}}$ is the energy projection of $e_{S_\Delta^p}$ into $\mathcal{U}_{0,\Delta}$, namely

$$\hat{e}^{\text{Dir}} = \pi_{\mathcal{U}_{0,\Delta}}\big(e_{S_\Delta^p}\big) \tag{5.1.19b}$$

and hence

$$\mathcal{E}^{\text{Dir}} \stackrel{\text{def}}{=} \|\hat{e}^{\text{Dir}}\|_{\mathcal{U}} = \big\|\pi_{\mathcal{U}_{0,\Delta}}\big(e_{S_\Delta^p}\big)\big\|_{\mathcal{U}} \le \|e_{S_\Delta^p}\|_{\mathcal{U}} \tag{5.1.19c}$$

which means that $\mathcal{E}^{\text{Dir}} = \|\hat{e}^{\text{Dir}}\|_{\mathcal{U}}$ is a *guaranteed lower estimator* of $\|e_{S_\Delta^p}\|_{\mathcal{U}}$. This can also be proven from

$$\|\hat{e}^{\text{Dir}}\|_{\mathcal{U}} = \frac{\mathcal{B}(\hat{e}^{\text{Dir}}, \hat{e}^{\text{Dir}})}{\|\hat{e}^{\text{Dir}}\|_{\mathcal{U}}} = \frac{\mathcal{R}(\hat{e}^{\text{Dir}})}{\|\hat{e}^{\text{Dir}}\|_{\mathcal{U}}} \le \sup_{v \in \mathcal{U}} \frac{\mathcal{R}(v)}{\|v\|_{\mathcal{U}}} = \sup_{v \in \mathcal{U}} \frac{\mathcal{B}\big(e_{S_\Delta^p}, v\big)}{\|v\|_{\mathcal{U}}} = \|e_{S_\Delta^p}\|_{\mathcal{U}}. \tag{5.1.19d}$$

Remark 5.1.4 Note that for any $\hat{e} \in \mathcal{U}$, we have

$$\underline{\mathcal{E}}(\hat{e}) \stackrel{\text{def}}{=} \frac{\mathcal{R}(\hat{e})}{\|\hat{e}\|_{\mathcal{U}}} \le \sup_{v \in \mathcal{U}} \frac{\mathcal{R}(v)}{\|v\|_{\mathcal{U}}} = \|\mathcal{R}\|_{L(\mathcal{U},\mathbb{R})} = \|e_{S_\Delta^p}\|_{\mathcal{U}} \tag{5.1.20}$$

and $\underline{\mathcal{E}}(\hat{e})$ is a guaranteed lower bound for the energy norm of the error $\|e_{S_\Delta^p}\|_{\mathcal{U}}$. However, if $\hat{e}$ is chosen arbitrarily, $\underline{\mathcal{E}}(\hat{e})$ may be zero or even negative, and hence useless as a lower estimator.

The Dirichlet indicator function $\hat{e}^{\text{Dir}}$ can be determined elementwise, as the energy projection of $e_{S_\Delta^p}$ into $\mathcal{U}_0(I_j^\Delta)$, and we will let

$$\hat{e}^{\text{Dir}}_{I_j^\Delta} \stackrel{\text{def}}{=} \hat{e}^{\text{Dir}}|_{I_j^\Delta} = \pi_{\mathcal{U}_0(I_j^\Delta)}\big(e_{S_\Delta^p}\big). \tag{5.1.21a}$$

We will call $\hat{e}^{\mathrm{Dir}}_{I^\Delta_j}$ the *Dirichlet element error indicator function.* It follows that $\hat{e}^{\mathrm{Dir}}_{I^\Delta_j}$ is the solution of the element residual problem:

Find $\hat{e}^{\mathrm{Dir}}_{I^\Delta_j} \in \mathcal{U}_0(I^\Delta_j)$ such that

$$\mathcal{B}_{I^\Delta_j}(\hat{e}^{\mathrm{Dir}}_{I^\Delta_j}, v) = \int_{I^\Delta_j} r_{I^\Delta_j} v \quad \forall v \in \mathcal{U}_0(I^\Delta_j). \tag{5.1.21b}$$

This is the variational formulation of the Dirichlet problem

$$-\frac{d}{dx}\left(AE\frac{d}{dx}(\hat{e}^{\mathrm{Dir}}_{I^\Delta_j})\right) + c\hat{e}^{\mathrm{Dir}}_{I^\Delta_j} = r_{I^\Delta_j}, \qquad \hat{e}^{\mathrm{Dir}}_{I^\Delta_j}(x^\Delta_{j-1}) = \hat{e}^{\mathrm{Dir}}_{I^\Delta_j}(x^\Delta_j) = 0. \tag{5.1.21c}$$

We will call this problem the *element residual problem with homogeneous Dirichlet boundary conditions.* We note that

$$\eta^{\mathrm{Dir}}_{I^\Delta_j} \stackrel{\mathrm{def}}{=} \|\hat{e}^{\mathrm{Dir}}_{I^\Delta_j}\|_{\mathcal{U}(I^\Delta_j)} \leq \|e_{S^p_\Delta}\|_{\mathcal{U}(I^\Delta_j)} \tag{5.1.22a}$$

and

$$\mathcal{E}^{\mathrm{Dir}} = \|\hat{e}^{\mathrm{Dir}}\|_{\mathcal{U}} = \sqrt{\sum_{j=1}^{N(\Delta)} \|\hat{e}^{\mathrm{Dir}}_{I^\Delta_j}\|^2_{\mathcal{U}(I^\Delta_j)}} \leq \|e_{S^p_\Delta}\|_{\mathcal{U}}. \tag{5.1.22b}$$

We will call $\mathcal{E}^{\mathrm{Dir}}$ the *Dirichlet estimator*, and $\eta^{\mathrm{Dir}}_{I^\Delta_j}$ the *Dirichlet element error indicator.*

Remark 5.1.5 The Dirichlet estimator $\mathcal{E}^{\mathrm{Dir}}$ is a reliable lower estimator of $\|e_{S^p_\Delta}\|_{\mathcal{U}}$ *without any assumptions on the pollution.*

Remark 5.1.6 In general, $e_{S^p(I^\Delta_j)}(x^\Delta_k) \neq 0$, $k = j-1, j$, and hence $\hat{e}^{\mathrm{Dir}}_{I^\Delta_j} \not\equiv e_{S^p_\Delta}|_{I^\Delta_j}$.

Remark 5.1.7 Another way of introducing the Dirichlet estimator $\mathcal{E}^{\mathrm{Dir}}$ is by constructing the recovered solution $\hat{u}^{\mathrm{Dir}}_{I^\Delta_j}$ as the solution of the Dirichlet problem

$$-\frac{d}{dx}\left(AE\frac{d}{dx}(\hat{u}^{\mathrm{Dir}}_{I^\Delta_j})\right) + c\hat{u}^{\mathrm{Dir}}_{I^\Delta_j} = f \tag{5.1.23a}$$

$$\hat{u}^{\mathrm{Dir}}_{I^\Delta_j}(x^\Delta_{j-1}) = u_{S^p_\Delta}(x^\Delta_{j-1}), \qquad \hat{u}^{\mathrm{Dir}}_{I^\Delta_j}(x^\Delta_j) = u_{S^p_\Delta}(x^\Delta_j) \tag{5.1.23b}$$

which can also be written in the variational form:

Find $\hat{u}^{\mathrm{Dir}}_{I^\Delta_j} \in u_{S^p_\Delta}|_{I^\Delta_j} \oplus \mathcal{U}_0(I^\Delta_j)$ such that

$$\mathcal{B}_{I^\Delta_j}(\hat{u}^{\mathrm{Dir}}_{I^\Delta_j}, v) = \int_{I^\Delta_j} fv \quad \forall v \in \mathcal{U}_0(I^\Delta_j). \tag{5.1.23c}$$

We then have

$$\hat{e}^{\mathrm{Dir}}_{I^\Delta_j} = \hat{u}^{\mathrm{Dir}}_{I^\Delta_j} - u_{S^p_\Delta}\big|_{I^\Delta_j}. \tag{5.1.23d}$$

In the case that *the pollution is negligible*, namely,

$$\max(e^{\Delta}_{j-1}, e^{\Delta}_{j}) \le \mathscr{C}\big(h_{I^\Delta_j}\big)^{p-1} \|e_{S^p_\Delta}\|_{L^\infty(I^\Delta_j)} \tag{5.1.24a}$$

it can be proven that (see Section 5.2)

$$e_{S^p_\Delta}\big|_{I^\Delta_j} \approx (1 + \mathscr{C}h^\Delta)\hat{e}^{\mathrm{Dir}}_{I^\Delta_j} \tag{5.1.24b}$$

and

$$\|u_{EX} - \hat{u}^{\mathrm{Dir}}_{I^\Delta_j}\|_{L^\infty(I^\Delta_j)} \le \mathscr{C}\big(h_{I^\Delta_j}\big)^{2p} \tag{5.1.24c}$$

while

$$\|u_{EX} - u_{S^p_\Delta}\|_{L^\infty(I^\Delta_j)} \approx \|\hat{u}^{\mathrm{Dir}}_{I^\Delta_j} - u_{S^p_\Delta}\|_{L^\infty(I^\Delta_j)} \approx \mathscr{C}\big(h_{I^\Delta_j}\big)^{p+1}. \tag{5.1.24d}$$

Hence for $p > 1$, and *negligible pollution*, $\hat{u}^{\mathrm{Dir}}_{I^\Delta_j}$ has better accuracy than the finite element solution $u_{S^p_\Delta}$. However, in the cases that the pollution cannot be neglected, $\hat{u}^{\mathrm{Dir}}_{I^\Delta_j}$ has practically the same accuracy as the finite element solution $u_{S^p_\Delta}|_{I^\Delta_j}$.

Let us underline that the error indicators $\eta^{\mathrm{Dir}}_{I^\Delta_j}$ and the estimator $\mathscr{E}^{\mathrm{Dir}}$, are the *exact* indicators and estimator, respectively. In the implementation, $\eta^{\mathrm{Dir}}_{I^\Delta_j}$ and $\mathscr{E}^{\mathrm{Dir}}$ are often computed by employing finite element approximations of $\hat{e}^{\mathrm{Dir}}_{I^\Delta_j}$, and this leads to under-estimation of the indicators $\eta^{\mathrm{Dir}}_{I^\Delta_j}$, and the estimator $\mathscr{E}^{\mathrm{Dir}}$. In practice, most often the p version is used, and $\hat{e}^{\mathrm{Dir}}_{I^\Delta_j}$ is approximated by $\pi_{S^{p+r}_0(I^\Delta_j)}(\hat{e}^{\mathrm{Dir}}_{I^\Delta_j})$, its energy projection into $S^{p+r}_0(I^\Delta_j) \equiv \mathscr{P}^{p+r}_0(I^\Delta_j)$, and we have

$$\eta^{\mathrm{Dir}}_{I^\Delta_j, S^{p+r}_\Delta} \overset{\mathrm{def}}{=} \|\pi_{S^{p+r}_0(I^\Delta_j)}\big(\hat{e}^{\mathrm{Dir}}_{I^\Delta_j}\big)\|_{\mathscr{U}(I^\Delta_j)} \le \eta^{\mathrm{Dir}}_{I^\Delta_j} = \|\hat{e}^{\mathrm{Dir}}_{I^\Delta_j}\|_{\mathscr{U}(I^\Delta_j)} \tag{5.1.25a}$$

and

$$\mathscr{E}^{\mathrm{Dir}}_{S^{p+r}_\Delta} \overset{\mathrm{def}}{=} \sqrt{\sum_{j=1}^{N(\Delta)} (\eta^{\mathrm{Dir}}_{I^\Delta_j, S^{p+r}_\Delta})^2} \le \mathscr{E}^{\mathrm{Dir}} = \sqrt{\sum_{j=1}^{N(\Delta)} \big(\eta^{\mathrm{Dir}}_{I^\Delta_j}\big)^2} \le \|e_{S^p_\Delta}\|_{\mathscr{U}}. \tag{5.1.25b}$$

We will call $\mathscr{E}^{\mathrm{Dir}}_{S^{p+r}_\Delta}$, the *computed* Dirichlet estimator. Below, we will often refer to $\mathscr{E}^{\mathrm{Dir}}_{S^{p+r}_\Delta}$ by the estimator *ERDPp* + *r*, which is an abbreviation of the *Element Residual estimator of the Dirichlet type which employs approximations of the element indicators by Polynomials of degree* $p + k$.

Another approximation of $\mathscr{E}^{\mathrm{Dir}}$ may be obtained by employing an explicit estimator of $\|\hat{e}^{\mathrm{Dir}}_{I^\Delta_j}\|_{\mathscr{U}(I^\Delta_j)}$ in terms of the element L^2 norms of the interior residuals $\|r_{I^\Delta_j}\|_{L^2(I^\Delta_j)}$. We have

$$\|\hat{e}^{\mathrm{Dir}}_{I^\Delta_j}\|_{\mathscr{U}(I^\Delta_j)} = \sup_{v \in \mathscr{U}_0(I^\Delta_j)} \frac{\mathscr{B}\big(\hat{e}^{\mathrm{Dir}}_{I^\Delta_j}, v\big)}{\|v\|_{\mathscr{U}(I^\Delta_j)}} = \sup_{v \in \mathscr{U}_0(I^\Delta_j)} \frac{\int_{I^\Delta_j} r_{I^\Delta_j} v}{\|v\|_{\mathscr{U}(I^\Delta_j)}} \le \mathscr{C}h_{I^\Delta_j}\|r_{I^\Delta_j}\|_{L^2(I^\Delta_j)} \tag{5.1.26a}$$

where $\mathscr{C}$ is a computable constant. It follows that

$$\mathscr{E}^{\mathrm{Dir}} \leq \mathscr{E}^{\mathrm{Dir}}_{EXPL} \stackrel{\mathrm{def}}{=} \mathscr{C} \sqrt{\sum_{j=1}^{N(\Delta)} h^2_{I^\Delta_j} \left\| r_{I^\Delta_j} \right\|^2_{L^2(I^\Delta_j)}} \tag{5.1.26b}$$

and hence we cannot guarantee that $\mathscr{E}^{\mathrm{Dir}}_{EXPL}$ is a lower estimate for $\|e_{S^p_\Delta}\|_{\mathcal{U}}$.

Let us now give the construction of $\bar{\mathscr{E}}$. We will start by splitting the global residuum $\mathscr{R} \in L(\mathcal{U}, \mathbb{R})$ into *element residuums* $\mathscr{R}^{EQ}_{I^\Delta_j} \in L(\mathcal{U}(I^\Delta_j), \mathbb{R})$, namely

$$\mathscr{R}(v) = \sum_{j=1}^{N(\Delta)} \mathscr{R}^{EQ}_{I^\Delta_j}(v|_{I^\Delta_j}) = \sum_{j=1}^{N(\Delta)} \int_{I^\Delta_j} \Re^{EQ}_{I^\Delta_j} v \tag{5.1.27a}$$

where $\mathscr{R}^{EQ}_{I^\Delta_j} \in L(\mathcal{U}(I^\Delta_j), \mathbb{R})$ is given by

$$\mathscr{R}^{EQ}_{I^\Delta_j}(v) = \int_{I^\Delta_j} \Re^{EQ}_{I^\Delta_j} v = \int_{I^\Delta_j} r_{I^\Delta_j} v + J^L_j v(x^\Delta_j) + J^R_{j-1} v(x^\Delta_{j-1}) \quad \forall v \in \mathcal{U}(I^\Delta_j) \tag{5.1.27b}$$

and

$$\Re^{EQ}_{I^\Delta_j} \stackrel{\mathrm{def}}{=} r_{I^\Delta_j} + J^L_j \delta(\cdot; x^{\Delta,-}_j) + J^R_{j-1} \delta(\cdot; x^{\Delta,+}_{j-1}) \tag{5.1.27c}$$

is the *residual load* corresponding to $\mathscr{R}^{EQ}_{I^\Delta_j}$. From the condition of the exact splitting (5.1.27a), it follows that J^L_j and J^R_j must satisfy the condition

$$J^\Delta_j = J^L_j + J^R_j. \tag{5.1.28}$$

We will call J^L_j, J^R_j an *exact splitting* of the jump J^Δ_j. Below, we will determine J^L_j, J^R_j, such that they satisfy the exact splitting condition (5.1.28), and the orthogonality condition

$$\mathscr{R}^{EQ}_{I^\Delta_j}(v) = \int_{I^\Delta_j} r_{I^\Delta_j} v + J^L_j v(x^\Delta_j) + J^R_{j-1} v(x^\Delta_{j-1}) = 0 \quad \forall v \in S^p(I^\Delta_j). \tag{5.1.29}$$

We will call (5.1.29) the *equilibrium condition*, and $\mathscr{R}^{EQ}_{I^\Delta_j}$ the *equilibrated residuum* for I^Δ_j.

Before we discuss the determination of J^L_j and J^R_j such that they satisfy (5.1.28) and (5.1.29), let us give the construction of $\bar{\mathscr{E}}$ the upper bound for $\|e_{S^p_\Delta}\|_{\mathcal{U}}$. We have

$$\|e_{S^p_\Delta}\|_{\mathcal{U}} = \sup_{v \in \mathcal{U}} \frac{\sum_{j=1}^{N(\Delta)} \mathscr{R}^{EQ}_{I^\Delta_j}(v|_{I^\Delta_j})}{\|v\|_{\mathcal{U}}} \leq \sup_{v \in \mathcal{U}} \sum_{j=1}^{N(\Delta)} \left(\frac{|\mathscr{R}^{EQ}_{I^\Delta_j}(v|_{I^\Delta_j})|}{\|v\|_{\mathcal{U}(I^\Delta_j)}} \frac{\|v\|_{\mathcal{U}(I^\Delta_j)}}{\|v\|_{\mathcal{U}}} \right). \tag{5.1.30a}$$

Employing the Cauchy–Schwarz inequality for the $\mathbb{R}^{N(\Delta)}$ inner product, and exchanging the supremum with the sum, we get

$$\|e_{S^p_\Delta}\|_{\mathcal{U}} \leq \sqrt{\sum_{j=1}^{N(\Delta)} \left(\sup_{v \in \mathcal{U}(I^\Delta_j)} \frac{\left|\mathscr{R}^{EQ}_{I^\Delta_j}(v)\right|}{\|v\|_{\mathcal{U}(I^\Delta_j)}} \right)^2} = \sqrt{\sum_{j=1}^{N(\Delta)} \|\mathscr{R}^{EQ}_{I^\Delta_j}\|^2_{L(\mathcal{U}(I^\Delta_j),\mathbb{R})}}. \tag{5.1.30b}$$

The upper estimator $\bar{\mathscr{E}}$ can now be constructed by employing that

$$\|\mathscr{R}^{EQ}_{I_j^\Delta}\|_{L(\mathscr{U}(I_j^\Delta),\mathbb{R})} = \sup_{v\in\mathscr{U}(I_j^\Delta)} \frac{\mathscr{R}^{EQ}_{I_j^\Delta}(v)}{\|v\|_{\mathscr{U}(I_j^\Delta)}} = \sup_{v\in\mathscr{U}(I_j^\Delta)} \frac{\mathscr{B}_{I_j^\Delta}(\hat{e}^{\text{Neum}}_{I_j^\Delta}, v)}{\|v\|_{\mathscr{U}(I_j^\Delta)}}$$
$$= \|\hat{e}^{\text{Neum}}_{I_j^\Delta}\|_{\mathscr{U}(I_j^\Delta)} \tag{5.1.31}$$

where $\hat{e}^{\text{Neum}}_{I_j^\Delta}$ is defined as the solution of the following problem:

Find $\hat{e}^{\text{Neum}}_{I_j^\Delta} \in \mathscr{U}(I_j^\Delta)$ such that

$$\mathscr{B}_{I_j^\Delta}(\hat{e}^{\text{Neum}}_{I_j^\Delta}, v) = \mathscr{R}^{EQ}_{I_j^\Delta}(v) \quad \forall v \in \mathscr{U}(I_j^\Delta). \tag{5.1.32a}$$

For all the elements except the first one, $\hat{e}^{\text{Neum}}_{I_j^\Delta}$ is the solution of the following Neumann problem

$$-\frac{d}{dx}\left(AE\frac{d}{dx}(\hat{e}^{\text{Neum}}_{I_j^\Delta})\right) + c\hat{e}^{\text{Neum}}_{I_j^\Delta} = r_{I_j^\Delta} \tag{5.1.32b}$$

$$AE(x^{\Delta,+}_{j-1})\frac{d}{dx}(\hat{e}^{\text{Neum}}_{I_j^\Delta})(x^\Delta_{j-1}) = -J^R_{j-1}, \qquad AE(x^{\Delta,-}_j)\frac{d}{dx}(\hat{e}^{\text{Neum}}_{I_j^\Delta})(x^\Delta_j) = J^L_j \tag{5.1.32c}$$

for $j = 2, \ldots, N(\Delta)$. The boundary value problem for $\hat{e}^{\text{Neum}}_{I_1^\Delta}$ is obtained from (5.1.32b), (5.1.32c) by employing $j = 1$, and by replacing the Neumann boundary condition at the left end point by the Dirichlet boundary condition, $\hat{e}^{\text{Neum}}_{I_1^\Delta}(x_0^\Delta) = 0$. We will call $\hat{e}^{\text{Neum}}_{I_j^\Delta}$ the *Neumann element error indicator function*. Note that in the case that $c \equiv 0$, the *consistency (equilibrium) condition*

$$\int_{I_j^\Delta} r_{I_j^\Delta} + J^L_j + J^R_{j-1} = \mathscr{R}^{EQ}_{I_j^\Delta}(1) = 0 \tag{5.1.32d}$$

is necessary for the existence of $\hat{e}^{\text{Neum}}_{I_j^\Delta}$ for $j = 2, \ldots, N(\Delta)$. This condition follows from the orthogonality condition $\mathscr{R}^{EQ}_{I_j^\Delta}(v) = 0$, $v \in S^p(I_j^\Delta)$, by noting that for $j = 2, \ldots, N(\Delta)$, $v \equiv 1 \in S^p(I_j^\Delta)$. In the case $c \equiv 0$, $\hat{e}^{\text{Neum}}_{I_j^\Delta}$, $j = 2, \ldots, N(\Delta)$, is unique up to a constant, which does not influence the value of the norm $\|\hat{e}^{\text{Neum}}_{I_j^\Delta}\|_{\mathscr{U}(I_j^\Delta)}$. The value of the constant will be chosen such that

$$\int_{I_j^\Delta} \hat{e}^{\text{Neum}}_{I_j^\Delta} = 0. \tag{5.1.32e}$$

Having established the existence of $\hat{e}^{\text{Neum}}_{I_j^\Delta}$, and employing (5.1.31) in (5.1.30b), we obtain

$$\|e_{S^p_\Delta}\|_{\mathscr{U}} \le \mathscr{E}^{\text{Neum}} \stackrel{\text{def}}{=} \sqrt{\sum_{j=1}^{N(\Delta)} \|\hat{e}^{\text{Neum}}_{I_j^\Delta}\|^2_{\mathscr{U}(I_j^\Delta)}} = \sqrt{\sum_{j=1}^{N(\Delta)} (\eta^{\text{Neum}}_{I_j^\Delta})^2} \tag{5.1.33}$$

where we employed the notation

$$\eta_{I_j^\Delta}^{\text{Neum}} = \left\| \hat{e}_{I_j^\Delta}^{\text{Neum}} \right\|_{\mathcal{U}(I_j^\Delta)}. \tag{5.1.34}$$

We will call $\eta_{I_j^\Delta}^{\text{Neum}}$ the *exact Neumann element error indicator*, and $\mathscr{E}^{\text{Neum}}$ the *exact Neumann estimator*. Note that the exact Neumann estimator $\mathscr{E}^{\text{Neum}}$ is a guaranteed upper estimator for $\|e_{S_\Delta^p}\|_{\mathcal{U}}$.

Remark 5.1.8 In the case that $\int_{I_j^\Delta} c > 0$, the equilibrium condition $\mathscr{R}_{I_j^\Delta}^{EQ}(1) = 0$ is not necessary for the existence of $\hat{e}_{I_j^\Delta}^{\text{Neum}}$, and hence any splitting of the jumps J_k^Δ may be used. However, the stabilizing effect of c diminishes as ch^2 tends to zero, we may have gross overestimation $\|e_{S_\Delta^p}\|_{\mathcal{U}} \ll \mathscr{E}^{\text{Neum}}$, if the jumps are split arbitrarily without satisfying the element equilibrium conditions.

Remark 5.1.9 In the case $c \equiv 0$, the Neumann estimator $\mathscr{E}^{\text{Neum}}$ can be written in the form

$$\mathscr{E}^{\text{Neum}} = \left\| \frac{1}{\sqrt{AE}} (\hat{P}_{\Delta,p}^{\text{Neum}} - P_{S_\Delta^p}) \right\|_{L^2(I)} \tag{5.1.35a}$$

where $\hat{P}_{\Delta,p}^{\text{Neum}}$ is defined elementwise by

$$\hat{P}_{\Delta,p}^{\text{Neum}}|_{I_j^\Delta} = \hat{P}_{I_j^\Delta}^{\text{Neum}} \overset{\text{def}}{=} P_{S^p(I_j^\Delta)} + AE\frac{d}{dx}(\hat{e}_{I_j^\Delta}^{\text{Neum}}). \tag{5.1.35b}$$

Note that $\hat{P}_{\Delta,p}^{\text{Neum}}$ satisfies the differential equation and the force boundary condition

$$\frac{d}{dx}(\hat{P}_{\Delta,p}^{\text{Neum}}) + f \equiv 0, \qquad \hat{P}_{\Delta,p}^{\text{Neum}}(x_N^\Delta) = a_2 \tag{5.1.35c}$$

and hence it is *statically admissible*. Note that although $P_{S^p(I_j^\Delta)}(x_j^\Delta) \neq P_{S^p(I_{j+1}^\Delta)}(x_j^\Delta)$, the function $\hat{P}_{\Delta,p}^{\text{Neum}}$ is continuous on I, and hence (5.1.35c) has meaning. When the pollution is negligible, $\hat{P}_{\Delta,p}^{\text{Neum}}$ is a superconvergent approximation of the exact axial force P_{EX} in I_j^Δ, namely,

$$\left\| \frac{1}{\sqrt{AE}} (\hat{P}_{\Delta,p}^{\text{Neum}} - P_{EX}) \right\|_{L^\infty(I_j^\Delta)} \leq \mathscr{C}\left(h_{I_j^\Delta}\right)^{p+1} \tag{5.1.36a}$$

and we have the estimate

$$|||P_{EX} - P_{S_\Delta^p}|||_{I_j^\Delta} \approx |||\hat{P}_{\Delta,p}^{\text{Neum}} - P_{S_\Delta^p}|||_{I_j^\Delta} \tag{5.1.36b}$$

in any norm $|||\cdot|||_{I_j^\Delta}$. The idea of the construction of an upper estimate of the error measured in the energy norm in the form (5.1.35a), based on a statically admissible axial force is the main theme of the classical method of *hypercircle* of [Synge, 1957]. The method of hypercircle can also be extended to the case $c \not\equiv 0$; see [Aubin and Burchard, 1970]. Nevertheless, above we proved the upper bound

$$\|e_{S_\Delta^p}\|_{\mathcal{U}} = \|\mathscr{R}\|_{L(\mathcal{U},\mathbb{R})} \leq \mathscr{E}^{\text{Neum}} \tag{5.1.37a}$$

directly by employing that

$$\|\mathcal{R}\|_{L(\mathcal{U},\mathbb{R})} \leq \sqrt{\sum_{j=1}^{N(\Delta)} \|\mathcal{R}^{EQ}_{I_j^\Delta}\|^2_{L(\mathcal{U}(I_j^\Delta),\mathbb{R})}} = \sqrt{\sum_{j=1}^{N(\Delta)} \|\hat{e}^{\text{Neum}}_{I_j^\Delta}\|^2_{\mathcal{U}(I_j^\Delta)}} = \mathcal{E}^{\text{Neum}} \tag{5.1.37b}$$

and we did not use the method of hypercircle. The exact splitting of the residuals into equilibrated element sets, $\mathcal{R} = \sum_{j=1}^{N(\Delta)} \mathcal{R}^{EQ}_{I_j^\Delta}$, with $\mathcal{R}^{EQ}_{I_j^\Delta}(1) = 0$, for $j = 2, \ldots, N(\Delta)$, was essential for the argument. As we will see below, such a splitting is possible because of the the orthogonality

$$\mathcal{R}(v) = \mathcal{B}(e_{S^p_\Delta}, v) = 0 \quad \forall v \in S^p_\Delta. \tag{5.1.37c}$$

As in the case of $\mathcal{E}^{\text{Dir}}$, we must also consider the effect of the approximation of the indicators. In practice, most often the p method is used and we let

$$\eta^{\text{Neum}}_{I_j^\Delta, S^{p+r}_\Delta} \stackrel{\text{def}}{=} \|\pi_{S^{p+r}(I_j^\Delta)}(\hat{e}^{\text{Neum}}_{I_j^\Delta})\|_{\mathcal{U}(I_j^\Delta)} \tag{5.1.38a}$$

and

$$\mathcal{E}^{\text{Neum}}_{S^{p+r}_\Delta} \stackrel{\text{def}}{=} \sqrt{\sum_{j=1}^{N(\Delta)} \left(\eta^{\text{Neum}}_{I_j^\Delta, S^{p+r}(I_j^\Delta)}\right)^2}. \tag{5.1.38b}$$

We will call $\mathcal{E}^{\text{Neum}}_{S^{p+r}_\Delta}$ the *computed Neumann* estimator. Below, we will often refer to $\mathcal{E}^{\text{Neum}}_{S^{p+r}_\Delta}$ as the *ERNp* + *k* estimate, namely, the estimate which employs the *Element Residual estimator of the Neumann type which employs approximation of the element indicators by Polynomials of degree* $p + k$. It follows that

$$\eta^{\text{Neum}}_{I_j^\Delta, S^{p+r}(I_j^\Delta)} \stackrel{\text{def}}{=} \|\pi_{S^{p+r}(I_j^\Delta)}(\hat{e}^{\text{Neum}}_{I_j^\Delta})\|_{\mathcal{U}(I_j^\Delta)} \leq \eta^{\text{Neum}}_{I_j^\Delta} = \|\hat{e}^{\text{Neum}}_{I_j^\Delta}\|_{\mathcal{U}(I_j^\Delta)} \tag{5.1.38c}$$

and hence

$$\mathcal{E}^{\text{Neum}}_{S^{p+r}_\Delta} \stackrel{\text{def}}{=} \sqrt{\sum_{j=1}^{N(\Delta)} \left(\eta^{\text{Neum}}_{I_j^\Delta, S^{p+r}(I_j^\Delta)}\right)^2} \leq \mathcal{E}^{\text{Neum}} = \sqrt{\sum_{j=1}^{N(\Delta)} \left(\eta^{\text{Neum}}_{I_j^\Delta}\right)^2} \tag{5.1.38d}$$

and hence the computed estimator $\mathcal{E}^{\text{Neum}}_{S^{p+r}_\Delta}$ is not guaranteed to be an upper estimator for $\|e_{S^p_\Delta}\|_{\mathcal{U}}$.

Another alternative is to estimate the element indicators $\eta^{\text{Neum}}_{I_j^\Delta}$ from above. Using (5.1.32a), (5.1.32d), and employing the Cauchy–Schwarz inequality, we obtain

$$\eta^{\text{Neum}}_{I_j^\Delta} = \sup_{\substack{v \in \mathcal{U}(I_j^\Delta) \\ \int_{I_j^\Delta} v = 0}} \frac{1}{\|v\|_{\mathcal{U}(I_j^\Delta)}} \left| \int_{I_j^\Delta} r_{I_j^\Delta} v + J_j^L v(x_j^\Delta) + J_{j-1}^R v(x_{j-1}^\Delta) \right| \leq \eta^{\text{Neum}}_{I_j^\Delta, EXPL} \tag{5.1.39a}$$

where

$$\eta^{\text{Neum}}_{I_j^\Delta,EXPL} \stackrel{\text{def}}{=} \mathcal{C}\sqrt{h^2_{I_j^\Delta}\|r_{I_j^\Delta}\|^2_{L^2(I_j^\Delta)} + h_{I_j^\Delta}((J_j^L)^2 + (J_{j-1}^R)^2)} \tag{5.1.39b}$$

where we used that for any $v \in \mathcal{U}(I_j^\Delta)$, with $\int_{I_j^\Delta} v = 0$, we have

$$\|v\|^2_{L^2(I_j^\Delta)} \leq \mathcal{C}h^2_{I_j^\Delta}\|v\|^2_{\mathcal{U}(I_j^\Delta)} \tag{5.1.39c}$$

and

$$|v(x_j^\Delta)|^2 \leq \mathcal{C}h_{I_j^\Delta}\|v\|^2_{\mathcal{U}(I_j^\Delta)}. \tag{5.1.39d}$$

It follows that

$$\|e_{S^p_\Delta}\|_{\mathcal{U}} \leq \mathcal{E}^{\text{Neum}} \leq \mathcal{E}^{\text{Neum}}_{EXPL} \stackrel{\text{def}}{=} \sqrt{\sum_{j=1}^{N(\Delta)}(\eta^{\text{Neum}}_{I_j^\Delta,EXPL})^2} \tag{5.1.39e}$$

and hence $\mathcal{E}^{\text{Neum}}_{EXPL}$ is a guaranteed upper estimate for $\|e_{S^p_\Delta}\|_{\mathcal{U}}$, *provided that the constant $\mathcal{C}$ is selected properly.* Recalling the above mentioned dichotomy of elements of odd and even degree, for p odd (respectively p even) we may use only the second (respectively first) term in (5.1.39b) with properly adjusted constant $\mathcal{C}$. For elements of degree $p > 1$, $\mathcal{E}^{\text{Neum}}_{EXPL}$ may grossly overestimate $\|e_{S^p_\Delta}\|_{\mathcal{U}}$. An improved explicit estimate can be also constructed by taking into account the orthogonalities

$$\int_{I_j^\Delta}(r_{I_j^\Delta}v + J_j^L v(x_j^\Delta) + J_{j-1}^R v(x_{j-1}^\Delta)) = 0 \quad \forall v \in S^p(I_j^\Delta). \tag{5.1.39f}$$

Let us now address the relative size of the contributions from the interior residual and the jumps for the case considered in Examples 5.1.1 and 5.1.2.

Example 5.1.3 We will consider the case of uniform mesh. In the case of linear elements ($p = 1$), we have

$$J_{j-1}^R = \int_{I_j^\Delta}\sin\left(\frac{\pi}{2}x\right)\frac{x - x_{j-1}^\Delta}{h_{I_j^\Delta}} = \frac{1}{2}J_{j-1}^\Delta(1 + O(h^2)) \tag{5.1.40a}$$

and similarly

$$J_j^L = \tfrac{1}{2}J_j^\Delta(1 + O(h^2)). \tag{5.1.40b}$$

It follows that

$$h^2\int_{I_j^\Delta}(r_{I_j^\Delta})^2 \approx h((J_{j-1}^R)^2 + (J_j^L)^2) \tag{5.1.40)}$$

which indicates that for $p = 1$, both the contributions of the interior residuals and the jumps are of the same order, and hence either term, $h_{I_j^\Delta}\sqrt{\int_{I_j^\Delta}(r_{I_j^\Delta})^2}$, or

$\sqrt{h_{I_j^\Delta}}\sqrt{(J_{j-1}^R)^2 + (J_j^L)^2}$, or their sum can be employed as an explicit element error indicator.

In the case of quadratic elements ($p = 2$) we have

$$-J_{j-1}^R = h\cos\left(\frac{\pi}{4}h\right) + \frac{h}{2}\sin\left(\frac{\pi}{2}x_{j-1/2}\right) + \frac{h}{2}h\frac{\sin^2((\pi/8)h)}{((\pi/8)h)^2}\sin\left(\frac{\pi}{2}x_{j-1/2}\right) = O(h^3) \quad (5.1.41a)$$

and similarly $J_j^L = O(h^3)$, and hence

$$h((J_{j-1}^R)^2 + (J_j^L)^2) \approx O(h^7) \ll h^2\int_{I_j^\Delta}(r_{I_j^\Delta})^2 \approx O(h^5). \quad (5.1.41b)$$

This means that for $p = 2$, the contribution of the jumps is negligible asymptotically, and it is sufficient to use only the first term $h_{I_j^\Delta}\sqrt{\int_{I_j^\Delta}(r_{I_j^\Delta})^2}$, as the explicit element error indicator. The same result holds for any p even ($p = 2, 4, 6, \ldots$).

Remark 5.1.10 A guaranteed computable upper estimate may also be obtained by noting that

$$\eta_{I_j^\Delta}^{\text{Neum}} = \sqrt{\left\|\pi_{S^{p+r}(I_j^\Delta)}\left(\hat{e}_{I_j^\Delta}^{\text{Neum}}\right)\right\|_{\mathcal{U}(I_j^\Delta)}^2 + \left\|\hat{e}_{I_j^\Delta}^{\text{Neum}} - \pi_{S^{p+r}(I_j^\Delta)}\left(\hat{e}_{I_j^\Delta}^{\text{Neum}}\right)\right\|_{\mathcal{U}(I_j^\Delta)}^2} \quad (5.1.42)$$

where we used the orthogonality property of the projection operator $\pi_{S^{p+r}(I_j^\Delta)}$. Hence we can obtain an upper estimate of the exact indicator $\eta_{I_j^\Delta}^{\text{Neum}}$ by augmenting the computed indicator $\eta_{I_j^\Delta,S^{p+r}(I_j^\Delta)}^{\text{Neum}} = \|\pi_{S^{p+r}(I_j^\Delta)}\ (\hat{e}_{I_j^\Delta}^{\text{Neum}})\|_{\mathcal{U}(I_j^\Delta)}$, by an explicit estimate of the energy norm of the error in the computed error indicator function $\|\hat{e}_{I_j^\Delta}^{\text{Neum}} - \pi_{S^{p+r}(I_j^\Delta)}(\hat{e}_{I_j^\Delta}^{\text{Neum}})\|_{\mathcal{U}(I_j^\Delta)}$. As we will see in Chapter 6, this approach often gives good results because $\pi_{S^{p+r}(I_j^\Delta)}(\hat{e}_{I_j^\Delta}^{\text{Neum}})$ is already a good approximation of $\hat{e}_{I_j^\Delta}^{\text{Neum}}$, and the second term where crude overestimation may occur could be small.

Let us now show that *there exists a splitting* $J_j^L + J_j^R = J_j^\Delta$ *which is such that the equilibrium condition*

$$\mathcal{R}_{I_j^\Delta}^{EQ}(v) = 0 \quad \forall v \in S^p(I_j^\Delta) \quad (5.1.43a)$$

holds in each element I_j^Δ. Let

$$P_j^L \stackrel{\text{def}}{=} \mathcal{B}_{I_j^\Delta}(u_{S_\Delta^p}, \varphi_j^\Delta) - \int_{I_j^\Delta} f\varphi_j^\Delta \quad (5.1.43b)$$

and

$$P_{j-1}^R \stackrel{\text{def}}{=} -\mathcal{B}_{I_j^\Delta}(u_{S_\Delta^p}, \varphi_{j-1}^\Delta) + \int_{I_j^\Delta} f\varphi_{j-1}^\Delta \quad (5.1.43c)$$

where φ_j^Δ denotes the hat function for node j, given in (3.2.5). Note that P_{j-1}^R and P_j^L are the end reactions which satisfy the discrete equation in element I_j^Δ, namely,

$$\mathcal{B}_{I_j^\Delta}(u_{S_\Delta^p}, v) = \int_{I_j^\Delta} fv + P_j^L v(x_j^\Delta) - P_{j-1}^R v(x_{j-1}^\Delta) \quad \forall v \in S^p(I_j^\Delta). \tag{5.1.43d}$$

Now let

$$J_j^L \stackrel{\text{def}}{=} P_j^L - P_{S_\Delta^p}(x_j^{\Delta,-}) \tag{5.1.44a}$$

and

$$J_j^R \stackrel{\text{def}}{=} -P_j^R + P_{S_\Delta^p}(x_j^{\Delta,+}) \tag{5.1.44b}$$

where $P_{S_\Delta^p}(x_j^{\Delta,\pm})$ were defined above in (5.1.4c); see also Fig. 5.1.1, where the 'free-body' diagrams of the elements are sketched.

We will now prove that the split jumps J_j^L, J_j^R given by (5.1.44a) and (5.1.44b) satisfy the exact splitting conditions

$$J_j^L + J_j^R = J_j^\Delta \quad j = 1, \ldots, N(\Delta) - 1 \tag{5.1.44c}$$

and the orthogonality conditions

$$\mathcal{R}_{I_j^\Delta}^{EQ}(v) = 0 \quad \forall v \in S^p(I_j^\Delta) \quad j = 1, \ldots, N(\Delta). \tag{5.1.44d}$$

Integrating by parts we get

$$\mathcal{B}_{I_j^\Delta}(u_{S_\Delta^p}, \varphi_{j-1}^\Delta) = -P_{S_\Delta^p}(x_{j-1}^{\Delta,+}) + \int_{I_j^\Delta} (\mathfrak{D}u_{S_\Delta^p})\varphi_{j-1}^\Delta \tag{5.1.45a}$$

where we employed the notation

$$\mathfrak{D}u_{S_\Delta^p} = -\frac{d}{dx}\left(AE\frac{d}{dx}(u_{S_\Delta^p})\right) + cu_{S_\Delta^p} \tag{5.1.45b}$$

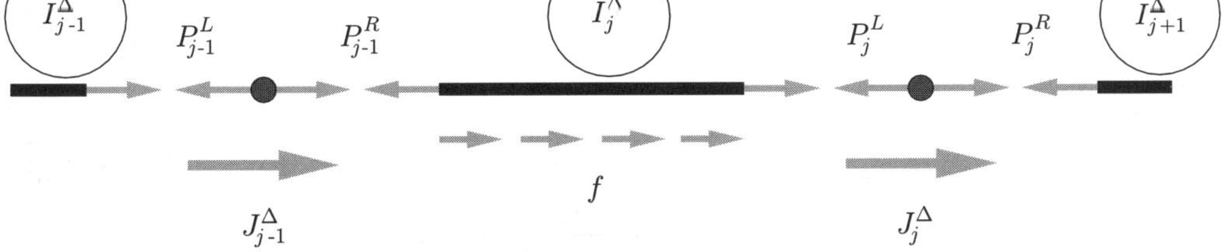

FIG. 5.1.1. A free-body diagram of the element I_j^Δ, and the nodes x_{j-1}^Δ and x_j^Δ connected to it. Note that the element I_j^Δ is in equilibrium under the action of the end forces $-P_{j-1}^R$, P_j^L and the distributed load f. The $(j-1)$th (respectively the jth) node is in equilibrium under the action of $-P_{j-1}^R$, P_{j-1}^L and the jump J_{j-1}^Δ (respectively P_j^L, P_j^R and the jump J_j^Δ)

and similarly

$$\mathcal{B}_{I_j^\Delta}(u_{S_\Delta^p}, \varphi_j^\Delta) = P_{S_\Delta^p}(x_j^{\Delta,-}) + \int_{I_j^\Delta} (\mathfrak{D}u_{S_\Delta^p})\varphi_j^\Delta. \tag{5.1.45c}$$

It follows that

$$J_{j-1}^R = -\int_{I_j^\Delta} (f - \mathfrak{D}u_{S_\Delta^p})\varphi_{j-1}^\Delta = -\int_{I_j^\Delta} r_{I_j^\Delta}\varphi_{j-1}^\Delta. \tag{5.1.46a}$$

Following similar steps, we get

$$J_j^L = -\int_{I_j^\Delta} r_{I_j^\Delta}\varphi_j^\Delta. \tag{5.1.46b}$$

Adding, we get

$$J_j^L + J_{j-1}^R = -\int_{I_j^\Delta} r_{I_j^\Delta}(\varphi_{j-1}^\Delta + \varphi_j^\Delta) = -\int_{I_j^\Delta} r_{I_j^\Delta} \tag{5.1.47a}$$

where we have used that $\varphi_{j-1}^\Delta + \varphi_j^\Delta \equiv 1$, and we have proven that the splitting (5.1.44a,b) satisfies the equilibrium condition

$$\mathcal{R}_{I_j^\Delta}^{EQ}(1) = \int_{I_j^\Delta} r_{I_j^\Delta} + J_j^L + J_{j-1}^R = 0. \tag{5.1.47b}$$

Noting that we have

$$\mathcal{R}_{I_j^\Delta}^{EQ}(v) = \int_{I_j^\Delta} r_{I_j^\Delta} v = \mathcal{R}(v) = 0 \quad \forall v \in S_0^p(I_j^\Delta) \tag{5.1.47c}$$

we also get

$$\mathcal{R}_{I_j^\Delta}^{EQ}(v) = 0 \quad \forall v \in S^p(I_j^\Delta). \tag{5.1.47d}$$

Further, recalling the orthogonality

$$\mathcal{R}(\varphi_j^\Delta) = \mathcal{B}(e_{S_\Delta^p}, \varphi_j^\Delta) = 0 \tag{5.1.48a}$$

we get

$$0 = \mathcal{B}(e_{S_\Delta^p}, \varphi_j^\Delta) = \mathcal{R}(\varphi_j^\Delta) = \underbrace{\int_{I_j^\Delta} r_{I_j^\Delta}\varphi_j^\Delta}_{-J_j^L} + \underbrace{\int_{I_{j+1}^\Delta} r_{I_{j+1}^\Delta}\varphi_j^\Delta}_{-J_j^R} + J_j^\Delta \tag{5.1.48b}$$

and hence we have proven that J_j^L, J_j^R, satisfy the exact splitting condition $J_j^L + J_j^R = J_j^\Delta$.

Remark 5.1.11 In the special case that essential boundary conditions are applied at both ends of the domain, the splitting $J_j^L + J_j^R = J_j^\Delta$ is unique up to an arbitrary constant, i.e., we may let

$$J_{j-1}^R = \int_{I_j^\Delta} r_{I_j^\Delta} \varphi_{j-1}^\Delta + \mathscr{C}, \qquad J_j^L = -\int_{I_j^\Delta} r_{I_j^\Delta} \varphi_j^\Delta + \mathscr{C} \tag{5.1.49}$$

where $\mathscr{C}$ is an arbitrary constant, the same for all the elements.

Remark 5.1.12 The orthogonality $\mathscr{B}(e_{S_\Delta^p}, \varphi_j^\Delta) = 0$, where φ_j^Δ is the hat function for the jth node, was essential in proving the splitting condition $J_j^L + J_j^R = J_j^\Delta$. This is also the case in the two-dimensional case.

Remark 5.1.13 Note that the term splitting of J_j^Δ is often understood (e.g., [Ainsworth and Oden, 1993, 1997, 2001]) in the sense that

$$J_j^R = \alpha_j J_j^\Delta, \qquad J_j^L = -\beta_j J_j^\Delta, \qquad \alpha_j + \beta_j = 1. \tag{5.1.50}$$

The splitting, given in (5.1.44), is more general because we could have $J_j^\Delta = 0$ while $J_j^R \neq 0$, $J_j^L \neq 0$.

In Section 4.7 we introduced a general approach for the asymptotic analysis of the finite element solution. In two dimensions, the following three types of asymptotics are of main interest:

(a) in the interior for smooth solutions

(b) at a straight boundary for smooth solutions

(c) at a corner point for solutions with algebraic singularities.

It follows that the asymptotic behavior of the error estimates must be analyzed for, at least, these three cases. We will now elaborate on the case (a), namely, the asymptotics in the interior of a locally uniform mesh for smooth solutions, and we will present a one-dimensional version of the two-dimensional analysis given in [Babuška, Strouboulis, and Upadhyay, 1994]. The cases (b) and (c) can be dealt with analogously.

Let the assumptions of Theorem 4.7.1 hold. Employing that

$$u_{S_\Delta^p} = u_{EX} - e_{S_\Delta^p} = u_{EX} - \bar{\Psi}_{S_\Delta^p}^{\bar{x},H} + \bar{\Psi}_{S_\Delta^p}^{\bar{x},H} - e_{S_\Delta^p} \tag{5.1.51a}$$

where we recall that (see (4.7.20a))

$$\bar{\Psi}_{S_\Delta^p}^{\bar{x},H} = \overline{\rho_{S_\Delta^p}^{\bar{x},H} - \pi^0_{S_{PER}(I_j^\Delta)}\left(\rho_{S_\Delta^p}^{\bar{x},H}\right)} \tag{5.1.51b}$$

and

$$\rho_{S_\Delta^p}^{\bar{x},H} = U_{EX}^{\bar{x},(p+1)} - \mathcal{I}_\Delta^p U_{EX}^{\bar{x},(p+1)} \tag{5.1.51c}$$

we get

$$u_{S_\Delta^p} = (u_{EX} - U_{EX}^{\bar{x},(p+1)}) + \mathcal{I}_\Delta^p U_{EX}^{\bar{x},(p+1)} + \pi^0_{S_{PER}(I_j^\Delta)}\left(\rho_{S_\Delta^p}^{\bar{x},H}\right) + \left(\bar{\Psi}_{S_\Delta^p}^{\bar{x},H} - e_{S_\Delta^p}\right). \tag{5.1.51d}$$

Recalling the definition of the *asymptotic finite element solution*

$$u_{S^p_\Delta} \approx \tilde{U}_{S^p_\Delta} \stackrel{\text{def}}{=} \mathcal{I}^p_\Delta U^{\bar{x},(p+1)}_{EX} + \pi^0_{S_{PER}(I^\Delta_j)}\big(\rho^{\bar{x},H}_{S^p_\Delta}\big) \tag{5.1.52a}$$

we get

$$\xi_{S^p_\Delta} = (u_{EX} - U^{\bar{x},(p+1)}_{EX}) + \big(\bar{\Psi}^{\bar{x},H}_{S^p_\Delta} - e_{S^p_\Delta}\big) = u_{S^p_\Delta} - \tilde{U}_{S^p_\Delta} \tag{5.1.52b}$$

and hence we have proven that

$$u_{S^p_\Delta} = \tilde{U}_{S^p_\Delta} + \xi_{S^p_\Delta} \tag{5.1.53a}$$

with

$$\Big\|\frac{\partial}{\partial x}(\xi_{S^p_\Delta})\Big\|_{L^2(S(\bar{x},\gamma H))} \le \mathscr{C}h^{p+\nu} \le \mathscr{C}h^{\nu}\,\big\|\bar{\Psi}^{\bar{x},H}_{S^p_\Delta}\big\|_{L^2(S(\bar{x},\gamma H))} \tag{5.1.53b}$$

where the second inequality follows from Theorem 4.7.1. When we write $\tilde{U}_{S^p_\Delta}$ and $\xi_{S^p_\Delta}$ below we will mean their restriction in $S(\bar{x},\gamma H)$.

Recalling now the orthogonality conditions

$$\mathscr{B}\big(e_{S^p_\Delta}, v\big) = \mathscr{B}\big(\bar{\Psi}^{\bar{x},H}_{S^p_\Delta}, v\big) = 0 \quad \forall v \in \overset{0}{S^p_\Delta}(S(\bar{x},\gamma H)) \tag{5.1.54a}$$

we get

$$\mathscr{B}\big(e_{S^p_\Delta} - \bar{\Psi}^{\bar{x},H}_{S^p_\Delta}, v\big) = 0 \quad \forall v \in \overset{0}{S^p_\Delta}(S(\bar{x},\gamma H)) \tag{5.1.54b}$$

we get that $\xi_{S^p_\Delta} \in S^p_\Delta(S(\bar{x},\gamma H))$ satisfies the condition

$$\mathscr{B}\big(\xi_{S^p_\Delta}, v\big) = \mathscr{B}(u_{EX} - U^{\bar{x},(p+1)}_{EX}, v) = \int_{S(\bar{x},\gamma H)} (f - \mathfrak{D}U^{\bar{x},(p+1)}_{EX})v \tag{5.1.54c}$$

for all $v \in \overset{0}{S^p_\Delta}(S(\bar{x},\gamma H))$. Similarly $\tilde{U}_{S^p_\Delta} \in S^p_\Delta(S(\bar{x},\gamma H))$ satisfies the condition

$$\mathscr{B}\big(\tilde{U}_{S^p_\Delta}, v\big) = \int_{S(\bar{x},\gamma H)} \mathfrak{D}U^{\bar{x},(p+1)}_{EX} v \quad \forall v \in \overset{0}{S^p_\Delta}(S(\bar{x},\gamma H)) \tag{5.1.55}$$

and $u_{S^p_\Delta}|_{S(\bar{x},\gamma H)} \in S^p_\Delta(S(\bar{x},\gamma H))$ satisfies the condition

$$\mathscr{B}\big(u_{S^p_\Delta}, v\big) = \int_{S(\bar{x},\gamma H)} fv \quad \forall v \in \overset{0}{S^p_\Delta}(S(\bar{x},\gamma H)). \tag{5.1.56}$$

This leads us to the notion of *admissible pair.*

We will say the pair $(\varphi_{S^p_\Delta}, g)$, with $\varphi_{S^p_\Delta} \in S^p_\Delta(S(\bar{x},\gamma H))$ and $g \in L^2(S(\bar{x},\gamma H))$, is *admissible* if

$$\mathscr{B}\big(\varphi_{S^p_\Delta}, v\big) = \int_{S(\bar{x},\gamma H)} gv \quad \forall v \in \overset{0}{S^p_\Delta}(S(\bar{x},\gamma H)). \tag{5.1.57}$$

It follows that the pairs $(u_{S_\Delta^p}, f)$, $(\tilde{U}_{S_\Delta^p}, \mathfrak{D}U_{EX}^{\bar{x},(p+1)})$, and $(\xi_{S_\Delta^p}, f - \mathfrak{D}U_{EX}^{\bar{x},(p+1)})$ are admissible.

Note that *if* $(\varphi_{S_\Delta^p}, g)$ *is an admissible pair in* $S(\bar{x}, \gamma H)$, we can define $\mathscr{E}_{S(\bar{x},\gamma H)}(\varphi_{S_\Delta^p}, g)$, which will be the error estimator.

We will make the following assumptions about the error estimator $\mathscr{E}_{S(\bar{x},\gamma H)}$:

Assumption A. For any admissible pair $(\varphi_{S_\Delta^p}, g)$, we have

$$\mathscr{E}_{S(\bar{x},\gamma H)}(\varphi_{S_\Delta^p}, g) \le \mathscr{C}\sqrt{\|\varphi_{S_\Delta^p}\|^2_{\mathfrak{U}(S(\bar{x},\gamma H))} + h^2 \|g\|^2_{L^2(S(\bar{x},\gamma H))}}. \tag{5.1.58}$$

Below we will often use the abbreviation $\mathscr{E}^2 \equiv \mathscr{E}^2_{S(\bar{x},\gamma H)}$.

Assumption A bounds the growth of $\mathscr{E}^2(\varphi_{S_\Delta^p}, g)$.

Assumption B. For any admissible pairs $(v_{S_\Delta^p}, s)$, $(\varphi_{S_\Delta^p}, g)$, we have

$$\left|\mathscr{E}^2(v_{S_\Delta^p} + \varphi_{S_\Delta^p}, s+g) - \mathscr{E}^2(v_{S_\Delta^p}, s)\right| \le \mathscr{C}\left[\sigma\mathscr{E}^2(v_{S_\Delta^p}, s) + \left(1 + \frac{1}{\sigma}\right)\mathscr{E}^2(\varphi_{S_\Delta^p}, g)\right] \tag{5.1.59}$$

for any σ, $0 < \sigma < 1$, with $\mathscr{C}$ independent of σ, $v_{S_\Delta^p}$, $\varphi_{S_\Delta^p}$, s, and g.

Assumption B is an assumption of *stability of the error estimator.*

Assumption C. For any $U \in \mathscr{P}^p(S(\bar{x}, H))$ we have

$$\mathscr{E}_{S(\bar{x},H)}(U, \mathfrak{D}U) = 0. \tag{5.1.60}$$

Assumption C states that, in the case $e_{S_\Delta^p} \equiv 0$, we also have $\mathscr{E} = 0$.

We will now prove that if $\mathscr{E}$ satisfies Assumptions A, B, and C, it can be analyzed from one cell.

Theorem 5.1.1 *Let Assumptions 0, I, and II of Section 4.7 hold, assume $\mathscr{E}$ satisfies Assumptions A, B, C, and that*

$$0 < \delta \le \kappa_{I_j^\Delta}(\tilde{U}_{S_\Delta^p}, F_{EX}^{(p+1)}) \overset{\text{def}}{=} \frac{\mathscr{E}_{I_j^\Delta}(\tilde{U}_{S_\Delta^p}, F_{EX}^{(p+1)})}{\|\Psi_{S_\Delta^p}^{\bar{x},H}\|_{\mathfrak{U}(I_j^\Delta)}} \tag{5.1.61a}$$

where we let $F_{EX}^{(p+1)} \overset{\text{def}}{=} \mathfrak{D}U_{EX}^{\bar{x},(p+1)}$. *Then there exists a* $\nu > 0$ *such that*

$$\kappa^2_{S(\bar{x},\gamma H)}(u_{S_\Delta^p}, f) \overset{\text{def}}{=} \frac{\mathscr{E}^2_{S(\bar{x},\gamma H)}(u_{S_\Delta^p}, f)}{\|e_{S_\Delta^p}\|^2_{\mathfrak{U}(S(\bar{x},\gamma H))}} = \kappa^2_{I_j^\Delta}(\tilde{U}_{S_\Delta^p}, F_{EX}^{(p+1)})(1 + \lambda\mathscr{C}h^\nu) \tag{5.1.61b}$$

where we used the notation $\kappa^2(\cdot) = (\kappa(\cdot))^2$.

Proof Recalling that

$$u_{S_\Delta^p} = \tilde{U}_{S_\Delta^p} + \xi_{S_\Delta^p} \tag{5.1.62a}$$

and letting

$$\psi \overset{\text{def}}{=} f - F_{EX}^{(p+1)} = \mathfrak{D}(u_{EX} - U_{EX}^{\bar{x},(p+1)}) \tag{5.1.62b}$$

we have

$$\mathscr{E}^2(u_{S^p_\Delta}, f) - \mathscr{E}^2(\tilde{U}_{S^p_\Delta}, F^{(p+1)}_{EX}) = \mathscr{E}^2(\tilde{U}_{S^p_\Delta} + \xi_{S^p_\Delta}, F^{(p+1)}_{EX} + \psi) - \mathscr{E}^2(\tilde{U}_{S^p_\Delta}, F^{(p+1)}_{EX}) \tag{5.1.62c}$$

where we used the notation $\mathscr{E}^2(\cdot) = (\mathscr{E}_{S(\bar{x},\gamma H)}(\cdot))^2$, and employing Assumption B, we get

$$|\mathscr{E}^2(u_{S^p_\Delta}, f) - \mathscr{E}^2(\tilde{U}_{S^p_\Delta}, F^{(p+1)}_{EX})| \le \mathscr{C}\Big|\sigma\mathscr{E}^2(\tilde{U}_{S^p_\Delta}, F^{(p+1)}_{EX}) + \Big(1 + \frac{1}{\sigma}\Big)\mathscr{E}^2(\xi_{S^p_\Delta}, \psi)\Big|. \tag{5.1.62d}$$

Let us now estimate $\mathscr{E}^2(\xi_{S^p_\Delta}, \psi)$. From Assumption A we have

$$\mathscr{E}^2_{S(\bar{x},\gamma H)}(\xi_{S^p_\Delta}, \psi) \le \mathscr{C}(\|\xi_{S^p_\Delta}\|^2_{\mathscr{U}(S(\bar{x},\gamma H))} + h^2\,\|\psi\|^2_{L^2(S(\bar{x},\gamma H))}). \tag{5.1.63a}$$

From Assumption 0 it follows that

$$u_{EX} - U^{\bar{x},(p+1)}_{EX} \approx \frac{1}{(p+2)!}\frac{d^{p+2}u_{EX}}{dx^{p+2}}(\bar{x})(x-\bar{x})^{p+2} \tag{5.1.63b}$$

and we get

$$\|\psi\|_{L^2(S(\bar{x},\gamma H))} = \|\mathfrak{D}(u_{EX} - U^{\bar{x},(p+1)}_{EX})\|_{L^2(S(\bar{x},\gamma H))} \le \mathscr{C}H^{p+1/2}. \tag{5.1.63c}$$

We also have

$$\|\xi_{S^p_\Delta}\|_{\mathscr{U}(S(\bar{x},\gamma H))} \le \mathscr{C}\left\|\frac{\partial}{\partial x}(\xi_{S^p_\Delta})\right\|_{L^2(S(\bar{x},\gamma H))} \le \mathscr{C}h^{p+\nu}. \tag{5.1.63d}$$

Employing these results in (5.1.63a) we obtain

$$\mathscr{E}^2_{S(\bar{x},\gamma H)}(\xi_{S^p_\Delta}, \psi) \le \mathscr{C}[(h^{p+\nu})^2 + h^2(H^{p+1})^2] \le \mathscr{C}[h^{2+2(1+p)\beta} + h^{2(p+\nu)}] \tag{5.1.63e}$$

where $p/(p+1) < \beta < 1$, and the relation between h and H is given in Theorem 4.7.1. It follows that for sufficiently small h, we have

$$\mathscr{E}^2_{S(\bar{x},\gamma H)}(\xi_{S^p_\Delta}, \varphi) \le \mathscr{C}h^{2(p+\nu)}. \tag{5.1.63f}$$

Employing now this result into (5.1.62d), and selecting $\sigma = h^\nu$, we get

$$|\mathscr{E}^2(u_{S^p_\Delta}, f) - \mathscr{E}^2(\tilde{U}_{S^p_\Delta}, F^{(p+1)}_{EX})| \le \mathscr{C}(h^\nu\mathscr{E}^2(\tilde{U}_{S^p_\Delta}, F^{(p+1)}_{EX}) + h^{2p+\nu}) \tag{5.1.64a}$$

and thus we have proven

$$\mathscr{E}^2_{S(\bar{x},\gamma H)}(u_{S^p_\Delta}, f) = \mathscr{E}^2_{S(\bar{x},\gamma H)}(\tilde{U}_{S^p_\Delta}, F^{(p+1)}_{EX})(1 + \mathscr{C}h^\nu) + \mathscr{C}h^{2p+\nu} \tag{5.1.64b}$$

from which it follows that

$$\kappa^2_{S(\bar{x},\gamma H)}\big(u_{S^p_\Delta}, f\big) = \frac{\mathscr{E}^2_{S(\bar{x},\gamma H)}\big(u_{S^p_\Delta}, f\big)}{\|e_{S^p_\Delta}\|^2_{\mathscr{U}(S(\bar{x},\gamma H))}}$$

$$= \frac{\mathscr{E}^2\big(\tilde{U}_{S^p_\Delta}, F^{(p+1)}_{EX}\big)(1 + \mathscr{C}h^\nu) + \mathscr{C}h^{2p+\nu}}{\|e_{S^p_\Delta}\|^2_{\mathscr{U}(S(\bar{x},\gamma H))}}. \tag{5.1.64c}$$

Recalling (4.7.36a), we have

$$\|e_{S^p_\Delta}\|_{\mathscr{U}(S(\bar{x},\gamma H))} = (1 + \mathscr{C}h^{\bar{\nu}})\|\bar{\Psi}^{\bar{x},H}_{S^p_\Delta}\|_{\mathscr{U}(S(\bar{x},\gamma H))} \tag{5.1.64d}$$

so that we get

$$\kappa^2_{S(\bar{x},\gamma H)}\big(u_{S^p_\Delta}, f\big) = \frac{\mathscr{E}^2\big(\tilde{U}_{S^p_\Delta}, F^{(p+1)}_{EX}\big)}{\|\bar{\Psi}^{\bar{x},H}_{S^p_\Delta}\|^2_{\mathscr{U}(S(\bar{x},\gamma H))}}(1 + \lambda\mathscr{C}h^{\bar{\nu}}). \tag{5.1.64e}$$

The proof of the Theorem is concluded by noting that

$$\mathscr{E}^2_{I^\Delta_j}\big(\tilde{U}_{S^p_\Delta}, F^{(p+1)}_{EX}\big) = \mathscr{E}^2_{I^\Delta_k}\big(\tilde{U}_{S^p_\Delta}, F^{(p+1)}_{EX}\big), \quad I^\Delta_j, I^\Delta_k \subset S(\bar{x},\gamma H) \tag{5.1.65}$$

and recalling that $\bar{\Psi}^{\bar{x},H}_{S^p_\Delta}$ is periodic. Note that (5.1.65) is a consequence of Assumption C. □

We have proven that

$$\kappa_{S(\bar{x},\gamma H)}\big(u_{S^p_\Delta}, f\big) \approx \kappa_{I^\Delta_j}\big(\tilde{U}_{S^p_\Delta}, F^{(p+1)}_{EX}\big) \tag{5.1.66}$$

where $\approx$ is an asymptotic equality and h tends to zero, and for $H \approx \mathscr{C}h^\gamma$, $0 < \gamma < 1$. Although we proved this for $AE \equiv$ *constant*, $c \equiv 0$, it also holds for sufficiently smooth AE. The asymptotic equality (5.1.66) is the basis of *a constructive approach for determining the asymptotic value of the effectivity index in the interior of any locally uniform mesh which satisfies Assumptions I and II of Section 4.7, for any estimator satisfying Assumptions* A, B, *and* C.

Remark 5.1.14 For elements adjacent to the boundary the effect of the boundary layer must also be included.

In this section, we have proven that the error, $e_{S^p_\Delta} \in \mathscr{U}$, is the exact solution of the residual equation

$$\mathscr{B}\big(e_{S^p_\Delta}, v\big) = \mathscr{R}(v) = \mathscr{L}(v) - \mathscr{B}\big(u_{S^p_\Delta}, v\big) \quad \forall v \in \mathscr{U} \tag{5.1.67a}$$

from which it follows that

$$\|e_{S^p_\Delta}\|_{\mathscr{U}} = \|\mathscr{R}\|_{L(\mathscr{U},\mathbb{R})} \tag{5.1.67b}$$

and hence $\|e_{S^p_\Delta}\|_{\mathcal{U}}$ may be estimated by estimating $\|\mathcal{R}\|_{L(\mathcal{U},\mathbb{R})}$. We obtained the lower estimator

$$\mathcal{E}^{\mathrm{Dir}}_{S^{p+r}_\Delta} \overset{\mathrm{def}}{=} \sqrt{\sum_{j=1}^{N(\Delta)} \left\|\pi_{S_0^{p+r}}\left(\hat{e}^{\mathrm{Dir}}_{I_j^\Delta}\right)\right\|^2_{\mathcal{U}(I_j^\Delta)}} \leq \mathcal{E}^{\mathrm{Dir}} = \sqrt{\sum_{j=1}^{N(\Delta)} \left\|\hat{e}^{\mathrm{Dir}}_{I_j^\Delta}\right\|^2_{\mathcal{U}(I_j^\Delta)}} \leq \|e_{S^p_\Delta}\|_{\mathcal{U}} \quad (5.1.68a)$$

where $\hat{e}^{\mathrm{Dir}}_{I_j^\Delta}$ is the Dirichlet element error indicator function, which is the solution of the problem:

Find $\hat{e}^{\mathrm{Dir}}_{I_j^\Delta} \in \mathcal{U}_0(I_j^\Delta)$ such that

$$\mathcal{B}_{I_j^\Delta}\left(\hat{e}^{\mathrm{Dir}}_{I_j^\Delta}, v\right) = \int_{I_j^\Delta} r_{I_j^\Delta} v \quad \forall v \in \mathcal{U}_0(I_j^\Delta). \quad (5.1.68b)$$

We call $\mathcal{E}^{\mathrm{Dir}}$ (respectively $\mathcal{E}^{\mathrm{Dir}}_{S^p_\Delta}$) the exact (respectively the computed) Dirichlet estimator.

We also constructed an upper estimator by splitting the residuum

$$\mathcal{R}(v) = \sum_{j=1}^{N(\Delta)} \mathcal{R}^{EQ}_{I_j^\Delta}\left(v\big|_{I_j^\Delta}\right) \quad \forall v \in \mathcal{U} \quad (5.1.69a)$$

into equilibrated element residua $\mathcal{R}^{EQ}_{I_j^\Delta}$, with the property that

$$\mathcal{R}^{EQ}_{I_j^\Delta}(v) = 0 \quad \forall v \in S^p(I_j^\Delta). \quad (5.1.69b)$$

It follows that

$$\|e_{S^p_\Delta}\|_{\mathcal{U}} = \|\mathcal{R}\|_{L(\mathcal{U},\mathbb{R})} \leq \mathcal{E}^{\mathrm{Neum}} = \sqrt{\sum_{j=1}^{N(\Delta)} \left\|\mathcal{R}^{EQ}_{I_j^\Delta}\right\|^2_{L(\mathcal{U}(I_j^\Delta),\mathbb{R})}}$$
$$= \sqrt{\sum_{j=1}^{N(\Delta)} \left\|\hat{e}^{\mathrm{Neum}}_{I_j^\Delta}\right\|^2_{\mathcal{U}(I_j^\Delta)}} \quad (5.1.69c)$$

where $\hat{e}^{\mathrm{Neum}}_{I_j^\Delta}$ is the solution of the element residual problem with Neumann boundary conditions:

Find $\hat{e}^{\mathrm{Neum}}_{I_j^\Delta} \in \mathcal{U}(I_j^\Delta)$ such that

$$\mathcal{B}_{I_j^\Delta}\left(\hat{e}^{\mathrm{Neum}}_{I_j^\Delta}, v\right) = \mathcal{R}^{EQ}_{I_j^\Delta}(v) \quad \forall v \in \mathcal{U}(I_j^\Delta) \quad (5.1.69d)$$

where we call $\mathcal{E}^{\mathrm{Neum}}$ is the exact Neumann estimator. We also have the computed Neumann estimator $\mathcal{E}^{\mathrm{Neum}}_{S^{p+r}_\Delta}$ which satisfies

$$\mathcal{E}^{\mathrm{Neum}}_{S^{p+r}_\Delta} \overset{\mathrm{def}}{=} \sqrt{\sum_{j=1}^{N} \left\|\pi_{S^{p+r}(I_j^\Delta)}\left(\hat{e}^{\mathrm{Neum}}_{I_j^\Delta}\right)\right\|^2_{\mathcal{U}(I_j^\Delta)}} \leq \mathcal{E}^{\mathrm{Neum}} \quad (5.1.69e)$$

and hence is not necessarily an upper estimator for $\|e_{S^p_\Delta}\|_{\mathcal{U}}$.

We also addressed the asymptotic analysis of the effectivity-index of the estimators. A more detailed analysis of the properties of the estimates $\mathscr{E}^{\text{Dir}}$ and $\mathscr{E}^{\text{Neum}}$ will be given in Sections 5.2 and 5.3, respectively.

Problems

P5.1.1. The model problem defines a linear operator $\mathscr{T} : \mathfrak{F} \mapsto \mathscr{S}$ which maps $\mathfrak{F}$, the class of admissible loads, into $\mathscr{S}$, the corresponding class of admissible solutions, namely,

$$u_{EX} = \mathscr{T}(\mathfrak{f}) \tag{5.1.70a}$$

where $\mathfrak{f}$ is the load (which includes both f and a_2). It follows that

$$e_{S_\Delta^p} = \mathscr{T}(\mathfrak{R}) \tag{5.1.70b}$$

where $\mathfrak{R}$ is the residual load. Let $\|\cdot\|_{\mathfrak{F}}$ and $\|\cdot\|_{\mathscr{S}}$ be the norms associated with the class $\mathfrak{F}$ and $\mathscr{S}$, respectively, and let $\|\cdot\|_{L(\mathfrak{F},\mathscr{S})}$: $L(\mathfrak{F},\mathscr{S}) \mapsto \mathbb{R}_0^+$ be the operator norm. Prove the two-sided estimate

$$\|\mathscr{T}\|^0_{L(\mathfrak{F},\mathscr{S})}\|\mathfrak{R}\|_{\mathfrak{F}} \leq \|e_{S_\Delta^p}\|_{\mathscr{S}} \leq \|\mathscr{T}\|_{L(\mathfrak{F},\mathscr{S})}\|\mathfrak{R}\|_{\mathfrak{F}} \tag{5.1.71a}$$

where

$$\|\mathscr{T}\|^0_{L(\mathfrak{F},\mathscr{S})} \stackrel{\text{def}}{=} \inf_{\substack{\mathfrak{f}\in\mathfrak{F} \\ \mathfrak{f}\neq 0}} \frac{\|\mathscr{T}(\mathfrak{f})\|_{\mathscr{S}}}{\|\mathfrak{f}\|_{\mathfrak{F}}}. \tag{5.1.71b}$$

In the problems that follow, we will use the notation

$$\lambda(\mathfrak{F},\mathscr{S}) = \frac{\|\mathscr{T}\|_{L(\mathfrak{F},\mathscr{S})}}{\|\mathscr{T}\|^0_{L(\mathfrak{F},\mathscr{S})}}. \tag{5.1.71c}$$

P5.1.2. Apply P5.1.1 to the regular model problem with $AE \equiv 1$, $c \equiv 0$, $a_2 = 0$, in the case that

$$\mathfrak{R}(x) - \sum_{k=5}^{10} a_k \sin\left(\frac{2k+1}{2}\pi\frac{x}{l}\right), \quad 0 < x < l \tag{5.1.72}$$

and $\mathscr{R}(v) = \int_0^l \mathfrak{R}v$, for $\mathscr{S} = \mathscr{U} = H^1(I)$, and $\mathfrak{F} = \mathscr{U}^{-1} = H^{-1}(I)$. Prove that $\mathscr{R}$ is bounded over $\mathscr{U}$ and compute $\|\mathscr{R}\|_{L(\mathscr{U},\mathbb{R})}$. Let $e = \mathscr{T}(\mathfrak{R})$ and determine an explicit expression for e. Compute the values of $\|\mathscr{T}\|_{L(\mathscr{U}^{-1},\mathscr{U})}$, $\|\mathscr{T}\|^0_{L(\mathscr{U}^{-1},\mathscr{U})}$, and and the value of the ratio $\lambda(\mathscr{U}^{-1},\mathscr{U})$.

P5.1.3. Repeat problem P5.1.2 in the case that

$$\mathscr{R}(v) = \int_0^l \mathfrak{R}v + 3v(l). \tag{5.1.73}$$

P5.1.4. Repeat problem P5.1.2 for $\mathfrak{F} = L^2(I)$ and $\mathcal{S} = H^2(I)$. Compute $\|\mathfrak{R}\|_{L^2(I)}$ and $\|\mathcal{T}(\mathfrak{R})\|_{H^2(I)}$ as a function of a_k and the values of $\|\mathcal{T}\|_{L(L^2(I),H^2(I))}$, $\|\mathcal{T}\|^0_{L(L^2(I),H^2(I))}$ and the ratio $\lambda(L^2(I), H^2(I))$.

P5.1.5. Repeat P5.1.4 letting $\mathfrak{F} = H^{-2}_{(0}(I)$ and $\mathcal{S} = L^2(I)$.

P5.1.6. For the model problem given in Example 5.1.1 compute $\|\mathfrak{R}\|_{H^{-2}_{(0}(I)}$ and

$$\|e_{S^1_\Delta}\|_{L^2(I)} = \|\mathcal{T}(\mathfrak{R})\|_{L^2(I)} \tag{5.1.74a}$$

as a function of h and show that

$$\|e_{S^1_\Delta}\|_{L^2(I)} \leq \mathscr{C}\|\mathfrak{R}\|_{H^{-2}_{(0}(I)} \tag{5.1.74b}$$

where $\mathscr{C}$ is a constant independent of h. Compute $\mathscr{C}$ and compare it with the value of $\|\mathcal{T}\|_{L(H^{-2}_{(0}(I),L^2(I))}$.

P5.1.7. Prove the guaranteed lower bound

$$\|e_{S^p_\Delta}\|_{\mathcal{U}} \geq \frac{\mathcal{R}\Big(\mathcal{A}^*_u(u_{S^p_\Delta})\Big)}{\big\|\mathcal{A}^*_u(u_{S^p_\Delta})\big\|_{\mathcal{U}}} \tag{5.1.75}$$

where $\mathcal{A}^*_u$ is any continuous recovery of the value.

5.1* The residuum and the construction of lower and upper estimators for the energy norm of the error: the two-dimensional case

We will now discuss the construction of element residual upper and lower estimators

$$\mathscr{E}^{\text{Dir}} \leq \|e_{S^p_\Delta}\|_{\mathcal{U}} \leq \mathscr{E}^{\text{Neum}} \tag{5.1*.1}$$

in the two-dimensional case.

As in the one-dimensional case, we will determine the Dirichlet error indicator function, $\hat{e}^{\text{Dir}}_\tau$, as the solution of the Dirichlet element residual problem:

Find $\hat{e}^{\text{Dir}}_\tau \in \mathcal{U}_0(\tau) \stackrel{\text{def}}{=} \{v \in \mathcal{U}(\tau) \mid v|_{\partial\tau} \equiv 0\}$ such that

$$\mathcal{B}_\tau(\hat{e}^{\text{Dir}}_\tau, v) = \int_\tau r_\tau v \quad \forall v \in \mathcal{U}_0(\tau). \tag{5.1*.2a}$$

Analogously as in the one-dimensional case, we have

$$\eta^{\text{Dir}}_\tau \stackrel{\text{def}}{=} \|\hat{e}^{\text{Dir}}_\tau\|_{\mathcal{U}(\tau)} \leq \|e_{S^p_\Delta}\|_{\mathcal{U}(\tau)} \tag{5.1*.2b}$$

where η^{Dir}_τ is the Dirichlet element error indicator, and

$$\mathscr{E}^{\text{Dir}} \stackrel{\text{def}}{=} \sqrt{\sum_{\tau\in\Delta} \|\hat{e}^{\text{Dir}}_\tau\|^2_{\mathcal{U}(\tau)}} \leq \|e_{S^p_\Delta}\|_{\mathcal{U}} \tag{5.1*.2c}$$

for $\mathscr{E}^{\mathrm{Dir}}$, the Dirichlet error estimator. Hence, as in the one-dimensional case, the Dirichlet error estimator $\mathscr{E}^{\mathrm{Dir}}$ is a guaranteed lower bound for the energy norm of the error $\|e_{S^p_\Delta}\|_{\mathcal{U}}$. *However, in contrast to the one-dimensional case,* $\mathscr{E}^{\mathrm{Dir}}$ *may be much smaller than* $\|e_{S^p_\Delta}\|_{\mathcal{U}}$, *and hence useless as an estimator.* Consider, for example, the finite element approximation of the solution of the Laplace equation ($\Delta u_{EX} = 0$), on a mesh of linear triangles ($p = 1$). In this case, we have

$$r_\tau = \Delta u_{S^1_\Delta} \equiv 0 \quad \Longrightarrow \quad \hat{e}^{\mathrm{Dir}}_\tau \equiv 0 \tag{5.1*.3a}$$

for all the elements τ in the mesh Δ, and hence

$$\mathscr{E}^{\mathrm{Dir}} = 0 \tag{5.1*.3b}$$

which is a guaranteed but useless lower bound. This observation holds, more generally, *for elements of odd order* ($p = 1, 3, 5, \ldots$). For elements of even order ($p = 2, 4, 6, \ldots$), and meshes of rectangles, it has been proven [Yu, 1991b] that $\mathscr{E}^{\mathrm{Dir}}$ is reliable, analogously as in the one-dimensional case; see Section 5.2*.

As in the one-dimensional case, the upper estimator $\mathscr{E}^{\mathrm{Neum}}$ will be based on a splitting of the global residuum $\mathscr{R}$ into equilibrated element residua $\mathscr{R}^{EQ}_\tau$, namely

$$\mathscr{R}(v) = \sum_{\tau\in\Delta} \mathscr{R}^{EQ}_\tau(v|_\tau) \tag{5.1*.4a}$$

with the property that

$$\mathscr{R}^{EQ}_\tau(\mathit{1}) = 0. \tag{5.1*.4b}$$

As in the one-dimensional case, we have

$$\mathscr{R}(v) = \mathscr{L}(v) - \mathscr{B}\big(u_{S^p_\Delta}, v\big) = \mathscr{L}(v) - \sum_{\tau\in\Delta} \mathscr{B}_\tau\big(u_{S^p_\Delta}, v\big) \tag{5.1*.5a}$$

and using the integration by parts

$$\mathscr{B}_\tau\big(u_{S^p_\Delta}, v\big) = \int_\tau \mathfrak{D} u_{S^p_\Delta} v + \oint_{\partial\tau} \mathbf{q}_{S^p_\Delta}\cdot \mathbf{n} v + \int_{\partial\tau\cap\Gamma_{\mathrm{N}}} \alpha u_{S^p_\Delta} v \tag{5.1*.5b}$$

where

$$\mathfrak{D} u_{S^p_\Delta} = \sum_{i,j=1}^{2} -\frac{\partial}{\partial x_i}\Big(a_{ij}\frac{\partial}{\partial x_j}\big(u_{S^p_\Delta}\big)\Big) + c u_{S^p_\Delta} \tag{5.1*.5c}$$

and

$$\mathbf{q}_{S^p_\Delta}\cdot \mathbf{n} = \sum_{i,j=1}^{2} n_i a_{ij}\frac{\partial}{\partial x_j}\big(u_{S^p_\Delta}\big) \tag{5.1*.5d}$$

we can express $\mathcal{R}$ in the form

$$\mathcal{R}(v) = \sum_{\tau\in\Delta}\left(\int_\tau r_\tau^\Delta v - \int_{\partial\tau}(\mathbf{q}_{S_\Delta^p}\cdot\mathbf{n}_\tau)v - \int_{\partial\tau\cap\Gamma_{\mathrm{N}}}\alpha u_{S_\Delta^p} v\right) \tag{5.1*.5e}$$

where

$$r_\tau^\Delta = f - \mathfrak{D}u_{S_\Delta^p} \tag{5.1*.6}$$

is the interior residual for the element τ. Further, rearranging the terms on the element boundaries, we obtain

$$\sum_{\tau\in\Delta}\left(\int_{\partial\tau}(\mathbf{q}_{S_\Delta^p}\cdot\mathbf{n}_\tau)v + \int_{\partial\tau\cap\Gamma_{\mathrm{N}}}\alpha u_{S_\Delta^p} v\right) = \sum_{\epsilon\in\mathcal{E}(\Delta)}\int_\epsilon J_\epsilon^\Delta v \tag{5.1*.7a}$$

where

$$J_\epsilon^\Delta = \begin{cases}(\mathbf{q}_{S_\Delta^p}|_{\tau^*} - \mathbf{q}_{S_\Delta^p}|_\tau)\cdot\mathbf{n}_\tau, & \epsilon = \partial\tau\cap\partial\tau^* \not\subseteq \partial\Omega\\ g - \alpha u_{S_\Delta^p} - \mathbf{q}_{S_\Delta^p}\cdot\mathbf{n}_{\Gamma_{\mathrm{N}}}, & \epsilon\subseteq\Gamma_{\mathrm{N}}\\ 0, & \epsilon\subseteq\Gamma_{\mathrm{D}}\end{cases} \tag{5.1*.7b}$$

is the jump on the edge ϵ, and $\mathbf{n}_\tau$ is the exterior unit normal for the element τ, and, as in the one-dimensional case, we get

$$\mathcal{R}(v) = \sum_{\tau\in\Delta}\int_\tau r_\tau^\Delta v + \sum_{\epsilon\in\mathcal{E}(\Delta)}\int_\epsilon J_\epsilon^\Delta v. \tag{5.1*.7c}$$

Letting now

$$\mathcal{R}_\tau^{EQ}(v) \stackrel{\text{def}}{=} \int_\tau r_\tau^\Delta v + \sum_{\epsilon\subseteq\partial\tau}\int_\epsilon J_\epsilon^\tau v \tag{5.1*.8a}$$

where J_ϵ^τ is the part of the jump J_ϵ^Δ assigned to the element τ, from the requirement

$$\mathcal{R}(v) = \sum_{\tau\in\Delta}\mathcal{R}_\tau^{EQ}(v|_\tau) \tag{5.1*.8b}$$

we obtain the exact splitting condition

$$J_\epsilon^\tau + J_\epsilon^{\tau^*} = J_\epsilon^\Delta \tag{5.1*.8c}$$

for every interior edge, $\epsilon = \partial\tau\cap\partial\tau^*$. As in the one-dimensional case, it is possible to construct the split jumps, J_ϵ^τ and $J_\epsilon^{\tau^*}$, to satisfy (5.1.8c) and such that each $\mathcal{R}_\tau^{EQ}$ satisfies the orthogonality condition

$$\mathcal{R}_\tau^{EQ}(v) = 0 \quad \forall v\in S_{\partial\tau_{\mathrm{D}}}^p(\tau). \tag{5.1*.8d}$$

As in the one-dimensional case, we will determine the Neumann element residual indicator function $\hat{e}_\tau^{\text{Neum}}$ to be the solution of the element residual problem:

Find $\hat{e}_\tau^{\text{Neum}} \in \mathcal{U}_{\partial\tau_\text{D}}(\tau)$ such that

$$\mathcal{B}_\tau(\hat{e}_\tau^{\text{Neum}}, v) = \mathcal{R}_\tau^{EQ}(v) \quad \forall v \in \mathcal{U}_{\partial\tau_\text{D}}(\tau) \tag{5.1*.9a}$$

where

$$\mathcal{U}_{\partial\tau_\text{D}}(\tau) \stackrel{\text{def}}{=} \{v \in \mathcal{U}(\tau) \mid v|_{\partial\tau_\text{D}} = 0\} \tag{5.1*.9b}$$

where in the elements with $\partial\tau_\text{D} = \emptyset$, the consistency condition

$$\mathcal{R}_\tau^{EQ}(1) = \int_\tau r_\tau + \sum_{\epsilon\subseteq\partial\tau} \int_\epsilon J_\epsilon^\tau = 0 \tag{5.1*.9c}$$

follows from the orthogonality (5.1*.8d), and hence $\hat{e}_\tau^{\text{Neum}}$ exists for every element. Following exactly the same steps as in the one-dimensional case, we obtain $\mathcal{E}^{\text{Neum}}$, the upper Neumann estimator

$$\|e_{S_\Delta^p}\|_{\mathcal{U}} \leq \mathcal{E}^{\text{Neum}} \stackrel{\text{def}}{=} \sqrt{\sum_{\tau\in\Delta} \|\mathcal{R}_\tau^{EQ}\|^2_{L(\mathcal{U}(\tau),\mathbb{R})}} = \sqrt{\sum_{\tau\in\Delta} \|\hat{e}_\tau^{\text{Neum}}\|^2_{\mathcal{U}(\tau)}} \tag{5.1*.9d}$$

which is a guaranteed upper estimator for $\|e_{S_\Delta^p}\|_{\mathcal{U}}$.

The construction of the split jumps J_ϵ^τ is done analogously as in the one dimensional case. Let us, for simplicity, consider the case of linear elements ($p = 1$), and the Poisson equation

$$-\Delta u|_\Omega = f, \qquad u|_{\Gamma_\text{N}} = 0, \qquad \left.\frac{\partial u}{\partial n}\right|_{\Gamma_\text{N}} = g. \tag{5.1*.10a}$$

As in the one-dimensional case, we will let

$$J_\epsilon^\tau = P_\epsilon^\tau + \frac{\partial}{\partial n_\tau}(u_{S_\Delta^1}|_\tau) \tag{5.1*.10b}$$

where P_ϵ^τ is the reaction for the edge ϵ, which will be constructed as an affine function over ϵ, namely

$$P_\epsilon^\tau = \underbrace{\left(\int_\epsilon P_\epsilon^\tau \psi_1^\epsilon\right)}_{\mathcal{P}_{\epsilon,1}^\tau} \underbrace{\frac{2}{|\epsilon|}(2\psi_1^\epsilon - \psi_2^\epsilon)}_{\lambda_1^\epsilon} + \underbrace{\left(\int_\epsilon P_\epsilon^\tau \psi_2^\epsilon\right)}_{\mathcal{P}_{\epsilon,2}^\tau} \underbrace{\frac{2}{|\epsilon|}(2\psi_2^\epsilon - \psi_1^\epsilon)}_{\lambda_2^\epsilon} \tag{5.1*.10c}$$

where $\psi_i^\epsilon \in \mathcal{P}^1(\epsilon)$, $i = 1, 2$, are the linear shape functions for the edge ϵ, and $\lambda_i^\epsilon \in \mathcal{P}^1(\epsilon)$, $i = 1, 2$ are the affine functions which are conjugate to ψ_i^ϵ in the sense

$$\int_\epsilon \psi_i^\epsilon \lambda_j^\epsilon = \delta_{ij} \tag{5.1*.10d}$$

where δ_{ij} is Krönecker's delta. The orthogonality conditions (5.1*.8d) can now be written in the form

$$\mathcal{R}_\tau^{EQ}(\varphi^\Delta_{j(k,\tau)}) = \int_\tau (f + \Delta u_{S^p_\Delta})\varphi^\Delta_{j(k,\tau)} + \sum_{\epsilon\subseteq\partial\tau}\int_\epsilon \Big(P^\tau_\epsilon + \frac{\partial}{\partial n_\tau}(u_{S^1_\Delta}|_\tau)\Big)\varphi^\Delta_{j(k,\tau)} = 0 \tag{5.1*.11a}$$

where $j(k,\tau)$ is the global index assigned to the kth vertex of the element τ, and $\varphi^\Delta_{j(k,\tau)}$ the piecewise linear hat function assigned to this vertex. Integrating by parts we get the system of linear equations

$$\sum_{\epsilon\subseteq\partial\tau}\int_\tau P^\tau_\epsilon \varphi^\Delta_{j(k,\tau)}|_\epsilon = \mathcal{B}_\tau(u_{S^1_\Delta}, \varphi^\Delta_{j(k,\tau)}) - \int_\tau f\varphi^\Delta_{j(k,\tau)} \tag{5.1*.11b}$$

for $k = 1, 2, 3$, for all $\tau \in \Delta$, with unknowns

$$\mathcal{P}^\tau_{\epsilon,l} \overset{\text{def}}{=} \int_\epsilon P^\tau_\epsilon \psi^\epsilon_l \tag{5.1*.11c}$$

where for each $\epsilon = \partial\tau \cap \partial\tau^*$ we have $\mathcal{P}^\tau_{\epsilon,l} = \mathcal{P}^{\tau^*}_{\epsilon,l}$.

A systematic approach for solving the linear system (5.1*.11b) was proposed in [Ladeveze and Leguillon, 1983], [Ladeveze and Maunder, 1996], [Bank and Weiser, 1985], and [Ainsworth and Oden, 1993], and is based on the observation that the subsystem of linear equations (5.1*.11b) associated with each vertex of the mesh can be solved. Let X denote an interior vertex of the mesh, e.g., the vertex shown in Fig. 5.1*.1, and denote by τ_l^X the lth element which is connected to the vertex X, and by $\epsilon^X_{k(1,\tau_l^X)}$, $\epsilon^X_{k(2,\tau_l^X)}$, the edges of τ_l^X with an end point at X. Then the unknowns associated with the vertex X are

$$\mathcal{P}^{\tau_l^X}_{\epsilon^X_{k(m,\tau_l^X)}} \overset{\text{def}}{=} \int_{\epsilon^X_{k(m,\tau_l^X)}} P^{\tau_l^X}_{\epsilon^X_{k(m,\tau_l^X)}} \varphi^\Delta_X|_{\epsilon^X_{k(m,\tau_l^X)}}, \quad m = 1, 2 \tag{5.1*.12a}$$

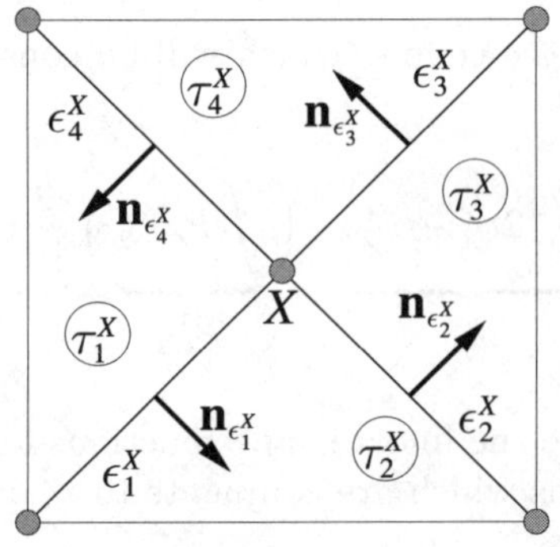

FIG. 5.1*.1. Examples of interior vertices with the enumeration of the edges and vertices used to set up the linear system for the unknowns $\mathcal{P}^\tau_{\epsilon,l}$ associated with each vertex

and letting $\varphi^{\Delta}_{j(k,\tau)} = \varphi^{\Delta}_{X}$ into (5.1*.11b) we obtain

$$\mathcal{P}^{\tau_l^X}_{\epsilon^X_{k(2,\tau_l^X)}} - \mathcal{P}^{\tau_l^X}_{\epsilon^X_{k(1,\tau_l^X)}} = \mathcal{B}_{\tau_l^X}\left(u_{S^1_\Delta}, \varphi^{\Delta}_{X}\right) - \int_{\tau_l^X} f\varphi^{\Delta}_{X}, \quad l = 1, \dots, ne(X) \qquad (5.1^*.12b)$$

where $ne(X)$ is the number of elements (and edges) connected to vertex X. The matrix of the linear system of equations (5.1*.12b) has exactly one zero eigenvalue with corresponding eigenvector

$$\underbrace{[1, \dots, 1]}_{ne(X)}^{\mathrm{T}},$$

and the right-hand side of (5.1*.12b) is orthogonal to this eigenvector (see also Section 5.3*), because of the orthogonality condition satisfied by the error

$$\sum_{l=1}^{ne(X)} \left(\mathcal{B}_{\tau_l^X}\left(u_{S^1_\Delta}, \varphi^{\Delta}_{X}\right) - \int_{\tau_l^X} f\varphi^{\Delta}_{X}\right) = \mathcal{B}\left(u_{S^1_\Delta}, \varphi^{\Delta}_{X}\right) - \mathcal{L}(\varphi^{\Delta}_{X})$$
$$= \mathcal{B}\left(e_{S^1_\Delta}, \varphi^{\Delta}_{X}\right) = 0. \qquad (5.1^*.12c)$$

It follows that the linear system (5.1*.12b) has a solution which is unique up to an arbitrary constant vector

$$\mathcal{C}_X \underbrace{[1, 1, \dots, 1]}_{ne(X)}^{\mathrm{T}},$$

which can be fixed in various ways, as discussed in Section 5.3*. The same approach can be used to obtain the unknowns associated with the vertices at the boundary of the domain. It follows that *the existence of* P^{τ}_{ϵ}, *for the edge* $\epsilon = (X_1, X_2)$, *relies on the orthogonality condition* $\mathcal{B}(e_{S^1_\Delta}, \varphi^{\Delta}_{X}) = 0$, for $X = X_1$, X_2, and P^{τ}_{ϵ} depends on the choice of the constants $\mathcal{C}_{X_1}$ and $\mathcal{C}_{X_2}$. The construction of $\mathcal{R}^{EQ}_{\tau}$ on a particular element τ utilizes the finite element solution $u_{S^p_\Delta}$ from all the elements connected to the vertices of τ. For a detailed description of the construction of $\mathcal{R}^{EQ}_{\tau}$, see Example 5.3*.1 below.

Remark 5.1*.1 As in the one-dimensional case, the construction of the estimator $\mathcal{E}^{\mathrm{Neum}}$ also follows from the classical method of the hypercircle. For example, in the case $c \equiv 0$,

$$\hat{\mathbf{q}}^{\mathrm{Neum}}_{\Delta,p} \stackrel{\mathrm{def}}{=} \mathbf{q}_{S^p_\Delta}|_{\tau} + \mathbf{q}(\hat{e}^{\mathrm{Neum}}_{\tau}) \qquad (5.1^*.13a)$$

is a *statically admissible* or *admissible flux*, because we have

$$\boldsymbol{\nabla} \cdot \hat{\mathbf{q}}^{\mathrm{Neum}}_{\Delta,p}|_{\tau} = 0 \quad \forall \tau \in \Delta \qquad (5.1^*.13b)$$

and

$$\hat{\mathbf{q}}^{\mathrm{Neum}}_{\Delta,p} \cdot \mathbf{n}_{\epsilon} = 0 \quad \forall \epsilon \in \mathcal{E}_{\mathrm{int}}(\Delta), \qquad \hat{\mathbf{q}}^{\mathrm{Neum}}_{\Delta,p} \cdot \mathbf{n}_{\Gamma_N}|_{\Gamma_N} = g \qquad (5.1^*.13c)$$

where $\mathbf{n}_\epsilon$ is the unit normal associated with the edge ϵ, and $\mathbf{n}_{\Gamma_N}$ is the exterior unit normal on Γ_N, and from the main result in the theory of the hypercircle we obtain

$$\|e_{S^p_\Delta}\|_{\mathcal{U}} \leq \mathscr{E}^{\mathrm{Neum}} = \left\|\mathbf{A}^{-1/2}\big(\hat{\mathbf{q}}^{\mathrm{Neum}}_{\Delta,p} - \mathbf{q}_{S^p_\Delta}\big)\right\|_{L^2(\Omega)}. \tag{5.1*.13d}$$

As in the one-dimensional case, the error indicator function $\hat{e}^{\mathrm{Neum}}_\tau$ is most often approximated by the finite element method and the corresponding computed estimator fails to be a guaranteed upper bound for $\|e_{S^p_\Delta}\|_{\mathcal{U}}$. Most often the p version is employed, and we have

$$\eta^{\mathrm{Neum}}_{\tau,S^{p+r}(\tau)} = \|\pi_{S^{p+r}(\tau)}(\hat{e}^{\mathrm{Neum}}_\tau)\|_{\mathcal{U}(\tau)} \leq \eta^{\mathrm{Neum}}_\tau = \|\hat{e}^{\mathrm{Neum}}_\tau\|_{\mathcal{U}(\tau)} \tag{5.1*.14a}$$

where $\eta^{\mathrm{Neum}}_\tau$ is the *exact*, and $\eta^{\mathrm{Neum}}_{\tau,S^{p+r}(\tau)}$ is the *computed* Neumann element error indicator, and

$$\mathscr{E}^{\mathrm{Neum}}_{S^{p+r}_\Delta} \stackrel{\mathrm{def}}{=} \sqrt{\sum_{\tau\in\Delta}(\eta^{\mathrm{Neum}}_{\tau,S^{p+r}(\tau)})^2} \leq \mathscr{E}^{\mathrm{Neum}} = \sqrt{\sum_{\tau\in\Delta}(\eta^{\mathrm{Neum}}_\tau)^2} \tag{5.1*.14b}$$

where $\mathscr{E}^{\mathrm{Neum}}$ is the *exact*, and $\mathscr{E}^{\mathrm{Neum}}_{S^{p+r}_\Delta}$ the *computed* Neumann estimator.

As in the one-dimensional case, an alternative approach is to construct an explicit upper estimator for $\eta^{\mathrm{Neum}}_\tau$, namely,

$$\|\hat{e}^{\mathrm{Neum}}_\tau\|_{\mathcal{U}(\tau)} = \sup_{v\in\mathcal{U}(\tau)} \frac{\mathscr{R}^{EQ}_\tau(v)}{\|v\|_{\mathcal{U}(\tau)}} \leq \eta^{\mathrm{EXPL}}_\tau$$

$$\stackrel{\mathrm{def}}{=} \mathscr{C}\sqrt{|\tau|\int_\tau r_\tau^2 + \frac{1}{2}\sum_{\epsilon\subseteq\partial\tau}|\epsilon|\int_\epsilon (J^\tau_\epsilon)^2} \tag{5.1*.15a}$$

and we get the explicit upper estimate

$$\|e_{S^p_\Delta}\|_{\mathcal{U}} \leq \mathscr{E}^{\mathrm{EXPL}} = \sqrt{\sum_{\tau\in\Delta}(\eta^{\mathrm{EXPL}}_\tau)^2}$$

$$= \mathscr{C}\sqrt{\sum_{\tau\in\Delta}\left(|\tau|\int_\tau r_\tau^2 + \frac{1}{2}\sum_{\epsilon\subseteq\partial\tau}|\epsilon|\int_\epsilon (J^\tau_\epsilon)^2\right)} \tag{5.1*.15b}$$

where the determination of $\mathscr{C}$ is critical. This estimate can be rather crude. It can also be restated in terms of the equilibrated jumps using a different value for the constant. Further, by the dichotomy principle, in the case of linear elements ($p = 1$) it is also possible to construct an upper estimator which employs only the jumps, namely

$$\|e_{S^1_\Delta}\|_{\mathcal{U}} \leq \mathscr{C}'\sqrt{\sum_{\tau\in\Delta}\left(\frac{1}{2}\sum_{\epsilon\subseteq\partial\tau}|\epsilon|\int_\epsilon (J^\tau_\epsilon)^2\right)}. \tag{5.1*.15c}$$

As in the one-dimensional case, we have

$$\eta^{\mathrm{Neum}}_\tau = \sqrt{(\eta^{\mathrm{Neum}}_{\tau,S^{p+r}(\tau)})^2 + \|\hat{e}^{\mathrm{Neum}}_\tau - \pi_{S^{p+r}(\tau)}(\hat{e}^{\mathrm{Neum}}_\tau)\|^2_{\mathcal{U}(\tau)}} \tag{5.1*.16}$$

and we can obtain a *computable* upper estimator of $\eta_{\mathcal{T}}^{\text{Neum}}$ by adding to the squares of the computed indicators $\eta_{\tau,S^{p+r}(\tau)}^{\text{Neum}}$, an explicit upper estimator for the energy of the error in the computed error indicator $\|\hat{e}_{\tau}^{\text{Neum}} - \pi_{S^{p+r}(\tau)}(\hat{e}_{\tau}^{\text{Neum}})\|_{\mathcal{U}(\tau)}^2$. We will elaborate on this approach in Chapter 6.

Assumptions A, B, C in Section 5.1 can be directly extended to the two-dimensional setting, and are satisfied for the $\mathscr{E}^{\text{Dir}}$, $\mathscr{E}^{\text{Neum}}$, and $\mathscr{E}^{\text{EXPL}}$, and the two-dimensional analogue of Theorem 5.1.1 holds for these estimators. It follows that $\mathscr{E}^{\text{Dir}}$, $\mathscr{E}^{\text{Neum}}$, and $\mathscr{E}^{\text{EXPL}}$ can be analyzed by employing asymptotic analyses of the effectivity index for interior, boundary elements and elements at the corners of the domain, similarly as in the one-dimensional case. The effectivity index for the explicit estimator $\mathscr{E}^{\text{EXPL}}$, will be analyzed in Section 5.7*.

5.2 Analysis of the Dirichlet element residual estimator

In this section we present a detailed analysis of the Dirichlet element residual estimator.

We start by recalling the definition of the Dirichlet element residual error indicator function $\hat{e}_{I_j^{\Delta}}^{\text{Dir}}$ as the solution of the Dirichlet element residual problem:

Find $\hat{e}_{I_j^{\Delta}}^{\text{Dir}} \in \mathcal{U}_0(I_j^{\Delta}) = \{v \in \mathcal{U}(I_j^{\Delta}) \mid v(x_{j-1}^{\Delta}) = v(x_j^{\Delta}) = 0\}$ such that

$$\mathscr{B}_{I_j^{\Delta}}(\hat{e}_{I_j^{\Delta}}^{\text{Dir}}, v) = \int_{I_j^{\Delta}} r_{I_j^{\Delta}} v, \quad \forall v \in \mathcal{U}_0(I_j^{\Delta}). \tag{5.2.1}$$

Let us first prove that $\hat{e}_{I_j^{\Delta}}^{\text{Dir}}$ exists and it is unique. In Chapter 2 we proved the existence and uniqueness of solutions of the model problem (2.2.8) under the rather general assumptions

$$f \in \mathcal{U}^{-1}, \qquad 0 < \beta_1 \leq AE(x) \leq \beta_2 < \infty, \qquad 0 \leq c(x) \leq \gamma < \infty, \qquad x \in I. \tag{5.2.2a}$$

Here the existence and uniqueness of $\hat{e}_{I_j^{\Delta}}^{\text{Dir}}$ will be proven assuming that, in addition to (5.2.2a), we also have

$$f \in L^2(I), \qquad \left\|\frac{d}{dx}(AE)\right\|_{L^{\infty}(I_j^{\Delta})} \leq K < \infty \quad j = 1, \ldots, N(\Delta). \tag{5.2.2b}$$

Theorem 5.2.1 *Assume that f, AE, c satisfy (5.2.2a) and (5.2.2b). Then*

$$r_{I_j^{\Delta}} \in L^2(I_j^{\Delta}) \tag{5.2.3a}$$

and

$$\mathscr{B}_{I_j^{\Delta}}(\hat{e}_{I_j^{\Delta}}^{\text{Dir}}, v) = \mathscr{B}_{I_j^{\Delta}}(e_{S_{\Delta}^p}, v), \quad \forall v \in \mathcal{U}_0(I_j^{\Delta}) \tag{5.2.3b}$$

for $j = 1, \ldots, N(\Delta)$, and there exists a unique $\hat{e}_{I_j^{\Delta}}^{\text{Dir}}$ which is the solution of (5.2.1).

Proof From the assumptions (5.2.2a) and (5.2.2b) it follows that $u_{EX} \in H^2(I_j^\Delta)$, and hence

$$e_{S_\Delta^p}|_{I_j^\Delta} \in H^2(I_j^\Delta) \tag{5.2.4a}$$

and we have

$$r_{I_j^\Delta} = -\frac{d}{dx}\Big(AE\frac{d}{dx}(e_{S_\Delta^p})\Big) + ce_{S_\Delta^p} \in L^2(I_j^\Delta). \tag{5.2.4b}$$

For any $v \in \mathcal{U}_0(I_j^\Delta)$, we have

$$\mathcal{B}_{I_j^\Delta}(e_{S_\Delta^p}, v) = \mathcal{B}_{I_j^\Delta}(u_{EX}, v) - \mathcal{B}_{I_j^\Delta}(u_{S_\Delta^p}, v) = \int_{I_j^\Delta} fv - \mathcal{B}_{I_j^\Delta}(u_{S_\Delta^p}, v). \tag{5.2.5a}$$

Further, the integration by parts

$$\int_{I_j^\Delta} AE\frac{d}{dx}(u_{S_\Delta^p})\frac{dv}{dx} = -\int_{I_j^\Delta} \frac{d}{dx}\Big(AE\frac{d}{dx}(u_{S_\Delta^p})\Big)v \quad \forall v \in \mathcal{U}_0(I_j^\Delta) \tag{5.2.5b}$$

is justified because $u_{S_\Delta^p}|_{I_j^\Delta} \in \mathcal{P}^p(I_j^\Delta)$. Employing (5.2.5b) into (5.2.5a) we get

$$\mathcal{B}_{I_j^\Delta}(e_{S_\Delta^p}, v) = \int_{I_j^\Delta} r_{I_j^\Delta} v \quad \forall v \in \mathcal{U}_0(I_j^\Delta). \tag{5.2.5c}$$

Hence we have proven that

$$\mathcal{B}_{I_j^\Delta}(\hat{e}_{I_j^\Delta}^{\text{Dir}}, v) = \int_{I_j^\Delta} r_{I_j^\Delta} v = \mathcal{B}_{I_j^\Delta}(e_{S_\Delta^p}, v) \quad \forall v \in \mathcal{U}_0(I_j^\Delta). \tag{5.2.5d}$$

It follows that $\hat{e}_{I_j^\Delta}^{\text{Dir}} \in \mathcal{U}_0(I_j^\Delta)$ is the energy projection of $e_{S_\Delta^p}|_{I_j^\Delta} \in \mathcal{U}(I_j^\Delta)$ into $\mathcal{U}_0(I_j^\Delta)$, namely,

$$\hat{e}_{I_j^\Delta}^{\text{Dir}} = \pi_{\mathcal{U}_0(I_j^\Delta)}(e_{S_\Delta^p}) \tag{5.2.5e}$$

and the existence and uniqueness of $\hat{e}_{I_j^\Delta}^{\text{Dir}}$ has been proven. □

Remark 5.2.1 The variational problem (5.2.1) can also be written in the equivalent form:

Find

$$\tilde{e}_{I_j^\Delta}^{\text{Dir}} \in \tilde{\mathcal{U}}_0(I_j^\Delta) \stackrel{\text{def}}{=} \{v \in \mathcal{U} \mid \exists w \in \mathcal{U}_0(I_j^\Delta)\colon v|_{I_j^\Delta} \equiv w, \text{ and } v|_{I-I_j^\Delta} \equiv 0\} \tag{5.2.6a}$$

such that

$$\mathscr{B}\big(\tilde{\hat{e}}^{\mathrm{Dir}}_{I^\Delta_j}, v\big) = \mathscr{R}(v) \quad \forall v \in \tilde{\mathscr{U}}_0(I^\Delta_j). \tag{5.2.6b}$$

It follows that for the existence and uniqueness of $\tilde{\hat{e}}^{\mathrm{Dir}}_{I^\Delta_j} \in \tilde{\mathscr{U}}_0(I^\Delta_j)$ it is sufficient to assume that the input data satisfy only (5.2.2a).

Remark 5.2.2 In general, $\hat{e}^{\mathrm{Dir}}_{I^\Delta_j} \not\equiv e_{S^p_\Delta}|_{I^\Delta_j}$. In the special case where the nodal values are exact, we have $\hat{e}^{\mathrm{Dir}}_{I^\Delta_j} \equiv e_{S^p_\Delta}|_{I^\Delta_j}$.

In Section 5.1, we have already introduced $\eta^{\mathrm{Dir}}_{I^\Delta_j}$, the Dirichlet element error indicator, and $\mathscr{E}^{\mathrm{Dir}}$, the Dirichlet error estimator

$$\eta^{\mathrm{Dir}}_{I^\Delta_j} \stackrel{\mathrm{def}}{=} \big\|\hat{e}^{\mathrm{Dir}}_{I^\Delta_j}\big\|_{\mathscr{U}(I^\Delta_j)}, \qquad \mathscr{E}^{\mathrm{Dir}} \stackrel{\mathrm{def}}{=} \sqrt{\sum_{j=1}^{N(\Delta)} \big(\eta^{\mathrm{Dir}}_{I^\Delta_j}\big)^2}. \tag{5.2.7}$$

We will now prove the equivalence $\mathscr{E}^{\mathrm{Dir}} \approx \|e_{S^p_\Delta}\|_{\mathscr{U}}$.

Theorem 5.2.2 *Assume that the input data satisfy assumptions (5.2.2a). Then,*

$$\mathscr{E}^{\mathrm{Dir}} \leq \big\|e_{S^p_\Delta}\big\|_{\mathscr{U}} \leq \mathcal{K}^{\mathrm{Dir}}\mathscr{E}^{\mathrm{Dir}} \tag{5.2.8}$$

where $\mathcal{K}^{\mathrm{Dir}}$ depends only on AE and c.

Proof The lower bound has been proven in Section 5.1. However, for completeness we will repeat the proof here. We have

$$(\mathscr{E}^{\mathrm{Dir}})^2 = \sum_{j=1}^{N(\Delta)} \mathscr{B}_{I^\Delta_j}\big(\hat{e}^{\mathrm{Dir}}_{I^\Delta_j}, \hat{e}^{\mathrm{Dir}}_{I^\Delta_j}\big) = \sum_{j=1}^{N(\Delta)} \int_{I^\Delta_j} r_{I^\Delta_j}\hat{e}^{\mathrm{Dir}}_{I^\Delta_j} = \mathscr{R}(\hat{e}^{\mathrm{Dir}}) \tag{5.2.9a}$$

and recalling that

$$\mathscr{R}(\hat{e}^{\mathrm{Dir}}) = \mathscr{B}\big(e_{S^p_\Delta}, \hat{e}^{\mathrm{Dir}}\big) \tag{5.2.9b}$$

we obtain

$$(\mathscr{E}^{\mathrm{Dir}})^2 = \mathscr{R}(\hat{e}^{\mathrm{Dir}}) = \mathscr{B}\big(e_{S^p_\Delta}, \hat{e}^{\mathrm{Dir}}\big) = \sum_{j=1}^{N(\Delta)} \mathscr{B}_{I^\Delta_j}\big(e_{S^p_\Delta}, \hat{e}^{\mathrm{Dir}}_{I^\Delta_j}\big). \tag{5.2.9c}$$

Using the Cauchy–Schwarz inequality for the energy, and the $\mathbb{R}^{N(\Delta)}$, inner products, we get

$$(\mathscr{E}^{\mathrm{Dir}})^2 \leq \sum_{j=1}^{N(\Delta)} \big\|e_{S^p_\Delta}\big\|_{\mathscr{U}(I^\Delta_j)}\big\|\hat{e}^{\mathrm{Dir}}_{I^\Delta_j}\big\|_{\mathscr{U}(I^\Delta_j)} \leq \big\|e_{S^p_\Delta}\big\|_{\mathscr{U}}\mathscr{E}^{\mathrm{Dir}} \tag{5.2.9d}$$

which proves the lower bound.

We will now prove the upper bound, $\|e_{S^p_\Delta}\|_{\mathscr{U}} \leq \mathcal{K}^{\mathrm{Dir}}\mathscr{E}^{\mathrm{Dir}}$, for the case of linear elements ($p = 1$). The proof for higher order elements ($p > 1$) follows similar arguments.

Let $v \in \mathcal{U}$ be arbitrary. Then letting,

$$v = \mathcal{I}_{\Delta}^{1} v + (v - \mathcal{I}_{\Delta}^{1} v) \tag{5.2.10a}$$

and noting that $\mathcal{I}_{\Delta}^{1} v \in S_{\Delta}^{1}$, and hence

$$\mathcal{R}(\mathcal{I}_{\Delta}^{1} v) = 0 \tag{5.2.10b}$$

we get

$$|\mathcal{R}(v)| = |\mathcal{R}(v - \mathcal{I}_{\Delta}^{1} v)| = \left| \sum_{j=1}^{N(\Delta)} \mathcal{B}_{I_j^{\Delta}} \left(\hat{e}_{I_j^{\Delta}}^{\text{Dir}}, v - \mathcal{I}_{I_j^{\Delta}}^{1} v \right) \right| \tag{5.2.10c}$$

and employing the Cauchy–Schwarz inequality for the energy and the $\mathbb{R}^{N(\Delta)}$ inner products we get

$$|\mathcal{R}(v)| \leq \sum_{j=1}^{N(\Delta)} \|\hat{e}_{I_j^{\Delta}}^{\text{Dir}}\|_{\mathcal{U}(I_j^{\Delta})} \|v - \mathcal{I}_j^{1} v\|_{\mathcal{U}(I_j^{\Delta})} \leq \sqrt{\sum_{j=1}^{N(\Delta)} \|\hat{e}_{I_j^{\Delta}}^{\text{Dir}}\|_{\mathcal{U}(I_j^{\Delta})}^{2}} \, \|v - \mathcal{I}_{\Delta}^{1} v\|_{\mathcal{U}} \tag{5.2.10d}$$

and hence we have proven that

$$|\mathcal{R}(v)| \leq \mathcal{E}^{\text{Dir}} \, \|v - \mathcal{I}_{\Delta}^{1} v\|_{\mathcal{U}}. \tag{5.2.10e}$$

To complete the proof of the upper bound, we will utilize Lemma 5.2.1, which is proven below, and which states that: There exists a constant $\mathcal{K}$ such that

$$\|v - \mathcal{I}_{\Delta}^{1} v\|_{\mathcal{U}} \leq \mathcal{K} \|v\|_{\mathcal{U}} \quad \forall v \in \mathcal{U}. \tag{5.2.11a}$$

Using this result into (5.2.10e) we get

$$|\mathcal{R}(v)| \leq \mathcal{E}^{\text{Dir}} \|v - \mathcal{I}_{\Delta}^{1} v\|_{\mathcal{U}} \leq \mathcal{K} \mathcal{E}^{\text{Dir}} \|v\|_{\mathcal{U}} \tag{5.2.11b}$$

and hence

$$\|e_{S_{\Delta}^{p}}\|_{\mathcal{U}} = \sup_{v \in \mathcal{U}} \frac{|\mathcal{B}(e_{S_{\Delta}^{p}}, v)|}{\|v\|_{\mathcal{U}}} = \sup_{v \in \mathcal{U}} \frac{|\mathcal{R}(v)|}{\|v\|_{\mathcal{U}}} \leq \mathcal{K} \mathcal{E}^{\text{Dir}} \tag{5.2.11c}$$

which proves the upper bound with $\mathcal{K}^{\text{Dir}} = \mathcal{K}$. □

Let us now show that $\mathcal{K}$ exists.

Lemma 5.2.1 *Under the assumption that*

$$0 < \beta_1 \leq AE(x) \leq \beta_2 < \infty, \quad 0 \leq c(x) \leq \gamma < \infty \qquad x \in I \tag{5.2.12a}$$

there exists a $\mathcal{K}$ *depending only on* β_1, β_2, *and* γ, *such that*

$$\|v - \mathcal{I}_{\Delta}^{1} v\|_{\mathcal{U}} \leq \mathcal{K} \|v\|_{\mathcal{U}} \quad \forall v \in \mathcal{U}. \tag{5.2.12b}$$

Proof Let us first consider the case with $AE \equiv 1$, $c \equiv 0$. Noting that

$$\frac{d}{dx}(\mathcal{I}^1_{I^\Delta_j} v) = \frac{1}{h_{I^\Delta_j}}\Big(v(x^\Delta_j) - v(x^\Delta_{j-1})\Big) \equiv constant \tag{5.2.13a}$$

and that

$$\int_{I^\Delta_j} \frac{d}{dx}(v - \mathcal{I}^1_{I^\Delta_j} v) = (v - \mathcal{I}^1_j v)(x^\Delta_j) - (v - \mathcal{I}^1_j v)(x^\Delta_{j-1}) = 0 \tag{5.2.13b}$$

it follows that

$$\int_{I^\Delta_j} \frac{d}{dx}(\mathcal{I}^1_{I^\Delta_j} v)\frac{d}{dx}(v - \mathcal{I}^1_{I^\Delta_j} v) = \frac{d}{dx}(\mathcal{I}^1_{I^\Delta_j} v)\int_{I^\Delta_j} \frac{d}{dx}(v - \mathcal{I}^1_{I^\Delta_j} v) = 0 \tag{5.2.13c}$$

and we get

$$\int_{I^\Delta_j}\Big(\frac{dv}{dx}\Big)^2 = \int_{I^\Delta_j}\Big(\frac{d}{dx}(v - \mathcal{I}^1_{I^\Delta_j} v)\Big)^2 + \int_{I^\Delta_j}\Big(\frac{d}{dx}(\mathcal{I}^1_{I^\Delta_j} v)\Big)^2. \tag{5.2.13d}$$

Omitting the second term in the right-hand side, we get

$$\int_{I^\Delta_j}\Big(\frac{d}{dx}(v - \mathcal{I}^1_{I^\Delta_j} v)\Big)^2 \le \int_{I^\Delta_j}\Big(\frac{dv}{dx}\Big)^2. \tag{5.2.13e}$$

This proves that, in the particular case $AE \equiv 1$, $c \equiv 0$, we have $\mathcal{K} = 1$.

In the general case, we have

$$\int_{I^\Delta_j} AE\Big(\frac{d}{dx}(v - \mathcal{I}^1_{I^\Delta_j} v)\Big)^2 \le \beta_{2,I^\Delta_j}\int_{I^\Delta_j}\Big(\frac{d}{dx}(v - \mathcal{I}^1_{I^\Delta_j} v)\Big)^2 \le \beta_2\int_{I^\Delta_j}\Big(\frac{dv}{dx}\Big)^2 \tag{5.2.14a}$$

where we used the notations

$$\beta_{2,I^\Delta_j} \stackrel{\text{def}}{=} \|AE\|_{L^\infty(I^\Delta_j)} \le \beta_2 \stackrel{\text{def}}{=} \|AE\|_{L^\infty(I)}. \tag{5.2.14b}$$

Further, using that

$$\beta_1 \stackrel{\text{def}}{=} \min_{x\in I} AE(x) \le \beta_{1,I^\Delta_j} \stackrel{\text{def}}{=} \min_{x\in I^\Delta_j}(AE(x)) \tag{5.2.14c)}$$

we get

$$\int_{I^\Delta_j}\Big(\frac{dv}{dx}\Big)^2 \le \frac{1}{\beta_{1,I^\Delta_j}}\int_{I^\Delta_j} AE\Big(\frac{dv}{dx}\Big)^2 \le \frac{1}{\beta_1}\int_{I^\Delta_j} AE\Big(\frac{dv}{dx}\Big)^2 \tag{5.2.14d}$$

and hence

$$\int_{I^\Delta_j} AE\Big(\frac{d}{dx}(v - \mathcal{I}^1_{I^\Delta_j} v)\Big)^2 \le \frac{\beta_{2,I^\Delta_j}}{\beta_{1,I^\Delta_j}}\int_{I^\Delta_j} AE\Big(\frac{dv}{dx}\Big)^2 \le \frac{\beta_2}{\beta_1}\|v\|^2_{\mathcal{U}(I^\Delta_j)}. \tag{5.2.14e}$$

Also, because $(v - \mathcal{I}^1_{I_j^\Delta} v)(x^\Delta_{j-1}) = 0$ we have

$$\left|(v - \mathcal{I}^1_{I_j^\Delta} v)(x)\right| \le \sqrt{x - x^\Delta_{j-1}} \sqrt{\int_{I_j^\Delta} \left(\frac{d}{dx}(v - \mathcal{I}^1_{I_j^\Delta} v)\right)^2} \le \sqrt{h_{I_j^\Delta}} \sqrt{\int_{I_j^\Delta} \left(\frac{dv}{dx}\right)^2}. \tag{5.2.15a}$$

Squaring both sides, multiplying by c, integrating over I_j^Δ, and employing that

$$\gamma_{I_j^\Delta} \stackrel{\text{def}}{=} \|c\|_{L^\infty(I_j^\Delta)} \le \gamma \stackrel{\text{def}}{=} \|c\|_{L^\infty(I)} \tag{5.2.15b}$$

we get

$$\int_{I_j^\Delta} c\left(v - \mathcal{I}^1_{I_j^\Delta} v\right)^2 \le \left(h_{I_j^\Delta}\right)^2 \gamma_{I_j^\Delta} \int_{I_j^\Delta} \left(\frac{d}{dx}(v - \mathcal{I}^1_{I_j^\Delta} v)\right)^2 \le \left(h_{I_j^\Delta}\right)^2 \gamma \int_{I_j^\Delta} \left(\frac{dv}{dx}\right)^2 \tag{5.2.15c}$$

and using (5.2.14c) we get

$$\int_{I_j^\Delta} c\left(v - \mathcal{I}^1_{I_j^\Delta} v\right)^2 \le \left(h_{I_j^\Delta}\right)^2 \frac{\gamma_{I_j^\Delta}}{\beta_{1,I_j^\Delta}} \int_{I_j^\Delta} AE\left(\frac{dv}{dx}\right)^2 \le \left(h_{I_j^\Delta}\right)^2 \frac{\gamma}{\beta_1} \|v\|^2_{\mathcal{U}}. \tag{5.2.15d}$$

Combining (5.2.14e) and (5.2.15d), we obtain

$$\int_{I_j^\Delta} AE\left(\frac{d}{dx}(v - \mathcal{I}^1_{I_j^\Delta} v)\right)^2 + \int_{I_j^\Delta} c\left(v - \mathcal{I}^1_{I_j^\Delta} v\right)^2 \le \left(\frac{\beta_2}{\beta_1} + \frac{\gamma}{\beta_1}(h^\Delta)^2\right) \|v\|^2_{\mathcal{U}(I_j^\Delta)} \tag{5.2.16a}$$

and hence

$$\|v - \mathcal{I}^1_\Delta v\|_{\mathcal{U}(I_j^\Delta)} \le \sqrt{\frac{\beta_2}{\beta_1} + \frac{\gamma}{\beta_1}(h^\Delta)^2}\, \|v\|_{\mathcal{U}(I_j^\Delta)}. \tag{5.2.16b}$$

Hence we have proven (5.2.12b) with $\mathcal{K}$ given by

$$\mathcal{K} = \sqrt{\frac{\beta_2}{\beta_1} + \frac{\gamma}{\beta_1}(h^\Delta)^2} = \frac{\sqrt{\beta_2}}{\sqrt{\beta_1}} \sqrt{1 + \frac{\gamma}{\beta_2}(h^\Delta)^2} \approx \frac{\sqrt{\beta_2}}{\sqrt{\beta_1}}\left(1 + \frac{1}{2}\frac{\gamma}{\beta_2}(h^\Delta)^2\right). \tag{5.2.16c}$$

Note that $\mathcal{K}$ may be replaced by $\bar{\mathcal{K}}$, where

$$\mathcal{K} \le \overline{\mathcal{K}} = \frac{\sqrt{\beta_2}}{\sqrt{\beta_1}} \sqrt{1 + \frac{\gamma}{\beta_2}}. \tag{5.2.16d}$$

□

Remark 5.2.3 Let us underline that the estimate

$$\mathcal{E}^{\text{Dir}} \le \|e_{S^p_\Delta}\|_{\mathcal{U}} \le \mathcal{K}^{\text{Dir}}\, \mathcal{E}^{\text{Dir}} \tag{5.2.17}$$

holds under the minimal assumptions (5.2.2a), and hence also in the case that AE is a very rough function, and the convergence is arbitrarily slow.

The Dirichlet estimator $\mathcal{E}^{\text{Dir}}$ and its equivalence with $\|e_{S^p_\Delta}\|_{\mathcal{U}}$ were first addressed in [Babuška and Rheinboldt, 1978b].

The estimate $\mathscr{E}^{\text{Dir}} \le \|e_{S^p_\Delta}\|_{\mathcal{U}} \le \mathcal{K}^{\text{Dir}}\mathscr{E}^{\text{Dir}}$ *was proven without making any assumptions about the pollution error.* Let us now give an example which illustrate how the estimate works in cases of high pollution.

Example 5.2.1 Let us, once more, consider the cases given in Examples 4.4.1 and 4.4.2. In these examples the pollution error dominates the local error in all the elements except the first element, which is the biggest element in Example 4.4.1, and the element next to the singular point in Example 4.4.2.

Let us first discuss a case from Example 4.4.1. Employing the approximation $\hat{e}^{\text{Dir}}_{I^\Delta_j} \approx \pi_{\mathscr{P}^9_0}(\hat{e}^{\text{Dir}}_{I^\Delta_j})$, and the mesh of $N(\Delta) = 81$ elements employed in Table 4.4.1, we find that the effectivity index of the error indicator in the first element is

$$\kappa^{\text{Dir}}_{I^\Delta_1} = \frac{\|\hat{e}^{\text{Dir}}_{I^\Delta_1}\|_{\mathcal{U}(I^\Delta_1)}}{\|e_{S^1_\Delta}\|_{\mathcal{U}(I^\Delta_1)}} = \frac{0.04068628}{0.04069639} = 0.9998 \tag{5.2.18a}$$

while the effectivity index of the estimate in the rest of the mesh is

$$\kappa^{\text{Dir}}_{I-I^\Delta_1} = \frac{\sqrt{\sum_{j=2}^{N(\Delta)} \|\hat{e}^{\text{Dir}}_{I^\Delta_j}\|^2_{\mathcal{U}(I^\Delta_j)}}}{\|e_{S^1_\Delta}\|_{\mathcal{U}(I-I^\Delta_1)}} = \frac{\sqrt{(\mathscr{E}^{\text{Dir}})^2 - \|\hat{e}^{\text{Dir}}_{I^\Delta_1}\|^2_{\mathcal{U}(I^\Delta_1)}}}{\|e_{S^1_\Delta}\|_{\mathcal{U}(I-I^\Delta_1)}}$$

$$= \frac{0.00036817}{0.00098633} = 0.3733. \tag{5.2.18b}$$

Nevertheless, because

$$\|e_{S^1_\Delta}\|_{\mathcal{U}(I-I^\Delta_1)} = 0.00098633 \ll \|e_{S^1_\Delta}\|_{\mathcal{U}(I^\Delta_1)} = 0.04069639 \tag{5.2.18c}$$

we have

$$\kappa^{\text{Dir}} = \frac{\mathscr{E}^{\text{Dir}}}{\|e_{S^1_\Delta}\|_{\mathcal{U}}} = \frac{0.04068795}{0.04070834} = 0.9995 \approx \kappa^{\text{Dir}}_{I^\Delta_1} = \frac{\|\hat{e}^{\text{Dir}}_{I^\Delta_1}\|_{\mathcal{U}(I^\Delta_1)}}{\|e_{S^1_\Delta}\|_{\mathcal{U}(I^\Delta_1)}} = 0.9998. \tag{5.2.18d}$$

We also have $\beta_2 = \max_{x \in I}(1+x) = 2$, $\beta_1 = \min_{x \in I}(1+x) = 1$, and $\gamma = 0$, and hence

$$\mathcal{K}^{\text{Dir}} = \frac{\sqrt{\beta_2}}{\sqrt{\beta_1}} = \sqrt{2} \tag{5.2.18e}$$

and we have

$$\mathscr{E}^{\text{Dir}} = 0.04068795 \le \|e_{S^1_\Delta}\|_{\mathcal{U}} = 0.04070834 \le \mathcal{K}^{\text{Dir}}\mathscr{E}^{\text{Dir}} = 0.05754145. \tag{5.2.18f}$$

In summary, we have $\kappa^{\text{Dir}}_{I^\Delta_1} \approx 1$, while, because of the pollution, we have $\kappa^{\text{Dir}}_{I-I^\Delta_1} \ll 1$. Nevertheless, because the major contribution to the energy norm of the error is due to the first element, namely, $\|e_{S^p_\Delta}\|_{\mathcal{U}} \approx \|e_{S^p_\Delta}\|_{\mathcal{U}(I^\Delta_1)}$, we also have $\kappa^{\text{Dir}} \approx \kappa^{\text{Dir}}_{I^\Delta_1} \approx 1$.

Let us also consider the case of Example 4.4.2 with $\vartheta = \frac{3}{4}$, $N(\Delta) = 16$, $\gamma = 1.0$, $\alpha_2/(1-\vartheta) = 10$, i.e., the case given in Table 4.4.4. Similarly, we have

$$\kappa^{\text{Dir}}_{I^\Delta_1} = \frac{\|\hat{e}^{\text{Dir}}_{I^\Delta_1}\|_{\mathcal{U}(I^\Delta_1)}}{\|e_{S^1_\Delta}\|_{\mathcal{U}(I^\Delta_1)}} = \frac{0.07619}{0.09974} = 0.7634 \tag{5.2.19a}$$

while

$$\kappa^{\text{Dir}}_{I-I_1^\Delta} = \frac{\sqrt{\sum_{j=2}^{N(\Delta)} \|\hat{e}^{\text{Dir}}_{I_j^\Delta}\|^2_{\mathcal{U}(I_j^\Delta)}}}{\|e_{S^1_\Delta}\|_{\mathcal{U}(I-I_1^\Delta)}} = \frac{\sqrt{(\mathcal{E}^{\text{Dir}})^2 - \|\hat{e}^{\text{Dir}}_{I_1^\Delta}\|^2_{\mathcal{U}(I_1^\Delta)}}}{\|e_{S^1_\Delta}\|_{\mathcal{U}(I-I_1^\Delta)}}$$
$$= \frac{0.01330}{0.04575} = 0.2907. \tag{5.2.19b}$$

Nevertheless, because

$$\|e_{S^1_\Delta}\|_{\mathcal{U}(I-I_1^\Delta)} = 0.04575 < \|e_{S^1_\Delta}\|_{\mathcal{U}(I_1^\Delta)} = 0.09974 \tag{5.2.19c}$$

we have

$$\kappa^{\text{Dir}} = \frac{\mathcal{E}^{\text{Dir}}}{\|e_{S^1_\Delta}\|_{\mathcal{U}}} = \frac{0.07730}{0.1097} = 0.7046 \approx \kappa^{\text{Dir}}_{I_1^\Delta} = \frac{\|\hat{e}^{\text{Dir}}_{I_1^\Delta}\|_{\mathcal{U}(I_1^\Delta)}}{\|e_{S^1_\Delta}\|_{\mathcal{U}(I_1^\Delta)}} = 0.7634. \tag{5.2.19d}$$

Using

$$\mathcal{K}^{\text{Dir}} = \sqrt{2}\sqrt{1 + \frac{1}{1-\frac{3}{4}}} = \sqrt{10} \tag{5.2.19e}$$

which is derived in Remark 5.2.4 below, we get

$$\mathcal{E}^{\text{Dir}} = 0.07730 \le \|e_{S^1_\Delta}\|_{\mathcal{U}} = 0.1097 \le \mathcal{K}^{\text{Dir}}\mathcal{E}^{\text{Dir}} = 0.24444. \tag{5.2.19f}$$

Hence in both cases we have $\|e_{S^p_\Delta}\|_{\mathcal{U}} \approx \|e_{S^p_\Delta}\|_{\mathcal{U}(I_1^\Delta)}$ and $\kappa^{\text{Dir}} \approx \kappa^{\text{Dir}}_{I_1^\Delta}$, and *the pollution does not affect the asymptotic quality of the estimator for the global energy norm.*

Let us elaborate further on the equivalence between $\|e_{S^p_\Delta}\|_{\mathcal{U}}$ and $\mathcal{E}^{\text{Dir}}$.

Corollary 5.2.1 *Under the assumption (5.2.2a), we have*

$$\mathcal{E}^{\text{Dir}} \le \|e_{S^p_\Delta}\|_{\mathcal{U}} \le (\mathcal{K}^{\text{Dir}}_0 + \mathcal{K}^{\text{Dir}}_1(h^\Delta)^2)\mathcal{E}^{\text{Dir}} = \left(\frac{\sqrt{\beta_2}}{\sqrt{\beta_1}} + \frac{1}{2}\frac{\gamma}{\sqrt{\beta_1\beta_2}}(h^\Delta)^2\right)\mathcal{E}^{\text{Dir}} \tag{5.2.20}$$

asymptotically, as h^Δ tends to zero. Further, under the additional assumption

$$\left\|\frac{d}{dx}(AE)\right\|_{L^\infty(I)} = K < \infty. \tag{5.2.21a}$$

we also have

$$\mathcal{E}^{\text{Dir}} \le \|e_{S^p_\Delta}\|_{\mathcal{U}} \le \left(1 + \frac{1}{2}\frac{K}{\beta_1}h^\Delta + \frac{1}{2}\frac{\gamma}{\beta_1}(h^\Delta)^2\right)\mathcal{E}^{\text{Dir}}. \tag{5.2.21b}$$

Proof A more careful proof of Lemma 5.2.1 gives

$$\|v - \mathcal{I}^1_\Delta v\|^2_{\mathcal{U}} \le \max_{j=1,\dots,N(\Delta)}\left(\frac{\beta_{2,I_j^\Delta}}{\beta_{1,I_j^\Delta}} + \frac{\gamma_{I_j^\Delta}}{\beta_{1,I_j^\Delta}}\left(h_{I_j^\Delta}\right)^2\right)\|v\|^2_{\mathcal{U}} \tag{5.2.22a}$$

and hence

$$\|v - \mathcal{I}_\Delta^1 v\| \le \sqrt{\frac{\beta_2}{\beta_1} + \frac{\gamma}{\beta_1}(h^\Delta)^2}\; \|v\|_{\mathcal{U}} \tag{5.2.22b}$$

and, for sufficiently small h^Δ, (5.2.20) follows. Further, under the assumption (5.2.21a), we have

$$\begin{aligned}\beta_{2,I_j^\Delta} - \beta_{1,I_j^\Delta} &= \max_{x\in I_j^\Delta}(AE(x)) - \min_{x\in I_j^\Delta}(AE(x))\\ &\le \left\|\frac{d}{dx}(AE)\right\|_{L^\infty(I_j^\Delta)} h_{I_j^\Delta} \le K_{I_j^\Delta} h_{I_j^\Delta}\end{aligned} \tag{5.2.22c}$$

where we let $K_{I_j^\Delta} \stackrel{\text{def}}{=} \|(d/dx)(AE)\|_{L^\infty(I_j^\Delta)}$, and hence

$$\frac{\beta_{2,I_j^\Delta}}{\beta_{1,I_j^\Delta}} \le 1 + \frac{K_{I_j^\Delta}}{\beta_{1,I_j^\Delta}} h_{I_j^\Delta} \le 1 + \frac{K}{\beta_1} h_{I_j^\Delta}. \tag{5.2.22d}$$

Using this result into (5.2.22a), we get

$$\mathcal{K} = \sqrt{\max_{j=1,\ldots,N(\Delta)} \left(\frac{\beta_{2,I_j^\Delta}}{\beta_{1,I_j^\Delta}} + \frac{\gamma_{I_j^\Delta}}{\beta_{1,I_j^\Delta}}\left(h_{I_j^\Delta}\right)^2\right)} \le \sqrt{1 + \frac{K}{\beta_1}h^\Delta + \gamma\beta_1(h^\Delta)^2} \tag{5.2.22e}$$

and hence

$$\mathcal{K} \approx 1 + \frac{1}{2}\left(\frac{K}{\beta_1}h^\Delta + \frac{\gamma}{\beta_1}(h^\Delta)^2\right) \tag{5.2.22f}$$

which proves (5.2.21b). □

Similar results can also be shown for the estimator $\mathcal{R}(\mathcal{A}_u^*(u_{S_\Delta^p}))$, where $\mathcal{A}_u^*$ is any continuous recovery of the value.

Remark 5.2.4 *The $\mathcal{K}_1^{\text{Dir}}$ term expresses the effect of the pollution.* In the cases that there is no pollution (e.g., $AE \equiv$ *constant* and $c \equiv 0$), we have $\mathcal{K}_0^{\text{Dir}} = 1$ and $\mathcal{K}_1^{\text{Dir}} = 0$.

Remark 5.2.5 The proof of the upper estimate $\|e_{S_\Delta^p}\|_{\mathcal{U}} \le \mathcal{K}^{\text{Dir}}\mathcal{E}^{\text{Dir}}$ in the case of elements of degree $p > 1$, can be obtained by employing the splitting $v = \mathcal{I}_\Delta^p v + (v - \mathcal{I}_\Delta^p v)$. Employing additional assumptions of smoothness of the input data, we also get $\mathcal{K}^{\text{Dir}} \approx \mathcal{K}_0^{\text{Dir}} + \mathcal{K}_1^{\text{Dir}}(h^\Delta)^{2p}$.

The above derivation of the upper bound $\|e_{S_\Delta^p}\|_{\mathcal{U}} \le \mathcal{K}^{\text{Dir}}\mathcal{E}^{\text{Dir}}$ was based on the assumption that $0 < \beta_1 \le AE$, which does not hold in the case of the singular model problem in which $AE(x) \equiv x^\vartheta$, and $AE(0) = 0$. We will now prove a more general version of Lemma 5.2.1, which holds both for the regular and the singular model problems.

Lemma 5.2.2 *Assume that the input data satisfy (5.2.2a) with $0 \le \beta_1$, where the equality $\beta_1 = 0$ may be achieved at several points, and that we have*

$$0 < \int_a^b AE, \qquad \int_a^b \frac{1}{AE} < \infty \tag{5.2.23a}$$

for any $a < b$. We then have

$$\|v - \mathcal{I}^1_{I_j^\Delta} v\|_{\mathcal{U}} \le \sqrt{2}\sqrt{1 + \max_{j=1,\dots,N(\Delta)} \left(\frac{1}{(h_{I_j^\Delta})^2} \int_{I_j^\Delta} AE \int_{I_j^\Delta} \frac{1}{AE} \right)} \|v\|_{\mathcal{U}}. \tag{5.2.23b}$$

Proof We will give the proof for the case $c \equiv 0$. We have

$$\int_{I_j^\Delta} AE\Big(\frac{d}{dx}(v - \mathcal{I}^1_{I_j^\Delta} v)\Big)^2 \le 2\int_{I_j^\Delta} AE\Big(\frac{dv}{dx}\Big)^2 + 2\int_{I_j^\Delta} AE\Big(\frac{d}{dx}(\mathcal{I}^1_{I_j^\Delta} v)\Big)^2. \tag{5.2.24}$$

To bound the second term, we note that for any $x \in I_j^\Delta$, we have

$$\Big|\frac{d}{dx}(\mathcal{I}^1_{I_j^\Delta} v)(x)\Big| = \frac{1}{h_{I_j^\Delta}}\Big|\int_{I_j^\Delta} \frac{dv}{dx}\Big| = \frac{1}{h_{I_j^\Delta}}\Big|\int_{I_j^\Delta} \frac{1}{\sqrt{AE}}\sqrt{AE}\frac{dv}{dx}\Big| \tag{5.2.25a}$$

and, using the Cauchy–Schwarz inequality for the L^2 inner product, we get

$$\Big|\frac{d}{dx}(\mathcal{I}^1_{I_j^\Delta} v)(x)\Big| \le \frac{1}{h_{I_j^\Delta}}\sqrt{\int_{I_j^\Delta} \frac{1}{AE}}\sqrt{\int_{I_j^\Delta} AE\Big(\frac{dv}{dx}\Big)^2}. \tag{5.2.25b}$$

Here we utilized assumption (5.2.23a) which states that $1/AE$ is integrable in any interval $(a, b) \subset I$. Squaring, multiplying both sides by AE, and integrating over I_j^Δ, we get

$$\int_{I_j^\Delta} AE\Big(\frac{d}{dx}(\mathcal{I}^1_{I_j^\Delta} v)\Big)^2 \le \frac{1}{(h_{I_j^\Delta})^2}\Big(\int_{I_j^\Delta} AE\Big)\Big(\int_{I_j^\Delta} \frac{1}{AE}\Big)\int_{I_j^\Delta} AE\Big(\frac{dv}{dx}\Big)^2. \tag{5.2.25c}$$

Employing (5.2.25c) in (5.2.24), we get

$$\int_{I_j^\Delta} AE\Big(\frac{d}{dx}(v - \mathcal{I}^1_{I_j^\Delta} v)\Big)^2 \le 2\Big(1 + \frac{1}{(h_{I_j^\Delta})^2}\Big(\int_{I_j^\Delta} AE\Big)\Big(\int_{I_j^\Delta} \frac{1}{AE}\Big)\Big)\int_{I_j^\Delta} AE\Big(\frac{dv}{dx}\Big)^2. \tag{5.2.26}$$

Hence we have proven (5.2.23b) for the special case $c \equiv 0$. □

Remark 5.2.6 Let us now apply Lemma 5.2.2 in the case of the singular model problem in which $AE \equiv x^\vartheta$. It follows that

$$\mathcal{K} = \sqrt{2}\sqrt{1 + \max_{j=1,\dots,N(\Delta)} \left(\frac{1}{(h_{I_j^\Delta})^2} \Big(\int_{I_j^\Delta} x^\vartheta\Big)\Big(\int_{I_j^\Delta} x^{-\vartheta}\Big)\right)}. \tag{5.2.27a}$$

Further, in the case that the mesh Δ is uniform, we have

$$\mathcal{K} = \sqrt{2}\sqrt{1 + \max_{j=1,\dots,N(\Delta)} \left(\frac{1}{1-\vartheta^2} \frac{1-(1-1/j)^{1+\vartheta}}{1/j} \frac{1-(1-1/j)^{1-\vartheta}}{1/j} \right)} \tag{5.2.27b}$$

and using the inequalities given in the proof of Lemma 3.4.1, we get

$$\mathcal{K} \leq \sqrt{2}\sqrt{1 + \frac{1}{1-\vartheta}} \tag{5.2.27c}$$

and hence $\mathcal{K}$ exists for any $0 \leq \vartheta < 1$. Note that for $\vartheta = 0$, (5.2.27c) gives $\mathcal{K} \leq 2$, instead of the optimal value $\mathcal{K} = 1$, and hence the value of $\mathcal{K}$ given by (5.2.27a,b) is not optimal.

Above we have proven the upper bound

$$\|e_{S^p_\Delta}\|_{\mathcal{U}} \leq \mathcal{K}\mathcal{E}^{\mathrm{Dir}} = \mathcal{K}\sqrt{\sum_{j=1}^{N(\Delta)} \left(\eta^{\mathrm{Dir}}_{I^\Delta_j}\right)^2} \tag{5.2.28a}$$

where $\mathcal{E}^{\mathrm{Dir}}$ is the exact Dirichlet estimator. As we already discussed in Section 5.1, in practice $\hat{e}^{\mathrm{Dir}}_{I^\Delta_j}$ is replaced by its projection $\pi_{S'}(\hat{e}^{\mathrm{Dir}}_{I^\Delta_j})$, where $S^p(I^\Delta_j) \subset S' \subseteq \mathcal{U}_0(I^\Delta_j)$, and we get the computed Dirichlet element error indicator

$$\eta^{\mathrm{Dir}}_{I^\Delta_j,S} = \|\pi_S(\hat{e}^{\mathrm{Dir}}_{I^\Delta_j})\|_{\mathcal{U}(I^\Delta_j)} \leq \eta^{\mathrm{Dir}}_{I^\Delta_j} \tag{5.2.28b}$$

and the computed Dirichlet estimator

$$\mathcal{E}^{\mathrm{Dir}}_S = \sqrt{\sum_{j=1}^{N(\Delta)} \|\pi_S(\hat{e}^{\mathrm{Dir}}_{I^\Delta_j})\|^2_{\mathcal{U}(I^\Delta_j)}} \leq \mathcal{E}^{\mathrm{Dir}} \tag{5.2.28c}$$

and hence $\mathcal{K}\mathcal{E}^{\mathrm{Dir}}_S$ is not a guaranteed upper estimator. A guaranteed computable upper estimator may be obtained by employing the identity

$$\eta^{\mathrm{Dir}}_{I^\Delta_j} = \sqrt{\left(\eta^{\mathrm{Dir}}_{I^\Delta_j,S}\right)^2 + \|\hat{e}^{\mathrm{Dir}}_{I^\Delta_j} - \pi_S(\hat{e}^{\mathrm{Dir}}_{I^\Delta_j})\|^2_{\mathcal{U}(I^\Delta_j)}} \tag{5.2.28d}$$

and by employing an explicit upper estimator for $\|\hat{e}^{\mathrm{Dir}}_{I^\Delta_j} - \pi_S(\hat{e}^{\mathrm{Dir}}_{I^\Delta_j})\|_{\mathcal{U}(I^\Delta_j)}$, the energy norm of the error in the computed element error indicator function.

In practice, we often employ $S = B^{p+k} \stackrel{\mathrm{def}}{=} S^{p+k}_0 - S^p_0$, which is the span of all internal element shape-functions of degree $p+k$, and we get the computed estimator $\mathcal{E}^{\mathrm{Dir}}_{B^{p+k}}$, which will be denoted by *ERDBp* + *k*. In the one-dimensional case, we have $\mathcal{E}^{\mathrm{Dir}}_{B^{p+k}} \equiv \mathcal{E}^{\mathrm{Dir}}_{S^{p+k}_0}$.

Let us now give an example which illustrates the effect of the approximation of the error-indicator functions.

Example 5.2.2 Let us consider the regular model problem with $AE \equiv 1$, $c \equiv 0$, $u_{EX} \equiv x^{3/4}$, and the error in the finite element solution for a radical mesh of $N(\Delta) = 5$ quadratic elements ($p = 2$) with $\gamma = 2$. In this case the nodal values are exact and hence the nodal errors are zero, $e_{S^2_\Delta}(x^\Delta_j) = 0$, $j = 1, \ldots, N(\Delta)$, and it follows that the exact indicators $\hat{e}^{\mathrm{Dir}}_{I^\Delta_j}$ coincide with the error $e_{S^2_\Delta}|_{I^\Delta_j}$, namely, $\hat{e}^{\mathrm{Dir}}_{I^\Delta_j} \equiv e_{S^2_\Delta}|_{I^\Delta_j}$, $j = 1, \ldots, N(\Delta) = 5$. Figure 5.2.1 shows the graphs of $|(d/dx)(e_{S^2_\Delta})|$, the absolute value of the derivative of the exact error, and $|(d/dx)(\hat{e}^{\mathrm{Dir}}_{I^\Delta_j} - \pi_{B^{p+k}}(\hat{e}^{\mathrm{Dir}}_{I^\Delta_j}))|$, the absolute value of the derivative of the error in the element error-indicator functions for $k = 1$ and $k = 6$, in the second, third

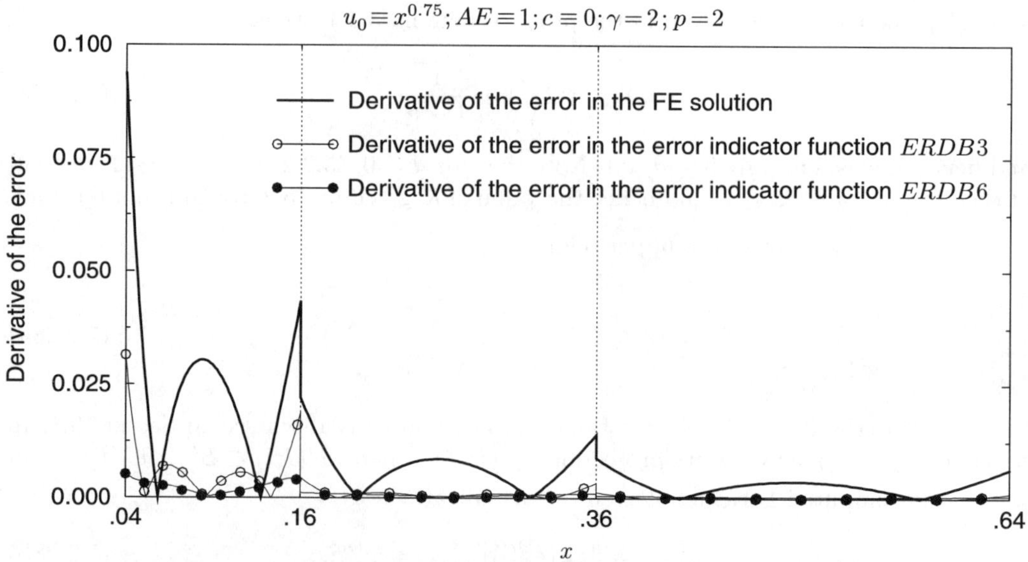

FIG. 5.2.1. Element residual error-indicator function for the element residual problem with homogeneous Dirichlet boundary conditions. Input data: $AE \equiv 1$, $c \equiv 0$, $u_{EX} \equiv x^{3/4}$. Mesh data: Radical mesh with $\gamma = 2$, $N(\Delta) = 5$, quadratic elements ($p = 2$). Graphs of $|(d/dx)(\hat{e}^{\text{Dir}}_{I^\Delta_j} - \pi_{B^{2+k}}(\hat{e}^{\text{Dir}}_{I^\Delta_j}))|$, the absolute value of the derivative of the error in the error indicator function for $k = 1$ and $k = 6$ in the second, third and fourth element from the singular point. The absolute value of the derivative of the error, $|(d/dx)(e_{S^2_\Delta})|$, is given for comparison of the magnitudes of the two errors

and fourth element of a radical mesh ($\gamma = 2$) of quadratic elements ($p = 2$). It can be seen that in the elements where the solution is smooth (e.g., the third and fourth element) we have $(d/dx)(\hat{e}^{\text{Dir}}_{I^\Delta_j}) \approx (d/dx)(\pi_{B^{2+k}}(\hat{e}^{\text{Dir}}_{I^\Delta_j}))$ for $k = 1$, while in the elements where the solution is unsmooth (e.g., the second element) it is necessary to employ a sufficiently high k in order to get an accurate approximation of the error-indicator function.

As we have already discussed in Section 5.1, another approach for obtaining an upper estimator is to construct explicit upper bounds for the error indicators

$$\eta^{\text{Dir}}_{I^\Delta_j} = \|\hat{e}^{\text{Dir}}_{I^\Delta_j}\|_{\mathcal{U}(I^\Delta_j)} \leq \eta^{\text{Dir}}_{I^\Delta_j, EXPL} = \mathscr{C}_{I^\Delta_j} h^\Delta_{I^\Delta_j} \|r_{I^\Delta_j}\|_{L^2(I^\Delta_j)} \tag{5.2.29}$$

where $\mathscr{C}_{I^\Delta_j}$ is an a-priori computable constant independent of $h_{I^\Delta_j}$. Here, of course, the main problem is the determination of $\mathscr{C}_{I^\Delta_j}$.

As an example, we will determine $\mathscr{C}_{I^\Delta_j}$ in the case of linear elements ($p = 1$), for $AE|_{I^\Delta_j} \equiv$ *constant*, and $c \equiv 0$. In this case we have

$$-AE|_{I^\Delta_j} \frac{d^2}{dx^2}(\hat{e}^{\text{Dir}}_{I^\Delta_j}) = r_{I^\Delta_j}(x), \quad x \in I^\Delta_j \qquad \hat{e}^{\text{Dir}}_{I^\Delta_j}(x^\Delta_k) = 0, \quad k = j-1, j \tag{5.2.30a}$$

and taking into account that the nodal errors are zero, $e_{S^p_\Delta}(x^\Delta_k) = 0$, $k = j-1, j$, we obtain

$$\hat{e}^{\rm Dir}_{I^\Delta_j} = e_{S^p_\Delta}\big|_{I^\Delta_j} = \sum_{k=1}^{\infty} a_k \sin\left(\frac{k\pi}{h_{I^\Delta_j}}(x - x^\Delta_{j-1})\right) \tag{5.2.30b}$$

and

$$r_{I^\Delta_j} = -AE|_{I^\Delta_j} \frac{d^2}{dx^2}\left(e_{S^p_\Delta}\big|_{I^\Delta_j}\right) = \sum_{k=1}^{\infty} b_k \sin\left(\frac{k\pi}{h_{I^\Delta_j}}(x - x^\Delta_{j-1})\right) \tag{5.2.30c}$$

where

$$b_k = AE|_{I^\Delta_j}\left(\frac{k\pi}{h_{I^\Delta_j}}\right)^2 a_k \tag{5.2.30d}$$

and hence

$$\hat{e}^{\rm Dir}_{I^\Delta_j} = e_{S^p_\Delta}\big|_{I^\Delta_j} = \frac{1}{AE|_{I^\Delta_j}} \sum_{k=1}^{\infty} b_k \left(\frac{h_{I^\Delta_j}}{k\pi}\right)^2 \sin\left(\frac{k\pi}{h_{I^\Delta_j}}(x - x^\Delta_{j-1})\right). \tag{5.2.30e}$$

It follows that

$$\|r_{I^\Delta_j}\|^2_{L^2(I^\Delta_j)} = \int_{I^\Delta_j} (r_{I^\Delta_j})^2 = \frac{h_{I^\Delta_j}}{2} \sum_{k=1}^{\infty} b_k^2 \tag{5.2.31a}$$

and

$$\|\hat{e}^{\rm Dir}_{I^\Delta_j}\|^2_{\mathfrak{U}(I^\Delta_j)} = \int_{I^\Delta_j} AE\Big(\frac{d}{dx}(\hat{e}^{\rm Dir}_{I^\Delta_j})\Big)^2 = \frac{h_{I^\Delta_j}}{2} \frac{1}{AE|_{I^\Delta_j}} \sum_{k=1}^{\infty} b_k^2 \left(\frac{h_{I^\Delta_j}}{k\pi}\right)^2 \tag{5.2.31b}$$

and using the inequality

$$\left(\frac{\pi}{h_{I^\Delta_j}}\right)^2 \sum_{k=1}^{\infty} \left(\frac{h_j}{k\pi}\right)^2 b_k^2 = \sum_{k=1}^{\infty} \frac{1}{k^2} b_k^2 \le \sum_{k=1}^{\infty} b_k^2 \tag{5.2.31c}$$

we get

$$\sqrt{\int_{I^\Delta_j} AE\Big(\frac{d}{dx}(\hat{e}^{\rm Dir}_{I^\Delta_j})\Big)^2} \le \frac{1}{\sqrt{AE|_{I^\Delta_j}}} \frac{h_{I^\Delta_j}}{\pi} \sqrt{\int_{I^\Delta_j} (r_{I^\Delta_j})^2}. \tag{5.2.31d}$$

Thus we have proven that

$$\eta^{\rm Dir}_{I^\Delta_j} = \|\hat{e}^{\rm Dir}_{I^\Delta_j}\|_{\mathfrak{U}(I^\Delta_j)} \le \frac{1}{\pi\sqrt{AE|_{I^\Delta_j}}} h_{I^\Delta_j} \|r_{I^\Delta_j}\|_{L^2(I^\Delta_j)} \tag{5.2.32a}$$

and using the upper bound

$$\|e_{S^p_\Delta}\|_{\mathcal{U}} \le \mathcal{K}\sqrt{\sum_{j=1}^{N(\Delta)} \left(\eta^{\mathrm{Dir}}_{I^\Delta_j}\right)^2} \tag{5.2.32b}$$

we get the explicit upper bound

$$\|e_{S^p_\Delta}\|_{\mathcal{U}} \le \frac{\mathcal{K}}{\pi}\frac{1}{\min_{j=1,\ldots,N(\Delta)}\sqrt{AE|_{I^\Delta_j}}}\sqrt{\sum_{j=1}^{N(\Delta)} \left(h_{I^\Delta_j}\right)^2 \|r_{I^\Delta_j}\|^2_{L^2(I^\Delta_j)}}. \tag{5.2.32c}$$

We have thus shown that in the case that $AE|_{I^\Delta_j} \equiv$ *constant*, $c \equiv 0$, and $p = 1$, we have

$$\mathcal{C}_{I^\Delta_j} = \frac{1}{\pi\sqrt{AE|_{I^\Delta_j}}} \tag{5.2.33a}$$

which is achieved for

$$r_{I^\Delta_j} \equiv \sin\left(\frac{\pi}{h_{I^\Delta_j}}(x - x^\Delta_j)\right). \tag{5.2.33b}$$

Nevertheless, as is, for example, shown in P5.2.4, $\mathcal{C}_{I^\Delta_j}$ may be pessimistic for the concrete case of interest. Let us also note that the above value of $\mathcal{C}_{I^\Delta_j}$ (respectively the function $r_{I^\Delta_j}$) is the first eigenvalue (respectively eigenfunction) of the problem

$$\left(AE|_{I^\Delta_j}\frac{d^2u}{dx^2} - \lambda u\right)\Big|_{I^\Delta_j} = 0, \qquad u(x^\Delta_{j-1}) = u(x^\Delta_j) = 0. \tag{5.2.33c}$$

In the special case that $AE|_{I^\Delta_j} \equiv$ *constant*, $c \equiv 0$, $p = 1$, and f is smooth, we can also employ the approximation

$$r_{I^\Delta_j} = f|_{I^\Delta_j} \approx f(x_{j-1/2}) \tag{5.2.34a}$$

and

$$\eta^{\mathrm{Dir}}_{I^\Delta_j} = \|\hat{e}^{\mathrm{Dir}}_{I^\Delta_j}\|_{\mathcal{U}(I^\Delta_j)} \approx \frac{h_{I^\Delta_j}}{\sqrt{12}}\frac{1}{\sqrt{AE|_{I^\Delta_j}}}\|f\|_{L^2(I^\Delta_j)}. \tag{5.2.34b}$$

Let us now summarize the above results.

Theorem 5.2.3 *Under the assumptions (5.2.2), there exists a constant $\mathcal{C}$ such that*

$$\|e_{S^p_\Delta}\|_{\mathcal{U}} \le \mathcal{C}\sqrt{\sum_{j=1}^{N(\Delta)} \left(h_{I^\Delta_j}\right)^2 \|r_{I^\Delta_j}\|^2_{L^2(I^\Delta_j)}}. \tag{5.2.35}$$

Proof Above we have proven (5.2.35) by employing the global upper estimate

$$\|e_{S^p_\Delta}\|_{\mathcal{U}} \leq \mathcal{K}\sqrt{\sum_{j=1}^{N(\Delta)} \left(\eta^{\mathrm{Dir}}_{I^\Delta_j}\right)^2} \tag{5.2.36}$$

which was proven in Theorem 5.2.2, and the elementwise upper estimate

$$\eta^{\mathrm{Dir}}_{I^\Delta_j} \leq \mathscr{C}_{I^\Delta_j} h_{I^\Delta_j} \|r^\Delta_j\|_{L^2(I^\Delta_j)}. \tag{5.2.37}$$

We will now prove (5.2.35) directly.

From the definition of the energy norm via duality, the definition of the residuum, and the orthogonality condition, $\mathcal{R}(\mathcal{I}^p_\Delta v) = 0$, for all $v \in \mathcal{U}$, we obtain

$$\|e_{S^p_\Delta}\|_{\mathcal{U}} = \sup_{v\in\mathcal{U}} \frac{\mathcal{B}\left(e_{S^p_\Delta}, v\right)}{\|v\|_{\mathcal{U}}} = \sup_{v\in\mathcal{U}} \frac{\mathcal{R}(v)}{\|v\|_{\mathcal{U}}} = \sup_{v\in\mathcal{U}} \frac{\mathcal{R}(v - \mathcal{I}^p_\Delta v)}{\|v\|_{\mathcal{U}}}. \tag{5.2.38a}$$

Noting that

$$\mathcal{R}(v - \mathcal{I}^p_\Delta v) = \sum_{j=1}^{N(\Delta)} \int_{I^\Delta_j} r_{I^\Delta_j}(v - \mathcal{I}^p_\Delta v) \tag{5.2.38b}$$

and using the Cauchy–Schwarz inequality for the L^2 inner product, we obtain

$$|\mathcal{R}(v - \mathcal{I}^p_\Delta v)| \leq \sqrt{\sum_{j=1}^{N(\Delta)} \left(h_{I^\Delta_j}\right)^2 \int_{I^\Delta_j} \left(r_{I^\Delta_j}\right)^2} \sqrt{\sum_{j=1}^{N(\Delta)} \frac{1}{\left(h_{I^\Delta_j}\right)^2} \int_{I^\Delta_j} (v - \mathcal{I}^p_\Delta v)^2}. \tag{5.2.38c}$$

Employing the inequality

$$\int_{I^\Delta_j} (v - \mathcal{I}^p_\Delta v)^2 \leq \mathscr{C}\left(h_{I^\Delta_j}\right)^2 \|v\|^2_{\mathcal{U}(I^\Delta_j)} \tag{5.2.38d}$$

which follows from (5.2.15a), we get

$$|\mathcal{R}(v - \mathcal{I}^1_\Delta v)| \leq \mathscr{C}\sqrt{\sum_{j=1}^{N(\Delta)} \left(h_{I^\Delta_j}\right)^2 \int_{I^\Delta_j} \left(r_{I^\Delta_j}\right)^2} \sqrt{\sum_{j=1}^{N(\Delta)} \|v\|^2_{\mathcal{U}(I^\Delta_j)}} \tag{5.2.38c}$$

and hence

$$\|e_{S^p_\Delta}\|_{\mathcal{U}} = \sup_{v\in\mathcal{U}} \frac{|\mathcal{R}(v - \mathcal{I}^1_\Delta v)|}{\|v\|_{\mathcal{U}(I^\Delta_j)}} \leq \mathscr{C}\sqrt{\sum_{j=1}^{N(\Delta)} \left(h_{I^\Delta_j}\right)^2 \int_{I^\Delta_j} \left(r_{I^\Delta_j}\right)^2}. \tag{5.2.38f}$$

□

Remark 5.2.7 For elements of degree $p > 1$ the estimate (5.2.35) can be improved by taking into account the orthogonalities $\int_{I^\Delta_j} r_{I^\Delta_j} v = 0$, for all $v \in S^p_0(I^\Delta_j)$.

The proof of the explicit estimate (5.2.35) was based on the identity

$$\mathcal{R}(v - \mathcal{I}^p_\Delta v) = \sum_{j=1}^{N(\Delta)} \int_{I_j^\Delta} r_{I_j^\Delta}(v - \mathcal{I}^p_\Delta v). \tag{5.2.39}$$

Note that we took $v \in \mathcal{U}$, and the interpolant $\mathcal{I}^p_\Delta v$ may not have good meaning in higher dimensions, because a function v with finite norm $\|v\|_{\mathcal{U}}$ may not have well-defined point values. We can overcome this difficulty by employing the mean interpolation operator $\hat{\mathcal{I}}^1_\Delta$ which is defined by

$$\hat{\mathcal{I}}^1_\Delta v \stackrel{\text{def}}{=} \sum_{j=1}^{N(\Delta)} \frac{\int_{\omega_j^\Delta} v\varphi_j^\Delta}{\int_{\omega_j^\Delta} \varphi_j^\Delta} \varphi_j^\Delta \tag{5.2.40}$$

where φ_j^Δ is the piecewise linear hat function for the jth node, and $\omega_j^\Delta = \operatorname{supp}(\varphi_j^\Delta)$. Note that $\hat{\mathcal{I}}^1_\Delta u$ has good meaning for all $u \in \mathcal{U}$ in higher dimensions, while this is not the case for the classical interpolant $\mathcal{I}^1_\Delta u$. Then we have:

Lemma 5.2.3

$$\sum_{j=1}^{N(\Delta)} \left(\frac{1}{2h_{I_j^\Delta}} \sum_{k=j-1}^{j} ((v - \hat{\mathcal{I}}^1_\Delta v)(x_k^\Delta))^2 + \frac{1}{\left(h_{I_j^\Delta}\right)^2} \int_{I_j^\Delta} (v - \hat{\mathcal{I}}^1_\Delta v)^2 \right) \leq \mathcal{C}(\mu)\|v\|^2_{\mathcal{U}} \tag{5.2.41}$$

where $\omega_{I_j^\Delta} \stackrel{\text{def}}{=} \omega_{j-1}^\Delta \cup \omega_j^\Delta$, *and* $1/\mu < h_{I_j^\Delta}/h_{I_{j-1}^\Delta} < \mu$.

Proof The proof of (5.2.41) follows from the inequality

$$\frac{1}{2h_{I_j^\Delta}} \sum_{k=j-1}^{j} ((v - \hat{\mathcal{I}}_\Delta v)(x_k^\Delta))^2 + \frac{1}{\left(h_{I_j^\Delta}\right)^2} \int_{I_j^\Delta} (v - \hat{\mathcal{I}}_\Delta v)^2 \leq \mathcal{C}(\mu)\|v\|^2_{\mathcal{U}(\omega_{I_j^\Delta})}. \tag{5.2.42}$$

Employing a scaling argument, it follows that, in order to prove (5.2.42), it is sufficient to prove the inequality

$$\frac{1}{2}(((\tilde{v} - \tilde{\hat{\mathcal{I}}}^1\tilde{v})(\tilde{x}_{-1}^\Delta))^2 + ((\tilde{v} - \tilde{\hat{\mathcal{I}}}^1\tilde{v})(\tilde{x}_0^\Delta))^2) + \int_{\tilde{I}_0^\Delta} (\tilde{v} - \tilde{\hat{\mathcal{I}}}^1\tilde{v})^2 \leq \tilde{\mathcal{C}}\|\tilde{v}\|^2_{\mathcal{U}(\tilde{\omega})} \tag{5.2.43a}$$

where we let $\tilde{\omega} = \tilde{I}_{-1} \cup \tilde{I}_0 \cup \tilde{I}_1$, under the assumption

$$\tilde{\mathcal{C}}_1(\mu)(\tilde{h}_{-1} + \tilde{h}_0 + \tilde{h}_1) \leq \tilde{h}_0 \leq \tilde{\mathcal{C}}_2(\mu)(\tilde{h}_{-1} + \tilde{h}_0 + \tilde{h}_1). \tag{5.2.43b}$$

We will not elaborate on the proof of (5.2.43a). It follows directly from the construction of the interpolation operator $\tilde{\hat{\mathcal{I}}}^1$. Summing (5.2.42) for $j = 1, \ldots, N(\Delta)$, we get

$$\sum_{j=1}^{N(\Delta)} \left(\frac{1}{2h_{I_j^\Delta}} \sum_{k=j-1}^{j} ((v - \hat{\mathcal{I}}^1_\Delta v)(x_k))^2 + \frac{1}{\left(h_{I_j^\Delta}\right)^2} \int_{I_j^\Delta} (v - \hat{\mathcal{I}}^1_\Delta v)^2 \right) \leq \mathcal{C} \sum_{j=1}^{N(\Delta)} \|v\|^2_{\mathcal{U}(\omega_{I_j^\Delta})}. \tag{5.2.44a}$$

Noting that

$$\sum_{j=1}^{N(\Delta)} \|v\|^2_{\mathcal{U}(\omega_{I_j^\Delta})} \le \mathscr{C}'\|v\|^2_{\mathcal{U}} \tag{5.2.44b}$$

where, in the one-dimensional case, $\mathscr{C}' = 2$, the right-hand side of (5.2.44a) can be replaced by $\mathscr{C}\mathscr{C}'\|v\|_{\mathcal{U}}$, and (5.2.41) has been proven. □

Let us also give a proof of the explicit upper bound $\|e_{S^p_\Delta}\|_{\mathcal{U}} \le \mathscr{C}\mathscr{E}^{EXPL}$, based on the preceding Lemma.

Employing the orthogonality $\mathscr{R}(\hat{\mathcal{I}}^1_\Delta v) = 0$, we have

$$|\mathscr{R}(v)| = |\mathscr{R}(v - \hat{\mathcal{I}}^1_\Delta v)| = \left| \sum_{j=1}^{N(\Delta)} \int_{I_j^\Delta} r_{I_j^\Delta}(v - \hat{\mathcal{I}}^1_\Delta v) + \sum_{j=0}^{N(\Delta)} J_j^\Delta (v - \hat{\mathcal{I}}^1_\Delta v)(x_j^\Delta) \right| \tag{5.2.45a}$$

and rearranging the sum $\sum_{j=0}^{N(\Delta)} J_j^\Delta (v - \hat{\mathcal{I}}^1_\Delta v)(x_j^\Delta)$ into the elements, we get

$$\begin{aligned} \mathscr{R}(v) &= \mathscr{R}(v - \hat{\mathcal{I}}^1_\Delta v) \\ &= \sum_{j=1}^{N(\Delta)} \left(\int_{I_j^\Delta} r_{I_j^\Delta}(v - \hat{\mathcal{I}}^1_\Delta v) + \sum_{k=j-1}^{j} \frac{1}{2} \check{J}_k^\Delta (v - \hat{\mathcal{I}}^1_\Delta v)(x_k^\Delta) \right) \end{aligned} \tag{5.2.45b}$$

where we used the notation

$$\check{J}_0^\Delta = 0, \qquad \check{J}_j^\Delta = J_j^\Delta, \quad j = 1, \ldots, N(\Delta) - 1, \qquad \check{J}_N^\Delta = 2J_N^\Delta. \tag{5.2.45c}$$

Using the Cauchy–Schwarz inequality we get

$$\begin{aligned} |\mathscr{R}(v)| \le& \sqrt{\sum_{j=1}^{N} (\eta^{EXPL}_{I_j^\Delta})^2} \\ &\times \sqrt{\sum_{j=1}^{N} \left(\frac{1}{2h_{I_j^\Delta}} \sum_{k=j-1}^{j} ((v - \hat{\mathcal{I}}^1_\Delta v)(x_k))^2 + \frac{1}{\left(h_{I_j^\Delta}\right)^2} \int_{I_j^\Delta} (v - \hat{\mathcal{I}}^1_\Delta v)^2 \right)} \end{aligned} \tag{5.2.46a}$$

where we let

$$\eta^{EXPL}_{I_j^\Delta} = \sqrt{\frac{h_{I_j^\Delta}}{2}((\check{J}^\Delta_{j-1})^2 + (\check{J}^\Delta_j)^2) + (h_{I_j^\Delta})^2 \int_{I_j^\Delta} r^2_{I_j^\Delta}}. \tag{5.2.46b}$$

Employing now the interpolation estimate (5.2.41), we get

$$\begin{aligned} \|e_{S^p_\Delta}\|_{\mathcal{U}} &= \sup_{v \in \mathcal{U}} \frac{|\mathscr{R}(v)|}{\|v\|_{\mathcal{U}}} \le \bar{\mathscr{C}} \sqrt{\sum_{j=1}^{N} \left(\frac{h_{I_j^\Delta}}{2}((\check{J}^\Delta_{j-1})^2 + (\check{J}^\Delta_j)^2) + \left(h_{I_j^\Delta}\right)^2 \int_{I_j^\Delta} r^2_{I_j^\Delta} \right)} \\ &= \bar{\mathscr{C}}\mathscr{E}^{EXPL} \end{aligned} \tag{5.2.47}$$

where $\bar{\mathscr{C}}$ is computable, and depends on the parameter μ, which reflects the uniformity of the mesh.

Remark 5.2.8 Note that the explicit estimate (5.2.47) includes also the jumps J_j^{Δ} and has a complete analogue in two dimensions. In contrast, the estimate

$$\|e_{S_\Delta^p}\|_{\mathscr{U}} \leq \mathscr{C}\sqrt{\sum_{j=1}^{N(\Delta)} \left(h_{I_j^\Delta}\right)^2 \int_{I_j^\Delta} \left(r_{I_j^\Delta}\right)^2} \tag{5.2.48}$$

holds only in the one-dimensional case.

Remark 5.2.9 Recall that the Dirichlet estimator $\mathscr{E}^{\mathrm{Dir}}$ is given by

$$\mathscr{E}^{\mathrm{Dir}} = \sqrt{(\mathscr{E}_S^{\mathrm{Dir}})^2 + \sum_{j=1}^{N(\Delta)} \|\hat{e}_{I_j^\Delta}^{\mathrm{Dir}} - \pi_S(\hat{e}_{I_j^\Delta}^{\mathrm{Dir}})\|_{\mathscr{U}(I_j^\Delta)}^2} \tag{5.2.49a}$$

where $\mathscr{E}_S^{\mathrm{Dir}}$ is the computed Dirichlet estimator. Employing now the explicit upper bound

$$\|\hat{e}_{I_j^\Delta}^{\mathrm{Dir}} - \pi_S(\hat{e}_{I_j^\Delta}^{\mathrm{Dir}})\|_{\mathscr{U}(I_j^\Delta)} \leq \frac{1}{\pi\sqrt{\beta_{1,I_j^\Delta}}} h_{I_j^\Delta} \|r_{I_j^\Delta} - \mathfrak{D}(\pi_S(\hat{e}_{I_j^\Delta}^{\mathrm{Dir}}))\|_{\mathscr{U}(I_j^\Delta)} \tag{5.2.49b}$$

we get the computable upper bound

$$\|e_{S_\Delta^p}\|_{\mathscr{U}} \leq \mathcal{K}^{\mathrm{Dir}} \sqrt{(\mathscr{E}_S^{\mathrm{Dir}})^2 + \pi^2 \sum_{j=1}^{N(\Delta)} \frac{\left(h_{I_j^\Delta}\right)^2}{\beta_{1,I_j^\Delta}} \|r_{I_j^\Delta} - \mathfrak{D}(\pi_S(\hat{e}_{I_j^\Delta}^{\mathrm{Dir}}))\|_{\mathscr{U}(I_j^\Delta)}^2}. \tag{5.2.49c}$$

Above we have proven the equivalence $\mathscr{E}^{\mathrm{Dir}} \leq \|e_{S_\Delta^p}\|_{\mathscr{U}} \leq \mathcal{K}^{\mathrm{Dir}} \mathscr{E}^{\mathrm{Dir}}$. Let us now examine the relationship between $\eta_{I_j^\Delta}^{\mathrm{Dir}}$ and $\|e_{S_\Delta^p}\|_{\mathscr{U}(I_j^\Delta)}$.

Theorem 5.2.4 *Assume that the input data satisfy (5.2.2a). Then we have*

$$\eta_{I_j^\Delta}^{\mathrm{Dir}} = \|\hat{e}_{I_j^\Delta}^{\mathrm{Dir}}\|_{\mathscr{U}(I_j^\Delta)} \leq \|e_{S_\Delta^p}\|_{\mathscr{U}(I_j^\Delta)}. \tag{5.2.50}$$

Proof In general, we have $\hat{e}_{I_j^\Delta}^{\mathrm{Dir}} \not\equiv e_{S_\Delta^p}|_{I_j^\Delta}$. Noting that

$$\mathscr{B}_{I_j^\Delta}(e_{S_\Delta^p} - \hat{e}_{I_j^\Delta}^{\mathrm{Dir}}, v) = \mathscr{B}_{I_j^\Delta}(e_{S_\Delta^p}, v) - \mathscr{B}_{I_j^\Delta}(\hat{e}_{I_j^\Delta}^{\mathrm{Dir}}, v) = 0 \quad \forall v \in \mathscr{U}_0(I_j^\Delta) \tag{5.2.51a}$$

we obtain

$$\|e_{S_\Delta^p}\|_{\mathscr{U}(I_j^\Delta)}^2 = \|\hat{e}_{I_j^\Delta}^{\mathrm{Dir}}\|_{\mathscr{U}(I_j^\Delta)}^2 + \|e_{S_\Delta^p} - \hat{e}_{I_j^\Delta}^{\mathrm{Dir}}\|_{\mathscr{U}(I_j^\Delta)}^2 \tag{5.2.51b}$$

and omitting the second term, we get (5.2.50). □

The elementwise lower estimate $\eta_{I_j^\Delta}^{\mathrm{Dir}} \leq \|e_{S_\Delta^p}\|_{\mathscr{U}(I_j^\Delta)}$, and the global upper estimate $\|e_{S_\Delta^p}\|_{\mathscr{U}} \leq \mathcal{K}^{\mathrm{Dir}} \mathscr{E}^{\mathrm{Dir}}$, hold under assumptions (5.2.2a), which are the *minimal* assumptions needed to guarantee the existence of solutions of the model problem *and without any assumptions about the mesh.*

Let us now show that, under the additional assumption that the pollution is negligible, the element error-indicators $\eta_{I_j^\Delta}^{\mathrm{Dir}}$ are *asymptotically exact*, namely,

$$\|e_{S_\Delta^p}\|_{\mathcal{U}(I_j^\Delta)} \approx (1 + O(h^\Delta))\eta_{I_j^\Delta}^{\mathrm{Dir}}. \tag{5.2.52}$$

Theorem 5.2.5 *Assume that (5.2.2a) holds, and that there exists an $\alpha > 0$ such that*

$$\max\left(|e_j^\Delta + e_{j-1}^\Delta|, \frac{|e_j^\Delta - e_{j-1}^\Delta|}{h_{I_j^\Delta}}\right) \le \mathscr{C}(h_{I_j^\Delta})^{\alpha-1/2} \min_{v\in S^p(I_j^\Delta)} \|u_{EX} - v\|_{\mathcal{U}(I_j^\Delta)}. \tag{5.2.53a}$$

Then, for sufficiently small $h_{I_j^\Delta}$, we have

$$\|e_{S_\Delta^p}\|_{\mathcal{U}(I_j^\Delta)} \le \left(1 + \mathscr{C}(h_{I_j^\Delta})^{2\alpha}\right)\|\hat{e}_{I_j^\Delta}^{\mathrm{Dir}}\|_{\mathcal{U}(I_j^\Delta)}. \tag{5.2.53b}$$

Proof Recalling that the definition of the best approximation

$$\|\tilde{e}_{S^p(I_j^\Delta)}^{loc}\|_{\mathcal{U}(I_j^\Delta)} \stackrel{\mathrm{def}}{=} \min_{v\in S^p(I_j^\Delta)} \|u_{EX} - v\|_{\mathcal{U}(I_j^\Delta)} \le \|e_{S_\Delta^p}\|_{\mathcal{U}(I_j^\Delta)} \tag{5.2.54a}$$

we have

$$\|\tilde{e}_{S^p(I_j^\Delta)}^{loc}\|_{\mathcal{U}(I_j^\Delta)} \le \|e_{S_\Delta^p}\|_{\mathcal{U}(I_j^\Delta)} = \sqrt{\|\hat{e}_{I_j^\Delta}^{\mathrm{Dir}}\|^2_{\mathcal{U}(I_j^\Delta)} + \|e_{S_\Delta^p} - \hat{e}_{I_j^\Delta}^{\mathrm{Dir}}\|^2_{\mathcal{U}(I_j^\Delta)}}. \tag{5.2.54b}$$

From the orthogonality condition, $\mathscr{B}_{I_j^\Delta}(e_{S_\Delta^p} - \hat{e}_{I_j^\Delta}^{\mathrm{Dir}}, v) = 0$, for any $v \in \mathcal{U}_0(I_j^\Delta)$, it follows that

$$\|e_{S_\Delta^p} - \hat{e}_{I_j^\Delta}^{\mathrm{Dir}}\|^2_{\mathcal{U}(I_j^\Delta)} \le \|e_{S_\Delta^p} - \hat{e}_{I_j^\Delta}^{\mathrm{Dir}} + v\|^2_{\mathcal{U}(I_j^\Delta)} \quad \forall v \in \mathcal{U}_0(I_j^\Delta) \tag{5.2.55a}$$

and choosing $v = \mathcal{I}^1_{I_j^\Delta} e_{S_\Delta^p} - e_{S_\Delta^p} - \hat{e}_{I_j^\Delta}^{\mathrm{Dir}} \in \mathcal{U}_0(I_j^\Delta)$, we get

$$\|e_{S_\Delta^p} - \hat{e}_{I_j^\Delta}^{\mathrm{Dir}}\|_{\mathcal{U}(I_j^\Delta)} \le \|\mathcal{I}^1_{I_j^\Delta} e_{S_\Delta^p}\|_{\mathcal{U}(I_j^\Delta)}. \tag{5.2.55b}$$

Further, we have

$$\begin{aligned}\|\mathcal{I}^1_{I_j^\Delta} e_{S_\Delta^p}\|_{\mathcal{U}(I_j^\Delta)} = \int_{I_j^\Delta} \Bigg(AE\left(\frac{e_j^\Delta - e_{j-1}^\Delta}{h_{I_j^\Delta}}\right)^2 \\ + c\left(\frac{e_j^\Delta + e_{j-1}^\Delta}{2} + \frac{e_j^\Delta - e_{j-1}^\Delta}{h_{I_j^\Delta}}(x - x_{j+1/2})\right)^2\Bigg)\end{aligned} \tag{5.2.55c}$$

and using the notation

$$\rho_1^{I_j^\Delta} \stackrel{\mathrm{def}}{=} \tfrac{1}{2}(e_j^\Delta + e_{j-1}^\Delta), \qquad \rho_2^{I_j^\Delta} \stackrel{\mathrm{def}}{=} \frac{1}{h_{I_j^\Delta}}(e_j^\Delta - e_{j-1}^\Delta) \tag{5.2.55d}$$

we get

$$\|\mathcal{I}^1_{I_j^\Delta} e_{S^p_\Delta}\|_{\mathcal{U}(I_j^\Delta)} \le \mathcal{C}\sqrt{h_{I_j^\Delta}}\max(|\rho_1^{I_j^\Delta}|, |\rho_2^{I_j^\Delta}|) \le \mathcal{C}(h_{I_j^\Delta})^\alpha \|\tilde{e}^{loc}_{S^p(I_j^\Delta)}\|_{\mathcal{U}(I_j^\Delta)} \tag{5.2.55e}$$

where the second inequality follows from the assumption (5.2.53a). Combining the above, we get

$$\|e_{S^p_\Delta} - \hat{e}^{\mathrm{Dir}}_{I_j^\Delta}\|_{\mathcal{U}(I_j^\Delta)} \le \|\mathcal{I}^1_{I_j^\Delta} e_{S^p_\Delta}\|_{\mathcal{U}(I_j^\Delta)} \le \mathcal{C}(h_{I_j^\Delta})^\alpha \|\tilde{e}^{loc}_{S^p(I_j^\Delta)}\|_{\mathcal{U}(I_j^\Delta)}. \tag{5.2.55f}$$

Further, using (5.2.54b), we get

$$\|e_{S^p_\Delta} - \hat{e}^{\mathrm{Dir}}_{I_j^\Delta}\|_{\mathcal{U}(I_j^\Delta)} \le \mathcal{C}(h_{I_j^\Delta})^\alpha \sqrt{\|\hat{e}^{\mathrm{Dir}}_{I_j^\Delta}\|^2_{\mathcal{U}(I_j^\Delta)} + \|e_{S^p_\Delta} - \hat{e}^{\mathrm{Dir}}_{I_j^\Delta}\|^2_{\mathcal{U}(I_j^\Delta)}} \tag{5.2.56a}$$

from which it follows that

$$\|e_{S^p_\Delta} - \hat{e}^{\mathrm{Dir}}_{I_j^\Delta}\|_{\mathcal{U}(I_j^\Delta)} \le \frac{\mathcal{C}(h_{I_j^\Delta})^\alpha}{1+\frac{1}{2}\mathcal{C}(h_{I_j^\Delta})^{2\alpha}} \|\hat{e}^{\mathrm{Dir}}_{I_j^\Delta}\|_{\mathcal{U}(I_j^\Delta)} \le \mathcal{C}(h_{I_j^\Delta})^\alpha \|\hat{e}^{\mathrm{Dir}}_{I_j^\Delta}\|_{\mathcal{U}(I_j^\Delta)}. \tag{5.2.56b}$$

Using (5.2.56b) into the identity

$$\|e_{S^p_\Delta}\|^2_{\mathcal{U}(I_j^\Delta)} = \sqrt{\|\hat{e}^{\mathrm{Dir}}_{I_j^\Delta}\|^2_{\mathcal{U}(I_j^\Delta)} + \|e_{S^p_\Delta} - \hat{e}^{\mathrm{Dir}}_{I_j^\Delta}\|^2_{\mathcal{U}(I_j^\Delta)}} \tag{5.2.56c}$$

(5.2.53b) follows. □

Let us now combine the results of Theorem 5.2.4 and Theorem 5.2.5.

Corollary 5.2.2 *Assume that the input data satisfy (5.2.2a), and that the meshes Δ are such that (5.2.53a) holds. Then, for sufficiently small $h_{I_j^\Delta}$, we have*

$$\|\hat{e}^{\mathrm{Dir}}_{I_j^\Delta}\|_{\mathcal{U}(I_j^\Delta)} \le \|e_{S^p_\Delta}\|_{\mathcal{U}(I_j^\Delta)} \le \left(1+\mathcal{C}(h_{I_j^\Delta})^{2\alpha}\right)\|\hat{e}^{\mathrm{Dir}}_{I_j^\Delta}\|_{\mathcal{U}(I_j^\Delta)} \tag{5.2.57a}$$

and hence

$$\left(1-\mathcal{C}(h_{I_j^\Delta})^{2\alpha}\right) \le \kappa^{\mathrm{Dir}}_{I_j^\Delta} = \frac{\|\hat{e}^{\mathrm{Dir}}_{I_j^\Delta}\|_{\mathcal{U}(I_j^\Delta)}}{\|e_{S^p_\Delta}\|_{\mathcal{U}(I_j^\Delta)}} \le 1. \tag{5.2.57b}$$

It follows that when the pollution is negligible, the error indicators $\eta^{\mathrm{Dir}}_{I_j^\Delta}$ are asymptotically exact, namely, $\eta^{\mathrm{Dir}}_{I_j^\Delta} \approx \|e_{S^p_\Delta}\|_{\mathcal{U}(I_j^\Delta)}$, as h^Δ tends to zero.

Let us underline that the above corollary holds for the exact indicators $\eta^{\mathrm{Dir}}_{I_j^\Delta}$, and for the computed error indicators $\eta^{\mathrm{Dir}}_{I_j^\Delta,S}$ if the error $\sqrt{(\eta^{\mathrm{Dir}}_{I_j^\Delta})^2 - (\eta^{\mathrm{Dir}}_{I_j^\Delta,S})^2}$ is of higher order than h^α.

Remark 5.2.10 Let us underline that *in many cases the pollution error is not negligible*, and we may have $\eta^{\mathrm{Dir}}_{I_j^\Delta} \ll \|e_{S^p_\Delta}\|_{\mathcal{U}(I_j^\Delta)}$. This is why $\eta^{\mathrm{Dir}}_{I_j^\Delta}$ is called an indicator and not an estimator for $\|e_{S^p_\Delta}\|_{\mathcal{U}(I_j^\Delta)}$. On the other hand, the estimator $\mathcal{E}^{\mathrm{Dir}}$ is a reliable estimate of $\|e_{S^p_\Delta}\|_{\mathcal{U}}$, even if there is severe pollution in the majority of the elements.

Remark 5.2.11 In many cases an adaptive procedure creates a sequence of meshes for which the pollution is negligible asymptotically, and in that case the element error indicators $\eta^{\mathrm{Dir}}_{I_j^\Delta}$ can also be understood as estimates of $\|e_{S^p_\Delta}\|_{\mathcal{U}(I_j^\Delta)}$.

A question that will arise later, in connection with the problem of a-posteriori estimation of the pollution error, is about the relation between $\eta^{\mathrm{Dir}}_{I_j^\Delta}$ and $\min_{v\in S^p(I_j^\Delta)}\|u_{EX}-v\|_{\mathcal{U}(I_j^\Delta)}$, the energy norm of the error in the best approximation. As we have proven above, *when the pollution is negligible*

$$\eta^{\mathrm{Dir}}_{I_j^\Delta} = \left(1+\mathcal{C}\left(h_{I_j^\Delta}\right)^{2\alpha}\right) \min_{v\in S^p(I_j^\Delta)} \|u_{EX}-v\|_{\mathcal{U}(I_j^\Delta)} \tag{5.2.58a}$$

with $\alpha > 0$. In the more general case that the pollution cannot be neglected, one would think that we should expect that

$$\mathcal{C}_1\eta^{\mathrm{Dir}}_{I_j^\Delta} \le \min_{v\in S^p(I_j^\Delta)} \|u_{EX}-v\|_{\mathcal{U}(I_j^\Delta)} \le \mathcal{C}_2\eta^{\mathrm{Dir}}_{I_j^\Delta} \tag{5.2.58b}$$

with $\mathcal{C}_1$, $\mathcal{C}_2$ of comparable magnitude close to one. However, as it can be seen from the following counterexample, this is not the case.

Counterexample 5.2.1. Let us consider the case $AE \equiv 1$, $c \equiv 1$, $p = 1$, and let us assume that in the neighborhood of I_j^Δ the exact solution is approximated with negligible accuracy by the quadratic monomial, namely

$$u_{EX}(x) \approx (x - x^\Delta_{j-1/2})^2. \tag{5.2.59a}$$

It follows that

$$\mathcal{I}^1_{I_j^\Delta} u_{EX} \equiv \tfrac{1}{4}\left(h_{I_j^\Delta}\right)^2 \tag{5.2.59b}$$

and, if the meshes Δ are such that the pollution is negligible, we also have

$$u_{S^1_\Delta}|_{I_j^\Delta} \approx \mathcal{I}^1_{I_j^\Delta} u_{EX} \equiv \tfrac{1}{4}\left(h_{I_j^\Delta}\right)^2 \tag{5.2.59c}$$

and hence

$$\eta^{\mathrm{Dir}}_{I_j^\Delta} \approx \mathcal{C}\left(h_{I_j^\Delta}\right)^{3/2} \approx \min_{v\in S^p(I_j^\Delta)} \|u_{EX}-v\|_{\mathcal{U}(I_j^\Delta)}. \tag{5.2.59d}$$

Let us now assume that, for the employed meshes, *the pollution is not negligible and that the finite element solution $u_{S^1_\Delta}|_{I_j^\Delta}$, is shifted away from $\mathcal{I}^1_{I_j^\Delta} u_{EX}$ by a constant $\mathcal{C}_0$,* namely

$$u_{S^1_\Delta}|_{I_j^\Delta} = \mathcal{I}^1_{I_j^\Delta} u_{EX} + \mathcal{C}_0 = \tfrac{1}{4}\left(h_{I_j^\Delta}\right)^2 + \mathcal{C}_0 \tag{5.2.60a}$$

where $\mathscr{C}_0$ is independent of $h_{I_j^\Delta}$. This can happen when the meshes are very nonuniform such as, for example, the meshes employed in Example 4.4.1 with $h_{I_1^\Delta} = h_0$ fixed, while $h_{I_j^\Delta} = h = (1-h_0)/N(\Delta) \to 0$, $j = 2, \ldots, N(\Delta)$. Then, the interior residual $r_{I_j^\Delta}$ is

$$r_{I_j^\Delta}(x) = e_{S_\Delta^1}(x) - \frac{d^2}{dx^2}\big(e_{S_\Delta^1}\big)(x) = \Big(\big(x - x_{j-1/2}^\Delta\big)^2 - \frac{1}{4}\big(h_{I_j^\Delta}\big)^2 - \mathscr{C}_0\Big) - 2. \qquad (5.2.60b)$$

Letting $\mathscr{C}_0 = -2$, and recalling that

$$\eta_{I_j^\Delta}^{\mathrm{Dir}} \le \mathscr{C}_{I_j^\Delta} h_{I_j^\Delta} \sqrt{\int_{I_j^\Delta} \big(r_{I_j^\Delta}\big)^2} \qquad (5.2.60c)$$

we obtain

$$\eta_{I_j^\Delta}^{\mathrm{Dir}} \le \mathscr{C}_{I_j^\Delta} h_{I_j^\Delta} \sqrt{\int_{I_j^\Delta} \Big((x - x_{j-1/2}^\Delta)^2 - \frac{1}{4}\big(h_{I_j^\Delta}\big)^2\Big)^2} \approx \mathscr{C}\big(h_{I_j^\Delta}\big)^{5/2} \qquad (5.2.60d)$$

while

$$\min_{v \in S^p(I_j^\Delta)} \|u_{EX} - v\|_{\mathscr{U}(I_j^\Delta)} \approx \mathscr{C}\big(h_{I_j^\Delta}\big)^{3/2}. \qquad (5.2.60e)$$

Hence we have constructed a case for which

$$\eta_{I_j^\Delta}^{\mathrm{Dir}} \approx \mathscr{C}\big(h_{I_j^\Delta}\big)^{5/2} \ll \min_{v \in S^p(I_j^\Delta)} \|u_{EX} - v\|_{\mathscr{U}(I_j^\Delta)} \approx \mathscr{C}\big(h_{I_j^\Delta}\big)^{3/2}. \qquad (5.2.60f)$$

This means that the element error indicators $\eta_{I_j^\Delta}^{\mathrm{Dir}}$, can be much smaller (by an order of magnitude in $h_{I_j^\Delta}$) than $\min_{v \in S^p(I_j^\Delta)} \|u_{EX} - v\|_{\mathscr{U}(I_j^\Delta)}$, the energy norm of the error in the best approximation. Hence, unless some additional assumptions are imposed, the equivalence (5.2.58b) does not hold.

The construction of the above counterexample was based on the assumption that $u_{S_\Delta^1}|_{I_j^\Delta} - \mathcal{I}_{I_j^\Delta}^1 u_{EX} \equiv \mathscr{C}_0$, with $\mathscr{C}_0$ independent of the mesh. This cannot happen for quasi-uniform meshes, which are the meshes most often employed in practice.

Let us now prove that $\mathscr{E}^{\mathrm{Dir}}$ satisfies Assumptions A, B, C and hence it can be analyzed by employing Theorem 5.5.1.

Theorem 5.2.6 *The estimator $\mathscr{E}^{\mathrm{Dir}}$ satisfies Assumptions A, B, C given in Section 5.1.*

Proof Below we will write $\hat{e}_{I_j^\Delta}^{\mathrm{Dir}} = \hat{e}_{I_j^\Delta}^{\mathrm{Dir}}(u_{S_\Delta^p}, f)$ to denote the dependence of $\hat{e}_{I_j^\Delta}^{\mathrm{Dir}}$ on the admissible pair $(u_{S_\Delta^p}, f)$.

Assumption A: Recall that $\hat{e}_{I_j^\Delta}^{\mathrm{Dir}}$ is the solution of the variational problem:

Find $\hat{e}_{I_j^\Delta}^{\mathrm{Dir}} \in \mathscr{U}_0(I_j^\Delta)$ such that

$$\mathscr{B}_{I_j^\Delta}\big(\hat{e}_{I_j^\Delta}^{\mathrm{Dir}}, v\big) = \int_{I_j^\Delta} fv - \mathscr{B}_{I_j^\Delta}\big(u_{S_\Delta^p}, v\big) = \int_{I_j^\Delta} r_{I_j^\Delta} v \quad \forall v \in \mathscr{U}_0(I_j^\Delta) \qquad (5.2.61a)$$

where $r_{I_j^\Delta} = r_{I_j^\Delta}(u_{S_\Delta^p}, f)$ is the interior residual corresponding to the admissible pair $(u_{S_\Delta^p}, f)$. Letting $v = \hat{e}_{I_j^\Delta}^{\mathrm{Dir}} \in \mathcal{U}_0(I_j^\Delta)$, we get

$$\|\hat{e}_{I_j^\Delta}^{\mathrm{Dir}}\|_{\mathcal{U}(I_j^\Delta)}^2 = \int_{I_j^\Delta} f\hat{e}_{I_j^\Delta}^{\mathrm{Dir}} - \mathcal{B}_{I_j^\Delta}\left(u_{S_\Delta^p}, \hat{e}_{I_j^\Delta}^{\mathrm{Dir}}\right) \tag{5.2.61b}$$

and employing the Cauchy–Schwarz inequality we get

$$\|\hat{e}_{I_j^\Delta}^{\mathrm{Dir}}\|_{\mathcal{U}(I_j^\Delta)}^2 \leq \|f\|_{L^2(I_j^\Delta)}\|\hat{e}_{I_j^\Delta}^{\mathrm{Dir}}\|_{L^2(I_j^\Delta)} + \|u_{S_\Delta^p}\|_{\mathcal{U}(I_j^\Delta)}\|\hat{e}_{I_j^\Delta}^{\mathrm{Dir}}\|_{\mathcal{U}(I_j^\Delta)}. \tag{5.2.61c}$$

Recalling that $\hat{e}_{I_j^\Delta}^{\mathrm{Dir}}(x_{j-1}^\Delta) = 0$, we have

$$\|\hat{e}_{I_j^\Delta}^{\mathrm{Dir}}\|_{L^2(I_j^\Delta)} \leq \mathcal{C} h_{I_j^\Delta}\|\hat{e}_{I_j^\Delta}^{\mathrm{Dir}}\|_{\mathcal{U}(I_j^\Delta)} \tag{5.2.61d}$$

and we get

$$\eta_{I_j^\Delta}^{\mathrm{Dir}}(u_{S_\Delta^p}, f) = \|\hat{e}_{I_j^\Delta}^{\mathrm{Dir}}(u_{S_\Delta^p}, f)\|_{\mathcal{U}(I_j^\Delta)} \leq \mathcal{C}\sqrt{(h_{I_j^\Delta})^2\|f\|_{L^2(I_j^\Delta)}^2 + \|u_{S_\Delta^p}\|_{\mathcal{U}(I_j^\Delta)}^2} \tag{5.2.61e}$$

which proves that $\mathcal{E}^{\mathrm{Dir}}$ satisfies Assumption A.

Assumption B: Let $(v_{S_\Delta^p}, s)$ and $(\varphi_{S_\Delta^p}, g)$ be admissible pairs; then

$$(w_{S_\Delta^p}, t) = (v_{S_\Delta^p} + \varphi_{S_\Delta^p}, s + g) \tag{5.2.62a}$$

is also an admissible pair, and we have

$$\hat{e}_{I_j^\Delta}^{\mathrm{Dir}}(w_{S_\Delta^p}, t) = \hat{e}_{I_j^\Delta}^{\mathrm{Dir}}(v_{S_\Delta^p}, s) + \hat{e}_{I_j^\Delta}^{\mathrm{Dir}}(\varphi_{S_\Delta^p}, g). \tag{5.2.62b}$$

It follows that

$$\begin{aligned}\mathcal{E}^2(w_{S_\Delta^p}, t) &= \mathcal{E}^2(v_{S_\Delta^p}, s) + \mathcal{E}^2(\varphi_{S_\Delta^p}, g)\\ &\quad + 2\sum_{j=j_1}^{j_2} \mathcal{B}_{I_j^\Delta}\left(\hat{e}_{I_j^\Delta}^{\mathrm{Dir}}(v_{S_\Delta^p}, s), \hat{e}_{I_j^\Delta}^{\mathrm{Dir}}(\varphi_{S_\Delta^p}, g)\right)\end{aligned} \tag{5.2.62c}$$

where we employed the notation

$$\mathcal{E}^2(v_{S_\Delta^p}, s) \stackrel{\mathrm{def}}{=} \left(\mathcal{E}_{S(\bar{x},\gamma H)}^{\mathrm{Dir}}(v_{S_\Delta^p}, s)\right)^2 = \sum_{j=j_1}^{j_2} \|\hat{e}_{I_j^\Delta}^{\mathrm{Dir}}(v_{S_\Delta^p}, s)\|_{\mathcal{U}(I_j^\Delta)}^2. \tag{5.2.62d}$$

From the properties of the inner product, it follows that

$$\left|2\mathcal{B}_{I_j^\Delta}\left(\hat{e}_{I_j^\Delta}^{\mathrm{Dir}}(v_{S_\Delta^p}, s), \hat{e}_{I_j^\Delta}^{\mathrm{Dir}}(\varphi_{S_\Delta^p}, g)\right)\right| \leq \alpha\|\hat{e}_{I_j^\Delta}^{\mathrm{Dir}}(v_{S_\Delta^p}, s)\|_{\mathcal{U}(I_j^\Delta)}^2 + \frac{1}{\alpha}\|\hat{e}_{I_j^\Delta}^{\mathrm{Dir}}(\varphi_{S_\Delta^p}, g)\|_{\mathcal{U}(I_j^\Delta)}^2 \tag{5.2.62e}$$

for any α, $0 < \alpha < 1$. Using (5.2.62e) in (5.2.62c) we get

$$\left|\mathscr{E}^2(v_{S^p_\Delta} + \varphi_{S^p_\Delta}, s+g) - \mathscr{E}^2(v_{S^p_\Delta}, g)\right| \le \mathscr{E}^2(\varphi_{S^p_\Delta}, g) + \alpha\mathscr{E}^2(v_{S^p_\Delta}, s) + \frac{1}{\alpha}\mathscr{E}^2(\varphi_{S^p_\Delta}, g) \qquad (5.2.62f)$$

and Assumption B has been proven.

Assumption C: This assumption is clearly satisfied because for any $w \in S^p_\Delta$, we have

$$r_{I_j^\Delta}(\pi_{S^p(I_j^\Delta)}(w), \mathfrak{D}w|_{I_j^\Delta}) \equiv 0 \quad \Longrightarrow \quad \hat{e}^{\mathrm{Dir}}_{I_j^\Delta}(\pi_{S^p(I_j^\Delta)}(w), \mathfrak{D}w|_{I_j^\Delta}) \equiv 0. \qquad (5.2.63)$$

□

In this section, we analyzed the estimator $\mathscr{E}^{\mathrm{Dir}}$ based on the element-residual problems with homogeneous Dirichlet boundary conditions. The conclusions of this analysis are:

1. The quantity $\mathscr{E}^{\mathrm{Dir}}$ is a lower estimator for $\|e_{S^p_\Delta}\|_{\mathcal{U}}$. Further, it is possible to determine a constant $\mathcal{K}^{\mathrm{Dir}}$, such that we get the upper estimate $\|e_{S^p_\Delta}\|_{\mathcal{U}} \le \mathcal{K}^{\mathrm{Dir}}\mathscr{E}^{\mathrm{Dir}}$.
2. The element error-indicator $\eta^{\mathrm{Dir}}_{I_j^\Delta}$ is a lower estimator for the energy norm of the error in each element, namely, $\eta^{\mathrm{Dir}}_{I_j^\Delta} \le \|e_{S^p_\Delta}\|_{\mathcal{U}(I_j^\Delta)}$.
3. When the pollution error is negligible, we practically have asymptotical exactness, $\eta^{\mathrm{Dir}}_{I_j^\Delta} \approx \|e_{S^p_\Delta}\|_{\mathcal{U}(I_j^\Delta)}$, in each element.
4. The estimator $\mathscr{E}^{\mathrm{Dir}}$ satisfies Assumptions A, B, and C of Section 5.1.

Problems

P5.2.1. Consider the singular model problem with $f \equiv 1$, $l = 1$, $\alpha_2 = a_2 = 0$, and corresponding exact solution

$$u_{EX}(x) = \frac{x^{2-\vartheta}}{2-\vartheta} - \frac{x^{1-\vartheta}}{1-\vartheta}. \qquad (5.2.64a)$$

Let Δ be the two element mesh with $I_1^\Delta = (0, h_{I_1^\Delta})$ and $I_2^\Delta = (h_{I_1^\Delta}, 1)$. Let

$$\mathrm{I}\left(\|u_{EX}\|^2_{\mathcal{U}(I_j^\Delta)}; \{\tilde{x}_l, \tilde{w}_l\}_{l=1}^{nintp}\right) = \sum_{l=1}^{nintp} \tilde{w}_l \left(F_{I_j^\Delta}(\tilde{x}_l)\right)^{\vartheta} \left(\frac{d}{dx}(u_{EX})(F_{I_j^\Delta}(\tilde{x}_l))\right)^2 \left(\frac{h_{I_j^\Delta}}{2}\right) \qquad (5.2.64b)$$

be the value of the energy in the element I_j^Δ, computed by numerical integration where $\tilde{x}_l$, $\tilde{w}_l$, $l = 1, \ldots, nintp$, are the integration points and the corresponding weights defined over the master element $\tilde{I} = (-1, 1)$, and $nintp$ denotes the number of integration points. Estimate the minimal value of $nintp$ for the Gauss–Legendre integration rule required to compute $\|u_{EX}\|^2_{\mathcal{U}(I_j^\Delta)}$, in the first and the second element ($j = 1$ and 2), with 1% accuracy, namely

$$\left|\|u_{EX}\|^2_{\mathcal{U}(I_j^\Delta)} - \mathrm{I}\left(\|u_{EX}\|^2_{\mathcal{U}(I_j^\Delta)}; \{\tilde{x}_l, \tilde{w}_l\}_{l=1}^{nintp}\right)\right| \le 0.01\|u_{EX}\|^2_{\mathcal{U}(I_j^\Delta)}. \qquad (5.2.64c)$$

Show that for ϑ close to $\frac{1}{2}$, it is practically impossible to achieve the desired accuracy in I_1^Δ.

P5.2.2. Consider the regular model problem with $AE \equiv 1$, $c \equiv 0$, $\alpha_2 = a_2 = 0$, $f \equiv \sin(\pi x/2l)$. Let Δ be a uniform mesh of linear elements ($p = 1$). Show that for the particular choice of f we have

$$\lim_{h\to 0} \frac{\eta^{\text{Dir}}_{I_j^\Delta}}{h_{I_j^\Delta}\|r_{I_j^\Delta}\|_{L^2(I_j^\Delta)}} \neq \frac{1}{\pi}. \tag{5.2.65a}$$

Nevertheless, for every mesh Δ there exists an f, which depends on Δ, such that

$$\eta^{\text{Dir}}_{I_j^\Delta} = \frac{1}{\pi} h_{I_j^\Delta}\|r_{I_j^\Delta}\|_{L^2(I_j^\Delta)}. \tag{5.2.65b}$$

P5.2.3. Consider the regular model problem employed in P5.2.2 and assume that the finite element solution $u_{S^p_\Delta}$ is computed using a uniform mesh Δ of quadratic elements ($p = 2$). Compute the $\lim_{h\to 0}(\eta^{\text{Dir}}_{I_j^\Delta}/h_{I_j^\Delta}\|r_{I_j^\Delta}\|_{L^2(I_j^\Delta)})$.

P5.2.4. Let $AE \equiv 1$, $c \equiv 0$, and $r_{I_j^\Delta} = e^{-\gamma(x - x^\Delta_{j-1})}$. Determine the $\lim_{h\to 0}(\eta^{\text{Dir}}_{I_j^\Delta}/h_{I_j^\Delta}\|r_{I_j^\Delta}\|_{L^2(I_j^\Delta)})$, and compare it to the value $1/\pi$.

5.2* Analysis of the Dirichlet element residual estimator in two dimensions

We will now analyze the estimator $\mathscr{E}^{\text{Dir}}$ in the two-dimensional case. As in the one-dimensional case, we will employ the Dirichlet element-residual problems:

Find $\hat{e}^{\text{Dir}}_\tau \in \mathscr{U}_0(\tau)$ such that

$$\mathscr{B}_\tau(\hat{e}^{\text{Dir}}_\tau, v) = \int_\tau r_\tau v = \int_\tau \underbrace{(f + \mathfrak{D}u_{S^p_\Delta})}_{r_\tau} v \quad \forall v \in \mathscr{U}_0(\tau) \tag{5.2*.1}$$

where $\mathfrak{D}$ is the differential operator for the two-dimensional problem under consideration.

Analogously as in the one-dimensional case, the existence and uniqueness of $\hat{e}^{\text{Dir}}_\tau$ can be proven for all admissible input data, namely:

Theorem 5.2*.1 *A unique $\hat{e}^{\text{Dir}}_\tau$ exists for all admissible classes of input data for which u_{EX} exists.*

Proof As in Remark 5.2.1, letting

$$\tilde{\mathscr{U}}_0(\tau) \stackrel{\text{def}}{=} \{v \in \mathscr{U} \mid \exists w \in \mathscr{U}_0(\tau)\colon v|_\tau \equiv w, \text{ and } v|_{\Omega-\tau} \equiv 0\} \tag{5.2*.2a}$$

we may reformulate the element residual problem (5.2*.1) as the global problem:

Find $\tilde{\hat{e}}^{\text{Dir}}_\tau \in \tilde{\mathscr{U}}_0(\tau)$ such that

$$\mathscr{B}(\tilde{\hat{e}}^{\text{Dir}}_\tau, v) = \mathscr{R}(v) \quad \forall v \in \tilde{\mathscr{U}}_0(\tau). \tag{5.2*.2b}$$

Now, because $\mathscr{R}$ is bounded on $\mathscr{U}$, it is also bounded on $\tilde{\mathscr{U}}_0(\tau) \subset \mathscr{U}$, and hence $\tilde{\hat{e}}^{\text{Dir}}_\tau$ exists and it is unique, and we have $\hat{e}^{\text{Dir}}_\tau = \tilde{\hat{e}}^{\text{Dir}}_\tau|_\tau$. □

As in the one-dimensional case, we will denote by η_τ^{Dir} the Dirichlet element error indicator, and by $\mathscr{E}^{\mathrm{Dir}}$ the Dirichlet estimator, namely

$$\eta_\tau^{\mathrm{Dir}} = \|\hat{e}_\tau^{\mathrm{Dir}}\|_{\mathcal{U}(\tau)}, \qquad \mathscr{E}^{\mathrm{Dir}} = \sqrt{\sum_{\tau\in\Delta} (\eta_\tau^{\mathrm{Dir}})^2}. \tag{5.2*.3}$$

As in the one-dimensional case, $\mathscr{E}^{\mathrm{Dir}}$ is a lower estimator for $\|e_{S_\Delta^p}\|_{\mathcal{U}}$.

Theorem 5.2*.2 *For all admissible classes of input data we have*

$$\mathscr{E}^{\mathrm{Dir}} \le \|e_{S_\Delta^p}\|_{\mathcal{U}}. \tag{5.2*.4}$$

Proof As in the one-dimensional case, letting $\hat{e}^{\mathrm{Dir}} \in \mathcal{U}$ be defined by $\hat{e}^{\mathrm{Dir}}|_\tau = \hat{e}_\tau^{\mathrm{Dir}}$, $\tau \in \Delta$, we have

$$\mathscr{E}^{\mathrm{Dir}} = \|\hat{e}^{\mathrm{Dir}}\|_{\mathcal{U}} = \frac{\mathcal{R}(\hat{e}^{\mathrm{Dir}})}{\|\hat{e}^{\mathrm{Dir}}\|_{\mathcal{U}}} \le \sup_{v\in\mathcal{U}} \frac{\mathcal{R}(v)}{\|v\|_{\mathcal{U}}} = \|e_{S_\Delta^p}\|_{\mathcal{U}}. \tag{5.2*.5}$$

□

As we have already discussed in Section 5.1*, *in two or higher dimensions* $\mathscr{E}^{\mathrm{Dir}}$ *may be too small.* For example, for the piecewise linear approximation of the Laplacian on a mesh of triangles; $\mathfrak{D} = -\Delta$, $p = 1$, $f \equiv 0$, we get $\mathscr{E}^{\mathrm{Dir}} = 0$. *Hence* $\mathscr{E}^{\mathrm{Dir}}$ *may be worthless as an estimate.* However, for elements of even degree ($p = 2, 4, \ldots$), and meshes of rectangles, there exists a constant $\mathcal{K}^{\mathrm{Dir}}$, depending only on the class of input data, such that

$$\|e_{S_\Delta^p}\|_{\mathcal{U}} \le \mathcal{K}^{\mathrm{Dir}} \mathscr{E}^{\mathrm{Dir}} \tag{5.2*.6}$$

and $\mathscr{E}^{\mathrm{Dir}}$ has good meaning as an estimator of $\|e_{S_\Delta^p}\|_{\mathcal{U}}$ (see [Babuška and Yu, 1987] and [Yu, 1991b], and [Baehmann, Shephard and Flaherty, 1992]).

We will now prove a local version of the upper bound (5.2*.6) by employing the asymptotic analysis of Section 4. For concreteness we will consider here the Poisson equation $\Delta u = f$.

Theorem 5.2*.3 *Assume that* Δ *are locally uniform meshes of squares,* S_Δ^p *is the finite element space obtained by employing bi-p element spaces, and that Assumptions A, B, C given in Section 5.1 hold. Then for* $p \ge 2$*, even, there exists a constant* $\mathcal{K}(p)$ *such that*

$$\|e_{S_\Delta^p}\|_{\mathcal{U}(\tau)} \le \mathcal{K}(p)\mathscr{E}^{\mathrm{Dir}}(\tau) \tag{5.2*.7}$$

in the asymptotic sense of Section 4.7

Proof Noting that bi-p elements reproduce all monomials $x_1^l x_2^m$, where $0 \le l, m \le p$, to analyze the $(p+1)$ asymptotics of the error it is sufficient to let

$$U_{EX}^{\bar{x},(p+1)} = a_1(x_1 - \bar{x}_1)^{p+1} + a_2(x_2 - \bar{x}_2)^{p+1}. \tag{5.2*.8a}$$

The asymptotic finite element solution is

$$\tilde{U}_{S_\Delta^p}|_\tau = a_1\mathcal{I}_\tau^p((x_1 - \bar{x}_1)^{p+1}) + a_2\mathcal{I}_\tau^p((x_2 - \bar{x}_2)^{p+1}) \tag{5.2*.8b}$$

and the corresponding error

$$\Psi_{S_\Delta^p}^{\bar{x},H}|_\tau = a_1\Psi_1^{(p+1)} + a_2\Psi_2^{(p+1)} \tag{5.2*.8c}$$

where

$$\Psi_i^{(p+1)} = x_i^{p+1} - \mathcal{I}_\tau^p x_i^{p+1}. \tag{5.2*.8d}$$

Note that $\Psi_i^{(p+1)}$ varies only in the ith direction, and is often referred to as the ith tunnel function.

It follows that the asymptotic form of the error indicator $\hat{e}_\tau^{\mathrm{Dir}}$ is

$$\hat{e}_\tau^{\mathrm{Dir}} = a_1\Phi_1^{(p+1)} + a_2\Phi_2^{(p+1)} \tag{5.2*.9a}$$

where $\Phi_i^{(p+1)}$ is the solution of the element problem:

Find $\Phi_i^{(p+1)} \in \mathcal{U}_0(\tau)$ such that

$$\mathcal{B}_\tau(\Phi_i^{(p+1)}, v) = \mathcal{B}_\tau(\Psi_i^{(p+1)}, v) \quad \forall v \in \mathcal{U}_0(\tau). \tag{5.2*.9b}$$

Now let

$$\mathcal{K}(p) = \sqrt{\max_{a_1,a_2} \frac{\|a_1\Psi_1^{(p+1)} + a_2\Psi_2^{(p+1)}\|^2_{\mathcal{U}(\tau)}}{\|a_1\Phi_1^{(p+1)} + a_2\Phi_2^{(p+1)}\|^2_{\mathcal{U}(\tau)}}}. \tag{5.2*.10}$$

For p even, $\mathcal{K}(p)$ is determined by solving a generalized eigenvalue problem. For p odd, $\mathcal{K}(p)$ does not exist (see P5.2*.1). □

Remark 5.2*.1 In the case of p even we can construct an explicit estimator in terms of the element interior residuals, namely

$$\|e_{S_\Delta^p}\|_{\mathcal{U}} \le \mathcal{E}^{\mathrm{EXPL}} = \mathcal{C}\sqrt{\sum_{\tau\in\Delta}\int_\tau (r_\tau)^2}. \tag{5.2*.11}$$

See [Yu, 1991b]. In the case of p odd, we can construct an explicit estimator which employs only the jumps

$$\|e_{S_\Delta^p}\|_{\mathcal{U}} \le \mathcal{E}^{\mathrm{EXPL}} = \mathcal{C}\sqrt{\sum_{\epsilon\in\mathcal{E}(\Delta)}\int_\epsilon (J_\epsilon^\Delta)^2}. \tag{5.2*.12}$$

This has been proven in [Yu, 1991a] for rectangular elements. For $p = 1$ the explicit estimator $\mathcal{E}^{\mathrm{EXPL}}$ is addressed in [Carstensen and Verfürth, 1999] and [Kunert and Verfürth, 1999] by employing the arguments introduced in [Babuška and Miller, 1981, 1987]; see also [Rodriguez, 1994]. The problem of the determination of the constant $\mathcal{C}$ has been addressed for special meshes of linear triangles by [Carstensen and Funken, 2000a], where it was shown that the constant needed to obtain the guaranteed upper estimator (5.2*.12) leads to gross overestimation ($\kappa \approx 70$) in many problems of practical interest.

Remark 5.2*.2 In the case of the Laplacian $\Delta u_{EX} \equiv 0$, meshes of square elements, and p even, we can also construct an estimator based on the idea of dimensional splitting. Let

$$r_{\tau,i} = \frac{\partial^2}{\partial x_i^2}(u_{S^p_\Delta}|_\tau) \quad i = 1, 2 \tag{5.2*.13a}$$

and let $\hat{e}^{\mathrm{Dir}}_{\tau,i}$ be the solution of the problem:

Find $\hat{e}^{\mathrm{Dir}}_{\tau,i} \in \mathcal{U}_{0,i}(\tau)$ such that

$$\mathcal{B}_{\tau,i}(\hat{e}^{\mathrm{Dir}}_{\tau,i}, v) = \int_{\tau,i} r_{\tau,i} v \quad \forall v \in \mathcal{U}_{0,i}(\tau). \tag{5.2*.13b}$$

We then have

$$\|e_{S^p_\Delta}\|_{\mathcal{U}} \approx \mathcal{E}^{\mathrm{Dir}}_{split} \stackrel{\mathrm{def}}{=} \sqrt{\sum_{\tau\in\Delta}\left(\sum_{i=1}^{2} \|\hat{e}^{\mathrm{Dir}}_{\tau,i}\|^2_{\mathcal{U}_{0,i}(\tau)}\right)}. \tag{5.2*.13c}$$

Remark 5.2*.3 The above estimator can be extended to mapped meshes of squares.

Remark 5.2*.4 The Dirichlet estimators $\mathcal{E}^{\mathrm{Dir}}$ and $\mathcal{E}^{\mathrm{Dir}}_{split}$ were implemented in NFEARS, a research program for the adaptive solution of nonlinear elliptic equations which was developed at the University of Maryland in the 1980s. The NFEARS was based on a tree data structure which allows local refinements of the elements without imposing a constraint on the relative refinement of neighboring elements, and employs biquadratic elements. The fact that the Dirichlet estimator does not require the computation of the jumps makes it very suitable for NFEARS because, for the employed data structure, the determination of the elements connected to an edge is rather cumbersome.

We will now give an example which illustrates the accuracy of the above Dirichlet and explicit estimators in a concrete case.

Example 5.2*.1 Let us consider the Neumann problem for the Laplace equation in the rectangular domain $\Omega = (-1, 1) \times (0, 1)$ with boundary condition $\partial u/\partial n = g = \nabla u_{EX} \cdot \mathbf{n}_\Gamma$ where

$$u_{EX}(r, \theta) = \sqrt{r} \sin\left(\frac{\theta}{2}\right) \tag{5.2*.14}$$

where (r, θ) denotes the polar coordinates with pole at $O' = (0, -a)$, $a = 0.1$, as shown in Fig. 5.2*.1. Note that u_{EX} has an algebraic singularity at O' outside Ω, and it is smooth in Ω and its vicinity, the domain Ω is convex, and Neumann boundary conditions are applied on the entire boundary. It can be proven that for quasi-uniform meshes of degree $p \leq 2$ the pollution is negligible, and hence the element error indicators η_τ can be interpreted as estimators of the element energy norms of the error $\|e_{S^p_\Delta}\|_{\mathcal{U}(\tau)}$. Here we computed the finite element solution $u_{S^p_\Delta}$, using the eight-, fourteen-, and twenty-element mesh of biquadratic elements ($p = 2$) shown in Figs 5.2*.2a, 5.2*.2b and 5.2*.2c, respectively. Figure 5.2*.2 (respectively Fig. 5.2*.3) gives the element effectivity indices for the indicator η^{Dir}_τ (respectively for the indicator

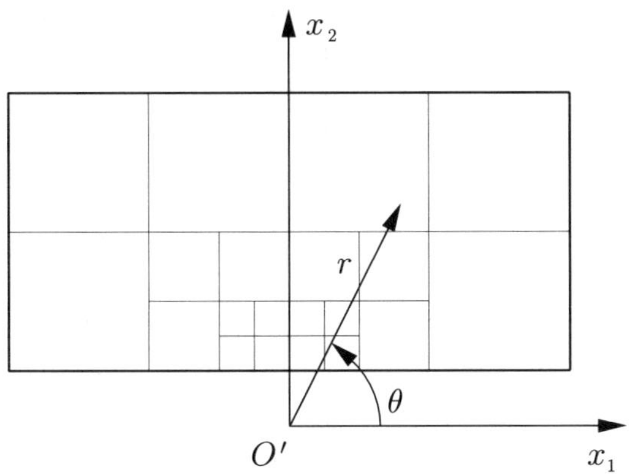

FIG. 5.2*.1. The domain with the polar coordinate system employed in the definition of the exact solution, and the 20-element mesh employed in Example 5.2*.1

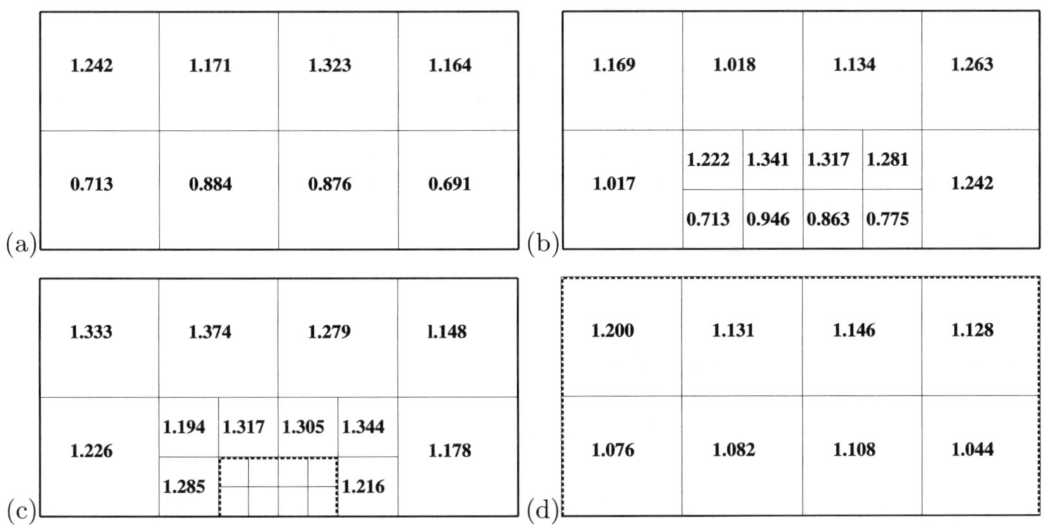

FIG. 5.2*.2. An example which illustrates the accuracy of the element error indicators η_τ^{Dir}, which employ only the interior residuals. Input data: Laplace's equation in $\Omega = (-1, 1) \times (0, 1)$, Neumann boundary conditions, $u_{EX} \equiv \sqrt{r}\sin(\theta/2)$ for the polar coordinate system shown in Fig. 5.2*.1. Mesh data: Mesh of biquadratic squares ($p = 2$) with one to two irregular connections. Element effectivity indices for the element residual error indicator with homogeneous Dirichlet boundary conditions (bubble-type indicator) for the mesh of: (a) 8 elements (global effectivity index $\kappa = 0.880$); (b) 14 elements ($\kappa = 0.897$); (c) 20 elements ($\kappa = 1.091$); (d) the 8-element patch in the subdomain $(-\frac{1}{4}, \frac{1}{4}) \times (0, \frac{1}{4})$ depicted with dashed-line perigram in the mesh of 20 elements shown in (c)

$\eta_{\tau,split}^{\mathrm{Dir}}$). Note that as the mesh is refined, the element effectivity indices get closer to one. Thus, for elements of degree $p = 2$ (and, more generally, of even degree $p = 2, 4, \ldots$) the indicators η_τ^{Dir} and $\eta_{\tau,split}^{\mathrm{Dir}}$ are sufficiently accurate even for relatively coarse meshes.

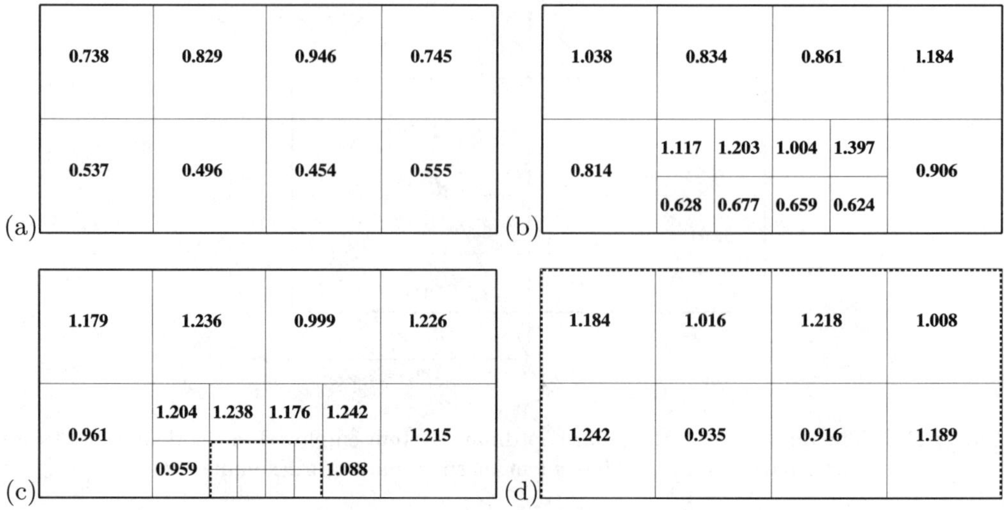

FIG. 5.2*.3. An example which illustrates the accuracy of the element error indicators $\eta^{\mathrm{Dir}}_{\tau,split}$, which employ only the interior residuals. Input data: Laplace's equation in $\Omega = (-1, 1) \times (0, 1)$, Neumann boundary conditions, $u_{EX} \equiv \sqrt{r}\sin(\theta/2)$ for the polar coordinate system shown in Fig. 5.2*.1. Mesh data: Mesh of biquadratic squares ($p = 2$) with one to two irregular connections. Element effectivity indices for the element residual indicator which employs the tunnel functions for the mesh of: (a) 8 elements (global effectivity index $\kappa = 0.469$); (b) 14 elements ($\kappa = 0.676$); (c) 20 elements ($\kappa = 0.934$); (d) the 8-element patch in the subdomain $(-\frac{1}{4}, \frac{1}{4}) \times (0, \frac{1}{4})$ depicted with dashed-line perigram in the mesh of 20 elements shown in (c)

Let us also give corresponding results for elements of odd degree (bilinear, $p = 1$, and bicubic, $p = 3$) for the explicit indicator

$$\eta^{\mathrm{Jumps}}_{\tau} = \frac{h}{4\sqrt{2p+1}}\sqrt{\sum_{\epsilon\subseteq\partial\tau} (J^{\tau}_{\epsilon})^2}. \qquad (5.2^*.15)$$

The element effectivity indices for this indicator and the four meshes are given in Fig. 5.2*.4 for $p = 1$, and Fig. 5.2*.5 for $p = 3$. We see that the explicit indicators $\eta^{\mathrm{Jumps}}_{\tau}$ have reasonable effectivity indices even in coarse meshes.

Remark 5.2*.5 The explicit estimator $\mathscr{E}^{\mathrm{EXPL}} = \mathscr{C}\sqrt{\sum_{\tau\in\Delta} |\tau| \int_{\tau} (r_{\tau})^2}$ is asymptotically exact for meshes of even bi-p squares. Here $\mathscr{C}$ is a constant obtained from the local asymptotic analysis in the interior of the mesh; see [Yu, 1991b].

Remark 5.2*.6 The dichotomy was proven only for meshes of quadrilaterals in [Babuška and Yu, 1987] and [Yu, 1991a,b]. Nevertheless, numerical experiences indicate that it also holds for meshes of triangles; see [Baehmann *et al.*, 1992].

Problems

P5.2*.1. Prove that $\mathcal{K}(p)$, defined in (5.2*.10), does not exist for p odd.

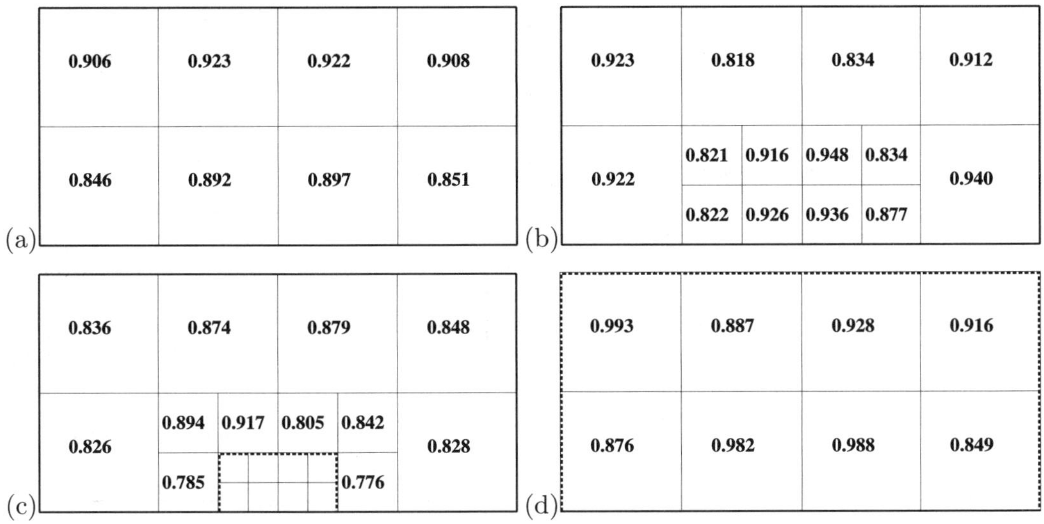

FIG. 5.2*.4. An example which illustrates the accuracy of the element error indicators η_τ^{Jumps}, which employ only the jumps. Input data: Laplace's equation in $\Omega = (-1, 1) \times (0, 1)$, Neumann boundary conditions, $u_{EX} \equiv \sqrt{r}\sin(\theta/2)$ for the polar coordinate system shown in Fig. 5.2*.1. Mesh data: Mesh of bilinear squares ($p = 1$) with one to two irregular connections. Element effectivity indices for the element error indicator which employs the jumps η_τ^{Jumps} (given in (5.2*.12)) for the mesh of: (a) 8 elements (global effectivity index $\kappa = 0.894$); (b) 14 elements ($\kappa = 0.928$); (c) 20 elements ($\kappa = 0.985$); (d) the 8-element patch in the subdomain $(-\frac{1}{4}, \frac{1}{4}) \times (0, \frac{1}{4})$ depicted with dashed-line perigram in the mesh of 20 elements shown in (c)

5.3 Analysis of the Neumann element residual estimator

We will now analyze the Neumann element residual error indicator. Recall that for $j = 2, \ldots, N$, $\hat{e}_{I_j^\Delta}^{\text{Neum}}$ is the solution of the Neumann element residual problem:

Find $\hat{e}_{I_j^\Delta}^{\text{Neum}} \in \mathcal{U}(I_j^\Delta)$ such that

$$\mathcal{B}_{I_j^\Delta}(\hat{e}_{I_j^\Delta}^{\text{Neum}}, v) = \mathcal{R}_{I_j^\Delta}^{EQ}(v) \stackrel{\text{def}}{=} \int_{I_j^\Delta} r_{I_j^\Delta} v + J_j^L v(x_j) + J_{j-1}^R v(x_{j-1}) \quad \forall v \in \mathcal{U}(I_j^\Delta) \tag{5.3.1a}$$

where

$$J_{j-1}^R = -\int_{I_j^\Delta} r_{I_j^\Delta} \varphi_{j-1}^\Delta, \qquad J_j^L = -\int_{I_j^\Delta} r_{I_j^\Delta} \varphi_j^\Delta \tag{5.3.1b}$$

where φ_j^Δ is the piecewise linear hat function corresponding to the jth node. In the cases that $c \equiv 0$, we impose the condition $\int_{I_j^\Delta} \hat{e}_{I_j^\Delta}^{\text{Neum}} = 0$.

The error indicator function $\hat{e}_{I_1^\Delta}^{\text{Neum}}$ will be constructed to satisfy $\hat{e}_{I_1^\Delta}^{\text{Neum}}(x_0^\Delta) = 0$, the same Dirichlet condition, as the exact error, $e_{S_\Delta^p}(x_0^\Delta) = 0$, and it is the solution of the

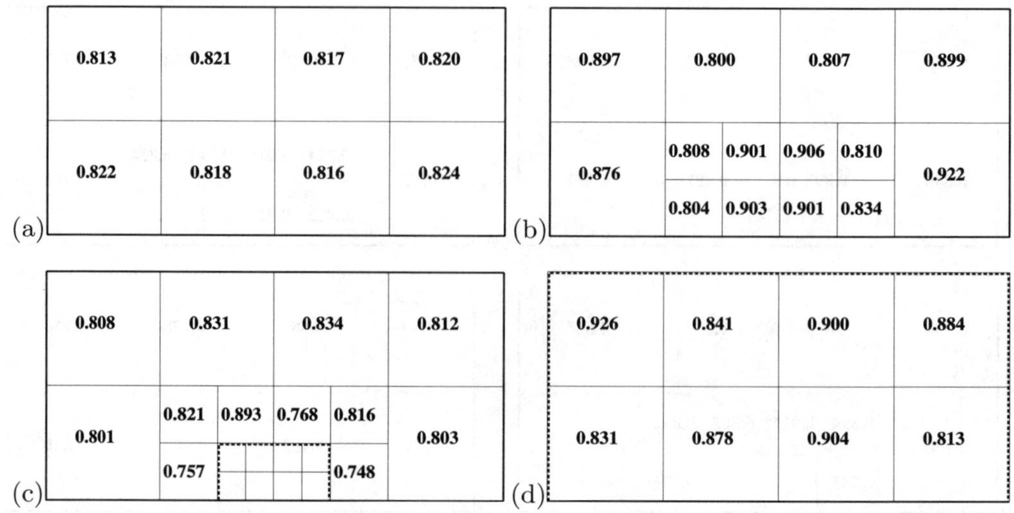

FIG. 5.2*.5. An example which illustrates the accuracy of the element error indicators η_τ^{Jumps}, which employ only the jumps. Input data: Laplace's equation in $\Omega = (-1,1)\times(0,1)$, Neumann boundary conditions, $u_{EX} \equiv \sqrt{r}\sin(\theta/2)$ for the polar coordinate system shown in Fig. 5.2*.1. Mesh data: Mesh of bicubic squares ($p=3$) with one to two irregular connections. Element effectivity indices for the element error indicator which employs the jumps η_τ^{Jumps} (given in (5.2*.12)) for the mesh of: (a) 8 elements (global effectivity index $\kappa = 0.817$); (b) 14 elements ($\kappa = 0.902$); (c) 20 elements ($\kappa = 0.887$); (d) the 8-element patch in the subdomain $(-\frac{1}{4},\frac{1}{4})\times(0,\frac{1}{4})$ depicted with dashed-line perigram in the mesh of 20 elements shown in (c)

mixed boundary value problem:

Find $\hat{e}^{\text{Neum}}_{I_1^\Delta} \in \mathcal{U}_{(0}(I_1^\Delta) \overset{\text{def}}{=} \{w \in \mathcal{U}(I_1^\Delta) \mid w(x_0^\Delta) = 0\}$ such that

$$\mathcal{B}_{I_1^\Delta}\left(\hat{e}^{\text{Neum}}_{I_1^\Delta}, v\right) = \int_{I_1^\Delta} r_{I_1^\Delta} v + J_1^L v(x_1^\Delta) \quad \forall v \in \mathcal{U}_{(0}(I_1^\Delta) \tag{5.3.1c}$$

Let us now discuss the existence and uniqueness of $\hat{e}^{\text{Neum}}_{I_j^\Delta}$.

Theorem 5.3.1 *Assume that the input data satisfy (5.2.2a) and (5.2.2b). Then,* $\hat{e}^{\text{Neum}}_{I_1^\Delta}$ *exists and it is unique. The solution* $\hat{e}^{\text{Neum}}_{I_j^\Delta}$*, for* $j = 2, \ldots, N(\Delta)$*, exists and, in the case that* $c \not\equiv 0$*, it is also unique. In the case that* $c \equiv 0$*,* $\hat{e}^{\text{Neum}}_{I_j^\Delta}$*, for* $j = 2, \ldots, N-1$*, is unique up to a constant which is determined by imposing the condition* $\int_{I_j^\Delta} \hat{e}^{\text{Neum}}_{I_j^\Delta} = 0$*. If* $\alpha_2 > 0$*,* $\hat{e}^{\text{Neum}}_{I_N^\Delta}$ *is unique, while if* $\alpha_2 = 0$*,* $\hat{e}^{\text{Neum}}_{I_N^\Delta}$ *is unique up to a constant.*

Proof To prove the existence of $\hat{e}^{\text{Neum}}_{I_j^\Delta}$, it is sufficient to show that $\mathcal{R}^{EQ}_{I_j^\Delta}$ is bounded. Let us first consider the case of interior elements, $j = 2, \ldots, N(\Delta)-1$, and $c \equiv 0$. From the construction of J_j^L, J_{j-1}^R, it follows that

$$\mathcal{R}^{EQ}_{I_j^\Delta}(1) = J_j^L + J_{j-1}^R + \int_{I_j^\Delta} r_{I_j^\Delta} = -\int_{I_j^\Delta} r_{I_j^\Delta}\varphi_j^\Delta - \int_{I_j^\Delta} r_{I_j^\Delta}\varphi_{j-1}^\Delta + \int_{I_j^\Delta} r_{I_j^\Delta} = 0 \tag{5.3.2}$$

where we used that $\varphi_{j-1}^{\Delta} + \varphi_j^{\Delta} \equiv 1$. This is the consistency condition, which is essential for proving the boundedness of $\mathcal{R}_{I_j^{\Delta}}^{EQ}$. Given any $v \in \mathcal{U}(I_j^{\Delta})$, let

$$w = v - v(x_{j-1}^{\Delta}) \tag{5.3.3a}$$

and note that

$$\mathcal{R}_{I_j^{\Delta}}^{EQ}(v) = \mathcal{R}_{I_j^{\Delta}}^{EQ}(w) + \mathcal{R}_{I_j^{\Delta}}^{EQ}(v(x_{j-1}^{\Delta})) = \mathcal{R}_{I_j^{\Delta}}^{EQ}(w) = J_j^L w(x_j^{\Delta}) + \int_{I_j^{\Delta}} r_{I_j^{\Delta}} w \tag{5.3.3b}$$

where we used that $\mathcal{R}_{I_j^{\Delta}}^{EQ}(v(x_{j-1}^{\Delta})) = 0$, and $w(x_{j-1}^{\Delta}) = 0$. It follows that

$$\left|\mathcal{R}_{I_j^{\Delta}}^{EQ}(v)\right| = \left|J_j^L w(x_j^{\Delta}) + \int_{I_j^{\Delta}} r_{I_j^{\Delta}} w\right| \le \left(|J_j^L| + \int_{I_j^{\Delta}} |r_{I_j^{\Delta}}|\right) \|w\|_{L^\infty(I_j^{\Delta})}. \tag{5.3.3c}$$

Recalling that $J_j^L = -\int_{I_j^{\Delta}} r_{I_j^{\Delta}} \varphi_j^{\Delta}$, we have

$$|J_j^L| = \left|\int_{I_j^{\Delta}} r_{I_j^{\Delta}} \varphi_j^{\Delta}\right| \le \left(\int_{I_j^{\Delta}} |r_{I_j^{\Delta}}|\right) \|\varphi_j^{\Delta}\|_{L^\infty(I_j^{\Delta})} = \int_{I_j^{\Delta}} |r_{I_j^{\Delta}}| \tag{5.3.3d}$$

where we employed that $\|\varphi_j^{\Delta}\|_{L^\infty(I_j^{\Delta})} = 1$, and, using this result in (5.3.3c), we obtain

$$\left|\mathcal{R}_{I_j^{\Delta}}^{EQ}(v)\right| \le 2\int_{I_j^{\Delta}} |r_{I_j^{\Delta}}| \|v - v(x_{j-1}^{\Delta})\|_{L^\infty(I_j^{\Delta})}. \tag{5.3.3e}$$

To finish the proof, we need to bound $\|v - v(x_{j-1}^{\Delta})\|_{L^\infty(I_j^{\Delta})}$ in terms of $\|v\|_{\mathcal{U}(I_j^{\Delta})}$. We note that

$$\begin{aligned}|v(x) - v(x_{j-1}^{\Delta})| &= \left|\int_{x_{j-1}^{\Delta}}^{x} \frac{d}{dx}(v - v(x_{j-1}^{\Delta}))\right| \\ &= \left|\int_{x_{j-1}^{\Delta}}^{x} \frac{dv}{dx}\right| = \left|\int_{x_{j-1}^{\Delta}}^{x} \frac{1}{\sqrt{AE}}\sqrt{AE}\frac{dv}{dx}\right|\end{aligned} \tag{5.3.4a}$$

and using the Cauchy–Schwarz inequality for the L^2 inner product, we obtain

$$\|v - v(x_{j-1}^{\Delta})\|_{L^\infty(I_j^{\Delta})} \le \sqrt{\int_{x_{j-1}^{\Delta}}^{x} \frac{1}{AE}}\sqrt{\int_{x_{j-1}^{\Delta}}^{x} AE\Big(\frac{dv}{dx}\Big)^2} \le \sqrt{\int_{I_j^{\Delta}} \frac{1}{AE}}\, \|v\|_{\mathcal{U}(I_j^{\Delta})}. \tag{5.3.4b}$$

Employing this result in (5.3.3e) we get

$$\left|\mathcal{R}_{I_j^{\Delta}}^{EQ}(v)\right| \le 2\sqrt{\int_{I_j^{\Delta}} \frac{1}{AE}} \int_{I_j^{\Delta}} |r_{I_j^{\Delta}}| \|v\|_{\mathcal{U}(I_j^{\Delta})}. \tag{5.3.5a}$$

Hence we have proven that $\mathcal{R}_{I_j^{\Delta}}^{EQ}$ is bounded on $\mathcal{U}(I_j^{\Delta})$ with norm

$$\|\mathcal{R}_{I_j^{\Delta}}^{EQ}\|_{L(\mathcal{U}(I_j^{\Delta}),\mathbb{R})} = \sup_{v \in \mathcal{U}(I_j^{\Delta})} \frac{\left|\mathcal{R}_{I_j^{\Delta}}^{EQ}(v)\right|}{\|v\|_{\mathcal{U}(I_j^{\Delta})}} \le 2\sqrt{\int_{I_j^{\Delta}} \frac{1}{AE}} \int_{I_j^{\Delta}} |r_{I_j^{\Delta}}|. \tag{5.3.5b}$$

Now, because

$$\|\hat{e}^{\text{Neum}}_{I_j^\Delta}\|_{\mathcal{U}(I_j^\Delta)} = \sup_{v\in\mathcal{U}(I_j^\Delta)} \frac{|\mathcal{B}_{I_j^\Delta}(\hat{e}^{\text{Neum}}_{I_j^\Delta}, v)|}{\|v\|_{\mathcal{U}(I_j^\Delta)}}$$

$$= \sup_{v\in\mathcal{U}(I_j^\Delta)} \frac{|\mathcal{R}^{EQ}_{I_j^\Delta}(v)|}{\|v\|_{\mathcal{U}(I_j^\Delta)}} = \|\mathcal{R}^{EQ}_{I_j^\Delta}\|_{L(\mathcal{U}(I_j^\Delta),\mathbb{R})} \tag{5.3.6a}$$

it immediately follows that

$$\|\hat{e}^{\text{Neum}}_{I_j^\Delta}\|_{\mathcal{U}(I_j^\Delta)} = \|\mathcal{R}^{EQ}_{I_j^\Delta}\|_{L(\mathcal{U}(I_j^\Delta),\mathbb{R})} \le 2\sqrt{\int_{I_j^\Delta} \frac{1}{AE}} \int_{I_j^\Delta} |r_{I_j^\Delta}|. \tag{5.3.6b}$$

This bound holds under the assumption that $1/AE$ is integrable on any arbitrary subdomain, namely,

$$0 < \int_a^b \frac{1}{AE} < \infty, \quad 0 < a < b < 1. \tag{5.3.6c}$$

In the case of the regular model problem, we also have

$$0 < \int_{I_j^\Delta} \frac{1}{AE} \le \frac{1}{\beta_1} h_{I_j^\Delta} \tag{5.3.6d}$$

and hence

$$\|\hat{e}^{\text{Neum}}_{I_j^\Delta}\|_{\mathcal{U}(I_j^\Delta)} \le \frac{2}{\sqrt{\beta_1}}\sqrt{h_{I_j^\Delta}} \int_{I_j^\Delta} |r_{I_j^\Delta}| \le \frac{2}{\sqrt{\beta_1}} h_{I_j^\Delta} \sqrt{\int_{I_j^\Delta} (r_{I_j^\Delta})^2}. \tag{5.3.6e}$$

The boundedness of $\mathcal{R}^{EQ}_{I_1^\Delta}$ follows by employing similar steps as in Lemma 2.2.1, and $\hat{e}^{\text{Neum}}_{I_1^\Delta}$ exists and it is always unique. In the case $c \not\equiv 0$ the consistency condition (5.3.2) is not necessary for ensuring the boundedness of $\mathcal{R}^{EQ}_{I_j^\Delta}$, and $\hat{e}^{\text{Neum}}_{I_j^\Delta}$ exists and is unique in all the elements. Similarly the consistency condition (5.3.2) is not needed for proving the boundedness of $\mathcal{R}^{EQ}_{I_N^\Delta}$, with the exception of the case $\alpha_2 = 0$, $c \equiv 0$, in which it must be satisfied and $\hat{e}^{\text{Neum}}_{I_N^\Delta}$ is unique up to a constant. □

As we already mentioned in Section 5.1, we will use the notations

$$\eta^{\text{Neum}}_{I_j^\Delta} = \|\hat{e}^{\text{Neum}}_{I_j^\Delta}\|_{\mathcal{U}(I_j^\Delta)}, \qquad \mathcal{E}^{\text{Neum}} = \sqrt{\sum_{j=1}^{N(\Delta)} \left(\eta^{\text{Neum}}_{I_j^\Delta}\right)^2} \tag{5.3.7}$$

for the Neumann element error indicator, and the Neumann estimator, respectively.

We will now prove the equivalence

$$\mathcal{C}_L^{\text{Neum}} \mathcal{E}^{\text{Neum}} \le \|e_{S_\Delta^p}\|_{\mathcal{U}} \le \mathcal{E}^{\text{Neum}}. \tag{5.3.8}$$

The upper bound $\|e_{S^p_\Delta}\|_{\mathcal{U}} \leq \mathscr{E}^{\text{Neum}}$ has already been proven in Section 5.1. The lower bound

$$\mathscr{C}_L^{\text{Neum}}\mathscr{E}^{\text{Neum}} \leq \|e_{S^p_\Delta}\|_{\mathcal{U}} \tag{5.3.9a}$$

will be proven below by showing that

$$\mathscr{E}^{\text{Neum}} = \mathscr{E}^{\text{Dir}}(1 + O(h^\Delta)). \tag{5.3.9b}$$

The proof will be based on an estimate of the difference $(d/dx)(\hat{e}^{\text{Neum}}_{I^\Delta_j} - \hat{e}^{\text{Dir}}_{I^\Delta_j})(x^\Delta_k)$, $k = j-1, j$, which will be derived by employing the following lemma:

Lemma 5.3.1 *The Dirichlet error indicator function $\hat{e}^{\text{Dir}}_{I^\Delta_j}$, satisfies*

$$AE(x^{\Delta,+}_{j-1})\frac{d}{dx}(\hat{e}^{\text{Dir}}_{I^\Delta_j})(x^{\Delta,+}_{j-1}) = \int_{I^\Delta_j} r_{I^\Delta_j}\hat{\varphi}^\Delta_{j-1} \tag{5.3.10a}$$

and

$$AE(x^{\Delta,-}_{j})\frac{d}{dx}(\hat{e}^{\text{Dir}}_{I^\Delta_j})(x^{\Delta,-}_{j}) = -\int_{I^\Delta_j} r_{I^\Delta_j}\hat{\varphi}^\Delta_{j} \tag{5.3.10b}$$

where $\hat{\varphi}^\Delta_j \in \mathcal{U}$ is defined elementwise as the solution of the following Dirichlet problems:
Find $\hat{\varphi}^\Delta_j|_{I^\Delta_k} \in \mathcal{U}(I^\Delta_k)$ which satisfies the boundary conditions

$$\hat{\varphi}^\Delta_j(x^\Delta_{k-1}) = \varphi^\Delta_j(x^\Delta_{k-1}) = \delta_{j(k-1)}, \qquad \hat{\varphi}^\Delta_j(x^\Delta_{k}) = \varphi^\Delta_j(x^\Delta_{k}) = \delta_{jk} \tag{5.3.11a}$$

and is such that

$$\mathscr{B}_{I^\Delta_k}(\hat{\varphi}^\Delta_j, v) = 0 \quad \forall v \in \mathcal{U}_0(I^\Delta_k) \tag{5.3.11b}$$

where φ^Δ_j is the piecewise linear hat function, and δ_{jk} is Kronecker's delta.

Proof From (5.3.11), it follows that $\hat{\varphi}^\Delta_j \equiv 0$ outside $I^\Delta_j \cup I^\Delta_{j+1}$, and that $\hat{\varphi}^\Delta_{j-1}|_{I^\Delta_j}$ and $\hat{\varphi}^\Delta_j|_{I^\Delta_j}$ are linearly independent solutions of the homogeneous differential equation in I^Δ_j. Using that $\hat{\varphi}^\Delta_{j-1}(x^\Delta_{j-1}) = 1$, and $\hat{\varphi}^\Delta_{j-1}(x^\Delta_j) = 0$, we obtain

$$\int_{I^\Delta_j} -\frac{d}{dx}\Big(AE\frac{d}{dx}(\hat{e}^{\text{Dir}}_{I^\Delta_j})\hat{\varphi}^\Delta_{j-1}\Big) = AE(x^{\Delta,+}_{j-1})\frac{d}{dx}(\hat{e}^{\text{Dir}}_{I^\Delta_j})(x^{\Delta,+}_{j-1}) \tag{5.3.12a}$$

and employing the integration by parts

$$\begin{aligned}\int_{I^\Delta_j} -\frac{d}{dx}\Big(AE\frac{d}{dx}(\hat{e}^{\text{Dir}}_{I^\Delta_j})\Big)\hat{\varphi}^\Delta_{j-1} &= \int_{I^\Delta_j} -\Big(\frac{d}{dx}AE\frac{d}{dx}(\hat{e}^{\text{Dir}}_{I^\Delta_j})\hat{\varphi}^\Delta_{j-1}\Big) \\ &\quad + \int_{I^\Delta_j} AE\frac{d}{dx}(\hat{e}^{\text{Dir}}_{I^\Delta_j})\frac{d}{dx}(\hat{\varphi}^\Delta_{j-1})\end{aligned} \tag{5.3.12b}$$

we get that

$$\int_{I_j^\Delta} -\frac{d}{dx}\Big(AE\frac{d}{dx}(\hat{e}^{\mathrm{Dir}}_{I_j^\Delta})\Big)\hat{\varphi}^\Delta_{j-1} = AE(x^{\Delta,+}_{j-1})\frac{d}{dx}(\hat{e}^{\mathrm{Dir}}_{I_j^\Delta})(x^{\Delta,+}_{j-1}) + \int_{I_j^\Delta} AE\frac{d}{dx}(\hat{e}^{\mathrm{Dir}}_{I_j^\Delta})\frac{d}{dx}(\hat{\varphi}^\Delta_{j-1}). \tag{5.3.12c}$$

Adding $\int_{I_j^\Delta} c\hat{e}^{\mathrm{Dir}}_{I_j^\Delta}\hat{\varphi}^\Delta_{j-1}$ to both sides, and recalling that

$$r_{I_j^\Delta} = \mathfrak{D}\hat{e}^{\mathrm{Dir}}_{I_j^\Delta} = -\frac{d}{dx}\Big(AE\frac{d}{dx}(\hat{e}^{\mathrm{Dir}}_{I_j^\Delta})\Big) + c\hat{e}^{\mathrm{Dir}}_{I_j^\Delta} \tag{5.3.12d}$$

we get

$$\int_{I_j^\Delta} r_{I_j^\Delta}\hat{\varphi}^\Delta_{j-1} = AE(x^{\Delta,+}_{j-1})\frac{d}{dx}(\hat{e}^{\mathrm{Dir}}_{I_j^\Delta})(x^{\Delta,+}_{j-1}) + \mathcal{B}_{I_j^\Delta}(\hat{\varphi}^\Delta_{j-1}, \hat{e}^{\mathrm{Dir}}_{I_j^\Delta}). \tag{5.3.12e}$$

Further, because $\hat{e}^{\mathrm{Dir}}_{I_j^\Delta} \in \mathcal{U}_0(I_j^\Delta)$, we have

$$\mathcal{B}_{I_j^\Delta}(\hat{\varphi}^\Delta_{j-1}, \hat{e}^{\mathrm{Dir}}_{I_j^\Delta}) = 0 \tag{5.3.12f}$$

and (5.3.10a) follows from (5.3.12e). Following similar steps we obtain (5.3.10b). □

We are now ready to estimate $(d/dx)(\hat{e}^{\mathrm{Neum}}_{I_j^\Delta} - \hat{e}^{\mathrm{Dir}}_{I_j^\Delta})(x^\Delta_k)$, for $k = j, j-1$. We have:

Lemma 5.3.2 *Let*

$$\delta J^R_{j-1} \stackrel{\mathrm{def}}{=} J^R_{j-1} + AE(x^{\Delta,+}_{j-1})\frac{d}{dx}(\hat{e}^{\mathrm{Dir}}_{I_j^\Delta})(x^\Delta_{j-1}) = -AE(x^{\Delta,+}_{j-1})\frac{d}{dx}(\hat{e}^{\mathrm{Neum}}_{I_j^\Delta} - \hat{e}^{\mathrm{Dir}}_{I_j^\Delta})(x^{\Delta,+}_{j-1}) \tag{5.3.13a}$$

and

$$\delta J^L_j \stackrel{\mathrm{def}}{=} J^L_j - AE(x^{\Delta,-}_j)\frac{d}{dx}(\hat{e}^{\mathrm{Dir}}_{I_j^\Delta})(x^\Delta_j) = AE(x^{\Delta,-}_{j-1})\frac{d}{dx}(\hat{e}^{\mathrm{Neum}}_{I_j^\Delta} - \hat{e}^{\mathrm{Dir}}_{I_j^\Delta})(x^{\Delta,-}_{j-1}). \tag{5.3.13b}$$

Under the assumptions (5.2.2a) and (5.2.2b), we have

$$\max(|\delta J^R_{j-1}|, |\delta J^L_j|) \le \mathcal{C}\sqrt{h_{I_j^\Delta}}\,\|\hat{e}^{\mathrm{Dir}}_{I_j^\Delta}\|_{\mathcal{U}(I_j^\Delta)} \tag{5.3.13c}$$

with $\mathcal{C}$ *depending only on* β_{1,I_j^Δ}, β_{2,I_j^Δ}, $\gamma_{I_j^\Delta}$, *and* $K_{I_j^\Delta}$.

Proof Recalling that

$$J^R_{j-1} = -\int_{I_j^\Delta} r_{I_j^\Delta}\varphi^\Delta_{j-1} \tag{5.3.14a}$$

and

$$AE(x_{j-1}^{\Delta,+})\frac{d}{dx}(\hat{e}_{I_j^\Delta}^{\mathrm{Dir}})(x_{j-1}^{\Delta,+}) = \int_{I_j^\Delta} r_{I_j^\Delta}\hat{\varphi}_{j-1}^\Delta \tag{5.3.14b}$$

which was proven in Lemma 5.3.1, we get

$$\delta J_{j-1}^R = -\int_{I_j^\Delta} r_{I_j^\Delta}(\varphi_{j-1}^\Delta - \hat{\varphi}_{j-1}^\Delta) = -\mathcal{B}_{I_j^\Delta}(\hat{e}_{I_j^\Delta}^{\mathrm{Dir}}, \varphi_{j-1}^\Delta - \hat{\varphi}_{j-1}^\Delta) \tag{5.3.14c}$$

where we used that

$$\varphi_{j-1}^\Delta - \hat{\varphi}_{j-1}^\Delta \in \mathcal{U}_0(I_j^\Delta) \tag{5.3.14d}$$

and hence

$$\mathcal{B}_{I_j^\Delta}(\hat{e}_{I_j^\Delta}^{\mathrm{Dir}}, \varphi_{j-1}^\Delta - \hat{\varphi}_{j-1}^\Delta) = \int_{I_j^\Delta} r_{I_j^\Delta}(\varphi_{j-1}^\Delta - \hat{\varphi}_{j-1}^\Delta). \tag{5.3.14e}$$

Using the Cauchy–Schwarz inequality in (5.3.14c) we get

$$|\delta J_{j-1}^R| = |\mathcal{B}_{I_j^\Delta}(\hat{e}_{I_j^\Delta}^{\mathrm{Dir}}, \varphi_{j-1}^\Delta - \hat{\varphi}_{j-1}^\Delta)| \le \|\hat{e}_{I_j^\Delta}^{\mathrm{Dir}}\|_{\mathcal{U}(I_j^\Delta)}\|\varphi_{j-1}^\Delta - \hat{\varphi}_{j-1}^\Delta\|_{\mathcal{U}(I_j^\Delta)}. \tag{5.3.14f}$$

Similarly, we have

$$\delta J_j^L = -\int_{I_j^\Delta} r_{I_j^\Delta}(\varphi_j^\Delta - \hat{\varphi}_j^\Delta) = -\mathcal{B}_{I_j^\Delta}(\hat{e}_{I_j^\Delta}^{\mathrm{Dir}}, \varphi_j^\Delta - \hat{\varphi}_j^\Delta) \tag{5.3.15a}$$

and

$$|\delta J_j^L| = |\mathcal{B}_{I_j^\Delta}(\hat{e}_{I_j^\Delta}^{\mathrm{Dir}}, \varphi_j^\Delta - \hat{\varphi}_j^\Delta)| \le \|\hat{e}_{I_j^\Delta}^{\mathrm{Dir}}\|_{\mathcal{U}(I_j^\Delta)}\|\varphi_j^\Delta - \hat{\varphi}_j^\Delta\|_{\mathcal{U}(I_j^\Delta)}. \tag{5.3.15b}$$

The proof of the Lemma follows from (5.3.14f) and (5.3.15b) by employing that

$$\max\left(\|\varphi_{j-1}^\Delta - \hat{\varphi}_{j-1}^\Delta\|_{\mathcal{U}(I_j^\Delta)}, \|\varphi_j^\Delta - \hat{\varphi}_j^\Delta\|_{\mathcal{U}(I_j^\Delta)}\right) \le \mathcal{C}\sqrt{h_{I_j^\Delta}} \tag{5.3.16}$$

which will be proven in Lemma 5.3.3. □

Lemma 5.3.3 *Let φ_j^Δ be the piecewise linear hat function for node j, and $\hat{\varphi}_j^\Delta$ the function defined in (5.3.11) above. Under the assumption that the input data satisfy (5.2.2a) and (5.2.2b), we have*

$$\|\varphi_j^\Delta - \hat{\varphi}_j^\Delta\|_{\mathcal{U}(I_j^\Delta)} \le \mathcal{C}\sqrt{h_{I_j^\Delta}} \tag{5.3.17}$$

with $\mathcal{C}$ depending on β_{1,I_j^Δ}, β_{2,I_j^Δ}, $\gamma_{I_j^\Delta}$, and $K_{I_j^\Delta}$.

Proof In the special case $AE \equiv$ *constant*, $c \equiv 0$, we have $\hat{\varphi}_j^\Delta \equiv \varphi_j^\Delta$, and $\delta\varphi_j^\Delta \equiv 0$, and hence (5.3.17) holds with $\mathcal{C} = 0$.

In the cases, where either $AE \not\equiv$ *constant* and/or $c \not\equiv 0$, we have

$$|\mathcal{B}_{I_j^\Delta}(\varphi_{j-1}^\Delta, v)| \le AE(\bar{x})\left|\int_{I_j^\Delta} \frac{d}{dx}(\varphi_{j-1}^\Delta)\frac{dv}{dx}\right| + \left|\int_{I_j^\Delta}\Big((AE - AE(\bar{x}))\frac{d}{dx}(\varphi_{j-1}^\Delta)\frac{dv}{dx} + c\varphi_{j-1}^\Delta v\Big)\right|. \tag{5.3.18a}$$

Recalling that $(d/dx)(\varphi_j^\Delta) = 1/h_{I_j^\Delta}$, we obtain

$$\int_{I_j^\Delta} \frac{d}{dx}(\varphi_j^\Delta)\frac{dv}{dx} = \frac{1}{h_{I_j^\Delta}}\int_{I_j^\Delta}\frac{dv}{dx} = \frac{1}{h_{I_j^\Delta}}(v(x_j^\Delta) - v(x_{j-1}^\Delta)) = 0 \quad \forall v \in \mathcal{U}_0(I_j^\Delta) \tag{5.3.18b}$$

and the first term in the right-hand side of (5.3.18a) is zero. To estimate the second term we employ the assumptions

$$\left\|\frac{d}{dx}(AE)\right\|_{L^\infty(I_j^\Delta)} \le K_{I_j^\Delta} < \infty, \qquad \|c\|_{L^\infty(I_j^\Delta)} \le \gamma_{I_j^\Delta} < \infty \tag{5.3.18c}$$

and we get

$$\left|\int_{I_j^\Delta}\Big((AE - AE(\bar{x}))\frac{d}{dx}(\varphi_j^\Delta)\frac{dv}{dx} + c\varphi_j^\Delta v\Big)\right| \le \max\left(K_{I_j^\Delta}, \gamma_{I_j^\Delta}\right)\int_{I_j^\Delta}\Big(\left|\frac{dv}{dx}\right| + |v|\Big). \tag{5.3.18d}$$

Hence we have proven that

$$|\mathcal{B}_{I_j^\Delta}(\varphi_j^\Delta, v)| \le \max\left(K_{I_j^\Delta}, \gamma_{I_j^\Delta}\right)\int_{I_j^\Delta}\Big(\left|\frac{dv}{dx}\right| + |v|\Big). \tag{5.3.18e}$$

Recalling that

$$\mathcal{B}_{I_j^\Delta}(\hat{\varphi}_j^\Delta, v) = 0 \quad \forall v \in \mathcal{U}_0(I_j^\Delta) \tag{5.3.19a}$$

we obtain

$$\mathcal{B}_{I_j^\Delta}(\varphi_j^\Delta - \hat{\varphi}_j^\Delta, v) = \mathcal{B}_{I_j^\Delta}(\varphi_j^\Delta, v) \quad \forall v \in \mathcal{U}_0(I_j^\Delta) \tag{5.3.19b}$$

and employing (5.3.18e), we get

$$|\mathcal{B}_{I_j^\Delta}(\varphi_j^\Delta - \hat{\varphi}_j^\Delta, v)| \le \max\left(K_{I_j^\Delta}, \gamma_{I_j^\Delta}\right)\int_{I_j^\Delta}\Big(\left|\frac{dv}{dx}\right| + |v|\Big) \quad \forall v \in \mathcal{U}_0(I_j^\Delta) \tag{5.3.19c}$$

Then using that

$$\int_{I_j^\Delta}\Big(\left|\frac{dv}{dx}\right| + |v|\Big) \le \mathcal{C}\sqrt{h_{I_j^\Delta}}\,\|v\|_{\mathcal{U}(I_j^\Delta)} \quad \forall v \in \mathcal{U}_0(I_j^\Delta) \tag{5.3.19d}$$

we get

$$\mathcal{B}_{I_j^\Delta}(\varphi_j^\Delta - \hat{\varphi}_j^\Delta, v) \le \mathcal{C} \max\left(K_{I_j^\Delta}, \gamma_{I_j^\Delta}\right) \sqrt{h_{I_j^\Delta}}\, \|v\|_{\mathcal{U}(I_j^\Delta)} \quad \forall v \in \mathcal{U}_0(I_j^\Delta). \tag{5.3.19e}$$

Now, letting $v = \varphi_j^\Delta - \hat{\varphi}_j^\Delta \in \mathcal{U}_0(I_j^\Delta)$, we obtain

$$\|\varphi_j^\Delta - \hat{\varphi}_j^\Delta\|_{\mathcal{U}(I_j^\Delta)} \le \mathcal{C} \max\left(K_{I_j^\Delta}, \gamma_{I_j^\Delta}\right) \sqrt{h_{I_j^\Delta}} \tag{5.3.19f}$$

which proves the Lemma. □

We are now ready to prove that $\eta_{I_j^\Delta}^{\text{Neum}}$ is bounded from above by $\eta_{I_j^\Delta}^{\text{Dir}}$, asymptotically.

Theorem 5.3.2 *Under the assumption that the input data satisfy (5.2.2a–b), for sufficiently small $h_{I_j^\Delta}$, we have*

$$\eta_{I_j^\Delta}^{\text{Neum}} \le (1 + \mathcal{C} h_{I_j^\Delta}) \eta_{I_j^\Delta}^{\text{Dir}} \tag{5.3.20}$$

with $\mathcal{C}$ depending on β_{1,I_j^Δ}, β_{2,I_j^Δ}, $\gamma_{I_j^\Delta}$, and $K_{I_j^\Delta}$.

Proof From the triangle inequality we have

$$\|\hat{e}_{I_j^\Delta}^{\text{Neum}}\|_{\mathcal{U}(I_j^\Delta)} \le \|\hat{e}_{I_j^\Delta}^{\text{Dir}}\|_{\mathcal{U}(I_j^\Delta)} + \|\hat{e}_{I_j^\Delta}^{\text{Neum}} - \hat{e}_{I_j^\Delta}^{\text{Dir}}\|_{\mathcal{U}(I_j^\Delta)} \tag{5.3.21a}$$

and hence in order to prove (5.3.20), it is sufficient to show that

$$\|\hat{e}_{I_j^\Delta}^{\text{Neum}} - \hat{e}_{I_j^\Delta}^{\text{Dir}}\|_{\mathcal{U}(I_j^\Delta)} \le \mathcal{C} h_{I_j^\Delta} \|\hat{e}_{I_j^\Delta}^{\text{Dir}}\|_{\mathcal{U}(I_j^\Delta)}. \tag{5.3.21b}$$

We note that, for $j = 2, \ldots, N(\Delta)$, the difference $\hat{e}_{I_j^\Delta}^{\text{Neum}} - \hat{e}_{I_j^\Delta}^{\text{Dir}}$ is the solution of the Neumann problem:

Find $\hat{e}_{I_j^\Delta}^{\text{Neum}} - \hat{e}_{I_j^\Delta}^{\text{Dir}} \in \mathcal{U}(I_j^\Delta)$ such that

$$\mathcal{B}_{I_j^\Delta}\left(\hat{e}_{I_j^\Delta}^{\text{Neum}} - \hat{e}_{I_j^\Delta}^{\text{Dir}}, v\right) = \mathcal{L}_{I_j^\Delta}^{\text{ND}}(v) \stackrel{\text{def}}{=} \delta J_j^L v(x_j^\Delta) - \delta J_{j-1}^R v(x_{j-1}^\Delta) \quad \forall\, v \in \mathcal{U}(I_j^\Delta). \tag{5.3.22a}$$

In the first element, the difference $\hat{e}_{I_j^\Delta}^{\text{Neum}} - \hat{e}_{I_j^\Delta}^{\text{Dir}} \in \mathcal{U}_{(0}(I_1^\Delta)$, and satisfies (5.3.22b) for every $v \in \mathcal{U}_{(0}(I_1^\Delta)$.

It follows that

$$\|\hat{e}_{I_j^\Delta}^{\text{Neum}} - \hat{e}_{I_j^\Delta}^{\text{Dir}}\|_{\mathcal{U}(I_j^\Delta)} = \|\mathcal{L}_{I_j^\Delta}^{\text{ND}}\|_{L(\mathcal{U}(I_j^\Delta),\mathbb{R})} \tag{5.3.22b}$$

for $j = 1, \ldots, N(\Delta)$, where for $j = 1$, $\mathcal{U}(I_j^\Delta)$ must be replaced by $\mathcal{U}_{(0}(I_j^\Delta)$. Hence, to prove (5.3.21b) it is sufficient to show that

$$\|\mathcal{L}_{I_j^\Delta}^{\text{ND}}\|_{L(\mathcal{U}(I_j^\Delta),\mathbb{R})} \le \mathcal{C} h_{I_j^\Delta} \|\hat{e}_{I_j^\Delta}^{\text{Dir}}\|_{\mathcal{U}(I_j^\Delta)}. \tag{5.3.22c}$$

We will prove this by employing the splitting

$$\mathscr{L}^{\mathrm{ND}}_{I_j^\Delta}(v) = \underbrace{(\delta J_j^L + \delta J_{j-1}^R) v(x_j^\Delta)}_{\mathscr{L}^{\mathrm{ND}}_{I_j^\Delta,1}(v)} - \underbrace{\delta J_{j-1}^R \big(v(x_j^\Delta) - v(x_{j-1}^\Delta)\big)}_{\mathscr{L}^{\mathrm{ND}}_{I_j^\Delta,2}(v)} \tag{5.3.23a}$$

and by showing that

$$\max\left(\|\mathscr{L}^{\mathrm{ND}}_{I_j^\Delta,1}\|_{L(\mathscr{U}(I_j^\Delta),\mathbb{R})}, \|\mathscr{L}^{\mathrm{ND}}_{I_j^\Delta,2}\|_{L(\mathscr{U}(I_j^\Delta),\mathbb{R})}\right) \le \mathscr{C} h_{I_j^\Delta} \|\hat{e}^{\mathrm{Dir}}_{I_j^\Delta}\|_{\mathscr{U}(I_j^\Delta)}. \tag{5.3.23b}$$

Let us first bound $\|\mathscr{L}^{\mathrm{ND}}_{I_j^\Delta,1}\|_{L(\mathscr{U}(I_j^\Delta),\mathbb{R})}$. We note that

$$|\mathscr{L}^{\mathrm{ND}}_{I_j^\Delta,1}(v)| = |(\delta J_j^L + \delta J_{j-1}^R) v(x_j^\Delta)| \le |\delta J_j^L + \delta J_{j-1}^R|\, |v(x_j^\Delta)| \tag{5.3.24}$$

and hence we need to estimate $|\delta J_j^L + \delta J_{j-1}^R|$, and to bound $|v(x_j^\Delta)|$ in terms of $\|v\|_{\mathscr{U}(I_j^\Delta)}$.

Letting $v \equiv 1$ in (5.3.22a) we get

$$\mathscr{L}^{\mathrm{ND}}_{I_j^\Delta}(1) = \delta J_j^L + \delta J_{j-1}^R = \mathscr{B}_{I_j^\Delta}\big(\hat{e}^{\mathrm{Neum}}_{I_j^\Delta} + \hat{e}^{\mathrm{Dir}}_{I_j^\Delta}, 1\big) \tag{5.3.25a}$$

and recalling the consistency condition

$$\mathscr{B}_{I_j^\Delta}\big(\hat{e}^{\mathrm{Neum}}_{I_j^\Delta}, 1\big) = \mathscr{R}^{EQ}_{I_j^\Delta}(1) = 0 \tag{5.3.25b}$$

we obtain

$$\delta J_j^L + \delta J_{j-1}^R = -\mathscr{B}_{I_j^\Delta}\big(\hat{e}^{\mathrm{Dir}}_{I_j^\Delta}, 1\big) = -\int_{I_j^\Delta} c\,\hat{e}^{\mathrm{Dir}}_{I_j^\Delta}. \tag{5.3.25c}$$

It follows that in the case $c|_{I_j^\Delta} \equiv 0$, we have

$$\delta J_j^L + \delta J_{j-1}^R = 0 \quad \Longrightarrow \quad \mathscr{L}^{\mathrm{ND}}_{I_j^\Delta,1} \equiv 0. \tag{5.3.26}$$

In the case that $c|_{I_j^\Delta} \not\equiv 0$, we have $\int_{I_j^\Delta} c > 0$, and (5.3.25c) gives

$$|\delta J_j^L + \delta J_{j-1}^R| = \left|\int_{I_j^\Delta} c\,\hat{e}^{\mathrm{Dir}}_{I_j^\Delta}\right| \le \left(\int_{I_j^\Delta} c\right) \|\hat{e}^{\mathrm{Dir}}_{I_j^\Delta}\|_{L^\infty(I_j^\Delta)}. \tag{5.3.27a}$$

We also have

$$\|\hat{e}^{\mathrm{Dir}}_{I_j^\Delta}\|_{L^\infty(I_j^\Delta)} \le \mathscr{C}\sqrt{h_{I_j^\Delta}}\, \|\hat{e}^{\mathrm{Dir}}_{I_j^\Delta}\|_{\mathscr{U}(I_j^\Delta)} \tag{5.3.27b}$$

where we used that $\hat{e}^{\mathrm{Dir}}_{I_j^\Delta}(x_{j-1}^\Delta) = 0$, and hence

$$|\delta J_j^L + \delta J_{j-1}^R| \le \left(\int_{I_j^\Delta} c\right) \mathscr{C} \frac{1}{\sqrt{\beta_{1,j}}} \sqrt{h_{I_j^\Delta}}\, \|\hat{e}^{\mathrm{Dir}}_{I_j^\Delta}\|_{\mathscr{U}(I_j^\Delta)} \tag{5.3.27c}$$

which is the desired estimate for $|\delta J_j^L + \delta J_{j-1}^R|$, with the explicit dependence of the constant on $\int_{I_j^\Delta} c$.

We will now prove that

$$|v(x_j^\Delta)| \le \mathscr{C}\|v\|_{\mathscr{U}(I_j^\Delta)} \tag{5.3.28}$$

where, as in (5.3.27c), we will give the explicit dependence of $\mathscr{C}$ on $\int_{I_j^\Delta} c$. We have

$$\begin{aligned}|v(x_j^\Delta)| &= \left(\int_{I_j^\Delta} c\right)^{-1}\left|\int_{I_j^\Delta} cv(x_j^\Delta)\right| \\ &\le \left(\int_{I_j^\Delta} c\right)^{-1}\left(\left|\int_{I_j^\Delta} cv\right| + \left|\int_{I_j^\Delta} c(v - v(x_j^\Delta))\right|\right)\end{aligned} \tag{5.3.29a}$$

and hence

$$|v(x_j^\Delta)| \le \left(\int_{I_j^\Delta} c\right)^{-1}\left(\sqrt{\int_{I_j^\Delta} c}\sqrt{\int_{I_j^\Delta} cv^2} + \left(\int_{I_j^\Delta} c\right)\|v - v(x_j^\Delta)\|_{L^\infty(I_j^\Delta)}\right) \tag{5.3.29b}$$

and because

$$\sqrt{\int_{I_j^\Delta} cv^2} \le \sqrt{\int_{I_j^\Delta} AE\left(\frac{dv}{dx}\right) + cv^2} \le \|v\|_{\mathscr{U}(I_j^\Delta)} \tag{5.3.29c}$$

we get

$$|v(x_j^\Delta)| \le \left(\sqrt{\int_{I_j^\Delta} c}\right)^{-1}\left(\|v\|_{\mathscr{U}(I_j^\Delta)} + \sqrt{\int_{I_j^\Delta} c}\|v - v(x_j^\Delta)\|_{L^\infty(I_j^\Delta)}\right). \tag{5.3.29d}$$

Employing that

$$\|v - v(x_j^\Delta)\|_{L^\infty(I_j^\Delta)} \le \mathscr{C}\frac{1}{\sqrt{\beta_{1,I_j^\Delta}}}\sqrt{h_{I_j^\Delta}}\|v\|_{\mathscr{U}(I_j^\Delta)} \tag{5.3.29e}$$

and

$$\sqrt{\int_{I_j^\Delta} c} \le \sqrt{\gamma_{I_j^\Delta}}\sqrt{h_{I_j^\Delta}} \tag{5.3.29f}$$

we obtain

$$|v(x_j^\Delta)| \le \left(\sqrt{\int_{I_j^\Delta} c}\right)^{-1}\left(\|v\|_{\mathscr{U}(I_j^\Delta)} + \mathscr{C}\frac{\sqrt{\gamma_{I_j^\Delta}}}{\sqrt{\beta_{1,j}}}h_{I_j^\Delta}\|v\|_{\mathscr{U}(I_j^\Delta)}\right) \tag{5.3.29g}$$

and hence we have proven that

$$|v(x_j^\Delta)| \le \left(1 + \mathscr{C}\frac{\sqrt{\gamma_{I_j^\Delta}}}{\sqrt{\beta_{1,j}}}h_{I_j^\Delta}\right)\left(\sqrt{\int_{I_j^\Delta} c}\right)^{-1}\|v\|_{\mathscr{U}(I_j^\Delta)}. \tag{5.3.29h}$$

Employing (5.3.27c) and (5.3.29h) in

$$\left|\mathcal{L}^{\mathrm{ND}}_{I_j^{\Delta},1}(v)\right| \leq |\delta J_j^L + \delta J_{j-1}^R|\,|v(x_j^{\Delta})| \tag{5.3.30a}$$

we get

$$\left|\mathcal{L}^{\mathrm{ND}}_{I_j^{\Delta},1}(v)\right| \leq \mathscr{C}\sqrt{\int_{I_j^{\Delta}} c\frac{1}{\sqrt{\beta_{1,j}}}\left(1+\mathscr{C}\frac{\sqrt{\gamma_j}}{\sqrt{\beta_{1,j}}}h_{I_j^{\Delta}}\right)}\sqrt{h_{I_j^{\Delta}}}\,\|\hat{e}^{\mathrm{Dir}}_{I_j^{\Delta}}\|_{\mathcal{U}(I_j^{\Delta})}\|v\|_{\mathcal{U}(I_j^{\Delta})} \tag{5.3.30b}$$

and using that $\sqrt{\int_{I_j^{\Delta}} c} \leq \sqrt{\gamma_{I_j^{\Delta}}}\sqrt{h_{I_j^{\Delta}}}$, we also obtain

$$\left|\mathcal{L}^{\mathrm{ND}}_{I_j^{\Delta},1}(v)\right| \leq \mathscr{C}\frac{\sqrt{\gamma_{I_j^{\Delta}}}}{\sqrt{\beta_{1,I_j^{\Delta}}}}\left(1+\mathscr{C}\frac{\sqrt{\gamma_j}}{\sqrt{\beta_{1,j}}}h_{I_j^{\Delta}}\right)h_{I_j^{\Delta}}\,\|\hat{e}^{\mathrm{Dir}}_{I_j^{\Delta}}\|_{\mathcal{U}(I_j^{\Delta})}\|v\|_{\mathcal{U}(I_j^{\Delta})}. \tag{5.3.30c}$$

Further, for sufficiently small $h_{I_j^{\Delta}}$, the second term in the parenthesis may be neglected, and we have

$$\left|\mathcal{L}^{\mathrm{ND}}_{I_j^{\Delta},1}(v)\right| \leq \mathscr{C}\frac{\sqrt{\gamma_{I_j^{\Delta}}}}{\sqrt{\beta_{1,I_j^{\Delta}}}}h_{I_j^{\Delta}}\,\|\hat{e}^{\mathrm{Dir}}_{I_j^{\Delta}}\|_{\mathcal{U}(I_j^{\Delta})}\|v\|_{\mathcal{U}(I_j^{\Delta})} \tag{5.3.30d}$$

which proves that

$$\left\|\mathcal{L}^{\mathrm{ND}}_{I_j^{\Delta},1}\right\|_{L(\mathcal{U}(I_j^{\Delta}),\mathbb{R})} \leq \mathscr{C}\frac{\sqrt{\gamma_j}}{\sqrt{\beta_{1,j}}}h_{I_j^{\Delta}}\,\|\hat{e}^{\mathrm{Dir}}_{I_j^{\Delta}}\|_{\mathcal{U}(I_j^{\Delta})} \tag{5.3.30e}$$

for $j = 2,\ldots,N(\Delta)$. The proof, with obvious modifications, also holds for $\|\mathcal{L}^{\mathrm{ND}}_{I_1^{\Delta},1}\|_{L(\mathcal{U}_{(0}(I_j^{\Delta}),\mathbb{R})}$.

Following similar steps, we will bound $\|\mathcal{L}^{\mathrm{ND}}_{I_j^{\Delta},2}\|_{L(\mathcal{U}(I_j^{\Delta}),\mathbb{R})}$. We have

$$\left|\mathcal{L}^{\mathrm{ND}}_{I_j^{\Delta},2}(v)\right| \leq |\delta J_{j-1}^R|\,|v(x_j^{\Delta}) - v(x_{j-1}^{\Delta})| \tag{5.3.31}$$

and we will bound $|\delta J_{j-1}^R|$ in terms of $\|\hat{e}^{\mathrm{Dir}}_{I_j^{\Delta}}\|_{\mathcal{U}(I_j^{\Delta})}$, and $|v(x_j^{\Delta}) - v(x_{j-1}^{\Delta})|$ in terms of $\|v\|_{\mathcal{U}(I_j^{\Delta})}$. We have

$$\delta J_{j-1}^R = \mathscr{B}_{I_j^{\Delta}}\left(\hat{e}^{\mathrm{Dir}}_{I_j^{\Delta}} - \hat{e}^{\mathrm{Neum}}_{I_j^{\Delta}}, \varphi_{j-1}^{\Delta}\right) \tag{5.3.32a}$$

and recalling that

$$\mathscr{B}_{I_j^{\Delta}}\left(\hat{e}^{\mathrm{Neum}}_{I_j^{\Delta}}, \varphi_{j-1}^{\Delta}\right) = \mathscr{R}^{EQ}_{I_j^{\Delta}}(\varphi_{j-1}^{\Delta}) = 0 \tag{5.3.32b}$$

δJ_{j-1}^R can be written as

$$\delta J_{j-1}^R = \mathscr{B}_{I_j^{\Delta}}\left(\hat{e}^{\mathrm{Dir}}_{I_j^{\Delta}}, \varphi_{j-1}^{\Delta}\right). \tag{5.3.32c}$$

Employing the definition of $\hat{\varphi}^{\Delta}_{j-1}$, and that $\hat{e}^{\mathrm{Dir}}_{I^{\Delta}_j} \in \mathcal{U}_0(I^{\Delta}_j)$, we get

$$\mathcal{B}_{I^{\Delta}_j}\big(\hat{e}^{\mathrm{Dir}}_{I^{\Delta}_j}, \hat{\varphi}^{\Delta}_{j-1}\big) = \mathcal{B}_{I^{\Delta}_j}\big(\hat{\varphi}^{\Delta}_{j-1}, \hat{e}^{\mathrm{Dir}}_{I^{\Delta}_j}\big) = 0 \tag{5.3.32d}$$

and hence we have

$$\delta J^R_{j-1} = \mathcal{B}_{I^{\Delta}_j}\big(\hat{e}^{\mathrm{Dir}}_{I^{\Delta}_j}, \varphi^{\Delta}_{j-1}\big) - \mathcal{B}_{I^{\Delta}_j}\big(\hat{e}^{\mathrm{Dir}}_{I^{\Delta}_j}, \hat{\varphi}^{\Delta}_{j-1}\big) = \mathcal{B}_{I^{\Delta}_j}\big(\hat{e}^{\mathrm{Dir}}_{I^{\Delta}_j}, \varphi^{\Delta}_{j-1} - \hat{\varphi}^{\Delta}_{j-1}\big) \tag{5.3.32e}$$

and employing the Cauchy–Schwarz inequality we get

$$|\delta J^R_{j-1}| \leq \|\hat{e}^{\mathrm{Dir}}_{I^{\Delta}_j}\|_{\mathcal{U}(I^{\Delta}_j)} \|\varphi^{\Delta}_{j-1} - \hat{\varphi}^{\Delta}_{j-1}\|_{\mathcal{U}(I^{\Delta}_j)} \leq \mathcal{C}\sqrt{h_{I^{\Delta}_j}}\,\|\hat{e}^{\mathrm{Dir}}_{I^{\Delta}_j}\|_{\mathcal{U}(I^{\Delta}_j)} \tag{5.3.32f}$$

where we used that

$$\|\varphi^{\Delta}_{j-1} - \hat{\varphi}^{\Delta}_{j-1}\|_{\mathcal{U}(I^{\Delta}_j)} \leq \mathcal{C}\max(K_{I^{\Delta}_j}, \gamma_{I^{\Delta}_j})\sqrt{h_{I^{\Delta}_j}} \tag{5.3.32g}$$

which was proven in Lemma 5.3.3 above. We also have

$$|v(x^{\Delta}_j) - v(x^{\Delta}_{j-1})| \leq \mathcal{C}\sqrt{h_{I^{\Delta}_j}}\,\|v\|_{\mathcal{U}(I^{\Delta}_j)}. \tag{5.3.33}$$

It follows that

$$\begin{aligned} \big|\mathcal{L}^{\mathrm{ND}}_{I^{\Delta}_j,2}(v)\big| &\leq |\delta J^R_{j-1}|\,|v(x^{\Delta}_j) - v(x^{\Delta}_{j-1})| \\ &\leq \mathcal{C}\sqrt{h_{I^{\Delta}_j}}\,\|\hat{e}^{\mathrm{Dir}}_{I^{\Delta}_j}\|_{\mathcal{U}(I^{\Delta}_j)}\,\mathcal{C}\sqrt{h_{I^{\Delta}_j}}\,\|v\|_{\mathcal{U}(I^{\Delta}_j)} \end{aligned} \tag{5.3.34a}$$

and hence we have proven that

$$\big\|\mathcal{L}^{\mathrm{ND}}_{I^{\Delta}_j,2}\big\|_{L(\mathcal{U}(I^{\Delta}_j),\mathbb{R})} \leq \mathcal{C}h_{I^{\Delta}_j}\,\|\hat{e}^{\mathrm{Dir}}_{I^{\Delta}_j}\|_{\mathcal{U}(I^{\Delta}_j)} \tag{5.3.34b}$$

with $\mathcal{C}$ depending only on β_{1,I^{Δ}_j}, β_{2,I^{Δ}_j}, $K_{I^{\Delta}_j}$, and $\gamma_{I^{\Delta}_j}$.

Using the triangle inequality

$$\big\|\mathcal{L}^{\mathrm{ND}}_{I^{\Delta}_j,1} + \mathcal{L}^{\mathrm{ND}}_{I^{\Delta}_j,2}\big\|_{L(\mathcal{U}(I^{\Delta}_j),\mathbb{R})} \leq \big\|\mathcal{L}^{\mathrm{ND}}_{I^{\Delta}_j,1}\big\|_{L(\mathcal{U}(I^{\Delta}_j),\mathbb{R})} + \big\|\mathcal{L}^{\mathrm{ND}}_{I^{\Delta}_j,2}\big\|_{L(\mathcal{U}(I^{\Delta}_j),\mathbb{R})} \tag{5.3.35a}$$

we get

$$\|\ddot{e}^{\mathrm{Dir}}_{I^{\Delta}_j} - \dot{e}^{\mathrm{Neum}}_{I^{\Delta}_j}\|_{\mathcal{U}(I^{\Delta}_j)} \leq \mathcal{C}h_{I^{\Delta}_j}\,\|\hat{e}^{\mathrm{Dir}}_{I^{\Delta}_j}\|_{\mathcal{U}(I^{\Delta}_j)}. \tag{5.3.35b}$$

This bound holds for all the elements, including the first and the last element. □

We are now ready to prove the equivalence $\mathcal{E}^{\mathrm{Neum}} \approx \|e_{S^p_{\Delta}}\|_{\mathcal{U}}$.

Theorem 5.3.3 *Assume that the input data satisfy assumptions (5.2.2a) and (5.2.2b). Then, for sufficiently small h^{Δ}, we have*

$$(1 - \mathcal{K}^{\mathrm{Neum}}h^{\Delta})\mathcal{E}^{\mathrm{Neum}} \leq \|e_{S^p_{\Delta}}\|_{\mathcal{U}} \leq \mathcal{E}^{\mathrm{Neum}} \tag{5.3.36}$$

where $\mathcal{K}^{\mathrm{Neum}}$ depends only on β_1, β_2, K, and γ.

Proof From Theorem 5.3.2 we have

$$\frac{\eta^{\mathrm{Neum}}_{I^{\Delta}_j}}{1+\mathscr{C}h_{I^{\Delta}_j}} \le \eta^{\mathrm{Dir}}_{I^{\Delta}_j} \tag{5.3.37a}$$

and squaring and summing over the elements, we get

$$\sqrt{\sum_{j=1}^{N(\Delta)} \frac{\left(\eta^{\mathrm{Neum}}_{I^{\Delta}_j}\right)^2}{\left(1+\mathscr{C}h_{I^{\Delta}_j}\right)^2}} \le \sqrt{\sum_{j=1}^{N(\Delta)} \left(\eta^{\mathrm{Dir}}_{I^{\Delta}_j}\right)^2} = \mathscr{E}^{\mathrm{Dir}} \le \|e_{S^p_\Delta}\|_{\mathscr{U}}. \tag{5.3.37b}$$

Using that $h^\Delta \le h_{I^{\Delta}_j}$, $j = 1, \ldots, N(\Delta)$, we get

$$\frac{1}{(1+\mathscr{C}h^\Delta)}\sqrt{\sum_{j=1}^{N(\Delta)} \left(\eta^{\mathrm{Neum}}_{I^{\Delta}_j}\right)^2} \le \|e_{S^p_\Delta}\|_{\mathscr{U}} \tag{5.3.37c}$$

and choosing $\mathcal{K}^{\mathrm{Neum}}$ such that

$$(1-\mathcal{K}^{\mathrm{Neum}}h^\Delta) \le \frac{1}{(1+\mathscr{C}h^\Delta)} \tag{5.3.37d}$$

the lower bound follows.

The upper bound has been already proven in Section 5.1. Let us now give a more elementary version of this proof. Recalling that

$$\mathscr{R}(v) = \sum_{j=1}^{N(\Delta)} \mathscr{R}^{EQ}_{I^{\Delta}_j}(v|_{I^{\Delta}_j}) = \sum_{j=1}^{N(\Delta)} \mathscr{B}_{I^{\Delta}_j}\left(\hat{e}^{\mathrm{Neum}}_{I^{\Delta}_j}, v|_{I^{\Delta}_j}\right) \quad \forall v \in \mathscr{U} \tag{5.3.38a}$$

and using the Cauchy–Schwarz inequality for the energy and the $\mathbb{R}^N$ inner products, we get

$$\begin{aligned} |\mathscr{R}(v)| &= \left| \sum_{j=1}^{N(\Delta)} \mathscr{B}_{I^{\Delta}_j}\left(\hat{e}^{\mathrm{Neum}}_{I^{\Delta}_j}, v|_{I^{\Delta}_j}\right)\right| \\ &\le \sum_{j=1}^{N(\Delta)} \|\hat{e}^{\mathrm{Neum}}_{I^{\Delta}_j}\|_{\mathscr{U}(I^{\Delta}_j)} \|v\|_{\mathscr{U}(I^{\Delta}_j)} \le \mathscr{E}^{\mathrm{Neum}}\|v\|_{\mathscr{U}} \end{aligned} \tag{5.3.38b}$$

for any $v \in \mathscr{U}$, and employing $v = e_{S^p_\Delta}$, we get

$$\|e_{S^p_\Delta}\|^2_{\mathscr{U}} = \mathscr{B}(e_{S^p_\Delta}, e_{S^p_\Delta}) = \mathscr{R}(e_{S^p_\Delta}) \le \mathscr{E}^{\mathrm{Neum}}\|e_{S^p_\Delta}\|_{\mathscr{U}} \tag{5.3.38c}$$

and the upper bound follows. □

Remark 5.3.1 The upper bound, $\|e_{S^p_\Delta}\|_{\mathscr{U}} \le \mathscr{E}^{\mathrm{Neum}}$, requires only assumption (5.2.2a).

Note that although we have the global upper estimate

$$\|e_{S^p_\Delta}\|_{\mathcal{U}} \leq \mathcal{E}^{\text{Neum}} = \sqrt{\sum_{j=1}^{N(\Delta)} \|\hat{e}^{\text{Neum}}_{I^\Delta_j}\|^2_{\mathcal{U}(I^\Delta_j)}} \tag{5.3.39a}$$

the element error indicators $\eta^{\text{Neum}}_{I^\Delta_j}$ may underestimate the element energy norms of the error $\|e_{S^p_\Delta}\|_{\mathcal{U}(I^\Delta_j)}$, namely

$$\eta^{\text{Neum}}_{I^\Delta_j} = \|\hat{e}^{\text{Neum}}_{I^\Delta_j}\|_{\mathcal{U}(I^\Delta_j)} \leq \|e_{S^p_\Delta}\|_{\mathcal{U}(I^\Delta_j)}. \tag{5.3.39b}$$

Let us illustrate this by an example.

Example 5.3.1 Let us consider the cases analyzed in Examples 4.4.1 and 4.4.2, where there is severe pollution in all the elements except I^Δ_1. In the case of Example 4.4.1 when a mesh of $N(\Delta) = 81$ elements is employed, we have

$$\frac{\sqrt{(\mathcal{E}^{\text{Neum}})^2 - \left(\eta^{\text{Neum}}_{I^\Delta_1}\right)^2}}{\|e_{S^1_\Delta}\|_{\mathcal{U}(I-I^\Delta_1)}} = 0.37 \ll \frac{\mathcal{E}^{\text{Neum}}}{\|e_{S^1_\Delta}\|_{\mathcal{U}}} = 1.00 \approx \frac{\eta^{\text{Neum}}_{I^\Delta_1}}{\|e_{S^1_\Delta}\|_{\mathcal{U}(I^\Delta_1)}} = 1.00. \tag{5.3.40a}$$

Similarly in the case considered in Example 4.4.2 with $\vartheta = \frac{3}{4}$, a uniform mesh of $N(\Delta) =$ 16 elements, and $\alpha_2/(1-\vartheta) = 10$ (see Table 4.4.4) we have

$$\frac{\sqrt{(\mathcal{E}^{\text{Neum}})^2 - \left(\eta^{\text{Neum}}_{I^\Delta_1}\right)^2}}{\|e_{S^1_\Delta}\|_{\mathcal{U}(I-I^\Delta_1)}} = 0.29 \ll \frac{\mathcal{E}^{\text{Neum}}}{\|e_{S^1_\Delta}\|_{\mathcal{U}}} = 0.94 \approx \frac{\eta^{\text{Neum}}_{I^\Delta_1}}{\|e_{S^1_\Delta}\|_{\mathcal{U}(I^\Delta_1)}} = 1.02. \tag{5.3.40b}$$

Hence in both cases we have severe underestimation in all the elements except the first one, namely $\sqrt{(\mathcal{E}^{\text{Neum}})^2 - (\eta^{\text{Neum}}_{I^\Delta_1})^2} \ll \|e_{S^1_\Delta}\|_{\mathcal{U}(I-I^\Delta_1)}$. However, because the energy norm of the error is governed by the first element, $\|e_{S^p_\Delta}\|_{\mathcal{U}} \approx \|e_{S^p_\Delta}\|_{\mathcal{U}(I^\Delta_1)}$, and we have $\|e_{S^p_\Delta}\|_{\mathcal{U}(I^\Delta_1)} \leq \eta^{\text{Neum}}_{I^\Delta_1}$, we get the global upper estimate $\|e_{S^p_\Delta}\|_{\mathcal{U}} \leq \mathcal{E}^{\text{Neum}}$, where the effectivity index is practically the same as the effectivity index of the indicator in the first element.

Remark 5.3.2 When the pollution is negligible the element error indicators $\|\hat{e}^{\text{Dir}}_{I^\Delta_j}\|_{\mathcal{U}(I^\Delta_j)}$ are asymptotically exact, and hence we have

$$\|\hat{e}^{\text{Neum}}_{I^\Delta_j}\|_{\mathcal{U}(I^\Delta_j)} \leq \left(1 + \mathcal{C}h_{I^\Delta_j}\right)\|\hat{e}^{\text{Dir}}_{I^\Delta_j}\|_{\mathcal{U}(I^\Delta_j)} \leq \left(1 + \mathcal{C}h_{I^\Delta_j}\right)\|e_{S^p_\Delta}\|_{\mathcal{U}(I^\Delta_j)} \tag{5.3.41}$$

as $h_{I^\Delta_j}$ tends to zero.

As we have already noted in Section 5.1, $\eta^{\text{Neum}}_{I^\Delta_j}$ is the exact Neumann indicator and $\mathcal{E}^{\text{Neum}}$ the exact Neumann estimator. As we have already seen above, most often in practice, $\hat{e}^{\text{Neum}}_{I^\Delta_j}$ is approximated by the p version of the finite element method and we obtain the computed Neumann indicator

$$\eta^{\text{Neum}}_{I^\Delta_j, S^{p+k}(I^\Delta_j)} \stackrel{\text{def}}{=} \|\pi_{S^{p+k}(I^\Delta_j)}\left(\hat{e}^{\text{Neum}}_{I^\Delta_j}\right)\|_{\mathcal{U}(I^\Delta_j)} \tag{5.3.42}$$

and the corresponding computed estimator $\mathscr{E}^{\text{Neum}}_{S^{p+k}_\Delta}$, which is smaller than the exact estimator, $\mathscr{E}^{\text{Neum}}$, and hence it is not a guaranteed upper estimator for $\|e_{S^p_\Delta}\|_{\mathcal{U}}$.

Using similar notation as in Section 5.1, we will refer to the $\eta^{\text{Neum}}_{I^\Delta_j, S^{p+k}}$ (respectively $\eta^{\text{Neum}}_{I^\Delta_j, B^{p+k}}$) and the corresponding estimator by the name $ERNp+k$ (respectively $ERNBp+k$).

Another approach is to construct an explicit upper estimate $\eta^{\text{EXPL}}_{I^\Delta_j}$ for the exact indicator $\eta^{\text{Neum}}_{I^\Delta_j}$, in terms of the L^2 norms of the residual loads J^R_{j-1}, J^L_j, and $r_{I^\Delta_j}$. As we have already discussed in Section 5.1, the derivation of the explicit upper estimate for $\eta^{\text{Neum}}_{I^\Delta_j}$ starts from the observation

$$\eta^{\text{Neum}}_{I^\Delta_j} = \|\hat{e}^{\text{Neum}}_{I^\Delta_j}\|_{\mathcal{U}(I^\Delta_j)} = \|\mathcal{R}^{EQ}_{I^\Delta_j}\|_{L(\mathcal{U}(I^\Delta_j),\mathbb{R})} \tag{5.3.43a}$$

and proceeds with the construction of an explicit upper estimate for $\|\mathcal{R}^{EQ}_{I^\Delta_j}\|_{L(\mathcal{U}(I^\Delta_j),\mathbb{R})}$. Noting that

$$\begin{aligned}|\mathcal{R}^{EQ}_{I^\Delta_j}(v)| &= |\mathcal{R}^{EQ}_{I^\Delta_j}(v - v(x^\Delta_{j-1}))| \\ &\leq \left|\int_{I^\Delta_j} r_{I^\Delta_j}(v - v(x^\Delta_{j-1}))\right| + |J^L_j|\,|v(x^\Delta_j) - v(x^\Delta_{j-1})|\end{aligned} \tag{5.3.43b}$$

and hence

$$\begin{aligned}|\mathcal{R}^{EQ}_{I^\Delta_j}(v)| \leq &\int_{I^\Delta_j} |r_{I^\Delta_j}| \|v - v(x^\Delta_{j-1})\|_{L^\infty(I^\Delta_j)} \\ &+ \sqrt{(J^L_j)^2 + (J^R_{j-1})^2}\,|v(x^\Delta_j) - v(x^\Delta_{j-1})|\end{aligned} \tag{5.3.43c}$$

and recalling that

$$\|v - v(x^\Delta_{j-1})\|_{L^\infty(I^\Delta_j)} \leq \mathscr{C}\frac{1}{\sqrt{\beta_{1,I^\Delta_j}}}\sqrt{h_{I^\Delta_j}}\,\|v\|_{\mathcal{U}(I^\Delta_j)} \tag{5.3.43d}$$

we get

$$\begin{aligned}&\|\mathcal{R}^{EQ}_{I^\Delta_j}\|_{L(\mathcal{U}(I^\Delta_j),\mathbb{R})} \\ &\leq \mathscr{C}\frac{1}{\sqrt{\beta_{1,I^\Delta_j}}}\left(h_{I^\Delta_j}\sqrt{\int_{I^\Delta_j}\left(r_{I^\Delta_j}\right)^2} + \sqrt{h_{I^\Delta_j}}\sqrt{(J^L_j)^2 + (J^R_{j-1})^2}\right).\end{aligned} \tag{5.3.43e}$$

It follows that

$$\eta^{\text{Neum}}_{I^\Delta_j} \leq \eta^{\text{EXPL}}_{I^\Delta_j} \stackrel{\text{def}}{=} \mathscr{C}\frac{1}{\sqrt{\beta_{1,I^\Delta_j}}}\left(h_{I^\Delta_j}\sqrt{\int_{I^\Delta_j}\left(r_{I^\Delta_j}\right)^2} + \sqrt{h_{I^\Delta_j}}\sqrt{(J^L_j)^2 + (J^R_{j-1})^2}\right) \tag{5.3.44}$$

where $\mathscr{C}$ must be estimated from above. For elements of degree $p > 1$, a sharper estimate may be derived starting from the identity

$$|\mathscr{R}^{EQ}_{I^\Delta_j}(v)| = |\mathscr{R}^{EQ}_{I^\Delta_j}(v - \pi_{S^p(I^\Delta_j)}(v))|. \quad (5.3.45)$$

Remark 5.3.3 The explicit element error indicator $\eta^{\mathrm{EXPL}}_{I^\Delta_j}$ and the corresponding estimator can also be proven by employing a generalized interpolant, like, for example, $\hat{\mathcal{I}}^1_\Delta v$, which was defined in Section 5.2. Such a generalized interpolant is of interest because it can also be used in proving the explicit estimate (5.3.44) in higher dimensions.

Let us now give an example which illustrates the effect of the approximate determination of the error-indicator functions.

Example 5.3.2 Let us consider the regular model problem with $AE \equiv 1$, $c \equiv 0$, $u_{EX} \equiv x^{3/4}$, and, as before, let us compute the finite element solution by employing a radical mesh with $N(\Delta) = 5$ elements and $\gamma = 2$. Figure 5.3.1a compares the graphs of the absolute value of the errors, $|(d/dx)(\hat{e}^{\mathrm{Neum}}_{I^\Delta_j} - \pi_{\mathscr{P}^{p+k}}(\hat{e}^{\mathrm{Neum}}_{I^\Delta_j}))|$, for $k = 3$ and $k = 5$, in the second, third, and fourth element. From the comparison of these errors with the graph of the absolute value of the error in the derivative of the finite element solution $|(d/dx)(e_{S^p_\Delta})|$ in the corresponding elements, it can be seen that the error in the derivative of the error indicator is rather insignificant for a sufficiently high k which depends on the local smoothness of the exact solution. Let us remark that for the considered model problem, the nodal values are exact and we have $(d/dx)(\hat{e}^{\mathrm{Neum}}_{I^\Delta_j}) \equiv (d/dx)(e_{S^p_\Delta}|_{I^\Delta_j}) \equiv (d/dx)(\hat{e}^{\mathrm{Dir}}_{I^\Delta_j})$. In Fig. 5.3.1b we compare the errors $|(d/dx)(e_{S^p_\Delta}|_{I^\Delta_j} - \pi_{\mathscr{P}^{p+5}}(\hat{e}^{\mathrm{Neum}}_{I^\Delta_j}))|$ and $|(d/dx)(e_{S^p_\Delta}|_{I^\Delta_j} - \pi_{B^{p+5}}(\hat{e}^{\mathrm{Dir}}_{I^\Delta_j}))|$. Note that $(d/dx)(\pi_{\mathscr{P}^{p+5}}(\hat{e}^{\mathrm{Neum}}_{I^\Delta_j}))$ is a better approximation of $(d/dx)(e_{S^p_\Delta}|_{I^\Delta_j})$ than $(d/dx)(\pi_{B^{p+5}}(\hat{e}^{\mathrm{Neum}}_{I^\Delta_j}))$.

The estimator $\mathscr{E}^{\mathrm{Neum}}$ satisfies Assumptions A, B, C, and hence it can be analyzed using the asymptotic setting of Section 4.7. We have

Theorem 5.3.4 *Assumptions A, B, C of Section 5.1 hold for the estimator* $\mathscr{E}^{\mathrm{Neum}}$.

Proof Recall that $J^R_{j-1} = -\int_{I^\Delta_j} r_{I^\Delta_j}\varphi_{j-1}$ and $J^L_j = -\int_{I^\Delta_j} r_{I^\Delta_j}\varphi_j$. It follows that $\hat{e}^{\mathrm{Neum}}_{I^\Delta_j}$ is completely determined by the residual $r_{I^\Delta_j}$, and Assumptions A, B, and C can be proven for $\mathscr{E}^{\mathrm{Neum}}$ as in the case of $\mathscr{E}^{\mathrm{Dir}}$. □

Let us now summarize the results of Sections 5.2 and 5.3. In both cases we have to employ an element residual problem of the form:

Find $\hat{e}^{BC}_{I^\Delta_j} \in \mathscr{U}_{BC}(I^\Delta_j)$, such that

$$\mathscr{B}(\hat{e}^{BC}_{I^\Delta_j}, v) = \mathscr{R}^{EQ}_{I^\Delta_j}(v) \quad \forall v \in \mathscr{U}^{BC}(I^\Delta_j) \quad (5.3.46a)$$

with $\mathscr{U}^{BC} = \mathscr{U}_0$ when Dirichlet boundary conditions apply, and $\mathscr{U}^{BC} = \mathscr{U}$ when Neumann boundary conditions apply. In each case we define the exact indicator and

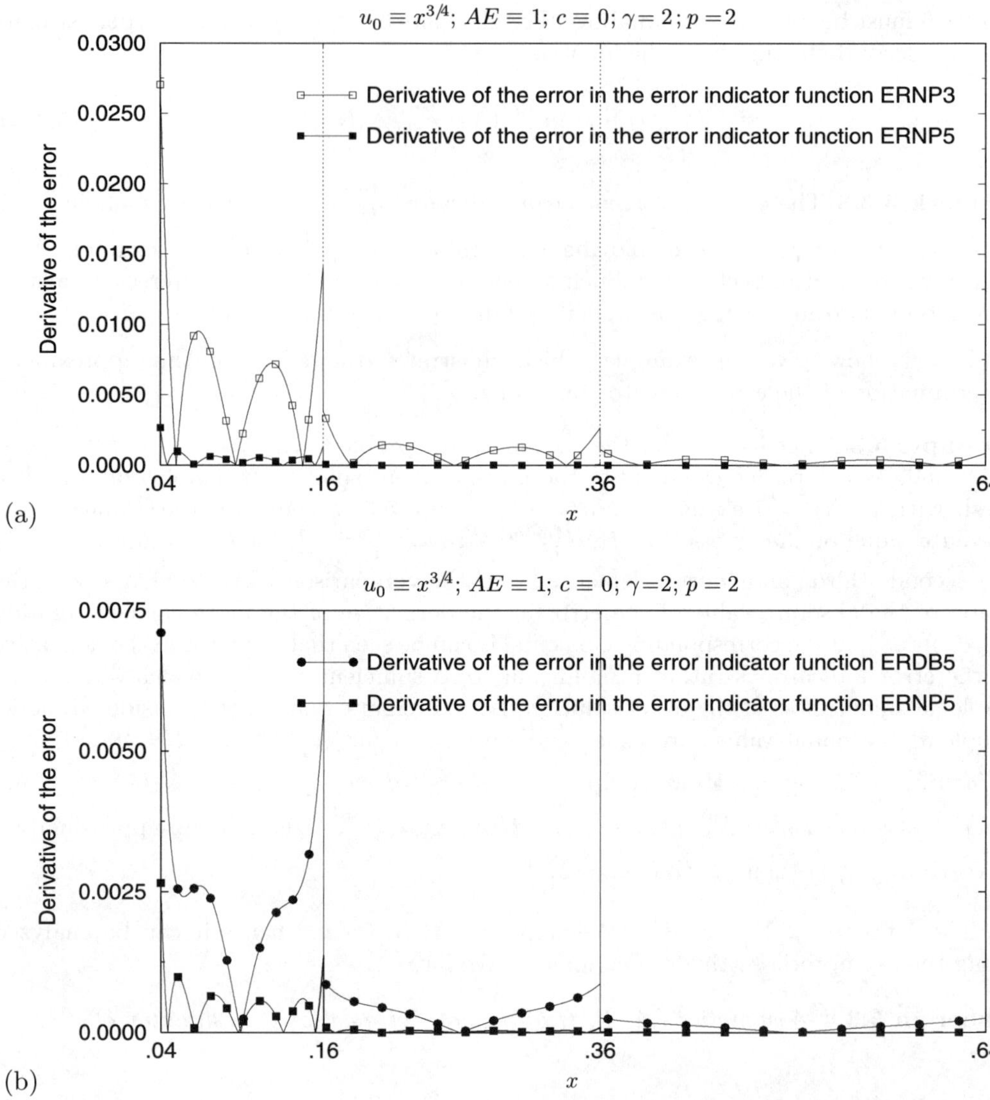

FIG. 5.3.1. Element residual error indicator function for the element residual problem with Neumann boundary conditions. Input data: $AE \equiv 1$, $c \equiv 0$, $u_0 \equiv x^{3/4}$. Mesh data: Radical mesh with $\gamma = 2$, $N(\Delta) = 5$, quadratic elements ($p = 2$). (a) Graphs of the absolute value of the derivative of the error in the error indicator function $|(d/dx)(\hat{e}^{\text{Neum}}_{I^\Delta_j} - \pi_{\mathcal{P}^{p+k}}(\hat{e}^{\text{Neum}}_j))|$ for $k = 3$ and $k = 5$. (b) Graphs of the absolute value of the error in the error indicator function $|(d/dx)(\hat{e}^{\text{Neum}}_{I^\Delta_j} - \pi_{\mathcal{P}^{p+5}}(\hat{e}^{\text{Neum}}_j))|$ and $|(d/dx)(\hat{e}^{\text{Dir}}_{I^\Delta_j} - \pi_{B^{p+5}}(\hat{e}^{\text{Dir}}_{I^\Delta_j}))|$, in the second, third, and fourth element from the singular point. Note that for the case considered the nodal values are exact, and it follows that $(d/dx)(\hat{e}^{\text{Neum}}_{I^\Delta_j}) \equiv (d/dx)(e_{S^p_\Delta})|_{I^\Delta_j}$ and

$$(d/dx)(\pi_{\mathcal{P}^{p+k}}(\hat{e}^{\text{Neum}}_{I^\Delta_j})) \equiv (d/dx)(\pi_{\mathcal{P}^{p+k}_0}(\hat{e}^{\text{Dir}}_{I^\Delta_j}))$$

the corresponding estimator by

$$\eta_{I_j^\Delta}^{BC} = \|\hat{e}_{I_j^\Delta}^{BC}\|_{\mathcal{U}(I_j^\Delta)}, \qquad \mathscr{E}^{BC} = \sqrt{\sum_{j=1}^{N(\Delta)} \left(\eta_{I_j^\Delta}^{BC}\right)^2} \tag{5.3.46b}$$

and we obtain the two-sided estimates

$$\mathscr{E}^{\text{Dir}} \leq \|e_{S_\Delta^p}\| \leq \mathcal{K}^{\text{Dir}}\,\mathscr{E}^{\text{Dir}} \tag{5.3.46c}$$

and

$$(1 - \mathcal{K}^{\text{Neum}} h^\Delta)\mathscr{E}^{\text{Neum}} \leq \|e_{S_\Delta^p}\| \leq \mathscr{E}^{\text{Neum}} \tag{5.3.46d}$$

where $\mathcal{K}^{\text{Dir}}$ and $\mathcal{K}^{\text{Neum}}$ depend only on the input data AE and c. We may also employ $\mathscr{E}^{\text{Dir}}$ and $\mathscr{E}^{\text{Neum}}$, simultaneously, to obtain the two-sided estimate

$$\mathscr{E}^{\text{Dir}} \leq \|e_{S_\Delta^p}\|_{\mathcal{U}} \leq \mathscr{E}^{\text{Neum}} \tag{5.3.46e}$$

which is, in general, sharper than the two-sided estimates (5.3.46c,d).

We also introduced the computed indicators and the corresponding computed estimators

$$\eta_{I_j^\Delta,S}^{BC} = \|\pi_S(\hat{e}_{I_j^\Delta}^{BC})\|_{\mathcal{U}(I_j^\Delta)}, \qquad \mathscr{E}_S^{BC} = \sqrt{\sum_{j=1}^{N(\Delta)} \left(\eta_{I_j^\Delta,S}^{BC}\right)^2} \tag{5.3.46f}$$

where $S = S^{p+k}(I_j^\Delta)$ or $B^{p+k}(I_j^\Delta)$, etc., and we noted that

$$\eta_{I_j^\Delta,S}^{BC} \leq \eta_{I_j^\Delta}^{BC}, \qquad \mathscr{E}_S^{BC} \leq \mathscr{E}^{BC} \tag{5.3.46g}$$

and hence the upper bounds in (5.3.46c,d) are not guaranteed for the computed estimators. We can still obtain an upper computed estimator by employing the identity

$$\mathscr{E}^{BC} = \sqrt{(\mathscr{E}_S^{BC})^2 + \sum_{j=1}^{N(\Delta)} \|\hat{e}_{I_j^\Delta}^{BC} - \pi_S(\hat{e}_{I_j^\Delta}^{BC})\|_{\mathcal{U}(I_j^\Delta)}^2} \tag{5.3.46h}$$

and by using an explicit upper estimator

$$\|\hat{e}_{I_j^\Delta}^{BC} - \pi_S(\hat{e}_{I_j^\Delta}^{BC})\|_{\mathcal{U}(I_j^\Delta)} \leq \mathscr{E}_{\hat{e}_{I_j^\Delta}^{BC}}^{EXPL}. \tag{5.3.46i}$$

We will elaborate on this approach in Chapter 6.

Let us emphasize that $\mathscr{E}^{\text{Neum}}$ (respectively $\mathscr{E}^{\text{Dir}}$) is a guaranteed upper (respectively lower) estimator for the global energy norm of the error without any assumptions about the pollution. On the other hand, the element error indicators $\eta_{I_j^\Delta}^{\text{Neum}}$ are reliable estimates of the element energy norms of the error only if the pollution error is negligible relative to the local error. Adaptive construction of the mesh often leads to control of the pollution error over the entire mesh and, in this case, the effectivity of the element error indicators improves. However, as we have seen in Chapter 4, in general, one cannot assess the magnitude of the pollution a priori, and an a-posteriori estimation of the pollution error is also needed.

Problems

P5.3.1. Consider the regular model problem with $AE \equiv 1$, $c > 0$, $f \equiv 1$. For an interior element I_j^Δ, compare $\eta^{\mathrm{Dir}}_{I_j^\Delta}$ and $\eta^{\mathrm{Neum}}_{I_j^\Delta}$ under the assumption that the nodal values of the finite element solution are exact.

P5.3.2. Repeat P5.3.1 assuming that $e_{S^p_\Delta}(x_j^\Delta) = e_{S^p_\Delta}(x_{j+1}^\Delta) = \epsilon$.

5.3* Analysis of the Neumann element residual estimator in two dimensions

We will now analyze the Neumann estimator $\mathscr{E}^{\mathrm{Neum}}$ in the two-dimensional case. As in the one-dimensional case, in the elements with $\partial\tau \cap \Gamma_{\mathrm{D}} = \emptyset$ the Neumann element residual error indicators $\hat{e}_\tau^{\mathrm{Neum}}$ are the solutions of the Neumann problems:

Find $\hat{e}_\tau^{\mathrm{Neum}} \in \mathscr{U}(\tau)$ such that

$$\mathscr{B}_\tau(\hat{e}_\tau^{\mathrm{Neum}}, v) = \mathscr{R}_\tau^{EQ}(v) = \int_\tau r_\tau^\Delta v + \sum_{\epsilon \subseteq \partial\tau} \int_\epsilon J_\epsilon^\tau v \quad \forall v \in \mathscr{U}(\tau). \tag{5.3*.1a}$$

For the elements which intersect the Dirichlet boundary $\hat{e}_\tau^{\mathrm{Neum}}$ is the solution of the mixed boundary value problem which is obtained by replacing $\mathscr{U}(\tau)$ by

$$\mathscr{U}_{\partial\tau_{\mathrm{D}},0}(\tau) \stackrel{\mathrm{def}}{=} \{v \in \mathscr{U}(\tau) \mid v|_{\partial\tau_{\mathrm{D}}} \equiv 0\} \tag{5.3*.1b}$$

where $\partial\tau_{\mathrm{D}} \stackrel{\mathrm{def}}{=} \partial\tau \cap \Gamma_{\mathrm{D}}$ is assumed to consist of one or more edges. Here we recall that

$$r_\tau^\Delta = f|_\tau - \mathfrak{D}u_{S^p_\Delta}|_\tau \tag{5.3*.2a}$$

is the interior residual in the element τ, and

$$J_\epsilon^\tau = \tfrac{1}{2} J_\epsilon^\Delta + \theta_\epsilon^\tau \tag{5.3*.2b}$$

is the jump assigned to the element τ on the edge ϵ, where J_ϵ^Δ, θ_ϵ^τ are, respectively, the jump and the *edge correction* function for the edge $\epsilon \subseteq \partial\tau$. For example, in the case that $\mathfrak{D}\cdot = \boldsymbol{\nabla} \cdot (\mathbf{K}\boldsymbol{\nabla}\cdot)$, we have

$$J_\epsilon^\Delta = \begin{cases} \left((\mathbf{K}\boldsymbol{\nabla} u_{S^p_\Delta})\big|_{\tau^*} - (\mathbf{K}\boldsymbol{\nabla} u_{S^p_\Delta})\big|_\tau\right) \cdot \mathbf{n}_\epsilon, & \epsilon = \partial\tau \cap \partial\tau^* \\ g - (\mathbf{K}\boldsymbol{\nabla} u_{S^p_\Delta})\big|_\tau \cdot \mathbf{n}_\epsilon, & \epsilon = \partial\tau \cap \Gamma_{\mathrm{N}} \end{cases} \tag{5.3*.2c}$$

and $J_\epsilon^\Delta = 0$, for $\epsilon = \partial\tau \cap \Gamma_{\mathrm{D}}$, where $\mathbf{n}_\epsilon$ is the exterior unit normal on $\epsilon \subseteq \partial\tau$.

The edge corrections θ_ϵ^τ will be specified in terms of edge functions θ_ϵ, namely,

$$\theta_\epsilon^\tau = sg(\epsilon, \tau)\theta_\epsilon \tag{5.3*.3a}$$

and

$$\theta_\epsilon^{\tau^*} = sg(\epsilon, \tau^*)\theta_\epsilon = -sg(\epsilon, \tau)\theta_\epsilon \tag{5.3*.3b}$$

for each interior edge, $\epsilon = \partial\tau \cap \partial\tau^*$, where $sg(\epsilon, \tau) = +1$ or -1, and

$$sg(\epsilon, \tau^*) = -sg(\epsilon, \tau) \tag{5.3*.3c}$$

and $\theta_\epsilon^\tau \equiv 0$, for each edge on the boundary, $\epsilon \subseteq \partial\Omega$. It follows that for each interior edge, $\epsilon = \partial\tau \cap \partial\tau^*$, we have

$$\theta_\epsilon^\tau + \theta_\epsilon^{\tau^*} = 0 \tag{5.3*.4a}$$

and hence we have an exact splitting of the jumps

$$J_\epsilon^\tau + J_\epsilon^{\tau^*} = J_\epsilon^\Delta \tag{5.3*.4b}$$

and an exact splitting of the global residuum

$$\mathcal{R}(v) = \sum_{\tau\in\Delta} \mathcal{R}_\tau^{EQ}(v|_\tau) \quad \forall v \in \mathcal{U}. \tag{5.3*.5}$$

As we have already seen in Section 5.1*, the edge functions θ_ϵ can be constructed such that $\mathcal{R}_\tau^{EQ}$ is orthogonal to

$$S_{\partial\tau_{\mathrm{D}}}^p(\tau) \stackrel{\text{def}}{=} \{v \in S^p(\tau) \mid v|_{\partial\tau_{\mathrm{D}}} \equiv 0\} \tag{5.3*.6a}$$

the restriction of the finite element space in the element τ, namely,

$$0 = \mathcal{R}_\tau^{EQ}(v) = \mathcal{R}_\tau^{UNEQ}(v) + \sum_{\substack{\epsilon\subseteq\partial\tau \\ \epsilon\subseteq\partial\Omega}} \int_\epsilon sg(\epsilon,\tau)\theta_\epsilon v \quad \forall v \in S_{\partial\tau_{\mathrm{D}}}^p(\tau) \tag{5.3*.6b}$$

where

$$\mathcal{R}_\tau^{UNEQ}(v) \stackrel{\text{def}}{=} \int_\tau r_\tau^\Delta v + \sum_{\substack{\epsilon\subseteq\partial\tau \\ \epsilon\not\subseteq\partial\Omega}} \int_\epsilon \frac{1}{2} J_\epsilon^\Delta v + \sum_{\epsilon\subseteq\partial\tau\cap\partial\Omega} \int_\epsilon J_\epsilon^\Delta v \tag{5.3*.6c}$$

is the *unequilibrated* element residuum corresponding to one half splitting of the jumps in the interior edges.

Below we will show that many constructions of θ_ϵ such that they satisfy (5.3*.6b) are possible. Let us first address the existence and uniqueness of $\hat{e}_\tau^{\mathrm{Neum}}$.

Theorem 5.3*.1 *Consider the case of the model problem*

$$(-\boldsymbol{\nabla}\cdot(\mathbf{K}\boldsymbol{\nabla}u) + cu)|_\Omega = f, \qquad u|_{\Gamma_{\mathrm{D}}} \equiv 0, \qquad \left.\frac{\partial u}{\partial n}\right|_{\Gamma_{\mathrm{N}}} = g \tag{5.3*.7a}$$

and assume that the input data f, c, g, and K_{ij}, are elementwise smooth, namely,

$$f|_\tau \in L^2(\tau), \qquad c|_\tau \in L^2(\tau), \qquad g|_{\partial\tau\cap\Gamma_{\mathrm{N}}} \in L^2(\tau) \tag{5.3*.7b}$$

and

$$\max_{k=1,2} \left\| \frac{\partial}{\partial x_k}(K_{ij}) \right\|_{L^\infty(\tau)} \le K < \infty. \tag{5.3*.7c}$$

Then $\hat{e}_\tau^{\mathrm{Neum}}$ exists, and in the cases that either $c \not\equiv 0$, or $\partial\tau_{\mathrm{D}} = \epsilon \subseteq \Gamma_{\mathrm{D}}$, it is also unique.

Proof The proof of existence follows from the proof of existence of solutions for the model problem. The only difference is that the prescribed flux boundary condition is in terms of the split jumps J_ϵ^τ, and the load in terms of the interior residual r_τ^Δ, and assumption (5.3*.7c) is needed to guarantee that they have good meaning. Let us address the case that $\partial\tau_{\mathrm{D}} = \emptyset$, and the problem (5.3*.1) is a pure Neumann problem. In this case $1 \in S^p(\tau)$, and letting $v \equiv 1$ in the orthogonality condition (5.3*.6b) we get

$$\mathcal{R}_\tau^{EQ}(1) = \int_\tau r_\tau^\Delta + \sum_{\epsilon \subseteq \partial\tau} \int_\epsilon \left(\frac{1}{2} J_\epsilon^\Delta + \theta_\epsilon^\tau\right) = 0 \tag{5.3*.8}$$

which is the *consistency* condition required for the existence of $\hat{e}_\tau^{\mathrm{Neum}}$, when $c|_\tau \equiv 0$, in which case the solution is unique up to a constant. In the cases that $\partial\tau_{\mathrm{D}}$ is one or several edges of Γ_{D}, and/or $c \not\equiv 0$, (5.3*.8) is not needed for existence, and $\hat{e}_\tau^{\mathrm{Neum}}$ exists and is unique. □

As in the one-dimensional case, we will denote by

$$\eta_\tau^{\mathrm{Neum}} = \|\hat{e}_\tau^{\mathrm{Neum}}\|_{\mathcal{U}(\tau)} \qquad \mathcal{E}^{\mathrm{Neum}} = \sqrt{\sum_{\tau\in\Delta} (\eta_\tau^{\mathrm{Neum}})^2} \tag{5.3*.10a}$$

respectively, the exact Neumann element error indicator, and the exact Neumann estimator.

As in the one-dimensional case, the exact estimator $\mathcal{E}^{\mathrm{Neum}}$ is an upper bound for the energy norm of the error.

Theorem 5.3*.2 *Consider the boundary value problem (5.3*.7a) with data which satisfy (5.3*.7b,c). Then we have*

$$\|e_{S_\Delta^p}\|_{\mathcal{U}} \le \mathcal{E}^{\mathrm{Neum}} \stackrel{\mathrm{def}}{=} \sqrt{\sum_{\tau\in\Delta} \|\hat{e}_\tau^{\mathrm{Neum}}\|^2_{\mathcal{U}(\tau)}}. \tag{5.3*.10b}$$

Proof The proof is exactly the same as in the one-dimensional case. □

Note that the proof of the upper bound, $\|e_{S_\Delta^p}\|_{\mathcal{U}} \le \mathcal{E}^{\mathrm{Neum}}$, is based on the assumption that there exists an exact splitting

$$\mathcal{R} = \sum_{\tau\in\Delta} \mathcal{R}_\tau^{EQ} \tag{5.3*.11a}$$

with

$$\mathcal{R}_\tau^{EQ}(v) = 0 \quad \forall v \in S^p_{\partial\tau_{\mathrm{D}}}(\tau), \ \ \forall\tau \in \Delta. \tag{5.3*.11b}$$

Let us now construct such a splitting. We will employ the orthogonality conditions

$$\mathcal{R}_\tau^{EQ,q}(\psi_i^\tau) = \mathcal{R}_\tau^{UNEQ}(\psi_i^\tau) + \sum_{\substack{\epsilon\subseteq\partial\tau \\ \epsilon\not\subseteq\partial\tau\cap\partial\Omega}} \int_\epsilon sg(\epsilon,\tau)\theta_\epsilon\psi_i^\tau = 0 \tag{5.3*.12}$$

for $i = 1, \ldots, N_\tau^q$, and for all $\tau \in \Delta$, where we suppress the degrees of freedom on $\partial\tau_{\mathrm{D}}$. Here ψ_i^τ, $i = 1, \ldots, N_\tau^q$, are the hierarchic shape functions of degree q, and N_τ^q denotes the total number of shape functions ($N_\tau^q = (q+1)(q+2)/2$ for triangles). The notation $\mathcal{R}_\tau^{EQ,q}$ underlines that $\mathcal{R}_\tau^{EQ,q}$ is orthogonal to $S_{\partial\tau_{\mathrm{D}}}^q(\tau)$, and $q \geq 1$ will be referred to as the order of the equilibration. As in the one-dimensional case, the orthogonality of $\mathcal{R}_\tau^{EQ}$ with respect to the interior element shape functions is automatically satisfied.

We will employ a polynomial representation of θ_ϵ, namely

$$\theta_\epsilon^{(q)} = \sum_{i=1}^{q+1} \theta_{\epsilon,i}\lambda_{\epsilon,i} = \theta_\epsilon^{(1)} + \sum_{i=3}^{q+1} \theta_{\epsilon,i}\lambda_{\epsilon,i}. \tag{5.3*.13a}$$

Here $\lambda_{\epsilon,1}$, $\lambda_{\epsilon,2} \in \mathcal{P}^1(\epsilon)$ are the affine edge functions which satisfy the orthogonalities

$$\int_\epsilon \lambda_{\epsilon,i}\psi_{\epsilon,j} = \delta_{ij} \tag{5.3*.13b}$$

where $\psi_{\epsilon,1}$, $\psi_{\epsilon,2} \in \mathcal{P}^1(\epsilon)$, are the standard linear shape functions over the edge ϵ. It follows that

$$\theta_{\epsilon,k} = \int_\epsilon \theta_\epsilon^{(1)}\psi_{\epsilon,k}, \quad k = 1, 2 \tag{5.3*.13c}$$

and $\lambda_{\epsilon,i} \in \mathcal{P}^{i-1}(\epsilon)$, for $i > 2$, are the edge polynomials which satisfy the orthogonality conditions

$$\int_\epsilon \lambda_{\epsilon,i}\varphi_{\epsilon,j}^\Delta|_\epsilon = \delta_{ij}, \quad j \leq i = 3, \ldots, q+1 \tag{5.3*.13d}$$

where $\varphi_{\epsilon,k}^\Delta$, $k = 1, \ldots, q+1$ are the global basis functions which do not vanish on the edge ϵ.

Remark 5.3*.1 We have

$$\lambda_{\epsilon,1} = \frac{2}{|\epsilon|}(2\psi_{\epsilon,1} - \psi_{\epsilon,2}), \qquad \lambda_{\epsilon,2} = \frac{2}{|\epsilon|}(2\psi_{\epsilon,2} - \psi_{\epsilon,1}) \tag{5.3*.14a}$$

and

$$\tilde{\lambda}_{\tilde{\epsilon},3}(\tilde{x}) = \frac{5\sqrt{6}}{4}(-1 + 3\tilde{x}^2), \qquad \tilde{\lambda}_{\tilde{\epsilon},4}(\tilde{x}) = \frac{7\sqrt{5}}{2\sqrt{2}}(-3\tilde{x} + 5\tilde{x}^3), \tag{5.3*.14b}$$

$$\tilde{\lambda}_{\tilde{\epsilon},5}(\tilde{x}) = \frac{9\sqrt{7}}{8\sqrt{2}}(3 - 30\tilde{x}^2 + 35\tilde{x}^4), \tag{5.3*.14c}$$

$$\tilde{\lambda}_{\tilde{\epsilon},6}(\tilde{x}) = \frac{3}{8\sqrt{2}}(165\tilde{x} - 770\tilde{x}^3 + 693\tilde{x}^5),$$

$$\tilde{\lambda}_{\tilde{\epsilon},7}(\tilde{x}) = \frac{13\sqrt{11}}{16\sqrt{2}}(-5 + 105\tilde{x}^2 - 315\tilde{x}^4 + 231\tilde{x}^6) \tag{5.3*.14d}$$

where $\tilde{x}$ is the master coordinate along the master edge $\tilde{\epsilon} = \tilde{I} = (-1, 1)$. Here it was assumed that the shape functions are constructed from the primitives of the Legendre polynomials.

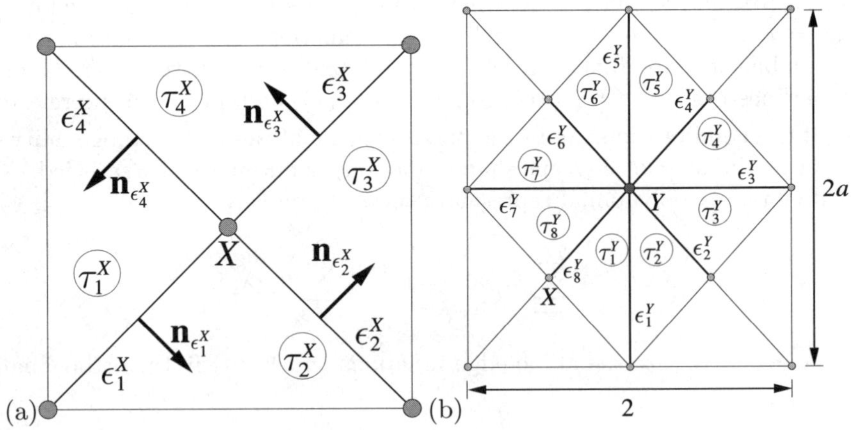

FIG. 5.3*.1. The basic cell of the criss-cross pattern with the enumeration of the edges and the elements connected to the two typical vertices of the pattern. (a) The vertex X with the elements τ_k^X and the edges ϵ_k^X, $k = 1, \ldots, 4$, connected to it. The directions of the edge unit normals $\mathbf{n}_{\epsilon_k^X}$ are chosen such that they rotate in the counterclockwise direction around the vertex X. (b) The vertex Y with the elements τ_k^Y and the edges ϵ_k^Y, $k = 1, \ldots, 8$, connected to it

Let us now give an example of how the representation (5.3*.13) can be used to simplify the determination of θ_ϵ such that they satisfy (5.3*.12).

Example 5.3*.1 (*Construction of the edge corrections for interior vertices*) Let X be an interior vertex of the mesh, and let τ_i^X and ϵ_i^X, $i = 1, \ldots, 4$, denote the elements and edges connected to the vertex X, as shown in Fig. 5.3*.1, and let $\nu(\epsilon_k^X)$ be the local number (1 or 2) of the vertex X with respect to the edge ϵ_k^X. Using these notations, the unknowns are

$$\theta_{\epsilon^X, \nu(\epsilon^X)} = \int_{\epsilon^X} \theta_{\epsilon^X} \varphi_X^\Delta|_{\epsilon^X} \tag{5.3*.15a}$$

and the orthogonality condition $\mathcal{R}_{\tau_k^X}^{EQ,1}(\varphi_X^\Delta|_{\tau_k^X}) = 0$, gives

$$\sum_{\epsilon \subseteq \partial\tau_k^X} \int_\epsilon \theta_\epsilon^{\tau_k^X} \varphi_X^\Delta|_{\tau_k^X} = -\mathcal{R}_{\tau_k^X}^{UNEQ}(\varphi_X^\Delta|_{\tau_k^X}) \tag{5.3*.15b}$$

for $k = 1, \ldots, 4$, which, employing (5.3*.15a), can be written as the following system of linear equations

$$\begin{bmatrix} 1 & 0 & 0 & -1 \\ -1 & 1 & 0 & 0 \\ 0 & -1 & 1 & 0 \\ 0 & 0 & -1 & 1 \end{bmatrix} \begin{Bmatrix} \theta_{\epsilon_1^X, \nu(\epsilon_1^X)} \\ \theta_{\epsilon_2^X, \nu(\epsilon_2^X)} \\ \theta_{\epsilon_3^X, \nu(\epsilon_3^X)} \\ \theta_{\epsilon_4^X, \nu(\epsilon_4^X)} \end{Bmatrix} = \begin{Bmatrix} -\mathcal{R}_{\tau_1^X}^{UNEQ}(\varphi_X^\Delta) \\ -\mathcal{R}_{\tau_2^X}^{UNEQ}(\varphi_X^\Delta) \\ -\mathcal{R}_{\tau_3^X}^{UNEQ}(\varphi_X^\Delta) \\ -\mathcal{R}_{\tau_4^X}^{UNEQ}(\varphi_X^\Delta) \end{Bmatrix} \tag{5.3*.15c}$$

or

$$\mathbf{M}_X\boldsymbol{\theta}_X = -\boldsymbol{\mathfrak{R}}_X^{UNEQ} \tag{5.3*.15d}$$

where $\mathbf{M}_X \in \mathbb{R}^{ne(X)\times ne(X)}$, $\boldsymbol{\theta}_X \in \mathbb{R}^{ne(X)}$, and $\boldsymbol{\mathfrak{R}}_X^{UNEQ} \in \mathbb{R}^{ne(X)}$ are, respectively, the matrix, the vector or unknowns, and the right-hand side of the system (5.3*.15c), and $ne(X)$ is the number of edges connected to vertex X (here $ne(X) = 4$). The matrix $\mathbf{M}_X$ is singular and has exactly one zero eigenvalue with eigenvector

$$\mathbf{V}_X = [\underbrace{1, 1, \ldots, 1}_{ne(X)}]^{\mathrm{T}}.$$

Recalling the orthogonality of the error $\mathscr{B}(e_{S^p_\Delta}, \varphi_X^\Delta) = 0$, we obtain

$$\sum_{k=1}^{ne(X)} \mathscr{R}_{\tau_k^X}^{UNEQ}(\varphi_X^\Delta|_{\tau_k^X}) = \mathscr{R}(\varphi_X^\Delta) = \mathscr{B}(e_{S^p_\Delta}, \varphi_X^\Delta) = 0 \tag{5.3*.16a}$$

which can also be written as

$$\mathbf{V}_X^{\mathrm{T}}\boldsymbol{\mathfrak{R}}_X^{UNEQ} = 0. \tag{5.3*.16b}$$

Thus $\boldsymbol{\mathfrak{R}}_X^{UNEQ}$ is orthogonal to $\mathbf{V}_X$, and hence the system of linear equations (5.3*.15d) has exactly a one-parameter family of solutions. Particular solutions can be obtained by letting (see [Bank and Weiser, 1985])

$$\theta^{\mathrm{BW}}_{\epsilon_n^X, \nu(\epsilon_n^X)} = 0, \quad n = ne(X) \tag{5.3*.17a}$$

or, by minimizing the weighted l^2 norm

$$\|\boldsymbol{\theta}_X\|_{l^2, \{w_k\}_{k=1}^{ne(X)}} \overset{\mathrm{def}}{=} \sqrt{\sum_{k=1}^{ne(X)} w_k \left(\theta_{\epsilon_k^X, \nu(\epsilon_k^X)}\right)^2} \tag{5.3*.17b}$$

where $\{w_k\}_{k=1}^{ne(X)}$ is a set of weights assigned to the edges which are connected to vertex X; for example, one may choose $w_k = 1$ or $w_k = |\epsilon_k^X|^{-2}$. This choice was proposed by [Ladeveze, 1977], [Ladeveze and Leguillon, 1983], and [Ladeveze and Maunder, 1996].

We constructed $\theta_{\epsilon_k,1}$, $k = 1, 2, 3, 4$ from the patch shown in Fig. 5.3*.1b. The values of $\theta_{\epsilon_k,2}$ are determined similarly using the appropriate patches. For example, in Fig. 5.3*.1b we have $\theta_{\epsilon_1^X,2} = \theta_{\epsilon_8^Y,1}$. The computation of $\mathscr{R}_\tau^{EQ,1}$ is based on the use of all patches of elements associated with the vertices of the element τ. Realizing that $\sum \varphi_X^\Delta = 1$, where the sum is taken over the vertices of the element τ, we get $\mathscr{R}_\tau^{EQ,1}(1) = 0$. Let us underline that this argument holds also for meshes of curvilinear elements.

Remark 5.3*.2 In the case of the elasticity problem, for the existence of solutions of the element residual problems, it is necessary to have

$$\mathscr{R}_\tau^{EQ}(\mathbf{v}) = 0, \quad \mathbf{v} = \mathbf{i}_x, \mathbf{i}_y, \; -y\mathbf{i}_x + x\mathbf{i}_y \tag{5.3*.18a}$$

where $\mathbf{i}_x$ and $\mathbf{i}_y$ are the unit-vectors along the x and the y coordinate directions, respectively. The conditions

$$\mathcal{R}_{\tau}^{EQ}(\mathbf{i}_x) = 0, \qquad \mathcal{R}_{\tau}^{EQ}(\mathbf{i}_y) = 0 \tag{5.3*.18b}$$

are always satisfied, while the condition

$$\mathcal{R}_{\tau}^{EQ}(-y\mathbf{i}_x + x\mathbf{i}_y) = 0 \tag{5.3*.18c}$$

which expresses the equilibrium of moments for the residuals assigned to the element τ, is satisfied by the construction of the edge splitting of the residuals, only in the case that the element transformation is such that the mapped shape functions reproduce the linear monomials, namely,

$$x = \sum x_i \varphi_{X_i}^{\Delta}, \qquad y = \sum y_i \varphi_{X_i}^{\Delta} \tag{5.3*.18d}$$

where the sums are taken over the vertices of the element τ. This is the case when subparametric or isoparametric element transformations are employed, but not in the case that a superparametric or a blending-function transformation is used.

After the coefficients $\theta_{\epsilon,1}$, $\theta_{\epsilon,2}$ have been obtained for the edge ϵ, the coefficients for the higher-order corrections, $\theta_{\epsilon,i}$, $i = 3, \ldots, q+1$, can be determined such that

$$\mathcal{R}_{\tau}^{EQ,q}(\varphi_{\epsilon,j}^{\Delta}|_{\tau}) = \mathcal{R}_{\tau}^{UNEQ}(\varphi_{\epsilon,j}^{\Delta}|_{\tau}) + \sum_{k=1}^{j} \theta_{\epsilon,k} \int_{\epsilon} \lambda_{\epsilon,k} \varphi_{\epsilon,j}^{\Delta}|_{\tau} = 0 \tag{5.3*.19a}$$

for $j = 3, \ldots, q+1$. This gives

$$\theta_{\epsilon,j} = -\mathcal{R}_{\tau}^{UNEQ}(\varphi_{\epsilon,j}^{\Delta}|_{\tau}) - \sum_{k=1}^{j-1} \theta_{\epsilon,k} \int_{\epsilon} \lambda_{\epsilon,k} \varphi_{\epsilon,j}^{\Delta}|_{\tau}, \quad j = 3, \ldots, q+1. \tag{5.3*.19b}$$

Note that we have

$$0 = \mathcal{R}(\varphi_{\epsilon,j}^{\Delta}) = \mathcal{R}_{\tau}^{EQ,q}(\varphi_{\epsilon,j}^{\Delta}|_{\tau}) + \mathcal{R}_{\tau^*}^{EQ,q}(\varphi_{\epsilon,j}^{\Delta}|_{\tau^*}), \quad \epsilon = \partial\tau \cap \partial\tau^* \tag{5.3*.19c}$$

and hence

$$\mathcal{R}_{\tau}^{EQ,q}(\varphi_{\epsilon,j}^{\Delta}|_{\tau}) = 0 \implies \mathcal{R}_{\tau^*}^{EQ}(\varphi_{\epsilon,j}^{\Delta}|_{\tau^*}) = 0 \tag{5.3*.19d}$$

for $j = 3, \ldots, q+1$. Thus, the interior degrees of freedom for the edge ϵ $\theta_{\epsilon,j}$, $j = 3, \ldots, q+1$ may be determined for either element τ or τ^* where $\epsilon = \partial\tau \cap \partial\tau^*$. Note that the determination of the higher-order terms $\theta_{\epsilon,j}$, $j = 3, \ldots, q+1$ is explicit, and hence its cost is negligible compared to the cost of determining $\theta_{\epsilon,1}$, $\theta_{\epsilon,2}$.

Remark 5.3*.3 It is also possible to construct the θ_{ϵ} such that they satisfy only the consistency condition

$$\mathcal{R}_{\tau}^{EQ,0}(1) = \int_{\tau} r_{\tau}^{\Delta} + \sum_{\epsilon \subseteq \partial\tau} \int_{\epsilon} \left(\frac{1}{2} J_{\epsilon} + sg(\epsilon,\tau)\theta_{\epsilon}\right) = 0. \tag{5.3*.20}$$

However, as we will see below, the reliability of $\mathscr{E}^{\text{Neum}}$, for elements of degree $p > 1$, improves when the orthogonality (5.3*.12) is enforced with $q = p$.

Remark 5.3*.4 The main difference between the one- and two-dimensional cases is that in two dimensions the split jumps J_ϵ^τ are not unique. We will underline the dependence of $\mathscr{E}^{\text{Neum}}$ on the particular choice of the θ_ϵ, by writing

$$\mathscr{E}^{\text{Neum}} = \mathscr{E}^{\text{Neum}}(\boldsymbol{\theta}_{\mathcal{E}_{int}}), \quad \boldsymbol{\theta}_{\mathcal{E}_{int}} \stackrel{\text{def}}{=} \left[\theta_\epsilon; \epsilon \in \mathcal{E}_{int}(\Delta)\right]^{\mathrm{T}}. \tag{5.3*.21a}$$

Depending on the particular choice of $\boldsymbol{\theta}_{\mathcal{E}_{int}}$, the estimate $\mathscr{E}^{\text{Neum}}$ could be much bigger than $\|e_{S_\Delta^p}\|_{\mathcal{U}}$. Above we employed polynomial edge corrections $\theta_\epsilon \in \mathscr{P}^p(\epsilon)$; nevertheless, various other representations are also possible. The *optimal* edge corrections $\boldsymbol{\theta}_{\mathcal{E}_{int}}^{opt}$ may be defined as the ones which which minimize the upper bound $\|e_{S_\Delta^p}\|_{\mathcal{U}} \leq \mathscr{E}^{\text{Neum}}$, namely

$$\|e_{S_\Delta^p}\|_{\mathcal{U}} \leq \mathscr{E}^{\text{Neum}}(\boldsymbol{\theta}_{\mathcal{E}_{int}}^{opt}) = \min_{\boldsymbol{\theta} \in \boldsymbol{\Theta}^{EQ,p}_{\boldsymbol{\mathscr{P}}^p_{\mathcal{E}_{int}}}} \mathscr{E}(\boldsymbol{\theta}_{\mathcal{E}_{int}}) \tag{5.3*.21b}$$

where $\boldsymbol{\Theta}^{EQ,p}_{\boldsymbol{\mathscr{P}}^p_{\mathcal{E}_{int}}}$ is the set of *admissible* edge corrections given by

$$\boldsymbol{\Theta}^{EQ,p}_{\boldsymbol{\mathscr{P}}^p_{\mathcal{E}_{int}}} = \left\{\boldsymbol{\theta}_\epsilon \in \boldsymbol{\mathscr{P}}^p_{\mathcal{E}_{int}} \mid \mathscr{R}_\tau^{UNEQ}(v) + \sum_{\epsilon \subseteq \partial\tau} \int_\epsilon \operatorname{sg}(\epsilon,\tau)\theta_\epsilon v = 0 \quad \forall v \in S^p(\tau),\ \forall \tau \in \Delta\right\} \tag{5.3*.21c}$$

where we let $\boldsymbol{\mathscr{P}}^p_{\mathcal{E}_{int}} \stackrel{\text{def}}{=} \prod_{\epsilon \in \mathcal{E}_{int}(\Delta)} \mathscr{P}^p(\epsilon)$. Note that $\boldsymbol{\theta}_{\mathcal{E}_{int}}^{opt}$ was defined above for the exact estimator. The same definition can also be used for a computed upper estimator. Let us note that the construction of $\boldsymbol{\theta}_{\mathcal{E}_{int}}^{opt}$ requires the solution of a global discrete problem and does not lead to a dramatic improvement of the estimator. A more effective approach for improving the accuracy of the estimator is to employ an iterative improvement via equilibration iterations, which will be discussed in Chapter VI.

Remark 5.3*.5 Let the *exact* corrections θ_ϵ^{EX} be defined by

$$\frac{\partial}{\partial n_\tau}(u_{S_\Delta^p})\Big|_\epsilon + \operatorname{sg}(\epsilon,\tau)\theta_\epsilon^{EX} = \frac{\partial}{\partial n_\tau}(e_{S_\Delta^p})\Big|_\epsilon. \tag{5.3*.22a}$$

We then have

$$\|e_{S_\Delta^p}\|_{\mathcal{U}} = \mathscr{E}(\boldsymbol{\theta}_{\mathcal{E}_{int}}^{EX}) = \min_{\boldsymbol{\theta} \in \boldsymbol{\Theta}^{EQ,0}_{\mathbf{L}^2_{\mathcal{E}_{int}}}} \mathscr{E}(\boldsymbol{\theta}_{\mathcal{E}_{int}}) \leq \mathscr{E}(\boldsymbol{\theta}_{\mathcal{E}_{int}}^{opt}) = \min_{\boldsymbol{\theta} \in \boldsymbol{\Theta}^{EQ,p}_{\boldsymbol{\mathscr{P}}^p_{\mathcal{E}_{int}}}} \mathscr{E}(\boldsymbol{\theta}) \tag{5.3*.22b}$$

where

$$\boldsymbol{\Theta}^{EQ,0}_{\mathbf{L}^2_{\mathcal{E}_{int}}} \stackrel{\text{def}}{=} \left\{\boldsymbol{\theta}_\epsilon \in \mathbf{L}^2_{\mathcal{E}_{int}} = \prod_{\epsilon \in \mathcal{E}_{int}(\Delta)} L^2(\epsilon) \mid \mathscr{R}_\tau^{UNEQ}(1) + \sum_{\epsilon \subseteq \partial\tau} \int_\epsilon \operatorname{sg}(\epsilon,\tau)\theta_\epsilon = 0, \quad \forall \tau \in \Delta\right\}. \tag{5.3*.22c}$$

In general, $\boldsymbol{\theta}_{\mathcal{E}_{int}}^{EX} \neq \boldsymbol{\theta}_{\mathcal{E}_{int}}^{opt}$.

Let us now give an example of the dependence of θ_ϵ on the vertex constants.

Example 5.3*.2 (*Edge corrections for various choices of the vertex constant*) Let us determine the asymptotic θ_ϵ for the criss-cross pattern and linear elements ($p = 1$). We will employ the notations shown in Fig. 5.3*.1. Letting $Q^{(2)}(x_1, x_2) = x_1 x_2$, the asymptotic finite element solution is (see also Section 4.7*)

$$\tilde{U}_{S^1_\Delta}^{PER(h)} = \mathcal{I}^1_\Delta Q^{(2)} + Z^1_{PER}. \tag{5.3*.23a}$$

There are two kinds of vertices in the criss-cross pattern: The vertex X which is connected to four edges, which is shown in Fig. 5.3*.1a, and the vertex Y which is connected to eight edges, which is shown in Fig. 5.3*.1b. Let us write down the system of equations (5.3*.15d) for these vertices.

For vertex X, the coefficients $\theta_{\epsilon_i^X, \nu(\epsilon_i^X)}$, $i = 1, \ldots, ne(X) = 4$ satisfy (5.3*.15c) with $\boldsymbol{\mathcal{R}}_X^{UNEQ} = \mathbf{0}$, and $\boldsymbol{\theta}_X = \mathbf{0}$ is the solution with minimal norm.

For the vertex Y we have

$$\begin{bmatrix} 1 & 0 & 0 & 0 & 0 & 0 & 0 & -1 \\ -1 & 1 & 0 & 0 & 0 & 0 & 0 & 0 \\ 0 & -1 & 1 & 0 & 0 & 0 & 0 & 0 \\ 0 & 0 & -1 & 1 & 0 & 0 & 0 & 0 \\ 0 & 0 & 0 & -1 & 1 & 0 & 0 & 0 \\ 0 & 0 & 0 & 0 & -1 & 1 & 0 & 0 \\ 0 & 0 & 0 & 0 & 0 & -1 & 1 & 0 \\ 0 & 0 & 0 & 0 & 0 & 0 & -1 & 1 \end{bmatrix} \begin{Bmatrix} \theta_{Y,1} \\ \theta_{Y,2} \\ \theta_{Y,3} \\ \theta_{Y,4} \\ \theta_{Y,5} \\ \theta_{Y,6} \\ \theta_{Y,7} \\ \theta_{Y,8} \end{Bmatrix} = \frac{1}{16}(1+a^2) \begin{Bmatrix} 1 \\ -1 \\ 1 \\ -1 \\ -1 \\ 1 \\ -1 \\ 1 \end{Bmatrix} \tag{5.3*.23b}$$

where we used the abbreviated notation $\theta_{Y,i} = \theta_{\epsilon_i^Y, \nu(\epsilon_i^Y)}$. Let us now give some choices of $\theta_{Y,i}$.

(a) *Bank–Weiser equilibration:* Let

$$\theta_{Y,n}^{\mathrm{BW}} = 0, \qquad \theta_{Y,n-i}^{\mathrm{BW}} = \sum_{j=1}^{i} \mathcal{R}_{\tau_{n-j+1}}^{UNEQ}(\varphi_Y^\Delta), \quad i = 1, \ldots, n-1 = 7 \tag{5.3*.24a}$$

where $n = ne(Y)$, and for our example

$$\boldsymbol{\theta}_Y^{\mathrm{BW}} = -\tfrac{1}{16}(1+a^2)[1, 0, 1, 0, -1, 0, -1, 0]^{\mathrm{T}}. \tag{5.3*.24b}$$

(b) *Ladeveze equilibrations:* Let $\boldsymbol{\theta}_Y^{\mathrm{Lad},1}$ be the solution of $\mathbf{M}_Y \boldsymbol{\theta}_Y = -\boldsymbol{\mathcal{R}}_Y^{UNEQ}$, with minimal l^2 norm, namely

$$\|\boldsymbol{\theta}_Y^{Lad,1}\|_{l^2}^2 = (\boldsymbol{\theta}_Y^{Lad,1})^{\mathrm{T}} \boldsymbol{\theta}_Y^{Lad,1} \stackrel{\mathrm{def}}{=} \min_{\substack{\boldsymbol{\theta}_Y \in \mathbb{R}^{ne(Y)} \\ \mathbf{M}_Y \boldsymbol{\theta}_Y = -\boldsymbol{\mathcal{R}}_Y^{UNEQ}}} \|\boldsymbol{\theta}_Y\|_{l^2}^2. \tag{5.3*.25a}$$

It follows that $\boldsymbol{\theta}_Y^{Lad,1}$ can be obtained as the stationary value of the Lagrangian $L : \mathbb{R}^{ne(Y)} \times \mathbb{R}^{ne(Y)} \mapsto \mathbb{R}$, given by

$$L(\boldsymbol{\theta}_Y, \boldsymbol{\Lambda}_Y) = \tfrac{1}{2} \boldsymbol{\theta}_Y^{\mathrm{T}} \boldsymbol{\theta}_Y + (\mathbf{M}_Y \boldsymbol{\theta}_Y + \boldsymbol{\mathcal{R}}_Y^{UNEQ})^{\mathrm{T}} \boldsymbol{\Lambda}_Y \tag{5.3*.25b}$$

and hence

$$0 = \frac{\partial L}{\partial \boldsymbol{\theta}_Y}(\boldsymbol{\theta}_Y^{\mathrm{Lad},1}, \boldsymbol{\Lambda}_Y^{\mathrm{Lad},1}) = \boldsymbol{\theta}_Y^{\mathrm{Lad},1} + \mathbf{M}_Y^{\mathrm{T}} \boldsymbol{\Lambda}_Y^{\mathrm{Lad},1}$$
$$\implies \quad \boldsymbol{\theta}_Y^{\mathrm{Lad},1} = -\mathbf{M}_Y^{\mathrm{T}} \boldsymbol{\Lambda}_Y^{\mathrm{Lad},1}. \tag{5.3*.25c}$$

Substituting $\boldsymbol{\theta}_Y^{\mathrm{Lad},1} = -\mathbf{M}_Y^{\mathrm{T}} \boldsymbol{\Lambda}_Y^{\mathrm{Lad},1}$, into the original system, $\mathbf{M}_Y \boldsymbol{\theta}_Y = -\boldsymbol{\mathfrak{R}}_Y^{UNEQ}$, it follows that the Lagrange multiplier $\boldsymbol{\Lambda}_Y^{\mathrm{Lad},1}$ satisfies

$$\mathbf{M}_Y \mathbf{M}_Y^{\mathrm{T}} \boldsymbol{\Lambda}_Y^{\mathrm{Lad},1} = \boldsymbol{\mathfrak{R}}_Y^{UNEQ}. \tag{5.3*.25d}$$

The matrix $\mathbf{M}_Y \mathbf{M}_Y^{\mathrm{T}} \in \mathbb{R}^{ne(Y) \times ne(Y)}$ has one zero eigenvalue with corresponding eigenvector

$$\mathbf{V}_Y = [\underbrace{1, 1, \ldots, 1}_{ne(Y)}]^{\mathrm{T}},$$

and hence $\boldsymbol{\Lambda}_Y^{\mathrm{Lad},1}$ is uniquely determined from (5.3*.25d) up to a multiple of $\mathbf{V}_Y$. Choosing $\boldsymbol{\Lambda}_Y^{\mathrm{Lad},1}$ to be orthogonal to $\mathbf{V}_Y$, namely,

$$\mathbf{V}_Y^{\mathrm{T}} \boldsymbol{\Lambda}_Y^{\mathrm{Lad},1} = 0, \tag{5.3*.25e}$$

$\boldsymbol{\Lambda}_Y^{\mathrm{Lad},1}$ can be determined by solving the linear system

$$(\mathbf{M}_Y \mathbf{M}_Y^{\mathrm{T}} + \mathbf{D}_Y) \boldsymbol{\Lambda}_Y^{\mathrm{Lad},1} = \boldsymbol{\mathfrak{R}}_Y^{UNEQ} \tag{5.3*.25f}$$

where

$$\mathbf{D}_Y = [\underbrace{\mathbf{V}_Y, \mathbf{V}_Y, \ldots, \mathbf{V}_Y}_{ne(Y)}]$$

is the matrix with columns equal to $\mathbf{V}_Y$. The entries of $\boldsymbol{\theta}_Y^{\mathrm{Lad},1}$ can be written explicitly in the form

$$\begin{aligned} \theta_{Y,n}^{\mathrm{Lad},1} &= -\frac{1}{n} \sum_{i-1}^{n} i \mathfrak{R}_{Y,i}^{UNEQ}, \\ \theta_{Y,n-j}^{\mathrm{Lad},1} &= \theta_{Y,n-j+1}^{\mathrm{Lad},1} + \mathfrak{R}_{Y,n-j+1}^{UNEQ}, \quad j = 1, \ldots, n-1. \end{aligned} \tag{5.3*.25g}$$

For our example we obtain

$$\boldsymbol{\Lambda}_Y^{\mathrm{Lad},1} = \frac{1+a^2}{16}[1, 0, 0, -1, -1, 0, 0, 1]^{\mathrm{T}} \tag{5.3*.25h}$$

and

$$\boldsymbol{\theta}_Y^{\mathrm{Lad},1} = -\frac{1+a^2}{16}[1, 0, 1, 0, -1, 0, -1, 0]^{\mathrm{T}}. \tag{5.3*.25i}$$

Let $\boldsymbol{\theta}_Y^{\mathrm{Lad},2}$ be the solution of $\mathbf{M}_Y\boldsymbol{\theta}_Y = -\mathfrak{R}_Y^{UNEQ}$, with minimal weighted l^2 norm $\|\cdot\|^2_{l^2,\{w_k\}_{k=1}^{ne(Y)}}$, with weights $w_k = |\epsilon_k^Y|^{-2}$, namely

$$\begin{aligned}\|\boldsymbol{\theta}_Y^{Lad,2}\|^2_{l^2,\{w_k\}_{k=1}^{ne(Y)}} &\stackrel{\mathrm{def}}{=} \sum_{k=1}^{ne(Y)} |\epsilon_k^Y|^{-2}(\theta_{\epsilon_k^Y,\nu(\epsilon_k^Y)})^2 \\ &= \min_{\substack{\boldsymbol{\theta}_Y\in\mathbb{R}^{ne(Y)}\\ \mathbf{M}_Y\boldsymbol{\theta}_Y=\mathfrak{R}_Y^{UNEQ}}} \|\boldsymbol{\theta}_Y\|^2_{l^2,\{|\epsilon_k^Y|^{-2}\}_{k=1}^{ne(Y)}}.\end{aligned} \tag{5.3*.26a}$$

Employing the method of Lagrange multipliers we obtain

$$\boldsymbol{\theta}_Y^{\mathrm{Lad},2} = -\mathbf{diag}(|\epsilon_i^Y|^2)\mathbf{M}_Y^{\mathrm{T}}\boldsymbol{\Lambda}_Y^{\mathrm{Lad},2} \tag{5.3*.26b}$$

and

$$(\mathbf{M}_Y\mathbf{diag}(|\epsilon_i^Y|^2)\mathbf{M}_Y^{\mathrm{T}} + \mathbf{D}_Y)\boldsymbol{\Lambda}_Y^{\mathrm{Lad},2} = \mathfrak{R}_Y^{UNEQ} \tag{5.3*.26c}$$

where $\mathbf{diag}(|\epsilon_i^Y|^2)$ denotes the diagonal matrix with ith diagonal element given by $|\epsilon_i^Y|^2$. For our example, we obtain

$$\theta_{Y,1}^{\mathrm{Lad},2} = -\theta_{Y,3}^{\mathrm{Lad},2} = \theta_{Y,5}^{\mathrm{Lad},2} = -\theta_{Y,7}^{\mathrm{Lad},2} = -\frac{1}{8}\frac{a^2(a^4+6a^2+5)}{(a^4+10a^2+1)} \tag{5.3*.26d}$$

$$\theta_{Y,2}^{\mathrm{Lad},2} = \theta_{Y,4}^{\mathrm{Lad},2} = \theta_{Y,6}^{\mathrm{Lad},2} = \theta_{Y,8}^{\mathrm{Lad},2} = -\frac{1}{16}\frac{(1+a^2)^2(a^2-1)}{(a^4+10a^2+1)}. \tag{5.3*.26e}$$

Hence *different choices of the free vertex parameter give different sets of* θ_ϵ.

Below we will use the notation $\hat{e}_\tau^{\mathrm{Neum},q}$, $\eta_\tau^{\mathrm{Neum},q}$, and $\mathscr{E}^{\mathrm{Neum},q}$ for the element residual indicators and estimator corresponding to the equilibrated residuals $\mathcal{R}_\tau^{EQ,q}$. When necessary we will also mention how the selection of the vertex constant was done, e.g., Bank–Weiser or Ladeveze equilibration.

As in the one-dimensional case, we will employ the computed indicator and estimator

$$\eta_{\tau,S^{p+k}}^{\mathrm{Neum},q} \stackrel{\mathrm{def}}{=} \left\|\pi_{S^{p+k}_{\partial\tau_{\mathrm{D}},0}(\tau)}(\hat{e}_\tau^{\mathrm{Neum},q})\right\|_{\mathcal{U}(\tau)}, \qquad \mathscr{E}_{S_\Delta^{p+k}}^{\mathrm{Neum},q} \stackrel{\mathrm{def}}{=} \sqrt{\sum_{\tau\in\Delta}(\eta_{\tau,S^{p+k}}^{\mathrm{Neum},q})^2} \tag{5.3*.27a}$$

and we will refer to this estimator by the notation $ERqPp+k$. Similarly, $\eta_{\tau,B^{p+k}}^{\mathrm{Neum}}$ and $\mathscr{E}_{B_\Delta^{p+k}}^{\mathrm{Neum}}$ will be referred to as the $ERqBp+k$ indicator and estimator, respectively, where we recall that $B^{p+k}(\tau)$ is the element space spanned by the edge and internal element shape functions of degree $p+1$ to $p+k$.

As we have already discussed in Section 5.1*, the computed estimator $\mathscr{E}_{S_\Delta^{p+k}}^{\mathrm{Neum}}$ is not a guaranteed upper estimator; nevertheless, may obtain a computable upper estimator by recalling the identity

$$\mathscr{E}^{\mathrm{Neum}} = \sqrt{\left(\mathscr{E}_{S_\Delta^{p+k}}^{\mathrm{Neum}}\right)^2 + \sum_{\tau\in\Delta}\|\hat{e}_\tau^{\mathrm{Neum}} - \pi_{S^{p+k}(\tau)}(\hat{e}_\tau^{\mathrm{Neum}})\|^2_{\mathcal{U}(\tau)}} \tag{5.3*.27b}$$

and by employing an explicit upper estimator

$$\|\hat{e}_\tau^{\mathrm{Neum}} - \pi_{S^{p+k}(\tau)}(\hat{e}_\tau^{\mathrm{Neum}})\|_{\mathcal{U}(\tau)} \le \mathcal{E}^{EXPL}_{\pi_{S^{p+k}(\tau)}(\hat{e}_\tau^{\mathrm{Neum}})}. \tag{5.3*.27c}$$

The same approach can also be used for $\mathcal{E}^{\mathrm{Neum}}_{B^{p+k}_\Delta}$. We will elaborate on this in Chapter 6.

Let us also note that it is also possible (see [Bank and Weiser, 1985]) to employ a residual estimator based on the following *discrete* element residual problem:

Find $\hat{e}^{UNEQ}_{\tau,B^{p+k}} \in B^{p+k}(\tau)$ such that

$$\mathcal{B}_\tau(\hat{e}^{UNEQ}_{\tau,B^{p+k}}, v) = \mathcal{R}^{UNEQ}_\tau(v) = \int_\tau r^\Delta_\tau v + \sum_{\epsilon \subseteq \partial\tau} \int_\epsilon \frac{1}{2} J^\tau_\epsilon v \quad \forall v \in B^{p+k}(\tau). \tag{5.3*.28a}$$

Note that this problem has a unique solution because $1 \notin B^{p+k}$, and hence the consistency condition $\mathcal{R}^{UNEQ}_\tau(1) = 0$ is not needed. The corresponding element error indicator and the estimator are

$$\eta^{UNEQ}_{\tau,B^{p+k}} = \|\hat{e}^{UNEQ}_{\tau,B^{p+k}}\|_{\mathcal{U}(\tau)}, \qquad \mathcal{E}^{UNEQ}_{B^{p+k}} = \sqrt{\sum_{\tau\in\Delta} (\eta^{UNEQ}_{\tau,B^{p+k}})^2}. \tag{5.3*.28b}$$

Let us also underline that *the discrete problem (5.3*.28a) does not have a continuous counterpart and it is not possible to use it to obtain an upper bound for the energy norm of the error.* We will refer to $\mathcal{E}^{UNEQ}_{B^{p+k}}$, as the *ER_Bp+k* (*E*lement *R*esidual problem with unequilibrated residual and *B*ubble solution of degree $p+k$) estimator. The asymptotic exactness of the *ER_B* estimator was addressed in [Duran *et al.*, 1991, 1992], [Ainsworth, 1994], and in [Duran and Rodriguez, 1992]. The estimator *ER_Bp + k* is closely related to many estimators used in the literature; for example, we refer to [Bornemann *et al.*, 1996] and [Deuflhard *et al.*, 1989].

It can be proven that the equilibrated Neumann error indicators satisfy Assumptions A, B, C of Section 5.1, and hence they can be analyzed by using the framework of Section 4.7*.

Theorem 5.3*.3 *Assumptions A, B, C of Section 5.1 are satisfied for all the error indicators introduced in Section 5.3*.*

Above we have shown that the edge corrections θ_ϵ depend on the particular choice of free vertex parameters. Let us now demonstrate that *the effectiveness of the element residual estimators also depend on the choice of the vertex parameters.*

Example 5.3*.3 (*Analysis of the asymptotic effectiveness of various versions of the element residual indicator as a function of the mesh aspect ratio*) Let us consider the criss-cross pattern with aspect ratio $1/a$, as shown in Fig. 5.3*.1 (see also Fig. 4.7*.3d), and let us analyze the asymptotic behavior of the effectivity index for the various element residual estimators introduced above, in the case of linear elements ($p = 1$), for the Laplacian ($-\Delta u = 0$). Letting

$$\tilde{Q}^{(2)} = \sum_{l=1}^{2} \alpha_l \tilde{Q}^{(2)}_l \tag{5.3*.29a}$$

where $\tilde{Q}_1^{(2)}(x_1, x_2) = x_1 x_2, \tilde{Q}_2^{(2)}(x_1, x_2) = x_1^2 - x_2^2$. The asymptotic error, written in the master basic cell, is given by

$$\tilde{\Psi}^{PER}_{\tilde{S}^1_{\tilde{\Delta}}} = \sum_{l=1}^{2} \alpha_l \tilde{\Psi}^{PER}_{\tilde{S}^1_{\tilde{\Delta}},l}, \qquad \hat{e}^{\mathrm{Neum},q}_{\tilde{\tau}} = \sum_{l=1}^{2} \alpha_l \hat{e}^{\mathrm{Neum},q}_{\tilde{\tau},l} \tag{5.3*.29b}$$

where $\tilde{\Psi}^{PER}_{\tilde{S}^1_{\tilde{\Delta}},l}$ is the asymptotic error introduced in section 4.7*, and $\hat{e}^{\mathrm{Neum},q}_{\tilde{\tau},l}$ is the indicator function corresponding to $\tilde{Q}_l^{(2)}$. The asymptotic value of the square of the cell effectivity index is given by

$$(\kappa^{EQ,q}(\boldsymbol{\alpha}))^2 = \frac{\sum_{\tilde{\tau}\in\tilde{\Delta}} \|\hat{e}^{\mathrm{Neum},q}_{\tilde{\tau}}(\boldsymbol{\alpha})\|^2_{\mathcal{U}(\tilde{\tau})}}{\sum_{\tilde{\tau}\in\tilde{\Delta}} \left\|\tilde{\Psi}^{PER}_{\tilde{S}^1_{\tilde{\Delta}}}(\boldsymbol{\alpha})\right\|^2_{\mathcal{U}(\tilde{\tau})}} = \frac{\sum_{\tilde{\tau}\in\tilde{\Delta}} \left\|\sum_{l=1}^2 \alpha_l \hat{e}^{\mathrm{Neum},q}_{\tilde{\tau},l}\right\|^2_{\mathcal{U}(\tilde{\tau})}}{\sum_{\tilde{\tau}\in\tilde{\Delta}} \left\|\sum_{l=1}^2 \alpha_l \tilde{\Psi}^{PER}_{\tilde{S}^1_{\tilde{\Delta}},l}\right\|^2_{\mathcal{U}(\tilde{\tau})}} \tag{5.3*.30a}$$

and its square can be written in the form

$$(\kappa^{EQ,q}(\boldsymbol{\alpha}))^2 = \frac{\sum_{l=1}^2 \sum_{m=1}^2 \alpha_l \alpha_m \sum_{\tilde{\tau}\in\tilde{\Delta}} \mathcal{B}_{\tilde{\tau}}(\hat{e}^{\mathrm{Neum},q}_{\tilde{\tau},l}, \hat{e}^{\mathrm{Neum},q}_{\tilde{\tau},m})}{\sum_{l=1}^2 \sum_{m=1}^2 \alpha_l \alpha_m \sum_{\tilde{\tau}\in\tilde{\Delta}} \mathcal{B}_{\tilde{\tau}}\big(\tilde{\Psi}^{PER}_{\tilde{S}^1_{\tilde{\Delta}},l}, \tilde{\Psi}^{PER}_{\tilde{S}^1_{\tilde{\Delta}},m}\big)} = \frac{\boldsymbol{\alpha}^{\mathrm{T}}\hat{\mathbf{M}}\boldsymbol{\alpha}}{\boldsymbol{\alpha}^{\mathrm{T}}\bar{\mathbf{M}}\boldsymbol{\alpha}}. \tag{5.3*.30b}$$

Here $\boldsymbol{\alpha} = [\alpha_1, \alpha_2]^{\mathrm{T}}$, and $\hat{\mathbf{M}}, \bar{\mathbf{M}} \in \mathbb{R}^{2\times 2}$ are the matrices with entries

$$\hat{M}_{lm} = \sum_{\tilde{\tau}\in\tilde{\Delta}} \mathcal{B}_{\tilde{\tau}}(\hat{e}^{\mathrm{Neum},q}_{\tilde{\tau},l}, \hat{e}^{\mathrm{Neum},q}_{\tilde{\tau},m}), \qquad \bar{M}_{lm} = \sum_{\tilde{\tau}\in\tilde{\Delta}} \mathcal{B}_{\tilde{\tau}}\big(\tilde{\Psi}^{PER}_{\tilde{S}^1_{\tilde{\Delta}},l}, \tilde{\Psi}^{PER}_{\tilde{S}^1_{\tilde{\Delta}},m}\big). \tag{5.3*.30c}$$

It follows that, the asymptotic bounds for the cell effectivity index $\kappa^{EQ,q}$, are given by

$$\begin{aligned} \mathcal{C}_L &= \sqrt{\min_{\boldsymbol{\alpha}\in\mathbb{R}^2}\left(\frac{\boldsymbol{\alpha}^{\mathrm{T}}\hat{\mathbf{M}}\boldsymbol{\alpha}}{\boldsymbol{\alpha}^{\mathrm{T}}\bar{\mathbf{M}}\boldsymbol{\alpha}}\right)} = \sqrt{\mu_{\min}} \le \kappa^{EQ,q} \le \mathcal{C}_U \\ &= \sqrt{\max_{\boldsymbol{\alpha}\in\mathbb{R}^2}\left(\frac{\boldsymbol{\alpha}^{\mathrm{T}}\hat{\mathbf{M}}\boldsymbol{\alpha}}{\boldsymbol{\alpha}^{\mathrm{T}}\bar{\mathbf{M}}\boldsymbol{\alpha}}\right)} = \sqrt{\mu_{\max}} \end{aligned} \tag{5.3*.30d}$$

where $\mu_{\min}$ and $\mu_{\max}$ are respectively the minimum and the maximum eigenvalue of the generalized eigenvalue problem

$$\hat{\mathbf{M}}\mathbf{w} = \mu\bar{\mathbf{M}}\mathbf{w}. \tag{5.3*.30e}$$

In the particular example considered here, the matrix $\bar{\mathbf{M}}$ and the matrices $\hat{\mathbf{M}}$ are diagonal for all the considered error indicators, the eigenvectors of the generalized eigenvalue problem are $\boldsymbol{\alpha}^{(1)} = [1, 0]^{\mathrm{T}}$, and $\boldsymbol{\alpha}^{(2)} = [0, 1]^{\mathrm{T}}$, and the bounds for the cell effectivity index, which are given in Tables 5.3*.1 and 5.3*.2, are the values corresponding to the basis monomials. For example, for the unequilibrated *ER_B2* error indicator the bounds for the cell effectivity index are

$$\mathcal{C}_L(a) = \begin{cases} 1, & a_1 \le a \le a_2 \\ 2\sqrt{2}\dfrac{a}{1+a^2}, & \text{otherwise} \end{cases} \tag{5.3*.30f}$$

and

$$\mathscr{C}_U(a) = \begin{cases} 2\sqrt{2}\dfrac{a}{1+a^2}, & a_1 \le a \le a_2 \\ 1, & \text{otherwise} \end{cases} \tag{5.3*.30g}$$

where $a_1 = \sqrt{2} - 1$ and $a_2 = \sqrt{2} + 1$.

TABLE 5.3*.1. The effectivity index for the estimate of the energy norm in the basic cell of the criss-cross pattern for the *ER1B2*, the *ER1P2* and the *ER_B2* error indicators as a function of a, where $1/a$ is the aspect ratio as shown in Fig. 5.3*.2, for various constructions of the edge corrections. Input data: Laplace equation, exact solution $Q^{(2)} \equiv x_1 x_2$. The graphs of the effectivity index as a function of a, in the cases for which it is not identically equal to one, are shown in Fig. 5.3*.3. Note that, for this mesh pattern and solution, the B–W and Lad, 1 equilibration give identical results, but this should not be expected, in general (see, for example, the results in Table 5.3*.2)

Equilibration	*ER1B2*	*ER1P2*
Bank–Weiser	1	$\sqrt{6}\dfrac{1+a^2}{(3a^4+10a^2+3)^{1/2}}$
Ladeveze, 1	1	$\sqrt{6}\dfrac{1+a^2}{(3a^4+10a^2+3)^{1/2}}$
Ladeveze, 2	$\dfrac{\sqrt{5a^8+80a^6+406a^4+80a^2+5}}{2(a^4+10a^2+1)}$	$\sqrt{\dfrac{3(1+a^2)^2(3a^8+64a^6+442a^4+64a^2+3)}{2(3a^4+10a^2+3)(a^4+10a^2+1)^2}}$
Unequilibrated	1	—

TABLE 5.3*.2. The effectivity index for the estimate of the energy norm in the basic cell of the criss-cross pattern for the *ER1B2*, the *ER1P2*, and the *ER_B2* error indicators as a function of a, where $1/a$ is the aspect ratio as shown in Fig. 5.3*.2, for various constructions of the edge corrections. Input data: Laplace equation, exact solution $Q^{(2)} \equiv x_1^2 - x_2^2$. The graphs of the effectivity index, in the cases for which it is not identically equal to $\sqrt{3}$, are shown in Fig. 5.3*.4. Note that for this mesh pattern and solution the Lad, 1 and Lad, 2 equilibration gave identical results, but this should not be expected, in general (see, for example, the results in Table 5.3*.1). Note that for the *ER1B2* error indicator (respectively the *ER1P2* error indicator) the effectivity index converges to $1/\sqrt{2}$ (respectively to $\sqrt{3}$) as either a or $1/a$ tends to zero. On the other hand, the effectivity index for the unequilibrated *ER_B2* error indicator converges to zero as either a or $1/a$ tends to zero

Equilibration	*ER1B2*	*ER1P2*
Bank–Weiser	$\dfrac{1}{\sqrt{2}}\dfrac{\sqrt{a^4+10a^2+1}}{1+a^2}$	$\sqrt{3}\sqrt{\dfrac{3a^4+14a^2+3}{3a^4+10a^2+3}}$
Ladeveze, 1	$\dfrac{1}{\sqrt{2}}\dfrac{\sqrt{a^4+6a^2+1}}{a^2+1}$	$\sqrt{3}$
Ladeveze, 2	$\dfrac{1}{\sqrt{2}}\dfrac{\sqrt{a^4+6a^2+1}}{a^2+1}$	$\sqrt{3}$
Unequilibrated	$2\sqrt{2}\dfrac{a}{a^2+1}$	—

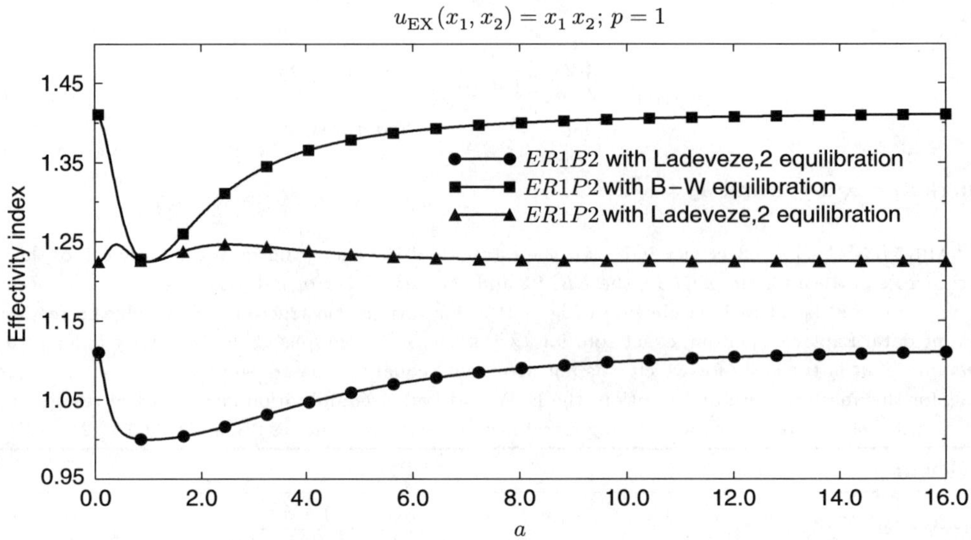

FIG. 5.3*.3. Graphs of the effectivity index for the estimate of the energy norm in the basic cell of the criss-cross pattern for the asymptotic finite element solution corresponding to the exact solution $Q^{(2)} \equiv x_1x_2$ as a function of a, $\frac{1}{16} \leq a \leq 16$. Here we give the graphs of the expressions given in Table 5.3*.1 for the cases that the effectivity index is not identically equal to one. Note that in this case the $ER1P2$ error indicator is more effective when the Lad, 2 equilibration is employed instead of either the Lad, 1 equilibration or the B–W equilibration

Let us now discuss the effectivity indices for the basis monomials. Table 5.3*.1 gives the effectivity index for the energy norm in the basic cell for the monomial $Q^{(2)} \equiv x_1x_2$ as a function of a for the three different types of equilibration (B–W, Lad, 1, and Lad, 2) for the error indicators $\eta_{\tau,B^2}^{\text{Neum},1}$ ($ER1B2$), $\eta_{\tau,S^2}^{\text{Neum},1}$ ($ER1P2$), and η_{τ,B^2}^{UNEQ} (ER_B2). Figure 5.3*.3 shows the corresponding graphs of the effectivity index as a function of the aspect ratio for $\frac{1}{16} \leq a \leq 16$. Note that in the case of the $ER1B2$ indicator the effectivity index for the B–W and Lad, 1 equilibration is equal to one, irrespective of the value of the aspect ratio. Note also that the effectivity index for the ER_B2 indicator is equal to one. The $ER1P2$ indicator is not exact, for any of the three equilibrations employed, and overestimates the error in the mesh patch. The best version of the $ER1P2$ indicator is obtained using Lad, 2 equilibration. Table 5.3*.2 reports the effectivity index as a function of a, and Fig. 5.3*.4 gives the corresponding graphs for $\frac{1}{16} \leq a \leq 16$ for the monomial $Q^{(2)} \equiv x_1^2 - x_2^2$. From these results we conclude that *the ER1B2 error indicator with Lad, 1 equilibration is the most effective for the class of cases considered (i.e. the criss-cross mesh pattern with linear elements* $(p = 1)$ *and aspect ratio* $\frac{1}{16} \leq 1/a \leq 16$).

In the preceding example we have seen the influence of the vertex parameters on the sensitivity of the various residual estimators. Let us now recall the optimal θ_ϵ defined in Remark 5.3*.3. Recall that for any interior edge $\epsilon = (\mathbf{X}, \mathbf{Y})$, we have

$$\theta_\epsilon = \theta_{\epsilon,\nu(\epsilon,X)}\lambda_{\nu(\epsilon,X)} + \theta_{\epsilon,\nu(\epsilon,Y)}\lambda_{\nu(\epsilon,Y)} \tag{5.3*.31a}$$

and

$$\theta_{\epsilon,\nu(\epsilon,X)} = \theta^{\text{BW}}_{\epsilon,\nu(\epsilon,X)} + \mathscr{C}_X \tag{5.3*.31b}$$

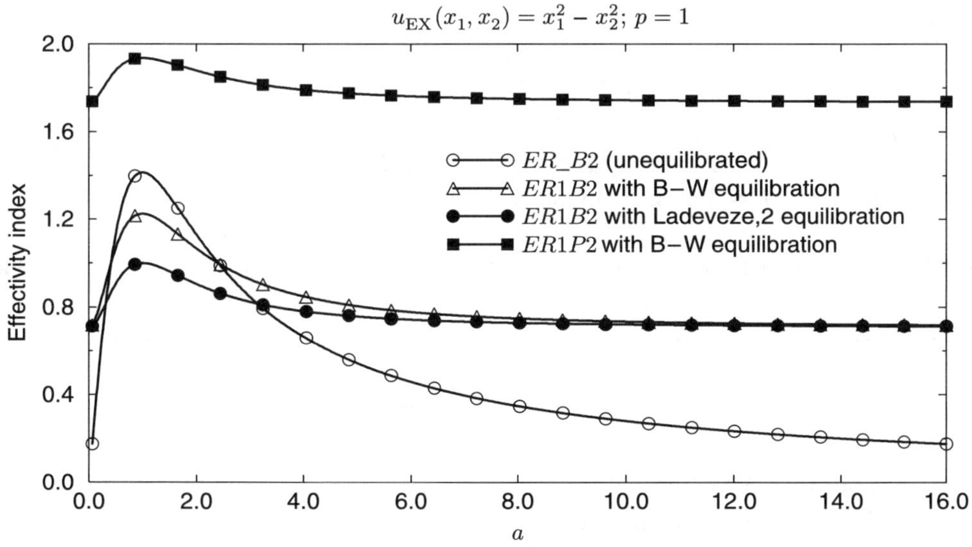

FIG. 5.3*.4. Graphs of the effectivity index for the estimate of the energy norm in the basic cell of the criss-cross pattern for the asymptotic finite element solution corresponding to the exact solution $Q^{(2)} \equiv x_1^2 - x_2^2$ as a function of a, $\frac{1}{16} \leq a \leq 16$. Here we give the graphs of the expressions given in Table 5.3*.2 for the cases that the effectivity index is not identically equal to $\sqrt{3}$. Note that in this case the *ER1B2* error indicator with Lad, 2 (or Lad, 1) equilibration is the most effective. Note also that the unequilibrated error indicator *ER_B2* has the poorest quality

and similarly for $\theta_{\epsilon,\nu(\epsilon,Y)}$. It follows that $\mathscr{E}^{\text{Neum}} = \mathscr{E}^{\text{Neum}}(\mathbf{\mathscr{C}}^{vert})$ is a quadratic form of the vertex parameters, where $\mathbf{\mathscr{C}}^{vert}$ is the vector of the vertex parameters, and following Remark 5.3*.3, we may define the optimal vertex parameters $\mathbf{\mathscr{C}}^{vert}_{opt}$, as the ones which minimize the value of $\mathscr{E}^{\text{Neum},q}$, namely

$$\|e_{S^p_\Delta}\|_{\mathscr{U}} \leq \mathscr{E}^{\text{Neum},q}(\mathbf{\mathscr{C}}^{vert}_{opt}) = \min_{\mathbf{\mathscr{C}}^{vert} \in \mathbb{R}^{nvi}} \mathscr{E}^{\text{Neum},q}(\mathbf{\mathscr{C}}^{vert}) \tag{5.3*.31c}$$

where nvi is the number of interior vertices.

Let us now give an example where the optimal vertex parameters are employed.

Example 5.3*.4 (*The asymptotic analysis of the computed residual estimator based on the optimal vertex parameters*) Let us consider the asymptotic case in a uniform mesh of linear elements ($p = 1$) in the criss-cross pattern, which was also considered in Examples 5.3*.2 and 5.3*.3. Letting

$$\boldsymbol{\theta}_X = \boldsymbol{\theta}_X^{\text{BW}} + \mathscr{C}_X \mathbf{V}_X, \qquad \boldsymbol{\theta}_Y = \boldsymbol{\theta}_Y^{\text{BW}} + \mathscr{C}_Y \mathbf{V}_Y \tag{5.3*.32a}$$

where $\mathscr{C}_X$, $\mathscr{C}_Y$ are the free parameters for the vertices X, Y, respectively, and

$$\mathbf{V}_X = [\ \underbrace{1, \ldots, 1}_{ne(X)=4}\ ]^{\text{T}}, \qquad \mathbf{V}_Y = [\ \underbrace{1, \ldots, 1}_{ne(Y)=8}\ ]^{\text{T}} \tag{5.3*.32b}$$

we have

$$\theta_\epsilon(\mathscr{C}_X, \mathscr{C}_Y) = \theta_\epsilon^{\mathrm{BW}} + \mathscr{C}_{\epsilon,1}\lambda_{\epsilon,1} + \mathscr{C}_{\epsilon,2}\lambda_{\epsilon,2} \tag{5.3*.33a}$$

where $\mathscr{C}_{\epsilon,k} = \mathscr{C}_X$ if $(\epsilon, k) \equiv X$, and $\mathscr{C}_{\epsilon,k} = \mathscr{C}_Y$ if $(\epsilon, k) \equiv Y$, where (ϵ, k) is the kth end point of the edge ϵ, $k = 1, 2$. Employing the notations

$$\hat{e}_{\tau,X}^{\mathrm{Neum}} \stackrel{\mathrm{def}}{=} \hat{e}_\tau^{\mathrm{Neum}}(1,0) - \hat{e}_\tau^{\mathrm{Neum}}(0,0), \qquad \hat{e}_{\tau,Y}^{\mathrm{Neum}} \stackrel{\mathrm{def}}{=} \hat{e}_\tau^{\mathrm{Neum}}(0,1) - \hat{e}_\tau^{\mathrm{Neum}}(0,0) \tag{5.3*.33b}$$

and $\hat{e}_\tau^{\mathrm{Neum,BW}} \stackrel{\mathrm{def}}{=} \hat{e}_\tau^{\mathrm{Neum}}(0,0)$, we obtain

$$\hat{e}_\tau^{\mathrm{Neum}}(\mathscr{C}_X, \mathscr{C}_Y) = \hat{e}_\tau^{\mathrm{Neum,BW}} + \mathscr{C}_X e_{\tau,X}^{\mathrm{Neum}} + \mathscr{C}_Y e_{\tau,Y}^{\mathrm{Neum}} \tag{5.3*.33c}$$

where we recall that $\hat{e}_\tau^{\mathrm{Neum}}(\mathscr{C}_X, \mathscr{C}_Y)$ is the solution of the variational problem:
Find $\hat{e}_\tau^{\mathrm{Neum}}(\mathscr{C}_X, \mathscr{C}_Y) \in \mathscr{U}_{\partial\tau_{\mathrm{D}}}(\tau)$ such that

$$\mathscr{B}_\tau(\hat{e}_\tau^{\mathrm{Neum}}(\mathscr{C}_X, \mathscr{C}_Y), v) = \sum_{\epsilon \subseteq \partial\tau} \int_\epsilon \left(\frac{1}{2} J_\epsilon + \theta_\epsilon^\tau(\mathscr{C}_X, \mathscr{C}_Y)\right) \quad \forall v \in \mathscr{U}(\tau). \tag{5.3*.33d}$$

The optimal constants $\mathscr{C}_X^{opt}$, $\mathscr{C}_Y^{opt}$ are the ones which minimize $\mathscr{E}^{\mathrm{Neum}}(\mathscr{C}_X, \mathscr{C}_Y)$, namely

$$\mathscr{E}^2(\mathscr{C}_X^{opt}, \mathscr{C}_Y^{opt}) = \min_{\mathscr{C}_X, \mathscr{C}_Y} \mathscr{E}^2(\mathscr{C}_X, \mathscr{C}_Y) \tag{5.3*.34a}$$

where we used the notation $\mathscr{E}^2(\mathscr{C}_X, \mathscr{C}_Y) = (\mathscr{E}^{\mathrm{Neum}}(\mathscr{C}_X, \mathscr{C}_Y))^2$, and are determined by solving the linear system

$$\frac{\partial \mathscr{E}^2}{\partial \mathscr{C}_X}(\mathscr{C}_X^{opt}, \mathscr{C}_Y^{opt}) = 0, \qquad \frac{\partial \mathscr{E}^2}{\partial \mathscr{C}_Y}(\mathscr{C}_X^{opt}, \mathscr{C}_Y^{opt}) = 0 \tag{5.3*.34b}$$

which can be written as

$$\begin{bmatrix} \sum_{i=1}^{4} \mathscr{B}_\tau(\hat{e}_{\tau,X}^{\mathrm{Neum}}, \hat{e}_{\tau,X}^{\mathrm{Neum}}) & \sum_{i=1}^{4} \mathscr{B}_\tau(\hat{e}_{\tau,X}^{\mathrm{Neum}}, \hat{e}_{\tau,Y}^{\mathrm{Neum}}) \\ \sum_{i=1}^{4} \mathscr{B}_\tau(\hat{e}_{\tau,X}^{\mathrm{Neum}}, \hat{e}_{\tau,Y}^{\mathrm{Neum}}) & \sum_{i=1}^{4} \mathscr{B}_\tau(\hat{e}_{\tau,Y}^{\mathrm{Neum}}, \hat{e}_{\tau,Y}^{\mathrm{Neum}}) \end{bmatrix} \begin{Bmatrix} \mathscr{C}_X^{opt} \\ \mathscr{C}_Y^{opt} \end{Bmatrix}$$
$$= \begin{Bmatrix} -\sum_{i=1}^{4} \mathscr{B}_\tau(\hat{e}_{\tau,X}^{\mathrm{Neum}}, \hat{e}_\tau^{\mathrm{Neum,BW}}) \\ -\sum_{i=1}^{4} \mathscr{B}_\tau(\hat{e}_{\tau,X}^{\mathrm{Neum}}, \hat{e}_\tau^{\mathrm{Neum,BW}}) \end{Bmatrix}. \tag{5.3*.34c}$$

Let us now denote by $\mathscr{C}^{opt}_{X,i}$, $\mathscr{C}^{opt}_{Y,i}$, the optimal vertex parameters corresponding to the ith harmonic monomial, $Q^{(2)}_i$. For $i = 1$, $Q^{(2)}_1(x_1, x_2) = x_1 x_2$, we have

$$\mathscr{C}^{opt}_{X,1} = -\mathscr{C}^{opt}_{Y,1} = \frac{a^2 - 1}{512}, \qquad \kappa^{opt}_1 = \sqrt{\frac{3}{2}} \tag{5.3*.35a}$$

while the effectivity index corresponding to $\hat{e}^{\text{Neum,BW}}_\tau$ is

$$\kappa^{\text{BW}}_1 = \sqrt{6}\frac{a^2 + 1}{\sqrt{3a^4 + 10a^2 + 3}}. \tag{5.3*.35b}$$

Similarly, for $i = 2$, $Q^{(2)}_2(x_1, x_2) = x_1^2 - x_2^2$ we have

$$\mathscr{C}^{opt}_{X,2} = -\mathscr{C}^{opt}_{Y,2} = \frac{a}{64}, \qquad \kappa^{opt}_2 = \sqrt{3}. \tag{5.3*.36a}$$

Letting $\mathscr{C}_X = \mathscr{C}_Y = 0$, we get

$$\kappa^{\text{BW}}_2 = \sqrt{3}\sqrt{\frac{3a^4 + 14a^2 + 3}{3a^4 + 10a^2 + 3}}. \tag{5.3*.36b}$$

In the more general case, we let $Q^{(2)} = \sum_{l=1}^{2} \alpha_l Q_l$, and we obtain

$$\begin{aligned}\mathscr{C}^{opt}_L &= \min_{\alpha_1,\alpha_2} \kappa^{opt}(\alpha_1, \alpha_2) = \sqrt{\frac{3}{2}} \le \kappa^{opt}(\alpha_1, \alpha_2) \le \mathscr{C}^{opt}_U \\ &= \max_{\alpha_1,\alpha_2} \kappa^{opt}(\alpha_1, \alpha_2) = \sqrt{3}.\end{aligned} \tag{5.3*.37}$$

It is interesting to note that in this example the bounds $\mathscr{C}^{opt}_L$, $\mathscr{C}^{opt}_U$ are achieved by $Q^{(2)}_1$, $Q^{(2)}_2$, respectively, and they do not depend on the aspect ratio.

Remark 5.3*.6 In the above example we obtained the optimal values for the vertex constants for a mesh patch in the asymptotic setting. The same approach may be used for the entire mesh in a concrete case of interest.

Above we have given the construction of the edge corrections θ_ϵ for *regular meshes* with the property that every inter-element edge is an entire edge for both elements sharing it. Let us now explain how the above constructions can be extended to meshes with one to two irregular connections.

Example 5.3*.5 Let us consider the mesh patch of bilinear elements ($p = 1$) shown in Fig. 5.3*.5, and let φ^Δ_X be the basis function corresponding to the vertex X, and τ^X_i, $i = 1, \ldots, 6$, ϵ^X_i, $i = 1, \ldots, 8$, be the elements and the edges which belong to the interior of the support of φ^Δ_X. As before, we will denote by $\psi_{\epsilon,i}$ the linear shape functions which are the restrictions on the edge ϵ of the vertex basis functions associated with the edge.

For example, $\psi_{\epsilon^X_{5,6},1}$, $\psi_{\epsilon^X_{5,6},2}$ are the linear shape functions associated with the edge $\epsilon^X_{5,6} = \epsilon^X_5 \cup \epsilon^X_6$, as shown in Fig. 5.3*.5b. Note that $\epsilon^X_{5,6}$ is an edge with an irregular node and $\psi_{\epsilon^X_{5,6},1}|_{\epsilon^X_5}$ (respectively $\psi_{\epsilon^X_{5,6},2}|_{\epsilon^X_6}$) is not the standard one-dimensional shape function over the edge ϵ^X_6 (respectively ϵ^X_5). Let $\lambda_{\epsilon^X_k,j}$, $j = 1, 2$, be the conjugate linear

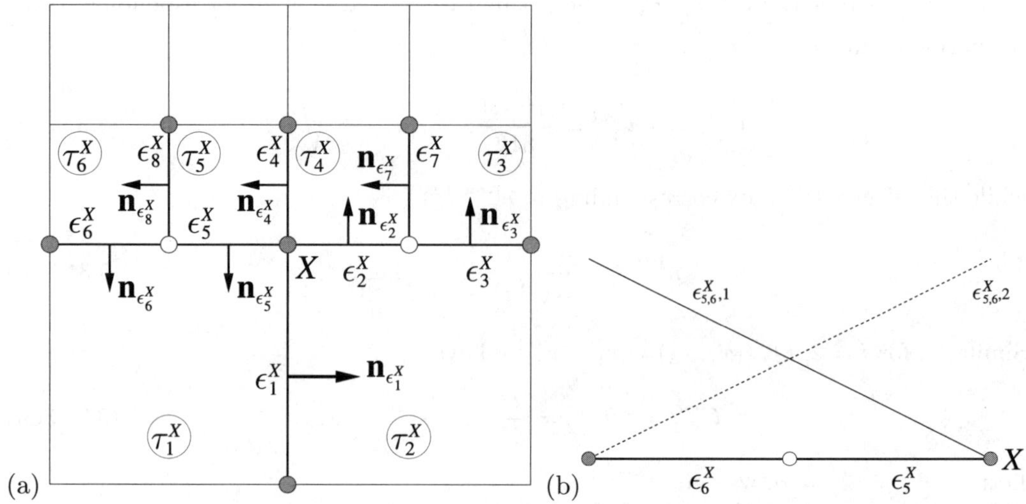

FIG. 5.3*.5. Construction of the edge corrections for meshes with irregular connections: An example of a vertex mesh patch. (a) The elements τ_k^X, $k = 1, \ldots, 6$, and the edges ϵ_k^X, $k = 1, \ldots, 8$, which belong to the interior of the support of the elementwise bilinear basis function φ_X^{Δ}. (b) The edges $\epsilon_5^X = \partial\tau_1^X \cap \partial\tau_5^X$, $\epsilon_6^X = \partial\tau_1^X \cap \partial\tau_6^X$, and the graphs of the linear edge basis functions $\psi_{\epsilon_{5,6}^X, j}$, $j = 1, 2$, associated with the edge $\epsilon_{5,6}^X = \epsilon_5 \cup \epsilon_6$

basis functions over ϵ_k^X, which satisfy $\int_{\epsilon_k^X} \lambda_{\epsilon_k^X, j} \psi_{\epsilon, i}|_{\epsilon_k^X} = \delta_{ij}$, where either $\epsilon = \epsilon_k^X$ is a *regular edge* shared by only two elements and $\psi_{\epsilon,i}$ are the standard linear shape functions for the edge ϵ_k^X, or ϵ is an *irregular edge*, i.e., an edge with an irregular node like for example $\epsilon = \epsilon_{5,6}^X$ or $\epsilon = \epsilon_{2,3}^X$ (for example in Fig. 5.3*.5 the edges ϵ_1^X, ϵ_7^X, ϵ_4^X, ϵ_8^X are regular edges, and $\epsilon_{5,6}^X = \epsilon_5^X \cup \epsilon_6^X$, and $\epsilon_{2,3}^X = \epsilon_2^X \cup \epsilon_3^X$ are irregular edges).

For example, on the edge $\epsilon_5^X \subseteq \epsilon_{5,6}^X$ we have

$$\begin{aligned} \lambda_{\epsilon_5^X,1} &= \frac{1}{|\epsilon_5^X|}\big(4\psi_{\epsilon_{5,6}^X,1} - 8\psi_{\epsilon_{5,6}^X,2}\big), \\ \lambda_{\epsilon_5^X,2} &= \frac{1}{|\epsilon_5^X|}\Big(-\frac{2}{3}\psi_{\epsilon_{5,6}^X,1} + \frac{7}{3}\psi_{\epsilon_{5,6}^X,2}\Big) \end{aligned} \tag{5.3*.38a}$$

and on the edge $\epsilon_6^X \subseteq \epsilon_{5,6}^X$ we have

$$\begin{aligned} \lambda_{\epsilon_6^X,1} &= \frac{1}{|\epsilon_6^X|}\Big(\frac{7}{3}\psi_{\epsilon_{5,6}^X,1} - \frac{2}{3}\psi_{\epsilon_{5,6}^X,2}\Big), \\ \lambda_{\epsilon_6^X,2} &= \frac{1}{|\epsilon_6^X|}\big(-8\psi_{\epsilon_{5,6}^X,1} + 4\psi_{\epsilon_{5,6}^X,2}\big). \end{aligned} \tag{5.3*.38b}$$

As before, we will construct affine edge corrections θ_ϵ by employing the orthogonalities

$$\sum_{\epsilon \subseteq \partial\tau_k^X} \int_\epsilon \theta_\epsilon^{\tau_k^X} \varphi_X^{\Delta}|_{\tau_k^X} = -\mathscr{R}_{\tau_k^X}^{UNEQ}(\varphi_X^{\Delta}|_{\tau_k^X}), \quad k = 1, \ldots, ne(X) = 6 \tag{5.3*.39a}$$

which give

$$\begin{bmatrix} 1 & 0 & 0 & 0 & -1 & -1 & 0 & 0 \\ -1 & 1 & 1 & 0 & 0 & 0 & 0 & 0 \\ 0 & 0 & -1 & 0 & 0 & 0 & 1 & 0 \\ 0 & -1 & 0 & 1 & 0 & 0 & -1 & 0 \\ 0 & 0 & 0 & -1 & 1 & 0 & 0 & 1 \\ 0 & 0 & 0 & 0 & 0 & 1 & 0 & -1 \end{bmatrix} \begin{Bmatrix} \theta_{\epsilon_1^X,\nu(\epsilon_1^X)} \\ \theta_{\epsilon_2^X,\nu(\epsilon_2^X)} \\ \theta_{\epsilon_3^X,\nu(\epsilon_3^X)} \\ \theta_{\epsilon_4^X,\nu(\epsilon_4^X)} \\ \theta_{\epsilon_5^X,\nu(\epsilon_5^X)} \\ \theta_{\epsilon_6^X,\nu(\epsilon_6^X)} \\ \theta_{\epsilon_7^X,\nu(\epsilon_7^X)} \\ \theta_{\epsilon_8^X,\nu(\epsilon_8^X)} \end{Bmatrix} = \begin{Bmatrix} -\mathcal{R}_{\tau_1^X}^{UNEQ}(\varphi_X^{\Delta}) \\ -\mathcal{R}_{\tau_2^X}^{UNEQ}(\varphi_X^{\Delta}) \\ -\mathcal{R}_{\tau_3^X}^{UNEQ}(\varphi_X^{\Delta}) \\ -\mathcal{R}_{\tau_4^X}^{UNEQ}(\varphi_X^{\Delta}) \\ -\mathcal{R}_{\tau_5^X}^{UNEQ}(\varphi_X^{\Delta}) \\ -\mathcal{R}_{\tau_6^X}^{UNEQ}(\varphi_X^{\Delta}) \end{Bmatrix} \tag{5.3*.39b}$$

or, in abbreviated form, $\mathbf{M}_X\boldsymbol{\theta}_X = -\boldsymbol{\mathcal{R}}_X^{UNEQ}$. Note that, in this case, $\mathbf{M}_X$ is a rectangular matrix, $\mathbf{M}_X \in \mathbb{R}^{ne(X)\times nedges(X)}$, where $nedges(X)$ is the number of edges which belong to the support of φ_X^{Δ}. In this example $ne(X) = 6$ and $nedges(X) = 8$, and $\mathbf{M}_X \in \mathbb{R}^{6\times 8}$. The null space of $\mathbf{M}_X^{\mathrm{T}}$ is spanned by the vector

$$\mathbf{V}_X = [\underbrace{1, 1, \dots, 1}_{ne(X)}]^{\mathrm{T}},$$

and, as before, we have $\mathbf{V}_X^{\mathrm{T}}\boldsymbol{\mathcal{R}}_X^{UNEQ} = 0$, and the linear system admits a $nedges(X) - ne(X) + 1$ parameter family of solutions. As before, a solution $\boldsymbol{\theta}_X$ may be chosen by either fixing the values of $nedges(X) - ne(X) + 1$ components of $\boldsymbol{\theta}_X$, or by employing the method of Lagrange multipliers to determine $\boldsymbol{\theta}_X$ with minimal weighted l^2 norm.

The higher order edge corrections are constructed similarly as in the case of regular meshes. For example, to obtain the higher order corrections for the edge ϵ_5^X of Fig. 5.3*.5, we construct the conjugate basis function $\lambda_{\epsilon_5^X,i}$, over the edge ϵ_5^X such that

$$\int_{\epsilon_5^X} \lambda_{\epsilon_5^X,i}\varphi_{\epsilon_{5,6}^X,j}|_{\epsilon_5^X} = \delta_{ij}, \quad i \geq j \tag{5.3*.40a}$$

where $\varphi_{\epsilon_{5,6}^X,j}$ are the global basis functions which do not vanish on $\epsilon_{5,6}^X \supset \epsilon_5^X$. For example, in the case that $\varphi_{\epsilon_{5,6}^X,j}|_{\epsilon_{5,6}^X}$ are the primitives of the Legendre polynomials, we obtain

$$\tilde{\lambda}_{\tilde{\epsilon}_5,3}(\tilde{x}) = 20\sqrt{6}(1 + 6\tilde{x} + 6\tilde{x}^2), \tag{5.3*.40b}$$

$$\tilde{\lambda}_{\tilde{\epsilon}_5,4}(\tilde{x}) = 56\sqrt{10}(1 + 12\tilde{x} + 30\tilde{x}^2 + 20\tilde{x}^3), \tag{5.3*.40c}$$

$$\tilde{\lambda}_{\tilde{\epsilon}_5,5}(\tilde{x}) = 144\sqrt{14}(1 + 20\tilde{x} + 90\tilde{x}^2 + 140\tilde{x}^3 + 70\tilde{x}^4), \tag{5.3*.40d}$$

$$\tilde{\lambda}_{\tilde{\epsilon}_5,6}(\tilde{x}) = 1056\sqrt{2}(1 + 30\tilde{x} + 210\tilde{x}^2 + 560\tilde{x}^3 + 630\tilde{x}^4 + 252\tilde{x}^5), \tag{5.3*.40e}$$

$$\tilde{\lambda}_{\tilde{\epsilon}_5,7}(\tilde{x}) = 832\sqrt{22}(1 + 42\tilde{x} + 42\tilde{x}^2 + 168\tilde{x}^3 + 3150\tilde{x}^4 + 2184\tilde{x}^5 + 924\tilde{x}^6) \tag{5.3*.40f}$$

where $\tilde{x} \in \tilde{\epsilon}_5 = (-1, 0)$, $\tilde{\epsilon}_5$ is the master edge corresponding to edge $\epsilon_5^X \subset \epsilon_{5,6}^X$, and $\epsilon_{5,6}^X$ is mapped to $\tilde{\epsilon} = (-1, 1)$ with the vertex X mapped to $\tilde{x} = 1$. The conjugate basis

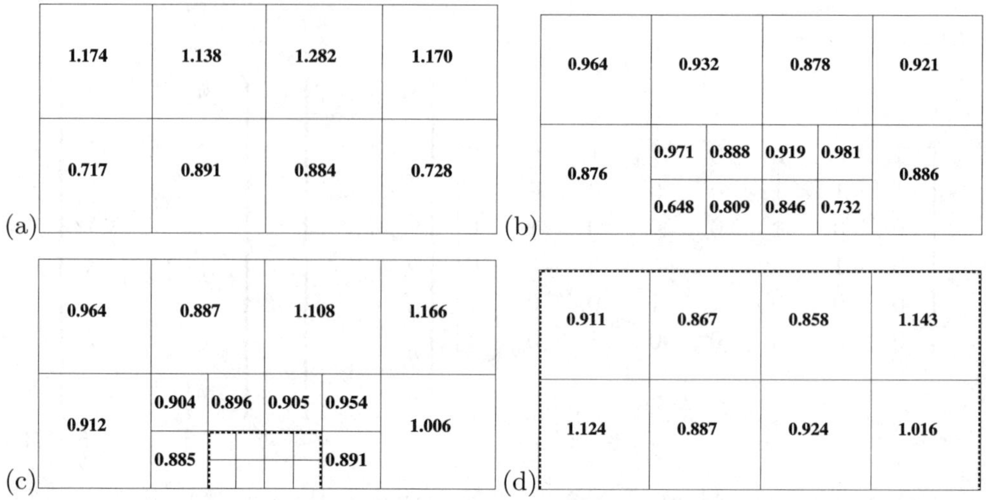

FIG. 5.3*.6. An example which illustrates the accuracy of the element error indicators. Input data: Laplace's equation in $\Omega = (-1, 1) \times (0, 1)$, Neumann boundary conditions, $u_0 \equiv \sqrt{r}\sin(\theta/2)$ for the polar coordinate system shown in Fig. 5.2*.1. Mesh data: Mesh of biquadratic squares ($p = 2$) with one to two irregular connections. Element effectivity indices for the *ERN2B3* estimator for the mesh of: (a) 8 elements (global effectivity index $\kappa = 0.886$); (b) 14 elements ($\kappa = 0.821$); (c) 20 elements ($\kappa = 0.913$); (d) the 8-element patch in the subdomain $(-\frac{1}{4}, \frac{1}{4}) \times (0, \frac{1}{4})$ depicted with dashed perigram in the mesh of 20 elements shown in (c)

function for the edge ϵ_6^X can be written in terms of the master coordinate $\tilde{x} \in \tilde{\epsilon}_6 = (0, 1)$ by employing $\lambda_{\tilde{\epsilon}_6,i}(\tilde{x}) = \lambda_{\tilde{\epsilon}_5,i}(-\tilde{x})$, $i = 3, \ldots, 7$. Using this basis we obtain

$$\theta_{\epsilon_5^X} = \sum_{i=1}^{q+1} \left(\mathcal{R}_{\tau_6^X}(\phi_{\epsilon,i}) - \sum_{k=1}^{i-1} \theta_{\epsilon_5^X,k} \int_{\epsilon_5^X} \lambda_{\epsilon_5^X,k} \phi_{\epsilon,i}|_{\tau_6^X} \right) \lambda_{\epsilon_5^X,i}. \tag{5.3*.40g}$$

Let us now give an example which gives us an idea of the accuracy of the various element residual error indicators for meshes with one to two irregular connections.

Example 5.3*.6 Let us, once more, employ the Neumann problem for the Laplace equation, the domain and the meshes that were also used in Example 5.2*.1. Figure 5.3*.6 (respectively Fig. 5.3*.7) gives the element effectivity indices for the $ERN2B2{+}1$ (respectively $ERN2P2+1$) error indicators, and Fig. 5.3*.8 shows the effectivity indices for the $ERN2B2+\infty \approx ERN2B2+5$ error indicators. Note that the $ERN2B2+1$ estimator is the most effective from the three estimators considered.

As we discussed above, for the construction of the Neumann residual estimator it is sufficient to construct θ_ϵ to satisfy only the consistency condition

$$\mathcal{R}_\tau(1) + \sum_{\epsilon \subseteq \partial\tau} \int_\epsilon \theta_\epsilon^\tau = 0 \tag{5.3*.41}$$

and no other orthogonalities. We will refer to such a construction as *zeroth-order equilibration.* Let us now give such a construction of θ_ϵ, which was proposed in [Ainsworth

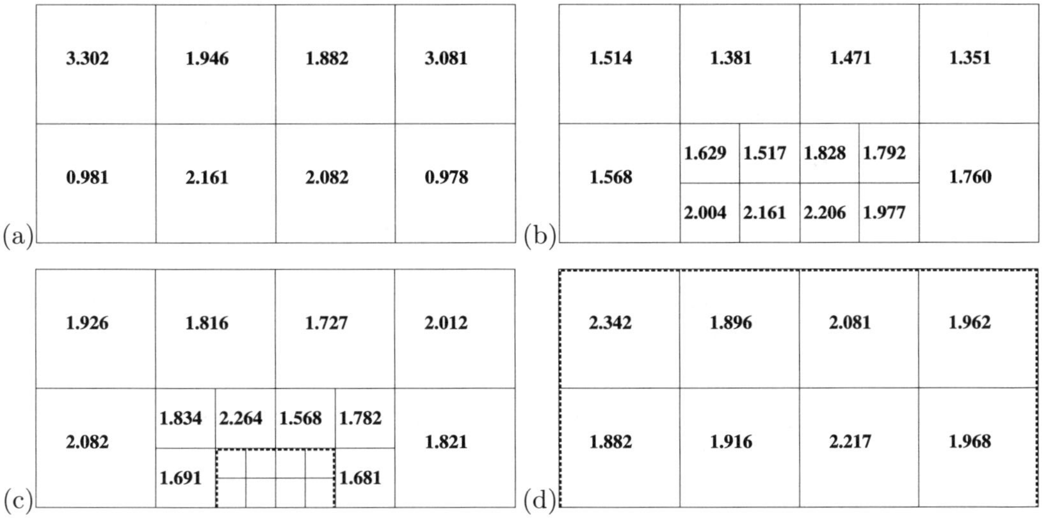

FIG. 5.3*.7. An example which illustrates the accuracy of the element error indicators. Input data: Laplace's equation in $\Omega = (-1, 1) \times (0, 1)$, Neumann boundary conditions, $u_0 \equiv \sqrt{r}\sin(\theta/2)$ for the polar coordinate system shown in Fig. 5.2*.1. Mesh data: Mesh of biquadratic squares ($p = 2$) with one to two irregular connections. Element effectivity indices for the *ERN2P3* estimator for the mesh of: (a) 8 elements (global effectivity index $\kappa = 2.087$); (b) 14 elements ($\kappa = 2.182$); (c) 20 elements ($\kappa = 2.204$); (d) the 8-element patch in the subdomain $(-\frac{1}{4}, \frac{1}{4}) \times (0, \frac{1}{4})$ depicted with dashed perigram in the mesh of 20 elements shown in (c). It can be seen that the error indictors overestimate the error in the majority of the elements

and Oden, 1993]. Letting

$$\theta_\epsilon = (\theta_{\epsilon,1}\lambda_{\epsilon,1} + \theta_{\epsilon,2}\lambda_{\epsilon,2})J_\epsilon \tag{5.3*.42a}$$

into (5.3*.41) we get

$$\sum_{i=1}^{3} \mathfrak{R}_\tau(\phi_{X_i^\tau}) + \sum_{\epsilon \subseteq \partial\tau} \int_\epsilon (\theta_{\epsilon,1}J_\epsilon\lambda_{\epsilon,1} + \theta_{\epsilon,2}J_\epsilon\lambda_{\epsilon,2}) = 0 \tag{5.3*.42b}$$

where X_i^τ, $i = 1, 2, 3$, are the vertices of the element and $\lambda_{\epsilon,1}$, $\lambda_{\epsilon,2}$ are the linear shape functions for the edge ϵ. Equation (5.3*.42b) can also be written as

$$\mathfrak{R}_\tau(\phi_{X_i^\tau}) + \int_{\epsilon_i^\tau} \theta_{\epsilon,\nu(\epsilon,i)}\lambda_{\epsilon,\nu(\epsilon,i)}J_\epsilon\eta_\epsilon^\tau + \int_{\epsilon^\tau_{\text{mod }(i+2,3)}} \theta_{\epsilon,\nu(\epsilon,i)}\lambda_{\epsilon,\nu(\epsilon,i)}J_\epsilon\eta_\epsilon^\tau = 0 \tag{5.3*.42c}$$

for $i = 1, 2, 3$, where $\nu(\epsilon, i) = 1$ or 2 depending on the degree of freedom of θ_ϵ on the edge associated with the vertex i; $\eta_\epsilon^\tau = +1$ or -1, depending on the global positive direction of the edge ϵ; $\epsilon_i^\tau = (X_i^\tau, X^\tau_{\text{mod }(i+1,3)})$ is the ith edge of the element τ. Letting

$$\mu_{\epsilon,\nu(\epsilon,i)} = \theta_{\epsilon,\nu(\epsilon,i)} \int_\epsilon J_\epsilon\lambda_{\epsilon,\nu(\epsilon,i)} \tag{5.3*.42d}$$

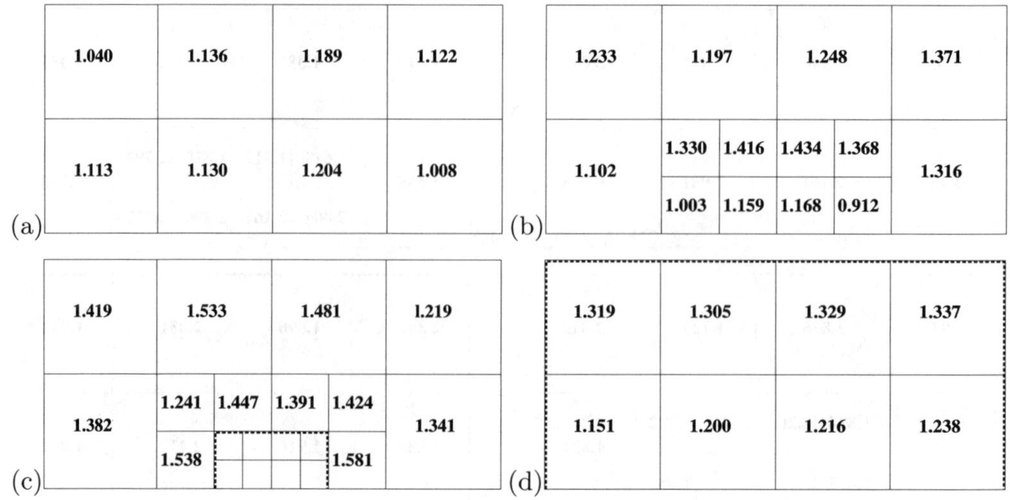

FIG. 5.3*.8. An example which illustrates the accuracy of the element error indicators. Input data: Laplace's equation in $\Omega = (-1, 1) \times (0, 1)$, Neumann boundary conditions, $u_0 \equiv \sqrt{r}\sin(\theta/2)$ for the polar coordinate system shown in Fig. 5.2*.1. Mesh data: Mesh of biquadratic squares ($p = 2$) with one to two irregular connections. Element effectivity indices for the $ERN2B7$ estimator for the mesh of: (a) 8 elements (global effectivity index $\kappa = 1.164$); (b) 14 elements ($\kappa = 1.166$); (c) 20 elements ($\kappa = 1.208$); (d) the 8-element patch in the subdomain $(-\frac{1}{4}, \frac{1}{4}) \times (0, \frac{1}{4})$ depicted with dashed perigram in the mesh of 20 elements shown in (c).

we can solve for $\mu_{\epsilon,1}$, $\mu_{\epsilon,2}$ for an edge ϵ in exactly the same way as we did for $\theta_{\epsilon,1}$, $\theta_{\epsilon,2}$ in Example 5.3*.2 above. The coefficients $\theta_{\epsilon,\nu(\epsilon,i)}$ are then obtained by dividing $\mu_{\epsilon,\nu(\epsilon,i)}$ by $\int_\epsilon J_\epsilon \lambda_{\epsilon,\nu(\epsilon,i)}$; in the case that the jump J_ϵ is zero the coefficient $\theta_{\epsilon,\nu(\epsilon,i)}$ is set to zero. It should be noted, however, that the above construction of the zeroth-order equilibration is not possible in some cases.

5.4 Subdomain residual estimates

In this section we will discuss the *subdomain residual estimator* which was first introduced in [Babuška and Rheinboldt, 1978b,c]; see also [Babuška and Miller, 1981, 1987].

Let

$$\omega_j^\Delta = \operatorname{supp} \varphi_j^\Delta \overset{\text{def}}{=} \{x \in I \mid \varphi_j^\Delta(x) > 0\} \tag{5.4.1a}$$

denote the jth *subdomain*, which is defined as the *support* of the hat function φ_j^Δ associated with the jth vertex of the mesh $j = 0, \ldots, N$. In particular, we have

$$\omega_0^\Delta = I_1^\Delta, \qquad \omega_j^\Delta = I_j^\Delta \cup I_{j+1}^\Delta, \quad 1 \le j \le N-1, \qquad \omega_N^\Delta = I_N^\Delta. \tag{5.4.1b}$$

We will introduce the subdomain residual error indicator functions $\hat{e}^{\text{Dir}}_{\omega_j^\Delta}$, $j = 0, \ldots, N-1$, as the solution of the following subdomain residual problems with zero Dirichlet boundary conditions:

Find $\hat{e}^{\mathrm{Dir}}_{\omega_j^\Delta} \in \mathcal{U}_0(\omega_j^\Delta)$, such that

$$\mathcal{B}_{\omega_j^\Delta}\big(\hat{e}^{\mathrm{Dir}}_{\omega_j^\Delta}, v\big) = \mathcal{R}_{\omega_j^\Delta}(v) \stackrel{\mathrm{def}}{=} \sum_{I_j^\Delta \subseteq \omega_j^\Delta} \int_{I_j^\Delta} r_{I_j^\Delta} v + J_j^\Delta v(x_j^\Delta) \quad \forall v \in \mathcal{U}_0(\omega_j^\Delta). \tag{5.4.1c}$$

The indicator function $\hat{e}^{\mathrm{Dir}}_{\omega_N^\Delta}$ will be determined as the solution of the following subdomain residual problem with mixed boundary conditions:

Find $\hat{e}^{\mathrm{Dir}}_{\omega_N^\Delta} \in \mathcal{U}_{(0}(\omega_N^\Delta)$ such that

$$\mathcal{B}_{\omega_N^\Delta}\big(\hat{e}^{\mathrm{Dir}}_{\omega_N^\Delta}, v\big) = \mathcal{R}_{\omega_N^\Delta}(v) = \int_{I_N^\Delta} r_{I_N^\Delta} v + J_N^\Delta v(x_N^\Delta) \quad \forall v \in \mathcal{U}_{(0}(\omega_N^\Delta). \tag{5.4.1d}$$

Let us now address the existence and uniqueness of the subdomain-residual indicator functions $\hat{e}^{\mathrm{Dir}}_{\omega_j^\Delta}$.

Theorem 5.4.1 *The solutions $\hat{e}^{\mathrm{Dir}}_{\omega_j^\Delta}$ of (5.4.1c,d) exist and are unique under the minimal assumptions which guarantee the existence of solutions of the model problem (2.2.8).*

Proof Let $\tilde{v} \in \mathcal{U}$ be the extension of $v \in \mathcal{U}_0(\omega_j^\Delta)$ by zero outside ω_j^Δ. We then have

$$\big|\mathcal{R}_{\omega_j^\Delta}(v)\big| = \big|\mathcal{R}(\tilde{v})\big| = |\mathcal{B}(e_{S_\Delta^p}, \tilde{v})| \le \|e_{S_\Delta^p}\|_{\mathcal{U}} \|\tilde{v}\|_{\mathcal{U}} = \|e_{S_\Delta^p}\|_{\mathcal{U}} \|v\|_{\mathcal{U}(\omega_j^\Delta)}. \tag{5.4.2}$$

Hence $\mathcal{R}_{\omega_j^\Delta}$ is bounded on $\mathcal{U}_0(\omega_j^\Delta)$, and $\hat{e}^{\mathrm{Dir}}_{\omega_j^\Delta}$ exists and is unique, for $j = 0, \ldots, N$. □

The subdomain residual estimator for the global energy norm is constructed as the square root of the sum of the energies of the solutions of the subdomain residual problems, namely:

$$\eta^{\mathrm{Dir}}_{\omega_j^\Delta} = \|\hat{e}^{\mathrm{Dir}}_{\omega_j^\Delta}\|_{\mathcal{U}(\omega_j^\Delta)}, \qquad \mathcal{E}^{\mathrm{Dir}}_{\mathrm{subd}} \stackrel{\mathrm{def}}{=} \sqrt{\sum_{j=0}^{N(\Delta)} \big(\eta^{\mathrm{Dir}}_{\omega_j^\Delta}\big)^2}. \tag{5.4.3}$$

Remark 5.4.1 The subdomain residual error indicator functions $\hat{e}^{\mathrm{Dir}}_{\omega_j^\Delta}$ exist under the minimal assumptions needed for the existence of solutions of the model problem. In contrast, the existence of the Neumann element residual indicator functions $\hat{e}^{\mathrm{Neum}}_{I_j^\Delta}$ requires an additional assumption of smoothness of AE. There exist problems of practical interest, e.g. problems in highly heterogeneous materials for which AE is a very rough function, for which the Neumann element residual estimator $\mathcal{E}^{\mathrm{Neum}}$ has no meaning. On the other hand, the subdomain residual estimator $\mathcal{E}^{\mathrm{Dir}}_{\mathrm{subd}}$ is always meaningful.

Remark 5.4.2 Note that although both estimators $\mathcal{E}^{\mathrm{Dir}}$ and $\mathcal{E}^{\mathrm{Dir}}_{\mathrm{subd}}$ employ indicators which are the solutions of homogeneous Dirichlet residual problems, they are essentially different. The estimator $\mathcal{E}^{\mathrm{Dir}}_{\mathrm{subd}}$ does take into account the jumps J_ϵ^Δ at the edges of the mesh, while the estimator $\mathcal{E}^{\mathrm{Dir}}$ does not. This leads to different behavior of $\mathcal{E}^{\mathrm{Dir}}$ and $\mathcal{E}^{\mathrm{Dir}}_{\mathrm{subd}}$.

We will now prove the equivalence

$$\mathscr{C}_L^{\text{subd}}\mathscr{E}_{\text{subd}}^{\text{Dir}} \leq \|e_{S^p_\Delta}\|_{\mathscr{U}} \leq \mathscr{C}_U^{\text{subd}}\,\mathscr{E}_{\text{subd}}^{\text{Dir}}. \tag{5.4.4}$$

Theorem 5.4.2 *Under the minimal assumptions which guarantee the existence of solutions of (2.2.8), we have*

$$\frac{1}{\sqrt{2}}\mathscr{E}_{\text{subd}}^{\text{Dir}} \leq \|e_{S^p_\Delta}\|_{\mathscr{U}} \leq \sqrt{2}\sqrt{2\frac{\beta_2}{\beta_1}\Big(1+\frac{1}{\pi^2}\Big)+\frac{\gamma}{\pi^2\beta_1}}\,\mathscr{E}_{\text{subd}}^{\text{Dir}}. \tag{5.4.5}$$

Proof Recalling the orthogonality of the error, $\mathscr{B}(e_{S^p_\Delta}, v) = 0$, for any $v \in S^p_\Delta$, and that $\sum_{j=0}^{N(\Delta)} \varphi_j^\Delta \equiv 1$, we obtain

$$\|e_{S^p_\Delta}\|_{\mathscr{U}}^2 = \mathscr{B}(e_{S^p_\Delta}, e_{S^p_\Delta}) = \mathscr{B}(e_{S^p_\Delta}, e_{S^p_\Delta} - v) = \mathscr{B}\left(e_{S^p_\Delta}, \sum_{j=0}^{N(\Delta)} \varphi_j^\Delta(e_{S^p_\Delta} - v)\right) \tag{5.4.6a}$$

for all $v \in S^p_\Delta$. Using now that $\varphi_j^\Delta(e_{S^p_\Delta} - v)|_{\omega_j^\Delta} \in \mathscr{U}_0(\omega_j^\Delta)$ and $\varphi_N^\Delta(e_{S^p_\Delta} - v)|_{\omega_N^\Delta} \in \mathscr{U}_{(0}(\omega_N^\Delta)$, we get

$$\mathscr{B}\big(e_{S^p_\Delta}, \varphi_j^\Delta(e_{S^p_\Delta} - v)\big) = \mathscr{R}_{\omega_j^\Delta}\big(\varphi_j^\Delta(e_{S^p_\Delta} - v)\big) = \mathscr{B}_{\omega_j^\Delta}\big(\hat{e}^{\text{Dir}}_{\omega_j^\Delta}, \varphi_j^\Delta(e_{S^p_\Delta} - v)\big) \tag{5.4.6b}$$

we obtain

$$\|e_{S^p_\Delta}\|_{\mathscr{U}}^2 = \sum_{j=0}^{N(\Delta)} \mathscr{B}\big(e_{S^p_\Delta}, \varphi_j^\Delta(e_{S^p_\Delta} - v)\big) = \sum_{j=0}^{N(\Delta)} \mathscr{B}_{\omega_j^\Delta}\big(\hat{e}^{\text{Dir}}_{\omega_j^\Delta}, \varphi_j^\Delta(e_{S^p_\Delta} - v)\big)$$
$$\forall v \in S^p_\Delta \tag{5.4.6c}$$

and employing the Cauchy–Schwarz inequality we get

$$\|e_{S^p_\Delta}\|_{\mathscr{U}}^2 \leq \sqrt{\sum_{j=0}^{N(\Delta)} \|\hat{e}^{\text{Dir}}_{\omega_j^\Delta}\|^2_{\mathscr{U}(\omega_j^\Delta)}}\sqrt{\sum_{j=0}^{N(\Delta)} \|\varphi_j^\Delta(e_{S^p_\Delta} - v)\|^2_{\mathscr{U}(\omega_j^\Delta)}} \quad \forall v \in S^p_\Delta. \tag{5.4.6d}$$

Employing now Lemma 5.4.1, which will be proven below, and states that

$$\min_{v\in S^p_\Delta} \sum_{j=0}^{N(\Delta)} \|\varphi_j^\Delta(e_{S^p_\Delta} - v)\|^2_{\mathscr{U}(\omega_j^\Delta)} \leq 2\Big(2\frac{\beta_2}{\beta_1}\Big(1+\frac{1}{\pi^2}\Big)+\frac{\gamma}{\pi^2\beta_1}\Big)\|e_{S^p_\Delta}\|^2_{\mathscr{U}}$$
$$\forall v \in S^p_\Delta \tag{5.4.6e}$$

we get the upper estimate

$$\|e_{S^p_\Delta}\|_{\mathscr{U}} \leq \sqrt{2}\sqrt{2\frac{\beta_2}{\beta_1}\Big(1+\frac{1}{\pi^2}\Big)+\frac{\gamma}{\pi^2\beta_1}}\mathscr{E}_{\text{subd}}^{\text{Dir}}. \tag{5.4.6f}$$

The proof of the lower estimate follows from the identity

$$(\mathscr{E}_{subd}^{\text{Dir}})^2 = \sum_{j=0}^{N(\Delta)} \|\hat{e}_{\omega_j^\Delta}^{\text{Dir}}\|_{\mathcal{U}(\omega_j^\Delta)}^2 = \sum_{j=0}^{N(\Delta)} \mathscr{B}_{\omega_j^\Delta}(\hat{e}_{\omega_j^\Delta}^{\text{Dir}}, \hat{e}_{\omega_j^\Delta}^{\text{Dir}}) = \sum_{j=0}^{N(\Delta)} \mathscr{B}_{\omega_j^\Delta}(e_{S_\Delta^p}, \hat{e}_{\omega_j^\Delta}^{\text{Dir}}) \tag{5.4.7a}$$

where we used the identity

$$\mathscr{B}_{\omega_j^\Delta}(\hat{e}_{\omega_j^\Delta}^{\text{Dir}}, \hat{e}_{\omega_j^\Delta}^{\text{Dir}}) = \mathscr{R}_{\omega_j^\Delta}(\hat{e}_{\omega_j^\Delta}^{\text{Dir}}) = \mathscr{B}_{\omega_j^\Delta}(e_{S_\Delta^p}, \hat{e}_{\omega_j^\Delta}^{\text{Dir}}). \tag{5.4.7b}$$

Employing the Cauchy–Schwarz inequality we obtain

$$(\mathscr{E}_{subd}^{\text{Dir}})^2 = \mathscr{B}_{\omega_j^\Delta}\left(e_{S_\Delta^p}, \sum_{j=0}^{N(\Delta)} \hat{e}_{\omega_j^\Delta}^{\text{Dir}}\right) \le \|e_{S_\Delta^p}\|_{\mathcal{U}} \left\| \sum_{j=0}^{N(\Delta)} \hat{e}_{\omega_j^\Delta}^{\text{Dir}} \right\|_{\mathcal{U}} \tag{5.4.7c}$$

and using that

$$\left\| \sum_{j=0}^{N(\Delta)} \hat{e}_{\omega_j^\Delta}^{\text{Dir}} \right\|_{\mathcal{U}} \le \sqrt{2} \sqrt{\sum_{j=0}^{N(\Delta)} \|\hat{e}_{\omega_j^\Delta}^{\text{Dir}}\|_{\mathcal{U}(\omega_j^\Delta)}^2} \tag{5.4.7d}$$

which will be proven in Lemma 5.4.2 below, we get

$$(\mathscr{E}_{subd}^{\text{Dir}})^2 \le \|e_{S_\Delta^p}\|_{\mathcal{U}} \sqrt{2} \underbrace{\sqrt{\sum_{j=0}^{N(\Delta)} \|\hat{e}_{\omega_j^\Delta}^{\text{Dir}}\|_{\mathcal{U}(\omega_j^\Delta)}^2}}_{\mathscr{E}_{subd}^{\text{Dir}}} \tag{5.4.7e}$$

and hence we have proven that

$$\mathscr{E}_{subd}^{\text{Dir}} = \sqrt{\sum_{j=0}^{N(\Delta)} \|\hat{e}_{\omega_j^\Delta}^{\text{Dir}}\|_{\mathcal{U}(\omega_j^\Delta)}^2} \le \sqrt{2}\|e_{S_\Delta^p}\|_{\mathcal{U}}. \tag{5.4.7f}$$

□

The effect of the approximate determination of the indicator $\hat{e}_{\omega_j^\Delta}^{\text{Dir}}$, has to also be addressed following similar steps as in Section 5.3.

Let us now prove the lemmas employed in the above proof.

Lemma 5.4.1 *Let* φ_j^Δ, $j = 1, \ldots, N(\Delta)$, *denote the piecewise linear hat functions for the mesh* Δ, *and let* $\omega_j^\Delta = \operatorname{supp}(\varphi_j^\Delta)$. *Then, we have*

$$\min_{v \in S_\Delta^p} \sum_{j=0}^{N(\Delta)} \|\varphi_j^\Delta (w - v)\|_{\mathcal{U}(\omega_j^\Delta)}^2 \le 2\left(2\frac{\beta_2}{\beta_1}\left(1 + \frac{1}{\pi^2}\right) + \frac{\gamma}{\pi^2 \beta_1}\right)\|w\|_{\mathcal{U}}^2 \quad \forall w \in \mathcal{U}. \tag{5.4.8}$$

Proof From the definition of the energy norm we have

$$\|\varphi_j^\Delta v\|_{\mathcal{U}(\omega_j^\Delta)}^2 = \left\| \sqrt{AE}\frac{d}{dx}(\varphi_j^\Delta v)\right\|_{L^2(\omega_j^\Delta)}^2 + \|\sqrt{c}\varphi_j^\Delta v\|_{L^2(\omega_j^\Delta)}^2. \tag{5.4.9}$$

The first term in the right-hand side of (5.4.9) will be estimated using that

$$\left\|\sqrt{AE}\frac{d}{dx}(\varphi_j^\Delta v)\right\|^2_{L^2(\omega_j^\Delta)} \leq 2\beta_2\left(\left\|\varphi_j^\Delta \frac{dv}{dx}\right\|^2_{L^2(\omega_j^\Delta)} + \left\|v\frac{d}{dx}(\varphi_j^\Delta)\right\|^2_{L^2(\omega_j^\Delta)}\right) \tag{5.4.10a}$$

where we used that $\|\sqrt{AE}\|^2_{L^\infty(\omega_j^\Delta)} \leq \beta_2$. We have

$$\left\|\varphi_j^\Delta \frac{dv}{dx}\right\|^2_{L^2(\omega_j^\Delta)} \leq \|\varphi_j^\Delta\|^2_{L^\infty(\omega_j^\Delta)}\left\|\frac{dv}{dx}\right\|^2_{L^2(\omega_j^\Delta)} \leq \left\|\frac{dv}{dx}\right\|^2_{L^2(\omega_j^\Delta)} \tag{5.4.10b}$$

where we used that $\|\varphi_j^\Delta\|_{L^\infty(\omega_j^\Delta)} = 1$, and

$$\left\|v\frac{d}{dx}(\varphi_j^\Delta)\right\|^2_{L^2(\omega_j^\Delta)} \leq \sum_{k=j}^{j+1}\left\|\frac{d}{dx}(\varphi_j^\Delta)\right\|^2_{L^\infty(I_k^\Delta)}\|v\|^2_{L^2(I_k^\Delta)} \leq \sum_{k=j}^{j+1}\frac{\|v\|^2_{L^2(I_k^\Delta)}}{\left(h_{I_k^\Delta}\right)^2} \tag{5.4.10c}$$

where we used that $\|(d/dx)(\varphi_j^\Delta)\|_{L^\infty(I_k^\Delta)} = 1/h_k^\Delta$.

The second term in the right-hand side of (5.4.9) will be estimated by

$$\|\sqrt{c}\varphi_j^\Delta v\|^2_{L^2(\omega_j^\Delta)} \leq \|c\|_{L^\infty(\omega_j^\Delta)}\|\varphi_j^\Delta\|^2_{L^\infty(\omega_j^\Delta)}\|v\|^2_{L^2(\omega_j^\Delta)} \leq \gamma\|v\|^2_{L^2(\omega_j^\Delta)} \tag{5.4.11}$$

where we used that $\|c\|_{L^\infty(\omega_j^\Delta)} \leq \gamma$, and that $\|\varphi_j^\Delta\|_{L^\infty(\omega_j^\Delta)} = 1$.

Combining the above we get

$$\|\varphi_j^\Delta v\|^2_{\mathcal{U}(\omega_j^\Delta)} \leq 2\beta_2\left(\left\|\frac{dv}{dx}\right\|^2_{L^2(\omega_j^\Delta)} + \sum_{k=j}^{j+1}\frac{\|v\|^2_{L^2(I_k^\Delta)}}{\left(h_{I_k^\Delta}\right)^2}\right) + \gamma\|v\|^2_{L^2(\omega_j^\Delta)} \tag{5.4.12a}$$

which can also be written as

$$\|\varphi_j^\Delta v\|^2_{\mathcal{U}(\omega_j^\Delta)} \leq 2\beta_2\sum_{k=j}^{j+1}\left(\left\|\frac{dv}{dx}\right\|^2_{L^2(\omega_k^\Delta)} + \left(\frac{1}{\left(h_{I_k^\Delta}\right)^2} + \frac{\gamma}{\beta_2}\right)\|v\|^2_{L^2(\omega_k^\Delta)}\right). \tag{5.4.12b}$$

Recalling now that *in the one-dimensional case*

$$w \in \mathcal{U} \quad \Longrightarrow \quad w \in C^0(I) \tag{5.4.13a}$$

and the interpolant $\mathcal{I}_\Delta^1 w$ is well defined, letting

$$v = w - \mathcal{I}_\Delta^1 w \tag{5.4.13b}$$

in (5.4.12b) and employing now the interpolation error estimate

$$\|w - \mathcal{I}_\Delta^1 w\|_{L^2(I_k^\Delta)} \leq \frac{1}{\pi}h_{I_k^\Delta}\left\|\frac{dw}{dx}\right\|_{L^2(I_k^\Delta)} \leq \frac{1}{\pi}\frac{1}{\sqrt{\beta_1}}h_{I_k^\Delta}\|w\|_{\mathcal{U}(I_k^\Delta)} \tag{5.4.14}$$

and the inequality

$$\left\|\frac{d}{dx}(w - \mathcal{I}_\Delta^1 w)\right\|_{L^2(I_k^\Delta)} \leq \left\|\frac{dw}{dx}\right\|_{L^2(I_k^\Delta)} \leq \frac{1}{\sqrt{\beta_1}}\|w\|_{\mathcal{U}(I_k^\Delta)} \tag{5.4.15}$$

we get

$$\|\varphi_j^\Delta(w-\mathcal{I}_\Delta^1 w)\|^2_{\mathcal{U}(\omega_j^\Delta)} \le 2\frac{\beta_2}{\beta_1}\sum_{k=j}^{j+1}\left(1+h^2_{I_k^\Delta}\left(\frac{1}{\left(h_{I_k^\Delta}\right)^2}+\frac{\gamma}{2\beta_2}\right)\frac{1}{\pi^2}\right)\|w\|^2_{\mathcal{U}(I_k^\Delta)} \qquad (5.4.16a)$$

that is

$$\|\varphi_j^\Delta(w-\mathcal{I}_\Delta^1 w)\|^2_{\mathcal{U}(\omega_j^\Delta)} \le \left(2\frac{\beta_2}{\beta_1}\left(1+\frac{1}{\pi^2}\right)+\frac{\gamma}{\pi^2\beta_1}\right)\sum_{k=j}^{j+1}\|w\|^2_{\mathcal{U}(I_k^\Delta)}. \qquad (5.4.16b)$$

Using that $\|w\|^2_{\mathcal{U}(\omega_j^\Delta)} = \sum_{k=j}^{j+1}\|w\|^2_{\mathcal{U}(I_k^\Delta)}$, and summing over the subdomains, we get

$$\sum_{j=0}^{N}\|\varphi_j^\Delta(w-\mathcal{I}_\Delta^1 w)\|^2_{\mathcal{U}(\omega_j^\Delta)} \le \left(2\frac{\beta_2}{\beta_1}\left(1+\frac{1}{\pi^2}\right)+\frac{\gamma}{\pi^2\beta_1}\right)\sum_{j=0}^{N}\|w\|^2_{\mathcal{U}(\omega_j^\Delta)}. \qquad (5.4.16c)$$

Now, because any element belongs *at most* in two subdomains we have

$$\sum_{j=0}^{N(\Delta)}\|w\|^2_{\mathcal{U}(\omega_j^\Delta)} = \sum_{j=0}^{N(\Delta)}\left(\sum_{k=j}^{j+1}\|w\|^2_{\mathcal{U}(I_k)}\right) \le 2\|w\|^2_{\mathcal{U}} \qquad (5.4.16d)$$

where we employed the definition $\|w\|_{\mathcal{U}(I_0^\Delta)} = 0$, and $\|w\|_{\mathcal{U}(I_{N+1}^\Delta)} = 0$, and the proof of the lemma is complete. □

Remark 5.4.3 Under the additional assumption $\|(d/dx)(AE)\|_{L^\infty(I)} \le K$, it follows that

$$\frac{\beta_2}{\beta_1} = 1 + O(h^\Delta) \qquad (5.4.17a)$$

and, for sufficiently small h^Δ, (5.4.8) can be written in the asymptotic form

$$\min_{v\in S_\Delta^p}\sum_{j=0}^{N(\Delta)}\|\varphi_j^\Delta(w-v)\|^2_{\mathcal{U}(\omega_j^\Delta)} \le 2\left(1+\frac{1}{\pi^2}\right)(1+O(h^\Delta))\|w\|^2_{\mathcal{U}}. \qquad (5.4.17b)$$

We will now prove the second lemma employed in the proof of Theorem 5.4.2.

Lemma 5.4.2 *For any $w_j \in \mathcal{U}_0(\omega_j^\Delta)$, $j = 0,\ldots,N-1$, and $w_N \in \mathcal{U}_{(0}(\omega_N^\Delta)$, we have*

$$\left\|\sum_{j=0}^{N(\Delta)}\tilde{w}_j\right\|^2_{\mathcal{U}} \le 2\sum_{j=0}^{N(\Delta)}\|w_j\|^2_{\mathcal{U}(\omega_j^\Delta)} \qquad (5.4.18)$$

where $\tilde{w}_j \in \mathcal{U}$ is the extension of w_j outside ω_j^Δ by zero.

Proof Noting

$$\left(\sum_{j=0}^{N(\Delta)}\tilde{w}_j\right)\Bigg|_{I_k^\Delta} = w_{k-1}|_{I_k^\Delta} + w_k|_{I_k^\Delta} \qquad (5.4.19a)$$

for $k = 2, \ldots, N(\Delta) - 1$, we have

$$\left\| \sum_{j=0}^{N(\Delta)} \tilde{w}_j \right\|^2_{\mathcal{U}(I_k^\Delta)} \leq 2\left(\|w_{k-1}\|^2_{\mathcal{U}(I_k^\Delta)} + \|w_k\|^2_{\mathcal{U}(I_k^\Delta)}\right) \tag{5.4.19b}$$

and hence

$$\sum_{k=1}^{N(\Delta)} \left\| \sum_{j=0}^{N(\Delta)} \tilde{w}_j \right\|^2_{\mathcal{U}(I_k^\Delta)} \leq 2 \sum_{k=1}^{N(\Delta)} \left(\|w_{k-1}\|^2_{\mathcal{U}(I_k^\Delta)} + \|w_k\|^2_{\mathcal{U}(I_k^\Delta)}\right)$$

$$= 2 \sum_{j=0}^{N(\Delta)} \|w_j\|^2_{\mathcal{U}(\omega_j)} \tag{5.4.19c}$$

which is (5.4.18). □

We also have:

Theorem 5.4.3 *Assumptions A, B, C of Section 5.1 hold for the estimator* $\mathcal{E}^{\mathrm{Dir}}_{subd}$.

Remark 5.4.4 The subdomain residual estimates given above were based on the *partition of unity* generated by the nodal hat functions, φ_j^Δ, $j = 0, \ldots, n$. A more general version of the estimate can be constructed by employing any partition of unity

$$\sum_{j=0}^{n} \varphi_j \equiv 1, \quad \varphi_j \geq 0, \qquad \omega_j = \mathrm{supp}(\varphi_j), \;\; j = 0, \ldots, n \tag{5.4.20a}$$

which satisfies the following properties:

1. The partition of unity has bounded *overlap index*

$$\mathrm{overlap}(\{\varphi_j\}_{j=0}^n) \stackrel{\mathrm{def}}{=} \max_{k=1,\ldots,n} (\mathrm{card}\{j \mid \omega_i \cap \omega_j \neq \emptyset, \; j = 1, \ldots, n\}) \leq M \tag{5.4.20b}$$

with M independent of the partition of unity. Here card{ } is the cardinal number of the set { }. The overlap index is the maximal number of intersecting subdomains in the partition of unity.

2. The relative size of any two intersecting partition-of-unity subdomains ω_i, ω_j, $\omega_i \cap \omega_j \neq \emptyset$, is bounded

$$\frac{1}{\delta} \leq \frac{\mathrm{diam}(\omega_i)}{\mathrm{diam}(\omega_j)} \leq \delta \tag{5.4.20c}$$

with δ independent of the partition of unity.

3. The partition-of-unity functions φ_j are bounded,

$$\|\varphi_j\|_{L^\infty(\omega_j)} \leq \alpha \tag{5.4.20d}$$

and their derivatives are bounded by the inverse of the diameter of ω_j

$$\left\| \frac{d}{dx}(\varphi_j) \right\|_{L^\infty(\omega_j)} \leq \frac{\alpha}{\mathrm{diam}(\omega_j)}. \tag{5.4.20e}$$

The partition of unity generated by the nodal hat functions φ_j^{Δ}, has overlap index equal to two, can be made to satisfy the condition of local uniformity of the supports (5.4.20c) by employing locally uniform meshes, and satisfies conditions (5.4.20d) and (5.4.20e) because we have

$$\|\varphi_j^{\Delta}\|_{L^\infty(\omega_j^{\Delta})} = 1, \qquad \left\|\frac{d}{dx}(\varphi_j^{\Delta})\right\|_{L^\infty(\omega_j^{\Delta})} = \frac{1}{\min\left(h_{I_j^{\Delta}}, h_{I_{j+1}^{\Delta}}\right)}. \tag{5.4.21}$$

Remark 5.4.5 The subdomain residual estimates can be directly generalized to the higher-dimensional case. The construction of the partition of unity functions is completely analogous with the one-dimensional case. The proof of Lemma 5.4.1 in higher dimensions is more technical because $e_{S^p_\Delta} \in \mathcal{U}$ *does not imply* $e_{S^p_\Delta} \in C^0(\Omega)$, *and the interpolant* $\mathcal{I}^1_\Delta e_{S^p_\Delta}$ *may not have meaning.*

Remark 5.4.6 The above subdomain residual estimator employs the indicators

$$\eta^{\text{Dir}}_{\omega_j^{\Delta}} = \|\hat{e}^{\text{Dir}}_{\omega_j^{\Delta}}\|_{\mathcal{U}(\omega_j^{\Delta})} = \sup_{v \in \mathcal{U}_0(\omega_j^{\Delta})} \frac{1}{\|v\|_{\mathcal{U}(\omega_j^{\Delta})}} \Big(\sum_{I_j^{\Delta} \subseteq \omega_j^{\Delta}} \int_{I_j^{\Delta}} r_{I_j^{\Delta}} v + J_j^{\Delta} v(x_j^{\Delta}) \Big) \tag{5.4.22a}$$

for $j = 0, \ldots, N-1$, and similarly for $j = N$. [Carstensen and Funken, 1999] analyzed a modified two-dimensional version of this estimator based on the indicators

$$\begin{aligned} \eta^{\text{Dir}}_{\omega_j^{\Delta}, \varphi_j} &= \|\hat{e}^{\text{Dir}}_{\omega_j^{\Delta}, \varphi_j}\|_{\mathcal{U}(\omega_j^{\Delta}), \varphi_j} \\ &= \sup_{v \in \mathcal{U}_0(\omega_j^{\Delta})} \frac{1}{\|v\|_{\mathcal{U}(\omega_j^{\Delta}), \varphi_j}} \Big(\sum_{I_j^{\Delta} \subseteq \omega_j^{\Delta}} \int_{I_j^{\Delta}} \varphi_j r_{I_j^{\Delta}} v + J_j^{\Delta} \varphi_j(x_j^{\Delta}) v(x_j^{\Delta}) \Big) \end{aligned} \tag{5.4.22b}$$

where $\|\cdot\|_{\mathcal{U}(\omega_j^{\Delta}), \varphi_j}$ is the weighted energy norm defined

$$\|v\|_{\mathcal{U}(\omega_j^{\Delta})} = \sqrt{\sum_{I_j^{\Delta} \subseteq \omega_j^{\Delta}} \int_{I_j^{\Delta}} \varphi_j \Big(AE\Big(\frac{dv}{dx}\Big)^2 + cv^2 \Big)} \tag{5.4.22c}$$

for $j = 0, \ldots, N-1$, and analogously for $j = N$, and $\hat{e}^{\text{Dir}}_{\omega_j^{\Delta}, \varphi_j}$ is the solution of the subdomain residual problem (5.4.1c,d) with the residuals $r_{I_j^{\Delta}}$, J_j^{Δ}, replaced by $\varphi_j r_{I_j^{\Delta}}$ and $J_j^{\Delta} \varphi_j$. Carstensen and Funken showed in [Carstensen and Funken, 2000b] that the modified estimator is slightly more effective than the original one because it avoids the influence of the overlap in the constants. Problems of this type are related, for example, to the results from the dimensional reduction method; see [Babuška and Schwab, 1996], [Babuška and Vogelius, 1981].

Remark 5.4.7 The subdomain residual estimator performs well with respect to the parameter γh^2 which measures the importance of the zeroth-order term in the differential equation. This fact was exploited by [Verfürth, 1988] to obtain an explicit estimator with uniform accuracy for the entire range of γh^2. Ainsworth and Babuška [Ainsworth and Babuška, 1999] showed that the estimator $\mathscr{E}^{\text{Neum}}$ grossly overestimates for large γh^2, and that this overestimation is due to the employed splitting of the jumps and may be controlled by modifying the sets of functions for which the element orthogonalities are enforced.

Above we have proven the equivalence $\|e_{S^p_\Delta}\|_{\mathcal{U}} \approx \mathcal{E}^{\mathrm{Dir}}_{subd}$, under the minimal assumptions which guarantee the existence of solutions of the model problem. We will prove that for locally quasi-uniform meshes, and for AE, c, f, piecewise polynomials over the meshes, $\mathcal{E}^{\mathrm{Dir}}_{subd}$, $\mathcal{E}^{\mathrm{EXPL}}$, $\mathcal{E}^{\mathrm{Neum}}$, and $\mathcal{E}^{\mathrm{Dir}}$ are equivalent. We have:

Theorem 5.4.4 *Assume that the meshes Δ are locally μ quasi-uniform, namely*

$$\frac{1}{\mu} \leq \frac{h_{I^\Delta_j}}{h_{I^\Delta_{j+1}}} \leq \mu \tag{5.4.23a}$$

for $j = 0, \ldots, N(\Delta)-1$, for a fixed $\mu \geq 1$, and that AE, c, and f are piecewise polynomials over the meshes Δ, namely

$$AE|_{I^\Delta_j} \in \mathcal{P}^{q_j}(I^\Delta_j), \qquad c|_{I^\Delta_j} \in \mathcal{P}^{r_j}(I^\Delta_j), \qquad f|_{I^\Delta_j} \in \mathcal{P}^{s_j}(I^\Delta_j) \tag{5.4.23b}$$

for $j = 1, \ldots, N(\Delta)$, for some integers $q_j \leq q$, $r_j \leq r$, $s_j \leq s$, for $j = 1, \ldots, N(\Delta)$. Then

$$\eta^{\mathrm{EXPL}}_{\omega^\Delta_j} \stackrel{\mathrm{def}}{=} \sqrt{\sum_{k=j}^{j+1} (h^\Delta_k)^2 \int_{I^\Delta_k} (r_{I^\Delta_k})^2 + \frac{1}{2}(h_j + h_{j+1})(J^\Delta_j)^2} \leq \mathcal{C}\|\hat{e}^{\mathrm{Dir}}_{\omega^\Delta_j}\|_{\mathcal{U}(\omega^\Delta_j)} \tag{5.4.23c}$$

with $\mathcal{C}$ depending only on μ, and on β_1, β_2, γ, and on q, r, s.

Proof Let $F_{\omega^\Delta_j} : \hat{\omega} \stackrel{\mathrm{def}}{=} (-1,1) \mapsto \omega^\Delta_j = (x^\Delta_{j-1}, x^\Delta_{j+1})$, be the piecewise affine mapping given by

$$F_{\omega^\Delta_j}(\hat{x}) = \begin{cases} x^\Delta_j + (x^\Delta_j - x^\Delta_{j-1})\hat{x}, & \hat{x} \in \hat{I}_1 \stackrel{\mathrm{def}}{=} (-1,0) \\ x^\Delta_j + (x^\Delta_{j+1} - x^\Delta_j)\hat{x}, & \hat{x} \in \hat{I}_2 \stackrel{\mathrm{def}}{=} (0,1). \end{cases} \tag{5.4.24}$$

We then have $I^\Delta_j = F_{\omega^\Delta_j}(\hat{I}_1)$ and $I^\Delta_{j+1} = F_{\omega^\Delta_j}(\hat{I}_2)$, and the subdomain residual problem can be written in the master patch $\hat{\omega}$, namely:

Find $\hat{e}^{\mathrm{Dir}}_{\hat{\omega}} \in \mathcal{U}_0(\hat{\omega})$ such that

$$\mathcal{B}_{\hat{\omega}}(\hat{e}^{\mathrm{Dir}}_{\hat{\omega}}, v) = \mathcal{R}_{\hat{\omega}}(v) \stackrel{\mathrm{def}}{=} \sum_{k=1}^{2} \int_{\hat{I}_k} \hat{r}_{\hat{I}_k} v + \hat{J}v(0) \quad \forall v \in \mathcal{U}_0(\hat{\omega}) \tag{5.4.25a}$$

where

$$\hat{r}_{\hat{I}_k}(\hat{x}) \stackrel{\mathrm{def}}{=} r_{I_k}(F_{\omega^\Delta_j}(\hat{x})) \frac{d}{dx}(F_{\omega^\Delta_j})(\hat{x}), \quad \hat{J} = J^\Delta_j. \tag{5.4.25b}$$

Then (5.4.23c) can be restated as

$$\mathcal{C}\eta^{\mathrm{EXPL}}_{\hat{\omega}} = \mathcal{C}\sqrt{\sum_{k=1}^{2} \int_{\hat{I}_k} (\hat{r}_{\hat{I}_k})^2 + (\hat{J})^2} \leq \|\hat{e}^{\mathrm{Dir}}_{\hat{\omega}}\|_{\mathcal{U}(\hat{\omega})} \tag{5.4.25c}$$

where we incorporated $\mathcal{C}$ into the indicator $\eta^{\mathrm{EXPL}}_{\hat{\omega}}$.

We will now prove (5.4.25c) by contradiction. Assume that (5.4.25c) is false. Then there exists a sequence $\{(\hat{r}_{\hat{I}_1,n}, \hat{r}_{\hat{I}_2,n}, \hat{J}_n)\}_{n=1}^{\infty}$ which is such that

$$\|\hat{e}_{\hat{\omega},n}^{\mathrm{Dir}}\|_{\mathfrak{U}(\hat{\omega})} \leq \epsilon_n < \eta_{\hat{\omega},n}^{\mathrm{EXPL}} = \sqrt{\sum_{k=1}^{2} \int_{\hat{I}_k} (\hat{r}_{\hat{I}_k,n})^2 + (\hat{J}_n)^2} = 1 \tag{5.4.26}$$

with $\lim_{n\to\infty} \epsilon_n = 0$, where $\hat{e}_{\hat{\omega},n}^{\mathrm{Dir}}$ is the solution of the subdomain residual problem loaded by $\hat{r}_{\hat{I}_1,n}$, $\hat{r}_{\hat{I}_2,n}$, $\hat{J}_n$.

From the assumptions about the data AE, c, and f, it follows that $\hat{r}_{\hat{I}_k,n}$ and $\hat{J}_n$ depend on the finite number of parameters which determine $u_{S_\Delta^p}|_{\omega_j^\Delta}$, $AE|_{\omega_j^\Delta}$, $c|_{\omega_j^\Delta}$, and $f|_{\omega_j^\Delta}$, and thus every $(\hat{r}_{\hat{I}_1,n}, \hat{r}_{\hat{I}_2,n}, \hat{J}_n)$ can be indexed by an element of $\mathbb{R}^M$, where $M = p+1+\sum_{k=j}^{j+1}(q_k+r_k+s_k)+1$. Hence the sequence $\{(\hat{r}_{\hat{I}_1,n}, \hat{r}_{\hat{I}_2,n}, \hat{J}_n)\}_{n=1}^{\infty}$ can be rewritten as a sequence in $\mathbb{R}^M$, which is bounded because

$$\sum_{k=1}^{2} \int_{\hat{I}_k} (\hat{r}_{\hat{I}_k,n})^2 + (\hat{J}_n)^2 = 1 \tag{5.4.27}$$

and hence has a convergent subsequence, the limit of which we will denote by $(\hat{r}_{\hat{I}_1,\infty}, \hat{r}_{\hat{I}_2,\infty}, \hat{J}_\infty)$. Replacing the original sequence by its convergent subsequence we have

$$\lim_{n\to\infty} \|\hat{r}_{\hat{I}_k,n} - \hat{r}_{\hat{I}_k,\infty}\|_{L^2(\hat{I}_k)} = 0, \quad k = 1,2, \qquad \lim_{n\to\infty} |\hat{J}_n - \hat{J}_\infty| = 0. \tag{5.4.28}$$

Taking now the limit on both sides of assumption (5.4.26), we get

$$\|\hat{e}_{\hat{\omega},\infty}^{\mathrm{Dir}}\|_{\mathfrak{U}(\hat{\omega})}^2 = 0 < (\eta_{\hat{\omega},\infty}^{\mathrm{EXPL}})^2 = \sum_{k=1}^{\infty} \int_{\hat{I}_k} (\hat{r}_{\hat{I}_k,\infty})^2 + (\hat{J}_\infty)^2 = 1 \tag{5.4.29}$$

where $\hat{e}_{\hat{\omega},\infty}^{\mathrm{Dir}}$ denotes the solution of the subdomain residual problem loaded by $\hat{r}_{\hat{I}_1,\infty}$, $\hat{r}_{\hat{I}_2,\infty}$, and $\hat{J}_\infty$. It follows that

$$\|\mathcal{R}_{\hat{\omega},\infty}\|_{L(\mathfrak{U}_0(\hat{\omega}),\mathbb{R})} = \sup_{v\in\mathfrak{U}_0(\hat{\omega})} \frac{\sum_{k=1}^{2} \int_{\hat{I}_k} \hat{r}_{\hat{I}_k,\infty} v + \hat{J}_\infty v(0)}{\|v\|_{\mathfrak{U}(\hat{\omega})}} = \|\hat{e}_{\hat{\omega},\infty}^{\mathrm{Dir}}\|_{\mathfrak{U}(\hat{\omega})} = 0. \tag{5.4.30a}$$

However, this is impossible because

$$\sum_{k=1}^{2} \int_{\hat{I}_k} (\hat{r}_{\hat{I}_k,\infty})^2 + (\hat{J}_\infty)^2 = 1 \tag{5.4.30b}$$

and hence at least one of $\int_{\hat{I}_1} (\hat{r}_{\hat{I}_1,\infty})^2$, $\int_{\hat{I}_2} (\hat{r}_{\hat{I}_2,\infty})^2$, and $(\hat{J}_\infty)^2$ is nonzero, and there exist $v \in \mathfrak{U}_0(\hat{\omega})$ such that

$$0 < |\mathcal{R}_{\hat{\omega},\infty}(v)|. \tag{5.4.30c}$$

The proof of (5.4.23c) for the end patches ω_0^Δ and ω_N^Δ, follows from the same argument. □

Let us also prove the reverse inequality, $\|\hat{e}^{\mathrm{Dir}}_{\omega_j^\Delta}\|_{\mathcal{U}(\omega_j^\Delta)} \le \mathscr{C}\eta^{\mathrm{EXPL}}_{\omega_j^\Delta}$.

Theorem 5.4.5 *Under the assumptions of Theorem 5.4.4, we have*

$$\|\hat{e}^{\mathrm{Dir}}_{\omega_j^\Delta}\|_{\mathcal{U}(\omega_j^\Delta)} \le \mathscr{C}\eta^{\mathrm{EXPL}}_{\omega_j^\Delta}. \tag{5.4.31}$$

Proof The proof of (5.4.31) is obtained by rescaling ω_j^Δ into $\hat{\omega}$, as in the proof of Theorem 5.4.4, and employing that

$$\|\hat{e}^{\mathrm{Dir}}_{\hat{\omega}}\|_{\mathcal{U}(\hat{\omega})} = \|\mathscr{R}_{\hat{\omega}}\|_{L(\mathcal{U}_0(\hat{\omega}),\mathbb{R})} \le \sqrt{\sum_{k=1}^{2}\int_{\hat{I}_k}(\hat{r}_{\hat{I}_k})^2 + (\hat{J})^2 \frac{\sqrt{\|v\|^2_{\mathcal{U}(\hat{\omega})} + (v(0))^2}}{\|v\|_{\mathcal{U}(\hat{\omega})}}} \tag{5.4.32a}$$

and by noting that there exists a constant $\mathscr{C}$, such that

$$|v(0)| \le \mathscr{C}\|v\|_{\mathcal{U}(\hat{\omega})}, \quad \forall v \in \mathcal{U}_0(\hat{\omega}). \tag{5.4.32b}$$

□

Combining Theorems 5.4.2, 5.4.4, and 5.4.5 we obtain the equivalence

$$\mathscr{E}^{\mathrm{Dir}}_{\mathrm{subd}} \approx \mathscr{E}^{\mathrm{EXPL}} = \sqrt{\sum_{j=0}^{N}\left(\eta^{\mathrm{EXPL}}_{\omega_j^\Delta}\right)^2} \approx \|e_{S^p_\Delta}\|_{\mathcal{U}}. \tag{5.4.33}$$

Hence, we do not distinguish between the various versions of the explicit estimator $\mathscr{E}^{\mathrm{EXPL}}$ when we are only interested in their equivalence with the energy norm of the error.

Theorem 5.4.6 *Under the same assumptions as in Theorem 5.4.4 we have*

$$\mathscr{C}\mathscr{E}^{\mathrm{EXPL}} \le \mathscr{E}^{\mathrm{Dir}}_{\mathrm{subd}} \le \mathscr{C}\mathscr{E}^{\mathrm{EXPL}} \tag{5.4.34a}$$

and

$$\mathscr{C}\mathscr{E}^{\mathrm{EXPL}} \le \|e_{S^p_\Delta}\|_{\mathcal{U}} \le \mathscr{C}\mathscr{E}^{\mathrm{EXPL}} \tag{5.4.34b}$$

where the constants $\mathscr{C}$ *depend only on* μ, *on* β_1, β_2, γ, *and on* q, r, s.

Proof Combining Theorems 5.4.4 and 5.4.5 we obtain

$$\|\hat{e}^{\mathrm{Dir}}_{\omega_j^\Delta}\|_{\mathcal{U}(\omega_j^\Delta)} \approx \eta^{\mathrm{EXPL}}_{\omega_j^\Delta} = \sqrt{\sum_{k=j}^{j+1}\left(h_{I_k^\Delta}\right)^2\int_{I_k^\Delta}(r_{I_k^\Delta})^2 + \frac{1}{2}\left(h_{I_j^\Delta} + h_{I_{j+1}^\Delta}\right)(J_j^\Delta)^2} \tag{5.4.35}$$

and summing over all the subdomains, we get (5.4.34a). Then (5.4.34b) follows from (5.4.34a) and Theorem 5.4.2. □

Remark 5.4.8 Theorems 5.4.4–5.4.6 also hold in the case of the modified subdomain residual estimator.

We will also prove that

$$\mathscr{E}^{\mathrm{Dir}}_{\mathrm{subd}} \approx \mathscr{E}^{\mathrm{Neum}} \approx \mathscr{E}^{\mathrm{Dir}} \approx \|e_{S^p_\Delta}\|_{\mathcal{U}}. \tag{5.4.36}$$

Theorem 5.4.7 *Under the assumptions of Theorem 5.4.4, we have*

$$\mathscr{C}\mathscr{E}^{\mathrm{Dir}}_{\mathrm{subd}} \leq \mathscr{E}^{\mathrm{BC}} \leq \mathscr{C}\,\mathscr{E}^{\mathrm{Dir}}_{\mathrm{subd}} \tag{5.4.37}$$

where BC= Dir, or BC= Neum.

Proof Let us first consider the $\mathscr{E}^{\mathrm{Neum}}$ estimator. Recalling that

$$\eta^{\mathrm{Neum}}_{I^{\Delta}_j} = \|\hat{e}^{\mathrm{Neum}}_{I^{\Delta}_j}\|_{\mathscr{U}(I^{\Delta}_j)} = \sup_{v\in\mathscr{U}(I^{\Delta}_j)} \frac{\mathscr{R}^{EQ}_{I^{\Delta}_j}(v)}{\|v\|_{\mathscr{U}(I^{\Delta}_j)}} \leq \mathscr{C}\eta^{\mathrm{EXPL}}_{I^{\Delta}_j} \tag{5.4.38a}$$

and squaring and summing over the elements we get

$$\mathscr{E}^{\mathrm{Neum}} \leq \mathscr{E}^{\mathrm{EXPL}} \leq \mathscr{C}\sqrt{\sum_{j=1}^{N}\left(\eta^{\mathrm{EXPL}}_{I^{\Delta}_j}\right)^2} \leq \sqrt{\sum_{j=0}^{N(\Delta)}\left(\eta^{\mathrm{EXPL}}_{\omega^{\Delta}_j}\right)^2}. \tag{5.4.38b}$$

On the other hand, from Theorem 5.4.6, we have

$$\mathscr{C}\mathscr{E}^{\mathrm{EXPL}} \leq \|e_{S^p_{\Delta}}\|_{\mathscr{U}} \tag{5.4.38c}$$

and hence we have

$$\mathscr{C}\mathscr{E}^{\mathrm{EXPL}} \leq \|e_{S^p_{\Delta}}\|_{\mathscr{U}} \leq \mathscr{E}^{\mathrm{Neum}} \leq \mathscr{E}^{\mathrm{EXPL}} \tag{5.4.38d}$$

which proves that

$$\mathscr{E}^{\mathrm{EXPL}} \approx \mathscr{E}^{\mathrm{Neum}}. \tag{5.4.38e}$$

Recalling now (5.4.33)

$$\|e_{S^p_{\Delta}}\|_{\mathscr{U}} \approx \mathscr{E}^{\mathrm{EXPL}} \approx \mathscr{E}^{\mathrm{Dir}}_{\mathrm{subd}} \tag{5.4.38f}$$

we obtain

$$\mathscr{E}^{\mathrm{Dir}}_{\mathrm{subd}} \approx \|e_{S^p_{\Delta}}\|_{\mathscr{U}} \approx \mathscr{E}^{\mathrm{Neum}}. \tag{5.4.38g}$$

For the $\mathscr{E}^{\mathrm{Dir}}$ estimator we have

$$\|\hat{e}^{\mathrm{Dir}}_j\|_{\mathscr{U}(I^{\Delta}_j)} = \sup_{v\in\mathscr{U}_0(I^{\Delta}_j)} \frac{\int_{I^{\Delta}_j} r_{I^{\Delta}_j} v}{\|v\|_{\mathscr{U}(I^{\Delta}_j)}} \leq \mathscr{C}h_{I^{\Delta}_j}\|r_{I^{\Delta}_j}\|_{L^2(I^{\Delta}_j)} \leq \mathscr{C}\eta^{\mathrm{EXPL}}_{I^{\Delta}_j} \tag{5.4.39a}$$

and hence we have

$$\mathscr{E}^{\mathrm{Dir}} \leq \mathscr{C}\,\mathscr{E}^{\mathrm{EXPL}}. \tag{5.4.39b}$$

The lower bound

$$\mathscr{C}\mathscr{E}^{\mathrm{EXPL}} \leq \mathscr{E}^{\mathrm{Dir}} \tag{5.4.40a}$$

follows from the observation that from

$$\mathcal{R}(\varphi_j^\Delta) = \int_{I_j^\Delta} r_{I_j^\Delta}\varphi_j^\Delta + \int_{I_{j+1}^\Delta} r_{I_{j+1}^\Delta}\varphi_j^\Delta + J_j^\Delta = 0 \tag{5.4.40b}$$

and we obtain

$$|J_j^\Delta| = \left| \int_{I_j^\Delta} r_{I_j^\Delta}\varphi_j^\Delta + \int_{I_{j+1}^\Delta} r_{I_{j+1}^\Delta}\varphi_j^\Delta \right| \le \int_{I_j^\Delta} |r_{I_j^\Delta}| + \int_{I_{j+1}^\Delta} |r_{I_{j+1}^\Delta}| \tag{5.4.40c}$$

and hence

$$|J_j^\Delta| \le \sqrt{h_{I_j^\Delta}}\, \|r_{I_j^\Delta}\|_{L^2(I_j^\Delta)} + \sqrt{h_{I_{j+1}^\Delta}}\, \|r_{I_{j+1}^\Delta}\|_{L^2(I_{j+1}^\Delta)}. \tag{5.4.40d}$$

Using (5.4.40d) into the definition of $\eta_{\omega_j^\Delta}^{\mathrm{EXPL}}$, we get

$$\left(\eta_{\omega_j^\Delta}^{\mathrm{EXPL}}\right)^2 \le \mathscr{C}\left(\left(h_{I_j^\Delta}\right)^2 \|r_{I_j^\Delta}\|_{L^2(I_j^\Delta)}^2 + \left(h_{I_{j+1}^\Delta}\right)^2 \|r_{I_{j+1}^\Delta}\|_{L^2(I_{j+1}^\Delta)}^2\right). \tag{5.4.40e}$$

Further, because AE, c, and f are piecewise polynomials, there exists a $\mathscr{C} > 0$, such that

$$\mathscr{C} h_{I_j^\Delta} \|r_{I_j^\Delta}\|_{L^2(I_j^\Delta)} \le \|r_{I_j^\Delta}\|_{L(\mathscr{U}_0(I_j^\Delta),\mathbb{R})} = \|\hat{e}_{I_j^\Delta}^{\mathrm{Dir}}\|_{\mathscr{U}(I_j^\Delta)} \tag{5.4.40f}$$

and hence

$$\left(\eta_{I_j^\Delta}^{\mathrm{EXPL}}\right)^2 \le \mathscr{C} \sum_{k=j}^{j+1} \|\hat{e}_{I_k^\Delta}^{\mathrm{Dir}}\|_{\mathscr{U}(I_k^\Delta)}^2 \tag{5.4.40g}$$

and summing over the elements we get

$$\mathscr{E}^{\mathrm{EXPL}} \le \mathscr{C}\mathscr{E}^{\mathrm{Dir}}. \tag{5.4.40h}$$

□

Note that the *equivalence*

$$\mathscr{E}_{\mathrm{subd}}^{\mathrm{Dir}} \approx \mathscr{E}^{\mathrm{Neum}} \approx \mathscr{E}^{\mathrm{Dir}} \approx \mathscr{E}^{\mathrm{EXPL}} \approx \|e_{S_\Delta^p}\|_{\mathscr{U}} \tag{5.4.41}$$

holds without any assumptions on the size of h^Δ.

Remark 5.4.9 Note that the estimate

$$\mathscr{C}\eta_{\omega_j^\Delta}^{\mathrm{EXPL}} \le \|\hat{e}_{\omega_j^\Delta}^{\mathrm{Dir}}\|_{\mathscr{U}(\omega_j^\Delta)} \tag{5.4.42}$$

was proven under the assumption that f is piecewise polynomial over the mesh. It is easy to see that this assumption is necessary. For example, if $f \in \mathscr{U}^{-1}$, but $f \notin L^2(I)$, we could have bounded $\|\hat{e}_{\omega_j^\Delta}^{\mathrm{Dir}}\|_{\mathscr{U}(\omega_j^\Delta)}$, while $\|f\|_{L^2(I)}$ and $\eta_{\omega_j^\Delta}^{\mathrm{EXPL}}$ are meaningless.

Remark 5.4.10 Theorems 5.4.4–5.4.7 can also be extended to the more general case that f is sufficiently smooth. This can be done by replacing f in each element by $\mathcal{I}_{I_j^\Delta}^{s_j} f$, its local finite element interpolant of a sufficiently high order s_j. Then, all the estimates must be corrected to account for the interpolation error $\sqrt{\sum_{j=1}^{N(\Delta)} \|f - \mathcal{I}_{I_j^\Delta}^{s_j} f\|^2_{L^2(I_j^\Delta)}}$.

Remark 5.4.11 In Theorem 5.3.2 the equivalence $\mathscr{E}^{\text{Neum}} \approx \mathscr{E}^{\text{Dir}}$ was proven without the assumptions that the meshes Δ are locally quasi-uniform, and AE, c, and f are piecewise polynomial. However *the proof of Theorem 5.3.2 cannot be extended to higher dimensions.* On the other hand, the proof of Theorems 5.4.4–5.4.7 has a direct analogue in higher dimensions.

Remark 5.4.12 The practical value of the above equivalences between the various residual estimators and the energy norm of the error is rather limited, because the values of the equivalence constants determined by going through the proofs are rather pessimisstic for practical use.

Remark 5.4.13 Employing the recovery

$$\mathcal{A}_u^{\text{Dir}}(u_{S^p_\Delta}) \stackrel{\text{def}}{=} u_{S^p_\Delta} + \sum_{j=1}^{N} \hat{e}^{\text{Dir}}_{\omega_j^\Delta} \tag{5.4.43a}$$

we also obtain the guaranteed lower bound

$$\|e_{S^p_\Delta}\|_{\mathcal{U}} \geq \frac{\mathscr{R}\big(\mathcal{A}_u^{\text{Dir}}(u_{S^p_\Delta})\big)}{\big\|\mathcal{A}_u^{\text{Dir}}(u_{S^p_\Delta})\big\|_{\mathcal{U}}}. \tag{5.4.43b}$$

In this section we have introduced the subdomain error estimator and have shown that it is an upper and lower estimator with specified constants. We have also proven that for locally quasi-uniform meshes and piecewise polynomial input data, the explicit Neumann and the Dirichlet residual estimators are equivalent to the subdomain residual estimator, and hence, to the energy norm of the error.

Problems

P5.4.1. Show that the results of this section hold for any partition of unity $\{\varphi_j\}_{j=1}^M$ which is such that

$$\|\varphi_j\|_{L^\infty} \leq K_1, \qquad \left\|\frac{d}{dx}(\varphi_j)\right\|_{L^\infty} \leq \frac{K_2}{|\omega_j|} \tag{5.4.44}$$

where ω_j is the support of φ_j.

P5.4.2. Let φ_j be any functions $j = 0, \dots, M$, and let $\psi_j(x) = (\sum_{i=0}^M \varphi_i(x))\varphi_j(x)$. Under what conditions can $\{\psi_j\}_{j=1}^M$ be used in the construction of the subdomain residual estimate?

5.4* Subdomain residual estimates in two dimensions

Let us now discuss the two-dimensional version of the subdomain residual estimate. As in the one-dimensional case, we will employ partition of unity corresponding to the hat functions φ_X^Δ, corresponding to the vertices X of the mesh, the corresponding subdomains

$\omega_X^\Delta = \operatorname{supp}(\varphi_X^\Delta)$, and the subdomain residual problems with zero Dirichlet boundary conditions:

Find $\hat{e}^{\mathrm{Dir}}_{\omega_X^\Delta} \in \mathcal{U}_0(\omega_X^\Delta)$ such that

$$\mathcal{B}_{\omega_X^\Delta}(\hat{e}^{\mathrm{Dir}}_{\omega_X^\Delta}, v) = \mathcal{R}_{\omega_X^\Delta}(v) = \sum_{\tau \subseteq \omega_X^\Delta} \left(\int_\tau r_\tau v + \sum_{\substack{\epsilon \subseteq \partial\tau \\ \epsilon \cap \partial\omega_X^\Delta = \emptyset}} \int_\epsilon J_\epsilon v \right)$$

$$\forall v \in \mathcal{U}_0(\omega_X^\Delta). \tag{5.4*.1a}$$

This problem will be used for the subdomains ω_X which do not intersect the Neumann boundary, namely, $\partial\omega_X^\Delta \cap \Gamma_{\mathrm{N}} = \emptyset$. For the subdomains ω_X with a finite length intersection of the Neumann boundary $\partial\omega_{X,N}^\Delta = \partial\omega_X^\Delta \cap \Gamma_{\mathrm{N}} \neq \emptyset$, we will impose zero Dirichlet boundary conditions only on $\partial\omega_{X,D}^\Delta = \partial\omega_X^\Delta - \partial\omega_{X,N}^\Delta$, namely:

Find $\hat{e}^{\mathrm{Dir}}_{\omega_X^\Delta} \in \mathcal{U}_{\partial\omega_{X,D}^\Delta}(\omega_X^\Delta) \overset{\mathrm{def}}{=} \{v \in \mathcal{U}(\omega_X^\Delta) \mid v|_{\partial\omega_{X,D}^\Delta} \equiv 0\}$ such that

$$\mathcal{B}_{\omega_X^\Delta}(\hat{e}^{\mathrm{Dir}}_{\omega_X^\Delta}, v) = \mathcal{R}_{\omega_X^\Delta}(v) = \sum_{\tau \subseteq \omega_X^\Delta} \left(\int_\tau r_\tau v + \sum_{\substack{\epsilon \subseteq \partial\tau \\ \epsilon \not\subseteq \Gamma_{\mathrm{N}}}} \frac{1}{2} \int_\epsilon J_\epsilon v \right)$$

$$\forall v \in \mathcal{U}_{\partial\omega_{X,D}^\Delta}(\omega_X^\Delta). \tag{5.4*.1b}$$

The existence and uniqueness of $\hat{e}^{\mathrm{Dir}}_{\omega_X^\Delta}$ can be proven exactly as in the one-dimensional case.

Theorem 5.4*.1 *The solutions $\hat{e}^{\mathrm{Dir}}_{\omega_X^\Delta}$ of (5.4.1a) and (5.4.1b) exist and are unique under the minimal assumptions which guarantee the existence of solutions for the two-dimensional model problem.*

As in the one-dimensional case, we will employ the notation

$$\eta^{\mathrm{Dir}}_{\omega_X^\Delta} = \|\hat{e}^{\mathrm{Dir}}_{\omega_X^\Delta}\|_{\mathcal{U}(\omega_X^\Delta)}, \qquad \mathcal{E}^{\mathrm{Dir}}_{subd} \overset{\mathrm{def}}{=} \sqrt{\sum_{X \in \mathcal{N}(\Delta)} (\eta^{\mathrm{Dir}}_{\omega_X^\Delta})^2} \tag{5.4*.2}$$

respectively, for the subdomain residual error indicators and the error estimator, and we will prove the equivalence $\mathcal{E}^{\mathrm{Dir}}_{subd} \approx \|e_{S^p_\Delta}\|_{\mathcal{U}}$.

Theorem 5.4*.2 *Under the assumption that the input data satisfy the minimal conditions needed for the existence of solutions to the model problem, we have*

$$\mathcal{C}_L \mathcal{E}^{\mathrm{Dir}}_{subd} \le \|e_{S^p_\Delta}\|_{\mathcal{U}} \le \mathcal{C}_U \mathcal{E}^{\mathrm{Dir}}_{subd} \tag{5.4*.3}$$

with $\mathcal{C}_L$ and $\mathcal{C}_U$ depending only on the class of meshes, for example, by the minimal angle condition, the type of element basis, and the differential equation.

Proof The proof follows exactly the same steps as in the one-dimensional case. Employing the orthogonality of the error, we get

$$\|e_{S^p_\Delta}\|^2_{\mathcal{U}} = \mathcal{B}(e_{S^p_\Delta}, e_{S^p_\Delta} - v) \quad \forall v \in S^p_\Delta \tag{5.4*.4a}$$

and taking into account that $\sum_{X\in\mathcal{N}(\Delta)} \varphi_X^\Delta \equiv 1$, we get

$$\|e_{S_\Delta^p}\|_{\mathcal{U}}^2 = \mathcal{B}\left(e_{S_\Delta^p}, \left(\sum_{X\in\mathcal{N}(\Delta)} \varphi_X^\Delta\right)(e_{S_\Delta^p} - v)\right)$$

$$= \sum_{X\in\mathcal{N}(\Delta)} \mathcal{B}_{\omega_X^\Delta}\left(e_{S_\Delta^p}, \varphi_X^\Delta(e_{S_\Delta^p} - v)\right) \tag{5.4*.4b}$$

for any $v \in S_\Delta^p$. Now, because

$$\varphi_X^\Delta(e_{S_\Delta^p} - v)|_{\omega_X^\Delta} \in \mathcal{U}_{\partial\omega_{X,D}^\Delta}(\omega_X^\Delta) \tag{5.4*.4c}$$

we have

$$\mathcal{B}_{\omega_X^\Delta}\left(e_{S_\Delta^p}, \varphi_X^\Delta(e_{S_\Delta^p} - v)\right) = \mathcal{R}_{\omega_X^\Delta}(\varphi_X^\Delta(e_{S_\Delta^p} - v)) = \mathcal{B}_{\omega_X^\Delta}\left(\hat{e}_{\omega_X^\Delta}^{\mathrm{Dir}}, \varphi_X^\Delta(e_{S_\Delta^p} - v)\right) \tag{5.4*.4d}$$

for any $v \in S_\Delta^p$, and we get

$$\|e_{S_\Delta^p}\|_{\mathcal{U}}^2 = \sum_{X\in\mathcal{N}(\Delta)} \mathcal{B}_{\omega_X^\Delta}\left(\hat{e}_{\omega_X^\Delta}^{\mathrm{Dir}}, \varphi_X^\Delta(e_{S_\Delta^p} - v)\right) \quad \forall v \in S_\Delta^p. \tag{5.4*.4e}$$

Employing the Cauchy–Schwarz inequality for the energy and the $\mathbb{R}^{nv}$ inner products, where nv is the number of vertices in the mesh Δ, we get

$$\|e_{S_\Delta^p}\|_{\mathcal{U}} \le \sqrt{\sum_{X\in\mathcal{N}(\Delta)} \|\hat{e}_{\omega_X^\Delta}^{\mathrm{Dir}}\|_{\mathcal{U}(\omega_X^\Delta)}^2} \sqrt{\sum_{X\in\mathcal{N}(\Delta)} \|\varphi_X^\Delta(e_{S_\Delta^p} - v)\|_{\mathcal{U}(\omega_X^\Delta)}^2} \tag{5.4*.4f}$$

and employing Lemma 5.4*.1 given below to obtain

$$\min_{v\in S_\Delta^p} \sqrt{\sum_{X\in\mathcal{N}(\Delta)} \|\varphi_X^\Delta(e_{S_\Delta^p} - v)\|_{\mathcal{U}(\omega_X^\Delta)}^2} \le \mathcal{K}\|e_{S_\Delta^p}\|_{\mathcal{U}} \tag{5.4*.5}$$

we get the upper bound

$$\|e_{S_\Delta^p}\|_{\mathcal{U}} \le \mathcal{K}\sqrt{\sum_{X\in\mathcal{N}(\Delta)} \|\hat{e}_{\omega_X^\Delta}^{\mathrm{Dir}}\|_{\mathcal{U}(\omega_X^\Delta)}^2}. \tag{5.4*.6}$$

As in the one-dimensional case, the lower bound can be proven by employing that

$$(\mathcal{E}_{subd}^{\mathrm{Dir}})^2 = \sum_{X\in\mathcal{N}(\Delta)} \mathcal{B}_{\omega_X^\Delta}\left(\hat{e}_{\omega_X^\Delta}^{\mathrm{Dir}}, \hat{e}_{\omega_X^\Delta}^{\mathrm{Dir}}\right) = \sum_{X\in\mathcal{N}(\Delta)} \mathcal{B}_{\omega_X^\Delta}\left(e_{S_\Delta^p}, \hat{e}_{\omega_X^\Delta}^{\mathrm{Dir}}\right) \tag{5.4*.7a}$$

where we have used that

$$\mathcal{B}_{\omega_X^\Delta}\left(\hat{e}_{\omega_X^\Delta}^{\mathrm{Dir}}, \hat{e}_{\omega_X^\Delta}^{\mathrm{Dir}}\right) = \mathcal{R}_{\omega_X^\Delta}(e_{S_\Delta^p}) = \mathcal{B}_{\omega_X^\Delta}\left(e_{S_\Delta^p}, \hat{e}_{\omega_X^\Delta}^{\mathrm{Dir}}\right). \tag{5.4*.7b}$$

Extending each $\hat{e}^{\mathrm{Dir}}_{\omega^{\Delta}_{X}}$ by zero outside ω^{Δ}_{X}, and using the Cauchy–Schwarz inequality we get

$$(\mathcal{E}^{\mathrm{Dir}}_{subd})^2 = \mathcal{B}\left(e_{S^p_\Delta}, \sum_{X\in\mathcal{N}(\Delta)} \widetilde{\hat{e}^{\mathrm{Dir}}_{\omega^{\Delta}_{X}}}\right) \le \|e_{S^p_\Delta}\|_{\mathcal{U}} \left\| \sum_{X\in\mathcal{N}(\Delta)} \hat{e}^{\mathrm{Dir}}_{\omega^{\Delta}_{X}} \right\|_{\mathcal{U}} \tag{5.4*.7c}$$

where $\widetilde{\hat{e}^{\mathrm{Dir}}_{\omega^{\Delta}_{X}}} \in \mathcal{U}$ is the extension of $\hat{e}^{\mathrm{Dir}}_{\omega^{\Delta}_{X}}$ by zero, and employing that

$$\left\| \sum_{X\in\mathcal{N}(\Delta)} \widetilde{\hat{e}^{\mathrm{Dir}}_{\omega^{\Delta}_{X}}} \right\|^2_{\mathcal{U}} \le M \sum_{X\in\mathcal{N}(\Delta)} \|\hat{e}^{\mathrm{Dir}}_{\omega^{\Delta}_{X}}\|^2_{\mathcal{U}(\omega^{\Delta}_{X})} \tag{5.4*.8}$$

which follows from Lemma 5.4.2 given below, we get

$$(\mathcal{E}^{\mathrm{Dir}}_{subd})^2 \le \|e_{S^p_\Delta}\|_{\mathcal{U}} \left\| \sum_{X\in\mathcal{N}(\Delta)} \widetilde{\hat{e}^{\mathrm{Dir}}_{\omega^{\Delta}_{X}}} \right\|_{\mathcal{U}} \le \|e_{S^p_\Delta}\|_{\mathcal{U}} \sqrt{M} \sqrt{\sum_{X\in\mathcal{N}(\Delta)} \|\hat{e}^{\mathrm{Dir}}_{\omega^{\Delta}_{X}}\|^2_{\mathcal{U}(\omega^{\Delta}_{X})}} \tag{5.4*.9}$$

which proves the lower bound with $\mathcal{C}_L = 1/\sqrt{M}$. □

Let us note that the effect of the approximate determination of the indicators must also be addressed.

Let us now state, without proving them, the lemmas on which the above proof is based.

Lemma 5.4*.1 *We have*

$$\min_{v\in S^p_\Delta} \sqrt{\sum_{X\in\mathcal{N}(\Delta)} \|\varphi^{\Delta}_{X}(w-v)\|^2_{\mathcal{U}(\omega^{\Delta}_{X})}} \le \mathcal{K}\|w\|_{\mathcal{U}} \quad \forall w \in \mathcal{U} \tag{5.4*.10}$$

where $\mathcal{N}(\Delta)$ is the set of vertices for the mesh Δ, and $\mathcal{K}$ depends on the coefficients of the differential equation, the type of meshes (triangles, quadrilaterals with irregular connections, etc.), and the element basis $\tilde{S}^p$ ($\tilde{\mathcal{P}}^p$ for triangles, bi-p or trunc space for quadrilaterals, etc.). □

Lemma 5.4*.2 *For every $X \in \mathcal{N}(\Delta)$, let $w_X \in \mathcal{U}_{\partial\omega^{\Delta}_{X,D}}(\omega^{\Delta}_{X})$ be an arbitrary function, and let $\tilde{w}_X$ denote its extension by zero outside ω^{Δ}_{X}. We then have*

$$\left\| \sum_{X\in\mathcal{N}(\Delta)} \tilde{w}_X \right\|^2_{\mathcal{U}} \le M \sum_{X\in\mathcal{N}(\Delta)} \|w_X\|^2_{\mathcal{U}(\omega^{\Delta}_{X})} \tag{5.4*.11}$$

where M is the overlap index for the partition of unity $\{\varphi^{\Delta}_{X}\}_{X\in\mathcal{N}(\Delta)}$ ($M = 3$ for triangles and $M = 4$ for quadrilaterals). □

Remark 5.4*.1 As we have already mentioned in Section 5.1, the estimator $\mathcal{E}^{\mathrm{Dir}}_{subd}$ is rather insensitive to the magnitude of the zeroth-order term. This observation was used by [Verfürth, 1998] to construct an explicit estimator which is robust for the entire range of the parameter γh^2, which measures the effect of the zeroth order term.

Remark 5.4*.2 As in the one-dimensional case, we may also employ the weighted subdomain residual indicator

$$\eta^{\mathrm{Dir}}_{\omega^{\Delta}_{X},\varphi^{\Delta}_{X}} = \|\hat{e}^{\mathrm{Dir}}_{\omega^{\Delta}_{X},\varphi^{\Delta}_{X}}\|_{\mathcal{U}(\omega^{\Delta}_{X})}$$
$$= \sup_{v\in\mathcal{U}_0(\omega^{\Delta}_{X})} \frac{\sum_{\tau\subseteq\omega^{\Delta}_{X}}\int_\tau r_\tau\varphi^{\Delta}_{X}v + \sum_{\substack{\epsilon\subseteq\partial\tau\\ \epsilon\not\subseteq\Gamma_{\mathrm{N}}}}\int_\epsilon J_\epsilon\varphi^{\Delta}_{X}v}{\|v\|_{\mathcal{U}(\omega^{\Delta}_{X}),\varphi^{\Delta}_{X}}} \tag{5.4*.12a}$$

where $\|\cdot\|_{\mathcal{U}_0(\omega^{\Delta}_{X}),\varphi^{\Delta}_{X}}$ is the φ^{Δ}_{X}-weighted energy norm in ω^{Δ}_{X}, e.g., for the case of the Laplacian

$$\|v\|_{\mathcal{U}_0(\omega^{\Delta}_{X}),\varphi^{\Delta}_{X}} = \sqrt{\sum_{\tau\in\omega^{\Delta}_{X}}\int_\tau \varphi_X|\boldsymbol{\nabla}u|^2}. \tag{5.4*.12b}$$

The weighted subdomain residual estimator

$$\mathscr{E}^{\mathrm{Dir}}_{subd,\{\varphi^{\Delta}_{X}\}_{X\in\mathcal{N}(\Delta)}} \stackrel{\mathrm{def}}{=} \sqrt{\sum_{X\in\mathcal{N}(\Delta)}\left(\eta^{\mathrm{Dir}}_{\omega^{\Delta}_{X},\varphi^{\Delta}_{X}}\right)^2} \tag{5.4*.12c}$$

was analyzed by [Carstensen and Funken, 2000b]. Another version of the weighted subdomain residual estimator was recently proposed in [Morin *et al.*, 2001].

Remark 5.4*.3 The subdomain residual estimate was introduced in [Babuška and Rheinboldt, 1978c], and further analyzed in [Babuška and Miller, 1981] and [Babuška and Miller, 1987]. More recently, it was employed for the a-posteriori estimation of the error in the partition-of-unity method (PUM) [Melenk and Babuška, 1996]. The proofs of Lemmas 5.4*.1 and 5.4*.2 are simple and can be found in the references mentioned in this Remark.

We also have:

Theorem 5.4*.3 *Assumptions A, B, C of Section 5.1* hold for the estimator* $\mathscr{E}^{\mathrm{Dir}}_{subd}$.

Let us, once more, underline that *the subdomain residual estimator* $\mathscr{E}^{\mathrm{Dir}}_{subd}$ *can be used in problems with very rough input data, for which the estimators* $\mathscr{E}^{\mathrm{Neum}}$ *and* $\mathscr{E}^{\mathrm{EXPL}}$ *may have no meaning.*

Let us now address the relation between the estimators $\mathscr{E}^{\mathrm{Dir}}_{subd}$, $\mathscr{E}^{\mathrm{Neum}}$, and $\mathscr{E}^{EXPL}$. As in the one-dimensional case, we will assume that the meshes Δ are locally quasi-uniform, namely, that

$$\frac{1}{\mu} \le \frac{h_\tau}{h_{\tau'}} \le \mu \quad \forall\tau,\tau'\in\Delta\colon\ \partial\tau\cap\partial\tau'\neq\emptyset \tag{5.4*.13a}$$

for a fixed $\mu\ge 1$, and the elements satisfy a regularity condition, e.g., the minimal-angle condition, and that the coefficients and the right-hand side of the partial differential equation, and the boundary conditions, can be expressed in terms of a finite number of parameters, e.g., they can be represented by polynomials in each element and on each edge of the boundary, for example, in the case of the Poisson equation we have

$$f|_\tau \in \mathscr{P}^{s_\tau}(\tau) \quad \forall\tau\in\Delta, \qquad g|_\epsilon \in \mathscr{P}^{s_\epsilon}(\epsilon) \quad \forall\epsilon\subseteq\Gamma_{\mathrm{N}}. \tag{5.4*.13b}$$

We have:

Theorem 5.4*.4 *For finite element solutions of the Poisson equation, under the assumptions (5.4*.13a) and (5.4*.13b), we have*

$$\mathscr{C}(\eta^{\text{EXPL}}_{\omega^{\Delta}_X})^2 = \mathscr{C}\Big(\sum_{\tau\subseteq\omega^{\Delta}_X} (h_\tau)^2 \int_\tau (r^{\Delta}_\tau)^2 + \sum_{\epsilon\subseteq\omega^{\Delta}_X\cup\partial\omega^{\Delta}_{X,N}} h_\epsilon \int_\epsilon (J^{\Delta}_\epsilon)^2 \Big)$$
$$\leq \|\hat{e}^{\text{Dir}}_{\omega^{\Delta}_X}\|^2_{\mathfrak{U}(\omega^{\Delta}_X)}. \tag{5.4*.14}$$

Proof The proof is analogous to the proof of the one-dimensional case given in Theorem 5.4.4. Each patch ω^{Δ}_X will be scaled to a corresponding master patch. From assumption (5.4*.13a), it follows that there is a finite number of elements connected to each vertex X, and, therefore, we need to employ only a finite number of master patches. The rest of the proof follows exactly the same arguments as the proof of Theorem 5.4.4. □

As in the one-dimensional case, the reverse inequality also holds.

Theorem 5.4*.5 *Under the assumptions of Theorem 5.4*.4, we have*

$$\|\hat{e}^{\text{Dir}}_{\omega^{\Delta}_X}\|^2_{\mathfrak{U}(\omega^{\Delta}_X)} \leq \mathscr{C}(\eta^{\text{EXPL}}_{\omega^{\Delta}_X})^2$$
$$= \mathscr{C}\Big(\sum_{\tau\subseteq\omega^{\Delta}_X} (h_\tau)^2 \int_\tau (r^{\Delta}_\tau)^2 + \sum_{\epsilon\subseteq\omega^{\Delta}_X\cup\partial\omega^{\Delta}_{X,N}} h_\epsilon \int_\epsilon (J^{\Delta}_\epsilon)^2 \Big). \tag{5.4*.15}$$

Proof The proof is exactly analogous as the proof of Theorem 5.4.5. □

Combining (5.4*.14) and (5.4*.15) we get the equivalence

$$\|\hat{e}^{\text{Dir}}_{\omega^{\Delta}_X}\|_{\mathfrak{U}(\omega^{\Delta}_X)} \approx \eta^{\text{EXPL}}_{\omega^{\Delta}_X} = \sqrt{\sum_{\tau\subseteq\omega^{\Delta}_X} (h_\tau)^2 \int_\tau (r^{\Delta}_\tau)^2 + \sum_{\epsilon\subseteq\omega^{\Delta}_X\cup\partial\omega^{\Delta}_{X,N}} h_\epsilon \int_\epsilon (J^{\Delta}_\epsilon)^2} \tag{5.4*.16a}$$

and hence, by summing over all the subdomains, we obtain

$$\mathscr{E}^{\text{EXPL}} \approx \mathscr{E}^{\text{Dir}}_{subd}. \tag{5.4*.16b}$$

Hence, as in the one-dimensional case, we have:

Theorem 5.4*.6 *Consider the setting of Theorem 5.4*.4. Then, we have*

$$\mathscr{C}_1\mathscr{E}^{\text{EXPL}} \leq \mathscr{E}^{\text{Dir}}_{subd} \leq \mathscr{C}_2\mathscr{E}^{\text{EXPL}} \tag{5.4*.17a}$$

and

$$\mathscr{C}_3\mathscr{E}^{\text{EXPL}} \leq \|e_{S^p_\Delta}\|_{\mathfrak{U}} \leq \mathscr{C}_4\mathscr{E}^{\text{EXPL}}. \tag{5.4*.17b}$$

Proof The equivalence $\mathscr{E}^{\text{Dir}}_{subd} \approx \mathscr{E}^{\text{EXPL}}$ has already been proven. The equivalence $\mathscr{E}^{\text{EXPL}} \approx \|e_{S^p_\Delta}\|_{\mathfrak{U}}$ follows from $\mathscr{E}^{\text{EXPL}} \approx \mathscr{E}^{\text{Dir}}_{subd}$ and $\mathscr{E}^{\text{Dir}}_{subd} \approx \|e_{S^p_\Delta}\|_{\mathfrak{U}}$. □

As in the one-dimensional case, we also have $\mathscr{E}^{\text{Dir}}_{subd} \approx \mathscr{E}^{\text{Neum}}$.

Theorem 5.4*.7 *Under the assumptions of Theorem 5.4*.4 we have*

$$\mathscr{C}_5\,\mathscr{E}^{\text{Dir}}_{subd} \leq \mathscr{E}^{\text{Neum}} \leq \mathscr{C}_6\,\mathscr{E}^{\text{Dir}}_{subd}. \tag{5.4*.18}$$

Proof As in the one-dimensional case we have

$$\eta_\tau^{\text{Neum}} = \|\hat{e}_\tau^{\text{Neum}}\|_{\mathscr{U}(\tau)} = \sup_{v\in\mathscr{U}(\tau)} \frac{|\mathscr{R}_\tau^{EQ}(v)|}{\|v\|_{\mathscr{U}(\tau)}} \leq \mathscr{C}\eta_\tau^{\text{EXPL}} \tag{5.4*.19a}$$

and hence

$$\mathscr{E}^{\text{Neum}} \leq \mathscr{C}\,\mathscr{E}^{\text{EXPL}}. \tag{5.4*.19b}$$

We also have

$$\mathscr{C}\mathscr{E}^{\text{EXPL}} \leq \|e_{S^p_\Delta}\|_{\mathscr{U}} \leq \mathscr{E}^{\text{Neum}} \tag{5.4*.19c}$$

and hence $\mathscr{E}^{\text{EXPL}} \approx \mathscr{E}^{\text{Neum}}$. □

Remark 5.4*.4 In general $\mathscr{E}^{\text{Dir}}_{subd} \not\approx \mathscr{E}^{\text{Dir}}$. Consider, for example, the case of the finite element approximation of the Laplacian on a mesh of linear ($p = 1$) triangles for which we have $\mathscr{E}^{\text{Dir}} = 0$.

Remark 5.4*.5 Theorems 5.4*.2, 5.4*.4–5.4*.7 hold for all meshes under consideration, including meshes with irregular connections.

Let us now give an example which illustrates the accuracy of the subdomain residual element error indicators.

Example 5.4*.1 We employed the model problem and the meshes given in Example 5.2*.1. In order to be able to compare the subdomain residual error indicators with the indicators corresponding to the other element residual estimators, we will let

$$\eta_\tau^{subd} = \sqrt{\frac{1}{4}\sum_{\omega_X^\Delta\cap\tau\neq\emptyset} \|\hat{e}^{\text{Dir}}_{\omega_X^\Delta}\|^2_{\mathscr{U}(\tau)}}. \tag{5.4*.20}$$

The element effectivity indices for the three meshes employed in the computation are given in Fig. 5.4*.1.

Remark 5.4*.6 [Carstensen and Funken, 2000a] employed the weighted subdomain residual estimator to derive an explicit upper estimator with patch error indicators

$$\eta^{\text{EXPL}}_{\omega_X^\Delta,\varphi_X^\Delta} = \mathscr{C}_1(\omega_X^\Delta)\sqrt{\sum_{\substack{\tau\in\Delta\\ \tau\subseteq\omega_X^\Delta}}\int_\tau r_\tau^2\varphi_X^\Delta} + \mathscr{C}_2(\omega_X^\Delta)\sqrt{\sum_{\substack{\tau\in\Delta\\ \tau\subseteq\omega_X^\Delta}}\sum_{\epsilon\subseteq\partial\tau}\frac{1}{2}\int_\epsilon J_\epsilon^2\varphi_X^\Delta} \tag{5.4*.21}$$

where $\mathscr{C}_1(\omega_X^\Delta)$, $\mathscr{C}_2(\omega_X^\Delta)$ are constants depending on the topology of the mesh in ω_X^Δ.

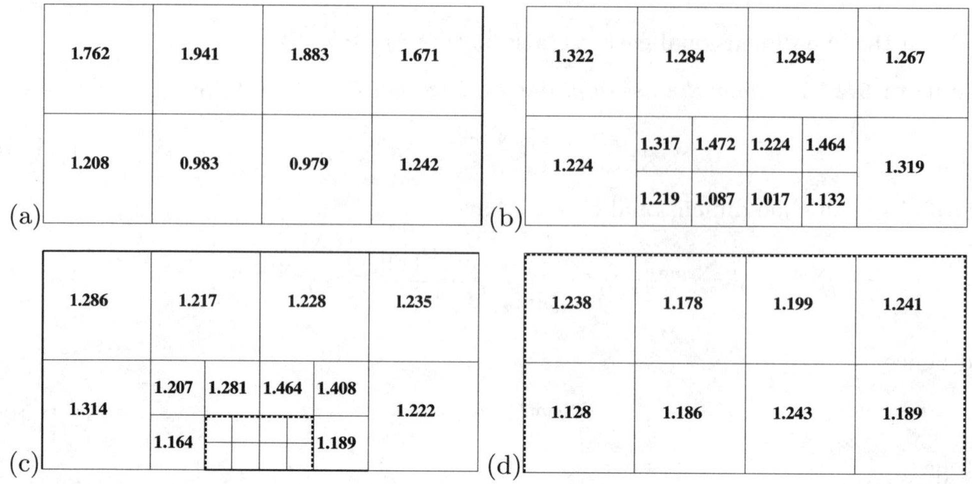

FIG. 5.4*.1. An example which illustrates the accuracy of the element error indicators. Input data: Laplace's equation in $\Omega = (-1,1) \times (0,1)$, Neumann boundary conditions, $u_0 \equiv \sqrt{r}\sin(\theta/2)$ for the polar coordinate system shown in Fig. 5.2*.1. Mesh data: Mesh of biquadratic squares with one to two irregular connections. Element effectivity indices for the subdomain residual error indicators for the mesh of: (a) 8 elements (global effectivity index $\kappa = 0.984$); (b) 14 elements ($\kappa = 1.104$); (c) 20 elements ($\kappa = 1.194$); (d) the 8-element patch in the subdomain $(-\frac{1}{4}, \frac{1}{4}) \times (0, \frac{1}{4})$ depicted with dashed perigram in the mesh of 20 elements

Remark 5.4*.7 The theoretical analysis of the subdomain residual estimator was first given in [Babuška and Miller, 1981, 1987], where its equivalence with the explicit estimator was proven.

Remark 5.4*.8 The explicit estimator was widely used in the adaptive program FEARS which was developed in the early 1980s at the University of Maryland and employed bilinear elements; see [Mesztenyi and Sczymczak, 1982], [Babuška and Rheinboldt, 1980], and [Babuška, 1981].

Remark 5.4*.9 The equivalence between the various estimators was addressed in various papers like for example in [Rank and Zienkiewicz, 1987], [Babuška and Rodriguez, 1993], [Rodriguez, 1993], [Ainsworth and Craig, 1992], [Ainsworth and Oden, 1993], [Carstensen and Bartel, 1999].

Remark 5.4*.10 Similarly as in the one-dimensional case, we can construct the guaranteed lower bound

$$\|e_{S^p_\Delta}\|_{\mathcal{U}} \geq \frac{\mathcal{R}\big(\mathcal{A}^{\mathrm{Dir}}_u(u_{S^p_\Delta})\big)}{\big\|\mathcal{A}^{\mathrm{Dir}}_u(u_{S^p_\Delta})\big\|_{\mathcal{U}}} \tag{5.4*.22a}$$

where $\mathcal{A}^{\mathrm{Dir}}_u(u_{S^p_\Delta})$ is constructed by adding the solutions of the subdomain residual problems to the finite element solution, namely

$$\mathcal{A}^{\mathrm{Dir}}_u(u_{S^p_\Delta}) = u_{S^p_\Delta} + \sum_{X\in\mathcal{N}(\Delta)} \tilde{e}^{\mathrm{Dir}}_{\omega^\Delta_X}. \tag{5.4*.22b}$$

In this section we introduced the subdomain residual estimator and discussed its equivalence with the explicit and the Neumann residual estimators. We also underlined that the subdomain residual estimator is robust with respect to the zeroth-order term and hence it is reliable for singularly perturbed equations of the form $-\varepsilon\Delta u + u = f$.

Problems

P5.4*.1. Let Δ be a uniform square mesh on $\Omega = (-1,1)\times(-1,1)$. Consider the problem $-\Delta u = f$ in $\partial\Omega$, $u = 0$ on $\partial\Omega$, and bilinear elements. Compute $\mathscr{C}_U$ and $\mathscr{C}_L$ for the subdomain estimate.

P5.4*.2. Prove Lemmas 5.4*.1 and 5.4*.2.

5.5 Analysis of the explicit element residual estimator

In this section we will analyze the explicit element error indicator

$$\eta_{I_j^\Delta}^{\mathrm{EXPL}} \stackrel{\mathrm{def}}{=} \sqrt{\left(h_{I_j^\Delta}\right)^2 \|r_{I_j^\Delta}\|_{L^2(I_j^\Delta)}^2 + \frac{h_{I_j^\Delta}}{2}(J_{j-1}^2 + J_j^2)} \tag{5.5.1a}$$

and the corresponding explicit estimator

$$\mathscr{E}^{\mathrm{EXPL}} \stackrel{\mathrm{def}}{=} \sqrt{\sum_{j=1}^{N(\Delta)} \left(\eta_{I_j^\Delta}^{\mathrm{EXPL}}\right)^2}. \tag{5.5.1b}$$

Remark 5.5.1 The estimator $\mathscr{E}^{\mathrm{EXPL}}$ was the first a-posteriori estimator in the finite element method; it was proposed in [Babuška and Rheinboldt, 1978a].

In Section 5.4 we have already proven that

$$\mathscr{E}^{\mathrm{EXPL}} \approx \mathscr{E}_{subd}^{\mathrm{Dir}} \approx \|e_{S_\Delta^p}\|_{\mathscr{U}} \tag{5.5.1c}$$

by employing the equivalences $\eta_{\omega_j^\Delta}^{\mathrm{EXPL}} \approx \|\hat{e}_{\omega_j^\Delta}^{\mathrm{Dir}}\|_{\mathscr{U}(\omega_j^\Delta)}$ and $\mathscr{E}_{subd}^{\mathrm{Dir}} \approx \|e_{S_\Delta^p}\|_{\mathscr{U}}$. In this section we will show that it is possible to prove the equivalence

$$\mathscr{E}^{\mathrm{EXPL}} \approx \|e_{S_\Delta^p}\|_{\mathscr{U}} \tag{5.5.1d}$$

directly, in an asymptotic sense. The analysis of the one-dimensional case is presented as an analogue of the two-dimensional case, which was given in [Babuška *et al.*, 1992]. The main idea is to prove that

$$\mathscr{E}^{\mathrm{EXPL}} \approx \|e_{S_\Delta^p}^{S_\Delta^{p+k}}\|_{\mathscr{U}} \tag{5.5.2a}$$

where

$$e_{S_\Delta^p}^{S_\Delta^{p+k}} \stackrel{\mathrm{def}}{=} u_{S_\Delta^{p+k}} - u_{S_\Delta^p}. \tag{5.5.2b}$$

Then, employing the triangle inequality

$$\|e_{S^p_\Delta}\|_{\mathcal{U}} = \|u_{EX} - u_{S^p_\Delta}\|_{\mathcal{U}} \le \|\underbrace{u_{EX} - u_{S^{p+k}_\Delta}}_{e_{S^{p+k}_\Delta}}\|_{\mathcal{U}} + \|\underbrace{u_{S^{p+k}_\Delta} - u_{S^p_\Delta}}_{e^{S^{p+k}_\Delta}_{S^p_\Delta}}\|_{\mathcal{U}} \tag{5.5.2c}$$

and choosing k sufficiently high such that $\|e_{S^{p+k}_\Delta}\|_{\mathcal{U}}$ is neglibible with respect to $\|e^{S^{p+k}_\Delta}_{S^p_\Delta}\|_{\mathcal{U}}$, we obtain

$$\|e_{S^p_\Delta}\|_{\mathcal{U}} = \|e^{S^{p+k}_\Delta}_{S^p_\Delta}\|_{\mathcal{U}}(1 + o(1)) \tag{5.5.2d}$$

from which it follows that

$$\kappa^{\text{EXPL}} = \frac{\mathscr{E}^{\text{EXPL}}}{\|e_{S^p_\Delta}\|_{\mathcal{U}}} = \frac{\mathscr{E}^{\text{EXPL}}}{\|e^{S^{p+k}_\Delta}_{S^p_\Delta}\|_{\mathcal{U}}}\big(1 + o(1)\big). \tag{5.5.2e}$$

Let us now derive a representation for $\|e^{S^{p+k}_\Delta}_{S^p_\Delta}\|_{\mathcal{U}}$ in terms of the residuals. We have

$$\|e^{S^{p+k}_\Delta}_{S^p_\Delta}\|^2_{\mathcal{U}} = \mathscr{B}\Big(\overbrace{\underbrace{u_{S^{p+k}_\Delta} - u_{EX}}_{-e_{S^{p+k}_\Delta}} + \underbrace{u_{EX} - u_{S^p_\Delta}}_{e_{S^p_\Delta}}}^{e^{S^{p+k}_\Delta}_{S^p_\Delta}}, e^{S^{p+k}_\Delta}_{S^p_\Delta}\Big) \tag{5.5.3a}$$

and employing the orthogonality condition

$$\mathscr{B}\big(e_{S^{p+k}_\Delta}, e^{S^{p+k}_\Delta}_{S^p_\Delta}\big) = 0 \tag{5.5.3b}$$

we get

$$\|e^{S^{p+k}_\Delta}_{S^p_\Delta}\|^2_{\mathcal{U}} = \mathscr{B}\big(e_{S^p_\Delta}, e^{S^{p+k}_\Delta}_{S^p_\Delta}\big). \tag{5.5.3c}$$

Further, employing the orthogonality condition

$$\mathscr{B}(e_{S^p_\Delta}, \chi) = 0, \quad \chi \in S^p_\Delta \tag{5.5.4a}$$

we get

$$\|e^{S^{p+k}_\Delta}_{S^p_\Delta}\|^2_{\mathcal{U}} = \mathscr{B}\big(e_{S^p_\Delta}, e^{S^{p+k}_\Delta}_{S^p_\Delta} - \chi\big) = \mathscr{R}\big(e^{S^{p+k}_\Delta}_{S^p_\Delta} - \chi\big) \tag{5.5.4b}$$

and employing the splitting of the residuum

$$\mathscr{R}(v) = \sum_{s=1}^{N(\Delta)} \mathscr{R}^{UNEQ}_{I^\Delta_s}(v|_{I^\Delta_s}) \quad \forall v \in \mathcal{U} \tag{5.5.4c}$$

where we recall that

$$\mathcal{R}_{I_s^\Delta}^{UNEQ}(v) \stackrel{\text{def}}{=} \int_{I_s^\Delta} r_{I_s^\Delta} v + \frac{1}{2}(J_{s-1}^\Delta v(x_{s-1}^\Delta) + J_s^\Delta v(x_s^\Delta)), \quad v \in \mathcal{U}(I_j^\Delta) \tag{5.5.4d}$$

we get

$$\|e_{S_\Delta^p}^{S_\Delta^{p+k}}\|_{\mathcal{U}}^2 = \sum_{s=1}^{N(\Delta)} \mathcal{R}_{I_s^\Delta}^{UNEQ}((e_{S_\Delta^p}^{S_\Delta^{p+k}} - \chi)|_{I_s^\Delta}) \quad \forall \chi \in S_\Delta^p. \tag{5.5.4e}$$

Further, because

$$e_{S_\Delta^p}^{S_\Delta^{p+k}} - \chi \in S_\Delta^{p+k} \tag{5.5.5a}$$

and assuming that $f \in L^2(I)$, we have

$$\int_{I_s^\Delta} r_{I_s^\Delta}(e_{S_\Delta^p}^{S_\Delta^{p+k}} - \chi) = \int_{I_s^\Delta} \pi_{S^{p+k}(I_s^\Delta)}^{L^2}(r_{I_s^\Delta})(e_{S_\Delta^p}^{S_\Delta^{p+k}} - \chi) \tag{5.5.5b}$$

where $\pi_{S^{p+k}(I_s^\Delta)}^{L^2}$ is the L^2 projection operator into $S^{p+k}(I_s^\Delta)$, and we obtain

$$\mathcal{R}_{I_s^\Delta}^{UNEQ}(v) = \mathcal{R}_{I_s^\Delta, S_\Delta^{p+k}}^{UNEQ}(v) \stackrel{\text{def}}{=} \int_{I_s^\Delta} \pi_{S^{p+k}(I_s^\Delta)}^{L^2}(r_{I_s^\Delta})v + \frac{1}{2}\sum_{t=s-1}^{s} J_t^\Delta v(x_t^\Delta) \tag{5.5.5c}$$

for all $v \in S^{p+k}(I_s^\Delta)$. Further, letting $v = e_{S_\Delta^p}^{S_\Delta^{p+k}} - \chi$, and using the Cauchy–Schwarz inequality, as discussed in the proof of Theorem 5.5.1 below, we get

$$\|e_{S_\Delta^p}^{S_\Delta^{p+k}}\|_{\mathcal{U}} \le \mathscr{C}\mathscr{E}_{S_\Delta^{p+k}}^{\text{EXPL}} \stackrel{\text{def}}{=} \mathscr{C}\sqrt{\sum_{j=1}^{N(\Delta)} \left(\eta_{I_j^\Delta, S^{p+k}(I_j^\Delta)}^{\text{EXPL}}\right)^2} \tag{5.5.6a}$$

where

$$\eta_{I_j^\Delta, S^{p+k}(I_j^\Delta)}^{\text{EXPL}} \stackrel{\text{def}}{=} \sqrt{(h_{I_j^\Delta})^2 \|\pi_{S^{p+k}(I_j^\Delta)}^{L^2}(r_{I_j^\Delta})\|_{L^2(I_j^\Delta)}^2 + \tfrac{1}{2}h_{I_j^\Delta}(J_{j-1}^2 + J_j^2)}. \tag{5.5.6b}$$

Below we will often write $\eta_{I_j^\Delta}^{\text{EXPL}}$ (respectively $\mathscr{E}^{\text{EXPL}}$) instead of $\eta_{j,S^{p+k}(I_j^\Delta)}^{\text{EXPL}}$ (respectively $\mathscr{E}_{S_\Delta^{p+k}}^{\text{EXPL}}$), for simplicity.

Remark 5.5.2 As we have already discussed above, *the explicit estimate given in (5.5.6) has meaning under the assumption that $f \in L^2(I)$. In the more general case where $f \in \mathcal{U}^{-1}$, but $f \notin L^2(I)$, the constant $\mathscr{C}$ in (5.5.6a) does not exist, and the explicit estimate $\mathscr{E}^{\text{EXPL}}$ is meaningless.*

Let us now now elaborate on the determination of the constant $\mathscr{C}$.

Theorem 5.5.1 *(Upper bound) Under the assumption*

$$f \in L^2(I) \tag{5.5.7a}$$

we have

$$\|e_{S_\Delta^p}^{S_\Delta^{p+k}}\|_{\mathfrak{U}} \leq \sqrt{\sum_{s=1}^{N(\Delta)} \mathscr{C}_{I_s^\Delta}^2(p,k)\big(\eta_{s,S^{p+k}(I_s^\Delta)}^{\mathrm{EXPL}}\big)^2} \leq \mathscr{C}(p,k)\mathscr{E}_{S_\Delta^{p+k}}^{\mathrm{EXPL}} \tag{5.5.7b}$$

where $\mathscr{C}_{I_s^\Delta}(p,k)$ and $\mathscr{C}(p,k)$ are given, respectively, in (5.5.10a) and (5.5.12) below.

Proof Let $v, w \in L^2(I_s^\Delta)$, $\mathbf{a}, \mathbf{b} \in \mathbb{R}^2$, and $(v;\mathbf{a}), (w;\mathbf{b}) \in L^2(I_s^\Delta) \times \mathbb{R}^2$, and let

$$\langle(v;\mathbf{a}),(w;\mathbf{b})\rangle_{I_s^\Delta} \stackrel{\mathrm{def}}{=} \int_{I_s^\Delta} vw + \frac{1}{2}\mathbf{a}^{\mathrm{T}}\mathbf{b} \tag{5.5.8a}$$

which is an inner product over $L^2(I_s^\Delta) \times \mathbb{R}^2$, with corresponding norm

$$\|(v;\mathbf{a})\|_{I_s^\Delta,(h_{I_s^\Delta})^{-1}} = \sqrt{\frac{1}{(h_{I_s^\Delta})^2}\int_{I_s^\Delta} v^2 + \frac{1}{2h_{I_s^\Delta}}\mathbf{a}^{\mathrm{T}}\mathbf{a}} \tag{5.5.8b}$$

and we have

$$\langle(v;\mathbf{a}),(w;\mathbf{b})\rangle_{I_s^\Delta} = \left\langle (h_{I_s^\Delta} v; \sqrt{h_{I_s^\Delta}}\mathbf{a}), \left(\frac{1}{h_{I_s^\Delta}} w; \frac{1}{\sqrt{h_{I_s^\Delta}}}\mathbf{b}\right)\right\rangle_{I_s^\Delta} \tag{5.5.8c}$$

and hence

$$\langle(v;\mathbf{a}),(w;\mathbf{b})\rangle_{I_s^\Delta} \leq \|(v,\mathbf{a})\|_{I_s^\Delta,h_{I_s^\Delta}} \|(w;\mathbf{b})\|_{I_s^\Delta,(h_{I_s^\Delta})^{-1}}. \tag{5.5.8d}$$

Using these notations we have

$$\mathscr{R}_{I_s^\Delta,S_\Delta^{p+k}}^{UNEQ}(v) = \big\langle\big(\pi_{S^{p+k}(I_s^\Delta)}^{L^2}(r_{I_s^\Delta}); \mathbf{J}_s^\Delta\big), (v;\mathbf{v}_s^\Delta)\big\rangle_{I_s^\Delta,h_{I_s^\Delta}} \tag{5.5.8e}$$

where $\mathbf{J}_s^\Delta = [J_{s-1}^\Delta, J_s^\Delta]^{\mathrm{T}}$ and $\mathbf{v}_s^\Delta = [v(x_{s-1}^\Delta), v(x_s^\Delta)]^{\mathrm{T}}$, and

$$\|e_{S_\Delta^p}^{S_\Delta^{p+k}}\|_{\mathfrak{U}}^2 = \sum_{s=1}^{N(\Delta)} \mathscr{R}_{I_s^\Delta,S_\Delta^{p+k}}^{UNEQ}\big(e_{S_\Delta^p}^{S_\Delta^{p+k}} - \chi\big) \tag{5.5.9a}$$

can be written in the form

$$\|e_{S_\Delta^p}^{S_\Delta^{p+k}}\|_{\mathfrak{U}}^2 = \sum_{s=1}^{N(\Delta)} \big\langle\big(\pi_{S^{p+k}(I_s^\Delta)}^{L^2} r_{I_s^\Delta}; \mathbf{J}_s^\Delta\big), \big(e_{S_\Delta^p}^{S_\Delta^{p+k}} - \chi; \big(\mathbf{e}_{S_\Delta^p}^{S_\Delta^{p+1}} - \boldsymbol{\chi}\big)_s^\Delta\big)\big\rangle_{I_s^\Delta,h_{I_s^\Delta}} \tag{5.5.9b}$$

for any $\chi \in S_\Delta^p$, and employing the inequality (5.5.8d), we get

$$\|e_{S_\Delta^p}^{S_\Delta^{p+k}}\|_{\mathfrak{U}}^2 \leq \sum_{s=1}^{N(\Delta)} \underbrace{\|(\pi_{S^{p+k}(I_s^\Delta)}^{L^2} r_{I_s^\Delta}; \mathbf{J}_s^\Delta)\|_{I_s^\Delta,h_{I_s^\Delta}}}_{\eta_{I_s^\Delta}^{\mathrm{EXPL}}} \|e_{S_\Delta^p}^{S_\Delta^{p+1}} - \chi\|_{I_s^\Delta,(h_{I_s^\Delta})^{-1}} \tag{5.5.9c}$$

where we used the simplified notation

$$\|v\|_{I_s^\Delta,(h_{I_s^\Delta})^{-1}} \stackrel{\text{def}}{=} \|(v;\mathbf{v}_s^\Delta)\|_{I_s^\Delta,(h_{I_s^\Delta})^{-1}}. \tag{5.5.9d}$$

Further, given any $v \in S_\Delta^{p+k}$, let $\mathcal{A}^p_{I_s^\Delta,h_{I_s^\Delta}} v$, be the solution of the minimization problem:

Find $\mathcal{A}^p_{I_s^\Delta,h_{I_s^\Delta}} v$ such that $\mathcal{A}^p_{I_s^\Delta,h_{I_s^\Delta}} v(x_t^\Delta) = v(x_t^\Delta)$, $t = s-1, s$, and

$$\|\mathcal{A}^p_{I_s^\Delta,h_{I_s^\Delta}} v - v\|_{I_s^\Delta,(h_{I_s^\Delta})^{-1}} = \min_{\substack{\psi\in S_\Delta^p \\ \psi(x_{s-1}^\Delta)=v(x_{s-1}^\Delta) \\ \psi(x_s^\Delta)=v(x_s^\Delta)}} \|\psi - v\|_{I_s^\Delta,(h_{I_s^\Delta})^{-1}}. \tag{5.5.10a}$$

Letting

$$\begin{aligned}\mathscr{C}_s(p,k) &\stackrel{\text{def}}{=} \sup_{v\in S^{p+k}(I_s^\Delta)} \frac{\|v - \mathcal{A}^p_{I_s^\Delta,h_{I_s^\Delta}} v\|_{I_s^\Delta,(h_{I_s^\Delta})^{-1}}}{\|v\|_{\mathcal{U}(I_s^\Delta)}} \\ &= \sup_{\tilde{v}\in\tilde{S}^{p+k}} \frac{\|\tilde{v} - \tilde{A}_{\tilde{I},2}\tilde{v}\|_{\tilde{I},2^{-1}}}{\|\tilde{v}\|_{\tilde{\mathcal{U}}(\tilde{I})}}\end{aligned} \tag{5.5.10b}$$

we have

$$\|e_{S_\Delta^p}^{S_\Delta^{p+k}} - \mathcal{A}^p_{I_s^\Delta,h_{I_s^\Delta}} e_{S_\Delta^p}^{S_\Delta^{p+k}}\|_{I_s^\Delta,(h_{I_s^\Delta})^{-1}} \le \mathscr{C}_{I_s^\Delta}(p,k)\|e_{S_\Delta^p}^{S_\Delta^{p+k}}\|_{\mathcal{U}(I_s^\Delta)} \tag{5.5.10c}$$

and employing $\chi = \mathcal{A}^p_{I_s^\Delta,h_{I_s^\Delta}} e_{S_\Delta^p}^{S_\Delta^{p+k}}$ into (5.5.9c), we obtain

$$\|e_{S_\Delta^p}^{S_\Delta^{p+k}}\|_{\mathcal{U}}^2 \le \sum_{s=1}^{N(\Delta)} \eta_{I_s^\Delta}^{\text{EXPL}} \mathscr{C}_{I_s^\Delta}(p,k)\|e_{S_\Delta^p}^{S_\Delta^{p+k}}\|_{\mathcal{U}(I_s^\Delta)} \tag{5.5.11a}$$

and we get

$$\|e_{S_\Delta^p}^{S_\Delta^{p+k}}\|_{\mathcal{U}}^2 \le \sqrt{\sum_{s=1}^{N(\Delta)} \mathscr{C}_{I_s^\Delta}^2(p,k)(\eta_s^{\text{EXPL}})^2}\,\|e_{S_\Delta^p}^{S_\Delta^{p+k}}\|_{\mathcal{U}} \tag{5.5.11b}$$

which proves the first inequality in (5.5.7b). The second inequality follows by employing

$$\mathscr{C}(p,k) = \max_{s=1,\dots,N(\Delta)} \mathscr{C}_{I_s^\Delta}(p,k) \le \mathscr{C}(\beta_1,\beta_2,\gamma) < \infty. \tag{5.5.12}$$

□

Remark 5.5.3 As we will see in Section 5.5*, the constant $\mathscr{C}_{I_s^\Delta}(p,k)$ can be determined as an eigenvalue of a matrix, and in the one-dimensional case considered here $\mathscr{C}_{I_s^\Delta}(p,k)$ is bounded independently of k. This reflects the fact that in one dimension we have $\mathcal{U}(I_s^\Delta) \subseteq C^0(I_s^\Delta)$. In higher dimensions, $\mathscr{C}_{I_s^\Delta}(p,k)$ grows with k.

Let us now give a proof of the lower bound $\mathscr{C}\mathscr{E}^{\text{EXPL}} \le \|e_{S_\Delta^p}^{S_\Delta^{p+k}}\|_{\mathcal{U}}$, which is the one-dimensional analogue of the proof given by [Babuška *et al.*, 1992] for the two-dimensional case.

Theorem 5.5.2 *(Lower bound) Under the assumption $f \in L^2(I)$, we have*

$$\frac{1}{\tilde{\mathscr{C}}(p,k)} \mathscr{E}^{\mathrm{EXPL}} \leq \left\| e_{S^p_\Delta}^{S^{p+k}_\Delta} \right\|_{\mathscr{U}} \tag{5.5.13}$$

where $\tilde{\mathscr{C}}(p,k)$ is given in (5.5.19) below.

Proof We have

$$\sum_{s=1}^{N(\Delta)} \left(\eta^{\mathrm{EXPL}}_{I^\Delta_s}\right)^2 = \sum_{s=1}^{N(\Delta)} \left(h_{I^\Delta_s}\right)^2 \int_{I^\Delta_s} \left(\pi^{L^2}_{S^{p+k}(I^\Delta_s)}(r_{I^\Delta_s})\right)^2 + \sum_{s=1}^{N(\Delta)} \frac{1}{2} h_{I^\Delta_s}\left((J^\Delta_{s-1})^2 + (J^\Delta_s)^2\right) \tag{5.5.14a}$$

or, using the notations introduced in the proof of Theorem 5.5.1, we have

$$\sum_{s=1}^{N(\Delta)} \left(\eta^{\mathrm{EXPL}}_{I^\Delta_s}\right)^2 = \sum_{s=1}^{N(\Delta)} \left\langle \left(h_{I^\Delta_s} \pi^{L^2}_{S^{p+k}(I^\Delta_s)}(r_{I^\Delta_s}); \sqrt{h_{I^\Delta_s}} \mathbf{J}^\Delta_s\right), \left(h_{I^\Delta_s} \pi^{L^2}_{S^{p+k}(I^\Delta_s)}(r_{I^\Delta_s}); \sqrt{h_{I^\Delta_s}} \mathbf{J}^\Delta_s\right) \right\rangle_{I^\Delta_s}. \tag{5.5.14b}$$

Let $w \in \mathscr{U}$ be such that it satisfies

$$\int_{I^\Delta_s} \left(\pi^{L^2}_{S^{p+k}(I^\Delta_s)}(r_{I^\Delta_s})\right) w = \left(h_{I^\Delta_s}\right)^2 \int_{I^\Delta_s} \left(\pi^{L^2}_{S^{p+k}(I^\Delta_s)}(r_{I^\Delta_s})\right)^2 \tag{5.5.15a}$$

and

$$w(x^\Delta_l) = J^\Delta_l \left(h_{I^\Delta_l} + h_{I^\Delta_{l-1}}\right), \quad l = s, s-1 \tag{5.5.15b}$$

or in the alternative notation

$$\left(\int_{I^\Delta_s} \pi^{L^2}_{S^{p+k}(I^\Delta_s)}(r_{I^\Delta_s}) w; \mathbf{w}^\Delta_s \right) = \left((h_{I^\Delta_s})^2 \|r_{I^\Delta_s}\|^2_{L^2(I^\Delta_s)}; (h\mathbf{J})^\Delta_s \right) \tag{5.5.15c}$$

where

$$(h\mathbf{J})^\Delta_s \stackrel{\mathrm{def}}{=} \left((h_{I^\Delta_{s-1}} + h_{I^\Delta_s}) J^\Delta_{s-1}, (h_{I^\Delta_s} + h_{I^\Delta_{s+1}}) J^\Delta_s \right)^{\mathrm{T}} \tag{5.5.15d}$$

for $s = 1, \ldots, N(\Delta)$. Using such a w, (5.5.14) can be rewritten as

$$\sum_{s=1}^{N(\Delta)} \left(\eta^{\mathrm{EXPL}}_{I^\Delta_s}\right)^2 = \sum_{s=1}^{N(\Delta)} \int_{I^\Delta_s} \pi^{L^2}_{S^{p+k}(I^\Delta_s)}(r_{I^\Delta_s}) w + \frac{1}{2} \sum_{s=1}^{N(\Delta)} \left(J^\Delta_{s-1} w(x^\Delta_{s-1}) + J^\Delta_s w(x^\Delta_s)\right) \tag{5.5.16a}$$

or in the alternative notation

$$\sum_{s=1}^{N(\Delta)} \left(\eta_{I_s^\Delta}^{\mathrm{EXPL}}\right)^2 = \sum_{s=1}^{N(\Delta)} \Big\langle \left(\pi^{L^2}_{S^{p+k}(I_s^\Delta)}(r_{I_s^\Delta}); (J_{s-1}^\Delta, J_s^\Delta)^{\mathrm{T}}\right), \left(w; (w(x_{s-1}^\Delta), w(x_s^\Delta))^{\mathrm{T}}\right)\Big\rangle \tag{5.5.16b}$$

and integrating by parts we get

$$\sum_{s=1}^{N(\Delta)} \left(\eta_{I_s^\Delta}^{\mathrm{EXPL}}\right)^2 = \mathcal{B}\left(e_{S_\Delta^p}^{S_\Delta^{p+k}}, w\right) \le \|e_{S_\Delta^p}^{S_\Delta^{p+k}}\|_{\mathcal{U}} \|w\|_{\mathcal{U}}. \tag{5.5.16c}$$

To complete the proof we will select the w which satisfies (5.5.15) and has minimal the energy norm. In particular, given any $\rho_s \in S^{p+k}(I_s^\Delta)$, and $\alpha_{I_s^\Delta}$, $\beta_{I_s^\Delta} \in \mathbb{R}$, we will let

$$W_{I_s^\Delta}(\rho_{I_s^\Delta}, \alpha_{I_s^\Delta}, \beta_{I_s^\Delta}) = \Big\{ w \in \mathcal{U}(I_s^\Delta) \Big| \int_{I_s^\Delta} w\rho_{I_s^\Delta} = \left(h_{I_s^\Delta}\right)^2 \int_{I_s^\Delta} (\rho_{I_s^\Delta})^2, \; w(x_{s-1}^\Delta) = \alpha_{I_s^\Delta} h_{I_s^\Delta},\; w(x_s^\Delta) = \beta_{I_s^\Delta} h_{I_s^\Delta} \Big\}, \tag{5.5.17a}$$

or in the alternative notation

$$W_{I_s^\Delta}(\rho_{I_s^\Delta}, \alpha_{I_s^\Delta}, \beta_{I_s^\Delta}) = \Big\{ w \in \mathcal{U}(I_s^\Delta) \Big| \left(\int_{I_s^\Delta} w\rho_{I_s^\Delta}; (w(x_{s-1}^\Delta), w(x_s^\Delta))^{\mathrm{T}} \right) = \left(\left(h_{I_s^\Delta}\right)^2 \int_{I_s^\Delta} (\rho_{I_s^\Delta})^2; \left((h_{I_{s-1}^\Delta} + h_{I_s^\Delta})\alpha_{I_s^\Delta}, (h_{I_s^\Delta} + h_{I_{s+1}^\Delta})\beta_{I_s^\Delta}\right)^{\mathrm{T}} \right) \Big\} \tag{5.5.17b}$$

and we will denote by $\mathfrak{w}_{I_s^\Delta}(\rho_{I_s^\Delta}, \alpha_{I_s^\Delta}, \beta_{I_s^\Delta})$ the solution of the minimization problem:

Find $\mathfrak{w}_{I_s^\Delta}(\rho_{I_s^\Delta}, \alpha_{I_s^\Delta}, \beta_{I_s^\Delta}) \in W_{I_s^\Delta}(\rho_{I_s^\Delta}, \alpha_{I_s^\Delta}, \beta_{I_s^\Delta})$ such that

$$\|\mathfrak{w}_{I_s^\Delta}(\rho_{I_s^\Delta}, \alpha_{I_s^\Delta}, \beta_{I_s^\Delta})\|_{\mathcal{U}(I_s^\Delta)} = \min_{w \in W_{I_s^\Delta}(\rho_{I_s^\Delta}, \alpha_{I_s^\Delta}, \beta_{I_s^\Delta})} \|w\|_{\mathcal{U}(I_s^\Delta)}. \tag{5.5.18}$$

The function $\mathfrak{w}(\rho_{I_s^\Delta}, \alpha_{I_s^\Delta}, \beta_{I_s^\Delta})$ exists and is unique. Note that $\rho_{I_s^\Delta}$ depends on $p+k+1$ parameters, and hence $\mathfrak{w}$ depends on $p+k+1+2$ parameters and $h_{I_s^\Delta}$. Now we will let

$$\tilde{\mathcal{C}}_{I_s^\Delta}(p+k) = \sup_{\substack{\alpha_{I_s^\Delta}, \beta_{I_s^\Delta} \in \mathbb{R} \\ \rho_{I_s^\Delta} \in S^{p+k}(I_s^\Delta)}} \frac{\|\mathfrak{w}_{I_s^\Delta}(\rho_{I_s^\Delta}, \alpha_{I_s^\Delta}, \beta_{I_s^\Delta})\|_{\mathcal{U}(I_s^\Delta)}}{\sqrt{\left(h_{I_s^\Delta}\right)^2 \int_{I_s^\Delta} \rho_s^2 + (h_{I_s^\Delta}/2)(\alpha_s^2 + \beta_s^2)}} \tag{5.5.19a}$$

or in the alternative notation

$$\tilde{\mathcal{C}}_{I_s^\Delta}(p+k) = \sup_{(\rho_{I_s^\Delta}, \alpha_{I_s^\Delta}, \beta_{I_s^\Delta}) \in S^{p+k}(I_s^\Delta) \times \mathbb{R}^2} \frac{\|\mathfrak{w}_{I_s^\Delta}(\rho_{I_s^\Delta}, \alpha_{I_s^\Delta}, \beta_{I_s^\Delta})\|_{\mathcal{U}(I_s^\Delta)}}{\left\|\left(\rho_s; [\alpha_{I_s^\Delta}, \beta_{I_s^\Delta}]^{\mathrm{T}}\right)\right\|_{I_s^\Delta, h_{I_s^\Delta}}} \tag{5.5.19b}$$

and

$$\tilde{\mathscr{C}}(p,k) = \max_{1\le s\le N(\Delta)} \tilde{\mathscr{C}}_{I_s^\Delta}(p+k). \tag{5.5.19c}$$

By scaling (5.5.19b) to the master element $\tilde{I}$, it can be seen that $\tilde{\mathscr{C}}_{I_s^\Delta}(p+k)$ is bounded from below and above independently of $h_{I_s^\Delta}$, with the bounds depending on β_1, β_2, γ. The computation of $\tilde{\mathscr{C}}_{I_s^\Delta}(p+k)$ will be given in Section 5.5*.

From the construction of $\tilde{\mathscr{C}}_{I_s^\Delta}(p+k)$ it follows that

$$\|\mathfrak{w}_{I_s^\Delta}\|_{\mathscr{U}(I_s^\Delta)} \le \tilde{\mathscr{C}}_{I_s^\Delta}(p+k)\eta_{I_s^\Delta}^{\text{EXPL}}. \tag{5.5.20a}$$

Now letting $\mathfrak{w} \in \mathscr{U}$ be such that $\mathfrak{w}|_{I_s^\Delta} = \mathfrak{w}_{I_s^\Delta}$, $s = 1,\dots,N(\Delta)$, we obtain

$$\|\mathfrak{w}\|_{\mathscr{U}}^2 = \sum_{s=1}^{N(\Delta)} \|\mathfrak{w}\|_{\mathscr{U}(I_s^\Delta)}^2 \le \sum_{s=1}^{N(\Delta)} \tilde{\mathscr{C}}_{I_s^\Delta}^2 (\eta_{I_s^\Delta}^{\text{EXPL}})^2 \le \tilde{\mathscr{C}}^2 \sum_{s=1}^{N(\Delta)} (\eta_{I_s^\Delta}^{\text{EXPL}})^2 \tag{5.5.20b}$$

where we omitted the arguments from $\tilde{\mathscr{C}}_{I_s^\Delta}(p+k)$ and $\tilde{\mathscr{C}}(p,k)$, and employing (5.5.16c) we get

$$\|\mathfrak{w}\|_{\mathscr{U}}^2 \le \tilde{\mathscr{C}}^2 \sum_{s=1}^{N(\Delta)} (\eta_{I_s^\Delta}^{\text{EXPL}})^2 = \tilde{\mathscr{C}}^2 \mathscr{B}(e_{S_\Delta^p}^{S_\Delta^{p+k}}, \mathfrak{w}) \le \tilde{\mathscr{C}}^2 \|e_{S_\Delta^p}^{S_\Delta^{p+k}}\|_{\mathscr{U}} \|\mathfrak{w}\|_{\mathscr{U}} \tag{5.5.21}$$

which completes the proof of the lower bound. □

We also have:

Theorem 5.5.3 *The estimator $\mathscr{E}^{\text{EXPL}}$ satisfies Assumptions A, B, and C of Section 5.1.*

Let us explain the purpose of the above approach. In Section 5.4 we have proven the equivalence $\mathscr{E}^{\text{EXPL}} \approx \|e_{S_\Delta^p}\|_{\mathscr{U}}$, namely, that

$$\mathscr{C}_L \mathscr{E}^{\text{EXPL}} \le \|e_{S_\Delta^p}\|_{\mathscr{U}} \le \mathscr{C}_U \mathscr{E}^{\text{EXPL}} \tag{5.5.22}$$

where $\mathscr{C}_L$, $\mathscr{C}_U$ *depend on the geometry of the nodal patches and are bounded. However the constants* $\mathscr{C}_L$, $\mathscr{C}_U$ *are not computable.* In this section we have proven the equivalence $\mathscr{E}^{\text{EXPL}} \approx \|e_{S_\Delta^p}^{S_\Delta^{p+k}}\|_{\mathscr{U}}$, namely, that

$$\mathscr{C}_L(p,k)\mathscr{E}^{\text{EXPL}} \le \|e_{S_\Delta^p}^{S_\Delta^{p+k}}\|_{\mathscr{U}} \le \mathscr{C}_U(p,k)\,\mathscr{E}^{\text{EXPL}} \tag{5.5.23a}$$

with constants $\mathscr{C}_L(p,k)$, $\mathscr{C}_U(p,k)$, which can be computed from one element. As we will see in Section 5.5*, in the higher-dimensional case we have

$$\lim_{k\to\infty} \mathscr{C}_L(p,k) = 0, \qquad \lim_{k\to\infty} \mathscr{C}_U(p,k) = \infty \tag{5.5.23b}$$

and hence the constants $\mathscr{C}_L$, $\mathscr{C}_U$ of (5.5.22) cannot be obtained as limits of the constants $\mathscr{C}_L(p,k)$, $\mathscr{C}_U(p,k)$ of (5.5.23a).

Remark 5.5.4 The main idea above was the replacement of u_{EX} by its finite element approximation $u_{S^{p+k}_{\Delta}}$. This leads to underestimation of the true error. Of course, $\|u_{EX} - u_{S^{p+k}_{\Delta}}\|_{\mathcal{U}}$ can be made to be negligible by selecting a sufficiently high k, but the problem is that $\mathscr{C}_L(p,k)$ and $\mathscr{C}_U(p,k)$ lose practical meaning for large k. Nevertheless, the constants $\mathscr{C}_L(p,k)$ and $\mathscr{C}_U(p,k)$ give us insight about the theoretical bounds of $\mathscr{C}_L$, $\mathscr{C}_U$ of (5.5.22), which have to be pessimistic by nature. This indicates that another approach is needed. We will discuss it in Chapter 6.

Remark 5.5.5 The same approach can be also used to obtain constants $\mathscr{C}_L(p,k)$ and $\mathscr{C}_U(p,k)$ for the other explicit estimators introduced in Section 5.4.

In this section we have given an asymptotic approach for analyzing the equivalence between the explicit estimator and the energy norm of the error. The main purpose was to give the one-dimensional analogue of the two-dimensional case given in Section 5.4*.

Problems

P5.5.1. Derive the eigenvalue problem for the computation of $\tilde{\mathscr{C}}_{I^{\Delta}_{s}}(p+k)$: (a) for the special case $AE \equiv 1$ and $c \equiv 0$; (b) for general AE and c.

P5.5.2. Derive the boundary value problem satisfied by $\mathfrak{w}$ in the special case $AE \equiv 1$, $c \equiv 0$.

5.5* Analysis of the explicit element residual estimator in two dimensions

Here we will prove the equivalence

$$\mathscr{C}_L(p,k;\alpha)\mathscr{E}^{\mathrm{EXPL}} \le \left\|e^{S^{p+k}_{\Delta}}_{S^{p}_{\Delta}}\right\|_{\mathcal{U}} \le \mathscr{C}_U(p,k;\alpha)\mathscr{E}^{\mathrm{EXPL}} \tag{5.5*.1}$$

where α is the minimal interior angle in the meshes employed, and the constants $\mathscr{C}_L(p,k;\alpha)$, $\mathscr{C}_U(p,k;\alpha)$ can be computed from one element.

More specifically, we will prove (5.5*.1) for the explicit estimator

$$\mathscr{E}^{\mathrm{EXPL}}_{S^{p+k}_{\Delta}} \stackrel{\mathrm{def}}{=} \sqrt{\sum_{\tau\in\Delta}(\eta^{\mathrm{EXPL}}_{\tau,S^{p+k}(\tau)})^2} \tag{5.5*.2a}$$

where

$$\eta^{\mathrm{EXPL}}_{\tau,S^{p+k}(\tau)} \stackrel{\mathrm{def}}{=} \sqrt{|\tau|\int_{\tau}(\pi^{L^2}_{S^{p+k}(\tau)}(r_\tau))^2 + \frac{1}{2}\sum_{\epsilon\in\mathcal{E}(\tau)}|\epsilon|\int_{\epsilon}(\pi^{L^2}_{S^{p+k}(\epsilon)}(J_\epsilon))^2}. \tag{5.5*.2b}$$

Here $\pi^{L^2}_{S^{p+k}(\tau)}$: $L^2(\tau) \mapsto S^{p+k}(\tau)$ is the $L^2(\tau)$ projection:

Given $f \in L^2(\tau)$, find $\pi^{L^2}_{S^{p+k}(\tau)}f \in S^{p+k}(\tau)$ such that

$$\int_{\tau}\pi^{L^2}_{S^{p+k}(\tau)}fv = \int_{\tau}fv \quad \forall v \in S^{p+k}(\tau). \tag{5.5*.2c}$$

Similarly, $\pi^{L^2}_{S^{p+k}(\epsilon)}$: $L^2(\epsilon) \mapsto S^{p+k}(\epsilon)$ we denote the $L^2(\epsilon)$ projection defined by:

Given $J \in L^2(\epsilon)$, find $\pi^{L^2}_{S^{p+k}(\epsilon)}(J) \in S^{p+k}(\epsilon)$ such that

$$\int_\epsilon \pi^{L^2}_{S^{p+k}(\epsilon)}(J)v = \int_\epsilon Jv \quad \forall v \in S^{p+k}(\epsilon). \tag{5.5*.2d}$$

As in the one-dimensional case, below we will often use the simplified notations η^{EXPL}_τ and $\mathscr{E}^{\text{EXPL}}$ instead of $\eta^{\text{EXPL}}_{\tau,S^{p+k}(\tau)}$ and $\mathscr{E}^{\text{EXPL}}_{S^{p+k}_\Delta}$.

Following [Babuška *et al.*, 1992] below, we will analyze the estimator $\mathscr{E}^{\text{EXPL}}$ in the case of triangular elements. The analysis of $\mathscr{E}^{\text{EXPL}}$ for quadrilateral elements, and more general differential operators follows similar steps.

Let us first prove the upper bound.

Theorem 5.5*.1 *(Upper bound) Under the assumptions*

$$f \in L^2(\Omega), \qquad g \in L^2(\Gamma_{\text{N}}) \tag{5.5*.3a}$$

we have

$$\|e^{S^{p+k}_\Delta}_{S^p_\Delta}\|_{\mathscr{U}} \le \sqrt{\sum_{\tau\in\Delta} \mathscr{C}^2_\tau(p,k;\alpha_\tau)(\eta^{\text{EXPL}}_{\tau,S^{p+k}(\tau)})^2} \le \mathscr{C}(p,k;\alpha^\Delta)\mathscr{E}^{\text{EXPL}}_{S^{p+k}_\Delta} \tag{5.5*.3b}$$

where α_τ is the minimal of the triangle τ, and $\alpha^\Delta \stackrel{\text{def}}{=}$ $\min_{\tau\in\Delta}\alpha_\tau$ *is the minimal angle of the mesh Δ.*

Proof As in the one-dimensional case we have

$$\|e^{S^{p+k}_\Delta}_{S^p_\Delta}\|^2_{\mathscr{U}} = \mathscr{B}(e^{S^{p+k}_\Delta}_{S^p_\Delta}, \underbrace{u_{S^{p+k}_\Delta} - u_{EX}}_{-e_{S^{p+k}_\Delta}} + \underbrace{u_{EX} - u_{S^p_\Delta}}_{e_{S^p_\Delta}}) \tag{5.5*.4a}$$

and employing the orthogonality condition

$$\mathscr{B}(e_{S^{p+k}_\Delta}, e^{S^{p+k}_\Delta}_{S^p_\Delta}) = 0 \tag{5.5*.4b}$$

and the symmetry of $\mathscr{B}$, we get

$$\|e^{S^{p+k}_\Delta}_{S^p_\Delta}\|^2_{\mathscr{U}} = \mathscr{B}(e^{S^{p+k}_\Delta}_{S^p_\Delta}, e_{S^p_\Delta}). \tag{5.5*.4c}$$

Using the orthogonality condition

$$\mathscr{B}(e_{S^p_\Delta}, \mathcal{I}^p_\Delta e^{S^{p+k}_\Delta}_{S^p_\Delta}) = 0 \tag{5.5*.4d}$$

and the symmetry of $\mathscr{B}$, we get

$$\|e^{S^{p+k}_\Delta}_{S^p_\Delta}\|^2_{\mathscr{U}} = \mathscr{B}(e_{S^p_\Delta}, e^{S^{p+k}_\Delta}_{S^p_\Delta} - \mathcal{I}^p_\Delta e^{S^{p+k}_\Delta}_{S^p_\Delta}) = \mathscr{R}(e^{S^{p+k}_\Delta}_{S^p_\Delta} - \mathcal{I}^p_\Delta e^{S^{p+k}_\Delta}_{S^p_\Delta}). \tag{5.5*.4e}$$

Note that $e_{S^p_\Delta}^{S^{p+k}_\Delta} \in S^{p+k}_\Delta \subseteq C^0(\Omega)$, *and hence the interpolant* $\mathcal{I}^p_\Delta e_{S^p_\Delta}^{S^{p+k}_\Delta}$ *is well defined.* As in the one-dimensional case, we will use the splitting

$$\mathcal{R}(v) = \sum_{\tau\in\Delta} \mathcal{R}^{UNEQ}_\tau(v|_\tau) = \sum_{\tau\in\Delta} \mathcal{R}^{UNEQ}_{\tau,S^{p+k}(\tau)}(v|_\tau) \quad \forall v \in S^{p+k}_\Delta \tag{5.5*.5a}$$

where

$$\mathcal{R}^{UNEQ}_{\tau,S^{p+k}(\tau)}(v) = \int_\tau \pi^{L^2}_{S^{p+k}(\tau)}(r^\Delta_\tau)v + \frac{1}{2}\sum_{\epsilon\in\mathcal{E}(\tau)} \int_\epsilon \pi^{L^2}_{S^{p+k}(\epsilon)}(J_\epsilon)v$$

$$\forall v \in S^{p+k}(\tau), \tag{5.5*.5b}$$

the inner product

$$\langle(v;\mathbf{a}),(w;\mathbf{b})\rangle_\tau \stackrel{\text{def}}{=} \sqrt{\int_\tau vw + \sum_{i=1}^{3}\frac{1}{2}\int_{\epsilon_i} a_i b_i} \tag{5.5*.5c}$$

where $(v;\mathbf{a}),(w;\mathbf{b}) \in L^2(\tau)\times(\prod_{i=1}^3 L^2(\epsilon_i))$, and the corresponding norm

$$\|(v;\mathbf{a})\|_{\tau,h^{-1}} = \sqrt{\frac{1}{|\tau|^2}\int_\tau v^2 + \sum_{\epsilon\in\mathcal{E}(\tau)}\frac{1}{2|\epsilon|^2}\int_\epsilon a_i^2}. \tag{5.5*.5d}$$

Using these notations (5.5*.5b) can be written in the form

$$\mathcal{R}^{UNEQ}_{\tau,S^{p+k}(\tau)}(v) = \langle(\pi_{S^{p+k}(\tau)}(r^\Delta_\tau);\mathbf{J}^\Delta_\epsilon),(v;\mathbf{v}^\Delta_\epsilon)\rangle_\tau \quad \forall v \in S^{p+k}(\tau) \tag{5.5*.6a}$$

and (5.5*.4e) gives

$$\|e_{S^p_\Delta}^{S^{p+k}_\Delta}\|^2_{\mathcal{U}} = \sum_{\tau\in\Delta}\langle(\pi_{S^{p+k}(\tau)}(r^\Delta_\tau);\mathbf{J}^\Delta_\epsilon),((Id-\mathcal{I}^p_\tau)e_{S^p_\Delta}^{S^{p+k}_\Delta};$$

$$(Id-\mathcal{I}^p_\tau)\mathbf{e}_{S^p_\Delta}^{S^{p+k}_\Delta}|_{\partial\tau})\rangle_\tau \tag{5.5*.6b}$$

where Id is the identity operator, and $\mathbf{c}_{S^p_\Delta}^{S^{p+k}_\Delta}|_{\partial\tau} \stackrel{\text{def}}{=} [c_{S^p_\Delta}^{S^{p+k}_\Delta}|_{\epsilon_1}, c_{S^p_\Delta}^{S^{p+k}_\Delta}|_{\epsilon_2}, c_{S^p_\Delta}^{S^{p+k}_\Delta}|_{\epsilon_3}]^{\mathrm{T}}$. Here we are considering the case of triangular elements, and we will use $S^{p+k}(\tau)$ interchangeably with $\mathcal{P}^{p+k}(\tau)$. Employing the Cauchy–Schwarz inequality

$$\langle(v;\mathbf{a}),(w;\mathbf{b})\rangle_\tau \le \|(|\tau|v;|\epsilon|\mathbf{a})\|_\tau \left\|\left(\frac{1}{|\tau|}w;\frac{1}{|\epsilon|}\mathbf{b}\right)\right\|_\tau \tag{5.5*.7a}$$

where we let

$$|\epsilon|\mathbf{a} \stackrel{\text{def}}{=} [|\epsilon_1|a_1,|\epsilon_2|a_2,|\epsilon_3|a_3]^{\mathrm{T}}, \qquad \frac{1}{|\epsilon|}\mathbf{b} \stackrel{\text{def}}{=} \left[\frac{1}{|\epsilon_1|}b_1,\frac{1}{|\epsilon_2|}b_2,\frac{1}{|\epsilon_3|}b_3\right]^{\mathrm{T}} \tag{5.5*.7b}$$

we get

$$\|e_{S^p_\Delta}^{S^{p+k}_\Delta}\|^2_{\mathcal{U}} \leq \sum_{\tau\in\Delta} \underbrace{\|(|\tau|\pi_{S^{p+k}(\tau)}(r^\Delta_\tau); |\epsilon|\mathbf{J}^\Delta_\epsilon)\|_\tau}_{\eta^{\text{EXPL}}_{\tau,S^p(\tau)}}$$

$$\times \left\| \left(\frac{1}{|\tau|}(Id - \mathcal{I}^p_\tau) e_{S^p_\Delta}^{S^{p+k}_\Delta}; (Id - \mathcal{I}^p_\tau) \frac{1}{|\epsilon|} \mathbf{e}_{S^p_\Delta}^{S^{p+k}_\Delta}|_{\partial\tau} \right) \right\|_\tau. \tag{5.5*.8}$$

Letting

$$\mathcal{C}_\tau(p,k) = \max_{v\in\mathcal{P}^{p+k}(\tau)} \frac{1}{\|v\|_{\mathcal{U}(\tau)}} \left\| \left(\frac{1}{|\tau|}(Id - \mathcal{I}^p_\tau)v; (Id - \mathcal{I}^p_\tau)\frac{1}{|\epsilon|}\mathbf{v}|_{\partial\tau} \right) \right\|_\tau \tag{5.5*.9}$$

we get

$$\|e_{S^p_\Delta}^{S^{p+k}_\Delta}\|^2_{\mathcal{U}} \leq \sum_{\tau\in\Delta} \eta^{\text{EXPL}}_\tau \mathcal{C}_\tau(p,k) \|e_{S^p_\Delta}^{S^{p+k}_\Delta}\|_{\mathcal{U}} \tag{5.5*.10a}$$

and hence

$$\|e_{S^p_\Delta}^{S^{p+k}_\Delta}\|_{\mathcal{U}} \leq \max_{\tau\in\Delta}(\mathcal{C}_\tau(p,k)) \sqrt{\sum_{\tau\in\Delta} (\eta^{\text{EXPL}}_\tau)^2}\, \|e_{S^p_\Delta}^{S^{p+k}_\Delta}\|_{\mathcal{U}} \tag{5.5*.10b}$$

which proves (5.5*.3b). □

Let us now address the determination of $\mathcal{C}_\tau(p,k)$ in the case of linear triangles ($p = 1$), and the Poisson equation

$$-\Delta u|_\Omega = f, \qquad u|_{\Gamma_\text{D}} \equiv 0, \qquad \left.\frac{\partial u}{\partial n}\right|_{\Gamma_\text{N}} = g \tag{5.5*.11a}$$

with f and g piecewise constant, namely

$$f|_\tau \in \mathcal{P}^0(\tau) \quad \forall\tau\in\Delta, \qquad g|_\epsilon \in \mathcal{P}^0(\epsilon) \quad \forall\epsilon \subseteq \Gamma_\text{N}. \tag{5.5*.11b}$$

In this case the indicator takes the form

$$\eta^{\text{EXPL}}_{\tau,S^{1+k}(\tau)} = \sqrt{|\tau|(f|_\tau)^2 + \frac{1}{2}\sum_{\epsilon\in\mathcal{E}(\tau)} |\epsilon|(J_\epsilon)^2} \tag{5.5*.11c}$$

which *does not depend on k*.

Let us first modify the proof of Theorem 5.5*.1 to take into account that f, g are piecewise constant.

Equation (5.5*.6b) can now be written as

$$\|e_{S^1_\Delta}^{S^{1+k}_\Delta}\|^2_{\mathcal{U}} = \sum_{\tau\in\Delta}\left(f|_\tau \int_\tau (Id - \mathcal{I}^1_\tau) e_{S^1_\Delta}^{S^{1+k}_\Delta} + \frac{1}{2}\sum_{\epsilon\in\mathcal{E}(\tau)} J^\Delta_\epsilon \int_\epsilon (Id - \mathcal{I}^1_\tau) e_{S^1_\Delta}^{S^{1+k}_\Delta} \right) \tag{5.5*.12a}$$

or

$$\|e_{S^1_\Delta}^{S^{1+k}_\Delta}\|_{\mathcal{U}}^2 = \sum_{\tau\in\Delta}\left\langle\left[|\tau||f|_\tau,\ \frac{|\epsilon_i|}{\sqrt{2}}J^\Delta_{\epsilon_i},\ i=1,2,3\right]^{\mathrm{T}},\ \left[\frac{1}{|\tau|}(Id-\mathcal{I}^1_\tau)e_{S^1_\Delta}^{S^{1+k}_\Delta},\right.\right.$$
$$\left.\left.\frac{1}{\sqrt{2}|\epsilon_i|}(Id-\mathcal{I}^1_\tau)e_{S^1_\Delta}^{S^{1+k}_\Delta}\Big|_{\epsilon_i},\ i=1,2,3\right]\right\rangle \tag{5.5*.12b}$$

and using the Cauchy–Schwarz inequality for the $\mathbb{R}^4$ inner product, we get

$$\|e_{S^1_\Delta}^{S^{1+k}_\Delta}\|_{\mathcal{U}}^2 \le \sum_{\tau\in\Delta}\underbrace{\left|\left[|\tau||f|_\tau,\ \frac{|\epsilon_i|}{\sqrt{2}}J^\Delta_{\epsilon_i},\ i=1,2,3\right]^{\mathrm{T}}\right|_{\mathbb{R}^4}}_{\eta_\tau^{\mathrm{EXPL}}}\left|\left[\frac{1}{|\tau|}\int_\tau(Id-\mathcal{I}^1_\tau)e_{S^1_\Delta}^{S^{1+k}_\Delta},\right.\right.$$
$$\left.\left.\frac{1}{\sqrt{2}|\epsilon_i|}\int_{\epsilon_i}(Id-\mathcal{I}^1_\tau)e_{S^1_\Delta}^{S^{1+k}_\Delta},\ i=1,2,3\right]\right|_{\mathbb{R}^4}. \tag{5.5*.12c}$$

Letting

$$\mathscr{C}_\tau(k) = \max_{v\in\mathscr{P}^{1+k}(\tau)}\frac{1}{\|v\|_{\mathcal{U}(\tau)}}$$
$$\times\left|\left[\frac{1}{|\tau|}\int_\tau(v-\mathcal{I}^1_\tau v),\ \frac{1}{\sqrt{2}|\epsilon_i|}\int_{\epsilon_i}(v-\mathcal{I}^1_\tau v),\ i=1,2,3\right]^{\mathrm{T}}\right|_{\mathbb{R}^4} \tag{5.5*.13a}$$

or explicitly

$$\mathscr{C}_\tau(k) = \max_{v\in\mathscr{P}^{1+k}(\tau)}\frac{1}{\|v\|_{\mathcal{U}(\tau)}}\sqrt{\frac{1}{|\tau|^2}\left(\int_\tau(v-\mathcal{I}^1_\tau v)\right)^2+\frac{1}{2|\epsilon_i|^2}\left(\int_{\epsilon_i}(v-\mathcal{I}^1_\tau v)\right)^2} \tag{5.5*.13b}$$

we get

$$\|e_{S^1_\Delta}^{S^{1+k}_\Delta}\|_{\mathcal{U}} \le \sqrt{\sum_{\tau\in\Delta}(\mathscr{C}_\tau(k))^2\left(\eta^{\mathrm{EXPL}}_{\tau,S^{1+k}(\tau)}\right)^2}. \tag{5.5*.13c}$$

Let us now determine $\mathscr{C}_\tau(k)$. Letting

$$z_0\in\mathscr{P}^{1+k}(\tau)\colon\ \int_\tau\nabla z_0\cdot\nabla v = \frac{1}{|\tau|}\int_\tau(v-\mathcal{I}^1_\tau v)\quad\forall v\in\mathscr{P}^{1+k}(\tau) \tag{5.5*.14a}$$

and

$$z_i\in\mathscr{P}^{1+k}(\tau)\colon\ \int_\tau\nabla z_i\cdot\nabla v = \frac{1}{\sqrt{2}|\epsilon_i^\tau|}\int_{\epsilon_i^\tau}(v-\mathcal{I}^1_\tau v)\quad\forall v\in\mathscr{P}^{1+k}(\tau) \tag{5.5*.14b}$$

for $i = 1, 2, 3$, we have

$$(\mathscr{C}_\tau(k))^2 = \max_{\substack{v \in \mathscr{P}^{1+k}(\tau) \\ v \not\equiv constant}} \frac{\sum_{i=0}^{3} \left(\int_\tau \nabla z_i \cdot \nabla v \right)^2}{\int_\tau |\nabla v|^2}$$

$$= \max_{\substack{v = \sum_{i=0}^{3} \beta_i z_i \\ v \not\equiv constant}} \frac{\sum_{i=0}^{3} \left(\int_\tau \nabla z_i \cdot \nabla v \right)^2}{\int_\tau |\nabla v|^2} \tag{5.5*.15a}$$

and hence

$$(\mathscr{C}_\tau(k))^2 = \max_{\substack{\boldsymbol{\beta} \in \mathbb{R}^3 \\ v = \sum_{k=0}^{3} \beta_k z_k \not\equiv constant}} \frac{\sum_{i=0}^{3} \left(\int_\tau \nabla z_i \cdot \nabla \left(\sum_{j=0}^{3} \beta_j z_j \right) \right)^2}{\int_\tau \left| \nabla \left(\sum_{k=0}^{3} \beta_k z_k \right) \right|^2}. \tag{5.5*.15b}$$

Now, employing that

$$\int_\tau \left| \nabla \left(\sum_{k=0}^{3} \beta_k z_k \right) \right|^2 = \sum_{l=0}^{3} \sum_{m=0}^{3} \left(\int_\tau \nabla z_l \cdot \nabla z_m \right) \beta_l \beta_m = \boldsymbol{\beta}^{\mathrm{T}} \mathbf{C}^\tau \boldsymbol{\beta} \tag{5.5*.16a}$$

where we let

$$C_{lm}^\tau \stackrel{\text{def}}{=} \int_\tau \nabla z_l \cdot \nabla z_m \tag{5.5*.16b}$$

and that

$$\begin{aligned} \sum_{i=0}^{3} \left(\int_\tau \nabla z_i \cdot \nabla \left(\sum_{j=0}^{3} \beta_j z_j \right) \right)^2 &= \sum_{i=0}^{3} \left(\sum_{j=0}^{3} \beta_j \left(\int_\tau \nabla z_i \cdot \nabla z_j \right) \right)^2 \\ &= \sum_{i=0}^{3} \sum_{j=0}^{3} \sum_{k=0}^{3} \beta_j \beta_k \left(\int_\tau \nabla z_i \cdot \nabla z_j \right) \left(\int_\tau \nabla z_i \cdot \nabla z_k \right) \end{aligned} \tag{5.5*.16c}$$

and hence

$$\sum_{i=0}^{3} \left(\int_\tau \nabla z_i \cdot \nabla \left(\sum_{j=0}^{3} \beta_j z_j \right) \right)^2 = \sum_{i=0}^{3} \sum_{j=0}^{3} \sum_{k=0}^{3} \beta_j C_{ij}^\tau C_{ik}^\tau \beta_k = \boldsymbol{\beta}^{\mathrm{T}} (\mathbf{C}^\tau)^{\mathrm{T}} \mathbf{C}^\tau \boldsymbol{\beta} \tag{5.5*.16d}$$

we obtain

$$\mathscr{C}_\tau(k) = \sqrt{\max_{\substack{\mathbf{v} \in \mathbb{R}^4 \\ \mathbf{v}^{\mathrm{T}} \mathbf{C}^\tau \mathbf{v} \neq \mathbf{0}}} \frac{\mathbf{v}^{\mathrm{T}} (\mathbf{C}^\tau)^{\mathrm{T}} \mathbf{C}^\tau \mathbf{v}}{\mathbf{v}^{\mathrm{T}} \mathbf{C}^\tau \mathbf{v}}} = \sqrt{\max_{\substack{\mathbf{v} \in \mathbb{R}^4 \\ \mathbf{v} \neq \mathbf{0}}} \frac{\mathbf{v}^{\mathrm{T}} \mathbf{C}^\tau \mathbf{v}}{\mathbf{v}^{\mathrm{T}} \mathbf{v}}}. \tag{5.5*.17}$$

TABLE 5.5*.1. The values of the constants $(\mathscr{C}'_\tau)^{-1}$ and $\mathscr{C}_\tau(k,\alpha)$

α	$(\mathscr{C}'_\tau)^{-1}$	$\mathscr{C}_\tau(k,\alpha)$			
		$k=1$	$k=3$	$k=5$	$k=7$
7.5°	0.051	2.390	3.306	3.660	3.988
15.0°	0.072	1.682	2.341	2.609	2.839
22.5°	0.087	1.363	1.918	2.156	2.343
30.0°	0.099	1.169	1.670	1.895	2.058
37.5°	0.108	1.035	1.508	1.727	1.876
45.0°	0.115	0.939	1.400	1.615	1.757
52.5°	0.119	0.875	1.334	1.547	1.684
60.0°	0.121	0.850	1.309	1.522	1.657

It follows that $\mathscr{C}_\tau(k)$ is the square root of the maximum eigenvalue of $\mathbf{C}^\tau = [\int_\tau \nabla z_i \cdot \nabla z_j,\ i,j=0,\ldots,3]$, and depends on the polynomial degree $1+k$ and the interior angles of the element τ, but not on h_τ. The computations of [Babuška *et al.*, 1992] for $k=1,\ldots,7$, have shown that

$$\frac{0.548\sqrt{\ln(1+k)}}{\sqrt{\sin(\alpha/2)}} \leq \mathscr{C}_\tau(k,\alpha) \leq \frac{0.813\sqrt{\ln(1+k)}}{\sqrt{\sin(\alpha/2)}}. \tag{5.5*.18a}$$

Table 5.5*.1 gives the values of the constant $\mathscr{C}_\tau(k,\alpha)$ for $k = 1,3,5,7$, and $\alpha = 7.5, 15, 22.5, 30, 37.5, 45, 52.5$, and 60 degrees. Note that $\mathscr{C}_\tau = \mathscr{C}(k;\alpha_1^\tau,\alpha_2^\tau,\alpha_3^\tau)$, where α_i^τ, $i=1,2,3$ are the interior angles of τ, and the dependence on the interior angles α_i^τ was simplified by determining the worst value

$$\mathscr{C}(k,\alpha) = \max_{\alpha_1^\tau,\alpha_2^\tau,\alpha_3^\tau \leq \alpha} \mathscr{C}(k;\alpha_1^\tau,\alpha_2^\tau,\alpha_3^\tau). \tag{5.5*.18a}$$

Note that

$$\mathscr{C}_\tau(k) \approx \sqrt{\ln(1+k)} \to \infty \quad \text{as } k \to \infty \tag{5.5*.19}$$

and it is not possible to recover the result $\|e_{S^p_\Delta}\|_{\mathscr{U}} \leq \mathscr{C}_U \mathscr{E}^{\mathrm{EXPL}}$ which was proven using a different approach in Theorem 5.4*.6.

We will now prove the lower bound.

Theorem 5.5*.2 *(Lower bound) Assume that $f \in L^2(\Omega)$ and $g \in L^2(\Gamma_{\mathrm{N}})$. Then there exists a constant $\mathscr{C}'$ such that*

$$\mathscr{C}'\mathscr{E}^{\mathrm{EXPL}}_{S^{p+k}_\Delta} \leq \|e^{S^{p+k}_\Delta}_{S^1_\Delta}\|_{\mathscr{U}} \tag{5.5*.20}$$

where $\mathscr{C}'$ depends on α^Δ, the minimal interior angle in the mesh.

As in the one-dimensional case, the proof of the lower bound is based on the construction of a special function $\mathfrak{w} \in \mathcal{U}$ with the property that

$$\underbrace{|\tau| \int_\tau (\pi^{L^2}_{S^{p+k}(\tau)}(r^\Delta_\tau))^2 + \frac{1}{2} \sum_{\epsilon \in \mathcal{E}(\tau)} |\epsilon| \int_\epsilon (\pi^{L^2}_{S^{p+k}(\epsilon)}(J_\epsilon))^2}_{\eta^2_{\tau,S^{p+k}(\tau)}} = \mathcal{R}^{UNEQ}_{\tau,S^{p+k}(\tau)}(\mathfrak{w}|_\tau). \tag{5.5*.21}$$

As in the one-dimensional case, we will construct $\mathfrak{w}$ elementwise by letting $\mathfrak{w}|_\tau = \mathfrak{w}_\tau$ where $\mathfrak{w}_\tau \in H^1(\tau)$ is such that

$$\int_\tau (\pi^{L^2}_{S^{p+k}(\tau)}(r^\Delta_\tau))\mathfrak{w}_\tau = |\tau| \int_\tau (\pi^{L^2}_{S^{p+k}(\tau)}(r^\Delta_\tau))^2 \tag{5.5*.22b}$$

and

$$\int_\epsilon (\pi^{L^2}_{S^{p+k}(\epsilon)}(J_\epsilon))\mathfrak{w}_\tau = k_\epsilon |\epsilon| \int_\epsilon (\pi^{L^2}_{S^{p+k}(\epsilon)}(J_\epsilon))^2 \quad \forall \epsilon \in \mathcal{E}(\tau) \tag{5.5*.22c}$$

and such that

$$\mathfrak{w}_\tau|_\epsilon = \mathfrak{w}_{\tau^*}|_\epsilon \quad \epsilon = \partial\tau \cap \partial\tau^*. \tag{5.5*.22d}$$

This construction is possible and we have

$$(\mathcal{E}^{\mathrm{EXPL}})^2 = \sum_{\tau \in \Delta} (\eta^{\mathrm{EXPL}}_\tau)^2 = \sum_{\tau \in \Delta} \mathcal{R}_\tau(\mathfrak{w}_\tau) = \mathcal{R}(\mathfrak{w}) = \mathcal{B}(e_{S^p_\Delta}^{S^{p+k}_\Delta}, \mathfrak{w}) \tag{5.5*.23a}$$

and using the Cauchy–Schwarz inequality we get

$$(\mathcal{E}^{\mathrm{EXPL}})^2 \le \|e_{S^1_\Delta / S^{1+k}_\Delta}\|_{\mathcal{U}} \|\mathfrak{w}\|_{\mathcal{U}}. \tag{5.5*.23b}$$

The rest of the proof is exactly analogous as in the one-dimensional case.

Let us now specialize the proof of the lower bound to the case of the Poisson equation with piecewise-constant f, g, and linear triangles ($p = 1$).

In this case we have

$$\mathcal{C}'_\tau = \sqrt{\max_{\mathbf{m} \in \mathbb{R}^4} \frac{\|v_\mathbf{m}\|^2_{\mathcal{U}(\tau)}}{m_0^2 + \frac{1}{2}\sum_{i=1}^3 m_i^2}} \tag{5.5*.24a}$$

where $v_\mathbf{m} \in H^1(\tau)$ is such that

$$\int_\tau v_\mathbf{m} = |\tau| m_0, \qquad \int_{\epsilon_i} v_\mathbf{m} = |\epsilon_i| m_i, \quad i = 1, 2, 3. \tag{5.5*.24b}$$

We also have $\mathfrak{w}_\tau \equiv v_{\bar{\mathbf{m}}}$ where $\bar{\mathbf{m}} \in \mathbb{R}^4$ is the vector with components

$$\bar{m}_0 = |\tau| f|_\tau, \qquad \bar{m}_i = k_{\epsilon_i} |\epsilon_i| J^\Delta_{\epsilon_i}, \quad i = 1, 2, 3. \tag{5.5*.24c}$$

By the construction of $\mathcal{C}'_\tau$, we have

$$\|v_{\bar{\mathbf{m}}}\|^2_{\mathfrak{U}(\tau)} \leq (\mathcal{C}'_\tau)^2 \left(|\tau|^2 (f|_\tau)^2 + \frac{1}{2} \sum_{i=1}^{3} k_{\epsilon_i} |\epsilon_i|^2 \left(J^\Delta_{\epsilon_i}\right)^2 \right) = (\mathcal{C}'_\tau)^2 (\eta^{\mathrm{EXPL}}_\tau)^2 \qquad (5.5^*.24\mathrm{d})$$

and hence

$$\|\mathfrak{w}_\tau\|^2_{\mathfrak{U}(\tau)} = \|v_{\bar{\mathbf{m}}}\|^2_{\mathfrak{U}(\tau)} \leq (\mathcal{C}'_\tau)^2 (\eta^{\mathrm{EXPL}}_\tau)^2 \qquad (5.5^*.24\mathrm{e})$$

and summing over the elements we get

$$\|\mathfrak{w}\|_{\mathfrak{U}} = \sqrt{\sum_{\tau\in\Delta} \|\mathfrak{w}_\tau\|^2_{\mathfrak{U}(\tau)}} \leq \sqrt{\sum_{\tau\in\Delta} (\mathcal{C}'_\tau)^2 (\eta^{\mathrm{EXPL}}_\tau)^2} \leq \max_{\tau\in\Delta} (\mathcal{C}'_\tau) \mathcal{E}^{\mathrm{EXPL}}. \qquad (5.5^*.25)$$

Using this result in (5.5*.23b), we get the proof of the lower bound with $\mathcal{C}' = \max_{\tau\in\Delta}(\mathcal{C}'_\tau)$.

Let us now elaborate on the computation of the $\mathcal{C}'_\tau$. Letting

$$\bar{\mathbf{m}} = [f|_\tau, k_{\epsilon_1} J_{\epsilon_1}, k_{\epsilon_2} J_{\epsilon_2}, k_{\epsilon_3} J_{\epsilon_3}]^{\mathrm{T}} \qquad (5.5^*.26\mathrm{a})$$

and

$$\mathbf{D} \stackrel{\mathrm{def}}{=} \tfrac{1}{\sqrt{2}} \mathbf{diag}(\sqrt{2}|\tau|, |\epsilon^\tau_1|, |\epsilon^\tau_2|, |\epsilon^\tau_3|) \qquad (5.5^*.26\mathrm{b})$$

we have

$$\begin{aligned}(\eta^{\mathrm{EXPL}}_\tau)^2 &= |\tau| \int_\tau (\pi^{L^2}_{S^{p+k}(\tau)}(r^\Delta_\tau))^2 + \frac{1}{2} \sum_{\epsilon\in\mathcal{E}(\tau)} k_\epsilon |\epsilon| \int_\epsilon (\pi^{L^2}_{S^{p+k}(\epsilon)}(J_\epsilon))^2 \\ &= \bar{\mathbf{m}}^{\mathrm{T}} \mathbf{D} \bar{\mathbf{m}}. \end{aligned} \qquad (5.5^*.26\mathrm{c})$$

Let $\psi_{\epsilon^\tau_i} \in H^1_0(\epsilon^\tau_i)$, $i = 1, 2, 3$, be given and

$$W^\tau_{\mathbf{m}} \stackrel{\mathrm{def}}{=} \left\{ w \in H^1(\tau) \,\middle|\, \int_\tau w = |\tau|^2 m_0,\ w|_{\epsilon^\tau_i} = \frac{|\epsilon^\tau_i| m_i}{c^\tau_i} \psi_{\epsilon^\tau_i},\ i = 1, 2, 3 \right\} \qquad (5.5^*.27\mathrm{a})$$

where $c^\tau_i = (1/|\epsilon^\tau_i|) \int_{\epsilon^\tau_i} \psi_{\epsilon^\tau_i}$ $i = 1, 2, 3$, and

$$W^\tau_{\mathbf{0}} = \left\{ w \in H^1(\tau) \,\middle|\, \int_\tau w = 0,\ w|_{\epsilon^\tau_i} \equiv 0,\ i = 1, 2, 3 \right\}. \qquad (5.5^*.27\mathrm{b})$$

Note that $W^\tau_{\mathbf{m}} = W^\tau_{\mathbf{m}}(\psi_{\epsilon^\tau_1}, \psi_{\epsilon^\tau_2}, \psi_{\epsilon^\tau_3})$ is an affine subspace of $H^1(\tau)$.

We will now let $w_{\mathbf{m}} \in W^\tau_{\mathbf{m}}$ be such that

$$\int_\tau |\boldsymbol{\nabla} w_{\mathbf{m}}|^2 = \min_{w\in W^\tau_{\mathbf{m}}} \int_\tau |\boldsymbol{\nabla} w|^2. \qquad (5.5^*.28\mathrm{a})$$

We then have

$$\int_\tau \nabla w_{\mathbf{m}} \cdot \nabla v = 0 \quad \forall v \in W_{\mathbf{0}}^\tau \quad \Longrightarrow \quad w_{\mathbf{m}} \in (W_{\mathbf{0}}^\tau)^\perp \tag{5.5*.28b}$$

where $(W_{\mathbf{0}}^\tau)^\perp$ is the orthogonal complement of $W_{\mathbf{0}}^\tau$ in $H^1(\tau)$. Note that $(W_{\mathbf{0}}^\tau)^\perp$ *is a four-dimensional space.* It follows that

$$\mathscr{C}'_\tau = \sqrt{\max_{\substack{\mathbf{m}\in\mathbb{R}^4\\ \mathbf{m}\neq\mathbf{0}}} \frac{\|v_{\bar{\mathbf{m}}}\|^2_{\mathfrak{U}(\tau)}}{m_0^2 + \frac{1}{2}\sum_{i=1}^3 m_i^2}} = \sqrt{\max_{\substack{\mathbf{m}\in\mathbb{R}^4\\ \mathbf{m}\neq\mathbf{0}}} \frac{\int_\tau |\nabla w_{\mathbf{m}}^\tau|^2}{\mathbf{m}^{\mathrm{T}}\mathbf{D}\mathbf{m}}}. \tag{5.5*.29}$$

Let us now construct a basis for $(W_{\mathbf{0}}^\tau)^\perp$. Let $w_0 \in H_0^1(\tau)$ be the exact solution of the homogeneous Dirichlet problem

$$-\Delta w_0|_\tau = 1, \qquad w_0|_{\partial\tau} = 0. \tag{5.5*.30a}$$

Then, integrating by parts, we get

$$\int_\tau \nabla w_0 \cdot \nabla v = \int_\tau v + \int_{\partial\tau} \frac{\partial}{\partial n}(w_0) v = 0 \quad \forall v \in W_{\mathbf{0}}^\tau \tag{5.5*.30b}$$

and hence $w_0 \in (W_{\mathbf{0}}^\tau)^\perp$. This follows from the observation that $v \in W_{\mathbf{0}}^\tau$ implies that $\int_\tau v = 0$, and $v|_{\epsilon_i^\tau} \equiv 0$, for $i = 1, 2, 3$.

We will also let $w_i \in H^1(\tau)$, $i = 1, 2, 3$, be the exact solution of

$$-\Delta w_i = 0, \quad \text{in } \tau, \qquad w_i|_{\epsilon_i^\tau} = \psi_{\epsilon_i^\tau}, \qquad w_i|_{\partial\tau - \epsilon_i^\tau} = 0. \tag{5.5*.31}$$

If follows that $w_i \in (W_{\mathbf{0}}^\tau)^\perp$. Moreover w_0, w_1, w_2, w_3 *are linearly independent by their construction at the boundary*, and hence they form a basis for $(W_{\mathbf{0}}^\tau)^\perp$, and any $\mathfrak{w}_{\mathbf{m}}^\tau \in (W_{\mathbf{0}}^\tau)^\perp$ can be written in the form

$$\mathfrak{w}_{\mathbf{m}}^\tau = c_{\mathbf{m}} w_0 + \sum_{i=1}^3 \frac{|\epsilon_i^\tau| m_i}{c_i^\tau} w_i \tag{5.5*.32a}$$

where $c_i^\tau = (1/|\epsilon_i^\tau|) \int_{\epsilon_i^\tau} \psi_{\epsilon_i^\tau}$. It follows that

$$\int_{\epsilon_i^\tau} \mathfrak{w}_{\mathbf{m}} = \frac{|\epsilon_i^\tau| m_i}{c_i^\tau} \int_{\epsilon_i^\tau} \psi_{\epsilon_i^\tau} = m_i. \tag{5.5*.32b}$$

The constant $c_{\mathbf{m}}$ is chosen to be such that

$$\int_\tau \mathfrak{w}_{\mathbf{m}}^\tau = |\tau|^2 m_0 \tag{5.5*.32c}$$

namely,

$$c_{\mathbf{m}} = \frac{1}{\int_\tau |\nabla w_0|^2} \left(|\tau| m_0 - \sum_{i=1}^3 \frac{|\epsilon_i^\tau| m_i}{c_i^\tau} \int_\tau w_i \right) \tag{5.5*.32d}$$

where we have used that $\int_\tau |\nabla w_0|^2 = \int_\tau w_0$.

Hence, for a given choice of edge functions $\psi_{\epsilon_1^\tau}$, $\psi_{\epsilon_2^\tau}$, $\psi_{\epsilon_3^\tau}$, the corresponding function $\mathfrak{w}_{\mathbf{m}} = \mathfrak{w}_{\mathbf{m}}(\psi_{\epsilon_1^\tau}, \psi_{\epsilon_2^\tau}, \psi_{\epsilon_3^\tau})$ can be constructed by employing (5.5*.32). The corresponding constant $\mathscr{C}'_\tau(\psi_{\epsilon_1^\tau}, \psi_{\epsilon_2^\tau}, \psi_{\epsilon_3^\tau})$ can then be determined from

$$\mathscr{C}'_\tau(\psi_{\epsilon_1^\tau}, \psi_{\epsilon_2^\tau}, \psi_{\epsilon_3^\tau}) = \sqrt{\max_{\substack{\mathbf{m}\in\mathbb{R}^4 \\ \mathbf{m}\neq \mathbf{0}}} \frac{\mathbf{r}^{\mathrm{T}}\tilde{\mathbf{D}}^2\mathbf{r}}{\mathbf{r}^{\mathrm{T}}\mathbf{D}^2\mathbf{r}}} \tag{5.5*.33a}$$

where $\tilde{\mathbf{D}}^2 = \tilde{\mathbf{D}}^2(\psi_{\epsilon_1^\tau}, \psi_{\epsilon_2^\tau}, \psi_{\epsilon_3^\tau})$ is the matrix corresponding to the quadratic form

$$\mathbf{m}^{\mathrm{T}}\tilde{\mathbf{D}}^2\mathbf{m} = \int_\tau |\boldsymbol{\nabla}\mathfrak{w}^\tau_{\mathbf{m}}|^2 \tag{5.5*.33b}$$

which can be written as

$$\mathbf{m}^{\mathrm{T}}\tilde{\mathbf{D}}^2\mathbf{m} = \int_\tau \left|\sum_{i=1}^{3} \frac{|\epsilon_i^\tau| m_i}{c_i^\tau} \boldsymbol{\nabla} w_i\right|^2 + \frac{1}{\int_\tau |\boldsymbol{\nabla} w_0|^2}\left(|\tau|^2 m_0 - \sum_{i=1}^{3} \frac{|\epsilon_i^\tau| m_i}{c_i^\tau}\int_\tau w_i\right)^2. \tag{5.5*.33c}$$

To eliminate the dependence on the choice of $\psi_{\epsilon_i^\tau}$, $i = 1, 2, 3$, we will employ

$$\mathscr{C}'_\tau = \max_{\substack{\psi_{\epsilon_i^\tau}, i=1,2,3 \\ \psi_{\epsilon_i^\tau}\in H_0^1(\epsilon_i^\tau)}} \mathscr{C}'_\tau(\psi_{\epsilon_1^\tau}, \psi_{\epsilon_2^\tau}, \psi_{\epsilon_3^\tau}). \tag{5.5*.34a}$$

[Babuška *et al.*, 1992a] shows that this maximum is practically achieved by the quadratic edge bubbles. Further, the dependence on the geometry may be simplified by letting

$$\mathscr{C}'(\alpha) \stackrel{\text{def}}{=} \max_{\alpha_1^\tau, \alpha_2^\tau, \alpha_3^\tau \le \alpha} \mathscr{C}(\alpha_1^\tau, \alpha_2^\tau, \alpha_3^\tau). \tag{5.5*.34b}$$

The computations of [Babuška *et al.*, 1992a] show that

$$\frac{3.45}{\sqrt{\sin(\alpha/2)}} \le (\mathscr{C}'(\alpha))^{-1} \le \frac{5.85}{\sqrt{\sin(\alpha/2)}}. \tag{5.5*.34c}$$

See also Table 5.5*.1.

Using the above results, we obtain for $k = 7$

$$\frac{1}{\mathscr{C}(p,k)} = 0.591\sqrt{\sin\left(\frac{\alpha}{2}\right)} \le \kappa^{\mathrm{EXPL}}_{S_\Delta^{p+k}} = \frac{\mathscr{E}^{\mathrm{EXPL}}_{S_\Delta^{p+k}}}{\left\|e^{S_\Delta^{p+k}}_{S_\Delta^p}\right\|_{\mathfrak{U}}} \le \frac{1}{\mathscr{C}'} = \frac{5.85}{\sqrt{\sin(\alpha/2)}}. \tag{5.5*.35a}$$

Let us now compare this result with the asymptotic bounds for the criss-cross mesh pattern

$$2\sqrt{6}\sqrt{\frac{a}{1+a^2}} \le \kappa^{\mathrm{EXPL}}_{S_\Delta^{p+1}} \le \sqrt{6}\sqrt{\frac{1+a^2}{a}} \tag{5.5*.35b}$$

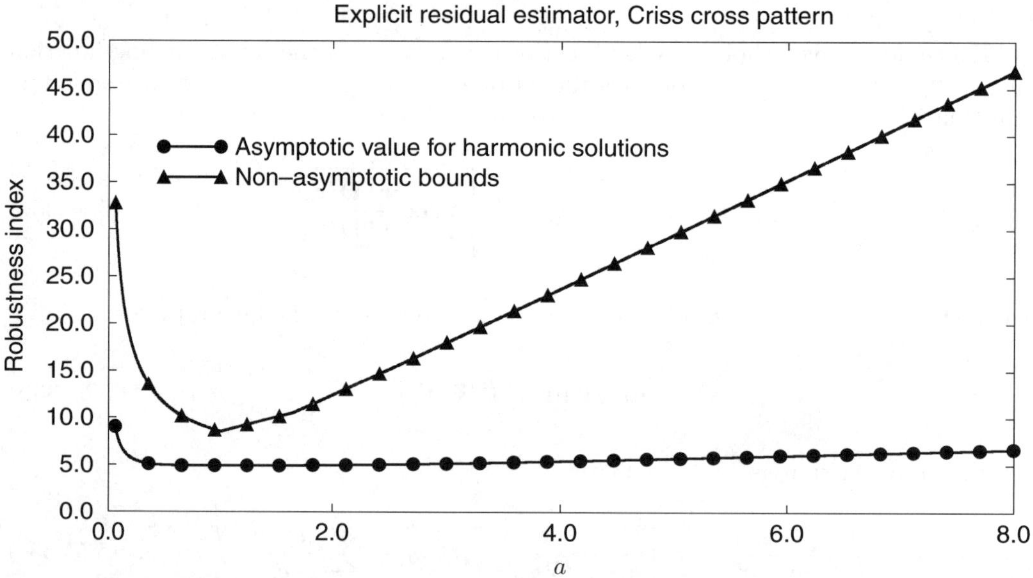

FIG. 5.5*.1. Comparison of the bounds resulting from the asymptotic analysis explained in Section 5.3*, and the analysis of [Babuška *et al.*, 1992a] for the explicit residual estimator. Input data: Laplace equation. Mesh data: criss-cross mesh pattern, linear elements ($p = 1$). Here we employed the constants in Table 5.5*.1 for $k = 7$. Note that in both analyses the value of the robustness index diverges to infinity as either a or $1/a$ tends to zero. However the robustness index obtained from the asymptotic analysis is almost constant in the practical range of the aspect ratio $1/a$

for the class of harmonic solutions, where a is the aspect ratio, which were obtained in Example 5.3*.2. A comparison of the bounds can be made by determining the *robustness index*

$$\mathcal{R}(a) \stackrel{\text{def}}{=} \max\left(\mathscr{C}_U(a) - 1, \frac{1}{\mathscr{C}_L(a)} - 1\right) \tag{5.5*.36}$$

for each case. Figure 5.5*.1 gives the graphs of $\mathcal{R}$ for $k = 7$. It can be seen that the analysis employed in this section is rather pessimistic for the asymptotic case. Nevertheless, it is important because it shows that the explicit estimator is less effective for small α.

Remark 5.5*.1 The above results are extremely pessimistic and have mainly theoretical value. The main corollary from these results is that the effectivity of the explicit estimator cannot be improved by multiplying it by an 'optimal' scaling constant, especially for meshes with high aspect ratios.

Remark 5.5*.2 The above analysis can be extended to meshes of quadrilaterals. In this case, however, two parameters are needed to characterize the mesh.

Remark 5.5*.3 [Babuška *et al.*, 1992] gives analysis of the explicit estimator for the elasticity problem, and showed that the dependence of the constants on the angle α is more pronounced than in the case of the Poisson equation.

Remark 5.5*.4 [Carstensen and Funken, 2000a] analyzes the explicit estimator

$$\mathcal{E}^{\mathrm{EXPL}} = \mathcal{C}_1 \sqrt{\sum_{\tau\in\Delta} h_\tau \|r_\tau^\Delta\|_{L^2(\tau)}^2} + \mathcal{C}_2 \sqrt{\sum_{\epsilon\in\mathcal{E}(\tau)} h_\epsilon \|J_\epsilon\|_{L^2(\epsilon)}^2} \tag{5.5*.37}$$

and determined the constants $\mathcal{C}_1$, $\mathcal{C}_2$, for which $\mathcal{E}^{\mathrm{EXPL}}$ is an upper estimator of $\|e_{S^p_\Delta}\|_{\mathcal{U}}$. For the uniform mesh of triangles arranged in the regular pattern with minimal angle $\pi/4$, it was found that $\mathcal{C}_1 = 1.18$ and $\mathcal{C}_2 = 12.71$. These constants were used to estimate the error in the uniform mesh finite element solution of the Laplacian in an L-shaped domain, and it was found that the effectivity index was $\kappa \approx 70$. It was also found that for the same example the effectivity of the weighted explicit estimator introduced in Section 5.4* was about $\kappa \approx 2.5$, and hence much better than the effectivity index $\kappa \approx 70$ of the original explicit estimator (5.5*.37); see [Carstensen, 2000], and [Carstensen and Funken, 1999]. Let us also note that [Carstensen and Funken, 2000b] showed that better values for $\mathcal{C}_1$ and $\mathcal{C}_2$ may be obtained by utilizing the relation between the subdomain residual and the explicit estimator.

Remark 5.5*.5 The ideas presented in this section are based on the *saturation* assumption, namely, that

$$\|e_{S^{p+k}_\Delta}\|_{\mathcal{U}} \le \eta \|e_{S^p_\Delta}\|_{\mathcal{U}} \tag{5.5*.38}$$

with $\eta < 1$, which means that the exact solution can be replaced by a higher order finite element solution. Using this assumption with *known* saturation coefficient η, it is possible to construct guaranteed error estimates; see e.g. [Bornemann *et al.*, 1996]. Note, however, that the value of η is practically never available, and hence such estimates have mostly academic value.

In this section we analyzed the behavior of the explicit estimator for the triangular elements with emphasis on the effects of the minimal angle condition, and we showed that the effectivity of the estimator deteriorates with decreasing angle.

5.6 A-posteriori error estimates based on local averaging

In Section 4.8 we introduced various recoveries $\mathcal{A}^*_{\mathcal{F}(u)}$ which construct a superconvergent approximation to the solution quantity $\mathcal{F}(u_{EX})$ by employing a local averaging of the computed finite element solution $u_{S^p_\Delta}$ or its derivatives $(d^k/dx^k)(u_{S^p_\Delta})$. Let us, for example, mention the recovery $\mathcal{A}^{*,ls}_{du/dx}$ which constructs a superconvergent approximation of $(d/dx)(u_{EX})$ by employing a least-squares fit of the values of $(d/dx)(u_{S^p_\Delta})$ at the mapped Gauss–Legendre points of order p in the element I_j^Δ and its neighbors.

Let $\mathcal{A}^*_{du/dx}$ be one of the operators for the superconvergent recovery of the derivative defined in Section 4.8. Then, under the Assumptions I–III of Section 4.7, we have

$$\left\| \frac{d}{dx}(u_{EX}) - \mathcal{A}^*_{du/dx}(u_{S^p_\Delta}) \right\|_{L^\infty(I_j^\Delta)} \approx \mathcal{C} h^{p+1} \ll \left\| \frac{d}{dx}(e_{S^p_\Delta}) \right\|_{L^\infty(I_j^\Delta)} \approx \mathcal{C} h^p \tag{5.6.1a}$$

and hence

$$\frac{d}{dx}(e_{S^p_\Delta}) = \frac{d}{dx}(u_{EX}) - \frac{d}{dx}(u_{S^p_\Delta}) \approx \mathcal{A}^*_{du/dx}(u_{S^p_\Delta}) - \frac{d}{dx}(u_{S^p_\Delta}). \tag{5.6.1b}$$

This observation motivates the introduction of the error indicator function

$$\hat{e}^{\mathcal{A}^*_{du/dx}} \stackrel{\text{def}}{=} \mathcal{A}^*_{du/dx}(u_{S^p_\Delta}) - \frac{d}{dx}(u_{S^p_\Delta}) \tag{5.6.2}$$

with the corresponding element error indicator $\eta_{I_j^\Delta}^{\mathcal{A}^*_{du/dx}}$ and estimator $\mathscr{E}^{\mathcal{A}^*_{du/dx}}$ given by

$$\eta_{I_j^\Delta}^{\mathcal{A}^*_{du/dx}} = \|\sqrt{AE}\hat{e}^{\mathcal{A}^*_{du/dx}}\|_{L^2(I_j^\Delta)}, \qquad \mathscr{E}^{\mathcal{A}^*_{du/dx}} = \sqrt{\sum_{j=1}^{N}(\eta_{I_j^\Delta}^{\mathcal{A}^*_{du/dx}})^2}. \tag{5.6.3}$$

Remark 5.6.1 The indicator $\eta_{I_j^\Delta}^{\mathcal{A}^*_{du/dx}}$ and the estimator $\mathscr{E}^{\mathcal{A}^*_{du/dx}}$ follow from the observation that

$$\|e_{S^p_\Delta}\|_{\mathscr{U}(I_j^\Delta)} = \sqrt{\left\|\sqrt{AE}\frac{d}{dx}(e_{S^p_\Delta})\right\|^2_{L^2(I_j^\Delta)} + \|\sqrt{c}e_{S^p_\Delta}\|^2_{L^2(I_j^\Delta)}} \tag{5.6.4}$$

and that the zeroth-order term $\|\sqrt{c}e_{S^p_\Delta}\|_{L^2(I_j^\Delta)}$ is negligible in comparison with the derivative term $\|\sqrt{AE}(d/dx)(e_{S^p_\Delta})\|_{L^2(I_j^\Delta)}$, asymptotically.

Another approach is to employ $\mathcal{A}^*_u$, a superconvergent recovery for the value, and to define the indicator

$$\hat{e}^{\mathcal{A}^*_u} \stackrel{\text{def}}{=} \mathcal{A}^*_u(u_{S^p_\Delta}) - u_{S^p_\Delta}. \tag{5.6.5a}$$

Then $(d/dx)(\mathcal{A}^*_u(\cdot))$ is a superconvergent recovery for the derivative, and hence

$$\frac{d}{dx}(e_{S^p_\Delta}) \approx \frac{d}{dx}(\hat{e}^{\mathcal{A}^*_u}) \tag{5.6.5b}$$

and we are led to the indicator

$$\eta_{I_j^\Delta}^{\mathcal{A}^*_u} = \sqrt{\left\|\sqrt{AE}\frac{d}{dx}(\hat{e}^{\mathcal{A}^*_u})\right\|^2_{L^2(I_j^\Delta)} + \|\sqrt{c}\hat{e}^{\mathcal{A}^*_u}\|^2_{L^2(I_j^\Delta)}} \tag{5.6.5c}$$

where the second term may be omitted, asymptotically.

Remark 5.6.2 The element residual estimators may also be understood as recovery estimators by letting

$$\mathcal{A}_u^{EST}(u_{S^p_\Delta}) \stackrel{\text{def}}{=} u_{S^p_\Delta} + \hat{e}^{EST} \tag{5.6.6}$$

where $EST =$ Dir, or Neum, and $\hat{e}^{EST}$ is defined elementwise.

Let us now give an example.

Example 5.6.1 We constructed the error indicator functions $\hat{e}^{\mathcal{A}^*_{du/dx}}$ and $(d/dx)(\hat{e}^{\mathcal{A}^*_u})$, for the one-dimensional example problem employed in the previous sections of this chapter. Figure 5.6.1a compares the error in these indicators. Figure 5.6.1b compares

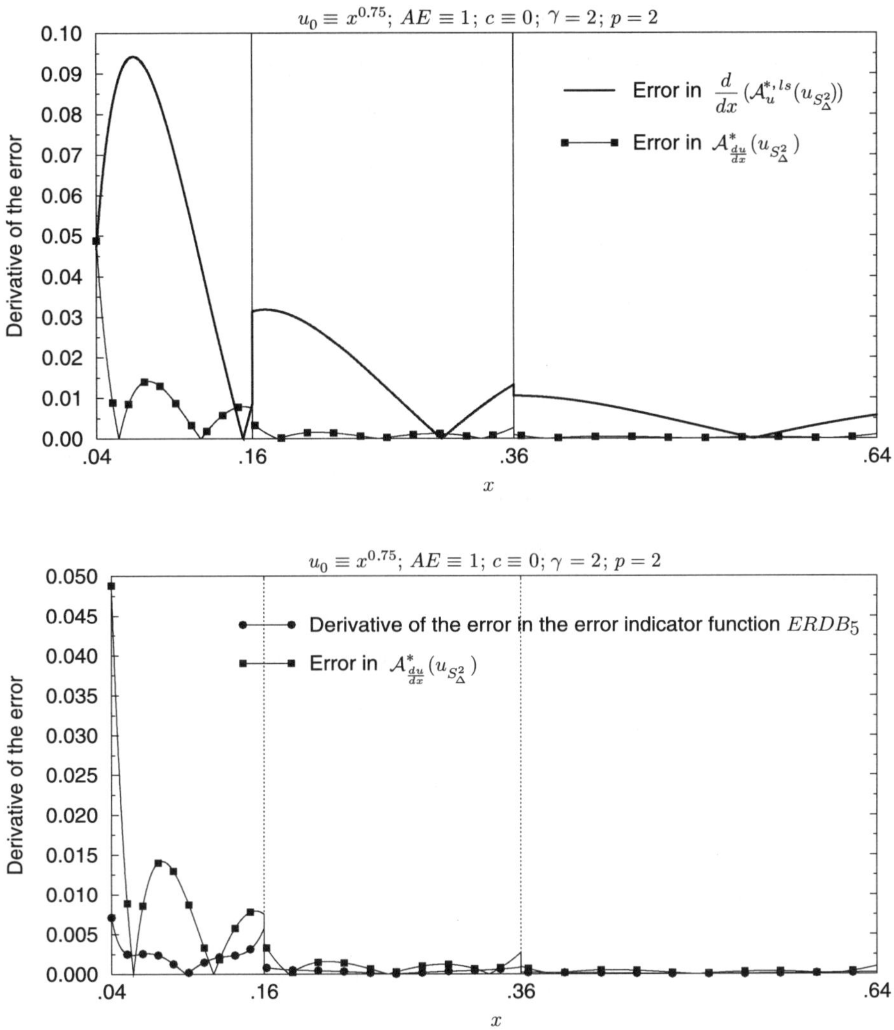

FIG. 5.6.1. Element error indicator functions based on superconvergent recovery. Input data: $AE \equiv 1$, $c \equiv 0$, $u_{EX} \equiv x^{3/4}$. Mesh data: Radical mesh with $\gamma = 2$, $N(\Delta) = 5$, quadratic elements ($p = 2$). (a) Graphs of the absolute values of the errors in the error indicator function $|(d/dx)(e_{S^2_\Delta}) - \hat{e}^{\mathcal{A}^*_{du/dx}}|$ for $\hat{e}^{\mathcal{A}^*_{du/dx}}$ where $\mathcal{A}^*_{du/dx} = \frac{1}{2}(\mathcal{A}^{*,int,L}_{du/dx} + \mathcal{A}^{*,int,R}_{du/dx})$, and of $|(d/dx)(e_{S^2_\Delta}) - (d/dx)(\hat{e}^{\mathcal{A}^*_{du/dx}})|$, where $\mathcal{A}^*_u = \mathcal{A}^{*,ls}_u$. (b) Graphs of the absolute values of the errors in the error indicator function $|(d/dx)(e_{S^2_\Delta}) - \hat{e}^{\mathcal{A}^*_{du/dx}}|$ where $\mathcal{A}^*_{du/dx}$ was given above and $|(d/dx)(e_{S^2_\Delta}) - (d/dx)(\pi_{B^{2+5}}(\hat{e}^{\text{Dir}}_j))|$

the error in $\hat{e}^{\mathcal{A}^*_{du/dx}}$ with the error in $(d/dx)(\pi_{B^{2+5}}(\hat{e}^{\text{Dir}}_{I^\Delta_j}))$. It can be seen that the accuracy of the indicators $\hat{e}^{\mathcal{A}^*_{du/dx}}$ and $(d/dx)(\hat{e}^{\mathcal{A}^*_u})$ is inferior to the accuracy of the computed element residual indicator $(d/dx)(\pi_{B^{2+5}}(\hat{e}^{\text{Dir}}_{I^\Delta_j}))$ in the first element which is adjacent to the singular point for the solution at $x = 0$. In the rest of the mesh, where the solution is smooth, the accuracy of the indicators $\hat{e}^{\mathcal{A}^*_{du/dx}}$ and $(d/dx)(\pi_{B^{2+5}}(\hat{e}^{\text{Dir}}_{I^\Delta_j}))$ is comparable and superior to the accuracy of the indicator $(d/dx)(\hat{e}^{\mathcal{A}^*_u})$.

Let us now state the main result about the error indicator $\hat{e}^{\mathcal{A}^*_{du/dx}}$.

Theorem 5.6.1 *Under Assumptions I–III of Theorem 4.6.3, we have*

$$\|e_{S^p_\Delta}\|_{\mathcal{U}(I^\Delta_j)} = (1 + \mathcal{C}\bar{h}_{I^\Delta_j}\theta)\eta^{\mathcal{A}^*_{du/dx}}_{I^\Delta_j} \tag{5.6.7}$$

where $\bar{h}_{I^\Delta_j} = \max_{k=j-1,j,j+1} h_{I^\Delta_k}$, $-1 \le \theta \le 1$, *and* $\mathcal{C}$ *depends on the ratios* $h_{I^\Delta_j}/h_{I^\Delta_{j-1}}$ *and* $h_{I^\Delta_j}/h_{I^\Delta_{j+1}}$.

Proof From Theorem 4.7.1 it follows that

$$\left\|\frac{d}{dx}(u_{EX}) - \mathcal{A}^*_{du/dx}(u_{S^p_\Delta})\right\|_{L^\infty(I^\Delta_j)} \le \mathcal{C}\bar{h}_{I^\Delta_j}\left\|\frac{d}{dx}(e_{S^p_\Delta})\right\|_{L^\infty(\omega^{(1)}_{I^\Delta_j})}. \tag{5.6.8a}$$

Noting that

$$\frac{d}{dx}(e_{S^p_\Delta}) - \hat{e}^{\mathcal{A}^*_{du/dx}} = \frac{d}{dx}(u_{EX}) - \mathcal{A}^*_{du/dx}(u_{S^p_\Delta}) \tag{5.6.8b}$$

(5.6.8a) can be rewritten as

$$\left\|\frac{d}{dx}(e_{S^p_\Delta}) - \hat{e}^{\mathcal{A}^*_{du/dx}}\right\|_{L^\infty(I^\Delta_j)} \le \mathcal{C}\bar{h}_{I^\Delta_j}\left\|\frac{d}{dx}(e_{S^p_\Delta})\right\|_{L^\infty(\omega^{(1)}_{I^\Delta_j})} \tag{5.6.8c}$$

from which it follows that

$$\frac{d}{dx}(e_{S^p_\Delta}) = (1 + \mathcal{C}\bar{h}_{I^\Delta_j}\theta)\hat{e}^{\mathcal{A}^*_{du/dx}} \tag{5.6.8d}$$

and hence

$$\eta^{\mathcal{A}^*_{du/dx}}_{I^\Delta_j} = \|\sqrt{AE}\,\hat{e}^{\mathcal{A}^*_{du/dx}}\|_{L^2(I^\Delta_j)} = \frac{1}{1 + \mathcal{C}\bar{h}_{I^\Delta_j}\theta}\left\|\sqrt{AE}\frac{d}{dx}(e_{S^p_\Delta})\right\|_{L^2(I^\Delta_j)}. \tag{5.6.8e}$$

Further, recalling that

$$\int_{I^\Delta_j} c(e_{S^p_\Delta})^2 \le (h_{I^\Delta_j})^2\frac{\gamma}{\beta_1}\left\|\sqrt{AE}\frac{d}{dx}(e_{S^p_\Delta})\right\|^2_{L^2(I^\Delta_j)} \tag{5.6.9a}$$

and employing (5.6.8e) we obtain

$$\int_{I^\Delta_j} c(e_{S^p_\Delta})^2 \le (h_{I^\Delta_j})^2\frac{\gamma}{\beta_1}(1 + \mathcal{C}h_{I^\Delta_j}\theta)\|\sqrt{AE}\,\hat{e}^{\mathcal{A}^*_{du/dx}}\|_{L^2(I^\Delta_j)} \tag{5.6.9b}$$

and employing that

$$\|e_{S^p_\Delta}\|_{\mathcal{U}(I^\Delta_j)} = \sqrt{\left\|\sqrt{AE}\frac{d}{dx}(e_{S^p_\Delta})\right\|_{L^2(I^\Delta_j)} + \|\sqrt{c}\, e_{S^p_\Delta}\|_{L^2(I^\Delta_j)}} \tag{5.6.9c}$$

we obtain

$$\|e_{S^p_\Delta}\|_{\mathcal{U}(I^\Delta_j)} \approx \sqrt{1 + (h_{I^\Delta_j})^2 \frac{\gamma}{\beta_1}} \left\|\sqrt{AE}\frac{d}{dx}(e_{S^p_\Delta})\right\|_{L^2(I^\Delta_j)} \tag{5.6.9d}$$

and hence

$$\|e_{S^p_\Delta}\|_{\mathcal{U}(I^\Delta_j)} \approx \sqrt{(1 + \mathscr{C}\bar{h}_{I^\Delta_j}\theta)\left(1 + (h_{I^\Delta_j})^2 \frac{\gamma}{\beta_1}\right)}\, \eta^{\mathcal{A}^*_{du/dx}}_{I^\Delta_j} \tag{5.6.9e}$$

and (5.6.7) follows. □

Remark 5.6.3 Under the assumption that the pollution in I^Δ_j is negligible, we have

$$\left\|\frac{d}{dx}(e_{S^p_\Delta})\right\|_{L^\infty(I^\Delta_j)} \approx \left\|\mathcal{A}^*_{du/dx}(u_{S^p_\Delta}) - \frac{d}{dx}(u_{S^p_\Delta})\right\|_{L^\infty(I^\Delta_j)}. \tag{5.6.10}$$

Remark 5.6.4 Another possibility is to employ the element error indicators

$$\eta^{\mathcal{A}^*_{du/dx}/\mathcal{A}^*_u}_{I^\Delta_j} = \sqrt{\|\sqrt{AE}\,\hat{e}^{\mathcal{A}^*_{du/dx}}\|^2_{L^2(I^\Delta_j)} + \|\sqrt{c}\,\hat{e}^{\mathcal{A}^*_u}\|^2_{L^2(I^\Delta_j)}}. \tag{5.6.11}$$

Remark 5.6.5 We can use any continuous recovery for the value $\mathcal{A}^*_u$ to construct a corresponding guaranteed lower estimate

$$\|e_{S^p_\Delta}\|_{\mathcal{U}} \geq \frac{\mathscr{R}\left(\mathcal{A}^{\mathrm{Dir}}_u(u_{S^p_\Delta})\right)}{\left\|\mathcal{A}^{\mathrm{Dir}}_u(u_{S^p_\Delta})\right\|_{\mathcal{U}}}. \tag{5.6.12}$$

The accuracy of this bound depends on the accuracy of the employed recovery $\mathcal{A}^*_u$.

Remark 5.6.6 The estimator based on the least-squares recovery of the derivatives was first introduced by [Zienkiewicz and Zhu, 1987] and [Zienkiewicz and Zhu, 1992a,b], and is known as the ZZ-SPR estimator.

Remark 5.6.7 The estimate $\mathscr{E}^{\mathcal{A}^*_{du/dx}}$ can be corrected to obtain upper and lower bounds for $\|e_{S^p_\Delta}\|_{\mathcal{U}}$. This will be addressed in Chapter 6.

Remark 5.6.8 Various possibilities of optimal averaging were addressed in [Babuška and Rodriguez, 1993].

In this section we outlined the construction of error estimators based on local least-square recoveries, and we have proven that for superconvergent recoveries and in the cases of negligible pollution the element error indicators are asymptotically exact.

5.6* A-posteriori estimates based on local averaging: the two-dimensional case

As in the one-dimensional case we will let

$$\eta_\tau^{\mathcal{A}^*_{\mathcal{F}(u)}} = \left\|\mathcal{A}^*_{\mathcal{F}(u)}(u_{S^p_\Delta}) - \mathcal{F}(u_{S^p_\Delta})\right\|_{*,\tau}, \qquad \mathscr{E}^{\mathcal{A}^*_{\mathcal{F}(u)}} = \sqrt{\sum_{\tau\in\Delta}(\eta_\tau^{\mathcal{A}^*_{\mathcal{F}(u)}})^2} \tag{5.6*.1}$$

where $\|\cdot\|_{*,\tau}$ is a norm such that $\eta_\tau^{\mathcal{A}^*_{\mathcal{F}(u)}}$ has the meaning of error indicator for $\|e_{S^p_\Delta}\|_{\mathscr{U}(\tau)}$. For example, in the case of the Poisson equation we may employ a recovery of the gradient $\mathcal{A}^*_{\nabla u}$, and we have

$$\eta_\tau^{\mathcal{A}^*_{\nabla u}} \stackrel{\text{def}}{=} \left\|\mathcal{A}^*_{\nabla u}(u_{S^p_\Delta}) - \nabla u_{S^p_\Delta}\right\|_{L^2(\tau)} \tag{5.6*.2a}$$

and hence $\|\cdot\|_{*,\tau}$, in this case is the L^2 norm for vector valued functions, or we may employ a recovery of the value, $\mathcal{A}^*_u$, and in this case we have

$$\eta_\tau^{\mathcal{A}^*_u} \stackrel{\text{def}}{=} \left\|\mathcal{A}^*_u(u_{S^p_\Delta}) - u_{S^p_\Delta}\right\|_{H^1(\tau)} = \left\|\nabla(\mathcal{A}^*_u(u_{S^p_\Delta})) - \nabla u_{S^p_\Delta}\right\|_{L^2(\tau)} \tag{5.6*.2b}$$

and in this case $\|\cdot\|_{*,\tau}$ is the H^1-seminorm $\|\cdot\|_{H^1(\tau)}$.

As we have already discussed in Section 4.8*, the major difference between the one and two-dimensional cases is that, with the exception of special meshes, in the two-dimensional case there are no superconvergent recoveries, and even in the case of negligible pollution, the indicators $\eta_\tau^{\mathcal{A}^*_{\mathcal{F}(u)}}$ are only asymptotically equivalent to the energy norm of the error. We have:

Theorem 5.6*.1 *Under Assumptions 0, I, and II of Section 4.7*, we have*

$$\mathscr{C}_L(1+\mathscr{C}h_\tau^{\bar{\mu}}) \le \kappa_\tau^{\mathcal{A}^*_{\mathcal{F}(u)}} \stackrel{\text{def}}{=} \frac{\eta_\tau^{\mathcal{A}^*_{\mathcal{F}(u)}}}{\|e_{S^p_\Delta}\|_{\mathscr{U}(\tau)}} \le \mathscr{C}_U(1+\mathscr{C}h_\tau^{\bar{\mu}}) \tag{5.6*.3}$$

where τ is an element in the interior of the mesh-patch $S(\bar{x}, H)$, as described in n Section 4.7, $0 < \bar{\mu} < 1$, and $\mathscr{C}_L$, $\mathscr{C}_U$ depend on the employed recovery $\mathcal{A}^*_{\mathcal{F}(u)}$ and on the definition of the indicators.* □

This result also holds for the error indicators given in the previous sections.

Remark 5.6*.1 In Chapter 6 we will show how to construct guaranteed upper and lower estimates of the error in $\mathcal{A}^*_{\mathcal{F}(u)}(u_{S^p_\Delta})$.

Let us now illustrate the accuracy of the indicators based on local least-squares recovery in a concrete case.

Example 5.6*.1 Let us consider the Neumann problem for the Laplace equation, the domain and the meshes that were employed in Example 5.2*.1, and let us report the element effectivity indices for the ZZ-SPR element error indicators with and without boundary patches. In Fig. 5.6*.1 (respectively Fig. 5.6*.2) we give the element effectivity

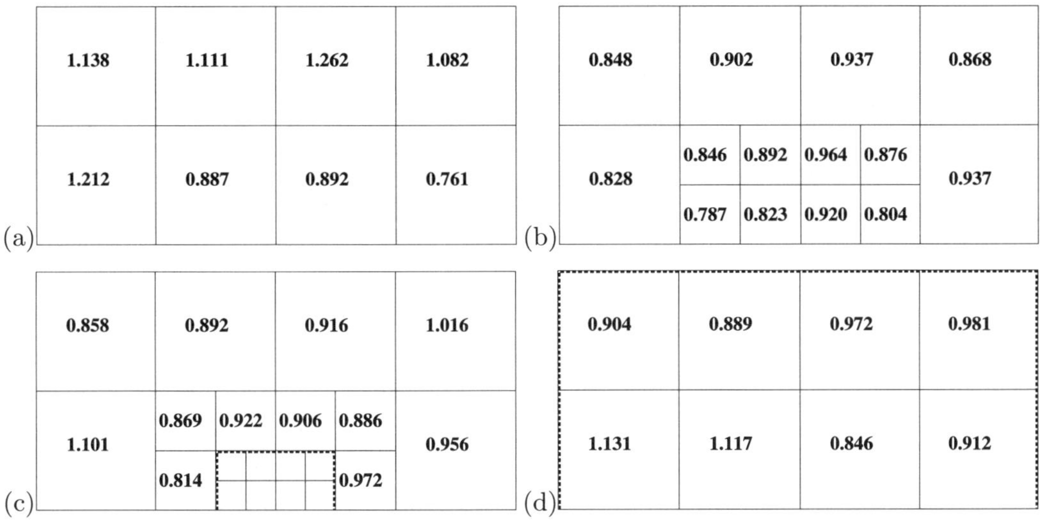

FIG. 5.6*.1. An example which illustrates the accuracy of the element error indicators. Input data: Laplace's equation in $\Omega = (-1, 1) \times (0, 1)$, Neumann boundary conditions, $u_{EX} \equiv \sqrt{r}\sin(\theta/2)$ for the polar coordinate system shown in Fig. 5.2*.1. Mesh data: Mesh of biquadratic squares with one to two irregular connections. Element effectivity indices for the ZZ-SPR element error indicators, with the boundary patches included in the recovery for the mesh of: (a) 8 elements (global effectivity index $\kappa = 0.890$); (b) 14 elements ($\kappa = 0.896$); (c) 20 elements ($\kappa = 0.906$); (d) the 8-element patch in the subdomain $(-\frac{1}{4}, \frac{1}{4}) \times (0, \frac{1}{4})$ in the mesh of 20 elements shown in (c)

indices for the ZZ-SPR with boundary patches (respectively the ZZ-SPR without boundary patches) for the example. For the precise definition of the ZZ-SPR with, or without boundary patches, see [Zienkiewicz and Zhu, 1994], and [Babuška, Strouboulis *et al.*, 1997]. From Figs. 5.6*.1 and 5.6*.2 it can be seen that the ZZ-SPR has very reasonable effectivity index both elemental and global index, but it can underestimate the error for biquadratic elements by 20–30% in this example. We will discuss more the effectivity of this estimator in the next section.

Remark 5.6*.2 Estimates of the types addressed in this section are very popular because of their simplicity, low cost, and good performance. They were introduced in [Zienkiewicz and Zhu, 1992a,b,c].

Remark 5.6*.3 For an interesting analysis of a-posteriori estimation based on averaging, we recommend [Carstensen and Bartels, 1999].

In this section we have outlined the construction of error estimators based on least-squares recovery in two dimensions.

5.7 The effectivity of the error indicators: the principles for their comparison and the detailed analysis of their quality

In the previous sections, we introduced two families of estimators, namely the residual estimators and the estimators based on local averaging. In each case, we determined

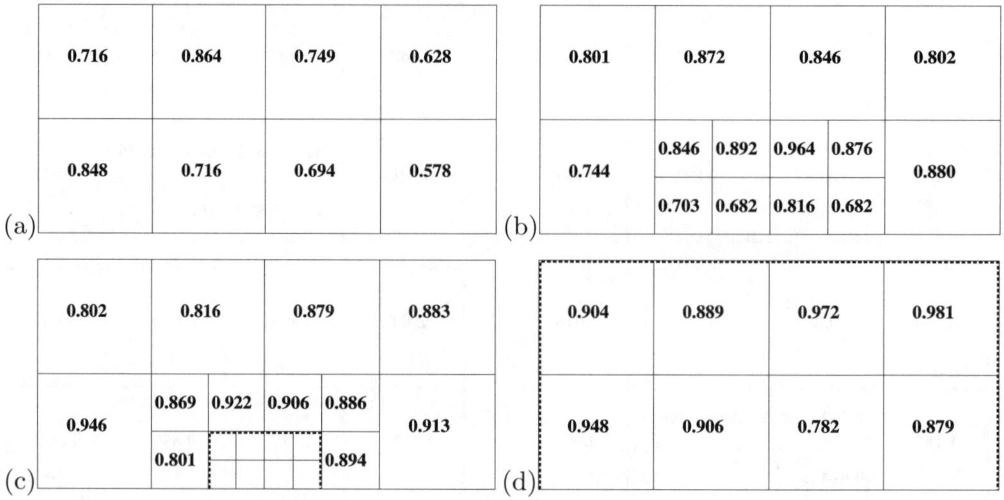

FIG. 5.6*.2. An example which illustrates the accuracy of the element error indicators. Input data: Laplace's equation in $\Omega = (-1,1) \times (0,1)$, Neumann boundary conditions, $u_{EX} \equiv \sqrt{r}\sin(\theta/2)$ for the polar coordinate system shown in Fig. 5.2*.1. Mesh data: Mesh of biquadratic squares with one to two irregular connections. Element effectivity indices for the ZZ-SPR element error indicators with extrapolation from the interior patches for the mesh of: (a) 8 elements (global effectivity index $\kappa = 0.703$); (b) 14 elements ($\kappa = 0.714$); (c) 20 elements ($\kappa = 0.827$); (d) the 8-element patch in the subdomain $(-\frac{1}{4}, \frac{1}{4}) \times (0, \frac{1}{4})$ in the mesh of 20 elements shown in (c)

element error indicators $\eta_{I_j^\Delta}^{EST}$, and the estimator $\mathscr{E}^{EST}$, in the form

$$\eta_{I_j^\Delta}^{EST} \stackrel{\text{def}}{=} \|\hat{u}_{I_j^\Delta}^{EST} - u_{S_\Delta^p}\|_{\mathscr{U}(I_j^\Delta)}, \qquad \mathscr{E}^{EST} = \sqrt{\sum_{j=1}^{N(\Delta)} \left(\eta_{I_j^\Delta}^{EST}\right)^2} \tag{5.7.1a}$$

where the more general definition

$$\eta_{I_j^\Delta}^{EST} = \left\|\mathcal{A}_{\mathcal{F}(u)}^*(u_{S_\Delta^p}) - \mathcal{F}(u_{S_\Delta^p})\right\|_{*,I_j^\Delta} \tag{5.7.1b}$$

is needed to include various indicators based on recovery. The accuracy of $\mathscr{E}^{EST}$ is measured by employing the *global effectivity index*

$$\kappa^{EST} \stackrel{\text{def}}{=} \frac{\mathscr{E}^{EST}}{\|e_{S_\Delta^p}\|_{\mathscr{U}}}. \tag{5.7.1c}$$

The element error indicators $\eta_{I_j^\Delta}^{EST}$, the estimator $\mathscr{E}^{EST}$, and the effectivity index κ^{EST} depend on the mesh Δ, the degree p of the elements, the model parameters AE, c, and the exact solution u_{EX}. To underline this, we will write

$$\eta_{I_j^\Delta}^{EST} = \eta_{I_j^\Delta}^{EST}(u_{EX}; (\Delta, p); (AE, c)) \tag{5.7.2a}$$

and similarly for $\mathscr{E}^{EST}$ and κ^{EST}. Equivalently we may employ the loads (f, a_2), instead of u_{EX}, in the arguments of $\eta_{I_j^\Delta}^{EST}$, namely,

$$\eta_{I_j^\Delta}^{EST} = \eta_{I_j^\Delta}^{EST}((\Delta, p); (AE, c); (f, a_2)) \tag{5.7.2b}$$

and similarly for $\mathscr{E}^{EST}$ and κ^{EST}. In the cases that *the pollution error in the element I_j^Δ is negligible*, it also makes sense to determine the effectivity index

$$\kappa_{I_j^\Delta}^{EST} = \frac{\eta_{I_j^\Delta}^{EST}}{\|e_{S_\Delta^p}\|_{\mathcal{U}(I_j^\Delta)}} \tag{5.7.3a}$$

for $\eta_{I_j^\Delta}^{EST}$. Recall that $\eta_{I_j^\Delta}^{EST}$ is based only on the local residuals and cannot account for the pollution in $\|e_{S_\Delta^p}\|_{\mathcal{U}(I_j^\Delta)}$. *This is the reason why $\eta_{I_j^\Delta}^{EST}$ is called an element error indicator and not an element error estimator.* The indicator $\eta_{I_j^\Delta}^{EST}$ can be understood as an estimator for the energy norm of the local error $\|e_{S_\Delta^p}^{loc}\|_{\mathcal{U}(I_j^\Delta)}$. Therefore the accuracy of $\eta_{I_j^\Delta}^{EST}$ can be measured by computing

$$\kappa_{I_j^\Delta}^{loc,IND} = \frac{\eta_{I_j^\Delta}^{EST}}{\left\|e_{S^p(I_j^\Delta)}^{loc}\right\|_{\mathcal{U}(I_j^\Delta)}} \tag{5.7.3b}$$

and it is desirable to have $\kappa_{I_j^\Delta}^{loc,IND} \approx 1$ in all elements I_j^Δ.

An estimator $\mathscr{E}^{EST}$ is called reliable if its effectivity index κ^{EST} is close to one for the exact solutions u_{EX}, the meshes Δ, and model parameters AE, c, of interest. The notion of the reliability is not quantitative because we did not specify how close to one the effectivity index has to be. Below we will quantify the performance characteristics of the estimator. In the special case that

$$\kappa^{EST}(u_{EX}; (\Delta, p); (AE, c)) \quad \to \quad 1 \quad \text{as } h \quad \to \quad 0 \tag{5.7.4}$$

the estimator $\mathscr{E}^{EST}$ is called *asymptotically exact.* Note that asymptotical exactness is a well-defined notion. In general, asymptotical exactness does not characterize the accuracy of an estimator because it occurs only for a rather restricted class of solutions and meshes which do not include the majority of solutions and meshes which occur in practical engineering problems in higher dimensions. The accuracy of an estimator $\mathscr{E}^{EST}$ can be measured by determining the range of κ^{EST} for the admissible classes of solutions, meshes and element degrees, and model parameters. More precisely:

Let $\mathcal{U}$, $\mathcal{M}$, $\mathcal{D}$ denote the admissible classes of solutions, meshes and element degrees and model parameters, respectively, and let

$$\underline{\kappa}^{EST}(\mathcal{U}; \mathcal{M}; \mathcal{D}) \stackrel{\text{def}}{=} \inf_{\substack{u_{EX} \in \mathcal{U} \\ (\Delta,p) \in \mathcal{M} \\ (AE,c) \in \mathcal{D}}} \kappa^{EST}(u_{EX}; (\Delta, p); (AE, c)) \tag{5.7.5a}$$

denote the greatest lower bound for κ^{EST}, and

$$\bar{\kappa}^{EST}(\mathcal{U};\mathcal{M};\mathcal{D}) \stackrel{\text{def}}{=} \sup_{\substack{u_{EX}\in\mathcal{U}\\(\Delta,p)\in\mathcal{M}\\(AE,c)\in\mathcal{D}}} \kappa^{EST}(u_{EX};(\Delta,p);(AE,c)) \tag{5.7.5b}$$

the least upper bound for κ for the classes $\mathcal{U}$, $\mathcal{M}$, $\mathcal{D}$, or

$$\underline{\kappa}^{EST}(\mathcal{M};\mathcal{D}) \stackrel{\text{def}}{=} \inf_{\substack{(\Delta,p)\in\mathcal{M}\\((AE,c),(f,a_2))\in\mathcal{D}}} \kappa^{EST}((\Delta,p);(AE,c);(f,a_2)) \tag{5.7.5c}$$

and analogously for $\bar{\kappa}^{EST}(\mathcal{M};\mathcal{D})$, where $\mathcal{D}$ is understood here as the class of the input data, the model parameters and the loads. The accuracy of the estimator $\mathscr{E}^{EST}$ for the classes $\mathcal{U}$, $\mathcal{M}$, $\mathcal{D}$, can then be measured in terms of the *robustness index*

$$\mathcal{R}^{EST}(\mathcal{U};\mathcal{M};\mathcal{D}) \stackrel{\text{def}}{=} \max\left(\bar{\kappa}^{EST}-1, \frac{1}{\underline{\kappa}^{EST}}-1\right). \tag{5.7.6}$$

A small value of the robustness index $\mathcal{R}^{EST}$ guarantees that $\mathscr{E}^{EST}$ is practically equal with $\|e_{S^p_\Delta}\|_{\mathscr{U}}$.

Remark 5.7.1 Note that the robustness index $\mathcal{R}(\mathcal{U};\mathcal{M};\mathcal{D})$ depends on the admissible sets $\mathcal{U}$, $\mathcal{M}$, $\mathcal{D}$, and always improves when smaller sets are considered. For the analysis to be useful, we must consider the largest possible sets of solutions $\mathcal{U}$ and model parameters $\mathcal{D}$ which can occur in the engineering problems of interest, and the class of meshes $\mathcal{M}$ which are constructed by the mesh generator employed in the computation. The accuracy of an estimator can be improved by restricting the class of admissible meshes to exclude the types of meshes for which the estimator performs poorly. However, *we have to allow for sufficiently general meshes in order to be able to discretize complex domains and to perform the computations as economically as possible.*

Remark 5.7.2 The robustness index $\mathcal{R}$ weighs equally the sub-intervals [0,1] and $[1,\infty)$ of the range $[0,\infty)$ of the effectivity index κ. For example, the cases $\underline{\kappa} = 1/1000$, $\bar{\kappa} = 1$ and $\underline{\kappa} = 1$, $\bar{\kappa} = 1000$ result in the same value of $\mathcal{R}$, as they should.

We will say that an estimator $\mathscr{E}$ is ϵ robust for the classes $\mathcal{U}$, $\mathcal{M}$, $\mathcal{D}$ if

$$\mathcal{R}(\mathcal{U},\mathcal{M},\mathcal{D}) \leq \epsilon. \tag{5.7.7}$$

In practice we would like to have at least $\epsilon \leq 0.5$.

We will say that $\mathscr{E}$ is $(\mathcal{U},\mathcal{M},\mathcal{D})$ *proper* if

$$0 < \underline{\kappa}(\mathcal{U};\mathcal{M};\mathcal{D}) \leq \bar{\kappa}(\mathcal{U};\mathcal{M};\mathcal{D}) < \infty \tag{5.7.8a}$$

or, equivalently, if

$$\mathcal{R}(\mathcal{U};\mathcal{M};\mathcal{D}) < \infty. \tag{5.7.8b}$$

For a family of meshes Δ_h, parameterized by its maximum element size h, we will say that $\mathscr{E}$ is $(\mathcal{U};\mathcal{M};\mathcal{D})$ *asymptotically exact* if

$$\lim_{h\to 0} \mathcal{R}(\mathcal{U};\mathcal{M};\mathcal{D}) = 0. \tag{5.7.9}$$

Let us summarize the results of the previous sections in the context of the above definitions.

Theorem 5.7.1 *Let* $\mathcal{U} \equiv \mathfrak{U}$, *and*

$$\mathcal{D} = \{(AE, c, f) \mid 0 < \beta_1 \le AE \le \beta_2,\ 0 \le c \le \gamma,\ f \in \mathfrak{U}^{-1}\}. \tag{5.7.10a}$$

Then

$$\mathscr{E}^{\mathrm{Dir}} \le \|e_{S^p_\Delta}\|_{\mathfrak{U}} \le \mathcal{K}^{\mathrm{Dir}}\,\mathscr{E}^{\mathrm{Dir}} \tag{5.7.10b}$$

and

$$\frac{1}{\mathcal{K}^{\mathrm{Dir}}} \le \kappa^{\mathrm{Dir}} \le 1 \tag{5.7.10c}$$

and for

$$\mathcal{D}' = \left\{(AE, c) \in \mathcal{D} \,\middle|\, \left\|\frac{d}{dx}(AE)\right\|_{L^\infty(I)} \le K,\ f \in L^2(I)\right\} \tag{5.7.10d}$$

we have

$$(1 - \mathcal{K}^{\mathrm{Neum}} h^\Delta)\mathscr{E}^{\mathrm{Neum}} \le \|e_{S^p_\Delta}\|_{\mathfrak{U}} \le \mathscr{E}^{\mathrm{Neum}} \tag{5.7.10e}$$

and

$$1 \le \kappa^{\mathrm{Neum}} \le \frac{1}{1 - \mathcal{K}^{\mathrm{Neum}} h^\Delta} \tag{5.7.10f}$$

where $\mathcal{K}^{\mathrm{Dir}}$ *depends only on* β_1, β_2, *and* γ, *and* $\mathcal{K}^{\mathrm{Neum}}$ *depends only on* β_1, β_2, γ, *and* K. *Further, we have*

$$\lim_{h^\Delta \to 0} \mathcal{R}^{EST}(\mathcal{U}; \mathcal{M}; \mathcal{D}') = 0 \tag{5.7.10g}$$

for both $EST = \mathrm{Dir}$ *and* $EST = \mathrm{Neum}$.

Proof From Theorem 5.2.2 we have

$$\mathscr{E}^{\mathrm{Dir}} \le \|e_{S^p_\Delta}\|_{\mathfrak{U}} \le \mathcal{K}^{\mathrm{Dir}}\,\mathscr{E}^{\mathrm{Dir}} = \frac{\sqrt{\beta_2}}{\sqrt{\beta_1}}\sqrt{1 + \frac{1}{2}\frac{\gamma}{\beta_2}(h^\Delta)^2}\,\mathscr{E}^{\mathrm{Dir}} \tag{5.7.11a}$$

and hence

$$\frac{1}{\mathcal{K}^{\mathrm{Dir}}} = \frac{\sqrt{\beta_1}}{\sqrt{\beta_2}}\frac{1}{\sqrt{1 + \frac{1}{2}(\gamma/\beta_2)l^2}} \le \kappa^{\mathrm{Dir}} = \frac{\mathscr{E}^{\mathrm{Dir}}}{\|e_{S^p_\Delta}\|_{\mathfrak{U}}} \le 1. \tag{5.7.11b}$$

Further, under the additional assumptions

$$\left\|\frac{d}{dx}(AE)\right\|_{L^\infty(I)} \le K < \infty, \quad f \in L^2(I) \tag{5.7.12a}$$

from Corollary 5.2.1 it follows that

$$\kappa^{\text{Dir}} = 1 - O(h^{\Delta}) \tag{5.7.12b}$$

and from Theorem 5.3.2 we have

$$(1 - \mathcal{K}^{\text{Neum}} h^{\Delta})\mathscr{E}^{\text{Neum}} \leq \|e_{S^p_\Delta}\|_{\mathfrak{U}} \leq \mathscr{E}^{\text{Neum}} \tag{5.7.13a}$$

and hence

$$1 \leq \kappa^{\text{Neum}} = \frac{\mathscr{E}^{\text{Neum}}}{\|e_{S^p_\Delta}\|_{\mathfrak{U}}} \leq \frac{1}{1 - \mathcal{K}^{\text{Neum}} h^{\Delta}} \tag{5.7.13b}$$

where $\mathcal{K}^{\text{Neum}}$ depends on K. □

Let us underline that *the above results were proven for the exact estimators.*

In Sections 5.1–5.6 we introduced several error indicators and analyzed their properties. Several of the results are asymptotic and are applicable if the mesh size h^{Δ} is sufficiently small. Although the asymptotic analysis is very valuable, the behavior of the finite element solution is often outside the asymptotic range for the required accuracies. This question is very important for practical computation and was addressed in [Strouboulis and Haque, 1992a,b] and in [Babuška, Strouboulis *et al.*, 1994a,b]. The major questions that arise are:

- How does the theory explain the behavior in the concrete setting?
- Which error indicator should be preferred?

We will now give examples of how the various estimators can be analyzed by computing the bounds for the effectivity index κ, and the robustness index $\mathcal{R}$ for various choices of the classes $\mathcal{U}$, $\mathcal{M}$, $\mathcal{D}$, which are relevant to practical computation. We will analyze the cases which have direct analogues in higher dimensions.

We will consider the class of *locally quasi-uniform* meshes of uniform polynomial degree p

$$\mathcal{M}(\beta; p) \stackrel{\text{def}}{=} \left\{ (\Delta, p) \,\middle|\, \frac{1}{\beta} \leq \frac{h_{I_i^{\Delta}}}{h_{I_{i+1}^{\Delta}}} \leq \beta,\ 1 \leq i \leq N(\Delta);\ p_i^{\Delta} = p \right\}. \tag{5.7.14}$$

The majority of the meshes employed in practical computations are quasi-uniform.

We will analyze the error indicators for the following cases:

(a) *The class of problems with negligible zeroth-order term* $\left(c(h^{\Delta})^2 \ll 1\right)$: For this class it is sufficient to analyze the special case

$$\mathcal{D}_1 \stackrel{\text{def}}{=} \{AE \equiv 1,\ c \equiv 0\} \tag{5.7.15}$$

and the conclusions will be valid for smooth AE, c, and sufficiently small h^{Δ}.

We will consider the following choices of exact solutions:

(a1) *The class of smooth solutions* $\mathcal{U}^{smth}_{\Delta,p,\epsilon}$: Given a mesh Δ, the degree p of the elements, and an $\epsilon > 0$, we will define $\mathcal{U}^{smth}_{\Delta,p,\epsilon}$ to be the set of functions $u \in \mathfrak{U}$, which are such that

$$0 < \min_{x \in I_j^{\Delta}} \left| \frac{d^{p+1}}{dx^{p+1}}(u)(x) \right| \tag{5.7.16a}$$

and

$$\max_{j=1,\dots,N(\Delta)} \left\| \frac{d^{p+2}}{dx^{p+2}}(u) \right\|_{L^\infty(I_j^\Delta)} \le \frac{\epsilon}{h_{I_j^\Delta}} \min_{j=1,\dots,N(\Delta)} \left(\min_{x\in I_j^\Delta} \left| \frac{d^{p+1}}{dx^{p+1}}(u) \right| \right). \tag{5.7.16b}$$

Note that *the smoothness of a function* $u \in \mathcal{U}^{smth}_{\Delta,p,\epsilon}$ *is relative to the employed mesh* Δ, *and the employed element degree* p. Hence a given u may or may not belong to $\mathcal{U}^{smth}_{\Delta,p,\epsilon}$, and hence may be characterized as smooth or unsmooth, depending on the mesh Δ and p. For example, consider $x^{0.51} \in \mathcal{U}^{smth}_{\Delta^\gamma,p,\epsilon}$, where Δ^γ is any radical mesh with $\gamma_{cr} = p/(\alpha - \frac{1}{2}) = 100p < \gamma$; the function $x^{0.51}$ is smooth with respect to these meshes.

(a2) *The class of unsmooth solutions* $\mathcal{U}^{unsm}_{\Delta,p,\epsilon}$: Given a mesh Δ and an $\epsilon > 0$, we will let

$$\mathcal{U}^{unsm}_{\Delta,p,\epsilon} = \left\{ u \in \mathfrak{U} \;\middle|\; \frac{\epsilon}{h_{I_j^\Delta}} \min_{j=1,\dots,N(\Delta)} \left(\min_{x\in I_j^\Delta} \left| \frac{d^{p+1}u}{dx^{p+1}}(x) \right| \right) \le \max_{j=1,\dots,N(\Delta)} \left\| \frac{d^{p+2}u}{dx^{p+2}} \right\|_{L^\infty(I_j^\Delta)} \right\}. \tag{5.7.17}$$

Once more the smoothness of the functions in $\mathcal{U}^{unsm}_{\Delta,p,\epsilon}$ is relative to Δ and p. For example, the function $\sin x$ is unsmooth if Δ is a sufficiently nonuniform mesh.

(a3) *The set of solutions* $\mathfrak{U}_{\alpha,d}$ *with an algebraic singularity at a distance* d *from the boundary*: We will let

$$\mathfrak{U}_{\alpha,d} = \left\{ u \in \mathfrak{U} \;\middle|\; \lim_{x\to 0}((x+d)^{-\alpha}u(x)) = \mathscr{C} \neq 0 \right\} \tag{5.7.18a}$$

with $\frac{1}{2} < \alpha \not\equiv \text{integer}, 0 \le d$.

As we already noted above, the functions in $\mathfrak{U}_{\alpha,d}$ belong to $\mathcal{U}^{smth}_{\Delta,p,\epsilon}$, for sufficiently refined meshes Δ depending on p, and are one-dimensional analogues of two-dimensional solutions in the neighborhood of corner-points ($d = 0$) and sharp fillets ($d > 0$). The goal of this analysis is to show the influence of the relation between the $(p+1)$ and $(p+2)$ derivatives, which expresses the smoothness of the exact solution on the performance of the estimator.

(b) *The class of problems with significant zeroth-order term* ($ch^2 \simeq O(1)$): To analyze this class of problems it is sufficient to analyze the case

$$\mathcal{D}_2 = \{AE \equiv 1,\ c \equiv c_0 \not\equiv 0\} \tag{5.7.19}$$

for the classes of solutions $\mathcal{U}^{smth}_{\Delta,p,\epsilon}$, which were defined above $\mathcal{U}^{bl}_{\Delta,p,\epsilon}$, which will be defined below and is related to the boundary-layer behavior. Cases where the zeroth-order term dominates occur, for example, when a higher-dimensional, e.g., 3D, problem is solved by the dimensional reduction method. As mentioned in [Ainsworth and Babuška, 1999], the Neumann estimator does not perform well for $ch^2 \ge 1$.

(c) *The singular model problem with* $AE \equiv x^\vartheta$, $c \equiv 0$: In this case we have

$$\mathcal{D}_3 = \{AE \equiv x^\vartheta,\ c \equiv 0\}. \tag{5.7.20}$$

This is a one-dimensional analogue of the behavior of the differential operator in the neighborhood of corner points in two dimensions.

Let us now list the error indicators which will analyzed below:

(a) ***ERDBp + k*** (***E***lement ***R***esidual estimator of the ***D***irichlet type which employs approximation of the indicators by ***B***ubble functions from the space (span) of monomials of degree $p+1$ to $\boldsymbol{p+k}$ which belong to S_0^{p+k}). This is the indicator given by

$$\eta^{\text{Dir}}_{I_j^\Delta, B^{p+k}} = \left\| \pi_{B^{p+k}} \left(\hat{e}^{\text{Dir}}_{I_j^\Delta} \right) \right\|_{\mathcal{U}(I_j^\Delta)}. \tag{5.7.21a}$$

(b) ***ERDPp + k*** (***E***lement ***R***esidual estimator of the ***D***irichlet type which employs approximation of the indicators by ***P***olynomial functions of degree $\boldsymbol{p+k}$). This is the indicator

$$\eta^{\text{Dir}}_{I_j^\Delta, S_0^{p+k}} = \left\| \pi_{S_0^{p+k}} \left(\hat{e}^{\text{Dir}}_{I_j^\Delta} \right) \right\|_{\mathcal{U}(I_j^\Delta)}. \tag{5.7.21b}$$

Note that because of the orthogonality of e^{Dir} we have $\eta^{\text{Dir}}_{I_j^\Delta, B^{p+k}} = \eta^{\text{Dir}}_{I_j^\Delta, S_0^{p+k}}$.

(c) ***ERNPp + k*** (***E***lement ***R***esidual estimator of the ***N***eumann type which employs approximation of the indicators by ***P***olynomial functions of degree $\boldsymbol{p+k}$). This indicator is

$$\eta^{\text{Neum}}_{I_j^\Delta, S^{p+k}} = \left\| \pi_{S^{p+k}} \left(\hat{e}^{\text{Neum}}_{I_j^\Delta} \right) \right\|_{\mathcal{U}(I_j^\Delta)}. \tag{5.7.21c}$$

(d) ***CRec2dPp*** (**2**-sided ***Rec***overy of the ***d***erivative using a ***C***ontinuous least-squares fit by ***P***olynomials of degree $\boldsymbol{p}$). This is the indicator based on the recovery $\mathcal{A}^*_{du/dx}$, and is given by

$$\eta^{\mathcal{A}^*_{du/dx}}_{I_j^\Delta} = \left\| \sqrt{AE} \Big(\mathcal{A}^*_{du/dx}(u_{S^p_\Delta}) - \frac{d}{dx}(u_{S^p_\Delta}) \Big) \right\|_{L^2(I_j^\Delta)}. \tag{5.7.21d}$$

Here we will employ for $\mathcal{A}^*_{du/dx}$ the C^0 recovery based on the continuous least-squares fit, defined by (4.7.6), with the right and left recovery operators obtained from (4.7.4) with $\|\cdot\|_{*,sup}$ replaced by $\|\cdot\|_{L^2(I_j^\Delta)}$.

(e) ***DRec2dPp*** (**2**-sided ***Rec***overy of the ***d***erivative using a ***D***iscrete least-squares fit by ***P***olynomials of degree $\boldsymbol{p}$). This is the $\eta^{\mathcal{A}^*_{du/dx}}_{I_j^\Delta}$ indicator of (5.7.21d) in the case that $\mathcal{A}^*_{du/dx}$ is the C^0 recovery based on least-squares fit, given by (4.7.6), which uses the discrete norm (4.7.4c) with weights equal to one.

(f) ***DRecvPp*** **+ 1** (***Rec***overy of the ***v***alue using a ***D***iscrete least-squares fit by ***P***olynomials of degree $\boldsymbol{p+1}$). This is the indicator

$$\eta^{\mathcal{A}^*_{du/dx}/\mathcal{A}^*_u}_{I_j^\Delta} = \sqrt{ \left\| \sqrt{AE} \frac{d}{dx} \left(\hat{e}^{\mathcal{A}^*_{du/dx}} \right) \right\|^2_{L^2(I_j^\Delta)} + \| \sqrt{c} \hat{e}^{\mathcal{A}^*_u} \|^2_{L^2(I_j^\Delta)} } \tag{5.7.21e}$$

defined in Section 5.6.

All these estimators have higher-dimensional analogues.

As we already mentioned above, the analysis of the error indicators will be done by determining the range of the local effectivity index of the indicators

$$\kappa_{I_j^\Delta}^{loc,EST} = \frac{\eta_{I_j^\Delta}^{EST}}{\|e_{S^p(I_j^\Delta)}^{loc}\|_{\mathcal{U}(I_j^\Delta)}}. \tag{5.7.22a}$$

For the class $\mathcal{D}_1$ given in (5.7.15), the nodal values of the finite element solution are exact, and we have

$$\kappa_{I_j^\Delta}^{loc,EST} = \kappa_{I_j^\Delta}^{EST} = \frac{\eta_{I_j^\Delta}^{EST}}{\|e_{S_\Delta^p}\|_{\mathcal{U}(I_j^\Delta)}}. \tag{5.7.22b}$$

Below we will omit the superscript *loc* from the local effectivity index $\kappa_{I_j^\Delta}^{loc,\,EST}$. In general, $\eta_{I_j^\Delta}^{EST}$ (e.g., *CRec2dPp*, *DRec2dPp*, and *DRecvPp+1*) depends on the error in the element I_j^Δ and the neighboring elements I_{j-1}^Δ, I_{j+1}^Δ, and we will write

$$\kappa_{I_j^\Delta}^{EST} = \kappa_{I_j^\Delta}^{EST}(u_{EX}; \tilde{h}_{j-1}^\Delta, \tilde{h}_{j+1}^\Delta; p) \tag{5.7.23a}$$

where

$$\tilde{h}_{I_{j-1}^\Delta} = \frac{h_{I_{j-1}^\Delta}}{h_{I_j^\Delta}}, \qquad \tilde{h}_{I_{j+1}^\Delta} = \frac{h_{I_{j+1}^\Delta}}{h_{I_j^\Delta}}. \tag{5.7.23b}$$

We will analyze an indicator *EST*, for the class $\mathcal{M}(\beta, p)$ and we will determine the lower bound for the effectivity index

$$\underline{\kappa}_{I_j^\Delta}^{EST} = \underline{\kappa}_{I_j^\Delta}^{EST}(u_{EX}; \beta; p) = \min_{\frac{1}{\beta} \le \tilde{h}_{I_{j-1}^\Delta}, \tilde{h}_{I_{j+1}^\Delta} \le \beta} \kappa_{I_j^\Delta}^{EST}(u_{EX}; \tilde{h}_{I_{j-1}^\Delta}, \tilde{h}_{I_{j+1}^\Delta}; p) \tag{5.7.23c}$$

and the upper bound $\bar{\kappa}_{I_j^\Delta}^{EST}$ defined as in (5.7.23c) by replacing min by max, and the corresponding robustness index

$$\mathcal{R}_{I_j^\Delta}^{EST}(u_{EX}; \beta; p) = \max\left(\bar{\kappa}_{I_j^\Delta}^{EST} - 1, \frac{1}{\underline{\kappa}_{I_j^\Delta}^{EST}} - 1\right). \tag{5.7.23d}$$

For the class $\mathcal{D}_2$ the pollution is practically negligible and we can analyze the effectivity index $\kappa_{I_j^\Delta}^{EST}$ by employing the error instead of the local error. Similarly for the class $\mathcal{D}_3$, the pollution in the first element I_1^Δ is negligible, and the effectivity index $\kappa_{I_1^\Delta}^{EST}$ can be analyzed by employing the error instead of the local error.

Let us now proceed with the analysis of the indicators.

(1) Regular model problem with $AE \equiv 1$, $c \equiv 0$.

(i) *Smooth solutions*: To analyze the robustness of the indicators for the class of smooth solutions it is sufficient to analyze them for $u_{EX} \in \mathcal{P}^{p+1}$, and because the finite element

TABLE 5.7.1. Quality of the *CRec2dPp* estimator for smooth solutions. Input data: $AE \equiv 1$, $c \equiv 0$, $\tilde{u}_{EX} \equiv \tilde{x}^{p+1}$; quasi-uniform meshes with $1/\beta \leq h_{I_i^\Delta}/h_{I_{i-1}^\Delta}, h_{I_i^\Delta}/h_{I_{i+1}^\Delta} \leq \beta$, $\beta = 2, 4, 8, 16, 32$, elements of degree p, $p = 1$, 2. The minimum and maximum value of the effectivity index and the value of the robustness index for $\beta = 2, 4, 8, 16, 32$. Note that as the nonuniformity of the meshes increases (i.e., as β increases) the estimator *CRec2dPp* has very poor quality even for smooth solutions ($\tilde{u}_{EX} \equiv \tilde{x}^{p+1}$, $1/\beta \leq h_{I_i^\Delta}/h_{I_{i-1}^\Delta}, h_{I_i^\Delta}/h_{I_{i+1}^\Delta} \leq \beta$)

β	$p = 1$			$p = 2$		
	$\underline{\kappa}$	$\bar{\kappa}$	$\mathcal{R}$	$\underline{\kappa}$	$\bar{\kappa}$	$\mathcal{R}$
2	0.8333	1.3333	0.3333	1.1714	1.6692	0.6691
4	0.6100	2.5600	1.5599	1.5334	6.3157	5.3157
8	0.3843	5.9259	4.9259	0.9231	62.9019	61.9019
16	0.2182	13.5087	12.5087	0.5889	388.0738	387.0738
32	0.1166	29.2672	28.2672	0.3116	1980.5516	1979.5516

solution reproduces exactly all monomials of degree less than or equal to p, it is sufficient to consider the monomial

$$u_{EX}(x) = (x - x^\Delta_{j-(1/2)})^{p+1}, \quad x \in I. \tag{5.7.24}$$

Note that all the estimators are exact in this case, except the *CRec2dPp* estimator which is not exact and its robustness depends on β. Table 5.7.1 reports the maximum and the minimum effectivity index $\bar{\kappa}_{I_j^\Delta}$, $\underline{\kappa}_{I_j^\Delta}$, and the robustness index $\mathcal{R}_{I_j^\Delta}$ for the *CRec2dPp* estimator for various values of β, for elements of degree $p = 1, 2$. Note that the robustness deteriorates as the non-uniformity of the mesh increases. For linear elements ($p = 1$), the minimum effectivity index $\underline{\kappa}_{I_j^\Delta}$ is achieved for $\tilde{h}_{I_{j-1}^\Delta} = \tilde{h}_{I_{j+1}^\Delta} = 1/\beta$ and the maximum effectivity index $\bar{\kappa}_{I_j^\Delta}$ is achieved for $\tilde{h}_{I_{j-1}^\Delta} = \tilde{h}_{I_{j+1}^\Delta} = \beta$. For quadratic elements ($p = 2$) $\underline{\kappa}_{I_j^\Delta}$ and $\bar{\kappa}_{I_j^\Delta}$ may not correspond to the extreme values of $\tilde{h}_{I_{j-1}^\Delta}$, $\tilde{h}_{I_{j+1}^\Delta}$. For instance, in the case $\beta = 8$, the minimum $\underline{\kappa}_{I_j^\Delta} = 0.8789$ is achieved for $\tilde{h}_{I_{j-1}^\Delta} = 0.75$, $\tilde{h}_{I_{j+1}^\Delta} = 0.125$. From the above results, we conclude that *the CRec2dPp estimator is not robust even for smooth solutions.*

(ii) *Unsmooth solutions*: To analyze the error indicators for this case it is sufficient to analyze their robustness for the $(p + 2)$ degree polynomial solution,

$$u_{EX}(x) = a_1(x - x^\Delta_{j-1/2})^{p+1} + a_2(x - x^\Delta_{j-1/2})^{p+2} \tag{5.7.25}$$

as a function of $a_1/a_2 = 1/a$, where $a = a_2/a_1$ characterizes the unsmoothness of the solution.

(a) *The ERDBp + 1 estimator*: Figure 5.7.1 shows the variation of the effectivity index $\kappa^{ERDBp+1}(a_1/a_2)$ for $p = 1, 2, 3, 4$. Note that as the solution becomes unsmooth ($a \to \infty$), κ tends to zero, and we have

$$\lim_{a \to \infty} \mathcal{R}^{ERDBp+1}(a, \beta, p) = \infty. \tag{5.7.26}$$

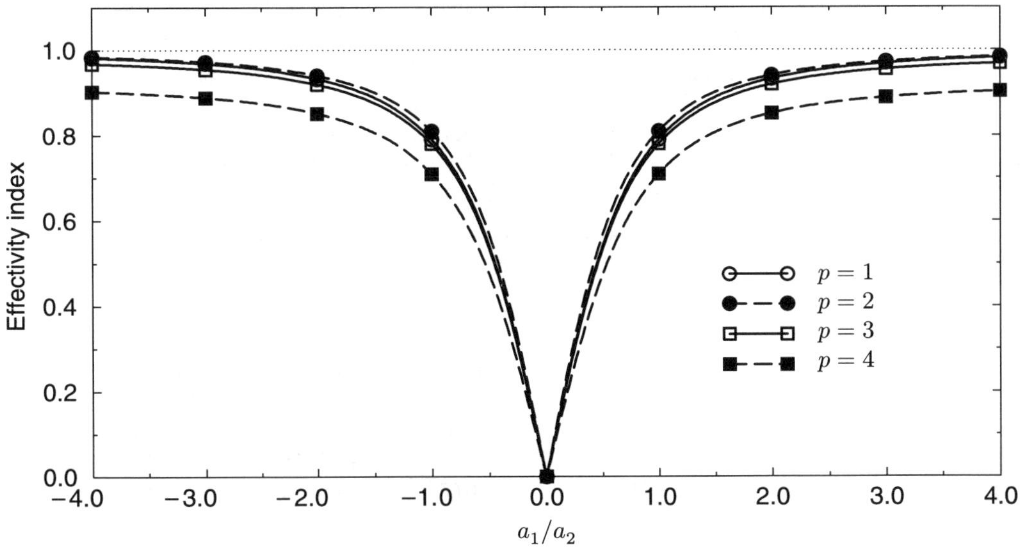

FIG. 5.7.1. Quality of the *ERDBp + 1* estimator for generally unsmooth solutions. Input data: $\tilde{u}_{EX} \equiv a_1\tilde{x}^{p+1} + a_2\tilde{x}^{p+2}$, $AE \equiv 1$, $c \equiv 0$, elements of degree p. Graphs of the variation of the effectivity index as a function of a_1/a_2 for $p = 1, 2, 3, 4$. Note that the effectivity index is equal to zero for $a_1/a_2 = 0$, which means that this estimator is not proper for the class of generally unsmooth solutions

Further, for fixed a, the robustness deteriorates with increasing p. This is because the $ERDPp+1$ indicator cannot detect the derivatives of the solution of order higher than $p+1$ from the residuals.

The reason for this behavior is that the estimator $ERDPp+1$ cannot 'see' the $p+2$ derivative, and is directly related to the assumption (4.6.26). This behavior can also occur in two dimensions in the cases that the solution is symmetric or antisymmetric. Similar conclusions also hold for the *ERDPp + 1* estimator. This situation can be remedied by employing the $ERDBp+k$ estimator with $k > 1$.

(b) *The DRec2dPp estimator*: Table 5.7.2 reports $\mathcal{R}^{DRec2dPp}(a_1/a_2, \beta)$, for various values of a_1/a_2 and β. Note that the robustness of the estimator deteriorates ($\mathcal{R}^{DRec2dPp}$ takes on large values) as the nonuniformity of the mesh increases, and that the maximum value of $\mathcal{R}^{DRec2dPp}$ is achieved for $a_1/a_2 = 0$. For linear elements ($p = 1$), the minimum value of the effectivity index $\underline{\kappa}_{I_j^\Delta}$ is achieved for $\tilde{h}_{I_{j-1}^\Delta} = \tilde{h}_{I_{j+1}^\Delta} = 1/\beta$, and the maximum value $\bar{\kappa}_{I_j^\Delta}^{DRec2dPp}$ for $\tilde{h}_{I_{j-1}^\Delta} = \tilde{h}_{I_{j+1}^\Delta} = \beta$. In the case of quadratic elements ($p = 2$), $\underline{\kappa}_{I_j^\Delta}^{DRec2dPp}$, $\bar{\kappa}_{I_j^\Delta}^{DRec2dPp}$ are not achieved for the extreme values of $\tilde{h}_{I_{j-1}^\Delta}$, $\tilde{h}_{I_{j+1}^\Delta}$; for instance for $a_1/a_2 = \frac{2}{3}$ and $\beta = 32$, the minimum effectivity index is achieved for $\tilde{h}_{I_{j-1}^\Delta} = 0.82, \tilde{h}_{I_{j+1}^\Delta} = 1/\beta$, and the maximum for $\tilde{h}_{I_{j-1}^\Delta} = \tilde{h}_{I_{j+1}^\Delta} = \beta$.

The *DRecvPp + 1* estimator has similar behavior.

Similar behavior should be expected for the two-dimensional versions of the estimators.

TABLE 5.7.2. Quality of the *DRec2dPp* estimator for generally unsmooth solutions. Input data: $AE \equiv 1$, $c \equiv 0$, $\tilde{u}_{EX} \equiv a_1\tilde{x}^{p+1} + a_2\tilde{x}^{p+2}$; quasi-uniform meshes with $1/\beta \le \tilde{h}_{I^{\Delta}_{i-1}}, \tilde{h}_{I^{\Delta}_{i+1}} \le \beta$; elements of degree p. The robustness index $\mathcal{R}^{DRec2dPp}$ is given for various values of β, a_1/a_2 and for $p = 1, 2$. Note that for quadratic elements $\mathcal{R}^{DRec2dPp}$ diverges to infinity as $a = a_2/a_1 \to \infty$ for any $\beta > 1$ (nonuniform mesh). For $p = 1$ the estimator is proper ($0 < \mathcal{R}^{DRec2dPp} < \infty$), but has poor robustness even in the case of the uniform mesh ($\beta = 1$) ($\mathcal{R}^{DRec2dPp}_{I^{\Delta}_i}(a_1/a_2, \beta, p;\ AE \equiv 1,\ c \equiv 0)$, $\tilde{u}_{EX} \equiv a_1\tilde{x}^{p+1} + a_2\tilde{x}^{p+2}$, $1/\beta \le \tilde{h}_{I^{\Delta}_{i-1}}, \tilde{h}_{I^{\Delta}_{i+1}} \le \beta$)

β	$a_1/a_2 = 0.0$	$a_1/a_2 = 1.0$	$a_1/a_2 = 2.0$	$a_1/a_2 = 3.0$	$a_1/a_2 = 4.0$
$p = 1$					
1	11.4164	6.3666	3.1920	1.8730	1.2288
2	13.6525	7.7425	4.0092	2.4371	1.6401
4	21.4787	12.5499	6.8564	4.4043	3.2055
8	38.8082	23.1770	13.1346	8.7465	6.5106
16	74.3058	44.9255	25.9681	17.7978	13.1525
32	145.7202	88.6642	51.7677	35.6572	26.7260
$p = 2$					
1	1.5675	0.6477	0.2581	0.1399	0.0817
2	∞	5.0034	2.7171	1.5835	1.1706
4	∞	38.2109	21.3338	14.3018	10.8931
8	∞	150.6382	84.5613	57.4172	44.0566
16	∞	594.2121	351.5182	238.1996	184.9659
32	∞	2442.0817	1642.1147	1002.4987	883.3018

(iii) *Solutions with an algebraic singularity*: To analyze the indicator for this case, it is sufficient to employ

$$u_{EX}(x) = (x + d)^{\alpha} \tag{5.7.27}$$

where $\alpha > 0.5$ and $d \ge 0$.

(a) *The ERDBp + k estimator*: For the considered case $AE \equiv$ *constant*, $c \equiv 0$, the effectivity index $\kappa^{ERDBp+k}_{I^{\Delta}_j}$ depends only on α and $d/h_{I^{\Delta}_j}$ and p. Table 5.7.3a and Table 5.7.3b report $\kappa^{ERDBp+k}(\alpha, d/h, p)$ for various values α and d/h. From Table 5.7.3a which reports $\kappa^{ERDBp+k}$ for linear elements ($p = 1$) for the cases that $k = 1$ and $k = 2$, it can be seen that the estimator severely underestimates as the exponent α approaches the critical value $\frac{1}{2}$, and as the singular point approaches the left end point of the element (as d/h approaches zero), while for $\alpha > 1$, $\kappa^{ERDBp+k}$ is practically equal to one. Table 5.7.3b reports $\kappa = \kappa(\alpha, d/h, p)$ for the *ERDBp + 1* estimator for $p = 2$ and $p = 3$. It can be seen that the estimator severely underestimates the error, more than in the case $p = 1$, as $\alpha \to \frac{1}{2}$, $d/h \to 0$. Figure 5.7.2 gives the graphs of $\mathcal{R}^{ERDBp+k}(\alpha, d/h = 0, p = 1)$ for $k = 1, 2, 3, 4,$ and 6. From these graphs it is clear that for α close to $\frac{1}{2}$ the *ERDBp + k* estimator is not robust even for high k's (e.g., for $k = 6$, $\alpha = 0.55$, we have $\mathcal{R}_1 \approx 1.5$). Figures 5.7.3a, and 5.7.3b give the graphs of $\kappa^{ERDBp+k}(\alpha, d/h = 0, p = 2)$ and $\mathcal{R}^{ERDBp\ +k}(\alpha, d/h = 0, p = 2)$. Very similar results

TABLE 5.7.3a. Quality of the *ERDBp* + k estimator for specially unsmooth solutions. Input data: $AE \equiv 1$, $c \equiv 0$, $u_{EX} \equiv (x+d)^\alpha$, $\alpha > 0.5$; linear elements ($p = 1$). The effectivity index $\kappa^{ERDBp+k}$ of the *ERDBp* + k estimator for various values of α, d/h, and $k = 1, 2$. Note that as the solution becomes more unsmooth (as $\alpha \to 0.5$, $d/h \to 0$) the estimator underestimates the error, and increasing k improves the performance ($\kappa^{ERDBp+k}(\alpha, d/h, p; AE \equiv 1,\ c \equiv 0)$, $u_{EX}(x) = (x+d)^\alpha$, $p = 1$)

α	$d/h = 0.0$		$d/h = 0.01$		$d/h = 0.1$	
	$k = 1$	$k = 2$	$k = 1$	$k = 2$	$k = 1$	$k = 2$
0.55	0.3533	0.4383	0.7598	0.8896	0.9123	0.9829
0.75	0.6998	0.8114	0.8337	0.9354	0.9362	0.9895
0.95	0.8426	0.9273	0.8935	0.9662	0.9562	0.9940
1.15	0.9186	0.9727	0.9378	0.9845	0.9724	0.9969
1.35	0.9610	0.9907	0.9678	0.9939	0.9845	0.9987
1.55	0.9843	0.9974	0.9864	0.9981	0.9929	0.9995
1.75	0.9959	0.9996	0.9963	0.9997	0.9979	0.9999
1.95	0.9999	1.0000	0.9999	1.0000	0.9999	1.0000
2.15	0.9989	1.0000	0.9989	1.0000	0.9993	1.0000
2.35	0.9945	0.9999	0.9948	0.9999	0.9966	1.0000

TABLE 5.7.3b. Quality of the *ERDBp* + k estimator for specially unsmooth solutions. Input data: $AE \equiv 1$, $c \equiv 0$, $u_{EX} \equiv (x+d)^\alpha$, $\alpha > 0.5$; elements of degree $p = 2, 3$. The effectivity index $\kappa^{ERDBp+k}$ of the *ERDBp* + k estimator for various values of α, d/h, and $k = 1$. Note that as the unsmoothness of the solution increases (as $\alpha \to 0.5$, $d/h \to 0$), the underestimation is more severe than in the case of linear elements ($\kappa^{ERDBp+k}(\alpha, d/h, p;\ AE \equiv 1, c \equiv 0)$, $u_{EX}(x) = (x+d)^\alpha$, $k = 1$)

α	$p = 2$			$p = 3$		
	$d/h = 0.0$	$d/h = 0.01$	$d/h = 0.1$	$d/h = 0.0$	$d/h = 0.01$	$d/h = 0.1$
0.55	0.2554	0.7243	0.8934	0.2086	0.7061	0.8830
0.75	0.5624	0.7768	0.9113	0.4807	0.7463	0.8971
0.95	0.7176	0.8246	0.9276	0.6328	0.7845	0.9105
1.15	0.8117	0.8663	0.9423	0.7318	0.8199	0.9230
1.35	0.8730	0.9012	0.9553	0.7996	0.8519	0.9345
1.55	0.9147	0.9296	0.9664	0.8484	0.8801	0.9451
1.75	0.9440	0.9518	0.9758	0.8849	0.9046	0.9547
1.95	0.9646	0.9687	0.9835	0.9130	0.9255	0.9632
2.15	0.9791	0.9811	0.9896	0.9349	0.9429	0.9708
2.35	0.9889	0.9898	0.9941	0.9522	0.9574	0.9774

are obtained for the *ERDPp* + k indicator. Hence, for the elements in close vicinity of the corners, the robustness deteriorates as α decreases and the use of high k may not be a practical way to improve the indicator. This should have been expected from the earlier results where we showed that the effectivity of the estimator deteriorates when

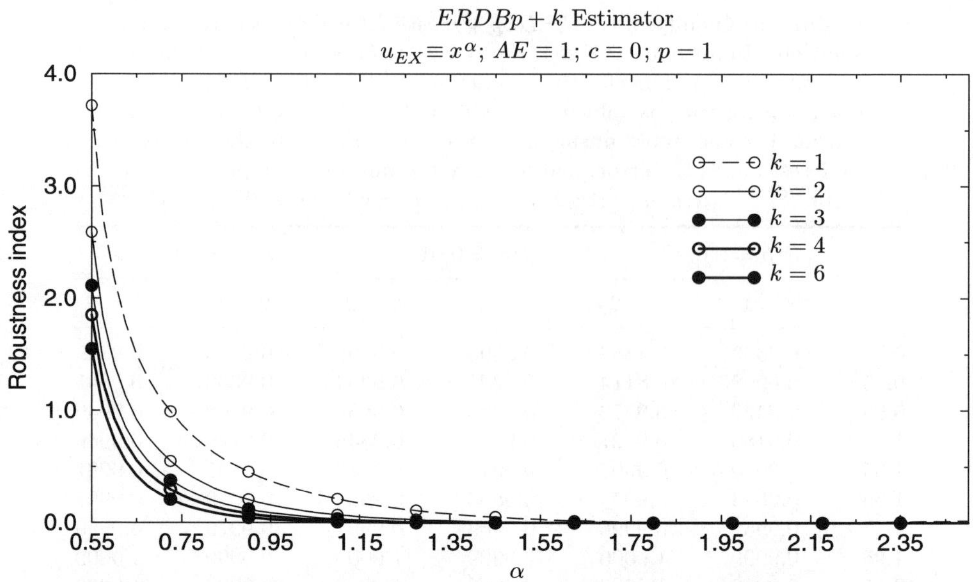

FIG. 5.7.2. Quality of the $ERDBp + k$ estimator for specially unsmooth solutions. Input data: $u_{EX} \equiv x^\alpha$, $AE \equiv 1$, $c \equiv 0$, linear elements ($p = 1$). Graphs of the robustness index $\mathcal{R}^{ERDBp+k}(\alpha, d/h = 0, p = 1)$, as a function of α, for $k = 1, \ldots, 6$. Note that for very strong singularities ($\alpha \leq 0.65$) even $k = 6$ is not sufficient to produce a robust estimator

the higher-order terms in the Taylor series expansion of the solution dominate the $p + 1$ term. Analogous conclusions also hold in the two-dimensional case.

(b) *The DRec2dPp estimator*: The indicator $\eta_{I_j^\Delta}^{DRec2dPp}$ depends on $\alpha, d/h, p$, and also on $\tilde{h}_{I_{j-1}^\Delta}, \tilde{h}_{I_{j+1}^\Delta}$. Here we will consider meshes which are uniform, or refined towards the singular point. Table 5.7.4 and Figs 5.7.4–5.7.7 report the results of the analysis in the element I_1^Δ, which is next to the singular point. Table 5.7.4 reports the values of $\mathcal{R}_{I_j^\Delta}^{DRec2dPp}(\alpha, \beta, p)$, in the first element for various α, β, for $p = 1$ and $p = 2$. Note that the robustness is poor for the entire range of α, except for the neighborhoods of $\alpha = p + 1$. In the case of linear elements ($p = 1$), we have

$$\kappa_{I_1^\Delta}^{DRec2dPp}(\alpha, h_{I_2^\Delta}/h_{I_1^\Delta}, p = 1) = \frac{(1 + (h_2^\Delta/h_1^\Delta))^{\alpha-1} - 1}{\sqrt{3}} \frac{\sqrt{2\alpha - 1}}{(\alpha - 1)} \frac{h_{I_1^\Delta}}{h_{I_2^\Delta}} \tag{5.7.28a}$$

and the minimum effectivity index $\underline{\kappa}_{I_1^\Delta}^{DRec2dPp}(\alpha)$, for $\frac{1}{2} < \alpha < 2$, is obtained by employing $h_{I_2^\Delta}/h_{I_1^\Delta} = \beta$, and for $\alpha \geq 2$, by letting $h_{I_2^\Delta}/h_{I_1^\Delta} = 1/\beta$, namely

$$\underline{\kappa}_{I_1^\Delta}^{DRec2dPp}(\alpha) = \begin{cases} \dfrac{(1 + \beta)^{\alpha-1} - 1}{\sqrt{3}} \dfrac{\sqrt{2\alpha - 1}}{(\alpha - 1)} \dfrac{1}{\beta}, & \dfrac{1}{2} < \alpha < 2 \\ \dfrac{(1 + 1/\beta)^{\alpha-1} - 1}{\sqrt{3}} \dfrac{\sqrt{2\alpha - 1}}{(\alpha - 1)} \beta, & 2 \leq \alpha. \end{cases} \tag{5.7.28b}$$

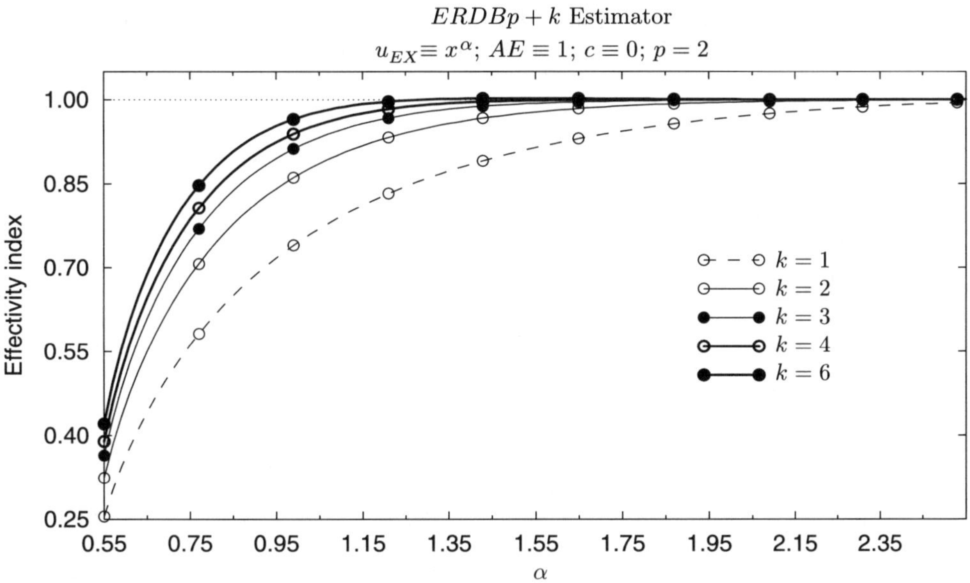

(a)

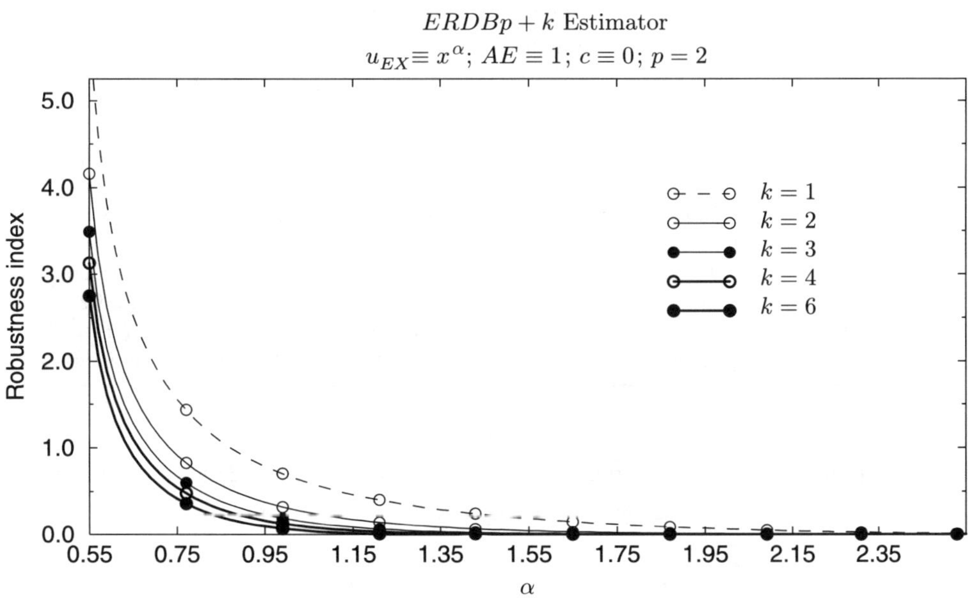

(b)

FIG. 5.7.3. Quality of the $ERDBp+k$ estimator for specially unsmooth solutions. Input data: $u_{EX} \equiv x^{\alpha}$, $AE \equiv 1$, $c \equiv 0$, quadratic elements ($p = 2$). (a) Graph of the effectivity index $\kappa^{ERDBp+k}(\alpha, d/h = 0, p = 2)$, and (b) graph of the robustness index $\mathcal{R}^{ERDBp+k}(\alpha, d/h = 0, p = 2)$, as functions of α, for $k = 1, 2, 3, 4$, and 6

TABLE 5.7.4. Quality of the *DRec2dPp* estimator for specially unsmooth solutions. Input data: $AE \equiv 1$, $c \equiv 0$, $u_{EX} \equiv x^{\alpha}$, $\alpha > 0.5$; quasi-uniform meshes with $1/\beta \leq h_{I^{\Delta}_{i+1}}/h_{I^{\Delta}_{i}} \leq \beta$; elements of degree $p = 1, 2$. The robustness index $\mathcal{R}^{DRec2dPp}_{I^{\Delta}_1}(\alpha, \beta, p)$ for various α and β, for $p = 1$ and 2. Note that the robustness deteriorates with increasing singularity ($\alpha \to 0.5$) and increasing nonuniformity of the mesh ($\beta \to \infty$) and that this deterioration is much more than in the case of the residual estimator *ERDPp + 1*. For $\beta \geq 2$, the robustness of the estimator is rather poor for practically the entire range of α ($\mathcal{R}^{DRec2dPp}_{I^{\Delta}_1}(\alpha, \beta, p;\ AE \equiv 1, c \equiv 0)$, $u_{EX} \equiv x^{\alpha}$, $1/\beta \leq \tilde{h}_{I^{\Delta}_2} \leq \beta$)

α	$\beta = 2$	$\beta = 4$	$\beta = 8$	$\beta = 16$	$\beta = 32$
$p = 1$					
0.55	15.2301	20.0033	30.1537	50.2021	89.1803
0.75	5.3696	7.0126	10.5082	17.1609	29.7938
0.95	3.3064	4.1196	6.0281	9.5521	15.8480
1.15	2.0806	2.6023	3.7422	5.6456	8.9274
1.35	1.3028	1.6439	2.2727	3.3009	4.9208
1.55	0.7830	0.9688	1.2942	1.7903	2.5071
1.75	0.3855	0.4672	0.6023	0.7930	1.0428
1.95	0.0693	0.0822	0.1024	0.1283	0.1591
2.15	0.2137	0.2597	0.3338	0.4338	0.5580
2.35	0.5366	0.6816	0.9315	1.3016	1.8132
$p = 2$					
0.55	20.6780	18.9520	18.0187	17.3476	17.3756
0.95	4.5189	4.2585	3.9296	4.1091	3.6860
1.35	2.7361	1.9246	2.4410	2.5988	2.5946
1.75	1.6255	1.4645	1.7544	2.4999	3.1489
2.15	1.0589	0.9653	1.8893	3.1388	5.0526
2.55	0.6214	0.5232	0.4883	0.5137	6.2625
2.95	0.0735	0.0730	0.0714	0.0573	1.1193
3.35	0.5769	1.7308	5.1149	13.0219	34.8780

The maximum effectivity index $\bar{\kappa}^{DRec2dPp}_{I^{\Delta}_1}(\alpha)$ for $\frac{1}{2} < \alpha < 2$, is obtained by letting $h_{I^{\Delta}_2}/h_{I^{\Delta}_1} = \beta$, and for $\alpha \geq 2$ by employing $h_{I^{\Delta}_2}/h_{I^{\Delta}_1} = 1/\beta$, namely

$$\bar{\kappa}^{DRec2dPp}_{I^{\Delta}_1}(\alpha) = \begin{cases} \dfrac{(1+1/\beta)^{\alpha-1} - 1}{\sqrt{3}} \dfrac{\sqrt{2\alpha - 1}}{(\alpha - 1)} \beta, & \dfrac{1}{2} < \alpha < 2 \\ \dfrac{(1+\beta)^{\alpha-1} - 1}{\sqrt{3}} \dfrac{\sqrt{2\alpha - 1}}{(\alpha - 1)} \dfrac{1}{\beta}, & 2 \leq \alpha. \end{cases} \tag{5.7.28c}$$

In the case of quadratic elements, the bounds $\underline{\kappa}^{DRec2dPp}_{I^{\Delta}_1}$, $\bar{\kappa}^{DRec2dPp}_{I^{\Delta}_1}$ do not necessarily correspond to $h_{I^{\Delta}_2}/h_{I^{\Delta}_1} = \beta$ or $1/\beta$. For instance for $\beta = 32$, $\alpha = 0.55$, the minimum $\underline{\kappa}^{DRec2dPp}_{I^{\Delta}_1}$ is achieved for $h_{I^{\Delta}_2}/h_{I^{\Delta}_1} = 0.7726$ and maximum $\bar{\kappa}^{DRec2dPp}_{I^{\Delta}_1}$ for

$h_{I_2^\Delta}/h_{I_1^\Delta} = 1/\beta$. For linear elements ($p = 1$), the *DRec2dPp* estimator underestimates for $\frac{1}{2} < \alpha < 2$, while it overestimates for $\alpha > 2$. The robustness of the estimator deteriorates with increasing β practically over the entire range of α.

Let us also analyze the indicator in the second element I_2^Δ. Table 5.7.5a (respectively Table 5.7.5b) gives the values of the robustness index $\mathcal{R}^{DRec2dPp}(\alpha, \beta, p)$ for $p = 1$

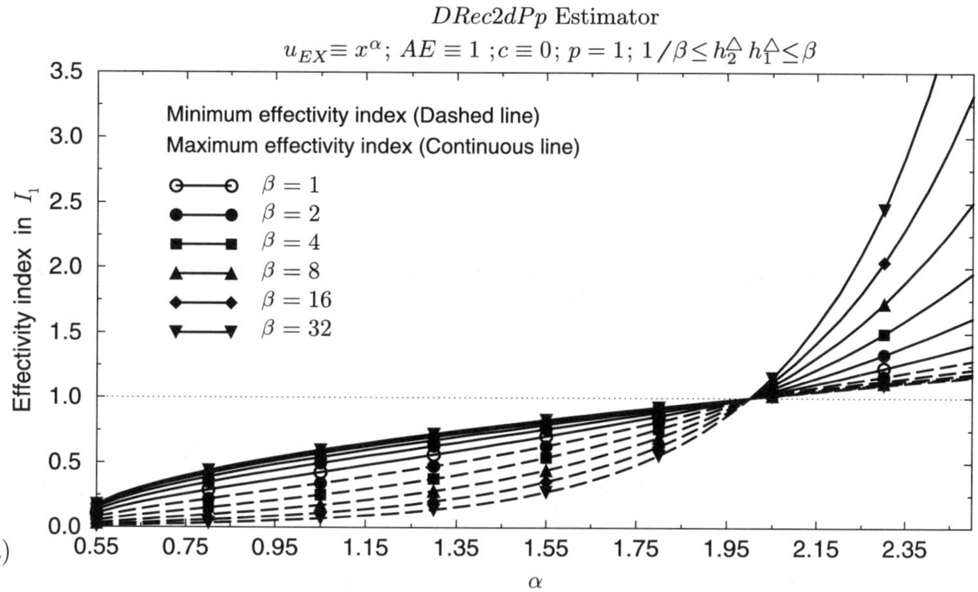

(a)

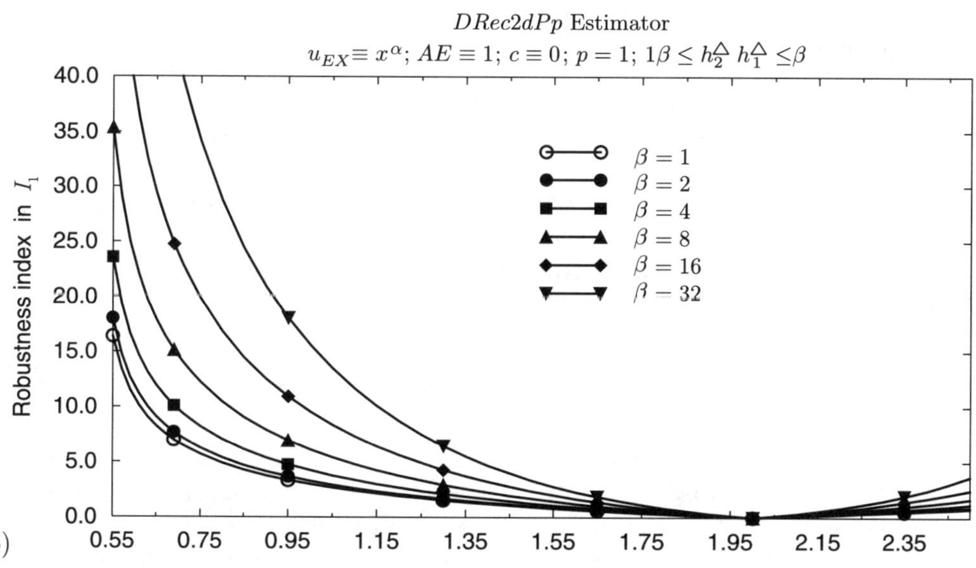

(b)

FIG. 5.7.4.

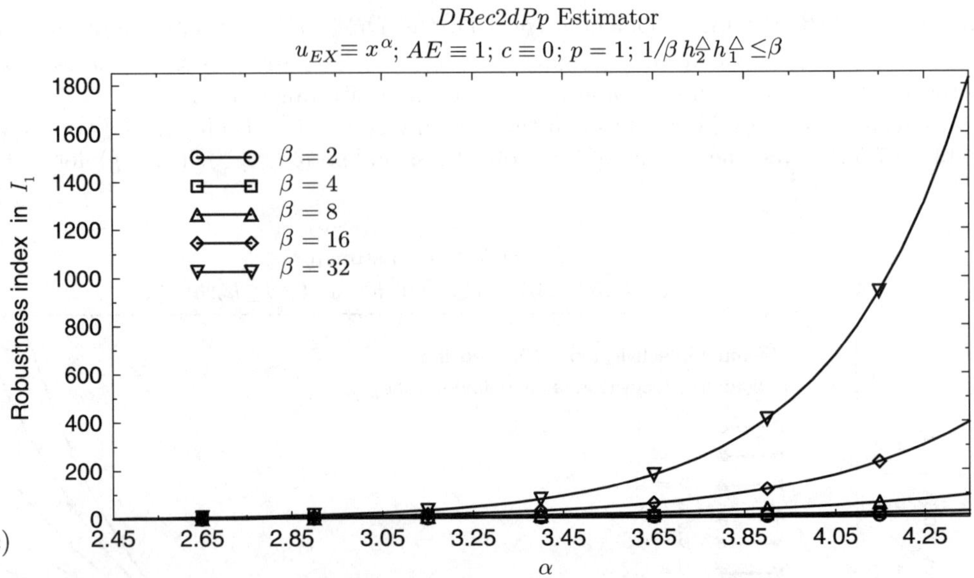

(c)

FIG. 5.7.4. Quality of the *DRec2dPp* estimator for specially unsmooth solutions. Input data: $u_{EX}\equiv x^{\alpha}$, $AE\equiv 1$, $c\equiv 0$, linear elements ($p=1$), quasi-uniform meshes with $1/\beta \leq h_{I_2^{\triangle}}/h_{I_1^{\triangle}} \leq \beta$. (a) Graph of the maximum and minimum effectivity index and (b) graph of the robustness index in the first element as a function of α for $0.55 \leq \alpha \leq 2.5$ and $\beta = 1, 2, 4, 8, 16, 32$. (c) Graph of the robustness index in the first element for $2.45 \leq \alpha \leq 4.35$, and $\beta = 2, 4, 8, 16, 32$

(respectively $p=2$). Note that $\mathcal{R}$ grows with β, for fixed α. As expected, the *DRec2dPp* indicator is exact for $\alpha = 2$ in the case of linear elements ($p=1$), and for $\alpha = 3$ in the case of quadratic elements ($p=2$).

From the above results it can be seen that, *for the solutions with algebraic singularities, the residual estimators are significantly more robust than the estimators based on recovery.* This can be explained by noting that for singular solutions, unless the mesh is properly constructed, e.g., to be sufficiently graded towards the singular point, the superconvergence effects on which the good performance of the recovery estimators is based, are not present. The same conclusions also hold in the two-dimensional case.

(2) Regular model problem with $AE\equiv 1$, $c\equiv c_0 > 0$, and $c_0h^2 \gg 1$. In this case all indicators depend on the mesh, and we have

$$\kappa_{I_j^{\triangle}}^{EST} = \kappa_{I_j^{\triangle}}^{EST}\big(u_{EX}, \tilde{h}_{I_{i-1}^{\triangle}}, \tilde{h}_{I_{i+1}^{\triangle}}, p\big). \qquad (5.7.29a)$$

(i) *Solutions with algebraic singularities*: Once more, it is sufficient to analyze the indicators for the solution $u_{EX}\equiv (x+d)^{\alpha}$. The effectivity index $\kappa_{I_1^{\triangle}}^{EST}$ practically depends on the mesh in the first few elements. Here we computed the finite element solution from a three-element patch, and we assessed the effect of the size of the patch by comparing the effectivity index computed using the finite element solution computed from the two-element patch which consists of the first element and its neighbor, and the three-element patch of the first three elements. In each case, we employed Neumann boundary conditions at the right end point of the patch.

(a) *The ERDBp + 2 estimator*: Table 5.7.6a (respectively Table 5.7.6b) reports the relative robustness index

$$\bar{\mathcal{R}}_{I_1^{\Delta}}^{ERDBp+2} = \frac{\mathcal{R}_{I_1^{\Delta}}^{ERDBp+2}\big(\alpha, d/h_{I_1^{\Delta}}, p, \bar{c}\big)}{\mathcal{R}\big(\alpha, d/h_{I_1^{\Delta}}, p, 0\big)} \tag{5.7.29b}$$

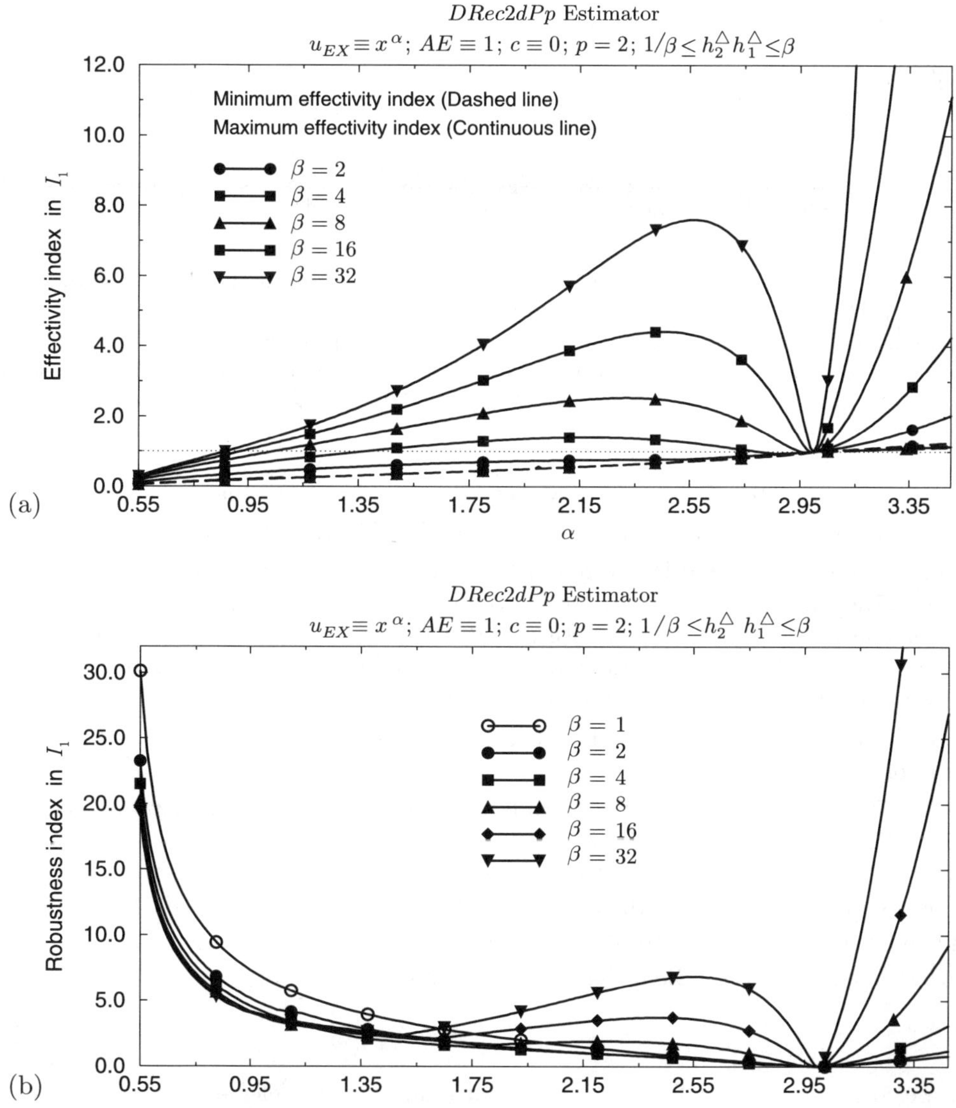

FIG. 5.7.5.

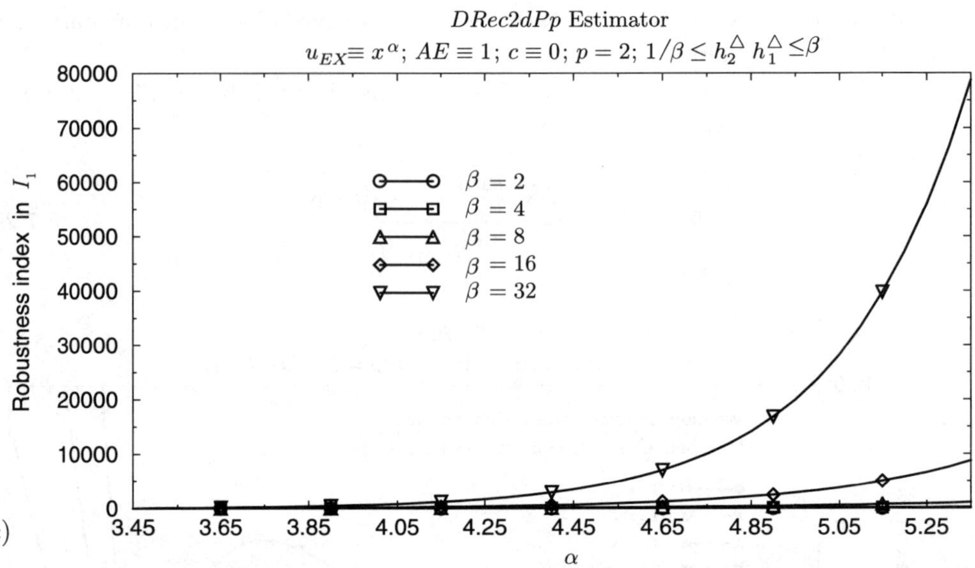

(c)

FIG. 5.7.5. Quality of the *DRec2dPp* estimator for specially unsmooth solutions. Input data: $u_{EX} \equiv x^\alpha$, $AE \equiv 1$, $c \equiv 0$, quadratic elements, quasi-uniform meshes with $1/\beta \leq h_{I_2^\Delta}/h_{I_1^\Delta} \leq \beta$. (a) Graphs of the maximum and minimum effectivity index in the first element as a function of α for $0.55 \leq \alpha \leq 3.5$ and $\beta = 2, 4, 8, 16, 32$. Graphs of the robustness index in the first element as a function of α for $\beta = 2, 4, 8, 16, 32$: (b) for $0.55 \leq \alpha \leq 3.45$; (c) for $3.45 \leq \alpha \leq 5.45$

for $d/h_{I_1^\Delta} = 0.1, 0.01$, for $p = 1$ (respectively $p = 2$). Note that as $\bar{c}$ increases, the robustness deteriorates. Table 5.7.6c reports $\bar{\mathcal{R}}_{I_1^\Delta}^{ERDBp+2}(\alpha, d/h_{I_1^\Delta}, p, \bar{c})$ for the case that the finite element solution is computed from the two-element patch. A comparison of Tables 5.7.6a and 5.7.6c shows that for $\alpha > 1.55$ and $\bar{c} \geq 10$, there is a significant difference in the values of the robustness index from the two- and three-element patch. Nevertheless, for strong singularities ($\alpha < 1$), the difference in the results obtained using the two- and three-element patch is insignificant.

(b) *The ERNp + 2 estimator*: The behavior is similar to that of the *ERDBp* + *2* estimator.

(c) *The DRec2dPp estimator*: This estimator is, in general, more robust with respect to c than $ERDPp + 2$ or $ERDBp + 2$.

Here we see the simultaneous effects of the singularity and the zeroth-order term. As it can be seen from Table 5.7.4, the presence of the zeroth-order term leads to an improvement in the accuracy of the recovery estimator.

(ii) *Boundary-layer solution*: Here we considered the solution

$$u_{EX} \equiv e^{-\mu x}, \qquad \mu = \sqrt{c_0} \tag{5.7.30}$$

and we analyzed the quality of the estimator in the first element. The results were computed by considering a three-element patch, for $\frac{1}{2} \leq h_{I_2^\Delta}/h_{I_1^\Delta}, h_{I_3^\Delta}/h_{I_1^\Delta} \leq 2$. Once more this case has larger importance in the relation to the analysis of the dimensional

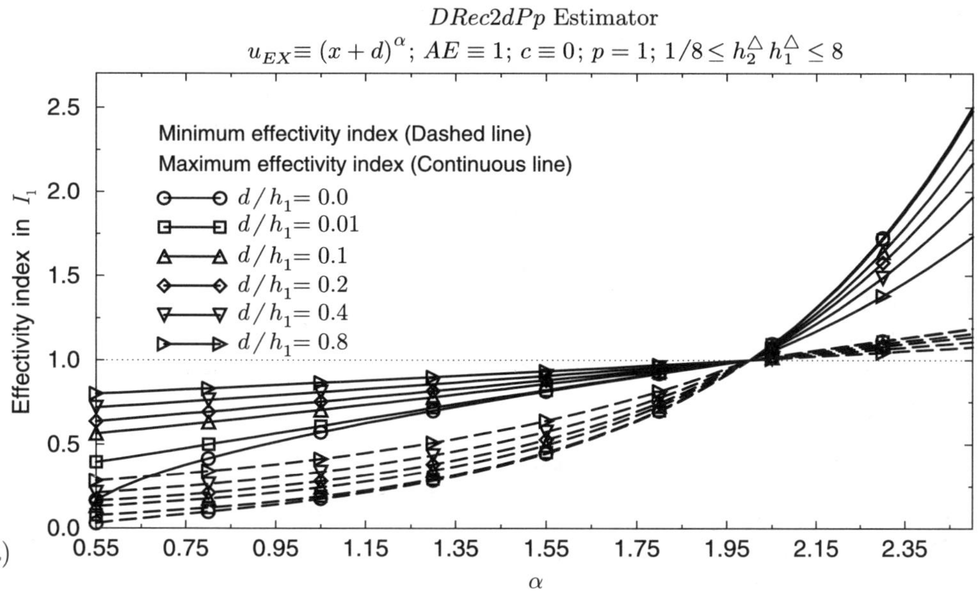

(a)

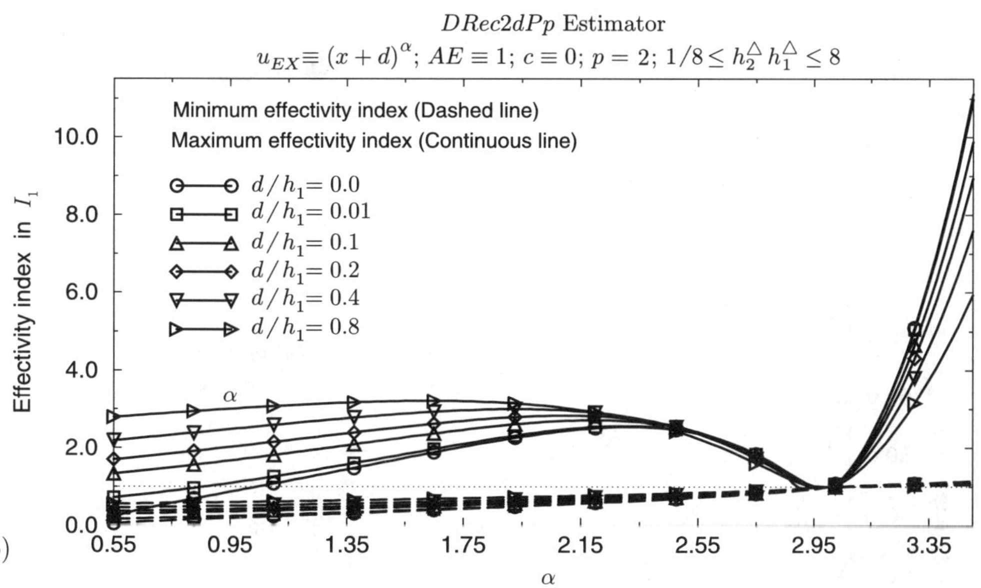

(b)

FIG. 5.7.6. Quality of the *DRec2dPp* estimator for specially unsmooth solutions. Input data: $u_{EX} \equiv (x+d)^{\alpha}$, $AE \equiv 1$, $c \equiv 0$, elements of degree p, quasi-uniform meshes with $\frac{1}{8} \leq h_{I_2^{\triangle}}/h_{I_1^{\triangle}} \leq 8$. Graphs of the maximum and the minimum effectivity index in the first element as a function of α for various values of d/h_1 for (a) linear elements ($p = 1$), $0.55 \leq \alpha \leq 2.50$, (b) quadratic elements ($p = 2$), $0.55 \leq \alpha \leq 3.5$

reduction approach used, e.g., in the derivation of the models of plates and shells, where the boundary-layer effects are then present, where we often have $\bar{c} > 100$.

(a) *The ERDBp + 1 estimator*: Table 5.7.7 gives the results for the *ERDBp + 1* estimator for various values of $\bar{c}$ and for elements of degree $p = 1, 2, 3$. Note that as $\bar{c} \to \infty$

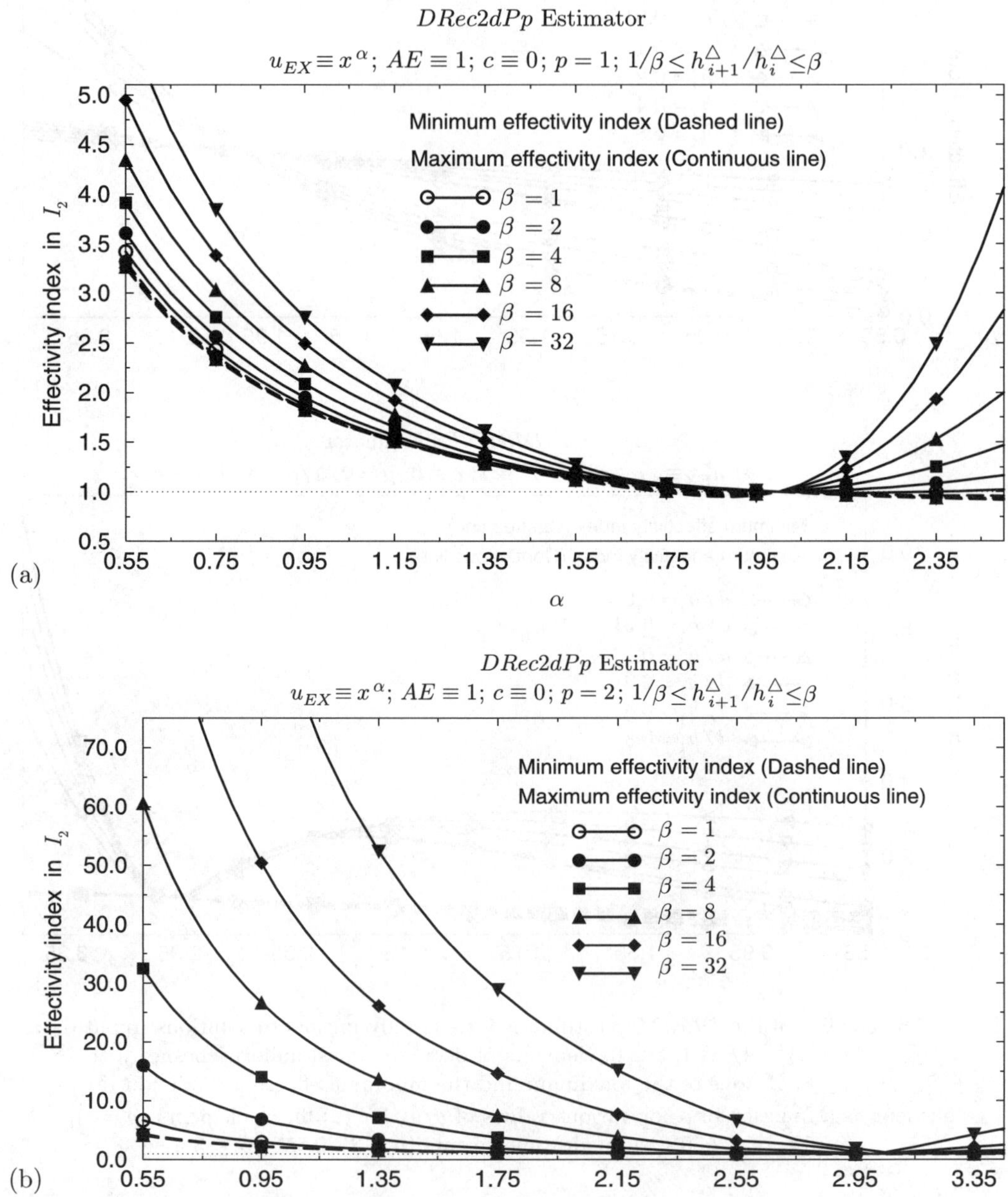

FIG. 5.7.7.

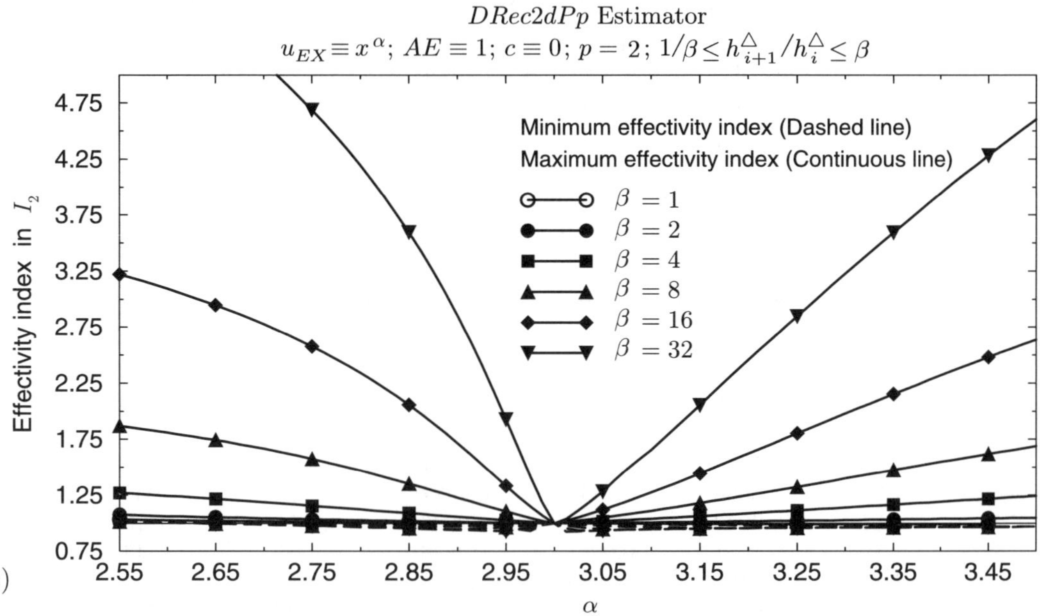

(c)

FIG. 5.7.7. Quality of the *DRec2dPp* estimator for specially unsmooth solutions. Input data: $u_{EX} \equiv x^{\alpha}$, $AE \equiv 1$, $c \equiv 0$, elements of degree p, quasi-uniform meshes with $1/\beta \leq h^{\Delta}_{i+1}/h^{\Delta}_{i} \leq \beta$. Graphs of the maximum and the minimum effectivity index in the second element as a function of the exponent α for $\beta = 1, 2, 4, 8, 16, 32$ for: (a) Linear elements ($p = 1$). (The graphs are shown for $0.55 \leq \alpha \leq 2.5$.) (b) Quadratic elements ($p = 2$). (The graphs are shown for $0.55 \leq \alpha \leq 3.5$.) (c) Quadratic elements ($p = 2$). (The graphs are shown for $2.55 \leq \alpha \leq 3.5$.) Note that the robustness of the estimator is poor practically for the entire range of α and deteriorates as the degree of the elements is increased from linear to quadratic

(i.e., $\mu \to \infty$) the estimator severely underestimates the error. The robustness improves if elements of higher degree are employed.

(b) *The ERNp + 1 estimator*: Table 5.7.8 gives the results for the *ERNp + 1* estimator for various values of $\bar{c}$ and elements of degree $p = 1, 2, 3$. From Table 5.7.8 it can be seen that, in contrast with the *ERDBp + 1* estimator, the *ERNp + 1* estimator overestimates for the entire range of $\bar{c}$. For $\bar{c} > 100$, the robustness of the *ERNp + 1* estimator is practically insensitive to $\bar{c}$.

(c) *The DRec2dPp estimator*: Table 5.7.9 gives the results for the *DRec2dPp* estimator. Note that the quality of the estimator deteriorates as $\bar{c} \to \infty$.

Above we have seen that all the estimators analyzed deteriorate as c takes large values. Analogous conclusions hold also in the higher-dimensional case. The problem of large c is encountered in the case of dimensional reduction which is employed in the analysis of problems in thin domains; see, e.g., [Babuška and Vogelius, 1981], [Babuška and Schwab, 1996]. [Verfürth, 1998] and [Ainsworth and Babuška, 1999] proposed estimators which are uniformly robust for the entire range of c.

TABLE 5.7.5a. Quality of the *DRec2dPp* estimator for specially unsmooth solutions. Input data: $AE \equiv 1$, $c \equiv 0$, $u_{EX} \equiv x^\alpha$, $\alpha > 0.5$; quasi-uniform meshes with $1/\beta \leq \tilde{h}_{I_1^\Delta}$, $\tilde{h}_{I_3^\Delta} \leq \beta$; linear elements ($p = 1$). The robustness index $\mathcal{R}_{I_2^\Delta}^{DRec2dPp}(\alpha, \beta, p = 1)$ is given for the second element as a function of α for various values of β. It can be seen that the robustness deteriorates with increasing strength of the singularity ($\alpha \to 0.5$) and increasing nonuniformity of the mesh ($\beta \to \infty$) ($\mathcal{R}_{I_2^\Delta}^{DRec2dPp}(\alpha, \beta, p;\ AE \equiv 1, c \equiv 0)$, $u_{EX} \equiv x^\alpha$, $1/\beta \leq \tilde{h}_{I_1^\Delta}$, $\tilde{h}_{I_3^\Delta} \leq \beta$, $p = 1$)

α	$\beta = 2$	$\beta = 4$	$\beta = 8$	$\beta = 16$	$\beta = 32$
0.55	4.9331	5.1934	5.6095	6.2058	7.0338
0.75	2.9297	3.1067	3.3712	3.7215	4.1738
0.95	1.7916	1.9172	2.0976	2.3181	2.5777
1.15	1.0754	1.1586	1.2811	1.4248	1.5792
1.35	0.5966	0.6349	0.7081	0.7993	0.8938
1.55	0.2946	0.2933	0.2986	0.3390	0.3919
1.75	0.1051	0.0941	0.0846	0.0782	0.0843
1.95	0.0106	0.0205	0.0319	0.0434	0.0544
2.15	0.0416	0.0954	0.1687	0.2626	0.3773
2.35	0.1246	0.3073	0.5896	0.9962	1.5490

TABLE 5.7.5b. Quality of the *DRec2dPp* estimator for specially unsmooth solutions. Input data: $AE \equiv 1$, $c \equiv 0$, $u_{EX} \equiv x^\alpha$, $\alpha > 0.5$; quasi-uniform meshes with $1/\beta \leq \tilde{h}_{I_1^\Delta}$, $\tilde{h}_{I_3^\Delta} \leq \beta$; elements of degree $p = 2$. The robustness index $\mathcal{R}_{I_2^\Delta}^{DRec2dPp}(\alpha, \beta, p = 2)$ is given for the second element as a function of α for various values of β. Note that the robustness deteriorates with increasing strength of the singularity ($\alpha \to 0.5$) and increasing nonuniformity of the mesh ($\beta \to \infty$) ($\mathcal{R}_{I_2^\Delta}^{DRec2dPp}(\alpha, \beta, p;\ AE \equiv 1, c \equiv 0)$, $u_{EX} \equiv x^\alpha$, $1/\beta \leq \tilde{h}_{I_1^\Delta}$, $\tilde{h}_{I_3^\Delta} \leq \beta$, $p = 2$)

α	$\beta = 2$	$\beta = 4$	$\beta = 8$
0.55	18.4235	34.6400	62.7274
0.95	7.1647	14.3258	26.8901
1.35	3.0672	6.7284	13.2766
1.75	1.2157	3.0918	6.8109
2.15	0.3619	1.1572	3.3062
2.55	0.0394	0.2836	0.8831
2.95	0.0094	0.0424	0.1402
3.35	0.0603	0.2023	0.5103

TABLE 5.7.6a. Quality of the $ERDBp+2$ estimator for specially unsmooth solutions: The effect of the zeroth-order term. Input data: $AE \equiv 1$, $c \equiv c_0 > 0$, $u_{EX} \equiv (x+d)^\alpha$, $\alpha > 0.5$; linear elements ($p = 1$); $\beta = 2$. The relative robustness index $\bar{\mathcal{R}}_{I_1^\Delta}^{ERDBp+2}(\alpha, d/h_{I_1^\Delta}, p, \bar{c})$ as a function of α, $\bar{c} = c_0 h_1^2$ for $d/h_1^\Delta = 0.1$ and $d/h_1^\Delta = 0.01$, determined by employing the finite element solution corresponding to a three-element patch. Note that the robustness of the estimator deteriorates as $\bar{c}$ increases. This deterioration is much more pronounced for smooth solutions ($\alpha > 2$)
$(\mathcal{R}_{I_1^\Delta}^{ERDBp+2}(\alpha, d/h_{I_1^\Delta}, p, \bar{c})/\mathcal{R}_{I_1^\Delta}^{ERDBp+2}(\alpha, d/h_{I_1^\Delta}, p, \bar{c} \equiv 0)$
$u_{EX} \equiv (x+d)^\alpha$, $\beta = 2$, $p = 1)$

α	$\bar{c} = 0.01$	$\bar{c} = 0.1$	$\bar{c} = 1.0$	$\bar{c} = 10.0$	$\bar{c} = 100.0$
$d/h_1^\Delta = 0.1$					
0.55	0.8786	0.9180	1.6646	8.1728	11.4062
0.75	0.9000	0.9784	2.4867	15.3678	21.8884
0.95	0.9221	1.1205	4.5749	33.1179	47.0187
1.15	0.9449	1.5347	10.5860	82.6424	113.7438
1.35	0.9686	3.1002	30.8648	244.7607	311.7031
1.55	0.9932	9.4166	101.4400	799.7236	869.6769
1.75	1.0186	26.5186	254.8944	2019.3707	1723.0067
1.95	1.0450	36.5908	297.1761	2447.1565	1532.9005
2.15	1.0724	37.2500	253.6006	2223.6255	974.7138
2.35	1.1010	34.8601	199.7577	1608.1735	561.6605
$d/h_1^\Delta = 0.01$					
0.55	0.8790	0.8899	0.9817	1.7002	2.0733
0.75	0.8899	0.9156	1.1042	2.6451	3.3555
0.95	0.9651	0.9538	1.4253	5.2592	7.0440
1.15	0.9792	1.0358	2.4681	13.7691	18.4959
1.35	1.0093	1.3450	6.6290	46.8319	60.3090
1.55	1.0286	3.1500	27.6884	210.4691	238.1449
1.75	1.0516	15.9431	152.4431	1185.9203	1047.7484
1.95	1.0951	36.0064	293.5276	2404.1470	1520.4960
2.15	1.1199	34.6707	233.5145	2085.3076	885.5652
2.35	1.1513	33.0357	185.7747	1488.6765	506.4828

(3) Singular model problem with $AE \equiv (x+d)^\vartheta$, $c = 0$

We will analyze the effectivity index of the indicators in the first element (i.e. the element next to the singularity) for the class of homogeneous solutions. For this it is sufficient to employ the solution

$$u_{EX} \equiv (x+d)^{1-\vartheta}. \tag{5.7.31}$$

Note that the nodal value $u_{S_\Delta^p}(x_1^\Delta)$ is practically independent of the mesh outside I_1^Δ, and that the effectivity index $\kappa_{I_1^\Delta}$ of the element residual estimators is independent of the mesh.

TABLE 5.7.6b. Quality of the *ERDBp + 2* estimator for specially unsmooth solutions: The effect of the zeroth-order term. Input data: $AE \equiv 1$, $c \equiv c_0 > 0$, $u_{EX} \equiv (x+d)^\alpha$, $\alpha > 0.5$; quadratic elements ($p = 2$); $\beta = 2$. The relative robustness index $\bar{\mathcal{R}}_{I_1^\Delta}^{ERDBp+2}(\alpha, d/h_{I_1^\Delta}, p, \bar{c})$ as a function of α and $\bar{c} = c_0 h_1^2$ for $d/h_1^\Delta = 0.1$ and $d/h_1^\Delta = 0.01$, determined by employing the finite element solution corresponding to a three-element patch. Note that the robustness deteriorates as $\bar{c}$ increases, but this deterioration is much less than the one observed in Table 5.5.6a for the case of linear elements $(\mathcal{R}_{I_1^\Delta}^{ERDBp+2}(\alpha, d/h_{I_1^\Delta}, p, \bar{c})/\mathcal{R}_{I_1^\Delta}^{ERDBp+2}(\alpha, d/h_{I_1^\Delta}, p, \bar{c} \equiv 0)$, $u_{EX} \equiv (x+d)^\alpha$, $\beta = 2$, $p = 2)$

α	$\bar{c} = 0.01$	$\bar{c} = 0.1$	$\bar{c} = 1.0$	$\bar{c} = 10.0$	$\bar{c} = 100.0$
$d/h_1^\Delta = 0.1$					
0.55	0.8756	0.8857	0.9460	1.5929	4.5014
0.75	0.8957	0.9074	0.9805	1.7785	5.5944
0.95	0.9164	0.9305	1.0211	1.9987	7.0377
1.15	0.9378	0.9556	1.0641	2.2285	8.8600
1.35	0.9599	0.9752	1.1036	2.3973	14.8160
1.55	0.9828	1.0035	1.1647	4.2382	29.5694
1.75	1.0063	1.0132	1.2099	8.6459	60.1579
1.95	1.0307	1.0377	1.3060	18.0992	128.0685
2.15	1.0560	1.1517	1.5162	39.1462	295.7212
2.35	1.0822	1.3073	3.0708	99.1376	773.4268
$d/h_1^\Delta = 0.01$					
0.55	0.8722	0.8797	0.9010	1.0595	1.6611
0.75	0.8908	0.8987	0.9243	1.1281	1.9243
0.95	0.9099	0.9181	0.9495	1.2190	2.3186
1.15	0.9296	0.9392	0.9789	1.3390	2.9078
1.35	0.9499	0.9616	1.0124	1.4879	3.7554
1.55	0.9708	0.9835	1.0502	1.6434	4.8785
1.75	0.9923	1.0047	1.0898	1.7104	11.3056
1.95	1.0146	1.0326	1.1401	4.1410	28.4668
2.15	1.0375	1.0445	1.2226	10.2894	73.6918
2.35	1.0611	1.1310	1.4549	27.8248	214.5131

(a) *The ERDBp+1 estimator*: Table 5.7.10 gives $\kappa_{I_1^\Delta}^{ERDBp+1} = \kappa_{I_1^\Delta}^{ERDBp+1}(\vartheta, d/h_{I_1^\Delta}, p)$ for $p = 1, 2, 3, 4$. Note that as the singularity becomes stronger, i.e., as $\vartheta \to 1$, $d/h_{I_1^\Delta} \to 0$, the estimator underestimates severely. As the strength of the singularity increases, the robustness of the estimators deteriorates and that $k = 2$ is not sufficient to get good robustness for the entire range of ϑ.

(b) *The ERNp+1 estimator*: Table 5.7.11 gives $\kappa_{I_1^\Delta} = \kappa_{I_1^\Delta}(\vartheta, d/h_{I_1^\Delta}, p)$ of the *ERNp+1* estimator for various values of ϑ, $d/h_{I_1^\Delta}$ and $p = 1, 2, 3, 4$. Once more, as the strength of the singularity increases (i.e., $\vartheta \to 1$, $d/h_{I_1^\Delta} \to 0$), the quality of the estimator deteriorates.

(c) *The DRec2dPp estimator*: The results for the *DRec2dPp* estimator are reported in Table 5.7.12. Note that $\kappa_{I_1^\Delta}^{DRec2dPp}$ depends on the ratio $h_{I_2^\Delta}/h_{I_1^\Delta}$; here we analyzed

TABLE 5.7.6c. Quality of the *ERDBp + 2* estimator for specially unsmooth solutions: The effect of the zeroth-order term. This table gives the values of the relative robustness index $\bar{\mathcal{R}}^{ERDBp+2}_{I_1^\Delta}(\alpha, d/h_{I_1^\Delta}, p, \bar{c})$ for data as in Table 5.7.6a, computed by employing the finite element solution from a two-element patch (instead of the three-element patch which was employed in Table 5.7.6a). By comparing Tables 5.7.6a and 5.7.6c, we see significant differences in the computed value of the robustness index for $\alpha > 1.55$ and $\bar{c} \geq 10$. Nevertheless, the results for strong singularities ($\alpha < 1$) are practically the same ($\mathcal{R}^{ERDBp+2}_{I_1^\Delta}(\alpha, d/h_{I_1^\Delta}, p, \bar{c})/\mathcal{R}^{ERDBp+2}_{I_1^\Delta}(\alpha, d/h_{I_1^\Delta}, p, \bar{c} \equiv 0)$, $u_{EX} \equiv (x+d)^\alpha$, $d/h_1^\Delta = 0.1$, $p = 1$)

α	$\bar{c} = 0.01$	$\bar{c} = 0.1$	$\bar{c} = 1.0$	$\bar{c} = 10.0$	$\bar{c} = 100.0$
0.55	0.8672	0.9060	1.6985	7.9643	11.3828
0.75	0.8838	0.9566	2.5465	14.6986	21.7106
0.95	0.9008	1.0721	4.6805	30.9618	46.5295
1.15	0.9181	1.4018	10.7669	75.1925	112.3358
1.35	0.9361	2.5247	30.9790	215.9152	310.2197
1.55	0.9543	7.9755	120.3541	816.9424	1069.1912
1.75	0.9731	28.6571	432.7798	2889.7976	3161.1218
1.95	0.9922	59.6096	846.5432	5749.5640	6627.4536
2.15	1.0121	57.2600	753.0685	5477.4043	6563.6558
2.35	1.0323	57.1560	700.3834	5501.5654	7672.9668

TABLE 5.7.7. Robustness index of the *ERDBp + 1* estimator. Input data: $AE \equiv 1$, $c \equiv c_0 > 0$, $u_{EX} \equiv e^{-\mu x}$, $\mu = \sqrt{c_0}$, $\beta = 2$; element of degree p ($p = 1, 2, 3$). The robustness index $\mathcal{R}^{ERDBp+1}_{I_1^\Delta}(\beta, p, \bar{c})$ for various values of $\bar{c}$. Note that as $\bar{c}$ tends to infinity the robustness index deteriorates but this deterioration is delayed as the degree p of the element is increased ($\mathcal{R}^{ERDBp+1}_{I_1^\Delta}(\beta, p, \bar{c})$, $u_{EX} \equiv e^{-\mu x}$, $\mu = \sqrt{c_0}$, $\beta = 2$)

$\bar{c}$	$p = 1$	$p = 2$	$p = 3$
0.01	0.0008	0.0001	0.0001
0.1	0.1117	0.0007	0.0003
1.0	0.0694	0.0064	0.0024
10.0	0.4512	0.0816	0.0446
100.0	1.7829	0.8319	0.4792

the values of $\kappa_{I_1^\Delta}$ for $\frac{1}{2} \leq h_{I_2^\Delta}/h_{I_1^\Delta} \leq 2$. Table 5.7.12 reports the robustness index $\mathcal{R}^{DRec2dPp}_{I_1^\Delta}(\vartheta, d/h_{I_1^\Delta}, p)$, for various values of ϑ, $d/h_{I_1^\Delta}$, and $p = 1, 2, 3, 4$. Note that $\mathcal{R}_{I_1^\Delta}$ deteriorates as the singularity becomes stronger, i.e., as $\vartheta \to 1$, $d/h_{I_1^\Delta} \to 0$, while for given ϑ and $d/h_{I_1^\Delta}$ the $\mathcal{R}_{I_1^\Delta}$ improves as the degree of the element p is increased.

TABLE 5.7.8. Robustness index of the *ERNp + 1* estimator. Input data: $AE \equiv 1$, $c \equiv c_0 > 0$, $u_{EX} \equiv e^{-\mu x}$, $\mu = \sqrt{c_0}$; element of degree p $(p = 1, 2, 3)$; $K = 1$; $\beta = 2$. The robustness index $\mathcal{R}_{I_1^\Delta}^{ERNp+1}(\beta, p, \bar{c})$ for various values of $\bar{c}$. By comparing this table with Table 5.7.7 it can be seen that the *ERDPp + 1* and *ERNp + 1* estimators have practically the same robustness ($\mathcal{R}_{I_1^\Delta}^{ERNp+1}(\beta, p, \bar{c})$, $u_{EX} \equiv e^{-\mu x}$, $\mu = \sqrt{c_0}$, $k = 1$, $\beta = 2$)

$\bar{c}$	$p = 1$	$p = 2$	$p = 3$
0.01	0.0008	0.0001	0.0000
0.1	0.0084	0.0007	0.0003
1.0	0.0713	0.0033	0.0022
10.0	0.6619	0.0544	0.0264
100.0	1.8342	0.6191	0.0571

TABLE 5.7.9. Robustness index of the *DRec2dPp* estimator. Input data: $AE \equiv 1$, $c \equiv c_0 > 0$, $u_{EX} \equiv e^{-\mu x}$, $\mu = \sqrt{c_0}$; element of degree p $(p = 1, 2, 3)$; $\beta = 2$. The robustness index $\mathcal{R}_{I_1^\Delta}^{DRec2dPp}(\beta, p, \bar{c})$ for various values of $\bar{c}$. By comparing this table with Tables 5.7.8 and 5.7.7 it can be seen that this estimator is much less robust than the residual estimators ($\mathcal{R}_{I_1^\Delta}^{DRec2dPp}(\beta, p, \bar{c})$, $u_{EX} \equiv e^{-\mu x}$, $\mu = \sqrt{c_0}$, $\beta = 2$)

$\bar{c}$	$p = 1$	$p = 2$	$p = 3$
0.01	0.1042	0.0096	0.0074
0.1	0.3291	0.2011	0.2793
1.0	0.9743	0.4399	0.9942
10.0	2.9962	0.8661	1.0371
100.0	7.3042	2.4788	1.1748

The effectivity index is not very sensitive to the value of ϑ and that the quality of the estimator improves as the singularity is moved outside the domain.

The purpose of this analysis is to show that the performance of the estimator is poor in the areas where the solution is singular and that it improves with the distance to the singularity (neglecting the pollution).

The above analyses, although one-dimensional, concentrated on the *cases which have direct analogues in two dimensions.* Various insights and conclusions could be made from the analyses of the one-dimensional cases. We have to also note that the

TABLE 5.7.10. Effectivity index of the *ERDBp + 1* estimator for the homogeneous solution of the singular model problem. Input data: $AE = (x+d)^{\vartheta}$, $c \equiv 0$, $u_{EX} \equiv (x+d)^{1-\vartheta}$, $0 < \vartheta < 1$; elements of degree p. The effectivity index $\kappa_{I_1^\Delta}^{ERDBp+1}(\vartheta, d/h_{I_1^\Delta}, p)$ as a function of ϑ for $d/h_{I_1^\Delta} = 0, 0.01, 0.1$, and for elements of degree $p = 1, 2, 3, 4$. Note that as the strength of the singularity increases, i.e., $\vartheta \to 1$, $d/h_{I_1^\Delta} \to 0$, the estimator underestimates severely $(\kappa_{I_1^\Delta}^{ERDBp+1}(\vartheta, d/h_{I_1^\Delta}, p)$, $u_{EX} \equiv (x+d)^{1-\vartheta})$

ϑ	$d/h_{I_1^\Delta} = 0.1$	$d/h_{I_1^\Delta} = 0.01$	$d/h_{I_1^\Delta} = 0.0$	$d/h_{I_1^\Delta} = 0.1$	$d/h_{I_1^\Delta} = 0.01$	$d/h_{I_1^\Delta} = 0.0$
		$p = 1$			$p = 2$	
0.15	0.7927	0.7337	0.6900	0.7909	0.6916	0.5960
0.25	0.7854	0.7103	0.6570	0.7894	0.6759	0.5737
0.35	0.7749	0.6819	0.6160	0.7867	0.6572	0.5419
0.45	0.7616	0.6497	0.5679	0.7829	0.6357	0.5033
0.55	0.7458	0.6144	0.5133	0.7779	0.6118	0.4584
0.65	0.7277	0.5769	0.4510	0.7715	0.5854	0.4058
0.75	0.7076	0.5382	0.3791	0.7637	0.5572	0.3435
0.85	0.6860	0.4987	0.2915	0.7543	0.5274	0.2660
0.95	0.6631	0.4591	0.1669	0.7436	0.4964	0.1533
		$p = 3$			$p = 4$	
0.15	0.7594	0.6376	0.5054	0.7686	0.6247	0.4843
0.25	0.7597	0.6265	0.4933	0.7697	0.6172	0.4610
0.35	0.7596	0.6134	0.4686	0.7705	0.6068	0.4342
0.45	0.7589	0.5978	0.4368	0.7711	0.5945	0.4034
0.55	0.7576	0.5804	0.3988	0.7713	0.5807	0.3681
0.65	0.7559	0.5612	0.3541	0.7712	0.5652	0.3270
0.75	0.7533	0.5403	0.3007	0.7708	0.5483	0.2779
0.85	0.7499	0.5178	0.2335	0.7699	0.5298	0.2161
0.95	0.7458	0.4939	0.1350	0.7686	0.5100	0.1251

higher-dimensional problem is much richer and the extrapolation of the results from one dimension has to be made with care.

5.7* The asymptotic analysis of the effectiveness of the element error indicators in two dimensions

As we have already discussed above, the element error indicators reflect only the local error. In the one-dimensional case it is often easier to construct cases where we know a-priori that the pollution is negligible. For example, in the case $AE \equiv$ *constant*, $c \equiv 0$, the error localizes in each element and there is no pollution for any mesh, while in the case of the singular model problem with $AE \equiv x^{\vartheta}$, $c \equiv 0$, the pollution is practically negligible in the first element. Hence in the one-dimensional case we were able to analyze the indicators by employing various example problems and by carefully interpreting the results. The approach of analyzing the indicators by using example problems does not lead to results with clear meaning in higher dimensions where pollution is always present in many elements, and a more precise approach is needed. Here, as we have already

TABLE 5.7.11. Effectivity index of the *ERNp + 1* estimator for the homogeneous solution of the singular model problem. Input data: $AE = (x+d)^{\vartheta}$, $c \equiv 0$, $u_{EX} \equiv (x+d)^{1-\vartheta}$, $0 < \vartheta < 1$; elements of degree p. The effectivity index $\kappa_{I_1^{\Delta}}^{ERNp+1}(\vartheta, d/h_{I_1^{\Delta}}, p)$ as a function of ϑ, for $d/h_{I_1^{\Delta}} = 0, 0.01, 0.1$, and for elements of degree $p = 1, 2, 3, 4$. Note that as the strength of the singularity increases, i.e., as $\vartheta \to 1$, $d/h_{I_1^{\Delta}} \to 0$, the estimator underestimates severely. The results are very similar with the corresponding results for the *ERDPp + 1* estimator given in Table 5.7.10 ($\kappa_{I_1^{\Delta}}^{ERNPp+1}(\vartheta, d/h_{I_1^{\Delta}}, p)$, $u_{EX} \equiv (x+d)^{1-\vartheta}$)

ϑ	$d/h_{I_1^{\Delta}} = 0.1$	$d/h_{I_1^{\Delta}} = 0.01$	$d/h_{I_1^{\Delta}} = 0.0$	$d/h_{I_1^{\Delta}} = 0.1$	$d/h_{I_1^{\Delta}} = 0.01$	$d/h_{I_1^{\Delta}} = 0.0$
		$p = 1$			$p = 2$	
0.15	0.7956	0.7387	0.6910	0.7814	0.6819	0.5879
0.25	0.7933	0.7231	0.6660	0.7805	0.6683	0.5681
0.35	0.7901	0.7050	0.6351	0.7791	0.6529	0.5398
0.45	0.7858	0.6845	0.5974	0.7771	0.6360	0.5056
0.55	0.7804	0.6616	0.5525	0.7746	0.6176	0.4653
0.65	0.7739	0.6364	0.4980	0.7715	0.5977	0.4172
0.75	0.7665	0.6091	0.4299	0.7677	0.5764	0.3584
0.85	0.7579	0.5800	0.3400	0.7633	0.5538	0.2820
0.95	0.7482	0.5491	0.2003	0.7582	0.5299	0.1654
		$p = 3$			$p = 4$	
0.15	0.7594	0.6379	0.5025	0.7590	0.6152	0.4442
0.25	0.7598	0.6275	0.4913	0.7602	0.6081	0.4424
0.35	0.7599	0.6152	0.4681	0.7611	0.5985	0.4240
0.45	0.7596	0.6016	0.4384	0.7619	0.5874	0.3978
0.55	0.7589	0.5869	0.4028	0.7623	0.5753	0.3655
0.65	0.7579	0.5709	0.3605	0.7625	0.5621	0.3269
0.75	0.7565	0.5538	0.3090	0.7624	0.5478	0.2799
0.85	0.7547	0.5357	0.2427	0.7621	0.5327	0.2196
0.95	0.7525	0.5166	0.1421	0.7615	0.5167	0.1284

done in the examples of Section 5.3*, we will show how to employ the asymptotic setting of Section 4.7* to analyze the various indicators. This setting allows us to analyze the indicators without any biases associated with the selection of example problems, and to obtain conclusions with clear meaning.

In this section we will give examples of the asymptotic analysis of the quality of the element error indicators for finite element approximation of the orthotropic Poisson equation, $\boldsymbol{\nabla} \cdot (-\mathbf{K}\boldsymbol{\nabla} u) = f$. As we have already seen in Section 4.7*, there are, at least, three types of asymptotics which are important for the analysis of the behavior of the finite element method in engineering practice, namely:

(i) in the interior of the mesh

(ii) adjacent to the boundary

(iii) adjacent to a corner point.

TABLE 5.7.12. Robustness index of the *DRec2dPp* estimator for the homogeneous solution of the singular model problem. Input data: $AE = (x+d)^{\vartheta}$, $c \equiv 0$, $u_{EX} \equiv (x+d)^{1-\vartheta}$, $0 < \vartheta < 1$; elements of degree p. The robustness index $\mathcal{R}^{DRec2dPp}_{I_1^{\Delta}}(\vartheta, d/h_{I_1^{\Delta}}, p)$ as a function of ϑ for $d/h_{I_1^{\Delta}} = 0, 0.01, 0.1$, and for elements of degree $p = 1, 2, 3, 4$. Note that the robustness index of the estimator deteriorates as the strength of the singularity is increased, i.e., as $\vartheta \to 1$, $d/h_{I_1^{\Delta}} \to 0$. By comparing this table with Tables 5.7.10 and 5.7.11 we conclude that the *DRec2dPp* estimator is significantly less robust than the *ERDPp + 1* and the *ERNp + 1* estimators ($\mathcal{R}^{DRec2dPp}_{I_1^{\Delta}}(\vartheta, d/h_{I_1^{\Delta}}, p)$, $u_{EX} \equiv (x+d)^{1-\vartheta}$)

ϑ	$d/h_{I_1^{\Delta}} = 0.1$	$d/h_{I_1^{\Delta}} = 0.01$	$d/h_{I_1^{\Delta}} = 0.0$	$d/h_{I_1^{\Delta}} = 0.1$	$d/h_{I_1^{\Delta}} = 0.01$	$d/h_{I_1^{\Delta}} = 0.0$
		$p = 1$			$p = 2$	
0.15	2.0504	3.1334	3.6958	1.6667	2.9286	3.9963
0.25	2.3600	3.6795	4.3868	1.7850	3.1963	4.3957
0.35	2.6866	4.2719	5.1971	1.9050	3.4817	4.9104
0.45	3.0309	4.9187	6.1711	2.0272	3.7884	5.5646
0.55	3.3940	5.6310	7.3955	2.1523	4.1216	6.4087
0.65	3.7773	6.4224	9.0405	2.2809	4.4865	7.5803
0.75	4.1824	7.3102	11.4952	2.4137	4.8897	9.3816
0.85	4.6113	8.3150	15.9464	2.5514	5.3385	12.7404
0.95	5.0665	9.4623	29.9183	2.6946	5.8413	23.5724
		$p = 3$			$p = 4$	
0.15	1.3048	2.1685	3.5217	1.1491	1.4897	2.8992
0.25	1.3289	2.3417	3.7731	1.1683	1.5444	3.0306
0.35	1.3541	2.5321	4.1866	1.1885	1.6247	3.3505
0.45	1.3802	2.7401	4.7334	1.2098	1.7688	3.8011
0.55	1.4076	2.9660	5.4663	1.2323	1.9239	4.4162
0.65	1.4363	3.2137	6.4886	1.2561	2.0916	5.2811
0.75	1.4663	3.4861	8.0699	1.2813	2.2734	6.6219
0.85	1.4980	3.7871	11.0286	1.3079	2.4715	9.1341
0.95	1.5670	4.1211	20.5739	1.3361	2.6877	17.2523

Let us give a list of the error indicators which will be addressed below. The notation is similar to that used in Section 5.7.

(a) *ERqPp + k*, $1 \leq q \leq p$: **E**lement **R**esidual estimator of the Neumann type with residuals equilibrated up to order q and approximation of the indicators by **P**olynomials of degree $p + k$.

(b) *ERqBp + k*, $1 \leq q \leq p$: **E**lement **R**esidual estimator of the Neumann type with residuals equilibrated up to order q and approximation of the indicators by **B**ubble polynomials of degree $p + k$.

A detailed description of these estimators can be found in Section 5.3*.

Below we will utilize various equilibrations introduced in Section 5.3*.

(c) *ER_Bp + k*: **E**lement **R**esidual estimator with unequilibrated residual and **B**ubble solution of degree $p + k$; see Section 5.3*.

(d) *Explicit*: Explicit **R**esidual estimator; see Section 5.5*.

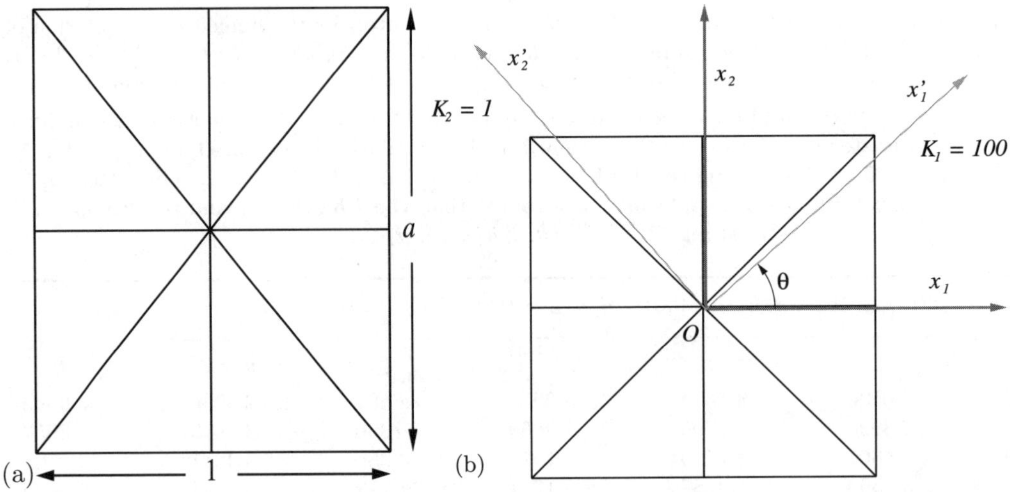

FIG. 5.7*.1. A typical basic mesh cell with: (a) Aspect ratio $1/a$; (b) The orientation of the principal material direction with respect to the mesh

(e) *ZZ*: Discrete Zienkiewicz Zhu indicator, which also will be referred to as *ZZ-SPR*; see Sections 4.8* and 5.3*.

(f) *ZZ*: Continuous Zienkiewicz Zhu estimator, which employs a continuous L^2 least-squares recovery in each patch.

We will determine the asymptotic quality of an error indicator for the master meshes $\tilde{T}$ from the class $\mathcal{T}$ which reflects the mesh topologies of interest with respect to the class of solutions $\mathcal{Q}$ of interest, and the class of coefficients of the differential equation $\mathcal{M}$ which are of interest. For the orthotropic Poisson equation

$$K_{11}\frac{\partial^2 u}{\partial x_1^2} + 2K_{12}\frac{\partial^2 u}{\partial x_1 x_2} + K_{22}\frac{\partial^2 u}{\partial x_2^2} = f \qquad (5.7^*.1a)$$

the class $\mathcal{M}$ is given by

$$\mathcal{M} = \{(K_1, K_2, \theta) \mid 0 < K_{\min} \le K_1/K_2 \le K_{\max} < \infty,\ \theta_{min} \le \theta \le \theta_{max}\} \qquad (5.7^*.1b)$$

where K_1, K_2 are the principal orthotropies, θ is the *grid-material orientation* as shown in Fig. 5.7*.1.

Let us first address the analysis of the indicators in the interior of the mesh. In this case the class of solutions $\mathcal{Q}$ is the class of monomials which satisfy the homogeneous differential equation, which we will call harmonic (i.e., monomials satisfying the homogeneous differential equation), or the class of general monomials of degree $p+1$, and the effectivity index depends only on the ratio K_1/K_2, the angle θ, the polynomial $Q^{(p+1)}$, and the mesh $\tilde{T}$, namely

$$\kappa^{EST} = \kappa^{EST}(Q^{(p+1)}; \tilde{T}; K_1, K_2, \theta). \qquad (5.7^*.2a)$$

We will analyze an indicator by determining the bounds

$$\mathscr{C}_L^{EST} = \min_{(K_1,K_2,\theta)\in\mathscr{M}} \min_{\tilde{T}\in\mathscr{T}} \min_{Q^{(p+1)}\in\mathfrak{Q}^{(p+1)}} \kappa^{EST}(Q^{(p+1)};\tilde{T};K_1,K_2,\theta) \tag{5.7*.2b}$$

and

$$\mathscr{C}_U^{EST} = \max_{(K_1,K_2,\theta)\in\mathscr{M}} \max_{\tilde{T}\in\mathscr{T}} \max_{Q^{(p+1)}\in\mathfrak{Q}^{(p+1)}} \kappa^{EST}(Q^{(p+1)};\tilde{T};K_1,K_2,\theta) \tag{5.7*.2c}$$

and the *robustness index*

$$\mathcal{R}^{EST}(\mathscr{T},\mathscr{M},\mathfrak{Q}^{(p+1)}) = \max\Big(\mathscr{C}_U^{EST}-1, \frac{1}{\mathscr{C}_L^{EST}}-1\Big). \tag{5.7*.2d}$$

Let us now explain how the robustness index is determined for interior mesh cells. Letting

$$\tilde{Q}^{(p+1)} = \sum_{i=1}^{s} a_i \tilde{Q}_i^{(p+1)} \tag{5.7*.3a}$$

where $\tilde{Q}_i^{(p+1)}$ is the ith basis monomial of degree $p+1$, and s is the number of basis monomials in the class $\mathfrak{Q}^{(p+1)}$, we have for the asymptotic error

$$\tilde{\Psi}^{(p+1)} \stackrel{\text{def}}{=} \sum_{i=1}^{s} a_i \tilde{\Psi}_i^{(p+1)} \tag{5.7*.3b}$$

where $\tilde{\Psi}_i^{(p+1)}$ is the asymptotic error corresponding to $\tilde{Q}_i^{(p+1)}$. Similarly, denoting by $\hat{\Psi}_i^{(p+1)}$ the asymptotic error indicator function corresponding to $Q_i^{(p+1)}$, we have for the error indicator function

$$\hat{\Psi}^{(p+1)} \stackrel{\text{def}}{=} \sum_{i,j=1}^{s} a_i \hat{\Psi}_i^{(p+1)}. \tag{5.7*.3c}$$

Denoting the master cell by $\tilde{c}$, we have

$$\begin{aligned} \|\tilde{\Psi}^{(p+1)}\|^2_{\mathcal{U}(\tilde{c})} &= \int_{\tilde{c}} \boldsymbol{\nabla}\tilde{\Psi}^{(p+1)}\cdot\boldsymbol{\nabla}\tilde{\Psi}^{(p+1)} = \sum_{i,j=1}^{p+1} a_i a_j \int_{\tilde{c}} \boldsymbol{\nabla}\tilde{\Psi}_i^{(p+1)}\cdot\boldsymbol{\nabla}\tilde{\Psi}_j^{(p+1)} \\ &= \mathbf{a}^T\mathbf{M}\mathbf{a} \end{aligned} \tag{5.7*.4a}$$

where $\mathbf{a}\in\mathbb{R}^{p+1}$, and $\mathbf{M}\in\mathbb{R}^{(p+1)\times(p+1)}$ with

$$M_{ij} = \int_{\tilde{c}} \boldsymbol{\nabla}\tilde{\Psi}_i^{(p+1)}\cdot\boldsymbol{\nabla}\tilde{\Psi}_j^{(p+1)}, \quad i,j=1,\dots,p+1. \tag{5.7*.4b}$$

The matrix $\mathbf{M}$ is symmetric and positive definite, and can be decomposed in the form

$$\mathbf{M} = \mathbf{F}^T\mathbf{F}, \qquad \mathbf{F} = \sqrt{\mathbf{M}} \stackrel{\text{def}}{=} \mathbf{Q}\sqrt{\mathbf{\Lambda}}\mathbf{Q}^T \tag{5.7*.4c}$$

where $\mathbf{Q}$ is the orthogonal matrix with the columns the normalized eigenvectors of $\mathbf{M}$ and $\sqrt{\mathbf{\Lambda}}$ is the diagonal matrix with the square roots of the corresponding eigenvalues of $\mathbf{M}$ in its diagonal. Similarly we have for the error indicator

$$\mathscr{E}_{\tilde{T}}^2 = \sum_{\tilde{\tau}\in\tilde{T}} \int_{\tilde{\tau}} \boldsymbol{\nabla}\hat{\Psi}^{(p+1)} \cdot \boldsymbol{\nabla}\hat{\Psi}^{(p+1)} = \sum_{i,j=1}^{p+1} a_i a_j \sum_{\tilde{\tau}\in\tilde{T}} \int_{\tilde{\tau}} \boldsymbol{\nabla}\hat{\Psi}_i^{(p+1)} \cdot \boldsymbol{\nabla}\hat{\Psi}_j^{(p+1)}$$
$$= \mathbf{a}^T\hat{\mathbf{M}}\mathbf{a} \tag{5.7*.5a}$$

where $\tilde{T}$ is the master mesh over the master cell $\tilde{c}$, $\tilde{\tau}$ is an element of $\tilde{T}$ and

$$\hat{M}_{ij} = \sum_{\tilde{\tau}\in\tilde{T}} \int_{\tilde{\tau}} \boldsymbol{\nabla}\hat{\Psi}_i^{(p+1)} \cdot \boldsymbol{\nabla}\hat{\Psi}_j^{(p+1)}, \quad i,j = 1,\ldots,p+1. \tag{5.7*.5b}$$

Note that $\hat{\mathbf{M}}$ depends on the choice of indicator, while $\mathbf{M}$ does not. Next we express the effectivity index as the Rayleigh quotient

$$\kappa_{\tilde{T}} = \frac{\mathscr{E}_{\tilde{T}}}{|||\tilde{\Psi}^{(2)}|||_{\tilde{c}}} = \sqrt{\frac{\mathbf{a}^T\hat{\mathbf{M}}\mathbf{a}}{\mathbf{a}^T\mathbf{M}\mathbf{a}}} = \sqrt{\frac{\mathbf{a}^T\mathbf{F}^T}{|\mathbf{F}\mathbf{a}|}\mathbf{F}^{-T}\hat{\mathbf{M}}\mathbf{F}^{-1}\frac{\mathbf{F}\mathbf{a}}{|\mathbf{F}\mathbf{a}|}} = \sqrt{\mathbf{v}^T\mathbf{K}\mathbf{v}} \tag{5.7*.6a}$$

where we let $\mathbf{v} = \mathbf{F}\mathbf{a}/|\mathbf{F}\mathbf{a}|$, and $\mathbf{K} \stackrel{\text{def}}{=} \mathbf{F}^{-T}\hat{\mathbf{M}}\mathbf{F}^{-1}$. It follows that

$$\sqrt{\mu_1} \le \kappa_{\tilde{T}} \le \sqrt{\mu_{p+1}} \tag{5.7*.6b}$$

where $0 \le \mu_1 \le \mu_2 \le \cdots \le \mu_{p+1} < \infty$ are the eigenvalues of $\mathbf{K}$.

Let us now give an example.

Example 5.7*.1 (*Asymptotic robustness of the ERpB and ZZ-SPR error indicators for linear elements* ($p = 1$), *for the criss-cross and the chevron patterns*)
Let us first determine the asymptotic bounds for the criss-cross pattern, which is shown in Fig. 4.7*.3d. Employing the monomial basis

$$Q_1^{(2)}(\tilde{x}_1,\tilde{x}_2) = \tilde{x}_1^2, \qquad Q_2^{(2)}(\tilde{x}_1,\tilde{x}_2) = \tilde{x}_1\tilde{x}_2, \qquad Q_3^{(2)}(\tilde{x}_1,\tilde{x}_2) = \tilde{x}_2^2 \tag{5.7*.7a}$$

we obtain

$$\mathbf{M} = \begin{bmatrix} 0.2222 & 0.0000 & -0.1111 \\ 0.0000 & 0.0833 & 0.0000 \\ -0.1111 & 0.0000 & 0.2222 \end{bmatrix},$$
$$\hat{\mathbf{M}}^{ER1B} = \begin{bmatrix} 0.2129 & 0.0000 & -0.1209 \\ 0.0000 & 0.0833 & 0.0000 \\ -0.1209 & 0.0000 & 0.2129 \end{bmatrix} \tag{5.7*.7b}$$

and we get

$$\mu_1^{ER1B} = \tfrac{5}{6}, \qquad \mu_2^{ER1B} = \mu_3^{ER1B} = 1. \tag{5.7*.7c}$$

Also

$$\mathbf{w}_1^{ER1B} = \left[\frac{1}{\sqrt{2}}, 0, \frac{1}{\sqrt{2}}\right]^T, \qquad \mathbf{w}_2^{ER1B} = [0, 1, 0]^T,$$
$$\mathbf{w}_3^{ER1B} = \left[-\frac{1}{\sqrt{2}}, 0, \frac{1}{\sqrt{2}}\right]^T \tag{5.7*.7d}$$

and we obtain

$$\sqrt{\tfrac{5}{6}} \le \kappa_{\tilde{T}}^{ER1B} \le 1 \tag{5.7*.7e}$$

with the robustness index

$$\mathcal{R}^{ER1B} = \sqrt{\tfrac{6}{5}} - 1. \tag{5.7*.7f}$$

The *ZZ-SPR* estimator for the criss-cross pattern gives $\hat{\mathbf{M}}^{ZZ} = \mathbf{M}$, and hence $\kappa_{\tilde{T}}^{ZZ} = 1$, and

$$\mathcal{R}^{ZZ} = 0 \tag{5.7*.7g}$$

which means that the *ZZ-SPR estimator is asymptotically exact for the criss-cross pattern.* Let us underline that *the asymptotic exactness is a very special property.* For example, if we perturb the central node of the criss-cross pattern from the point (0.5, 0.5) to the point (0.4, 0.6), we get $0.999 \le \kappa_{\tilde{c}}^{ZZ} \le 1.017$, with the robustness index $\mathcal{R}^{ZZ} = 0.017$, and the *ZZ-SPR* estimator is no longer asymptotically exact.

Let us now assume that we are interested in the bounds of the effectivity index for the same estimators and the basic cell of the chevron pattern which is shown in Fig. 4.7*.3b. We have

$$\mathbf{M} = \begin{bmatrix} 0.0417 & 0.0000 & 0.0000 \\ 0.0000 & 0.0417 & 0.0000 \\ 0.0000 & 0.0000 & 0.0417 \end{bmatrix}, \qquad \hat{\mathbf{M}}^{ZZ} = \begin{bmatrix} 0.0417 & 0.0000 & 0.0000 \\ 0.0000 & 0.0366 & 0.0000 \\ 0.0000 & 0.0000 & 0.0425 \end{bmatrix} \tag{5.7*.8a}$$

and we get

$$\mu_1^{ZZ} = 0.878, \qquad \mu_2^{ZZ} = 1, \qquad \mu_3^{ZZ} = 1.020 \tag{5.7*.8b}$$

with

$$\mathbf{w}_1^{ZZ} = [0, 1, 0]^T, \qquad \mathbf{w}_2^{ZZ} = [0, 0, 1]^T, \qquad \mathbf{w}_3^{ZZ} = [1, 0, 0]^T \tag{5.7*.8c}$$

and

$$\sqrt{0.878} \le \kappa_{\tilde{T}}^{ZZ} \le \sqrt{1.020} \tag{5.7*.8d}$$

with robustness index

$$\mathcal{R}^{ZZ} = \max\left(\sqrt{1.020} - 1, \frac{1}{\sqrt{0.878}} - 1\right) = 0.067. \tag{5.7*.8e}$$

Therefore the *ZZ-SPR* estimator is not asymptotically exact for the chevron pattern. For the *ERpB* estimator we have $\hat{\mathbf{M}}^{ERpB} = \mathbf{M}$ and hence, the *ERpB* estimator is asymptotically exact for the chevron pattern.

Note that *the robustness of an estimator depends on the mesh pattern.* The best estimator is the one with the smallest robustness index over the set of the mesh patterns which can occur in the meshes of interest. For example, if the meshes of interest include only the chevron and the criss-cross patterns and linear elements ($p = 1$), then, between the *ZZ-SPR* and *ER2B* estimators we would prefer the *ZZ-SPR* estimator for interior mesh cells.

Let us now assume that we are interested in the case $f \equiv 0$, i.e., the case of the Laplacian. Employing the quadratic harmonic monomials

$$\tilde{Q}^{(2)} = a_1 \tilde{Q}^{(2)}_{harm,1} + a_2 \tilde{Q}^{(2)}_{harm,2} \tag{5.7*.9a}$$

where

$$\tilde{Q}^{(2)}_{harm,1}(\tilde{x}_1, \tilde{x}_2) = \tilde{x}_1^2 - \tilde{x}_2^2, \qquad \tilde{Q}^{(2)}_{harm,2}(\tilde{x}_1, \tilde{x}_2) = 2\tilde{x}_1\tilde{x}_2 \tag{5.7*.9b}$$

we obtain for the chevron pattern

$$\mathbf{M} = \begin{bmatrix} 0.0208 & -2.77(-5) \\ -2.77(-5) & 0.0416 \end{bmatrix}, \qquad \hat{\mathbf{M}}^{ZZ} = \begin{bmatrix} 0.0210 & 0.0000 \\ 0.0000 & 0.0365 \end{bmatrix} \tag{5.7*.10a}$$

and we get

$$\sqrt{0.878} \le \kappa_{\tilde{T}}^{ZZ} \le \sqrt{1.007} \tag{5.7*.10b}$$

with robustness index

$$\mathcal{R}^{ZZ} = \max\left(\sqrt{1.007} - 1, \frac{1}{\sqrt{0.878}} - 1\right) = 0.067. \tag{5.7*.10c}$$

Hence, *when the class of solutions is restricted to the homogeneous class, the robustness of the estimator improves.* Above we reported the robustness of an indicator over an entire cell. Of course, we may also determine the robustness over a single element.

Let us now summarize a study of the effect of the mesh pattern and the aspect ratio on the robustness of the error estimators.

Example 5.7*.2 (Analysis of the robustness of various estimators for interior mesh cells.) We analyzed the asymptotic robustness of the estimators for linear ($p = 1$) and quadratic ($p = 2$) elements for the four basic patterns of mesh shown in Fig. 4.7*.3. We will analyze the estimators for the Laplacian by employing the harmonic polynomials

$$Q^{(p+1)}(x_1, x_2) = a_1 \mathrm{Re}((x_1 + ix_2)^{p+1}) + a_2 \mathrm{Im}((x_1 + ix_2)^{p+1}) \tag{5.7*.11a}$$

and for the Poisson equation by employing the general polynomials

$$Q^{(p+1)}(x_1, x_2) = \sum_{i,j=0}^{p+1} a_{ij} x_1^i x_2^j, \quad i + j = p + 1, \ i, j \le 0. \tag{5.7*.11b}$$

In the case of the Laplacian, the bounds of the effectivity index can be determined in closed form. Tables 5.7*.1–5.7*.4 list the formulae for $\mathscr{C}_L$ and $\mathscr{C}_U$ for the various estimators as functions of the aspect ratio a, which is defined in Fig. 5.7*.1. We used the above formulae for $\mathscr{C}_L$ and $\mathscr{C}_U$ to graph the robustness index; these are shown in Figs 5.7*.2–5.7*.9. We also determined the robustness indices of the estimators for general polynomials; their graphs are given in Figs 5.7*.10–5.7*.17.

From the results in Figs 5.7*.2–5.7*.17 we make the following conclusions:

1. The Continuous and Discrete ZZ error estimators are insensitive to the mesh geometry, they are among the most reliable error estimators, and they have practically the same robustness.
2. ER_B is very reliable for the regular pattern, while for the Union-Jack and criss-cross patterns, it is very sensitive to the aspect ratio of the mesh, and unreliable for large or small a. For the chevron pattern, ER_B is reliable for $a \leq 1$, but unreliable for $a \gg 1$.
3. Generally the ERqBp + k estimator is more reliable than the ERqPp + k estimator, for each one of the three equilibrations employed.

TABLE 5.7*.1. The upper and lower bounds, $\mathscr{C}_U$ and $\mathscr{C}_L$, of the effectivity index for the estimate of the energy norm in the basic cell of the periodic meshes for the Explicit and the Discrete ZZ error estimators as functions of a, where $1/a$ is the aspect ratio of the mesh for linear elements ($p = 1$)

	Explicit	Discrete ZZ
Regular	$\mathscr{C}_U = \sqrt{6}\sqrt{\frac{a^4+3a^2+1+\sqrt{a^8-a^4+1}}{a(a^2+1)}}$	$\mathscr{C}_U = 1$
	$\mathscr{C}_L = \sqrt{6}\sqrt{\frac{a^4+3a^2+1-\sqrt{a^8-a^4+1}}{a(a^2+1)}}$	$\mathscr{C}_L = 1$
Chevron	$\mathscr{C}_U = \sqrt{6}\sqrt{\frac{a^4+2a^2+2}{a(a^2+1)}}$	$\mathscr{C}_U = \frac{1}{7}\sqrt{\frac{50a^2+49}{a^2+1}}$
	$\mathscr{C}_L = 2\sqrt{6}\sqrt{\frac{a}{a^2+1}}$	$\mathscr{C}_L = \frac{1}{7}\sqrt{\frac{37a^2+49}{a^2+1}}$
Union Jack	$\mathscr{C}_U = \sqrt{6}\sqrt{\frac{a^2+1}{a}}$	$\mathscr{C}_U = 1$
	$\mathscr{C}_L = 2\sqrt{6}\sqrt{\frac{a}{a^2+1}}$	$\mathscr{C}_L = 1$
Criss cross	$\mathscr{C}_U = \sqrt{6}\sqrt{\frac{a^2+1}{a}}$	$\mathscr{C}_U = 1$
	$\mathscr{C}_L = 2\sqrt{6}\sqrt{\frac{a}{a^2+1}}$	$\mathscr{C}_L = 1$

TABLE 5.7*.2. The upper and lower bounds, $\mathscr{C}_U$ and $\mathscr{C}_L$, of the effectivity index for the estimate of the energy norm in the basic cell of the periodic meshes for the Continuous ZZ and the ER_B error estimators as functions of a, where $1/a$ is the aspect ratio of the mesh for linear elements ($p = 1$)

	Continuous ZZ	ER_B
Regular	$\mathscr{C}_U = 1$	$\mathscr{C}_U = 1$
	$\mathscr{C}_L = 1$	$\mathscr{C}_L = 1$
Chevron	$\mathscr{C}_U = \frac{1}{37}\sqrt{\frac{1370a^2 + 1369}{a^2+1}}$	$\mathscr{C}_U = \frac{\sqrt{2}}{2}\frac{\sqrt{a^6 + 4a^4 + 4a^2 + 2}}{a^2+1}$
	$\mathscr{C}_L = \frac{1}{37}\sqrt{\frac{949a^2 + 1369}{a^2+1}}$	$\mathscr{C}_L = 1$
Union Jack	$\mathscr{C}_U = 1$	$\mathscr{C}_U = \frac{\sqrt{2}}{2}\frac{a^2+1}{a}$
	$\mathscr{C}_L = 1$	$\mathscr{C}_L = 1$
Criss cross	$\mathscr{C}_U = 1$	$\mathscr{C}_U = \begin{cases} 2\sqrt{2}\frac{a}{a^2+1}, & \text{if } \sqrt{2}-1 \le a \le \sqrt{2}+1, \\ 1, & \text{otherwise.} \end{cases}$
	$\mathscr{C}_L = 1$	$\mathscr{C}_L = \begin{cases} 1, & \text{if } \sqrt{2}-1 \le a \le \sqrt{2}+1, \\ 2\sqrt{2}\frac{a}{a^2+1}, & \text{otherwise.} \end{cases}$

4. ERqBp+k(Lad2) is among the most reliable error estimators for all four of the mesh patterns. On the other hand, ERqPp + k(Lad2) is very unreliable for some cases (e.g., for the criss-cross pattern; nevertheless, its reliability can be improved by using quadratic equilibration; see Figs 5.7*.22–5.7*.25 and Figs 5.7*.30–5.7*.33 below).
5. Both ERqBp + k and ERqPp + k, with Bank–Weiser equilibration and Ladeveze equilibration without weight, are sensitive to mesh geometry. For example, for the Union-Jack pattern, both ERqBp + k and ERqPp + k with Bank–Weiser or Lad1 equilibration are unreliable for $a \ll 1$. Nevertheless ERqPp + k(Lad1) will be much improved by using quadratic equilibration; see Figs 5.7*.22–5.7*.25 and Figs 5.7*.30–5.7*.33.
6. The Explicit error estimator is very sensitive to the mesh geometry. Generally, it is unreliable, although it is reasonably reliable for $a \approx 1$.

Let us underline that the above conclusions hold, asymptotically, for mesh cells in the interior of the mesh. Although the above asymptotic analysis is an important tool for selecting an estimator, it is not the only one. See, for example, Chapter 6, where the problem of guaranteed a-posteriori estimation is addressed.

In order to analyze the effect of the order of the equilibration, we computed the robustness index for the estimators with quadratic equilibration, in order to investigate the linear and high order equilibrations.

TABLE 5.7*.3. The upper and lower bounds, $\mathscr{C}_U$ and $\mathscr{C}_L$, of the effectivity index for the estimate of the energy norm in the basic cell of the periodic meshes for the $ER1B2$ error estimators with three types of equilibration: (a) Bank–Weiser, (b) Ladeveze without weight, and (c) Ladeveze with weight. $\mathscr{C}_L$ and $\mathscr{C}_U$ are given as functions of a, where $1/a$ is the aspect ratio of the mesh for linear elements ($p = 1$)

(a) *ER1B2 (Bank–Weiser)*

Regular $\mathscr{C}_U = 1$

$\mathscr{C}_L = 1$

Chevron $\mathscr{C}_U = 1$

$\mathscr{C}_L = 1$

Union Jack $\mathscr{C}_U = \dfrac{\sqrt{2}}{2}\dfrac{\sqrt{2a^2+1}}{a}$

$\mathscr{C}_L = 1$

Criss cross
$$\mathscr{C}_U = \begin{cases} \dfrac{\sqrt{2}}{2}\dfrac{\sqrt{a^4+10a^2+1}}{a^2+1}, & \text{if } \sqrt{3-2\sqrt{2}} \le a \le \sqrt{3+2\sqrt{2}}, \\ 1, & \text{otherwise.} \end{cases}$$
$$\mathscr{C}_L = \begin{cases} 1, & \text{if } \sqrt{3-2\sqrt{2}} \le a \le \sqrt{3+2\sqrt{2}}, \\ \dfrac{\sqrt{2}}{2}\dfrac{\sqrt{a^4+10a^2+1}}{a^2+1}, & \text{otherwise.} \end{cases}$$

(b) *ER1B2 (Ladeveze 1)*

Regular $\mathscr{C}_U = 1$

$\mathscr{C}_L = 1$

Chevron $\mathscr{C}_U = 1$

$\mathscr{C}_L = 1$

Union Jack $\mathscr{C}_U = \dfrac{\sqrt{2}}{4}\dfrac{\sqrt{a^4+6a^2+1}}{a}$

$\mathscr{C}_L = 1$

Criss cross $\mathscr{C}_U = 1$

$\mathscr{C}_L = \dfrac{\sqrt{2}}{2}\dfrac{\sqrt{a^4+6a^2+1}}{a^2+1}$

(c) *ER1B2 (Ladeveze 2)*

Regular $\mathscr{C}_U = 1$

$\mathscr{C}_L = 1$

Chevron $\mathscr{C}_U = 1$

$\mathscr{C}_L = 1$

Union Jack $\mathscr{C}_U = \dfrac{\sqrt{2}}{2}\sqrt{\dfrac{2a^8+17a^6+34a^4+17a^2+2}{a^8+8a^6+18a^4+8a^2+1}}$

$\mathscr{C}_L = 1$

Criss cross $\mathscr{C}_U = \dfrac{1}{2}\sqrt{\dfrac{5a^8+80a^6+406a^4+80a^2+5}{a^8+20a^6+102a^4+20a^2+1}}$

$\mathscr{C}_L = \dfrac{\sqrt{2}}{2}\dfrac{\sqrt{a^4+6a^2+1}}{a^2+1}$

From the results in Figs 5.7*.18–5.7*.33 we make the following conclusions:

1. The main factor which affects the robustness of the equilibrated element residual estimators is the type of equilibration. The best robustness is obtained for the Ladeveze, Type 2 equilibration.
2. In some cases, a significant improvement of the robustness may be obtained by employing the maximal order of equilibration, $q = p$. For example, in the case of Ladeveze equilibration for the Union-Jack and the criss-cross patterns

Let us also analyze the effect of the orthotropy. The conclusions are similar to the ones obtained from the analysis of the aspect ratio. We will analyze the robustness of the

TABLE 5.7*.4. The upper and lower bounds, $\mathscr{C}_U$ and $\mathscr{C}_L$, of the effectivity index for the estimate of the energy norm in the basic cell of the periodic meshes for the ER1P2 error estimators with three equilibrations: (a) Bank–Weiser, (b) Ladeveze without weight, and (c) Ladeveze with weight. $\mathscr{C}_L$ and $\mathscr{C}_U$ are given as functions of a, where $1/a$ is the aspect ratio of the mesh for linear elements ($p = 1$)

(a)	*ER1P2 (Bank–Weiser)*	
	Regular	$\mathscr{C}_U = \sqrt{3}$
		$\mathscr{C}_L = \dfrac{\sqrt{6}}{2}$
	Chevron	$\mathscr{C}_U = \sqrt{3}$
		$\mathscr{C}_L = \dfrac{\sqrt{6}}{2}$
	Union Jack	$\mathscr{C}_U = \dfrac{\sqrt{3}}{2}\dfrac{\sqrt{4a^2+1}}{a}$
		$\mathscr{C}_L = \dfrac{\sqrt{6}}{2}$
	Criss cross	$\mathscr{C}_U = \dfrac{\sqrt{6}}{2}\sqrt{\dfrac{5a^4+18a^2+5+\sqrt{a^8+28a^6+86a^4+28a^2+1}}{3a^4+10a^2+3}}$
		$\mathscr{C}_L = \dfrac{\sqrt{6}}{2}\sqrt{\dfrac{5a^4+18a^2+5-\sqrt{a^8+28a^6+86a^4+28a^2+1}}{3a^4+10a^2+3}}$
(b)	*ER1P2 (Ladeveze 1)*	
	Regular	$\mathscr{C}_U = \sqrt{3}$
		$\mathscr{C}_L = \dfrac{\sqrt{6}}{2}$
	Chevron	$\mathscr{C}_U = \sqrt{3}$
		$\mathscr{C}_L = \dfrac{\sqrt{6}}{2}$
	Union Jack	$\mathscr{C}_U = \dfrac{\sqrt{3}}{4}\dfrac{\sqrt{a^4+14a^2+1}}{a}$
		$\mathscr{C}_L = \dfrac{\sqrt{6}}{2}$
	Criss cross	$\mathscr{C}_U = \sqrt{3}$
		$\mathscr{C}_L = \sqrt{6}\dfrac{a^2+1}{\sqrt{3a^4+10a^2+3}}$

TABLE 5.7*.4. (*Continued.*)

(c)	*ER1P2 (Ladeveze 2)*	
	Regular	$\mathscr{C}_U = \sqrt{3}$
		$\mathscr{C}_L = \frac{\sqrt{6}}{2}$
	Chevron	$\mathscr{C}_U = \sqrt{3}$
		$\mathscr{C}_L = \frac{\sqrt{6}}{2}$
	Union Jack	$\mathscr{C}_U = \frac{\sqrt{3}}{2}\sqrt{\frac{4a^8 + 33a^6 + 70a^4 + 33a^2 + 4}{a^8 + 8a^6 + 18a^4 + 8a^2 + 1}}$
		$\mathscr{C}_L = \frac{\sqrt{6}}{2}$
	Criss cross	$\mathscr{C}_U = \sqrt{3}$
		$\mathscr{C}_L = \frac{\sqrt{6}}{2}\sqrt{\frac{3a^{12} + 70a^{10} + 573a^8 + 1012a^6 + 573a^4 + 70a^2 + 3}{3a^{12} + 70a^{10} + 509a^8 + 1140a^6 + 509a^4 + 70a^2 + 3}}$

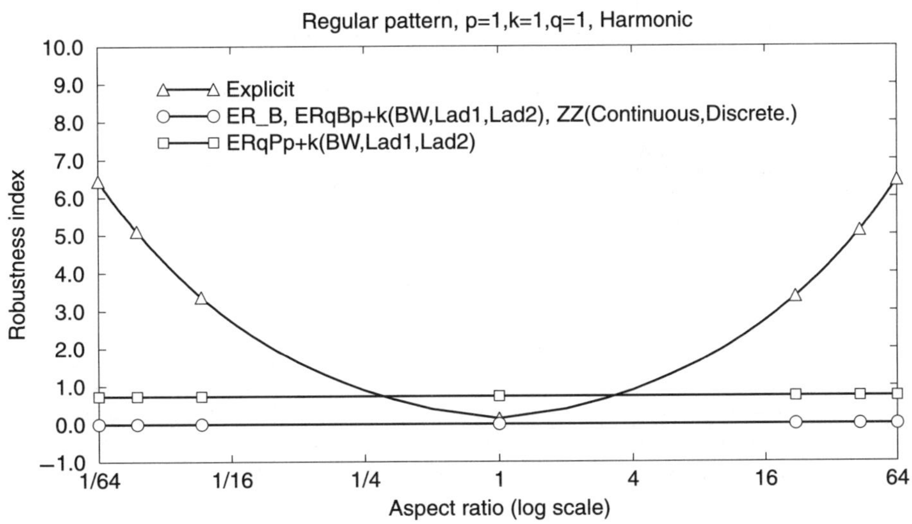

FIG. 5.7*.2. *Regular pattern*: ZZ Continuous, ZZ Discrete, ER_B, and ERqBp + k are reliable; in fact, they are exact. ERqPp + k is reliable. Explicit is not reliable. For this case, there is no difference between the three equilibrations. All error estimators are symmetric with respect to a and $1/a$

estimators as a function of the angle of grid-material orientation. Here we will present the results for linear elements $(p+1)$ and harmonic polynomials (Figs 5.7*.34–5.7*.37).

Let us summarize the conclusions obtained from the analysis of robustness with respect to the orthotropy (Figs 5.7*.34–5.7*.37):

1. The ERqBp + k estimator with Ladeveze equilibration of Type 2 is the most robust among the residual estimators.
2. The most robust estimators are the Continuous and the Discrete ZZ.

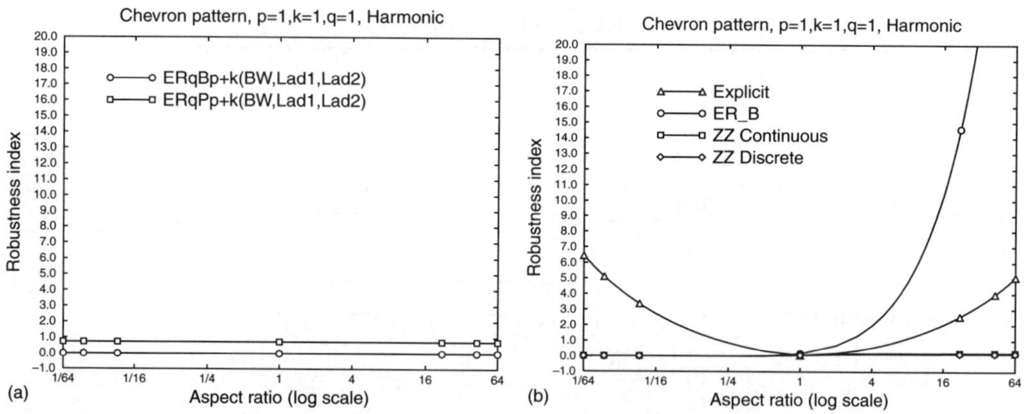

FIG. 5.7*.3. *Chevron pattern*: ERqBp + k are exact. ZZ Continuous and ZZ Discrete tend to be exact as a tends to zero. For large a, the Continuous and the Discrete ZZ are among the most reliable error estimators. ERqPp + k are also reliable. ER_B is reliable for $a \leq 1$, but it deteriorates for large a. The Explicit is reasonable for $a \approx 1$ but deteriorates for $a \to 0$ and $a \to \infty$. There is no difference between the three equilibrations for this case

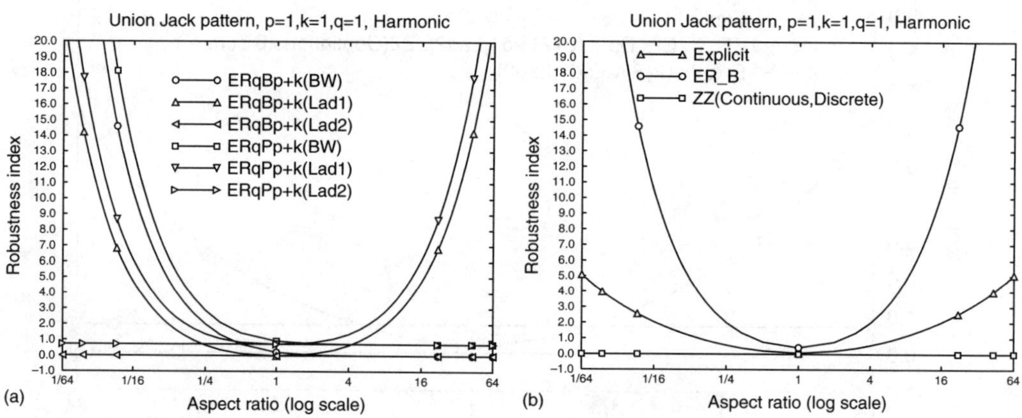

FIG. 5.7*.4. *Union-Jack pattern*: ZZ Continuous, ZZ Discrete, and ERqBp + k(Lad2) are exact. ERqBp + k(BW) is reliable when $a \to \infty$, but deteriorates as $a \to 0$. ERqPp + k(Lad2) is reliable. ERqPp + k(BW) are reliable when $a \to \infty$, but it deteriorates when $a \to 0$. The Explicit, ER_B, ERqBp + k(Lad1), and ERqPp + k(Lad1) are reliable for $a \approx 1$, but deteriorate as $a \to 0$ or $a \to \infty$. Note that the ERqBp + k(BW) and ERqPp + k(BW) are not symmetric with respect to a and $1/a$, even though this mesh (Union Jack) is symmetric. This is due to the way the equilibrated fluxes are constructed

Let us now explain how the above asymptotic analyses can be extended to general meshes which are employed in practice; see also [Babuška, Strouboulis *et al.*, 1994a].

Example 5.7*.3 (*Extension of the asymptotic analysis to general meshes*)
From the analysis given above and numerical experience, we know that the effectivity index for any estimator in the cell ω_0^h depends practically on the geometry of the mesh

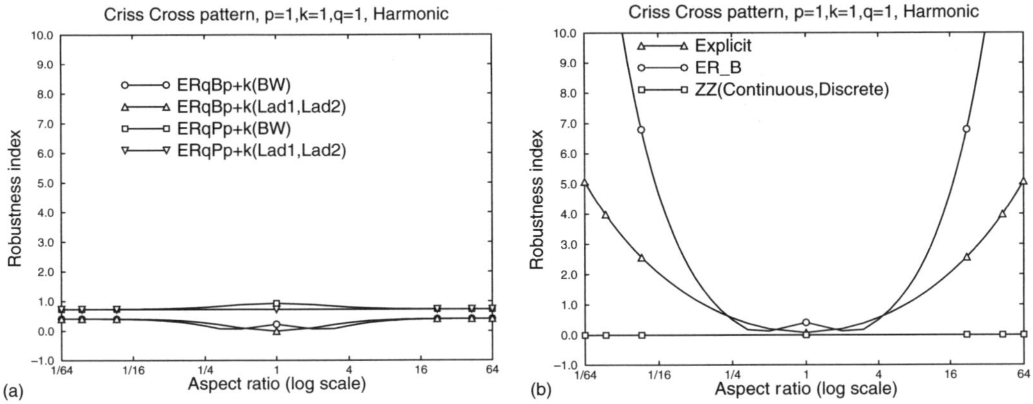

FIG. 5.7*.5. *Criss-cross pattern*: ZZ Continuous and ZZ Discrete are exact. ERqBp + k and ERqPp + k are reliable. Explicit and ER_B are reliable for $a \approx 1$, but deteriorate as $a \to 0$ or $a \to \infty$. All error estimators are symmetric with respect to a and $1/a$, including the two error estimators with Bank–Weiser equilibration (ERqBp + k(BW) and ERqPp + k(BW))

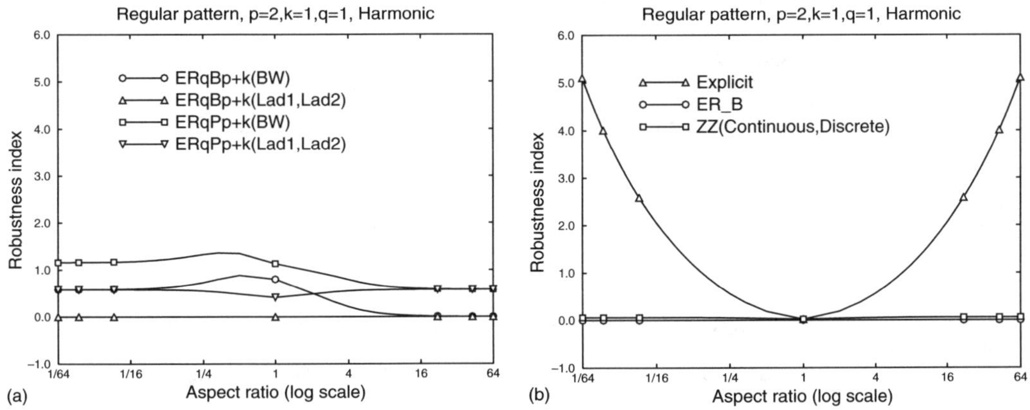

FIG. 5.7*.6. *Regular pattern*: ER_B, ERqBp + k(Lad1), and ERqBp + k(Lad2) are exact. ERqBp + k(BW) is reliable for the entire range of a. The Continuous and the Discrete ZZ are very reliable. The Explicit is reliable for $a \approx 1$, but deteriorates as $a \to 0$ or $a \to \infty$. All the estimators are symmetric about $a = 1$ except ERqBp + k(BW); this is due to the construction of the equilibration

in the patch ω_s^h, where s is the number of mesh layers around ω_0^h which are included in ω_s^h. Further, because we are interested only in the first two significant digits of the robustness index, it is sufficient to employ $s = 1$ or 2. Given a mesh T_h (respectively class of meshes $\mathcal{T}$) constructed by a mesh generator, there are patches ω_0^h/ω_s^h of various mesh geometries. The robustness index can be analyzed for all the different patches and its value for the mesh can be obtained as the maximum over all the patches. Let us illustrate this approach for the mesh shown in Fig. 5.7*.38. From this mesh, five mesh patterns were extracted, as shown in Fig. 5.7*.39 where the mesh cells ω_0^h are shaded gray and

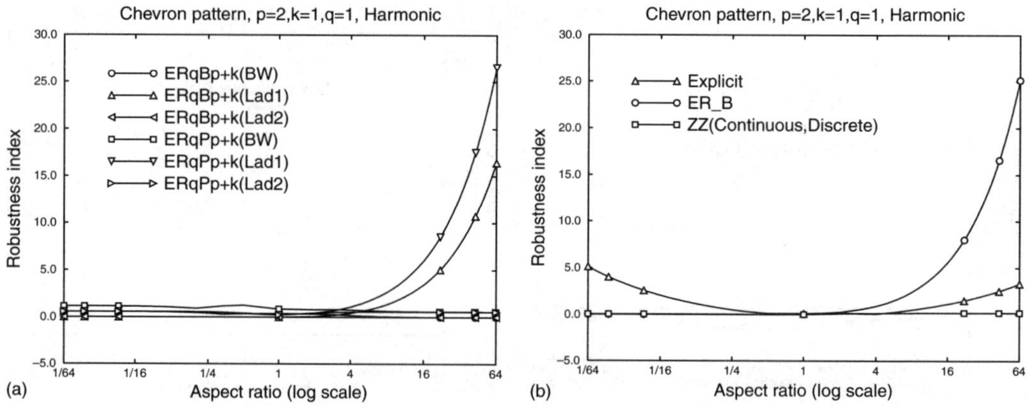

FIG. 5.7*.7. *Chevron pattern*: ZZ Continuous, ZZ Discrete, ERqBp + k(BW), and ERqBp + k(Lad2) are the most reliable. ERqPp + k(BW) and ERqPp + k(Lad2) are also reliable. ER_B, ERqBp + k(Lad1), and ERqPp + k(Lad1) are reliable for $a \leq 1$, but not as $a \to \infty$. The Explicit is reliable only for $a \approx 1$

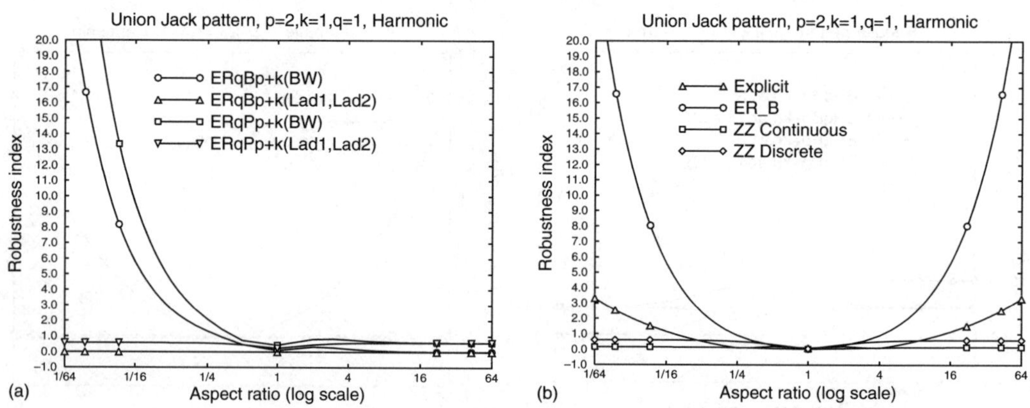

FIG. 5.7*.8. *Union-Jack pattern*: ERqBp + k(Lad1, Lad2) are exact. The Continuous and the Discrete ZZ, and ERqPp + k(Lad1, Lad2) are reliable. The two estimators with Bank–Weiser equilibration, ERqBp + k(BW) and ERqPp + k(BW), are not reliable for $a \to 0$, while they are reliable as $a \to \infty$. The Explicit and ER_B are reliable only for $a \approx 1$. All the estimators are symmetric with respect to $a = 1$, except the ones employing Bank–Weiser equilibration

the boundary of the patch ω_s^h, for $s = 3$, is shown with thick perigram. The robustness index of any error estimator in each mesh cell ω_0^h/ω_s^h was determined by employing the periodization approach discussed in Section 4.7*, namely:

(1) *Completion of the mesh cell to a periodic mesh cell*
Translate and scale the mesh patch ω_s^h so that its image $\tilde{\omega}_s^h \subseteq \tilde{c} = [-1, 1] \times [-1, 1]$ (see Fig. 5.7*.40). Then employ a mesh generator to complete the mesh in $\tilde{\omega}_s^h$ into a *periodic mesh* $\tilde{T}$ which covers $\tilde{c}$ (see Fig. 5.7*.40b).

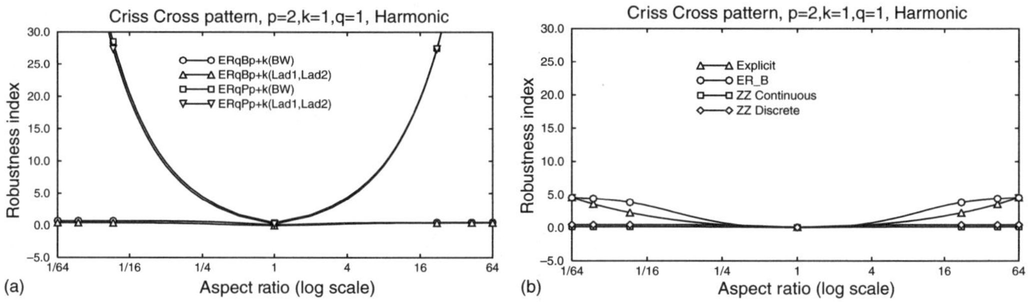

FIG. 5.7*.9. *Criss-cross pattern*: Continuous ZZ, Discrete ZZ, and ERqBp + k(BW, Lad1, Lad2) are the most reliable. Explicit and ER_B are reliable only for $a \approx 1$. ERqPp + k is unreliable

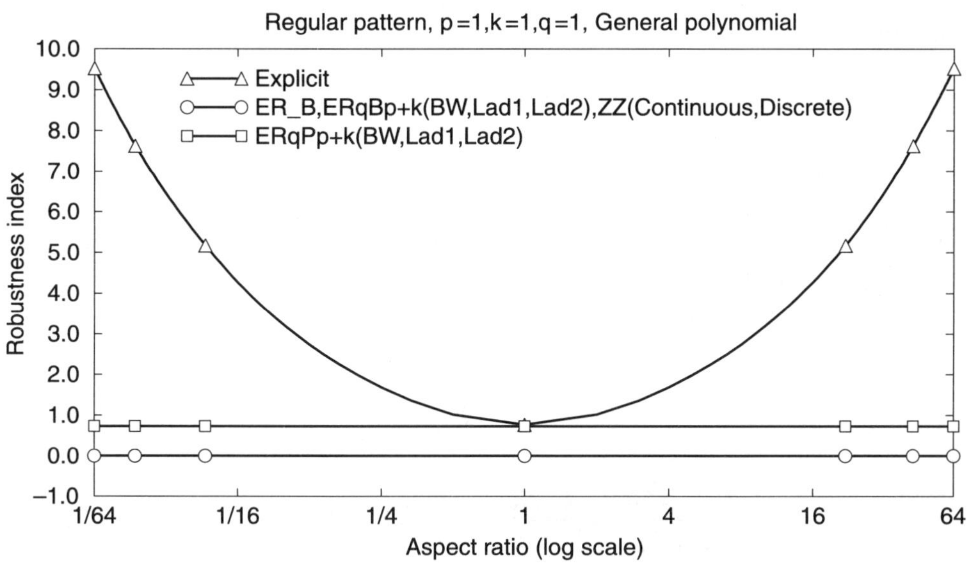

FIG. 5.7*.10. *Regular pattern*: ER_B, Continuous ZZ, Discrete ZZ, and ERqBp + k(BW, Lad1, Lad2) are exact. ERqPp + k(BW, Lad1, Lad2) are reliable, while the Explicit is reliable only for $a \approx 1$. All error estimators are symmetric with respect to a and $1/a$

(2) *Construction of the asymptotic finite element solution*
Obtain the asymptotic finite element solution for the mesh patch ω_s^h by solving the periodic boundary value problem over the periodic mesh cell, *which essentially means that, for all practical purposes, the local finite element solution is the same as the finite element solution for the infinite periodic mesh shown in Fig. 5.7*.40c.* The approach described above for patchwise uniform meshes is then employed to determine the robustness of the error estimators for the general meshes.

Tables 5.7*.5 and 5.7*.6 give the values of $\mathscr{C}_L^{\omega_0^h}$, $\mathscr{C}_U^{\omega_0^h}$, and $\mathcal{R}^s_{\omega_0^h}$ (where s represents the number of layers of elements around the patch of elements of interest. Here $s = 3$ was

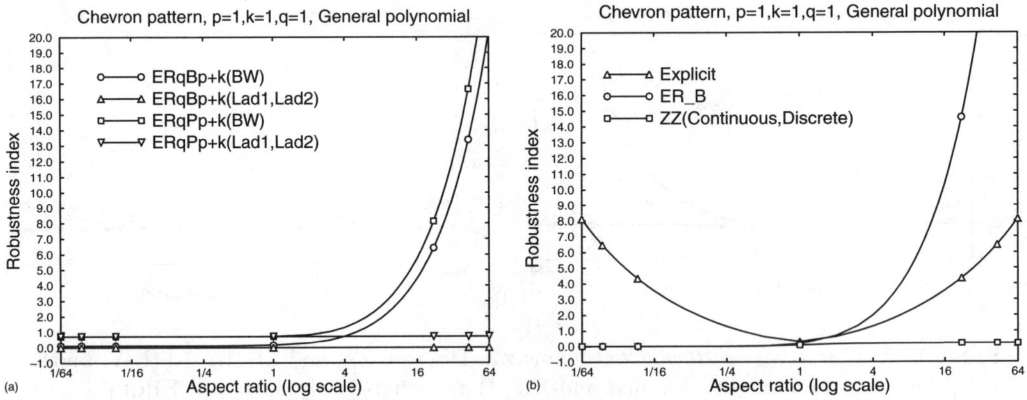

FIG. 5.7*.11. *Chevron pattern*: ERqBp + k(Lad1, Lad2) are exact. The Continuous and Discrete ZZ are also among the most reliable. ERqPp + k(Lad1, Lad2) are also reliable, while the ER_B and Explicit are reliable only for $a \approx 1$. ERqBp + k(BW) and ERqPp + k(BW) are unreliable as $a \to \infty$

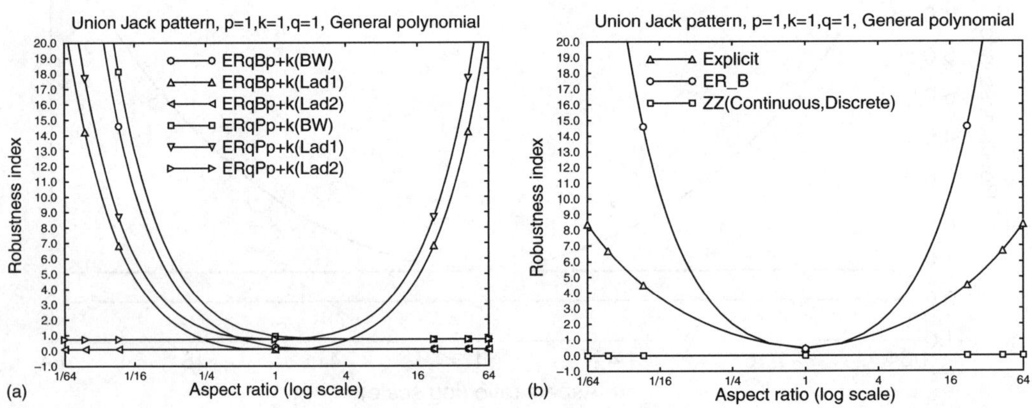

FIG. 5.7*.12. *Union-Jack pattern*: ZZ Continuous and ZZ Discrete are exact. ERqBp + k(Lad2) is very reliable. ERqBp + k(BW) and ERqPp + k(BW) are unreliable for $a \ll 1$. ERqPp + k(Lad2) is reliable. Explicit, ER_B, ERqBp + k(Lad1), and ERqPp + k(Lad1) are reliable only for $a \approx 1$

chosen) for the *ER_B*, *ER1B* and the discrete ZZ estimators for the Laplace equation and linear ($p = 1$) or quadratic ($p = 2$) elements, respectively. These are compared with the values of the effectivity indices $\bar{C}_L^{\omega_0^h}$, $\bar{C}_U^{\omega_0^h}$, and the corresponding $\bar{\mathcal{R}}$ obtained from the finite element solution of a Dirichlet boundary value problem using the grids shown in Fig. 5.7*.38 (the *big mesh*) with data consistent with the harmonic solutions which give the extremal values $\mathscr{C}_L^{\omega_0^h}$, $\mathscr{C}_U^{\omega_0^h}$ of the effectivity index $\kappa_{\omega_0^h}$ in the periodic super-patch calculations. The periodic super-patches employed for the various patterns were constructed using a mesh generator, as shown in Fig. 5.7*.40.

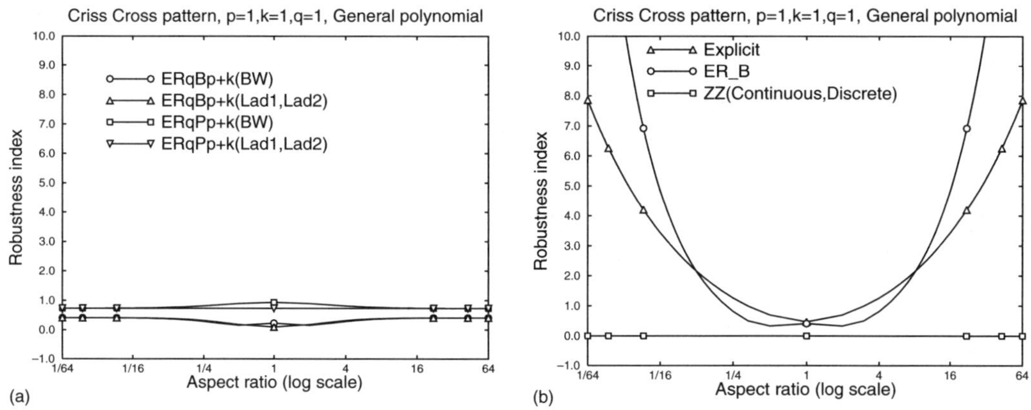

FIG. 5.7*.13. *Criss-cross pattern*: ZZ Continuous and ZZ Discrete are exact. ERqBp + k(BW, Lad1, Lad2) and ERqPp + k(BW, Lad1, Lad2) are reliable, while Explicit and ER_B are reliable only for $a \approx 1$

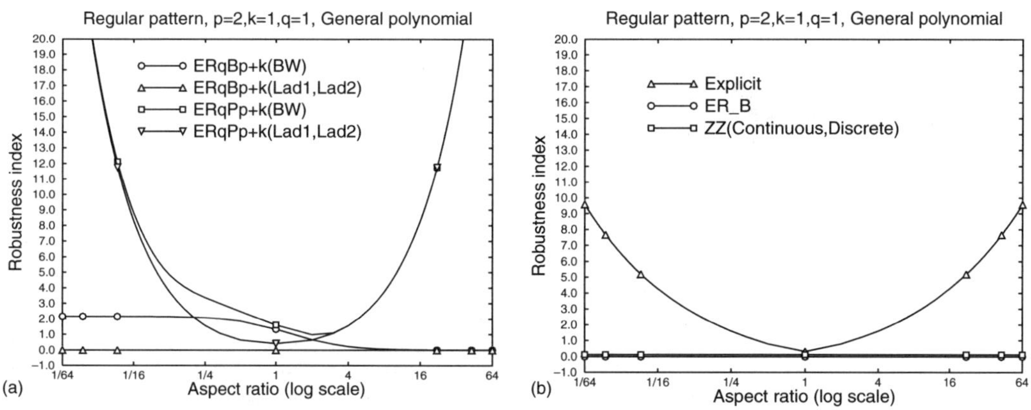

FIG. 5.7*.14. *Regular pattern*: ZZ Continuous, ZZ Discrete, ER_B, and ERqBp + k(Lad1, Lad2) are the most reliable. ERqBp + k(BW) becomes reliable as $a \to \infty$, while Explicit and ERqPp + k(BW, Lad1, Lad2) are not reliable

From the results for the bounds of the effectivity index and the robustness index we can make the following conclusions:

(1) The employment of $s = 3$ is sufficient to obtain the first two significant digits of the bounds of the effectivity index and the robustness index.
(2) The *ER_B* is not robust for the general meshes.
(3) The equilibration of the flux increases the robustness of the element residual estimators.
(4) The Discrete-*ZZ* estimator is the most robust.

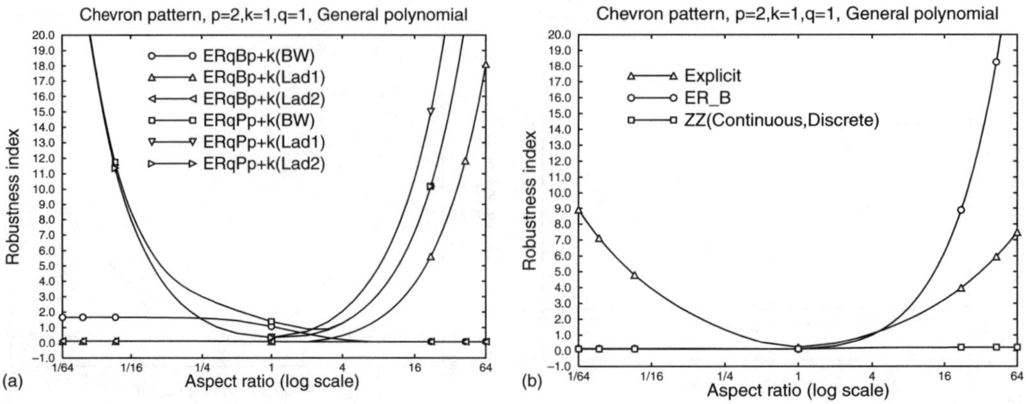

FIG. 5.7*.15. *Chevron pattern*: ZZ Continuous, ZZ Discrete, and ERqBp + k(Lad2) are the most reliable. ERqBp + k(BW) becomes reliable as $a \to \infty$, and ERqBp + k(Lad1) as $a \to 0$. ERqPp + k(BW, Lad1, Lad2) are not reliable. ER_B is reliable as $a \to 0$, but unreliable for $a \to \infty$. Explicit is unreliable

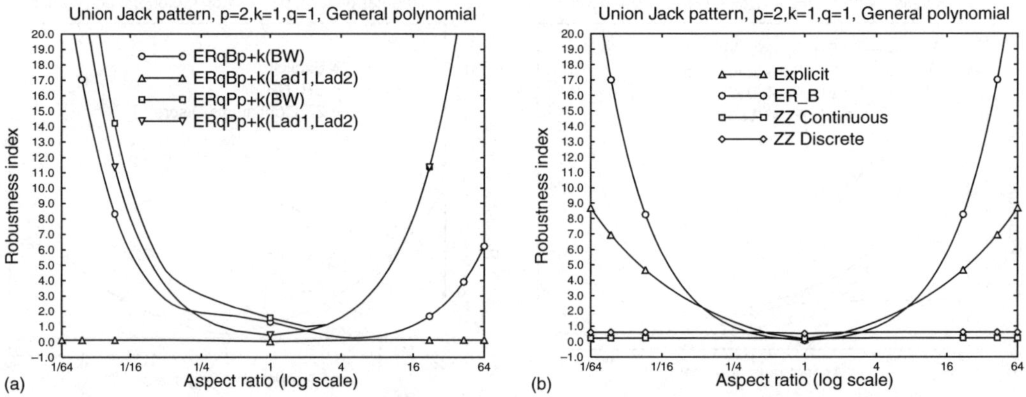

FIG. 5.7*.16. *Union-Jack pattern*: ERqBp + k(Lad1, Lad2) are exact. ZZ Continuous and ZZ Discrete are reliable. ER_B is reliable only for $a \approx 1$. ERqBp + k(BW), ERqPp + k(BW, Lad1, Lad2), and Explicit are unreliable

Above we have given examples of the asymptotic analysis of error indicators in the interior of the mesh. Let us now consider the case of elements which are adjacent to the boundary. The differences between the analysis for interior mesh cells and mesh cells adjacent to the boundary are:

(a) *The definition of the error indicators*: The definition of the error indicators for elements adjacent to the boundary differs from the definition in the interior of the mesh because the grid terminates at the boundary and the definitions of the error indicators have to be modified to utilize the available information. For example, the *ZZ* estimator employs averaging over nodal patches. Since the nodal patches at the boundary are incomplete, one approach (see [Zienkiewicz and Zhu, 1992a,b]) is to

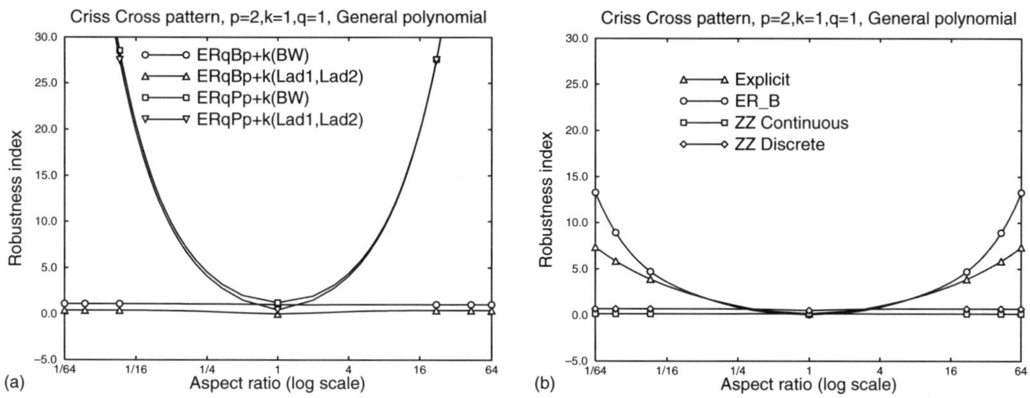

FIG. 5.7*.17. *Criss-cross pattern*: ZZ Continuous, ZZ Discrete, and ERqBp + k(Lad1, Lad2) are the most reliable. ERqBp + k(BW) is also reliable. Explicit and ER_B are reliable only for $a \approx 1$. ERqPp + k(BW, Lad1, Lad2) are unreliable

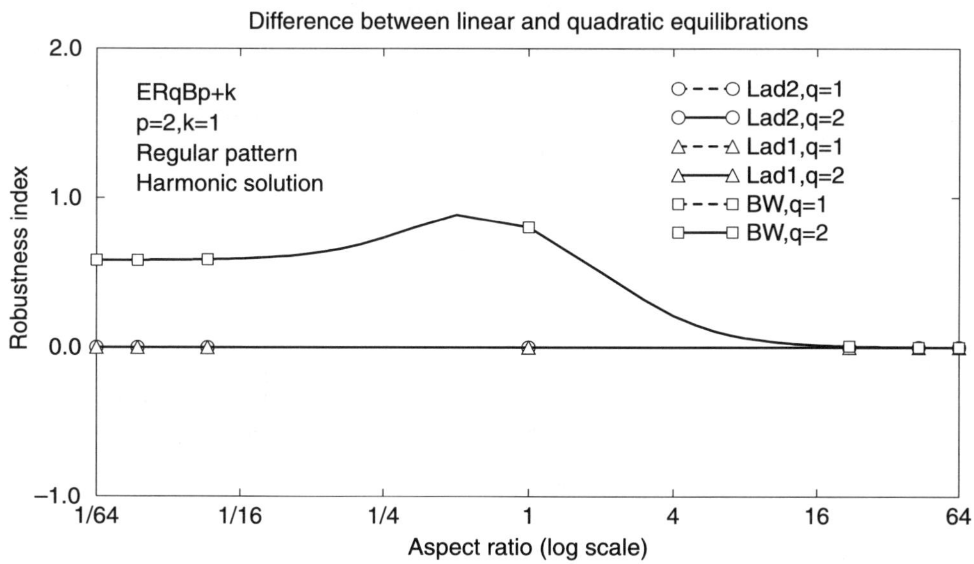

FIG. 5.7*.18. *The effect of the type and the order of equilibration on the robustness of the residual estimators*: Regular pattern, ERqBp + k estimator, with $p = 2$, $k = 1$, harmonic polynomial solutions. Note that, in this case, the robustness depends only on the type of equilibration

extrapolate the recovered stresses from the interior nodal patches to the elements adjacent to the boundary.

(b) *The boundary layer in the finite element solution:* As we have already seen in Section 4.7*, the asymptotic finite element solution for the interior mesh cells, is valid up to the boundary only if it satisfies the boundary conditions. If it does not, then it must be adjusted by adding a *boundary-layer solution.*

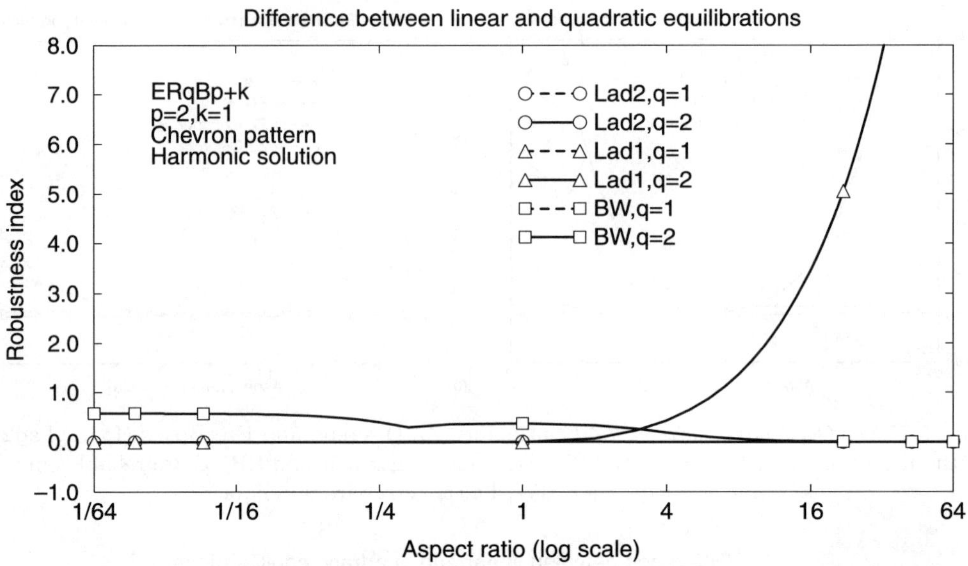

FIG. 5.7*.19. *The effect of the type and the order of equilibration on the robustness of the residual estimators*: Chevron pattern, ERqBp + k estimator, with $p = 2$, $k = 1$, harmonic polynomial solutions. Note that, in this case, the robustness depends only on the type of equilibration

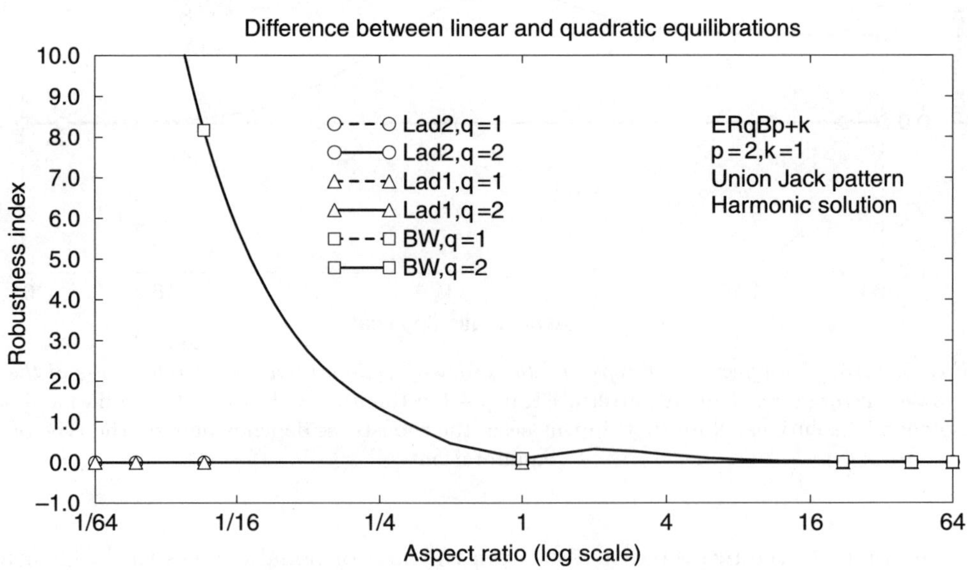

FIG. 5.7*.20. *The effect of the type and the order of equilibration on the robustness of the residual estimators*: Union-Jack pattern, ERqBp + k estimator, with $p = 2$, $k = 1$, harmonic polynomial solutions. Note that, in this case, the robustness depends only on the type of equilibration

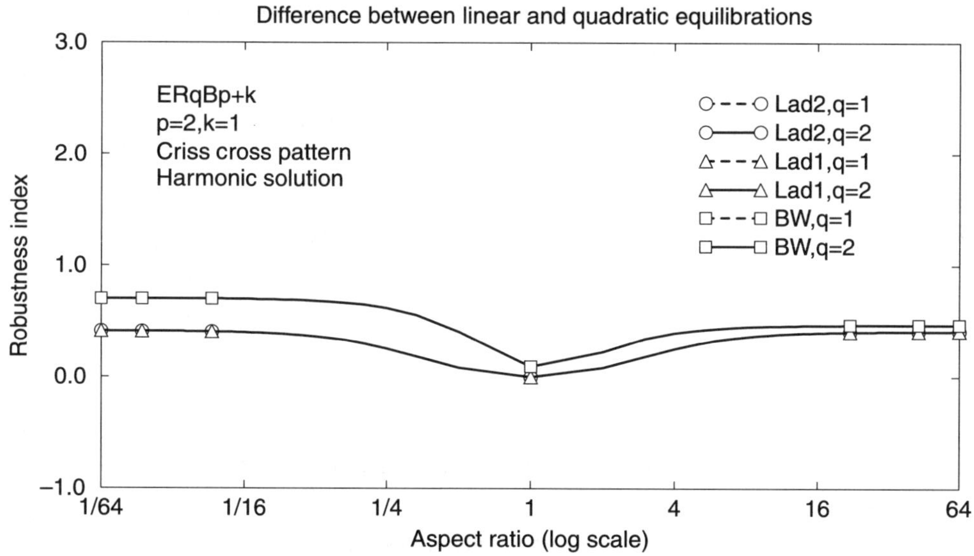

FIG. 5.7*.21. *The effect of the type and the order of equilibration on the robustness of the residual estimators*: Criss-cross pattern, ERqBp + k estimator, with $p = 2$, $k = 1$, harmonic polynomial solutions. Note that, in this case, the robustness depends only on the type of equilibration

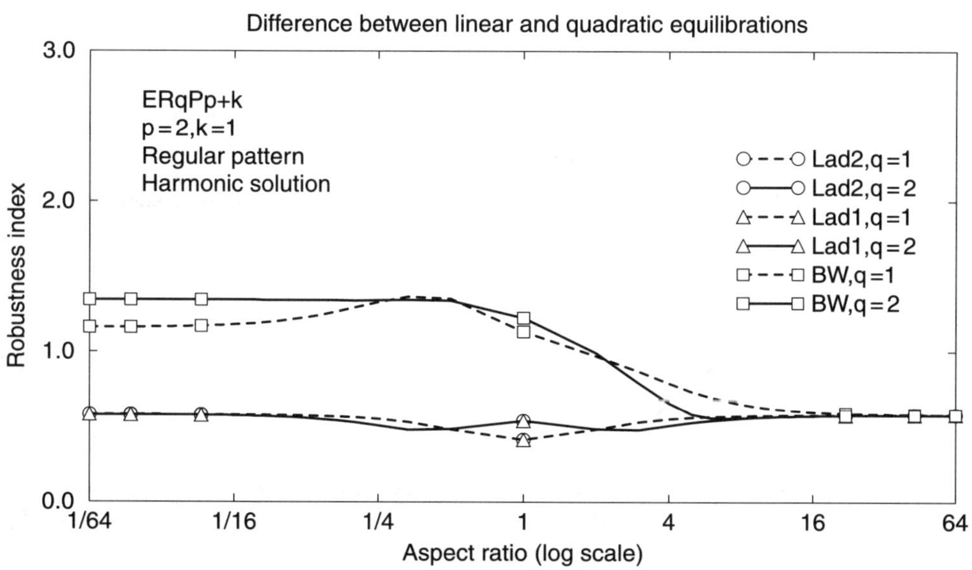

FIG. 5.7*.22. *The effect of the type and the order of equilibration on the robustness of the residual estimators*: Regular pattern, ERqPp + k estimator, with $p = 2$, $k = 1$, harmonic solutions. Note that the overall robustness depends primarily on the type of equilibration

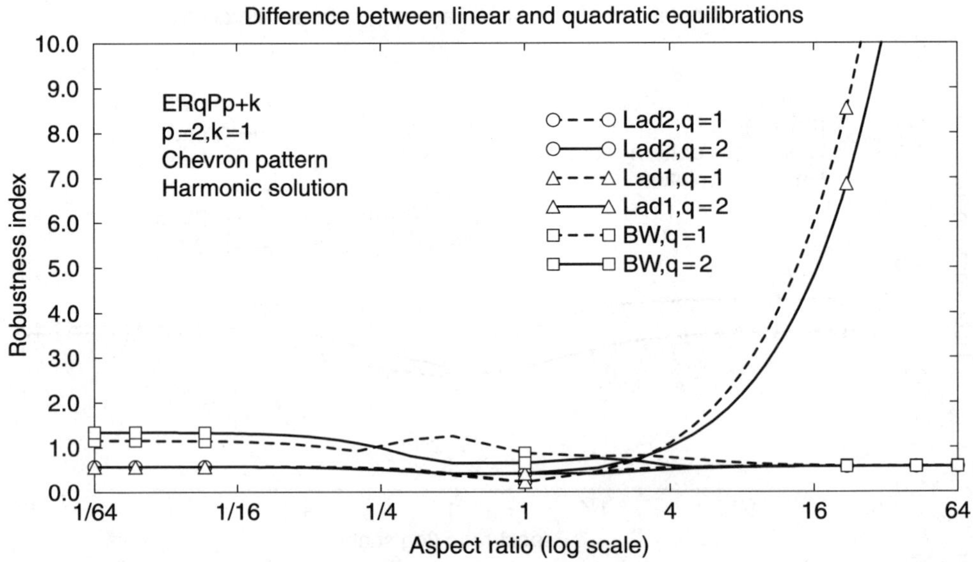

FIG. 5.7*.23. *The effect of the type and the order of equilibration on the robustness of the residual estimators*: Chevron pattern, ERqPp + k estimator, with $p = 2$, $k = 1$, harmonic solutions. Note that the overall robustness depends primarily on the type of equilibration.

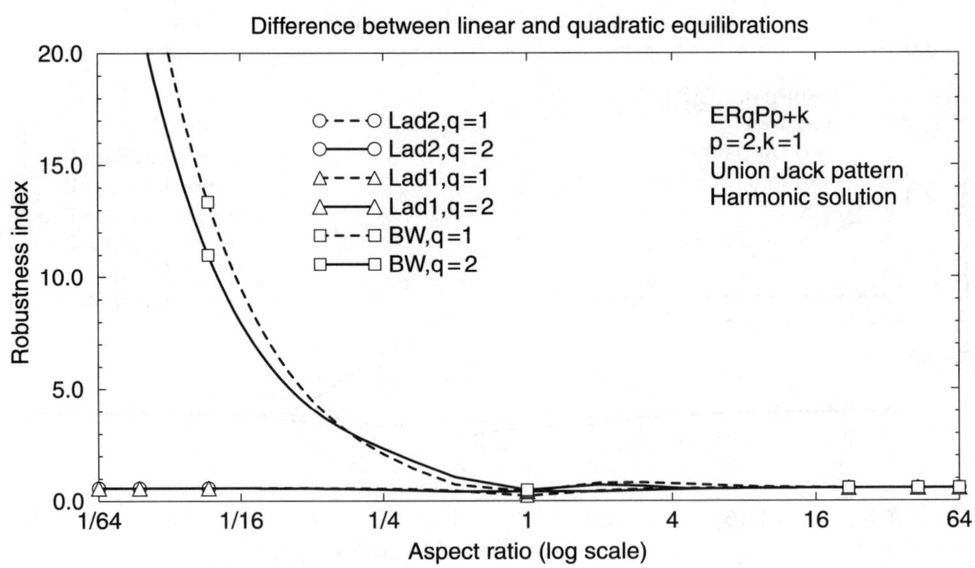

FIG. 5.7*.24. *The effect of the type and the order of equilibration on the robustness of the residual estimators*: Union-Jack pattern, ERqPp + k estimator, with $p = 2$, $k = 1$, harmonic solutions. Note that the overall robustness depends primarily on the type of equilibration

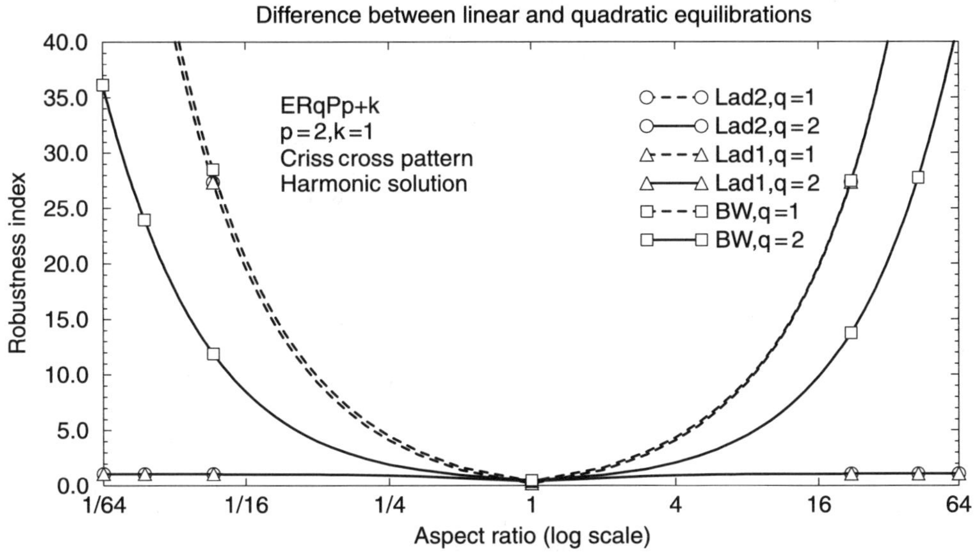

FIG. 5.7*.25. *The effect of the type and the order of equilibration on the robustness of the residual estimators*: Criss-cross pattern, ERqPp + k estimator, with $p = 2$, $k = 1$, harmonic solutions. Note that the overall robustness depends primarily on the type of equilibration

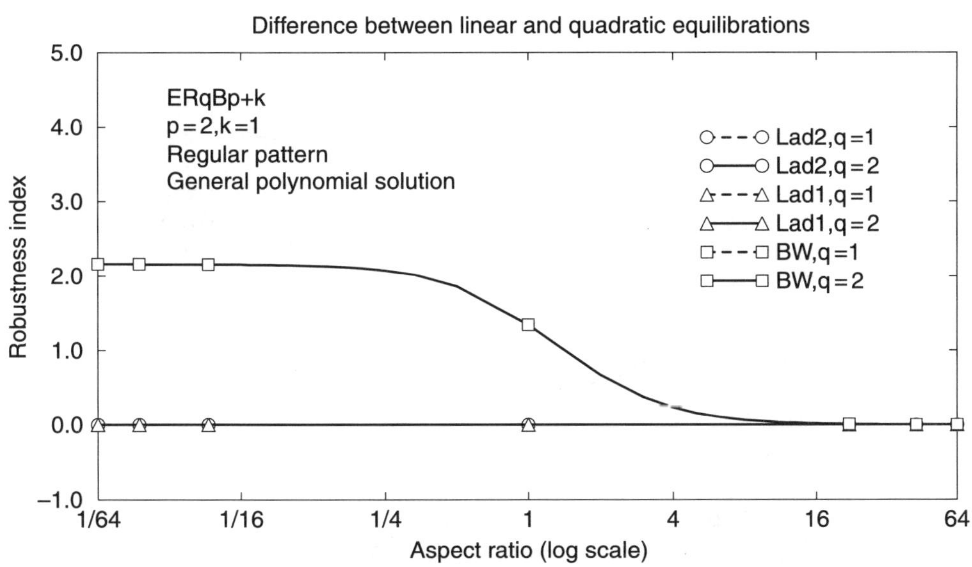

FIG. 5.7*.26. *The effect of the type and the order of equilibration on the robustness of the residual estimators*: Regular pattern, ERqBp + k estimator, with $p = 2$, $k = 1$, general polynomial solutions. Note that, in this case, the robustness depends only on the type of equilibration

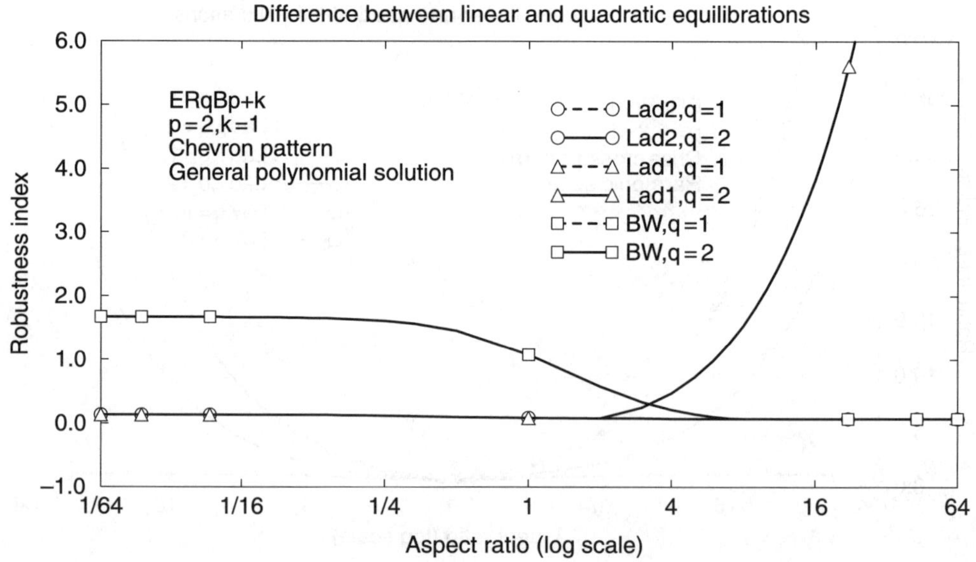

FIG. 5.7*.27. *The effect of the type and the order of equilibration on the robustness of the residual estimators*: Chevron pattern, ERqBp + k estimator, with $p = 2$, $k = 1$, general polynomial solutions. Note that, in this case, the robustness depends only on the type of equilibration

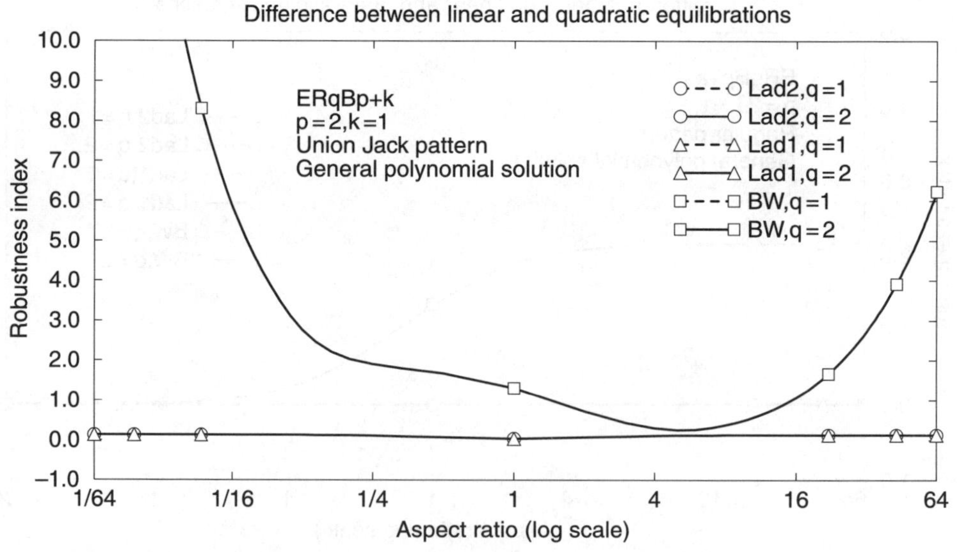

FIG. 5.7*.28. *The effect of the type and the order of equilibration on the robustness of the residual estimators*: Union-Jack pattern, ERqBp + k estimator, with $p = 2$, $k = 1$, general polynomial solutions. Note that, in this case, the robustness depends only on the type of equilibration

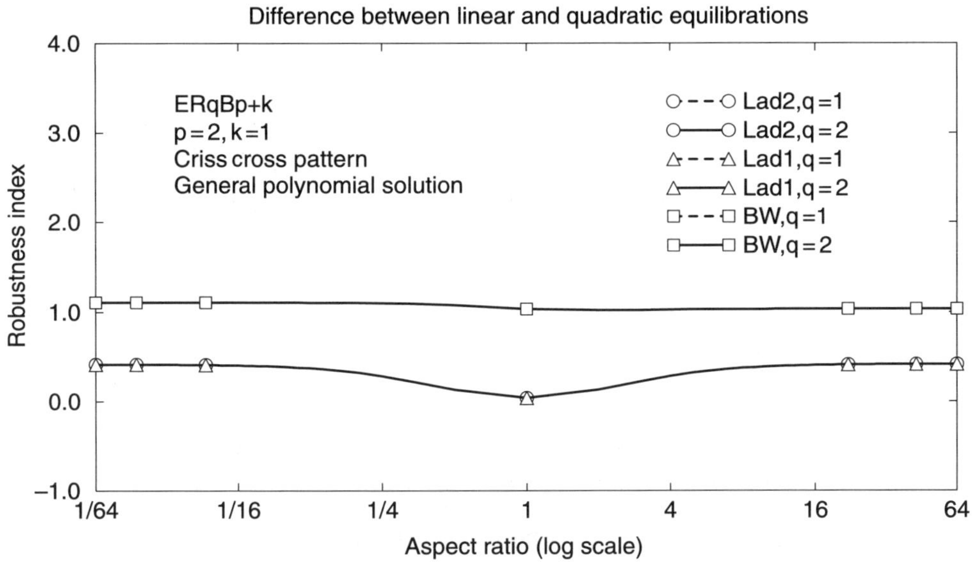

FIG. 5.7*.29. *The effect of the type and the order of equilibration on the robustness of the residual estimators*: Criss-cross pattern, ERqBp + k estimator, with $p = 2$, $k = 1$, general polynomial solutions. Note that, in this case, the robustness depends only on the type of equilibration

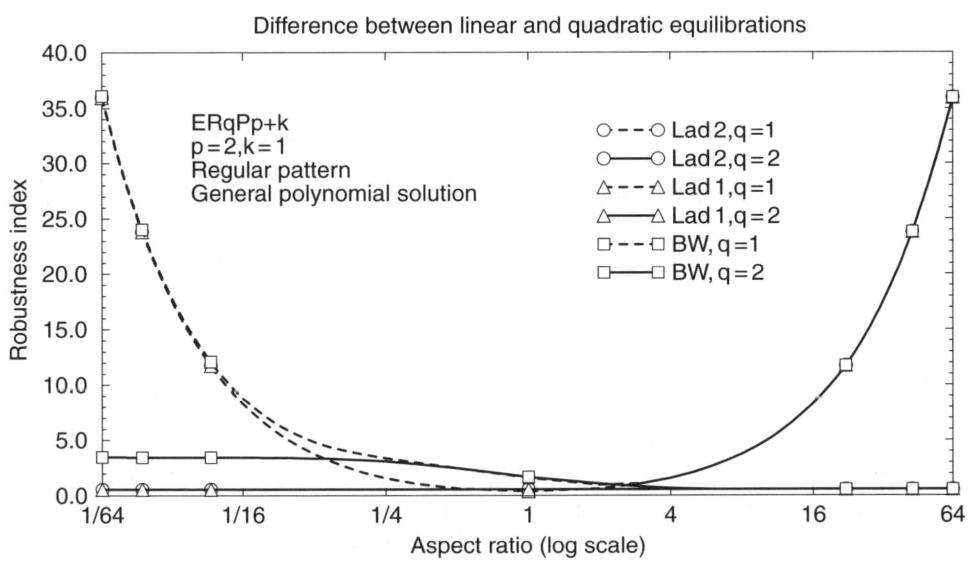

FIG. 5.7*.30. *The effect of the type and the order of equilibration on the robustness of the residual estimators*: Regular pattern, ERqPp + k estimator, with $p = 2$, $k = 1$, general solutions. Note that the overall robustness depends primarily on the type of equilibration

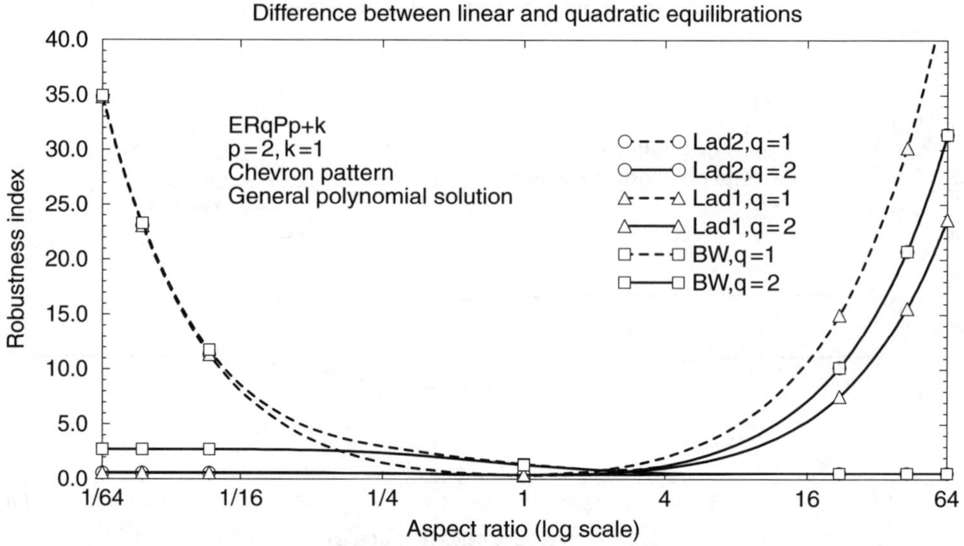

FIG. 5.7*.31. *The effect of the type and the order of equilibration on the robustness of the residual estimators*: Chevron pattern, ERqPp + k estimator, with $p = 2$, $k = 1$, general solutions. Note that the overall robustness depends primarily on the type of equilibration

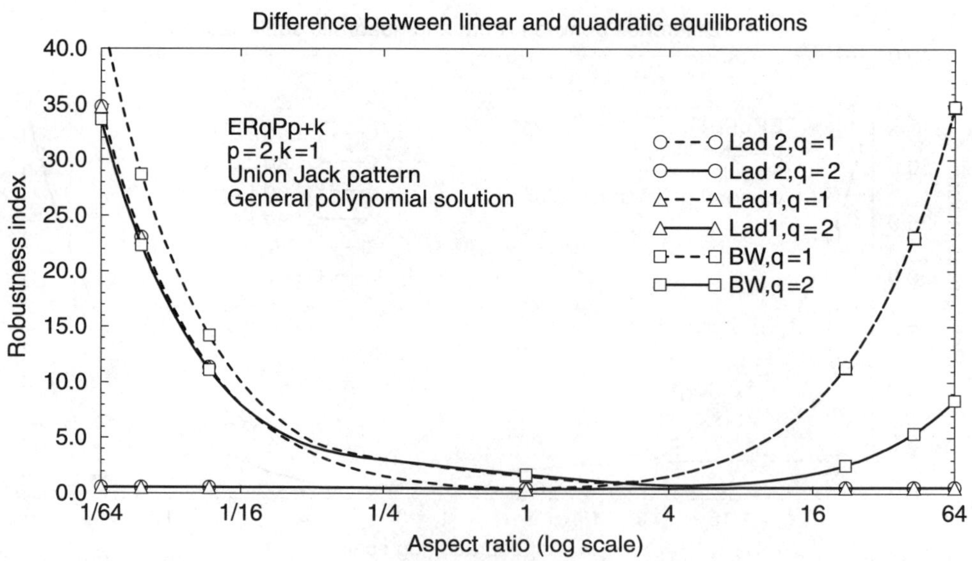

FIG. 5.7*.32. *The effect of the type and the order of equilibration on the robustness of the residual estimators*: Union-Jack pattern, ERqPp + k estimator, with $p = 2$, $k = 1$, general solutions. Note that the overall robustness depends primarily on the type of equilibration

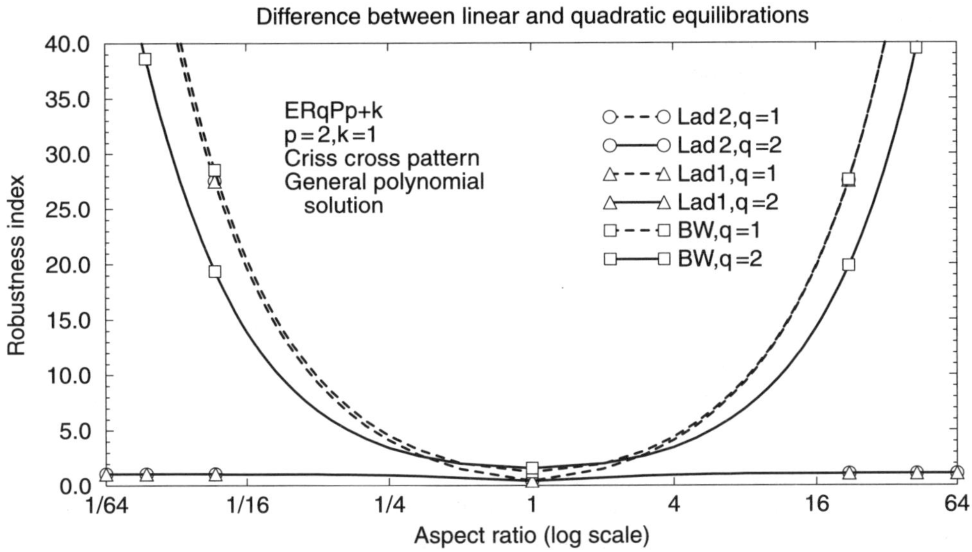

FIG. 5.7*.33. *The effect of the type and the order of equilibration on the robustness of the residual estimators*: Criss-cross pattern, ERqPp + k estimator, with $p = 2$, $k = 1$, general solutions. Note that the overall robustness depends primarily on the type of equilibration

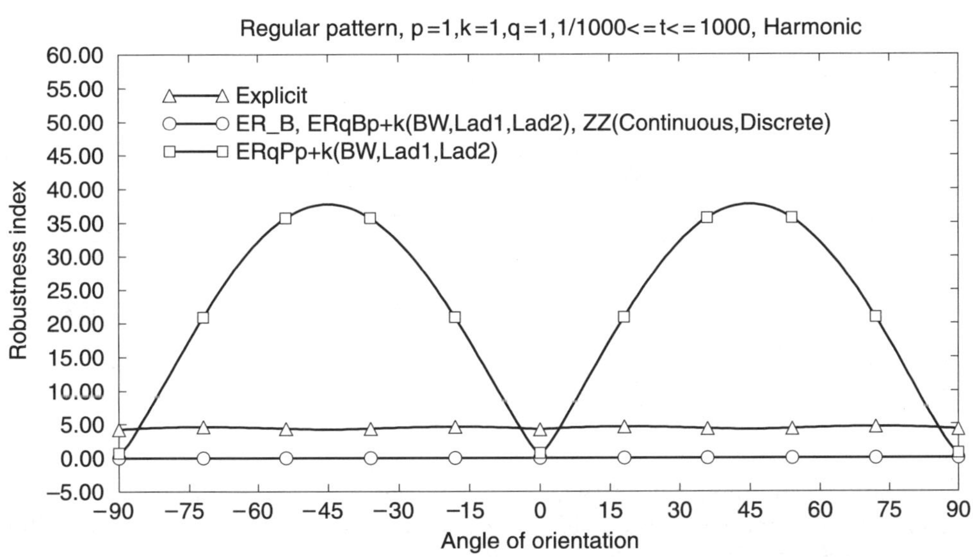

FIG. 5.7*.34. *Analysis of the robustness of the error estimators for the orthotropic Laplacian*: Regular pattern. Note that ER_B, ERqBp + k, ZZ Continuous and ZZ Discrete are exact, Explicit is insensitive to the angle of orientation, while ERqPp + k is very sensitive to the angle of orientation

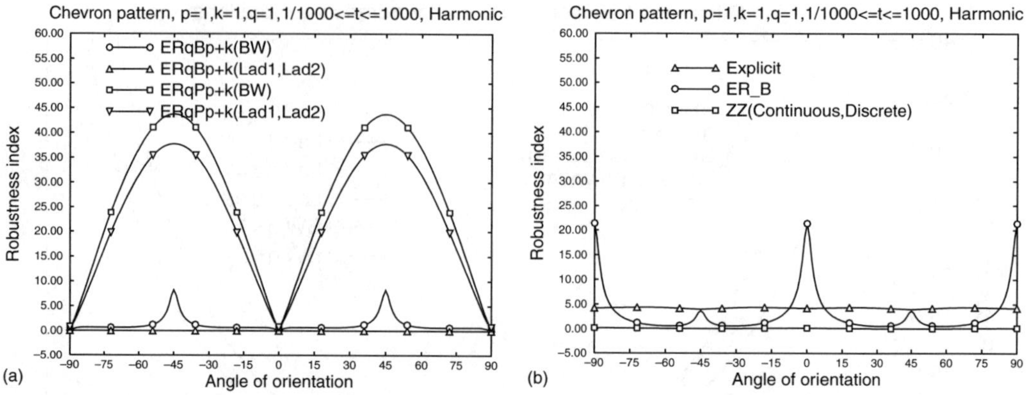

FIG. 5.7*.35. *Analysis of the robustness of the error estimators for the orthotropic Laplacian*: Chevron pattern. Note that Continuous ZZ, Discrete ZZ, and ERqBp + k are among the most robust, while Explicit is insensitive to the angle of orientation. ER_B and ERqPp + k are very sensitive to the angle of orientation, although they give good results in the special cases $\theta = 0, \pm 90°$

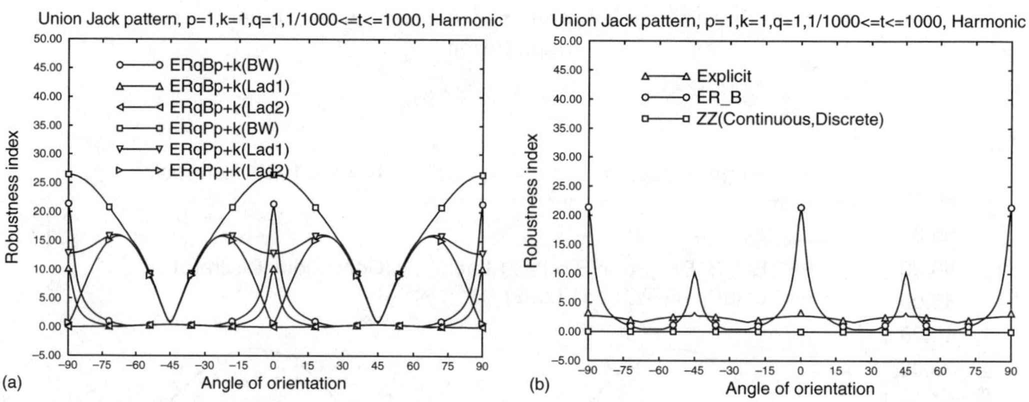

FIG. 5.7*.36. *Analysis of the robustness of the error estimators for the orthotropic Laplacian*: Union-Jack pattern. Note that Continuous ZZ, Discrete ZZ and ERqBp + k (Lad2) are the most robust. Explicit is not sensitive to the angle of orientation. All other estimators are not robust

In [Babuška, Strouboulis *et al.*, 1997] we gave an extensive analysis of the robustness of the error indicators for elements adjacent to the boundary. From this study we concluded that *the main factor which affects the quality of the error indicators at the boundary is the difference between their definitions for interior elements and those adjacent to the boundary, and the boundary layer in the asymptotic solution does not have a significant effect on the robustness of the indicators.*

Let us now summarize the results of a systematic analysis of the error indicators for mesh cells adjacent to the boundary; for the theoretical setting, see [Babuška, Strouboulis

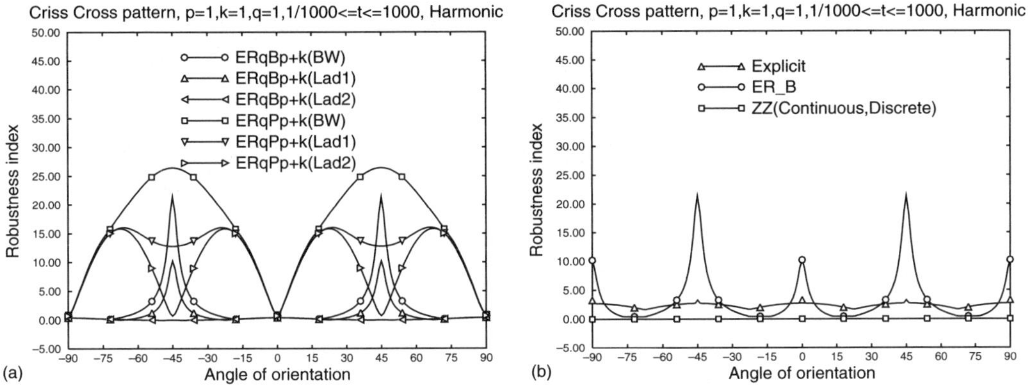

FIG. 5.7*.37. *Analysis of the robustness of the error estimators for the orthotropic Laplacian*: Criss-cross pattern. Note that Continuous ZZ, Discrete ZZ, and ERqBp + k (Lad2) are the most robust for this case. Explicit is not sensitive to the angle of orientation. ER_B, ERqPp + k, and ERqBp + k (BW,Lad1) are sensitive to the angle of orientation and hence not robust

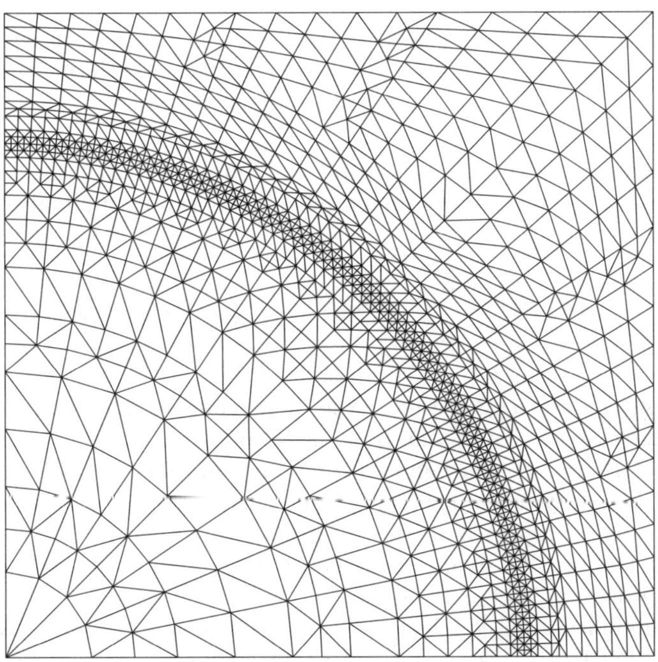

FIG. 5.7*.38. An example of a general mesh

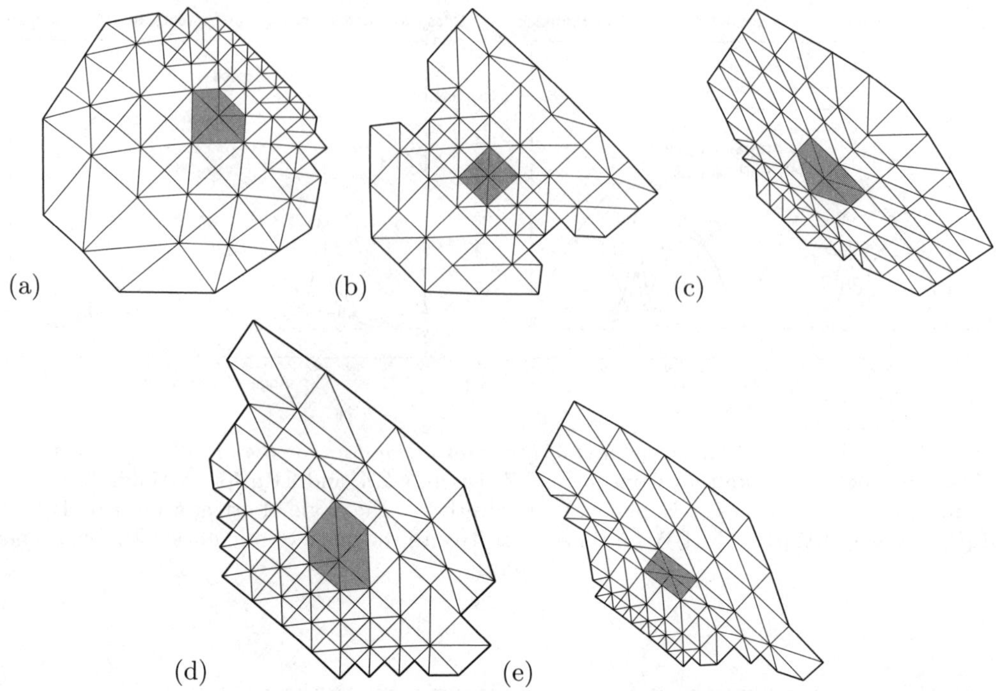

FIG. 5.7*.39. Typical patches of elements in the interior of the mesh shown in Fig. 5.7*.38. (a) Pattern 1, (b) Pattern 2, (c) Pattern 3, (d) Pattern 4, (e) Pattern 5

et al., 1997]. We analyzed the estimators for the four mesh patterns, for linear ($p = 1$) and quadratic ($p = 2$) elements, by employing the general polynomial solution.

$$u_{EX}(x_1, x_2) = \sum_{i,j=0}^{p+1} a_{ij} x_1^i x_2^j, \quad i + j = p + 1, \;\; i, j \leq 0 \qquad (5.7^*.12)$$

where a_{ij} are such that u_{EX} satisfies the homogeneous Dirichlet boundary condition or the homogeneous Neumann boundary condition at the boundary $x_2 = 0$. Tables 5.7*.7–5.7*.10 report the values of $\mathscr{C}_L$ and $\mathscr{C}_U$ for the first layer of elements adjacent to the boundary as a function of the aspect ratio for $p = 1$ and $p = 2$ in the cases of homogeneous Dirichlet and Neumann boundary conditions, for the ERqBp + k, ERqPp + k, and Discrete-ZZ estimators. The corresponding results for the robustness index are given in Figs 5.7*.41–5.7*.56.

Let us now summarize our conclusions from the asymptotic analysis of the robustness of the Discrete ZZ, ERqBp + k, and ERqPp + k estimators with respect to the aspect ratio in the first layer adjacent to the boundary.

(1) The Discrete ZZ estimator has the best overall robustness, although in some cases ERqBp + k is the best estimator.

(2) The ERqPp + k is the least robust of the three estimators.

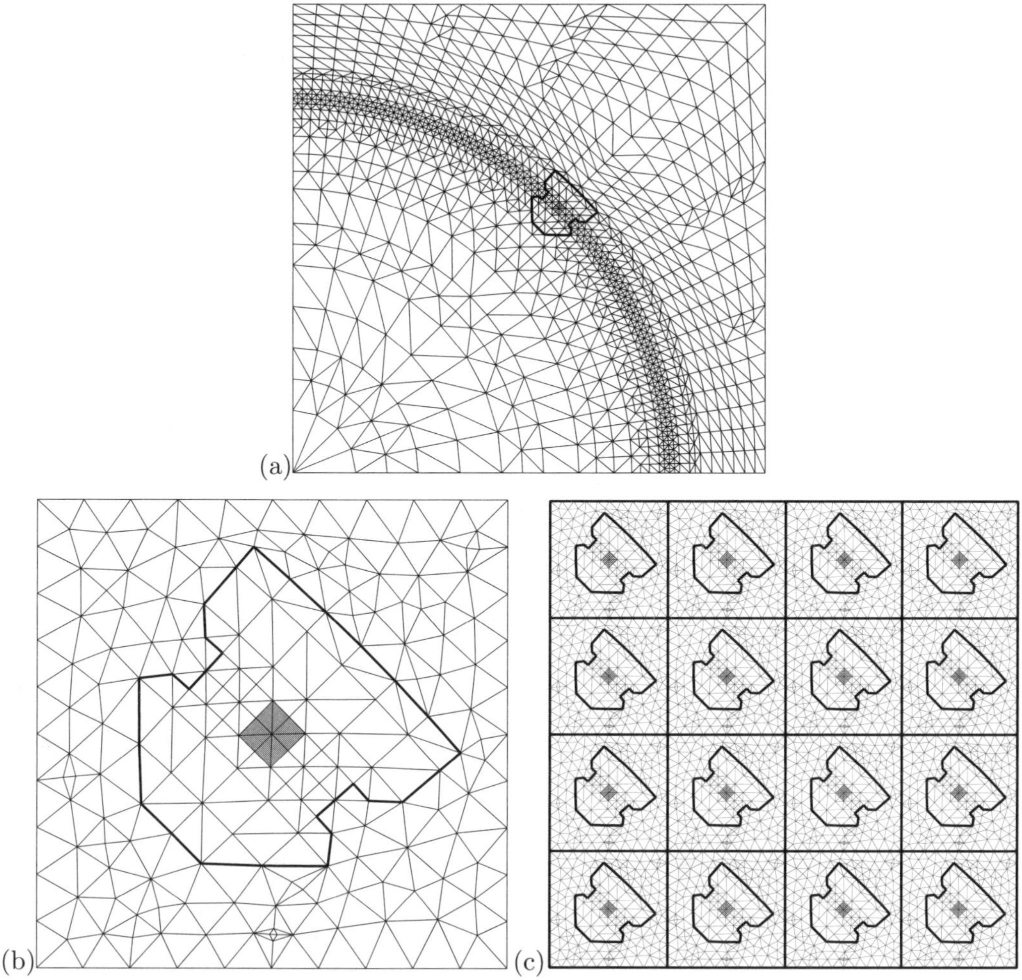

FIG. 5.7*.40. An example of the application of the method of periodization: (a) A general mesh with the mesh patch of interest shown shaded gray. (b) The periodic cell created by taking three layers of elements around the patch of interest and by creating a mesh in the remaining region of the square using a mesh generator. (c) The periodic mesh in the infinite domain

(3) All three estimators have reasonable robustness in the range of realistic aspect ratio, which is $\frac{1}{4} \le a \le 4$.

(4) The ZZ estimator is recommended to be used as an error indicator because it has low direct cost. Although it is also reasonable as an error estimator, it does not lead to a guaranteed upped or lower bound. For stopping criteria based on the desirable accuracy of the computed outputs, guaranteed estimators are needed, like for example, the estimators constructed in Chapter 6.

Let us also analyze the effect of the order of equilibration on the robustness of the estimator with respect to the aspect ratio in the first layer of elements adjacent to the

TABLE 5.7*.5. Asymptotic robustness of the estimators for general meshes: Laplace equation, quadratic harmonic polynomial solution, linear elements ($p = 1$). The mesh cell of Fig. 5.7*.1 and mesh patterns of Fig. 5.7*.38 are employed in the computations (Laplace equation, Linear elements)

Pattern	Periodic problem			Dirichlet BVP		
	$\mathcal{C}_L^{\omega_0^h}$	$\mathcal{C}_U^{\omega_0^h}$	$\mathcal{R}^3_{\omega_0^h}$	$\bar{\mathcal{C}}_L^{\omega_0^h}$	$\bar{\mathcal{C}}_U^{\omega_0^h}$	$\bar{\mathcal{R}}^3_{\omega_0^h}$
ER_B estimator						
1	1.08	1.31	0.39	1.08	1.31	0.39
2	1.04	1.42	0.46	1.03	1.41	0.44
3	1.00	1.01	0.01	1.00	1.01	0.02
4	1.00	1.13	0.13	0.98	1.12	0.13
5	0.96	1.70	0.74	0.95	1.70	0.75
ERpB estimator						
1	0.99	1.00	0.01	0.99	1.00	0.01
2	1.00	1.00	0.00	0.98	1.00	0.03
3	0.99	1.03	0.04	0.99	1.03	0.04
4	0.98	1.02	0.04	0.96	1.01	0.05
5	0.92	1.07	0.15	0.92	1.07	0.16
Discrete ZZ estimator						
1	1.00	1.01	0.02	1.00	1.01	0.01
2	1.00	1.04	0.04	0.99	1.02	0.03
3	0.99	0.99	0.02	0.99	1.00	0.01
4	0.96	1.01	0.05	0.94	1.00	0.06
5	0.98	1.01	0.03	0.98	1.01	0.03

TABLE 5.7*.6. Asymptotic robustness of the estimators for general meshes: Laplace equation, cubic harmonic polynomial solution, quadratic elements ($p = 2$). The mesh cell of Fig. 5.7*.1 and mesh patterns of Fig. 5.7*.38 are employed in the computations (Laplace equation, Quadratic elements)

Pattern	Periodic problem			Dirichlet BVP		
	$\mathcal{C}_L^{\omega_0^h}$	$\mathcal{C}_U^{\omega_0^h}$	$\mathcal{R}^3_{\omega_0^h}$	$\bar{\mathcal{C}}_L^{\omega_0^h}$	$\bar{\mathcal{C}}_U^{\omega_0^h}$	$\bar{\mathcal{R}}^3_{\omega_0^h}$
ERpB estimator						
1	0.96	1.03	0.07	0.96	1.03	0.07
2	0.98	1.01	0.03	0.98	1.01	0.03
3	1.01	1.02	0.03	1.01	1.02	0.03
4	0.97	1.01	0.04	0.97	1.01	0.04
5	0.72	0.89	0.51	0.72	0.89	0.51
Discrete ZZ estimator						
1	0.97	1.00	0.04	0.97	1.00	0.04
2	0.92	0.94	0.16	0.92	0.93	0.17
3	1.00	1.06	0.06	1.00	1.06	0.06
4	1.01	1.04	0.05	1.01	1.04	0.05
5	0.88	0.93	0.22	0.88	0.93	0.22

TABLE 5.7*.7. Range of the effectivity index in the first layer as a function of the aspect ratio for the four patterns. Poisson's equation with homogeneous Dirichlet boundary condition. Linear element ($p = 1$), First layer ($p = 1, k = 1, q = 1$). $ERqBp + k$, $ERqPp + k$, and discrete ZZ error estimators (Poisson's equation, Dirichlet boundary, First layer, $p = 1, k = 1, q = 1$)

Aspect ratio a	Regular		Chevron		Union Jack		Criss cross	
	$\mathscr{C}_L$	$\mathscr{C}_U$	$\mathscr{C}_L$	$\mathscr{C}_U$	$\mathscr{C}_L$	$\mathscr{C}_U$	$\mathscr{C}_L$	$\mathscr{C}_U$
$ERqBp + k$								
1/64	1.0000	1.0000	0.9244	1.0000	0.9129	1.0000	0.9895	1.1175
1/32	1.0000	1.0000	0.9496	1.0000	0.9130	1.0001	0.9896	1.1160
1/16	1.0000	1.0001	0.9862	1.0001	0.9134	1.0005	0.9897	1.1102
1/8	1.0000	1.0003	0.9993	1.0003	0.9151	1.0018	0.9901	1.0909
1/4	1.0000	1.0011	1.0000	1.0011	0.9219	1.0052	0.9923	1.0486
1	1.0000	1.0115	1.0000	1.0219	0.9711	1.0219	0.9763	1.0016
4	1.0000	1.0486	1.0000	1.3834	0.9975	1.3872	0.7621	1.0370
8	1.0000	1.0572	1.0000	2.1437	0.9994	2.1446	0.7224	1.0717
16	1.0000	1.0598	1.0000	3.9073	0.9998	3.9074	0.7110	1.0879
32	1.0000	1.0604	1.0000	7.6115	1.0000	7.6115	0.7081	1.0928
64	1.0000	1.0606	1.0000	15.1196	1.0000	15.1196	0.7074	1.0941
$ERqPp + k$								
1/64	1.0001	1.7319	1.0332	1.7319	1.0002	1.7320	1.0006	1.2247
1/32	1.0003	1.7316	1.1152	1.7316	1.0008	1.7317	1.0026	1.2247
1/16	1.0014	1.7304	1.2754	1.7304	1.0034	1.7308	1.0101	1.2247
1/8	1.0054	1.7257	1.3831	1.7257	1.0126	1.7270	1.0385	1.2258
1/4	1.0200	1.7094	1.3961	1.7093	1.0398	1.7120	1.1308	1.2337
1	1.1300	1.6016	1.3086	1.6112	1.1643	1.6112	1.1962	1.5512
4	1.2139	1.5118	1.2345	1.8986	1.2193	1.9057	1.3120	1.7160
8	1.2220	1.5029	1.2273	2.7556	1.2234	2.7573	1.7279	1.8865
16	1.2240	1.5007	1.2254	4.8570	1.2244	4.8573	1.7310	3.4338
32	1.2246	1.5002	1.2249	9.3590	1.2247	9.3590	1.7318	6.7153
64	1.2247	1.5000	1.2248	18.5362	1.2247	18.5362	1.7320	13.3574
Discrete ZZ								
1/64	1.0000	1.0000	1.0000	1.0038	1.0000	1.0002	1.0000	1.0000
1/32	1.0000	1.0000	0.9999	1.0113	1.0000	1.0005	0.9999	1.0000
1/16	1.0000	1.0000	0.9996	1.0181	0.9998	1.0013	0.9997	1.0001
1/8	1.0000	1.0000	0.9984	1.0151	0.9994	1.0021	0.9990	1.0003
1/4	1.0000	1.0000	0.9939	1.0134	0.9976	1.0028	0.9962	1.0013
1	1.0000	1.0000	0.9470	1.0134	0.9798	1.0033	0.9704	1.0111
4	1.0000	1.0000	0.8977	1.0134	0.9616	1.0015	0.9460	1.0207
8	1.0000	1.0000	0.8927	1.0134	0.9598	1.0013	0.9437	1.0216
16	1.0000	1.0000	0.8913	1.0134	0.9593	1.0013	0.9430	1.0219
32	1.0000	1.0000	0.8910	1.0134	0.9592	1.0013	0.9429	1.0220
64	1.0000	1.0000	0.8909	1.0134	0.9592	1.0013	0.9428	1.0220

TABLE 5.7*.8. Range of the effectivity index in the first layer as a function of the aspect ratio for the four patterns. Poisson's equation with homogeneous Dirichlet boundary condition. Quadratic element ($p = 2$), First layer ($p = 2, k = 1, q = 2$), and quadratic equilibration ($q = 2$). $ERqBp + k$, $ERqPp + k$, and discrete ZZ error estimators (Poisson's equation, Dirichlet boundary, First layer, $p = 2, k = 1, q = 2$)

Aspect ratio a	Regular		Chevron		Union Jack		Criss cross	
	$\mathscr{C}_L$	$\mathscr{C}_U$	$\mathscr{C}_L$	$\mathscr{C}_U$	$\mathscr{C}_L$	$\mathscr{C}_U$	$\mathscr{C}_L$	$\mathscr{C}_U$
$ERqBp + k$								
1/64	0.8650	1.0000	0.8649	0.9708	0.8649	2.0420	0.9745	1.0000
1/32	0.9818	1.0000	0.8806	0.9855	0.8807	1.3304	0.9970	1.0000
1/16	0.9997	1.0001	0.8811	0.9996	0.8816	1.0819	0.9973	1.0000
1/8	1.0000	1.0006	0.8835	1.0009	0.8852	1.0116	0.9950	1.0000
1/4	1.0000	1.0028	0.8915	1.0043	0.8973	1.0000	0.9848	1.0025
1	1.0000	1.0352	0.9235	1.0477	0.9544	1.0362	0.9618	1.0000
4	1.0000	1.0583	0.9351	1.1934	0.9745	1.1891	0.7803	1.0000
8	1.0000	1.0601	0.9372	1.5214	0.9758	1.5204	0.7289	1.0000
16	1.0000	1.0605	0.9378	2.4229	0.9762	2.4227	0.7128	1.0000
32	1.0000	1.0606	0.9380	4.4825	0.9763	4.4825	0.7086	1.0000
64	1.0000	1.0607	0.9381	8.7739	0.9763	8.7739	0.7075	1.0000
$ERqPp + k$								
1/64	1.0000	1.5810	1.0002	1.5349	1.0002	2.7769	1.0331	1.4393
1/32	1.0000	1.5806	1.0007	1.5576	1.0008	1.9262	1.0392	1.4395
1/16	1.0001	1.5801	1.0026	1.5753	1.0033	1.6450	1.0401	1.4404
1/8	1.0003	1.5776	1.0098	1.5711	1.0129	1.5625	1.0419	1.4433
1/4	1.0016	1.5694	1.0345	1.5478	1.0469	1.5280	1.0513	1.4475
1	1.0691	1.5158	1.1602	1.4487	1.2189	1.4508	1.1905	1.4032
4	1.2535	1.4650	1.2488	1.6177	1.2832	1.6161	1.3220	1.9569
8	1.2813	1.4596	1.2759	2.0412	1.2891	2.0406	1.3374	2.0471
16	1.2885	1.4581	1.2866	3.1805	1.2905	3.1804	1.3415	3.3525
32	1.2904	1.4578	1.2898	5.8159	1.2909	5.8159	1.3426	6.3826
64	1.2909	1.4578	1.2908	11.3420	1.2910	11.3420	1.3429	12.5988
Discrete ZZ								
1/64	0.9145	1.0018	0.9141	1.0076	1.0411	1.4914	1.0968	1.7991
1/32	0.9139	1.0027	0.9284	1.0078	0.9961	1.4550	1.0893	1.8293
1/16	0.9134	1.0031	0.9486	1.0085	0.9665	1.4445	1.0846	1.8351
1/8	0.9133	1.0023	0.9589	1.0099	0.9647	1.4458	1.0808	1.8305
1/4	0.9131	0.9994	0.9611	1.0121	0.9642	1.4565	1.0681	1.8118
1	0.9112	0.9945	0.9529	1.0081	0.9843	1.5578	0.9765	1.6464
4	0.9093	0.9993	0.9073	0.9950	0.9937	1.6441	1.0195	1.7211
8	0.9092	0.9998	0.8988	0.9986	0.9981	1.6519	1.0453	1.7408
16	0.9091	1.0000	0.8963	0.9996	0.9995	1.6540	1.0535	1.7462
32	0.9091	1.0000	0.8956	0.9999	0.9999	1.6545	1.0556	1.7476
64	0.9091	1.0000	0.8954	1.0000	1.0000	1.6546	1.0562	1.7480

TABLE 5.7*.9. Range of the effectivity index in the first layer as a function of the aspect ratio for the four patterns. Poisson's equation with homogeneous Neumann boundary condition. Linear element ($p = 1$), First layer ($p = 1, k = 1, q = 1$). $ERqBp + k$, $ERqPp + k$, and discrete ZZ error estimators (Poisson's equation, Neumann boundary, First layer, $p = 1, k = 1, q = 1$)

Aspect ratio a	Regular		Chevron		Union Jack		Criss cross	
	$\mathscr{C}_L$	$\mathscr{C}_U$	$\mathscr{C}_L$	$\mathscr{C}_U$	$\mathscr{C}_L$	$\mathscr{C}_U$	$\mathscr{C}_L$	$\mathscr{C}_U$
$ERqBp + k$								
1/64	1.0000	1.0000	0.9105	1.0000	0.9129	1.0000	0.7073	1.0000
1/32	1.0000	1.0000	0.9046	1.0000	0.9129	1.0000	0.7080	1.0000
1/16	1.0000	1.0000	0.8909	1.0000	0.9129	1.0000	0.7108	1.0000
1/8	1.0000	1.0001	0.8852	1.0000	0.9129	1.0000	0.7212	1.0000
1/4	1.0000	1.0003	0.9126	1.0000	0.9129	1.0000	0.7575	1.0000
1	1.0000	1.0031	0.9720	1.0000	0.9129	1.0000	0.9129	1.0000
4	1.0000	1.0234	0.9837	1.0000	0.9129	1.0000	0.7575	1.0000
8	1.0000	1.0286	0.9845	1.0000	0.9129	1.0000	0.7212	1.0000
16	1.0000	1.0302	0.9847	1.0000	0.9129	1.0000	0.7108	1.0000
32	1.0000	1.0306	0.9847	1.0000	0.9129	1.0000	0.7080	1.0000
64	1.0000	1.0307	0.9847	1.0000	0.9129	1.0000	0.7073	1.0000
$ERqPp + k$								
1/64	1.1726	1.3837	0.9995	1.2247	1.0000	1.2247	1.0000	1.7320
1/32	1.1727	1.3835	0.9988	1.2247	1.0000	1.2247	1.0000	1.7321
1/16	1.1731	1.3831	1.0018	1.2247	1.0000	1.2247	1.0000	1.7321
1/8	1.1746	1.3823	1.0343	1.2247	1.0000	1.2247	1.0000	1.7321
1/4	1.1787	1.3620	1.1215	1.2247	1.0000	1.2247	1.0000	1.7321
1	1.1937	1.3508	1.2247	1.3012	1.0000	1.2247	1.0000	1.7321
4	1.2188	1.2864	1.2247	1.3446	1.0000	1.2247	1.0000	1.7321
8	1.2230	1.2779	1.2247	1.3474	1.0000	1.2247	1.0000	1.7321
16	1.2243	1.2755	1.2247	1.3482	1.0000	1.2247	1.0000	1.7321
32	1.2246	1.2750	1.2247	1.3483	1.0000	1.2247	1.0000	1.7321
64	1.2247	1.2748	1.2247	1.3484	1.0000	1.2247	1.0000	1.7320
Discrete ZZ								
1/64	1.0000	1.0000	0.9997	1.0008	0.9998	1.0002	0.9994	1.0006
1/32	1.0000	1.0000	0.9997	1.0023	0.9997	1.0003	0.9988	1.0012
1/16	1.0000	1.0000	0.9996	1.0064	0.9994	1.0006	0.9975	1.0024
1/8	1.0000	1.0000	0.9995	1.0140	0.9987	1.0012	0.9950	1.0046
1/4	1.0000	1.0000	0.9992	1.0216	0.9975	1.0021	0.9897	1.0089
1	1.0000	1.0000	0.9982	1.0190	0.9953	1.0032	0.9609	1.0215
4	1.0000	1.0000	0.9996	1.0140	0.9990	1.0017	0.9435	1.0056
8	1.0000	1.0000	0.9999	1.0135	0.9997	1.0014	0.9429	1.0016
16	1.0000	1.0000	1.0000	1.0134	0.9999	1.0013	0.9428	1.0004
32	1.0000	1.0000	1.0000	1.0134	1.0000	1.0013	0.9428	1.0001
64	1.0000	1.0000	1.0000	1.0134	1.0000	1.0013	0.9428	1.0000

TABLE 5.7*.10. Range of the effectivity index in the first layer as a function of the aspect ratio for the four patterns. Poisson's equation with homogeneous Neumann boundary condition. Quadratic element ($p = 2$), First layer ($p = 2, k = 1, q = 2$), and quadratic equilibration ($q = 2$). $ERqBp + k$, $ERqPp + k$, and discrete ZZ error estimators (Poisson's equation, Neumann boundary, First layer, $p = 2, k = 1, q = 2$)

Aspect ratio a	Regular		Chevron		Union Jack		Criss cross	
	$\mathscr{C}_L$	$\mathscr{C}_U$	$\mathscr{C}_L$	$\mathscr{C}_U$	$\mathscr{C}_L$	$\mathscr{C}_U$	$\mathscr{C}_L$	$\mathscr{C}_U$
$ERqBp + k$								
1/64	0.9124	1.0635	0.8569	1.0606	0.8570	2.7488	0.5600	1.0000
1/32	0.9043	1.0672	0.8590	1.0604	0.8592	1.6609	0.5629	1.0001
1/16	0.9069	1.0684	0.8664	1.0598	0.8671	1.2513	0.5742	1.0004
1/8	0.9293	1.0632	0.8792	1.0580	0.8818	1.1298	0.6142	1.0015
1/4	0.9590	1.0533	0.8921	1.0560	0.9005	1.0986	0.7233	1.0051
1	0.9958	1.0710	0.9221	1.0750	0.9527	1.0812	0.9725	1.0238
4	0.9895	1.1005	0.9889	1.2413	0.9945	1.2456	0.7820	1.0740
8	0.9888	1.1028	0.9972	1.7121	0.9986	1.7712	0.7297	1.0772
16	0.9887	1.1034	0.9993	2.9133	0.9996	3.0874	0.7131	1.0781
32	0.9886	1.1036	0.9998	5.5418	0.9999	5.9271	0.7086	1.0783
64	0.9886	1.1036	1.0000	10.9366	1.0000	11.7273	0.7075	1.0784
$ERqPp + k$								
1/64	1.0020	1.7078	1.0042	10.9598	1.0043	11.4348	1.0954	2.3901
1/32	1.0065	1.7077	1.0141	5.6687	1.0145	5.8962	1.0954	2.3872
1/16	1.0149	1.7074	1.0365	3.1882	1.0379	3.2824	1.0954	2.3757
1/8	1.0223	1.7074	1.0694	2.1703	1.0744	2.1915	1.0958	2.3319
1/4	1.0189	1.7098	1.1010	1.8491	1.1174	1.8440	1.1033	2.1895
1	1.0311	1.6491	1.1452	1.6709	1.2079	1.6455	1.3619	1.5760
4	1.2381	1.6383	1.2667	2.0149	1.2819	2.0344	1.2419	1.9499
8	1.2775	1.6469	1.2846	2.5553	1.2887	2.6560	1.2454	2.6501
16	1.2877	1.6498	1.2894	3.9786	1.2904	4.2189	1.2494	4.5978
32	1.2902	1.6506	1.2906	7.2718	1.2908	7.7773	1.2508	8.7923
64	1.2909	1.6508	1.2910	14.1789	1.2910	15.2039	1.2511	17.3743
Discrete ZZ								
1/64	0.9302	0.9641	0.9063	1.0050	1.0411	1.5793	0.9016	1.5327
1/32	0.9365	0.9652	0.9050	1.0051	1.0410	1.5792	0.9018	1.5313
1/16	0.9400	0.9682	0.9040	1.0052	1.0407	1.5787	0.9027	1.5263
1/8	0.9397	0.9730	0.9118	1.0049	1.0394	1.5769	0.9062	1.5128
1/4	0.9390	0.9793	0.9290	1.0044	1.0345	1.5697	0.9183	1.5385
1	0.9356	0.9928	0.9596	1.0100	0.9951	1.5202	1.0026	1.7367
4	0.9347	0.9999	0.9702	1.0371	0.9884	1.6364	1.0243	1.8978
8	0.9347	1.0006	0.9708	1.0429	0.9914	1.6498	1.0460	1.9130
16	0.9347	1.0007	0.9710	1.0446	0.9923	1.6534	1.0536	1.9170
32	0.9347	1.0008	0.9710	1.0450	0.9926	1.6544	1.0557	1.9180
64	0.9347	1.0008	0.9710	1.0451	0.9927	1.6546	1.0562	1.9183

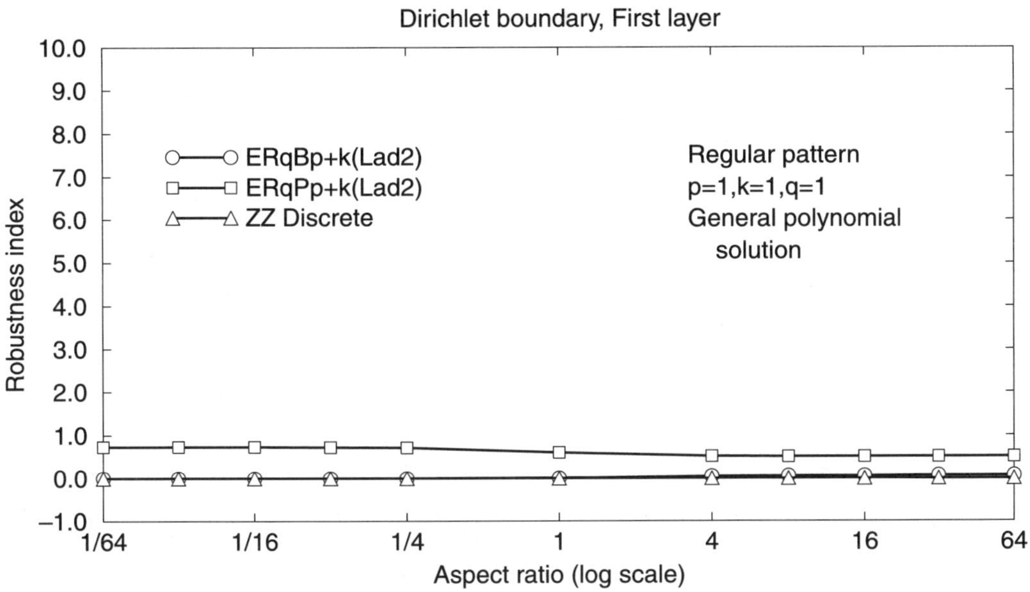

FIG. 5.7*.41. *The effect of the aspect ratio on the robustness of the estimators in the first layer of elements adjacent to the boundary*: Homogeneous Dirichlet boundary condition, linear elements ($p = 1$). Regular pattern. Note that all three estimators are robust for any aspect ratio for this pattern

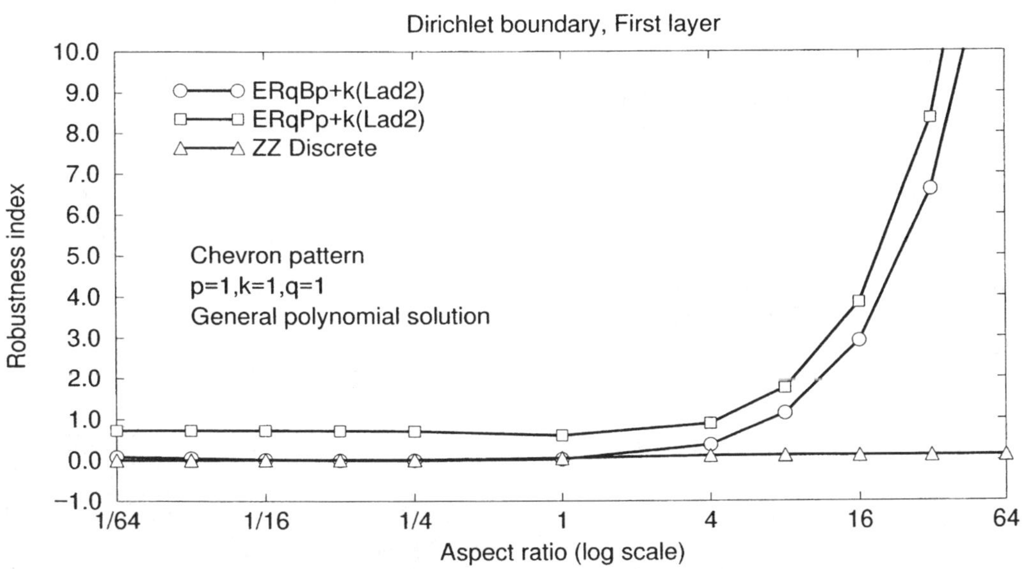

FIG. 5.7*.42. *The effect of the aspect ratio on the robustness of the estimators in the first layer of elements adjacent to the boundary*: Homogeneous Dirichlet boundary condition, linear elements ($p = 1$). Chevron pattern. Note that the Discrete ZZ is practically exact, while the ERqBp + k and ERqPp + k are not robust for large a

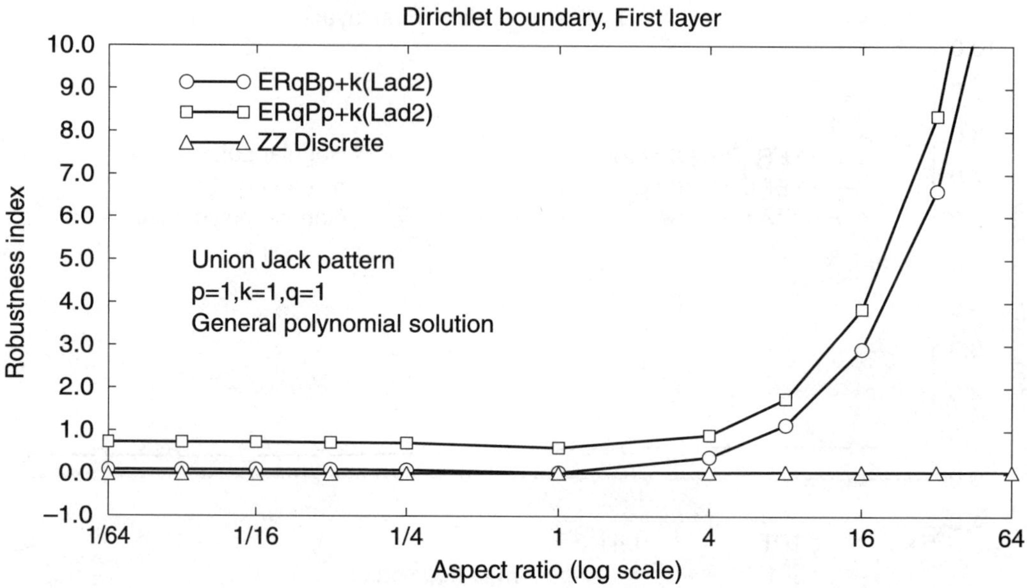

FIG. 5.7*.43. *The effect of the aspect ratio on the robustness of the estimators in the first layer of elements adjacent to the boundary*: Homogeneous Dirichlet boundary condition, linear elements ($p = 1$). Union-Jack pattern. Note that the Discrete ZZ is practically exact, while the ERqBp + k and ERqPp + k are not robust for large a

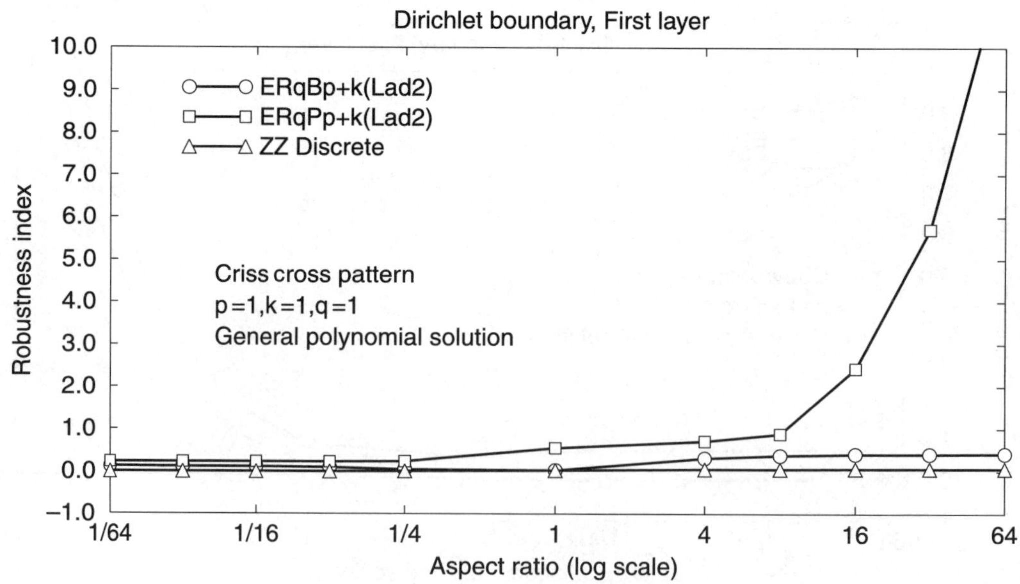

FIG. 5.7*.44. *The effect of the aspect ratio on the robustness of the estimators in the first layer of elements adjacent to the boundary*: Homogeneous Dirichlet boundary condition, linear elements ($p = 1$). Criss-cross pattern. Note that ERqBp + k and Discrete ZZ are robust, while ERqPp + k is not robust

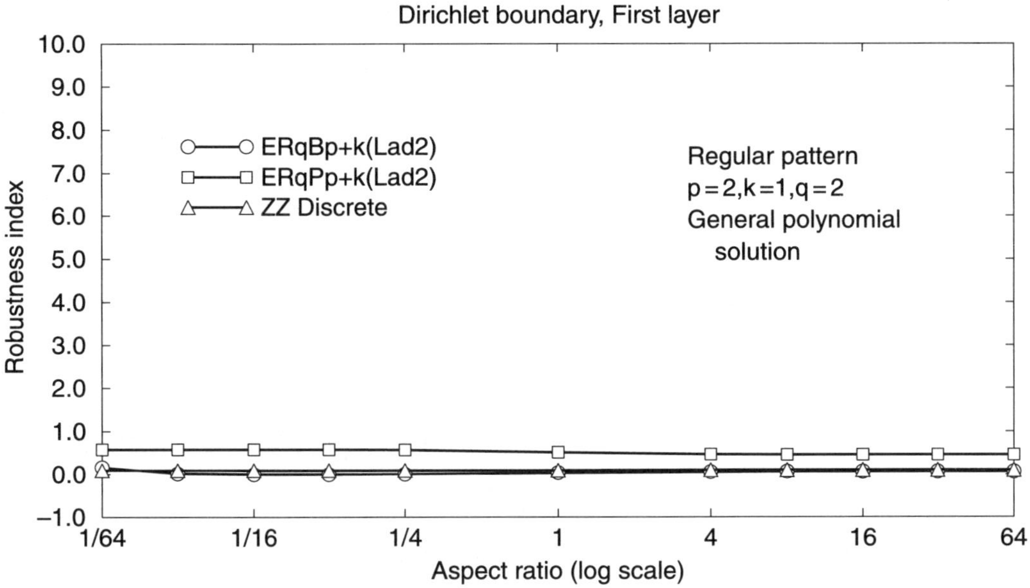

FIG. 5.7*.45. *The effect of the aspect ratio on the robustness of the estimators in the first layer of elements adjacent to the boundary*: Homogeneous Dirichlet boundary condition, quadratic elements ($p = 2$). Regular pattern. As in the case $p = 1$, all three estimators are robust for this pattern

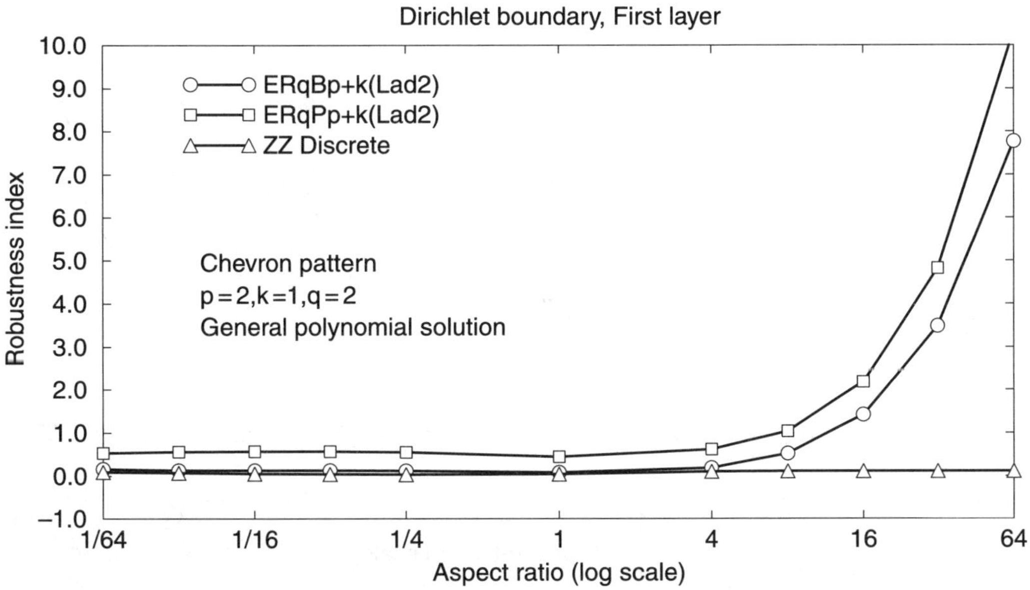

FIG. 5.7*.46. *The effect of the aspect ratio on the robustness of the estimators in the first layer of elements adjacent to the boundary*: Homogeneous Dirichlet boundary condition, quadratic elements ($p = 2$). Chevron pattern. As in the case $p = 1$, only the Discrete ZZ is robust for this pattern

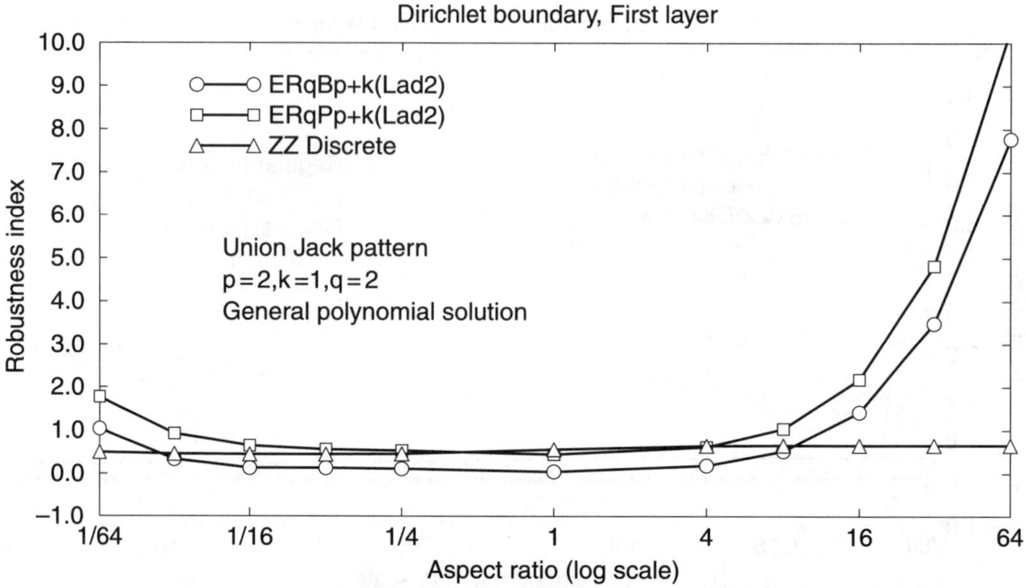

FIG. 5.7*.47. *The effect of the aspect ratio on the robustness of the estimators in the first layer of elements adjacent to the boundary*: Homogeneous Dirichlet boundary condition, quadratic elements ($p = 2$). Union-Jack pattern. As in the case $p = 1$, only the Discrete ZZ is robust for any aspect ratio for this pattern. Note also that for $\frac{1}{4} \leq a \leq 4$ the ERqBp + k is the most robust of the three estimators

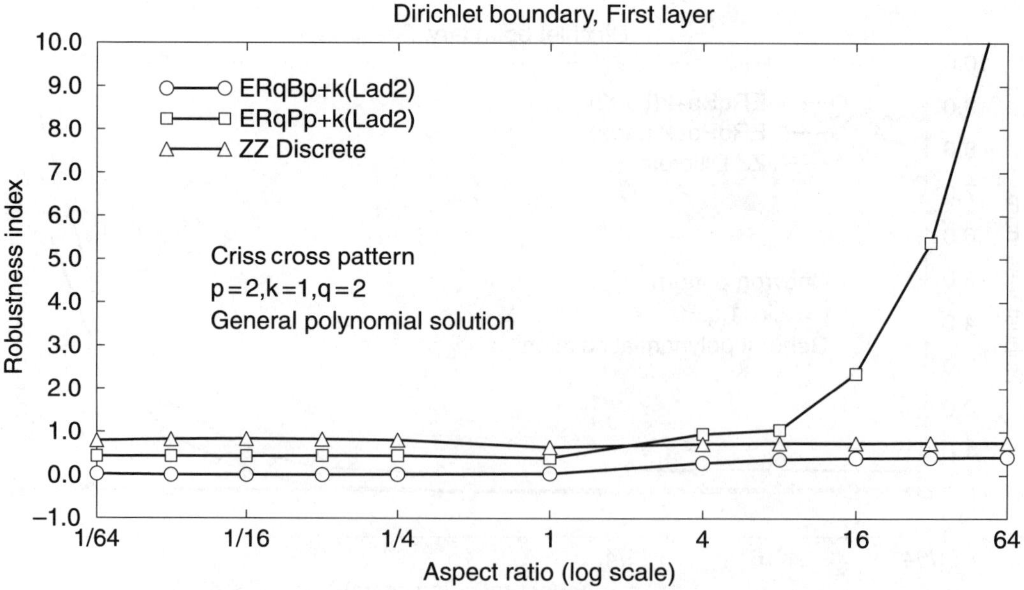

FIG. 5.7*.48. *The effect of the aspect ratio on the robustness of the estimators in the first layer of elements adjacent to the boundary*: Homogeneous Dirichlet boundary condition, quadratic elements ($p = 2$). Criss-cross pattern. Both the Discrete ZZ and ERqBp + k are robust for any aspect ratio a, while the ERqPp + k is not robust for large a. Note also that for this pattern the ERqBp + k is the most robust of the three estimators

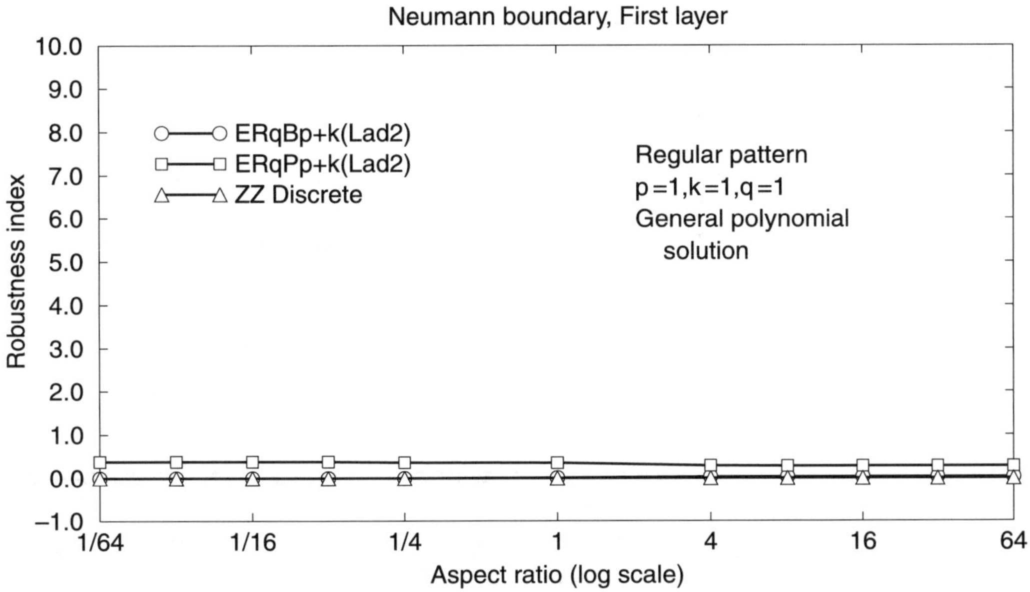

FIG. 5.7*.49. *The effect of the aspect ratio on the robustness of the estimators in the first layer of elements adjacent to the boundary*: Homogeneous Neumann boundary condition, linear elements ($p = 1$). Regular pattern. Note that all three estimators are robust with respect to the aspect ratio for this pattern

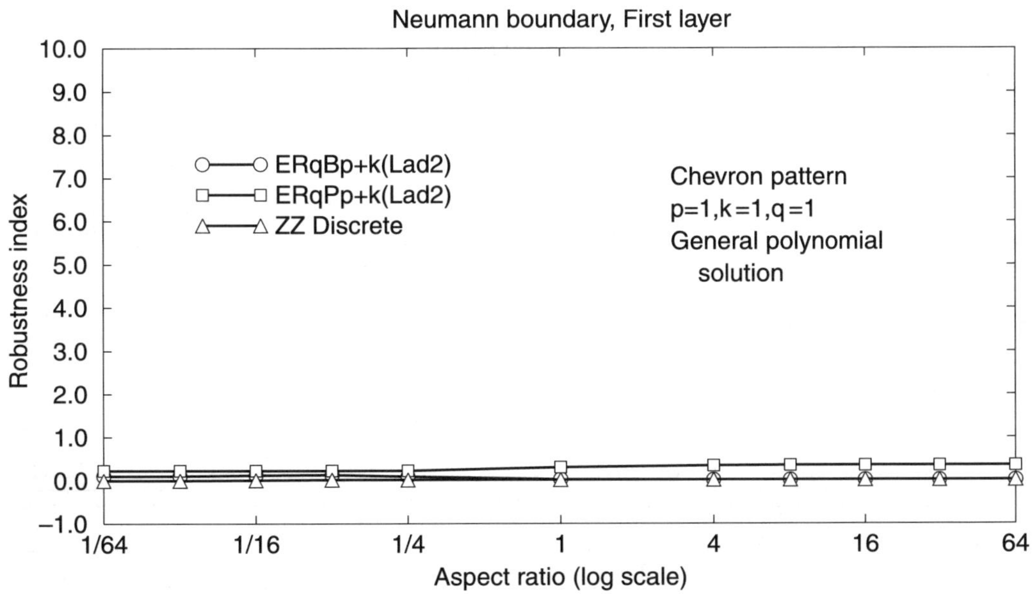

FIG. 5.7*.50. *The effect of the aspect ratio on the robustness of the estimators in the first layer of elements adjacent to the boundary*: Homogeneous Neumann boundary condition, linear elements ($p = 1$). Chevron pattern. Note that all three estimators are robust with respect to the aspect ratio for this pattern

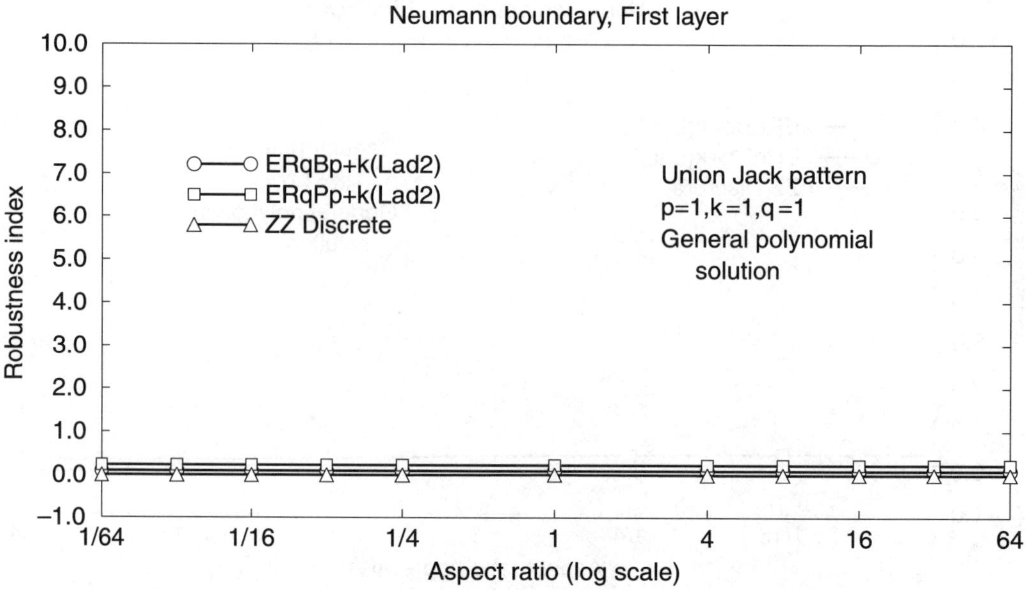

FIG. 5.7*.51. *The effect of the aspect ratio on the robustness of the estimators in the first layer of elements adjacent to the boundary*: Homogeneous Neumann boundary condition, linear elements ($p = 1$). Union-Jack pattern. Note that all three estimators are robust with respect to the aspect ratio for this pattern

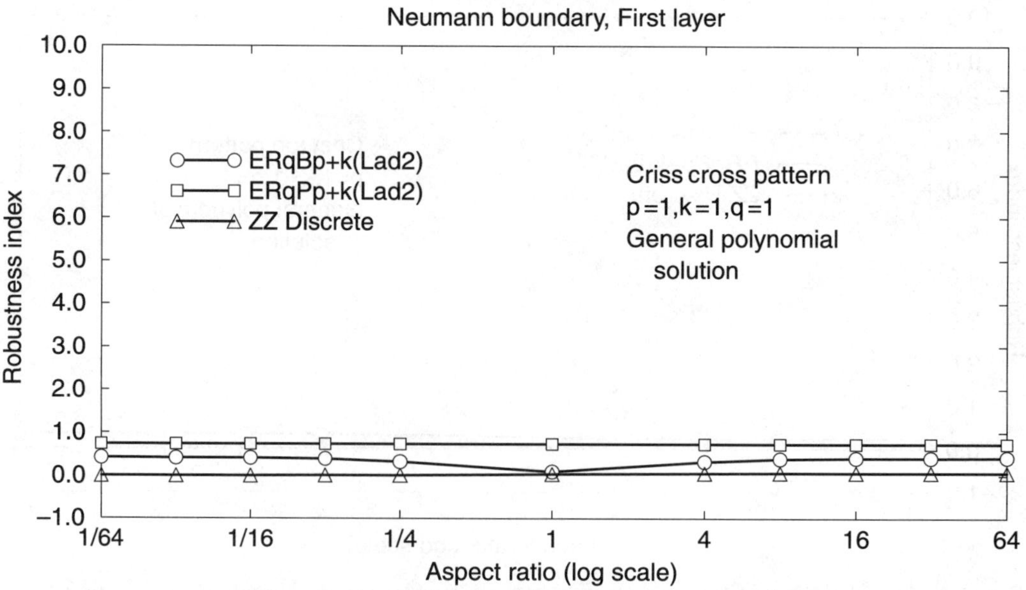

FIG. 5.7*.52. *The effect of the aspect ratio on the robustness of the estimators in the first layer of elements adjacent to the boundary*: Homogeneous Neumann boundary condition, linear elements ($p = 1$). Criss-cross pattern. The Discrete ZZ is the most robust estimator

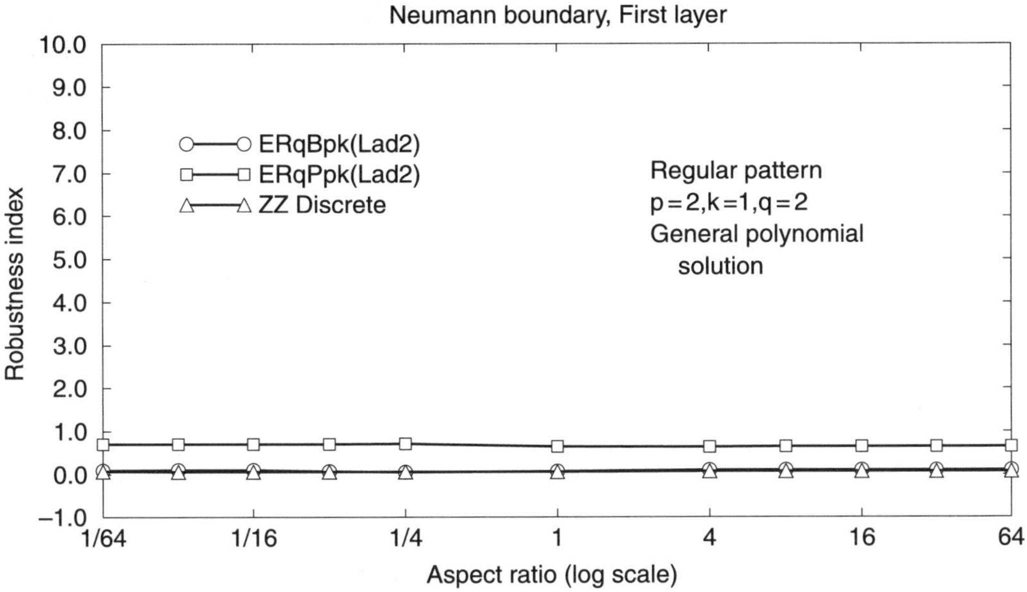

FIG. 5.7*.53. *The effect of the aspect ratio on the robustness of the estimators in the first layer of elements adjacent to the boundary*: Homogeneous Neumann boundary condition, quadratic elements ($p = 2$). Regular pattern. Note that all three estimators are robust with respect to the aspect ratio for this pattern

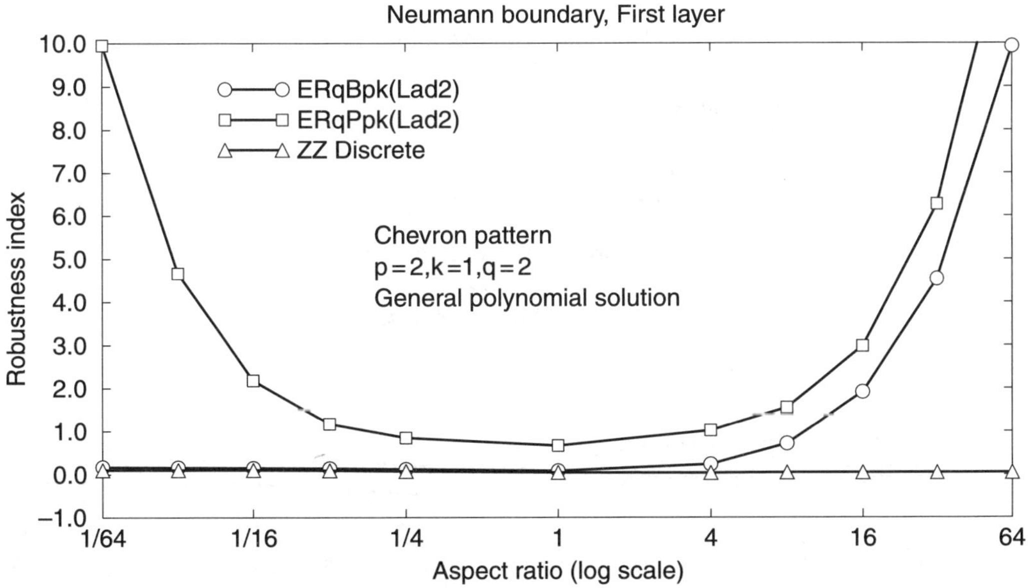

FIG. 5.7*.54. *The effect of the aspect ratio on the robustness of the estimators in the first layer of elements adjacent to the boundary*: Homogeneous Neumann boundary condition, quadratic elements ($p = 2$). Chevron pattern. Note that only the Discrete ZZ is robust with respect to the aspect ratio for this pattern

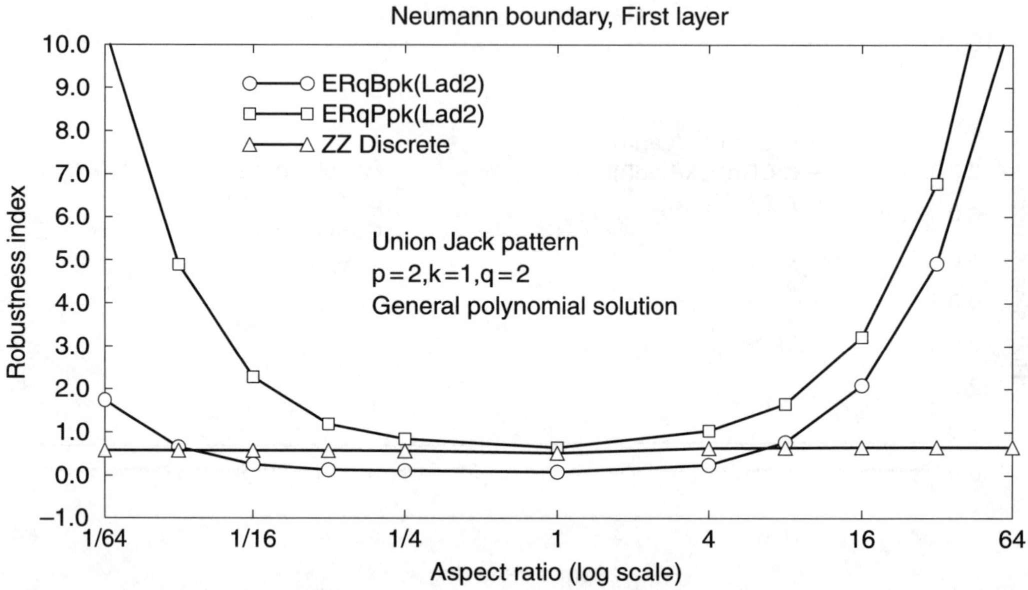

FIG. 5.7*.55. *The effect of the aspect ratio on the robustness of the estimators in the first layer of elements adjacent to the boundary*: Homogeneous Neumann boundary condition, quadratic elements ($p = 2$). Union-Jack pattern. Note that only the Discrete ZZ is robust for the entire range of the aspect ratio for this pattern. However, for the range $\frac{1}{4} \leq a \leq 4$, the ERqBp + k is the most robust estimator

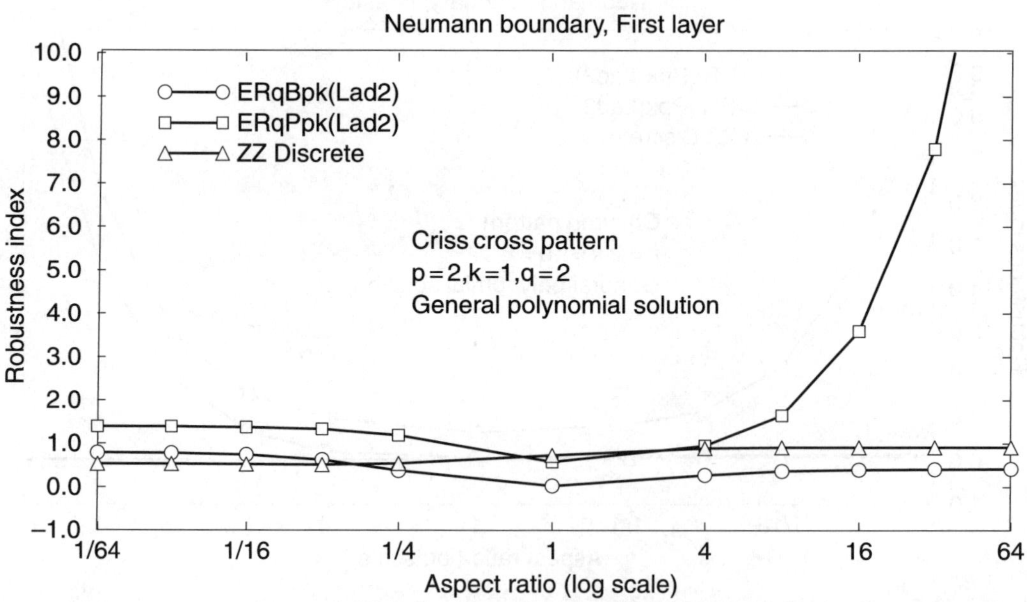

FIG. 5.7*.56. *The effect of the aspect ratio on the robustness of the estimators in the first layer of elements adjacent to the boundary*: Homogeneous Neumann boundary condition, quadratic elements ($p = 2$). Criss-cross pattern. Note that only the Discrete ZZ is robust for the entire range of the aspect ratio for this pattern. However, for the range $\frac{1}{4} \leq a \leq 4$, the ERqBp + k is the most robust estimator

boundary. Figure 5.7*.57 (respectively Fig. 5.7*.58) shows the results for the robustness of the ERqPp + k estimator with Type 2 Ladeveze equilibration for quadratic elements with linear ($q = 1$) and quadratic ($q = 2$) equilibration for the first mesh layer adjacent to the homogeneous Dirichlet (respectively Neumann) boundary for the four mesh patterns.

From the results on the effect of the order of equilibration (q) on the robustness of the ERqPp + k estimators we can make the following conclusions:

(1) The ERpPp + k estimator with Ladeveze equilibration of Type 2 is significantly more robust than the ERqPp + k estimator with $0 \leq q \leq p$. Here we analyzed the case $p = 2$, but similar results could be also obtained for other values of p ($p = 3$, 4, etc.).

(2) In cases where we are interested in meshes with high a or $1/a$, even ERpPp + k may not be sufficiently robust, depending on the mesh pattern.

Local error indicators are used in adaptive procedures, the goal of which is to construct nearly optimal meshes. Such meshes are obtained by achieving equidistribution of an indicator, and are not particularly sensitive to its definition. So, it is not critical which indicator is used in the adaptive procedure.

Another use of the estimators is in the *stopping criterion*, namely, to decide if the desired accuracy has been achieved. For this purpose, it is important to employ only *guaranteed* estimators, like the ones which will be addressed in the next chapter.

Many of the constructions presented in this and the next chapter can also be applied to important classes of nonlinear and transient problems.

In this section we presented detailed numerical analysis of the asymptotic performance of various estimators. Such a detailed analysis was necessary for the justification of our conclusions. We analyzed the robustness of various estimators which are employed in practice. The results have asymptotic character and were presented for elements with straight edges. The extension of the asymptotic approach, with some modifications to curved elements, is also possible, but the results are different, especially for the *ZZ* indicator. Let us, once more, underline that the *analysis of the estimators is local and was done in a careful way to avoid inclusion of the pollution effect, which cannot be accounted for by local indicators.* The a-posteriori estimation of the pollution will be addressed in the next chapter.

5.8 Historical and bibliographical comments

Although the term *a-posteriori error estimate* was likely used for the first time in [Ostrowski, 1940], the idea was already employed long before 1940. A classical example is the estimation of the error in a partial sum of an infinite series in terms of the first term not included in the sum, which can be made rigorous by employing a saturation assumption. Another example are the two-sided estimates for approximate solutions of elliptic partial differential equations constructed by the method of *orthogonal projections* by [Zaremba, 1909] (see Section 6.1, and Remark 6.1*.10), and by the Treftz method [Treftz, 1926], and were later extended by [Mikhlin, 1946]. The methods employed in the construction of the two-sided estimates are related to the primal and dual formulation of the problems in mechanics, which were employed in the construction of error bounds by [Synge, 1957]; see also [Zeidler, 1990] for an abstract setting.

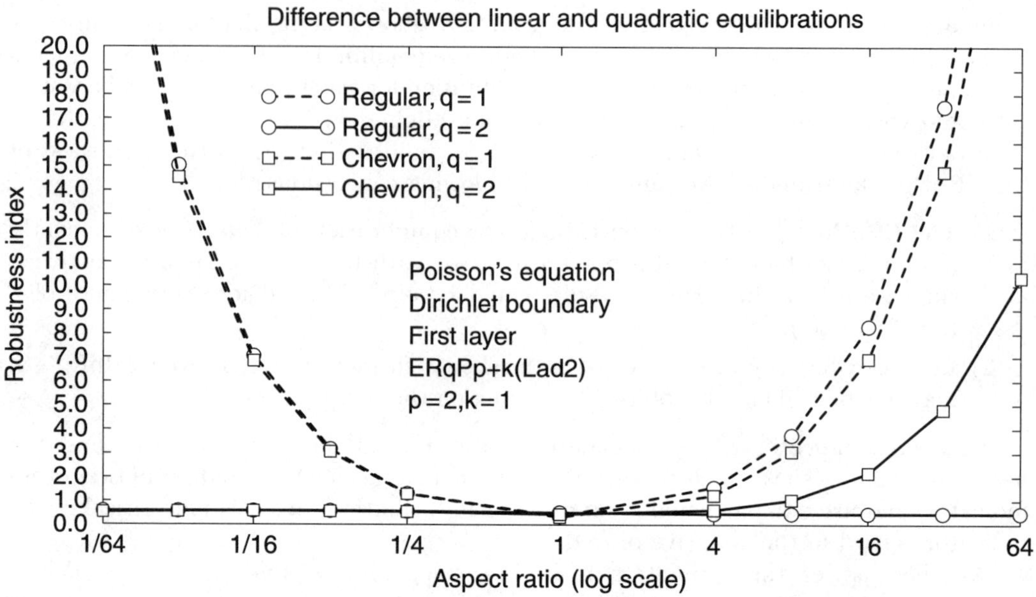

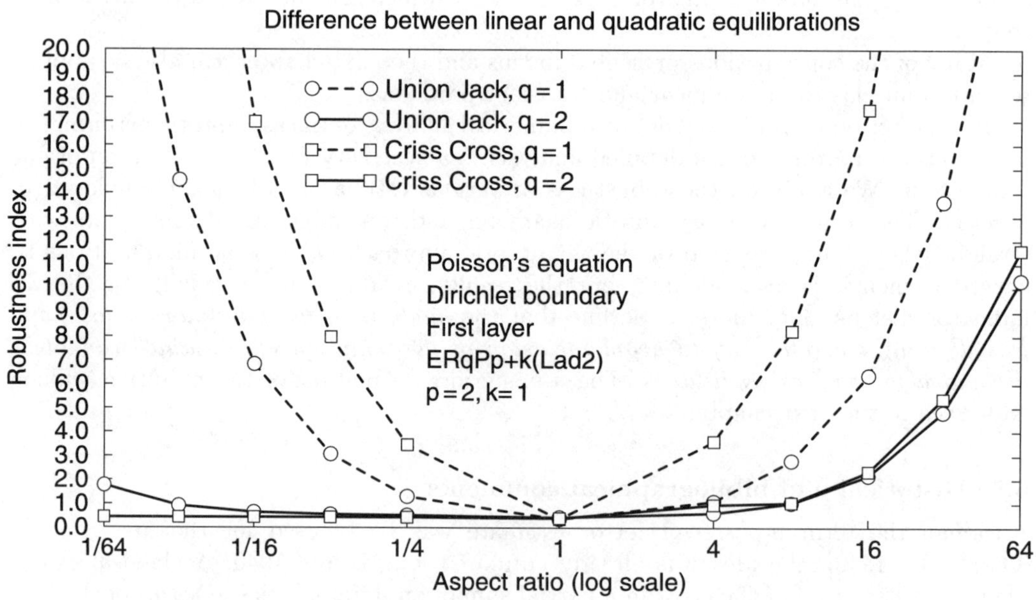

FIG. 5.7*.57. *The effect of the order of equilibration on the robustness of the equilibrated element residual estimator with respect to the aspect ratio in the first layer adjacent to the boundary*: Homogeneous Dirichlet boundary condition, quadratic ($p = 2$) elements, ERpPp + k estimator, $k = 1$. Note that the overall robustness with respect to the aspect ratio dramatically improves if quadratic ($q = p = 2$) equilibration is used. The estimator ERqPp + k, with $q < p$, is not robust with respect to the aspect ratio

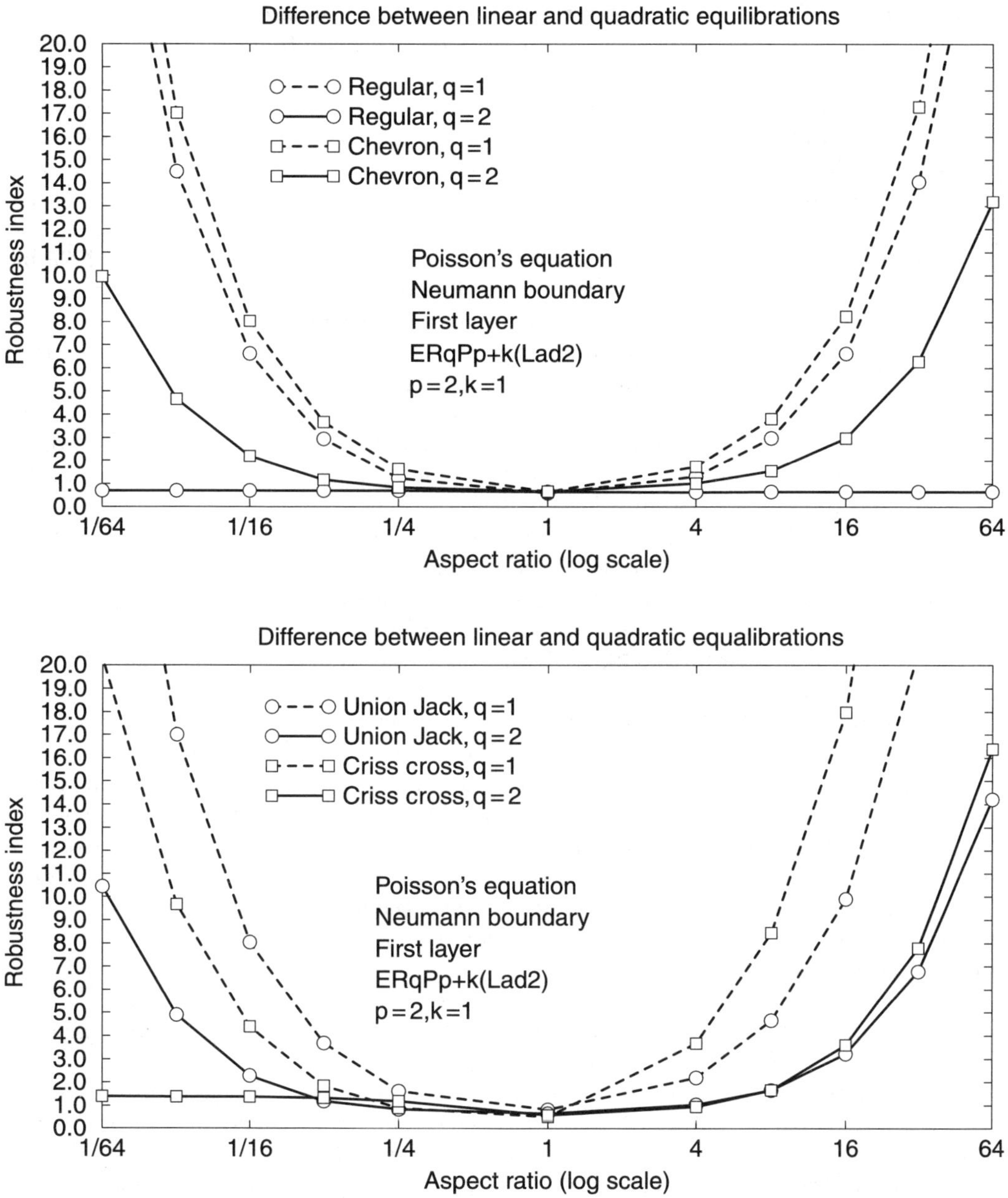

FIG. 5.7*.58. *The effect of the order of equilibration on the robustness of the element residual estimator in the first layer adjacent to the boundary*: Homogeneous Neumann boundary condition, quadratic ($p = 2$) elements, ERpPp + k estimator, $k = 1$. Note that the overall robustness with respect to the aspect ratio improves if quadratic ($q = p = 2$) equilibration is used. However, even the ERpPp + k estimator is not sufficiently robust for the chevron and the Union-Jack patterns

Other classical approaches of a-posteriori estimation often used, for example, in finite difference methods, are based on the comparison of solutions obtained using two meshes and Richardson extrapolation [Richardson, 1910]. This approach is widely used for the assessment of the error in the numerical integration of ordinary differential equations. Another version of this approach is based on the comparison of approximate solutions of different order and can be used in the p-version of the finite element method (see, for example [Szabo abd Babuška, 1991]) and in the numerical solution of ordinary differential equations (see [Fehlberg, 1970]).

A-posteriori error estimates in the finite element method were introduced in 1978 by Babuška and Rheinboldt, particularly the explicit and the subdomain residual estimator. The adaptive code FEARS employing meshes of curvilinear quadrilaterals with capability for 1 to n irregular connections, and bilinear $p = 1$ elements for two-dimensional elasticity problems was written in 1978–1979 at the University of Maryland. FEARS used the explicit estimator with an asymptotic constant which lead to the effectivity index in the range 0.5–2.00 in all test computations. The explicit indicators were used for adaptive refinement and the stopping criterion [Mesztenyi and Szymczak, 1982], [Babuška, 1981], [Babuška and Rheinboldt, 1978b].

Since the late 1980s the a-posteriori error estimation and adaptive procedures have become to be widely researched for various norms and various types of differential equation. In the last few years various a-posteriori error estimations are also included in commercial codes. There are many papers in the literature. Among others, we refer to [Verfürth, 1996], [Verfürth, 1997], [Eriksson *et al.*, 1995], [Johnson and Hansbo, 1992], [Johnson, 1993], [Eriksson and Johnson, 1988], [Strouboulis and Haque, 1992a,b], [Ainsworth and Oden, 2000], [Ainsworth and Oden, 1997], [Oden *et al.*, 1989], [Bank and Smith, 1993], [Ainsworth, 1996], [Bank and Weiser, 1985], [Bornemann, Erdmann, and Kornhuber, 1996], [Carstensen, 1999], [Rodriguez, 1994], [Ainsworth, 1994], [Ainsworth and Craig, 1992], [Babuška, Plank, and Rodriguez, 1992], [Hugger, 2001], [Rannacher, 1998], and [Appel *et al.*, 2000]. All of these works are related to the ideas and results addressed in this book.

Various papers addressing the asymptotic exactness of the estimators for various special meshes have also appeared, e.g., [Durán *et al.*, 1991] [Durán *et al.*, 1992] [Durán and Rodriguez, 1992] [Yu, 1991a,b].

For recent results on L^∞ a-posteriori error estimates on the elemental level, under assumption of negligible pollution, we refer to [Hoffmann *et al.*, 1999]; see also [Schatz, 1998, 2001].

A-posteriori estimates can be based on the difference between the computed finite element solution and an 'enhanced' solution based on superconvergence or post processing based on local averaging. Superconvergence was discussed in Chapter 4 where also its history was mentioned. The construction of various 'enhanced' solutions was addressed in the engineering literature long time ago. As an example let us mention [Stein and Ahmad, 1977], where 'enhanced' stresses were constructed. Comparison of the enhanced stresses with the stresses coming from the finite element solution could be used for a-posteriori estimation. Further, recovered continuous solutions can also be employed for obtaining guaranteed lower bounds.

Finally, let us underline that the notion of an a-posteriori error estimate is used in the literature in *a very loose way* and it *does not mean* a true, i.e., guaranteed error estimate.

The guaranteed upper and lower estimates will be discussed in Chapter 6. Development of such estimates is a very recent accomplishment and is the subject of ongoing research.

In this chapter we addressed the estimation for the global energy norm of the error. The error indicators may also be used as estimates of the error in an element or a region in cases that the pollution does not play an important role. A-posteriori error estimation of the pollution will be addressed in Chapter 6.

References

M. Ainsworth and A. Craig (1992) A posteriori error estimators in the finite element method. *Numerische Mathematik*, **60**, 429–463.

M. Ainsworth and J.T. Oden (1993) A unified approach to a-posteriori error estimation using element residual methods. *Numerische Mathematik*, **65**, 23–50.

M. Ainsworth (1994) The performance of the Bank–Weiser's error estimator for quadrilateral finite elements. *Numerical Methods for Partial Differential Equations*, **10**, 609–623.

M. Ainsworth (1996) The influence and selection of subspaces for a-posteriori estimators. *Numerische Mathematik*, **73**, 399–418.

M. Ainsworth and J.T. Oden (1997) A-posteriori error estimation in finite element analysis. *Computer Methods in Applied Mechanics and Engineering: Computational Mechanics Advances*, **142**, 1–88.

M. Ainsworth and I. Babuška (1999) Reliable and robust a-posteriori error estimation for singularly perturbed reaction diffusion problems. *SIAM Journal on Numerical Analysis*, **36**, 331–353.

M. Ainsworth and J.T. Oden (2000) *A-posteriori error estimation in finite element analysis*. John Wiley and Sons, New York.

T. Appel, M. Berzins, P.K. Jimack, G. Kunert, A. Planks, I. Tsukerman and M. Walkley (2000) Mesh shape and anisotropic elements: theory and practice. In: J.R. Whiteman (ed.) *The Mathematics of Finite Elements and Applications X*. Elsevier, Amsterdam, pp. 367–376.

J.P. Aubin and H.G. Burchard (1971) Some aspects of the method of the hypercircle applied to elliptic variational problems. In: B. Hubbard (ed.) *SYNSPADE 1970*. Academic Press, New York, pp. 1–67.

I. Babuška and W.C. Rheinboldt (1978a) A-posteriori error estimates for the finite element method. *International Journal for Numerical Methods in Engineering*, **12**, 1597–1615.

I. Babuška and W.C. Rheinboldt (1978b) A-posteriori bounds and adaptive procedures for the finite element method. In: *Recent Advances in Engineering Science, Proceedings of 15th Annual Meeting, Society of Engineering Science*. University of Florida, Division of Continuing education, pp. 413–423.

I. Babuška and W.C. Rheinboldt (1978c) Error estimates for adaptive finite element computation. *SIAM Journal on Numerical Analysis*, **15**, 736–754.

I. Babuška and W.C. Rheinboldt (1980) Reliable error estimation and mesh adaptation for the finite element method. In: J.T. Oden (ed.) *Computational Methods in Nonlinear Mechanics*. North Holland, Amsterdam, pp. 67–108.

I. Babuška (1981) FEARS, the adaptive solver with an error estimation. The design theory and experience. In: R. Vichnevetsky and R.S. Stepleman (eds) *Advances of computer Methods for Partial Differential Equations IV*, *Proceedings of the Fourth IMACS International Symposium on Computer Methods for Partial Differential Equations, June 30–July 2*, Lehigh University Bethlehem, PA.

I. Babuška and A. Miller (1981) *A-posteriori error estimates and adaptive techniques for the finite element method, Technical Note BN-968*. Institute for Physical Science and Technology, University of Maryland.

I. Babuška and M. Vogelius (1981) On a dimensional reduction method. I The optimal selection of basis functions. II Some approximation theoretical results. *Mathematics of Computation*, **37**, 31–45, 47–68.

I. Babuška and D. Yu (1987) Asymptotically exact a posteriori error estimator for biquadratic elements. *Finite Elements in Analysis and Design*, **3**, 341–354.

I. Babuška and A. Miller (1987) A feedback finite element method with a posteriori error estimation: Part I. The finite element method and some basic properties of the a posteriori error estimator. *Computer Methods in Applied Mechanics and Engineering*, **61**, 1–40.

I. Babuška, R. Durán, and R. Rodriguez (1992) Analysis of the efficiency of an a-posteriori error estimator for linear triangular finite elements. *SIAM Journal on Numerical Analysis*, **29**, 947–964.

I. Babuška, L. Plank and R. Rodriguez (1992) Quality assessment of the a-posteriori error estimation in finite elements. *Finite Elements in Analysis and Design*, **11**, 285–306.

I. Babuška and R. Rodriguez (1993) The problem of the selection of an a-posteriori error indicator based on smoothening techniques. *International Journal for Numerical Methods in Engineering*, **36**, 539–567.

I. Babuška, T. Strouboulis, C.S. Upadhyay, S.K. Gangaraj, and K. Copps (1994a) An objective criterion for assessing the reliability of a-posteriori estimators in finite element computations. *IACM Bulletin*, pp. 27–37.

I. Babuška, T. Strouboulis, C.S. Upadhyay, S.K. Gangaraj, and K. Copps (1994b) Validation of a posteriori error estimators by numerical approach. *International Journal for Numerical Methods in Engineering*, **37**, 1073–1123.

I. Babuška, T. Strouboulis, and C.S. Upadhyay (1994) A model study of the quality of a posteriori error estimators for linear elliptic problems. Error estimation in the interior of patchwise uniform grids of triangles. *Computer Methods in Applied Mechanics and Engineering*, **114**, 307–378.

I. Babuška and C. Schwab (1996) A posteriori error estimation for hierarchic models of elliptic boundary value problems on thin domains. *SIAM Journal on Numerical Analysis*, **33**, 221–246.

I. Babuška, T. Strouboulis, and C.S. Upadhyay (1997) A model study of the quality of a-posteriori error estimators for finite element solutions of linear elliptic problems with particular reference to the behavior near the boundary. *International Journal for Numerical Methods in Engineering*, **40**, 2521–2577.

P.L. Baehmann, M.S. Shephard, and J.E. Flaherty (1992) A-posteriori error estimation for triangular quadratic elements using interior residuals. *International Journal for Numerical Methods in Engineering*, **34**, 979–996.

R.E. Bank and A. Weiser (1985) Some a posteriori error estimators for elliptic partial differential equations. *Mathematics of Computation*, **44**, 283–301.

R.E. Bank and R.K. Smith (1993) A posteriori error estimates based on hierarchical bases. *SIAM Journal on Numerical Analysis*, **30**, 921–935.

F. Bornemann, B. Erdmann, and R. Kornhuber (1996) A-posteriori error estimates for elliptic problems in two and three dimensions. *SIAM Journal on Numerical Analysis*, **33**, 1188–1204.

C. Carstensen (1999) Quasi-interpolant and a-posteriori error analysis in Finite Element Analysis. *Mathematical Modelling and Numerical Analysis*, **33**, 1187–1202.

C. Carstensen and R. Verfürth (1999) Edge residuals dominate a-posteriori error estimates for low order finite element methods. *SIAM Journal on Numerical Analysis*, **36**, 1571–1587.

C. Carstensen and S. Bartels (1999) Each averaging technique yields reliable a-posteriori control in FEM on unstructured grids. Part I: Low order conforming, nonconforming and mixed FEM. *Berichtsreihe des Mathematischen Seminar Kiel*, 99–11.

C. Carstensen and S.A. Funken (2000a) Constants in Clement-interpolation error and residual based a-posteriori estimates in finite element methods. *Berichtsreihe des Mathematischen Seminar Kiel*, 97–11 (1997). *East-West Journal of Mathematical Analysis* (to appear).

C. Carstensen and S.A. Funken (2000b) Fully reliable localized error control in FEM. Berichtsreihe des Mathematischen Seminar Kiel, 97–12 (1997). *SIAM Journal on Scientific Computing* (to appear).

P. Deuflhard, P. Leinen, and H. Yserentant (1989) Concept of an adaptive hierarchic finite element code. *Impact Computational Science and Engineering*, **1**, 3–35.

R. Durán, M.A. Muschietti and R. Rodriguez (1991) On the asymptotic exactness of the error estimators for linear triangular elements. *Numerische Mathematik*, **59**, 107–127.

R. Durán, M.A. Muschietti and R. Rodriguez (1992) Asymptotically exact error estimators for rectangular finite elements. *SIAM Journal on Numerical Analysis*, **29**, 78–88.

R. Durán and R. Rodriguez (1992) On the asymptotic exactness of Bank–Weiser's estimator. *Numerische Mathematik*, **62**, 297–303.

K. Eriksson and C. Johnson (1988) An adaptive finite element method for linear elliptic problems. *Mathematics of Computation*, **50**, 361–383.

K. Eriksson, D. Estep, P. Hansbo, and C. Johnson (1995) Introduction to adaptive methods for differential equations. *Acta Numerica*, Cambridge University Press, pp. 105–158.

E. Fehlberg (1970) Klassische Runge-Kutta formeln vierter und niedrigerer Ordnung mit Schrittweiten Kontrolle und ihre Anwendung auf Wärmeleitung Probleme. *Computing*, **6**, 61–71.

W. Hoffmann, A.H. Schatz, L.B. Wahlbin, and G. Wittum (2000) Asymptotically exact a-posteriori estimators for the pointwise gradient error on each element in irregular meshes Part I: A smooth problem and globally quasi-uniform meshes. *Mathematics of Computation* (to appear).

J. Hugger (2001) A theory of local a-posteriori pointwise residual based estimation of the finite element error. *Journal of Computational and Applied Mathematics* (to appear).

C. Johnson and P. Hansbo (1992) Adaptive finite element methods in computational mechanics. *Computer Methods in Applied Mechanics and Engineering*, **101**, 143–181.

C. Johnson (1993) A new paradigm for adaptive finite element methods. In: J.R. Whiteman (ed.) *The Mathematics of Finite Elements and Applications VIII.* Wiley, Chichester, pp. 105–120.

G. Kunert and R. Verfürth (1999) *Edge residuals dominate a-posteriori error estimates for linear finite element methods on anisotropic trianglular and tetrahedral meshes*, Bericht Nr 256 Facultät fur Mathematik Ruhr-Universität Bochum.

P. Ladeveze and D. Leguillon (1983) Error estimate procedure in the finite element method and applications. *SIAM Journal on Numerical Analysis*, **20**, 485–509.

P. Ladeveze and E.A.W. Maunder (1996) A general method for recovering equilibrating element tractions. *Computer Methods in Applied Mechanics and Engineering*, **137**, 111–151.

J.M. Melenk and I. Babuška (1996) The partition of unity finite element method: Basic theory and applications. *Computer Methods in Applied Mechanics and Engineering*, **139**, 289–314.

C. Mesztenyi and W. Szymczak (1982) *FEARS User's Manual for UNIAC 1100.* Technical Note BN-991, IPST, University of Maryland, College Park.

S.G. Mikhlin (1964) *Variational Methods in Mathematical Physics.* Pergamon Press, Oxford.

P. Morin, R.H. Nochetto, and G. Siebert (2001) Local problems on stars: a-posteriori error estimation, convergence and performance (Preprint).

J.T. Oden, L. Demkowicz, W. Rachowicz and T.A. Westermann (1989) Toward a universal *h–p* adaptive finite element strategy: Part 2. A Posteriori Error Estimates. *Computer Methods in Applied Mechanics and Engineering*, **77**, 113–180.

A. Ostrowski (1940) Recherches sur la methode de Graeffe et les zéros des polynomes et des séries des Laurent. *Acta Mathematica*, **72**, 99–257.

E. Rank and O.C. Zienkiewicz (1987) A simple error estimator in the finite element method. *Communications in Applied Numerical Methods*, **3**, 243–249.

R. Rannacher (1998) Error control in finite element computations. Proc. Summer School, *Error Control and Adaptivity in Scientific Computing.* Antalya, Turkey, August 1998, Kluwer.

L.F. Richardson (1910) The approximate arithmetic solution by finite differences of physical problems involving differential equations with an application to stresses in a masonry dam. *Transactions of the Royal Soc.*, London, Ser. A, **210**, 307–357.

R. Rodriguez (1993) Some remarks on Zienkiewicz–Zhu estimator. *International Journal of Numerical Methods for Partial Differential Equations*, **10**, 625–635.

R. Rodriguez (1994) A-posteriori error analysis in the finite element method. In: M. Krizek *et al.* (eds), *Finite Element Methods, fifty years of the Courant Element*, pp. 389–397, Lecture Notes in Pure and Applied Mathematics, **164**, M. Dekker, New York.

A.H. Schatz (1998) Pointwise error estimates and asymptotic expansion error inequalities for the finite element method on irregular grids, Part 1: Global estimates. *Mathematics of Computation*, **67**, 877–899.

A.H. Schatz (2001) Pointwise error estimates and asymptotic expansion error inequalities for the finite element method on irregular grids, Part II: Interior estimates. *SIAM Journal on Numerical Analysis* (to appear).

E. Stein and R. Ahmad (1977) An equilibrium method for stress calculation using finite element displacement models. *Computer Methods in Applied Mechanics and Engineering*, **10**, 175–198.

T. Strouboulis and K.A. Haque (1992a) Recent experiences with error estimation and adaptivity, Part I: Review of error estimators for scalar elliptic problems. *Computer Methods in Applied Mechanics and Engineering*, **97**, 399–436.

T. Strouboulis and K.A. Haque (1992b) Recent experiences with error estimation and adaptivity, Part II: Error estimation for h-adaptive approximations on grids of triangles and quadrilaterals. *Computer Methods in Applied Mechanics and Engineering*, **100**, 359–430.

J.L. Synge (1957) *The hypercircle in mathematical physics.* Cambridge University Press.

B.A. Szabo and I. Babuška (1991) *Finite Element Analysis.* Wiley, New York.

E. Treftz (1926) Ein Gengenstück zum Ritzschen Verfahren. *Verhandlungen des 2 Internationalean Kongress für Technische Mechanik, Zurich 1926.*

R. Verfürth (1996) *A review of a-posteriori error estimation and adaptive mesh-refinement.* Wiley Teubner, New York.

R. Verfürth (1997) *A review of a-posteriori error estimation techniques for elasticity problems.* Bericht 223/1997 Facultät fur Mathematik, Ruhr University Bochum.

R. Verfürth (1998) Robust a-posteriori error estimators for a singularly perturbed reaction diffusion equation. *Numerische Mathematik*, **78**, 479–493.

D. Yu (1991a) Asymptotically exact a-posteriori error estimators for elements of bi-odd degree. *Chinese Journal of Numerical Mathematics and Applications*, **13**, 64–78.

D. Yu (1991b) Asymptotically exact a-posteriori error estimators for elements of bi-even degree. *Chinese Journal of Numerical Mathematics and Applications*, **13**, 82–90.

S. Zaremba (1909) Sur le principle de minimum. *Bull. Internat. l'Acad. d. Sciences de Cracovie, Ac. de Sciences Math. et Nat.*, **7**, 197–264.

E. Zeidler (1990) *Nonlinear Functional Analysis and its Applications II/A.* Springer, New York.

O.C. Zienkiewicz and J.Z. Zhu (1987) A simple error estimator and the adaptive procedure for practical engineering analysis. *International Journal for Numerical Methods in Engineering*, **24**, 337–357.

O.C. Zienkiewicz and J.Z. Zhu (1992a) The superconvergent patch recovery and a posteriori error estimates: Part 1: The recovery technique. *International Journal for Numerical Methods in Engineering*, **33**, 1331–1364.

O.C. Zienkiewicz and J.Z. Zhu (1992b) The superconvergent patch recovery and a posteriori error estimates: Part 2: Error estimates and adaptivity. *International Journal for Numerical Methods in Engineering*, **33**, 1365–1382.

O.C. Zienkiewicz and J.Z. Zhu (1994) *Superconvergence and superconvergent patch recovery, or the obvious which escaped attention*, CR/823/94, Department of Civil Engineering, University of Wales, Swansea, U.K., March 1994.

6

GUARANTEED A-POSTERIORI ERROR ESTIMATION AND A-POSTERIORI ESTIMATION OF THE POLLUTION ERROR

6.0 Introduction

In Chapter 5 we discussed various estimators for the energy norm of the error in the finite element solution. We discussed the Neumann element residual estimator, the subdomain residual estimator, the explicit residual estimator, the recovery estimators (e.g. the ZZ–SPR estimator, etc), and analyzed them in two ways:

(a) By establishing their equivalence with the energy norm of the error, namely, by proving the existence of positive constants $\mathscr{C}_L$, $\mathscr{C}_U$, such that

$$\mathscr{C}_L \mathscr{E} \leq \|e_{S^p_\Delta}\|_{\mathcal{U}} \leq \mathscr{C}_U \mathscr{E}. \tag{6.0.1}$$

The constants $\mathscr{C}_L$, and $\mathscr{C}_U$ may be estimated, by going through classical proofs of equivalence and estimating the various constants involved. The values of $\mathscr{C}_L$, $\mathscr{C}_U$ obtained by this approach however are often too pessimistic to be of any practical use, or are not exactly applicable to the cases of interest.

(b) By analyzing the *asymptotic range* of their robustness index

$$\mathcal{R}^{EST}_\omega = \max\left(\sup(\kappa^{EST,loc}_\omega) - 1, \frac{1}{\inf(\kappa^{EST,loc}_\omega)} - 1\right) \tag{6.0.2a}$$

where

$$\kappa^{EST,loc}_\omega \overset{\text{def}}{=} \frac{\mathscr{E}^{EST}_\omega}{\|e^{loc,\omega}_{S^p_\Delta}\|_{\mathcal{U}(\omega)}} \tag{6.0.2b}$$

is the effectivity index of the estimator with respect to the local error. This definition has good meaning because $\mathscr{E}^{EST}_\omega$ reflects only the energy norm of the local error $\|e^{loc,\omega}_{S^p_\Delta}\|_{\mathcal{U}(\omega)}$. In Section 5.7* we analyzed the range of $\mathcal{R}^{EST}_\omega$ with respect to the anisotropy of the mesh and the differential operator, for various estimators, for some of the asymptotics which are relevant to engineering computation, i.e. the asymptotics in the interior of the mesh, at smooth boundaries, in the neighborhood of corner points, etc. Using this asymptotic approach we were able to rank the estimators according to their robustness with respect to the local geometry, the distortion of the mesh, and the anisotropy of the differential operator, keeping in mind that the results of the asymptotic analysis are applicable for sufficiently refined meshes.

In practice we are interested to obtain computable upper and lower estimates for:

(a) The energy norm of the error, namely

$$\mathscr{E}^L \leq \|e_{S^p_\Delta}\|_{\mathscr{U}} = \|u_{EX} - u_{S^p_\Delta}\|_{\mathscr{U}} \leq \mathscr{E}^U. \tag{6.0.3}$$

(b) The error in output functionals $\mathscr{F}$, which express data of interest like for example the value of stress, displacement, heat flux at some points, or their averages on chosen subdomains, namely

$$\mathscr{E}^{L,\mathscr{F}} \leq e^{\mathscr{F}}_{S^p_\Delta} \stackrel{\text{def}}{=} \mathscr{F}(u_{EX}) - \mathscr{F}(u_{S^p_\Delta}) \leq \mathscr{E}^{U,\mathscr{F}}. \tag{6.0.4a}$$

Note that the error $e^{\mathscr{F}}_{S^p_\Delta}$ could be positive or negative, and the accuracy of $e^{\mathscr{F}}_{S^p_\Delta}$ is guaranteed by the bound

$$\left|e^{\mathscr{F}}_{S^p_\Delta}\right| \leq \max\left\{|\mathscr{E}^{L,\mathscr{F}}|, |\mathscr{E}^{U,\mathscr{F}}|\right\}. \tag{6.0.4b}$$

The upper bounds $\mathscr{E}^U$, and/or $\max\{|\mathscr{E}^{L,\mathscr{F}}|, |\mathscr{E}^{U,\mathscr{F}}|\}$, where $\mathscr{F}$ is the output or outputs of interest, are needed as a safe criterion for accepting the computed results. The lower bound $\mathscr{E}^L$ is needed as assurance that the upper bound $\mathscr{E}^U$ is not potentialy overly pessismistic, and hence it will not lead to unnecessary overkill if it is used to decide if the target accuracy has been achieved. When the effectivity of the upper bound relative to the lower bound $\mathscr{E}^U/\mathscr{E}^L$, is close to one then we are certain that both bounds are accurate.

Let us now list the desirable properties of the bounds.

1. *The bounds must be guaranteed for the exact error $e^{\mathscr{F}}_{u_{S^p_\Delta}}$ for any accuracy.* Let us note that some authors (see e.g. [Paraschivoiu, Peraire, and Patera, 1997]) construct upper bounds for the energy norm of the *truth-mesh error* $\|e^{S^{p+k}_{\Delta'}}_{S^p_\Delta}\|_{\mathscr{U}}$, where

$$e^{S^{p+k}_{\Delta'}}_{S^p_\Delta} \stackrel{\text{def}}{=} u_{S^{p+k}_{\Delta'}} - u_{S^p_\Delta} \tag{6.0.5a}$$

and the truth-mesh error in $\mathscr{F}$, $|e^{\mathscr{F},S^{p+k}_{\Delta'}}|$, which is given by

$$e^{\mathscr{F},S^{p+k}_{\Delta'}}_{S^p_\Delta} \stackrel{\text{def}}{=} \mathscr{F}(u_{S^{p+k}_{\Delta'}}) - \mathscr{F}(u_{S^p_\Delta}). \tag{6.0.5b}$$

However, it is possible to have

$$\left\|e^{S^{p+k}_{\Delta'}}_{S^p_\Delta}\right\|_{\mathscr{U}} \leq \mathscr{E}_{S^{p+k}_{\Delta'}} < \|e_{S^p_\Delta}\|_{\mathscr{U}} \tag{6.0.5c}$$

where $\mathscr{E}_{S^{p+k}_{\Delta'}}$, is the upper bound for the energy norm of the truth-mesh error $\|e^{S^{p+k}_{\Delta'}}_{S^p_\Delta}\|_{\mathscr{U}}$, and/or

$$\left|e^{\mathscr{F},S^{p+k}_{\Delta'}}_{S^p_\Delta}\right| \leq \max\left\{\left|\mathscr{E}^{L,\mathscr{F}}_{S^{p+k}_{\Delta'}}\right|, \left|\mathscr{E}^{U,\mathscr{F}}_{S^{p+k}_{\Delta'}}\right|\right\} < \left|e^{\mathscr{F}}_{S^p_\Delta}\right| \tag{6.0.5d}$$

where $\mathscr{E}^{L,\mathscr{F}}_{S^{p+k}_{\Delta'}}$, $\mathscr{E}^{U,\mathscr{F}}_{S^{p+k}_{\Delta'}}$, are the bounds for the error in $\mathscr{F}$ determined by using the truth-mesh error bound $\mathscr{E}_{S^{p+k}_{\Delta'}}$. A guaranteed upper estimator could be obtained from the upper estimate for the truth-mesh error if the saturation coefficient η was known (but it is not; see Remark 5.5*.5). Hence *the truth-mesh error bounds cannot be used (or should be used with caution) to obtain a safe stopping criterion, especially when the finite element approximation is crude and the error is high which is the case of most interest in practice.*

2. *It should be possible to construct the bounds also for the error in post-processed solutions* $\tilde{u}$, namely,

$$\mathscr{E}^L_{\tilde{u}} \leq \|u_{EX} - \tilde{u}\|_{\mathscr{U}} \leq \mathscr{E}^U_{\tilde{u}} \tag{6.0.6a}$$

or

$$\mathscr{E}^{L,\mathscr{F}}_{\tilde{u}} \leq \mathscr{F}(u_{EX}) - \mathscr{F}(\tilde{u}) \leq \mathscr{E}^{U,\mathscr{F}}_{\tilde{u}} \tag{6.0.6b}$$

because in most practical computations some kind of post-processing is used for extracting the data of interest from the computed finite element solution.

3. *The construction of the bounds should be such that their accuracy can be monitored and improved at a relatively small cost.* This is important because in many practical cases, we need

$$\|u_{EX} - \tilde{u}\|_{\mathscr{U}} \leq tol \tag{6.0.7a}$$

and/or

$$|e^{\mathscr{F}}_{S^p_\Delta}| \leq tol \tag{6.0.7b}$$

and for economy we also want $\|u_{EX} - \tilde{u}\|_{\mathscr{U}}$ and/or $|e^{\mathscr{F}}_{S^p_\Delta}| \approx tol$, where tol is the desired tolerance. By decreasing the size of the *reliability interval* $(\mathscr{E}^L_{\tilde{u}}, \mathscr{E}^U_{\tilde{u}})$ and/or $(\mathscr{E}^{L,\mathscr{F}}, \mathscr{E}^{U,\mathscr{F}})$, we may be able to avoid the computation of another finite element solution for a refined mesh, or we can be certain that a refined mesh solution is actually needed in order to achieve the target accuracy.

4. *The bounds should not depend on unspecified constants, or constants which are known for a rather restricted class of input data and meshes.* This is often the case for bounds constructed by following the steps in mathematical proofs of equivalence between various estimators and the error.

5. *The cost of the bounds should be reasonable, where what is reasonable depends on the goal of the computation.* As we have already discussed in Chapter 5, the cost is the sum of direct and indirect cost. For example, a guaranteed estimator which could help prevent a very costly accident (e.g. the Sleipner accident discussed in Chapter 1) is preferable to a less computationally expensive estimator which is not guaranteed and may not have been able to prevent the accident. What cost may be admissible as reasonable also depends on the implementation, and on the computer architecture.

6. *The bounds must be for the output of interest.* Often in the literature various residual measures which are equal to zero when the finite element solution coincides with the exact solution, are employed as estimators. This property only is not sufficient for a

quantity to be characterized as an estimator. In practice we are interested in quantitative control of the error in critical outputs and the estimators must be directly related to these outputs.

Let us underline that the results on guaranteed upper and lower bounds discussed in this chapter hold under the following assumptions:

- all integrals are evaluated exactly
- all systems of linear equations are solved exactly
- there are no roundoff errors.

Note that, in the mathematical literature, the term estimate most often means a guaranteed upper, or lower estimate. In Chapter 5 we used the term estimate in a broader sense, to mean a quantity which is close to the energy-norm of the error, although it may be over- or under estimating. Here we use the adjective guaranteed to underline that we know, a-priori that the estimate bounds the estimated quantity from above or below.

Following this introduction, in Section 6.1 we give the construction of guaranteed computable upper and lower bounds for the energy norm of the error in the finite element solution, or a post-processed solution, and in Section 6.2 we construct bounds for the error in the computed outputs. The construction of all the bounds is based on the computed Neumann element or patch residual estimator supplemented by the explicit residual estimator with known constants, and it is recursive, allowing for an inexpensive iterative improvement. We will also show that any estimator of the residual or recovery type (e.g. the ZZ–SPR) can be made to be guaranteed at the expense of additional computation.

6.1 Upper and lower bounds for the energy norm of the error

In this section we will construct bounds for $\|e_{S^p_\Delta}\|_{\mathcal{U}} = \|u_{EX} - u_{S^p_\Delta}\|_{\mathcal{U}}$, the energy norm of the error in the finite element solution $u_{S^p_\Delta}$, and for $\|u_{EX} - \tilde{u}_{S^{p+k}_{\Delta'}}\|_{\mathcal{U}}$, the energy norm of the error in post-processed solutions, $\tilde{u}_{S^{p+k}_{\Delta'}}$. The discussion below follows the work of [Babuška, Strouboulis *et al.*, 1999].

The construction of the upper bound, $\|e_{S^p_\Delta}\|_{\mathcal{U}} \leq \mathscr{E}^U$, will be based on the following results:

Theorem 6.1.1 *(The exact estimator, $\mathscr{E}^{\text{Neum}}$, and the computed estimator $\mathscr{E}^{\text{Neum}}_{S^{p+k}_{\Delta'}}$): Under the assumptions which guarantee the existence of $\mathscr{E}^{\text{Neum}}$, we have*

$$\left\| e^{S^{p+k}_{\Delta'}}_{S^p_\Delta} \right\|_{\mathcal{U}} \leq \|e_{S^p_\Delta}\|_{\mathcal{U}} \leq \mathscr{E}^{\text{Neum}}_{u_{S^p_\Delta}} \stackrel{\text{def}}{=} \sqrt{\sum_{j=1}^{N(\Delta)} \left\| \hat{e}^{\text{Neum}}_{I^\Delta_j, u_{S^p_\Delta}} \right\|^2_{\mathcal{U}(I^\Delta_j)}} \tag{6.1.1}$$

and

$$\left\| e^{S^{p+k}_{\Delta'}}_{S^p_\Delta} \right\|_{\mathcal{U}} \leq \mathscr{E}^{\text{Neum}}_{u_{S^p_\Delta}, S^{p+k}_{\Delta'}} \stackrel{\text{def}}{=} \sqrt{\sum_{j=1}^{N(\Delta)} \left\| \hat{e}^{\text{Neum}}_{I^\Delta_j, u_{S^p_\Delta}, S^{p+k}_{\Delta'}} \right\|^2_{\mathcal{U}(I^\Delta_j)}} \leq \mathscr{E}^{\text{Neum}}_{u_{S^p_\Delta}} \tag{6.1.2}$$

where $\hat{e}^{\text{Neum}}_{I^\Delta_j, u_{S^p_\Delta}}$ (respectively $\hat{e}^{\text{Neum}}_{I^\Delta_j, u_{S^p_\Delta}, S^{p+k}_{\Delta'}} \stackrel{\text{def}}{=} \pi_{S^{p+k}_{\Delta'}(I^\Delta_j)}(\hat{e}^{\text{Neum}}_{I^\Delta_j, u_{S^p_\Delta}})$) is the exact (respectively the computed) Neumann error indicator function, and $\mathscr{E}^{\text{Neum}}_{u_{S^p_\Delta}}$ (respectively $\mathscr{E}^{\text{Neum}}_{u_{S^p_\Delta}, S^{p+k}_{\Delta'}}$) is *the exact (respectively the computed) estimator for the energy norm of the error*

$$e_{S^p_\Delta} \stackrel{\text{def}}{=} u_{EX} - u_{S^p_\Delta}. \tag{6.1.3a}$$

Remark 6.1.1 The computed indicator function $\hat{e}^{\text{Neum}}_{I^\Delta_j, u_{S^p_\Delta}, S^{p+k}_{\Delta'}}$ is the finite element approximation of the exact indicator function $\hat{e}^{\text{Neum}}_{I^\Delta_j, u_{S^p_\Delta}}$, computed using the mesh Δ' on I^Δ_j, and finite elements of degree $p + k$.

Let us also recall that the definition of the *truth-mesh error*

$$e^{S^{p+k}_{\Delta'}}_{S^p_\Delta} \stackrel{\text{def}}{=} u_{S^{p+k}_{\Delta'}} - u_{S^p_\Delta} = \pi_{S^{p+k}_{\Delta'}}(e_{S^p_\Delta}). \tag{6.1.3b}$$

Here Δ' is a mesh which either coincides with Δ, or is obtained by its nested refinement. The restriction of Δ' in each element is the mesh utilized in the computation of the computed indicator $\hat{e}^{\text{Neum}}_{I^\Delta_j, u_{S^p_\Delta}, S^{p+k}_{\Delta'}}$. We also used the subscript $u_{S^p_\Delta}$ in $\hat{e}^{\text{Neum}}_{I^\Delta_j, u_{S^p_\Delta}}$, $\mathscr{E}^{\text{Neum}}_{u_{S^p_\Delta}}$, etc, to underline that these quantities are determined for this solution. Below, we will omit this subscript from the indicators and the estimators, unless it is needed for clarity.

Proof The proof follows directly from the definition of the energy norm by duality, the residual equation, and the definition of the error indicator functions. For the exact Neumann estimator $\mathscr{E}^{\text{Neum}}$, we have

$$\|e_{S^p_\Delta}\|_{\mathfrak{U}} = \sup_{v \in \mathfrak{U}} \frac{\mathscr{R}(v)}{\|v\|_{\mathfrak{U}}} \le \sqrt{\sum_{j=1}^{N(\Delta)} \|\mathscr{R}^{EQ}_{I^\Delta_j}\|_{L(\mathfrak{U}(I^\Delta_j),\mathbb{R})}} = \sqrt{\sum_{j=1}^{N(\Delta)} (\eta^{\text{Neum}}_{I^\Delta_j})^2} = \mathscr{E}^{\text{Neum}} \tag{6.1.4a}$$

where we employed that

$$\eta^{\text{Neum}}_{I^\Delta_j} \stackrel{\text{def}}{=} \|\hat{e}^{\text{Neum}}_{I^\Delta_j}\|_{\mathfrak{U}(I^\Delta_j)} = \|\mathscr{R}^{EQ}_{I^\Delta_j}\|_{L(\mathfrak{U}(I^\Delta_j),\mathbb{R})}. \tag{6.1.4b}$$

Similarly, for the computed Neumann estimator $\mathscr{E}^{\text{Neum}}_{S^{p+k}_{\Delta'}}$, we have

$$\begin{aligned}\left\|e^{S^{p+k}_{\Delta'}}_{S^p_\Delta}\right\|_{\mathfrak{U}} &= \sup_{v \in S^{p+k}_{\Delta'}} \frac{\mathscr{R}(v)}{\|v\|_{\mathfrak{U}}} \le \sqrt{\sum_{j=1}^{N(\Delta)} \|\mathscr{R}^{EQ}_{I^\Delta_j}\|_{L(S^{p+k}_{\Delta'}(I^\Delta_j),\mathbb{R})}} \\ &= \sqrt{\sum_{j=1}^{N(\Delta)} \left(\eta^{\text{Neum}}_{I^\Delta_j, S^{p+k}_{\Delta'}}\right)^2} = \mathscr{E}^{\text{Neum}}_{S^{p+k}_{\Delta'}}\end{aligned} \tag{6.1.5a}$$

where we employed

$$\begin{aligned}\eta^{\text{Neum}}_{I^\Delta_j, S^{p+k}_{\Delta'}} &\stackrel{\text{def}}{=} \|\hat{e}^{\text{Neum}}_{I^\Delta_j, S^{p+k}_{\Delta'}}\|_{\mathfrak{U}(I^\Delta_j)} \stackrel{\text{def}}{=} \|\pi_{S^{p+k}_{\Delta'}(I^\Delta_j)}(\hat{e}^{\text{Neum}}_{I^\Delta_j})\|_{\mathfrak{U}(I^\Delta_j)} \\ &= \|\mathscr{R}^{EQ}_{I^\Delta_j}\|_{L(S^{p+k}_{\Delta'}(I^\Delta_j),\mathbb{R})}.\end{aligned} \tag{6.1.5b}$$

The left inequality in (6.1.1) follows from (6.1.4). □

Hence, as we have already discussed in Section 5.3:

(a) The exact Neumann estimator $\mathscr{E}^{\text{Neum}}$, is an upper bound for the energy norm of the exact error $\|e_{S^p_\Delta}\|_{\mathcal{U}}$.

(b) The computed Neumann estimator $\mathscr{E}^{\text{Neum}}_{S^{p+k}_{\Delta'}}$, is an upper bound for the energy norm of the truth-mesh error $\|e^{S^{p+k}_{\Delta'}}_{S^p_\Delta}\|_{\mathcal{U}}$, but is not a guaranteed upper bound for $\|e_{S^p_\Delta}\|_{\mathcal{U}}$, i.e. there exists cases for which we have

$$\left\|e^{S^{p+k}_{\Delta'}}_{S^p_\Delta}\right\|_{\mathcal{U}} \leq \mathscr{E}^{\text{Neum}}_{S^{p+k}_{\Delta'}} < \|e_{S^p_\Delta}\|_{\mathcal{U}}. \tag{6.1.6}$$

This can be easily proven by a counterexample, e.g. Example 6.1.1.

The exact Neumann estimator $\mathscr{E}^{\text{Neum}}$, is an upper bound for $\|e_{S^p_\Delta}\|_{\mathcal{U}}$, but it is not computable. We will now construct a *computable* upper bound, $\mathscr{E}^{U}_{S^{p+k}_{\Delta'}}$, which is bigger than $\mathscr{E}^{\text{Neum}}$, namely,

$$\|e_{S^p_\Delta}\|_{\mathcal{U}} \leq \mathscr{E}^{\text{Neum}} \leq \mathscr{E}^{U}_{S^{p+k}_{\Delta'}} \tag{6.1.7}$$

by employing the following two lemmas:

Lemma 6.1.1 *(The relation between the exact estimator $\mathscr{E}^{\text{Neum}}$ and the computed estimator $\mathscr{E}^{\text{Neum}}_{S^{p+k}_{\Delta'}}$) Under the assumptions which guarantee the existence of the Neumann estimator $\mathscr{E}^{\text{Neum}}$, we have*

$$\mathscr{E}^{\text{Neum}} = \sqrt{\left(\mathscr{E}^{\text{Neum}}_{S^{p+k}_{\Delta'}}\right)^2 + \sum_{j=1}^{N(\Delta)} \left\|\hat{e}^{\text{Neum}}_{I^\Delta_j} - \hat{e}^{\text{Neum}}_{I^\Delta_j, S^{p+k}_{\Delta'}}\right\|^2_{\mathcal{U}(I^\Delta_j)}}. \tag{6.1.8}$$

Proof Employing the orthogonality

$$\mathscr{B}_{I^\Delta_j}\left(\hat{e}^{\text{Neum}}_{I^\Delta_j} - \hat{e}^{\text{Neum}}_{I^\Delta_j, S^{p+k}_{\Delta'}}, v\right) = 0 \quad \forall v \in S^{p+k}_{\Delta'}(I^\Delta_j) \tag{6.1.9a}$$

we obtain

$$\left(\eta^{\text{Neum}}_{I^\Delta_j}\right)^2 = \left(\eta^{\text{Neum}}_{I^\Delta_j, S^{p+k}_{\Delta'}}\right)^2 + \left\|\hat{e}^{\text{Neum}}_{I^\Delta_j} - \hat{e}^{\text{Neum}}_{I^\Delta_j, S^{p+k}_{\Delta'}}\right\|^2_{\mathcal{U}(I^\Delta_j)}. \tag{6.1.9b}$$

Summing (6.1.9b) over the elements, and taking the square root we get (6.1.8). □

Lemma 6.1.2 *(Explicit upper estimator for $\|\hat{e}^{\text{Neum}}_{I^\Delta_j} - \hat{e}^{\text{Neum}}_{I^\Delta_j, S^{p+k}_{\Delta'}}\|_{\mathcal{U}(I^\Delta_j)}$, the energy norm of the error in the computed error indicator $\hat{e}^{\text{Neum}}_{I^\Delta_j, S^{p+k}_{\Delta}}$) Under the assumptions which guarantee the existence of $\mathscr{E}^{\text{Neum}}$, we have*

$$\left\|\hat{e}^{Neum}_{I^\Delta_j} - \hat{e}^{\text{Neum}}_{I^\Delta_j, S^{p+k}_\Delta}\right\|_{\mathcal{U}(I^\Delta_j)} \leq \mathscr{E}^{EXPL}_{\hat{e}^{\text{Neum}}_{S^{p+k}_\Delta}/I^\Delta_j} \tag{6.1.10a}$$

with

$$\mathscr{E}^{EXPL}_{\hat{e}^{\text{Neum}}_{S^{p+k}_\Delta}/I^\Delta_j} = \mathscr{C}_1^{I^\Delta_j} \|r_{I^\Delta_j/\hat{e}^{\text{Neum}}_{S^{p+k}_\Delta}}\|_{L^2(I^\Delta_j)} + \mathscr{C}_2^{I^\Delta_j}\left(\left|J^R_{j-1/\hat{e}^{\text{Neum}}_{S^{p+k}_\Delta}}\right| + \left|J^L_{j/\hat{e}^{\text{Neum}}_{S^{p+k}_\Delta}}\right|\right) \tag{6.1.10b}$$

where $r_{I_j^\Delta/\hat{e}^{\text{Neum}}_{S_\Delta^{p+k}}}$, $J^R_{j-1/\hat{e}^{\text{Neum}}_{S_\Delta^{p+k}}}$, *and* $J^L_{j/\hat{e}^{\text{Neum}}_{S_\Delta^{p+k}}}$, *are the residuals corresponding to* $\hat{e}^{\text{Neum}}_{S_\Delta^{p+k}}$, *namely,*

$$r_{I_j^\Delta/\hat{e}^{\text{Neum}}_{S_\Delta^{p+k}}} \stackrel{\text{def}}{=} r_{I_j^\Delta} + \frac{d}{dx}\Big(AE\frac{d}{dx}(\hat{e}^{\text{Neum}}_{I_j^\Delta,S_\Delta^{p+k}})\Big) - c\,\hat{e}^{\text{Neum}}_{I_j^\Delta,S_\Delta^{p+k}} \tag{6.1.10c}$$

and

$$J^R_{j-1/\hat{e}^{\text{Neum}}_{S_\Delta^{p+k}}} \stackrel{\text{def}}{=} J^R_{j-1} + \Big(AE\frac{d}{dx}(\hat{e}^{\text{Neum}}_{I_j^\Delta,S_\Delta^{p+k}})\Big)(x_{j-1}^{\Delta,+}) \tag{6.1.10d}$$

and

$$J^L_{j/\hat{e}^{\text{Neum}}_{S_\Delta^{p+k}}} \stackrel{\text{def}}{=} J^L_j + \Big(AE\frac{d}{dx}(\hat{e}^{\text{Neum}}_{I_j^\Delta,S_\Delta^{p+k}})\Big)(x_j^{\Delta,-}) \tag{6.1.10e}$$

and

$$\mathscr{C}_1^{I_j^\Delta} = \frac{1}{\sqrt{\beta_{1,I_j^\Delta}}}\frac{h_{I_j^\Delta}}{\pi}, \qquad \mathscr{C}_2^{I_j^\Delta} = \frac{1}{\sqrt{\beta_{1,I_j^\Delta}}}\sqrt{\frac{h_{I_j^\Delta}}{3}} \tag{6.1.10f}$$

and $r_{I_j^\Delta}$, *is the interior element residual, and* J^R_{j-1}, J^L_j, *are the split jumps corresponding to* $u_{S_\Delta^p}$, *which were defined in Section 5.1, and* $\beta_{1,I_j^\Delta} \stackrel{\text{def}}{=} \inf_{x\in I_j^\Delta}(AE(x))$.

Proof Denoting by

$$e_{\hat{e}^{\text{Neum}}_{I_j^\Delta,S_\Delta^{p+k}}} \stackrel{\text{def}}{=} \hat{e}^{\text{Neum}}_{I_j^\Delta} - \hat{e}^{\text{Neum}}_{I_j^\Delta,S_\Delta^{p+k}} \tag{6.1.11a}$$

the error in $\hat{e}^{\text{Neum}}_{I_j^\Delta,S_\Delta^{p+k}}$, (6.1.10a) becomes

$$\|e_{\hat{e}^{\text{Neum}}_{I_j^\Delta,S_\Delta^{p+k}}}\|_{\mathfrak{U}(I_j^\Delta)} \le \mathscr{E}^{EXPL}_{\hat{e}^{\text{Neum}}_{S_\Delta^{p+k}}/I_j^\Delta}. \tag{6.1.11b}$$

Let us first prove (6.1.11b) for $j = 2,\ldots,N(\Delta)$. We will employ that $e_{\hat{e}^{\text{Neum}}_{I_j^\Delta,S_\Delta^{p+k}}}$, is the solution of the residual equation:

Find $e_{\hat{e}^{\text{Neum}}_{I_j^\Delta,S_\Delta^{p+k}}} \in \mathfrak{U}(I_j^\Delta)$ such that

$$\mathscr{B}_{I_j^\Delta}(e_{\hat{e}^{\text{Neum}}_{I_j^\Delta,S_\Delta^{p+k}}}, v) = \mathscr{R}^{EQ}_{I_j^\Delta/\hat{e}^{\text{Neum}}_{S_\Delta^{p+k}}}(v) \stackrel{\text{def}}{=} \mathscr{R}^{EQ}_{I_j^\Delta}(v) - \mathscr{B}_{I_j^\Delta}(\hat{e}^{\text{Neum}}_{I_j^\Delta,S_\Delta^{p+k}}, v) \tag{6.1.12a}$$

for all $v \in \mathfrak{U}(I_j^\Delta)$. Here $\mathscr{R}^{EQ}_{I_j^\Delta/\hat{e}^{\text{Neum}}_{S_\Delta^{p+k}}}$, is the residuum corresponding to $\hat{e}^{\text{Neum}}_{I_j^\Delta,S_\Delta^{p+k}}$. Integrating by parts the term $\mathscr{B}_{I_j^\Delta}(\hat{e}^{\text{Neum}}_{I_j^\Delta,S_\Delta^{p+k}}, v)$, we get

$$\mathscr{R}^{EQ}_{I_j^\Delta/\hat{e}^{\text{Neum}}_{S_\Delta^{p+k}}}(v) = \int_{I_j^\Delta} r_{I_j^\Delta/\hat{e}^{\text{Neum}}_{S_\Delta^{p+k}}} v + J^L_{j/\hat{e}^{\text{Neum}}_{S_\Delta^{p+k}}} v(x_j^\Delta) + J^R_{j-1/\hat{e}^{\text{Neum}}_{S_\Delta^{p+k}}} v(x_{j-1}^\Delta) \tag{6.1.12b}$$

where $v \in \mathcal{U}(I_j^\Delta)$. It follows that

$$\|e_{\hat{e}^{\text{Neum}}_{I_j^\Delta, S_\Delta^{p+k}}}\|_{\mathcal{U}(I_j^\Delta)} = \sup_{v\in\mathcal{U}(I_j^\Delta)} \frac{\mathcal{R}^{EQ}_{I_j^\Delta/\hat{e}^{\text{Neum}}_{S_\Delta^{p+k}}}(v)}{\|v\|_{\mathcal{U}(I_j^\Delta)}} \le \frac{1}{\sqrt{\beta_{1,I_j^\Delta}}} \sup_{v\in\mathcal{U}(I_j^\Delta)} \frac{\mathcal{R}^{EQ}_{I_j^\Delta/\hat{e}^{\text{Neum}}_{S_\Delta^{p+k}}}(v)}{\|dv/dx\|_{L^2(I_j^\Delta)}} \tag{6.1.13a}$$

where we used that

$$\sqrt{\beta_{1,I_j^\Delta}}\left\|\frac{dv}{dx}\right\|_{L^2(I_j^\Delta)} \le \left\|\sqrt{AE}\frac{dv}{dx}\right\|_{L^2(I_j^\Delta)} \le \|v\|_{\mathcal{U}(I_j^\Delta)} \quad \forall v \in \mathcal{U}(I_j^\Delta). \tag{6.1.13b}$$

Further, employing the equilibrium condition

$$\mathcal{R}^{EQ}_{I_j^\Delta/\hat{e}^{\text{Neum}}_{S_\Delta^{p+k}}}(1) = \underbrace{\mathcal{R}^{EQ}_{I_j^\Delta}(1)}_{0} - \underbrace{\mathcal{B}_{I_j^\Delta}\left(\hat{e}^{\text{Neum}}_{I_j^\Delta, S_\Delta^{p+k}}, 1\right)}_{0} = 0 \tag{6.1.14a}$$

into (6.1.13a) we obtain

$$\|e_{\hat{e}^{\text{Neum}}_{I_j^\Delta, S_\Delta^{p+k}}}\|_{\mathcal{U}(I_j^\Delta)} \le \frac{1}{\sqrt{\beta_{1,I_j^\Delta}}} \sup_{v\in\mathcal{U}(I_j^\Delta)} \frac{\mathcal{R}^{EQ}_{I_j^\Delta/\hat{e}^{\text{Neum}}_{S_\Delta^{p+k}}}\left(v - \int_{I_j^\Delta} v\right)}{\left\|(d/dx)\left(v - \int_{I_j^\Delta} v\right)\right\|_{L^2(I_j^\Delta)}} \tag{6.1.14b}$$

which can also be written as

$$\|e_{\hat{e}^{\text{Neum}}_{I_j^\Delta, S_\Delta^{p+k}}}\|_{\mathcal{U}(I_j^\Delta)} \le \frac{1}{\sqrt{\beta_{1,I_j^\Delta}}} \sup_{v\in\mathcal{U}_{\int_{I_j^\Delta}}(I_j^\Delta)} \frac{\mathcal{R}^{EQ}_{I_j^\Delta/\hat{e}^{\text{Neum}}_{S_\Delta^{p+k}}}(v)}{\|dv/dx\|_{L^2(I_j^\Delta)}} \tag{6.1.14c}$$

where we let

$$\mathcal{U}_{\int_{I_j^\Delta}}(I_j^\Delta) \stackrel{\text{def}}{=} \{v \in \mathcal{U}(I_j^\Delta) \mid \textstyle\int_{I_j^\Delta} v = 0\}. \tag{6.1.14d}$$

Employing now the Cauchy–Schwarz inequality, we obtain

$$\begin{aligned}\|e_{\hat{e}^{\text{Neum}}_{I_j^\Delta, S_\Delta^{p+k}}}\|_{\mathcal{U}(I_j^\Delta)} \le{}& \frac{1}{\sqrt{\beta_{1,I_j^\Delta}}}\|r_{I_j^\Delta/\hat{e}^{\text{Neum}}_{S_\Delta^{p+k}}}\|_{L^2(I_j^\Delta)} \sup_{v\in\mathcal{U}_{\int_{I_j^\Delta}}(I_j^\Delta)} \frac{\|v\|_{L^2(I_j^\Delta)}}{\|dv/dx\|_{L^2(I_j^\Delta)}}\\ &+ \frac{1}{\sqrt{\beta_{1,I_j^\Delta}}}\left(\left|J^L_{j/\hat{e}^{\text{Neum}}_{S_\Delta^{p+k}}}\right| + \left|J^R_{j-1/\hat{e}^{\text{Neum}}_{S_\Delta^{p+k}}}\right|\right)\\ &\times \sup_{v\in\mathcal{U}_{\int_{I_j^\Delta}}(I_j^\Delta)} \frac{\max(|v(x^\Delta_{j-1})|, |v(x^\Delta_j)|)}{\|dv/dx\|_{L^2(I_j^\Delta)}}\end{aligned} \tag{6.1.15a}$$

and hence, we have proven (6.1.10a), with

$$\mathcal{C}_1^{I_j^\Delta} = \frac{1}{\sqrt{\beta_{1,I_j^\Delta}}} \sup_{v\in\mathcal{U}_{\int_{I_j^\Delta}}(I_j^\Delta)} \frac{\|v\|_{L^2(I_j^\Delta)}}{\|dv/dx\|_{L^2(I_j^\Delta)}} \tag{6.1.15b}$$

and

$$\mathscr{C}_2^{I_j^\Delta} = \frac{1}{\sqrt{\beta_{1,I_j^\Delta}}} \sup_{v \in \mathscr{U}_{\int_{I_j^\Delta}}(I_j^\Delta)} \frac{\max(|v(x_{j-1}^\Delta)|, |v(x_j^\Delta)|)}{\|dv/dx\|_{L^2(I_j^\Delta)}}. \tag{6.1.15c}$$

Let us estimate $\mathscr{C}_1^{I_j^\Delta}$, and $\mathscr{C}_2^{I_j^\Delta}$. Any $v \in \mathscr{U}_{\int_{I_j^\Delta}}(I_j^\Delta)$ can be uniquely represented as the cosine series

$$v = \sum_{n=1}^{\infty} a_n \cos\left(n\pi \frac{x - x_{j-1}^\Delta}{h_{I_j^\Delta}}\right) \tag{6.1.16a}$$

and we have

$$\|v\|_{L^2(I_j^\Delta)}^2 = \frac{h_{I_j^\Delta}}{2} \sum_{n=1}^{\infty} a_n^2 \tag{6.1.16b}$$

and

$$\left\|\frac{dv}{dx}\right\|_{L^2(I_j^\Delta)}^2 = \frac{\pi^2}{2h_{I_j^\Delta}} \sum_{n=1}^{\infty} n^2 a_n^2 \tag{6.1.16c}$$

and hence

$$\sup_{v \in \mathscr{U}_{\int_{I_j^\Delta}}(I_j^\Delta)} \frac{\|v\|_{L^2(I_j^\Delta)}^2}{\|dv/dx\|_{L^2(I_j^\Delta)}^2} = \frac{h_{I_j^\Delta}}{\pi} \sup_{\{a_n\}} \frac{\sum_{n=1}^{\infty} a_n^2}{\sum_{n=1}^{\infty} n^2 a_n^2} = \frac{h_{I_j^\Delta}}{\pi}. \tag{6.1.16d}$$

It follows that

$$\mathscr{C}_1^{I_j^\Delta} = \frac{1}{\sqrt{\beta_{1,I_j^\Delta}}} \sup_{v \in \mathscr{U}_{\int_{I_j^\Delta}}(I_j^\Delta)} \frac{\|v\|_{L^2(I_j^\Delta)}}{\|dv/dx\|_{L^2(I_j^\Delta)}} = \frac{1}{\sqrt{\beta_{1,I_j^\Delta}}} \frac{h_{I_j^\Delta}}{\pi}. \tag{6.1.16e}$$

We also have

$$|v(x_j^\Delta)| = \left|\sum_{n=1}^{\infty} (-1)^n a_n\right| = \left|\sum_{n=1}^{\infty} (-1)^n \frac{1}{n} n a_n\right| \tag{6.1.17a}$$

and employing the Cauchy–Schwarz inequality we get

$$|v(x_j^\Delta)| \le \sqrt{\sum_{n=1}^{\infty} \frac{1}{n^2}} \sqrt{\sum_{n=1}^{\infty} n^2 a_n^2} = \frac{\pi}{\sqrt{6}} \sqrt{\sum_{n=1}^{\infty} n^2 a_n^2} \tag{6.1.17b}$$

where we used the identity $\sum_{n=1}^{\infty} 1/n^2 = \pi^2/6$. Employing now (6.1.15c), we obtain

$$|v(x_j^\Delta)| \le \sqrt{\frac{h_{I_j^\Delta}}{3}} \left\|\frac{dv}{dx}\right\|_{L^2(I_j^\Delta)} \tag{6.1.17c}$$

and noting that the same estimate also holds for $|v(x_{j-1}^{\Delta})|$, we obtain

$$\sup_{v \in \mathcal{U}_{\mathcal{J}_{I_j^\Delta}}(I_j^\Delta)} \frac{\max(|v(x_{j-1}^\Delta)|, |v(x_j^\Delta)|)}{\|dv/dx\|_{L^2(I_j^\Delta)}} \leq \sqrt{\frac{h_{I_j^\Delta}^\Delta}{3}} \tag{6.1.17d}$$

and we have proven that

$$\mathscr{C}_2^{I_j^\Delta} = \frac{1}{\sqrt{\beta_{1,I_j^\Delta}}} \sup_{v \in \mathcal{U}_{\mathcal{J}_{I_j^\Delta}}(I_j^\Delta)} \frac{\max(|v(x_{j-1}^\Delta)|, |v(x_j^\Delta)|)}{\|dv/dx\|_{L^2(I_j^\Delta)}} \leq \frac{1}{\sqrt{\beta_{1,I_j^\Delta}}} \sqrt{\frac{h_{I_j^\Delta}}{3}}. \tag{6.1.17e}$$

For $j = 1$, $e_{\hat{e}^{\text{Neum}}_{I_1^\Delta, S_\Delta^{p+k}}} \in \mathcal{U}_{(0}(I_1^\Delta)$, and satisfies (6.1.12a) for all $v \in \mathcal{U}_{(0}(I_1^\Delta)$, and $\mathscr{C}_1^{I_1^\Delta}$, $\mathscr{C}_2^{I_1^\Delta}$ are defined as in (6.1.15b), (6.1.15c) with the supremum taken over $\mathcal{U}_{(0}(I_1^\Delta)$, and can be estimated by employing that any $v \in \mathcal{U}_{(0}(I_1^\Delta)$ can be expanded in the sine series

$$v = \sum_{n=1}^{\infty} a_n \sin\left(\frac{2n-1}{2}\pi \frac{x}{h_{I_1^\Delta}}\right) \tag{6.1.18a}$$

and we have

$$\|v\|_{L^2(I_1^\Delta)}^2 = \frac{h_{I_1^\Delta}}{2} \sum_{n=1}^{\infty} a_n^2 \tag{6.1.18b}$$

and

$$\frac{\pi^2}{2h_{I_1^\Delta}} \sum_{n=1}^{\infty} n^2 a_n^2 \leq \frac{\pi^2}{2\,h_{I_1^\Delta}} \sum_{n=1}^{\infty} \left(\frac{2n-1}{2}\right)^2 a_n^2 = \left\|\frac{dv}{dx}\right\|_{L^2(I_1^\Delta)}^2 \tag{6.1.18c}$$

and

$$|v(h_{I_1^\Delta})| = \left|\sum_{n=1}^{\infty} a_n (-1)^{2k+1}\right|. \tag{6.1.18d}$$

Proceeding exactly as above, we obtain the upper bound with

$$\mathscr{C}_1^{I_1^\Delta} = \frac{1}{\sqrt{\beta_{1,I_j^\Delta}}} \frac{h_{I_j^\Delta}}{\pi}, \qquad \mathscr{C}_2^{I_1^\Delta} = \frac{1}{\sqrt{\beta_{1,I_j^\Delta}}} \sqrt{\frac{h_{I_j^\Delta}}{3}} \tag{6.1.18e}$$

and $J_0^{R/S_\Delta^{p+k}} = 0$. □

Remark 6.1.2 The constants $\mathscr{C}_1^{I_1^\Delta}$, and $\mathscr{C}_2^{I_1^\Delta}$, may be obtained by solving simple eigenvalue problems.

Remark 6.1.3 In Lemma 6.1.2 we considered the case that the computed indicators $\hat{e}^{\text{Neum}}_{I_j^\Delta, S_\Delta^{p+k}}$, are obtained by the p version, namely by employing a $p+k$ degree approximation of $\hat{e}^{\text{Neum}}_{I_j^\Delta}$.

Remark 6.1.4 Taking into account the orthogonalities

$$\mathcal{R}^{EQ}_{I_j^\Delta/\hat{e}^{\rm Neum}_{S_\Delta^{p+k}}}(v) = \underbrace{\mathcal{R}^{EQ}_{I_j^\Delta}(v)}_{0} - \underbrace{\mathcal{B}_{I_j^\Delta}(\hat{e}^{\rm Neum}_{I_j^\Delta,S_\Delta^{p+k}}, v)}_{0} = 0 \quad \forall v \in S^{p+k}(I_j^\Delta) \tag{6.1.19}$$

in (6.1.13a) we obtain

$$\|e_{\hat{e}^{\rm Neum}_{I_j^\Delta,S_\Delta^{p+k}}}\|_{\mathcal{U}(I_j^\Delta)} \leq \frac{1}{\sqrt{\beta_{1,I_j^\Delta}}} \sup_{v\in\mathcal{U}^{(p+k)}_{\int_{I_j^\Delta}}(I_j^\Delta)} \frac{\mathcal{R}^{EQ}_{I_j^\Delta/\hat{e}^{\rm Neum}_{S_\Delta^{p+k}}}(\tilde{v})}{\|d\tilde{v}/dx\|_{L^2(I_j^\Delta)}} \tag{6.1.20a}$$

where

$$\mathcal{U}^{(p+k)}_{\int_{I_j^\Delta}}(I_j^\Delta) = \{v \in \mathcal{U}_{\int_{I_j^\Delta}}(I_j^\Delta) \mid \mathcal{B}_{I_j^\Delta}(v,w) = 0 \; \forall w \in S^{p+k}(I_j^\Delta)\} \tag{6.1.20b}$$

It follows that the upper bound (6.1.10) also holds with the better constants

$$\mathcal{C}_1^{I_j^\Delta} = \frac{h_{I_j^\Delta}}{\sqrt{\beta_{1,I_j^\Delta}}}\tilde{\mathcal{C}}_1(p+k), \qquad \mathcal{C}_2^{I_j^\Delta} = \frac{\sqrt{h_{I_j^\Delta}}}{\sqrt{\beta_{1,I_j^\Delta}}}\tilde{\mathcal{C}}_2(p+k) \tag{6.1.21a}$$

where

$$\tilde{\mathcal{C}}_1(p+k) \stackrel{\rm def}{=} \sup_{v\in\tilde{\mathcal{U}}^{(p+k)}_{\int_{I_j^\Delta}}(\tilde{I})} \frac{\|\tilde{v}\|_{L^2(I_j^\Delta)}}{\|d\tilde{v}/dx\|_{L^2(\tilde{I})}} \tag{6.1.21b}$$

and

$$\tilde{\mathcal{C}}_2(p+k) \stackrel{\rm def}{=} \sup_{v\in\tilde{\mathcal{U}}^{(p+k)}_{\int_{I_j^\Delta}}(\tilde{I})} \frac{|\tilde{v}(0)|}{\|d\tilde{v}/dx\|_{L^2(\tilde{I})}} \tag{6.1.21c}$$

where

$$\tilde{\mathcal{U}}^{(p+k)}_{\tilde{I}} = \{v \mid \textstyle\int_{\tilde{I}}(dv/dx)(dw/dx) = 0 \; \forall v \in S^{p+k}(\tilde{I}), \; \int_{\tilde{I}} v = 0\} \tag{6.1.21d}$$

The constants $\tilde{\mathcal{C}}_1$, $\tilde{\mathcal{C}}_2$ can be computed numerically, for each value of $p+k$, and can be used to significantly improve the effectivity of the explicit estimate.

We are now ready to construct the computable upper bound for $\|e_{S_\Delta^p}\|_{\mathcal{U}}$.

Theorem 6.1.2 *(The computable upper Neumann estimator $\mathcal{E}^{{\rm Neum}/EXPL}_{u_{S_\Delta^p},S_\Delta^{p+k}}$) Under the assumptions which guarantee the existence of $\mathcal{E}^{\rm Neum}$, we have*

$$\|e_{S_\Delta^p}\|_{\mathcal{U}} \leq \mathcal{E}^{\rm Neum}_{u_{S_\Delta^p}} \leq \mathcal{E}^{{\rm Neum}/EXPL}_{u_{S_\Delta^p},S_\Delta^{p+k}} \stackrel{\rm def}{=} \sqrt{\left(\mathcal{E}^{\rm Neum}_{u_{S_\Delta^p},S_\Delta^{p+k}}\right)^2 + \sum_{j=1}^{N(\Delta)} \left(\mathcal{E}^{EXPL}_{u_{S_\Delta^p}/\hat{e}^{\rm Neum}_{S_\Delta^{p+k}}/I_j^\Delta}\right)^2}. \tag{6.1.22}$$

Proof It follows directly from Lemmas 6.1.1, and 6.1.2. □

TABLE 6.1.1. The coefficient AE, employed in Example 6.1.1

x	$AE(x)$
$0 \leq x \leq 0.15$	a_1
$0.15 \leq x \leq 0.25$	$\frac{a_1+a_2}{2} + \frac{-a_1+a_2}{2}\sin\left(\frac{\pi}{0.1}(x-0.2)\right)$
$0.25 \leq x \leq 0.35$	a_2
$0.35 \leq x \leq 0.45$	$\frac{a_1+a_2}{2} + \frac{a_1-a_2}{2}\sin\left(\frac{\pi}{0.1}(x-0.4)\right)$
$0.45 \leq x \leq 0.55$	a_1
$0.55 \leq x \leq 0.65$	$\frac{a_1+a_2}{2} + \frac{-a_1+a_2}{2}\sin\left(\frac{\pi}{0.1}(x-0.6)\right)$
$0.65 \leq x \leq 0.75$	a_2
$0.75 \leq x \leq 0.85$	$\frac{a_1+a_2}{2} + \frac{a_1-a_2}{2}\sin\left(\frac{\pi}{0.1}(x-0.8)\right)$
$0.85 \leq x \leq 1.0$	a_1

Everywhere below we will use the simpler notation $\mathscr{E}^{U}_{u_{S^p_\Delta}, S^{p+k}_\Delta}$, instead of the more descriptive $\mathscr{E}^{\mathrm{Neum}/EXPL}_{u_{S^p_\Delta}, S^{p+k}_\Delta}$, and we will omit the subscript $u_{S^p_\Delta}$.

Remark 6.1.5 Above we have assumed that the estimators are obtained by employing the p version of the finite element method to construct approximations of the solutions of the element residual problems. A more general construction which employs the hp version can be obtained by a recursive application of Theorem 6.1.2. Let Δ' be a mesh which is obtained by nested refinement of Δ. The computed estimator $\mathscr{E}^{\mathrm{Neum}}_{S^{p+k}_{\Delta'}}$ can be obtained by employing Theorem 6.1.2 for the estimation of the energy norm of the error in the error indicator $\|\hat{e}^{\mathrm{Neum}}_{I^\Delta_j} - \hat{e}^{\mathrm{Neum}}_{I^\Delta_j, S^{p+k}_{\Delta'}}\|_{\mathscr{U}(I^\Delta_j)}$. We will elaborate on this construction in Section 6.1*.

Let us now give an example problem which illustrates the effect of the unsmoothness of the coefficient AE on the accuracy of the upper bounds.

Example 6.1.1 (*The effect of the unsmoothness of AE on the accuracy of the upper bounds*) Let us consider the example problem

$$-\frac{d}{dx}\left(AE(x)\frac{d}{dx}(u_{EX})(x)\right) = 0, \quad 0 \leq x \leq 1 \tag{6.1.23a}$$

$$u_{EX}(0) = 0, \qquad AE(1)\frac{d}{dx}u_{EX}(1) = 1 - \alpha u_{EX}(1) \tag{6.1.23b}$$

with AE given in Table 6.1.1 and Fig. 6.1.1a, and α is a constant. The exact solution of (6.1.23) is

$$u_{EX}(x) = \frac{1}{1 + \alpha\int_0^1 (1/AE)} \int_0^x \frac{1}{AE} \tag{6.1.23c}$$

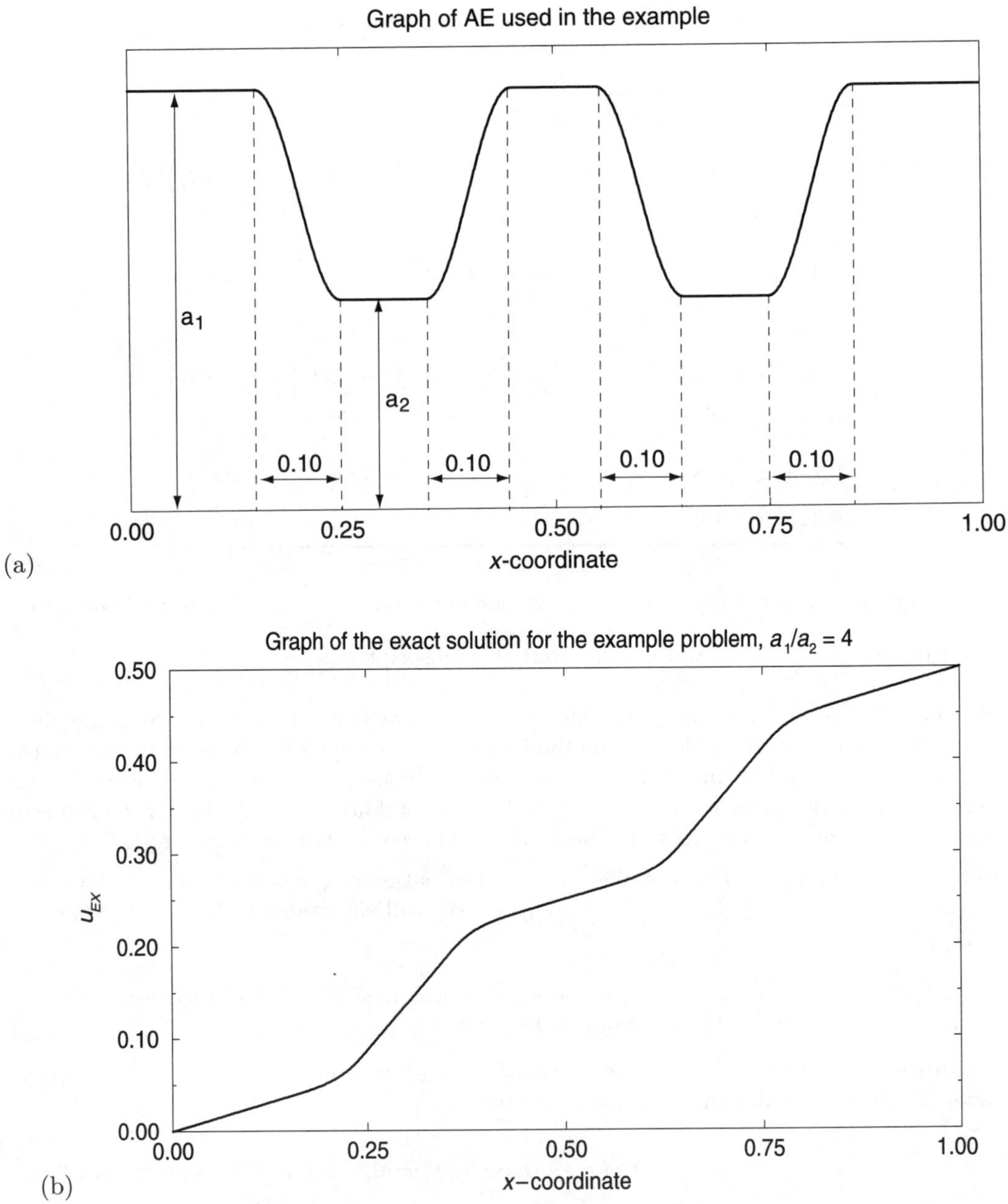

FIG. 6.1.1. *The one-dimensional example problem.* (a) The graph of the coefficient AE. (b) The graph of u_{EX} in the case that $a_1/a_2 = 4$ and $\alpha = 0$. Note that, as the ratio a_1/a_2, is increased AE and u_{EX} become rather unsmooth

and the graph for $\alpha = 0$ is shown in Fig. 6.1.1b. Similarly, the exact Neumann error indicator function is given by

$$\hat{e}^{\text{Neum}}_{I^{\Delta}_j}(x) = -u_{S^p_\Delta}(x) + \left(J^L_j + AE(x^\Delta_j)\frac{d}{dx}(u_{S^p_\Delta})(x^\Delta_j)\right)\int_{x^\Delta_{j-1}}^{x} \frac{1}{AE} + C_{I^\Delta_j} \qquad (6.1.24a)$$

where $C_{I_j^\Delta}$ is fixed from,

$$\hat{e}_{I_1^\Delta}^{\text{Neum}}(0) = 0 \tag{6.1.24b}$$

and $C_{I_j^\Delta}$, for $j = 2, ..., N(\Delta)$, by employing

$$\int_{I_j^\Delta} \hat{e}_{I_j^\Delta}^{\text{Neum}} = 0. \tag{6.1.24c}$$

Using (6.1.24a–c) the value of the exact estimator $\mathcal{E}^{\text{Neum}}$ can now be determined.

Figure 6.1.2a compares the exact relative error $E_{u_{S_\Delta^p}}^{rel} = \|e_{S_\Delta^p}\|_{\mathcal{U}}/\|u_{EX}\|_{\mathcal{U}}$, with the relative value of the upper bound $\mathcal{E}^{\text{Neum}}/\|u_{EX}\|_{\mathcal{U}}$, as a function of a_1/a_2 for $\alpha = 1, 10$, $p = 1, 2$, and $2 \leq a_1/a_2 \leq 10$. Figure 6.1.2b gives the values of the effectivity index $\kappa^{\text{Neum}} = \mathcal{E}^{\text{Neum}}/\|e_{S_\Delta^p}\|_{\mathcal{U}}$, as a function of a_1/a_2 for $\alpha = 1, 10$, and $p = 1, 2$. It can be seen that as the unsmoothness of AE is increased (as $a_1/a_2 \to \infty$) $\mathcal{E}^{\text{Neum}}$ severely overestimates $\|e_{S_\Delta^p}\|_{\mathcal{U}}$. *Let us also note that for* $\alpha = 0$, *we have* $\mathcal{E}^{\text{Neum}} = \|e_{S_\Delta^p}\|_{\mathcal{U}}$, *namely the effectivity index of the exact estimator is equal to one.*

Figure 6.1.3 (Table 6.1.2) shows the graphs (reports the values) of the effectivity index for the computed estimator $\kappa_{S_\Delta^{p+k}}^{\text{Neum}} \stackrel{\text{def}}{=} \mathcal{E}_{S_\Delta^{p+k}}^{\text{Neum}}/\|e_{S_\Delta^p}\|_{\mathcal{U}}$, and the effectivity index for the computable upper estimator $\kappa_{S_\Delta^{p+k}}^{U} \stackrel{\text{def}}{=} \mathcal{E}_{S_\Delta^{p+k}}^{U}/\|e_{S_\Delta^p}\|_{\mathcal{U}}$, for $\alpha = 0$ and $p = 1$. It can be seen that, *unless* $p + k$ *is taken sufficiently large, the computed estimator* $\mathcal{E}_{S_\Delta^{p+k}}^{\text{Neum}}$ *underestimates* $\|e_{S_\Delta^p}\|_{\mathcal{U}}$. On the other hand, the computable upper estimator $\mathcal{E}_{S_\Delta^{p+k}}^{U}$, always overestimates $\|e_{S_\Delta^p}\|_{\mathcal{U}}$, and its accuracy is in the acceptable range, and practically independent of k. The accuracy of the upper estimator $\mathcal{E}_{S_\Delta^{p+k}}^{U}$ is primarily influenced by the accuracy of the explicit estimator. Note the oscillatory behavior of the effectivity index with p in the case $a_1/a_2 = 10$.

Let us also note that AE has discontinuous second derivative in the interior of I_j^Δ, and hence we have $\hat{e}_{I_j^\Delta}^{\text{Neum}} \in B^{7/2}(I_j^\Delta)$, and we expect the computed effectivity index $\kappa_{S_\Delta^{p+k}}^{\text{Neum}}$, to converge to exact κ^{Neum} as $(p + k)^{-5/2}$.

Remark 6.1.6 In every case there exists a k, such that

$$\|e_{S_\Delta^p}\|_{\mathcal{U}} \approx \mathcal{E}_{S_\Delta^{p+k}}^{\text{Neum}} \leq \mathcal{E}^{\text{Neum}}. \tag{6.1.25}$$

This is because $\mathcal{E}^{\text{Neum}}$ is an upper bound for $\|e_{S_\Delta^p}\|_{\mathcal{U}}$, and k can be chosen such that $\mathcal{E}_{S_\Delta^{p+k}}^{\text{Neum}}$ underestimates $\mathcal{E}^{\text{Neum}}$ by the right amount so that it is close to $\|e_{S_\Delta^p}\|_{\mathcal{U}}$. In Chapter 5, we gave asymptotic analyses of the effect of the choice of k. However, these asymptotic analyses cannot give us guidelines for the selection of k for very coarse meshes which is the most interesting case in practice.

Above we have given the construction of a computable upper bound, $\|e_{S_\Delta^p}\|_{\mathcal{U}} \leq \mathcal{E}_{S_\Delta^{p+k}}^{U}$. We will now construct a computable lower bound $\mathcal{E}_{S_{\Delta'}^{p+k}}^{L} \leq \|e_{S_\Delta^p}\|_{\mathcal{U}}$, by employing the following lemma.

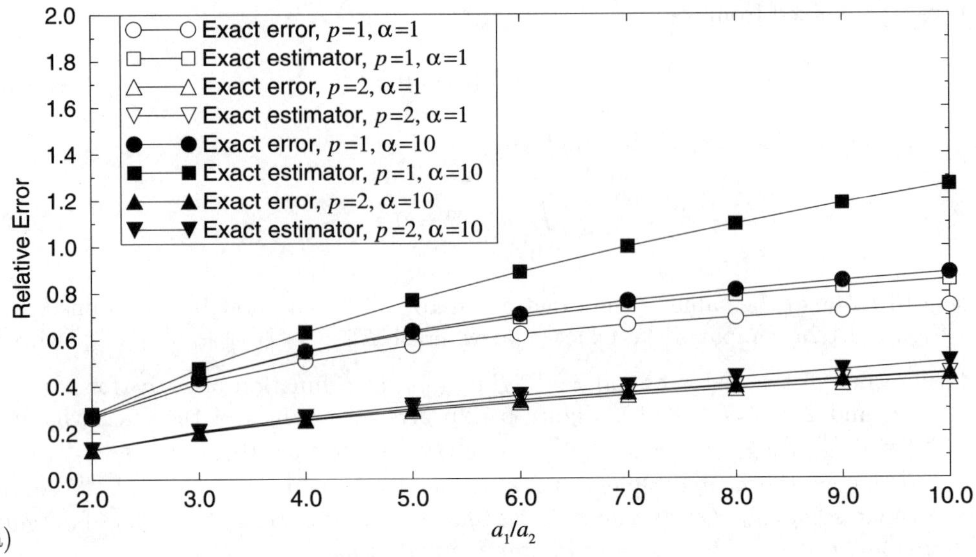

(a)

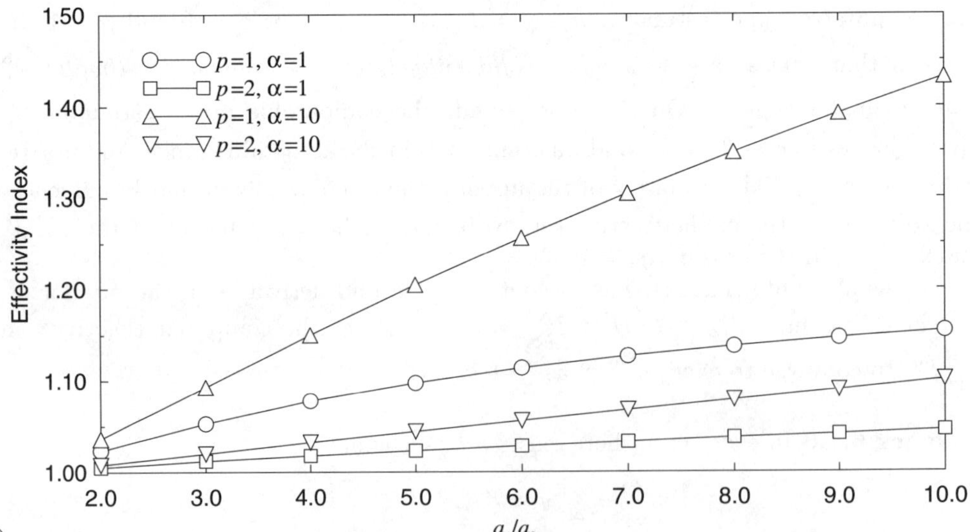

(b)

FIG. 6.1.2. *The influence of the unsmoothness of AE on the accuracy of the exact Neumann estimator* $\mathscr{E}^{\text{Neum}}$: $p = 1, 2, h = \frac{1}{4}, 2 \leq a_1/a_2 \leq 10, \alpha = 1, 10$. (a) Graph of the exact relative error $E^{rel}_{u_{S^p_\Delta}} = \|e_{S^p_\Delta}\|_{\mathfrak{u}}/\|u_{EX}\|_{\mathfrak{u}}$ and the relative value of the exact estimator $\mathscr{E}^{\text{Neum}}/\|u_{EX}\|_{\mathfrak{u}}$, for $p = 1$, and $p = 2$; (b) Graph of the effectivity index $\kappa^{\text{Neum}} = \mathscr{E}^{\text{Neum}}/\|e_{S^p_\Delta}\|_{\mathfrak{u}}$, for $p = 1$, and $p = 2$. It can be seen that the relative error and its overestimation by the estimator increases with the unsmoothness of AE (as $a_1/a_2 \to \infty$)

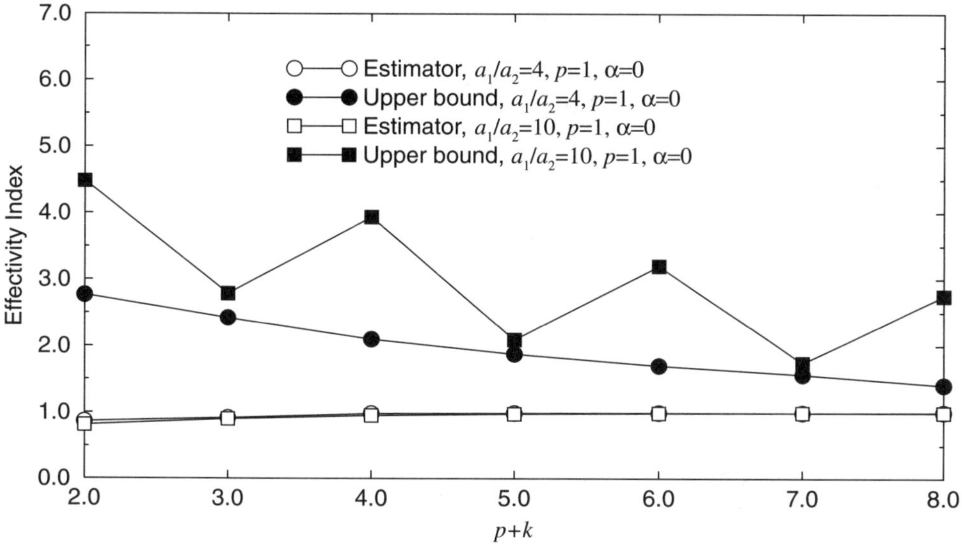

FIG. 6.1.3. *The influence of the degree* $p + k$ *on the accuracy of the computed Neumann estimator,* $\mathscr{E}^{\text{Neum}}_{S^{p+k}_{\Delta}}$*, and the computable upper bound* $\mathscr{E}^{U}_{S^{p+k}_{\Delta}}$: The effectivity indices $\kappa^{\text{Neum}}_{S^{p+k}_{\Delta}} = \mathscr{E}^{\text{Neum}}_{S^{p+k}_{\Delta}}/\|e_{S^p_{\Delta}}\|_{\mathscr{U}}$, and $\kappa^{U}_{S^{p+k}_{\Delta}} = \mathscr{E}^{U}_{S^{p+k}_{\Delta}}/\|e_{S^p_{\Delta}}\|_{\mathscr{U}}$ for $p = 1$, $p + k = 2, 3, \ldots, 8$, $a_1/a_2 = 4, 10$, and $\alpha = 0$. The numbers are listed in Table 6.1.3

TABLE 6.1.2. *The influence of the degree* $p + k$ *on the accuracy of the computed Neumann estimator.* The values of the effectivity indices $\kappa^{\text{Neum}}_{S^{p+k}_{\Delta}} = \mathscr{E}^{\text{Neum}}_{S^{p+k}_{\Delta}}/\|e_{S^p_{\Delta}}\|_{\mathscr{U}}$, and $\kappa^{U}_{S^{p+k}_{\Delta}} = \mathscr{E}^{U}_{S^{p+k}_{\Delta}}/\|e_{S^p_{\Delta}}\|_{\mathscr{U}}$, respectively for the computed and the computable upper estimator are given for the case $\alpha = 0$

$p+k$	$a_1/a_2 = 4$		$a_1/a_2 = 10$	
	$\mathscr{E}^{\text{Neum}}_{S^{p+k}_{\Delta}}/\|e_{S^p_{\Delta}}\|_{\mathscr{U}}$	$\mathscr{E}^{U}_{S^{p+k}_{\Delta}}/\|e_{S^p_{\Delta}}\|_{\mathscr{U}}$	$\mathscr{E}^{\text{Neum}}_{S^{p+k}_{\Delta}}/\|e_{S^p_{\Delta}}\|_{\mathscr{U}}$	$\mathscr{E}^{U}_{S^{p+k}_{\Delta}}/\|e_{S^p_{\Delta}}\|_{\mathscr{U}}$
2	0.86397	2.76791	0.81170	4.47722
3	0.92001	2.41519	0.89964	2.78811
4	0.98246	2.09476	0.95608	3.93608
5	0.98624	1.87283	0.97895	2.08852
6	0.99541	1.69928	0.98759	3.19866
7	0.99583	1.56729	0.99274	1.74128
8	0.99853	1.40848	0.99415	2.75101

Lemma 6.1.3 *(The basic identity) Under the assumptions which guarantee the existence of* $\mathscr{E}^{\text{Neum}}$*, we have*

$$\|e_{S^p_{\Delta}}\|^2_{\mathscr{U}} = \left(\mathscr{E}^{\text{Neum}}_{S^{p+k}_{\Delta'}}\right)^2 - \sum_{j=1}^{N(\Delta)} \|\tilde{e}_{S^{p+k}_{\Delta'}} - \hat{e}^{\text{Neum}}_{I^{\Delta}_j, S^{p+k}_{\Delta'}}\|^2_{\mathscr{U}(I^{\Delta}_j)} + \|e_{S^p_{\Delta}} - \tilde{e}_{S^{p+k}_{\Delta'}}\|^2_{\mathscr{U}} \tag{6.1.26a}$$

and

$$\left\| e_{S_\Delta^p}^{S_{\Delta'}^{p+k}} \right\|_{\mathcal{U}}^2 = \left(\mathcal{E}_{S_{\Delta'}^{p+k}}^{\text{Neum}}\right)^2 - \sum_{j=1}^{N(\Delta)} \left\| \tilde{e}_{S_{\Delta'}^{p+k}} - \hat{e}_{I_j^\Delta, S_{\Delta'}^{p+k}}^{\text{Neum}} \right\|_{\mathcal{U}(I_j^\Delta)}^2 + \left\| e_{S_\Delta^p}^{S_{\Delta'}^{p+k}} - \tilde{e}_{S_{\Delta'}^{p+k}} \right\|_{\mathcal{U}}^2 \quad (6.1.26b)$$

for any $\tilde{e}_{S_{\Delta'}^{p+k}} \in S_{\Delta'}^{p+k}(I) \subset \mathcal{U}$.

Proof Equation (6.1.26a) follows directly from the orthogonalities

$$\sum_{j=1}^{N(\Delta)} \mathcal{B}_{I_j^\Delta}\left(\hat{e}_{I_j^\Delta}^{\text{Neum}} - e_{S_\Delta^p}, v\right) = \sum_{j=1}^{N(\Delta)} \mathcal{R}_{I_j^\Delta}^{EQ}(v) - \mathcal{R}(v) = 0 \quad \forall v \in \mathcal{U} \quad (6.1.27a)$$

and

$$\sum_{j=1}^{N(\Delta)} \mathcal{B}_{I_j^\Delta}\left(\hat{e}_{I_j^\Delta}^{\text{Neum}} - \hat{e}_{I_j^\Delta, S_{\Delta'}^{p+k}}^{\text{Neum}}, v\right) = 0 \quad \forall v \in S_{\Delta'}^{p+k}. \quad (6.1.27b)$$

Equation (6.1.26b) follows from (6.1.26a) by employing that

$$\| e_{S_\Delta^p} - \tilde{e}_{S_{\Delta'}^{p+k}} \|_{\mathcal{U}}^2 = \left\| e_{S_\Delta^p} - e_{S_\Delta^p}^{S_{\Delta'}^{p+k}} \right\|_{\mathcal{U}}^2 + \left\| e_{S_\Delta^p}^{S_{\Delta'}^{p+k}} - \tilde{e}_{S_{\Delta'}^{p+k}} \right\|_{\mathcal{U}}^2 \quad (6.1.28a)$$

where we have used the orthogonality

$$\mathcal{B}\left(e_{S_\Delta^p} - e_{S_\Delta^p}^{S_{\Delta'}^{p+k}}, e_{S_\Delta^p}^{S_{\Delta'}^{p+k}} - \tilde{e}_{S_{\Delta'}^{p+k}}\right) = 0 \quad (6.1.28b)$$

and that

$$\left\| e_{S_\Delta^p} - e_{S_\Delta^p}^{S_{\Delta'}^{p+k}} \right\|_{\mathcal{U}}^2 = \| e_{S_\Delta^p} \|_{\mathcal{U}}^2 - \left\| e_{S_\Delta^p}^{S_{\Delta'}^{p+k}} \right\|_{\mathcal{U}}^2. \quad (6.1.28c)$$

□

Let us now state the lower bound.

Theorem 6.1.3 *(The lower estimator* $\mathcal{E}_{S_{\Delta'}^{p+k}}^{\text{Neum}/L}(\tilde{e}_{S_{\Delta'}^{p+k}})$*) Under the assumptions which guarantee the existence of* $\mathcal{E}^{\text{Neum}}$*, we have*

$$\mathcal{E}_{S_{\Delta'}^{p+k}}^{\text{Neum}/L}(\tilde{e}_{S_{\Delta'}^{p+k}}) \stackrel{\text{def}}{=} \sqrt{\left(\mathcal{E}_{S_{\Delta'}^{p+k}}^{\text{Neum}}\right)^2 - \sum_{j=1}^{N(\Delta)} \left\| \tilde{e}_{S_{\Delta'}^{p+k}} - \hat{e}_{I_j^\Delta, S_{\Delta'}^{p+k}}^{\text{Neum}} \right\|_{\mathcal{U}(I_j^\Delta)}^2}$$

$$\leq \left\| e_{S_\Delta^p}^{S_{\Delta'}^{p+k}} \right\|_{\mathcal{U}} \leq \| e_{S_\Delta^p} \|_{\mathcal{U}} \quad (6.1.29)$$

for any $\tilde{e}_{S_{\Delta'}^{p+k}} \in S_{\Delta'}^{p+k}$*, where in the cases that the quantity under the square root is less than zero, we replace it by zero.*

Proof The relation (6.1.29) follows from (6.1.26b) by omitting the last term in its right-hand side. □

Remark 6.1.7 Note that (6.1.29) holds for all positive integers k, and hence also for the exact estimator.

Below we will denote the lower bound $\mathcal{E}^{\mathrm{Neum}/L}_{S^{p+k}_{\Delta'}}$, by the simpler notation $\mathcal{E}^{L}_{S^{p+k}_{\Delta'}}$.

Remark 6.1.8 From (6.1.26b) it follows that the lower bound $\mathcal{E}^{L}_{S^{p+k}_{\Delta'}}$, achieves its maximum for $\tilde{e}_{S^{p+k}_{\Delta'}} \equiv e^{S^{p+k}_{\Delta'}}_{S^{p}_{\Delta}}$, and we have

$$\max_{\tilde{e}_{S^{p+k}_{\Delta'}} \in S^{p+k}_{\Delta'}} \mathcal{E}^{L}_{S^{p+k}_{\Delta'}}(\tilde{e}_{S^{p+k}_{\Delta'}}) = \mathcal{E}^{L}_{S^{p+k}_{\Delta'}}\left(e^{S^{p+k}_{\Delta'}}_{S^{p}_{\Delta}}\right) = \left\| e^{S^{p+k}_{\Delta'}}_{S^{p}_{\Delta}} \right\|_{\mathcal{U}} \leq \|e_{S^{p}_{\Delta}}\|_{\mathcal{U}}. \tag{6.1.30}$$

Let us now address the construction of $\tilde{e}_{S^{p+k}_{\Delta'}}$. Taking advantage that $\hat{e}^{\mathrm{Neum}}_{S^{p+k}_{\Delta'}}$, is already known, we will employ for $\tilde{e}_{S^{p+k}_{\Delta'}} \in \mathcal{U}$, a *smoothened* Neumann error indicator function

$$\tilde{e}^{\mathrm{Neum}}_{S^{p+k}_{\Delta'}}(\Phi_{\hat{e}^{\mathrm{Neum}}_{S^{p+k}_{\Delta'}}}) \stackrel{\mathrm{def}}{=} \hat{e}^{\mathrm{Neum}}_{S^{p+k}_{\Delta'}} + \Phi_{\hat{e}^{\mathrm{Neum}}_{S^{p+k}_{\Delta'}}} \tag{6.1.31a}$$

where $\Phi_{\hat{e}^{\mathrm{Neum}}_{S^{p+k}_{\Delta'}}}$ is a *gap function*, corresponding to $\hat{e}^{\mathrm{Neum}}_{S^{p+k}_{\Delta'}}$, namely a function from the set of *admissible gap functions*

$$S^{p+k,brok}_{\Delta',\{G^{\mathrm{Neum}}_{j,S^{p+k}_{\Delta'}}\}_{j=1}^{N(\Delta)-1}} \stackrel{\mathrm{def}}{=} \{v \mid v|_{I^{\Delta}_{k}} \in S^{p+k}_{\Delta'}(I^{\Delta}_{k}), [\![v]\!](x^{\Delta}_{j}) = -G^{\mathrm{Neum}}_{j,S^{p+k}_{\Delta'}}, j = 1, \ldots, N(\Delta)-1\} \tag{6.1.31b}$$

where

$$G^{\mathrm{Neum}}_{j,S^{p+k}_{\Delta'}} \stackrel{\mathrm{def}}{=} [\![\hat{e}^{\mathrm{Neum}}_{S^{p+k}_{\Delta'}}]\!](x^{\Delta}_{j}) = \hat{e}^{\mathrm{Neum}}_{S^{p+k}_{\Delta'}}(x^{\Delta,+}_{j}) - \hat{e}^{\mathrm{Neum}}_{S^{p+k}_{\Delta'}}(x^{\Delta,-}_{j}) \tag{6.1.31c}$$

is *the gap of the computed Neumann error indicator function* $\hat{e}^{\mathrm{Neum}}_{S^{p+k}_{\Delta'}}$ *at the* j*th node.*

Using these notations, Theorem 6.1.3 can now be rewritten in the following equivalent form.

Theorem 6.1.4 *(The lower estimator* $\tilde{\mathcal{E}}^{L}_{S^{p+k}_{\Delta'}}(\Phi_{\hat{e}^{\mathrm{Neum}}_{S^{p+k}_{\Delta'}}})$ *in terms of the gap function* $\Phi_{\hat{e}^{\mathrm{Neum}}_{S^{p+k}_{\Delta'}}}$*) Under the assumptions which guarantee the existence of* $\mathcal{E}^{\mathrm{Neum}}_{S^{p+k}_{\Delta'}}$, *we have*

$$\tilde{\mathcal{E}}^{L}_{S^{p+k}_{\Delta'}}(\Phi_{\hat{e}^{\mathrm{Neum}}_{S^{p+k}_{\Delta'}}}) \stackrel{\mathrm{def}}{=} \sqrt{\left(\mathcal{E}^{\mathrm{Neum}}_{S^{p+k}_{\Delta'}}\right)^2 - \sum_{j=1}^{N(\Delta)} \|\Phi_{\hat{e}^{\mathrm{Neum}}_{S^{p+k}_{\Delta'}}}\|^2_{\mathcal{U}(I^{\Delta}_{j})}} \leq \left\| e^{S^{p+k}_{\Delta'}}_{S^{p}_{\Delta}} \right\|_{\mathcal{U}} \leq \|e_{S^{p}_{\Delta}}\|_{\mathcal{U}} \tag{6.1.32}$$

for any $\Phi_{\hat{e}^{\mathrm{Neum}}_{S^{p+k}_{\Delta'}}} \in S^{p+k,brok}_{\Delta',\{G^{\mathrm{Neum}}_{j,S^{p+k}_{\Delta'}}\}_{j=1}^{N(\Delta)-1}}$.

From (6.1.32) it follows that the best choice of $\Phi_{\hat{e}^{\mathrm{Neum}}_{S^{p+k}_{\Delta'}}}$, *is the one which minimizes* $\sum_{j=1}^{N(\Delta)} \|\Phi_{\hat{e}^{\mathrm{Neum}}_{S^{p+k}_{\Delta'}}}\|^2_{\mathcal{U}(I_j^\Delta)}$. *We thus have:*

Corollary 6.1.1 *(The optimal gap function) Let the assumptions which guarantee the existence of* $\mathcal{E}^{\mathrm{Neum}}$ *hold, and given* $\bar{\alpha} \in \mathbb{R}^{N-1}$, *let*

$$\Phi^{\bar{\alpha}}_{\hat{e}^{\mathrm{Neum}}_{S^{p+k}_{\Delta'}}} \in S^{p+k,brok}_{\Delta',\{G^{\mathrm{Neum}}_{j,S^{p+k}_{\Delta'}}\}_{j=1}^{N(\Delta)-1}} : \quad \Phi^{\bar{\alpha}}_{\hat{e}^{\mathrm{Neum}}_{S^{p+k}_{\Delta'}}}\Big|_{I_j^\Delta} \stackrel{\mathrm{def}}{=} \Phi^{\bar{\alpha},I_j^\Delta}_{\hat{e}^{\mathrm{Neum}}_{S^{p+k}_{\Delta'}}} \tag{6.1.33a}$$

where $\Phi^{\bar{\alpha},I_j^\Delta}_{\hat{e}^{\mathrm{Neum}}_{S^{p+k}_{\Delta'}}}$ *is the solution of the following Dirichlet problem:*

Find $\Phi^{\bar{\alpha},I_j^\Delta}_{\hat{e}^{\mathrm{Neum}}_{S^{p+k}_{\Delta'}}} \in S^{p+k}_{\Delta'}(I_j^\Delta)$ *such that*

$$\Phi^{\bar{\alpha},I_j^\Delta}_{\hat{e}^{\mathrm{Neum}}_{S^{p+k}_{\Delta'}}}(x^\Delta_{j-1}) = \bar{\alpha}_{j-1} G^{\mathrm{Neum}}_{j-1,S^{p+k}_{\Delta'}}, \qquad \Phi^{\bar{\alpha},I_j^\Delta}_{\hat{e}^{\mathrm{Neum}}_{S^{p+k}_{\Delta'}}}(x^\Delta_{j}) = (1-\bar{\alpha}_{j-1}) G^{\mathrm{Neum}}_{j-1,S^{p+k}_{\Delta'}} \tag{6.1.33b}$$

such that

$$\mathcal{B}_{I_j^\Delta}\left(\Phi^{\bar{\alpha},I_j^\Delta}_{\hat{e}^{\mathrm{Neum}}_{S^{p+k}_{\Delta'}}}, v\right) = 0 \quad \forall v \in S^{p+k}_{\Delta',0}(I_j^\Delta) \stackrel{\mathrm{def}}{=} S^{p+k}_{\Delta'}(I_j^\Delta) \cap H^1_0(I_j^\Delta). \tag{6.1.33c}$$

We then have

$$\tilde{\mathcal{E}}^L_{S^{p+k}_{\Delta'}}\left(\Phi^{\bar{\alpha}^{opt}}_{\hat{e}^{\mathrm{Neum}}_{S^{p+k}_{\Delta'}}}\right) = \sqrt{\left(\mathcal{E}^{\mathrm{Neum}}_{S^{p+k}_{\Delta'}}\right)^2 - \min_{\bar{\alpha}\in\mathbb{R}^{N-1}} \sum_{j=1}^{N(\Delta)} \|\Phi^{\bar{\alpha}}_{\hat{e}^{\mathrm{Neum}}_{S^{p+k}_{\Delta'}}}\|^2_{\mathcal{U}(I_j^\Delta)}} = \left\|e^{S^{p+k}_{\Delta'}}_{S^p_\Delta}\right\|_{\mathcal{U}}. \tag{6.1.33d}$$

Proof It is sufficient to choose $\bar{\alpha}^{opt} \in \mathbb{R}^{N-1}$, such that we have

$$\Phi^{\bar{\alpha}^{opt}}_{\hat{e}^{\mathrm{Neum}}_{S^{p+k}_{\Delta'}}} = e^{S^{p+k}_{\Delta'}}_{S^p_\Delta} - \hat{e}^{\mathrm{Neum}}_{S^{p+k}_{\Delta'}} \tag{6.1.34}$$

and (6.1.33d) follows. □

Remark 6.1.9 In practice the choice of $\bar{\alpha}_j = \frac{1}{2}$, $j = 1, \ldots, N(\Delta)-1$, is used in absence of a better one. However, as we will see in Example 6.1.2, this choice is far from optimal and better values for the $\bar{\alpha}_j$ may be obtained by employing a few iterations of a minimization algorithm.

Remark 6.1.10 Lemma 6.1.3 can be also formulated for the exact estimator $\mathcal{E}^{\mathrm{Neum}}$. We have

$$\|e_{S^p_\Delta}\|^2_{\mathcal{U}} = (\mathcal{E}^{\mathrm{Neum}})^2 - \sum_{j=1}^{N(\Delta)} \|\tilde{e} - \hat{e}^{\mathrm{Neum}}_{I_j^\Delta}\|^2_{\mathcal{U}(I_j^\Delta)} + \|e_{S^p_\Delta} - \tilde{e}\|^2_{\mathcal{U}} \quad \forall \tilde{e} \in \mathcal{U}. \tag{6.1.35}$$

Omitting the last term in the right-hand side we obtain the lower bound

$$\mathcal{E}^L(\tilde{e}) \stackrel{\mathrm{def}}{=} \sqrt{(\mathcal{E}^{\mathrm{Neum}})^2 - \sum_{j=1}^{N(\Delta)} \|\tilde{e} - \hat{e}^{\mathrm{Neum}}_{I_j^\Delta, u_{S^p_\Delta}}\|^2_{\mathcal{U}(I_j^\Delta)}} \le \|e_{S^p_\Delta}\|_{\mathcal{U}} \tag{6.1.36}$$

for any $\tilde{e} \in \mathcal{U}$, or equivalently

$$\tilde{\mathcal{E}}^L(\Phi_{\hat{e}^{\text{Neum}}}) \stackrel{\text{def}}{=} \sqrt{(\mathcal{E}^{\text{Neum}})^2 - \sum_{j=1}^{N(\Delta)} \|\Phi_{\hat{e}^{\text{Neum}}}\|^2_{\mathcal{U}(I_j^\Delta)}} \leq \|e_{S_\Delta^p}\|_{\mathcal{U}} \tag{6.1.37a}$$

for any $\Phi_{\hat{e}^{\text{Neum}}} \in \mathcal{U}^{brok}_{\Delta,\{G_j^{\text{Neum}}\}_{j=1}^{N(\Delta)-1}}$, where

$$\mathcal{U}^{brok}_{\Delta,\{G_j^{\text{Neum}}\}_{j=1}^{N(\Delta)-1}} \stackrel{\text{def}}{=} \left\{v \mid v|_{I_k^\Delta} \in \mathcal{U}(I_k^\Delta), [\![v]\!](x_j^\Delta) = -G_j^{\text{Neum}}, j = 1, \dots, N(\Delta) - 1\right\} \tag{6.1.37b}$$

is the set of admissible gap functions for the exact Neumann error indicator $\hat{e}^{\text{Neum}}$, $\mathcal{U}(I_k^\Delta)$ is the restriction of $\mathcal{U}$ in I_k^Δ, and

$$G_j^{\text{Neum}} \stackrel{\text{def}}{=} [\![\hat{e}^{\text{Neum}}]\!](x_j^\Delta) = \hat{e}^{\text{Neum}}(x_j^{\Delta,+}) - \hat{e}^{\text{Neum}}(x_j^{\Delta,-}) \tag{6.1.37c}$$

is the gap of the exact Neumann error indicator function $\hat{e}^{\text{Neum}}$ at the jth node. We will call $\mathcal{E}^L(\tilde{e})$, or $\tilde{\mathcal{E}}^L(\Phi_{\hat{e}^{\text{Neum}}})$, the lower bound in terms of the exact estimator $\mathcal{E}^{\text{Neum}}$, or, in short, the *'exact' lower bound*.

Note that if the gaps are zero, we have $\Phi_{\hat{e}^{\text{Neum}}} = 0$, and

$$\mathcal{E}^{\text{Neum}} = \tilde{\mathcal{E}}^L(\Phi_{\hat{e}^{\text{Neum}}}) = \|e_{S_\Delta^p}\|_{\mathcal{U}} \tag{6.1.37d}$$

Remark 6.1.11 The accuracy of the 'exact' lower bound $\tilde{\mathcal{E}}^L(\Phi_{\hat{e}^{\text{Neum}}})$ depends on construction of the gap function $\Phi_{\hat{e}^{\text{Neum}}}$. For example, for the model problem, and the finite element approximation employed in Example 6.1.1, and we may construct the 'exact' lower bound $\tilde{\mathcal{E}}^L(\Phi_{\hat{e}^{\text{Neum}}})$, given in (6.1.37), for the gap functions $\Phi^{lin}_{\hat{e}^{\text{Neum}}}$ and $\Phi^{\alpha}_{\hat{e}^{\text{Neum}}}$, which are defined elementwise respectively by

$$\Phi^{lin}_{\hat{e}^{\text{Neum}}}\Big|_{I_j^\Delta} = -\frac{1}{2}G_{j-1}^{\text{Neum}}\frac{x_j^\Delta - x}{h} + \frac{1}{2}G_j^{\text{Neum}}\frac{x - x_{j-1}^\Delta}{h} \tag{6.1.38a}$$

where we let $G_0^{\text{Neum}} = G_N^{\text{Neum}} = 0$, and

$$\Phi^{\bar{\alpha},harm}_{\hat{e}^{\text{Neum}}}\Big|_{I_j^\Delta} = -\bar{\alpha}_{j-1}G_{j-1}^{\text{Neum}} + \left((1-\bar{\alpha}_j)G_j^{\text{Neum}} + \bar{\alpha}_{j-1}G_{j-1}^{\text{Neum}}\right) \times \left(\int_{I_j^\Delta}\frac{1}{AE}\right)\int_{x_{j-1}^\Delta}^{x}\frac{1}{AE}. \tag{6.1.38b}$$

Note that, the best choice is

$$\bar{\alpha}_k^{EX} \stackrel{\text{def}}{=} \frac{1}{G_k^{\text{Neum}}}\left(e_{S_\Delta^p}(x_k^\Delta) - \hat{e}^{\text{Neum}}(x_k^\Delta)\right) \tag{6.1.38c}$$

for $k = 1, ..., N(\Delta)$, for which we have

$$\tilde{\mathcal{E}}^L\left(\Phi^{\bar{\alpha}^{EX},harm}_{\hat{e}^{\text{Neum}}}\right) \stackrel{\text{def}}{=} \sqrt{(\mathcal{E}^{\text{Neum}})^2 - \sum_{j=1}^{N(\Delta)} \left\|\Phi^{\bar{\alpha}^{EX},harm}_{\hat{e}^{\text{Neum}}}\right\|^2_{\mathcal{U}(I_j^\Delta)}} = \|e_{S_\Delta^p}\|_{\mathcal{U}}. \tag{6.1.38d}$$

and we expect that

$$\tilde{\mathscr{E}}^L(\Phi^{lin}_{\hat{e}^{\rm Neum}}) \le \tilde{\mathscr{E}}^L(\Phi^{\{1/2\},harm}_{\hat{e}^{\rm Neum}}) \le \tilde{\mathscr{E}}(\Phi^{\bar{\alpha}^{EX},harm}_{\hat{e}^{\rm Neum}}) = \|e_{S^p_\Delta}\|_{\mathscr{U}}. \tag{6.1.38e}$$

Remark 6.1.12 Note that we determined the exact indicator $\hat{e}^{\rm Neum}_{I^\Delta_j}$, by imposing the condition $\int_{I^\Delta_j} \hat{e}^{\rm Neum}_{I^\Delta_j} = 0$. This choice is arbitrary and influences the gap function. More generally we could have imposed the condition $\int_{I^\Delta_j} \hat{e}^{\rm Neum}_{I^\Delta_j} = \mathscr{C}_{I^\Delta_j}$, where the constants $\mathscr{C}_{I^\Delta_j}$ may be chosen such that the corresponding lower bound $\tilde{\mathscr{E}}^L(\Phi_{\hat{e}^{\rm Neum}})$ is maximized.

We can also use Lemma 6.1.3 to obtain improved lower and upper bounds. For this we will need the following result:

Lemma 6.1.4 *(Lower and upper bounds for $\|u_{EX} - \tilde{u}^*_{S^{p+k}_{\Delta'}}\|_{\mathscr{U}}$, the energy norm of the error in any admissible finite element solution $\tilde{u}^*_{S^{p+k}_{\Delta'}}$). For any $\tilde{u}^*_{S^{p+k}_{\Delta'}} \in S^{p+k}_{\Delta'}$, which satisfies the orthogonality condition*

$$\mathscr{B}(u_{EX} - \tilde{u}^*_{S^{p+k}_{\Delta'}}, v) = 0 \quad \forall v \in S^p_\Delta \tag{6.1.39a}$$

we have

$$\mathscr{E}^L_{\tilde{u}^*_{S^{p+k}_{\Delta'}}, S^{p+k}_{\Delta'}}(\tilde{\tilde{e}}_{S^{p+k}_{\Delta'}}) \le \|u_{EX} - \tilde{u}^*_{S^{p+k}_{\Delta'}}\|_{\mathscr{U}} \le \mathscr{E}^U_{\tilde{u}^*_{S^{p+k}_{\Delta'}}, S^{p+k}_{\Delta'}} \tag{6.1.39b}$$

for any $\tilde{\tilde{e}}_{S^{p+k}_{\Delta'}} \in S^{p+k}_{\Delta'}$.

Proof The construction of the bounds for $\|u_{EX} - \tilde{u}^*_{S^{p+k}_{\Delta'}}\|_{\mathscr{U}}$ follows exactly the same steps as the construction of the bounds for $\|e_{S^p_\Delta}\|_{\mathscr{U}}$, which was given in Theorems 6.1.1–6.1.4. Let $\mathscr{R}_{\tilde{u}^*_{S^{p+k}_{\Delta'}}}$ denote the residuum corresponding to $\tilde{u}^*_{S^{p+k}_{\Delta'}}$, which is defined by

$$\mathscr{R}_{\tilde{u}^*_{S^{p+k}_{\Delta'}}}(v) \stackrel{\rm def}{=} \mathscr{L}(v) - \mathscr{B}(\tilde{u}^*_{S^{p+k}_{\Delta'}}, v) \quad \forall v \in \mathscr{U}. \tag{6.1.40a}$$

From the orthogonality (6.1.39a) we get that

$$\mathscr{R}_{\tilde{u}^*_{S^{p+k}_{\Delta'}}}(v) = \mathscr{B}(u_{EX} - \tilde{u}^*_{S^{p+k}_{\Delta'}}, v) = 0 \quad \forall v \in S^p_\Delta. \tag{6.1.40b}$$

It follows the splitting of the jumps can be constructed exactly as for the residuum $\mathscr{R} = \mathscr{R}_{u_{S^p_\Delta}}$, and the bounds $\mathscr{E}^L_{\tilde{u}^*_{S^{p+k}_{\Delta'}}, S^{p+k}_{\Delta'}}$ and $\mathscr{E}^U_{\tilde{u}^*_{S^{p+k}_{\Delta'}}, S^{p+k}_{\Delta'}}$ can be constructed exactly as the bounds $\mathscr{E}^L_{u_{S^p_\Delta}, S^{p+k}_{\Delta'}}$ and $\mathscr{E}^U_{u_{S^p_\Delta}, S^{p+k}_{\Delta'}}$ respectively. □

Lemma 6.1.4 was proven for *admissible solutions*, namely any $\tilde{u}^*_{S^{p+k}_{\Delta'}} \in S^{p+k}_{\Delta'}$, which satisfies the same orthogonality conditions as the finite element solution $u_{S^p_\Delta}$, namely (6.1.39a). Let us now extend Lemma 6.1.4 to any $\tilde{u}_{S^{p+k}_{\Delta'}} \in S^{p+k}_{\Delta'}$.

Lemma 6.1.5 *(Bounds for $\|u_{EX} - \tilde{u}_{S^{p+k}_{\Delta'}}\|_{\mathcal{U}}$, the energy norm of the error in any $\tilde{u}_{S^{p+k}_{\Delta'}} \in S^{p+k}_{\Delta'}$) For any $\tilde{u}_{S^{p+k}_{\Delta'}}, \tilde{e}_{S^{p+k}_{\Delta'}} \in S^{p+k}_{\Delta'}$, we have*

$$\hat{\mathscr{E}}^{L}_{\tilde{u}_{S^{p+k}_{\Delta'}}, S^{p+k}_{\Delta'}}(\tilde{e}_{S^{p+k}_{\Delta'}}) \leq \|u_{EX} - u_{S^{p+k}_{\Delta'}}\|_{\mathcal{U}} \leq \hat{\mathscr{E}}^{U}_{\tilde{u}_{S^{p+k}_{\Delta'}}, S^{p+k}_{\Delta'}} \tag{6.1.41a}$$

where

$$\hat{\mathscr{E}}^{L}_{\tilde{u}_{S^{p+k}_{\Delta'}}, S^{p+k}_{\Delta'}}(\tilde{e}_{S^{p+k}_{\Delta'}}) \stackrel{\text{def}}{=} \sqrt{\left\| q^{S^p_\Delta}_{\tilde{u}_{S^{p+k}_{\Delta'}}} \right\|^2_{\mathcal{U}} + \left(\mathscr{E}^{L}_{\tilde{u}^*_{S^{p+k}_{\Delta'}}, S^{p+k}_{\Delta'}}(\tilde{e}_{S^{p+k}_{\Delta'}})\right)^2} \tag{6.1.41b}$$

and

$$\hat{\mathscr{E}}^{U}_{\tilde{u}_{S^{p+k}_{\Delta'}}, S^{p+k}_{\Delta'}} \stackrel{\text{def}}{=} \sqrt{\left\| q^{S^p_\Delta}_{\tilde{u}_{S^{p+k}_{\Delta'}}} \right\|^2_{\mathcal{U}} + \left(\mathscr{E}^{U}_{\tilde{u}^*_{S^{p+k}_{\Delta'}}, S^{p+k}_{\Delta'}}\right)^2} \tag{6.1.41c}$$

where

$$\tilde{u}^*_{S^{p+k}_{\Delta'}} \stackrel{\text{def}}{=} \tilde{u}_{S^{p+k}_{\Delta'}} + q^{S^p_\Delta}_{\tilde{u}_{S^{p+k}_{\Delta'}}} \tag{6.1.42a}$$

and $q^{S^p_\Delta}_{\tilde{u}_{S^{p+k}_{\Delta'}}}$ is the solution of the discrete problem:

Find $q^{S^p_\Delta}_{\tilde{u}_{S^{p+k}_{\Delta'}}} \in S^p_\Delta$ such that

$$\mathscr{B}\left(q^{S^p_\Delta}_{\tilde{u}_{S^{p+k}_{\Delta'}}}, v\right) = \mathscr{R}_{\tilde{u}_{S^{p+k}_{\Delta'}}}(v) \stackrel{\text{def}}{=} \mathscr{L}(v) - \mathscr{B}(\tilde{u}_{S^{p+k}_{\Delta'}}, v) \quad \forall v \in S^p_\Delta. \tag{6.1.42b}$$

Proof The inequalities (6.1.41a) follow from the identity

$$\|u_{EX} - \tilde{u}_{S^{p+k}_{\Delta'}}\|^2_{\mathcal{U}} = \|u_{EX} - \tilde{u}^*_{S^{p+k}_{\Delta'}}\|^2_{\mathcal{U}} + \left\| q^{S^p_\Delta}_{\tilde{u}_{S^{p+k}_{\Delta'}}} \right\|^2_{\mathcal{U}} \tag{6.1.43}$$

and Lemma 6.1.4. □

Remark 6.1.13 We will refer to $q^{S^p_\Delta}_{\tilde{u}_{S^{p+k}_{\Delta'}}} \in S^p_\Delta$, as the *discrete equilibrium correction.* It is the correction which must be added to $\tilde{u}_{S^{p+k}_{\Delta'}} \in S^{p+k}_{\Delta'}$, to make it satisfy the orthogonality condition (6.1.39a).

Letting

$$\tilde{u}_{S^{p+k}_{\Delta'}} = u_{S^p_\Delta} + \tilde{e}_{S^{p+k}_{\Delta'}} \tag{6.1.44a}$$

Lemma 6.1.3 can be rewritten in the form

$$\|e_{S^p_\Delta}\|^2_{\mathcal{U}} = \left(\mathscr{E}^{\text{Neum}}_{S^{p+k}_{\Delta'}}\right)^2 - \sum_{j=1}^{N(\Delta)} \left\| \tilde{e}_{S^{p+k}_{\Delta'}} - \hat{e}^{\text{Neum}}_{I^\Delta_j, S^{p+k}_{\Delta'}} \right\|^2_{\mathcal{U}} + \|u_{EX} - \tilde{u}_{S^{p+k}_{\Delta'}}\|^2_{\mathcal{U}} \tag{6.1.44b}$$

for any $\tilde{u}_{S^{p+k}_{\Delta'}} \in S^{p+k}_{\Delta'}$. We will now construct improved bounds for $\|e_{S^p_\Delta}\|_{\mathcal{U}}$, by employing the bounds for $\|u_{EX} - \tilde{u}_{S^{p+k}_{\Delta'}}\|_{\mathcal{U}}$ proven in Lemma 6.1.5. We have:

Theorem 6.1.5 *(Improved bounds $\underline{\mathcal{E}}^L_{S^{p+k}_{\Delta'}}$, $\bar{\mathcal{E}}^U_{S^{p+k}_{\Delta'}}$ for $\|e_{S^p_\Delta}\|_{\mathcal{U}}$) Under the assumptions which guarantee the existence of $\mathcal{E}^{\text{Neum}}$, we have*

$$\mathcal{E}^L_{S^{p+k}_{\Delta'}}(\tilde{e}_{S^{p+k}_{\Delta'}}) \leq \underline{\mathcal{E}}^L_{S^{p+k}_{\Delta'}}(\tilde{e}_{S^{p+k}_{\Delta'}}, \tilde{\tilde{e}}_{S^{p+k}_{\Delta'}}) \leq \left\| e^{S^{p+k}_{\Delta'}}_{S^p_\Delta} \right\|_{\mathcal{U}} \leq \|e_{S^p_\Delta}\|_{\mathcal{U}} \leq \bar{\mathcal{E}}^U_{S^{p+k}_{\Delta'}}(\tilde{e}_{S^{p+k}_{\Delta'}}) \quad (6.1.45a)$$

for any $\tilde{e}_{S^{p+k}_{\Delta'}}$, $\tilde{\tilde{e}}_{S^{p+k}_{\Delta'}} \in S^{p+k}_{\Delta'}$, where

$$\underline{\mathcal{E}}^L_{S^{p+k}_{\Delta'}}(\tilde{e}_{S^{p+k}_{\Delta'}}, \tilde{\tilde{e}}_{S^{p+k}_{\Delta'}}) \stackrel{\text{def}}{=} \sqrt{\left(\mathcal{E}^L_{S^{p+k}_{\Delta'}}(\tilde{e}_{S^{p+k}_{\Delta'}})\right)^2 + \left(\hat{\mathcal{E}}^L_{\tilde{u}_{S^{p+k}_{\Delta'}}, S^{p+k}_{\Delta'}}(\tilde{\tilde{e}}_{S^{p+k}_{\Delta'}})\right)^2} \quad (6.1.45b)$$

and

$$\bar{\mathcal{E}}^U_{S^{p+k}_{\Delta'}}(\tilde{e}_{S^{p+k}_{\Delta'}}) \stackrel{\text{def}}{=} \sqrt{\left(\mathcal{E}^L_{S^{p+k}_{\Delta'}}(\tilde{e}_{S^{p+k}_{\Delta'}})\right)^2 + \left(\hat{\mathcal{E}}^U_{\tilde{u}_{S^{p+k}_{\Delta'}}, S^{p+k}_{\Delta'}}\right)^2} \quad (6.1.45c)$$

where we let

$$\tilde{u}_{S^{p+k}_{\Delta'}} \stackrel{\text{def}}{=} u_{S^p_\Delta} + \tilde{e}_{S^{p+k}_{\Delta'}} \quad (6.1.45d)$$

and $\hat{\mathcal{E}}^L_{\tilde{u}_{S^{p+k}_{\Delta'}}, S^{p+k}_{\Delta'}}(\tilde{\tilde{e}}_{S^{p+k}_{\Delta'}})$, $\hat{\mathcal{E}}^U_{\tilde{u}_{S^{p+k}_{\Delta'}}, S^{p+k}_{\Delta'}}$ are defined by (6.1.41b) and (6.1.41c) respectively.

Proof The proof follows directly from (6.1.44b), and Lemma 6.1.5.

□

A recursive application of the above gives a sequence of improved bounds $\underline{\mathcal{E}}^{L,(m)}_{S^{p+k}_{\Delta'}}$ and $\bar{\mathcal{E}}^{U,(m)}_{S^{p+k}_{\Delta'}}$, with $m = 1, 2, \ldots$, where $\underline{\mathcal{E}}^{L,(1)}_{S^{p+k}_{\Delta'}} = \underline{\mathcal{E}}^L_{S^{p+k}_{\Delta'}}$ and $\bar{\mathcal{E}}^{U,(1)}_{S^{p+k}_{\Delta'}} = \bar{\mathcal{E}}^U_{S^{p+k}_{\Delta'}}$.

Remark 6.1.14 The above approach for improving the estimators is based on the assumption that *a direct solver is used for determining the solution of the discrete finite element equations.* Because the stiffness matrix has been already factorized during the determination of the finite element solution $u_{S^p_\Delta}$ the cost of determining $q^{S^p_\Delta}_{u_{S^{p+k}_{\Delta'}}}$, is the same as the cost of processing an additional right-hand side, for the finite element equations which requires only $O(N)$ arithmetic operations, and hence it is negligible. Let us note that, in practice, for linear systems with up to 10^5 unknowns, direct solvers are used almost exclusively. The above approach can also be implemented using an iterative solver with a good preconditioner. In any case, a global solver is always needed to deal with the pollution.

Remark 6.1.15 Noting that

$$\tilde{\Phi}_{\hat{e}^{\text{Neum}}_{S^{p+k}_{\Delta'}}} = \tilde{e}_{S^{p+k}_{\Delta'}} - \hat{e}^{\text{Neum}}_{S^{p+k}_{\Delta'}} \quad (6.1.46a)$$

and that

$$\sum_{j=1}^{N(\Delta)} \mathcal{B}_{I_j^\Delta}\big(\hat{e}^{\text{Neum}}_{S^{p+k}_{\Delta'}}, v\big) = \sum_{j=1}^{N(\Delta)} \mathcal{R}^{EQ}_{I_j^\Delta}(v) = \mathcal{R}(v) = 0 \quad \forall v \in S^p_\Delta \tag{6.1.46b}$$

we obtain

$$\begin{aligned}\mathcal{R}_{\tilde{u}_{S^{p+k}_{\Delta'}}}(v) &= \mathcal{L}(v) - \mathcal{B}(\tilde{u}_{S^{p+k}_{\Delta'}}, v) = -\mathcal{B}(\tilde{e}_{S^{p+k}_{\Delta'}}, v) \\ &= -\sum_{j=1}^{N(\Delta)} \mathcal{B}_{I_j^\Delta}(\tilde{\Phi}_{\hat{e}^{\text{Neum}}_{S^{p+k}_{\Delta'}}}, v)\end{aligned} \tag{6.1.46c}$$

for all $v \in S^p_\Delta$. Using this result in (6.1.42b), it follows that $q^{S^p_\Delta}_{\tilde{u}_{S^{p+k}_{\Delta'}}}$ is the solution of the variational problem:

Find $q^{S^p_\Delta}_{\tilde{u}_{S^{p+k}_{\Delta'}}} \in S^p_\Delta$ such that

$$\mathcal{B}\Big(q^{S^p_\Delta}_{\tilde{u}_{S^{p+k}_{\Delta'}}}, v\Big) = -\sum_{j=1}^{N(\Delta)} \mathcal{B}_{I_j^\Delta}(\tilde{\Phi}_{\hat{e}^{\text{Neum}}_{S^{p+k}_{\Delta'}}}, v) \quad \forall v \in S^p_\Delta. \tag{6.1.46d}$$

Remark 6.1.16 Letting

$$\tilde{u} = u_{S^p_\Delta} + \tilde{e} \tag{6.1.47a}$$

the equation (6.1.35) can be rewritten in the form

$$\|e_{S^p_\Delta}\|^2_{\mathcal{U}} = (\mathcal{E}^{\text{Neum}})^2 - \sum_{j=1}^{N(\Delta)} \big\|\tilde{e} - e^{\text{Neum}}_{I_j^\Delta}\big\|^2_{\mathcal{U}(I_j^\Delta)} + \|u_{EX} - \tilde{u}\|^2_{\mathcal{U}} \tag{6.1.47b}$$

for any $\tilde{u} \in \mathcal{U}$, and following analogous steps as above, we obtain the improved bounds based on the exact estimator $\mathcal{E}^{\text{Neum}}$, namely

$$\mathcal{E}^L(\tilde{e}) \le \underline{\mathcal{E}}^L(\tilde{e}, \tilde{\tilde{e}}) \le \|e_{S^p_\Delta}\|_{\mathcal{U}} \le \bar{\mathcal{E}}^U(\tilde{e}) \quad \forall \tilde{e}, \tilde{\tilde{e}} \in \mathcal{U} \tag{6.1.48a}$$

where

$$\underline{\mathcal{E}}^L(\tilde{e}, \tilde{\tilde{e}}) \stackrel{\text{def}}{=} \sqrt{(\mathcal{E}^L(\tilde{e}))^2 + \|q^{S^p_\Delta}_{\tilde{u}}\|^2_{\mathcal{U}} + (\mathcal{E}^L_{\tilde{u}^*}(\tilde{\tilde{e}}))^2} \tag{6.1.48b}$$

is the 'exact' improved lower bound, and

$$\bar{\mathcal{E}}^U(\tilde{e}) \stackrel{\text{def}}{=} \sqrt{(\mathcal{E}^L(\tilde{e}))^2 + \|q^{S^p_\Delta}_{\tilde{u}}\|^2_{\mathcal{U}} + (\mathcal{E}^U_{\tilde{u}^*}(\tilde{e}))^2} \tag{6.1.48c}$$

is the 'exact' improved upper bound, where we let

$$\tilde{u}^* \stackrel{\text{def}}{=} \tilde{u} + q^{S^p_\Delta}_{\tilde{u}} \tag{6.1.49a}$$

and where $q_{\tilde{u}}^{S^p_\Delta}$ is discrete equilibrium correction for $\tilde{u}$, which is obtained as the solution of the discrete problem:

Find $q_{\tilde{u}}^{S^p_\Delta} \in S^p_\Delta$ suct that

$$\mathscr{B}(q_{\tilde{u}}^{S^p_\Delta}, v) = \mathscr{R}_{\tilde{u}}(v) \stackrel{\text{def}}{=} \mathscr{L}(v) - \mathscr{B}(\tilde{u}, v) = -\mathscr{B}(\tilde{e}, v) \quad \forall v \in S^p_\Delta. \tag{6.1.49b}$$

Let us now compare the improved bounds $\underline{\mathscr{E}}^L_{S^{p+k}_{\Delta'}}$, $\bar{\mathscr{E}}^U_{S^{p+k}_{\Delta'}}$ with the original ones $\mathscr{E}^L_{S^{p+k}_{\Delta'}}$, $\mathscr{E}^U_{S^{p+k}_{\Delta'}}$ in the setting of our example problem.

Example 6.1.2 (*The accuracy of the improved bounds*) Let us consider the model problem considered in Example 6.1.1 with $\alpha = 0$, and let $\Phi^{\bar{\alpha},harm}_{\hat{e}^{\text{Neum}}_{S^{p+k}_{\Delta'}}}$ be a gap function defined as in (6.1.33a)–(6.1.33c). It follows that

$$\sum_{j=1}^{N(\Delta)} \mathscr{B}_{I^\Delta_j}\big(\Phi^{\bar{\alpha},harm}_{\hat{e}^{\text{Neum}}_{S^{p+k}_{\Delta'}}}, v\big) = -\sum_{j=1}^{N(\Delta)} P_j\big(\Phi^{\bar{\alpha},harm}_{\hat{e}^{\text{Neum}}_{S^{p+k}_{\Delta'}}}\big) v(x^\Delta_j) \quad \forall v \in S^{p+k}_{\Delta'} \tag{6.1.50a}$$

where $P_j(\Phi^{\bar{\alpha},harm}_{\hat{e}^{\text{Neum}}_{S^{p+k}_{\Delta'}}})$ is the jump in the derivative of $\Phi^{\bar{\alpha},harm}_{\hat{e}^{\text{Neum}}_{S^{p+k}_{\Delta'}}}$ at the jth node. For example, for $\bar{\alpha} = \{\frac{1}{2}\}$ we have

$$P_j\left(\Phi^{\{1/2\},harm}_{\hat{e}^{\text{Neum}}_{S^{p+k}_{\Delta'}}}\right) = \frac{G^{\text{Neum}}_{j,S^{p+k}_{\Delta'}} + G^{\text{Neum}}_{j-1,S^{p+k}_{\Delta'}}}{2\int_{I^\Delta_j}(1/AE)} - \frac{G^{\text{Neum}}_{j+1,S^{p+k}_{\Delta'}} + G^{\text{Neum}}_{j,S^{p+k}_{\Delta'}}}{2\int_{I^\Delta_{j+1}}(1/AE)} \tag{6.1.50b}$$

for $j = 1, \ldots, N(\Delta) - 1$, and

$$P_N\left(\Phi^{\{1/2\},harm}_{\hat{e}^{\text{Neum}}_{S^{p+k}_{\Delta'}}}\right) = \frac{G^{\text{Neum}}_{N-1,S^{p+k}_{\Delta'}}}{2\int_{I^\Delta_N}(1/AE)}. \tag{6.1.50c}$$

It follows that $q^{S^p_\Delta}_{\tilde{u}_{S^{p+k}_{\Delta'}}} \in S^p_\Delta$, is the solution of the problem:

Find $q^{S^p_\Delta}_{\tilde{u}_{S^{p+k}_{\Delta'}}} \in S^p_\Delta$ such that

$$\mathscr{B}\left(q^{S^p_\Delta}_{\tilde{u}_{S^{p+k}_{\Delta'}}}, v\right) = \sum_{j=1}^{N(\Delta)} P_j\big(\Phi^{\bar{\alpha},harm}_{\hat{e}^{\text{Neum}}_{S^{p+k}_{\Delta'}}}\big) v(x^\Delta_j) \quad \forall v \in S^p_\Delta. \tag{6.1.50d}$$

Let us first consider the case of the bounds which corresponds to $k = 10$. Table 6.1.3 compares the effectivity indices for the original bounds $\kappa^L_{S^{p+k}_\Delta} = \mathscr{E}^L_{S^{p+k}_\Delta}/\|e_{S^p_\Delta}\|_{\mathfrak{U}}$, $\kappa^U_{S^{p+k}_\Delta} = \mathscr{E}^U_{S^{p+k}_\Delta}/\|e_{S^p_\Delta}\|_{\mathfrak{U}}$ and the improved bounds $\underline{\kappa}^L_{S^{p+k}_\Delta} = \underline{\mathscr{E}}^L_{S^{p+k}_\Delta}/\|e_{S^p_\Delta}\|_{\mathfrak{U}}$, $\bar{\kappa}^U_{S^{p+k}_\Delta} = \bar{\mathscr{E}}^U_{S^{p+k}_\Delta}/\|e_{S^p_\Delta}\|_{\mathfrak{U}}$ for $p = 1$, and 2. Note there is practically no improvement in the upper bound, while there is significant improvement in the lower bound.

Let us also examine the effect of the approximation of the indicators. Table 6.1.4 shows the values of the effectivity indices for the computed improved bounds $\underline{\kappa}^L_{S^{p+k}_\Delta} = \underline{\mathscr{E}}^L_{S^{p+k}_\Delta}/\|e_{S^p_\Delta}\|_{\mathfrak{U}}$, $\bar{\kappa}^U_{S^{p+k}_\Delta} = \bar{\mathscr{E}}^U_{S^{p+k}_\Delta}/\|e_{S^p_\Delta}\|_{\mathfrak{U}}$ as functions of the ratio a_1/a_2, for

TABLE 6.1.3. *Comparison of the original and the improved bounds based on the estimator* $\mathcal{E}^{\mathrm{Neum}}_{S^{p+k}_\Delta}$ with $k = 10$. The effectivity indices for the original bounds $\kappa^U_{S^{p+k}_\Delta} = \mathcal{E}^U_{S^{p+k}_\Delta}/\|e_{S^p_\Delta}\|_{\mathcal{U}}$, and the lower bound $\kappa^L_{S^{p+k}_\Delta} = \mathcal{E}^L_{S^{p+k}_\Delta}/\|e_{S^p_\Delta}\|_{\mathcal{U}}$ and the improved bounds $\bar{\kappa}^U_{S^{p+k}_\Delta} = \bar{\mathcal{E}}^U_{S^{p+k}_\Delta}/\|e_{S^p_\Delta}\|_{\mathcal{U}}$, $\underline{\kappa}^L_{S^{p+k}_\Delta} = \underline{\mathcal{E}}^L_{S^{p+k}_\Delta}/\|e_{S^p_\Delta}\|_{\mathcal{U}}$ for $\alpha = 0$, $h = \frac{1}{4}$ and $p = 1, 2$

a_1/a_2	$p = 1$				$p = 2$			
	$\kappa^U_{S^{p+k}_\Delta}$	$\kappa^L_{S^{p+k}_\Delta}$	$\bar{\kappa}^U_{S^{p+k}_\Delta}$	$\underline{\kappa}^L_{S^{p+k}_\Delta}$	$\kappa^U_{S^{p+k}_\Delta}$	$\kappa^L_{S^{p+k}_\Delta}$	$\bar{\kappa}^U_{S^{p+k}_\Delta}$	$\underline{\kappa}^L_{S^{p+k}_\Delta}$
2	1.047886	0.94937	1.047881	0.99965	1.139144	0.99183	1.139137	0.99975
3	1.110341	0.89564	1.110325	0.99704	1.413250	0.97962	1.413234	0.99928
4	1.149274	0.84439	1.149248	0.99109	1.781456	0.96678	1.781430	0.99858
5	1.178588	0.79742	1.178554	0.98216	2.150199	0.95371	2.150161	0.99769
6	1.227572	0.75423	1.227530	0.97104	2.472358	0.94054	2.472307	0.99664
7	1.318997	0.71405	1.318946	0.95840	2.728580	0.92731	2.728518	0.99544
8	1.448983	0.67620	1.448920	0.94474	2.922016	0.91404	2.921942	0.99410
9	1.613649	0.64011	1.613574	0.93040	3.060694	0.90074	3.060611	0.99261
10	1.805407	0.60532	1.805316	0.91563	3.152874	0.88743	3.152785	0.99096

TABLE 6.1.4. *The effect of the approximate solution of the element residual problems on the improved bounds.* The effectivity indices of the computed improved bounds $\bar{\kappa}^U_{S^{p+k}_\Delta} = \bar{\mathcal{E}}^U_{u_{S^p_\Delta}, S^{p+k}_\Delta}/\|e_{S^p_\Delta}\|_{\mathcal{U}}$, $\underline{\kappa}^L_{S^{p+k}_\Delta} = \underline{\mathcal{E}}^L_{u_{S^p_\Delta}, S^{p+k}_\Delta}/\|e_{S^p_\Delta}\|_{\mathcal{U}}$ for $p = 1$ and $p = 2$, as functions of a_1/a_2, for $k = 1, 2$ with $\alpha = 0$, $h = \frac{1}{4}$

a_1/a_2	$k = 1$		$k = 2$		$k = 1$		$k = 2$	
	$\bar{\kappa}^U_{S^{p+k}_\Delta}$	$\underline{\kappa}^L_{S^{p+k}_\Delta}$	$\bar{\kappa}^U_{S^{p+k}_\Delta}$	$\underline{\kappa}^L_{S^{p+k}_\Delta}$	$\bar{\kappa}^U_{S^{p+k}_\Delta}$	$\underline{\kappa}^L_{S^{p+k}_\Delta}$	$\bar{\kappa}^U_{S^{p+k}_\Delta}$	$\underline{\kappa}^L_{S^{p+k}_\Delta}$
	$p = 1$				$p = 2$			
2	2.13694	0.87732	1.98253	0.93968	3.70499	0.70198	2.68357	0.94335
3	2.45785	0.87043	2.24502	0.92637	4.21790	0.65050	3.17876	0.93989
4	2.76786	0.86060	2.41516	0.91475	4.47750	0.62774	3.78841	0.92856
5	3.07255	0.84889	2.52926	0.90363	4.59728	0.62170	4.41062	0.91539
6	3.37109	0.83608	2.60904	0.89263	4.63997	0.62479	4.96893	0.90253
7	3.66182	0.82272	2.66774	0.88165	4.64117	0.63255	5.45534	0.89074
8	3.94350	0.80917	2.71377	0.87066	4.62170	0.64242	5.87702	0.88022
9	4.21535	0.79567	2.75264	0.85966	4.59395	0.65301	6.24277	0.87093
10	4.47704	0.78235	2.78804	0.84865	4.56535	0.66354	6.59176	0.86275

$k = 1, 2$ and $p = 1, 2$. Note that bounds with reasonable effectivity are obtained for the entire range of a_1/a_2. Table 6.1.5 shows the convergence of the effectivity indices $\underline{\kappa}^{L,(m)}_{S^{p+k}_\Delta} \overset{\mathrm{def}}{=} \underline{\mathcal{E}}^{L,(m)}_{u_{S^p_\Delta}, S^{p+k}_\Delta}/\|e_{S^p_\Delta}\|_{\mathcal{U}}$, and $\bar{\kappa}^{U,(m)}_{u_{S^p_\Delta}, S^{p+k}_\Delta} = \bar{\mathcal{E}}^{U,(m)}_{S^{p+k}_\Delta}/\|e_{S^p_\Delta}\|_{\mathcal{U}}$, for $a_1/a_2 = 10$, $h = \frac{1}{4}$, $\alpha = 0$, and $k = 2, 4, 10$. In Table 6.1.5, $m = 1$ denotes the zeroth iteration. Note that there is a significant improvement in the lower bound in the first couple of iterations, while there is practically no improvement in the upper bound.

TABLE 6.1.5. *Iterative improvement of the bounds.* The effectivity indices of the improved bounds $\bar{\kappa}^{U,(m)}_{S^{p+k}_\Delta} = \bar{\mathscr{E}}^{U,(m)}_{u_{S^p_\Delta}, S^{p+k}_\Delta}/\|e_{S^p_\Delta}\|_{\mathcal{U}}$, $\underline{\kappa}^{L,(m)}_{S^{p+k}_\Delta} = \underline{\mathscr{E}}^{L,(m)}_{u_{S^p_\Delta}, S^{p+k}_\Delta}/\|e_{S^p_\Delta}\|_{\mathcal{U}}$ for (a) $p = 1$, (b) $p = 2$, for $a_1/a_2 = 10$, $h = \frac{1}{4}$, $\alpha = 0$, and $m = 1, \dots, 5$. Note that there is significant improvement in the lower bound in the first couple of iterations, while there is practically no improvement in the upper bound

m	$k = 2$		$k = 4$		$k = 10$	
	$\bar{\kappa}^{U,(m)}_{S^{p+k}_\Delta}$	$\underline{\kappa}^{L,(m)}_{S^{p+k}_\Delta}$	$\bar{\kappa}^{U,(m)}_{S^{p+k}_\Delta}$	$\underline{\kappa}^{L,(m)}_{S^{p+k}_\Delta}$	$\bar{\kappa}^{U,(m)}_{S^{p+k}_\Delta}$	$\underline{\kappa}^{L,(m)}_{S^{p+k}_\Delta}$
$p = 1$						
1	2.78811	0.55949	2.08852	0.60076	1.80541	0.60532
2	2.78804	0.84865	2.08845	0.90250	1.80532	0.91563
3	2.78803	0.89089	2.08843	0.96155	1.80529	0.97882
4	2.78802	0.89808	2.08842	0.97487	1.80527	0.99364
5	2.78802	0.89934	2.08841	0.97796	1.80527	0.99722
$p = 2$						
1	6.59194	0.78963	5.29539	0.86377	3.15287	0.88743
2	6.59176	0.86275	5.29529	0.95923	3.15278	0.99096
3	6.59175	0.86467	5.29527	0.96271	3.15277	0.99501
4	6.59174	0.86472	5.29527	0.96284	3.15277	0.99517
5	6.59174	0.86472	5.29527	0.96285	3.15277	0.99518

We will call the bounds $\underline{\mathscr{E}}^L_{S^{p+k}_{\Delta'}}(\tilde{e}_{S^{p+k}_{\Delta'}}, \tilde{\tilde{e}}_{S^{p+k}_{\Delta'}})$ and $\bar{\mathscr{E}}^L_{S^{p+k}_{\Delta'}}(\tilde{e}_{S^{p+k}_{\Delta'}})$ *the improved bounds based on orthogonality*, because they are based on Lemma 6.1.3 which follows from the orthogonalities satisfied by finite element solution, and the solutions of the element residual problems. We can also obtain improved bounds by employing the *triangle inequality*.

Theorem 6.1.6 *(Improved bounds $\underline{\mathscr{E}}^{L,tri}$, $\bar{\mathscr{E}}^{U,tri}$ based on the triangle inequality) Under the assumptions which guarantee the existence of $\mathscr{E}^{\mathrm{Neum}}$, we have*

$$\underline{\mathscr{E}}^{L,tri}_{\tilde{u}_{S^{p+k}_{\Delta'}}} \le \|e_{S^p_\Delta}\|_{\mathcal{U}} \le \bar{\mathscr{E}}^{U,tri}_{\tilde{u}_{S^{p+k}_{\Delta'}}}(\tilde{e}_{S^{p+k}_{\Delta'}}) \tag{6.1.51a}$$

for any $\tilde{u}_{S^{p+k}_{\Delta'}} \in S^{p+k}_{\Delta'}$, where we let

$$\underline{\mathscr{E}}^{L,tri}_{\tilde{u}_{S^{p+k}_{\Delta'}}} \stackrel{\mathrm{def}}{=} \left| \|\tilde{u}^*_{S^{p+k}_{\Delta'}} - u_{S^p_\Delta}\|_{\mathcal{U}} - \mathscr{E}^U_{\tilde{u}^*_{S^{p+k}_{\Delta'}}, S^{p+k}_{\Delta'}} \right| \tag{6.1.51b}$$

and

$$\bar{\mathscr{E}}^{U,tri}_{\tilde{u}_{S^{p+k}_{\Delta'}}} \stackrel{\mathrm{def}}{=} \|\tilde{u}^*_{S^{p+k}_{\Delta'}} - u_{S^p_\Delta}\|_{\mathcal{U}} + \mathscr{E}^U_{\tilde{u}^*_{S^{p+k}_{\Delta'}}, S^{p+k}_{\Delta'}} \tag{6.1.51c}$$

and $\tilde{u}^*_{S^{p+k}_{\Delta'}}$, is defined in Lemma 6.1.5.

Proof Noting that

$$e_{S^p_\Delta} = u_{EX} - u_{S^p_\Delta} = (u_{EX} - \tilde{u}^*_{S^{p+k}_{\Delta'}}) + (\tilde{u}^*_{S^{p+k}_{\Delta'}} - u_{S^p_\Delta}) \tag{6.1.52a}$$

and using the triangle inequality, we get

$$\left| \left\| \tilde{e}^{*,S^{p+k}_{\Delta'}}_{S^p_\Delta} \right\|_{\mathcal{U}} - \|u_{EX} - \tilde{u}^*_{S^{p+k}_{\Delta'}}\|_{\mathcal{U}} \right| \leq \|e_{S^p_\Delta}\|_{\mathcal{U}} \leq \left\| \tilde{e}^{*,S^{p+k}_{\Delta'}}_{S^p_\Delta} \right\|_{\mathcal{U}} + \|u_{EX} - \tilde{u}^*_{S^{p+k}_{\Delta'}}\|_{\mathcal{U}} \tag{6.1.52b}$$

where we let

$$\tilde{e}^{*,S^{p+k}_{\Delta'}}_{S^p_\Delta} \stackrel{\text{def}}{=} \tilde{u}^*_{S^{p+k}_{\Delta'}} - u_{S^p_\Delta}. \tag{6.1.52c}$$

Employing now the upper bound

$$\|u_{EX} - \tilde{u}^*_{S^{p+k}_{\Delta'}}\|_{\mathcal{U}} \leq \mathscr{E}^U_{\tilde{u}^*_{S^{p+k}_{\Delta'}}, S^{p+k}_{\Delta'}} \tag{6.1.53}$$

in (6.1.52b) we obtain (6.1.51). □

Remark 6.1.17 Following similar steps we also obtain the 'exact' improved bounds based on the triangle inequality

$$\underline{\mathscr{E}}^{L,tri}_{\tilde{u}} \stackrel{\text{def}}{=} \left| \|\tilde{u}^* - u_{S^p_\Delta}\|_{\mathcal{U}} - \mathscr{E}^{\text{Neum}}_{\tilde{u}} \right| \leq \|e_{S^p_\Delta}\|_{\mathcal{U}} \leq \bar{\mathscr{E}}^{U,tri}_{\tilde{u}} \stackrel{\text{def}}{=} \|\tilde{u}^* - u_{S^p_\Delta}\|_{\mathcal{U}} + \mathscr{E}^U_{\tilde{u}} \tag{6.1.54}$$

for any $\tilde{u} \in \mathcal{U}$, where $\tilde{u}^*$ is defined by (6.1.49a) and (6.1.49b).

Remark 6.1.18 The improved bounds $\mathscr{E}^{L,tri}_{\tilde{u}_{S^{p+k}_{\Delta'}}}$, $\mathscr{E}^{U,tri}_{\tilde{u}_{S^{p+k}_{\Delta'}}}$ depend on $\tilde{u}_{S^{p+k}_{\Delta'}}$, and because we employ

$$\tilde{u}_{S^{p+k}_{\Delta'}} = u_{S^p_\Delta} + \hat{e}^{\text{Neum}}_{S^{p+k}_{\Delta'}} + \Phi_{\hat{e}^{\text{Neum}}_{S^{p+k}_{\Delta'}}} \tag{6.1.55}$$

they also depend on $u_{S^p_\Delta}$.

Remark 6.1.19 In the above examples we have seen that the computable upper bound

$$\mathscr{E}^U_{S^{p+k}_{\Delta'}} = \sqrt{\left(\mathscr{E}^{\text{Neum}}_{S^{p+k}_{\Delta'}}\right)^2 + \sum_{j=1}^{N(\Delta)} \left(\mathscr{E}^{EXPL}_{\hat{e}_{S^{p+k}_{\Delta'}}/I^\Delta_j}\right)^2} \tag{6.1.56a}$$

may grossly overestimate $\|e_{S^p_\Delta}\|_{\mathcal{U}}$, although $\mathscr{E}^{\text{Neum}}_{S^{p+k}_\Delta}$ may be underestimating. This is because, in many cases, the explicit estimator $\mathscr{E}^{EXPL}_{\hat{e}_{S^{p+k}_{\Delta'}}/I^\Delta_j}$ grossly overestimates $\|\hat{e}^{\text{Neum}}_{I^\Delta_j} - \hat{e}^{\text{Neum}}_{I^\Delta_j, S^{p+k}_\Delta}\|_{\mathcal{U}}$. The effect of this overestimation on the quality of $\mathscr{E}^U_{S^{p+k}_\Delta}$ may be controlled by choosing a sufficiently high k such that we have

$$\mathscr{E}^{EXPL}_{\hat{e}_{S^{p+k}_{\Delta'}}/I^\Delta_j} \leq \delta \eta^{\text{Neum}}_{I^\Delta_j, S^{p+k}_{\Delta'}} \tag{6.1.56b}$$

where δ is a sufficiently small input parameter, e.g. $\delta = 0.05$.

Above we addressed the guaranteed error estimation for $\|e_{S^p_\Delta}\|_{\mathcal{U}}$, the energy norm of the error. In practice however we are most often interested in determining bounds for the relative error

$$E^{rel}_{u_{S^p_\Delta}} \stackrel{\text{def}}{=} \frac{\|e_{S^p_\Delta}\|_{\mathcal{U}}}{\|u_{EX}\|_{\mathcal{U}}} = \frac{\|u_{EX} - u_{S^p_\Delta}\|_{\mathcal{U}}}{\|u_{EX}\|_{\mathcal{U}}}. \tag{6.1.57}$$

Theorem 6.1.7 *(Computable upper and lower bounds $\mathscr{E}^{rel,L}_{S^{p+k}_{\Delta'}}$, $\mathscr{E}^{rel,U}_{S^{p+k}_{\Delta'}}$ for the relative error $E^{rel}_{u_{S^p_\Delta}}$) Under the assumptions which guarantee the existence of $\mathscr{E}^{\text{Neum}}$, and in the case that the Dirichlet conditions are homogeneous, we have*

$$\mathscr{E}^{rel,L}_{S^{p+k}_{\Delta'}}(\tilde{e}_{S^{p+k}_{\Delta'}}) \leq E^{rel}_{u_{S^p_\Delta}} \leq \mathscr{E}^{rel,U}_{S^{p+k}_{\Delta'}}(\tilde{e}_{S^{p+k}_{\Delta'}}) \tag{6.1.58a}$$

where

$$\mathscr{E}^{rel,L}_{S^{p+k}_{\Delta'}}(\tilde{e}_{S^{p+k}_{\Delta'}}) \stackrel{\text{def}}{=} \frac{\mathscr{E}^{L}_{S^{p+k}_{\Delta'}}(\tilde{e}_{S^{p+k}_{\Delta'}})}{\sqrt{\|u_{S^p_\Delta}\|^2_{\mathcal{U}} + (\mathscr{E}^{U}_{S^{p+k}_{\Delta'}})^2}} \tag{6.1.58b}$$

and

$$\mathscr{E}^{rel,U}_{S^{p+k}_{\Delta'}}(\tilde{e}_{S^{p+k}_{\Delta'}}) \stackrel{\text{def}}{=} \frac{\mathscr{E}^{U}_{S^{p+k}_{\Delta'}}}{\sqrt{\|u_{S^p_\Delta}\|^2_{\mathcal{U}} + (\mathscr{E}^{L}_{S^{p+k}_{\Delta'}}(\tilde{e}_{S^{p+k}_{\Delta'}}))^2}} \tag{6.1.58c}$$

Proof Because the Dirichlet boundary conditions are homogeneous, we have

$$\|u_{EX}\|_{\mathcal{U}} = \sqrt{\|u_{S^p_\Delta}\|^2_{\mathcal{U}} + \|e_{S^p_\Delta}\|^2_{\mathcal{U}}} \tag{6.1.59a}$$

and hence

$$E^{rel}_{u_{S^p_\Delta}} = \frac{\|e_{S^p_\Delta}\|_{\mathcal{U}}}{\|u_{EX}\|_{\mathcal{U}}} = \frac{\|e_{S^p_\Delta}\|_{\mathcal{U}}}{\sqrt{\|u_{S^p_\Delta}\|^2_{\mathcal{U}} + \|e_{S^p_\Delta}\|^2_{\mathcal{U}}}} \tag{6.1.59b}$$

and the proof is completed by employing the bounds

$$\mathscr{E}^{L}_{S^{p+k}_{\Delta'}}(\tilde{e}_{S^{p+k}_{\Delta'}}) \leq \|e_{S^p_\Delta}\|_{\mathcal{U}} \leq \mathscr{E}^{U}_{S^{p+k}_{\Delta'}}. \tag{6.1.59c}$$

□

Remark 6.1.20 Let us also address the case that the Dirichlet conditions are nonhomogeneous. Letting

$$u_{EX} = u^{S^p_\Delta}_{Dir} + \tilde{u}_{EX} \tag{6.1.60a}$$

where $u^{S^p_\Delta}_{Dir} \in S^p_\Delta$, is a function which satisfies the nonhomogeneous conditions (see (2.2.39b)), we have

$$e_{S^p_\Delta} = \tilde{e}_{S^p_\Delta} \stackrel{\text{def}}{=} \tilde{u}_{EX} - \tilde{u}_{S^p_\Delta} \tag{6.1.60b}$$

and

$$E^{rel}_{u_{S^p_\Delta}} = \frac{\|e_{S^p_\Delta}\|_\mathcal{U}}{\|u_{S^p_\Delta}\|_\mathcal{U}} = \frac{\|e_{S^p_\Delta}\|_\mathcal{U}}{\|\tilde{u}_{S^p_\Delta} + u^{S^p_\Delta}_{Dir}\|_\mathcal{U}}. \tag{6.1.60c}$$

Employing the triangle inequality for the denominator we obtain

$$\frac{\|e_{S^p_\Delta}\|_\mathcal{U}}{\|\tilde{u}_{S^p_\Delta}\|_\mathcal{U} + \|u^{S^p_\Delta}_{Dir}\|_\mathcal{U}} \le E^{rel}_{u_{S^p_\Delta}} \le \frac{\|e_{S^p_\Delta}\|_\mathcal{U}}{\big|\|\tilde{u}_{S^p_\Delta}\|_\mathcal{U} - \|u^{S^p_\Delta}_{Dir}\|_\mathcal{U}\big|} \tag{6.1.60d}$$

and using the bounds

$$\mathscr{E}^L_{S^{p+k}_{\Delta'}}(\tilde{e}_{S^{p+k}_{\Delta'}}) \le \|e_{S^p_\Delta}\|_\mathcal{U} \le \mathscr{E}^U_{S^{p+k}_{\Delta'}} \tag{6.1.60e}$$

we get

$$\mathscr{E}^{rel,L,tri}_{S^{p+k}_{\Delta'}}(\tilde{e}_{S^{p+k}_{\Delta'}}) \le E^{rel}_{u_{S^p_\Delta}} \le \mathscr{E}^{rel,U,tri}_{S^{p+k}_{\Delta'}} \tag{6.1.60f}$$

where we let

$$\mathscr{E}^{rel,L,tri}_{S^{p+k}_{\Delta'}}(\tilde{e}_{S^{p+k}_{\Delta'}}) = \frac{\|\tilde{u}_{S^p_\Delta}\|_\mathcal{U}}{\|\tilde{u}_{S^p_\Delta}\|_\mathcal{U} + \|\tilde{u}^{S^p_\Delta}_{Dir}\|_\mathcal{U}} \mathscr{E}^{rel,L}_{S^{p+k}_{\Delta'}}(\tilde{e}_{S^{p+k}_{\Delta'}}) \tag{6.1.60g}$$

and

$$\mathscr{E}^{rel,U,tri}_{S^{p+k}_{\Delta'}} = \frac{\|\tilde{u}_{S^p_\Delta}\|_\mathcal{U}}{\big|\|\tilde{u}_{S^p_\Delta}\|_\mathcal{U} - \|\tilde{u}^{S^p_\Delta}_{Dir}\|_\mathcal{U}\big|} \mathscr{E}^{rel,U}_{S^{p+k}_{\Delta'}}. \tag{6.1.60h}$$

Note that in the above derivation we assumed that $\|u_{S^p_\Delta}\|_\mathcal{U}$, which may not be satisfied in practice and hence the estimate (6.1.60h) could be very pessimistic.

Another approach is to employ the triangle inequality

$$\big|\|u_{S^p_\Delta}\|_\mathcal{U} - \|e_{S^p_\Delta}\|_\mathcal{U}\big| \le \|u_{EX}\|_\mathcal{U} \le \|u_{S^p_\Delta}\|_\mathcal{U} + \|e_{S^p_\Delta}\|_\mathcal{U} \tag{6.1.61a}$$

to obtain

$$\big|\|u_{S^p_\Delta}\|_\mathcal{U} - \mathscr{E}^U_{S^{p+k}_{\Delta'}}\big| \le \|u_{EX}\|_\mathcal{U} \le \|u_{S^p_\Delta}\|_\mathcal{U} + \mathscr{E}^U_{S^{p+k}_{\Delta'}} \tag{6.1.61b}$$

and employing (6.1.61b), and the bounds for $\|e_{S^p_\Delta}\|_\mathcal{U}$, in the definition of $E^{rel}_{u_{S^p_\Delta}}$, we get the bounds

$$\mathscr{E}^{rel,L,tri}_{S^{p+k}_{\Delta'}}(\tilde{e}_{S^{p+k}_{\Delta'}}) = \frac{\mathscr{E}^L_{S^{p+k}_{\Delta'}}(\tilde{e}_{S^{p+k}_{\Delta'}})}{\|u_{S^p_\Delta}\|_\mathcal{U} + \mathscr{E}^U_{S^{p+k}_{\Delta'}}} \le E^{rel}_{u_{S^p_\Delta}} \le \mathscr{E}^{rel,U,tri}_{S^{p+k}_{\Delta'}}.$$

$$= \frac{\mathscr{E}^U_{S^{p+k}_{\Delta'}}}{\big|\|u_{S^p_\Delta}\|_\mathcal{U} - \mathscr{E}^U_{S^{p+k}_{\Delta'}}\big|} \tag{6.1.61c*}$$

We will call the interval

$$R^{rel}_{S^{p+k}_{\Delta'}} \stackrel{\text{def}}{=} \left(\mathscr{E}^{rel,L}_{S^{p+k}_{\Delta'}}, \mathscr{E}^{rel,U}_{S^{p+k}_{\Delta'}}\right) \tag{6.1.62}$$

the *reliability interval* for the relative value of the energy norm of the error.

Above we constructed the lower bound $\mathscr{E}^L_{S^{p+k}_{\Delta'}}(\tilde{e}_{S^{p+k}_{\Delta'}})$, which was based on orthogonality. It is also possible to construct another lower bound based on duality.

Theorem 6.1.8 *(Lower bound based on duality) Under the minimal conditions which guarantee the existence of residuum $\mathscr{R}$, we have*

$$\mathscr{E}^L_{dual}(\tilde{e}_{S^{p+k}_{\Delta'}}) \stackrel{\text{def}}{=} \frac{\mathscr{R}(\tilde{e}_{S^{p+k}_{\Delta'}})}{\|\tilde{e}_{S^{p+k}_{\Delta'}}\|_{\mathscr{U}}} \leq \|e^{S^{p+k}_{\Delta'}}_{S^p_\Delta}\|_{\mathscr{U}} \leq \|e_{S^p_\Delta}\|_{\mathscr{U}} \quad \forall \tilde{e}_{S^{p+k}_{\Delta'}} \in S^{p+k}_{\Delta'}. \tag{6.1.63}$$

Proof For any $\tilde{e}_{S^{p+k}_{\Delta'}} \in S^{p+k}_{\Delta'}$, we have

$$\frac{\mathscr{R}(\tilde{e}_{S^{p+k}_{\Delta'}})}{\|\tilde{e}_{S^{p+k}_{\Delta'}}\|_{\mathscr{U}}} \leq \sup_{v \in S^{p+k}_{\Delta'}} \frac{\mathscr{R}(v)}{\|v\|_{\mathscr{U}}} = \left\| e^{S^{p+k}_{\Delta'}}_{S^p_\Delta} \right\|_{\mathscr{U}} \leq \|e_{S^p_\Delta}\|_{\mathscr{U}} \tag{6.1.64}$$

and (6.1.63) has been proven. □

Choosing for $\tilde{e}_{S^{p+k}_{\Delta'}}$, the smoothened Neumann error indicator function $\tilde{e}^{\text{Neum}}_{S^{p+k}_{\Delta'}}(\Phi_{\hat{e}^{\text{Neum}}_{S^{p+k}_{\Delta'}}})$, we obtain the lower bound from duality in terms of the gap function

$$\tilde{\mathscr{E}}^{\text{Neum}/L}_{dual}(\Phi_{\hat{e}^{\text{Neum}}_{S^{p+k}_{\Delta'}}}) = \frac{\mathscr{R}\big(\hat{e}^{\text{Neum}}_{S^{p+k}_{\Delta'}} + \Phi_{\hat{e}^{\text{Neum}}_{S^{p+k}_{\Delta'}}}\big)}{\|\hat{e}^{\text{Neum}}_{S^{p+k}_{\Delta'}} + \Phi_{\hat{e}^{\text{Neum}}_{S^{p+k}_{\Delta'}}}\|_{\mathscr{U}}} \leq \left\| e^{S^{p+k}_{\Delta'}}_{S^p_\Delta} \right\|_{\mathscr{U}} \leq \|e_{S^p_\Delta}\|_{\mathscr{U}}. \tag{6.1.65}$$

Below, for simplicity, we will write $\tilde{\mathscr{E}}^L_{dual}$ instead of $\tilde{\mathscr{E}}^{\text{Neum}/L}_{dual}$.

Remark 6.1.21 Let us now examine the relationship between the lower bounds $\tilde{\mathscr{E}}^L_{dual}(\Phi_{\hat{e}^{\text{Neum}}_{S^{p+k}_{\Delta'}}})$ and $\tilde{\mathscr{E}}^L_{S^{p+k}_{\Delta'}}(\Phi_{\hat{e}^{\text{Neum}}_{S^{p+k}_{\Delta'}}})$. Noting that

$$\mathscr{R}\big(\hat{e}^{\text{Neum}}_{S^{p+k}_{\Delta'}} + \Phi_{\hat{e}^{\text{Neum}}_{S^{p+k}_{\Delta'}}}\big) = \sum_{j=1}^{N(\Delta)} \mathscr{R}^{EQ}_{I^\Delta_j}\big(\hat{e}^{\text{Neum}}_{S^{p+k}_{\Delta'}} + \Phi_{\hat{e}^{\text{Neum}}_{S^{p+k}_{\Delta'}}}\big) \tag{6.1.66a}$$

and

$$\mathscr{R}^{EQ}_{I^\Delta_j}\big(\hat{e}^{\text{Neum}}_{S^{p+k}_{\Delta'}}\big) = \mathscr{B}_{I^\Delta_j}\big(\hat{e}^{\text{Neum}}_{S^{p+k}_{\Delta'}}, \hat{e}^{\text{Neum}}_{S^{p+k}_{\Delta'}}\big) = \|\hat{e}^{\text{Neum}}_{S^{p+k}_{\Delta'}}\|^2_{\mathscr{U}(I^\Delta_j)} \tag{6.1.66b}$$

and

$$\mathscr{R}^{EQ}_{I^\Delta_j}(\Phi_{\hat{e}^{\text{Neum}}_{S^{p+k}_{\Delta'}}}) = \mathscr{B}_{I^\Delta_j}\big(\hat{e}^{\text{Neum}}_{S^{p+k}_{\Delta'}}, \Phi_{\hat{e}^{\text{Neum}}_{S^{p+k}_{\Delta'}}}\big) \tag{6.1.66c}$$

we obtain that

$$\mathscr{R}\big(\hat{e}^{\text{Neum}}_{S^{p+k}_{\Delta'}} + \Phi_{\hat{e}^{\text{Neum}}_{S^{p+k}_{\Delta'}}}\big) = \big(\mathscr{E}^{\text{Neum}}_{S^{p+k}_{\Delta'}}\big)^2 + \sum_{j=1}^{N(\Delta)} \mathscr{B}_{I^\Delta_j}\big(\hat{e}^{\text{Neum}}_{S^{p+k}_{\Delta'}}, \Phi_{\hat{e}^{\text{Neum}}_{S^{p+k}_{\Delta'}}}\big). \tag{6.1.66d}$$

We also have

$$\|\hat{e}^{\text{Neum}}_{S^{p+k}_{\Delta'}} + \Phi_{\hat{e}^{\text{Neum}}_{S^{p+k}_{\Delta'}}}\|_{\mathcal{U}} = \sqrt{\left(\mathscr{E}^{\text{Neum}}_{S^{p+k}_{\Delta'}}\right)^2 + \sum_{j=1}^{N(\Delta)} \|\Phi_{\hat{e}^{\text{Neum}}_{S^{p+k}_{\Delta'}}}\|^2_{\mathcal{U}(I^{\Delta}_j)} + 2\sum_{j=1}^{N(\Delta)} \mathscr{B}_{I^{\Delta}_j}(\hat{e}^{\text{Neum}}_{S^{p+k}_{\Delta'}}, \Phi_{\hat{e}^{\text{Neum}}_{S^{p+k}_{\Delta'}}})} \tag{6.1.66e}$$

and hence

$$\tilde{\mathscr{E}}^L_{dual}(\Phi_{\hat{e}^{\text{Neum}}_{S^{p+k}_{\Delta'}}}) = \tilde{\mathscr{E}}^L(\Phi_{\hat{e}^{\text{Neum}}_{S^{p+k}_{\Delta'}}}) \frac{1 + \gamma^2(\hat{e}^{\text{Neum}}_{S^{p+k}_{\Delta'}}, \Phi_{\hat{e}^{\text{Neum}}_{S^{p+k}_{\Delta'}}})}{\sqrt{1 + 2\,\gamma^2(\hat{e}^{\text{Neum}}_{S^{p+k}_{\Delta'}}, \Phi_{\hat{e}^{\text{Neum}}_{S^{p+k}_{\Delta'}}})}} \tag{6.1.67a}$$

where we let

$$\gamma^2(\hat{e}^{\text{Neum}}_{S^{p+k}_{\Delta'}}, \Phi_{\hat{e}^{\text{Neum}}_{S^{p+k}_{\Delta'}}}) \stackrel{\text{def}}{=} \frac{\sum_{j=1}^{N(\Delta)} \|\Phi_{\hat{e}^{\text{Neum}}_{S^{p+k}_{\Delta'}}}\|^2_{\mathcal{U}(I^{\Delta}_j)} + \sum_{j=1}^{N(\Delta)} \mathscr{B}_{I^{\Delta}_j}(\hat{e}^{\text{Neum}}_{S^{p+k}_{\Delta'}}, \Phi_{\hat{e}^{\text{Neum}}_{S^{p+k}_{\Delta'}}})}{\left(\mathscr{E}^{\text{Neum}}_{S^{p+k}_{\Delta'}}\right)^2 - \sum_{j=1}^{N(\Delta)} \|\Phi_{\hat{e}^{\text{Neum}}_{S^{p+k}_{\Delta'}}}\|^2_{\mathcal{U}(I^{\Delta}_j)}}. \tag{6.1.67b}$$

Above we constructed computable upper and lower bounds $\mathscr{E}^U_{\tilde{u}_{S^{p+k}_{\Delta'}}, S^{p+k}_{\Delta'}}$ and $\mathscr{E}^L_{\tilde{u}_{S^{p+k}_{\Delta'}}, S^{p+k}_{\Delta'}}$ for the energy norm of the error in any $\tilde{u}_{S^{p+k}_{\Delta'}} \in S^{p+k}_{\Delta'}$, namely,

$$\mathscr{E}^L_{\tilde{u}_{S^{p+k}_{\Delta'}}, S^{p+k}_{\Delta'}} \le \|u_{EX} - \tilde{u}_{S^{p+k}_{\Delta'}}\|_{\mathcal{U}} \le \mathscr{E}^U_{\tilde{u}_{S^{p+k}_{\Delta'}}, S^{p+k}_{\Delta'}}. \tag{6.1.68a}$$

Letting $\tilde{u}_{S^{p+k}_{\Delta'}} \equiv u_{S^p_\Delta} \in S^p_\Delta \subseteq S^{p+k}_{\Delta'}$, we get the bounds for the energy norm of the error in the finite element solution $u_{S^p_\Delta}$, namely,

$$\mathscr{E}^L_{u_{S^p_\Delta}, S^{p+k}_{\Delta'}} \le \|e_{S^p_\Delta}\|_{\mathcal{U}} = \|u_{EX} - u_{S^p_\Delta}\|_{\mathcal{U}} \le \mathscr{E}^U_{u_{S^p_\Delta}, S^{p+k}_{\Delta'}} \tag{6.1.68b}$$

or by employing $\tilde{u}_{S^{p+k}_{\Delta'}} = \mathcal{A}^*_u(u_{S^p_\Delta}) \in S^{p+k}_{\Delta'}$, we get bounds for the energy norm of the error in the recovered solution $\mathcal{A}^*_u(u_{S^p_\Delta})$, namely

$$\mathscr{E}^L_{\mathcal{A}^*_u(u_{S^p_\Delta}), S^{p+k}_{\Delta'}} \le \|u_{EX} - \mathcal{A}^*_u(u_{S^p_\Delta})\|_{\mathcal{U}} \le \mathscr{E}^U_{\mathcal{A}^*_u(u_{S^p_\Delta}), S^{p+k}_{\Delta'}}. \tag{6.1.68c}$$

Recall that in Section 5.6 we employed

$$\mathscr{E}^{\mathcal{A}^*_u} \stackrel{\text{def}}{=} \|\mathcal{A}^*_u(u_{S^p_\Delta}) - u_{S^p_\Delta}\|_{\mathcal{U}} \tag{6.1.69}$$

as an estimator for $\|e_{S^p_\Delta}\|_{\mathcal{U}}$, the energy norm of the error in the finite element solution, and in Section 5.7 we have seen that $\mathscr{E}^{\mathcal{A}^*_u}$ often underestimates $\|e_{S^p_\Delta}\|_{\mathcal{U}}$. We will now construct guaranteed bounds for $\|e_{S^p_\Delta}\|_{\mathcal{U}}$, by correcting the estimator $\mathscr{E}^{\mathcal{A}^*_u}$.

Theorem 6.1.9 *(Guaranteed bounds based on continuous recoveries $\mathcal{A}_u^*$). Let $\mathcal{A}_u^*$ be a recovery operator characterized by the property*

$$\mathcal{A}_u^*(v) \in S_{\Delta'}^{p+k} \quad \forall v \in S_\Delta^p. \tag{6.1.70a}$$

We then have

$$\begin{aligned}\mathscr{E}_{u_{S_\Delta^p}}^{L,\mathcal{A}_u^*} &\overset{\text{def}}{=} \left|\mathscr{E}^{\mathcal{A}_u^*} - \mathscr{E}^U_{\mathcal{A}_u^*(u_{S_\Delta^p}),S_{\Delta'}^{p+k}}\right| \le \|e_{S_\Delta^p}\|_{\mathcal{U}} \le \mathscr{E}_{u_{S_\Delta^p}}^{U,\mathcal{A}_u^*} \\ &\overset{\text{def}}{=} \mathscr{E}^{\mathcal{A}_u^*} + \mathscr{E}^U_{\mathcal{A}_u^*(u_{S_\Delta^p}),S_{\Delta'}^{p+k}}.\end{aligned} \tag{6.1.69e}$$

Proof Noting that

$$e_{S_\Delta^p} = u_{EX} - \mathcal{A}_u^*(u_{S_\Delta^p}) + \mathcal{A}_u^*(u_{S_\Delta^p}) - u_{S_\Delta^p} \tag{6.1.71a}$$

taking the energy norm of both sides, and employing the triangle inequality, we obtain

$$\max\left(0, \mathscr{E}^{\mathcal{A}_u^*} - \|u_{EX} - \mathcal{A}_u^*(u_{S_\Delta^p})\|_{\mathcal{U}}\right) \le \|e_{S_\Delta^p}\|_{\mathcal{U}} \le \mathscr{E}^{\mathcal{A}_u^*} + \|u_{EX} - \mathcal{A}_u^*(u_{S_\Delta^p})\|_{\mathcal{U}} \tag{6.1.71b}$$

and employing the upper bound

$$\|u_{EX} - \mathcal{A}_u^*(u_{S_\Delta^p})\|_{\mathcal{U}} \le \mathscr{E}^U_{\mathcal{A}_u^*(u_{S_\Delta^p}),S_{\Delta'}^{p+k}} \tag{6.1.71c}$$

we get the bounds (6.1.70b). The bounds $\mathscr{E}^L_{\mathcal{A}_u^*(u_{S_\Delta^p}),S_{\Delta'}^{p+k}}$, $\mathscr{E}^U_{\mathcal{A}_u^*(u_{S_\Delta^p}),S_{\Delta'}^{p+k}}$ are the guaranteed bounds for $\|e_{S_\Delta^p}\|_{\mathcal{U}}$ based on the recovery estimator $\mathscr{E}^{\mathcal{A}_u^*}$. □

Remark 6.1.22 Above we introduced an iterative improvement of the error estimator. We did not prove the convergence of this iteration because we are interested in the reliability interval, the size of which can be controlled numerically.

There are several estimators which cannot be corrected as above. A typical example is the recovery estimator

$$\mathscr{E}^{\mathcal{A}_{du/dx}^*} \overset{\text{def}}{=} \sqrt{\sum_{j=1}^{N(\Delta)} \left\|\sqrt{AE}\left(\mathcal{A}_{du/dx}^*(u_{S_\Delta^p}) - \frac{d}{dx}(u_{S_\Delta^p}\right)\right\|^2_{L^2(I_j^\Delta)}} \tag{6.1.72a}$$

where $\mathcal{A}_{du/dx}^*$, may be constructed as discussed in Section 4.8. The problem is that *there may not exist a $\tilde{u}_{S_{\Delta'}^{p+k}} \in S_{\Delta'}^{p+k}$ such that*

$$\mathscr{E}^{\mathcal{A}_{du/dx}^*} = \|\tilde{u}_{S_{\Delta'}^{p+k}} - u\|_{\mathcal{U}} \tag{6.1.72b}$$

and hence Theorem 6.1.8 cannot be applied. Another example of an estimator is given by

$$\mathscr{E}^{\mathcal{A}_P^*} \overset{\text{def}}{=} \sqrt{\sum_{j=1}^{N(\Delta)} \left\|\mathcal{A}_P^*(u_{S_\Delta^p}) - AE\frac{d}{dx}(u_{S_\Delta^p})\right\|^2_{L^2(I_j^\Delta)}} \tag{6.1.73}$$

where $\mathcal{A}_P^*$ is a recovery for the axial force $P = AE(du/dx)$.

We will now show how we can construct guaranteed bounds based on the ZZ–SPR estimator $\mathscr{E}^{\mathcal{A}_P^*}$. Let us consider the regular model problem with $c \equiv 0$, $\alpha_2 \equiv 0$, namely:

Find $u \in \mathcal{U} \overset{\text{def}}{=} \{v \in L^2(I) \,|\, \|v\|_{\mathcal{U}} \overset{\text{def}}{=} \sqrt{\int_I AE(dv/dx)^2} < \infty,\ v(0) = 0\}$, such that

$$\mathscr{B}(u,v) = \int_I AE\frac{du}{dx}\frac{dv}{dx} = \mathscr{L}(v) = \int_I fv + a_2 v(1) \quad \forall v \in \mathcal{U}. \tag{6.1.74}$$

For simplicity, we will assume that $AE \in C^0(\bar{I})$. Let

$$\mathfrak{V} \overset{\text{def}}{=} \left\{ V \in L^2(I) \,\middle|\, \|V\|_{\mathfrak{V}} \overset{\text{def}}{=} \left\| \frac{1}{\sqrt{AE}} V \right\|_{L^2(I)} < \infty \right\}. \tag{6.1.75}$$

For any $V \in \mathfrak{V}$, we define its derivative $dV/dx \in \mathcal{U}^{-1}$, by

$$\int_I \frac{dV}{dx} v \overset{\text{def}}{=} -\int_I V\frac{dv}{dx} \quad \forall v \in \mathcal{U} \tag{6.1.76}$$

which means that we assume that $V(1) = 0$, weakly, i.e., if V is smooth on I, then $V(1) = 0$.

Remark 6.1.23 In two dimensions $\mathfrak{V}$ is a space of vectors $\mathbf{V}$ (gradient or fluxes) such that instead of dV/dx we have $div(\mathbf{V})$, and instead of $AE(dv/dx)$ we have $AE\nabla v$.

We will split $\mathfrak{V}$, into its subsets

$$\mathfrak{V}_1 \overset{\text{def}}{=} \left\{ V \in \mathfrak{V} \,\middle|\, \exists v \in \mathcal{U} :\ V = AE\frac{dv}{dx} \right\} \tag{6.1.77a}$$

and

$$\mathfrak{V}_2 \overset{\text{def}}{=} \left\{ V \in \mathfrak{V} \,\middle|\, \int_I V\frac{dv}{dx} = 0\ \forall v \in \mathcal{U} \right\}. \tag{6.1.77b}$$

The set $\mathfrak{V}_1$ is the subset of $\mathfrak{V}$ which consist of the functions which can be obtained as fluxes $AE(dv/dx)$, for a test function $v \in \mathcal{U}$. The set $\mathfrak{V}_2$ is the subset of $\mathfrak{V}$ which consists of functions V with zero derivative $(d/dx)(V) = 0$, which is understood in the generalized sense, and $V(1) = 0$, weakly. We have the following orthogonality:

Lemma 6.1.6 *Let $\mathfrak{V}$, $\mathfrak{V}_1$, and $\mathfrak{V}_2$, be defined as above. Then we have*

$$\int_I \frac{1}{AE} UV = 0 \quad \forall U \in \mathfrak{V}_1,\ V \in \mathfrak{V}_2. \tag{6.1.78}$$

Proof Because $U \in \mathfrak{V}_1$, there exists a $u \in \mathcal{U}$, such that

$$U = AE\frac{du}{dx} \tag{6.1.79a}$$

and hence

$$\int_I \frac{1}{AE} UV = \int_I \frac{du}{dx} V = 0 \quad \forall V \in \mathfrak{V}_2. \tag{6.1.79b}$$

□

We define $\mathcal{A}^*_P(u_{S^p_\Delta})$ similarly as $\mathcal{A}^*_{du/dx}$, which was introduced in Section 4.8, to be such that

$$\mathcal{A}^*_P(u_{S^p_\Delta}) \in S^p_\Delta \tag{6.1.80}$$

and we will employ the notations

$$\hat{e}^{\mathcal{A}^*_P} \stackrel{\text{def}}{=} \mathcal{A}^*_P(u_{S^p_\Delta}) - AE\frac{d}{dx}(u_{S^p_\Delta}) \tag{6.1.81}$$

for the ZZ error indicator function, and

$$e^{\mathcal{A}^*_P} \stackrel{\text{def}}{=} AE\frac{d}{dx}(u_{EX}) - \mathcal{A}^*_P(u_{S^p_\Delta}) = AE\frac{d}{dx}(e_{S^p_\Delta}) - \hat{e}^{\mathcal{A}^*_P} \tag{6.1.82}$$

for the error in $\mathcal{A}^*_P$. We then have

$$\frac{d}{dx}(e^{\mathcal{A}^*_P}) = \frac{d}{dx}\Big(AE\frac{d}{dx}(u_{EX}) - \mathcal{A}^*_P(u_{S^p_\Delta})\Big) = -f - \frac{d}{dx}(\mathcal{A}^*_P(u_{S^p_\Delta})). \tag{6.1.83}$$

Let $w^{\mathcal{A}^*_P}$, be the solution of the variational problem:

Find $w^{\mathcal{A}^*_P} \in \mathcal{U}$, such that

$$\mathcal{B}(w^{\mathcal{A}^*_P}, v) = \int_I -f - \frac{d}{dx}(\mathcal{A}^*_P(u_{S^p_\Delta}))v - (a_2 - \mathcal{A}^*_P(u_{S^p_\Delta})(1))v(1). \tag{6.1.84}$$

It follows that $w^{\mathcal{A}^*_P}$ is the solution of the boundary value problem

$$-\frac{d}{dx}\Big(AE\frac{d}{dx}(w^{\mathcal{A}^*_P})\Big) = \frac{d}{dx}(e^{\mathcal{A}^*_P}) = -f - \frac{d}{dx}(\mathcal{A}^*_P(u_{S^p_\Delta})) \tag{6.1.85a}$$

with the boundary conditions

$$w^{\mathcal{A}^*_P}(0) = 0, \qquad AE(1)\frac{d}{dx}(w^{\mathcal{A}^*_P})(1) = -a_2 + \mathcal{A}^*_P(u_{S^p_\Delta})(1). \tag{6.1.85b}$$

From (6.1.85a,b) it follows that

$$\frac{d}{dx}\Big(AE\frac{d}{dx}(w^{\mathcal{A}^*_P}) + e^{\mathcal{A}^*_P}\Big) = 0 \tag{6.1.86a}$$

and

$$\begin{aligned} 0 &= \Big(AE\frac{d}{dx}(w^{\mathcal{A}^*_P})\Big)(1) + a_2 - \mathcal{A}^*_P(u_{S^p_\Delta})(1) \\ &= \Big(AE\frac{d}{dx}(w^{\mathcal{A}^*_P})\Big)(1) + \Big(AE\frac{d}{dx}(u_{EX})\Big)(1) - \mathcal{A}^*_P(u_{S^p_\Delta})(1) \\ &= \Big(AE\frac{d}{dx}(w^{\mathcal{A}^*_P}) + e^{\mathcal{A}^*_P}\Big)(1) \end{aligned} \tag{6.1.86b}$$

hence

$$AE\frac{d}{dx}(w^{\mathcal{A}^*_P}) + e^{\mathcal{A}^*_P} \in \mathfrak{B}_2 \tag{6.1.86c}$$

and from Lemma 6.1.6, we get

$$\int_I \left(AE \frac{d}{dx}(w^{\mathcal{A}_P^*}) + e^{\mathcal{A}_P^*} \right) \frac{dv}{dx} = 0 \quad \forall v \in \mathcal{U} \tag{6.1.87a}$$

which can also be written as

$$\int_I \left(AE \frac{d}{dx}(w^{\mathcal{A}_P^*}) + AE \frac{d}{dx}(e_{S_\Delta^p}) - \hat{e}^{\mathcal{A}_P^*} \right) \frac{dv}{dx} = 0 \quad \forall v \in \mathcal{U} \tag{6.1.87b}$$

or

$$\mathcal{B}(w^{\mathcal{A}_P^*}, v) + \mathcal{B}(e_{S_\Delta^p}, v) - \mathcal{B}(g^{\mathcal{A}_P^*}, v) = 0 \quad \forall v \in \mathcal{U} \tag{6.1.87c}$$

where $g^{\mathcal{A}_P^*} \in \mathcal{U}$ such that

$$\mathcal{B}(g^{\mathcal{A}_P^*}, v) = \int_I \hat{e}^{\mathcal{A}_P^*} \frac{dv}{dx} = \int_I \frac{\hat{e}^{\mathcal{A}_P^*}}{\sqrt{AE}} \sqrt{AE} \frac{dv}{dx} \quad \forall v \in \mathcal{U}. \tag{6.1.87d}$$

Hence we get

$$-e_{S_\Delta^p} = w^{\mathcal{A}_P^*} - g^{\mathcal{A}_P^*}. \tag{6.1.88}$$

From (6.1.87d) it follows that

$$\|g^{\mathcal{A}_P^*}\|_{\mathcal{E}} \le \sqrt{\int_I \frac{1}{AE} (\hat{e}^{\mathcal{A}_P^*})^2} \stackrel{\text{def}}{=} \mathcal{E}^{\mathcal{A}_P^*} \tag{6.1.89a}$$

and hence we have from (6.1.88)

$$\|e_{S_\Delta^p}\|_{\mathcal{U}} \le \|w^{\mathcal{A}_P^*}\|_{\mathcal{U}} + \mathcal{E}^{\mathcal{A}_P^*} \tag{6.1.89b}$$

and

$$\|e_{S_\Delta^p}\|_{\mathcal{U}} \ge |\mathcal{E}^{\mathcal{A}_P^*} - \|w^{\mathcal{A}_P^*}\|_{\mathcal{U}}|. \tag{6.1.89c}$$

$\|w^{\mathcal{A}_P^*}\|_{\mathcal{U}}$ is not computable but we have

$$\|w^{\mathcal{A}_P^*}\|_{\mathcal{U}} = \sqrt{\left\| w_{S_\Delta^p}^{\mathcal{A}_P^*} \right\|_{\mathcal{U}}^2 + \left\| w^{\mathcal{A}_P^*} - w_{S_\Delta^p}^{\mathcal{A}_P^*} \right\|_{\mathcal{U}}^2} \tag{6.1.90a}$$

where $w_{S_\Delta^p}^{\mathcal{A}_P^*} \in S_\Delta^p$ is the finite element approximation of $w^{\mathcal{A}_P^*}$, which is computable by using the additional right-hand side given in (6.1.84) and

$$\left\| w^{\mathcal{A}_P^*} - w_{S_\Delta^p}^{\mathcal{A}_P^*} \right\|_{\mathcal{U}} \le \mathcal{E}^U_{w_{S_\Delta^p}^{\mathcal{A}_P^*}, S_{\Delta'}^{p+k}}. \tag{6.1.90b}$$

Hence from (6.1.89b,c) we get

$$\max\left(0, \mathscr{E}^{\mathcal{A}_P^*} - \sqrt{\left\|w_{S_\Delta^p}^{\mathcal{A}_P^*}\right\|_{\mathfrak{U}}^2 + \left(\mathscr{E}^U_{w_{S_\Delta^p}^{\mathcal{A}_P^*}, S_{\Delta'}^{p+k}}\right)^2}\right)$$

$$\leq \|e_{S_\Delta^p}\|_{\mathfrak{U}} \leq \mathscr{E}^{\mathcal{A}_P^*}. + \sqrt{\left\|w_{S_\Delta^p}^{\mathcal{A}_P^*}\right\|_{\mathfrak{U}}^2 + \left(\mathscr{E}^U_{w_{S_\Delta^p}^{\mathcal{A}_P^*}, S_{\Delta'}^{p+k}}\right)^2} \qquad (6.1.91)$$

We can improve the estimate by computation of $(w^{\mathcal{A}_P^*} - g^{\mathcal{A}_P^*})$ from (6.1.84) and (6.1.87d). It is easy to see that $(w^{\mathcal{A}_P^*} - g^{\mathcal{A}_P^*})_{S_\Delta^p} = 0$, and hence

$$\mathscr{E}^L_{(w^{\mathcal{A}_P^*} - g^{\mathcal{A}_P^*})_{S_\Delta^p}, S_{\Delta'}^{p+k}} \leq \|e_{S_\Delta^p}\|_{\mathfrak{U}} \leq \mathscr{E}^U_{(w^{\mathcal{A}_P^*} - g^{\mathcal{A}_P^*})_{S_\Delta^p}, S_{\Delta'}^{p+k}} \qquad (6.1.92)$$

are the guaranteed upper and lower estimates. Note that $\mathscr{E}^L_{(w^{\mathcal{A}_P^*} - g^{\mathcal{A}_P^*})_{S_\Delta^p}, S_{\Delta'}^{p+k}}$ and $\mathscr{E}^U_{(w^{\mathcal{A}_P^*} - g^{\mathcal{A}_P^*})_{S_\Delta^p}, S_{\Delta'}^{p+k}}$ are not zero, although $(w^{\mathcal{A}_P^*} - g^{\mathcal{A}_P^*})_{S_\Delta^p} = 0$. The estimate (6.1.91) is the guaranteed estimate of the error in the ZZ estimator. The left-hand side of (6.1.91) is very close to zero and hence the lower bound is very ineffective. On the right-hand side of (6.1.91) we have

$$\sqrt{\left\|w_{S_\Delta^p}^{\mathcal{A}_P^*}\right\|_{\mathfrak{U}}^2 + \left(\mathscr{E}^U_{w_{S_\Delta^p}^{\mathcal{A}_P^*}, S_{\Delta'}^{p+k}}\right)^2} \approx \mathscr{E}^{\mathcal{A}_P^*} \qquad (6.1.93)$$

and hence the guaranteed error estimate in the ZZ estimator leads to a doubling of the effectivity index. The estimate (6.1.92) is not the ZZ estimator but it uses the SPR postprocessing and leads to a better estimate than the ZZ estimator, although of course at higher expense.

Problems

P6.1.1. Construct an upper estimator for $\|e_{S_\Delta^p}\|_{\mathfrak{U}}$ which is based on the Dirichlet estimator $\mathscr{E}^{Dir}$.

Hint: Recalling the orthogonality

$$\mathscr{B}\left(e_{S_\Delta^p} - \hat{e}^{Dir}, \hat{e}^{Dir}\right) = 0 \qquad (6.1.94a)$$

we get

$$\|e_{S_\Delta^p}\|_{\mathfrak{U}} = \sqrt{\|\hat{e}^{Dir}\|_{\mathfrak{U}}^2 + \|e_{S_\Delta^p} - \hat{e}^{Dir}\|_{\mathfrak{U}}^2}. \qquad (6.1.94b)$$

Now let

$$e^{Dir} \stackrel{\text{def}}{=} e_{S_\Delta^p} - \hat{e}^{Dir} \qquad (6.1.95a)$$

and define $e^{Dir}_{S^p_\Delta}$ as the solution of the discrete problem:

Find $e^{Dir}_{S^p_\Delta} \in S^p_\Delta$ such that

$$\mathcal{B}\big(e^{Dir}_{S^p_\Delta}, v\big) = \mathcal{B}(e^{Dir}, v) = \sum_{j=1}^{N(\Delta)} [\![\hat{e}^{Dir}]\!](x_j^\Delta)\; v(x_j^\Delta) \quad \forall v \in S^p_\Delta. \tag{6.1.95b}$$

The determination of $e^{Dir}_{S^p_\Delta}$ is inexpensive, because it employs the stiffness matrix which has already been factorized during the determination of $u_{S^p_\Delta}$. Using now the explicit upper estimator discussed in Chapter 5, we obtain

$$\big\|e^{Dir} - e^{Dir}_{S^p_\Delta}\big\|_{\mathcal{U}} \le \mathcal{E}^{EXPL}_{e^{Dir}_{S^p_\Delta}} \tag{6.1.96}$$

and we get

$$\mathcal{E}^{Dir} \le \|e_{S^p_\Delta}\|_{\mathcal{U}} \le \sqrt{(\mathcal{E}^{Dir})^2 + \big\|e^{Dir}_{S^p_\Delta}\big\|^2_{\mathcal{U}} + \big(\mathcal{E}^{EXPL}_{e^{Dir}_{S^p_\Delta}}\big)^2}. \tag{6.1.97}$$

P6.1.2. Recall that in Theorem 5.2.2 we have proven the two-sided estimate

$$\mathcal{E}^{Dir} \le \|e_{S^p_\Delta}\|_{\mathcal{U}} \le \mathcal{K}^{Dir}\,\mathcal{E}^{Dir} \tag{6.1.98}$$

where $\mathcal{K}^{Dir}$ is a computable constant, and $\mathcal{E}^{Dir}$ is the exact Dirichlet estimator. Use this result to construct a computable two-sided estimate based on the computed Dirichlet estimator $\mathcal{E}^{Dir}_{S^{p+k}_{\Delta'}}$.

Hint: Note that

$$\|\hat{e}^{Dir}\|^2_{\mathcal{U}} = \big\|\hat{e}^{Dir}_{S^{p+k}_{\Delta'}}\big\|^2_{\mathcal{U}} + \big\|\hat{e}^{Dir} - \hat{e}^{Dir}_{S^{p+k}_{\Delta'}}\big\|^2_{\mathcal{U}} \tag{6.1.99a}$$

and employ an element by element explicit upper estimate

$$\big\|\hat{e}^{Dir} - \hat{e}^{Dir}_{S^{p+k}_{\Delta'}}\big\|^2_{\mathcal{U}(I_j^\Delta)} \le \mathcal{E}^{EXPL}_{\hat{e}^{Dir}_{S^{p+k}_{\Delta'}}/I_j^\Delta}. \tag{6.1.99b}$$

P6.1.3. Let $\check{e}^{\mathrm{Neum}}_{I_j^\Delta}$ denote the Neumann error indicator function which is the solution of the element residual problem:

Find $\check{e}^{\mathrm{Neum}}_{I_j^\Delta} \in \mathcal{U}(I_j^\Delta)$ such that

$$\int_{I_j^\Delta} \frac{d}{dx}\big(\check{e}^{\mathrm{Neum}}_{I_j^\Delta}\big)\frac{dv}{dx} = \mathcal{R}^{EQ}_{I_j^\Delta}(v) \quad \forall v \in \mathcal{U}(I_j^\Delta). \tag{6.1.100}$$

Show that $\check{e}^{\mathrm{Neum}}_{I_j^\Delta}$ exists up to an additive constant. Let

$$\check{\mathcal{E}}^{\mathrm{Neum}} \stackrel{\mathrm{def}}{=} \sqrt{\sum_{j=1}^{N(\Delta)} \Big\|\frac{d}{dx}\big(\check{e}^{\mathrm{Neum}}_{I_j^\Delta}\big)\Big\|^2_{L^2(I_j^\Delta)}} \tag{6.1.101a}$$

and

$$\breve{\mathscr{E}}^{\text{Neum}*} \stackrel{\text{def}}{=} \sqrt{\sum_{j=1}^{N(\Delta)} \left\| \frac{1}{\sqrt{\min\limits_{x\in I_j^\Delta} AE(x)}} \frac{d}{dx}\left(\breve{e}^{\text{Neum}}_{I_j^\Delta}\right) \right\|_{L^2(I_j^\Delta)}}. \tag{6.1.101b}$$

What can be proven about the relation between $\|e_{S^p_\Delta}\|_{\mathscr{U}}$, and $\breve{\mathscr{E}}^{\text{Neum}}$ (respectively $\breve{\mathscr{E}}^{\text{Neum}*}$)?

6.1* Upper and lower bounds for the energy norm of the error in two dimensions

We will now give the construction of the bounds for $\|e_{S^p_\Delta}\|_{\mathscr{U}}$ in two dimensions. We will follow the same steps as in Section 6.1, and we will focus mainly on the differences between the one- and two-dimensional cases.

We will give a general construction of the bounds based on *Neumann subdomain* or *patch residual problems.* We will let $\Delta_h \stackrel{\text{def}}{=} \{\tau_j^{\Delta_h}\}_{j=1}^{N(\Delta_h)}$ denote the finite element mesh, and $\Omega_h \stackrel{\text{def}}{=} \{\omega_j\}_{j=1}^{N(\Omega_h)}$ denote a partition of the domain Ω, into non-overlapping patches ω_j, each one of which consists of one or several elements, namely,

$$\bar{\Omega} = \bigcup_{\omega_j\in\Omega_h} \bar{\omega}_j, \quad \bar{\omega}_j = \bigcup_{\tau_k\subseteq\omega_j} \bar{\tau}_k, \quad \omega_j\cap\omega_k = \emptyset, \qquad j\neq k. \tag{6.1*.1}$$

The subscript h differentiates the partition Ω_h from the domain Ω. Examples of two possible partitions Ω_h of one of the the domains Ω, are shown in Fig. 6.1*.1. We will always employ curved quadrilateral elements and bi-p element shape functions.

The construction of the bounds will be based on the computed error indicators $\hat{e}^{\text{Neum}}_{\omega_j,S^{p+k}_{\Delta'}}$, which are the solutions of the following discrete *Neumann patch residual* problems:

Find $\hat{e}^{\text{Neum}}_{\omega_j,S^{p+k}_{\Delta'}} \in S^{p+k}_{\Delta',\Gamma_D}(\omega_j)$ such that

$$\mathscr{B}_{\omega_j}\left(\hat{e}^{\text{Neum}}_{\omega_j,S^{p+k}_{\Delta'}}, v\right) = \mathscr{R}^{EQ,p}_{\omega_j}(v) \stackrel{\text{def}}{=} \sum_{\tau_k\subseteq\omega_j} \mathscr{R}^{EQ,p}_{\tau_k}(v) \quad \forall v\in S^{p+k}_{\Delta',\Gamma_D}(\omega_j) \tag{6.1*.2a}$$

where $S^{p+k}_{\Delta',\Gamma_D}(\omega_j)$, is the restriction of the finite element space $S^{p+k}_{\Delta'}$ into ω_j, and the subscript Γ_D is used to underline that the functions in $S^{p+k}_{\Delta',\Gamma_D}(\omega_j)$ satisfy homogenous Dirichlet conditions on $\partial\omega_j\cap\Gamma_D$. In the cases that $\partial\omega_j\cap\Gamma_D = \emptyset$, and the zeroth order terms in the partial differential equation vanish, we fix $\hat{e}^{\text{Neum}}_{\omega_j,S^{p+k}_{\Delta'}}$, by imposing the additional condition

$$\int_{\omega_j} \hat{e}^{\text{Neum}}_{\omega_j,S^{p+k}_{\Delta'}} = 0. \tag{6.1*.2b}$$

Replacing $S^{p+k}_{\Delta',\Gamma_D}(\omega_j)$ by $\mathscr{U}_{\Gamma_D}(\omega_j)$, the restriction of $\mathscr{U}$ on ω_j, we obtain the definition of the corresponding *exact error indicator* function $\hat{e}^{\text{Neum}}_{\omega_j}$.

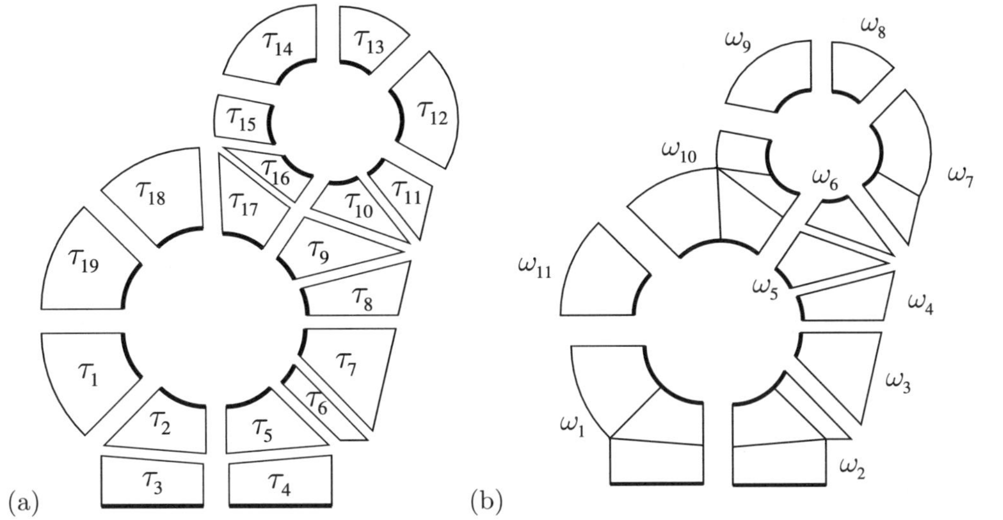

FIG. 6.1*.1. *Patch residual problems.* Example of partitions of a mesh into patches of elements. (a) The patches are identical with the elements; (b) The patches consist of the vertex patches of elements which are connected to the re-entrant corners, and the elements which remain after the vertex patches have been formed

Remark 6.1*.1 In two dimensions the major contribution to the error comes from the 'key' elements which are located at the re-entrant corners where the singularity of the solution is the strongest. The accuracy of the estimators in the key elements can be increased by employing patch residual problems in the 'key' patches, which consist of the elements which share the vertex at the re-entrant corner, as shown in Figure 6.1*.1b.

We have the following generalization of Theorem 6.1.1.

Theorem 6.1*.1 *(The exact estimator $\mathcal{E}^{\text{Neum}}_{\Omega_h}$ and the computed estimator $\mathcal{E}^{\text{Neum}}_{\Omega_h, S^{p+k}_{\Delta'}}$) Under the assumptions which guarantee the existence of $\mathcal{E}^{\text{Neum}}$, we have*

$$\left\| e_{S^p_\Delta}^{S^{p+k}_{\Delta'}} \right\|_{\mathcal{U}} \le \|e_{S^p_\Delta}\|_{\mathcal{U}} \le \mathcal{E}^{\text{Neum}}_{\Omega_h} \stackrel{\text{def}}{=} \sqrt{\sum_{\omega_j \in \Omega_h} \|\hat{e}^{\text{Neum}}_{\omega_j}\|^2_{\mathcal{U}(\omega_j)}} \qquad (6.1^*.3)$$

and

$$\left\| e_{S^p_\Delta}^{S^{p+k}_{\Delta'}} \right\|_{\mathcal{U}} \le \mathcal{E}^{\text{Neum}}_{\Omega_h, S^{p+k}_{\Delta'}} \stackrel{\text{def}}{=} \sqrt{\sum_{\omega_j \in \Omega_h} \left\| \hat{e}^{\text{Neum}}_{\omega_j, S^{p+k}_{\Delta'}} \right\|^2_{\mathcal{U}(\omega_j)}} \le \mathcal{E}^{\text{Neum}}_{\Omega_h}. \qquad (6.1^*.4)$$

Proof Replacing the sums over elements in the proof of Theorem 6.1.1, with sums over patches, we obtain

$$\|e_{S^p_\Delta}\|_{\mathcal{U}} = \sup_{v \in \mathcal{U}} \frac{\mathcal{R}(v)}{\|v\|_{\mathcal{U}}} \le \sqrt{\sum_{\omega_j \in \Omega_h} \|\mathcal{R}^{EQ,p}_{\omega_j}\|^2_{L(\mathcal{U}_{\Gamma_D}(\omega_j), \mathbb{R})}} = \mathcal{E}^{\text{Neum}}_{\Omega_h} \qquad (6.1^*.5a)$$

where we employed that

$$\eta_{\omega_j}^{\text{Neum}} \stackrel{\text{def}}{=} \|\hat{e}_{\omega_j}^{\text{Neum}}\|_{\mathcal{U}(\omega_j)} = \|\mathcal{R}_{\omega_j}^{EQ,p}\|_{L(\mathcal{U}_{\Gamma_D}(\omega_j),\mathbb{R})} \tag{6.1*.5b}$$

and similarly for the computed estimator $\mathcal{E}_{\Omega_h,S_{\Delta'}^{p+k}}^{\text{Neum}}$. □

Lemma 6.1.1 can be generalized as follows:

Lemma 6.1*.1 *(The relation between the exact estimator $\mathcal{E}_{\Omega_h}^{\text{Neum}}$ and the computed estimator $\mathcal{E}_{\Omega_h,S_{\Delta'}^{p+k}}^{\text{Neum}}$) Under the assumptions which guarantee the existence of the Neumann estimator $\mathcal{E}^{\text{Neum}}$, we have*

$$\mathcal{E}_{\Omega_h}^{\text{Neum}} = \sqrt{\left(\mathcal{E}_{\Omega_h,S_{\Delta'}^{p+k}}^{\text{Neum}}\right)^2 + \sum_{\omega_j\in\Omega_h} \|\hat{e}_{\omega_j}^{\text{Neum}} - \hat{e}_{\omega_j,S_{\Delta'}^{p+k}}^{\text{Neum}}\|_{\mathcal{U}}^2}. \tag{6.1*.6}$$

Proof It follows from the orthogonalities

$$\mathcal{B}_{\omega_j}\left(\hat{e}_{\omega_j}^{\text{Neum}} - \hat{e}_{\omega_j,S_{\Delta'}^{p+k}}^{\text{Neum}}, v\right) = 0 \quad \forall v \in S_{\Delta',\Gamma_D}^{p}(\omega_j). \tag{6.1*.7}$$

□

As in the one-dimensional case, we will construct computable upper bounds for $\|e_{S_\Delta^p}\|_{\mathcal{U}}$ by employing computable upper bounds for $\|\hat{e}_{\omega_j}^{\text{Neum}} - \hat{e}_{\omega_j,S_{\Delta'}^{p+k}}^{\text{Neum}}\|_{\mathcal{U}(\omega_j)}$, the energy norm of the error in the computed error indicator $\hat{e}_{\omega_j,S_{\Delta'}^{p+k}}^{\text{Neum}}$, into (6.1*.6). Employing the partition Ω_h which patches ω_j coincide with elements of the mesh Δ, letting $\Delta' = \Delta$, and considering the case of Poisson's equation, we have:

Lemma 6.1*.2 *(Explicit estimator for $\|\hat{e}_{\tau}^{\text{Neum}} - \hat{e}_{\tau,S_{\Delta}^{p+k}}^{\text{Neum}}\|_{\mathcal{U}(\tau)}$ the energy norm of the error in the computed error indicator $\hat{e}_{\tau,S_{\Delta}^{p+k}}^{\text{Neum}}$) Under the assumptions which guarantee the existence of $\mathcal{E}^{\text{Neum}}$, we have*

$$\|\hat{e}_{\tau}^{\text{Neum}} - \hat{e}_{\tau,S_{\Delta}^{p+k}}^{\text{Neum}}\|_{\mathcal{U}(\tau)} \le \mathcal{E}_{\hat{e}_{S_{\Delta}^{p+k}}^{\text{Neum}}/\tau}^{EXPL} \tag{6.1*.8a}$$

with

$$\mathcal{E}_{\hat{e}_{S_{\Delta}^{p+k}}^{\text{Neum}}/\tau}^{EXPL} = \mathcal{C}_1^{\tau}\|r_{\tau/\hat{e}_{S_{\Delta}^{p+k}}^{\text{Neum}}}\|_{L^2(\tau)} + \sum_{\epsilon\subseteq\partial\tau} \mathcal{C}_2^{\epsilon}\|J_{\epsilon/\hat{e}_{S_{\Delta}^{p+k}}^{\text{Neum}}}^{\tau}\|_{L^2(\epsilon)} \tag{6.1*.8b}$$

where $r_{\tau/\hat{e}_{S_{\Delta}^{p+k}}^{\text{Neum}}}$, is the interior residual, in the element τ, and $J_{\epsilon/\hat{e}_{S_{\Delta}^{p+k}}^{\text{Neum}}}^{\tau}$, is the jump on the edge ϵ of the element τ, corresponding to the residuum

$$\mathcal{R}_{\tau/\hat{e}_{S_{\Delta}^{p+k}}^{\text{Neum}}}^{EQ,p}(v) \stackrel{\text{def}}{=} \mathcal{R}_{\tau}^{EQ,p}(v) - \mathcal{B}_{\tau}(\hat{e}_{S_{\Delta}^{p+k}}^{\text{Neum}}, v) \quad \forall v \in \mathcal{U}(\tau) \tag{6.1*.8c}$$

namely,

$$\mathcal{R}_{\tau/\hat{e}_{S_{\Delta}^{p+k}}^{\text{Neum}}}^{EQ,p}(v) = \int_{\tau} r_{\tau/\hat{e}_{S_{\Delta}^{p+k}}^{\text{Neum}}}v + \sum_{\epsilon\subseteq\partial\tau}\int_{\epsilon} J_{\epsilon/\hat{e}_{S_{\Delta}^{p+k}}^{\text{Neum}}}^{\tau}v \quad \forall v \in \mathcal{U}(\tau). \tag{6.1*.8d}$$

In particular, in the case of the Poisson equation

$$-\nabla u\Big|_{\Omega} = f, \qquad \frac{\partial u}{\partial n}\Big|_{\partial\Omega} = g \tag{6.1*.9a}$$

we have

$$r_{\tau/\hat{e}^{\mathrm{Neum}}_{S^{p+k}_{\Delta}}} = r_{\tau} + \Delta \hat{e}^{\mathrm{Neum}}_{S^{p+k}_{\Delta}} \tag{6.1*.9b}$$

and

$$J^{\tau}_{\epsilon/\hat{e}^{\mathrm{Neum}}_{S^{p+k}_{\Delta}}} = J^{\tau}_{\epsilon} - \frac{\partial}{\partial n}\left(\hat{e}^{\mathrm{Neum}}_{S^{p+k}_{\Delta}}\right) \tag{6.1*.9c}$$

where r_{τ}, is the interior residual, and J^{τ}_{ϵ}, the split jump, which were defined in Section 5.1.*

The constants $\mathscr{C}^{\tau}_1$, and $\mathscr{C}^{\tau}_2$, are estimated by

$$\mathscr{C}^{\tau}_1 \le \frac{1}{\sqrt{\mathscr{K}_{\tau}}}\sqrt{\|det(\mathbf{J}_{\tau})\|_{L^{\infty}(\tau)}}\ \tilde{\mathscr{C}}_1 \tag{6.1*.10a}$$

and

$$\mathscr{C}^{\tau}_2 \le \frac{1}{\sqrt{\mathscr{K}_{\tau}}}\sqrt{\|det(\mathscr{J}_{\epsilon})\|_{L^{\infty}(\tau)}}\ \tilde{\mathscr{C}}_2 \tag{6.1*.10b}$$

where $\mathbf{J}_{\tau}$ (respectively $\mathscr{J}_{\epsilon}$) is the Jacobian matrix of the element (respectively edge) transformation,

$$\mathscr{K}_{\tau} = \min_{\mathbf{x}\in\tau}(det(\mathbf{J}_{\tau})(\mathbf{x})\lambda_{min}(\mathbf{J}^{-T}_{\tau}\mathbf{J}^{-1}_{\tau}(\mathbf{x}))) \tag{6.1*.10c}$$

where $\lambda_{min}(\mathbf{J}^{-T}_{\tau}\mathbf{J}^{-1}_{\tau})$, is the minimum eigenvalue of the matrix $\mathbf{J}^{-T}_{\tau}\mathbf{J}^{-1}_{\tau}$, and

$$\tilde{\mathscr{C}}_1 \overset{\mathrm{def}}{=} \sup_{\tilde{v}\in\tilde{\mathscr{U}}(\tilde{\tau})} \frac{\|\tilde{v}\|_{L^2(\tilde{\tau})}}{\|\tilde{v}\|_{\mathscr{U}(\tilde{\tau})}} = \frac{2}{\pi} \tag{6.1*.10d}$$

and

$$\tilde{\mathscr{C}}_2 \overset{\mathrm{def}}{=} \sup_{\tilde{v}\in\tilde{\mathscr{U}}(\tilde{\tau})} \frac{\|\tilde{v}\|_{L^2(\tilde{\epsilon})}}{\|\tilde{v}\|_{\mathscr{U}(\tilde{\tau})}} = \sqrt{\frac{2}{3}} \tag{6.1*.10e}$$

for $\partial\tau \cap \Gamma_D = \emptyset$, we have

$$\tilde{\mathscr{U}}(\tilde{\tau}) \overset{\mathrm{def}}{=} \{\tilde{v} \mid \|\tilde{v}\|_{\tilde{\mathscr{U}}(\tilde{\tau})} = \sqrt{\textstyle\int_{\tilde{\tau}} |\nabla\tilde{v}|^2} < \infty,\ \textstyle\int_{\tilde{\tau}} \tilde{v} = 0\}. \tag{6.1*.10f}$$

In the case that the intersection $\partial\tau \cap \Gamma_D \neq \emptyset$ has non zero length, the condition $\int_{\tilde{I}} \tilde{v} = 0$, is omitted.

Proof As in the one-dimensional case, we will denote the error in the computed indicator $\hat{e}^{\text{Neum}}_{S^{p+k}_\Delta}$ by

$$e_{\hat{e}^{\text{Neum}}_{S^{p+k}_\Delta}} \stackrel{\text{def}}{=} \hat{e}^{\text{Neum}} - \hat{e}^{\text{Neum}}_{S^{p+k}_\Delta}. \tag{6.1*.11a}$$

We then have

$$\mathcal{B}_\tau(e_{\hat{e}^{\text{Neum}}_{S^{p+k}_\Delta}}, v) = \mathcal{R}^{EQ,p}_{\tau/\hat{e}^{\text{Neum}}_\tau}(v) \stackrel{\text{def}}{=} \mathcal{R}^{EQ,p}_\tau(v) - \mathcal{B}_\tau\big(\hat{e}^{\text{Neum}}_{\tau,S^{p+k}_\Delta}, v\big)$$
$$\forall v \in \mathcal{U}_{\partial\tau_D}(\tau) \tag{6.1*.11b}$$

and

$$\mathcal{B}_\tau(e_{\hat{e}^{\text{Neum}}_{S^{p+k}_\Delta}}, v) = \mathcal{R}^{EQ,p}_{\tau,u_{S^p_\Delta}/\hat{e}^{\text{Neum}}_{S^{p+k}_\Delta}}(v) = 0 \quad \forall v \in S^{p+k}_{\Delta,\partial\tau_D}(\tau) \tag{6.1*.11c}$$

and for each element τ with $\partial\tau_D = \partial\tau \cap \Gamma_D = \emptyset$. We also have

$$\int_\tau e_{\hat{e}^{\text{Neum}}_{S^{p+k}_\Delta}} = \underbrace{\int_\tau \hat{e}^{\text{Neum}}}_{0} - \underbrace{\int_\tau \hat{e}^{\text{Neum}}_{S^{p+k}_\Delta}}_{0} = 0 \tag{6.1*.11d}$$

As in the one-dimensional case, we have

$$\|e_{\hat{e}^{\text{Neum}}_{S^{p+k}_\Delta}}\|_{\mathcal{U}(\tau)} = \sup_{\substack{v\in\mathcal{U} \\ \int_\tau v=0}} \frac{\mathcal{R}^{EQ,p}_{\tau/\hat{e}^{\text{Neum}}_{S^{p+k}_\Delta}}(v)}{\|v\|_{\mathcal{U}(\tau)}} \tag{6.1*.12a}$$

Integrating by parts the term $\mathcal{B}_\tau(\hat{e}^{\text{Neum}}_{S^{p+k}_\Delta}, v)$, we obtain

$$\mathcal{R}^{EQ,p}_{\tau/\hat{e}^{\text{Neum}}_{S^{p+k}_\Delta}}(v) = \int_\tau \big(r_\tau + \Delta\hat{e}^{\text{Neum}}_{S^{p+k}_\Delta}\big)v + \sum_{\varepsilon\subset\partial\tau}\int_\varepsilon \left(J^\tau_\varepsilon - \frac{\partial}{\partial n_\varepsilon}\hat{e}^{\text{Neum}}_{S^{p+k}_\Delta}\right)v \tag{6.1*.12b}$$

and employing the triangle, and the Cauchy–Schwarz inequality, we obtain

$$\mathcal{R}^{EQ,p}_{\tau/\hat{e}^{\text{Neum}}_{S^{p+k}_\Delta}}(v) \le \|r_\tau + \Delta\hat{e}^{\text{Neum}}_{S^{p+k}_\Delta}\|_{L^2(\tau)}\|v\|_{L^2(\tau)}$$
$$+ \sum_{\varepsilon\subset\partial\tau}\left\|J^\tau_\varepsilon - \frac{\partial}{\partial n_\varepsilon}\hat{e}^{\text{Neum}}_{S^{p+k}_\Delta}\right\|_{L^2(\varepsilon)}\|v\|_{L^2(\varepsilon)} \tag{6.1*.12c}$$

and substituting in (6.1*.12a) we get

$$\|e_{\hat{e}^{\text{Neum}}_{S^{p+k}_\Delta}}\|_{\mathcal{U}_\tau} \le \|r_\tau + \Delta\hat{e}^{\text{Neum}}_{S^{p+k}_\Delta}\|_{L^2(\tau)} \sup_{\substack{v\in\mathcal{U} \\ \int_\tau v=0}} \frac{\|v\|_{L^2(\tau)}}{\|v\|_{\mathcal{U}(\tau)}}$$
$$+ \sum_{\varepsilon\subset\partial\tau}\left\|J^\tau_\varepsilon - \frac{\partial}{\partial n_\varepsilon}\hat{e}^{\text{Neum}}_{S^{p+k}_\Delta}\right\|_{L^2(\varepsilon)} \sup_{\substack{v\in\mathcal{U} \\ \int_\tau v=0}} \frac{\|v\|_{L^2(\varepsilon)}}{\|v\|_{\mathcal{U}(\tau)}} \tag{6.1*.12d}$$

from which we obtain (6.1*.8b). The rest of the proof follows from the inequalities given in Lemma 6.1*.2a below. □

Lemma 6.1*.3 *For any $v \in \mathcal{U}(\tau)$, we have*

$$\|v\|_{L^2(\tau)} \leq \sqrt{\|det(\mathbf{J}_\tau)\|_{\mathbf{L}^\infty(\tau)}}\|\tilde{v}\|_{L^2(\tilde{\tau})} \tag{6.1*.13a}$$

and

$$\|v\|_{L^2(\varepsilon)} \leq \sqrt{\|det(\mathcal{J}_\varepsilon)\|_{L^\infty(\varepsilon)}}\,\|\tilde{v}\|_{L^2(\tilde{\varepsilon})} \tag{6.1*.13b}$$

and

$$\sqrt{\min_{\mathbf{x}\in\tau}(det(\mathbf{J}(\mathbf{x})\ \lambda_{min}(\mathbf{J}_\tau^{-T}\mathbf{J}_\tau^{-1})(\mathbf{x}))}\ \|\tilde{v}\|_{\mathcal{U}(\tilde{\tau})} \leq \|v\|_{\mathcal{U}(\tau)} \tag{6.1*.13c}$$

where $\tilde{\tau}$ (respectively $\tilde{\varepsilon}$) denotes the master square (respectively the edge of the master square).

Proof We have

$$\|v\|^2_{L^2(\tau)} = \int_\tau v^2 = \int_{\hat{\tau}} \tilde{v}^2\ det(\mathbf{J}_\tau) \leq \|det(\mathbf{J}_\tau)\|_{L^\infty(\tau)} \int_{\tilde{\tau}} \hat{v}^2. \tag{6.1*.14}$$

Further,

$$\begin{aligned}
\frac{\partial \hat{x}_1}{\partial x_1} &= \frac{1}{det(\mathbf{J}_\tau)}\frac{\partial x_2}{\partial \hat{x}_2}, & \frac{\partial \hat{x}_1}{\partial x_2} &= -\frac{1}{det(\mathbf{J}_\tau)}\frac{\partial x_1}{\partial \hat{x}_2} \\
\frac{\partial \hat{x}_2}{\partial x_1} &= -\frac{1}{det(\mathbf{J}_\tau)}\frac{\partial x_2}{\partial \hat{x}_1}, & \frac{\partial \hat{x}_2}{\partial x_2} &= \frac{1}{det(\mathbf{J}_\tau)}\frac{\partial x_1}{\partial \hat{x}_1}
\end{aligned} \tag{6.1*.15a}$$

and we obtain

$$\boldsymbol{\nabla} v \stackrel{\text{def}}{=} \begin{Bmatrix} \dfrac{\partial v}{\partial x_1} \\ \dfrac{\partial v}{\partial x_2} \end{Bmatrix} = \begin{bmatrix} \dfrac{\partial \tilde{x}_1}{\partial x_1} & \dfrac{\partial \tilde{x}_2}{\partial x_1} \\ \dfrac{\partial \tilde{x}_1}{\partial x_2} & \dfrac{\partial \tilde{x}_2}{\partial x_2} \end{bmatrix} \begin{Bmatrix} \dfrac{\partial \tilde{v}}{\partial \tilde{x}_1} \\ \dfrac{\partial \tilde{v}}{\partial \tilde{x}_2} \end{Bmatrix} = \mathbf{J}_\tau^{-1}\boldsymbol{\nabla}_{\tilde{\mathbf{x}}}\tilde{v} \tag{6.1*.15b}$$

where $\boldsymbol{\nabla}_{\tilde{\mathbf{x}}}\tilde{v} \stackrel{\text{def}}{=} \{\partial\tilde{v}/\partial\tilde{x}_1, \partial\tilde{v}/\partial\tilde{x}_2\}^{\mathrm{T}}$ and

$$\mathbf{J}_\tau^{-1} = \frac{1}{det(\mathbf{J}_\tau)} \begin{bmatrix} \dfrac{\partial \tilde{x}_2}{\partial x_2} & -\dfrac{\partial \tilde{x}_2}{\partial x_1} \\ -\dfrac{\partial \tilde{x}_1}{\partial x_2} & \dfrac{\partial \tilde{x}_1}{\partial x_1} \end{bmatrix} \tag{6.1*.15c}$$

is the inverse of the Jacobian matrix $\mathbf{J}_\tau$. By employing (6.1*.15a) and (6.1*.16b), we get

$$\begin{aligned}
\|v\|^2_{\mathcal{U}(\tau)} &= \int_\tau (\boldsymbol{\nabla} v)^{\mathrm{T}}(\boldsymbol{\nabla} v) = \int_{\tilde{\tau}} (\mathbf{J}_\tau^{-1}\boldsymbol{\nabla}_{\tilde{\mathbf{x}}}\tilde{v})^{\mathrm{T}}(\mathbf{J}_\tau^{-1}\boldsymbol{\nabla}_{\tilde{\mathbf{x}}}\tilde{v})\ det(\mathbf{J}_\tau) \\
&= \int_{\tilde{\tau}} (\boldsymbol{\nabla}_{\tilde{\mathbf{x}}}\tilde{v})^{\mathrm{T}}(\mathbf{J}_\tau)^{-\mathrm{T}}(\mathbf{J}_\tau)^{-1}(\boldsymbol{\nabla}_{\tilde{\mathbf{x}}}\tilde{v}) det(\mathbf{J}_\tau)
\end{aligned} \tag{6.1*.15d}$$

from which it follows that

$$\mathcal{K}_\tau \int_{\tilde{\tau}} (\boldsymbol{\nabla}_{\tilde{\mathbf{x}}}\tilde{v})^{\mathrm{T}}(\boldsymbol{\nabla}_{\tilde{\mathbf{x}}}\tilde{v}) \leq \int_{\tilde{\tau}} \lambda_{\min}(\mathbf{J}_\tau^{-\mathrm{T}}\mathbf{J}_\tau^{-1}) det(\mathbf{J}_\tau)(\boldsymbol{\nabla}_{\tilde{\mathbf{x}}}\tilde{v})^{\mathrm{T}}(\boldsymbol{\nabla}_{\tilde{\mathbf{x}}}\tilde{v}) \leq \|v\|^2_{\mathcal{U}(\tau)} \qquad (6.1^*.15e)$$

where $\lambda_{\min}((\mathbf{J}_\tau)^{-\mathrm{T}}(\mathbf{J}_\tau)^{-1})$ is the minimum eigenvalue of the matrix

$$\mathbf{J}_\tau^{-\mathrm{T}}\mathbf{J}_\tau^{-1} = \frac{1}{(det(\mathbf{J}_\tau))^2} \times \begin{bmatrix} \left(\frac{\partial x_1}{\partial \tilde{x}_2}\right)^2 + \left(\frac{\partial x_2}{\partial \tilde{x}_2}\right)^2 & -\left(\frac{\partial x_1}{\partial \tilde{x}_1}\right)\left(\frac{\partial x_1}{\partial \tilde{x}_2}\right) - \left(\frac{\partial x_2}{\partial \tilde{x}_1}\right)\left(\frac{\partial x_2}{\partial \tilde{x}_2}\right) \\ -\left(\frac{\partial x_1}{\partial \tilde{x}_1}\right)\left(\frac{\partial x_1}{\partial \tilde{x}_2}\right) - \left(\frac{\partial x_2}{\partial \tilde{x}_1}\right)\left(\frac{\partial x_2}{\partial \tilde{x}_2}\right) & \left(\frac{\partial x_1}{\partial \tilde{x}_1}\right)^2 + \left(\frac{\partial x_2}{\partial \tilde{x}_1}\right)^2 \end{bmatrix} \qquad (6.1^*.15f)$$

Hence

$$\sup_{\substack{v\in\mathcal{U} \\ \int_\tau v=0}} \frac{\|v\|_{L^2(\tau)}}{\|v\|_{\mathcal{U}(\tau)}} \leq \frac{\|det(\mathbf{J}_\tau)\|_{L^\infty(\tau)}}{\sqrt{\mathbf{J}_\tau}} \sup_{\tilde{v}\in\tilde{\mathcal{U}}(\tilde{\tau})} \frac{\|\tilde{v}\|_{L^2(\tilde{\tau})}}{\|\tilde{v}\|_{\mathcal{U}(\tilde{\tau})}} \qquad (6.1^*.16a)$$

$$\sup_{\substack{v\in\mathcal{U} \\ \int_\tau v=0}} \frac{\|v\|_{L^2(\varepsilon)}}{\|v\|_{\mathcal{U}(\tau)}} \leq \frac{\|det(\mathbf{J}_\tau)\|_{L^\infty(\varepsilon)}}{\sqrt{\mathbf{J}_\tau}} \sup_{\tilde{v}\in\tilde{\mathcal{U}}(\tilde{\tau})} \frac{\|\tilde{v}\|_{L^2(\tilde{\varepsilon})}}{\|\tilde{v}\|_{\mathcal{U}(\tilde{\tau})}} \qquad (6.1^*.16b)$$

where we used the fact that the mapped constant function is a constant function.

The constant $\tilde{\mathscr{C}}_1$, and $\tilde{\mathscr{C}}_2$ can be obtained as in one dimension. Because $\tilde{\tau}$ is a master square then any $\tilde{v} \in \tilde{\mathcal{U}}(\tau)$ can be written in the form

$$\tilde{v}(\tilde{x}_1, \tilde{x}_2) = \sum_{\substack{m,n=0 \\ m+n\neq 0}} a_{mn} \cos\left(\frac{m\pi}{2}(1+\tilde{x}_1)\right) \cos\left(\frac{n\pi}{2}(1+\tilde{x}_2)\right) \qquad (6.1^*.17a)$$

and we get as in one dimension

$$\tilde{\mathscr{C}}_1 = \frac{2}{\pi}, \qquad \tilde{\mathscr{C}}_2 = \sqrt{\frac{2}{3}} \qquad (6.1^*.17b)$$

If $\varepsilon \subset \Gamma_D$ then the proof is analogous and the constants $\tilde{\mathscr{C}}_1$, $\tilde{\mathscr{C}}_2$ can be used as in one dimension. $\tilde{\mathscr{C}}_1$, and $\tilde{\mathscr{C}}_2$ are the eigenvalues of the problem. □

Remark 6.1*.2 Analogously as in one dimension we can employ (6.1.11c) and the fact that we are using the mapped polynomials, we can replace $\tilde{\mathscr{C}}_1$, and $\tilde{\mathscr{C}}_2$ by $\tilde{\mathscr{C}}_1(p+k)$, and $\tilde{\mathscr{C}}_2(p+k)$ with

$$\tilde{\mathscr{C}}_1(p+k) = \sup_{\tilde{v}\in\tilde{\mathcal{U}}^{(p+k)}(\tilde{\tau})} \frac{\|\tilde{v}\|_{L^2(\tilde{\tau})}}{\|\tilde{v}\|_{\tilde{\mathcal{U}}(\tilde{\tau})}} \qquad (6.1^*.18a)$$

$$\tilde{\mathscr{C}}_2(p+k) = \sup_{\tilde{v}\in\tilde{\mathcal{U}}^{(p+k)}(\tilde{\tau})} \frac{\|\tilde{v}\|_{L^2(\tilde{\varepsilon})}}{\|\tilde{v}\|_{\tilde{\mathcal{U}}(\tilde{\tau})}} \qquad (6.1^*.18b)$$

where

$$\tilde{\mathcal{U}}^{(p+k)} = \left\{ \tilde{v} \in \tilde{\mathcal{U}}(\tilde{\tau}) \,\Big|\, \int_\tau \nabla\tilde{v}\cdot\nabla\tilde{w} = 0 \ \forall w \in S^{p+k} \right\} \tag{6.1*.18c}$$

$\tilde{\mathscr{C}}_1(p+k)$, and $\tilde{\mathscr{C}}_2(p+k)$ have to be computed numerically. For example we have $\tilde{\mathscr{C}}_1(4) = 0.1735$, $\tilde{\mathscr{C}}_2(4) = 0.3876$, and the use of the improved constants defined by (6.1*.18) can significantly improve the effectivity of the explicit estimator, and the overall estimate.

Analogous to the one-dimensional case, we obtain the following computable upper bound:

Theorem 6.1*.2 *(The computable upper estimator $\mathscr{E}^{\mathrm{Neum}/EXPL}_{S^{p+k}_\Delta}$) Under the assumptions which guarantee the existence of $\mathscr{E}^{\mathrm{Neum}}$, we have*

$$\|e_{S^p_\Delta}\|_{\mathcal{U}} \le \mathscr{E}^{\mathrm{Neum}} \le \mathscr{E}^{\mathrm{Neum}/EXPL}_{S^{p+k}_\Delta} \stackrel{\mathrm{def}}{=} \sqrt{\left(\mathscr{E}^{\mathrm{Neum}}_{S^{p+k}_\Delta}\right)^2 + \sum_{\tau\in\Delta}\left(\mathscr{E}^{EXPL}_{\hat{e}^{\mathrm{Neum}}_{S^{p+k}_\Delta}/\tau}\right)^2}. \tag{6.1*.19}$$

Remark 6.1*.3 The upper bound $\mathscr{E}^{\mathrm{Neum}/EXPL}_{S^{p+k}_\Delta}$, which will be also denoted by $\mathscr{E}^U_{S^{p+k}_\Delta}$, is based on the computed indicators $\hat{e}^{\mathrm{Neum}}_{\tau,S^{p+k}_\Delta}$, obtained by the p method. A more general estimator based on the computed indicators

$$\hat{e}^{\mathrm{Neum}}_{\tau,S^{p+k}_{\Delta_{h/2^n}}} = \pi_{S^{p+k}_{\Delta^\tau_{h/2^n},\partial\tau_D}}\left(\hat{e}^{\mathrm{Neum}}_\tau\right) \tag{6.1*.20a}$$

obtained by the hp method can also be constructed. Here π is the projection operator associated with the energy inner product, and $\Delta_{h/2^n}$ denotes a mesh which is obtained from Δ_h by employing n nested uniform or graded refinements, $S^{p+k}_{\Delta^\tau_{h/2^n},\partial\tau_D}$ is the restriction of the finite element space $S^{p+k}_{\Delta^\tau_{h/2^n},\partial\Gamma_D}$ on the element τ, and $\Delta^\tau_{h/2^n}$ is the restriction of the mesh $\Delta_{h/2^n}$ to the element τ. In this case, Lemma 6.1*.1 reads

$$\mathscr{E}^{\mathrm{Neum}}_{\Delta_h} = \sqrt{\left(\mathscr{E}^{\mathrm{Neum}}_{\Delta_h,S^{p+k}_{\Delta^\tau_{h/2^n}}}\right)^2 + \sum_{\tau\in\Delta_h}\left\|\hat{e}^{\mathrm{Neum}}_\tau - \hat{e}^{\mathrm{Neum}}_{\tau,S^{p+k}_{\Delta^\tau_{h/2^n}}}\right\|^2_{\mathcal{U}(\tau)}} \tag{6.1*.20b}$$

and an upper bound for $\|\hat{e}^{\mathrm{Neum}}_\tau - \hat{e}^{\mathrm{Neum}}_{\tau,S^{p+k}_{\Delta^\tau_{h/2^n}}}\|^2_{\mathcal{U}(\tau)}$ can be constructed by, employing Theorem 6.1*.2 to obtain

$$\left\|\hat{e}^{\mathrm{Neum}}_\tau - \hat{e}^{\mathrm{Neum}}_{\tau,S^{p+k}_{\Delta^\tau_{h/2^n}}}\right\|_{\mathcal{U}(\tau)} \le \mathscr{E}^{\mathrm{Neum}/EXPL}_{\hat{e}^{\mathrm{Neum}}_{S^{p+k}_{\Delta^\tau_{h/2^n}}},S^{p+k+l}_{\Delta_{h/2^n}}/\tau} \tag{6.1*.20c}$$

where $\mathscr{E}^{\mathrm{Neum}/EXPL}_{\hat{e}^{\mathrm{Neum}}_{S^{p+k}_{\Delta^\tau_{h/2^n}}},S^{p+k+l}_{\Delta_{h/2^n}}/\tau}$ is constructed analogously as $\mathscr{E}^{\mathrm{Neum}/EXPL}_{u_{S^p_\Delta},S^{p+k}_\Delta}$. Combining (6.1*.20b) and (6.1*.20c) we get the upper bound

$$\mathscr{E}^U_{S^{p+k}_{\Delta_{h/2^n}}/S^{p+k+l}_{\Delta_{h/2^n}}} = \sqrt{\left(\mathscr{E}^{\mathrm{Neum}}_{S^{p+k}_{\Delta_{h/2^n}}}\right)^2 + \sum_{\tau\in\Delta_h}\left(\mathscr{E}^{\mathrm{Neum}/EXPL}_{\hat{e}^{\mathrm{Neum}}_{S^{p+k}_{\Delta^\tau_{h/2^n}}},S^{p+k+l}_{\Delta_{h/2^n}}/\tau}\right)^2}. \tag{6.1*.20d}$$

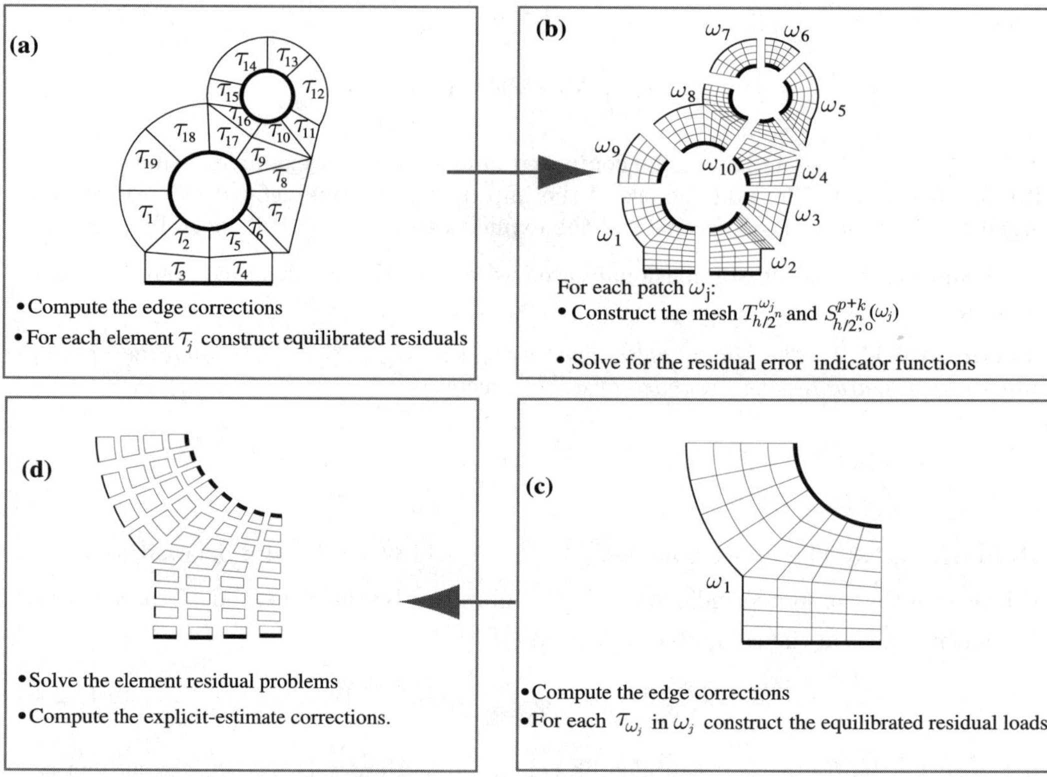

FIG. 6.1*.2. *Guaranteed upper bound for the energy norm of the error.* Schematic depiction of the steps involved in the construction of the guaranteed upper bound for $\|e_{S^p_\Delta}\|_{\mathcal{U}}$: (a) The mesh Δ_h which is employed in the computation of the finite element solution $u_{S^p_\Delta}$. (b) The partition of Δ_h into the patches ω_j and the meshes $\Delta^{\omega_j}_{h/2^n}$ in the computation of the error indicator functions. (c) The mesh $\Delta^{\omega_1}_{h/2^n}$ employed in the patch ω_1. (d) The elements $\tau' \in \Delta^{\omega_1}_{h/2^n}$ which are used for the solution of the element residual problems

Following similar steps as in the above remark, we can construct a generalization of Theorem 6.1*.2 to the case that $\Omega_h \not\equiv \Delta_h$, in which the patches do not coincide with the elements, as is schematically described in Fig. 6.1*.2. We have:

Theorem 6.1*.3 *Under the assumptions which guarantee the existence of* $\mathscr{E}^{\mathrm{Neum}}_{\Omega_h}$, *we have*

$$\|e_{S^p_\Delta}\|_{\mathcal{U}} \le \mathscr{E}^{\mathrm{Neum}}_{\Omega_h} \le \mathscr{E}^{U}_{\Omega_h, S^{p+k}_\Delta} \stackrel{\mathrm{def}}{=} \sqrt{\left(\mathscr{E}^{\mathrm{Neum}}_{\Omega_h, S^{p+k}_\Delta}\right)^2 + \sum_{\omega \in \Omega_h} \left(\mathscr{E}^{U}_{\hat{e}^{\mathrm{Neum}}_{S^{p+k}_{\Delta\omega}} / S^{p+k+l}_{\Delta\omega} / \omega}\right)^2}. \quad (6.1^*.21)$$

Remark 6.1*.4 The bound $\mathscr{E}^{U}_{\Omega_h, S^{p+k}_\Delta}$ given in (6.1*.25) is based on the computed indicators $\hat{e}^{\mathrm{Neum}}_{S^{p+k}_\Delta/\omega}$ obtained using the p version. A more general bound which employs the computed indicator $\hat{e}^{\mathrm{Neum}}_{\omega, S^{p+k}_{\Delta_{h/2^n}}}$ based on the hp version, can be obtained similarly using the explicit correction in each element in the last iteration.

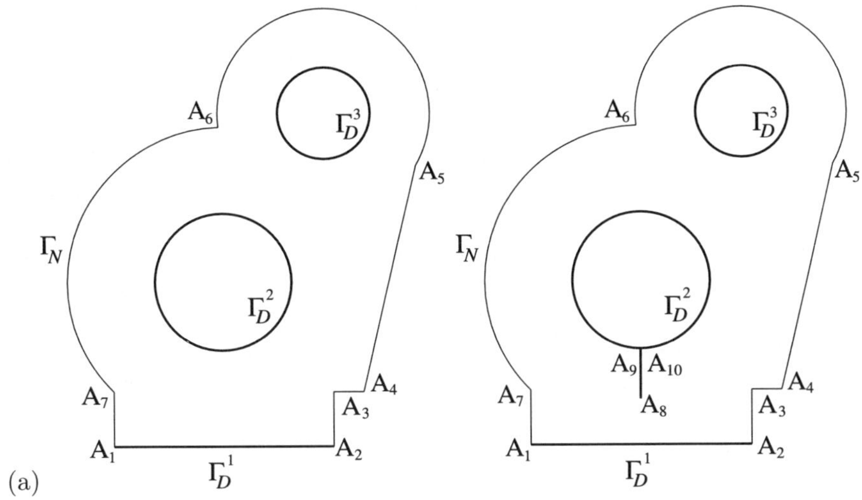

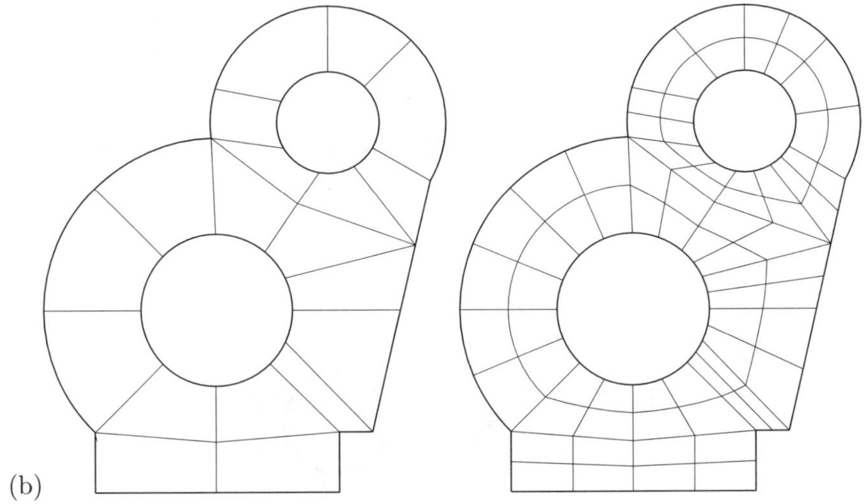

FIG. 6.1*.3. *The two-dimensional example problem.* (a) The domains Ω, with the Dirichlet, and Neumann parts of the boundary $\Gamma_D = \Gamma_D^1 \cup \Gamma_D^2 \cup \Gamma_D^3$, and Γ_N. We will call the domain on the left Domain I, and the domain on the right Domain II. (b) The meshes Δ_h employed in the construction of the finite element solution. We will call the mesh on the left Mesh I and the mesh on the right Mesh II, which was obtained by a uniform refinement of Mesh I. Note that both meshes are rather coarse and essentially governed only by the geometry of the domain, which is typical in practice

Let us now give an example which illustrates the above constructions.

Example 6.1*.1 (*The upper bound for the energy norm of the exact error*) Let us consider the boundary value problem for the Laplacian

$$-\Delta u_{EX}\big|_{\Omega} = 0 \qquad (6.1^*.22a)$$

with the boundary conditions

$$u_{EX}\big|_{\Gamma_D^1 \cup \Gamma_D^3} = 0, \qquad u_{EX}\big|_{\Gamma_D^2} = 1, \qquad \frac{\partial}{\partial n}(u_{EX})\Big|_{\Gamma_N} = 0 \tag{6.1*.22b}$$

where Ω is one of the two domains (Domain I, or Domain II) shown in Figure 6.1*.3a, with boundary $\Gamma = \partial\Omega = \Gamma_D \cup \Gamma_N$, and $\Gamma_D = \Gamma_D^1 \cup \Gamma_D^2 \cup \Gamma_D^3$. The solution u_{EX} has algebraic singularities at the vertices A_i, $i = 1, \ldots, N_v$ with $N_v = 7$ for Domain I, and $N_v = 10$ for Domain II, and is smooth in the rest of the domain. The exponent (strength) of the algebraic singularity at each vertex depends on the interior angle and the types of the boundary conditions applied on the edges emanating from the vertex. The major algebraic singularities are at A_3, A_5, A_6, A_7 for Domain I (respectively A_3, A_5, A_6, A_7, and A_8 for Domain II) the vertices of the re-entrant corners of the boundary $\partial\Omega$. There is no singularity at A_1, and A_2 (and also at A_8, A_{10} for Domain II), while there is only a weak singularity at A_4.

We will denote by $u_{S^p_{\Delta_h}}$, the finite element solution for the meshes Δ^{Ω_h}, namely Mesh I and Mesh II shown in Fig. 6.1*.3b. Note that unlike the one-dimensional example problem employed in Section 6.1, we do not have a closed form analytic solution for the two-dimensional example problem, and we will use the approximation

$$e_{S^p_\Delta} \approx u_{S^{p'}_{\Delta_{ovk}}} - u_{S^p_\Delta} \tag{6.1*.23a}$$

where $u_{S^{p'}_{\Delta_{ovk}}}$ is an 'overkill' solution, which has much higher accuracy than $u_{S^p_\Delta}$. In our example, we employed the overkill mesh Δ_h^{ovk}, which is obtained from Mesh I mesh by

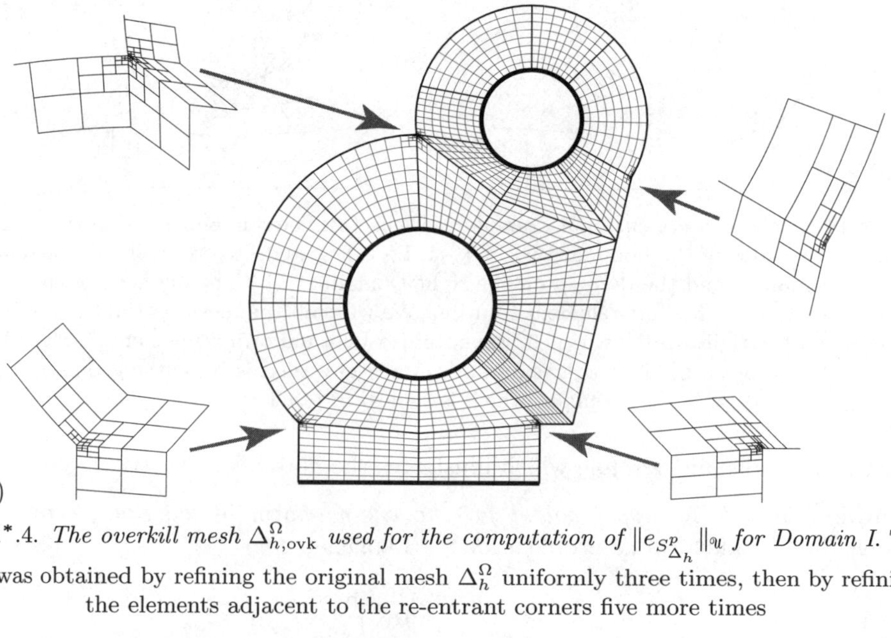

(a)

FIG. 6.1*.4. *The overkill mesh $\Delta^{\Omega}_{h,\mathrm{ovk}}$ used for the computation of $\|e_{S^p_{\Delta_h}}\|_{\mathfrak{U}}$ for Domain I.* This mesh was obtained by refining the original mesh Δ_h^{Ω} uniformly three times, then by refining the elements adjacent to the re-entrant corners five more times

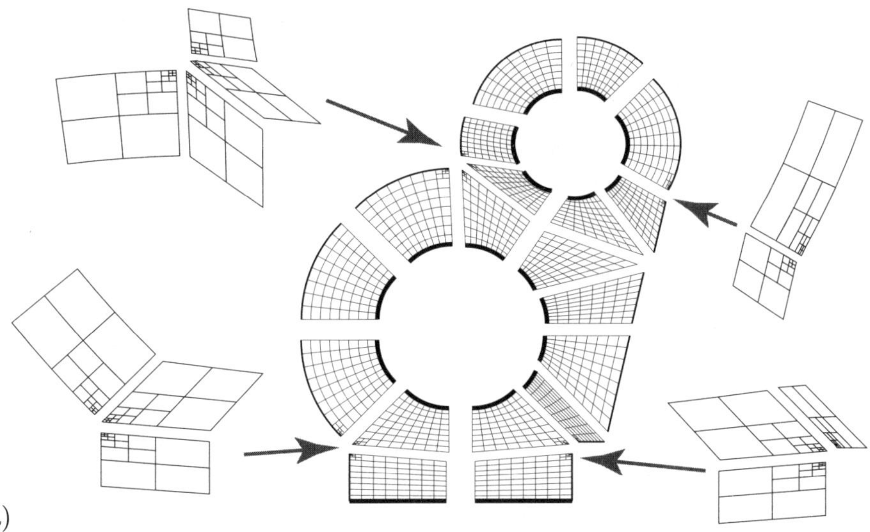
(a)

FIG. 6.1*.5. *The overkill meshes employed for the computation of the exact element error indicators $\eta_{\omega_j}^{\mathrm{EX}}$ for Domain I.* These meshes are the restrictions of the overkill mesh of Fig. 6.1*.4 in the elements of the original coarse mesh

refining the mesh three times uniformly, and the elements with a vertex at a re-entrant corner five additional times and bi-p finite element basis of degree $p' = 8$. The overkill mesh for Domain I is shown in Figure 6.1*.4. In each case, *we ensured that the overkill solution and its strain energy have sufficient accuracy for the purpose of evaluating the error for the employed approximations.* For example, for the finite element approximation of the solution of the model problem in Domain I, using linear ($p = 1$) elements on Mesh I, for the employed overkill approximation, we found that

$$\frac{\|e_{S^1_{\Delta_h}}\|_{\mathcal{U}}}{\|u_{S^8_{\Delta_h^{ovk}}}\|_{\mathcal{U}}} \approx \frac{\|u_{S^8_{\Delta_h^{ovk}}} - u_{S^1_{\Delta_h}}\|_{\mathcal{U}}}{\|u_{S^8_{\Delta_h^{ovk}}}\|_{\mathcal{U}}} = \frac{\sqrt{\left|\|u_{S^8_{\Delta_h^{ovk}}}\|^2_{\mathcal{U}} - \|u_{S^1_{\Delta_h}}\|^2_{\mathcal{U}}\right|}}{\|u_{S^8_{\Delta_h^{ovk}}}\|_{\mathcal{U}}} = 29.49\% \qquad (6.1^*.23\text{b})$$

which is 0.042% accurate. Note that in the last equality in (6.1*.23b) we used Remark 3.1.7. We also approximated the exact upper bound $\mathscr{E}^{\mathrm{Neum}}_{\Delta_h}$, by solving the element residual problems using the overkill meshes which are obtained as restrictions of the overkill mesh Δ_h^{ovk}, on the elements of Δ_h, as is, for example shown in Fig. 6.1*.5 for Domain I, and $p' = 8$. For the Mesh I, and $p = 1$, finite element approximation of the solution of the model problem in Domain I, the exact upper bound is

$$\frac{\|e_{S^1_{\Delta_h}}\|_{\mathcal{U}}}{\|u_{S^8_{\Delta_h^{ovk}}}\|_{\mathcal{U}}} = 29.49\% \le \frac{\mathscr{E}^{\mathrm{Neum}}_{\Delta_h}}{\|u_{S^8_{\Delta_h^{ovk}}}\|_{\mathcal{U}}} = 36.18\%. \qquad (6.1^*.23\text{c})$$

TABLE 6.1*.1. *The relative error in the finite element solution, and the effectivity index of the exact estimator.* The solution $u_{S^8_{\Delta_h^{ovk}}}$ is done using $p = 8$ on the mesh Δ_h^{ovk} which is obtained from Mesh I by refining all elements three times uniformly and by subsequent nested refinement of the elements with a vertex at a re-entrant corner five additional times. The exact Neumann estimator $\mathscr{E}^{\text{Neum}}_{\Delta_h}$, is computed using $p + k = 8$, and the element overkill meshes obtained by restricting the global overkill mesh Δ_h^{ovk} into the elements of the mesh Δ_h

p	Mesh I				Mesh II			
	$\|u_{S^p_\Delta}\|_\mathcal{U}$	$\|u_{S^8_{\Delta_h^{ovk}}}\|_\mathcal{U}$	$\frac{\|e_{S^p_{\Delta_h}}\|_\mathcal{U}}{\|u_{S^8_{\Delta_h^{ovk}}}\|_\mathcal{U}}$	$\frac{\mathscr{E}^{\text{Neum}}_{\Delta_h}}{\|e_{S^8_{\Delta_h^{ovk}}}\|_\mathcal{U}}$	$\|u_{S^p_\Delta}\|$	$\|u_{S^8_{\Delta_h^{ovk}}}\|_\mathcal{U}$	$\frac{\|e_{S^p_{\Delta_h}}\|_\mathcal{U}}{\|u_{S^8_{\Delta_h^{ovk}}}\|_\mathcal{U}}$	$\frac{\mathscr{E}^{\text{Neum}}_{\Delta_h}}{\|e_{S^8_{\Delta_h^{ovk}}}\|_\mathcal{U}}$
Domain I								
1	2.029275	1.946395	29.492%	1.2268	1.970682	1.946395	15.847%	1.2197
2	1.953125	1.946395	8.323%	1.3555	1.948106	1.946395	4.194%	1.4446
Domain II								
1	2.292093	2.091418	44.845 %	1.2222	2.165211	2.091418	26.798%	1.5269
2	2.127869	2.091418	18.751 %	1.2154	2.109951	2.091418	13.342%	1.4909

Table 6.1*.1 gives the complete results for the exact relative error, and the relative error predicted by the exact estimator $\mathscr{E}^{\text{Neum}}_{\Delta_h}$, for the bilinear ($p = 1$), and the biquadratic ($p = 2$) finite element solutions of the model problem in Domain I, and Domain II, using Mesh I, and Mesh II shown in Fig. 6.1*.3b.

Figure 6.1*.6 shows the graphs of the effectivity indices for the exact and the computed Neumann estimator, $\kappa^{\text{Neum}} \stackrel{\text{def}}{=} \mathscr{E}^{\text{Neum}}_{\Delta_h}/\|e_{S^p_{\Delta_h}}\|_\mathcal{U}$, and $\kappa^{\text{Neum}}_{S^{p+k}_{\Delta_h}} \stackrel{\text{def}}{=} \mathscr{E}^{\text{Neum}}_{\Delta_h, S^{p+k}_\Delta}/\|e_{S^p_{\Delta_h}}\|_\mathcal{U}$ for $p + k = 3, 4, \ldots 8$, for $p = 1$, and $p = 2$ for the finite element approximations of the solutions of the model problem in Domain I, and Domain II computed using Mesh I and Mesh II. Table 6.1*.2 lists the values of the effectivity indices for the computed Neumann estimator $\mathscr{E}^{\text{Neum}}_{\Delta_h, S^{p+k}_\Delta}$. Note that in some cases like, for example, Domain II with Mesh I, $p = 2$, and $p + k = 3$, and $p + k = 4$, we have

$$\left\| e^{S^{p+k}_{\Delta_h}}_{S^p_{\Delta_h}} \right\|_\mathcal{U} < \mathscr{E}^{\text{Neum}}_{\Delta_h, S^{p+k}_\Delta} < \|e_{S^p_{\Delta_h}}\|_\mathcal{U}. \tag{6.1*.24}$$

Thus, the computed Neumann estimator $\mathscr{E}^{\text{Neum}}_{\Delta_h, S^{p+k}_\Delta}$ may underestimate the energy norm of the exact error $\|e_{S^p_\Delta}\|_\mathcal{U}$.

Figure 6.1*.7 compares the graphs of the effectivity indices $\kappa^U_{S^{p+k}_{\Delta_h}} \stackrel{\text{def}}{=} (\mathscr{E}^U_{\Delta_h, S^{p+k}_{\Delta_h}})/\|e_{S^p_{\Delta_h}}\|_\mathcal{U}$ and $\kappa^{\text{Neum}} = \mathscr{E}^{\text{Neum}}_{\Delta_h}/\|e_{S^p_{\Delta_h}}\|_\mathcal{U}$, for $p + k = 3, 4, \ldots, 8$ for $p = 2$. We note that as expected we have

$$\|e_{S^p_{\Delta_h}}\|_\mathcal{U} < \mathscr{E}^{\text{Neum}}_{\Delta_h} < \mathscr{E}^U_{\Delta_h, S^{p+k}_{\Delta_h}}. \tag{6.1*.25}$$

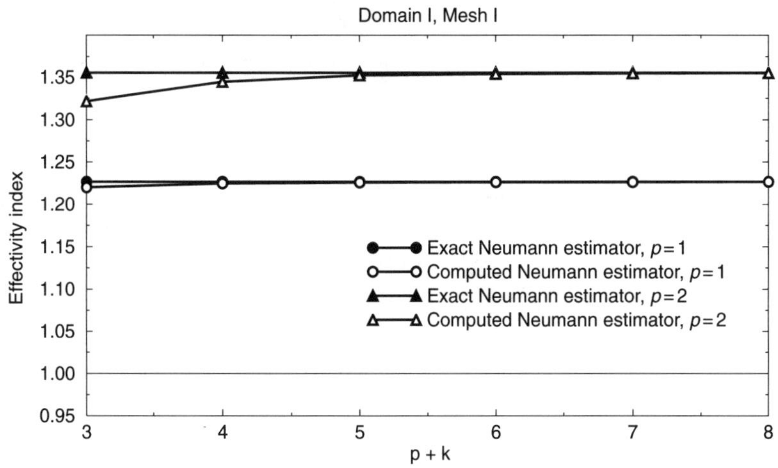
Domain I, Mesh I
Effectivity index
1.35
1.30
1.25
1.20
1.15
1.10
1.05
1.00
0.95
3
4
5
6
7
8
p + k
Exact Neumann estimator, p=1
Computed Neumann estimator, p=1
Exact Neumann estimator, p=2
Computed Neumann estimator, p=2

(a)

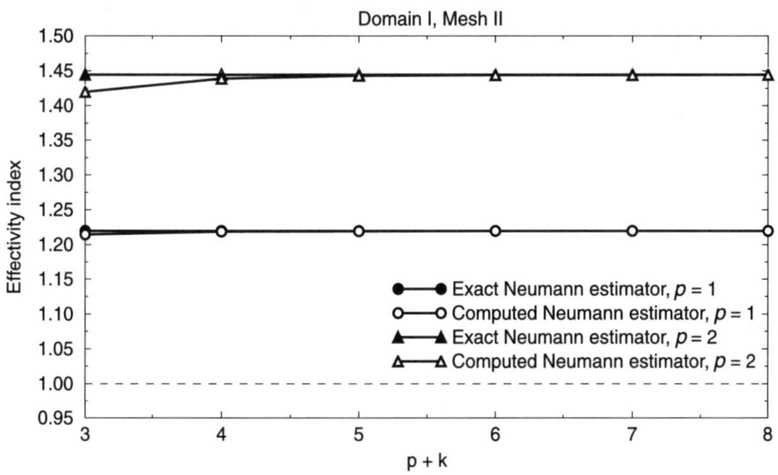
Domain I, Mesh II
Effectivity index
1.50
1.45
1.40
1.35
1.30
1.25
1.20
1.15
1.10
1.05
1.00
0.95
3
4
5
6
7
8
p + k
Exact Neumann estimator, p = 1
Computed Neumann estimator, p = 1
Exact Neumann estimator, p = 2
Computed Neumann estimator, p = 2

(b)

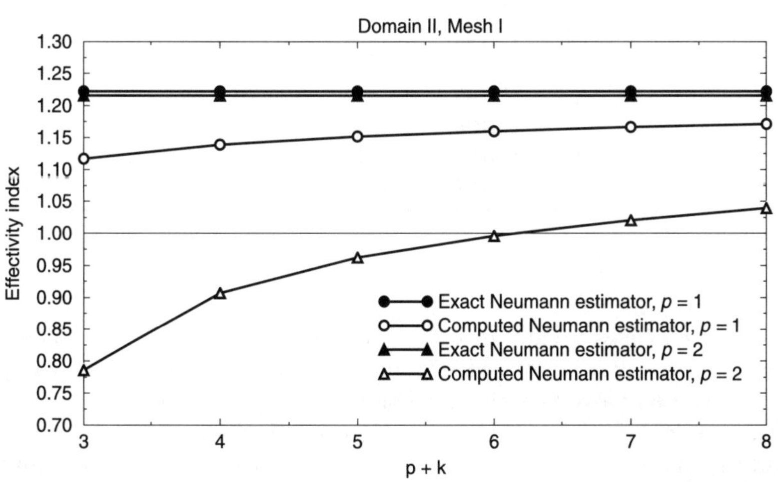
Domain II, Mesh I
Effectivity index
1.30
1.25
1.20
1.15
1.10
1.05
1.00
0.95
0.90
0.85
0.80
0.75
0.70
3
4
5
6
7
8
p + k
Exact Neumann estimator, p = 1
Computed Neumann estimator, p = 1
Exact Neumann estimator, p = 2
Computed Neumann estimator, p = 2

(c)

FIG. 6.1*.6.

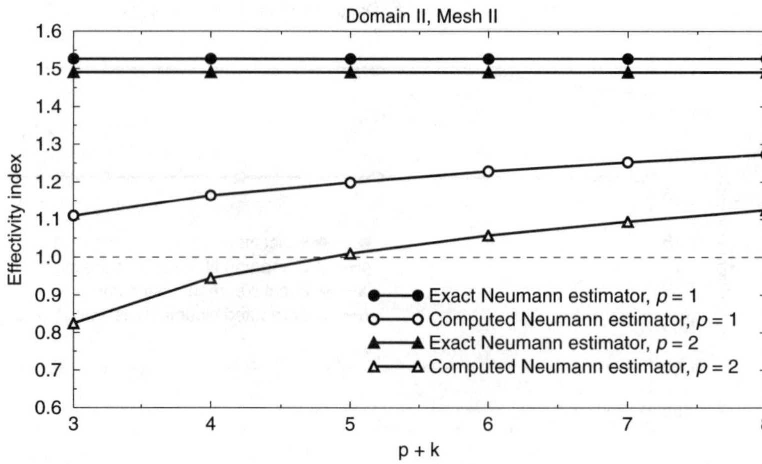

(d)

FIG. 6.1*.6. *The upper bound for energy norm of the error.* Approximate solution of the element residual problems using the p method. Graphs of the effectivity index of the exact Neumann estimator $\kappa^{\text{Neum}} = \mathscr{E}^{\text{Neum}}_{\Delta_h}/\|e_{S^p_{\Delta_h}}\|_{\mathscr{U}}$, the computed Neumann estimator $\kappa^{\text{Neum}}_{S^{p+k}_{\Delta_h}} \stackrel{\text{def}}{=} \mathscr{E}^{\text{Neum}}_{\Delta_h, S^{p+k}_{\Delta_h}}/\|e_{S^p_{\Delta_h}}\|_{\mathscr{U}}$, versus $p + k = 3, 4, \ldots, 8$ for $p = 1, 2$ for (a) Domain I, Mesh I. (b) Domain I, Mesh II (c) Domain II, Mesh I (d) Domain II, Mesh II

TABLE 6.1*.2. *Effectivity index of the computed Neumann estimator.* The values of the effectivity index $\kappa^{Neum}_{S^{p+k}_{\Delta_h}}$, for the computed Neumann estimator $\mathscr{E}^{\text{Neum}}_{\Delta_h, S^{p+k}_{\Delta}}$, for various values of k, for the finite element solutions of Domain I, and Domain II

$p+k$	Mesh I		Mesh II		Mesh I		Mesh II	
	$p = 1$	$p = 2$	$p = 1$	$p = 2$	$p = 1$	$p = 2$	$p = 1$	$p = 2$
	Domain I				Domain II			
3	1.2203	1.3217	1.2144	1.4198	1.1172	0.7859	1.1099	0.8246
4	1.2248	1.3446	1.2185	1.4391	1.1388	0.9070	1.1640	0.9449
5	1.2261	1.3524	1.2193	1.4432	1.1518	0.9625	1.1994	1.0105
6	1.2264	1.3538	1.2195	1.4440	1.1602	0.9964	1.2288	1.0578
7	1.2266	1.3547	1.2196	1.4443	1.1665	1.0209	1.2523	1.0943
8	1.2267	1.3549	1.2197	1.4445	1.1714	1.0394	1.2730	1.1247

Note that in all cases we get an upper estimate, the accuracy of which may improve or deteriorate as $p + k$ is increased depending on the smoothness of the problem, the degree p, and the employed mesh Δ_h. For example from Figure 6.1*.7 it can be seen that for the model problem in Domain I (respectively Domain II) the effectivity of the computable upper estimator $\mathscr{E}^{U}_{\Delta_h, S^{p+k}_{\Delta_h}}$ improves (respectively deteriorates) with k. This is because, when the solution is unsmooth, and the pure p version is used, the

L^2-norm of the residuals corresponding to $\hat{e}^{\mathrm{Neum}}_{S^{p+k}_{\Delta_h}}$, in each element do not converge to zero.

Remark 6.1*.5 Above we employed the p version for solving the element residual problems. We can also use the hp version, employing, for example, the meshes like the ones shown in Fig. 6.1*.8.

Remark 6.1*.6 The bounds depend on the particular choice of the edge corrections (Bank–Weiser, Ladeveze of Type 1 or 2, etc.), and the order of the equilibration. In the results given in this section, we employed linear Ladeveze Type 2 equilibration.

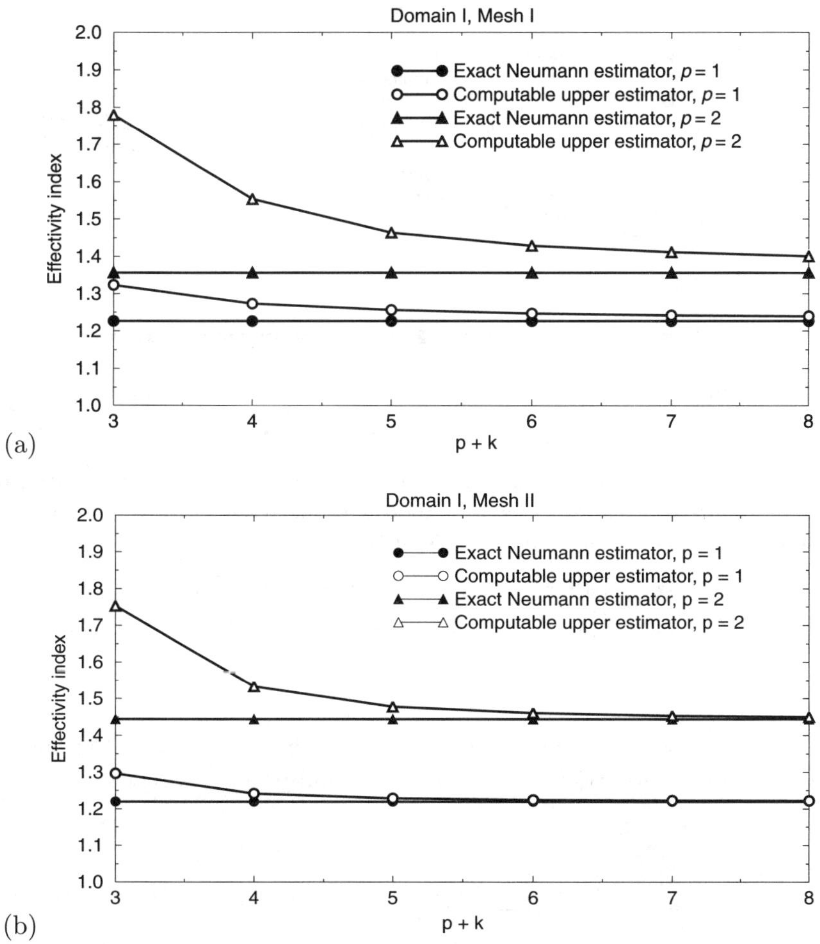

FIG. 6.1*.7. (Continued)

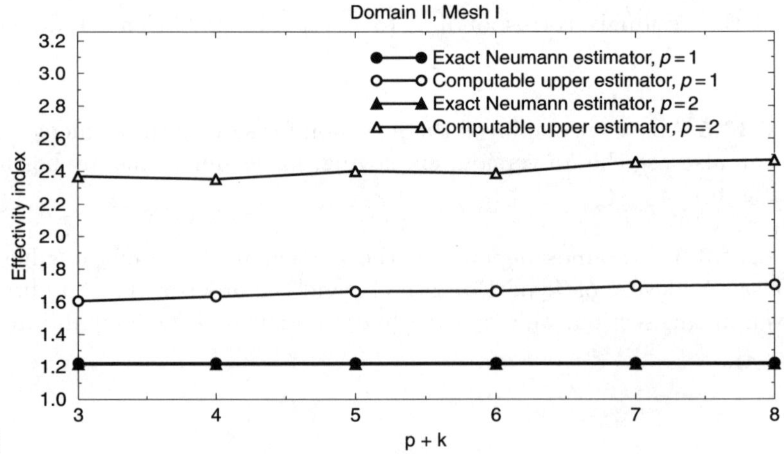

(c)

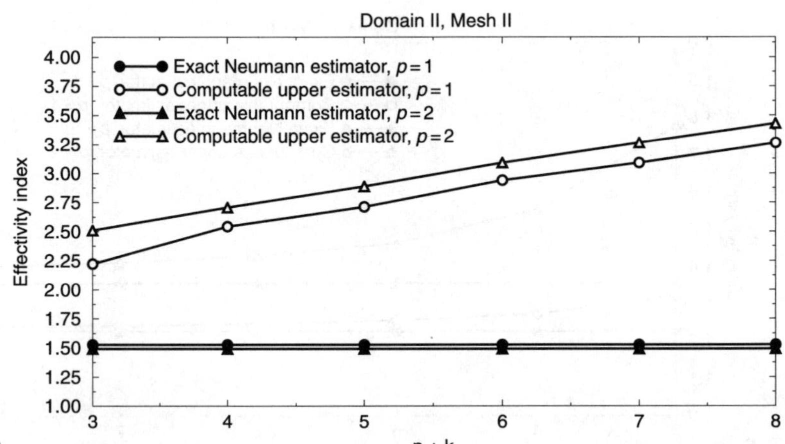

(d)

FIG. 6.1*.7. *The upper bound for energy norm of the error.* Approximate solutions of the element residual problems using the p method. Graphs of the effectivity indices for the computable upper estimator, and the exact Neumann estimator, $\kappa^U_{S^{p+k}_{\Delta_h}} = \mathscr{E}^U_{\Delta_h, S^{p+k}_{\Delta_h}} / \|e_{S^p_{\Delta_h}}\|_{\mathfrak{U}}$, and $\kappa^{\text{Neum}} = \mathscr{E}^{\text{Neum}}_{\Delta_h} / \|e_{S^p_{\Delta_h}}\|_{\mathfrak{U}}$ for $p+k = 3, 4, \ldots, 8$, for $p = 1$ and $p = 2$ for (a) Domain I, Mesh I (b) Domain I, Mesh II (c) Domain II, Mesh I (d) Domain II, Mesh II

Let us now address the lower bound. The generalized version of the basic identity, written in terms of the solutions of the patch residual problems, reads

$$\left\| e^{S^{p+k}_{\Delta'}}_{S^p_\Delta} \right\|^2_{\mathfrak{U}} = \left(\mathscr{E}^{\text{Neum}}_{\Omega_h, S^{p+k}_{\Delta'}}\right)^2 - \sum_{\omega_j \in \Omega_h} \left\| \tilde{e}_{S^{p+k}_{\Delta'}} - \hat{e}^{\text{Neum}}_{\omega_j, S^{p+k}_{\Delta'}} \right\|^2_{\mathfrak{U}(\omega_j)} + \left\| e^{S^{p+k}_{\Delta'}}_{S^p_\Delta} - \tilde{e}_{S^{p+k}_{\Delta'}} \right\|^2_{\mathfrak{U}} \tag{6.1*.26}$$

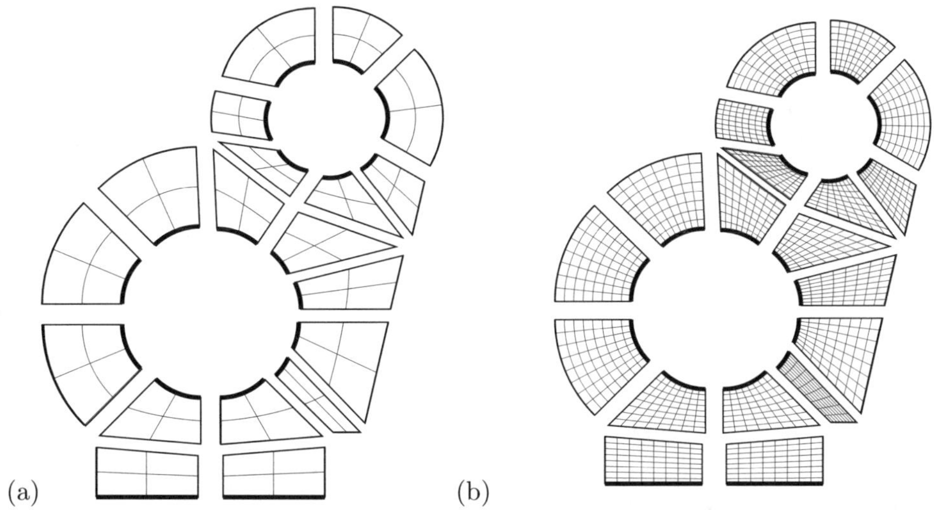

FIG. 6.1*.8. *Element residual problems.* The meshes $\Delta^{\tau}_{h/2^n}$, where τ are the elements of the original mesh Δ_h: (a) $n = 1$, (b) $n = 3$

for any $\tilde{e}_{S^{p+k}_{\Delta'}} \in S^{p+k}_{\Delta'}$. Omitting the last term we get the lower bound based on orthogonality. Further, as in the one-dimensional case, we can also construct a lower bound based on duality. We have:

Theorem 6.1*.4 *(The lower bounds $\mathscr{E}^L_{\Omega_h,S^{p+k}_{\Delta'}}$ and $\mathscr{E}^L_{dual}$) Under the assumptions which guarantee the existence of $\mathscr{E}^{\text{Neum}}_{\Omega_h}$ we have*

$$\mathscr{E}^L_{\Omega_h,S^{p+k}_{\Delta'}}(\tilde{e}_{S^{p+k}_{\Delta'}}) \stackrel{\text{def}}{=} \sqrt{\left(\mathscr{E}^{\text{Neum}}_{\Omega_h,S^{p+k}_{\Delta'}}\right)^2 - \sum_{\omega_j \in \Omega_h} \|\Phi_{\hat{e}^{\text{Neum}}_{\Omega_h,S^{p+k}_{\Delta'}}}(\tilde{e}_{S^{p+k}_{\Delta'}})\|^2_{\mathscr{U}(\omega_j)}}$$

$$\leq \left\| e^{S^{p+k}_{\Delta'}}_{S^p_{\Delta}} \right\|_{\mathscr{U}} \tag{6.1*.27a}$$

where

$$\Phi_{\hat{e}^{\text{Neum}}_{\Omega_h,S^{p+k}_{\Delta'}}}(\tilde{e}_{S^{p+k}_{\Delta'}}) = \tilde{e}_{S^{p+k}_{\Delta'}} - \hat{e}^{\text{Neum}}_{\Omega_h,S^{p+k}_{\Delta'}} \tag{6.1*.27b}$$

is the gap function which corresponds to $\tilde{e}_{S^{p+k}_{\Delta'}}$.

Further, under the minimal assumptions which guarantee the existence of the residuum $\mathscr{R}$, we have

$$\mathscr{E}^L_{dual}(\tilde{e}_{S^{p+k}_{\Delta'}}) \stackrel{\text{def}}{=} \frac{\mathscr{R}(\tilde{e}_{S^{p+k}_{\Delta'}})}{\|\tilde{e}_{S^{p+k}_{\Delta'}}\|_{\mathscr{U}}} \leq \left\| e^{S^{p+k}_{\Delta'}}_{S^p_{\Delta}} \right\|_{\mathscr{U}} \leq \|e_{S^p_{\Delta}}\|_{\mathscr{U}}. \tag{6.1*.28}$$

where $\tilde{e}_{S^{p+k}_{\Delta'}} = \hat{e}^{\text{Neum}}_{\Omega_h,S^{p+k}_{\Delta'}} + \Phi_{\hat{e}^{\text{Neum}}_{\Omega_h,S^{p+k}_{\Delta'}}}$.

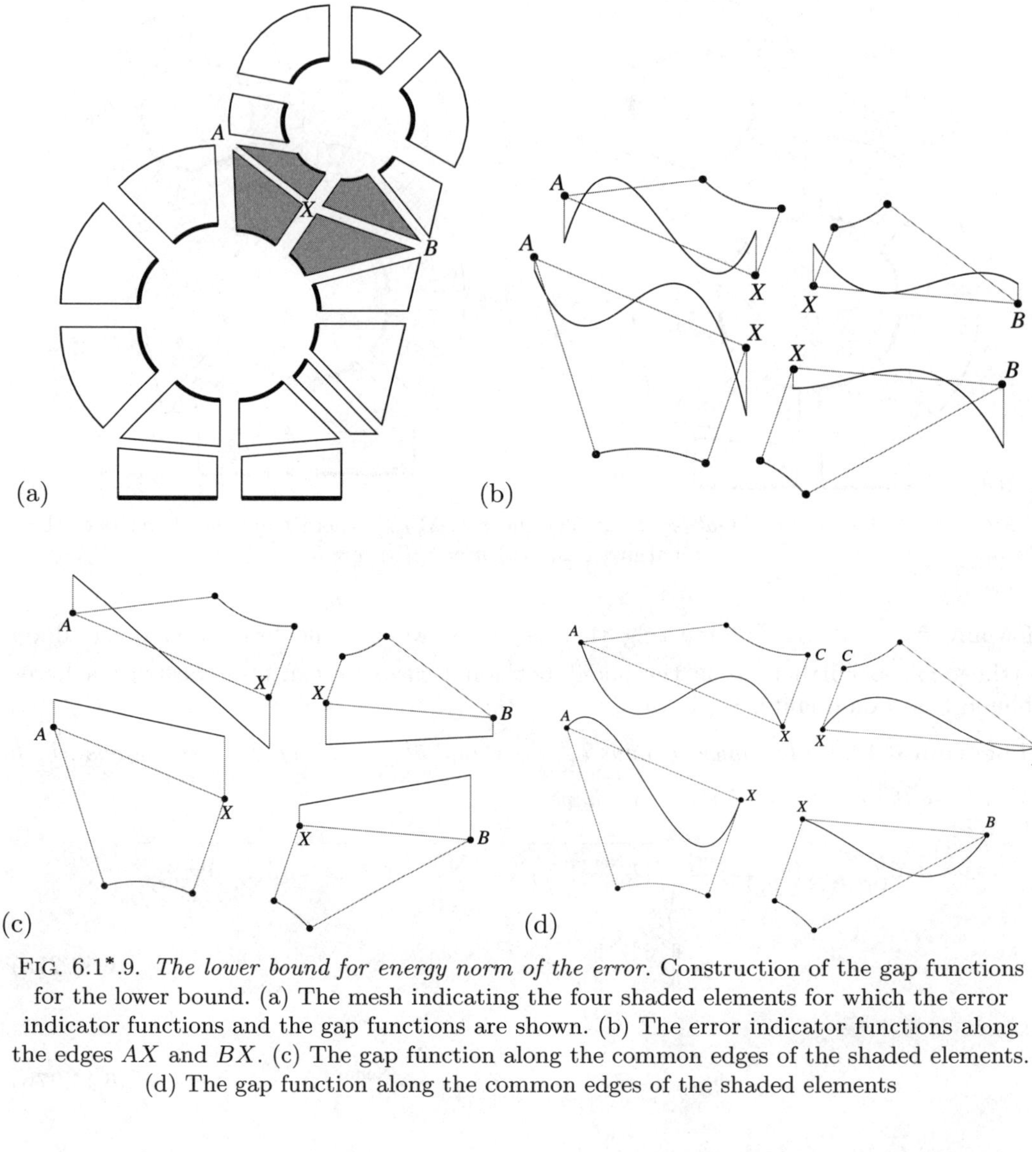

FIG. 6.1*.9. *The lower bound for energy norm of the error.* Construction of the gap functions for the lower bound. (a) The mesh indicating the four shaded elements for which the error indicator functions and the gap functions are shown. (b) The error indicator functions along the edges AX and BX. (c) The gap function along the common edges of the shaded elements. (d) The gap function along the common edges of the shaded elements

Let us now give one particular method of construction of the gap function $\Phi_{\hat{e}^{\mathrm{Neum}}}$. Let $\mathcal{X}_{\Delta_h}$ denote the set of vertices in the mesh. For any vertex $X \in \mathcal{X}_{\Delta_h}$, note that the vertex values $\hat{e}^{\mathrm{Neum}}(X)$ are well defined because we have

$$\hat{e}^{\mathrm{Neum}}|_{\omega_j} = \hat{e}^{\mathrm{EX}}_{\omega_j} \in H^{1+\delta}(\omega_j) \tag{6.1*.29}$$

where $\delta > 0$, and hence $\hat{e}^{\mathrm{Neum}}_{\omega_j}$ is continuous at the vertices of ω_j. Figure 6.1*.9b depicts the error indicator functions along the common edges of the shaded elements shown in Figure 6.1*.9a and illustrates the gap at the vertex node X. We define the *vertex gap*

$G^{\text{Neum}}_{X,\tau}$, at the vertex X of the element τ,

$$G^{\text{Neum}}_{X,\tau} \overset{\text{def}}{=} \hat{e}^{\text{Neum}}\big|_{\tau}(X) - \frac{1}{n_X}\sum_{j=1}^{n_X} \hat{e}^{\text{Neum}}\Big|_{\tau_j^X}(X) \tag{6.1*.30a}$$

where $\{\tau_j^X\}_{j=1}^{n_X}$ is the set of elements which are connected to X, and we will construct $\Phi^{(1)}_{\hat{e}\text{Neum}}$, by interpolating the vertex values, namely

$$\Phi^{(1)}_{\hat{e}\text{Neum}}\big|_{\tau} = -\sum_{k=1}^{4} G^{\text{Neum}}_{X,\tau}\varphi_{X_k^\tau}\Big|_{\tau} \tag{6.1*.30b}$$

where $\varphi_{X_k^\tau}$, is the mapped bilinear basis function corresponding to the vertex X_k^τ. The function $\hat{e}^{\text{Neum}} + \Phi^{(1)}_{\hat{e}\text{Neum}}$, is continuous at the vertices but still has gaps across the edges. Finally we let

$$\Phi^{\text{Neum}} = \Phi^{(1)}_{\hat{e}\text{Neum}} + \Phi^{(2)}_{\hat{e}\text{Neum}} \tag{6.1*.30c}$$

where

$$\begin{aligned} \Phi^{\text{Neum}(2)} &\in \mathcal{U}_{[\![\hat{e}^{\text{Neum}}+\Phi^{(1)}_{\hat{e}\text{Neum}}]\!]} \\ &\overset{\text{def}}{=} \left\{ v\big|_{\tau} \in \mathcal{U}_0(\tau) \;\middle|\; v|_{\varepsilon} = -\tfrac{1}{2}[\![\hat{e}^{\text{Neum}} + \Phi^{(1)}_{\hat{e}\text{Neum}}]\!]_{\varepsilon}\, \varepsilon \subset \partial\tau \right\}. \end{aligned} \tag{6.1*.30d}$$

As in the one-dimensional case $\Phi^{(2)}_{\hat{e}\text{Neum}}$, may be constructed to satisfy the homogenous differential equation in each element, although this is not necessary. Figure 6.1*.9d depicts the gap functions $\Phi^{(2)}_{\hat{e}\text{Neum}}$, along the common edges of the shaded elements shown in Fig. 6.1*.9a. As in the one-dimensional case $\Phi^{(2)}_{\hat{e}\text{Neum}}$, may be constructed to satisfy the homogenous differential equation in each element, although this is not necessary. Here we employed an equal splitting of the gaps at the vertices, and the interelement edges. A more general version can be obtained by employing a weighted splitting of the vertex and edge gaps.

Let us now examine the accuracy of the lower estimators $\mathcal{E}^L_{\Omega_h, S^{p+k}_{\Delta'}}$, and $\mathcal{E}^L_{dual}$, in the case of our example problem.

TABLE 6.1*.3. *The lower estimators for the energy norm of the error.* The estimators based on the $p+2$ solutions of the element residual problems. Comparison the effectivity indices of the lower bounds based on duality, and orthogonality. Note that the lower bound based on duality is overall better, than the lower bound based on orthogonality

	Mesh I		Mesh II		Mesh I		Mesh II	
	$p=1$	$p=2$	$p=1$	$p=2$	$p=1$	$p=2$	$p=1$	$p=2$
	Domain I				Domain II			
Orthogonality	0.779	0	0.829	0	0.803	0.749	0.853	0.756
Duality	0.929	0.517	0.894	0.395	0.882	0.759	0.658	0.417

Example 6.1*.2 (*The lower estimators for the energy norm of the error*) Table 6.1*.4 compares the effectivity indices for the two types of lower bounds based on the computed indicators with $k = 8$. Note that the lower bound based on duality is overall better, than the lower bound based on orthogonality. Table 6.1*.5 illustrates the effect of k. Note that there is a little improvement as k is increased.

Working similarly as in the one dimensional case, we obtain sequences of improved computable bounds

$$\mathscr{E}^{L,(m)}_{\Omega_h, S^{p+k}_{\Delta_{h/2^n}}} \leq \|e_{S^p_\Delta}\|_{\mathscr{U}} \leq \mathscr{E}^{U,(m)}_{\Omega_h, S^{p+k}_{\Delta_{h/2^n}}} . \tag{6.1*.31}$$

Let us now examine the accuracy of the improved bounds $\mathscr{E}^{L,(m)}_{\Omega_h, S^{p+k}_{\Delta_{h/2^n}}}$, $\mathscr{E}^{U,(m)}_{\Omega_h, S^{p+k}_{\Delta_{h/2^n}}}$ in the case of our example problem.

Example 6.1*.3 (*Iterative improvement of the bounds*) Figure 6.1*.10 shows the effectivity index of the estimators $\mathscr{E}^{L,(m)}_{\Delta_h, S^{p+k}_{\Delta_{h/2^n}}}$, and $\mathscr{E}^{U,(m)}_{\Delta_h, S^{p+k}_{\Delta_{h/2^n}}}$, for the error in the finite element solutions of the model problem in Domain I, and Domain II, using Mesh I, and Mesh II and elements of degree $p = 1, 2$, for various values of k. It can be seen that there is no improvement in the bounds after one iteration, and that further iterations may lead to divergence of the upper bound, and the effect of k in the improved bounds is not significant. Nevertheless, the first iteration often leads to significant improvement of the bounds.

Remark 6.1*.7 Note that in our example problem we employed a rather crude mesh. When a more refined mesh is employed, the accuracy of both bound improves.

Remark 6.1*.8 Above we have constructed the bounds based on the computed estimator $\mathscr{E}^{\text{Neum}}_{S^{p+k}_\Delta}$. In Chapter 5, we have seen that the computed estimator $\mathscr{E}^{\text{Neum}}_{B^{p+k}_\Delta}$, where B^{p+k}_Δ is the bubble space defined on the mesh Δ, has better effectivity that $\mathscr{E}^{\text{Neum}}_{S^{p+k}_\Delta}$, asymptotically. Following similar steps as above we can also construct computable bounds based on the $\hat{e}^{\text{Neum}}_{B^{p+k}_\Delta}$ indicators.

Following similar steps as in the one-dimensional case, we can construct a guaranteed estimator based on the ZZ–SPR estimator. Let us employ, as our model problem, the

TABLE 6.1*.4. *The effect of k on the accuracy of the lower bounds.* The effectivity indices of the lower bounds for $p + k = 4, 5, 6$, and 8 for the finite element solutions of the model problem for Domain I and Domain II using Mesh I and Mesh II and $p = 1, 2$

k	Mesh I		Mesh II		Mesh I		Mesh II	
	$p = 1$	$p = 2$	$p = 1$	$p = 2$	$p = 1$	$p = 2$	$p = 1$	$p = 2$
	Domain I				Domain II			
2	0.780	0	0.830	0	0.803	0.749	0.853	0.756
4	0.780	0.095	0.830	0	0.831	0.837	0.838	0.804
6	0.780	0.061	0.830	0	0.836	0.862	0.758	0.763

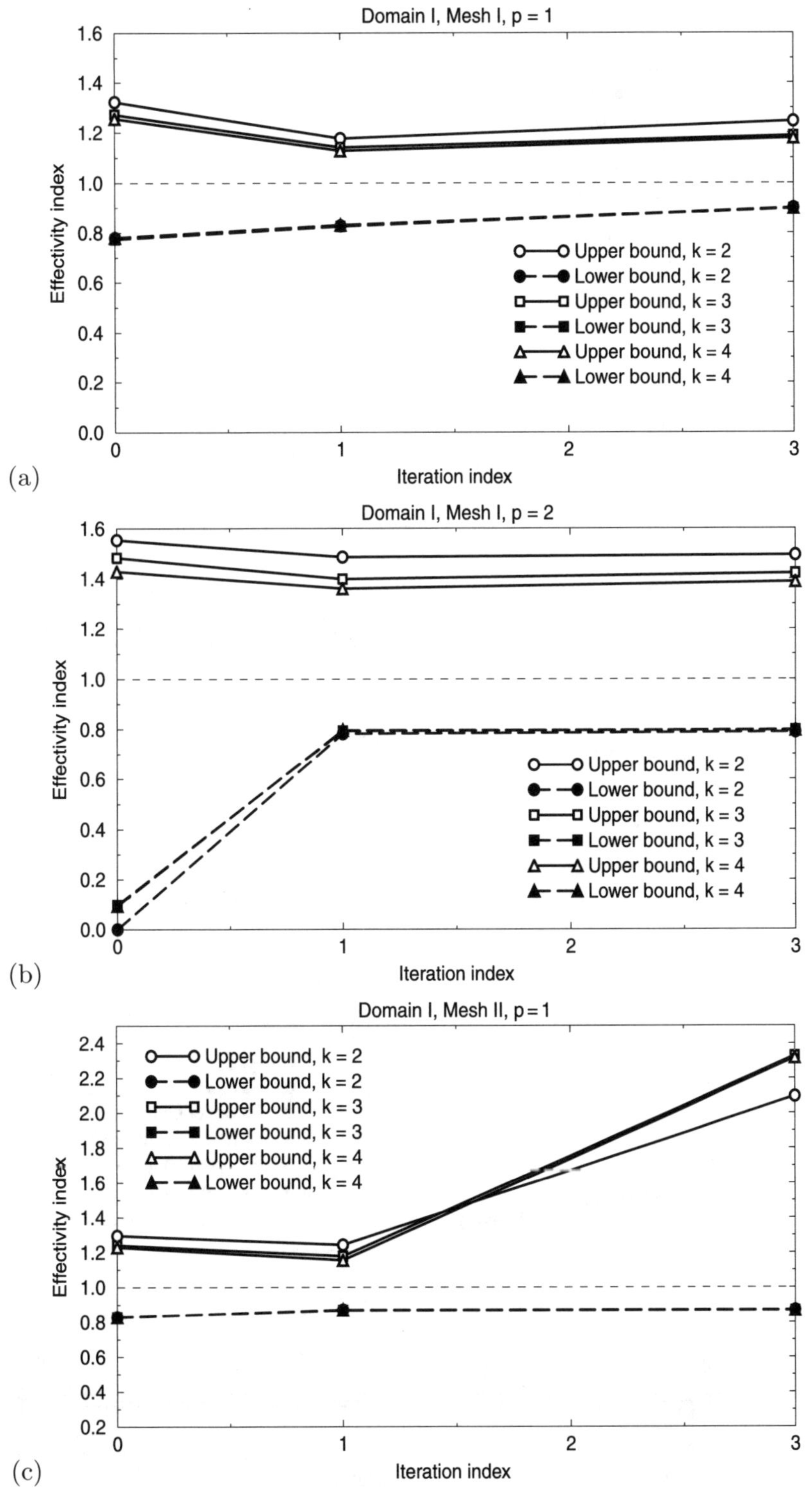
Domain I, Mesh I, p = 1
Effectivity index
Iteration index
Upper bound, k = 2
Lower bound, k = 2
Upper bound, k = 3
Lower bound, k = 3
Upper bound, k = 4
Lower bound, k = 4
(a)
Domain I, Mesh I, p = 2
Effectivity index
Iteration index
Upper bound, k = 2
Lower bound, k = 2
Upper bound, k = 3
Lower bound, k = 3
Upper bound, k = 4
Lower bound, k = 4
(b)
Domain I, Mesh II, p = 1
Effectivity index
Iteration index
Upper bound, k = 2
Lower bound, k = 2
Upper bound, k = 3
Lower bound, k = 3
Upper bound, k = 4
Lower bound, k = 4
(c)

FIG. 6.1*.10.

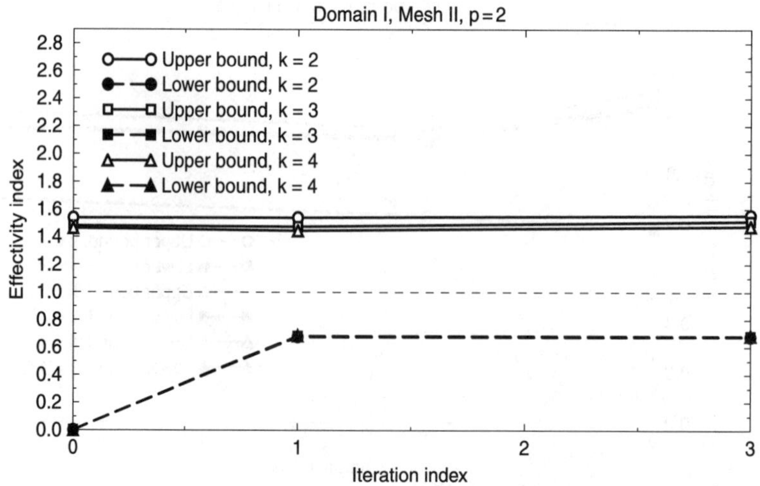
Domain I, Mesh II, p=2
Upper bound, k = 2
Lower bound, k = 2
Upper bound, k = 3
Lower bound, k = 3
Upper bound, k = 4
Lower bound, k = 4
Effectivity index
Iteration index

(d)

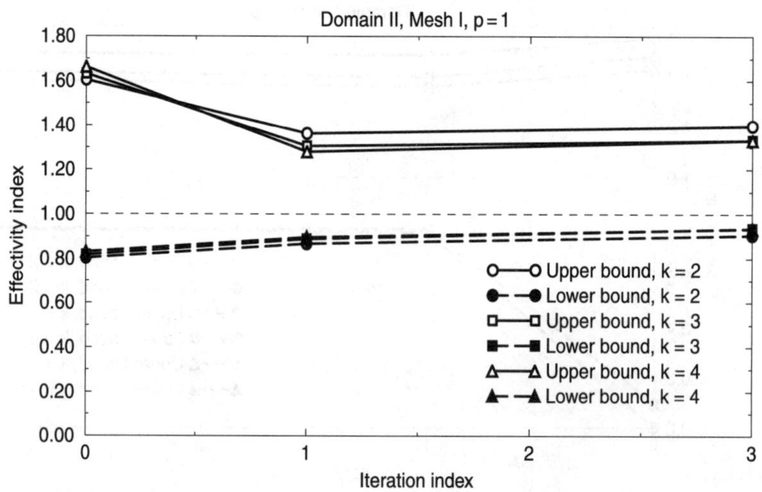
Domain II, Mesh I, p=1
Upper bound, k = 2
Lower bound, k = 2
Upper bound, k = 3
Lower bound, k = 3
Upper bound, k = 4
Lower bound, k = 4
Effectivity index
Iteration index

(e)

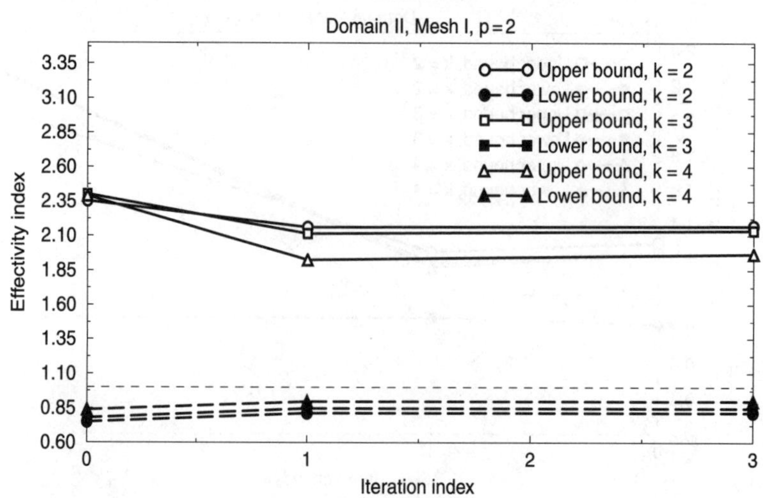
Domain II, Mesh I, p=2
Upper bound, k = 2
Lower bound, k = 2
Upper bound, k = 3
Lower bound, k = 3
Upper bound, k = 4
Lower bound, k = 4
Effectivity index
Iteration index

(f)

FIG. 6.1*.10.

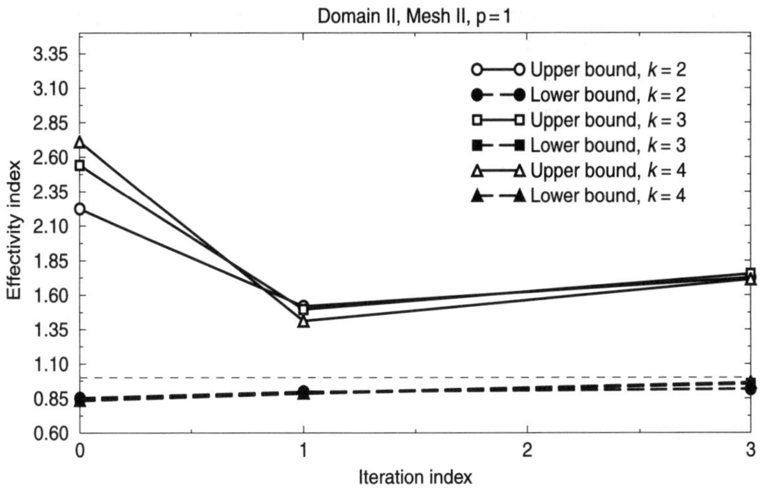

(g)

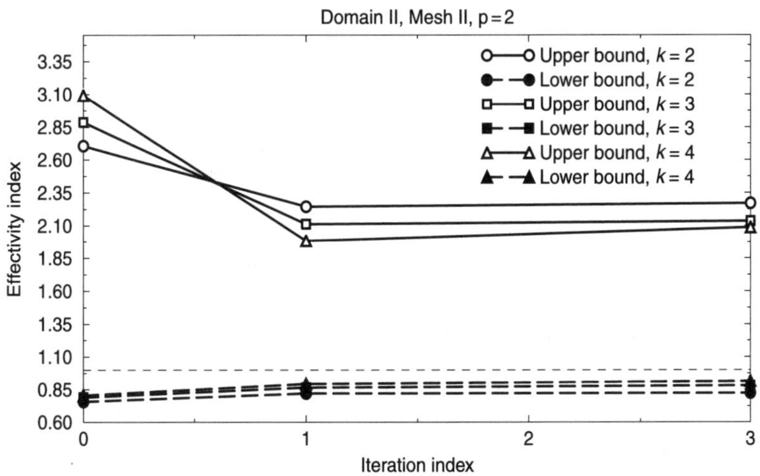

(h)

FIG. 6.1*.10. *Iterative improvement of the bounds.* The effectivity index of the estimators $\mathscr{E}^{L,(m)}_{\Delta_h,S^{p+k}_{\Delta_h}}$, and $\mathscr{E}^{U,(m)}_{\Delta_h,S^{p+k}_{\Delta_h}}$. For Domain I: (a) Mesh I, $p = 1$ (b) Mesh I, $p = 2$ (c) Mesh II, $p = 1$ (d) Mesh II, $p = 2$. For Domain II: (a) Mesh I, $p = 1$ (b) Mesh I, $p = 2$ (c) Mesh II, $p = 1$ (d) Mesh II, $p = 2$. Note that the first iteration often leads to significantly improved bounds

homogenous Dirichlet problem for the Poisson equation

$$-\Delta u\big|_{\Omega} = f, \qquad u\big|_{\Gamma_{\mathrm{D}}} = 0, \qquad \frac{\partial u}{\partial n}\bigg|_{\Gamma_{\mathrm{N}}} = g. \tag{6.1*.32}$$

We will let

$$\mathfrak{V} \stackrel{\text{def}}{=} \{\mathbf{V} = (V_1, V_2) \mid \|\mathbf{V}\|_{\mathfrak{V}} \stackrel{\text{def}}{=} \textstyle\int_{\Omega} \mathbf{V}\boldsymbol{\cdot}\mathbf{V} < \infty\} = L^2(\Omega) \times L^2(\Omega) \tag{6.1*.33}$$

be the space of square integrable vector-valued functions, and for any $\mathbf{V} \in \mathfrak{V}$ we will define its divergence $div\mathbf{V} = \boldsymbol{\nabla} \cdot \mathbf{V} \in \mathcal{U}^{-1}$, by

$$\int_\Omega v\boldsymbol{\nabla}\cdot\mathbf{V} = -\int_\Omega \mathbf{V}\cdot\boldsymbol{\nabla} v \quad \forall v \in \mathcal{U} = \mathcal{U}_0(\Omega). \tag{6.1*.34}$$

As in the one dimensional case we will split $\mathfrak{V}$, into the subsets

$$\mathfrak{V}_1 = \{\mathbf{V} \in \mathfrak{V} \mid \ \exists v \in \mathcal{U} : \ \mathbf{V} = \boldsymbol{\nabla} v\} \tag{6.1*.35a}$$

and

$$\mathfrak{V}_2 = \{\mathbf{V} \in \mathfrak{V} \mid \ \textstyle\int_\Omega \mathbf{V}\cdot\boldsymbol{\nabla} v = 0 \ \forall v \in \mathcal{U}\} \tag{6.1*.35b}$$

and we have

$$\int_\Omega \mathbf{U}\cdot\mathbf{V} = 0 \quad \forall \mathbf{U} \in \mathfrak{V}_1, \mathbf{V} \in \mathfrak{V}_2 \tag{6.1*.36}$$

which means, as in one dimension, that we are assuming that on Γ_{N} we have $\mathbf{V}\cdot\mathbf{n} = 0$ weakly. Letting

$$\hat{\mathbf{e}}^{\mathcal{A}^*_{\nabla u}} \stackrel{\text{def}}{=} \mathcal{A}^*_{\nabla u}(u_{S^p_\Delta}) - \boldsymbol{\nabla} u_{S^p_\Delta} \tag{6.1*.37}$$

for the ZZ–SPR error indicators, and

$$\mathbf{e}^{\mathcal{A}^*_{\nabla u}} \stackrel{\text{def}}{=} \boldsymbol{\nabla} u_{EX} - \mathcal{A}^*_{\nabla u}(u_{S^p_\Delta}) = \boldsymbol{\nabla} e_{S^p_\Delta} - \hat{\mathbf{e}}^{\mathcal{A}^*_{\nabla u}} \tag{6.1*.38}$$

for the error in the recovered flux, we get

$$\boldsymbol{\nabla}\cdot\mathbf{e}^{\mathcal{A}^*_{\nabla u}} = -f - \boldsymbol{\nabla}\cdot\mathcal{A}^*_{\nabla u}(u_{S^p_\Delta}). \tag{6.1*.39}$$

Assuming now that $\mathcal{A}^*_{\nabla u}$ is a continuous recovery as discussed in Section 4.8, we have $\mathcal{A}^*_{\nabla u}(u_{S^p_\Delta}) \in S^p_\Delta \times S^p_\Delta$, and hence

$$\boldsymbol{\nabla}\cdot\mathcal{A}^*_{\nabla u}(u_{S^p_\Delta}) \in L^2(\Omega). \tag{6.1*.40}$$

As in the one-dimensional case, we will construct the function $w^{\mathcal{A}^*_{\nabla u}}$, as the solution of the variational problem:

Find $w^{\mathcal{A}^*_{\nabla u}} \in \mathcal{U}$ such that

$$\begin{aligned} \mathcal{B}(w^{\mathcal{A}^*_{\nabla u}}, v) &= \int_\Omega (-f - \boldsymbol{\nabla}\cdot(\mathcal{A}^*_{\nabla u}(u_{S^p_\Delta})))\, v - \oint_{\partial\Omega_{\mathrm{N}}} (g - \mathcal{A}^*_{\nabla u}(u_{S^p_\Delta})\cdot\mathbf{n})\, v \qquad (6.1^*.41)\\ &= \int_\Omega \boldsymbol{\nabla}\cdot\mathbf{e}^{\mathcal{A}^*_{\nabla u}} - \oint_{\partial\Omega_{\mathrm{N}}} (g - \mathcal{A}^*_{\nabla u}(u_{S^p_\Delta})\cdot\mathbf{n})\, v \quad \forall v \in \mathcal{U} \end{aligned}$$

and we get

$$-\boldsymbol{\nabla}\cdot\boldsymbol{\nabla} w^{\mathcal{A}^*_{\nabla u}} = \boldsymbol{\nabla}\cdot\mathbf{e}^{\mathcal{A}^*_{\nabla u}} \tag{6.1*.42}$$

with the boundary conditions

$$\frac{\partial}{\partial n}(w^{\mathcal{A}^*_{\nabla u}}) = -(g - \mathcal{A}^*_{\nabla u}(u_{S^p_\Delta}) \cdot \mathbf{n}) \quad \text{on } \partial\Omega_{\mathrm{N}}.$$

Hence

$$\nabla \cdot (\nabla w^{\mathcal{A}^*_{\nabla u}} + \mathbf{e}^{\mathcal{A}^*_{\nabla u}}) = 0 \tag{6.1*.43a}$$

and

$$\begin{aligned} 0 &= \frac{\partial}{\partial n}(w^{\mathcal{A}^*_{\nabla u}}) + (g - \mathcal{A}^*_{\nabla u}(u_{S^p_\Delta}) \cdot \mathbf{n}) \\ &= \frac{\partial}{\partial n}(w^{\mathcal{A}^*_{\nabla u}}) + \frac{\partial}{\partial n}(u_{EX}) - \mathcal{A}^*_{\nabla u}(u_{S^p_\Delta}) \cdot \mathbf{n} \\ &= (\nabla w^{\mathcal{A}^*_{\nabla u}} + \nabla u_{EX} - \mathcal{A}^*_{\nabla u}(u_{S^p_\Delta}))\mathbf{n} \\ &= (\nabla w^{\mathcal{A}^*_{\nabla u}} + \mathbf{e}^{\mathcal{A}^*_{\nabla u}})\mathbf{n}. \end{aligned} \tag{6.1*.43b}$$

This shows that

$$\nabla w^{\mathcal{A}^*_{\nabla u}} + \mathbf{e}^{\mathcal{A}^*_{\nabla u}} \in \mathcal{B}_2 \tag{6.1*.48}$$

and from (6.1*.35) we get

$$\int_\Omega (\nabla w^{\mathcal{A}^*_{\nabla u}} + \mathbf{e}^{\mathcal{A}^*_{\nabla u}})\nabla v = 0 \quad \forall v \in \mathcal{U} \tag{6.1*.44a}$$

which can also be written as

$$\int_\Omega (\nabla w^{\mathcal{A}^*_{\nabla u}} + \nabla e_{S^p_\Delta} - \hat{\mathbf{e}}^{\mathcal{A}^*_{\nabla u}})\nabla v = 0 \quad \forall v \in \mathcal{U} \tag{6.1*.44b}$$

or

$$\mathcal{B}(\nabla w^{\mathcal{A}^*_{\nabla u}}, v) + \mathcal{B}(e_{S^p_\Delta}, v) - \mathcal{B}(g^{\mathcal{A}^*_p}, v) = 0 \quad \forall v \in \mathcal{U} \tag{6.1*.44c}$$

where $g^{\mathcal{A}^*_p} \in \mathcal{U}$ is such that

$$\mathcal{B}(g^{\mathcal{A}^*_p}, v) = \int_\Omega \hat{\mathbf{e}}^{\mathcal{A}^*_{\nabla u}} \nabla v. \tag{6.1*.45}$$

Hence

$$-e_{S^p_\Delta} = w^{\mathcal{A}^*_{\nabla u}} - g^{\mathcal{A}^*_p} \tag{6.1*.46}$$

and

$$\|g\|_{\mathcal{U}} \le \|\hat{\mathbf{e}}^{\mathcal{A}^*_{\nabla u}}\|_{L^2(\Omega)} = \sqrt{\|\hat{e}_1^{\mathcal{A}^*_{\nabla u}}\|^2_{L^2(\Omega)} + \|\hat{e}_2^{\mathcal{A}^*_{\nabla u}}\|^2_{L^2(\Omega)}} = \mathcal{E}^{\mathcal{A}^*_{\nabla u}} \tag{6.1*.47}$$

where $\hat{\mathbf{e}}^{\mathcal{A}^*_{\nabla u}} = (\hat{e}_1^{\mathcal{A}^*_{\nabla u}}, \hat{e}_2^{\mathcal{A}^*_{\nabla u}})^{\mathrm{T}}$. We have therefore

$$\|g\|_{\mathcal{U}} \le \mathcal{E}^{\mathcal{A}^*_{\nabla u}} \tag{6.1*.48}$$

and as in the one-dimensional case we get

$$|\mathcal{E}^{\mathcal{A}^*_{\nabla u}} - \|w^{\mathcal{A}^*_{\nabla u}}\|_{\mathcal{U}}| \le \|e_{S^p_\Delta}\|_{\mathcal{U}} \le \mathcal{E}^{\mathcal{A}^*_{\nabla u}} + \|w^{\mathcal{A}^*_{\nabla u}}\|_{\mathcal{U}}. \tag{6.1*.49}$$

We now estimate $\|w^{\mathcal{A}^*_{\nabla u}}\|_{\mathcal{U}}$ using

$$\sqrt{\left\|w^{\mathcal{A}^*_{\nabla u}}_{S^p_\Delta}\right\|^2_{\mathcal{U}} + \left(\mathcal{E}^L_{w^{\mathcal{A}^*_{\nabla u}}_{S^p_\Delta}, S^{p+k}_{\Delta'}}\right)^2} \le \|w^{\mathcal{A}^*_{\nabla u}}\|_{\mathcal{U}} \le \sqrt{\left\|w^{\mathcal{A}^*_{\nabla u}}_{S^p_\Delta}\right\|^2_{\mathcal{U}} + \left(\mathcal{E}^U_{w^{\mathcal{A}^*_{\nabla u}}_{S^p_\Delta}, S^{p+k}_{\Delta'}}\right)^2} \tag{6.1*.50}$$

and get desirable results. As in one dimension we have

$$\|e_{S^p_\Delta}\|_{\mathcal{U}} = \|w^{\mathcal{A}^*_{\nabla u}} - g^{\mathcal{A}^*_{\nabla u}}\|_{\mathcal{U}}.$$

We have $\|(w^{\mathcal{A}^*_{\nabla u}} - g^{\mathcal{A}^*_{\nabla u}})_{S^p_\Delta}\| = 0$ and, as in one dimension, we have

$$\mathcal{E}^L_{(w^{\mathcal{A}^*_{\nabla u}} - g^{\mathcal{A}^*_{\nabla u}}), S^{p+k}_{\Delta'}} \le \|e_{S^p_\Delta}\|_{\mathcal{U}} \le \mathcal{E}^U_{(w^{\mathcal{A}^*_{\nabla u}} - g^{\mathcal{A}^*_{\nabla u}}), S^{p+k}_{\Delta'}}. \tag{6.1*.51}$$

As in one dimension (6.1*.50) leads to the guaranteed a-posteriori estimate of the ZZ–SPR estimator and (6.1*.51) is the guaranteed estimator utilizing the recovery technique.

Remark 6.1*.9 The approach employed above for constructing guaranteed bounds based on the ZZ–SPR estimator is an application of the method of orthogonal projections; see [Weil, 1940], [Mikhlin, 1964], and [Vishik, 1949].

Remark 6.1*.10 Let us address briefly the cost of the computations of the estimators. The finite element computation has essentially three parts:

- computation of the stiffness matrix and right-hand side
- solution of the system of linear equations
- error estimation and postprocessing.

The cost of the computation of the stiffness matrix is $\approx C_1 N$, where N is the number of elements and C_1 strongly depends on how many integration points are used, especially for curved elements, and on the implementation.

In two dimensions the cost of elimination in practice is of order $\approx C_2 N^\alpha$ with $\alpha \approx 2$ up to $N \approx 10^5$.

Computation of the residual error estimation has the following essential parts:

- equilibration
- computation of the stiffness matrix and load vectors for the Neumann problem
- computation of indicator and its norm
- construction of the gap-function
- iterative improvement of the estimators.

All these parts are local and hence the cost is $\approx C_3 N$. Nevertheless, the constant C_3 can be large if the implementation is not sophisticated. Let us mention that by 'naive' implementation we can get $C_3 N > C_2 N^\alpha$ for N of practical interest.

If the load f is smooth and the elements are straight or curvilinear but not too distorted, then usually Gauss integration of degree $p+1$ (i.e., in square $(p+1)^2$) integration points are sufficient, although an adaptive integration would be desirable. Let us underline that the numerical quadrature is a major cost in the error estimation.

The cost of the error estimation can be reduced when the finite element mesh is not optimal and when the error is large only in a few elements. This is, for example, the case for uniform meshes and domains with unsmooth boundary. In this case we use the estimator $\mathscr{E}^{U,(m)}_{\Delta_h, S^{p+k}_{\Delta_h}}$ for $k > 0$ larger for these few elements and $k = 1$ for the others. By this we still can get guaranteed upper and lower estimates. Of course, the cost of the estimate (respectively its accuracy) has to be compared to the cost of the finite element solution on a refined mesh.

Remark 6.1*.11 Similar results can be obtained in the case of the elasticity problem. In this case however we must also address the effect of the residual moment in each element.

In this section we have shown that the results presented in Section 6.1 in one dimension have complete analogue in two dimensions. We demonstrated that for the large class of problems the estimator which satisfies all the desirable properties of an estimator listed in the introduction to this section exist and are computationally effective if properly implemented.

Problems

P6.1*.1. Repeat the derivation of the guaranteed estimator based on the ZZ–SPR estimator in the case that the model problem is

$$\boldsymbol{\nabla}\cdot(\mathbf{K}\boldsymbol{\nabla}u)\big|_{\Omega} = f, \qquad u|_{\Omega} = 0 \tag{6.1*.52a}$$

or

$$\Delta u|_{\Omega} = 0, \qquad \frac{\partial u}{\partial n}\bigg|_{\partial\Omega} = g \tag{6.1*.52b}$$

with g such that $\int_{\partial\Omega} g = 0$.

6.2 Upper and lower bounds for the error in the outputs, and a-posteriori estimation of the pollution

In this section we will address the construction of computable bounds

$$\mathscr{E}^{L,\mathscr{F}} \le e^{\mathscr{F}}_{S^p_\Delta} = \mathscr{F}(u_{EX}) - \mathscr{F}(u_{S^p_\Delta}) \le \mathscr{E}^{U,\mathscr{F}} \tag{6.2.1a}$$

where $\mathscr{F} : \mathscr{U} \mapsto \mathbb{R}$, is a functional which expresses the output of interest, the construction of estimators $\mathscr{E}^{\mathscr{F}}_{S^{p+k}_{\Delta'}}$, with asymptotic meaning, namely,

$$\left|e^{\mathscr{F}}_{S^p_\Delta}\right| \approx \mathscr{E}^{\mathscr{F}}_{S^{p+k}_{\Delta'}} \tag{6.2.1b}$$

and the a-posteriori estimation of the pollution error.

In practice, we are mostly interested in outputs which can be expressed as weighted averages of the solution, its derivative, or the flux over small subdomains, like for example

$$\mathscr{F}(u) = \frac{1}{b-a}\int_a^b wq(u) \tag{6.2.2a}$$

where $(a, b) \subseteq \bar{I}$, w, is a weight function, and $q(u) = u$, $(d^k u)/(dx^k)$, or $AE(du/dx)$, etc; or outputs which can be expressed as point values

$$\mathcal{F}(u) = q(u)(\bar{x}). \tag{6.2.2b}$$

We will assume that $\mathcal{F}$ is *linear*, namely that

$$\mathcal{F}(\alpha u + \beta v) = \alpha \mathcal{F}(u) + \beta \mathcal{F}(v) \tag{6.2.3}$$

for any u, $v \in \mathcal{U}$, and α, $\beta \in \mathbb{R}$, and that either

(a) $\mathcal{F}$ is bounded on the entire energy space $\mathcal{U}$, namely

$$|\mathcal{F}(v)| \leq \mathcal{C} \|v\|_{\mathcal{U}} \quad \forall v \in \mathcal{U} \tag{6.2.4}$$

or

(b) $\mathcal{F}$ is bounded on a subset of the energy space, $\mathcal{U}^{\mathcal{F}} \subset \mathcal{U}$, namely

$$|\mathcal{F}(v)| \leq \mathcal{C} \|v\|_{\mathcal{U}} \quad \forall v \in \mathcal{U}^{\mathcal{F}} \tag{6.2.5a}$$

and the exact solution of interest belongs in this subset, namely,

$$u_{EX} \in \mathcal{U}^{\mathcal{F}}. \tag{6.2.5b}$$

For example,

$$\mathcal{F}(u) = u(\bar{x}) \tag{6.2.6a}$$

and

$$\mathcal{F}(u) = \frac{1}{x_2 - x_1} \int_{x_1}^{x_2} AE \, \frac{du}{dx} \tag{6.2.6b}$$

are bounded in $\mathcal{U}$, while

$$\mathcal{F}(u) = AE \frac{du}{dx}(\bar{x}) \tag{6.2.6c}$$

is not. However, we have

$$|\mathcal{F}(v)| \leq \mathcal{C} \|v\|_{\mathcal{U}} \quad \forall v \in \mathcal{U}^{\mathcal{F}} \stackrel{\text{def}}{=} \left\{ v \in \mathcal{U} \,\middle|\, \left\| AE \frac{dv}{dx} \right\|_{L^\infty(I)} \leq \infty \right\} \tag{6.2.7a}$$

and in the case that we know that the input data, AE, c, and f, satisfy

$$\left\| \frac{d}{dx}(AE) \right\|_{L^\infty(I)} \leq A, \qquad \|c\|_{L^\infty(I)} \leq B, \qquad \|f\|_{L^\infty(I)} \leq C \tag{6.2.7b}$$

from Theorem 2.3.1, we also know that

$$\left\| AE \frac{d}{dx}(u_{EX}) \right\|_{L^\infty(I)} \leq D \tag{6.2.7c}$$

and hence $u_{EX} \in \mathcal{U}_{\mathcal{F}}$, and $\mathcal{F}(u_{EX})$ has good meaning.

Remark 6.2.1 Let us underline that in order to guarantee that $u_{EX} \in \mathcal{U}^{\mathcal{F}}$, we have to know some information about the smoothness of the input data.

Let us first address the case that $\mathcal{F}$ is bounded on the entire energy space $\mathcal{U}$, namely,

$$|\mathcal{F}(v)| \leq \mathcal{C}\|v\|_{\mathcal{U}} \quad \forall v \in \mathcal{U}. \tag{6.2.8}$$

We will let $H^{\mathcal{F}}$, denote the solution of the variational problem:

Find $H^{\mathcal{F}} \in \mathcal{U}$ such that

$$\mathcal{B}(H^{\mathcal{F}}, v) = \mathcal{F}(v) \quad \forall v \in \mathcal{U}. \tag{6.2.9a}$$

Note that $H^{\mathcal{F}}$ exists because $\mathcal{F}$ is bounded in $\mathcal{U}$ and is known as the *influence, dual,* or *extraction* function associated with $\mathcal{F}$. We will let $H^{\mathcal{F}}_{S^p_\Delta}$ denote the finite element approximation of $H^{\mathcal{F}}$, namely the solution of the discrete variational problem:

Find $H^{\mathcal{F}}_{S^p_\Delta} \in S^p_\Delta$ such that

$$\mathcal{B}\big(H^{\mathcal{F}}_{S^p_\Delta}, v\big) = \mathcal{F}(v) \quad \forall v \in S^p_\Delta. \tag{6.2.9b}$$

We then have

$$\mathcal{F}(u_{EX}) = \mathcal{B}(H^{\mathcal{F}}, u_{EX}) \tag{6.2.10a}$$

and

$$\mathcal{F}(u_{S^p_\Delta}) = \mathcal{B}\big(H^{\mathcal{F}}_{S^p_\Delta}, u_{S^p_\Delta}\big) \tag{6.2.10b}$$

and subtracting

$$e^{\mathcal{F}}_{S^p_\Delta} = \mathcal{F}(u_{EX}) - \mathcal{F}(u_{S^p_\Delta}) = \mathcal{B}(H^{\mathcal{F}}, u_{EX}) - \mathcal{B}\big(H^{\mathcal{F}}_{S^p_\Delta}, u_{S^p_\Delta}\big). \tag{6.2.10c}$$

Taking into account the orthogonalities satisfied by the finite element solutions $H^{\mathcal{F}}_{S^p_\Delta}$, and $u_{S^p_\Delta}$, we obtain:

Theorem 6.2.1 *Let $\mathcal{F}$ be bounded in $\mathcal{U}$, and let $H^{\mathcal{F}}$, be defined by (6.2.9a). We then have*

$$e^{\mathcal{F}}_{S^p_\Delta} = \mathcal{B}\big(H^{\mathcal{F}} - H^{\mathcal{F}}_{S^p_\Delta}, e_{S^p_\Delta}\big) \tag{6.2.11a}$$

and

$$\big|e^{\mathcal{F}}_{S^p_\Delta}\big| \leq \big\|H^{\mathcal{F}} - H^{\mathcal{F}}_{S^p_\Delta}\big\|_{\mathcal{U}} \|e_{S^p_\Delta}\|_{\mathcal{U}}. \tag{6.2.11b}$$

Proof It is the same as the proof of Corollary 4.2.1. □

Remark 6.2.2 From (6.2.11b) it follows that $|e^{\mathcal{F}}_{S^p_\Delta}|$, is smaller than $\|e_{S^p_\Delta}\|_{\mathcal{U}}$. For example, as we have already seen in Section 4.1, in the case of the regular model problem with input data

$$AE \equiv 1, \qquad c \equiv 1, \qquad \alpha_2 = 0, \qquad a_2 = 0 \tag{6.2.12a}$$

and smooth f, we have

$$\|e_{S^p_\Delta}\|_{\mathcal{U}} \le \mathscr{C}(h^\Delta)^p \tag{6.2.12b}$$

and hence

$$|e^{\mathscr{F}}_{S^p_\Delta}| \le \|H^{\mathscr{F}} - H^{\mathscr{F}}_{S^p_\Delta}\|_{\mathcal{U}}\|e_{S^p_\Delta}\|_{\mathcal{U}} \le \mathscr{C}(h^\Delta)^p\|H^{\mathscr{F}} - H^{\mathscr{F}}_{S^p_\Delta}\|_{\mathcal{U}}. \tag{6.2.12c}$$

For example, in the case that

$$\mathscr{F}(v) = v(x^\Delta_j) \tag{6.2.12d}$$

where x^Δ_j is a nodal point, we have $H^{\mathscr{F}} \equiv G(\cdot;x^\Delta_j)$, where $G(\cdot;x^\Delta_j)$ is the Green's function, and as we have seen in Section 4.1, we have

$$\|H^{\mathscr{F}} - H^{\mathscr{F}}_{S^p_\Delta}\|_{\mathcal{U}} = \|G(\cdot;x^\Delta_j) - G_{S^p_\Delta}(\cdot;x^\Delta_j)\|_{\mathcal{U}} \le \mathscr{C}(h^\Delta)^p \tag{6.2.12e}$$

where we let $G_{S^p_\Delta}(\cdot;x^\Delta_j)$, denote the finite element approximation of $G(\cdot;x^\Delta_j)$, and we get

$$|e^{\mathscr{F}}_{S^p_\Delta}| = |u_{EX}(x^\Delta_j) - u_{S^p_\Delta}(x^\Delta_j)| \le \mathscr{C}(h^\Delta)^{2p} \approx \mathscr{C}\|e_{S^p_\Delta}\|^2_{\mathcal{U}}. \tag{6.2.12f}$$

The construction of the bounds for $e^{\mathscr{F}}_{S^p_\Delta}$, will be based on the following identity.

Lemma 6.2.1 *For any u, $v \in \mathcal{U}$, and any $s \in \mathbb{R}$, we have*

$$\mathscr{B}(u,v) = \tfrac{1}{4}(\|s^{-1}u + su\|^2_{\mathcal{U}} - \|s^{-1}u - su\|^2_{\mathcal{U}}). \tag{6.2.13}$$

Proof Recall that

$$\mathscr{B}(u,v) = \tfrac{1}{4}(\|u + v\|^2_{\mathcal{U}} - \|u - v\|^2_{\mathcal{U}}) \tag{6.2.14}$$

and replace u, by $s^{-1}u$, and v, by sv. □

Let us now introduce a slight change in our notation for the error. We will let

$$e_{u_{S^p_\Delta}} = u_{EX} - u_{S^p_\Delta} \tag{6.2.15a}$$

denote the error in $u_{S^p_\Delta}$, instead of $e_{S^p_\Delta}$. Using this notation we can now write

$$e_{H^{\mathscr{F}}_{S^p_\Delta}} = H^{\mathscr{F}} - H^{\mathscr{F}}_{S^p_\Delta}. \tag{6.2.15b}$$

It follows that

$$e_{s^{-1}u_{S^p_\Delta} \pm sH^{\mathscr{F}}_{S^p_\Delta}} = s^{-1}e_{u_{S^p_\Delta}} \pm se_{H^{\mathscr{F}}_{S^p_\Delta}}. \tag{6.2.15c}$$

Employing these notations, and Lemma 6.2.1 we obtain:

Corollary 6.2.1 *For any bounded functional $\mathscr{F}$, and any $s \in \mathbb{R}$, we have*

$$e^{\mathscr{F}}_{S^p_\Delta} = \tfrac{1}{4}(\|e_{s^{-1}u_{S^p_\Delta} + sH^{\mathscr{F}}_{S^p_\Delta}}\|^2_{\mathcal{U}} - \|e_{s^{-1}u_{S^p_\Delta} - sH^{\mathscr{F}}_{S^p_\Delta}}\|^2_{\mathcal{U}}). \tag{6.2.16}$$

Proof Recalling that

$$e^{\mathcal{F}}_{S^p_\Delta} = \mathcal{B}(H^{\mathcal{F}}, e_{S^p_\Delta}) = \mathcal{B}(H^{\mathcal{F}} - H^{\mathcal{F}}_{S^p_\Delta}, e_{S^p_\Delta}) = \mathcal{B}(e_{H^{\mathcal{F}}_{S^p_\Delta}}, e_{u_{S^p_\Delta}}) \tag{6.2.17a}$$

and employing Lemma 6.2.1 with $u = e_{S^p_\Delta}$, and $v = e_{H^{\mathcal{F}}_{S^p_\Delta}}$, we obtain

$$e^{\mathcal{F}}_{S^p_\Delta} = \mathcal{B}(e_{u_{S^p_\Delta}}, e_{H^{\mathcal{F}}_{S^p_\Delta}}) = \tfrac{1}{4}(\|s^{-1}e_{u_{S^p_\Delta}} + se_{H^{\mathcal{F}}_{S^p_\Delta}}\|^2_{\mathcal{U}} - \|s^{-1}e_{u_{S^p_\Delta}} - se_{H^{\mathcal{F}}_{S^p_\Delta}}\|^2_{\mathcal{U}}) \tag{6.2.17b}$$

which is (6.2.17). □

The bounds for $e^{\mathcal{F}}_{S^p_\Delta}$, follow immediately from (6.2.16), and the bounds for the energy norm of the error $\|e_{s^{-1}u_{S^p_\Delta} - s\ H^{\mathcal{F}}_{S^p_\Delta}}\|_{\mathcal{U}}$. We have:

Theorem 6.2.2 *Under the assumption that $\mathcal{F}$ is bounded in $\mathcal{U}$, for any $s \in \mathbb{R}$, we have*

$$\mathcal{E}^{L,\mathcal{F},s}_{S^{p+k}_{\Delta'}} \le e^{\mathcal{F}}_{S^p_\Delta} \le \mathcal{E}^{U,\mathcal{F},s}_{S^{p+k}_{\Delta'}} \tag{6.2.18a}$$

where

$$\mathcal{E}^{L,\mathcal{F},s}_{S^{p+k}_{\Delta'}} \stackrel{\text{def}}{=} \tfrac{1}{4}\Big(\big(\mathcal{E}^{L}_{\bar{u}^{\mathcal{F},s,+}_{S^p_\Delta}, S^{p+k}_{\Delta'}}\big)^2 - \big(\mathcal{E}^{U}_{\bar{u}^{\mathcal{F},s,-}_{S^p_\Delta}, S^{p+k}_{\Delta'}}\big)^2\Big) \tag{6.2.18b}$$

and

$$\mathcal{E}^{U,\mathcal{F},s}_{S^{p+k}_{\Delta'}} \stackrel{\text{def}}{=} \tfrac{1}{4}\Big(\big(\mathcal{E}^{U}_{\bar{u}^{\mathcal{F},s,+}_{S^p_\Delta}, S^{p+k}_{\Delta'}}\big)^2 - \big(\mathcal{E}^{L}_{\bar{u}^{\mathcal{F},s,-}_{S^p_\Delta}, S^{p+k}_{\Delta'}}\big)^2\Big) \tag{6.2.18c}$$

where we let

$$\bar{u}^{\mathcal{F},s,\pm}_{S^p_\Delta} \stackrel{\text{def}}{=} s^{-1}u_{S^p_\Delta} \pm sH^{\mathcal{F}}_{S^p_\Delta}. \tag{6.2.18d}$$

We also have

$$|e^{\mathcal{F}}_{S^p_\Delta}| \le \mathcal{E}^{U}_{U_{S^p_\Delta}}\ \mathcal{E}^{U}_{H^{\mathcal{F}}_{S^p_\Delta}}. \tag{6.2.18e}$$

Proof Using the notation (6.2.19d), the identity (6.2.17) can be rewritten as

$$e^{\mathcal{F}}_{S^p_\Delta} = \tfrac{1}{4}(\|e_{\bar{u}^{\mathcal{F},s,+}_{S^p_\Delta}}\|^2_{\mathcal{U}} - \|e_{\bar{u}^{\mathcal{F},s,-}_{S^p_\Delta}}\|^2_{\mathcal{U}}) \tag{6.2.19a}$$

and employing the bounds

$$\mathcal{E}^{L}_{\bar{u}^{\mathcal{F},s,\pm}_{S^p_\Delta}, S^{p+k}_{\Delta'}} \le \|e_{\bar{u}^{\mathcal{F},s,\pm}_{S^p_\Delta}}\|_{\mathcal{U}} \le \mathcal{E}^{U}_{\bar{u}^{\mathcal{F},s,\pm}_{S^p_\Delta}, S^{p+k}_{\Delta'}} \tag{6.2.19b}$$

(6.2.18) follows. The bound (6.2.18e) is obtained from (6.2.17a), and the Cauchy-Schwarz inequality. □

Using $\mathcal{E}^{L,(m)}_{S^{p+k}_{\Delta'}}$, $\mathcal{E}^{U,(m)}_{S^p_\Delta}$, the mth iterate of the improved bounds for $\|e_{S^p_\Delta}\|_{\mathcal{U}}$, we obtain $\mathcal{E}^{L,\mathcal{F},(m)}_{S^{p+k}_{\Delta'}}$, $\mathcal{E}^{U,\mathcal{F},(m)}_{S^p_\Delta}$, the corresponding improved bounds for $e^{\mathcal{F}}_{S^p_\Delta}$.

Remark 6.2.3 Note that the bounds $\mathscr{E}^{L,\mathscr{F},s}_{S^{p+k}_{\Delta'}}$, $\mathscr{E}^{U,\mathscr{F},s}_{S^{p+k}_{\Delta'}}$, depend on s, and an optimal value s_{opt} may be determined by minimizing the difference $\mathscr{E}^{U,\mathscr{F},s}_{S^{p+k}_{\Delta'}} - \mathscr{E}^{L,\mathscr{F},s}_{S^{p+k}_{\Delta'}}$.

Let us now illustrate the accuracy of the bounds for $e^{\mathscr{F}}_{S^p_\Delta}$, by a numerical example.

Example 6.2.1 Let us consider the one-dimensional model problem employed in Section 6.1, and let

$$\mathscr{F}(u) = u(\bar{x}) \tag{6.2.20a}$$

with $\bar{x} = \frac{1}{2}$. Note that

$$|u(\bar{x})| \leq \left| \int_0^{\bar{x}} \frac{du}{dx} \right| \leq \bar{x} \left\| \frac{du}{dx} \right\|_{L^2(I)} \leq \mathscr{C}\|u\|_{\mathscr{U}} \tag{6.2.20b}$$

and hence $\mathscr{F}$ is bounded. We will be interested in the effectivity indices

$$\kappa^{L,\mathscr{F},(m)}_{S^{p+k}_\Delta} \stackrel{\text{def}}{=} \frac{\mathscr{E}^{L,\mathscr{F},(m)}_{S^{p+k}_\Delta}}{\mathscr{F}(e_{S^p_\Delta})}, \qquad \kappa^{U,\mathscr{F},(m)}_{S^{p+k}_\Delta} \stackrel{\text{def}}{=} \frac{\mathscr{E}^{U,\mathscr{F},(m)}_{S^{p+k}_\Delta}}{\mathscr{F}(e_{S^p_\Delta})} \tag{6.2.20c}$$

for the mth iterate of the lower and the upper bound for

$$e^{\mathscr{F}}_{S^p_\Delta} = \mathscr{F}(e_{S^p_\Delta}) = e_{S^p_\Delta}\left(\tfrac{1}{2}\right) = u_{EX}\left(\tfrac{1}{2}\right) - u_{S^p_\Delta}\left(\tfrac{1}{2}\right). \tag{6.2.20d}$$

Table 6.2.1 and 6.2.2 report the values of the bounds and the effectivity indices $\kappa^{L,\mathscr{F},(m)}_{S^{p+k}_{\Delta'}}$, and $\kappa^{U,\mathscr{F},(m)}_{S^{p+k}_{\Delta'}}$ for $p = 1$, for $a_1/a_2 = 4$, and $a_1/a_2 = 10$ In both cases we employed $k = 2$, and $h = \frac{1}{4}$. Note that the effectivity indices deteriorate as the unsmoothness of the solution is increased. Here we used the scaling factor $s = 1$. The quality of the estimator depends on the choice of the scaling factor s. We will discuss the effect of the scaling factor s in two dimension (see (6.2*.5b) and Table 6.2*.3). The poor effectivity index is caused by subtracting inaccurate estimates in (6.2.19). The subtraction leads to poor quality of the estimates in comparison to the quality of the estimates for the energy norm. Other methods to get an estimate for $\mathscr{E}^{\mathscr{F}}$ can be used, which will be discussed later.

Let us now address the case that $\mathscr{F}$ is bounded only on a subset $\mathscr{U}^{\mathscr{F}} \subset \mathscr{U}$, and $u_{EX} \in \mathscr{U}$. The main idea of our approach is to construct an *extraction functional* $E^{\mathscr{F}} : \mathscr{U} \mapsto \mathbb{R}$, which is such that

$$E^{\mathscr{F}}(u_{EX}) = \mathscr{F}(u_{EX}) \tag{6.2.21a}$$

and it is given in the form

$$E^{\mathscr{F}}(u) = \mathscr{L}_{\mathscr{F}}(u) + \mathscr{M}_{\mathscr{F}}(f, a_2) \quad \forall u \in \mathscr{U}^{\mathscr{F}} \tag{6.2.21b}$$

where f, a_2 are the loads corresponding to u_{EX} and $\mathscr{L}_{\mathscr{F}} : \mathscr{U} \mapsto \mathbb{R}$, is linear, and bounded in $\mathscr{U}$, namely,

$$|\mathscr{L}_{\mathscr{F}}(v)| \leq \mathscr{C}\|v\|_{\mathscr{U}} \quad \forall v \in \mathscr{U} \tag{6.2.21c}$$

TABLE 6.2.1. *Upper and lower bounds for the error in* $u(\frac{1}{2})$ *using the ERNPp + k indicators.* Convergence of the effectivity indices $\kappa^{L,\mathcal{F},(m)}_{S^{p+k}_{\Delta}}$, $\kappa^{U,\mathcal{F},(m)}_{S^{p+k}_{\Delta}}$, of the improved bounds $\mathcal{E}^{L,\mathcal{F},(m)}_{S^{p+k}_{\Delta}}$, $\mathcal{E}^{U,\mathcal{F},(m)}_{S^{p+k}_{\Delta}}$, for $h = \frac{1}{4}$, $k = 2$, $\bar{x} = \frac{1}{2}$, $s = 1$ and $a_1/a_2 = 4$ in the case $p = 1$ and $p = 2$. The numbers in the brackets are the corresponding effectivity indices

m	$\mathcal{E}^{L}_{\bar{u}^{\mathcal{F},s,+}_{S^p_\Delta}, S^{p+k}_\Delta}$	$\mathcal{E}^{U}_{\bar{u}^{\mathcal{F},s,+}_{S^p_\Delta}, S^{p+k}_\Delta}$	$\mathcal{E}^{L}_{\bar{u}^{\mathcal{F},s,-}_{S^p_\Delta}, S^{p+k}_\Delta}$	$\mathcal{E}^{U}_{\bar{u}^{\mathcal{F},s,-}_{S^p_\Delta}, S^{p+k}_\Delta}$	$\kappa^{L,\mathcal{F},(m)}_{S^{p+k}_\Delta}$	$\kappa^{U,\mathcal{F},(m)}_{S^{p+k}_\Delta}$
$u_{EX}(\frac{1}{2}) - u_{S^p_\Delta}(\frac{1}{2}) = 0.06283422475, p = 1$						
1	0.44512	1.359003	0.19507	0.601441	−0.65092	7.196859
	(0.79413)	(2.424585)	(0.77821)	(2.399357)		
2	0.51299	1.359000	0.22900	0.601440	−0.39218	7.139584
	(0.91522)	(2.424578)	(0.91355)	(2.399354)		
3	0.51557	1.358999	0.23052	0.601440	−0.38161	7.136787
	(0.91983)	(2.424577)	(0.91963)	(2.399353)		
4	0.51568	1.358999	0.23060	0.601440	−0.38117	7.136649
	(0.92002)	(2.424577)	(0.91993)	(2.399353)		
5	0.51569	1.358999	0.23060	0.601440	−0.38116	7.136642
	(0.92003)	(2.424577)	(0.91994)	(2.399353)		
$u_{EX}(\frac{1}{2}) - u_{S^p_\Delta}(\frac{1}{2}) = 0.01594214021, p = 2$						
1	0.25489	1.072225	0.11448	0.476358	−2.53963	17.823222
	(0.90280)	(3.797760)	(0.90665)	(3.772768)		
2	0.26215	1.072224	0.11725	0.476358	−2.48075	17.813119
	(0.92852)	(3.797758)	(0.92863)	(3.772767)		
3	0.26218	1.072224	0.11726	0.476358	−2.48053	17.813084
	(0.92862)	(3.797758)	(0.92870)	(3.772767)		
4	0.26218	1.072224	0.11726	0.476358	−2.48053	17.813084
	(0.92862)	(3.797758)	(0.92870)	(3.772767)		
5	0.26218	1.072224	0.11726	0.476358	−2.48053	17.813084
	(0.92862)	(3.797758)	(0.92870)	(3.772767)		

and $\mathcal{M}_{\mathcal{F}} : \mathcal{D} \mapsto \mathbb{R}$, where $\mathcal{D}$ is the class of admissible loads, and

$$|\mathcal{M}_{\mathcal{F}}(f, a_2)| < \infty \quad \forall (f; a_2) \in \mathcal{D}. \tag{6.2.21d}$$

Let us illustrate this idea of the extraction functional by an example.

Example 6.2.2 Let us consider the regular model problem with

$$\left\| \frac{d^{q'}}{dx^{q'}}(AE) \right\|_{L^\infty(I)} < A, \qquad \left\| \frac{d^{r'} c}{dx^{r'}} \right\|_{L^\infty(I)} < B, \qquad \left\| \frac{d^{t'} f}{dx^{t'}} \right\|_{L^\infty(I)} < C \tag{6.2.22a}$$

for all integers q', r', t', such that

$$0 \le q' \le q, \qquad 0 \le r' \le r, \qquad 0 \le t' \le t \tag{6.2.22b}$$

and let us be interested in the output

$$\mathcal{F}(u_{EX}) = AE(0) \frac{d}{dx}(u_{EX})(0) \tag{6.2.23}$$

TABLE 6.2.2. *Upper and lower bounds for the error in $u(\frac{1}{2})$ using the ERNPp + k indicators.* Convergence of the effectivity indices $\kappa^{L,\mathcal{F},(m)}_{S^{p+k}_\Delta}$, $\kappa^{U,\mathcal{F},(m)}_{S^{p+k}_\Delta}$, of the improved bounds $\mathcal{E}^{L,\mathcal{F},(m)}_{S^{p+k}_\Delta}$, $\mathcal{E}^{U,\mathcal{F},(m)}_{S^{p+k}_\Delta}$, for $h=\frac{1}{4}$, $k=2$, $x=\frac{1}{2}$, $s=1$ and $a_1/a_2=10$ in the case $p=1$ and $p=2$. The numbers in the brackets are the corresponding effectivity indices

m	$\mathcal{E}^L_{\bar{u}^{\mathcal{F},s,+}_{S^p_\Delta},S^{p+k}_\Delta}$	$\mathcal{E}^U_{\bar{u}^{\mathcal{F},s,+}_{S^p_\Delta},S^{p+k}_\Delta}$	$\mathcal{E}^L_{\bar{u}^{\mathcal{F},s,-}_{S^p_\Delta},S^{p+k}_\Delta}$	$\mathcal{E}^U_{\bar{u}^{\mathcal{F},s,-}_{S^p_\Delta},S^{p+k}_\Delta}$	$\kappa^{L,\mathcal{F},(m)}_{S^{p+k}_\Delta}$	$\kappa^{U,\mathcal{F},(m)}_{S^{p+k}_\Delta}$
$u_{EX}(\frac{1}{2})-u_{S^p_\Delta}(\frac{1}{2})=0.0984099222, p=1$						
1	0.39856	2.022819	0.18012	0.822165	−1.31365	10.312360
	(0.56819)	(2.883716)	(0.57416)	(2.620835)		
2	0.59744	2.022807	0.26784	0.822159	−0.81043	10.212412
	(0.85170)	(2.883700)	(0.85381)	(2.620815)		
3	0.62548	2.022803	0.27985	0.822157	−0.72330	10.195657
	(0.89168)	(2.883694)	(0.89209)	(2.620808)		
4	0.63011	2.022801	0.28180	0.822156	−0.70851	10.192859
	(0.89828)	(2.883691)	(0.89830)	(2.620805)		
5	0.63090	2.022801	0.28213	0.822155	−0.70599	10.192385
	(0.89941)	(2.883690)	(0.89934)	(2.620803)		
$u_{EX}(\frac{1}{2})-u_{S^p_\Delta}(\frac{1}{2})=0.0335905999, p=2$						
1	0.32353	2.738698	0.14608	1.179827	−9.58096	55.663854
	(0.78944)	(6.682674)	(0.79702)	(6.437385)		
2	0.35356	2.738690	0.15819	1.179825	−9.42960	55.636117
	(0.86272)	(6.682655)	(0.86310)	(6.437374)		
3	0.35436	2.738689	0.15848	1.179825	−9.42960	55.635382
	(0.86466)	(6.682652)	(0.86469)	(6.437374)		
4	0.35438	2.738689	0.15849	1.179825	−9.42528	55.635356
	(0.86472)	(6.682652)	(0.86473)	(6.437374)		
5	0.35438	2.738689	0.15849	1.179825	−9.42527	55.635355
	(0.86472)	(6.682652)	(0.86473)	(6.437374)		

which exists for the input data AE, c, and f, which satisfy (6.2.22a,b). Noting that

$$\left[AE\frac{d}{dx}(u_{EX})v\right]_{x=0}^{x=1} = \int_0^1 fv + \mathcal{B}(u_{EX}, v) \quad \forall v \in \mathcal{U} \tag{6.2.24a}$$

and employing $v=\psi_0$, where ψ_0 is any function which satisfies

$$\psi_0(0)=1, \qquad \psi_0(1)=0 \tag{6.2.24b}$$

and

$$\int_0^1 AE\Big(\frac{d}{dx}(\psi_0)\Big)^2 + c\psi_0^2 + \alpha_2(\psi(1))^2 < \infty \tag{6.2.24c}$$

we get

$$AE(0)\frac{d}{dx}(u_{EX})(0) = -\int_0^l f\psi_0 - \mathcal{B}(u_{EX}, \psi_0). \tag{6.2.24d}$$

Letting now

$$E^{\mathcal{F}}(u) \overset{\text{def}}{=} \int_0^l f\psi_0 - \mathcal{B}(u,\psi_0) = \mathcal{M}_{\mathcal{F}}(f) + \mathcal{L}_{\mathcal{F}}(u) \quad \forall u \in \mathcal{U} \tag{6.2.25a}$$

with

$$\mathcal{M}_{\mathcal{F}}(f) \overset{\text{def}}{=} -\int_0^l f\psi_0 \tag{6.2.25b}$$

and

$$\mathcal{L}_{\mathcal{F}}(u) \overset{\text{def}}{=} -\mathcal{B}(u,\psi_0) \quad \forall u \in \mathcal{U} \tag{6.2.25c}$$

we obtain

$$\mathcal{F}(u_{EX}) = AE(0)\frac{d}{dx}(u_{EX})(0) = E^{\mathcal{F}}(u_{EX}) \tag{6.2.25d}$$

and hence

$$\begin{aligned} \mathcal{F}(u_{EX}) - E^{\mathcal{F}}(u_{S^p_\Delta}) &= \mathcal{L}_{\mathcal{F}}(u_{EX}) - \mathcal{L}_{\mathcal{F}}(u_{S^p_\Delta}) = -\mathcal{B}(e_{S^p_\Delta},\psi_0) \\ &= \mathcal{B}(e_{S^p_\Delta}, H^{\mathcal{L}_{\mathcal{F}}}) \end{aligned} \tag{6.2.26a}$$

where $H^{\mathcal{L}_{\mathcal{F}}}$, is the solution of the variational problem:

Find $H^{\mathcal{L}_{\mathcal{F}}} \in \mathcal{U}$ such that

$$\mathcal{B}(H^{\mathcal{L}_{\mathcal{F}}}, v) = \mathcal{L}^{\mathcal{F}}(v) = -\mathcal{B}(\psi_0, v) \quad \forall v \in \mathcal{U}. \tag{6.2.26b}$$

Note that $\psi_0 \notin \mathcal{U}$, and hence

$$H^{\mathcal{L}_{\mathcal{F}}} \not\equiv -\psi_0. \tag{6.2.26c}$$

It follows that

$$\begin{aligned} \left|AE(0)\frac{d}{dx}(u_{EX})(0) - E^{\mathcal{F}}(u_{S^p_\Delta})\right| &\le \left|\mathcal{B}\left(e_{S^p_\Delta}, H^{\mathcal{L}_{\mathcal{F}}} - H^{\mathcal{L}_{\mathcal{F}}}_{S^p_\Delta}\right)\right| \\ &\le \|e_{S^p_\Delta}\|_{\mathcal{U}} \left\|H^{\mathcal{L}_{\mathcal{F}}} - H^{\mathcal{L}_{\mathcal{F}}}_{S^p_\Delta}\right\|_{\mathcal{U}} \end{aligned} \tag{6.2.27}$$

where we denote by $H^{\mathcal{L}_{\mathcal{F}}}_{S^p_\Delta} \in S^p_\Delta$, the finite element approximation of $H^{\mathcal{L}_{\mathcal{F}}}$. Employing now assumption (6.2.22a,b), about the smoothness of AE, c, and f, in Theorem 2.3.1, we obtain

$$\left\|\frac{d^j}{dx^j}(u_{EX})\right\|_{L^\infty(I)} \le D, \quad 0 \le j \le \min(r+2, t+2, q+1) \tag{6.2.28a}$$

and assuming that r, t, q, are such that

$$s \ge p-1 \tag{6.2.28b}$$

we get

$$\|e_{S^p_\Delta}\|_{\mathcal{U}} \le \mathscr{C}(h^\Delta)^p. \tag{6.2.28c}$$

Choosing ψ_0, to be sufficiently smooth, e.g. $\psi_0 \equiv 1 - x$, we also get

$$\left\| H^{\mathscr{L}_{\mathscr{F}}} - H^{\mathscr{L}_{\mathscr{F}}}_{S^p_\Delta} \right\|_{\mathcal{U}} \le \mathscr{C}\ (h^\Delta)^p \tag{6.2.28d}$$

and from (6.2.27) we obtain

$$\left| AE(0)\frac{d}{dx}(u_{EX})(0) - E^{\mathscr{F}}(u_{S^p_\Delta}) \right| \le \mathscr{C}(h^\Delta)^p \approx \mathscr{C}\|e_{S^p_\Delta}\|^2_{\mathcal{U}}. \tag{6.2.29}$$

Hence we have constructed a functional $E^{\mathscr{F}} : \mathcal{U} \mapsto \mathbb{R}$, such that for the class of the input data which satisfy (6.2.22). $E^{\mathscr{F}}(u_{S^p_\Delta})$, converges to $AE(0)(d/dx)(u_{EX})(0)$, as $\|e_{S^p_\Delta}\|^2_{\mathcal{U}}$, the square of the energy norm of the error.

Remark 6.2.4 The estimate

$$|\mathscr{F}(u_{EX}) - E^{\mathscr{F}}(u_{S^p_\Delta})| \approx \mathscr{C}\left\| H^{\mathscr{L}_{\mathscr{F}}} - H^{\mathscr{L}_{\mathscr{F}}}_{S^p_\Delta} \right\|_{\mathcal{U}} \|e_{S^p_\Delta}\|_{\mathcal{U}} \tag{6.2.30}$$

shows that, by a proper construction of $\mathscr{L}_{\mathscr{F}}$, i.e. by a proper choice of ψ_0, it is possible to obtain the output $\mathscr{F}(u_{EX})$ with good accuracy by employing the finite element solution $u_{S^p_\Delta}$, on a rather coarse mesh.

Remark 6.2.5 Let us note that in general, we have

$$\mathscr{F}(u_{EX}) - \mathscr{F}(u_{S^p_\Delta}) \ne \mathscr{F}(u_{EX}) - E^{\mathscr{F}}(u_{S^p_\Delta}). \tag{6.2.31a}$$

In fact for sufficiently smooth input data and proper choice of ψ_0, we have

$$|\mathscr{F}(u_{EX}) - E^{\mathscr{F}}(u_{S^p_\Delta})| \le \mathscr{C}h^{2p} \approx \mathscr{C}\|e_{S^p_\Delta}\|^2_{\mathcal{U}} \tag{6.2.31b}$$

while

$$|\mathscr{F}(u_{EX}) - \mathscr{F}(u_{S^p_\Delta})| \approx \mathscr{C}h^p \tag{6.2.31c}$$

and hence $E^{\mathscr{F}}(u_{S^p_\Delta})$, is more accurate than $\mathscr{F}(u_{S^p_\Delta})$. However, if we are interested for the bounds for $\mathscr{F}(u_{S^p_\Delta}) = (d/dx)(e_{S^p_\Delta})(0)$, the error in $(d/dx)(u_{S^p_\Delta})(0)$, we may employ that

$$\frac{d}{dx}(e_{S^p_\Delta})(0) = \left(\frac{d}{dx}(u_{EX})(0) - E^{\mathscr{F}}(u_{S^p_\Delta})\right) + \left(E^{\mathscr{F}}(u_{S^p_\Delta}) - \frac{d}{dx}(u_{S^p_\Delta})(0)\right). \tag{6.2.31d}$$

See also Remark 6.2*.1.

The approach illustrated in Example 6.2.2 is known as *extraction* of the output of interest. Let us formulate the extraction in a more general setting.

Theorem 6.2.3 *Let* $\mathscr{F} : \mathcal{U} \mapsto \mathbb{R}$, *and assume that*

$$u_{EX} \in \mathcal{U}^{\mathscr{F}} \stackrel{\text{def}}{=} \{u \in \mathcal{U} |\ |\mathscr{F}(u)| \le \mathscr{C}\|u\|_{\mathcal{U}}\} \tag{6.2.32a}$$

and let $E^{\mathcal{F}} : \mathcal{U} \mapsto \mathbb{R}$, be such that

$$\mathcal{F}(u_{EX}) = E^{\mathcal{F}}(u_{EX}) \tag{6.2.32b}$$

where

$$E^{\mathcal{F}}(u) \stackrel{\text{def}}{=} \mathcal{L}_{\mathcal{F}}(u) + \mathcal{M}_{\mathcal{F}}(f, a_2), \quad u \in \mathcal{U} \tag{6.2.33a}$$

with $\mathcal{L}_{\mathcal{F}} : \mathcal{U} \mapsto \mathbb{R}$, linear, and bounded in $\mathcal{U}$, namely,

$$|\mathcal{L}_{\mathcal{F}}(v)| \leq \mathcal{C}_1 \|v\|_{\mathcal{U}} \quad \forall v \in \mathcal{U} \tag{6.2.33b}$$

and $\mathcal{M}_{\mathcal{F}} : \mathcal{D} \mapsto \mathbb{R}$, bounded

$$|\mathcal{M}_{\mathcal{F}}(f, a_2)| < \infty \quad \forall (f, a_2) \in \mathcal{D} \tag{6.2.33c}$$

where $\mathcal{D}$ is the class of admissible data which correspond to solutions from $\mathcal{U}^{\mathcal{F}}$. Then we have

$$|\mathcal{F}(u_{EX}) - \mathcal{L}_{\mathcal{F}}(u_{S^p_\Delta})| \leq \|e_{S^p_\Delta}\|_{\mathcal{U}} \left\| H^{\mathcal{L}_{\mathcal{F}}} - H^{\mathcal{L}_{\mathcal{F}}}_{S^p_\Delta} \right\|_{\mathcal{U}} \tag{6.2.34a}$$

where $H^{\mathcal{L}_{\mathcal{F}}}$ is the solution of the problem:

Find $H^{\mathcal{L}_{\mathcal{F}}} \in \mathcal{U}$, such that

$$\mathcal{B}(H^{\mathcal{L}_{\mathcal{F}}}, v) = \mathcal{L}_{\mathcal{F}}(v) \quad \forall v \in \mathcal{U} \tag{6.2.34b}$$

and $H^{\mathcal{L}_{\mathcal{F}}}_{S^p_\Delta} \in S^p_\Delta$, is the finite element approximation of $H^{\mathcal{L}_{\mathcal{F}}}$.

Proof It follows exactly the same steps as Example 6.2.2. □

Above we have constructed guaranteed upper and lower estimates $\mathcal{E}^{U,\mathcal{F}}_{S^{p+k}_{\Delta'}}$ and $\mathcal{E}^{L,\mathcal{F}}_{S^{p+k}_{\Delta'}}$ for $e^{\mathcal{F}}_{S^p_\Delta}$, the error in the output $\mathcal{F}$. Let us now to construct estimators which are not guaranteed, but have reasonable asymptotic properties. Recalling that

$$e^{\mathcal{F}}_{S^p_\Delta} = \sum_{j=1}^{N(\Delta)} \mathcal{B}_{I_j^\Delta}(e_{u_{S^p_\Delta}}, e_{H^{\mathcal{F}}_{S^p_\Delta}}) \tag{6.2.35a}$$

and employing that

$$e_{u_{S^p_\Delta}} = \hat{e}_{u_{S^p_\Delta}} + \underbrace{e_{u_{S^p_\Delta}} - \hat{e}_{u_{S^p_\Delta}}}_{\delta\hat{e}_{u_{S^p_\Delta}}} = \hat{e}_{u_{S^p_\Delta}} + \delta\hat{e}_{u_{S^p_\Delta}} \tag{6.2.35b}$$

where $\hat{e}_{u_{S^p_\Delta}}$, is an error indicator for $e_{u_{S^p_\Delta}}$, e.g. the Dirichlet indicator $\hat{e}^{Dir}_{u_{S^p_\Delta}}$, or the Neumann indicator $\hat{e}^{\text{Neum}}_{u_{S^p_\Delta}}$, or a similar recovery estimator, and similarly

$$e_{H^{\mathcal{F}}_{S^p_\Delta}} = \hat{e}_{H^{\mathcal{F}}_{S^p_\Delta}} + \underbrace{e_{H^{\mathcal{F}}_{S^p_\Delta}} - \hat{e}_{H^{\mathcal{F}}_{S^p_\Delta}}}_{\delta\hat{e}_{H^{\mathcal{F}}_{S^p_\Delta}}} = \hat{e}_{H^{\mathcal{F}}_{S^p_\Delta}} + \delta\hat{e}_{H^{\mathcal{F}}_{S^p_\Delta}} \tag{6.2.35c}$$

we get

$$e^{\mathcal{F}}_{S^p_\Delta} = \underbrace{\sum_{j=1}^{N(\Delta)} \mathcal{B}_{I_j^\Delta}(\hat{e}_{u_{S^p_\Delta}} + \hat{e}_{H^{\mathcal{F}}_{S^p_\Delta}})}_{e^{\mathcal{F},1}_{S^p_\Delta}} + \underbrace{\sum_{j=1}^{N(\Delta)} \mathcal{B}_{I_j^\Delta}(\hat{e}_{u_{S^p_\Delta}}, \delta\hat{e}_{H^{\mathcal{F}}_{S^p_\Delta}})}_{e^{\mathcal{F},2}_{S^p_\Delta}}$$

$$+ \underbrace{\sum_{j=1}^{N(\Delta)} \mathcal{B}_{I_j^\Delta}(\delta\hat{e}_{u_{S^p_\Delta}}, \hat{e}_{H^{\mathcal{F}}_{S^p_\Delta}})}_{e^{\mathcal{F},3}_{S^p_\Delta}} + \underbrace{\sum_{j=1}^{N(\Delta)} \mathcal{B}_{I_j^\Delta}(\delta\hat{e}_{u_{S^p_\Delta}}, \delta\hat{e}_{\hat{H}^{\mathcal{F}}_{S^p_\Delta}})}_{e^{\mathcal{F},4}_{S^p_\Delta}}. \tag{6.2.35d}$$

Neglecting the terms $e^{\mathcal{F},2}_{S^p_\Delta}$, $e^{\mathcal{F},3}_{S^p_\Delta}$, and $e^{\mathcal{F},4}_{S^p_\Delta}$, as higher order, and taking the absolute value we get the estimator

$$\left|e^{\mathcal{F}}_{S^p_\Delta}\right| \approx \mathcal{E}^{\mathcal{F},1} \stackrel{\text{def}}{=} \left|e^{\mathcal{F},1}_{S^p_\Delta}\right| \stackrel{\text{def}}{=} \left|\sum_{j=1}^{N(\Delta)} \mathcal{B}_{I_j^\Delta}(\hat{e}_{u_{S^p_\Delta}}, \hat{e}_{H^{\mathcal{F}}_{S^p_\Delta}})\right|. \tag{6.2.36a}$$

The estimator $\mathcal{E}^{\mathcal{F},1}$, may underestimate $|e^{\mathcal{F}}_{S^p_\Delta}|$, because we have neglected the terms $e^{\mathcal{F},2}_{S^p_\Delta}$, $e^{\mathcal{F},3}_{S^p_\Delta}$, and $e^{\mathcal{F},4}_{S^p_\Delta}$, and because of cancellations in the sum $\sum_{j=1}^{N(\Delta)} \mathcal{B}_{I_j^\Delta}(\hat{e}_{u_{S^p_\Delta}}, \hat{e}_{H^{\mathcal{F}}_{S^p_\Delta}})$. The cancellation can be avoided by employing the estimator

$$|e^{\mathcal{F}}_{S^p_\Delta}| \approx \mathcal{E}^{\mathcal{F},2} \stackrel{\text{def}}{=} \sum_{j=1}^{N(\Delta)} |\mathcal{B}_{I_j^\Delta}(\hat{e}_{u_{S^p_\Delta}}, \hat{e}_{H^{\mathcal{F}}_{S^p_\Delta}}|. \tag{6.2.36b}$$

In the case that we can compute only the element error indicators $\|\hat{e}_{u_{S^p_\Delta}}\|_{\mathcal{U}}$, $\|\hat{e}_{H^{\mathcal{F}}_{S^p_\Delta}}\|_{\mathcal{U}}$, but not the inner products $\mathcal{B}_{I_j^\Delta}(\hat{e}_{u_{S^p_\Delta}}, \hat{e}_{H^{\mathcal{F}}_{S^p_\Delta}})$, of the indicator functions we may instead employ the estimator

$$|e^{\mathcal{F}}_{S^p_\Delta} \approx \mathcal{E}^{\mathcal{F},3}| \stackrel{\text{def}}{=} \sum_{j=1}^{N(\Delta)} \|\hat{e}_{u_{S^p_\Delta}}\|_{\mathcal{U}} \|\hat{e}_{H^{\mathcal{F}}_{S^p_\Delta}}\|_{\mathcal{U}}. \tag{6.2.36c}$$

A even safer estimator may be obtained by employing the Cauchy–Schwarz inequality, namely

$$\left|e^{\mathcal{F}}_{S^p_\Delta}\right| \approx \mathcal{E}^{\mathcal{F},4} \stackrel{\text{def}}{=} \sqrt{\sum_{j=1}^{N(\Delta)} \|\hat{e}_{u_{S^p_\Delta}}\|^2_{\mathcal{U}(I_j^\Delta)}} \sqrt{\sum_{j=1}^{N(\Delta)} \|\hat{e}_{H^{\mathcal{F}}_{S^p_\Delta}}\|^2_{\mathcal{U}(I_j^\Delta)}} = \mathcal{E}_{u_{S^p_\Delta}} \mathcal{E}_{H^{\mathcal{F}}_{S^p_\Delta}}. \tag{6.2.36d}$$

The estimators $\mathcal{E}^{\mathcal{F},1}$, $\mathcal{E}^{\mathcal{F},2}$, $\mathcal{E}^{\mathcal{F},3}$, and $\mathcal{E}^{\mathcal{F},4}$ *are not* guaranteed upper estimators for $e^{\mathcal{F}}_{S^p_\Delta}$. Nevertheless because we have

$$\mathcal{E}^{\mathcal{F},1} \leq \mathcal{E}^{\mathcal{F},2} \leq \mathcal{E}^{\mathcal{F},3} \leq \mathcal{E}^{\mathcal{F},4} \tag{6.2.36e}$$

and $\mathcal{E}^{\mathcal{F},2}$, $\mathcal{E}^{\mathcal{F},3}$, and $\mathcal{E}^{\mathcal{F},4}$ are very often upper estimators. As we have already seen in (6.2.18e), a guaranteed upper estimator for $\left|e^{\mathcal{F}}_{S^p_\Delta}\right|$ is given by $\mathcal{E}^U_{u_{S^p_\Delta}} \mathcal{E}^U_{H^{\mathcal{F}}_{S^p_\Delta}}$.

Let us now estimate the terms $e^{\mathcal{F},i}_{S^p_\Delta}$, $i = 2, 3, 4$, which were omitted from the above estimators, in the case that the Dirichlet error indicators $\hat{e}^{Dir}$ are used. We have:

Lemma 6.2.2 *Under the assumption that $\mathcal{F}$ is bounded in $\mathcal{U}$, we have*

$$\left|e^{\mathcal{F},i}_{S^p_\Delta}\right| \leq \mathcal{C}\mathcal{E}^{Dir}_{u_{S^p_\Delta}}\mathcal{E}^{Dir}_{H^{\mathcal{F}}_{S^p_\Delta}}\epsilon \tag{6.2.37}$$

for $i = 2, 3, 4$, where $\epsilon \to 0$, as $h^\Delta \to 0$ and $\mathcal{C}$ is independent of the mesh Δ.

Proof Using Theorem 4.3.2, we obtain

$$\|e_{S^p_\Delta} - \hat{e}^{Dir}\|_{\mathcal{U}(I^\Delta_j)} \leq \mathcal{C}\sqrt{h^\Delta_j}\sqrt{(\rho_1^{I^\Delta_j})^2 + (\rho_2^{I^\Delta_j})^2} \tag{6.2.38a}$$

where

$$\rho_1^{I^\Delta_j} = \left(e^\Delta_j + e^\Delta_{j-1}\right), \qquad \rho_2^{I^\Delta_j} = \frac{e^\Delta_j - e^\Delta_{j-1}}{h_{I^\Delta_j}}. \tag{6.2.38b}$$

In Section 4.4 we have proven that

$$|\rho_2^{I^\Delta_j}| \leq \sum_{i=1}^{N(\Delta)} \|e_{S^p_\Delta}\|_{\mathcal{U}(I^\Delta_i)}\|G'_j - \mathcal{A}^p_{I^\Delta_i}G'_j\|_{\mathcal{U}(I^\Delta_i)} \tag{6.2.38c}$$

and employing the asymptotic estimates for $\|e_{S^p_\Delta}\|_{\mathcal{U}(I^\Delta_j)}$, and $\|G'_j - \mathcal{A}^p_{I^\Delta_i}G'_j\|_{\mathcal{U}(I^\Delta_j)}$, we get

$$\|e_{S^p_\Delta} - \hat{e}^{Dir}\|^2_{\mathcal{U}(I^\Delta_j)} \leq \mathcal{C}h_{I^\Delta_j}\left(\underbrace{\left(\sum_{i=1}^{N(\Delta)}\left(\eta^{Dir}_{I^\Delta_j}\right)^2\right)}_{\mathcal{E}^{Dir}}\left(\sum_{i=1}^{N(\Delta)} h^{2\mu+1}_{I^\Delta_j}\right) + h^{2\mu-1}_{I^\Delta_j}\right) \tag{6.2.38d}$$

with

$$\mu = \min(p, r, q+1) \geq 1. \tag{6.2.38e}$$

It follows that

$$\sum_{j-1}^{N(\Delta)} \|e_{S^p_\Delta} - \hat{e}^{Dir}\|^2_{\mathcal{U}(I^\Delta_j)} \leq \mathcal{C}\mathcal{E}^{Dir}\sum_{i=1}^{N(\Delta)} h^{2\mu}_{I^\Delta_j} \leq (\mathcal{E}^{Dir})^2\epsilon^2 \tag{6.2.38f}$$

where

$$\epsilon^2 \leq (h^\Delta)^{2\mu-1}\mathcal{C}. \tag{6.2.38g}$$

Employing now the Cauchy–Schwarz inequality we obtain

$$\left|e^{\mathcal{F}}_{S^p_\Delta}\right| = \sum_{j=1}^{N} \|e_{H^{\mathcal{F}}_{S^p_\Delta}} - \hat{e}^{Dir}_{H^{\mathcal{F}}_{S^p_\Delta}}\|_{\mathcal{U}}\|e_{u_{S^p_\Delta}} - \hat{e}^{Dir}_{u_{S^p_\Delta}}\|_{\mathcal{U}} \leq \mathcal{C}\mathcal{E}^{Dir}_H\mathcal{E}^{Dir}_u\epsilon \tag{6.2.39}$$

where $\epsilon \to 0$, as $h^\Delta \to 0$. □

Taking into account that

$$\mathcal{B}_{I_j^\Delta}\left(\hat{e}^{Dir}_{H^{\mathcal{F}}_{S^p_\Delta}}, v\right) = 0 \quad \forall v \in S^p_0(I_j^\Delta) \tag{6.2.40a}$$

and following similar steps in the proof of Lemma 6.2.2 we get

$$\left|\sum_{j=1}^{N(\Delta)} \mathcal{B}_{I_j^\Delta}\left(\hat{e}^{Dir}_{H^{\mathcal{F}}_{S^p_\Delta}}, e_{u_{S^p_\Delta}} - \hat{e}^{Dir}_{u_{S^p_\Delta}}\right)\right| \le \mathcal{C}\mathcal{E}^{Dir}_H \mathcal{E}^{Dir}_u \epsilon. \tag{6.2.40b}$$

Similarly we obtain

$$\left|\sum_{j=1}^{N(\Delta)} \mathcal{B}_{I_j^\Delta}\left(\hat{e}^{Dir}_{u_{S^p_\Delta}}, e_{H^{\mathcal{F}}_{S^p_\Delta}} - \hat{e}^{Dir}_{H^{\mathcal{F}}_{S^p_\Delta}}\right)\right| \le \mathcal{C}\mathcal{E}^{Dir}_H \mathcal{E}^{Dir}_u \epsilon. \tag{6.2.41}$$

Employing now these we obtain:

Theorem 6.2.4 *Under the assumption*

$$\mathcal{C}\mathcal{E}_{u_{S^p_\Delta}} \mathcal{E}_{H^{\mathcal{F}}_{S^p_\Delta}} \le \left|\sum_{j=1}^{N(\Delta)} \mathcal{B}_{I_j^\Delta}\left(\hat{e}_{u_{S^p_\Delta}}, \hat{e}_{H^{\mathcal{F}}_{S^p_\Delta}}\right)\right| \tag{6.2.42a}$$

the terms $e^{\mathcal{F},k}_{S^p_\Delta}$, $k = 2, 3, 4$, *are negligible, and we have*

$$e^{\mathcal{F}}_{S^p_\Delta} \approx \mathcal{E}^{\mathcal{F},1} \tag{6.2.42b}$$

asymptotically as h^Δ *tends to zero.*

Remark 6.2.6 In the cases that $e^{\mathcal{F}}_{S^p_\Delta} = 0$, assumption (6.2.42a) does not hold. However this case is not of interest, and assumption (6.2.42a) can be satisfied in most cases of practical interest.

Remark 6.2.7 Theorem 6.1.3 also holds when the Neumann indicators are employed. In this case, the terms $e^{\mathcal{F},2}_{S^p_\Delta}$ and $e^{\mathcal{F},3}_{S^p_\Delta}$ are of the same magnitude only when the equilibration is done for all polynomials up to degree p.

Remark 6.2.8 Theorem 6.1.3 holds also for the recovery estimators, e.g. the ZZ–SPR estimator.

Let us also address the a-posteriori estimation of the pollution error. In Section 4.3 we defined the pollution error $e^{glob}_{S^p(I_j^\Delta)}$, in an element I_j^Δ, as the difference between the finite element solution $u_{S^p_\Delta}|_{I_j^\Delta}$ and the local finite element solution $\mathcal{A}^p_{I_j^\Delta} u_{EX}$ and we partitioned $e^{glob}_{S^p(I_j^\Delta)}$ into two components namely

$$e^{glob}_{S^p(I_j^\Delta)} \stackrel{\text{def}}{=} \mathcal{A}^p_{I_j^\Delta} u_{EX} - u_{S^p_\Delta}|_{I_j^\Delta} = e^{glob}_{S^1(I_j^\Delta)} + e^{glob}_{S^p_0(I_j^\Delta)} \tag{6.2.43a}$$

where $e^{glob}_{S^1(I^\Delta_j)} \in S^1(I^\Delta_j)$ is the affine component of the pollution error in I^Δ_j given by

$$e^{glob}_{S^1(I^\Delta_j)}(x) = \rho_1^{I^\Delta_j} + \rho_2^{I^\Delta_j}\left(x - x^\Delta_{j-1/2}\right) \tag{6.2.43b}$$

with

$$\rho_1^{I^\Delta_j} \stackrel{\text{def}}{=} \frac{e^\Delta_j + e^\Delta_{j-1}}{2}, \qquad \rho_2^{I^\Delta_j} \stackrel{\text{def}}{=} \frac{e^\Delta_j - e^\Delta_{j-1}}{h_{I^\Delta_j}} \tag{6.2.43c}$$

and $e^{glob}_{S^p_0(I^\Delta_j)} \in S^p_0(I^\Delta_j)$, is the bubble part of the pollution error in I^Δ_j. Noting that

$$\mathcal{B}_{I^\Delta_j}\left(e^{glob}_{S^p_0(I^\Delta_j)}, v\right) = -\mathcal{B}_{I^\Delta_j}\left(e^{glob}_{S^1(I^\Delta_j)}, v\right) \quad \forall v \in S^p_0(I^\Delta_j) \tag{6.2.43d}$$

we obtained

$$\left\|e^{glob}_{S^p_0(I^\Delta_j)}\right\|_{\mathcal{U}(I^\Delta_j)} \leq \left\|e^{glob}_{S^1(I^\Delta_j)}\right\|_{\mathcal{U}(I^\Delta_j)}. \tag{6.2.43e}$$

It follows that in order to estimate the pollution error $\|e^{glob}_{S^p(I^\Delta_j)}\|_{\mathcal{U}}$ it is sufficient to estimate $\|e^{glob}_{S^1(I^\Delta_j)}\|_{\mathcal{U}}$, which is accomplished by estimating $\rho_1^{I^\Delta_j}$, and $\rho_2^{I^\Delta_j}$.

The a-posteriori estimation of $\rho_1^{I^\Delta_j}$, and $\rho_2^{I^\Delta_j}$, will be based on the identities

$$\rho_1^{I^\Delta_j} = \tfrac{1}{2}\left(e_{S^p_\Delta}(x^\Delta_j) + e_{S^p_\Delta}(x^\Delta_{j-1})\right) = \mathcal{B}(e_{S^p_\Delta}, G_{I^\Delta_j}) = \mathcal{R}(G_{I^\Delta_j}) \tag{6.2.44a}$$

where

$$G_{I^\Delta_j} \stackrel{\text{def}}{=} \tfrac{1}{2}\left(G(\cdot; x^\Delta_j) + G(\cdot; x^\Delta_{j-1})\right) \tag{6.2.44b}$$

where $G(\cdot;\bar{x})$ is the Green's function, and

$$\rho_2^{I^\Delta_j} = \frac{1}{h_{I^\Delta_j}}\left(e_{S^p_\Delta}(x^\Delta_j) - e_{S^p_\Delta}(x^\Delta_{j-1})\right) = \mathcal{B}(e_{S^p_\Delta}, G'_{I^\Delta_j}) = \mathcal{R}(G'_{I^\Delta_j}) \tag{6.2.45a}$$

where

$$G'_{I^\Delta_j} = \frac{1}{h_{I^\Delta_j}}\left(G(\cdot; x^\Delta_j) - G(\cdot; x^\Delta_{j-1})\right). \tag{6.2.45b}$$

Note that, in general, we do not know the functions $G_{I^\Delta_j}$, $G'_{I^\Delta_j}$ and hence $\rho_1^{I^\Delta_j}$, $\rho_2^{I^\Delta_j}$ cannot be computed by evaluating $\mathcal{R}(G_{I^\Delta_j})$, $\mathcal{R}(G'_{I^\Delta_j})$. Employing the splitting of the residuum

$$\mathcal{R} = \sum_{j=1}^{N(\Delta)} \mathcal{R}^{EQ}_{I^\Delta_j} \tag{6.2.46a}$$

and the definition of the Neumann error indicator functions

$$\mathcal{B}\left(\hat{e}_{I_j^\Delta}^{\text{Neum}}, v\right) = \mathcal{R}_{I_j^\Delta}^{EQ}(v) \quad \forall v \in \mathcal{U}(I_j^\Delta) \tag{6.2.46b}$$

we obtain

$$\begin{aligned} \mathcal{R}(G'_{I_j^\Delta}) &= \sum_{k=1}^{N(\Delta)} \mathcal{R}_{I_k^\Delta}^{EQ}(G'_{I_j^\Delta}) = \sum_{k=1}^{N(\Delta)} \mathcal{B}_{I_k^\Delta}\left(\hat{e}_{I_k^\Delta}^{\text{Neum}}, G'_{I_j^\Delta}\right) \\ &= \sum_{k=1}^{N} \mathcal{B}_{I_k^\Delta}\left(\hat{e}_{I_k^\Delta}^{\text{Neum}}, G'_{I_j^\Delta} - \mathcal{B}^p_{I_k^\Delta} G'_{I_j^\Delta}\right) \end{aligned} \tag{6.2.46c}$$

where $\mathcal{B}^p_{I_k^\Delta} G'_{I_j^\Delta} \in S^p(I_j^\Delta)$, is the unconstrained best approximation of $G'_{I_j^\Delta}|_{I_k^\Delta}$ defined in Section 4.3.

Employing the notation $\hat{e}_{I_k^\Delta}^{\text{Neum}}(f, u_{S^p_\Delta})$, to indicate the dependence of the error indicator function $\hat{e}_{I_k^\Delta}^{\text{Neum}}$ on the right-hand side f and the finite element solution $u_{S^p_\Delta}$, letting $G'_{I_j^\Delta, S^p_\Delta}$ denote the finite element approximation of $G'_{I_j^\Delta}$ from S^p_Δ, and

$$\delta_{I_j^\Delta} \stackrel{\text{def}}{=} \tfrac{1}{2}\left(\delta(\cdot; x_j^\Delta) + \delta(\cdot; x_{j-1}^\Delta)\right) \tag{6.2.47a}$$

and

$$\delta'_{I_j^\Delta} \stackrel{\text{def}}{=} \frac{1}{h_{I_j^\Delta}}\left(\delta(\cdot; x_j^\Delta) - \delta(\cdot; x_{j-1}^\Delta)\right) \tag{6.2.47b}$$

we obtain

$$\rho_2^{I_j^\Delta} = \mathcal{R}(G'_{I_j^\Delta}) = \sum_{k=1}^{N(\Delta)} \mathcal{B}_{I_k^\Delta}\left(\hat{e}_{I_k^\Delta}^{\text{Neum}}(f, u_{S^p_\Delta}), \hat{e}_{I_k^\Delta}^{\text{Neum}}(\delta'_{I_j^\Delta}, G'_{I_j^\Delta, S^p_\Delta})\right) + E \tag{6.2.48a}$$

where

$$E = \sum_{k=1}^{N(\Delta)} \mathcal{B}_{I_k^\Delta}\left(\hat{e}_{I_k^\Delta}^{\text{Neum}}(f, u_{S^p_\Delta}), (G'_{I_j^\Delta} - \mathcal{B}^p_k G'_{I_j^\Delta}) - \hat{e}_{I_k^\Delta}^{\text{Neum}}(\delta'_{I_j^\Delta}, G'_{I_j^\Delta, S^p_\Delta})\right). \tag{6.2.48b}$$

Assuming that there exists an $\alpha > 0$, such that

$$\begin{aligned} &\left\|(G'_{I_j^\Delta} - \mathcal{B}^p_{I_j^\Delta} G'_{I_j^\Delta}) - \hat{e}_k^{\text{Neum}}(\delta'_{I_j^\Delta}, G'_{I_j^\Delta, S^p_\Delta})\right\|_{\mathcal{U}(I_k^\Delta)} \\ &\leq \mathcal{C}(h^\Delta)^\alpha \eta_{I_k^\Delta}^{\text{Neum}}(\delta'_{I_j^\Delta}, G'_{I_j^\Delta, S^p_\Delta}) \end{aligned} \tag{6.2.48c}$$

and employing the Cauchy–Schwarz inequality in (6.2.51b), we get

$$|E| \leq \mathcal{C}(h^\Delta)^\alpha \sum_{k=1}^{N(\Delta)} \eta_{I_k^\Delta}^{\text{Neum}}(f, u_{S^p_\Delta}) \eta_{I_k^\Delta}^{\text{Neum}}(\delta'_{I_j^\Delta}, G'_{I_j^\Delta, S^p_\Delta}). \tag{6.2.48d}$$

Remark 6.2.9 Note that *assumption* (6.2.48) *states that the error indicators are close to the error in the best approximation in each element*, and requires certain restrictions on the mesh, as we have already seen the Counterexample 5.2.1.

Neglecting E, we obtain the estimator

$$\hat{\rho}_2^{I_j^\Delta} = \sum_{k=1}^{N(\Delta)} \mathcal{B}_{I_k^\Delta}\left(\hat{e}_{I_k^\Delta}^{\text{Neum}}(f, u_{S_\Delta^p}), \hat{e}_{I_k^\Delta}^{\text{Neum}}(\delta'_{I_j^\Delta}, G'_{I_j^\Delta, S_\Delta^p})\right) \tag{6.2.49}$$

where $\hat{e}_{I_k^\Delta}^{\text{Neum}}(\delta'_{I_j^\Delta}, G'_{I_j^\Delta, S_\Delta^p})$ is the element residual error indicator function corresponding to $G'_{I_j^\Delta, S_\Delta^p}$. Following similar steps, we get

$$\hat{\rho}_1^{I_j^\Delta} = \sum_{k=1}^{N(\Delta)} \mathcal{B}_{I_k^\Delta}\left(\hat{e}_{I_k^\Delta}^{\text{Neum}}(f, u_{S_\Delta^p}), \hat{e}_{I_k^\Delta}^{\text{Neum}}(\delta_{I_j^\Delta}, G_{I_j^\Delta, S_\Delta^p})\right) \tag{6.2.50a}$$

which is an estimator for $\rho_1^{I_j^\Delta}$. We will call $\hat{\rho}_1^{I_j^\Delta}$, and $\hat{\rho}_2^{I_j^\Delta}$, the pollution estimators. Then, $e_{S^1(I_j^\Delta)}^{glob}$ may be estimated by

$$\hat{e}_{S^1(I_j^\Delta)}^{glob}(x) = \hat{\rho}_{j,1} + \hat{\rho}_{j,2}\left(x - x_{j-1/2}^\Delta\right) \tag{6.2.50b}$$

and $e_{S_0^p(I_j^\Delta)}^{glob}$, may be estimated by $\hat{e}_{S_0^p(I_j^\Delta)}^{glob}$, which is determined by solving the following discrete problem:

Find $\hat{e}_{S_0^p(I_j^\Delta)}^{glob} \in S_0^p(I_j^\Delta)$, such that

$$\mathcal{B}_{I_j^\Delta}\left(\hat{e}_{S_0^p(I_j^\Delta)}^{glob}, v\right) = -\mathcal{B}_{I_j^\Delta}\left(\hat{e}_{S^1(I_j^\Delta)}^{glob}, v\right) \quad \forall v \in S_0^p(I_j^\Delta). \tag{6.2.50c}$$

Taking the absolute values we obtain $|\hat{\rho}_2^{I_j^\Delta}|$, $|\hat{\rho}_1^{I_j^\Delta}|$, which are estimators for $|\rho_2^{I_j^\Delta}|$ and $|\rho_2^{I_j^\Delta}|$, respectively. Once more, to avoid cancellation in the sums we may employ the pollution estimator

$$|\tilde{\rho}_2^{I_j^\Delta}| = \sum_{k=1}^{N(\Delta)} \left|\mathcal{B}_{I_k^\Delta}\left(\hat{e}_{I_k^\Delta}^{\text{Neum}}(f, u_{S_\Delta^p}), \hat{e}_{I_k^\Delta}^{\text{Neum}}(\delta'_{I_j^\Delta}, G'_{I_j^\Delta, S_\Delta^p})\right)\right|. \tag{6.2.51a}$$

We will call the contributions

$$\xi_{I_k^\Delta}^{\text{Neum}}(\rho_2^{I_j^\Delta}) \stackrel{\text{def}}{=} \left|\mathcal{B}_{I_k^\Delta}\left(\hat{e}_{I_k^\Delta}^{\text{Neum}}(f, u_{S_\Delta^p}), \hat{e}_{I_k^\Delta}^{\text{Neum}}(\delta'_{I_j^\Delta}, G'_{I_j^\Delta, S_\Delta^p})\right)\right| \tag{6.2.51b}$$

the pollution indicators corresponding to the estimate $\tilde{\rho}_2^{I_j^\Delta}$. Employing the Cauchy–Schwarz inequality

$$\begin{aligned} &\left|\mathcal{B}_{I_k^\Delta}\left(\hat{e}_{I_k^\Delta}^{\text{Neum}}(f, u_{S_\Delta^p}), \hat{e}_{I_k^\Delta}^{\text{Neum}}(\delta'_{I_j^\Delta}, G'_{I_j^\Delta, S_\Delta^p})\right)\right| \\ &\quad \leq \eta_{I_k^\Delta}^{\text{Neum}}(f, u_{S_\Delta^p}) \eta_{I_k^\Delta}(\delta'_{I_j^\Delta}, G'_{I_j^\Delta, S_\Delta^p}) \end{aligned} \tag{6.2.51c}$$

we get the pollution estimate

$$|\bar{\rho}_2^{I_j^\Delta}| = \sum_{k=1}^{N(\Delta)} \eta_{I_k^\Delta}^{\text{Neum}}(f, u_{S_\Delta^p})\eta_{I_k^\Delta}^{\text{Neum}}(\delta'_{I_j^\Delta}, G'_{I_j^\Delta, S_\Delta^p}). \tag{6.2.51d}$$

Further, employing the Cauchy–Schwarz inequality for the $\mathbb{R}^{N(\Delta)}$ inner product, we get

$$|\breve{\rho}_2^{I_j^\Delta}| = \mathscr{E}^{\text{Neum}}(f, u_{S_\Delta^p})\mathscr{E}^{\text{Neum}}(\delta'_{I_j^\Delta}, G'_{I_j^\Delta, S_\Delta^p}). \tag{6.2.51e}$$

Note that

$$|\hat{\rho}_2^{I_j^\Delta}| \le |\tilde{\rho}_2^{I_j^\Delta}| \le |\bar{\rho}_2^{I_j^\Delta}| \le |\breve{\rho}_2^{I_j^\Delta}|. \tag{6.2.51f}$$

The estimators $|\hat{\rho}_1^{I_j^\Delta}|$, $|\tilde{\rho}_1^{I_j^\Delta}|$, $|\bar{\rho}_1^{I_j^\Delta}|$, and $|\breve{\rho}_1^{I_j^\Delta}|$, are defined analogously.

Let us now analyze the above pollution estimates using an example based on the singular model problem.

Example 6.2.3 Let us consider the model problem

$$-\frac{d}{dx}\Big(AE(x)\frac{du}{dx}(x)\Big) = 0, \quad x \in I \tag{6.2.52a}$$

$$u(0) = 0, \qquad AE(1)\frac{d}{dx}u(1) + \alpha u(1) = 1 \tag{6.2.52b}$$

where

$$AE(x) = (x+d)^{\vartheta}, \quad 0 \le \vartheta < 1 \tag{6.2.52c}$$

and $d \ge 0$. The exact solution of this problem is

$$u_{EX}(x) = \frac{1}{(1-\vartheta) + (1+d)^{1-\vartheta}\alpha - d^{1-\vartheta}}\Big((x+d)^{1-\vartheta} - \frac{d^{1-\vartheta}}{1-\vartheta}\Big). \tag{6.2.52d}$$

As we have already seen in Example 4.4.3, when uniform meshes are used, and $d = 0$, the pollution is the dominant component of the error in all the elements except the first one. We will analyze the estimate for the derivative of the pollution error in the last element and we will be interested in the values of the effectivity index

$$\hat{\kappa}_{2,I_j^\Delta} = \kappa(|\hat{\rho}_2^{I_j^\Delta}|) = \frac{|\hat{\rho}_2^{I_j^\Delta}|}{|\rho_2^{I_j^\Delta}|}. \tag{6.2.53a}$$

Note that the effectivity index $\kappa(\hat{\rho}_2^{I_j^\Delta})$ depends on the input data ϑ, d, α, the mesh Δ, and the degree p of the elements. In the majority of the cases below we will use the computed indicator $\hat{e}_{S_\Delta^{p+k}}^{\text{Neum}}$, instead of the exact indicator $\hat{e}^{\text{Neum}}$, and the effectivity index $\kappa(|\hat{\rho}_2^{I_j^\Delta}|)$ will also depend on the value of k. We will write

$$\hat{\kappa}_{2,I_j^\Delta} = \hat{\kappa}_{2,I_j^\Delta}(\vartheta, d, \alpha; \Delta, p; p+k). \tag{6.2.53b}$$

The goal of this example is to analyze the variation of the effectivity index with respect to ϑ, d, α, p, k, and the radical meshes Δ_N^γ, with nodes at $x_j^{\Delta_N^\gamma} = (j/N)^\gamma$. In all the results below we employ $\alpha = 10(1-\vartheta)$.

We first considered a *uniform mesh of eight elements* ($\gamma = 1, N = 8$), and studied the dependence of $\hat{\kappa}_{2,I_N^\Delta}$ on ϑ, and d, for the cases of linear ($p = 1$) and quadratic ($p = 2$) elements. Figure 6.2.2a (respectively Fig. 6.2.2b) shows the variation of $\hat{\kappa}_{2,I_N^\Delta}$ as a function of ϑ, for $d = 0$, and linear elements ($p = 1$) (respectively quadratic elements ($p = 2$)), where $\hat{\rho}_2^{I_N^\Delta}$ was computed by employing the ERNPp+k element error indicators. From the graphs it can be seen that very large values of k (e.g., $k = 12$) are needed in order to obtain an effectivity index $\hat{\kappa}_{2,I_N^\Delta} \approx 0.8$. Figure 6.2.3a (respectively Fig. 6.2.3b) shows the graphs of the variation of the effectivity index $\hat{\kappa}_{2,I_N^\Delta}$ with ϑ for various values of d/h, and $p = 1$ (respectively $p = 2$), in the case that $\hat{\rho}_2^{I_N^\Delta}$ is computed by employing the ERNPp + 6 element error indicators. Note that as the singularity is moved farther to the left of the point $x = 0$ (i.e., as d/h increases), we get $\hat{\kappa}_{2,I_N^\Delta} \approx 1$, for the entire range of ϑ. We also analyzed the effect of the degree p of the elements on the accuracy of $\hat{\rho}_2^{I_N^\Delta}$. Figure 6.2.4 shows $\hat{\kappa}_{2,I_N^\Delta}$ as a function of ϑ, $0.05 \le \vartheta \le 0.90$, for $p = 1, 2, 3, 4$ in the case that $d = 0$. Note that for fixed values of ϑ, as p increases $\hat{\kappa}_{2,I_N^\Delta}$ converges to zero.

We also considered the radical mesh $x_j^\Delta = (j/N)^\gamma$, with $N = 8$, and analyzed the effect of the parameter γ on the quality of the pollution estimator. Figure 6.2.5 shows the graph of $\hat{\rho}_2^{I_N^\Delta}$, as a function of ϑ for $\gamma = 0.75$, 1.0, 2.0, 3.0, and 4.0, linear elements ($p = 1$) and $d = 0$ where $\hat{\rho}_2^{I_N^\Delta}$ was computed using the ERNPp + 6 element error indicators. Note that the effectivity index $\hat{\kappa}_{2,I_N^\Delta}$ is not very sensitive to the choice of γ. Figure 6.2.6 gives $\hat{\kappa}_{2,I_N^\Delta}$ as a function of γ for $\vartheta = 0.75$, and $d = 0$, when $\hat{\rho}_2^{I_N^\Delta}$ is computed by employing the ERNPp + k error indicators with $k = 1$, 3, 6, 9, 12. Note that only in the case of $k = 12$, we obtain a reliable estimate of $\rho_2^{I_N^\Delta}$ with $0.8 \le \hat{\kappa}_{2,I_N^\Delta} \le 1.1$ for the entire range of γ.

Let us now show that the quality of the pollution estimate depends on the accuracy of the error indicator in the element closest to the singular point. Figure 6.2.7 gives the percent contribution of the first element and the rest of the mesh in the computed value of $\hat{\rho}_2^{I_N^\Delta}$ for the case $p = 1, N = 8$, and $\vartheta = 0.8$ and 0.9. Note that the percent contribution of the first element to the pollution estimate $\hat{\rho}_2^{I_N^\Delta}$ is over 60% for the entire range of ϑ. *Hence it is essential to compute the error indicator functions* $\hat{e}_{I_i^\Delta}^{\mathrm{Neum}}(f, u_{S_\Delta^p})$ *and* $\hat{e}_{I_i^\Delta}^{\mathrm{Neum}}(\delta'_{I_N^\Delta}, G'_{I_N^\Delta, S_\Delta^p})$ *with high accuracy in the element next to the singular point.*

Let us now describe two extraction methods which may be employed to increase the accuracy of the estimate in the first element.

(a) *Extraction method for the one-dimensional problem:*
Consider the singular problem on $I = (0, 1)$ with $0 < \vartheta < 1$:

$$-\frac{d}{dx}\left(x^\vartheta \frac{du}{dx}\right) = 0, \quad 0 < x < 1 \tag{6.2.54a}$$

$$u(0) = 0, \quad \left(x^\vartheta \frac{du}{dx}\right)(1) = a_2 \tag{6.2.54b}$$

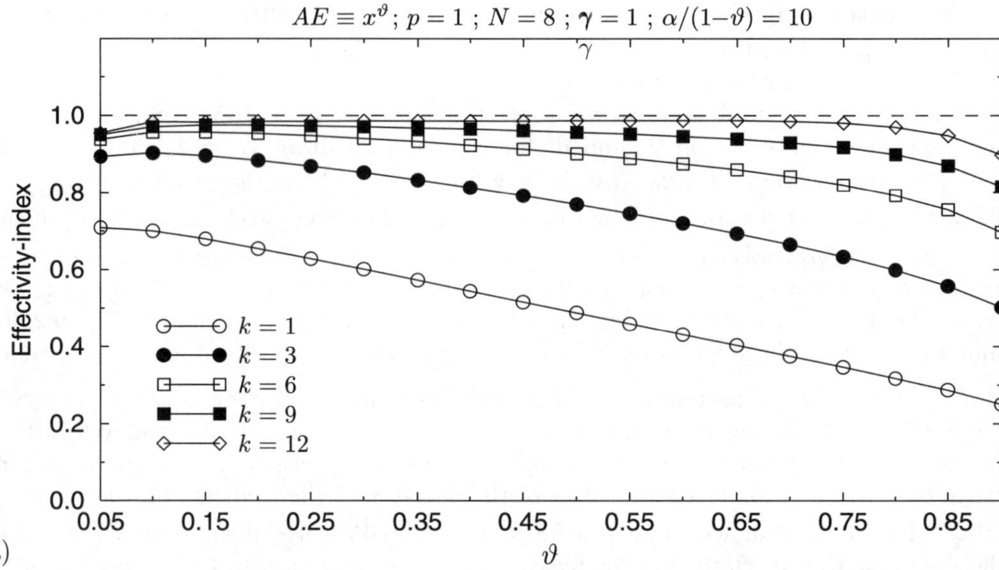

(a)

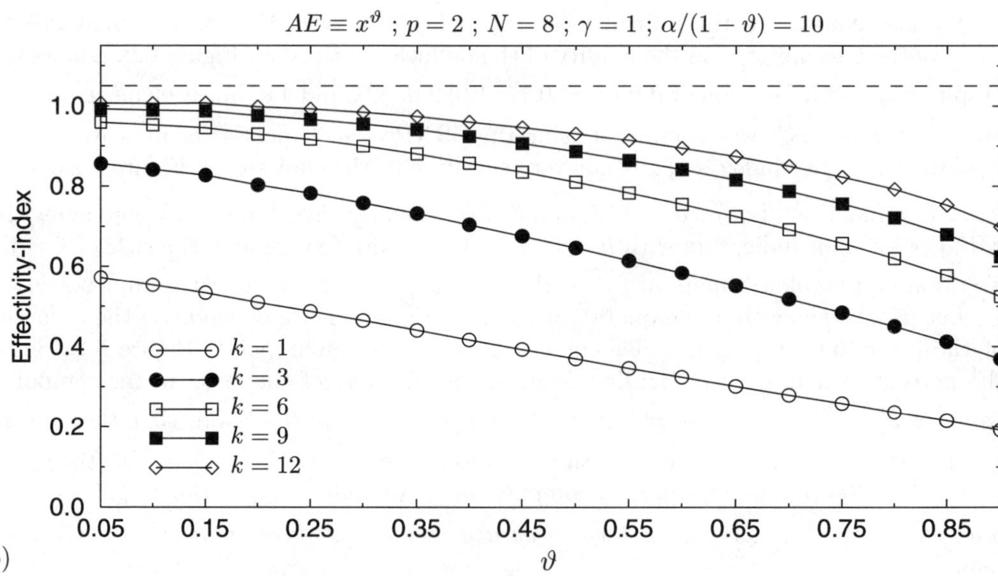

(b)

FIG. 6.2.2. *Analysis of the quality of the a-posteriori estimate of the pollution error.* The effect of the order of the polynomial approximation employed in the computation of the element error indicators. Input data: $d = 0$, $0.05 < \vartheta < 0.90$, $\alpha/(1-\vartheta) = 10$. Mesh data: Uniform mesh ($\gamma = 1$) of eight elements ($N = 8$) of degree p. The value of $\hat{\rho}_2^{I_N^{\Delta}}$ was computed using the ERNPp + k element error indicators. Graphs of the effectivity index $\kappa(\hat{\rho}_2^{I_N^{\Delta}})$ for the pollution estimator in the last element as a function of ϑ, for $k = 1, 3, 6, 9, 12$. (a) Linear elements ($p = 1$); (b) Quadratic elements ($p = 2$). *Note that as the strength of the singularity increases* ($\vartheta \to 1$) *very big values of k may be needed to get the effectivity index close to one*

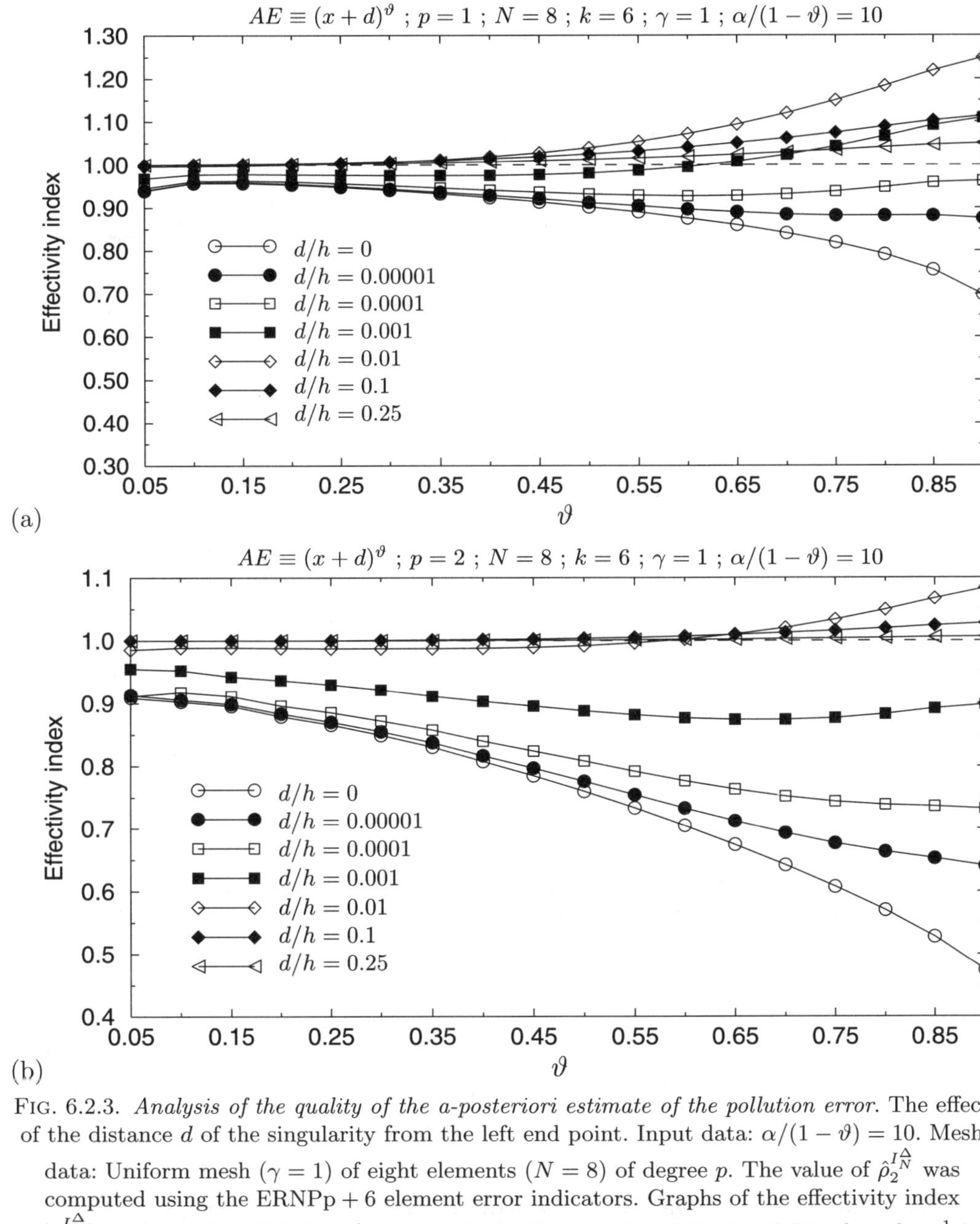

FIG. 6.2.3. *Analysis of the quality of the a-posteriori estimate of the pollution error.* The effect of the distance d of the singularity from the left end point. Input data: $\alpha/(1-\vartheta) = 10$. Mesh data: Uniform mesh ($\gamma = 1$) of eight elements ($N = 8$) of degree p. The value of $\hat{\rho}_2^{I_N^{\Delta}}$ was computed using the ERNPp + 6 element error indicators. Graphs of the effectivity index $\kappa(\hat{\rho}_2^{I_N^{\Delta}})$ as a function of ϑ, for $d/h = 0, 0.00001$, 0.0001, 0.001, 0.01, 0.1, 0.25, where $h = \frac{1}{8}$ is the mesh size. (a) Linear elements ($p = 1$); (b) Quadratic elements ($p = 2$). *Note that as d/h increases (i.e., the singularity is moved farther to the left of $x = 0$), we obtain a reliable estimate of the pollution error for the entire range of ϑ*

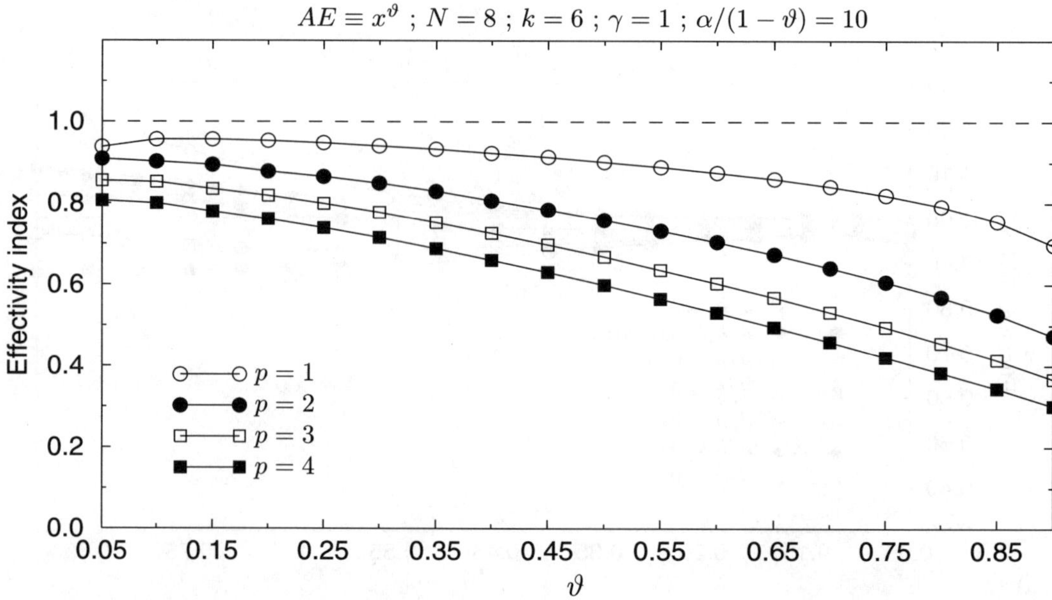

FIG. 6.2.4. *Analysis of the quality of the a-posteriori estimate of the pollution error.* The effect of the degree p of the elements. Input data: $d = 0$, $0.05 < \vartheta < 0.90$, $\alpha/(1-\vartheta) = 10$. Mesh data: Uniform mesh ($\gamma = 1$) of eight elements ($N = 8$) of degree p. The value of $\hat{\rho}_2^{I_N^{\Delta}}$ was computed using the ERNPp+6 element error indicators. Graphs of the effectivity index $\hat{\kappa}_{2,I_N^{\Delta}}$ as a function of ϑ, for $p = 1, 2, 3, 4$. *Note that for higher-order elements, we get severe underestimation as ϑ approaches one*

with the exact solution

$$u_{EX} = \mathscr{C} x^{1-\vartheta}. \tag{6.2.54c}$$

Assume that we know the character of u_{EX} given in (6.2.54c) but we do not know the constant $\mathscr{C}$ that we wish to determine. Then we use $u_{EX} \approx \mathscr{C} x^{1-\vartheta}$ for the error computation in the first element. (In two dimensions, $\mathscr{C}$ is the so-called stress intensity factor which is important in engineering.) Let

$$v(x) = \begin{cases} 1, & \text{for } 0 \le x \le \dfrac{\bar{x}}{2}, \quad \bar{x} < 1 \\ (\bar{x} - x)\left(\left(\dfrac{2}{\bar{x}}\right)^2 - \dfrac{2}{3}\left(\dfrac{2}{\bar{x}}\right)^3 (\bar{x} - x)\right) & \text{for } \dfrac{\bar{x}}{2} < x < \bar{x} \\ 0 & \text{for } x > \dfrac{\bar{x}}{2} \end{cases} \tag{6.2.55a}$$

Note that $v(x)$ has two continuous derivatives in I. Let $\varphi(x) = 1$, then

$$-\frac{d}{dx}\left(x^{\vartheta} \frac{d}{dx}\varphi\right) = 0 \tag{6.2.55b}$$

i.e., (x) satisfies the differential equation, and define

$$w = \varphi v. \tag{6.2.55c}$$

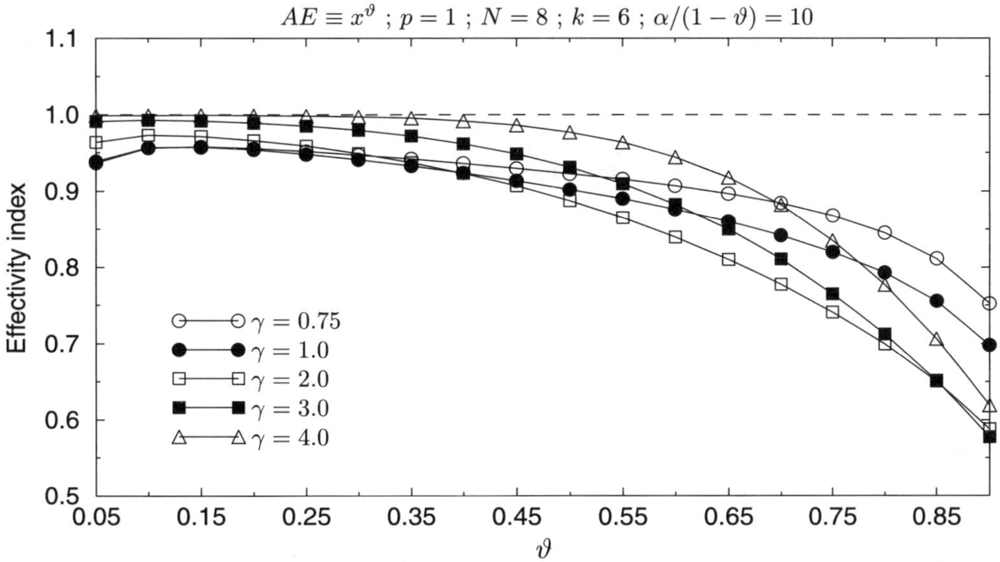

FIG. 6.2.5. *Analysis of the quality of the a-posteriori estimate of the pollution error.* The effect of the grading of the mesh. Input data: $d = 0$, $0.05 < \vartheta < 0.90$, $\alpha/(1-\vartheta) = 10$. Mesh data: Radical mesh $x_j^{\Delta} = (j/8)^{\gamma}$, $j = 0, 1, \ldots, 8$ of eight elements ($N = 8$) of degree $p = 1$. The value of $\hat{\rho}_2^{I_N^{\Delta}}$ was computed using the ERNPp+7 element error indicators. Graphs of the effectivity index $\kappa(|\hat{\rho}_2^{I_N^{\Delta}}|)$ as a function of ϑ, for $\gamma = 0.75, 1.0, 2.0, 4.0$. *Note that the quality of the pollution estimator is only mildly affected by the grading of the mesh*

Then we have

$$-\frac{d}{dx}\Big(x^{\vartheta}\frac{dw}{dx}\varphi\Big) = g \quad \text{on } I \tag{6.2.55d}$$

with $g = 0$ on $0 < x \leq \bar{x}/2$, $\bar{x} \leq x < 1$, and g is well defined on $\bar{x}/2 < x < \bar{x}$. Then we have

$$\int_0^1 u_{EX}\Big(-\frac{d}{dx}x^{\vartheta}\frac{dw}{dx}\Big) = \int_0^1 \Big(-\frac{d}{dx}x^{\vartheta}\frac{d}{dx}(u_{EX})\Big)w - \Big(u_{EX}x^{\vartheta}\frac{dw}{dx}\Big)(0)$$
$$+ \Big(wx^{\vartheta}\frac{d}{dx}(u_{EX})\Big)(0) = \Big(x^{\vartheta}\frac{d}{dx}(u_{EX})\Big)(0) = (1-\vartheta)\mathscr{C}. \tag{6.2.55e}$$

Hence

$$\mathscr{C} = \frac{1}{1-\vartheta}\int_{\bar{x}/2}^{x} u_{EX}g \tag{6.2.55f}$$

and $\mathscr{C}$ is a bounded functional which value can be approximately computed with guaranteed error estimation as explained above.

Function v is a cut-off function. Hence this method is called the cut off method for determination of the stress intensity factor (see also (6.2.51a,b)).

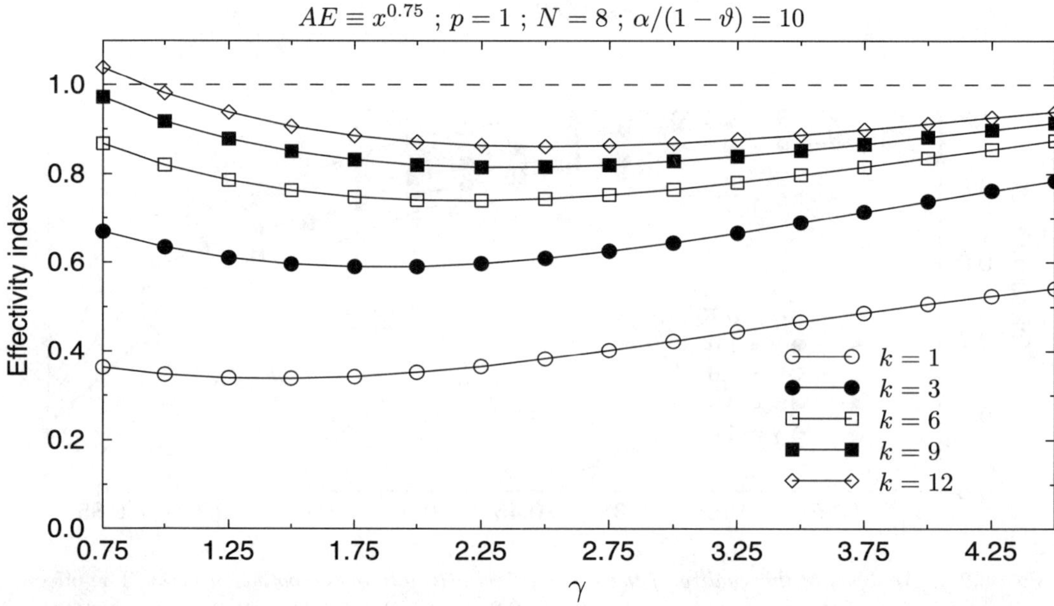

FIG. 6.2.6. *Analysis of the quality of the a-posteriori estimate of the pollution error.* The effect of the grading of the mesh and the order of the polynomial approximation employed in the computation of the element error indicators. Input data: $d = 0$, $\vartheta = 0.75$, $c \equiv 0$, $\alpha/(1-\vartheta) = 10$. Mesh data: Radical mesh $x_j^{\Delta} = (j/8)^{\gamma}$, $j = 0, 1, \ldots, 8$ of eight elements ($N = 8$) of degree $p = 1$. The value of $|\hat{\rho}_2^{I_N^{\Delta}}|$ was computed using the ERNPp + k element error indicators. Graphs of the effectivity index for the pollution estimator in the last element as a function of γ ($0.75 < \gamma < 4.5$), for $k = 1, 3, 6, 9, 12$. *Note that by selecting sufficiently high k, we obtain a reliable estimate of the pollution error in the last element, for the entire range of γ*

Another possibility is to compute

$$0 = \int_0^{\bar{x}/2} u_{EX}\Big(-\frac{d}{dx}x^{\vartheta}\frac{dw}{dx}\Big) = -\Big(x^{\vartheta}\frac{d}{dx}(u_{EX})\Big)\Big(\frac{\bar{x}}{2}\Big) + \Big(x^{\vartheta}\frac{d}{dx}(u_{EX})\Big)(0)$$

$$= -\Big(x^{\vartheta}\frac{d}{dx}(u_{EX})\Big)\Big(\frac{\bar{x}}{2}\Big) + (1-\vartheta)\mathscr{C}. \qquad (6.2.56\text{a})$$

Hence

$$\mathscr{C} = \frac{1}{1-\vartheta}\Big(\frac{\bar{x}}{2}\Big)^{\vartheta}\frac{d}{dx}(u_{EX})\Big(\frac{\bar{x}}{2}\Big). \qquad (6.2.56\text{b})$$

Now $\mathscr{C}$ is once more the value of a functional, but this time an unbounded one.

This is an analog of the contour integral method for determination of the stress intensity factor. Nevertheless, the functional is unbounded and we have to reduce it to a bounded one. See (6.2*.15a). Having approximated the function u_{EX} by $\tilde{\mathscr{C}}u_{EX}$, we use it for computation of the error in the first element.

(b) *Approximation of the error indicator functions in the first element using a geometric subgrid*: Let us now employ the p version on a geometric subgrid in order to

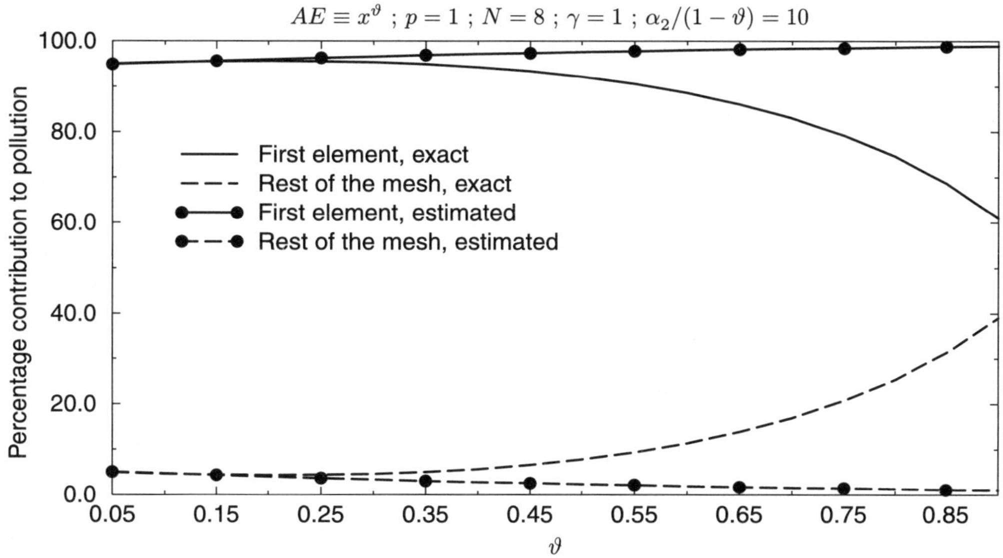

FIG. 6.2.7. *Analysis of the quality of the a-posteriori estimate of the pollution error.* The contribution of the first element and the rest of the mesh in the estimate $\hat{\rho}_2^{I_N^\Delta}$ compared with their corresponding exact values. Input data: $d = 0$, $c \equiv 0$, $\alpha/(1-\vartheta) = 10$. Mesh data: Uniform mesh ($\gamma = 1$) of eight elements ($N = 8$) of degree $p = 1$. The percent contribution of the first element $(\mathcal{B}_{I_1^\Delta}(\hat{e}^{\mathrm{Neum}}_{I_1^\Delta}(f, u_{S^p_\Delta}), \hat{e}^{\mathrm{Neum}}_{I_1^\Delta}(\delta'_N, G'_{I_N^\Delta, S^p_\Delta}))/\rho_2^{I_N^\Delta})100\%$ and the rest of the mesh $(\sum_{i=2}^{N(\Delta)} \mathcal{B}_{I_i^\Delta}(\hat{e}^{\mathrm{Neum}}_{I_i^\Delta}(f, u_{S^p_\Delta}), \hat{e}^{\mathrm{Neum}}_{I_i^\Delta}(\delta'_N, G'_{N,S^p_\Delta}))/\hat{\rho}_2^{I_N^\Delta})100\%$ compared with $(\mathcal{B}_{I_1^\Delta}(u_{EX} - u_{S^p_\Delta}, G'_{I_N^\Delta} - G'_{I_N^\Delta, S^p_\Delta})/\rho_2^{I_N^\Delta})$ 100% and $(\sum_{i=2}^{N(\Delta)} \mathcal{B}_{I_i^\Delta}(u_{EX} - u_{S^p_\Delta}, G'_{I_N^\Delta} - G'_{I^\Delta_{I_N^\Delta}, S^p_\Delta})/\hat{\rho}_2^{I_N^\Delta})$ 100% respectively

approximate $\hat{e}^{\mathrm{Neum}}_{I_1^\Delta}$. Let Δ^{I_1} be a grid with nodes $0 = x_0^{\Delta^{I_1}} < x_1^{\Delta^{I_1}} < x_2^{\Delta^{I_1}} < \cdots < x_{M-1}^{\Delta^{I_1}} < x_M^{\Delta^{I_1}} = x_1^\Delta$, which partitions I_1 into M sub-intervals (sub-elements) $I_i^{\Delta^{I_1}} = (x_{i-1}^{\Delta^{I_1}}, x_i^{\Delta^{I_1}})$, let

$$S^{p+k}_{\Delta^{I_1}} = \{w \mid w|_{I_i^{\Delta^{I_1}}} \in \mathcal{P}^{p+k}(I_i^{\Delta^{I_1}}),\ w(0) = 0\} \tag{6.2.57}$$

and let us employ the approximations $\hat{e}^{\mathrm{Neum}}_{I_1^\Delta, S^{p+k}_{\Delta^{I_1}}}(f, u_{S^p_\Delta})$, and $\hat{e}^{\mathrm{Neum}}_{I_1^\Delta, S^{p+k}_{\Delta^{I_1}}}(\delta'_N, G'_{N,S^p_\Delta})$, of the exact indicator functions $\hat{e}^{\mathrm{Neum}}_{I_1^\Delta}(f, u_{S^p_\Delta})$, and $\hat{e}^{\mathrm{Neum}}_{I_1^\Delta}(\delta'_{I_N^\Delta}, G'_{I_N^\Delta, S^p_\Delta})$, where $\Delta^{I_1} = \Delta^{I_1}_{(q,M)}$, is the geometric subgrid with nodes $x_j^{\Delta^{I_1}_{(q,M)}} = x_1^\Delta q^{M-j}$ for $j = 1, 2, \ldots M$, where q denotes the factor of the the geometric progression. Figure 6.2.8 (respectively Fig. 6.2.9) compares the values of the effectivity index $\hat{\kappa}_{2, I_N^\Delta}$, for the case of linear elements ($p = 1$) (respectively the case of quadratic elements ($p = 2$)) which are computed by employing extraction in the first element, with the values obtained by employing the computed indicators $\hat{e}^{\mathrm{Neum}}_{I_1^\Delta, S^{p+k}_{\Delta^{I_1}_{q,M}}}$, with $q = 0.15$.

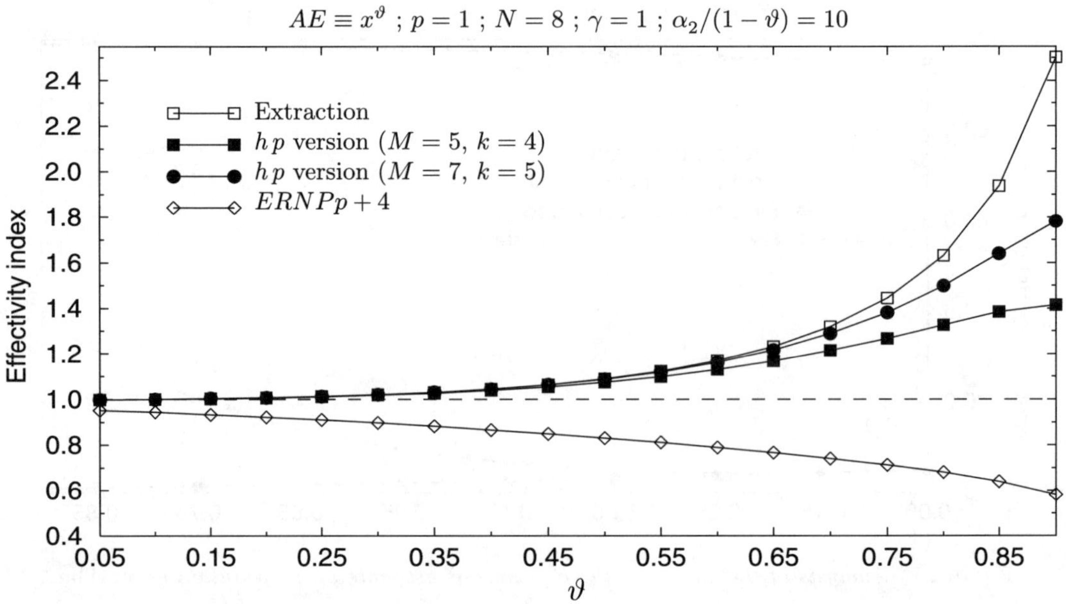

FIG. 6.2.8. *Analysis of the quality of the a-posteriori estimate of the pollution error.* The effect of the accuracy in the approximation of the error indicator in the first element. Input data: $d = 0$, $\alpha/(1-\vartheta) = 10$. Mesh data: Uniform mesh ($\gamma = 1$) of eight elements ($N = 8$) of degree $p = 1$. The value of $\hat{\rho}_2^{I_N^\Delta}$ was computed using the ERNPp + 4 element error indicators in all the elements, except the first element. In the first element, we considered the following cases: (i) extraction: $\hat{e}_{I_1^\Delta}(u_{S_\Delta^p}) = \hat{u}_{\text{EX}} - u_{S_\Delta^p}$, $\hat{e}_{I_1^\Delta}(G'_{I_N^\Delta, S_\Delta^p}) = G'_{I_N^\Delta} - G'_{I_N^\Delta, S_\Delta^p}$; (ii) $S^{p+k}_{\Delta^{I_1}_{(q,M)}}$ approximations of $\hat{e}^{\text{Neum}}_{I_1^\Delta}(f, u_{S_\Delta^p})$ and $\hat{e}^{\text{Neum}}_{I_1^\Delta}(\delta'_{I_N^\Delta}, G'_{I_N^\Delta, S_\Delta^p})$ with geometric progression ratio $q = 0.15$ and $M = 5$, $k = 4$ and $M = 7$, $k = 5$; (iii) ERNPp + 4 element error indicators. Graphs of the effectivity index $\hat{\kappa}_{2,I_N^\Delta}$ as a function of ϑ for various choices of the local error indicators for the first element

From the above numerical results, we note that:

1. The mesh Δ, and the polynomial degree p, do not have significant influence on the accuracy of the pollution estimator.
2. *The main factor which influences the accuracy of the pollution estimator is the accuracy in the approximation of the element error indicators in the element I_1^Δ.*

The above approach for the a-posteriori estimation of the pollution error cannot be generalized to higher dimensions. Let us now give a modified version of the above approach which can be directly extended to the multidimensional case. We will first recall another definition of the pollution error which is meaningful in higher dimensions. Let $\omega_j^\Delta = \bigcup_{k=j-1}^{j+1} I_k^\Delta$, and let $V_1^{\omega_j^\Delta} \in \mathcal{U}$ be such that

$$\mathcal{B}(V_1^{\omega_j^\Delta}, v) = \mathcal{R}_{\omega_j^\Delta}(v) \stackrel{\text{def}}{=} \sum_{k=j-1}^{j+1} \mathcal{R}^{EQ}_{I_k^\Delta}(v) \quad \forall v \in \mathcal{U} \tag{6.2.58a}$$

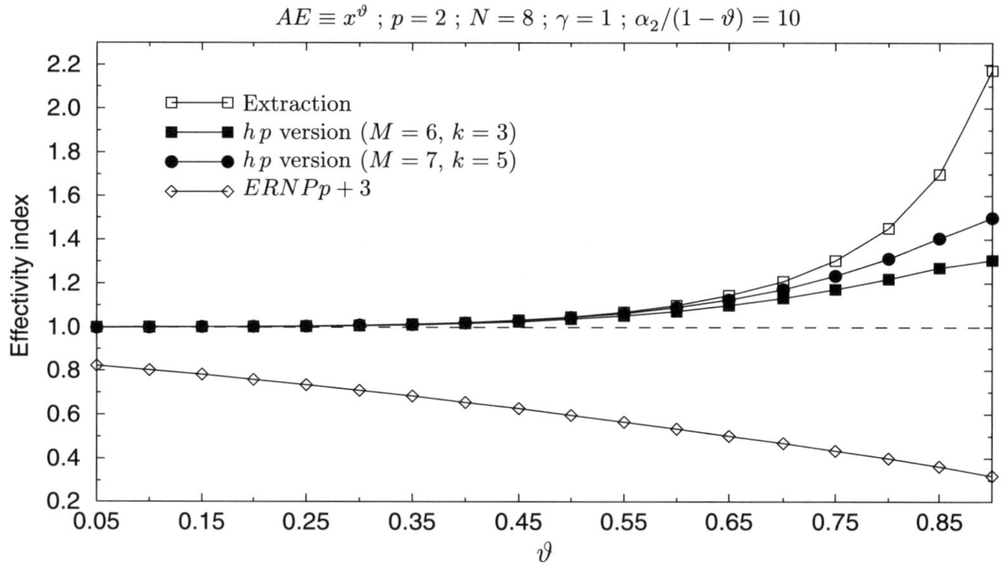

FIG. 6.2.9. *Analysis of the quality of the a-posteriori estimate of the pollution error.* The effect of the accuracy in the computation of the error indicator in the first element. Input data: $AE \equiv x^{\vartheta}$, $c \equiv 0$, $\alpha/(1-\vartheta) = 10$. Mesh data: Uniform mesh ($\gamma = 1$) of eight elements ($N = 8$) of degree $p = 2$. The value of $\hat{\rho}_2^{I_N^\Delta}$ was computed using the ERNPp+3 element error indicators in all the elements, except in the first element. In the first element, we considered the following cases: (i) extraction: $\hat{e}_{I_1^\Delta}(u_{S_\Delta^p}) = \hat{u}_{\mathrm{EX}} - u_{S_\Delta^p}$, $\hat{e}_{I_1^\Delta}(G'_{I_N^\Delta, S_\Delta^p}) = G'_{I_N^\Delta} - G'_{I_N^\Delta, S_\Delta^p}$; (ii) $S^{p+k}_{\Delta^{I_1}_{(q,M)}}$ approximations of $\hat{e}^{\mathrm{Neum}}_{I_1^\Delta}(f, u_{S_\Delta^p})$ and $\hat{e}^{\mathrm{Neum}}_{I_1^\Delta}(\delta'_{I_N^\Delta}, G'_{I_N^\Delta, S_\Delta^p})$ with geometric progression ratio $q = 0.15$ and $M = 6$, $k = 3$ and $M = 7$, $k = 5$; (iii) ERNPp+3 element error indicator. Graphs of the effectivity index $\hat{\kappa}_{2,I_N^\Delta}$ as a function of ϑ for various choices of the local error indicators for the first element

and let $V_2^{\omega_j^\Delta} \in \mathcal{U}$, be such that

$$\mathcal{B}(V_2^{\omega_j^\Delta}, v) = \mathcal{R}(v) - \mathcal{R}_{\omega_j^\Delta}(v) = \sum_{\substack{k=1\\ k\neq j-1,j,j+1}}^{N(\Delta)} \mathcal{R}^{EQ}_{I_k^\Delta}(v) \quad \forall v \in \mathcal{U}. \tag{6.2.58b}$$

Note that we have

$$\mathcal{B}(e_{S_\Delta^p}, v) = \mathcal{R}(v) = \mathcal{R}_{\omega_j^\Delta}(v) + (\mathcal{R}(v) - \mathcal{R}_{\omega_j^\Delta}(v)) \tag{6.2.58c}$$

and hence

$$e_{S_\Delta^p} = V_1^{\omega_j^\Delta} + V_2^{\omega_j^\Delta}. \tag{6.2.58d}$$

We will call $V_1^{\omega_j^\Delta}|_{I_j^\Delta}$ *the local error* and $V_2^{\omega_j^\Delta}|_{I_j^\Delta}$ *the pollution error* in I_j^Δ.

As we have already discussed in Section 4.3*, we get

$$\|e_{S_\Delta^p}\|^2_{\mathcal{U}(I_j^\Delta)} = \|V_1^{\omega_j^\Delta}\|^2_{\mathcal{U}(I_j^\Delta)} + \|V_2^{\omega_j^\Delta}\|^2_{\mathcal{U}(I_j^\Delta)} + 2\,\mathcal{B}_{I_j^\Delta}(V_1^{\omega_j^\Delta}, V_2^{\omega_j^\Delta}) \tag{6.2.59a}$$

and under rather mild assumptions about the meshes we have

$$\left|\mathcal{B}_{I_j^\Delta}(V_1^{\omega_j^\Delta}, V_2^{\omega_j^\Delta})\right| \le \mathcal{C}h^\alpha \|V_1^{\omega_j^\Delta}\|_{\mathcal{U}(I_j^\Delta)} \|V_2^{\omega_j^\Delta}\|_{\mathcal{U}(I_j^\Delta)} \tag{6.2.59b}$$

for some $\alpha > 0$, and hence the term $\mathcal{B}_{I_j^\Delta}(V_1^{\omega_j^\Delta}, V_2^{\omega_j^\Delta})$ is practically negligible, and we can employ the approximation

$$\|e_{S_\Delta^p}\|^2_{\mathcal{U}(I_j^\Delta)} \approx \|V_1^{\omega_j^\Delta}\|^2_{\mathcal{U}(I_j^\Delta)} + \|V_2^{\omega_j^\Delta}\|^2_{\mathcal{U}(I_j^\Delta)}. \tag{6.2.59c}$$

We also have

$$\|V_1^{\omega_j^\Delta}\|_{\mathcal{U}(I_j^\Delta)} \approx \eta_{I_j^\Delta}^{\text{Neum}} \tag{6.2.59d}$$

and hence

$$\|e_{S_\Delta^p}\|^2_{\mathcal{U}(I_j^\Delta)} \approx \left(\eta_{I_j^\Delta}^{\text{Neum}}\right)^2 + \|V_2^{\omega_j^\Delta}\|^2_{\mathcal{U}(I_j^\Delta)}. \tag{6.2.59e}$$

An a-posteriori error estimate for $\|V_2^{\omega_j^\Delta}\|_{\mathcal{U}(I_j^\Delta)}$, can be constructed by approximating $V_2^{\omega_j^\Delta}|_{I_j^\Delta}$, by its linear interpolant

$$V_2^{\omega_j^\Delta}|_{I_j^\Delta}(x) \approx \mathcal{I}_j^1 V_2^{\omega_j^\Delta}(x) = \rho_1(V_2^{\omega_j^\Delta}) + \rho_2(V_2^{\omega_j^\Delta})(x - x_{j-1/2}^\Delta), \quad x \in I_j^\Delta \tag{6.2.60a}$$

where we let

$$\rho_1(V_2^{\omega_j^\Delta}) \stackrel{\text{def}}{=} \tfrac{1}{2}(V_2^{\omega_j^\Delta}(x_j^\Delta) + V_2^{\omega_j^\Delta}(x_{j-1}^\Delta)) \tag{6.2.60b}$$

and

$$\rho_2(V_2^{\omega_j^\Delta}) \stackrel{\text{def}}{=} \frac{1}{h_{I_j^\Delta}}(V_2^{\omega_j^\Delta}(x_j^\Delta) - V_2^{\omega_j^\Delta}(x_{j-1}^\Delta)). \tag{6.2.60c}$$

Working as before we construct the estimate for $\rho_2(V_2^{\omega_j^\Delta})$

$$|\hat{\rho}_2(V_2^{\omega_j^\Delta})| = \left| \sum_{\substack{k=1 \\ k \ne j-1,j,j+1}}^{N(\Delta)} \mathcal{B}_{I_k^\Delta}\left(\hat{e}_{I_k^\Delta}^{\text{Neum}}(f, u_{S_\Delta^p}), \hat{e}_k^{\text{Neum}}(\delta'_{I_j^\Delta}, G'_{I_j^\Delta, S_\Delta^p})\right) \right| \tag{6.2.60d}$$

and analogously for $|\hat{\rho}_1(V_2^{\omega_j^\Delta})|$. As before, to prevent cancellations in the sum, we may define another estimator which sums the absolute values of the pollution indicators, we may estimate the pollution indicators by employing the Cauchy–Schwarz inequality, etc.

Remark 6.2.10 Note that in the estimate given in (6.2.60d) the terms $j-1$, j, and $j+1$ have been omitted from the sum. This precludes large errors in the computation of the indicators $\hat{e}_{I_k^\Delta}^{\text{Neum}}(\delta'_{I_j^\Delta}, G'_{I_j^\Delta, S_\Delta^p})$ from being included in the estimate. This is very important in higher dimensions where the Green's function $G'_{I_j^\Delta}$ does not belong to the energy space ($G'_{I_j^\Delta} \notin \mathcal{U}$), and the error $G'_{I_j^\Delta} - G'_{I_j^\Delta, S_\Delta^p}$, has unbounded energy in the elements in ω_j^Δ.

In this section we gave the construction of upper and lower bounds for the quantity of interest. We also constructed estimators which are not guaranteed, but are easier to compute and have good asymptotic properties. Finally, we constructed a-posteriori estimates for the pollution error.

Problems

P6.2.1. Consider the regular problem with $AE \equiv 1$, $c \equiv 1$, $\alpha_2 = a_2 = 0$, and $f = 1$. Given point $\bar{x}$, let the value of interest be $\mathcal{F}(u) = (d^2u/dx^2)(\bar{x})$, where $\bar{x}$ is a nodal point. Design an extraction functional $E^{\mathcal{F}}$ such that

$$|\mathcal{F}(u_{EX}) - E^{\mathcal{F}}(u_{S_\Delta^p})| \leq \mathcal{C}(h^\Delta)^{2p}. \tag{6.2.61}$$

P6.2.2. Repeat P6.2.1 for $\mathcal{F}(u) = (d^3u/dx^3)(\bar{x})$.

P6.2.3. Assume that $\bar{x} = \frac{1}{2}$, and let Δ consist of two elements of degree 1 of equal size. Compute the lower and upper bounds of $\mathcal{F}(u_{EX}) - E^{\mathcal{F}}(u_{S_\Delta^p})$, in P6.2.1.

P6.2.4. Repeat P6.2.3, for $\mathcal{F}(u) = (d^3u/dx^3)(\frac{1}{2})$.

6.2* Upper and lower bounds for the error in the outputs, and a-posteriori estimation of the pollution error in two dimensions

The construction of the bounds and the estimators for the error in the outputs, and the estimators for the pollution error in two dimensions is done analogously as in the one-dimensional case. Here, for simplicity, we will focus the presentation to the case of the mixed boundary value problem for the Laplacian

$$\Delta u\big|_\Omega = 0, \qquad u\big|_{\Gamma_D} = \bar{u}, \qquad \frac{\partial u}{\partial n}\Big|_{\Gamma_N} = g \tag{6.2*.1}$$

Let us first address the case of outputs which correspond to a functional $\mathcal{F}$, which is bounded in $\mathcal{U}$. As in the one-dimensional case, we have

$$\mathcal{E}^{L,\mathcal{F},s}_{S^{p+k}_{\Delta'}} \leq e^{\mathcal{F}}_{S^p_\Delta} \leq \mathcal{E}^{U,\mathcal{F},s}_{S^{p+k}_{\Delta'}} \tag{6.2*.2a}$$

where

$$\mathcal{E}^{L,\mathcal{F},s}_{S^{p+k}_{\Delta'}} = \frac{1}{4}\left(\left(\mathcal{E}^{L}_{\bar{u}^{\mathcal{F},s,+}_{S^p_\Delta}, S^{p+k}_{\Delta'}}\right)^2 - \left(\mathcal{E}^{U}_{\bar{u}^{\mathcal{F},s,-}_{S^p_\Delta}, S^{p+k}_{\Delta'}}\right)^2\right) \tag{6.2*.2b}$$

and

$$\mathcal{E}^{U,\mathcal{F},s}_{S^{p+k}_{\Delta'}} = \frac{1}{4}\left(\left(\mathcal{E}^{U}_{\bar{u}^{\mathcal{F},s,+}_{S^p_\Delta}, S^{p+k}_{\Delta'}}\right)^2 - \left(\mathcal{E}^{L}_{\bar{u}^{\mathcal{F},s,-}_{S^p_\Delta}, S^{p+k}_{\Delta'}}\right)^2\right) \tag{6.2*.2c}$$

and

$$\bar{u}^{\mathcal{F},s,\pm}_{S^p_\Delta} = s^{-1}u_{S^p_\Delta} \pm sH^{\mathcal{F}}_{S^p_\Delta}. \tag{6.2*.2d}$$

Let us now illustrate the accuracy of the bounds $\mathscr{E}^{L,\mathscr{F},s}_{S^{p+k}_{\Delta'}}$, $\mathscr{E}^{U,\mathscr{F},s}_{S^{p+k}_{\Delta'}}$, and their dependence on s.

Example 6.2*.1 (*Bounds for the average value over a subdomain*) Let u_{EX}, be the solution of the example problem in the Domain I which we employed in Example 6.1*.1, and let $u_{S^p_\Delta}$ be its finite element approximation determined using Mesh I, Mesh II, shown in Fig. 6.1*.3b or Mesh III, which is shown in Fig. 6.2*.1b, and biquadratic elements ($p = 2$), and let us be interested in the average value of the solution in a subdomain $\omega \subset \Omega$, namely

$$\mathscr{F}(u) = \frac{1}{|\omega|} \int_\omega u. \tag{6.2*.3a}$$

Taking into account that Dirichlet conditions are imposed on a finite portion of the boundary $\partial\Omega$, we have

$$|\mathscr{F}(u)| = \frac{1}{|\omega|} \left| \int_\omega u \right| \leq \frac{1}{\sqrt{|\omega|}} \|u\|_{L^2(\omega)} \leq \frac{1}{\sqrt{|\omega|}} \|u\|_{L^2(\Omega)} \leq \frac{\mathscr{C}}{\sqrt{|\omega|}} \|u\|_{\mathscr{U}} \tag{6.2*.3b}$$

and hence $\mathscr{F}$ is bounded for any ω, with positive area $|\omega|$. Below we employed the subdomain ω, which is shown in Fig. 6.2*.1b.

Table 6.2*.1 gives the values of the relative error in the computed average values over ω and the effectivity indices of the computable guaranteed bounds $\mathscr{E}^{L,\mathscr{F}}_{S^{p+1}_\Delta}$, $\mathscr{E}^{U,\mathscr{F}}_{S^{p+1}_\Delta}$ obtained

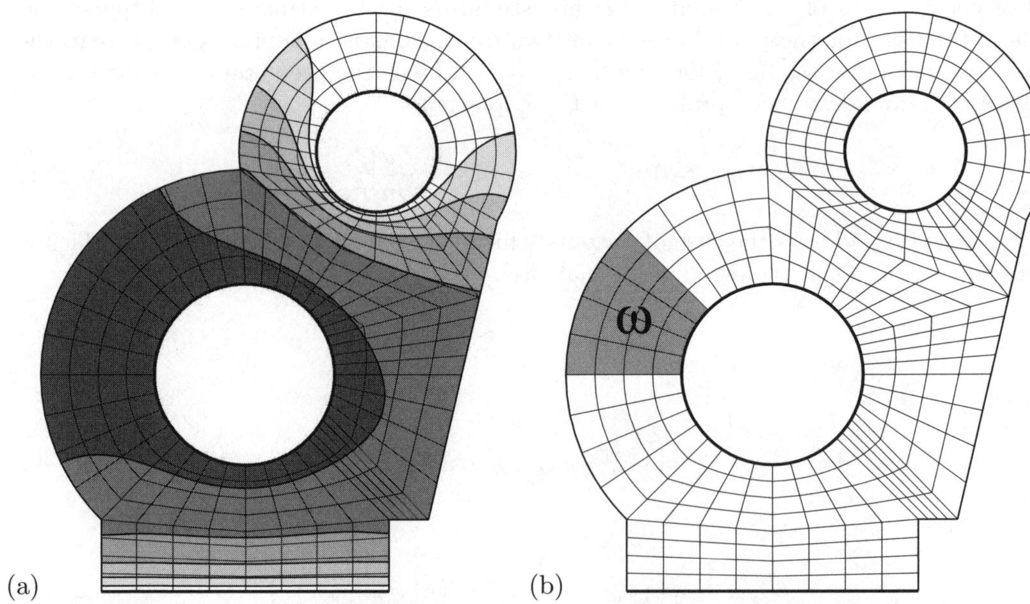

FIG. 6.2*.1. *Bounds for the error in the average value over a subdomain.* (a) The regions of $u_{S^p_\Delta}$ computed using Mesh III, where the darker colors indicate higher values. (b) Mesh III, which is obtained by two uniform refinements of Mesh I, and the subdomain ω where the average value is needed

TABLE 6.2*.1. *Bounds for the error in the average value over a subdomain.* The relative error in the average value over the subdomain ω, computed using Mesh I, II, and III, and biquadratic elements ($p = 2$) and the effectivity index for the computable bounds. The bounds were determined by employing the optimal value of the scaling factor s for the zeroth iteration. Note that we can have negative effectivity indices

	Mesh I		Mesh II		Mesh III	
$\dfrac{\int_\omega e_{S^p_\Delta}}{\int_\omega u_{S^8_{\Delta_h^{ovk}}}}$	0.1385 %		0.0444 %		0.0166 %	
Iteration index	$\dfrac{\mathcal{E}^{U,\mathcal{F}}}{\mathcal{F}(e_{S^p_\Delta})}$	$\dfrac{\mathcal{E}^{L,\mathcal{F}}}{\mathcal{F}(e_{S^p_\Delta})}$	$\dfrac{\mathcal{E}^{U,\mathcal{F}}}{\mathcal{F}(e_{S^p_\Delta})}$	$\dfrac{\mathcal{E}^{L,\mathcal{F}}}{\mathcal{F}(e_{S^p_\Delta})}$	$\dfrac{\mathcal{E}^{U,\mathcal{F}}}{\mathcal{F}(e_{S^p_\Delta})}$	$\dfrac{\mathcal{E}^{L,\mathcal{F}}}{\mathcal{F}(e_{S^p_\Delta})}$
0	7.923	−4.240	5.304	−2.994	5.173	−1.394
1	3.781	−0.409	2.868	0.002	3.014	0.004

using the optimal value of the scaling factor which is determined form the condition (6.2*.5a) below. The relatively poor effectivity of the bounds is due to cancellations in (6.2*.2b, c), which was first reported in [Babuška and Miller, 1983c]. Better effectivity indices may be obtained by improving the estimator, for example by replacing the element residual by the patch residual estimator, in the elements with major contribution to the bounds, namely the elements with a vertex at the tip of a re-entrant corner, and the elements in the neighborhood of ω where the function $H^{\mathcal{F}}$ is rather unsmooth.

Let us also give an example for which u_{EX}, and $H^{\mathcal{F}}$ are smoother and better effectivity indices may be obtained.

Example 6.2*.2 (*Bounds for the average value on a Neumann boundary.*) In order to illustrate the point about the effectivity of the bounds and its dependence on the smoothness of u_{EX} and $H^{\mathcal{F}}$, we employed the model problem in the smoothened Domain I shown in Figure 6.2*.2, where we applied homogenous Neumann boundary conditions on the outer boundary Γ_N, and the bounded functional

$$\mathcal{F}(u) = \int_{\Gamma_N} u\, ds \tag{6.2*.4}$$

namely the average value on the outer boundary.

Table 6.2*.2. gives the relative errors for the extracted value and the effectivity index of the computable upper, and lower bound $\mathcal{E}^{U,\mathcal{F}}$, and $\mathcal{E}^{L,\mathcal{F}}$, for the error in the average value over the outer boundary, for $p = 1, 2$, for the finite element solution of the model problem in the smoothened Domain I, and II. Note that in the case of smoothened Domain I for which both u_{EX}, and $H^{\mathcal{F}}$, are smooth functions and $p = 2$, bounds with improved effectivity are obtained after one iteration.

Let us address the dependence of the bounds on the scaling factor. For the exact bounds, we have

$$(\mathcal{E}^U)^2 - (\mathcal{E}^L)^2 = \sum_{\tau \in \Delta} \|\Phi_{\hat{e}^{\text{Neum}}}\|^2_{\mathcal{U}(\tau)} \tag{6.2*.4a}$$

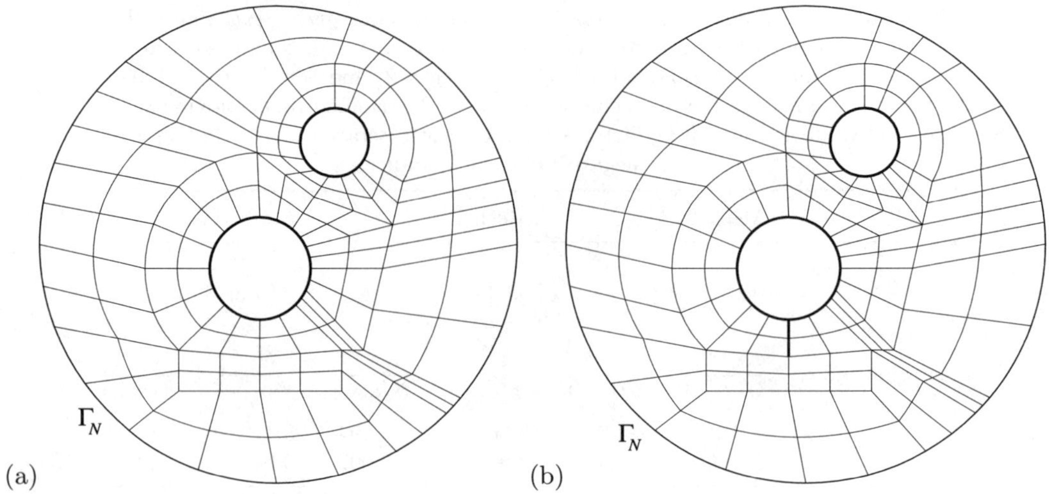

FIG. 6.2*.2. *Bounds for the error in the average value on the Neumann boundary.* (a) The smoothened Domain I, and (b) the smoothened Domain II, are shown with the mesh employed in the computation. Homogenous Neumann boundary conditions were applied on the outer boundary

TABLE 6.2*.2. *Bounds for the error in the average value on the Neumann boundary.* The relative error in the average value on the outer boundary, and the effectivity indices of the computable bounds of the error in the average value on the outer boundary, determined for the smoothened domains and the meshes shown in Figure 6.2*.2, by employing $k = 2$. The bounds were computed by employing the optimal value of the scaling factor s for the zeroth iteration

	Smoothened Domain I				Smoothened Domain II			
	$p = 1$		$p = 2$		$p = 1$		$p = 2$	
$\dfrac{\int_{\Gamma_N} e_{S^p_\Delta}}{\int_{\Gamma_N} u_{S^8_{\Delta^{ovk}_h}}}$	-7.6493×10^{-4}		-1.5569×10^{-5}		-2.1420×10^{-3}		-1.0161×10^{-3}	
Iteration index	$\dfrac{\mathscr{E}^{U,\mathscr{F}}}{\mathscr{F}(e_{S^p_\Delta})}$	$\dfrac{\mathscr{E}^{L,\mathscr{F}}}{\mathscr{F}(e_{S^p_\Delta})}$	$\dfrac{\mathscr{E}^{U,\mathscr{F}}}{\mathscr{F}(e_{S^p_\Delta})}$	$\dfrac{\mathscr{E}^{L,\mathscr{F}}}{\mathscr{F}(e_{S^p_\Delta})}$	$\dfrac{\mathscr{E}^{U,\mathscr{F}}}{\mathscr{F}(e_{S^p_\Delta})}$	$\dfrac{\mathscr{E}^{L,\mathscr{F}}}{\mathscr{F}(e_{S^p_\Delta})}$	$\dfrac{\mathscr{E}^{U,\mathscr{F}}}{\mathscr{F}(e_{S^p_\Delta})}$	$\dfrac{\mathscr{E}^{L,\mathscr{F}}}{\mathscr{F}(e_{S^p_\Delta})}$
0	−9.381	9.334	−13.352	13.303	−16.370	19.859	−2.097	9.132
1	−2.866	5.814	0.508	1.594	−4.250	8.497	−0.159	6.011

and hence

$$\mathscr{E}^{U,\mathscr{F},s} - \mathscr{E}^{L,\mathscr{F},s} = \frac{1}{4} \sum_{\tau \in \Delta} \left(\|\Phi_{\hat{e}^{\mathrm{Neum}}_{\tilde{u}^{\mathscr{F},s,+}_{S^p_\Delta}}}\|^2_{\mathcal{U}(\tau)} + \|\Phi_{\hat{e}^{\mathrm{Neum}}_{\tilde{u}^{\mathscr{F},s,-}_{S^p_\Delta}}}\|^2_{\mathcal{U}(\tau)} \right) \tag{6.2*.4b}$$

and expanding we get

$$\mathscr{E}^{U,\mathscr{F},s} - \mathscr{E}^{L,\mathscr{F},s} = \frac{1}{4} \Big(s^{-2} \sum_{\tau \in \Delta} \|\Phi_{\hat{e}^{\mathrm{Neum}}_{u_{S^p_\Delta}}}\|^2_{\mathcal{U}(\tau)} + s^2 \sum_{\tau \in \Delta} \|\Phi_{\hat{e}^{\mathrm{Neum}}_{H^{\mathscr{F}}_{S^p_\Delta}}}\|^2_{\mathcal{U}(\tau)} \Big). \tag{6.2*.4c}$$

The optimal value s^{opt}, can now be obtained from the condition

$$\frac{d}{ds}(\mathscr{E}^{U,\mathscr{F},s} - \mathscr{E}^{L,\mathscr{F},s})(s^{opt}) = 0 \tag{6.2*.5a}$$

which gives

$$s^{opt} = \left(\frac{\sum_{\tau\in\Delta} \left\| \Phi_{\hat{e}^{\mathrm{Neum}}_{u_{S^p_\Delta}}} \right\|^2_{\mathscr{U}(\tau)}}{\sum_{\tau\in\Delta} \left\| \Phi_{\hat{e}^{\mathrm{Neum}}_{H^{\mathscr{F}}_{S^p_\Delta}}} \right\|^2_{\mathscr{U}(\tau)}} \right)^{1/4} \tag{6.2*.5b}$$

and

$$\mathscr{E}^{U,\mathscr{F},s^{opt}} - \mathscr{E}^{U,\mathscr{F},s^{opt}} = \frac{1}{2}\sqrt{\sum_{\tau\in\Delta} \left\| \Phi_{\hat{e}^{\mathrm{Neum}}_{u_{S^p_\Delta}}} \right\|^2_{\mathscr{U}(\tau)}} \sqrt{\sum_{\tau\in\Delta} \left\| \Phi_{\hat{e}^{\mathrm{Neum}}_{H^{\mathscr{F}}_{S^p_\Delta}}} \right\|^2_{\mathscr{U}(\tau)}}. \tag{6.2*.5c}$$

Note that the optimal scaling factor s^{opt}, for the exact bounds and the corresponding minimum difference in the bounds depend only on the norms of the gap functions.

Let us give an example which illustrates the influence of s on the accuracy of the computable bounds.

Example 6.2*.3 We determined the effectivity of the computable bounds for the case considered in Example 6.2*.1 for various values of the scaling factor s. Table 6.2*.3 gives the effectivity indices of the relative difference in the first iterate of the computable bounds for the error in the average value over ω

$$\delta\mathscr{E}^{\mathscr{F}} \stackrel{\mathrm{def}}{=} \frac{1}{\mathscr{F}(e_{S^p_\Delta})}\left(\mathscr{E}^{U,\mathscr{F}}_{S^{p+k}_\Delta} - \mathscr{E}^{L,\mathscr{F}}_{S^{p+k}_\Delta}\right) \tag{6.2*.6}$$

for $0.01 \le s \le 10$, computed using Mesh I, II, and III. It can be seen that the choice of the scaling factor s is critical for the accuracy of the computable bounds.

TABLE 6.2*.3. *Influence of the scaling factor on the effectiveness of the computable bounds.* Values of $(1/\mathscr{F}(e_{S^p_\Delta}))(\mathscr{E}^{U,\mathscr{F}}_{S^{p+1}_\Delta} - \mathscr{E}^{L,\mathscr{F}}_{S^{p+1}_\Delta})$, for several values of s in the interval $0.01 \le s \le 10.0$. Note that the choice of the scaling factor s is critical for the accuracy of the bounds

s	Mesh I	Mesh II	Mesh III
0.01	44777	42247	45666
0.1	447.79	422.54	456.77
1.0	5.349	4.694	5.061
1.5	3.923	2.933	3.139
2.0	4.539	2.932	3.114
5.0	21.515	11.889	12.509
10.0	85.382	46.921	49.350

Let us now consider the case that the output of interest corresponds to an unbounded functional, like, for example, the value of the ith component of the gradient of $\bar{\mathbf{x}}$, namely

$$\mathcal{F}(u) = q_i^{\bar{\mathbf{x}}}(u) \stackrel{\text{def}}{=} \frac{\partial u}{\partial x_i}(\bar{\mathbf{x}}). \tag{6.2*.7}$$

As in the one-dimensional case, we will construct an *extraction functional*

$$E^{\mathcal{F}} = \mathcal{L}_{\mathcal{F}} + \mathcal{M}_{\mathcal{F}} \tag{6.2*.8a}$$

such that we have

$$E^{\mathcal{F}}(u_{EX}) = \mathcal{F}(u_{EX}) \tag{6.2*.8b}$$

with $\mathcal{L}_{\mathcal{F}} : \mathcal{U} \mapsto \mathbb{R}$ bounded in $\mathcal{U}$

$$\left|\mathcal{L}_{\mathcal{F}}(v)\right| \leq \mathcal{C}\ \|v\|_{\mathcal{U}} \qquad \forall v \in \mathcal{U} \tag{6.2*.8c}$$

and $\mathcal{M}_{\mathcal{F}} : \mathcal{D} \mapsto \mathbb{R}$, is computable from the input data, and is bounded on $\mathcal{D}$

$$|\mathcal{M}_{\mathcal{F}}(\bar{u}, g)| < \infty \quad \forall(\bar{u}, g) \in \mathcal{D}. \tag{6.2*.8d}$$

It follows the error in the extracted value $E^{\mathcal{F}}(u_{S^p_\Delta})$, is the same as $e^{\mathcal{L}_{\mathcal{F}}}_{S^p_\Delta} = \mathcal{L}_{\mathcal{F}}(e_{S^p_\Delta})$, the error in $\mathcal{L}_{\mathcal{F}}(u_{S^p_\Delta})$ namely

$$\mathcal{F}(u_{EX}) - E^{\mathcal{F}}(u_{S^p_\Delta}) = \mathcal{L}_{\mathcal{F}}(u_{EX}) - \mathcal{L}_{\mathcal{F}}(u_{S^p_\Delta}) = \mathcal{L}_{\mathcal{F}}(e_{S^p_\Delta}) = \mathcal{B}(e_{u_{S^p_\Delta}}, e_{H^{\mathcal{L}_{\mathcal{F}}}_{S^p_\Delta}}) \tag{6.2*.9a}$$

where $H^{\mathcal{L}_{\mathcal{F}}}_{S^p_\Delta} \in S^p_\Delta$, is the finite element approximation of the function $H^{\mathcal{L}_{\mathcal{F}}}$, which is the solution of the auxilliary problem:

Find $H^{\mathcal{L}_{\mathcal{F}}} \in \mathcal{U}$ such that

$$\mathcal{B}(H^{\mathcal{L}_{\mathcal{F}}}, v) = \mathcal{L}_{\mathcal{F}}(v) \quad \forall v \in \mathcal{U}. \tag{6.2*.9b}$$

Hence the bounds for the error in the extraction $\mathcal{F}(u_{EX}) - E^{\mathcal{F}}(u_{S^p_\Delta})$, are the same as the bounds for $e^{\mathcal{L}_{\mathcal{F}}}_{S^p_\Delta}$, namely

$$\mathcal{E}^{L,\mathcal{L}_{\mathcal{F}}}_{S^{p+k}_{\Delta'}} \leq \mathcal{F}(u_{EX}) - E^{\mathcal{F}}(u_{S^p_\Delta}) = e^{\mathcal{L}_{\mathcal{F}}}_{S^p_\Delta} \leq \mathcal{E}^{U,\mathcal{L}_{\mathcal{F}}}_{S^{p+k}_{\Delta'}}. \tag{6.2*.10}$$

Let us now elaborate on possible constructions of an extraction functional $E^{q_1^{\mathbf{x}}}$, for the output

$$q_1^{\mathbf{x}}(u_{EX}) \stackrel{\text{def}}{=} \frac{\partial}{\partial x_1}(u_{EX})(\bar{\mathbf{x}}) \tag{6.2*.11}$$

where $\bar{\mathbf{x}}$, is an interior point of Ω, as shown in Figure 6.2*.3. We will employ the following classical result.

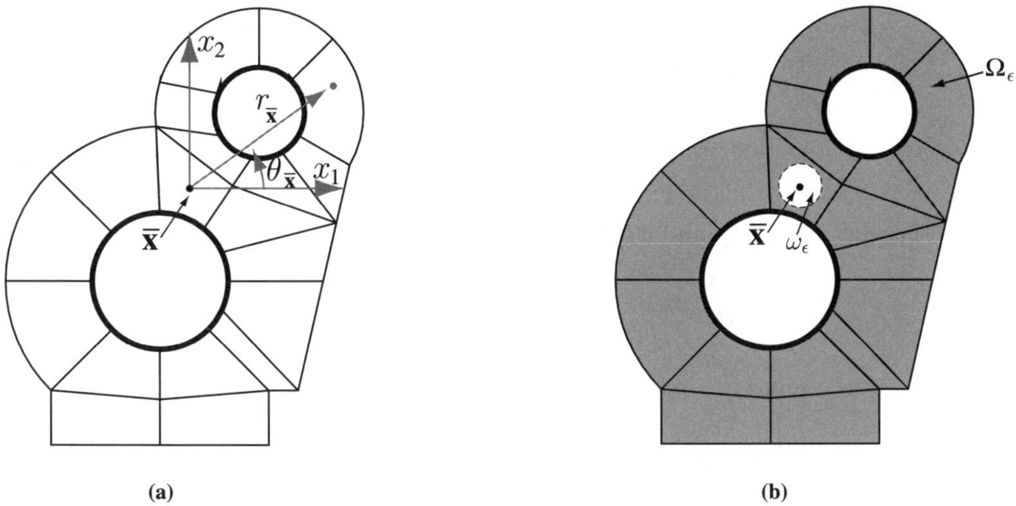

FIG. 6.2*.3. *The extraction of the derivative at a point.* (a) The point of interest $\bar{\mathbf{x}}$ and the polar coordinates $(r_{\bar{\mathbf{x}}}, \theta_{\bar{\mathbf{x}}})$ relative to $\bar{\mathbf{x}}$. (b) The circular disk ω_ϵ, of radius ϵ, around $\bar{\mathbf{x}}$

Lemma 6.2*.1 *Let $\bar{\mathbf{x}}$ be an interior point of Ω, and let*

$$w_0^{\bar{\mathbf{x}}} = \frac{1}{2\pi}\frac{\cos(\theta_{\bar{\mathbf{x}}})}{r_{\bar{\mathbf{x}}}} \tag{6.2*.12a}$$

where $r_{\bar{\mathbf{x}}}$, $\theta_{\bar{\mathbf{x}}}$, are polar coordinates centered at $\bar{\mathbf{x}}$. For any sufficiently smooth function u_{EX} which is harmonic in Ω, namely

$$\Delta u_{EX}\big|_{\Omega} \equiv 0 \tag{6.2*.12b}$$

we have

$$\frac{\partial}{\partial x_1}(u_{EX}(\bar{\mathbf{x}})) = \int_{\partial\Omega}\Big(\frac{\partial}{\partial n}(u_{EX})\ w_0^{\bar{\mathbf{x}}} - u_{EX}\frac{\partial}{\partial n}(w_0^{\bar{\mathbf{x}}})\Big). \tag{6.2*.12c}$$

Proof Let $\omega_\varepsilon^{\bar{\mathbf{x}}}$, be circular disk, centered at $\bar{\mathbf{x}}$, with ε sufficiently small such that its boundary $\partial\omega_\varepsilon^{\bar{\mathbf{x}}}$, does not intersect the boundary of the domain $\partial\Omega$, as is, for example, shown in Figure 6.2*.3b. Employing integration by parts, we obtain

$$\begin{aligned}\int_{\Omega-\omega_\varepsilon^{\bar{\mathbf{x}}}}&(u_{EX}\Delta w_0^{\bar{\mathbf{x}}} - w_0^{\bar{\mathbf{x}}}\Delta u_{EX})\\ &= \int_{\partial(\Omega-\omega_\varepsilon^{\bar{\mathbf{x}}})}\Big(\frac{\partial}{\partial n}(u_{EX})w_0^{\bar{\mathbf{x}}} - u_{EX}\frac{\partial}{\partial n}(w_0^{\bar{\mathbf{x}}})\Big)\end{aligned} \tag{6.2*.13a}$$

where n is the exterior normal of $\partial\big(\Omega - \omega_\varepsilon^{\bar{\mathbf{x}}}\big)$. Taking into account both u_{EX}, and $w_0^{\bar{\mathbf{x}}}$, are harmonic in $\Omega - \omega_\varepsilon^{\bar{\mathbf{x}}}$, namely

$$\Delta u_{EX}\big|_{\Omega-\omega_\varepsilon^{\bar{\mathbf{x}}}} = 0, \qquad \Delta w_0^{\bar{\mathbf{x}}}\big|_{\Omega-\omega_\varepsilon^{\bar{\mathbf{x}}}} = 0 \tag{6.2*.13b}$$

we get

$$\int_{\partial\omega_\varepsilon^{\bar{\mathbf{x}}}} \Big(w_0^{\bar{\mathbf{x}}} \frac{\partial}{\partial n}(u_{EX}) - u_{EX} \frac{\partial}{\partial n}(w_0^{\bar{\mathbf{x}}}) \Big)$$
$$= \int_{\partial\Omega} \Big(u_{EX} \frac{\partial}{\partial n}(w_0^{\bar{\mathbf{x}}}) - w_0^{\bar{\mathbf{x}}} \frac{\partial}{\partial n}(u_{EX}) \Big). \tag{6.2*.13c}$$

Taking the limit as $\varepsilon \to 0$, and employing that

$$\lim_{\varepsilon\to 0} \int_{\partial\omega_\varepsilon^{\bar{\mathbf{x}}}} w_0^{\bar{\mathbf{x}}} \frac{\partial}{\partial n}(u_{EX})$$
$$= \frac{1}{2\pi} \lim_{\varepsilon\to 0} \int_0^{2\pi} \cos\theta_{\bar{\mathbf{x}}} \Big(\frac{\partial}{\partial x_1}(u_{EX}) \cos\theta_{\bar{\mathbf{x}}} + \frac{\partial}{\partial x_2}(u_{EX}) \sin\theta_{\bar{\mathbf{x}}} \Big)\, d\theta_{\bar{\mathbf{x}}}$$
$$= \frac{1}{2} \frac{\partial}{\partial x_1}(u_{EX})(\bar{\mathbf{x}}) \tag{6.2*.14a}$$

and

$$\lim_{\varepsilon\to 0} \int_{\partial\omega_\varepsilon^{\bar{\mathbf{x}}}} u_{EX} \frac{\partial}{\partial n}(w_0^{\bar{\mathbf{x}}}) = \lim_{\varepsilon\to 0} \int_0^{2\pi} u_{EX} \Big(-\frac{1}{2\pi} \frac{\cos(\theta_{\bar{\mathbf{x}}})}{\varepsilon^2} \Big) \varepsilon\, d\theta_{\bar{\mathbf{x}}}$$
$$= -\frac{1}{2} \frac{\partial}{\partial x_1}(u_{EX})(\bar{\mathbf{x}}) \tag{6.2*.14b}$$

(6.2*.12c) follows. □

From the identity (6.2*.12c), we obtain

$$q_1^{\bar{\mathbf{x}}}(u_{EX}) = \frac{\partial}{\partial x_1}(u_{EX}(\bar{\mathbf{x}})) = \mathscr{L}_{q_1^{\bar{\mathbf{x}}}}^{unbnd}(u_{EX}) + \mathscr{M}_{q_1^{\bar{\mathbf{x}}}}(\bar{u}, g) \tag{6.2*.15a}$$

with

$$\mathscr{L}_{q_1^{\bar{\mathbf{x}}}}^{unbnd}(v) \stackrel{\text{def}}{=} \int_{\Gamma_D} w_0^{\bar{\mathbf{x}}} \frac{\partial v}{\partial n} - \int_{\Gamma_N} v \frac{\partial}{\partial n}(w_0^{\bar{\mathbf{x}}}) \tag{6.2*.15b}$$

and

$$\mathscr{M}_{q_1^{\bar{\mathbf{x}}}}(\bar{u}, g) \stackrel{\text{def}}{=} \int_{\Gamma_N} g w_0^{\bar{\mathbf{x}}} - \int_{\Gamma_N} \bar{u} \frac{\partial}{\partial n}(w_0^{\bar{\mathbf{x}}}) \tag{6.2*.15c}$$

where $\bar{u}$, g denote the boundary data. Note however that this is not the right splitting needed for the extraction because the functional

$$\mathscr{G}(v) \stackrel{\text{def}}{=} \int_{\Gamma_D} w_0^{\bar{\mathbf{x}}} \frac{\partial v}{\partial n}, \quad \forall v \in \mathscr{U} \tag{6.2*.15d}$$

is not bounded in $\mathscr{U}$, and hence also $\mathscr{L}_{q_1^{\bar{\mathbf{x}}}}^{unbnd}$ is not bounded in $\mathscr{U}$.

In order to replace $\mathscr{L}_{q_1^{\bar{\mathbf{x}}}}^{unbnd}$, by a bounded functional $\mathscr{L}_{q_1^{\bar{\mathbf{x}}}}$, we will let

$$w^{\bar{\mathbf{x}}} \stackrel{\text{def}}{=} w_0^{\bar{\mathbf{x}}} + w_1^{\bar{\mathbf{x}}} \tag{6.2*.16a}$$

where $w_1^{\bar{\mathbf{x}}} \in \mathcal{U}$, is such that

$$w^{\bar{\mathbf{x}}}\big|_{\Gamma_D} = (w_0^{\bar{\mathbf{x}}} + w_1^{\bar{\mathbf{x}}})\big|_{\Gamma_D} = 0, \qquad w_1^{\bar{\mathbf{x}}}\big|_{\Gamma_N} = 0. \tag{6.2*.16b}$$

The function $w_1^{\bar{\mathbf{x}}}$, can be constructed by employing finite element and blending function interpolation on the employed finite element mesh Δ. Replacing $w_0^{\bar{\mathbf{x}}}$ by $w^{\bar{\mathbf{x}}}$ in the proof of Lemma 6.2*.1, noting that

$$\Delta w^{\bar{\mathbf{x}}}\big|_{\Omega} = \Delta w_1^{\bar{\mathbf{x}}} \tag{6.2*.16c}$$

and writing the integrals over Ω, by employing the partition $\bar{\Omega} = \bigcup_{\tau\in\Delta} \bar{\tau}$, where Δ is the employed mesh, we obtain

$$q_1^{\bar{\mathbf{x}}}(u_{EX}) = \sum_{\tau\in\Delta} \left(\int_\tau u_{EX} \Delta w_1^{\bar{\mathbf{x}}} + \int_{\partial\tau} \frac{\partial}{\partial n_\tau}(w^{\bar{\mathbf{x}}}) \right) - \int_{\Gamma_N} g w_0^{\bar{\mathbf{x}}} \tag{6.2*.17a}$$

where $\mathbf{n}_\tau$, is the exterior unit normal for the element τ. Employing the integration by parts

$$\int_\tau u_{EX} \Delta w_1^{\bar{\mathbf{x}}} = -\int_\tau \nabla w_1^{\bar{\mathbf{x}}} \cdot \nabla u_{EX} - \int_{\partial\tau} u_{EX} \frac{\partial}{\partial n_\tau}(w_1^{\bar{\mathbf{x}}}) \tag{6.2*.17b}$$

we get

$$\frac{\partial}{\partial x_1}(u_{EX})(\bar{\mathbf{x}}) = \sum_{\tau\in\Delta} \left(-\int_\tau \nabla w_1^{\bar{\mathbf{x}}} \cdot \nabla u_{EX} + \int_{\partial\tau} u_{EX} \frac{\partial}{\partial n_\tau} \underbrace{(w^{\bar{\mathbf{x}}} - w_1^{\bar{\mathbf{x}}})}_{w_0^{\bar{\mathbf{x}}}} \right) - \int_{\Gamma_N} g w_0^{\bar{\mathbf{x}}} \tag{6.2*.18a}$$

which can also be written as

$$\begin{aligned} q_1^{\bar{\mathbf{x}}}(u_{EX}) = &-\sum_{\tau\in\Delta} \int_\tau \nabla w_1^{\bar{\mathbf{x}}} \cdot \nabla u_{EX} + \int_{\Gamma_N} u_{EX} \frac{\partial}{\partial n}(w_0^{\bar{\mathbf{x}}}) \\ &+ \int_{\Gamma_D} \bar{u} \frac{\partial}{\partial n}(w_0^{\bar{\mathbf{x}}}) - \int_{\Gamma_N} g w_0^{\bar{\mathbf{x}}}. \end{aligned} \tag{6.2*.18b}$$

Hence we have proven the identity

$$q_1^{\bar{\mathbf{x}}}(u_{EX}) = \frac{\partial}{\partial x_1}(u_{EX})(\bar{\mathbf{x}}) = \mathcal{L}_{q_1^{\bar{\mathbf{x}}}}(u_{EX}) + \mathcal{M}_{q_1^{\bar{\mathbf{x}}}}(\bar{u}, g) \tag{6.2*.19a}$$

with

$$\mathcal{L}_{q_1^{\bar{\mathbf{x}}}}(v) \stackrel{\text{def}}{=} -\sum_{\tau\in\Delta} \int_\tau \nabla w_1^{\bar{\mathbf{x}}} \cdot \nabla u_{EX} + \int_{\Gamma_N} v \frac{\partial}{\partial n}(w_0^{\bar{\mathbf{x}}}), \quad \forall\, v \in \mathcal{U} \tag{6.2*.19b}$$

bounded in $\mathcal{U}$, and $\mathcal{M}_{q_1^{\bar{\mathbf{x}}}}$, is given by (6.2*.15c).

Remark 6.2*.1 The extraction formula for $q_2^{\bar{\mathbf{x}}} \stackrel{\text{def}}{=} (\partial/\partial x_2)(u_{EX})(\bar{\mathbf{x}})$ is obtained by employing the extraction function

$$w_0^{\bar{\mathbf{x}}} = \frac{1}{2\pi}\frac{\sin(\theta_{\bar{\mathbf{x}}})}{r_{\bar{\mathbf{x}}}}. \tag{6.2*.20}$$

Remark 6.2*.2 In the above discussion we assumed that $\bar{\mathbf{x}}$, is an interior point of Ω. Let us now consider the case that $\bar{\mathbf{x}}$ is a point on Dirichlet boundary. In the case that $\bar{\mathbf{x}}$ is a point on Dirichlet boundary we need to extract only the normal derivative $(\partial/\partial n)(u_{EX})$, because the tangential derivative $(\partial/\partial t)(u_{EX})$ is known. This can be done by employing the same extraction function as in the interior $w_0^{\bar{\mathbf{x}}} = (1/\pi)(\cos(\theta_{\bar{\mathbf{x}}})/r_{\bar{\mathbf{x}}})$, where $r_{\bar{\mathbf{x}}}$, $\theta_{\bar{\mathbf{x}}}$ are the polar coordinates shown in Figure 6.2*.4a. Similarly in the case that $\bar{\mathbf{x}}$ is a point on Neumann boundary, we need to extract only the tangential derivative $(\partial/\partial t)(u_{EX})$.

Let us now elaborate further on the construction of the extraction for the pointwise derivatives, $q_i^{\bar{\mathbf{x}}}$. The accuracy of the extraction depends on the smoothness of the input data. In order to present the ideas and computations in the most clear way, once more we will consider the model problem in the smoothened domains, like, for example, the problem

$$\Delta u|_{\hat{\Omega}} = 0, \qquad u|_{\Gamma_D^1} = \bar{u}, \qquad u|_{\Gamma_D^2} = 0, \qquad \left.\frac{\partial u}{\partial n}\right|_{\Gamma_N} = 0 \tag{6.2*.21}$$

where $\hat{\Omega}$, is the smoothened Domain I shown in Fig. 6.2*.5. Note that the boundary $\partial\hat{\Omega}$, of $\hat{\Omega}$, the smoothened Domain I is analytic, and $\bar{u}$ is a constant, and the exact solution u_{EX} is analytic on $\bar{\hat{\Omega}}$.

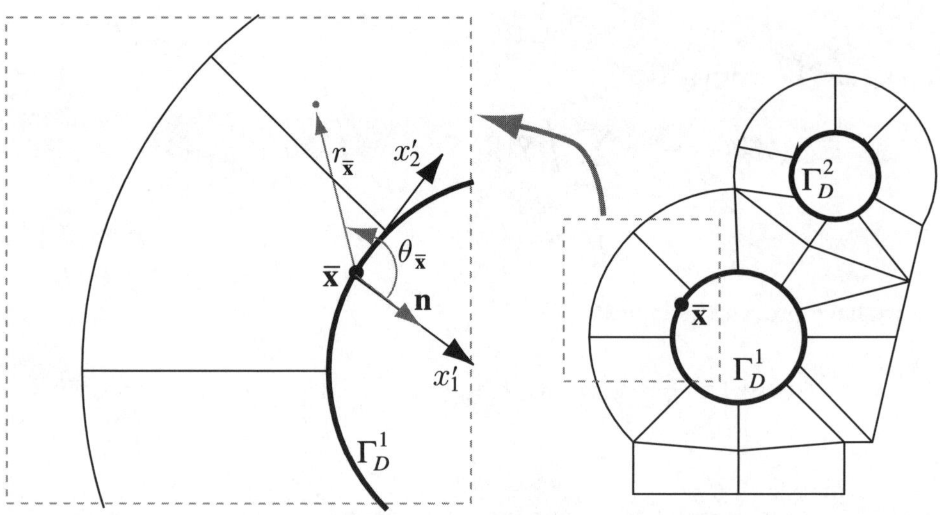

FIG. 6.2*.4. *Extraction of the derivatives at the boundary.* The orientation of the coordinate axes employed for extraction of the derivative at a point for the case that point lies on the Dirichlet boundary

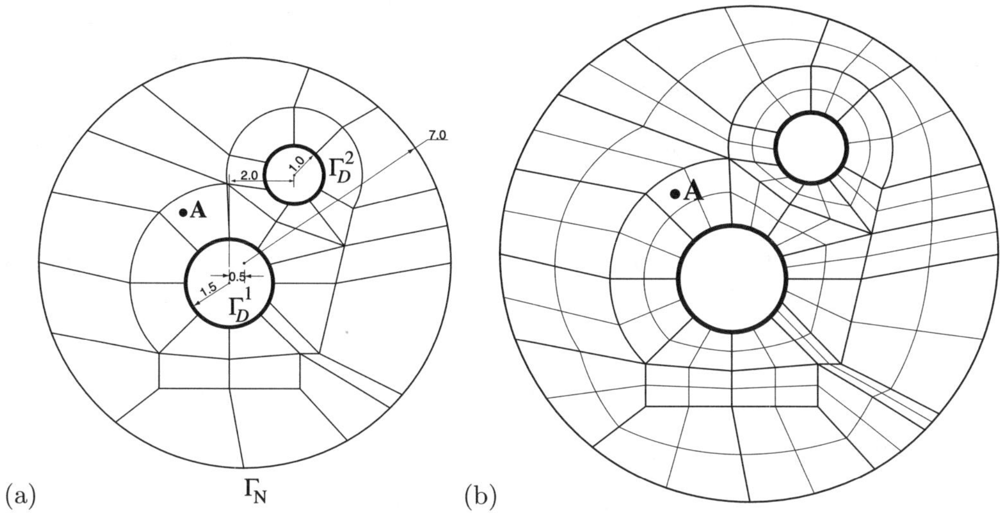

FIG. 6.2*.5. *Extraction of the pointwise derivatives of the solution.* (a) The smoothened domain $\hat{\Omega}$, the location of the point A, and the coarse mesh Δ_0. (b) The mesh Δ_2

Let us elaborate on the construction and the properties of the finite element solution. The exact solution u_{EX}, is the solution of the variational problem:

Find

$$u_{EX} \in \mathcal{U}_{\bar{u}} \stackrel{\text{def}}{=} \{u \mid \|u\|_{\mathcal{U}} \stackrel{\text{def}}{=} \sqrt{\mathcal{B}(u,u)} < \infty,\ u|_{\Gamma_D^1} = C,\ u|_{\Gamma_D^2} = 0\} \qquad (6.2^*.22a)$$

such that

$$\int_{\Omega} \nabla u_{EX} \cdot \nabla v = 0 \quad \forall\, v \in \mathcal{U} = \mathcal{U}_0 \qquad (6.2^*.22b)$$

where C is a constant. The finite element approximation, of u_{EX}, is the solution of the discrete problem:

Find

$$u_{S^p_{\Delta,\bar{u}}} \in S^p_{\Delta,\bar{u}} \stackrel{\text{def}}{=} \{v \in S^p_{\Delta} \mid v|_{\Gamma^1_{\mathrm{D}}} = \bar{u}, v|_{\Gamma^2_{\mathrm{D}}} = 0\} \qquad (6.2^*.23a)$$

such that

$$\int_{\Omega} \nabla u_{S^p_{\Delta,\bar{u}}} \cdot \nabla v = 0 \quad \forall v \in S^p_{\Delta,0} \subset \mathcal{U}. \qquad (6.2^*.23b)$$

Note that the set $S^p_{\Delta,\bar{u}}$ is non-empty because $\bar{u} = C = constant$. We will employ meshes $\Delta = \Delta_n$, $n = 0, 1, 2, \ldots$, where Δ_n is the mesh obtained by n successive refinements of the coarse mesh Δ_0, which is shown in Figure 6.2*.5a, and with each mesh Δ_n, we will associate the mesh size $h^{\Delta_n} = 1/2^n$. Recalling that u_{EX} is analytic on $\bar{\hat{\Omega}}$, we have

$$\|u_{EX} - u_{S^p_{\Delta,\bar{u}}}\|_{\mathcal{U}} \le \mathcal{C} h^p. \qquad (6.2^*.24)$$

Let $u^{(1)}_{S^p_{\Delta,\bar{u}}} \in S^p_{\Delta,\bar{u}} \subset \mathcal{U}_{\bar{u}}$, be the finite element function used to impose the Dirichlet boundary conditions on Γ_D, which is nonzero only in the elements τ which are adjacent to the Dirichlet boundary namely, $\partial\tau \cap \Gamma_D \neq \emptyset$. Letting

$$u^{(2)}_{S^p_{\Delta,0}} \stackrel{\text{def}}{=} u_{S^p_{\Delta,\bar{u}}} - u^{(1)}_{S^p_{\Delta,\bar{u}}} \in S^p_{\Delta,0} \tag{6.2*.25a}$$

we have

$$\int_\Omega \nabla u^{(2)}_{S^p_{\Delta,0}} \cdot \nabla v = -\int_\Omega \nabla u^{(1)}_{S^p_{\Delta,\bar{u}}} \cdot \nabla v \quad \forall\, v \in S^p_{\Delta,0}. \tag{6.2*.25b}$$

Similarly, we will let

$$u^{(2)}_{EX} \stackrel{\text{def}}{=} u_{EX} - u^{(1)}_{S^p_\Delta} \in \mathcal{U} \tag{6.2*.26a}$$

and we have

$$\int_\Omega \nabla u^{(2)}_{EX} \cdot \nabla v = -\int_\Omega u^{(1)}_{S^p_{\Delta,\bar{u}}} \cdot \nabla v \quad v \in \mathcal{U}. \tag{6.2*.26b}$$

Let us be interested in the output $q_1^{\mathbf{x}_A} = (\partial/\partial x_1)(u_{EX})(\mathbf{x}_A)$, where A is the interior point shown in Figure 6.2*.5. We will let

$$\Phi_A \stackrel{\text{def}}{=} w_0^{\mathbf{x}_A} = \frac{1}{2\pi}\frac{\cos(\theta_A)}{r_A} \tag{6.2*.27a}$$

and we will denote by φ^A_{EX}, the solution of the problem (6.2*.21) with Dirichlet boundary condition $\varphi^A_{EX}|_{\Gamma_D} = \Phi_A|_{\Gamma_D}$. It follows that φ^A_{EX} is analytic on $\bar{\hat{\Omega}}$, because $\partial\hat{\Omega}$, and $\Phi_A|_{\partial\hat{\Omega}}$, are analytic and it is the solution of the problem:

Find

$$\varphi^A_{EX} \in \mathcal{U}_{\Phi_A} \stackrel{\text{def}}{=} \{u \in L^2(\Omega) \mid \|u\|_{\mathcal{U}} < \infty,\ u|_{\Gamma_D} = \Phi_A|_{\Gamma_D}\} \tag{6.2*.27b}$$

such that

$$\int_\Omega \nabla \varphi^A_{EX} \cdot \nabla v = 0 \quad \forall v \in \mathcal{U}. \tag{6.2*.27c}$$

The Dirichlet boundary condition $u|_{\Gamma_D} = \Phi_A|_{\Gamma_D}$, can be imposed by employing the function

$$\varphi^{A,(1)}_{S^p_{\Delta,\Phi}} = {}^{(1)}\varphi^{A,(1)}_{S^p_\Delta} + {}^{(2)}\varphi^{A,(1)}_{S^p_{\Delta,\Phi}} \tag{6.2*.28a}$$

where ${}^{(1)}\varphi^{A,(1)}_{S^p_\Delta} \in S^p_\Delta$, is constructed by finite element interpolation by letting

$${}^{(1)}\varphi^{A,(1)}_{S^p_\Delta}|_{\Gamma_D} = \mathcal{I}^p_\Delta \Phi_A\big|_{\Gamma_D} \tag{6.2*.28b}$$

and by setting all the degrees of freedom which are not on Γ_D equal to zero. Similarly, the function ${}^{(2)}\varphi^{A,(1)}_{S^p_{\Delta,\Phi}}$, can be constructed by blending function interpolation to be such that

$${}^{(2)}\varphi^{A,(1)}_{S^p_{\Delta,\Phi}}|_{\Gamma_D} = \big(\Phi_A - {}^{(1)}\varphi^{A,(1)}_{S^p_\Delta}\big)|_{\Gamma_D} \tag{6.2*.28c}$$

and equal to zero on all the edges and vertices which are on Γ_D, and to be such that

$$\|{}^{(2)}\varphi^{A,(1)}_{S^p_{\Delta,\Phi}}\|_{\mathcal{U}} \leq \mathcal{C}h^p. \tag{6.2*.28d}$$

This is possible because Φ_A is smooth on Γ_D. Letting

$$\varphi^{A,(2)}_{EX} \stackrel{\text{def}}{=} \varphi^{A}_{EX} - \varphi^{A,(1)}_{S^p_{\Delta,\Phi_A}} \tag{6.2*.29a}$$

we have

$$\int_\Omega \nabla\varphi^{A,(2)}_{EX} \cdot \nabla v = -\int_\Omega \nabla\varphi^{A,(1)}_{S^p_{\Delta,\Phi_A}} \cdot \nabla v \quad \forall\, v \in \mathcal{U}. \tag{6.2*.29b}$$

The finite element approximation of φ^A_{EX}, is given by

$$\varphi^{A}_{S^p_{\Delta,\Phi}} \stackrel{\text{def}}{=} \varphi^{A,(1)}_{S^p_{\Delta,\Phi}} + \varphi^{A,(2)}_{S^p_{\Delta}} \tag{6.2*.30a}$$

where $\varphi^{A,(2)}_{S^p_\Delta}$ is the finite element approximation of $\varphi^{A,(2)}_{EX}$, and is obtained by solving the following discrete problem.

Find $\varphi^{A,(2)}_{S^p_\Delta} \in S^p_{\Delta,0}$, such that

$$\int_\Omega \nabla\varphi^{A,(2)}_{S^p_\Delta} \cdot \nabla v = -\int_\Omega \nabla\varphi^{A,(1)}_{S^p_{\Delta,\Phi}} \cdot \nabla v \quad \forall\, v \in S^p_{\Delta,0}. \tag{6.2*.30b}$$

It follows that

$$\|\varphi^{A}_{EX} - \varphi^{A}_{S^p_{\Delta,\Phi}}\|_{\mathcal{U}} = \|\varphi^{A,(2)}_{EX} - \varphi^{A,(2)}_{S^p_\Delta}\|_{\mathcal{U}} \leq \mathcal{C}h^p. \tag{6.2*.30c}$$

Let

$$\Psi_A \stackrel{\text{def}}{=} \Phi_A - \varphi^{A}_{S^p_{\Delta,\Phi}}. \tag{6.2*.31a}$$

We then have $\Psi_A\big|_{\Gamma_D} = 0$, and letting $\Omega_\varepsilon \stackrel{\text{def}}{=} \Omega - \bar{\omega}^\varepsilon_A$, and working as before we obtain

$$\begin{aligned}\int_\Gamma \frac{\partial}{\partial n}(\Psi_A)u_{EX} - \int_\Gamma \Psi_A \frac{\partial}{\partial n}(u_{EX}) \\ + \underbrace{\int_{\partial\omega^\varepsilon_A}\left(\frac{\partial}{\partial n}(\Psi_A)\, u_{EX} - \frac{\partial}{\partial n}(u_{EX})\Psi_A\right)}_{\frac{\partial}{\partial x_1}(u_{EX})(\mathbf{x}_A) + O(\varepsilon)} = -\int_\Omega u_{EX}\Delta\varphi^{A,(1)}_{S^p_{\Delta,\Phi}}\end{aligned} \tag{6.2*.31b}$$

where $\Delta\varphi^{(1)}_{S^p_{\Delta,\Phi_A}}$ is understood in a generalized sense, and has Dirac components on the interelement edges. Letting $\varepsilon \to 0$, we get

$$q_1^{\mathbf{x}_A}(u_{EX}) = \int_\Gamma \frac{\partial}{\partial n}(\Psi_A)u_{EX} - \int_\Gamma \Psi_A\, \frac{\partial}{\partial n}(u_{EX}) + \int_\Omega u_{EX}\Delta\varphi^{A,(1)}_{S^p_{\Delta,\Phi}}. \tag{6.2*.31c}$$

Using that $\Psi_A|_{\Gamma_D} = 0$, and integrating by parts of the last term, we get

$$q_1^{\mathbf{x}_A}(u_{EX}) = \int_\Gamma \frac{\partial \Phi_A}{\partial n} u_{EX} - \int_\Gamma \frac{\partial}{\partial n}(\varphi^{A,(1)}_{S^p_{\Delta,\Phi}}) u_{EX} - \int_{\Gamma_N} \Psi_A \frac{\partial}{\partial n}(u_{EX})$$
$$+ \int_\Gamma u_{EX} \frac{\partial}{\partial n}(\varphi^{A,(1)}_{S^p_{\Delta,\Phi}}) - \int_\Omega \boldsymbol{\nabla} u_{EX} \cdot \boldsymbol{\nabla} \varphi^{A,(1)}_{S^p_{\Delta,\Phi}} \tag{6.2*.32a}$$

and hence

$$q_1^{\mathbf{x}_A}(u_{EX}) = \int_{\Gamma_D^2} \frac{\partial}{\partial n}(\Phi_A)\bar{u} + \int_{\Gamma_N} \frac{\partial}{\partial n}(\Phi_A) u_{EX} - \int_{\Gamma_N} \Psi_A \frac{\partial}{\partial n}(u_{EX}).$$
$$- \int_\Omega \boldsymbol{\nabla} u_{EX} \cdot \boldsymbol{\nabla} \varphi^{A,(1)}_{S^p_{\Delta,\Phi}} \tag{6.2*.32b}$$

Further, employing that

$$\int_\Gamma \Psi_A \frac{\partial}{\partial n}(u_{EX}) = \int_{\Gamma_N} \underbrace{\left(\varphi^A_{EX} - \varphi^{A,(1)}_{S^p_{\Delta,\Phi}}\right)}_{\varphi^{A,(2)}_{EX}} \frac{\partial}{\partial n}(u_{EX}) = \int_\Omega \boldsymbol{\nabla}\varphi^{A,(2)}_{EX} \cdot \boldsymbol{\nabla} u_{EX} = 0 \tag{6.2*.32c}$$

we obtain

$$q_1^{\mathbf{x}_A}(u_{EX}) = \int_{\Gamma_D^2} \frac{\partial}{\partial n}(\Phi_A) C + \int_{\Gamma_N} \frac{\partial}{\partial n}(\Phi_A) u_{EX} - \int_\Omega \boldsymbol{\nabla} u_{EX} \cdot \boldsymbol{\nabla} \varphi^{A,(1)}_{S^p_{\Delta,\Phi}}. \tag{6.2*.32d}$$

Hence we have shown that

$$\frac{\partial}{\partial x_1}(u_{EX})(\mathbf{x}_A) = -\int_\Omega \boldsymbol{\nabla} u^{(2)}_{EX} \cdot \boldsymbol{\nabla} \varphi^{(1)}_{S^p_{\Delta,\Phi_A}} + \int_{\Gamma_N} u^{(2)}_{EX} \frac{\partial}{\partial n}(\Phi_A)$$
$$+ \int_{\Gamma_D^2} \bar{u} \frac{\partial}{\partial n}(\Phi_A) - \int_\Omega \boldsymbol{\nabla} u^{(1)}_{S^p_{\Delta,\bar{u}}} \cdot \boldsymbol{\nabla} \varphi^{(1)}_{S^p_{\Delta,\Phi_A}}. \tag{6.2*.33a}$$

Letting

$$\mathcal{L}_{q_1^{\mathbf{x}_A}}(v) \stackrel{\text{def}}{=} -\int_\Omega \boldsymbol{\nabla} v \cdot \boldsymbol{\nabla} \varphi^{A,(1)}_{S^p_{\Delta,\Phi}} + \int_{\Gamma_N} v \frac{\partial}{\partial n}(\Phi_A) \tag{6.2*.33b}$$

which is a linear bounded functional on the *entire* energy space $\mathcal{U}$, and

$$\mathcal{M}_{q_1^{\mathbf{x}_A}}(\bar{u}) \stackrel{\text{def}}{=} \int_{\Gamma_D^2} \frac{\partial}{\partial n}(\Phi_A)\bar{u} - \int_\Omega \boldsymbol{\nabla} u^{(1)}_{S^p_{\Delta,\bar{u}}} \cdot \boldsymbol{\nabla} \varphi^{A,(1)}_{S^p_{\Delta,\Phi}} \tag{6.2*.33c}$$

which depends *only* on the input data, because $\boldsymbol{\nabla} u^{(1)}_{S^p_{\Delta,\bar{u}}}$ is constructed *directly from* $\bar{u}$, it follows that

$$E^{q_1^{\mathbf{x}_A}}(v) = \mathcal{L}_{q_1^{\mathbf{x}_A}}(v) + \mathcal{M}_{q_1^{\mathbf{x}_A}}(\bar{u}) \quad \forall u \in \mathcal{U} \tag{6.2*.33d}$$

is an extraction functional for $q_1^{\mathbf{x}_A}(u_{EX})$.

Remark 6.2*.3 Note that in our example Γ_D^2, is a closed contour, and hence we have

$$\int_{\Gamma_D^2} \bar{u}\frac{\partial}{\partial n}(\Phi_A) = \bar{u} \in t_{\Gamma_D^2}\frac{\partial}{\partial n}(\Phi_A) = 0 \tag{6.2*.34a}$$

and we can write $\mathcal{M}_{q_1^{\times A}}$, in the form

$$\mathcal{M}_{q_1^{\times A}} = -\int_\Omega \nabla u^{(1)}_{S^p_{\Delta,\bar{u}}} \nabla\varphi^{(1)}_{S^p_{\Delta,\Phi_A}}. \tag{6.2*.34b}$$

Let us now analyze the error in $E^{q_1^{\times A}}(u_{S^p_\Delta})$. We have

$$q_1^{\times A}(u_{S^p_\Delta}) - E^{q_1^{\times A}} = \mathcal{L}_{q_1^{\times A}}(e_{S^p_\Delta}) = -\int_\Omega \nabla e_{S^p_\Delta}\cdot\nabla\varphi^{A,(1)}_{S^p_{\Delta,\Phi}} + \int_{\Gamma_N}\frac{\partial}{\partial n}(\Phi_A)\, e_{S^p_\Delta} \tag{6.2*.35a}$$

which can also be written in the form

$$\begin{aligned} q_1^{\times A}(u_{S^p_\Delta}) - E^{q_1^{\times A}} &= -\int_\Omega \nabla\big(u^{(2)}_{EX} - u^{(2)}_{S^p_{\Delta,0}}\big)\cdot\nabla\varphi^{A,(1)}_{S^p_{\Delta,\Phi}} \\ &\quad + \int_\Omega \nabla z^A_{EX}\cdot\nabla\big(u^{(2)}_{EX} - u^{(2)}_{S^p_{\Delta,\bar{u}}}\big) \end{aligned} \tag{6.2*.35b}$$

where z^A_{EX} be the solution of the variational problem:
Find $z^A_{EX} \in \mathcal{U}$ such that

$$\mathcal{B}(z^A_{EX}, v) = \int_{\Gamma_N}\frac{\partial}{\partial n}(\Phi_A)v \quad \forall\, v \in \mathcal{U}. \tag{6.2*.36a}$$

Letting $z_{S^p_{\Delta,0}} \in S^p_{\Delta,0}$, denote the finite element approximation of z^A_{EX}, we have

$$\|z^A_{EX} - z_{S^p_{\Delta,0}}\|_{\mathcal{U}} \le \mathcal{C}h^p \tag{6.2*.36b}$$

because z^A_{EX} is analytic in $\bar{\hat{\Omega}}$. Let us now estimate the error in the extracted value $q_1^{\times A}(u_{EX}) - E^{q_1^{\times A}}(u_{S^p_\Delta})$. The second term can be estimated by

$$\left|\int_\Omega \nabla z^A_{EX}\cdot\nabla\big(u^{(2)}_{EX} - u^{(2)}_{S^p_{\Delta,0}}\big)\right| = \left|\int_\Omega \nabla(z^A_{EX} - z^A_{S^p_{\Delta,0}})\cdot\nabla\big(u_{EX} - u_{S^p_{\Delta,\bar{u}}}\big)\right| \le \mathcal{C}h^{2p} \tag{6.2*.37a}$$

where we used that

$$u^{(2)}_{EX} - u^{(2)}_{S^p_{\Delta,\bar{u}}} = u_{EX} - u_{S^p_{\Delta,\bar{u}}} \in \mathcal{U} \tag{6.2*.37b}$$

and the Cauchy–Schwarz inequality. The first term can be written as

$$\int_\Omega \nabla e_{S^p_\Delta}\cdot\nabla\varphi^{A,(1)}_{S^p_{\Delta,\Phi}} = \int_\Omega \nabla e_{S^p_\Delta}\cdot\nabla\big(\varphi^A_{S^p_{\Delta,\Phi}} - \varphi^{A,(2)}_{S^p_{\Delta,\Phi}}\big) = \int_\Omega \nabla e_{S^p_\Delta}\cdot\nabla\varphi^A_{S^p_{\Delta,\Phi}} \tag{6.2*.38a}$$

and using that

$$\int_\Omega \nabla \varphi^A_{EX} \cdot \nabla e_{S^p_\Delta} = 0 \qquad (6.2^*.38\text{b})$$

we get

$$\left| \int_\Omega \nabla e_{S^p_\Delta} \cdot \nabla \varphi^{A,(1)}_{S^p_{\Delta,\Phi}} \right| = \left| \int_\Omega \nabla e_{S^p_\Delta} \cdot \nabla \big(\varphi^A_{S^p_{\Delta,\Phi}} - \varphi^A_{EX} \big) \right| \le \mathscr{C} h^{2p}. \qquad (6.2^*.38\text{c})$$

Hence we have proven that

$$\left| q_1^{\mathbf{x}_A} - E^{q_1^{\mathbf{x}_A}}(u_{EX}) \right| \le \mathscr{C} h^{2p}. \qquad (6.2^*.39)$$

Remark 6.2*.4 The derivation of the estimate

$$\left| q_1^{\mathbf{x}_A} - E^{q_1^{\mathbf{x}_A}}(u_{EX}) \right| \le \mathscr{C} h^{2p} \qquad (6.2^*.40)$$

was based on the assumption that A is sufficiently far from the boundary $\partial\Omega$, so that $\Phi_A|_\Omega$ and $(\partial/\partial n)(\Phi_A)|_\Omega$ are smooth. As the point A approaches the boundary of the domain $\partial\Omega$, the accuracy of the extraction is expected to deteriorate.

Remark 6.2*.5 Another important consideration is that as A approaches the boundary, the derivatives of $\Phi_A|_{\partial\Omega}$, and $(\partial/\partial n)(\Phi_A)|_{\partial\Omega}$, grow and in order to preserve the accuracy of the extraction, it is necessary to evaluate all integrals with sufficient accuracy, i.e., in an adaptive framework.

Noting that

$$q_1^{\mathbf{x}_A} - E^{q_1^{\mathbf{x}_A}}(u_{EX}) = \mathscr{B}\Big(\big(u_{EX} - u_{S^p_{\Delta,\bar{u}}}\big), \big(z^A_{EX} + \varphi^A_{EX}\big) - \big(z^A_{S^p_{\Delta,0}} + \varphi^A_{S^p_{\Delta,\Phi_A}}\big) \Big) \qquad (6.2^*.41\text{a})$$

and employing the bounds $\mathscr{E}^L_{\bar{u}^\pm_{S^p_\Delta,\bar{u}}}$, $\mathscr{E}^U_{\bar{u}^\pm_{S^p_\Delta,\bar{u}}}$, for the energy norm of the error in

$$\bar{u}^\pm_{S^p_\Delta,\bar{u}} \stackrel{\text{def}}{=} u_{S^p_{\Delta,\bar{u}}} \pm \big(z^A_{S^p_{\Delta,0}} + \varphi^A_{S^p_{\Delta,\Phi_A}} \big) \qquad (6.2^*.41\text{b})$$

we obtain the bounds for the error in the extraction.

Another related approach for extracting point values is the cut off function method which employs the identity

$$q_1^{\mathbf{x}_A} = -\int_\Omega \Delta(w^{\mathbf{x}_A} \Phi^{\mathbf{x}_A}) \qquad (6.2^*.42\text{a})$$

where $\Phi^{\mathbf{x}_A}$ is a sufficiently smooth cut off function, like for example (see [Szabo and Babuška, 1991])

$$\Phi^{\mathbf{x}_A} = \begin{cases} 1, & \text{if } \bar{r}^{\mathbf{x}_A} \le 0 \\ 1 - 10(\bar{r}^{\mathbf{x}_A})^3 + 15(\bar{r}^{\mathbf{x}_A})^4 - 6(\bar{r}^{\mathbf{x}_A})^5, & \text{if } 0 \le \bar{r}^{\mathbf{x}_A} \le 1 \\ 0 & \text{if } \bar{r}^{\mathbf{x}_A} \ge 1 \end{cases} \qquad (6.2^*.42\text{b})$$

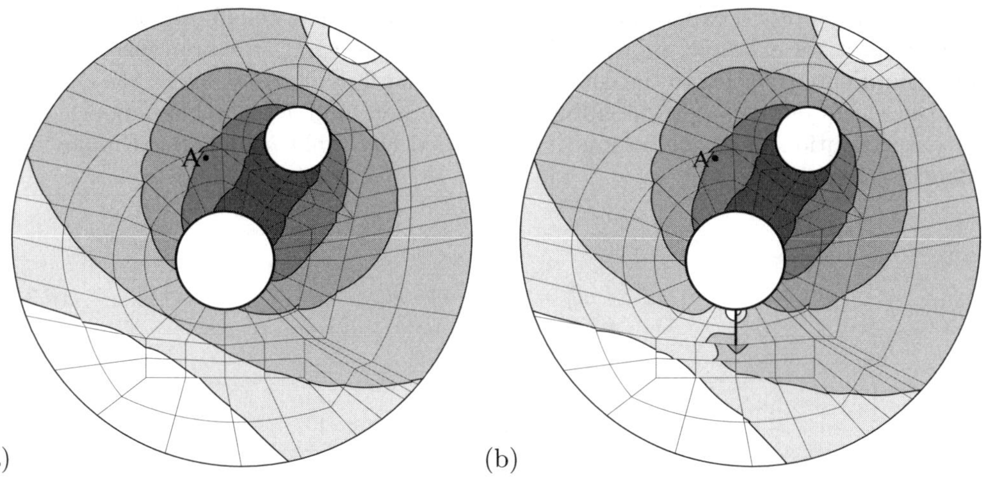

(a) (b)

FIG. 6.2*.6. *Contours of the modulus of the gradient.* (a) Smoothened Domain I, (b) Smoothened Domain II

Let us now give an example which compares the accuracy of the extracted values with the computed values directly from the finite element solution.

Example 6.2*.4 (*The rate of convergence of the extracted derivatives at interior points*) Let us employ the model problem in the smoothened Domain I, and II, and let us compare the convergence of the error in the extracted value of the x_1-derivative at point A obtained using the cut off function method, namely (6.2*.42), with the error in the derivative $(\partial/\partial x_1)(u_{S^p_\Delta})$ computed directly from the finite element solution, and, the error in the postprocessed derivative $(\partial/\partial x_1)(u_{S^p_\Delta})+(\partial/\partial x_1)(\hat{e}^{\text{Neum}}_{S^{p+k}_\Delta})$ obtained by adding the residual estimator to the finite element solution. Figure 6.2*.6 shows the location of the point in the domain, and the contours of the modulus of the gradient of the finite element solution. Figure 6.2*.7a (respectively Fig. 6.2*.7b) gives the convergence graphs for $p = 1$, and $p = 2$, for the smoothened Domain I, and II shown in Fig. 6.2*.2. Note that the derivative $q_1^{\mathbf{x}_A}$, extracted by the cut off function method has superior accuracy for all the employed meshes, and converges with twice the rate as the energy norm of the error of the finite element solution. Note also that, for coarse meshes, the derivative of the finite element solution $(\partial/\partial x_1)(u_{S^p_\Delta})(\mathbf{x}_A)$, may be more accurate than the postprocessed derivative $(\partial/\partial x_1)(u_{S^p_\Delta}) + (\partial/\partial x_1)(\hat{e}^{\text{Neum}}_{S^{p+k}_\Delta}))(\mathbf{x}_A)$.

Remark 6.2*.6 Similar results may be obtained by employing the extraction method outlined in (6.2*.21)–(6.2*.38).

We will also address the extraction of the intensity factors at the re-entrant corners. Let

$$u_{EX} = \sum_{i=1}^{N} A_i^{\bar{\mathbf{x}}}(u_{EX}) r_{\bar{\mathbf{x}}}^{\alpha_i} \Psi_i(\theta_{\bar{\mathbf{x}}}) + \tilde{u}_{EX} \tag{6.2*.43}$$

denote the asymptotic expansion of the exact solution in the neighborhood of the corner point $\bar{\mathbf{x}}$, as is, for example, shown in Fig. 6.2*.8, where $r_{\bar{\mathbf{x}}}, \theta_{\bar{\mathbf{x}}}$, are appropriately defined polar coordinates relative to the corner point, α_i and Ψ_i are, respectively, the eigenvalues and the eigenfunctions corresponding to the corner, and $\tilde{u}_{EX}$ is the smooth part of the exact solution. The coefficients A_i are called the *intensity factors.* Employing the extraction function

$$w_0^{\bar{\mathbf{x}}} = \alpha_1 r_{\bar{\mathbf{x}}}^{-\alpha_1} \Psi_1(\theta_{\bar{\mathbf{x}}}) \tag{6.2*.44}$$

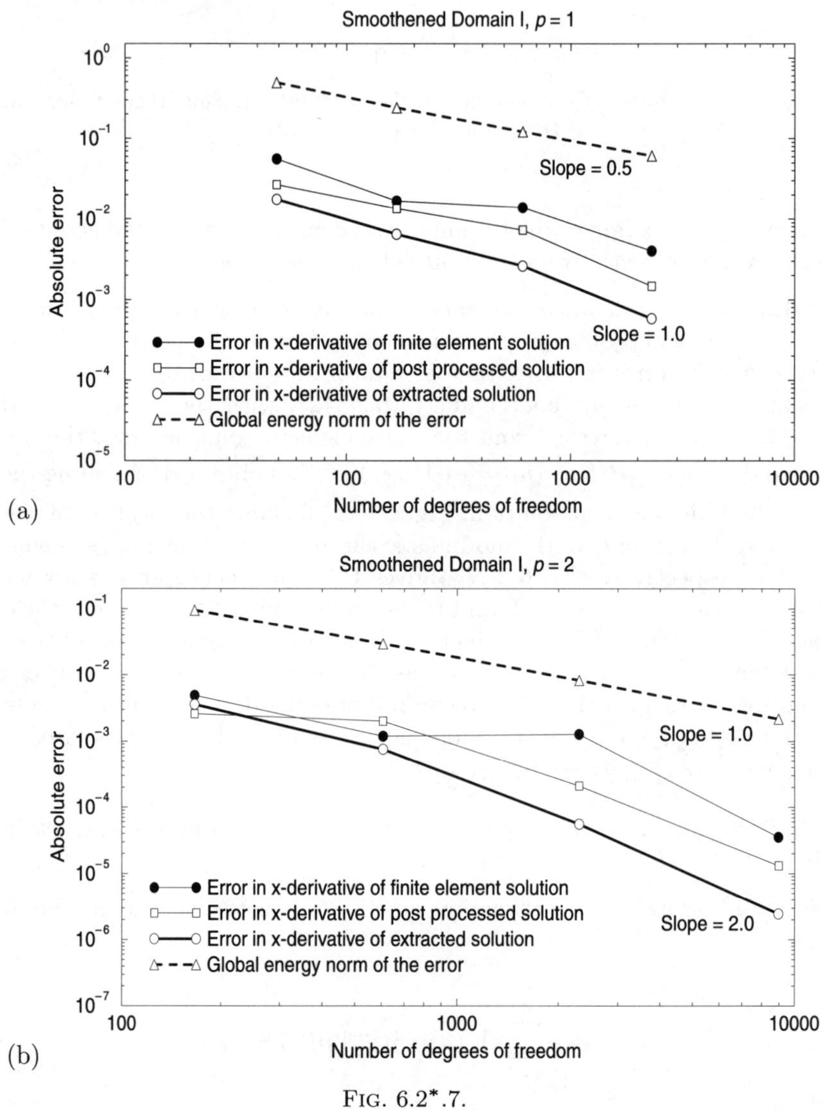

FIG. 6.2*.7.

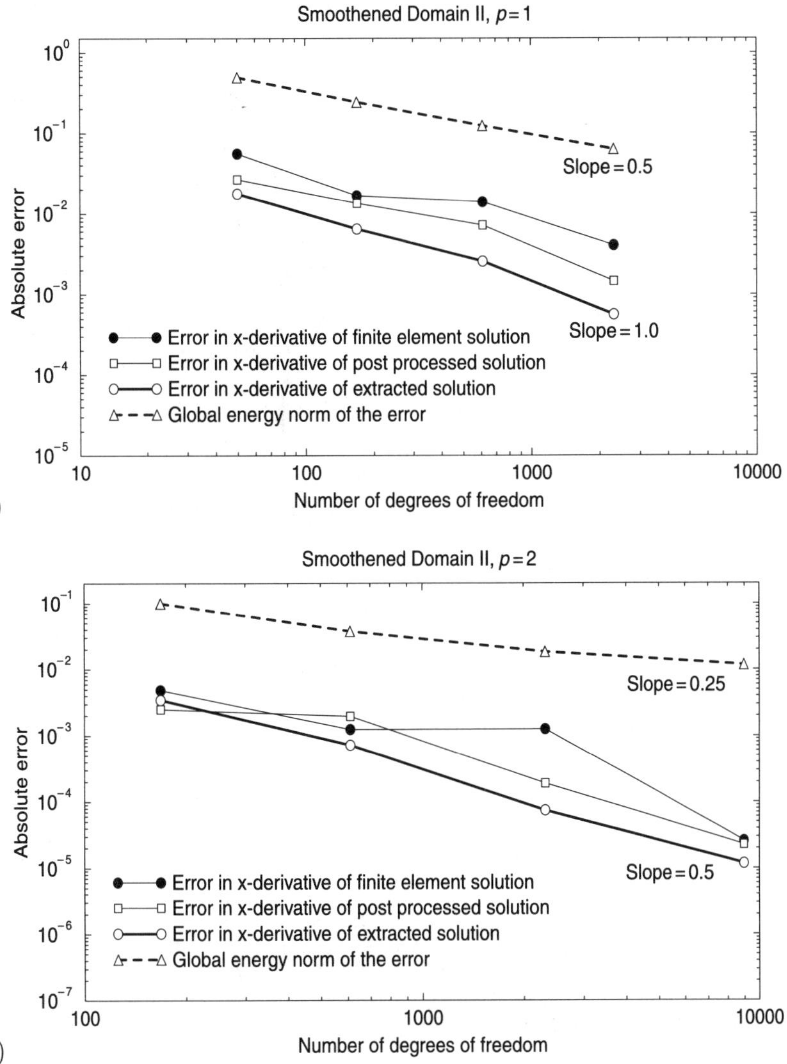

(c)

(d)

FIG. 6.2*.7. *Convergence of the x_1-derivative of the error at point A.* Smoothened Domain I (a) $p = 1$ and (b) $p = 2$. Smoothened Domain II (c) $p = 1$ and (d) $p = 2$. Note that in the case of linear elements ($p - 1$), and the case of Domain II the energy norm of the error has not achieved its expected asymptotic rate $N^{-1/4}$. If the mesh is refined further, the asymptotic rate is achieved, as it happens in the case of quadratic ($p = 2$) elements

and working as above, we obtain the identity

$$A^{\bar{\mathbf{x}}}_{1,EX} = -\sum_{\tau \in T_h} \mathscr{B}_\tau(w^{\bar{\mathbf{x}}}_1, u_{\mathrm{EX}}) + \int_{\Gamma_N} u_{EX} \frac{\partial w^{\bar{\mathbf{x}}}_0}{\partial n_{\Gamma_N}} + \int_{\Gamma_D} \bar{u} \frac{\partial w^{\bar{\mathbf{x}}}_0}{\partial n_{\Gamma_N}} - \int_{\Gamma_N} g w^{\bar{\mathbf{x}}}_0 \qquad (6.2^*.45)$$

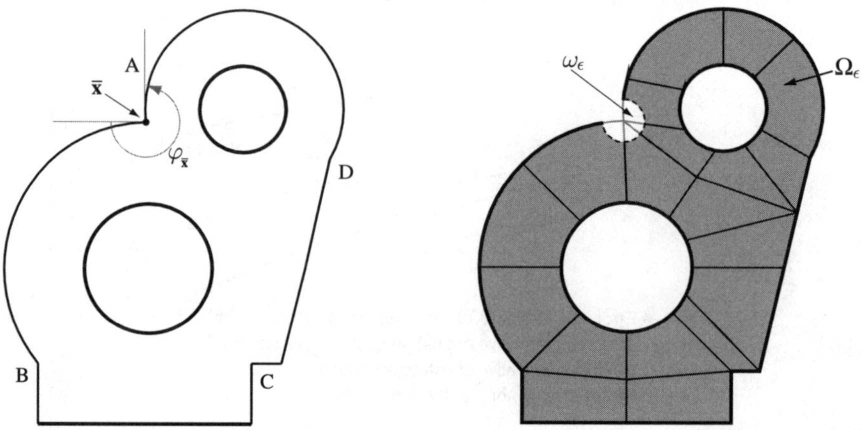

FIG. 6.2*.8. *The extraction of the flux intensity factor.* Notations used for the extraction function. The corner $\bar{\mathbf{x}}$, the angle $\varphi_{\bar{\mathbf{x}}}$ and the domain Ω_ϵ

where $w^{\bar{\mathbf{x}}} \stackrel{\text{def}}{=} w_0^{\bar{\mathbf{x}}} + w_1^{\bar{\mathbf{x}}}$, and $w_1^{\bar{\mathbf{x}}}$ such that $w_1^{\bar{\mathbf{x}}}|_{\Gamma_D} = -w_0^{\bar{\mathbf{x}}}|_{\Gamma_D}$. We thus have

$$A_1^{\bar{\mathbf{x}}}(u_{EX}) = \mathcal{L}_{A_1^{\bar{\mathbf{x}}}}(u_{EX}) + \mathcal{M}_{A_1^{\bar{\mathbf{x}}}}(\bar{u}; g) \tag{6.2*.46a}$$

where

$$\mathcal{L}_{A_1^{\bar{\mathbf{x}}}}(u) \stackrel{\text{def}}{=} -\sum_{\tau \in T_h} \mathcal{B}_\tau(w_1^{\bar{\mathbf{x}}}, u) + \int_{\Gamma_N} u \frac{\partial}{\partial n}(w_0^{\bar{\mathbf{x}}}) \tag{6.2*.46b}$$

and

$$\mathcal{M}_{A_1^{\bar{\mathbf{x}}}}(\bar{u}; g) \stackrel{\text{def}}{=} \int_{\Gamma_D} \bar{u} \frac{\partial}{\partial n_\tau}(w_0^{\bar{\mathbf{x}}}) - \int_{\Gamma_N} g w_0^{\bar{\mathbf{x}}} \tag{6.2*.46c}$$

with the extracted intensity factor given by

$$E_{A_1^{\bar{\mathbf{x}}}}(u_{S_\Delta^p}) = \mathcal{L}_{A_1^{\bar{\mathbf{x}}}}(u_{S_\Delta^p}) + \mathcal{M}_{A_1^{\bar{\mathbf{x}}}}(\bar{u}; g). \tag{6.2*.46d}$$

Below we will extract the intensity factor by employing the cut off function method (see [Szabo and Babuška, 1991]), which follows the identity,

$$\begin{aligned} A_{1,EX}^{\text{cutoff}} = & \int_{\omega_2^{\bar{\mathbf{x}}}} (\Delta W^{\bar{\mathbf{x}}}) u_{EX} + \int_{\Gamma_{\rho_1}} u_{EX} \frac{\partial}{\partial n}(w_0^{\bar{\mathbf{x}}}) - \int_{\Gamma_{\rho_2}} u_{EX} \frac{\partial}{\partial n}(w_0^{\bar{\mathbf{x}}}) \\ & + \int_{\Gamma^{\text{bound}}} u_{EX} \frac{\partial}{\partial n}(w_0^{\bar{\mathbf{x}}}) \end{aligned} \tag{6.2*.47}$$

where $W^{\bar{\mathbf{x}}} \stackrel{\text{def}}{=} w_0^{\bar{\mathbf{x}}}\, \Phi^{\bar{\mathbf{x}}}$, $\Gamma^{\text{bound}} \stackrel{\text{def}}{=} \partial\omega_2^{\bar{\mathbf{x}}} \cap \partial\Omega$, and $\Phi^{\bar{\mathbf{x}}}$ is a cut off function centered at the corner, for example the function given in (6.2*.42b). Note that here we have assumed that $\Gamma_{\rho_i} \cap \partial\Omega = \emptyset$, as shown in Fig. 6.2*.9.

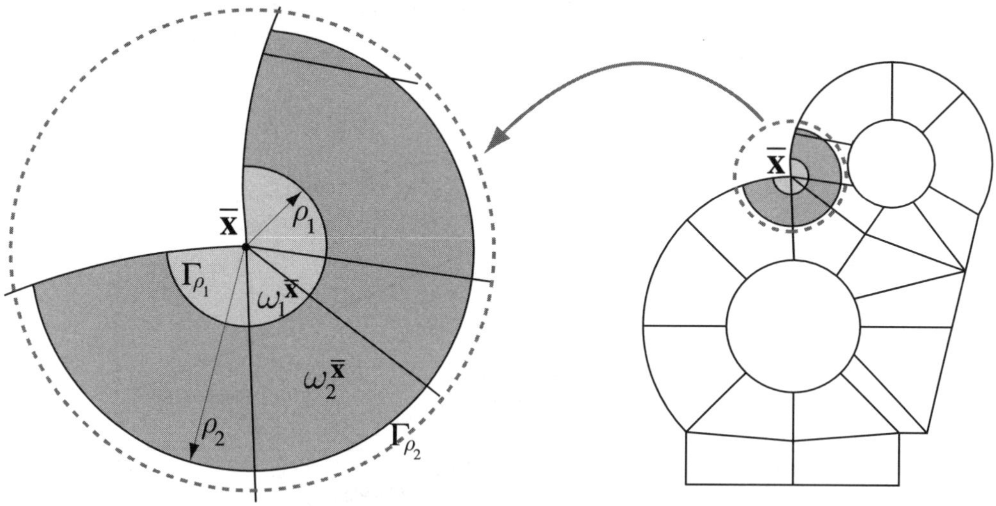

FIG. 6.2*.9. *The cut off function method for extraction of the flux intensity factor*. Illustration of the ring around the corner of interest showing the notation used for the cut off function method. The radii ρ_1 and ρ_2 with centers at the corner point $\bar{\mathbf{x}}$ form the ring with the corresponding boundaries Γ_{ρ_1}, Γ_{ρ_2}, respectively. The light gray region denotes $\omega_1^{\bar{\mathbf{x}}}$ and the dark gray shaded region denotes $\omega_2^{\bar{\mathbf{x}}}$. Note that Γ_{ρ_1} and Γ_{ρ_2} do not intersect with the boundary of the domain

Let us now examine the convergence of the error in the extracted intensity factor and the accuracy of its bounds.

Example 6.2*.5 (*Convergence and bounds of the error in the extracted intensity factor*) We considered the mixed boundary value problem for the Laplacian in the smoothened Domain II shown in Fig. 6.2*.2 and computed the intensity factor at the crack tip, and the computable bounds for its error. Figure 6.2*.10 show the convergence of the extracted value compared with the convergence of the energy of the error, for $p = 2$. Table 6.2*.4 gives the effectivity indices for the computable bounds for Mesh I, II, and III. As expected the effectivity of the bounds is poor due to the unsmoothness of the extraction function $H^{\mathcal{F}}$ for the cut off function method. Better results can be obtained by employing the extraction outlined in (6.2*.44)–(6.2*.47).

Remark 6.2*.7 The computable bounds for the errors in the outputs can be employed in an algorithm for adaptive control of the error, which is sketched in Fig. 6.2*.11. The main idea is to use the usual adaptivity procedure based on the equilibration of error indicators in combination with the stopping criterion. The goal is to obtain the guaranteed error in the range of the prescribed tolerance without over refinement. This includes the decision whether and when to use more expensive estimators rather than to refine the mesh. However, for this algorithm to be effective, bounds with better effectivity are needed. This is the subject of current research.

Remark 6.2*.8 Similar ideas can be used for the adaptive procedure for computation of the quantity of interest. Here we compute simultaneously the solution and the

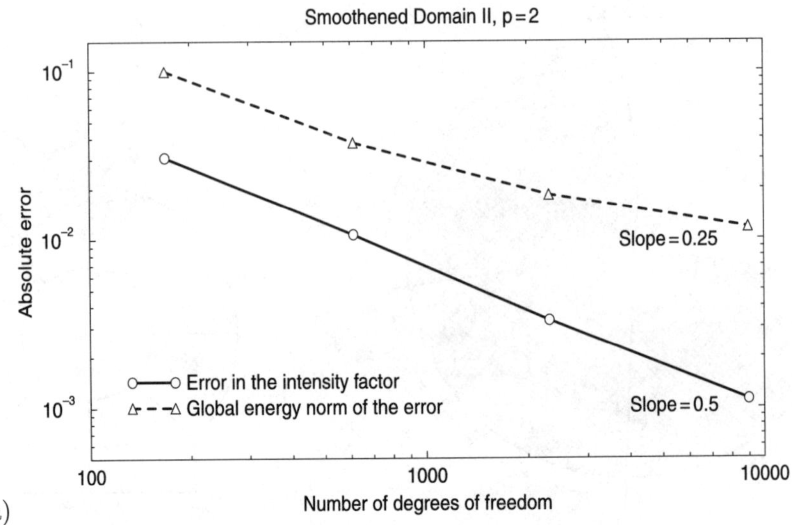

(a)

FIG. 6.2*.10. *Convergence of the error in the extracted intensity factor.* Graph of the global energy norm of the error and the error in the extracted intensity factors at the crack tip for the smoothened Domain II. As expected, the error in the extracted intensity factor converges with twice the rate of the global energy norm of the error

TABLE 6.2*.4. *Bounds for the error in the extracted value of the intensity factor.* The absolute error in the intensity factor computed using Mesh I, II, and III, and biquadratic elements and the effectivity indices for the computable bounds. The bounds were determined by employing the optimal value of the scaling factor s for the zeroth iteration

	Mesh I		Mesh II		Mesh III	
$A^{cutoff}_{1,EX} - A^{cutoff}_{(u_{S^p_\Delta})}$	3.0799×10^{-2}		1.0630×10^{-2}		3.2871×10^{-3}	
Iteration index	$\frac{\mathscr{E}^{U,\mathscr{F}}}{\mathscr{F}(e_{S^p_\Delta} z)}$	$\frac{\mathscr{E}^{L,\mathscr{F}}}{\mathscr{F}(e_{S^p_\Delta})}$	$\frac{\mathscr{E}^{U,\mathscr{F}}}{\mathscr{F}(e_{S^p_\Delta})}$	$\frac{\mathscr{E}^{L,\mathscr{F}}}{\mathscr{F}(e_{S^p_\Delta})}$	$\frac{\mathscr{E}^{U,\mathscr{F}}}{\mathscr{F}(e_{S^p_\Delta})}$	$\frac{\mathscr{E}^{L,\mathscr{F}}}{\mathscr{F}(e_{S^p_\Delta})}$
0	33.499	−25.158	15.192	−5.986	17.081	−2.776
1	24.370	−18.840	11.377	−3.001	7.842	−0.593

influence function on the same mesh and use one of the error indicators explained in Example 6.2*.6; see also [Babuška, Strouboulis *et al.*, 1995]. By this way we get much more economical meshes than we would get for computing the solution and the influence function on different meshes. If the goal is to compute few functionals then either we can compute every functional separately on different meshes. We could also compute them simultaneously using error indicators for the functionals together.

Let us now give an example of adaptive computation of an output.

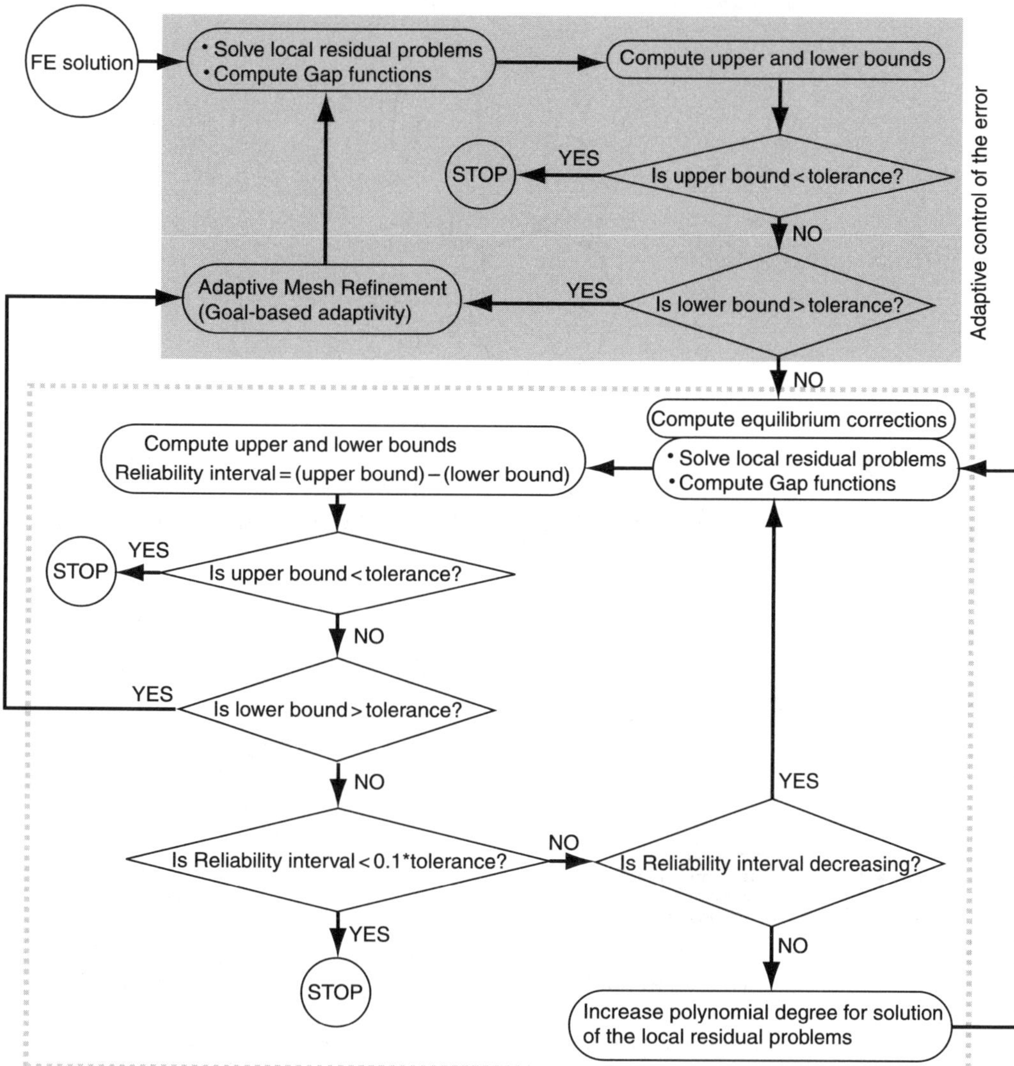

FIG. 6.2*.11. *An adaptive procedure for guaranteed error control based on the computable bounds*

Example 6.2*.6 *(Adaptive computation of the intensity factors).* Let us consider the mixed boundary value problem for the Laplacian

$$-\Delta u = 0 \quad \text{on } \Omega \tag{6.2*.48a}$$

$$\frac{\partial u}{\partial n} = \nabla(2x - y) \cdot \mathbf{n} \quad \text{on } \Gamma_1 \tag{6.2*.48b}$$

$$\frac{\partial u}{\partial n} = 0 \quad \text{on } \Gamma_2, \ldots, \Gamma_8 \tag{6.2*.48c}$$

$$u = 0 \quad \text{on } \Gamma_9, \Gamma_{10} \tag{6.2*.48d}$$

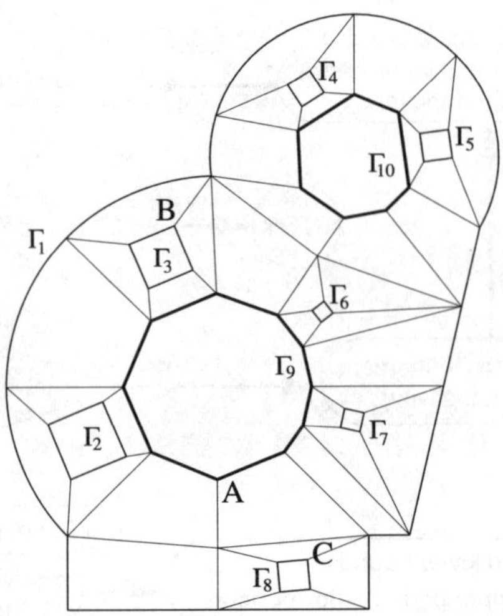

FIG. 6.2*.12. A complex domain with the corners A, B, and C where the intensity factor is desired

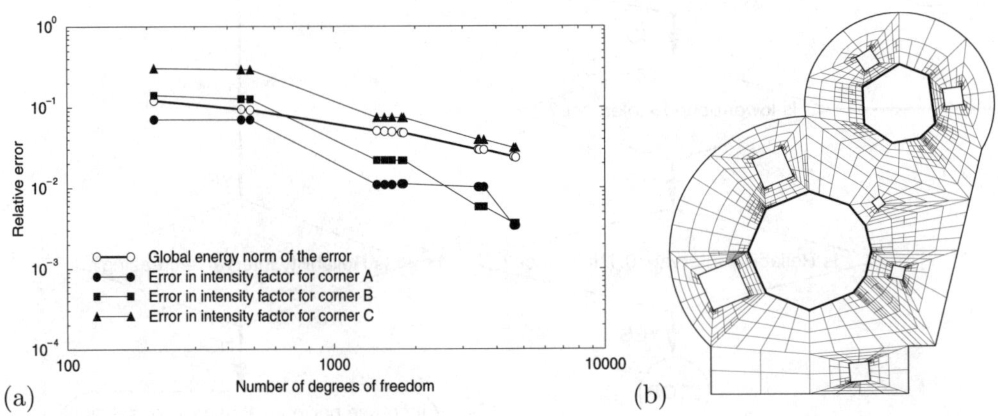

FIG. 6.2*.13. *Extraction of the intensity factors from a sequence of globally adaptive meshes.* (a) Convergence of the relative value of the global energy norm of the error and the error in the extracted intensity factor at corner A, B, and C for a sequence of globally adaptive meshes. (b) The last mesh in the employed adaptive sequence which achieves a tolerance of 2.36% in the global energy norm of the error

where Ω is the domain shown in Fig. 6.2*.12. We will be interested in controling the error in the intensity factor at corners A, B, and C. Figure 6.2*.13a shows the convergence of the relative error in the energy norm and the intensity factors computed using the cut off function approach for a sequence of globally adaptive meshes of biquadratic elements ($p = 2$), i.e. meshes which are constructed for the nearly optimal control of the global energy norm of the error, and Fig. 6.2*.13b shows the last mesh in the adaptive sequence which achieves a tolerance of 2.36% in the global energy norm of the error. Figure 6.2*.14a (respectively Figs 6.2*.14c,e) shows the convergence of the error in the extracted intensity

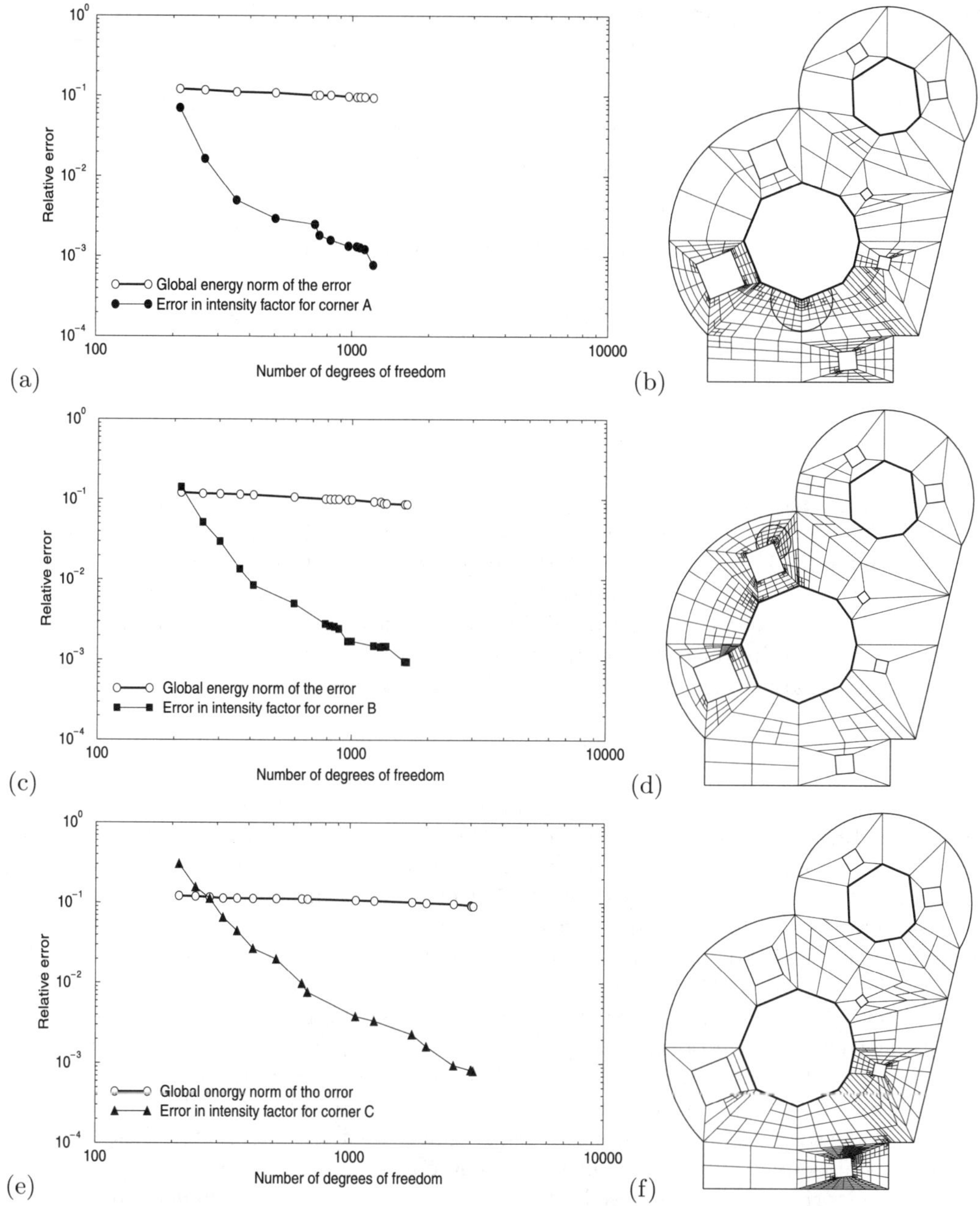

FIG. 6.2*.14. *Extraction of the intensity factors from sequences of goal adaptive meshes.* Convergence of the relative error in the extracted intensity factor for a sequence of goal adaptive meshes which controls only the error in the extracted intensity factor at (a) corner A, (c) corner B, and (e) corner C compared with the global energy norm of the error. The last goal adaptive mesh for (b) corner A which achieves a tolerance of 0.078%; (d) corner B which achieves a tolerance of 0.094%; (f) corner C which achieves a tolerance of 0.081%, in the extracted value of the corresponding intensity factor

factor at corner A (respectively corner B and C) compared with the energy norm of the error, for a sequence of goal adaptive meshes, namely the meshes which is nearly optimal for the extraction of the intensity factor at the corner A (respectively corner B and C). Figure 6.2*.14b (respectively Figs 6.2*.14d,f) shows the last mesh in the employed goal adaptive mesh sequence which achieves a tolerance of 0.078% for the extracted intensity factor at corner A (respectively 0.094% at corner B and 0.081% at corner C).

We note that:

a) Goal adaptive meshes can be much more economical than globally adaptive meshes for a desired accuracy.
b) Goal adaptive meshes can achieve very high accuracy in the goal, while at the same time the energy norm of the error in the finite element solution can be rather high.

Remark 6.2*.9 Different approaches for the extraction of the intensity factors lead to different adaptive sequences.

Remark 6.2*.10 The aim of the adaptive procedures is to construct meshes which are nearly optimal in the sense that they lead to the error in the prescribed tolerance range with minimal computational effort. All adaptive procedures are based on the same principle. They try to equidistribute the error for elements using some element error indicator. We discussed the structure of the optimal meshes in Section 3 and have shown that the equidistribution of the element estimators (indicators) is very effective. The error indicators are closely related to $p + k$ derivatives of the solution which governs the error.

There are various adaptive finite element codes. We mention for example the code PLTMG ([Bank, 1998]). There are books addressing the adaptive techniques, for example, [Verfürth, 1996]. There are many papers on the adaptive procedures, for example, [Babuška, 1976], [Babuška and Rheinboldt, 1978], [Babuška and Miller, 1984], [Johnson and Hansbo, 1992], [Johnson, 1993], [Becker and Rannacher, 1996], and [Carey, 1997]. The adaptive procedure strongly depends on the available mesh generator, for example, if it admits the use of constrains, anisotropic meshes, etc. The commonly used mesh generators produce isotropic meshes, but some anisotropic ones do exist, for example, [Buscaglia and Dari, 1997] and [Agoural *et al.*, 1999]. The main idea for the construction of anistropic meshes is to recover (for linear elements) the Hessian of the solution and construct triangles with optimal shape. The idea of computation of the error indicator function $\hat{e}_\tau^{\mathrm{Neum}}$ by using various shape functions which leads to information on anisotropy of the error was utilized in [Stein and Ohnimus, 1999]. Anisotropic meshes can be effective when the solution has boundary layers, singularities along edges, etc. Most adaptive appoaches attempt to control the error in some norm, e.g. the energy norm. There are also goal adaptive procedures based for the computation of the functionals, etc. Here as explained in Section 6.2, the error indicator is related to the scalar product of the solution and the function which generates the functional and the adaptive procedure attempts to equidistribute these indicators.

An analysis of adaptive procedures which proves that they lead to nearly optimal meshes for the energy norm of the error, without using any information about the exact solution, was presented for various versions of the procedure in [Babuška and Vogelius, 1984], for the one-dimensional case. In this work two essential results have been proven:

(a) The adaptive procedure which refines all elements τ according to the criterion $\eta_\tau \geq \Theta \max_{\tau \in \Delta} \eta_\tau$, $0 < \Theta \leq 1$ always converges.

(b) The adaptive procedure, under mild assumptions about the solution, leads to meshes with nearly minimized error, namely error $\|e_{S^p_\Delta}\|_{\mathcal{U}} \approx \inf \|e_{S^p_{\Delta'}}\|_{\mathcal{U}}$ where the infimum is taken over all meshes Δ' with the same number of degrees of freedom.

Similar results in higher dimensions were proven for (a) only in [Morin *et al.*, 2000] and [Dörfler, 1996]; for various strategies of refinement, see [Carey, 1997]. The analogue of the results in (b) is still open.

One of the major practical problem is the stopping criterion because the aim is to obtain the solution in the range of the prescribed accuracy (and not for given number of degrees of freedom). This stopping should be based on guaranteed a-posteriori error estimation. Experience shows that the quality of the mesh, i.e. the refinement at the right places is not very sensitive to the accuracy of the indicators. Further the equidistribution depends only on the ratio of the error indicators. Usually the adaptive procedures are not addressing the stopping criterion (related to the goal of the computation) which of course is a significant shortcoming of such procedures. Experience shows that it is possible to use cheap error indicators, e.g. unequilibrated indicator, but the error estimation and the construction of the reliabilty interval has to be accurate.

Essentially there are two basic strategies to construct the meshes with equidistributed indicators. The first one can be called the sequential, which is of the type addressed in [Babuška and Vogelius, 1984]; see also [Babuška *et al.*, 1986]. The other one is based on the construction of the mesh form its density derived from the recovered derivatives of the solution from the current finite element solution. Here 'in principle' it is possible to get directly the solution with prescribed accuracy. For the main idea of this type see [Babuška and Gui, 1986]. Unfortunately some of the experience indicates that this method when implemented is not so effective as it potentially would suggest.

Remark 6.2*.11 The accuracy in the computed value of the output depends on the accuracy of the finite element solutions for the original problem and the influence function, and their scalar product. Computation of the functionals a-posteriori bounds and the adaptive procedures were addressed in [Babuška and Miller, 1984]. Further, the problem was addressed in [Becker and Rannacher, 1996], [Rannacher, 2000], [Prudhomme and Oden, 1999], [Ainsworth and Oden, 2000], [Paraschivoiu *et al.*, 1997], [Datta, 1997].

Remark 6.2*.12 There are various methods for the computation of the stress intensity factor. We refer here to, for example, [Wahlbin, 1991].

Remark 6.2*.13 Let us also note that the extraction derived for the case of bounded functionals should not be used if the functional is not bounded. See, for example, [Becker and Rannacher, 1996], where the approach derived for bounded functionals is used for the extraction of pointwise derivatives and leads to meshes which are very refined in the neighborhood of the point of interest. A better approach seems to be is to replace the functional by a regularized one before attempting the extraction. Then high accuracy in the desired outputs can be achieved using rather coarse meshes.

Let us also address the a-posteriori estimation of the pollution. To facilitate the discussion, let us recall the definition of the local and the pollution error.

Given an element τ, we let

$$\omega_\tau^\Delta = \bigcup_{\substack{\tau' \in \Delta \\ \partial\tau' \cap \partial\tau \neq \emptyset}} \tau' \tag{6.2*.49}$$

denote the patch of elements which include τ, and one layer of elements around it, and we define $V_1^{\omega_\tau^\Delta}$, as the solution of the variational problem:

Find $V_1^{\omega_\tau^\Delta} \in \mathcal{U}$ such that

$$\mathcal{B}(V_1^{\omega_\tau^\Delta}, v) = \sum_{\substack{\tau' \in \Delta \\ \tau' \subseteq \omega_\tau^\Delta}} \mathcal{R}_{\tau'}^{EQ}(v) \quad \forall v \in \mathcal{U}. \tag{6.2*.50a}$$

We then let

$$V_2^{\omega_\tau^\Delta} \stackrel{\text{def}}{=} e_{S_\Delta^p} - V_1^{\omega_\tau^\Delta}. \tag{6.2*.50b}$$

It follows that $V_2^{\omega_\tau^\Delta}$, is the solution of the variational problem:

Find $V_2^{\omega_\tau^\Delta} \in \mathcal{U}$ such that

$$\mathcal{B}(V_2^{\omega_\tau^\Delta}, v) = \sum_{\substack{\tau' \in \Delta \\ \tau' \subseteq \Omega - \omega_\tau^\Delta}} \mathcal{R}_{\tau'}^{EQ}(v) \qquad \forall v \in \mathcal{U}. \tag{6.2*.50c}$$

The function $V_1^{\omega_\tau^\Delta}|_\tau$, is the local error in τ, and $V_2^{\omega_\tau^\Delta}|_\tau$, is the pollution error in τ. We have

$$\|e_{S_\Delta^p}\|^2_{\mathcal{U}(\tau)} \approx \|V_1^{\omega_\tau^\Delta}\|^2_{\mathcal{U}(\tau)} + \|V_2^{\omega_\tau^\Delta}\|^2_{\mathcal{U}(\tau)} \tag{6.2*.51a}$$

and

$$\|V_1^{\omega_\tau^\Delta}\|_{\mathcal{U}(\tau)} \approx \eta_\tau^{\text{Neum}} \tag{6.2*.51b}$$

and hence

$$\|e_{S_\Delta^p}\|^2_{\mathcal{U}(\tau)} \approx (\eta_\tau^{\text{Neum}})^2 + \|V_2^{\omega_\tau^\Delta}\|^2_{\mathcal{U}(\tau)} \tag{6.2*.51c}$$

asymptotically, as h^Δ tends to zero. Further, we have

$$\|V_2^{\omega_\tau^\Delta}\|_{\mathcal{U}(\tau)} \approx \sqrt{|\tau|}\,|\overline{\boldsymbol{\nabla} V_2^{\omega_\tau^\Delta}}| \stackrel{\text{def}}{=} \sqrt{|\tau|}\sqrt{\sum_{i=1}^{2}\Big(\frac{1}{|\tau|}\int_\tau \frac{\partial}{\partial x_i}(V_2^{\omega_\tau^\Delta})\Big)^2}. \tag{6.2*.51d}$$

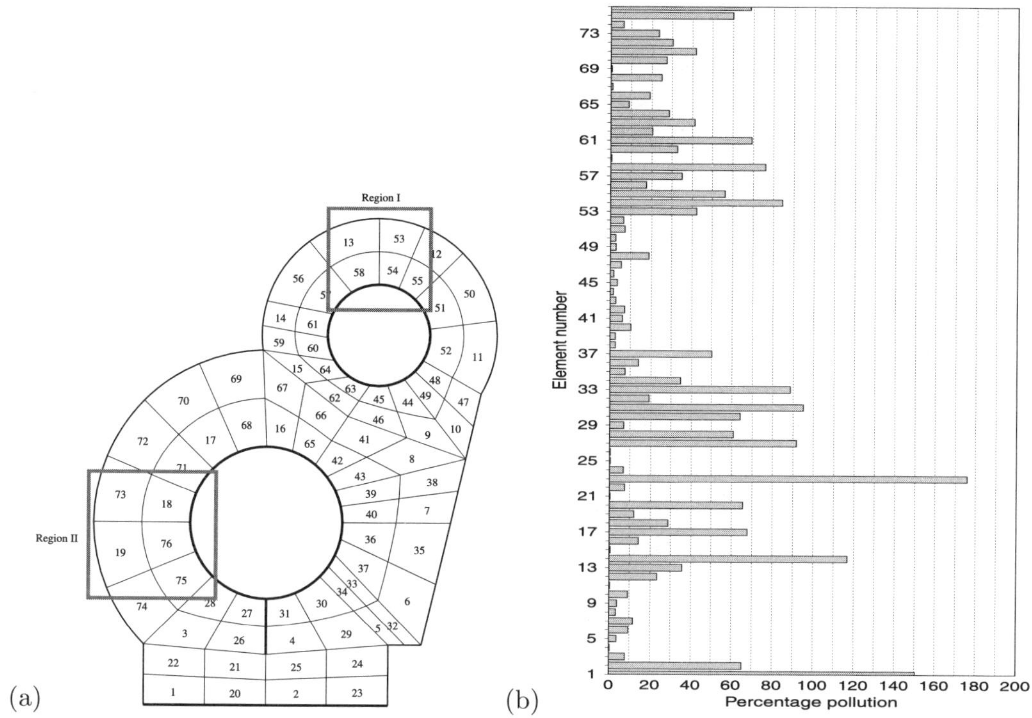

FIG. 6.2*.15. *The relative magnitude of the pollution and the local error.* (a) Domain II, Mesh II with the enumeration of the elements and the regions of interest. (b) Ratio $\|V_2^{\omega_\tau^\Delta}\|_{\mathcal{U}(\tau)}/\|V_1^{\omega_\tau^\Delta}\|_{\mathcal{U}(\tau)} * 100$ which measures the relative magnitude of the pollution to the local error in the element τ

Hence in order to estimate $\|V_2^{\omega_\tau^\Delta}\|_{\mathcal{U}(\tau)}$, it is sufficient to estimate $|\overline{\nabla V_2^{\omega_\tau^\Delta}}|$, the average gradient of the pollution error in the element τ.

Let us first give an example which illustrates the pollution error for our model problem.

Example 6.2*.7 (*The relative magnitude of the pollution and the local error*) We considered, once more the model problem on Domain II and its finite element approximation using biquadratic element in Mesh II. Figure 6.2*.15b gives the ratio $\|V_2^{\omega_\tau^\Delta}\|_{\mathcal{U}(\tau)}/\|V_1^{\omega_\tau^\Delta}\|_{\mathcal{U}(\tau)}$ for the elements in Mesh II shown in Fig. 6.2*.15a. Figure 6.2* 16 shows the shades of the modulus of the derivatives of the local error and the pollution error in the regions shown in Fig. 6.2*.15a. Note the oscillatory character of the local error in contrast with the pollution error.

Let us now address the a-posteriori estimation of the pollution error. We have

$$\frac{1}{|\tau|}\int_\tau \frac{\partial}{\partial x_i} V_2^{\omega_\tau^\Delta} = \sum_{\substack{\tau'\in\Delta \\ \tau'\subseteq\Omega-\omega_\tau^\Delta}} \mathcal{R}_{\tau'}^{EQ}(H^{q_i^\tau}) = \sum_{\substack{\tau'\in\Delta \\ \tau'\subseteq\Omega-\omega_\tau^\Delta}} \mathcal{B}_{\tau'}(\hat{e}^{\text{Neum}}_{u_{S^p_\Delta}}, H^{q_i^\tau}) \qquad (6.2^*.52a)$$

where $H^{q_i^\tau}$ is the solution of the variational problem.

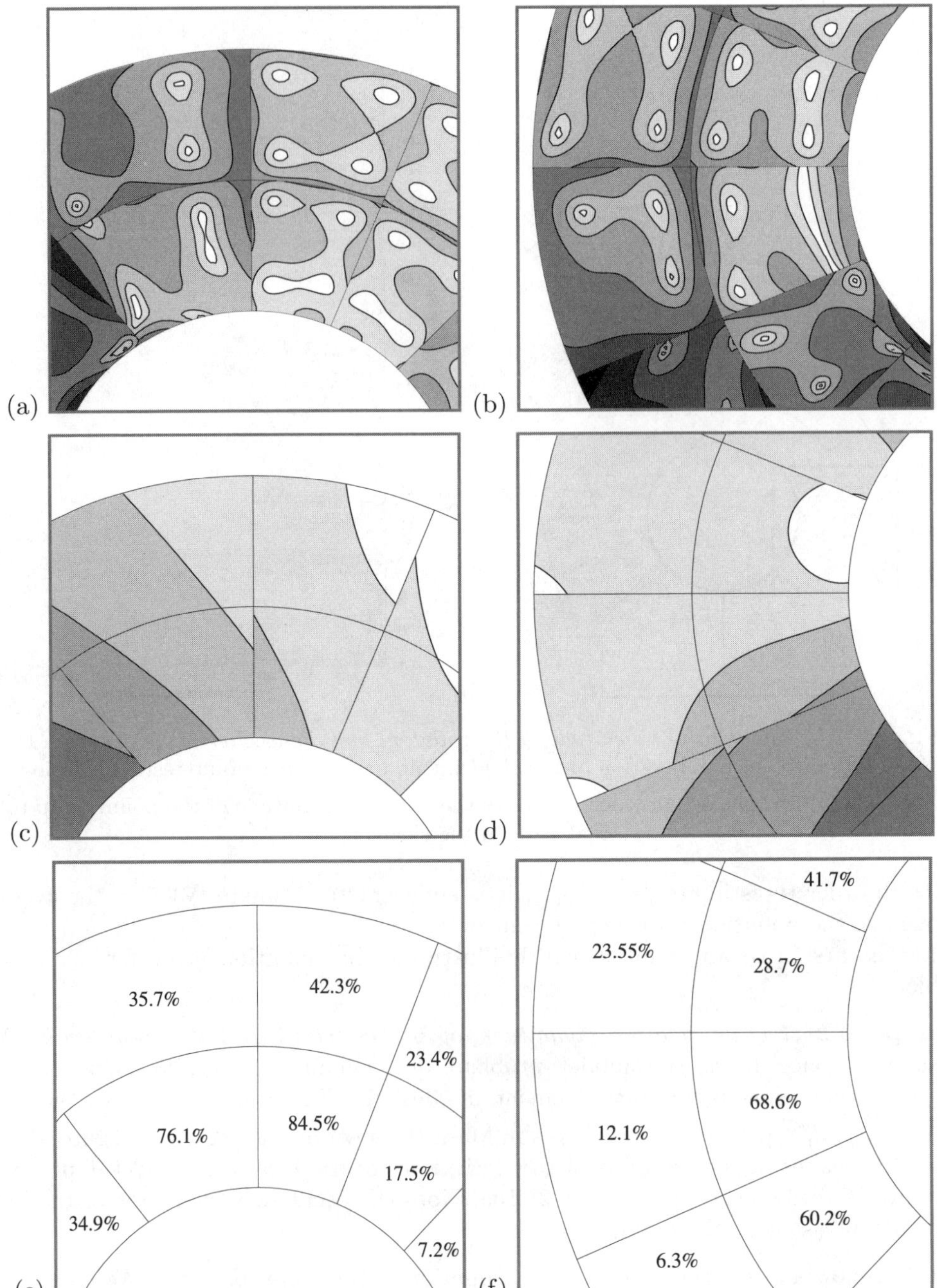

FIG. 6.2*.16. *The relative magnitude of the pollution and the local error.* The shades of the modulus of the derivatives of the local error in (a) Region I and (b) Region II. Shades of the of the modulus of the derivatives pollution error in (c) Region I and (d) Region II. Ratio $\|V_2^{\omega_\tau^\Delta}\|_{\mathcal{U}(\tau)}/\|V_1^{\omega_\tau^\Delta}\|_{\mathcal{U}(\tau)} * 100$ which measures the relative magnitude of the pollution to the local error in the element τ for the elements in (e) Region I and (f)Region II

Find $H^{q_i^\tau} \in \mathcal{U}$ such that

$$\mathcal{B}(H^{q_i^\tau}, v) = \sum_{\substack{\tau' \in \Delta \\ \tau' \subseteq \Omega - \omega_\tau^\Delta}} \mathcal{R}_{\tau'}^{EQ}(v) \quad \forall v \in \mathcal{U}. \tag{6.2*.52b}$$

It follows that

$$\frac{1}{|\tau|} \int_\tau \frac{\partial}{\partial x_i}(V_2^{\omega_\tau^\Delta}) = \sum_{\substack{\tau' \in \Delta \\ \tau' \subseteq \Omega - \omega_\tau^\Delta}} \mathcal{B}_{\tau'}(\hat{e}_{u_{S_\Delta^p}}^{\text{Neum}}, H^{q_i^\tau} - H_{S_\Delta^p}^{q_i^\tau}) \tag{6.2*.52c}$$

where $H_{S_\Delta^p}^{q_i^\tau} \in S_\Delta^p$, is the finite element approximation of $H^{q_i^\tau}$. As in the one-dimensional case we are led to the following estimator

$$\mathcal{E}_1^{q_i^\tau} \stackrel{\text{def}}{=} \left| \sum_{\substack{\tau' \in \Delta \\ \tau' \subseteq \Omega - \omega_\tau^\Delta}} \mathcal{B}_\tau(\hat{e}_{u_{S_\Delta^p}}^{\text{Neum}}, \hat{e}_{H_{S_\Delta^p}^{q_i^\tau}}^{\text{Neum}}) \right| \tag{6.2*.53a}$$

for $|(1/|\tau|) \int_\tau (\partial/\partial x_i)(V_2^{\omega_\tau^\Delta})|$, and by employing the Cauchy–Schwarz inequlity, we get the bigger estimator

$$\mathcal{E}_2^{q_i^\tau} = \sum_{\substack{\tau' \in \Delta \\ \tau' \subseteq \Omega - \omega_\tau^\Delta}} \eta_{\tau, u_{S_\Delta^p}}^{\text{Neum}} \eta_{\tau, H_{S_\Delta^p}^{q_i^\tau}}^{\text{Neum}} . \tag{6.2*.53b}$$

Remark 6.2*.14 We may also employ other indicator than the Neumann indicators in (6.2*.53a) and (6.2*.53b) to define the estimators of the pollution; see for example [Strouboulis, Babuška *et al.*, 1998, 2000].

Let us now give an example which shows the accuracy of the above estimator.

Example 6.2*.8 (*A-posteriori estimation of the pollution error*) We employed the pollution estimator (6.2*.58a) and the *ER*1*P*4 indicators to estimate the pollution errors for the model problem on Domain II and its finite element approximation using biquadratic element in Mesh II. Figure 6.2*.17a (respectively Fig. 6.2*.17b) reports the effectivity indices for the indicators for $p = 2$ for the elements in the regions shown in Fig. 6.2*.15a. Note that the effectivity indices are in the acceptable range. Improved estimates can be obtained by employing a special construction of the pollution indicators in the elements adjacent to the re-entrant corners, as described in [Strouboulis, Babuška *et al.*, 1998].

Remark 6.2*.15 As in the one-dimensional case we will call the contribution

$$\mu_{\tau'}^{q_i^\tau} \stackrel{\text{def}}{=} \mathcal{B}_{\tau'}(\hat{e}_{u_{S_\Delta^p}}^{\text{Neum}}, \hat{e}_{H_{S_\Delta^p}^{q_i^\tau}}^{\text{Neum}}) \tag{6.2*.54a}$$

of the elements the computed pollution indicators. Their accuracy can be assessed by comparing them with the corresponding exact values

$$\mu_{EX,\tau'}^{q_i^\tau} \stackrel{\text{def}}{=} \mathcal{B}_{\tau'}(\hat{e}_{u_{S_\Delta^p}}^{\text{Neum}}, H^{q_i^\tau} - H_{S_\Delta^p}^{q_i^\tau}) \tag{6.2*.54b}$$

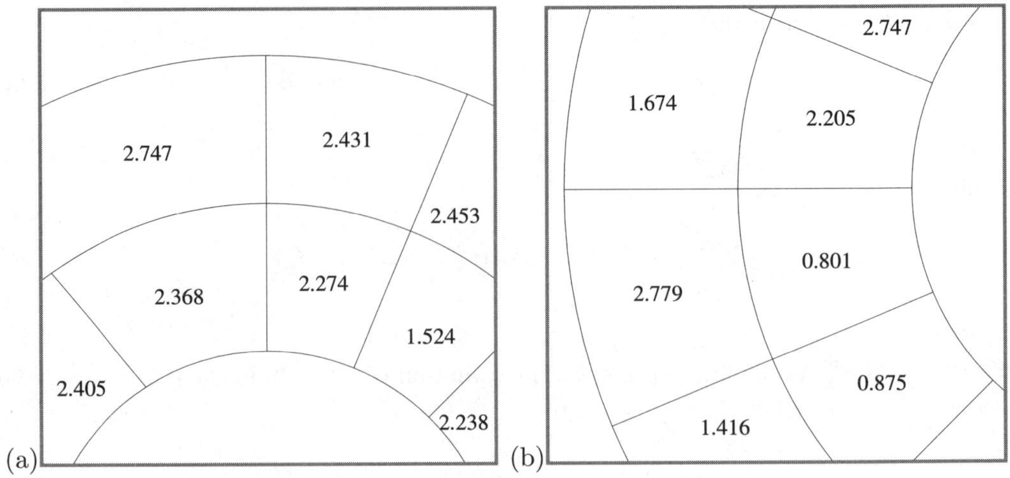

FIG. 6.2*.17. *A-posteriori estimation of the pollution error.* Element effectivity indices of the pollution estimator using (6.2*.58a) for the elements in (a) Region I and (b) Region II shown in Fig. 6.2*.21a. Here we employed the $ER1P4$ indicators

which, for example, may be computed by overkill. In [Strouboulis, Babuška *et al.*, 1998, 2000] we have analyzed the accuracy of the pollution indicators, and we have shown that their biggest values occur in the key elements which are the elements with a vertex at re-entrant corner. Hence the accuracy of the pollution estimate is influenced by the accuracy of the pollution indicators in the key elements. The pollution indicator are used to drive goal adaptive algorithms, for constructing meshes which are optimal with respect only to a local value, e.g. the derivative at a point, like for example the meshes employed in Example 6.2*.7.

Remark 6.2*.16 The pollution plays an essential role in the local analysis of the accuracy of the finite element method, for example, by local error estimators or recovered derivatives. This was discussed in [Babuška, Strouboulis *et al.*, 1994] and [Babuška, Strouboulis *et al.*, 1995] and [Babuška, Strouboulis *et al.*, 1998]. From the discussion the importance of the estimate of the pollution is clearly is visible.

In this section we addressed the construction of bounds for the outputs of interest, in the case that the output corresponds to a bounded or an unbounded functional. In the latter case we employed an extraction functional as in the one-dimensional case. We also addressed the a-posteriori estimation of the pollution, and illustrated its significance in problems with similar attributes as the problems encountered in practice.

Problems

P6.2*.1. Consider the problem

$$\Delta u\big|_{\Omega} = 0, \qquad \frac{\partial u}{\partial n}\Big|_{\partial\Omega} = g \tag{6.2*.55}$$

where $\Omega = (-1,1) \times (-1,1)$. Let the aim of the computation be $(\partial^2 u/\partial x^2)(0,0)$. Design the method for the computation and its a-posteriori estimate using uniform mesh and elements of degree $p = 2$. What asymptotic accuracy will we get?

6.3* Historical and bibliographical comments

Two-sided error estimation has a long history. There are essentially two approaches:

- the method of orthogonal projections
- the Treftz method.

The method of orthogonal projections was proposed by S. Zaremba [Zaremba, 1927]. The term 'method of orthogonal projections' was introduced by [Weil, 1940]. The method of orthogonal projections is also closely related to the Castigliano principle. The Treftz method was proposed by Treftz [Treftz, 1926]. Friedrichs [Friedrichs, 1929] has shown that the Treftz method is the Ritz method applied to the conjugate variational problem. The Treftz method was further developed in 1947 by Prager and Synge; see [Synge, 1957]; and subsequently by many others, e.g. [Aubin and Burchard, 1971]. For relevant discussions we also refer to [Mikhlin, 1964]. Physically the two-sided estimates are related to the use of the 'force' and 'displacement' method. We refer here to [de Veubeke, 1965].

The extraction function is related to the classical Riesz theorem in Hilbert spaces. The extraction function has been employed in structural mechanics for many years, like for example the dummy load method based on Castigliano's theorem. The first systematic use of the extraction function in the finite element computation was in [Babuška and Miller, 1984], which also addressed the a-posteriori estimation and the adaptive control of the error in the extracted outputs. The same approach has been used recently by many investigators, see e.g. [Rannacher, 1998]. The same technique can be extended to address the error in the computed outputs in finite element solutions of parabolic and hyperbolic problems.

Today many adaptive programs exist. The first adaptive program very likely was the program FEARS (see [Babuška, 1976], [Babuška and Rheinboldt, 1978], and [Mesztenyi and Szymczak, 1982]).

References

A. Agoural, K. Lipnikov and Y. Vasilevski (1999) Adaptive generation of quasi-optimal tetrahedral meshes. *East-West Journal of Numerical Mathematics*, **7**, 223–244.

M. Ainsworth and J.T. Oden (2000) *A-posteriori error estimation in Finite Element Analysis.* John Wiley, New York.

J.P. Aubin and H.G. Burchard (1971) Some aspects of the method of hypercircle applied to elliptic variational problems in numerical solution of partial differential equations Part II. In: B. Hubbard (ed.) SYNSPADE, Academic Press, New York, pp. 1–68.

I. Babuška (1976) The self-adaptive approach in the finite element method. In: J.R. Whiteman (ed.) *Mathematics of finite elements and applications II.* Academic Press, pp. 125–142.

I. Babuška and W.C. Rheinboldt (1978) On a system for adaptive parallel finite computations, *Proceedings 1978 Annual Conference Association for Computing Machinery*, pp. 480–489.

I. Babuška and M. Vogelius (1984) Feedback and adaptive finite element solution of one-dimensional boundary value problem. *Numerische Mathematik*, **44**, 75–102.

I. Babuška and A. Miller (1984) The post-processing approach in the finite element method. Part I: calculations of displacements, stresses and other higher derivatives of displacements. Part II: the calculation of stress intensity factors. Part III: a-posteriori error estimate and adaptive method. *International Journal for Numerical Methods in Engineering*, **20**, 1085–1109; 1111–1129; 2311–2325.

I. Babuška, O.C. Zienkiewicz, J. Gago, and E.R. de A Oliverira (eds) (1986) *Accuracy Estimates and Adaptive Refinements in Finite Element Computations.* John Wiley, New York.

I. Babuška and W. Gui (1986) Basic principles of feedback and adaptive approach in the finite element method. *Computer Methods in Applied Mechanics and Engineering*, **55**, 27–42.

I. Babuška, T. Strouboulis, A. Mathur and C.S. Upadhyay (1994) Pollution error in the *h* version of the finite element method and the local quality of a-posteriori error estimators. *Finite Elements in Analysis and Design*, **17**, 273–321.

I. Babuška, T. Strouboulis, C.S. Upadhyay and S.K. Gangaraj (1995) A Posteriori estimation and adaptive control of the pollution error in the *h* version of the finite element method. *International Journal for Numerical Methods in Engineering*, **38**, 4207–4235.

I. Babuška, T. Strouboulis, S.K. Gangaraj and C.S. Upadhyay (1997) Pollution error in the *h* version of the finite element method and the local quality of recovered derivatives. *Computer Methods in Applied Mechanics and Engineering*, **140**, 183–199.

I. Babuška, T. Strouboulis, S.K. Gangaraj, K. Copps, and D.K. Datta (1998) Practical aspects of a-posteriori estimation of the pollution error for reliable finite element analysis. *Computer Methods in Applied Mechanics and Engineering*, **66**, 626–664.

I. Babuška, T. Strouboulis, and S.K. Gangaraj (1999) Guaranteed computable bounds for the exact error in the finite element solution – Part I: One-dimensional model problem. *Computer Methods in Applied Mechanics and Engineering* **176**, 51–79.

R.E. Bank (1998) *PLTMG: a software package for solving elliptic partial differential equations, User's guide 7.0.* SIAM, Philadelphia.

R. Becker and R. Rannacher (1996) A feedback approach to error control in finite element methods: basic analysis and examples. *East-West Journal of Numerical Mathematics*, **4**, 237–264.

G.C. Buscaglia and E.A. Dari (1997) Anistropic mesh optimization and its application in adaptivity. *International Journal for Numerical Methods in Engineering*, **40**, 4119–4136.

G.F. Carey (1997) *Computational Grids Generation, Adaptation and Solution Strategies.* Taylor and Francis, London.

D.K. Datta (1997) *A posteriori estimation and adaptive control of the error in the quantities of interest.* M.S. Thesis, Texas A&M University, College Station, TX.

B.F. de Veubeke (1965) *Displacement and equilibrium models in the finite element method in stress analysis.* Zienkiewicz, Holister ed., Wiley.

W. Dörfler (1996) A convergent adaptive algorithm for Poisson equation. *SIAM Journal on Numerical Analysis*, **33**, 1106–1124.

K.O. Friedrichs (1929) Ein verfahren der variations rechnung des minimum eines integrals asl des maximum eines arderen ausdrucks darzustellen ger wiss Götingen, Nachrichten. *Mathematical Physics*, Kluwer, 13–20.

C. Johnson and P. Hansbo (1992) Adaptive finite element method in computational mechanics. *Computer Methods in Applied Mechanics and Engineering*, **101**, 143–181.

C. Johnson (1993) A new paradigm for adaptive finite element methods. In: J.R. Whiteman (ed.) *The mathematics of Finite Element and Applications*, **VIII**, Wiley, Chichester, pp. 105–120, 143–181.

C. Mesztenyi and W. Szymczak (1982) *FEARS user's manual for UNIAC 1100.* Technical Note BN 991, IPST University of Maryland, College Park.

S.G. Mikhlin (1964) *Variational methods in mathematical physics.* Pergamon Press.

P. Morin, R.H. Nochetto, and K.G. Siebert (2000) Data oscillation and convergence of adaptive FEM. *SIAM Journal on Numerical Analysis* (to appear).

M. Paraschivoiu, J. Peraire and A.T. Patera (1997) A-posteriori finite element bounds for linear functional outputs of elliptic partial differential equation. *Computer Methods in Applied Mechanics and Engineering*, **150**, 289–312.

S. Prudhomme and J.T. Oden (1999) On goal-oriented error estimation for elliptic problems: application to the control of pointwise errors. *Computer Methods in Applied Mechanics and Engineering*, **176**, 313–331.

E. Stein and S. Ohnimus (1999) Anisotropic discreteration and model-error estimation in solid mechanics by local Neumann problems. *Computer Methods in Applied Mechanics and Engineering*, **176**, 363–381.

R. Rannacher (1998) A posteriori error estimation in least-squares stabilized finite element schemes. *Computer Methods in Applied Mechanics and Engineering*, **166**, 99–114.

R. Rannacher (2000) The dual weighted residual method for error control and mesh adaptation in Finite Element Method. In: J.R. Whiteman (ed.) *Mathematics of Finite Elements and Applications*, **X**, Elsevier, Amsterdam, pp. 97–116.

T. Strouboulis, K. Copps and I. Babuška (2000) The generalized finite element method: an example of its implementation and illustration of its performance. *International Journal for Numerical Methods in Engineering*, **47**, 1401–1417.

J.L. Synge (1957) *The hypercircle in mathematical physics: a method for the approximate solution of boundary value problems.* Cambridge University Press, New York.

B.A. Szabo and I. Babuška (1991) *Finite element analysis.* John Wiley & Sons, Inc., New York.

E. Treftz (1926) Ein gegenstück zum Ritzschen Verfahren. *Proceeding of the Second International Congress of Applied Mechanics.* Zürich, pp. 131–137.

R. Verfürth (1996) *A review of a-posteriori error estimation and adaptive mesh-refinement technique.* Wiley, Teuber.

M.I. Vishik (1949) Method of orthogonal decomposition in the theory of elliptic differential equations. *Rossiĭskaya Akademiya Nauk. Matematicheskiĭ Sbornik. "Nauka", Moscow*, **25** (in Russian). (Translated in Sbornik. Mathematics. Presidium Russian Academy of Science, Moscow.)

L.B. Wahlbin (1991) Local behaviour in finite element methods. In: P.G. Ciarlet, J.L. Lions (eds) *Handbook of numerical analysis, Vol. II, Finite Element Methods, Part I.* North Holland, pp. 355–522.

H. Weil (1940) The method of orthogonal projections in potential theory. *Duke Mathematical Journal.* Duke University Press, Durham, NC, **7**.

S. Zaremba (1927) Sur un probleme toujours possible comprenant a' titre de ca particuliers, le probleme de Dirichlet et celui de Neumann. *Journal de Mathématiques Pures et Appliquées*, **6**, 127–163.

EPILOGUE

The finite element method, its h, p, hp and other versions are used in many fields of application, for linear and nonlinear problems for elliptic, parabolic, and hyperbolic differential equations, integral equations, etc. Rapid development of computers has allowed the computational analysis of problems of increasing complexity. Most often the goal of finite element computations is to replace experiments and to gain understanding of various effects and features which cannot be analyzed by experiments. Computations are relatively cheap and human cost is high.

The reliability of the model inclusive of all uncertainty (validation) and the quality of the computed results (verification) mostly by a-posteriori error estimation will be increasingly in the forefront of interest. The high human cost will increase the need for adaptive procedures in large scale problems. The criterion of the acceptance of the results will be essential. Some of these topics are addressed in an interesting book by Roache (see citations in Chapter I).

The goal of the present book was to present ideas and results related to the computational finite element analysis of elliptic equations. For computational analyses of other types of differential equations like, for example, hyperbolic equations, the state of the art is much less advanced, and further theoretical, computational and engineering work is needed. We hope that this book will further stimulate ideas which will lead to reliable quantitative analysis of engineering problems.

APPENDIX TO CHAPTER 2

A2.1.1 Normed linear spaces

A *normed linear space* X is a family of elements $u, v, \ldots$, which have the following properties:

1. If $u \in X$ and $v \in X$ then $(u+v) \in X$.
2. If $u \in X$ then $\alpha\, u \in X$.
3. $u + v = v + u$.
4. $u + (v + w) = (u + v) + w$.
5. There is a unique element in X, denoted by 0, such that $u + 0 = u$ for any $u \in X$.
6. Associated with every element $u \in X$ there exists a unique element $-u \in X$ such that $u + (-u) = 0$.
7. $\alpha(u + v) = \alpha u + \alpha v$.
8. $(\alpha + \beta)u = \alpha u + \beta u$.
9. $\alpha(\beta u) = (\alpha\beta)u$.
10. $1 \cdot u = u$.
11. $0 \cdot u = 0$.
12. With every $u \in X$ we associate a real number $\|u\|_X$, called *the norm*. The norm has the following properties:
    (a) $\|u\|_X \geq 0$.
    (b) $\|u\|_X = 0$ if and only if $u \equiv 0$.
    (c) $\|\alpha u\|_X = |\alpha|\|u\|_X$.
    (d) $\|u + v\|_X \leq \|u\|_X + \|v\|_X$. This is called the *triangle inequality*.
13. If X is complete, then X is a Banach space (see A2.2.2).

The *span* of $\{u_i\}$, $i = 1, \ldots, n$, $u_i \in X$, is the linear space of all elements of the form $u = \sum_{i=1}^{n} a_i u_i$, $a_i \in \mathbb{R}$.

A *seminorm* is a norm in X modulo $Y \subset X$, if $\|u\|_X$ satisfies 12a, 12c, and 12d in A2.1.1, and $\|u\|_X = 0 \Leftrightarrow u \in Y$. For example,

$$|u|_{H^k} \stackrel{\text{def}}{=} \sqrt{\sum_{\substack{k_1+k_2=k\\ k_1,k_2\geq 0}} (\partial^{k_1+k_2} u/\partial x_1^{k_1} \partial x_2^{k_2})^2}$$

is a seminorm in $H^k(\Omega)$ modulo polynomials of degree $k-1$.

A2.1.2 Linear functionals

Let X be a normed linear space and let $\mathcal{M} : X \mapsto \mathbb{R}$ be a functional. By functional we mean a mapping which associates with every $v \in X$ a real number $\mathcal{M}(v)$.

$\mathcal{M}$ is called a *linear functional* or *linear form* on X if it has the following properties:

1. $\mathcal{M}(v_1 + v_2) = \mathcal{M}(v_1) + \mathcal{M}(v_2)$.
2. $\mathcal{M}(\alpha v) = \alpha \mathcal{M}(v)$.

A functional $\mathcal{M} : X \mapsto \mathbb{R}$ is bounded if

3. $|\mathcal{M}(v)| \leq \mathcal{C}\|v\|_X$ with $\mathcal{C}$ independent of v. The smallest possible value of $\mathcal{C}$ is called the *norm* of M.

Often the terminology linear functional is employed for bounded linear functionals.

A2.1.3 Bilinear forms

Let X and Y be normed linear spaces and $\mathcal{B}(\cdot,\cdot)$ a map which associates with every $u \in X$ and $v \in Y$ a real number $\mathcal{B}(u, v)$.

$\mathcal{B}(\cdot,\cdot)$ is a *bilinear form* on $X \times Y$ if it has the following properties:

1. $\mathcal{B}(u_1 + u_2, v) = \mathcal{B}(u_1, v) + \mathcal{B}(u_2, v)$.
2. $\mathcal{B}(u, v_1 + v_2) = \mathcal{B}(u, v_1) + \mathcal{B}(u, v_2)$.
3. $\mathcal{B}(\alpha\, u, v) = \alpha\mathcal{B}(u, v)$, $\mathcal{B}(u, \beta v) = \beta\mathcal{B}(u, v)$.

A bilinear form $\mathcal{B} : X \times Y \mapsto \mathbb{R}$ is bounded if

4. $|\mathcal{B}(u, v)| \leq \mathcal{C}\|u\|_X\|v\|_Y$ with $\mathcal{C}$ independent of u and v. The smallest possible value of $\mathcal{C}$ is called the *norm* of $\mathcal{B}$.

A bilinear form $\mathcal{B} : X \times X \mapsto \mathbb{R}$ is *symmetric* if

5. $\mathcal{B}(u, v) = \mathcal{B}(v, u)$.

Often, the terminology bilinear form is employed only for bounded bilinear forms.

The space X is called a *trial space* and functions $u \in X$ are called *trial functions.* The space Y is called a *test space* and functions $v \in Y$ are called *test functions.* Note that, in general, $\mathcal{B}(u, v) \neq \mathcal{B}(v, u)$, i.e., $\mathcal{B}$ is not necessarily symmetric.

A2.1.4 The Hilbert space

Let X be a linear space which satisfies conditions 1–11 in A2.1.1, and let

$$(\cdot,\cdot) : X \times X \mapsto \mathbb{R}$$

be a map with the following properties:

1. $(x, y) = (y, x)$.
2. $(\lambda x_1 + \mu x_2, y) = \lambda(x_1, y) + \mu(x_2, y), \quad \lambda, \mu \in \mathbb{R}$.
3. $(x, x) \geq 0$, $(x, x) = 0 \Leftrightarrow x \equiv 0$.

The map $(\cdot,\cdot)$ with the above properties is called a *scalar* or *inner product* on X. Further, we have

$$|(x,y)| \leq \sqrt{(x,x)}\sqrt{(y,y)}$$

and

$$\|x\|_X \stackrel{\text{def}}{=} \sqrt{(x,x)}$$

is a norm on X. The linear space X, together with the inner product $(\cdot\,,\,\cdot)$, is an inner product space. If, in addition, we have:

4. If X is *complete*, then X is a *Hilbert* space (see A2.2.3).

A2.1.5 The energy space

Let X be a Hilbert space, $\mathcal{B} : X \times X \mapsto \mathbb{R}$ a symmetric bilinear form such that $\mathcal{B}(u,u) > 0$ if $u \not\equiv 0$. Then the functional $\|\cdot\|_{\mathcal{U}} : X \mapsto \mathbb{R}_0^+$ where

$$\|u\|_{\mathcal{U}} \stackrel{\text{def}}{=} \sqrt{\mathcal{B}(u,u)}$$

is a norm and the set

$$\mathcal{U} \stackrel{\text{def}}{=} \{u \in X \mid \|u\|_{\mathcal{U}} < \infty\}$$

is a Hilbert space. We will call $\mathcal{B}$ the *energy scalar product*, $\|\cdot\|_{\mathcal{U}}$ the *energy norm*, and $\mathcal{U}$ the *energy space*.

A2.2.1 Convergence in the space X

A sequence of functions $\{u_n\}_{n=1}^{\infty}$, $u_n \in X$, $n = 1, 2, \ldots$, *converges in the space* X to the function $u \in X$ if

$$\forall \epsilon > 0 \; \exists n_\epsilon \in \mathbb{N} : \quad \|u - u_n\|_X < \epsilon \quad \forall n > n_\epsilon .$$

A2.2.2 Cauchy sequence

A sequence $\{u_n\}_{n=1}^{\infty}$, $u_n \in X$, $n = 1, 2, \ldots$, is a *Cauchy sequence* if

$$\forall \epsilon > 0 \; \exists n_\epsilon \in \mathbb{N} : \quad \|u_n - u_m\|_X < \epsilon \quad \forall n, m > n_\epsilon .$$

A2.2.3 Completeness

X is *complete* if every Cauchy sequence $\{x_n\}_{n=1}^{\infty}$, $x_n \in X$ converges, i.e., there is $x \in X$ such that $\|x_n - x\|_X \to 0$ as $n \to \infty$.

A2.2.4 Denseness

A set $E \subseteq X$ is dense in X if for every $u \in X$ there exists a sequence $\{w_n\}_{n=1}^{\infty}$, $w_n \in E$, $n = 1, 2, \ldots$, such that

$$\|u - w_n\|_X \to 0 \quad \text{as } n \to \infty.$$

A2.3.1 The Lax–Milgram lemma

Let X be a Hilbert space and let $\mathscr{B} : X \times X \mapsto \mathbb{R}$ be a bounded symmetric bilinear form and assume that there exists a fixed $\alpha > 0$ such that

$$\mathscr{B}(u, u) \geq \alpha \|u\|_X^2 \quad \forall u \in X.$$

Then, given any linear form $\mathscr{L} : X \mapsto \mathbb{R}$, there exists a unique $u_{EX} \in X$ such that

$$\mathscr{B}(u_{EX}, v) = \mathscr{L}(v) \quad \forall v \in X.$$

A2.4.1 The L^p spaces

Let $\Omega \subseteq \mathbb{R}^2$ be a bounded open set (or $\Omega = (0, l) = I$ in the one-dimensional case). Then we define

$$\|f\|_{L^p(\Omega)} \stackrel{\text{def}}{=} \left(\int_\Omega |f|^p \right)^{1/p}, \quad 1 \leq p < \infty, \qquad \|f\|_{L^\infty(\Omega)} \stackrel{\text{def}}{=} \operatorname{ess.sup}_{x \in \Omega} |f(x)|.$$

The norms $\|\cdot\|_{L^p(\Omega)}$ are defined for functions for which the integral makes sense. Here we understand the integral in the Lebesgue sense. (Note that the Lebesgue integral is the same as the Riemann integral for 'reasonable' functions. However, it is possible to construct pathological functions for which the Riemann integral is not defined while the Lebesgue integral is well defined. Consider for example the function $q : I = (0, 1) \mapsto \{0, 1\}$ given by

$$q(x) = \begin{cases} 0, & x \text{ is irrational} \\ 1, & x \text{ is rational.} \end{cases}$$

The Lebesgue integral $\int_0^1 q = 0$ while the Riemann integral does not make sense.) It is possible to understand the $\|\cdot\|_{L^\infty(\Omega)}$ as the norm $\|\cdot\|_{L^p(\Omega)}$ as $p \to \infty$. Further, for a 'reasonable' function we have

$$\|f\|_{L^\infty(\Omega)} = \max_{x \in \Omega} |f(x)|$$

which we will always assume. (Note that for the pathological function q defined above we have $\|q\|_{L^\infty(I)} = 0$ while $\max_{x \in \Omega} |q(x)| = 1$.) The set of functions with finite $L^2(\Omega)$ norm form a Hilbert space which is denoted by $L^2(\Omega)$ with inner product $(\cdot, \cdot)_{L^2(\Omega)} : X \times X \mapsto \mathbb{R}$

$$(u, v)_{L^2(\Omega)} \stackrel{\text{def}}{=} \int_\Omega uv.$$

The set of functions with finite L^p norm form a complete linear space denoted by $L^p(\Omega)$.

A2.4.2 The Sobolev spaces

Let $\Omega \subset \mathbb{R}^2$ be a bounded set with piecewise analytic boundary (or $\Omega = (0, l) = I$ in the one-dimensional case). We will let

$$H^k(\Omega) \stackrel{\text{def}}{=} \left\{ v \in L^2(\Omega) \mid \|v\|_{H^k(\Omega)} < \infty \right\}$$

where

$$\|v\|_{H^k(\Omega)} \stackrel{\text{def}}{=} \sqrt{\sum_{0\le j\le k} \sum_{m_j+n_j=j} \left\| \frac{\partial^j u}{\partial x_1^{m_j} \partial x_2^{n_j}} \right\|^2_{L^2(\Omega)}}.$$

We will also let

$$|v|_{H^k(\Omega)} \stackrel{\text{def}}{=} \sqrt{\sum_{m+n=k} \left\| \frac{\partial^k u}{\partial x_1^{m} \partial x_2^{n}} \right\|^2_{L^2(\Omega)}}.$$

The set $H^k(\Omega)$ is a Hilbert space, $\|\cdot\|_{H^k(\Omega)}$ is a norm, and $|\cdot|_{H^k(\Omega)}$ is a seminorm modulo polynomials of degree $k-1$. Further we let $H^1_0(\Omega) \stackrel{\text{def}}{=} \{u \in H^1(\Omega) \mid u = 0 \text{ on } \partial\Omega\}$ or if $\Gamma_{\text{D}} \subset \partial\Omega$, we define $H^1_{\Gamma_{\text{D}}}(\Omega) \stackrel{\text{def}}{=} \{u \in H^1(\Omega) \mid u = 0 \text{ on } \Gamma_{\text{D}}\}$. If it is clear from the context what is Γ_{D}, we will sometimes write $H^1_{\Gamma_{\text{D}}}(\Omega) = H^1_0(\Omega)$. In the one-dimensional case we denote $H^1_{(0}(I) \stackrel{\text{def}}{=} \{u \in H^1(I) \mid u(0) = 0\}$ and $H^1_{0)}(I) \stackrel{\text{def}}{=} \{u \in H^1(I) \mid u(l) = 0\}$.

We will also let

$$W^k_p(\Omega) \stackrel{\text{def}}{=} \{v \in L^2(\Omega) \mid \|v\|_{W^k_p(\Omega)} < \infty\}$$

where

$$\|v\|_{W^k_p(\Omega)} \stackrel{\text{def}}{=} \left(\sum_{0\le j\le k} \sum_{m_j+n_j=j} \left\| \frac{\partial^j v}{\partial x_1^{m_j} \partial x_2^{n_j}} \right\|^p_{L^p(\Omega)} \right)^{1/p}, \quad 1 \le p < \infty$$

and

$$\|v\|_{W^k_\infty(\Omega)} \stackrel{\text{def}}{=} \sum_{0\le j\le k} \sum_{m_j+n_j=j} \left\| \frac{\partial^j v}{\partial x_1^{m_j} \partial x_2^{n_j}} \right\|_{L^\infty(\Omega)}$$

and the corresponding seminorms are defined analogously as above. The set $W^k_p(\Omega)$ is a Sobolev space. Note that $H^k(\Omega) = W^k_p(\Omega)$ is also a Sobolev space but $W^k_p(\Omega)$ for $p \neq 2$ is not a Hilbert space. If $v \in \mathbb{R}^n$, i.e., $v = (v_1, \ldots, v_n)^{\text{T}}$, then

$$\|v\|_{H^k(\Omega)} \stackrel{\text{def}}{=} \sqrt{\sum_{i=1}^{n} \|v_i\|^2_{H^k(\Omega)}}$$

$$|v|_{H^k(\Omega)} \stackrel{\text{def}}{=} \sqrt{\sum_{i=1}^{n} |v_i|^2_{H^k(\Omega)}}$$

$$\|v\|_{W^k_p(\Omega)} \stackrel{\text{def}}{=} \left(\sum_{i=1}^{n} \|v_i\|^p_{W^k_p(\Omega)} \right)^{1/p}$$

$$\|v\|_{W^k_\infty(\Omega)} \stackrel{\text{def}}{=} \max_{1\le i<n} \left(\|v_i\|_{W^k_\infty(\Omega)}\right).$$

A2.4.3 The C^k spaces

Let $\Omega \subset \mathbb{R}^2$ be a bounded set with piecewise analytic boundary (or $\Omega = (0, l) = I$ in the one-dimensional case). We will let $C^0(\Omega)$ denote the space of all continuous functions on $\bar{\Omega}$ (respectively on $\bar{I}$) with the norm $\|u\|_{C^0(\Omega)} = \max_{x \in \bar{\Omega}} |u(x)|$. Sometimes we will write $C(\Omega)$ instead of $C^0(\Omega)$. Further let

$$C^k(\Omega) \stackrel{\text{def}}{=} \{v \in C^0(\Omega) \mid \|v\|_{C^k(\Omega)} < \infty\}$$

where

$$\|v\|_{C^k(\Omega)} \stackrel{\text{def}}{=} \sum_{0 \le j \le k} \sum_{m_j + n_j = j} \left\| \frac{\partial^j u}{\partial x_1^{m_j} \partial x_2^{n_j}} \right\|_{C^0(\Omega)}.$$

We let also

$$|v|_{C^k(\Omega)} \stackrel{\text{def}}{=} \sum_{m+n=k} \left\| \frac{\partial^k u}{\partial x_1^m \partial x_2^n} \right\|_{C^0(\Omega)}.$$

Note that $C^0(\Omega) \subset L^\infty(\Omega)$ but $L^\infty(\Omega) \not\subset C^0(\Omega)$

$$C^\infty(\Omega) \stackrel{\text{def}}{=} \{v \in C^k(\Omega),\ k = 0, 1, \ldots\}.$$

If $v \in \mathbb{R}^n$, i.e., $v = (v_1, \ldots, v_n)^{\mathrm{T}}$, then

$$\|v\|_{C^k(\Omega)} \stackrel{\text{def}}{=} \max_{1 \le i < n} \|v_i\|_{C^k(\Omega)}.$$

A2.5.1 The symbols $O(|x|)$ and $o(|x|)$ as $|x| \to 0$

$$f(x) = O(|x|) \quad \Longleftrightarrow_{\text{df}} \quad |f(x)| \le \mathscr{C}|x| \quad \text{as} \quad |x| \to 0$$

where $\mathscr{C}$ is a constant independent of x.

$$f(x) = o(|x|) \quad \Longleftrightarrow_{\text{df}} \quad f(x) \to 0 \quad \text{as} \quad |x| \to 0.$$

For the basic notions of functional analysis mentioned above, we refer to (see Chapter 2 references) [Oden, 1979], [Mason, 1985], [Oden and Demkowicz, 1996], and many other textbooks. For the theory of Sobolev spaces we refer to [Adams, 1975].

A2.6.1 Kronecker's δ_{ij}

For i, j integers

$$\delta_{ij} = 1 \quad \text{for} \quad i = j$$

$$\delta_{ij} = 0 \quad \text{for} \quad i \ne j.$$

APPENDIX TO CHAPTER 3

A3.1 Special polynomials $P_n(x)$, $-1 < x < 1$

A3.1.1 The Legendre polynomials

$P_n(x)$ are solutions of the Legendre differential equation for $n = 0, 1, 2, \ldots$

$$(1 - x^2)P_n'' - 2xP_n' + n(n+1) = 0.$$

The explicit form is

$$P_n(x) = \frac{1}{2^n n!}\frac{d^n}{dx^n}(x^2 - 1)^n.$$

A3.1.2 The first four Legendre polynomials

$$\begin{aligned}
P_0(x) &= 1 \\
P_1(x) &= x \\
P_2(x) &= \tfrac{1}{2}(3x^2 - 1) \\
P_3(x) &= \tfrac{1}{2}(5x^3 - 3x) \\
P_4(x) &= \tfrac{1}{8}(35x^4 - 30x^2 + 3).
\end{aligned}$$

A3.1.3 Basic properties of $P_n(x)$

$$\int_{-1}^{+1} P_i(x)P_j(x) = \begin{cases} 2/(i+1) & \text{for } i = j \\ 0 & \text{for } i \neq j \end{cases}$$

i.e., the Legendre polynomials are orthogonal

$$\begin{aligned}
(n+1)P_{n+1}(x) &= (2n+1)xP_n(x) - nP_{n-1}(x) \\
(2n+1)P_n(x) &= P_{n+1}'(x) - P_{n-1}'(x).
\end{aligned}$$

A3.1.4 The Chebyshev polynomials $T_n(x)$, $-1 < x < 1$

$$T_n(x) = \cos(n \arccos x).$$

For more about the properties of the Legendre or Chebyshev polynomials, see [Gradshteyn and Ryzhik, 1980; Sections 8.91 and 8.94].

A3.2 The mappings

One dimension: Master element $\tilde{I}$, physical I.
Function F defined on $\tilde{I}$ and $G(x)$ defined on I_j^{Δ} such that $F(\tilde{x}) \in I$, $\forall \tilde{x} \in \tilde{I}$, $G(x) \in \tilde{I}$, $\forall x \in I$ and $G(F(\tilde{x})) = \tilde{x}$, $\forall \tilde{x} \in \tilde{I}$, $F(G(x)) = x$, $\forall x \in I$ and F and G have continuous first derivatives on $\bar{\tilde{I}}$ and $\bar{I}_j^{\Delta}$. They create smooth bijective mapping of the master element on the physical element.

Two dimensions: Master element $\tilde{\tau}$, physical $\tilde{\tau}_j^{\Delta}$

Function $F = (f_1, f_2)$ and $G = (g_1, g_2)$ such that $f_j(\tilde{x}_1, \tilde{x}_2) = x_j$, $j = 1, 2$, $\forall \tilde{x} = (\tilde{x}_1, \tilde{x}_2) \in \tilde{\tau}$, $x_j \in \tau$ and $g_j(x_1, x_2) = \tilde{x}_j$, $j = 1, 2$, $x = (x_1, x_2) \in \tau$, $\tilde{x}_j \in \tilde{\tau}$ and $f_j(g_1(x_1, x_2), g_2(x_1, x_2)) = x_j$, $g_j(f_1(\tilde{x}_1, \tilde{x}_2), f_2(\tilde{x}_1, \tilde{x}_2)) = \tilde{x}_j$, $j = 1, 2$ and f_j, g_j have continuous derivatives on $\bar{\tilde{\tau}}$ and $\bar{\tau}_j^{\Delta}$.

The mapping G is called the inverse of F and vice versa and will be written $G = F^{-1}$, $F = G^{-1}$.

A3.2.1 Jacobian of the mapping F in two dimensions

We denote

$$\nabla F = \begin{bmatrix} \dfrac{\partial f_1}{\partial x_1} & \dfrac{\partial f_1}{\partial x_2} \\ \dfrac{\partial f_2}{\partial x_1} & \dfrac{\partial f_2}{\partial x_2} \end{bmatrix}$$

and the Jacobian is the determinant of the matrix ∇F. If the mapping F is a smooth bijective mapping, then $|\det(\nabla F)| > 0$, $\forall \tilde{x} \in \bar{\tilde{\tau}}$ and $|\det(\nabla G)| > 0$, $\forall x \in \bar{\tau}_j^{\Delta}$ and

$$\det(\nabla F)(\tilde{x}) = \frac{1}{\det(\nabla G)(x)}$$

for $x = F(\tilde{x})$ and $\tilde{x} = G(x)$.

A3.2.2 The smooth bijective mapping of the master element $\tilde{\tau}$ onto the physical element τ

Let $\tilde{\tau}$ be the master element, τ be the physical element, and F be the smooth bijective mapping of $\tilde{\tau}$ onto τ. Let $\tilde{v}$ is a function on $\tilde{\tau}$, then we will write

$$v(x) = \tilde{v}(F^{-1}(x)) = \tilde{v} \circ F^{-1}$$

and $v(x)$ is defined on τ. We also have then

$$\tilde{v}(\tilde{x}) = v(F(\tilde{x})).$$

A3.3 The equivalent norms

Let X be a Banach or Hilbert space with norms $\|\cdot\|_1$ and $\|\cdot\|_2$. These norms will be called equivalent if there exists $0 < c_i < \infty$, $l = 1, 2$ such that

$$c_1 \|x\|_2 \le \|x\|_1 \le c_2 \|x\|_2 \quad \forall x \in X.$$

GLOSSARY OF SYMBOLS

1. General notation

$I \subseteq \mathbb{R}^1$, $\Omega \subseteq \mathbb{R}^2$	One, two dimensional domain
$\partial\Omega$	Boundary of Ω
$v(\cdot)$, $v(\cdot,\cdot)$	Functions of one, two variables
$v\|_I$	Restriction of v on I
$\mathscr{C}$	Constant, independent of the factors multiplying it
$\mathscr{U}$	Energy space
$\|\|\cdot\|\|_{\mathscr{U}}$	Energy norm
$\mathscr{B}$	Bilinear form associated with the energy space $\mathscr{U}$
$\mathscr{L}$	Linear functional associated with the variational formulation
$\mathfrak{D}$	Differential operator of the differential equation under consideration
u_{EX}	Exact solution of the problem under consideration
$\Pi(u)$	Total potential energy of u

2. Finite element method

Δ	Mesh
x_j^{Δ}	Nodal point of Δ
$I_j^{\Delta} \stackrel{\text{def}}{=} (x_{j-1}^{\Delta}, x_j^{\Delta})$	Element in one dimension
$\tilde{I}$	Master element in one dimension
$h_{I_j^{\Delta}} \stackrel{\text{def}}{=} x_j^{\Delta} - x_{j-1}^{\Delta}$	Size of I_j^{Δ}
τ	Element in two dimensions
$\tilde{\tau}$	Master element in two dimensions
$h_{\tau} \stackrel{\text{def}}{=} \max_{\mathbf{x},\mathbf{y}\in\tau} \|\mathbf{x}-\mathbf{y}\|_{\mathbb{R}^2}$	Size of τ
h^{Δ}	Maximal size of the elements in Δ
S_{Δ}^{p}	Finite element space of piecewise polynomials of degree p defined over Δ
$u_{S_{\Delta}^{p}}$	Galerkin solution from S_{Δ}^{p}
$\mathscr{P}^p(I_j^{\Delta})$	Space of polynomials of degree p defined over I_j^{Δ}
$\tilde{\mathscr{P}}^p(\tilde{\tau})$	Space of polynomials of total degree p defined over $\tilde{\tau}$
$\tilde{Q}^p$	Space of bi-p polynomials defined over the master square
$\tilde{Q}'^p$	Space of serendipity polynomials of degree p defined over the master square
F_{τ}	Bijective transformation which maps $\tilde{\tau}$ onto τ
$\mathcal{I}_{I_j^{\Delta}}^{p} u$	Polynomial interpolant of degree p over I_j^{Δ}
φ_j^{Δ}	Basis function defined over Δ
$\tilde{u}$, $\tilde{\mathcal{I}}^p\tilde{u}$	The function u transformed over the master element and its p-degree polynomial interpolant

Φ	Mesh density function
$\mathcal{A}^p_{I^\Delta_j} u_{EX}$	Best energy approximation of u_{EX} on I^Δ_j from $S^p(I^\Delta_j)$, which satisfies $\mathcal{A}^p_{I^\Delta_j} u_{EX}(x^\Delta_k) = u_{EX}(x^\Delta_k), \quad k = j-1, j$
$\mathcal{B}^p_{I^\Delta_j} u_{EX}$	Best energy approximation of u_{EX} on I^Δ_j from $S^p(I^\Delta_j)$
$e_{S^p_\Delta} \stackrel{\text{def}}{=} u_{EX} - u_{S^p_\Delta}$	Error in $u_{S^p_\Delta}$
$e^{glob}_{S^p_\Delta} \stackrel{\text{def}}{=} \mathcal{A}^p_\Delta u_{EX} - u_{S^p_\Delta}$	Global component of the error $e_{S^p_\Delta}$ in one dimension
$e^{loc}_{S^p_\Delta} \stackrel{\text{def}}{=} u_{EX} - \mathcal{A}^p_\Delta u_{S^p_\Delta}$	Local component of the error $e_{S^p_\Delta}$ in one dimension
$\mathcal{B}^p_{\omega_\tau} u_{EX}$	Best energy approximation of u_{EX} over the element patch ω_τ
$e^{loc}_{S^p(\tau)} \stackrel{\text{def}}{=} u_{EX}\vert_\tau - \mathcal{B}^p_{\omega_\tau} u_{EX}\vert_\tau$	Local component of the error $e_{S^p_\Delta}$ in the element τ
$e^{glob}_{S^p(\tau)} \stackrel{\text{def}}{=} \mathcal{B}^p_{\omega_\tau} u_{EX}\vert_\tau - u_{S^p_\Delta}\vert_\tau$	Global component of the error $e_{S^p_\Delta}$ in the element τ
$\Psi^{\bar{\mathbf{x}},H}_{S^p_\Delta}$	Asymptotic approximation of the error $e_{S^p_\Delta}$ in a locally uniform mesh
$x^{sup,\eta\%}_{\mathcal{F}(u)}$	Point of $\eta\%$ superconvergence for $\mathcal{F}(u)$ in the element τ
$x^{sup}_{\boldsymbol{\nabla} u}$	Point of superconvergence of $\boldsymbol{\nabla} u$

3. A-posteriori estimates

$\mathscr{E} \stackrel{\text{def}}{=} \sqrt{\sum_{\tau\in\Delta}(\eta_\tau)^2}$	Error estimator
η_τ	Error indicator for the element τ
$\kappa \stackrel{\text{def}}{=} \dfrac{\mathscr{E}}{\Vert e_{S^p_\Delta}\Vert_{\mathfrak{U}}}$	Effectivity index
$\hat{e}$	Error indicator function
$\mathfrak{R}^{EQ}_\tau$	Equilibrated residuum for τ
$\mathscr{E}_{S^{p+r}_\Delta}$	Estimator based on approximate indicators from S^{p+r}_Δ
S^{p+r}	Space of mapped polynomials of degree $p+r$ defined on an element as the span of all the element shape functions up to degree $p+r$
$B^{p+k} \stackrel{\text{def}}{=} S^{p+k} - S^p$	Space of bubble functions defined on an element as the span of the element shape functions of degrees $p+1$ through $p+k$

Estimators in one dimension:

$ERDBp+k$	Element Residual estimator of the Dirichlet type which employs approximate indicators from B^{p+k}
$ERDPp+k$	Element Residual estimator of the Dirichlet type which employs approximate indicators from S^{p+k}
$ERNPp+k$	Element Residual estimator of the Neumann type which employs approximate indicators from S^{p+k}

CRec2dPp	Estimator based on the 2 sided Recovery of the derivative using a Continuous least-squares fit by Polynomials of degree p
DRec2dPp	Estimator based on the 2 sided Recovery of the derivative using a Discrete least-squares fit by Polynomials of degree p
$DRecvPp+1$	Estimator based on the Recovery of the value using a Discrete least-squares fit by Polynomials of degree $p+1$

Estimators in two dimensions:

$ERqPp+k$	Element Residual estimator of Neumann type with residuals equilibrated up to order q which employs approximate indicators from S^{p+k}
$ERqBp+k$	Element Residual estimator of Neumann type with residuals equilibrated up to order q which employs approximate indicators from B^{p+k}
ER_Bp+k	Element Residual estimator using unequilibrated residual and indicators from B^{p+k}
Explicit	Explicit residual estimator
ZZD or *ZZ-SPR*	Zienkiewicz-Zhu estimator based on a discrete least-squares recovery
ZZ	Zienkiewicz-Zhu estimator based on a continuous least-squares recovery
$ERqBp+k$ (BW, Lad1, or Lad2)	Estimator $ERqBp+k$ using different types of residual equilibration
$\mathcal{R}$	Robustness index
$\mathscr{E}^{\text{Neum}}$	Exact Neumann estimator
$\mathscr{E}^{\text{Neum}}_{S^{p+k}_{\Delta'}}$	Computed Neumann estimator
$\mathscr{E}^{\text{EXPL}}_{\hat{e}^{\text{Neum}}_{S^p_\Delta}/I^\Delta_j}$	Explicit correction for the computed estimator $\mathscr{E}^{\text{Neum}}_{S^{p+k}_\Delta}$
$\mathscr{E}^{\text{U}}_{u_{S^p_\Delta},S^{p+k}_\Delta} = \mathscr{E}^{\text{Neum/EXPL}}_{u_{S^p_\Delta},S^{p+k}_\Delta}$	The guaranteed upper Neumann estimator
$\mathscr{E}^{\text{L}}_{S^{p+k}_{\Delta'}} = \mathscr{E}^{\text{Neum/L}}_{S^{p+k}_{\Delta'}}$	The guaranteed lower Neumann estimator
$\Phi_{\hat{e}^{\text{Neum}}_{S^{p+k}_{\Delta'}}}$	Gap function for $\hat{e}^{\text{Neum}}_{S^{p+k}_{\Delta'}}$
$\overline{\mathscr{E}}^{\text{U}}_{S^{p+k}_{\Delta'}}$, $\underline{\mathscr{E}}^{\text{L}}_{S^{p+k}_{\Delta'}}$	Improved upper and lower estimator
$\mathscr{E}^{\mathcal{A}^*_{\frac{du}{dx}}}$	Estimator based on the recovery $\mathcal{A}^*_{\frac{du}{dx}}$
$e^{\mathcal{F}}_{S^p_\Delta} = \mathcal{F}(u_{EX}) - \mathcal{F}(u_{S^p_\Delta})$	Error in $\mathcal{F}(u_{S^p_\Delta})$
$\mathscr{E}^{\text{L},\mathcal{F}}_{S^{p+k}_{\Delta'}}$, $\mathscr{E}^{\text{U},\mathcal{F}}_{S^{p+k}_{\Delta'}}$	Guaranteed lower and upper estimator for $e^{\mathcal{F}}_{S^p_\Delta}$

AUTHOR INDEX

SUBJECT INDEX